Commodities

Since a major source of income for many countries comes from exporting commodities, price discovery, and information transmission between commodity futures markets, are key issues for continued economic development. *Commodities: Fundamental Theory of Futures, Forwards and Derivatives Pricing, Second Edition* covers the fundamental theory of, and derivatives pricing for, major commodity markets, as well as the interaction between commodity prices, the real economy, and other financial markets.

After a thoroughly updated and extensive theoretical and practical introduction, this new edition of the book is divided into five parts – the last of which is entirely new material covering cutting-edge developments.

- Oil Products considers the structural changes in the demand and supply for hedging services that are increasingly determining the price of oil
- Other Commodities examines markets related to agricultural commodities, including natural gas, wine, soybeans, corn, gold, silver, copper, and other metals
- Commodity Prices and Financial Markets investigates the contemporary aspects of the financialization of commodities, including stocks, bonds, futures, currency markets, index products, and exchange traded funds
- Electricity Markets supplies an overview of the current and future modelling of electricity markets
- Contemporary Topics discuss rough volatility, order book trading, cryptocurrencies, text mining for price dynamics, and flash crashes.

Chapman & Hall/CRC Financial Mathematics Series

Aims and scope:
The field of financial mathematics forms an ever-expanding slice of the financial sector. This series aims to capture new developments and summarize what is known over the whole spectrum of this field. It will include a broad range of textbooks, reference works and handbooks that are meant to appeal to both academics and practitioners. The inclusion of numerical code and concrete real-world examples is highly encouraged.

Series Editors

M.A.H. Dempster
Statistical Laboratory
Centre for Mathematical Sciences
University of Cambridge, UK

Rama Cont
Mathematical Institute
University of Oxford, UK

Robert A. Jarrow
Lynch Professor of Investment Management
Johnson Graduate School of Management
Cornell University, USA

Dilip B. Madan
Robert H. Smith School of Business
University of Maryland, USA

Machine Learning for Factor Investing: R Version
Guillaume Coqueret, Tony Guida

Malliavin Calculus in Finance: Theory and Practice
Elisa Alos, David Garcia Lorite

Risk Measures and Insurance Solvency Benchmarks: Fixed-Probability Levels in Renewal Risk Models
Vsevolod K. Malinovskii

Financial Mathematics: A Comprehensive Treatment in Discrete Time, Second Edition
Giuseppe Campolieti, Roman N. Makarov

Pricing Models of Volatility Products and Exotic Variance Derivatives
Yue Kuen Kwok, Wendong Zheng

Quantitative Finance with Python: A Practical Guide to Investment Management, Trading, and Financial Engineering
Chris Kelliher

Stochastic Modelling of Big Data in Finance
Anatoliy Swishchuk

Introduction to Stochastic Finance with Market Examples, Second Edition
Nicolas Privault

Commodities: Fundamental Theory of Futures, Forwards, and Derivatives Pricing, Second Edition
M.A.H. Dempster, Ke Tang

For more information about this series please visit: https://www.crcpress.com/Chapman-and-HallCRC -Financial-Mathematics-Series/book *series/CHFINANCMTH*

Commodities

Fundamental Theory of Futures, Forwards, and Derivatives Pricing
Second Edition

Edited by
M.A.H. Dempster
University of Cambridge, UK

Ke Tang
Tsinghua University, People's Republic of China

CRC Press
Taylor & Francis Group
Boca Raton London New York

CRC Press is an imprint of the
Taylor & Francis Group, an **informa** business

A CHAPMAN & HALL BOOK

Second edition published 2023
by CRC Press
6000 Broken Sound Parkway NW, Suite 300, Boca Raton, FL 33487-2742

and by CRC Press
2 Park Square, Milton Park, Abingdon, Oxon, OX14 4RN

© 2023 selection and editorial matter, M.A.H. Dempster and Ke Tang; individual chapters, the contributors

First edition published by CRC Press 2016

CRC Press is an imprint of Taylor & Francis Group, LLC

Library of Congress Cataloging–in–Publication Data

Names: Dempster, M.A.H. (Michael Alan Howarth), 1938- editor. | Tang, Ke
(Professor of economics), editor.
Title: Commodities : fundamental theory of futures, forwards, and
derivatives pricing / [edited by] M.A.H. Dempster, University of
Cambridge, UK, Ke Tang, Tsinghua University, People's Republic of China.
Description: Second edition. | Boca Raton : C&H/CRC Press, 2023. | Series:
Chapman and Hall/CRC financial mathematics series | Includes
bibliographical references.
Identifiers: LCCN 2022024051 (print) | LCCN 2022024052 (ebook) | ISBN
9781032208176 (hardback) | ISBN 9781032208244 (paperback) | ISBN
9781003265399 (ebook)
Subjects: LCSH: Commodity futures. | Commodity exchanges.
Classification: LCC HG6046 .C5766 2023 (print) | LCC HG6046 (ebook) | DDC
332.64/4--dc23/eng/20220608
LC record available at https://lccn.loc.gov/2022024051
LC ebook record available at https://lccn.loc.gov/2022024052

ISBN: 978-1-032-20817-6 (hbk)
ISBN: 978-1-032-20824-4 (pbk)
ISBN: 978-1-003-26539-9 (ebk)

DOI: 10.1201/ 9781003265399

Typeset in Minion
by Deanta Global Publishing Services, Chennai, India

Contents

SECTION 3   Commodity Prices and Financial Markets

SECTION 4   Electricity Markets

SECTION 5   Contemporary Topics

About the Editors

M.A.H. Dempster is professor emeritus in the Statistical Laboratory, Centre for Mathematical Sciences, University of Cambridge. Educated at Toronto, Carnegie Mellon and Oxford Universities, he has taught and researched in leading universities on both sides of the Atlantic and is founding editor-in-chief of Quantitative Finance and the Oxford Handbooks in Finance. Consultant to many global financial institutions, corporations and governments, he is regularly involved in research presentation and executive education worldwide. He is the author of over 110 research articles in leading international journals and 14 books; his work has won several awards and he is an honorary fellow of the Stanford Center for Advanced Study in the Behavioral Sciences and the UK Institute of Actuaries, a foreign member of the Academia Lincei (Italian Academy) and managing director of Cambridge Systems Associates Limited, a financial analytics consultancy and software company.

Ke Tang is a professor in the Institute of Economics, School of Social Science, Tsinghua University, where he teaches courses in economics and finance. He earned his BA in engineering from Tsinghua University in 2000, Master of Financial Engineering from the University of California, Berkeley in 2004, and his doctoral degree in Finance from Cambridge University in 2008. His research has covered such topics as commodity markets, digital economy and fintech. He has published many papers in journals including *Journal of Finance, Review of Financial Studies, Annual Review of Financial Economics, Quantitative Finance*, etc. He is a frequent participant in various policy-related conferences held by institutions such as the United Nations Conference on Trade and Development, Organisation for Economic Co-operation and Development and the Food and Agriculture Organization of the United Nations. He currently serves as a managing editor of Quantitative Finance.

List of Contributors

Frédéric Abergel

Timur Aka

Lucia Baldi

Jonathan A. Batten

Mikkel Bennedsen

Fred E. Benth

Ilia Bouchouev

Chris Brooks

H.J. Cai

René Carmona

Jaime Casassus

Julien Chevallier

Troels S. Christensen

Cetin Ciner

Michael Coulon

Lei Cui

M.A.H. Dempster

Wenfeng Dong

Elyas Elyasiani

Jim Gatheral

Mathieu Gatumel

Michael Graham

Ben Hambly

Liyan Han

Eivind Helland

Freddy Higuera

Sam Howison

Ke Huang

Côme Huré

Florian Ielpo

Thibault Jaisson

Brett Johnson

Boda Kang

Jarno Kiviaho

Tino Kluge

Guowen Li

Jianping Li

Rong Liang

Brian M. Lucey	Mathieu Rosenbaum
Viktor Manahov	Sandro Sapio
Iqbal Mansur	Erik Schlögl
Massimiliano Marzo	Daniel Schwarz
Joseph McCarthy	J.E. Scott
Vicente Medina	Anna Sembos
Elena Medova	Gregorio Serna
Andrés García Mirantes	Kenichiro Shiraya
John M. Mulvey	Didier Sornette
Jussi Nikkinen	Peter G. Szilagyi
Babatunde Odusami	Akihiko Takahashi
Alexei G. Orlov	Ke Tang
Angel Pardo	Alexander Wehrli
N Hicks Pedrón	Eric Winnington
Massimo Peri	Daniela Vandone
Huyên Pham	Yanzhen Yao
Kay F. Pilz	Paolo Zagaglia
Javier Población	Xiaoqian Zhu
Marcel Prokopczuk	
Victor Rohde	

Introduction

Summary

For more than two decades, global commodity markets have experienced dramatic growth in terms of both prices and trading volumes. Starting in the early 2000s, key commodity prices grew at rates unprecedented in history and then fell sharply during the 2007–2008 financial crisis. The steady growth of US oil production from hydraulic fracturing of shale, "fracking", over the first decade of the 21st century and the slowdown of the growth of the Chinese economy caused oil and iron ore prices to halve spectacularly – in spite of adverse geopolitical events affecting supply. In both cases these events appear to have been caused by a mixture of supply and demand factors. However, from mid-2009, many commodity prices began to climb again, with oil prices increasing to over $100 a barrel. A related phenomenon over the post-crisis period is that most commodity prices have become extremely volatile, much more volatile than would be implied by variations in the fundamental balance of supply and demand. The past decade has seen a gradual increase in the financialization of commodities, likely related to the increased volatility of prices. A steadily increasing number of investors purchase commodity-based exchange traded funds or indices which treat commodities as an asset class for portfolio diversification and inflation hedging. Because of high energy prices, together with the necessity to rapidly drastically reduce carbon emissions, some food crops are used to produce bio-fuel, so that fluctuations of energy prices have also influenced food prices. All these developments in commodity markets have coincided with the fast growth of emerging market economies such as China, India and Brazil. The urbanization and expanding middle-class population of emerging economies signals a strong global demand for commodities in the long-term future. In fact, the major income of more than 70 developing countries comes from exporting commodities. Price discovery and information transmission between commodity forwards and futures markets are therefore key issues for the development of these economies. The linkages between different commodity markets and between commodity futures and other financial markets raise natural questions within a background of developing financialization and the use of bio-fuels. Closely related issues are the consequences of these developments for pollution control and global warming. Since the onset of the COVID-19 pandemic in 2020, economic growth in all countries and regions has become highly uncertain with very different and rapidly changing prospectives. It is generally agreed however that high population vaccination rates are very important for positive developments and that the future nature of human endeavour is changing forever. Some people even see a rerun of the World War One Spanish flu pandemic and the Roaring 20s to be followed by a horrendous financial market crash which will dwarf all previous market upsets and change humanity forever.

The chapters of this volume address topics relevant to an understanding of these issues. This second edition contains 33 chapters, one more than the original edition. Eight original chapters, which treated topics already well understood a decade ago, have been replaced by papers from mainly recent issues of *Quantitative Finance*, together with two from other journals. These are distributed by section as follows: two each in *Oil Products* and *Other Commodities*, one in *Commodity Prices and Financial Markets* and five on *Contemporary Topics*. The topics of contemporary interest discuss rough volatility, order book trading, cryptocurrencies, text mining for price dynamics and flash crashes. All chapters will be described in more detail below.

Introduction to Commodity Finance

Commodities have always been a part of our daily life and have now become an important asset class in financial markets. A *forward contract* between a buyer and a seller is an agreement for the seller to deliver a specified amount of a commodity to the buyer at an agreed price. The figure shows the nature of the first forward contracts employed in ancient Sumeria from 8,000 BC.* Clay tokens used to represent the amounts of different commodities, including livestock, were baked into a hollow clay vessel for security after marking them in the soft clay on its outside before firing. Contract disputes at delivery time were resolved by breaking the vessel to find inside the representative tokens of the previously agreed amounts.

The term *forward* comes from contract law around 317 BC in ancient Athens. Modern forward contracts date back to those for rice in 18th-century Japan and those for wheat and other grains in late 19th-century USA.

Most commodities, including gold, oil, natural gas, etc., are now traded by *futures contracts* in commodity futures exchanges which in current form date from the early 20th century for those set up for agricultural products in Chicago, USA. These *derivatives* are *standardized* contracts which involve a standardized amount of the underlying commodity to which they refer. (The contemporary importance of these first financial derivatives has been emphasized by the US government with its placement of the development of regulation for *all* derivatives under the 2017 post-crisis Dodd-Frank act in the hands of the Commodity Futures Trading Commission of the US Treasury.) To prevent *default* at contract maturity by an out-of-the-money speculative counterparty – a common event in 19th-century USA – prior to maturity the out-of-the-money party must periodically post *margin* with the commodity exchange which is passed to the in-the-money counterparty when the current market price for the futures contract is in their favour. This is termed *marking to market*. The usual measure of current trading volume for a specific commodity on a futures exchange is termed *open interest*, which refers to the number of standardized contracts, usually including both futures and options on futures, for the underlying commodity that are currently outstanding, i.e. which have not been settled in the immediately previous time period.

An *option* is a financial derivative contract which gives the holder the *right* but not the obligation to purchase (*call* option) or sell (*put* option) the underlying entity, here a commodity futures contract, at its maturity date (*European* option), or at any time up to its maturity (*American* option), at a fixed price termed the *strike price*. It is interesting to note that it was the empirical observation of agricultural prices by Kendall that led to the random walk hypothesis for price changes (Kendall and Bradford Hill, 1953), later made rigorous for futures contracts with stationary (explicitly not Brownian motion) prices by Samuelson (1965), leading to the efficient market hypothesis, geometric Brownian motion and the theory of option pricing by Black, Scholes and Merton (Black and Scholes, 1973; Merton, 1973).

There are broadly two classes of traders in commodity futures markets: *hedgers* and *speculators*. Hedgers, who are normally producers, seek to hedge their future physical position in a commodity by selling futures contracts, and speculators, pursuing speculative profits, take the buy side of these trades. Keynes (1923) argued that in order to attract speculators to take counterparty positions, hedgers need to offer *risk premia* to potential speculators by offering futures prices lower than they might otherwise be. These premia should increase with the maturity of the futures contract, leading to a *decreasing*

---

* We are indebted to Francesca Chiminello of Bloomberg for bringing our attention to the existence of these fascinating objects.

futures *term structure*, i.e. the variation of futures price with contract maturity. This phenomenon is referred to as *backwardization* and the theory is termed that of *normal backwardation*. After Keynes, many researchers have attempted to test this theory, including Rockwell (1967), Chang (1985) and others. The results are mixed.

One important feature of commodities is *storability*, i.e. all commodities except electricity can be stored, even agriculturals, at least for some time. The *theory of storage*, advanced by Kaldor (1939), Working (1949), Brennan (1958) and others, proposes that *inventory* serves as a buffer to offset demand or supply shocks in commodity markets. When the market-wide inventory is at a relatively low level, which corresponds to a high demand for the physical commodity and a futures term structure in backwardation, holders of scarce inventory should derive a larger benefit from holding the commodity. The marginal benefit of holding one unit of the commodity relative to holding the corresponding futures contract is thus defined to be the *convenience yield* of the commodity which corresponds to the dividend yield of holding a stock. The convenience yield represents the value of the option of using the stored physical commodity in production at any suitable time up to contract maturity. A larger convenience yield corresponds to a deeper backwardation of the futures term structure. Gorton, Hayashi and Rouwenhorst (2013) found that convenience yield is negatively proportional to the level of inventory of a commodity. Routledge, Seppi and Spatt (2000) showed that the convenience yield relates to the probability of inventory stockout. Convenience yield is therefore generally regarded as a distinguishing feature of commodities as an asset class. It is also the key to *modelling* the futures term structure. Gibson and Schwartz (1990), Schwartz (1997), Casassus and Collin-Dufresne (2005) and others propose a mean reverting Ornstein-Uhlenbeck process for the convenience yield in the framework of futures term structure modelling. The implication is that rather than the single factor model of Black (1976), until recently widely used by industry for trading commodity futures, at least a two-factor model, which includes a mean reverting convenience yield, is needed to capture the term structure of futures contracts.

Since the early 2000s commodities have become a new asset class. Many institutions, including hedge funds, mutual funds and pension funds, have been investing in commodities through commodity futures indices, exchange traded funds (ETFs), and related structures. This process of *financialization* is addressed in Tang and Xiong (2012), where it is argued that financial market factors influence commodity prices beyond the well understood fundamental effects of supply and demand. For example, Tang and Xiong show that the return correlation between different commodities increased substantially from 2004 to 2008. Henderson, Pearson and Wang (2012) note that trades motivated for financial reasons lead to significant price movements of commodity futures.

At a more general level, the fundamental reference in the field of commodities is the monograph of Geman (2005), published over 15 years ago at about the beginning of the accelerated financialization of commodity markets. The nine chapters of Section 3 of this volume represent the latest thinking on this development, in particular regarding the increasing importance of oil and gas prices in equity markets.

A recent more theoretical monograph is that of Pirrong (2012) which describes a structural approach to storable commodity futures/forwards pricing in the spirit of Working, as opposed to the reduced form arbitrage-based approaches originated by Keynes and Kaldor in the 1930s. The *structural approach* applies dynamic programming to derive the implications of non-negative inventories for prices determined by a market trading equilibrium with traders' rational expectations. The spectacular lack of empirical success with this approach using daily data is not surprising in that this theory essentially ignores the role of *speculation* and hence implies an *increasing* term structure of futures prices which is termed *contango*.

Structure of the Volume

Having briefly set out the practical and theoretical background, we now turn to the volume contents. The remaining first edition chapters are largely based on papers selected from two 2012–2013 special issues of *Quantitative Finance* (QF) with the theme of commodity markets.* The ten new chapters of this second edition are mainly on QF recent issue papers.

Chapter Contributions

Section 1 Oil Products

Chapter 1 by Bouchouev and Johnson, presents selected results from a comprehensive study of the volatility risk premium (VRP) in the oil market arising from the normal backwardization of the risk yield curve as a result of oil producers' hedging forward delivery contracts. It introduces the smile of the VRP which represents the variation in profitability and risk of this systematic strategy across option moneyness and maturities. A structural break in VRP evolution over time, driven by behavioural changes among hedging producers and strategy securitization by financial institutions, is identified.

Chapters 2 and 3 treat three-factor affine Gaussian spot price models in state space form with latent factor dynamics whose parameters are extracted from futures term structure data using the EM algorithm and the Kalman filter. Chapter 2, by Dempster, Medova and Tang, considers the question of what are the drivers of the short, medium and long-term economic factors of these models? This is a question that has been treated by macro-economists with financial markets treated as residual to the macroeconomy (see e.g. Ang and Piazzesi, 2003). However, their approach is inappropriate for pricing and hedging market modelling.

Using a similar three-factor model, in Chapter 3 Shiraya and Takahashi study both NYMEX oil futures and London Metal Exchange copper forwards markets. Their paper demonstrates the pricing and hedging efficiency of a three-factor stochastic mean reverting Gaussian model of commodity prices using oil and copper futures and forward contracts. The model is estimated using NYMEX WTI (light sweet crude oil) and LME Copper futures prices and is shown to fit the data well and significantly outperform the simple one-factor models commonly used.

Dempster, Hicks Pedrón, Medova, Scott and Sembos apply stochastic programming modelling and solution techniques to planning problems for a consortium of oil companies in Chapter 4. A multiperiod supply, transformation and distribution scheduling problem – the Depot and Refinery Optimization Problem (DROP) – is formulated for strategic and tactical level planning of the consortium's activities. This deterministic model is used as a basis for implementing a stochastic programming formulation with uncertainty in the product demands and spot supply costs (DROPS), whose solution process utilizes the deterministic equivalent linear programming problem. They employ their STOCHGEN stochastic problem generator to create a decision scenario tree for the unfolding future as deterministic equivalent. To project random demands for oil products at different spatial locations into the future and to generate random fluctuations in their future prices/costs a stochastic input data simulator is developed and calibrated to historical industry data. The models are written in the modelling language XPRESS-MP and solved by the XPRESS suite of linear programming solvers. From the viewpoint of implementation of large-scale stochastic programming models this study involves decisions in both space and time and careful revision of the original deterministic formulation.

In Chapter 5, García Mirantes, Población and Serna empirically analyze the stochastic mean dynamics of producers' heating oil and gasoline refining margins and trace the implications for financial market hedging. Using weekly observations on refining margin from January 1997 to May 2009, they show that the prices of these products are not only cointegrated with crude prices but that they share common long-term dynamics. The implication is that refining margin risk derives only from short-term effects which can be hedged with crack spread options based on modelling the spread directly.

---

* We would like to thank Collette Teasdale, formerly of Taylor & Francis, who pointed out the timely relevance of a collection of papers on commodity markets for the first edition and made a substantial initial contribution to bringing this collection about.

Dempster, Medova and Tang investigate in Chapter 6 the cointegration of commodity prices in long run dynamic equilibrium. They assert that the spread between the prices of two such commodities should be modelled directly and propose a two-factor model for the spot spread which is used to develop pricing and hedging formulae for options on spot and futures. These claims are illustrated for the *location spread* between Brent blend and WTI crude oil, using NYMEX daily futures data from 1993 to 2004, and the refining *crack spread* between heating oil and WTI crude, using NYMEX daily futures data from 1984 to 2004. The spread option values implied are significantly different from those of standard models but are consistent with practical observations.

Section 2 Other Commodities

Section 2 begins with Chapter 7 which is reprinted from an important recent paper in *Energy Economics*. In it Bennedsen develops for the first time a *multivariate* rough volatility model of electricity spot prices which accounts for the most important stylized facts of these time series: seasonality, spikes, stochastic volatility and mean reversion. Empirical studies have recently found a fifth stylized fact, roughness (see Chapter 29) which is explicitly incorporated into the multi-variate price model which it generalizes.

In Chapter 8, Baldi, Peri and Vandone consider investment in the global consumer wine market using a threshold cointegration model and apparently the first trading use of the Mediabanca Global Wine Industry Share Price Index. Most previous wine market studies have been concerned with high-priced fine wines for connoisseurs. Several have demonstrated that the returns of regional wine portfolios are comparable to those of the relevant stock exchange. The emphasis in this chapter is different. It attempts to use global equity indices to identify investment opportunities in wine price indices by examining their different speeds of adjustment to long-run equilibrium. The goal is to allow traders to identify information inefficiencies which can be exploited to construct profitable investment strategies for buying and selling wine producers' stocks. The nonlinear cointegration technique employed allows the index price to adjust differently to large and small deviations from the long-run equilibrium through a threshold representing a desired profit level. Market indices are examined for a variety of new and old wine producing countries, including Australia, Chile, China, France and the USA using daily data from January 2001 through February 2009. The results confirm the existence of threshold cointegration between the wine and composite industry equity indices in all countries studied except Australia, where the cointegration relationship is linear. The implications for regional economic development, macroeconomic conditions, political system and financial market sophistication are considered.

Chapter 9, by Han, Liang and Tang, analyzes the cross-market effects on global price discovery for soybean futures on the Dalian Commodity Exchange (DCE) in China and the Chicago Board of Trade (CBOT) exchange in the US. They use daily data from March 2002 to September 2011 and employ structural vector autoregression and vector error correction econometric models for soybean futures in both trading and non-trading hours, to demonstrate for the first time that information transfers between the DCE and CBOT flow in *both* directions with similar price impact magnitudes. Contrary to earlier opposite findings, this shows that Dalian plays a prominent role in global soybean futures price discovery.

The next two chapters of this section of the volume concern precious and base metals. They mainly focus on gold and silver markets, but the last discusses trading base metals on the Shanghai commodity exchange.

In Chapter 10, Batten, Ciner, Lucey and Szilagyi study the structure of gold and silver spread returns in terms of the dynamics of the bi-variate relationship between gold and silver prices. They analyze the spread between the daily prices of near-month futures contracts for New York Commodity Exchange (COMEX) gold and silver trading on the NYMEX from January 1999 to December 2005. After first filtering the spread return process using ARMA techniques to remove the short-term memory effects observed in autocorrelations, the resulting residual series reveals time-varying nonlinearities which are identified as fractility using Hurst techniques. The authors go on to test the effectiveness of simple trading rules based on the Hurst coefficient and find these superior both to the simple moving average strategies commonly applied by traders and to buy and hold.

Chapter 11 turns to the impact of US dollar (USO) currency fluctuations on gold price through the crisis. In it, Zagaglia and Marzo study the dynamic relationships between the spot price of gold and the USO-EUR and USD-GBP exchange rates with the euro and the pound using daily data over the period from 13 October 2004 to 5 March 2010. They find that the onset of the turmoil in August 2007 had little effect on these variables until the Lehman bankruptcy in September 2008, and that the gold price has been more stable throughout the crisis than USO exchange rates. They conclude that gold prices are less affected by market uncertainties than the USO, implying that gold can be a valuable addition to currency portfolios. They suggest that holding gold and selected other currencies in a portfolio removes the risks of volatility shock transmission caused by fluctuations in USO rates, independent of the severity of market fluctuations.

The penultimate paper of Section 2, like the last paper of Section 1, considers daily statistical arbitrage trading, here of zinc contracts on the Shanghai Futures Exchange (SHFE). However, Cui, Huang and Cai use entirely different methods in Chapter 12 to develop trading strategies based on a threshold GARCH-wavelet neural network. Four months of 2010 tick data is used to establish the optimal training and trading thresholds and the new method is compared to strategies based on historical projection and a standard back propagation neural network. Robustness checks on the rates of return of the three examined methods are performed under different market conditions to show the clear superiority of the new method. An investigation of the effects of trading commission charges on returns shows that the method employed remains very profitable for current commission levels. It is noted that the profitable trading strategy employed is risky in that it has difficulty coping with sudden large price changes and should therefore be used in conjunction with a suitable risk management system (see e.g. Dempster and Leemans, 2006).

The last chapter of this section by Benth, Christensen and Rohde treats the multivariate continuous time modelling of wind indices and the hedging of wind risk.

Section 3 Commodity Prices and Financial Markets

The nine chapters of this third section of the volume investigate contemporary aspects of the financialization of commodities, beginning with the impact of oil prices on US equities in Chapters 14, 15 and 16, and then widening to a range of commodity index drivers in a search for commodities whose prices are uncorrelated with equity prices and so can diversify investors' portfolios. In later chapters this theme of investment portfolio commodity enhancement is further investigated from different viewpoints and shorter-term models for derivating pricing are considered.

In Chapter 14, Casassus and Higuera show that oil price shocks in terms of three-month futures returns are a strong predictor of US excess stock returns at short horizons. They use as their shock variable quarterly WTI oil futures returns on the NYMEX from Q2 1983 (when trading in these contracts began) through Q4 2009 and the corresponding US stock returns were obtained from the value-weighted index of the Chicago University Center for the Study of Security Prices (CRSP) over the same period. The authors demonstrate that oil price shocks are a non-persistent negative leading indicator of excess equity returns whose impact dies away after three quarters. They compare their futures-based oil price shock variable with other popular predictors based on quarterly data over a sample period which includes the crisis to suggest that oil price change is the only variable with (linear) forecasting power for stock returns which, as they demonstrate, is robust against the inclusion of other macro-economic and financial variables and out of sample tests. The authors also study the cross-section of expected stock returns using an oil price shock conditional CAPM model which is highly statistically significant and a better fit to the data than the alternative conditional and unconditional models considered, including the Fama-French three-factor model. The leading indicator for excess stock returns they propose has the practical advantage over others of being market based, observable (rather than involving an arbitrary and possibly changing technical construction) and readily available to investors.

McCarthy and Orlov perform in Chapter 15 a time-frequency analysis of the relationship between daily closing prices of near month futures contracts on WTI crude and the S&P500 using data from end March 1983 to beginning March 2011. The authors use frequency domain techniques – wavelets and cross-spectra – to examine the association between crude oil and equity prices in order to reveal insights into this relationship that are not apparent from a conventional time-domain approach. They find a *long-term positive* correspondence between contemporaneous oil and S&P500 prices in levels. This of course does not contradict the findings in the previous chapter of a *short-term negative* lagged relationship between oil price returns (shocks) and excess returns of the S&P500. Their results also confirm that, when analyzed separately, oil prices lead oil volume and S&P500 trading volume leads S&P500 prices. These findings persist over a large number of time scales and across a wide range of Fourier frequencies. Their use of frequency-domain techniques overcomes the simultaneity problems which have plagued some earlier investigations of the impact of oil prices on the US economy. Taken together, the practical implications for investors of this chapter and the previous one are profound.

The next chapter, Chapter 16, by Elyasiani, Mansur and Odusami, investigates the relationship between oil price changes and US stocks at the sectoral level with particular regard to volatility persistence which affects asset pricing to reinforce and enhance the cross-sectional findings in Chapter 14. A nonlinear double threshold FIGARCH specification of sectoral relationships is employed because it nests the GARCH and IGARCH models to allow the identification of the true extent of long memory in conditional return volatility at the sector level. The authors use daily one month futures returns from January 1991 to December 2006 as oil price shocks and daily sector equity index data from the CRSP and Kenneth French databases on ten US industry sectors broadly classified as oil substitutes (2), oil-related (2), oil consuming (5) and financial (1). They consider different conditional oil return volatility regimes, calm and volatile, to adjust the FIGARCH thresholds for a Fama-French model augmented by oil returns when oil return shocks become large enough to exceed existing thresholds. The symmetry and signs of these thresholds are seen to vary widely by industry, with those of the oil-related industries found to be the tightest, i.e. most sensitive to volatility. The decay patterns of industry responses to oil price shocks are also seen to vary widely, which call for different policy responses by oil producers, oil consumers and regulators to these shocks. These sector-specific and period-specific results are tested against simpler model alternatives which are found wanting. The chapter concludes that regime shifts and long-term structural changes due to oil price shocks have profound influence on how investors perceive long-term risk and form their expectations.

In a much shorter contribution, Chapter 17, Mulvey outlines the construction of an investible index long-short commodity fund which is meant to be used as a risk reducing overlay to a standard portfolio robust to all market conditions. After discussing the sources of alpha in commodity portfolios and introducing the *ulcer index* as the cumulative sum of the standard deviation of drawdown values, he sketches the outline of a commodity overlay portfolio designed to minimize the effects of drawdown risks and negative rollover risks when markets are in contango. Six established tactics are employed which carefully balance long and short positions in the overlay portfolio: long and short momentum, long and short futures curve, trend following and breakout. Each of these strategies is equally weighted in the portfolio and the long and short positions are balanced to be market neutral. However, noting that three market regimes, growth, transition and contraction, occur in historical respective proportions two thirds and one-sixth each over time, the portfolio performance can be enhanced by regime detection and a tilt of about 20% in the appropriate long or short direction. Regular periodic rebalancing is also mentioned as a source of portfolio return enhancement (see e.g. Dempster, Evstigneev and Schenk-Hoppe, 2011). Unfortunately, neither the commodities nor the back test results used in the implementation of these ideas by a major global exchange are presented in the chapter.

In Chapter 18, Brooks and Prokopczuk use a basic nonnegative mean-reverting square root process to model log spot prices to which they add stochastic volatility a la Heston and jumps in both drift and stochastic volatility. They estimate the parameters of this succession of models, including three separate stochastic volatility models, using Markov Chain Monte Carlo (MCMC), which is a Bayesian technique used here with non-informative priors. MCMC breaks down the high dimensional posterior parameter distribution of a model into its low-dimensional complete conditional parameter distributions and latent factors, e.g. volatilities, jump times and jump sizes, which may be sampled efficiently. The authors employ US daily spot price data for crude oil, gasoline, gold, silver, soybeans and wheat from 1985 to March 2010, obtained from the Commodity Research Bureau, and the S&P500 from Bloomberg for comparison. Individual commodity analysis identifies jumps due to, for example, OPEC increasing oil supply in 1986, the Gulf War in 1990, Hurricane Katrina in 2005 and the Lehman default in 2008. As expected, return correlations are found to be highest between commodities in the same market sector and, with some interesting exceptions, so are volatility and jump correlations, but there is little interaction found between commodity market sectors. The results demonstrate that return correlations for their models between commodities and the stock index are also low, but, more importantly for portfolios of securities with embedded contingent claims, that their volatility correlations are low as well, so that commodities may act as a volatility diversifier for such embedded derivative portfolios. Some evidence is found for simultaneous equity and oil or gold jumps, but these results vary with the model employed. The alternative spot price models are compared for each of the six commodities by pricing European calls and barrier options, and by the effectiveness of their delta hedges. Substantial differences are found across commodities and between models. The authors conclude that this indicates the heterogeneous nature of the broad commodity asset class, and they stress the importance of choosing the appropriate model for each individual commodity and purpose at hand.

Chapter 19, by Graham, Kiviaho and Nikkinen, investigates the difference between short-term and long-term dependencies between a classification of all global commodity and S&P500 returns. This difference is important to the dissimilar goals of short- and long-term equity investors. The authors employ a wavelet coherency methodology to simultaneously take into account time and frequency domains in a manner similar to Chapter 15 for oil price shocks. They address the question of which, if any, commodity returns can genuinely be considered uncorrelated with equity returns and thus be useful to investors in actually diversifying their portfolios. The study employs weekly observations from January 1999 and December 2009 inclusive on the S&P500 and the S&P GSCI (Goldman Sachs Commodity Index) and its ten subcategories: Energy, Light Energy, Non-Energy, Reduced Energy, Agriculture, Livestock, Petroleum, Industrial Metals, Precious Metal and Softs. This sample period allows an analysis of the effects of the onset of the financial crisis on the studied relationships and a distinction between short-term (high frequency) and long-term (low frequency) effects. The findings show negligible short-term effects of the crisis on commodity-equity relationships but stronger, although still quite weak, long-term effects, as is commonly asserted. The authors conclude that some important distinctions can be made for combining equity and the various segmented commodity indices in investors' portfolios. They note in particular that the diversification benefits of combining equity (S&P500) returns with precious metal (S&P GSCI Precious Metal) returns are robust for both short- and long-term investors across time.

Chapter 20, by Chevallier, Gatumel and Ielpo, points out that if commodity markets can be shown to be strongly related to the regimes of the business cycle as important for a long-short commodity overlay portfolio, then this evidence contradicts the widespread intuition that commodity markets are a strong source of diversification for a standard long only cash-bond-equity portfolio. Based on a Bloomberg daily data set from 1990 to 2012 for the US, EMU and China, the authors comprehensively investigate this question, by first evaluating the impact of news on an extensive set of commodity markets and then employing a Markov regime-switching model to analyse the effects of US business cycle regimes on global commodity markets as an asset class. They analyze news impacts using EGARCH regression techniques on markets for gold, silver, platinum, aluminium, copper, nickel, zinc, lead, WTI crude, Brent crude, gasoil, heating oil, natural gas, corn, wheat, coffee, sugar, cocoa, cotton, soybeans and rice to reveal that the response of commodity prices to economic surprises is strong during global downturns, but weak during expansion periods. They note that their results suggest that markets over-reacted to news during the 2008–2009 crisis, as they did in 2001, but showed a decreased sensitivity to news flow in H2 2009–2010. Unsurprisingly, they note an increased sensitivity to economic activity in China over the whole period. The chapter goes on to estimate a Markov business cycle regime-switching model with five regimes to the economies of the US, EMU and China. From an investor's viewpoint, the authors' findings may be summarized as follows. Investors should not add commodities to portfolios on the grounds of low correlation of a commodity's returns with those of a standard portfolio, unless they are at least are prepared for the long term, since there is a strong correlation of commodity markets with risky assets during economic downturns. Secondly, economic influences on commodity markets are complex and vary greatly for the US, Europe and China. Thirdly, a cyclical rotation amongst commodity markets over economic regimes is suggested; during strong growth industrial metals and energy should overweigh agricultural and precious metals in portfolios and the reverse should apply in downturns. Finally, the authors note that in "stalled" periods of the US economy, commodities generally outperform the S&P500 with the obvious implications for portfolio construction in this regime of the cycle.

Chapter 21, by Pilz and Schlögl, specifies a joint model of commodity price and interest rate risks analogous to the fixed income multi-currency LIBOR market model by directly addressing the issues arising in applying this model to commodities. WTI crude oil is used as the "foreign currency" and USO as the domestic, with the spot "exchange rate" as WTI near month futures price and the WTI futures term structure used for "foreign" forward rates. The difference between futures and the forward rates used in fixed income must be specifically taken into account, by converting the futures term structure to an appropriate forward rate term structure by a suitable convexity correction. The authors construct a procedure to calibrate the model to market data for interest rates and commodity futures prices. NYMEX futures prices for 5 May 2008 are used to give the underlying WTI "forward" data. Calibration of the model is to options on these futures prices and to caplets, caps and swaptions on US forward rates, together with three-month prior historically estimated correlations between interest rates and futures prices. Standard Kirk approximations are used to *simultaneously* estimate prices for calendar spread options on futures prices and rates which are favourably compared in-sample with Monte Carlo simulations from the calibrated model. The authors conclude that the main contribution of their model is in its calibration, which they feel is appropriate to other settings.

Chapter 22, by Dong and Kang, proposes a lattice-based method (trinomial trees) and a simulation-based method (least-squares Monte Carlo simulations) for pricing swing contracts with indexation. A typical gas sales agreement, also called a gas swing contract, is an agreement between a supplier and purchaser for the delivery of variable daily quantities of gas between specified minimum and maximum daily limits. The primary constraint of such agreements that makes them difficult to value is that the strike price is set based on the indexation principle, under which the strike price is called the index. Each month, the value of the index is determined by the weighted average price of certain energy products (e.g. crude oil) in the previous month. With the help of graphics processing unit (GPU) technology, the algorithms can be efficiently evaluated. The authors also provide a detailed analysis using several numerical examples of the indexation and how different model parameters will affect both the optimal value and optimal decisions.

Section 4 Electricity Markets

The six chapters of this fourth section of the volume give an overview of the current and future modelling of electricity markets. Chapters 23 and 24 model electricity prices and demonstrate their complexity, while Chapters 25 and 26 treat swing options, the major derivative product to mitigate electricity price risks. Finally, Chapter 27 discusses clean spread options which link electricity, fuels and emission allowances, while, in the spirit of Section 3, Chapter 28 considers European Union Allowances for carbon emissions as a possible asset class.

In Chapter 23, Sapio investigates the statistical modelling of day-ahead returns in the NordPool, APX and Powernext European electricity markets. He points out that in these markets 24 auctions are run simultaneously each day to set prices for each hour of the following day. The day-ahead prices are determined by uniform price auctions so that *all* power is purchased and sold for the next day at the market-clearing prices. Using daily data from these three principal markets, NordPool (Scandinavia) from 1997 to 2002 inclusive, APX (Netherlands) from 2001 to 2004 and Powernext (France) from 2002 to 2007, the author compares the goodness-of-fit to empirical densities over these periods for four alternative return densities, a-stable, normal inverse Gaussian (NIG), exponential power (EP) and asymmetric exponential power (AEP), using Kolmogorov-Smirnov and Cramer-von Mises statistics. The author finds that electricity returns display heavy tails across both markets and hourly auctions, however returns are defined (log, percentage or price change) and seasonality is accounted for. The tails in the APX market and in day-time auctions are fatter than in other markets and in night-time auctions, but are dampened by the logarithmic transformation (calling into question its widespread use in financial economics of all sorts). Skew is an essential of all the fitted distributions, but the a-stable and NIG distributions systematically outperform the other two in goodness-of-fit to the data, although no clear ranking between these two emerges. Since both of them are closed under convolutions, daily return fat tails will persist for returns over longer intervals. The fitted a-stable distributions have exponents between 1 and 2, so that only the empirical *average* returns exist, but the empirical *variance* does not converge, unless of course the NIG distribution provides the better fit to a specific data set. Some potentially interesting cross market comparisons are inconclusive, using percentage or price change returns, but log returns indicate that skew is negative in Powernext, but positive in APX. The study reveals some interesting intraday patterns that are not observable when daily returns are averaged. The author sees these results as a starting point for further work, particularly as regards policy for retail market liberalization and the carbon emissions allowance trading.

Helland, Aka and Winnington develop a multi-period model for electricity market spot prices and apply it to option pricing in Chapter 24. At the largest energy company in Switzerland, the authors have developed the model described in this chapter for the purposes of optimizing power plant production from pumped storage hydro parks, pricing complex options and structured products, and dynamically hedging production. They assert that a deep understanding of stochastic price dynamics is necessary for both risk management and derivative pricing, so that their model is built up in several steps and calibrated using spot price distributions, log returns of forward contracts and standard option contracts over a number of delivery periods. The base price level of their stochastic spot price model is a forward curve for each hour of the day which has been estimated using a regression model and European Exchange (EEX) data for Germany. The main explanatory variables of this model are calendar dates (weekdays, holidays, etc.) to obtain (intraday, weekly and yearly) seasonality effects, historical meteorological data (daily temperature, wind and precipitation) and long-term market expectations. Each hourly price forward curve is then scaled to equal the power futures contract (PFC) price at each maturity to obtain an arbitrage-free model prediction. Daily changes in each PFC are based on a log-normal mean-reverting (Ornstein-Uhlenbeck) weekly process for the front year forward base contract to reflect shifts in spot prices due to changes in long-term contracts. Similarly, a log-normal mean-reverting hourly process is used to model intra-day and day-ahead perturbations in spot prices. Further, two jump processes are added in order to model the positive and negative price jumps observed in the German electricity market. Based on the most recent PFCs and a calibrated set of weekly, hourly and jump processes, a set of price scenarios is generated by Monte Carlo simulation using a dynamic spot price model and which is then used to evaluate the price of options on futures available on the EEX. The fair prices of calls and puts for selected strikes are calculated from the scenarios and compared with observed prices for options on the EEX, or with comparable broker data. The stochastic spot price multi-period model is then re-calibrated so that the monthly, quarterly and annual contracts fall within their quoted bid-ask spreads. The result is a practical and flexible model which is actually used for all the purposes intended in an arbitrage-free manner.

In Chapter 25, Hambly, Howison and Kluge address the modelling of the price spikes occasionally observed in electricity market prices and use the result to price the swing options used by energy traders to manage the risks of these high prices. "Swing contracts" are a broad class of path-dependent options allowing the holder to exercise a certain right multiple times over a specified period, but exercising only one right at a time, with a "refraction period" lag between exercises, or per time interval like a day. For example, the right could be to receive a call or put option payoff, or a mixture of these, or one or the other with different strikes. Another popular feature allows the swing option holder to exercise a number, termed the "volume", of calls or puts at once, typically with bounds on the volume and the sum of all multiple exercises. Such contracts protect the purchaser against excessive rises in electricity prices and, bundled with forward contracts, they ensure a constant stream of energy at a pre-determined price. If the strike price of the call options in a swing contract are set at a forward price maturity, the swing contract allows flexibility in the volume the holder will receive at the fixed price at that date. Such contracts can either "swing up" or "swing down" the volume received at forward dates, and hence the name. To price these contracts the authors use a general payoff function and an electricity spot price model that is the exponential of the sum of a deterministic seasonality function, a mean reverting Ornstein-Uhlenbeck process and an independent mean-reverting pure jump process with a constant intensity and independent identically distributed jumps. The mean reversion speeds of the two stochastic components may be different, a useful feature for modelling NordPool contracts. Although the Scandinavian exchange and the UKPX and German EEX exchanges are mentioned, no explicit calibrations to forward curve data are given in the chapter. The authors derive the moment generating function, and various approximations to the density, of the logarithm of the spot price process at a maturity date. They first present semi-analytic formulae for path-independent options and approximations for call and put options on forward contracts, with and without a delivery period. Then they use a tree/grid method of Jaillet. et al. (2004) and an approximation to the conditional density of the spot price process to price swing option payoff structures in the presence of price spikes obtained from the pure jump component of their spot price model, for which some numerical results are given.

The theme of pricing electricity swing options with multiple exercise rights is continued in Chapter 26 by Kudryavtev and Zannette, using different models and methods. They model the spiking spot price process with a general Levy process which punctuates infinitesimally-frequent small jumps with occasional large spikes. The authors concentrate on methods and numerical procedures which they illustrate with a swing option involving a fixed number of exercises of a futures put option within a specified maturity and with a positive refraction time. This multiple stopping time problem can be reduced to a cascade of single stopping time problems using ideas of Carmona and Touzi (2008), who used Monte Carlo methods for their numerical solution. The authors evaluate this problem using two different deterministic numerical techniques. They propose a Wiener-Hopf factorization method which solves the system of integro-partial differential equations corresponding to the cascade and compare their approach to a more standard finite difference technique to solve the same system. Numerical results are given for pricing the swing option for both GBM and the CGMY Levy spot price processes which show the reliability and robustness of both solution methods. Although again no empirical results are given, the authors conclude that the Wiener-Hopf approach is a very precise and efficient method for pricing options in the presence of multiple jumps.

The final two chapters in this section of the volume point to critical future considerations for the power industry. In Chapter 27, Carmona, Coulon and Schwartz discuss the valuation of clean spread options which link electricity, emissions and fuels. Building on their own recent previous work, and that of their colleagues, for this purpose, the authors construct a *structural* model. They assert that such a model is necessary to capture the economic factors entering power production which are not captured by alternative, usually partial, reduced-form models. Specifically, they concentrate on electricity production and CO_2 emissions, defining a *clean spread option*, in the spirit of the more common *spark* (gas) and *dark* (coal) spread options, as the spread between the price at which one megawatt hour (MWh) of electricity can be sold in the market and the unit costs of production, including *both* gas (or coal) inputs and the market costs of carbon *allowances*. Market mechanisms for controlling CO_2 emissions by cap-and-trade schemes involving these traded allowances, whether they are voluntary or mandatory, have helped to price carbon emissions in Europe, the US and globally. For the past decade, power generators (and other emitters) in Europe are given each year a number of allowances out of the total number for all industry, termed the *cap*, which must be surrendered at the end of the year against measured emissions. If their allowances are insufficient, a firm must go to the EU Emission Trading Scheme (ETS) and purchase allowances from firms in surplus, speculators and possibly investors. The ETS involves both a spot market and a futures market which, as is usual for other futures markets, leads spot allowance price discovery and can be used to anticipate future over and undersupply of allowances. The value of clean spread options involving these allowances is expressed here in a formulation in which the demand for power and input fuel prices are the only factors with exogenous stochastic dynamics. The spot prices for power are based on a structural model for the *bid-stack*, in which bids to a central electricity delivery operator from generators are taken in bid price order, lowest (and least carbon emitting) first. The emission allowances are derivatives on a non-traded underlying. Their prices are computed using a forward backward stochastic differential system solved by applying a finite difference scheme to a multi-dimensional PDE. Monte Carlo techniques are used to price clean spread options on them. These methods are used by the authors to investigate four indicative case studies. First, they compare the prices of spark and dark spread options in a market with no emission regulation with those of a cap-and-trade system in which the cap is increasingly strict. Next they analyze the effects that different fuel-type merit schemes for production have on option prices, and in the third case study they favourably compare option prices obtained with their structural scheme with those obtained from two alternative reduced-form schemes. Finally, they favourably compare from a policy viewpoint a cap-and-trade scheme using their structural model with a *fixed* carbon tax. The authors note the importance of the currently increased global market volatility in the price factors used in their study for designers of an increasing number of carbon emission regulation schemes. They conclude that their four case studies address the important considerations needed to understand the complex joint dynamics of electricity, emissions and fuels.

Medina and Pardo, in the final chapter of this section of the volume, Chapter 28, conduct an empirical study of the properties of the futures prices of EU allowances (EAUs) traded on the EU Emission Trading Scheme, with a view to finding out if their stylized properties can be considered similar to financial assets, like either stocks and bonds or commodity futures. The answer is preliminary to addressing the theme of Section 3, whether or not EU allowances in the world's most developed cap-and-trade market (and those of other schemes) can be considered a diversifying asset for global investors. The authors make use of previous work on the stylized properties of financial assets (Cont, 2001) and futures (Gorton et al., 2013) using daily data on EAU futures from the European Climate Exchange (ECX). Given that Phase 1 of the EU cap-and-trade scheme from 2005 to 2007 was considered to be a pilot and learning scheme, they concentrate on 2008, 2009 and 2010 EUAs of Phase 2 (which could be carried forward into Phase 3 starting in 2013) and utilize futures data on them from April 2005 to the December date of the allowance maturity for each year. They also use the Euro Stoxx 50 index and its futures, and fixed income futures for short, medium and long-term bond indices from the EUREX market, and Brent oil futures from the International Petroleum Exchange (IPE). One-month Euribor is taken as the risk-free return over the data period and inflation is measured by the European Harmonized Index of Consumer Prices (HICP). Computing return histograms and using well known statistical methods for EUA futures returns, including GARCH modelling for their volatility, the authors found evidence for intermittency, short-term predictability, long-term memory (Taylor effect), volatility clustering, asymmetric volatility and positive relationships between volatility and trading volume and open interest. As a result, they feel that temporal dependencies should be modelled using ARMA-GARCH techniques and caution that the persistence of heavy tails for conditional returns implies that any unconditional return distributions should be fat-tailed. Unlike commodity futures returns, but like those of stocks and bonds, the study found negative asymmetry, positive correlation with equity index returns and higher volatility during trading sessions than when the allowance exchange is closed. Like commodity futures, the properties of inflation hedging and positive return correlations with bonds were found. The authors conclude that EUAs do *not* behave like either common commodity futures or financial assets, but have selected features of both, to form a new asset class with implications for volatility modelling, hedging use, portfolio analysis and cointegration of the carbon market with other economic and financial variables.

Section 5 Contemporary Topics

The five chapters of the final section of this volume treat topics of contemporary interest since around the 2007–2008 financial crisis.

In Chapter 29, Gatheral, Jaisson and Rosenbaum revisit the question of the smoothness of the volatility process using recent high frequency data to find that on any reasonable time scale log-volatility behaves essentially like fractional Brownian motion with Hurst exponent H of order 0.1. They term their fractional stochastic volatility model *Rough FSV* (RFSV) to emphasise that its Hurst exponent H is strictly less than 1/2 and show that based on time series data, with which it is remarkably consistent, it improves forecasts of realized volatility. Further, they note that while the RFSV model is *not* long memory, classical statistical procedures aimed at detecting volatility persistence tend to find the presence of long memory in data generated from it to explain why the long memory of volatility has been widely accepted as a stylized fact. A quantitative market microstructure-based foundation for their findings relates the roughness of volatility to high frequency trading and order splitting.

Abergel, Huré and Pham set out in Chapter 30 a microstructural modelling framework for studying optimal market making policies in a *FIFO* (first in first out) limit order book. Limit order, market order and cancel order arrivals in the order book are modelled with a point process whose intensities depend only on the state of the order book. These recently developed high-dimensional policy models are realistic from a micro-structure viewpoint. They consider a market maker which stands ready to buy and sell stock on a regular and continuous basis at a publicly quoted price and identify strategies that maximize their profit and loss penalized by their inventory. An extension of the methodology is proposed to solve market-making problems in which order arrivals are modelled by Hawkes processes with an exponential kernel. Markov decision processes and dynamic programming are used to characterize the solutions to the optimal market-making problem analytically. The chapter goes on to treat the numerical aspects of the high-dimensional trading problem using controlled randomization and quantization to compute the optimal strategy analytically. Several computational tests are performed on simulated data to illustrate its efficiency. Naive strategies with constant, symmetrical, asymmetrical and state dependent intensities are compared with the optimal strategy.

In the Editors' opinion, Manahov gives the best currently available exposition of cryptocurrencies in Chapter 31. The enormous rise of cryptocurrencies over the last few years has created one of the largest unregulated markets in the world. Millisecond data for five major cryptocurrencies – Bitcoin, Ethereum, Ripple, Litecoin and Dash – and two indices – Crypto Index (GRIX) and CC130 Crypto Currencies Index – is used to investigate the relationships between cryptocurrency liquidity, herding behaviour and profitability during periods of extreme price movements (EPMs). Cryptocurrency traders (CTs) are seen to facilitate EPMs and to demand liquidity during even the worst EPMs and the presence of herding behaviour across the entire data set is observed. Robustness checks indicate that herding behaviour follows a dynamic pattern that varies over time with decreasing magnitude. Novel evidence is provided of CTs' profitability after transactions costs and the likelihood of strong profitability-generation in the future is demonstrated.

Li, Li, Zhu and Yao in Chapter 32 identify, through a new text mining approach, the influential factors in six Chinese commodity futures markets, principally in Shanghai, Shenzen and Hong Kong, which deal in agricultural commodity, metal and energy futures. How to determine the numerous and complicated factors that affect the dynamics of commodity futures prices is still a great challenge. A new text mining method, termed Dependency Parsing-Sentence-Latent Dirichlet Allocation (DP-Sent-LDA), uses 49,501 news headlines about six Chinese commodity futures prices over the period 2011–2018 to identify the influential factors for agricultural commodity, metal and energy futures. The identified influential factors are not only those widely studied, but also some rarely mentioned in the existing literature. Regression analysis is conducted to validate the effectiveness of the influential factors found.

Market flash crashes have been a feature of major financial markets since the period around the 2007–2008 financial crisis. Many of these are classified by Wehrli and Sornette in Chapter 33, the final chapter of this second edition volume. They introduce and parsimoniously disentangle and quantify the time-varying share of high frequency financial price changes that are due to endogenous feedback processes and not exogenous impulses. They show how both flexible exogenous arrival intensities and a time-dependent feedback parameter can be estimated in a structural manner using an Expectation Maximization (EM) algorithm. The EM algorithm is used to investigate potential characteristic signatures of anomalous market regimes in the vicinity of flash crashes – events in which prices exhibit highly irregular and cascading dynamics. The study covers some of the most liquid financial markets: equity and bond futures, foreign exchange and cryptocurrencies in the US, UK, Japan and Switzerland. Systematically balancing the degrees of freedom of both exogenously driven processes and endogenous feedback variation, it is shown that the dynamics around such events are not universal. This highlights the usefulness of the approach: (1) post-mortem, for developing remedies and better future processes – for example, developing circuit breakers and latency floor designs – and potentially ex ante, for short-term forecasts during exogenously driven events. The proposed model is tested against a process with a refined treatment of exogenous clustering dynamics using an auto-regressive moving average (ARMA) point process.

References

Ang, A & M Piazzesi (2003). A no-arbitrage vector autoregression of term structure dynamics with macroeconomic and latent variables. *Journal of Monetary Economics* **50** 745–787.

Benth, FE, JS Benth & S Koekebakker (2008). *Stochastic Modelling of Electricity and Related Markets*. World Scientific, Singapore.

Benth, FE, VA Kholodnyi & P Laurence (2014). *Quantitative Energy Finance: Modelling, Pricing, and Hedging in Energy and Commodity Markets*. Springer, Berlin.

Black, F & M Scholes (1973). The pricing of options and corporate liabilities. *Journal of Political Economy* **81** 637–659.

Black, F (1976). The pricing of commodity contracts. *Journal of Financial Economics* 3167–3179.

Brennan, M (1958). The supply of storage. *American Economic Review* **48** 50–72.

Carmona, R & N Touzi (2008). Optimal multiple stopping and valuation of swing options. *Mathematical Finance* **18** 239–268.

Casassus, J & P Collin-Dufresne (2005). Stochastic convenience yield implied from commodity futures and interest rates. *Journal of Finance* **60** 2283–2331.

Chang, E (1985). Returns to speculators and the theory of normal backwardation. *Journal of Finance* **40** 193–208.

Clark, E, J-B Lesourd & R Thieblemont (2001). *International Commodity Markets: Physical and Derivative Markets*. Wiley, New York.

Cont, R (2001). Empirical properties of asset returns: Stylized facts and statistical issues. *Quantitative Finance* **1** 223–236.

Cox, JC & SA Ross (1976). The valuation of options for alternative stochastic processes. *Journal of Financial Economics* **3** 145–166.

Dempster, MAH & V Leemans (2006). An automated FX trading system using adaptive reinforcement learning. *Expert Systems with Applications* **30** 543–552.

Dempster, MAH, IV Evstigneev & KR Schenk-Hoppe (2011). Growing wealth with fixed mix strategies. In: *The Kelly Capital Growth Investment Criterion*, LC Maclean, EO Thorpe and WT Ziemba, eds. World Scientific, Singapore, 427–458.

Dempster, MAH & K Tang (2011). Estimating exponential affine models with correlated measurement errors: Applications to fixed income and commodities. *Journal of Banking and Finance* **35** 639–652.

Geman, H (2005). *Commodities and Commodity Derivatives: Modelling and Pricing for Agriculturals, Metals and Energy*. Wiley, New York.

Gibson, R & E Schwartz (1990). Stochastic convenience yield and the pricing of oil contingent claims. *Journal of Finance* **45** 959–976.

Gorton G, F Hayashi & K Rouwenhorst (2013). The fundamentals of commodity futures returns. *Review of Finance* **17** 35–105.

Henderson, BJ, ND Pearson & L Wang (2012). New evidence on the financialization of commodity markets. *Review of Financial Studies* **28**(5) 1285–1311.

Kaldor, N (1939). Speculation and economic stability. *The Review of Economic Studies* **7**(1) 1–27.

Kendall, MG & A Bradford Hill (1953). The analysis of economic time-series – Part I: Prices. *Journal of the Royal Statistics Society. Series A (General)* **41**(5) 11–34.

Keynes, JM (1923). *A Tract on Monetary Reform*. Good Press.

Jaillet, P, EI Ronn & S Tompaidis (2004). Valuation of commodity-based swing options. *Management Science* **50** 909–921.

Merton, RC (1973). An intertemporal capital asset pricing model. *Econometrica* **41**(5) 867–887.

Pirrong, C (2012). Clearing and collateral mandates: A new liquidity trap? *Journal of Applied Corporate Finance* **24**(1) 67–73.

Rockwell, GL (1967). *White Power*. Ragnarok Press.

Routledge, B, DJ Seppi & CS Spatt (2000). Equilibrium forward curves for commodities. *Journal of Finance* **55**(3) 1297–1338.

Samuelson, PA (1965). A theory of induced innovation along Kennedy-Weisäcker lines. *The Review of Economics and Statistics* **47**(4) 343–356.

Schwartz, ES (1997). The stochastic behavior of commodity prices. *Journal of Finance* **52**(3) 923–953.

Tang, K & W Xiong (2012). Index investment and financialization of commodities. NBER Working Paper 16385.

Working, H (1949). The theory of the price of storage. *American Economic Review* **39** 1254–1262.

1

Oil Products

1

The Volatility Risk Premium in the Oil Market

Ilia Bouchouev*
and Brett Johnson

1.1 Introduction*

Commodity options resemble insurance contracts. Option buyers pay the premium to protect themselves against the adverse price movement. Producers buy puts to ensure minimum acceptable return on investments in capital intensive projects; consumers buy calls to protect against price appreciation of raw materials essential to their manufacturing process or to the services they provide. Sellers of commodity options are dealers, institutional investors, and professional volatility traders acting as insurers in anticipation of reward for providing the service of risk absorption.

Like any other insurance product, writing commodity options carries asymmetric risks. Small frequent gains in the form of premium collected create a buffer to offset the impact of rare but potentially large losses. To keep providers motivated to stay in business with such asymmetric payoffs, prices must be set at a premium to the contract actuarial value, or its average historical payout. Despite similarities to the traditional insurance industry, commodity options have one important advantage. In the derivatives market, the asymmetric option risk can be partially offset by dynamically trading the futures contract, the instrument on which the option is written. This technique, which became known as delta-hedging, replaced actuarial pricing of options with the cost of their dynamic replication. Delta-hedging cannot eliminate the asymmetry of the payoff, but it transforms the strategy risk profile. The investment return on such a dynamically hedged short option strategy is referred to by traders as the volatility risk premium, or VRP.

The performance of the VRP strategy in equity index options has been documented in numerous articles, which we don't discuss here. In contrast, the studies of VRP in commodity options have so far

---

* Email: ib@pentinv.com

DOI: 10.1201/9781003265399-2

been much more limited. Among all commodities, the market for crude oil options stands out in terms of its depth, liquidity, and overall significance for investment and hedging portfolios. For over three decades, oil options have actively traded not only on the organized exchanges, but also in the over-the-counter (OTC) market, which provided end-users with the flexibility to tailor hedging strategies to the highly customized oil price exposure of their enterprises. The development of the OTC market spurred the rapid growth of standardized oil options on exchanges allowing dealers to spread out volatility risks across the broader group of financial speculators. Even though the oil options market is now one of the largest option markets in the world, its dynamics are still poorly understood by a broader audience.

It should not be surprising that the desire to manage the high volatility of oil prices resulted in the structural imbalance between the end-user demand for price protection in the form of oil options and volatility dealer willingness to provide such protection. To incentivize the dealers to participate in this business, option prices must contain some additional premium, or compensation to the option seller, for risking their capital. To extract this volatility risk premium, the structurally short option market-making portfolios are managed by dealers using proprietary delta-hedging techniques. It should be emphasized that delta-hedging is highly model-dependent and is by no means a risk-free process.

The existence of the structural imbalance and oil VRP has been known to practitioners since the inception of oil options trading. However, despite its broad acceptance in the oil industry, the academic literature on this topic has been remarkably scarce. Doran and Ronn (2008) documented negative VRP for oil and natural gas markets and estimated the market price of the volatility risk in the context of the parametric stochastic model that explained the higher implied volatilities observed in at-the-money (ATM) options relative to the realized volatility. Another study of WTI option prices across various moneyness and expirations by Trolle and Schwartz (2009) demonstrated that oil volatility is stochastic, and option prices cannot be spanned even by a very general multifactor model specification for the dynamics of futures prices. In other words, contrary to the traditional paradigm of delta-hedging, the volatility risk implicit in the prices of oil options cannot be fully hedged by trading only underlying futures. Option prices always carry something of their own which cannot be replicated with other factors.

To quantify the richness of energy options, many authors resorted to the theoretically elegant, but less practical, analysis of the so-called variance risk premia, following the approach commonly used in the equity markets. The variance risk premium is defined as the profitability of a long position in the variance swap which pays the difference between the realized variance and the fixed price of the swap. In the absence of arbitrage, the payoff of such swap can be approximately replicated using a strip of options with different strikes of the appropriate tenor. The fixed price of the swap, or the market implied variance, is the risk neutral expectation of the future realized variance, which is calculated using a well-known transformation of option prices. Such implied variance is then compared to the realized variance computed from the historical data to determine the variance risk premium. This approach does not rely on any assumption about the stochastic behavior of futures, and it is relatively straightforward to implement.

Trolle and Schwartz (2010) applied the variance swap approach to crude oil and natural gas markets, and Prokopczuk *et al.* (2017) extended it across all major commodity markets. Both studies highlighted the existence of large negative variance risk premia in the crude oil market. Kang and Pan (2015) developed the theoretical model based on the theory of hedging pressure that derives the negativity of the variance risk premium and connects it to the futures risk premium under the assumption that producers are dominant hedgers in both markets. They also attempted to use the variance risk premium to predict futures returns. However, this dynamic has likely changed with the arrival of financial investors and inflation hedgers, as explained in Bouchouev (2020). Unfortunately, the methodology of variance swaps remains largely a theoretical construct, as in the oil markets variance swaps don't actively trade. Even though the first oil volatility swap was introduced back in 2003,[*] unlike the equity markets, neither

---

[*] The first energy volatility swap was originated by the energy trader, Koch Industries, in a deal with the hedge fund, Centaurus, with its details described in Ockenden (2003).

volatility nor variance swaps have been embraced by the oil market. The only practical way to extract VRP is still by implementing the dynamic delta-hedging strategy which, of course, is no longer model-free, as it depends on parameters of the hedging strategy. The historical performance of selling dynamically delta hedged oil options has been analyzed by Bouchouev and Johnson (2021) across options for different moneyness and maturities. Following the idea of implied volatility smile, they introduced the concept of a VRP smile that highlights higher returns for selling out-of-the-money (OTM) options. These results are consistent with the work of Ellwanger (2017) who found that option market-implied measures of extreme oil events exceed the equivalent realized measures of such events, and with the recent presentation of Jacobs and Li (2021) who also calculated historical delta-hedged returns.

The present article extends the results of Bouchouev and Johnson (2021) in several dimensions. We attempt to cover the existing gap in the academic literature and provide a comprehensive summary of an oil VRP as an investment strategy, going beyond a simple calculation of risk premia and paying particular attention to risks and risk-adjusted returns. We address important data challenges, which are often circumvented in academic studies, and analyze various VRP portfolios across moneyness and maturities. The portfolio analysis, which highlights subtle but important diversification benefits, is necessary for the proper comparison of strategies across multiple time horizons. We also study the performance of optimized and more cost-effective VRP strategies that hedge less frequently, hedge only above certain delta thresholds, or hedge with deltas computed using different volatilities. Finally, the performance of VRP strategies is presented under different regimes driven by behavioral changes among large participants in the option market.

Whenever possible, we present our results using the terminology accepted in the marketplace. For consistency with market standards, but in contrast to many academic works, we define VRP as the profitability of an investment strategy of selling rather than buying delta-hedged options. This eliminates some confusion caused by the definition of VRP as the returns from buying an option, which tend to be negative, causing unnecessary presentation challenges when risk premia are said to become either more or less negative. Our definition of VRP also refers to the realized historical performance of the investment strategy, even though the same term is sometimes used in other studies in the context of unobservable expected returns.

Our paper is organized as follows. In Section 1.2, we present an overview of our underlying data, models and methodology. We address important challenges of data quality and explain our rationale for the selection of the lookback period. In contrast to methods relying on the aggregate implied variance, our moneyness specific analysis requires more granular data and we describe the process in more detail. We discuss the pros and cons of alternative approaches for maintaining option prices and implied volatilities for different moneyness, and we present our results using traders' preferred method of defining moneyness as a function of option deltas.

In Section 1.3, we study the performance of daily delta-hedged strategies across options with different moneyness and expiration dates. In this section, we first analyze statistical properties of the VRP returns on the trade level. We demonstrate higher profitability for selling OTM options and introduce the concept of VRP smile in the spirit of the implied volatility smile. We also construct the term-structure of the VRP. However, the risk-adjusted performance of VRP cannot be easily compared across multiple time horizons or on the trade level, due to the different number of open trades held by portfolios with different expiration dates. We construct the corresponding equity lines for continuously run VRP portfolios to which we can then consistently apply traditional metrics or the investment performance, such as the returns on risk, and the return on drawdowns. We compare and contrast the smile and the term-structure of VRP to the corresponding smile and term-structure of implied volatilities.

Section 1.4 can be viewed as the study of VRP robustness to model risk, where we discuss the performance sensitivity to some crucial parameters of the hedging strategy. The most common practical challenge faced by volatility traders is the choice of the hedging frequency. We provide transparency for the cost of hedging by tracking the average count of futures for each strategy and we show that on a cost-adjusted basis, one can optimize the performance of the strategy by hedging the portfolio less frequently. For completeness, we also present the results for the naked VRP strategy with no delta hedging, which, not surprisingly, is dominated by directional price risks. We discuss another practical variation of the hedging strategy: re-hedging the portfolio only if the futures price moves beyond a certain threshold. Both strategy modifications minimize transaction costs and result in a slight under-hedging of the strategy's gamma risk. Such strategies implicitly bet on short-term price mean-reversion, and are often utilized by volatility traders. Finally, we analyze the results of varying the volatility used to calculate the options' delta and find that more aggressive hedging using deltas computed with lower volatility can substantially improve VRP performance for OTM puts. These results should not be interpreted as universal guidance, as ultimately the optimal hedging strategy depends on many factors, including individual traders' short-term views and risk tolerance, execution capabilities, and the strategy's contribution to a broader portfolio.

In Section 1.5, we discuss the evolution of VRP over time, given some structural behavioral changes among large option hedgers. We highlight that the magnitude of VRP has structurally decreased from approximately 2014, which was driven by an increasing supply of oil insurance and the simultaneous reduction in demand for protection. The arrival of financial investors has caused a well-known structural break in the oil futures market which occurred around 2004, as documented, among others, by Tang and Xiong (2012), Hamilton and Wu (2014), and Bouchouev (2020). The financialization of the oil volatility market occurred approximately a decade later when the securitization of oil VRP in the form of investable indices has brought additional volatility selling from pension funds and other yield-seeking long-term investors. At the same time, the rapid growth of U.S. independent and highly leveraged shale producers led to significant changes in the structure of their hedging programs. Being cash-constrained, many independent producers have switched from buying outright price protection in the form of put options to volatility neutral structures financing the required premium by selling call options. Some producers even net sell oil volatility to the market, monetizing their real options to produce oil. The abundance of such natural call sellers in the options market made the upside tail risk of futures prices more contained and has resulted in a stronger performance of a VRP strategy composed of OTM calls relative to one that sells OTM puts.

We summarize our conclusions and discuss opportunities for further research in Section 1.6.

1.2 Data Description

We use daily Chicago Monetary Exchange (CME) settlements for WTI futures and option prices from January 1, 2000 until October 31, 2021. Even though WTI options were introduced in 1986, we chose to exclude the years prior to 2000 from our analysis due to concerns about insufficient liquidity and quality of the data especially for deep OTM options. The settlement prices are published for every option that has non-zero open interest. However, only a small portion of long-dated options trade daily, and settlement prices for OTM options provided by the CME are based on various interpolation algorithms using ATM and other nearby options traded on the same day.

Futures and options are listed with a monthly expiration schedule. We analyze strategies initiated at 20, 40, 60, 120, 180, and 240 business days to maturity (DTM) of the option, which corresponds to the first three, the sixth, the ninth and the twelfth nearby futures and options contracts. For our base set of strategies, each option is hedged daily using settlement futures prices, but in section 4 we also discuss strategies that hedge less frequently.

The listed oil options have an American exercise style. We use the industry standard Barone-Adesi and Whaley (1987) approximation for American options to back out implied volatilities. This approximation has proven to be very accurate for the oil markets, as the early exercise premium tends to be low, since oil volatility is substantially higher than the interest rate, and the remaining value of holding an in-the-money (ITM) option dominates any potential time value benefits from receiving the payoff sooner. The early exercise premium becomes material only for very long-dated deeply ITM options, which we don't consider here. Following the industry standard practice, we only use more liquid prices for OTM puts and OTM calls. Deep ITM options trade extremely infrequently and due to their high delta, they mostly trade in the form of so-called synthetic conversions which are based on the price of the corresponding OTM option and the put-call parity relationship.

To characterize options for different moneyness, we have studied the results for several alternative models for moneyness, as there is still no established consensus in the industry. The most straightforward definitions of moneyness are either as the spread between the strike price K, and the futures price F, or their ratio. These metrics have the advantage of transparency, as options with different strikes can be easily converted into their corresponding moneyness. This simple moneyness approach is particularly helpful when analyzing options for any fixed time horizon. However, the results become more difficult to compare across multiple expiry dates, as the range of available and liquid option strikes expands with increasing DTM. While the (0.8, 1.2) moneyness range is often adequate for short-maturity one-month options, for long-maturity twelve-month options, a much wider range of percentage moneyness, such as (0.50, 1.50), may be needed to capture all liquid options.

To overcome this problem, some normalization with respect to time to maturity is often applied. One approach is to use log-moneyness and normalize it by the square root of the annualized total variance as follows:

$$m = \frac{\ln\left(\dfrac{K}{F}\right)}{\sigma_{ATM}\sqrt{\tau}}$$

where σ_{ATM} is the ATM volatility of the option and τ is the time remaining to expiration date. In this case, the moneyness of the option is measured by the number of standard deviations that the option is away from the ATM value.

Practitioners, however, like to go one step further and convert the log-moneyness into the option delta. The rationale behind this approach is best illustrated using call deltas computed from the standard Black (1976) formula:

$$\Delta_{call} = e^{-r\tau}N\left(\frac{\ln\left(\dfrac{F}{K}\right)}{\sigma_K\sqrt{\tau}} + \frac{\sigma_K\sqrt{\tau}}{2}\right),$$

For puts:

$$\Delta_{put} = -e^{-r\tau}N\left(\frac{\ln\left(\dfrac{K}{F}\right)}{\sigma_K\sqrt{\tau}} - \frac{\sigma_K\sqrt{\tau}}{2}\right)$$

where σ_K is the volatility of the option with the strike K, and r is the risk-free interest rate. Deltas effectively convert time normalized log-moneyness into a more intuitive moneyness scale by applying the cumulative normal distribution function. For oil options, which have American exercise style, deltas cannot be expressed analytically, but in practice, they are very close to Black deltas and the rationale above still holds.

Defining option moneyness as a function of delta has proven to be very handy for volatility traders as the hedging ratio immediately follows from the definition of moneyness. This standardized metric is particularly helpful for analyzing volatilities across different asset classes. With the rapid growth in

TABLE 1.1 Average Premium Collected and Implied Volatility at Trade Entry

Entry DTM	Value	Delta								
		10P	20P	30P	40P	0S(*)	40C	30C	20C	10C
20	AvePrem ($/bbl)	0.3335	0.7418	1.2212	1.7883	2.3110	1.5308	1.0227	0.6044	0.2684
	AveImpliedVol	0.4172	0.3904	0.3767	0.3671	0.3602	0.3563	0.3542	0.3554	0.3685
40	AvePrem ($)	0.4894	1.0812	1.7823	2.5985	3.2660	2.0999	1.4028	0.8293	0.3650
	AveImpliedVol	0.4195	0.3912	0.3756	0.3647	0.3562	0.3509	0.3478	0.3484	0.3587
60	AvePrem ($)	0.5994	1.3291	2.1841	3.1730	3.9420	2.4905	1.6681	0.9847	0.4347
	AveImpliedVol	0.4078	0.3812	0.3651	0.3535	0.3446	0.3394	0.3360	0.3365	0.3452
120	AvePrem ($)	0.8630	1.8834	3.0814	4.4730	5.3630	3.2543	2.1742	1.2937	0.5793
	AveImpliedVol	0.3860	0.3618	0.3457	0.3324	0.3228	0.3156	0.3116	0.3115	0.3205
180	AvePrem ($)	1.0349	2.2763	3.7168	5.3653	6.3136	3.7232	2.4983	1.4984	0.6832
	AveImpliedVol	0.3700	0.3463	0.3293	0.3151	0.3065	0.2993	0.2951	0.2956	0.3047
240	AvePrem ($)	1.1693	2.5773	4.2019	6.0414	6.9914	4.0542	2.7076	1.6229	0.7358
	AveImpliedVol	0.3560	0.3330	0.3166	0.3019	0.2923	0.2855	0.2814	0.2829	0.2931

(*) Zero delta straddles (OS) premium is scaled by 0.50 for consistency with OTM puts and calls

financialization of commodity markets and the inclusion of oil into broader financial volatility portfolio, maintaining volatility by delta is becoming the preferred choice among oil traders, and for this reason, we chose it to be our primary metric. In this article, we use prices and volatilities for zero-delta straddle and 10, 20, 30, and 40 delta OTM puts and calls. We first construct implied volatility surfaces for given strikes and expirations along with deltas using the pricing model for American options and then iterate in the strike dimension to match the deltas in our selected grid.

To preserve the uniform scale in terms of delta, we prefer to use zero-delta straddle instead of ATM straddle ($K = F$) as the latter has positive delta driven by $\sigma_K \sqrt{\tau} / 2$ term in the Black formula, which is the artifact of the industry standard lognormal assumption. In practice, VRP results for zero-delta straddles and ATM straddles are nearly identical, especially for shorter term options. For longer-term options, the anomaly of log-normal deltas is usually corrected by traders using either the second-order term that captures the change in implied volatility with respect to futures, often called the skew delta, or applying more accurate distributional assumptions for the underlying stochastic process.[*] While we fully agree with the need to make delta corrections for longer-term options and vega trading, such adjustments are often highly customized by traders, and adding extra level of complexity would have moved us further away from the main objectives of this article. Therefore, to preserve the transparency, the moneyness throughout the paper is measured in terms of log-normal American deltas, which became the standard for communication among oil traders.

Table 1.1 provides the descriptive statistics of the data that shows the average option premia and implied volatilities for six maturities and nine different delta options. The prices and volatilities are taken at the initiation of each position and averaged across positions in the corresponding maturity/delta bucket. For zero-delta straddles, the option premia are halved for consistency with puts and calls, which is what will be subsequently used in our simulations. The option prices also correspond to average option premia collected at the initiation of the trade, which also defines the average maximum profit that can be achieved by selling the option.

Figure 1.1 presents implied volatilities versus moneyness, the graph known as the volatility smile.

---

[*] See, for example, Gatheral (2006), and Derman and Miller (2016).

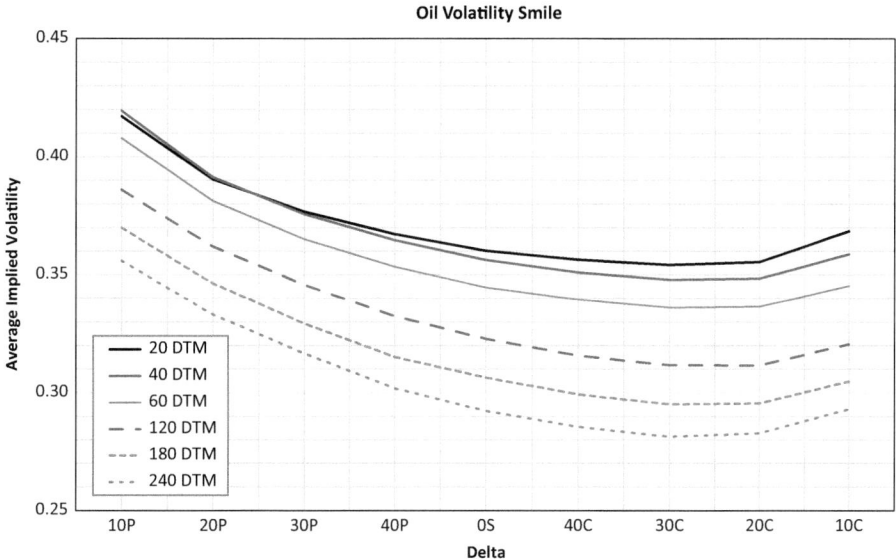

FIGURE 1.1 WTI average implied volatilities versus delta and maturity.

1.3 The Smile and Term-Structure of VRP

In this section we analyze the performance of a VRP strategy with daily delta-hedging across different moneyness and expirations. We can gain considerable clarity by splitting our analysis into two parts. In the first part, we focus on some key performance metrics for individual positions grouped by moneyness and time to maturity and characterize the probability distribution of profit and losses. In the second part, we extend the analysis to continuously run a systematic portfolio made up of such positions. The latter brings in subtle but important diversification benefits, as for any expiration longer than one month, the portfolio holds several positions at the same time. This allows us to compare the performance of different VRP portfolios consistently across multiple time horizons by applying traditional investment metrics for risk-adjusted returns.

1.3.1 VRP Position-Level Analysis

We simulate the historical performance of delta-hedged VRP strategies across various moneyness and maturities. Every option is delta-hedged daily using settlement prices for the underlying futures. Each moneyness and maturity combination comprises a distinct VRP strategy. Each strategy position involves selling 100 OTM put or call contracts consistent with the market standard size for a single transaction.* For zero-delta straddles, the strategy sells 50 put and 50 call contracts. Our first objective is to describe the average performance characteristics of the 262 positions comprising each VRP strategy (i.e., unique moneyness-maturity combination) between January 1, 2000 and October 31, 2021.

Table 1.2 summarizes the historical performance of 54 VRP strategies defined by selling 40, 30, 20 and 10 delta OTM puts and calls, as well as a zero-delta straddle, at 20, 40, 60, 120, 180, 240 days prior to the option expiry date with daily delta hedging. The average position profit-and-loss (P&L) is expressed in dollars, the metric commonly used by oil traders. The premium ratio retained is defined as the average position P&L scaled by the average premium collected. Premium retained is equivalent to the traditional definition of the return on the initial investment in an asset, taken with a negative sign to

---

* Each futures and option contract is for 1,000 barrels.

TABLE 1.2 Historical VRP Position Performance Versus Delta and Maturity (2000-2021)

Entry DTM	Value	Delta								
		10P	20P	30P	40P	0S	40C	30C	20C	10C
20	AveProfit ($)	8,357	9,212	6,749	7,727	8,946	11,370	14,505	15,118	12,379
	AvePremium ($)	33,347	74,182	122,122	178,828	231,096	153,084	102,267	60,442	26,839
	RatioRetained	0.2506	0.1242	0.0553	0.0432	0.0387	0.0743	0.1418	0.2501	0.4612
	AveFutures	71	128	169	201	174	213	177	128	65
40	AveProfit ($)	4,643	11,047	16,166	16,860	19,638	23,225	25,744	25,080	16,922
	AvePremium ($)	48,938	108,119	178,226	259,849	326,595	209,990	140,285	82,931	36,495
	RatioRetained	0.0949	0.1022	0.0907	0.0649	0.0601	0.1106	0.1835	0.3024	0.4637
	AveFutures	92	158	212	250	228	264	224	166	85
60	AveProfit ($)	1,886	14,802	20,186	24,339	30,144	32,260	31,441	26,454	19,675
	AvePremium ($)	59,942	132,910	218,406	317,296	394,201	249,045	166,812	98,472	43,468
	RatioRetained	0.0315	0.1114	0.0924	0.0767	0.0765	0.1295	0.1885	0.2686	0.4526
	AveFutures	105	179	239	280	262	301	265	199	105
120	AveProfit ($)	(9,889)	10,430	26,112	37,836	44,938	45,996	42,932	30,800	17,890
	AvePremium ($)	86,300	188,342	308,137	447,302	536,302	325,433	217,424	129,374	57,927
	RatioRetained	(0.1146)	0.0554	0.0847	0.0846	0.0838	0.1413	0.1975	0.2381	0.3088
	AveFutures	142	225	289	350	349	392	344	265	151
180	AveProfit ($)	(18,305)	6,532	25,213	36,621	42,777	44,385	37,137	25,781	11,953
	AvePremium ($)	103,494	227,631	371,676	536,533	631,362	372,322	249,830	149,835	68,324
	RatioRetained	(0.1769)	0.0287	0.0678	0.0683	0.0678	0.1192	0.1486	0.1721	0.1749
	AveFutures	166	263	341	410	408	441	391	307	192
240	AveProfit ($)	(25,860)	1,279	22,718	32,931	33,185	33,138	29,144	14,290	1,332
	AvePremium ($)	116,931	257,733	420,186	604,139	699,137	405,417	270,757	162,289	73,579
	RatioRetained	(0.2212)	0.0050	0.0541	0.0545	0.0475	0.0817	0.1076	0.0881	0.0181
	AveFutures	190	297	393	463	458	484	430	338	218

reflect the selling of an option, which makes our results directly comparable to other academic studies of investment returns.*

In this section we ignore the impact of transaction costs which will be discussed later, but we also report the average count of futures used for hedging the option, which will determine these transaction costs.

Figure 1.2 illustrates the premium retained for VRP strategies with different moneyness. For shorter-term options, a larger percentage of the premium can be collected by selling and delta-hedging OTM options. In other words, the investment returns are more negative for buying deep OTM options. The graph of the short-term VRP profitability versus moneyness looks like a smile which is more symmetric than the corresponding graph for implied volatility in Figure 1.1, as it appears to be more like a skew or a smirk.

The presence of a VRP smile reflects the insurance risk premium that end-users are willing to pay to ensure short-term protection provided by OTM puts and calls. While such protection is relatively inexpensive in dollar terms for hedgers to buy, it is most profitable for dealers to sell. This also means that the implied volatility smile cannot be explained solely by the distributional properties of the underlying

---

* We should emphasize that in the case of selling options, the return on the option premium is different from the return on cash investment. To calculate the latter, one must also incorporate the cost of the initial margin. This would be a rather difficult exercise as margin requirements are often adjusted by exchanges and clearing houses depending on prevalent volatilities and trade correlation with the rest of the traders' portfolio. Such analysis will inevitably include trader-specific cost of funding which is outside the scope of this article. Retaining the simple definition of investment returns from the long position, which is equivalent to the percentage of premium retained for the short position, allows us to directly compare our results to other studies.

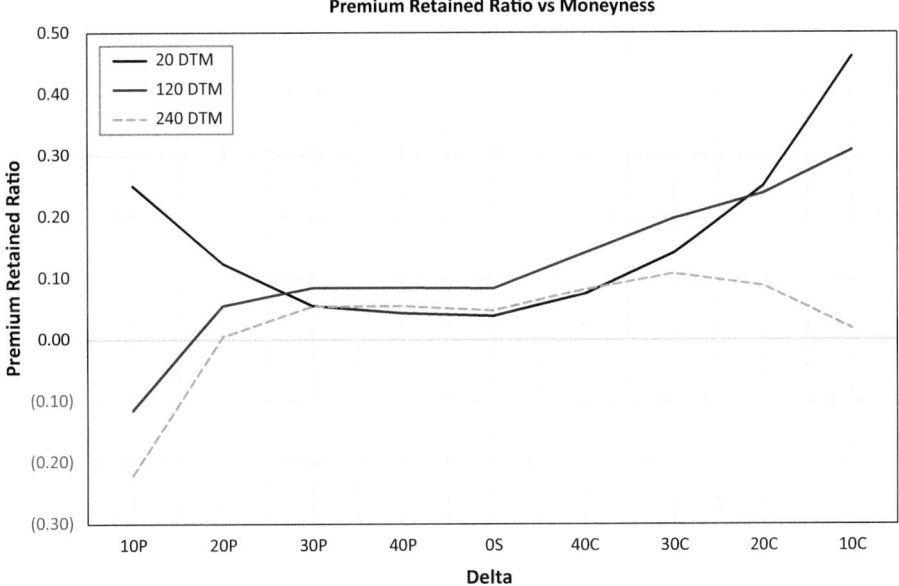

FIGURE 1.2 VRP smile, as the premium retained ratio versus moneyness.

futures prices as it also incorporates a risk premium. Alternatively, one can say that short-term options embed a jump risk premium which is not fully realized.

For medium-term expirations, the VRP smile flattens for the puts, but it retains its steepness for OTM calls. Such a shape of the VRP smile differs from the shape of the implied volatility smile, as it exhibits a strong put skew and sometimes leads to a false perception of puts being the most overpriced. As Figure 1.2 shows, puts are priced more fairly, and it is the risk premium for OTM calls that remains the highest, indicating that the market is structurally more concerned with unexpected upside spikes in oil prices. The premium for the calls is often driven by geopolitical uncertainty, which is rarely realized by being priced in the options market. For long-dated calls, the magnitude of the risk premium decreases, reflecting not only the impact of a larger denominator in the retained premium ratio, but, importantly, the lower dollar P&L as well. This often prompts traders to buy less overpriced longer-term options as a hedge against selling more overpriced shorter-term options. Such long-short investment portfolios can significantly improve VRP performance characteristics, but given their relative complexity, the proper analysis of such strategies is beyond the scope of the present chapter.

We should also highlight that despite the presence of the large put skew shown in Figure 1.1, selling medium and long-term deep OTM put options results in trading losses, or alternatively, buying them becomes profitable. However, these results should be interpreted with a degree of caution, because not all deep OTM options trade on a daily basis and settlement prices are often calculated using interpolated traded prices of other more liquid options. When such theoretically derived settlement prices for OTM options become relatively cheap, it becomes much more difficult to buy an option at the settlement price, as sellers often demand an additional excess premium, forcing the liquidity for these options to drop. This extra premium can remain invisible for a while until the new transaction occurs, but in the absence of new trades, it may not always be captured in interpolated prices for longer-dated OTM options. In other words, while it may appear that buying long-dated OTM put options is a profitable strategy, such a strategy can be difficult to execute in a systematic manner, since it requires consistent liquidity to be profitable.

Figure 1.3 displays the same results presented in the direction of time to maturity. For both OTM calls and OTM puts, such a term-structure of VRP, measured again by the premium retained ratio, resembles the term-structure of implied volatilities which steadily declines with increasing DTM. However, for zero-delta straddles the curve exhibits a pronounced hump indicating that selling medium-term

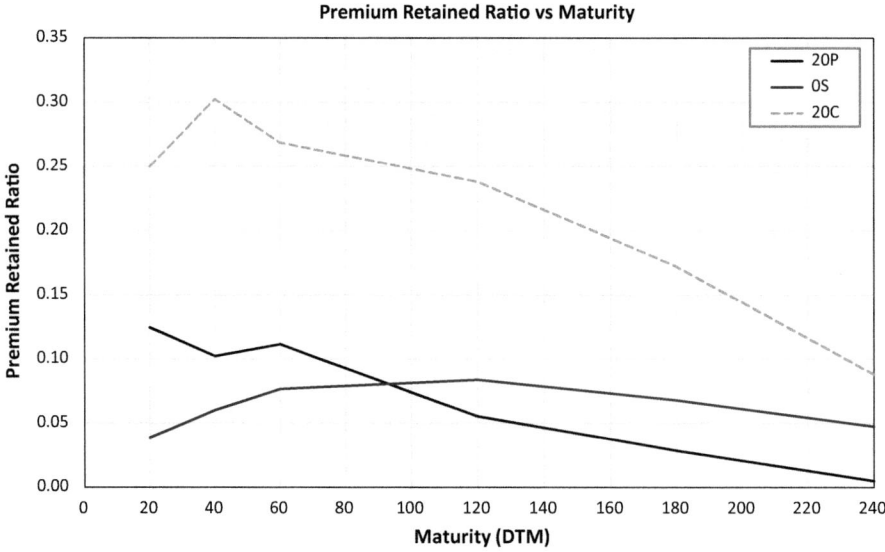

FIGURE 1.3 VRP term-structure, as the premium retained ratio versus maturity.

options is more profitable than selling either short-term options near ATM which have higher gamma or long-term options that maintain risk exposure for too long.

So far, we have only characterized the profitability or investment returns of VRP strategies, but we have not yet considered the corresponding risks. To analyze the uncertainty at the trade level, in Table 1.3 we present the summary statistics for the distribution of position P&L for each VRP strategy. As is typical for a short volatility insurance-like strategy, the distribution exhibits negative skewness. Skewness is generally most negative for OTM puts, which are less overpriced than OTM calls. Positive excess kurtosis results from the short gamma/volatility strategy, where frequent theta collection gains are countered by less frequent gamma losses resulting from significant price moves in the underlying futures. The result is fat tails relative to normally distributed returns. Excess kurtosis is generally higher for further OTM options, where the daily theta collection is lowest.

We define the position return on risk (PositionRoR) as the ratio of the mean of the distribution, which is the average position P&L, to its standard deviation, annualized by scaling with $\sqrt{252/DTM}$ to compare risk-adjusted performance across different expiry dates. However, we distinguish this metric from the traditional Sharpe ratio or the information ratio, both of which are more accurately estimated for continuously managed investment portfolios using annual samples. The annualization of risk-adjusted returns computed from lower frequency samples is prone to known statistical anomalies, and, for example, Lo (2002) highlighted the perils of simple time scaling of Sharpe ratios when portfolio returns are highly autocorrelated. Moreover, in our case the relative comparison of strategies across maturities becomes more complex due to multiple overlapping trades which remain open at the same time.

While PositionRoR provides useful guidance about VRP risk-adjusted performance at the individual position level, to properly characterize and compare systematic investment strategies, we analyze the portfolio of VRP trades typically implemented in practice.

1.3.2 VRP Portfolio-Level Analysis

To measure the relative performance of VRP strategies across different time horizons, we construct continuously run systematic strategies made up of corresponding VRP positions with the same moneyness and DTM at entry. Each shortest maturity (20 DTM) portfolio consists of only one position, as

TABLE 1.3 VRP Strategy Position P&L Moments and Risk Adjusted Returns

Entry DTM	Value	Delta								
		10P	20P	30P	40P	0S	40C	30C	20C	10C
20	Mean	8,357	9,212	6,749	7,727	8,946	11,370	14,505	15,118	12,379
	Stdev	42,304	57,209	61,880	68,356	74,959	71,880	64,186	58,431	38,369
	Skewness	(2.1842)	(0.9275)	(1.6909)	(1.4945)	(1.6901)	(1.5854)	(0.7478)	(0.2542)	(0.3387)
	Kurtosis	15.6393	9.9293	8.6272	6.3688	9.7332	9.9288	5.7263	6.3282	5.9776
	Mean/Stdev	0.1975	0.1610	0.1091	0.1130	0.1193	0.1582	0.2260	0.2587	0.3226
	PositionRoR	0.7012	0.5716	0.3871	0.4012	0.4236	0.5615	0.8022	0.9184	1.1452
40	Mean	4,643	11,047	16,166	16,860	19,638	23,225	25,744	25,080	16,922
	Stdev	61,603	72,907	76,658	84,339	90,664	87,054	79,169	64,308	46,055
	Skewness	(3.2199)	(1.1839)	(0.6298)	(0.6227)	(1.1953)	(0.6689)	(0.2524)	(0.1383)	0.2461
	Kurtosis	20.2597	6.8246	3.1288	3.8137	8.1517	5.7445	3.7710	3.4660	7.1273
	Mean/Stdev	0.0754	0.1515	0.2109	0.1999	0.2166	0.2668	0.3252	0.3900	0.3674
	PositionRoR	0.1892	0.3803	0.5293	0.5018	0.5437	0.6696	0.8162	0.9789	0.9223
60	Mean	1,886	14,802	20,186	24,339	30,144	32,260	31,441	26,454	19,675
	Stdev	75,059	83,783	85,726	98,281	101,924	98,664	91,775	75,865	53,029
	Skewness	(3.5203)	(1.8899)	(0.7641)	(0.6508)	(0.0460)	0.1894	0.5277	(0.0954)	0.3445
	Kurtosis	18.7074	8.9697	4.4346	5.0614	3.2773	1.7281	3.5462	2.6371	3.4825
	Mean/Stdev	0.0251	0.1767	0.2355	0.2476	0.2957	0.3270	0.3426	0.3487	0.3710
	PositionRoR	0.0515	0.3621	0.4826	0.5075	0.6061	0.6701	0.7021	0.7146	0.7604
120	Mean	(9,889)	10,430	26,112	37,836	44,938	45,996	42,932	30,800	17,890
	Stdev	111,255	122,008	139,728	153,959	152,534	149,263	136,474	109,249	74,043
	Skewness	(2.9828)	(1.4884)	(0.7173)	(0.2050)	0.0669	0.3583	0.4695	0.4301	(0.1309)
	Kurtosis	10.0738	5.0656	2.7528	2.7921	1.4638	1.2069	1.5743	2.5383	2.5662
	Mean/Stdev	(0.0889)	0.0855	0.1869	0.2458	0.2946	0.3082	0.3146	0.2819	0.2416
	PositionRoR	(0.1288)	0.1239	0.2708	0.3561	0.4269	0.4466	0.4559	0.4085	0.3501

(Continued)

TABLE 1.3 (CONTINUED) VRP Strategy Position P&L Moments and Risk Adjusted Returns

Entry DTM	Value	Delta								
		10P	20P	30P	40P	0S	40C	30C	20C	10C
180	Mean	(18,305)	6,532	25,213	36,621	42,777	44,385	37,137	25,781	11,953
	Stdev	133,697	157,954	180,565	198,822	206,286	197,140	178,757	145,107	98,207
	Skewness	(2.6240)	(1.8508)	(1.3169)	(0.8405)	(0.6885)	(0.4882)	(0.5387)	(0.5918)	(1.4988)
	Kurtosis	7.7134	5.5216	4.3623	3.5229	2.8925	2.7237	3.1465	3.8698	5.5818
	Mean/Stdev	(0.1369)	0.0414	0.1396	0.1842	0.2074	0.2251	0.2077	0.1777	0.1217
	PositionRoR	(0.1620)	0.0489	0.1652	0.2179	0.2454	0.2664	0.2458	0.2102	0.1440
240	Mean	(25,860)	1,279	22,718	32,931	33,185	33,138	29,144	14,290	1,332
	Stdev	146,706	192,384	227,306	246,762	250,297	2,46,075	230,831	193,505	135,682
	Skewness	(2.0798)	(1.5815)	(1.3246)	(1.3236)	(1.4009)	(1.4403)	(1.4453)	(1.8414)	(2.6022)
	Kurtosis	4.6866	4.4345	4.1327	4.1291	4.7098	5.2357	5.6685	7.2917	11.0706
	Mean/Stdev	(0.1763)	0.0067	0.0999	0.1335	0.1326	0.1347	0.1263	0.0739	0.0098
	PositionRoR	(0.1806)	0.0068	0.1024	0.1367	0.1359	0.1380	0.1294	0.0757	0.0101

the new trade is initiated within a few days of the expiration of the previous one. However, this is not the same for any other portfolio. The two-month (40 DTM) portfolio has two open trades with expiration twenty days apart, the three-month (60 DTM) portfolio has three open trades, and so on until the twelve-month (240 DTM) portfolio consists of twelve open trades. Only one option in each portfolio expires each month, so all portfolios are exposed to the same risk in the futures market at the expiration of each option. However, since a new option is added to the portfolio every month, the strikes of these options are spread out, providing some interesting diversification benefits for the portfolio of positions.

We construct continuous equity lines by cumulating daily P&L from all open positions within the portfolio. Table 1.4 presents traditional investment characteristics for each portfolio including the annualized P&L, its daily standard deviation, the return on risk (RoR),* and the return on maximum drawdown (RoD) which is often referred to as the return on stress risk.

In contrast to the previous analysis on the position level, where we have only studied the distribution of terminal P&L for individual VRP positions, here we look at the daily data for the strategy level equity lines. Using daily P&L fluctuations inevitably exposes that larger temporary drawdowns can occur throughout the life of the portfolio, even if some of such losses are subsequently recovered before the expiration of the portfolio positions.

Figures 1.4 and 1.5 illustrate portfolio RoR by moneyness and maturity. The shapes of these curves are similar to Figures 1.2 and 1.3, with the main difference coming from an improved performance for zero-delta straddles. Near ATM options, being highly exposed to large gamma risks, benefit the most from strike diversification. Such diversification contributes less to OTM options. On a risk-adjusted basis, OTM call portfolios generally outperform OTM put portfolios. This is likely explained by financial properties of the oil market and its correlation to equities which occasionally produces a negative spillover effect for oil, causing all asset class prices to fall rapidly. Unlike many other commodities, the return profile of oil futures is much closer to those observed in the equity markets, which have strong negative asymmetry, with frequent small gains countered by infrequent large losses. As a result, when large downside price gaps occur, spreading out put strikes provides less, rather than more, diversification, since they are more often simultaneously crossed by falling futures prices, which is the worst environment for selling put options, even on a delta-hedged basis.

Further insight can be gained by comparing portfolio level RoR to the position level RoR reported in Table 1.3. While both metrics exhibit significant similarities, the strike diversification effect resulting from overlapping positions improves portfolio RoR relative to position RoR in all cases, except for the shortest-maturity one-month portfolios, which only consist of a single open trade, and for deep OTM put options. For short-term portfolios, position RoR is slightly better than portfolio RoR as the former is computed using the position terminal P&L variance, while the portfolio metrics are constructed using daily equity price curves that often exhibit somewhat higher volatility. It is not surprising that many hedge funds and investment managers specializing in VRP strategies would rather report their results monthly to mask potentially large intra-month P&L fluctuations.

P&L fluctuations for a short gamma insurance-like strategy can be particularly nerve-racking for investors in VRP strategies, as they gradually accumulate theta gains in order to be able to sustain occasional significant gamma losses. Being able to absorb such losses is the essential prerequisite for VRP investors forcing them, who are forced to pay particular attention to anthe alternative metric of return on the maximum drawdown (RoD) as one of thea primary investment characteristics. As can be seen from Table 1.4, even though the best RoR areis generated forby the shortest one-month OTM call portfolio, RoD couldan be substantially improved by trading two- and three-month VRP call portfolios.

---

* Our definition of return on risk is the same as the information ratio with zero performance benchmark, or the Sharpe Ratio with zero risk-free interest rate, but both the numerator and the denominator in our case are computed in dollars, not in percent. Since we don't include the cost of capital and exchange margin in our calculations, we don't include any interest that could be received on such initial margin either. Overall, the impact of financing on the performance of VRP strategy is relatively minor.

TABLE 1.4 Portfolio Level VRP Strategy Performance

Entry DTM	Value	Delta								
		10P	20P	30P	40P	0S	40C	30C	20C	10C
20	AnnualizedPL ($)	100,644	110,946	81,285	93,061	107,743	136,944	174,701	182,074	149,091
	MaxDrawdown ($)	(778,885)	(830,918)	(1,070,249)	(1,274,355)	(1,658,357)	(1,314,051)	(1,174,683)	(959,191)	(491,929)
	DailyStdev ($)	13,241	14,401	15,519	16,812	17,453	16,797	15,122	12,549	8,272
	PortfolioRoR	0.4788	0.4853	0.3300	0.3487	0.3889	0.5136	0.7277	0.9140	1.1354
	PortfolioRoD	0.1292	0.1335	0.0759	0.0730	0.0650	0.1042	0.1487	0.1898	0.3031
40	AnnualizedPL ($)	55,915	133,047	194,696	203,053	236,517	279,718	310,052	302,062	203,805
	MaxDrawdown ($)	(1,708,293)	(1,657,691)	(1,469,025)	(1,786,199)	(1,715,953)	(1,432,181)	(1,034,110)	(755,840)	(429,220)
	DailyStdev ($)	18,613	21,460	23,281	24,676	25,694	24,418	22,082	18,794	12,212
	PortfolioRoR	0.1892	0.3906	0.5268	0.5184	0.5799	0.7216	0.8845	1.0125	1.0513
	PortfolioRoD	0.0327	0.0803	0.1325	0.1137	0.1378	0.1953	0.2998	0.3996	0.4748
60	AnnualizedPL ($)	22,714	178,270	243,120	293,133	363,042	388,532	378,671	318,600	236,967
	MaxDrawdown ($)	(2,046,489)	(1,935,929)	(1,637,852)	(1,919,368)	(1,636,007)	(1,422,527)	(1,141,153)	(1,037,781)	(584,951)
	DailyStdev ($)	23,785	27,308	29,578	31,711	31,729	31,142	28,723	25,287	18,269
	PortfolioRoR	0.0602	0.4112	0.5178	0.5823	0.7208	0.7859	0.8305	0.7937	0.8171
	PortfolioRoD	0.0111	0.0921	0.1484	0.1527	0.2219	0.2731	0.3318	0.3070	0.4051
120	AnnualizedPL ($)	(119,095)	125,614	314,486	455,685	541,222	553,971	517,058	370,950	215,465
	MaxDrawdown ($)	(5,353,575)	(4,210,027)	(4,336,461)	(4,337,404)	(4,204,970)	(3,966,196)	(3,540,122)	(3,112,739)	(2,429,636)
	DailyStdev ($)	49,557	51,635	53,322	55,453	55,278	54,020	51,102	45,742	34,868
	PortfolioRoR	(0.1514)	0.1532	0.3715	0.5177	0.6168	0.6460	0.6374	0.5109	0.3893
	PortfolioRoD	(0.0222)	0.0298	0.0725	0.1051	0.1287	0.1397	0.1461	0.1192	0.0887
180	AnnualizedPL ($)	(220,465)	78,674	303,654	441,056	515,201	534,561	447,264	310,496	143,953
	MaxDrawdown ($)	(7,835,172)	(6,339,726)	(6,896,521)	(7,305,140)	(7,471,086)	(7,069,798)	(6,523,592)	(5,823,212)	(5,071,743)
	DailyStdev ($)	70,333	77,809	82,809	84,572	84,074	81,981	78,076	67,735	50,616
	PortfolioRoR	(0.1975)	0.0637	0.2310	0.3285	0.3860	0.4108	0.3609	0.2888	0.1792
	PortfolioRoD	(0.0281)	0.0124	0.0440	0.0604	0.0690	0.0756	0.0686	0.0533	0.0284
240	AnnualizedPL ($)	(310,260)	15,351	272,566	395,100	398,152	397,583	349,664	171,455	15,986
	MaxDrawdown ($)	(10,302,199)	(8,495,851)	(9,546,062)	(10,364,112)	(10,826,782)	(11,003,556)	(10,730,515)	(10,065,658)	(7,550,030)
	DailyStdev ($)	93,448	103,900	112,322	114,716	116,005	112,305	104,666	93,791	109,085
	PortfolioRoR	(0.2091)	0.0093	0.1529	0.2170	0.2162	0.2230	0.2104	0.1152	0.0092
	PortfolioRoD	(0.0301)	0.0018	0.0286	0.0381	0.0368	0.0361	0.0326	0.0170	0.0021

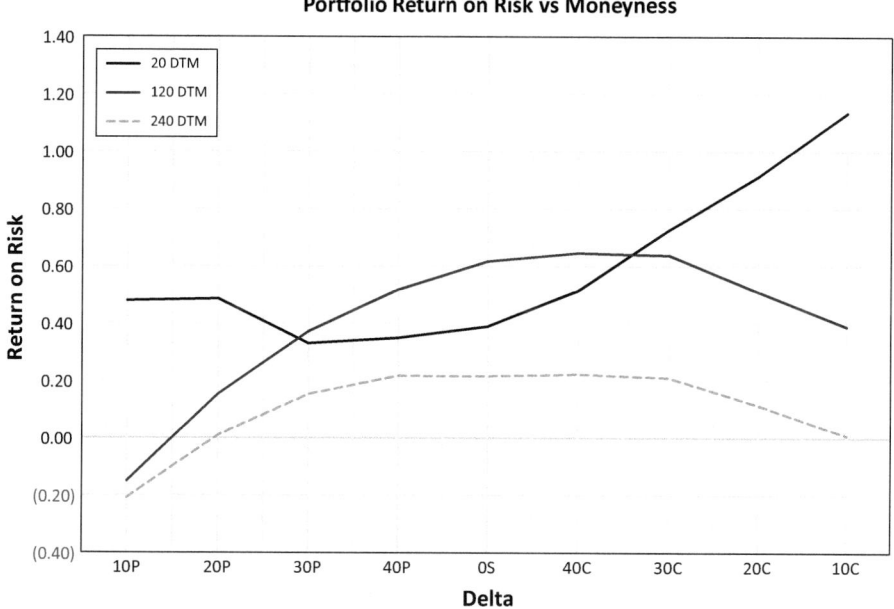

FIGURE 1.4 The return on risk (RoR) of VRP portfolios by moneyness.

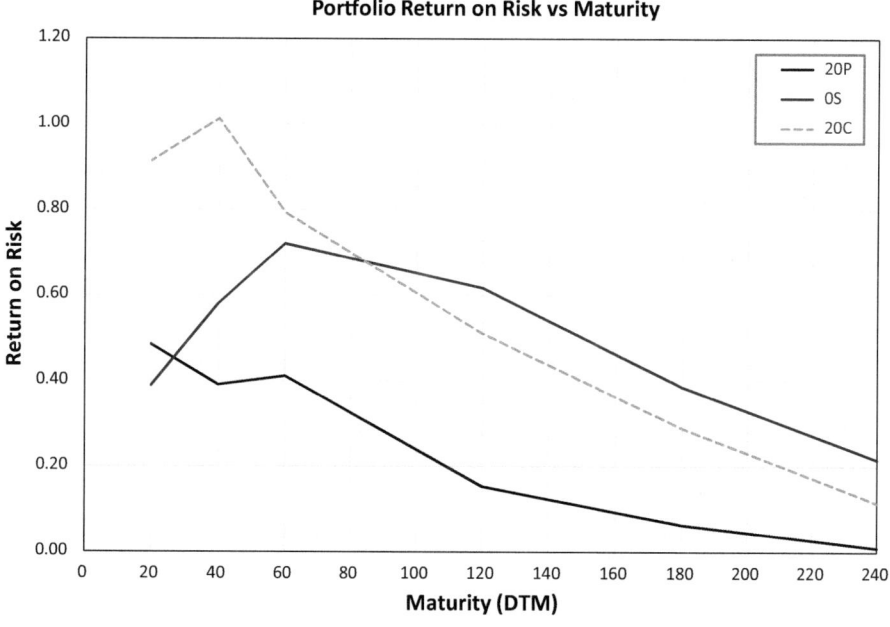

FIGURE 1.5 The return on risk (RoR) of VRP portfolios by maturity.

1.4 Delta-Hedging in Practice and Transaction Costs

In this section, we discuss some important practical aspects related to the implementation of a VRP investment strategy. As for any systematic strategy that trades relatively frequently, the question of bid-ask and transaction costs comes to the fore. In many academic studies, this topic is often sidestepped with simplistic assumptions about the execution slippage, the price impact of which is typically hidden within the final strategy output. We take a different approach, which provides users with more transparency regarding the cost of delta-hedging by explicitly tracking the quantity of futures required to hedge each configuration of the strategy. This allows one not only to see the true costs of dynamic hedging, but also to consider some practical alternatives for improving them. We also believe that hedging costs for VRP strategies vary among traders, who depend both on the execution tools available to them and on the particulars of the delta-hedging techniques they utilize.

One important feature of the oil futures market, that keeps the cost of delta-hedging relatively low, is the existence of a so-called trade-at-settlement (TAS) contract that allows traders to execute futures at a price to be determined later in the daily settlement window. This TAS contract trades actively for the nearest maturity futures throughout the day, and for the most part, the trader can execute any given number of futures with effectively a zero bid-ask spread. The TAS contract has proven to be handy for option traders who rebalance their portfolio at settlement prices. One challenge with hedging an options portfolio using a TAS contract arises from uncertainty in the quantity of futures required for rebalancing, as the settlement delta won't become known until the settlement price is known. In practice, traders dynamically adjust TAS orders throughout the day, based on updated estimates of the end-of-the-day delta, and then clean up the residual futures once the settlement price is published. This technique usually allows the trader to keep the average delta-hedging slippage down to fraction of one cent per barrel. For a short-term VRP strategy, one can use a rather conservative assumption of one cent per barrel for futures costs, which can then be easily multiplied by the reported futures count. However, liquidity of a TAS contract worsens with increasing DTM, and an alternative technique is to execute futures as close to their settlement prices as possible.

Since with the short-gamma strategy one always buys futures after they rally and sells futures after they drop, many traders have the natural temptation to cut some corners and hedge less, either by hedging less frequently or by trading fewer futures than are required by the portfolio's delta. The idea behind the deployment of such reduced hedging strategies is not only to minimize transaction costs, but also to profit from at least a partial price reversal following a large move on the previous day. Such delta-hedging techniques, with an embedded speculative bet on short-term price mean-reversion, are quite common among short-gamma traders, even though they embody significantly higher risks. The question is whether or not such risks are worth taking? To illustrate, we examine the performance of three popular hedging strategies in more detail.

The first strategy replaces daily rebalancing with hedging only every N days. Table 5 shows the results for this hedging strategy for several benchmark moneyness points and for $N = 1, 3, 5$ and 10. $N = 1$ corresponds to our base case of daily hedging. As above, we report the strategy statistics without transaction costs and show the futures count separately. For example, if the futures count is 100, then a transaction cost of one cent per barrel would reduce annualized P&L by $1,000, which can be easily scaled to reflect a trader's own assessment of execution slippage.

We pay particular attention to the performance of short-term OTM call strategies which have the best performance characteristics with daily hedging. It is not surprising that with less frequent hedging, the risks measured by daily portfolio standard deviation are increasing. More interestingly, for most hedging frequencies up to approximately $N = 5$, i.e., for hedging only once a week, profits are also increasing, effectively confirming additional gains from short-term price reversals. The increase in profits is sufficient for the strategy to maintain similar, or in some cases even higher, portfolio RoR. In other words, taking such risks appears to be well justified. In addition, the average position futures count for a weekly hedging strategy drops by approximately 40%-50% relative to daily hedging, substantially reducing transaction costs. Note that for strategies that wait longer than a week before rebalancing, such as for $N = 10$, increasing risks start dominating and the performance deteriorates significantly.

Somewhat inspired by the benefits from reduced delta-hedging, one can also wonder what are the implications of leaving options completely unhedged? We report the results of such an unhedged VRP strategy that does not trade any futures in Table 1.6. Clearly, this strategy won't be acceptable to most

TABLE 1.5 VRP Portfolio Performance Sensitivity to Hedging Frequency

Entry DTM	Value	Hedge Frequency											
		1			3			5			10		
		25P	0S	25C	25P	0S	25C	25P	0S	25C	25P	0S	25C
20	AnnualizedPL ($)	83,377	107,743	186,985	37,844	82,109	237,183	103,422	175,189	296,012	24,801	113,486	280,286
	MaxDrawdown ($)	(995,001)	(1,658,357)	(1,070,949)	(2,089,091)	(2,064,050)	(821,884)	(1,442,681)	(1,835,150)	(842,344)	(2,287,891)	(1,651,600)	(874,191)
	DailyStdev ($)	14,983	17,453	14,072	21,052	24,351	19,448	24,551	26,729	21,322	32,380	33,246	25,158
	PortfolioRoR	0.3506	0.3889	0.8370	0.1132	0.2124	0.7683	0.2654	0.4129	0.8745	0.0483	0.2150	0.7018
	AvePosFutures	149	174	153	107	109	105	87	81	84	66	45	65
40	AnnualizedPL ($)	168,989	236,517	301,932	160,887	284,177	380,200	193,729	327,418	424,231	120,531	282,206	398,821
	MaxDrawdown ($)	(1,613,421)	(1,715,953)	(883,850)	(2,048,783)	(1,161,158)	(1,237,052)	(1,871,284)	(1,361,670)	(1,154,414)	(3,013,281)	(1,520,737)	(1,458,675)
	DailyStdev ($)	22,563	25,694	20,771	28,559	32,260	27,011	34,068	35,856	29,182	41,107	43,370	34,226
	PortfolioRoR	0.4718	0.5799	0.9157	0.3549	0.5549	0.8867	0.3582	0.5752	0.9158	0.1847	0.4099	0.7341
	AvePosFutures	187	228	198	128	145	132	106	113	108	84	78	84
60	AnnualizedPL ($)	205,554	363,042	346,666	250,971	456,951	437,792	196,082	399,123	431,406	115,689	311,473	362,623
	MaxDrawdown ($)	(1,822,388)	(1,636,007)	(1,266,063)	(1,643,454)	(1,768,045)	(1,491,481)	(2,273,527)	(2,323,572)	(1,603,291)	(3,734,547)	(2,507,141)	(1,835,114)
	DailyStdev ($)	28,589	31,729	27,493	34,201	37,911	33,501	39,923	41,787	36,976	48,500	49,663	43,316
	PortfolioRoR	0.4529	0.7208	0.7943	0.4623	0.7593	0.8232	0.3094	0.6017	0.7350	0.1503	0.3951	0.5274
	AvePosFutures	211	262	236	141	166	155	118	134	128	95	99	100
120	AnnualizedPL ($)	225,584	541,222	445,464	277,663	602,849	529,728	248,814	609,470	569,460	66,107	433,291	492,909
	MaxDrawdown ($)	(4,291,257)	(4,204,970)	(3,310,935)	(4,046,015)	(4,166,870)	(3,202,401)	(4,882,057)	(4,625,190)	(3,360,714)	(6,903,246)	(5,147,059)	(3,728,735)
	DailyStdev ($)	52,123	55,278	48,958	62,757	61,816	54,799	69,901	67,709	59,488	82,789	77,932	66,793
	PortfolioRoR	0.2726	0.6168	0.5732	0.2787	0.6143	0.6089	0.2242	0.5670	0.6030	0.0503	0.3502	0.4649
	AvePosFutures	258	349	309	170	220	200	141	175	163	114	132	128
180	AnnualizedPL ($)	218,092	515,201	394,418	318,283	640,104	495,210	295,064	600,263	536,988	97,053	472,351	461,595
	MaxDrawdown ($)	(6,629,095)	(7,471,086)	(6,221,338)	(6,270,025)	(7,035,116)	(6,181,574)	(7,109,295)	(7,899,636)	(6,952,844)	(8,510,214)	(8,916,496)	(7,568,366)
	DailyStdev ($)	80,655	84,074	73,659	93,166	92,905	80,237	101,220	99,663	85,576	116,775	114,329	95,250
	PortfolioRoR	0.1703	0.3860	0.3373	0.2152	0.4340	0.3888	0.1836	0.3794	0.3953	0.0524	0.2603	0.3053
	AvePosFutures	303	408	352	195	252	225	160	202	183	129	152	143
240	AnnualizedPL ($)	174,221	398,152	269,221	325,302	517,272	372,071	264,854	455,321	375,151	65,177	204,234	269,418
	MaxDrawdown ($)	(8,960,182)	(10,826,782)	(10,494,962)	(8,572,752)	(10,494,808)	(10,329,548)	(9,229,062)	(11,725,266)	(11,381,168)	(10,836,351)	(13,260,748)	(12,316,693)
	DailyStdev ($)	109,399	116,005	99,664	122,007	125,639	106,729	130,312	133,971	113,246	147,792	151,002	124,925
	PortfolioRoR	0.1003	0.2162	0.1702	0.1680	0.2594	0.2196	0.1280	0.2141	0.2087	0.0278	0.0852	0.1359
	AvePosFutures	347	458	388	218	282	244	179	226	199	142	170	155

financial investors, as its risks, and particularly its drawdowns, are extremely large. It is important to highlight that the performance of any unhedged option strategy will always be dominated by directional bias, except for delta-neutral straddles. It is well known that since the beginning of oil financialization, systematic buying of short-dated oil futures has resulted in significant losses due to the prevalence of contango and negative roll yield.* Consistent with this directional bias, selling unhedged puts have also suffered large losses. Somewhat surprisingly, for longer maturity options where the roll-yield has far smaller impact, an unhedged VRP strategy has negative P&L for both puts and calls. This indicates the relative cheapness of all long-dated options and gives some appeal to buying and holding them until expiration as a long-term investment. This cheapness can be explained by leveraged producer hedges, as further discussed in the following section.

Another variant of the delta-hedging technique that we consider in this article is the strategy of intentionally leaving a residual unhedged long or short exposure above a threshold number of futures. This approach generally reduces the hedging cost while reducing the locked in losses when the underlying price reverts. Table 1.7 shows the results for hedging delta only over a threshold number of futures. A zero delta threshold corresponds to the base case of complete daily delta hedging. For non-zero delta threshold values, only the excess delta above or below the threshold is hedged and the portfolio has overnight exposure. The results, presented for this strategy in Table 1.7, qualitatively lead to similar conclusions as the hedge frequency analysis of Table 1.5. Both profits and risks increase slightly and the resulting RoR remains steady with a modest delta threshold, but the performance generally deteriorates as the hedging threshold increases. Reduction of futures trade count, and thus hedging cost, with an increased delta threshold provides an offset to performance deterioration that increases with threshold level.

Using hedging frequency and threshold strategies allows traders to reduce hedging, which benefits from short-term price reversals, but suffers losses if the trend continues. We now consider the third alternative strategy, where hedging deltas are calculated using implied volatility scaled by multiples ranging from 0.5 to 1.5. Table 1.8 summarizes the results. If implied volatility used for the delta calculation is increased, then hedging deltas for OTM options are reduced and such a strategy also leads to reduced hedging. The results for such a strategy with scaling factors above 1.0 lead to similar conclusions to the frequency and threshold results of Tables 1.5 and 1.7.

However, if the strategy is hedged using deltas computed with lower implied volatilities, the hedging deltas for OTM options increase. This more aggressive hedging strategy, which over-hedges standard Black deltas, benefits from price momentum and continuation of the trend. Table 1.8 shows that for a volatility scaling factor 0.5, such an over-hedged strategy leads to a substantial improvement in VRP performance for all OTM puts and zero-delta straddles. This result is consistent with the directional bias highlighted above in Table 1.6. Given large negative skewness in futures returns, the risks of selling OTM puts are mitigated by more aggressive hedging.

We should note that such hedging can be viewed as aggressive only when comparing it to hedging using standard log-normal deltas. Such a strategy, however, is somewhat inconsistent with the entire concept of VRP. Theoretically, the option payoff can only be perfectly replicated by a dynamic futures strategy if the hedging delta is calculated using the volatility realized during the life of the option, which, of course, is not known in advance. If options are deemed to be expensive and implied volatility is seen to be the upwardly biased expectation of future realized volatility, then why would a trader be hedging with a volatility that isnt even believed to be correct? In practice, traders often adjust the volatility input used for calculating the hedging delta more tactically, based on their short-term views about market behavior. If they perceive the market to be trendy, they can use lower input volatility and hedge more aggressively, but if the market is seen as range bounded, then either hedging using deltas computed with a higher volatility, or explicitly under-hedging using reduced frequency and threshold strategies, is a more appropriate strategy.

---

* See, for example, Bouchouev (2020).

TABLE 1.6 VRP portfolio Performance with No Delta Hedging

Entry DTM	Value		Delta								
		10P	20P	30P	40P	0S	40C	30C	20C	10C
20	AnnualizedPL ($)	27,480	(15,278)	(53,876)	(20,962)	42,691	142,916	213,432	264,403	218,204
	MaxDrawdown ($)	(3,956,976)	(5,913,959)	(8,767,951)	(10,461,935)	(2,917,469)	(4,106,679)	(2,086,847)	(1,116,999)	(593,984)
	DailyStdev ($)	35,513	49,961	62,554	74,602	41,351	60,817	47,686	33,321	16,639
	PortfolioRoR	0.0487	(0.0193)	(0.0543)	(0.0177)	0.0650	0.1480	0.2819	0.4999	0.8261
	AvePosFutures	0	0	0	0	0	0	0	0	0
40	AnnualizedPL ($)	(344,919)	(346,325)	(253,965)	(160,224)	2,986	174,841	285,908	338,827	264,492
	MaxDrawdown ($)	(12,084,345)	(15,445,961)	(17,081,786)	(19,457,179)	(6,843,069)	(7,413,993)	(5,434,184)	(3,380,714)	(1,617,933)
	DailyStdev ($)	71,570	96,462	119,608	141,332	72,266	109,462	85,153	58,351	29,250
	PortfolioRoR	(0.3036)	(0.2262)	(0.1338)	(0.0714)	0.0026	0.1006	0.2115	0.3658	0.5696
	AvePosFutures	0	0	0	0	0	0	0	0	0
60	AnnualizedPL ($)	(694,475)	(659,266)	(629,483)	(485,685)	(282,462)	(70,520)	116,110	222,433	269,268
	MaxDrawdown ($)	(21,255,580)	(25,532,862)	(30,202,562)	(32,732,498)	(13,140,506)	(13,625,604)	(10,538,499)	(7,101,325)	(2,462,706)
	DailyStdev ($)	113,978	149,578	181,512	210,618	108,231	165,776	131,543	92,362	49,948
	PortfolioRoR	(0.3838)	(0.2776)	(0.2185)	(0.1453)	(0.1644)	(0.0268)	0.0556	0.1517	0.3396
	AvePosFutures	0	0	0	0	0	0	0	0	0
120	AnnualizedPL ($)	(1,131,725)	(1,017,634)	(824,820)	(479,995)	(416,897)	(447,412)	(196,736)	(69,915)	126,735
	MaxDrawdown ($)	(38,139,726)	(41,721,056)	(44,405,079)	(45,338,784)	(25,421,696)	(40,516,755)	(31,325,949)	(21,619,935)	(10,270,557)
	DailyStdev ($)	230,621	297,718	353,785	405,744	214,758	333,377	271,478	205,014	124,923
	PortfolioRoR	(0.3091)	(0.2153)	(0.1469)	(0.0745)	(0.1223)	(0.0845)	(0.0457)	(0.0215)	0.0639
	AvePosFutures	0	0	0	0	0	0	0	0	0
180	AnnualizedPL ($)	(1,151,046)	(970,973)	(667,775)	(343,129)	(710,073)	(1,147,314)	(939,925)	(704,735)	(254,041)
	MaxDrawdown ($)	(44,582,170)	(49,476,456)	(50,550,970)	(52,880,225)	(33,987,534)	(69,860,877)	(57,900,757)	(44,349,983)	(25,514,392)
	DailyStdev ($)	321,246	417,202	502,058	582,257	319,877	506,251	423,439	329,818	210,730
	PortfolioRoR	(0.2257)	(0.1466)	(0.0838)	(0.0371)	(0.1398)	(0.1428)	(0.1398)	(0.1346)	(0.0759)
	AvePosFutures	0	0	0	0	0	0	0	0	0
240	AnnualizedPL ($)	(965,567)	(742,479)	(403,522)	(78,794)	(912,924)	(1,717,888)	(1,446,178)	(1,209,708)	(569,400)
	MaxDrawdown ($)	(44,778,970)	(53,165,800)	(56,319,888)	(63,807,925)	(44,987,478)	(100,516,916)	(84,973,783)	(65,885,950)	(40,905,550)
	DailyStdev ($)	405,510	527,625	635,957	741,120	415,135	667,980	564,616	451,387	317,292
	PortfolioRoR	(0.1500)	(0.0886)	(0.0400)	(0.0067)	(0.1385)	(0.1620)	(0.1614)	(0.1688)	(0.1130)
	AvePosFutures	0	0	0	0	0	0	0	0	0

TABLE 1.7 VRP Portfolio Performance Sensitivity to Hedging Delta Threshold

Entry DTM	Value	Hedge Delta Threshold											
		0			3			5			10		
		25P	0S	25C	25P	0S	25C	25P	0S	25C	25P	0S	25C
20	AnnualizedPL ($)	83,377	107,743	186,985	91,763	119,088	196,787	88,822	121,196	201,252	60,802	106,587	229,979
	MaxDrawdown ($)	(995,001)	(1,658,357)	(1,070,949)	(920,961)	(1,654,467)	(1,136,279)	(920,111)	(1,721,360)	(1,231,209)	(1,364,681)	(2,088,380)	(1,254,099)
	DailyStdev ($)	14,983	17,453	14,072	15,823	18,086	14,914	16,711	18,811	15,912	19,608	21,408	18,399
	PortfolioRoR	0.3506	0.3889	0.8370	0.3653	0.4148	0.8312	0.3348	0.4059	0.7968	0.1953	0.3136	0.7874
	AvePosFutures	149	174	153	117	133	119	106	116	107	92	88	90
40	AnnualizedPL ($)	168,989	236,517	301,932	166,577	249,888	309,907	140,971	245,662	311,333	76,158	207,476	313,078
	MaxDrawdown ($)	(1,613,421)	(1,715,953)	(883,850)	(1,531,411)	(1,681,103)	(1,027,610)	(1,704,411)	(1,806,643)	(1,284,854)	(2,350,710)	(2,266,444)	(2,096,175)
	DailyStdev ($)	22,563	25,694	20,771	24,257	27,276	22,596	26,128	28,940	24,370	31,436	33,498	29,886
	PortfolioRoR	0.4718	0.5799	0.9157	0.4326	0.5771	0.8640	0.3399	0.5347	0.8048	0.1526	0.3902	0.6599
	AvePosFutures	187	228	198	131	156	138	116	131	121	96	96	98
60	AnnualizedPL ($)	205,554	363,042	346,666	187,205	386,189	335,657	148,678	368,386	332,104	17,554	273,791	301,838
	MaxDrawdown ($)	(1,822,388)	(1,636,007)	(1,266,063)	(2,175,836)	(2,031,285)	(1,727,581)	(2,493,893)	(2,457,404)	(1,979,481)	(3,759,655)	(3,717,234)	(2,941,531)
	DailyStdev ($)	28,589	31,729	27,493	31,611	34,216	30,554	34,441	36,723	33,535	43,130	44,464	41,976
	PortfolioRoR	0.4529	0.7208	0.7943	0.3731	0.7110	0.6920	0.2719	0.6319	0.6238	0.0256	0.3879	0.4530
	AvePosFutures	211	262	236	137	164	155	119	136	133	98	98	106
120	AnnualizedPL ($)	225,584	541,222	445,464	203,104	565,733	450,297	127,972	533,656	415,840	(83,025)	367,217	343,094
	MaxDrawdown ($)	(4,291,257)	(4,204,970)	(3,310,935)	(5,352,097)	(5,117,060)	(3,949,725)	(6,317,677)	(5,897,519)	(4,658,185)	(8,992,368)	(8,610,089)	(7,274,395)
	DailyStdev ($)	52,123	55,278	48,958	57,774	60,717	55,614	63,358	65,737	61,903	81,142	82,464	80,000
	PortfolioRoR	0.2726	0.6168	0.5732	0.2215	0.5870	0.5101	0.1272	0.5114	0.4232	(0.0645)	0.2805	0.2702
	AvePosFutures	258	349	309	141	187	172	119	149	144	95	105	111
180	AnnualizedPL ($)	218,092	515,201	394,418	208,542	528,373	388,255	98,851	443,450	349,743	(182,107)	215,476	217,224
	MaxDrawdown ($)	(6,629,095)	(7,471,086)	(6,221,338)	(7,976,752)	(8,895,254)	(7,594,076)	(9,422,992)	(10,265,804)	(8,671,556)	(12,719,592)	(14,233,516)	(12,504,336)
	DailyStdev ($)	80,655	84,074	73,659	88,708	92,824	83,753	96,771	100,884	92,574	123,055	126,584	119,153
	PortfolioRoR	0.1703	0.3860	0.3373	0.1481	0.3586	0.2920	0.0643	0.2769	0.2380	(0.0932)	0.1072	0.1148
	AvePosFutures	303	408	352	147	197	173	122	155	142	96	107	108
240	AnnualizedPL ($)	174,221	398,152	269,221	160,874	401,434	258,271	32,513	275,054	209,143	(326,262)	(32,846)	22,899
	MaxDrawdown ($)	(8,960,182)	(10,826,782)	(10,494,962)	(10,507,113)	(12,372,852)	(12,135,973)	(12,253,464)	(14,436,262)	(13,707,783)	(16,431,360)	(19,305,162)	(18,652,813)
	DailyStdev ($)	109,399	116,005	99,664	119,520	127,288	113,149	129,974	137,872	125,015	163,410	171,678	160,751
	PortfolioRoR	0.1003	0.2162	0.1702	0.0848	0.1987	0.1438	0.0158	0.1257	0.1054	(0.1258)	(0.0121)	0.0090
	AvePosFutures	347	458	388	156	203	174	128	159	141	100	108	108

TABLE 1.8 VRP Portfolio Performance Sensitivity to Implied vol Fraction

Entry DTM	Value	Hedge Volatility								
		0.50			1.00			1.50		
		25P	0S	25C	25P	0S	25C	25P	0S	25C
20	AnnualizedPL ($)	118,487	91,011	139,217	83,377	107,743	186,985	55,322	107,925	229,713
	MaxDrawdown ($)	(1,009,032)	(1,778,582)	(1,080,360)	(995,001)	(1,658,357)	(1,070,949)	(1,334,651)	(1,888,547)	(1,327,538)
	DailyStdev ($)	20,317	22,180	19,057	14,983	17,453	14,072	18,162	20,116	19,355
	PortfolioRoR	0.3674	0.2585	0.4602	0.3506	0.3889	0.8370	0.1919	0.3380	0.7477
	AvePosFutures	166	228	169	149	174	153	138	141	145
40	AnnualizedPL ($)	239,228	279,110	173,620	168,989	236,517	301,932	86,614	220,091	355,560
	MaxDrawdown ($)	(1,715,491)	(1,608,123)	(1,764,689)	(1,613,421)	(1,715,953)	(883,850)	(1,908,101)	(2,269,094)	(2,346,345)
	DailyStdev ($)	30,068	31,613	28,672	22,563	25,694	20,771	28,201	31,678	31,030
	PortfolioRoR	0.5012	0.5562	0.3815	0.4718	0.5799	0.9157	0.1935	0.4377	0.7218
	AvePosFutures	210	295	223	187	228	198	170	185	184
60	AnnualizedPL ($)	408,841	515,880	99,687	205,554	363,042	346,666	32,930	254,698	410,422
	MaxDrawdown ($)	(1,693,446)	(1,619,966)	(3,014,201)	(1,822,388)	(1,636,007)	(1,266,063)	(3,266,435)	(4,009,194)	(3,240,282)
	DailyStdev ($)	38,853	37,852	37,357	28,589	31,729	27,493	37,856	42,477	43,832
	PortfolioRoR	0.6629	0.8585	0.1681	0.4529	0.7208	0.7943	0.0548	0.3777	0.5899
	AvePosFutures	237	338	273	211	262	236	190	216	216
120	AnnualizedPL ($)	735,077	745,077	129,897	225,584	541,222	445,464	(95,517)	377,198	523,397
	MaxDrawdown ($)	(2,966,496)	(2,940,257)	(7,189,469)	(4,291,257)	(4,204,970)	(3,310,935)	(8,246,578)	(9,094,396)	(6,746,592)
	DailyStdev ($)	66,872	65,771	64,428	52,123	55,278	48,958	73,711	81,047	87,563
	PortfolioRoR	0.6925	0.7136	0.1270	0.2726	0.6168	0.5732	(0.0816)	0.2932	0.3765
	AvePosFutures	278	463	350	258	349	309	233	284	280
180	AnnualizedPL ($)	708,876	855,102	213,044	218,092	515,201	394,418	(129,858)	269,740	365,588
	MaxDrawdown ($)	(5,183,866)	(3,957,361)	(9,201,131)	(6,629,095)	(7,471,086)	(6,221,338)	(12,081,482)	(14,539,114)	(12,252,531)
	DailyStdev ($)	99,723	91,973	91,453	80,655	84,074	73,659	110,951	127,874	134,578
	PortfolioRoR	0.4478	0.5857	0.1467	0.1703	0.3860	0.3373	(0.0737)	0.1329	0.1711
	AvePosFutures	336	546	383	303	408	352	268	331	321
240	AnnualizedPL ($)	677,109	773,042	146,953	174,221	398,152	269,221	(175,739)	128,309	218,723
	MaxDrawdown ($)	(8,073,365)	(4,861,492)	(12,360,342)	(8,960,182)	(10,826,782)	(10,494,962)	(15,542,144)	(19,833,442)	(18,473,296)
	DailyStdev ($)	135,963	120,580	117,247	109,399	116,005	99,664	147,043	176,425	185,118
	PortfolioRoR	0.3137	0.4039	0.0790	0.1003	0.2162	0.1702	(0.0753)	0.0458	0.0744
	AvePosFutures	395	614	413	347	458	388	301	372	353

1.5 Hedgers Behavior and VRP Regime Changes

We can gain additional insights by looking at the performance of VRP strategies over time. Many systematic risk premia strategies in energy markets are known to be sensitive to regimes, as described in Bouchouev and Zuo (2020). Energy markets constantly evolve, by adjusting for new fundamental drivers like the growth of shale extraction and structural factors resulting from market financialization. These factors lead to changing behavior among hedgers and speculators that impact the supply and demand for risk management services, and the resulting volatility risk premium. The VRP strategy is no exception. Figure 1.6 presents the equity history of a one-month VRP strategy comprising of OTM puts, OTM calls, and zero delta straddles, which highlights the presence of a structural break that separates two distinct regimes – or three, if we consider pre and post financial crisis markets as two regimes.

During the decade leading up to the financial crisis, this strategy generated impressive returns with RoR well above 1.0. Positive returns continued post-crisis as liquidity providers raised "insurance premiums" and then stopped providing liquidity altogether. However, by the beginning of 2014 the positive returns generated by the VRP strategy in oil markets were gone. Where did they go and why?

The strategy became a victim of its own success. As the business of passive commodity investments lost its allure, pressured by the prevalence of contango and punitive rolling costs, capital shifted towards more dynamic strategies designed to capture various systematic risk premia. To make it easier for investors, the VRP concept has also been packaged into investable indices, and the cumbersome task of daily delta-hedging has been effectively outsourced to index providers. Such indices allowed the large pools of capital held by pension funds and other institutional investors to access what used to be an obscure opportunity which previously could only be captured by oil specialists equipped with the right technology and risk management capabilities.

Another important factor behind the structural break in VRP is the evolution of hedging strategies by U.S. shale producers. Unlike traditional oil projects, shale is closer to mining operations in which constant drilling is required to keep production flowing. The shale business has been developed mostly by independent and highly leveraged producers whose access to capital provided by lending banks is often conditional on hedging the price risks. Thus while hedging for producers became nearly mandatory, their ability to pay the premium for the required insurance was limited. Instead, their hedging strategies shifted to more leveraged structures, such as costless collars which are net volatility neutral,

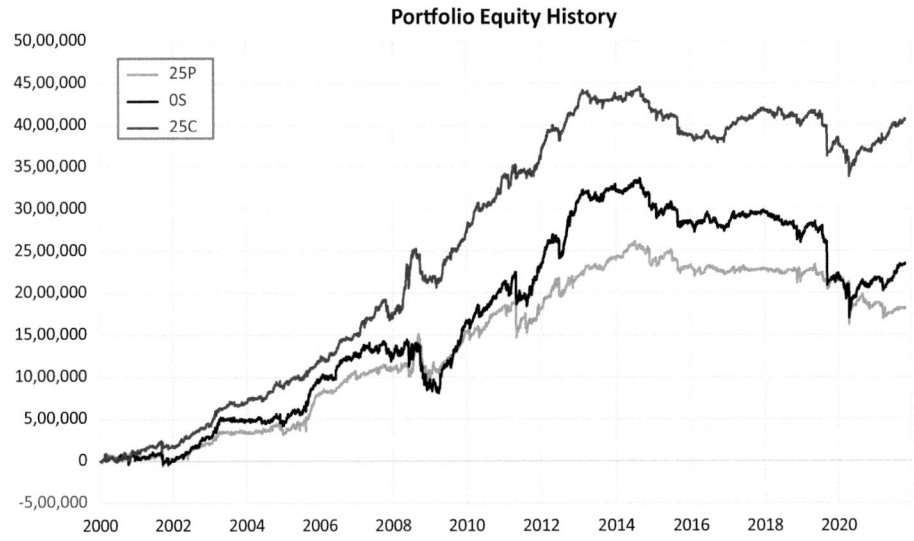

FIGURE 1.6   20-DTM VRP equity history for 25P, 25C and zero delta straddle.

and three-way collars where producers sell two options, an OTM put and an OTM call, to finance the purchase of an ATM put.

Finally, the impact of one large scale annual sovereign put buying program, consistently executed by the Finance Ministry of Mexico since 2002, gradually became more muted. This program, described in more detail in Bouchouev and Fattouh (2020), was designed to protect the country's export revenues which were heavily dependent on oil and quickly turned into the largest derivatives deal of each year. The hedge effectively treated put options as an insurance product with over one billion dollars spent annually on option premia. More recently, hedging volumes have been reduced as the country's overall oil production and exports decreased. In addition, to reduce costs in certain years, the strategy of buying outright puts has been replaced with buying cheaper put spreads with much smaller overall impact on volatility.

The impressive historical performance of the oil VRP strategy has attracted more providers of oil insurance willing to accept lower returns and take on larger risks. In addition, the natural buyers of the insurance, producers and consumers, demonstrated their own creativity by restructuring their approaches to hedging. Instead of buying relatively expensive outright insurance, these hedgers routinely bought an option and financed it by selling another option, thereby monetizing the real optionality embedded in their assets.

The business of extracting VRP via the traditional approach of selling ATM straddles is no longer attractive for financial investors. However, the aggregate premium has not entirely disappeared, rather it became somewhat spread out across various strikes and maturities, but not in a uniform manner. The opportunities to provide insurance-like products in commodity options has become more dynamic, with some options more expensive than others. To identify such trading opportunities, a trader needs more granular quantitative metrics for VRP, such as the ones introduced in this article.

1.6 Conclusions and Further Research

In this article, we present an in-depth analysis of the popular VRP investment strategy in the oil market. The strategy aims to extract a risk premium by selling and delta hedging WTI options that are widely perceived to be structurally overpriced, given the natural imbalance between buyers and sellers of optionality in the oil market. We analyze the strategy for options across different moneyness, as measured by the option's delta, and expiry dates, to conclude that the highest risk-adjusted returns can be obtained by selling and delta-hedging short-term OTM call options.

We then look at the practical implementation of the strategy, varying some key parameters of the delta-hedging model, such as hedging frequency, delta thresholds and hedging volatility. We found that such strategies not only reduce the transaction costs, but also bring in additional gains resulting from the short-term price reversals following particularly large price moves. We also study the performance of these strategies over time and argue that the structural break occurred around 2014. Following nearly two decades of strong performance, the simple VRP strategy returns flatlined. We attribute this structural change to the increase in the supply of volatility provided in the form of investable indices and to the simultaneous decrease in demand for volatility protection from producer hedgers. With the rapid growth of the U.S. shale industry, many producers opted to increase the leverage of their hedging programs by selling volatility to the market as a way to monetize the optionality embedded in their real assets.

Even though the simple directional volatility risk premium has largely evaporated as oil markets have become more efficient, the increasing complexity and diversity of hedging instruments has opened up new opportunities for investors in the form of more dynamic long-short volatility trading strategies. Such relative value strategies involve selling shorter-term options and hedging them by buying longer-term options, selling OTM calls hedged with some ratio of ATM options, or trading cross-energy volatility spreads between various energy products. The detailed analysis of such long-short strategies present interesting opportunities for further research, which could also be extended to studying more diversified oil volatility portfolios comprised of different systematic strategies. The well-known futures risk premia, such as carry, momentum, and value have yet to be applied to the oil volatility market.

Much more can also be done to improve the efficiency of delta-hedging volatility portfolios. In this article, we only considered daily delta-hedging, while practitioners often rebalance their VRP portfolios intraday. Intraday hedging using sophisticated statistical techniques, including methods of machine learning, can further improve the risk-adjusted returns on the VRP strategy in the oil market.

References

Barone-Adesi, G., and Whaley. R. E., Efficient Analytic Approximation of American Option Values, *J. Finance*, 1987, 42(2), 301–320.

Black, F., The pricing of commodity contracts, *J. Financ. Econ.*, 1976, 3, 167–179.

Bouchouev, I., From risk bearing to propheteering, *Quant. Finance*, 2020, 20(6), 887–894.

Bouchouev, I., and Fattouh, B., Can Russia and OPEC Draw any Lessons from Mexico's Oil Hedge?, *Oxford Institute for Energy Studies,* 2020, August.

Bouchouev, I., and Johnson, B., The Smile of the Volatility Risk Premia, *Global Commodities Applied Research Digest*, 2021, 6(2), Winter.

Bouchouev, I., and Zuo, L., Oil Risk Premia under Changing Regimes, 2020, *Global Commodities Applied Research Digest*, 2020, 5(2), Winter.

Dempster, M. A. H., Medova, E.A., and Tang, K., Determinants of Oil Futures Prices and Convenience Yields., *Quantitative Finance*, 2012, 12(12), 251–290.

Derman, E., and Miller, M., *The Volatility Smile*, 2016 (John Wiley & Sons, Inc., Hoboken, New Jersey).

Doran, J., and Ronn, E., Computing the market price of volatility risk in the energy commodity markets, 2008, *J. Bank. Finance*, 32, 2541–2552.

Ellwanger, R., On the Tail Risk Premium in the Oil Market, *Bank of Canada Working Paper*, 2017, 46.

Gatheral, J., *The Volatility Surface*, 2006 (John Wiley & Sons, Hoboken, New Jersey).

Hamilton, J, and Wu, J., Risk premia in crude oil futures market, *J. Int. Money Finance*, 2014, 42, 9–37.

Jacobs, K, and Li, B., Option Returns, Risk Premiums, and Demand Pressure in Energy Markets, *JP Morgan Center for Commodities Research Symposium*, 2021, August 17.

Kang, S. B, and Pan, X., Commodity Variance Risk Premia and Expected Futures Return: Evidence from the Crude Oil Market, *SSRN*, 2015.

Lo, A., The Statistics of Sharpe Ratio, *Financ. Anal. J.*, 2002, July/August, 36–52.

Ockenden, J., Koch smooths volatile waters, *Energy Risk*, 2003, November, 20–22.

Prokopczuk, M., Symeonidis, L., and Wese Simen, C., Variance risk in commodity markets, *J. Bank. Finance*, 2017, 81, 136–149.

Tang, K., and Xiong, W., Investment and financialization of commodities, *Financ. Anal. J.*, 2012, 68(6), 54–74.

Trolle, A., and Schwartz, E., Unspanned Stochastic Volatility and the Pricing of Commodity Derivatives, *Rev. Financ. Stud.*, 2009, 22(11), 4423–4461.

Trolle, A., and Schwartz, E., Variance Risk Premia in Energy Commodities, *J. Deriv.*, 2010, Spring, 15–32.

2

Determinants of Oil Futures Prices and Convenience Yields

M. A. H. Dempster,
Elena Medova
and Ke Tang

2.1 Introduction

Commodity futures exchanges arose in the late 19th and early 20th centuries to allow the forward contract mitigation of cyclical supply/demand imbalances between agricultural producers and consumers while limiting speculation through requiring the posting of margin on futures positions. In the 21st century, expanded to a wide range of commodities and related services and increasingly electronic, they continue to serve their original purpose. But they are also vital to the forecasting of spot prices by large resource producers for management evaluation of project alternatives and investment opportunities over very long-term horizons. Global resource firms may or may not hedge their physical activities in the futures markets, but increasingly they have come to see that sophisticated price forecasting is a prerequisite to the use of real option techniques for forward planning and risk management of ongoing operations.

However, for use in exploration, acquisition evaluation, or project development and risk management, senior management cannot be content with reduced form 'black box' price forecasting methods devoid of an economic understanding of the commodity markets involved. Focussing on crude oil prices, this paper attempts to meet these stringent managerial criteria by specifying a three-factor spot price model for oil and studying its relationship to different economic variables, which includes financial variables (such as SP500 returns, US dollar returns, etc.), business cycle variables (such as the business cycle coincident index), demand variables (such as inventory and the heating-crude oil spread) and trading

DOI: 10.1201/9781003265399-3

variables (such as futures open interest growth and hedging pressure). We hope to contribute to an understanding of the relationships between oil prices, physical inventory management, financial hedging and speculation. Although this paper treats the oil markets, the model treated here may be applied to a wider range of commodities, upon which we are currently engaged (see e.g. Dempster and Tang 2011).

Commodities are real assets and so their prices should be influenced by their supply and demand. The *theory of storage* (Kaldor 1939, Working 1949, Brennan 1958) sees optimal inventory management as the main determinant of commodity prices. But since commodities are traded through futures contracts, financial markets and trading behaviour will also influence the commodity prices and term structure, as noted by Keynes (1930) in his *theory of normal backwardation*. More recently, Bailey and Chan (1993) showed that financial factors such as the spread between BAA and AAA bonds can influence the convenience yield of many commodities. In this paper, we shall examine the impact both of supply, demand and business variables and of financial and trading factors on the movement and shape of the oil futures term structure.

Both producers and consumers wish to make forecasts for long-term planning and investment decisions since commodity prices represent their output revenues and input costs, respectively. Various authors have expressed differing opinions on the long-term evolution of commodity prices. Most (see, e.g., Cuddington and Urzua [1987] and Gersovitz and Paxson [1990]) believe that commodity prices are non-stationary since it is hard statistically to reject the most parsimonious geometric random walk model using historical time series data. Cashin *et al.* (2000) have shown that shocks to commodity prices are typically long lasting, while Grilli and Yang (1988) found that real primary commodity prices have a trend of about 0.5% a year using a dataset from 1900 to 1986. Schwartz and Smith (2000) use *geometric Brownian motion* (GBM) to model such long-term behaviour because of the ability of GBM to capture trend and the persistency of shocks. However, Bessembinder *et al.* (1995) discovered strong mean reversion in commodity log spot prices, suggesting that a *geometric Ornstein Uhlenbeck* (GOU) process might be more appropriate. Although not pointed out explicitly, Casassus and Collin-Dufresne (2005) use a mean-reverting process to model log spot commodity prices. Geman and Nguyen (2005) use a mean-reverting log spot price with stochastic mean and stochastic volatility to model soybean futures prices.* Many economists believe that commodity prices in the medium term are closely related to the business cycle (e.g., Fama and French 1988), which is usually considered to be a mean-reverting process. Short-term swings in commodity prices have substantial impacts for many speculators and short-term strategic investors and the short-term factors driving commodity prices are usually also considered to be mean-reverting (e.g., Schwartz and Smith 2000).

In this paper, we use a mean-reverting process to model the short-term factor in our three-factor (log) spot price model. However, it is not appropriate to model short-term influences using only a *single* mean-reverting factor, since one factor alone cannot adequately model the complicated short- to medium-term behaviour of commodity prices. This suggests that two mean-reverting factors—one short- and one medium-term—are needed to model price movements.[†] Our medium-term factor captures business cycles and long-term demand in the global economy, while a long-term GBM factor captures trend-related persistent shocks, such as technology growth, long-term supply through the discovery of new resources, etc. Intuitively, the time scale of the short-term factor is several months, that of the medium-term factor 1 to 2 years and that of the long-term factor decades or even longer. The three-factor model treated here nests several other models, including those of Gibson and Schwartz (1990) and Schwartz and Smith (2000), which are equivalent. Because of oil's importance to the global economy, and the ease of obtaining inventory data for it and relevant economic variables, we use crude oil futures to illustrate our model's development and to examine the various intuitions presented briefly above.

---

[*] Using soybean inventory data, Geman and Nguyen (2005) also show that soybean futures return volatility is negatively related to soybean inventory (or positively related to 'scarcity', the reciprocal of inventory), which is consistent with the theory of storage.

[†] The medium-term factor should obviously revert to its mean more slowly than the short-term factor.

After developing the model* in state space form, we use the Kalman filter to obtain the estimated historical paths of the three latent factors from observations of oil futures prices. We then perform a *structural vector auto-regression* (SVAR) analysis involving the three estimated factor paths and the historical paths of several economic variables, including financial, business-cycle, fundamental and trading variables. We find that financial variables mainly affect the long-term p factor and the business cycle variable influences mainly the medium-term y factor. The demand variables affect both x and y factors, and higher net demand results in both a higher short-term x factor (deeper short-end backwardation) and a higher y factor (deeper long-end backwardation). The trading variables also influence both x and y factors, and more intensive trading and stronger hedging pressure result in higher factor levels.

The paper is organized as follows. Section 2.2 examines several features of WTI crude oil prices and develops a detailed motivation for a three-factor spot price model. Section 2.3 presents such a model and explains its relationship to earlier models. Section 2.4 examines the relationship between the three estimated latent factors and several economic variables. Section 2.5 concludes.

2.2 Oil Price Features

In this section we characterize features of crude oil futures prices and their evolution.

2.2.1 Term Structure of Oil Futures Open Interest

Weekly oil futures prices and open interest for WTI crude oil (CL) traded on the New York Mercantile Exchange (NYMEX) were obtained from 1986.06 to 2010.12 from Pinnacle Data Corp. The times to maturity of these futures contracts range from several days to about 17 months[†] (the first to the seventeenth contract). Figure 2.1 shows futures term structure shapes commonly observed in the market with their observation dates. At the long end we can see both *contango* (top two diagrams) and *backwardation* (bottom two diagrams) term structures, while at the short end we see U and hump shapes. Thus the short end of the term structure appears to be more volatile and is not necessarily conformal with the long end.

Futures prices are discovered through trading, so to investigate their short-term behaviour in Figure 2.2 we plot the futures *open interest,* i.e. the total number of futures contracts that have not expired, or been fulfilled by delivery, against their times to maturity. The figure demonstrates that the main oil trading activities are concentrated in the one month futures contract. Its open interest is much larger than the value obtained from an exponential function fit to open interest at other maturities.[‡] This large open interest in the nearby contract indicates a uniquely high liquidity which can result in different behaviour of short-term oil futures prices. Furthermore, investors and hedgers all prefer to use short-term futures instead of long-term futures and thus contribute to this behaviour. For example, passive commodity index investors tend to invest in short-term rather than long-term futures (for details, refer to Tang and Xiong 2012). As shown by Culp and Miller (1995), Mellon and Parsons (1995) and Brennan and Crew (1997), hedgers employ short-term futures to hedge longer-term obligations.

2.2.2 Single-Factor Convenience Yield Models

In previous research, nearly all researchers have used a single factor to model *convenience yield*, which is the commodity equivalent of equity dividend yield representing the opportunity return to physical

---

* It is an affine futures term structure model for the log spot price in the $A_0(3)$ form of Dai and Singleton (2000).

[†] The open interest for futures with time to maturity longer than 17 months is very small (see Figure 2.4). Also, in the earlier part of our dataset, futures prices with maturity longer than this are not available.

[‡] We fit the open interest frequency by fitting $P_\tau = a \exp(b\tau)$ to it, where P_τ is the (time) average open interest for futures contracts with time to maturity.

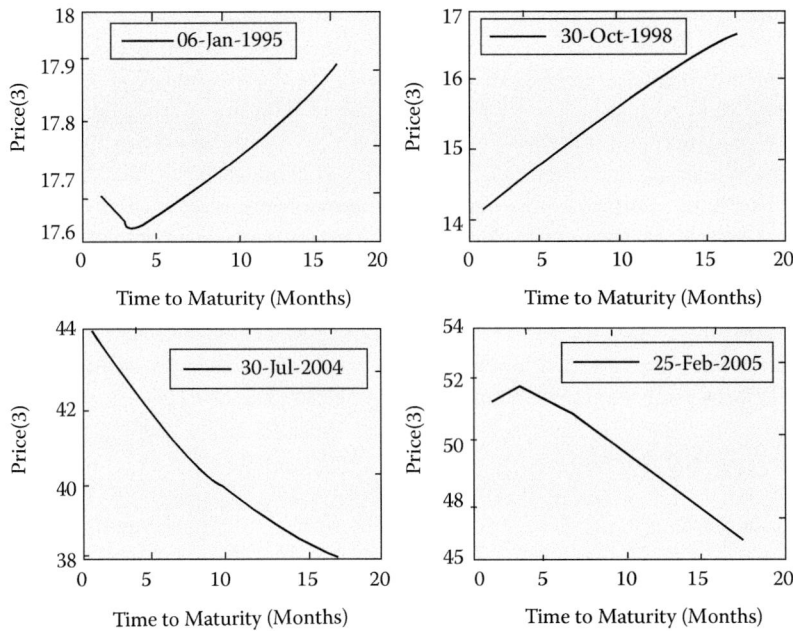

FIGURE 2.1   The oil futures term structure.

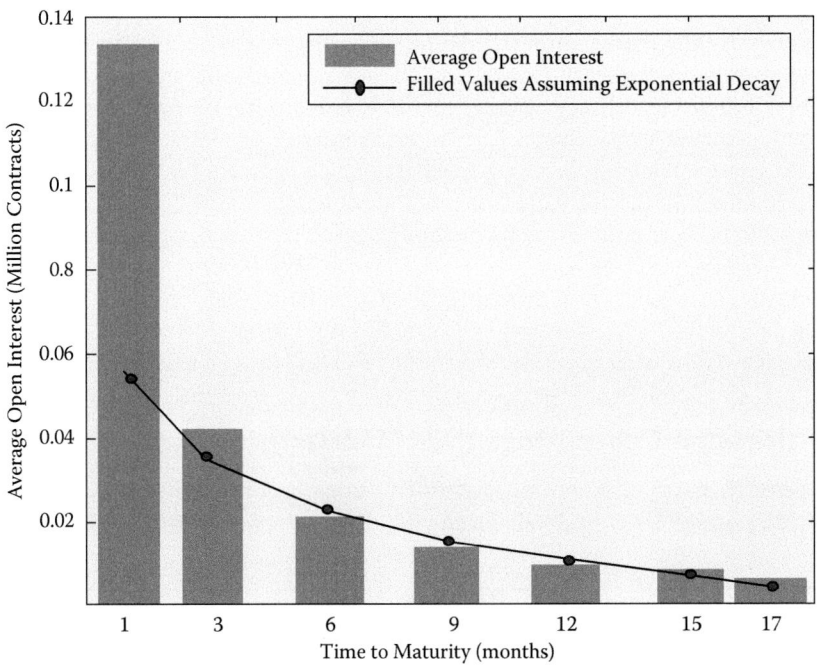

FIGURE 2.2   Average open interest vs. time to maturity.

TABLE 2.1　Model Errors for Futures with Different Times to Maturity

Maturity (Months)	Gibson–Schwartz Two-Factor Model	Schwartz–Smith Two-Factor Model	Schwartz Three-Factor Model
1	*0.043*	*0.042*	*0.045*
5	0.006	0.006	0.007
9	0.003	0.003	0.003
13	0	0	0
17	0.004	0.004	0.004

ownership of the commodity. Among such models, the Gibson–Schwartz (1990) two-factor model, the Schwartz–Smith (2000) two-factor model and the Schwartz three-factor model are commonly used. We use the Schwartz–Smith model here to investigate whether or not one factor is enough to model convenience yields.

2.2.2.1 Short Term Pricing Errors

Table 2.1 contains the log pricing errors from Schwartz (1997, p. 939) and Schwartz–Smith (2000, p. 903), where the short-term (1 month) futures prices have a noticeably larger error than the others.* This is consistent with the hypothesis that short-term futures price movements are often not conformal with longer-term futures price movements and thus a *separate* factor is needed to model short-term futures price movements.

2.2.2.2 Convenience Yields Inferred from the Schwartz–Smith (2000) Model

In the Gibson–Schwartz model, log futures prices are expressed as an affine combination of two latent factors: the log spot price and the convenience yield. Thus given parameter estimates for the model the convenience yield and spot price can be backed out using any two futures prices. We re-estimate the Gibson–Schwartz model using our own dataset and in Figure 2.3 plot the evolution of the model convenience yields inferred from 1 and 3 month futures and 15 and 17 month futures and their difference. It is clear that these two estimates of convenience yield are *strongly inconsistent*. The unconditional standard deviation of the convenience yield implied from 1 and 3 month futures is 26.3% per annum, while that from 15 and 17 months is 34.7% p.a. The unconditional standard deviation of the difference between these two convenience yields is surprisingly large, about 33.6% p.a.† The differences between these two convenience yield maturities should therefore not be overlooked, but should instead be modelled using a new short-term factor.

2.2.2.3 Principal Component Analysis on Convenience Yields

As we have seen, one factor is not enough to model the convenience yield; in this section we test how many factors are actually needed using a *principal component analysis* (PCA). Since convenience yield is not directly observable, we infer the implied convenience yield $\delta(t, T)$ from commodity futures prices and interest rates using

$$\delta(t,T) = r_r - \frac{\ln(F(t,T)) - \ln S_t}{T-t}$$

$$\approx r_t - \frac{\ln(F(t,T)) - \ln\left(F\left(t,T_0\right)\right)}{T-T_0},$$

(2.1)

---

* When we estimate the Schwartz-Smith (2000) model using our dataset we find a similar phenomenon.
† Note that the implied convenience yield during the financial crisis fluctuates widely.

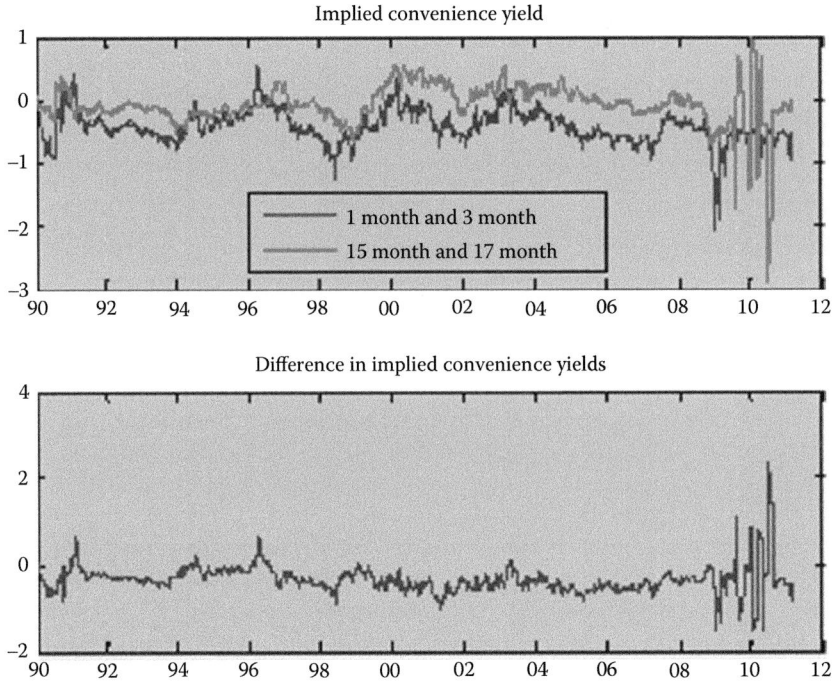

FIGURE 2.3   Implied convenience yields and their difference for the two-factor Gibson and Schwartz model.

TABLE 2.2   The Principal Components of the Implied Convenience Yield

Component	Variance Explained (%)
First	92.05
Second	5.99
Third	1.33
Fourth	0.39
Fifth	0.22
Sixth	0.02

where T_0 is the time to maturity of nearby futures contracts, T is the time to maturity of futures with a relatively longer horizon and r_t is the instantaneous rate corresponding to the three month LIBOR rate. We use nearby and 3, 6, 9, 12, 15, 17 month futures (i.e. $T = 3, 6, 9, 12, 15, 17$) to calculate six time series of implied convenience yield over our data period and then perform a PCA on them. Table 2.2 shows the variance explained by each factor. Clearly, two factors can explain more than 98% of the overall variance of the convenience yield, so that a two-factor model is good enough to catch its behaviour.

In the sequel we use two factors to model convenience yield, one short term and one for a longer medium-term horizon. The short-term factor should correct both the large pricing error of short-term contracts in two-factor models and the mismatch arising with these models of implied convenience yields which are backed out from short- and longer-maturity futures. Adding a long-term factor, the resulting three-factor model can capture the different shapes of the futures term structure shown in Figure 2.2.

2.3 Three-Factor Model Statement

We begin by modelling the dynamics of the log spot oil price **G** in terms of convenience yield using two factors.

2.3.1 Dynamics of Spot Prices

In the *market* (physical) *measure* the system is given by,*

$$dG_t = \left(r^f + \lambda_G - \delta_t - \gamma_t - \frac{1}{2}\sigma_G^2 \right)dt + \sigma_G d\mathbf{W}_G, \tag{2.2}$$

$$d\delta_t = k_\delta\left(\alpha - \delta_t \right)dt + \sigma_g d\mathbf{W}_\delta, \tag{2.3}$$

$$d\gamma_t = -k_\gamma\gamma_t dt + \sigma_\gamma d\mathbf{W}_\gamma, \tag{2.4}$$

$$Ed\mathbf{W}_G d\mathbf{W}_\delta = \rho_{\delta G}dt, Ed\mathbf{W}_\delta d\mathbf{W}_\gamma = \rho_{\delta\gamma}dt, Ed\mathbf{W}_G d\mathbf{W}_\gamma = \rho_{G\gamma}dt. \tag{2.5}$$

Here, at time t, $G_t := \ln(S_t)$ is the logarithm of the spot price, $\delta_t + \gamma_t$ is the spot (instantaneous) convenience yield with the medium-term δ_t and the short-term γ_t mean-reverting factors having long-run means α and 0 respectively in the market measure, λ_G is the *market price of risk* premium of the G process and $\mathbf{W}_G$, $\mathbf{W}_\delta$ and $\mathbf{W}_\gamma$ are Wiener processes with σ_G, σ_δ and σ_γ their corresponding volatilities.

In the *risk-neutral measure* this system becomes

$$dG_t = \left(r^f - \delta_t - \gamma_t - \frac{1}{2}\sigma_G^2 \right)dt + \sigma_G d\mathbf{W}_G^Q, \tag{2.6}$$

$$d\delta_t = k_\delta\left(\alpha - \delta_t - \lambda_\delta \right)dt + \sigma_\delta d\mathbf{W}_\delta^Q, \tag{2.7}$$

$$d\gamma_t = k_\gamma\left(-\gamma_t - \lambda_\gamma \right)dt + \sigma_\gamma d\mathbf{W}_\gamma^Q, \tag{2.8}$$

$$Ed\mathbf{W}_G^Q d\mathbf{W}_\delta^Q = \rho_{\delta G}dt, Ed\mathbf{W}_\delta^Q d\mathbf{W}_\gamma^Q = \rho_{\delta\gamma}dt,$$
$$Ed\mathbf{W}_G^Q d\mathbf{W}_\gamma^Q = \rho_{\Gamma\gamma}dt, \tag{2.9}$$

where $k_\delta\lambda_\delta$ and $k_\gamma\lambda_\gamma$ are, respectively, the market risk premia for the δ and γ processes.

Setting the γ factor identically equal to zero, (2.2) to (2.9) becomes the Gibson–Schwartz (1990) model so that our model is its extension, but with convenience yield decomposed into two parts, δ and γ, with different mean-reversion speeds.

Defining $x_t := (1/k_\delta)(\delta_t - \alpha)$, $\mathbf{y}_t := \gamma_t/k_\gamma$ and $\mathbf{p}_t := \mathbf{G}_t - \mathbf{x}_t - \mathbf{y}_t$ in the market measure, we have

$$d\mathbf{x}_t = \frac{1}{k_\gamma}d\delta_t = -k_\delta x_t dt + \frac{\sigma_\delta}{k_\delta}d\mathbf{W}_\delta, \tag{2.10}$$

$$d\mathbf{y}_t = \frac{1}{k_\gamma}d\gamma_t = -k_\gamma y_t dt + \frac{\sigma_\gamma}{k_\gamma}d\mathbf{W}_\gamma, \tag{2.11}$$

---

* Boldface is used throughout to denote random entities, here conditional.

$$d\mathbf{p}_t = d\mathbf{G}_t - \frac{d\delta_t}{k_.} - \frac{d\gamma_t}{k_{\ddagger}} = \left(r^f + \lambda_G - \alpha - \frac{1}{2}\sigma_G^2 \right)dt$$

$$+\sigma_G d\mathbf{W}_G - \frac{\sigma_\delta}{k_\delta}d\mathbf{W}_\delta - \frac{\sigma_\gamma}{k_\gamma}d\mathbf{W}_\gamma. \tag{2.12}$$

Setting $k_x := k_\delta$, $k_y := k_\gamma$, $\sigma_x := \sigma_\delta/k_\delta$, $\sigma_y := \sigma_\gamma/k_\gamma$, $\lambda_x := \lambda_\delta/k_\delta$, $\lambda_y := \lambda_\gamma/k_\gamma$, $\lambda_p := \lambda_G - \lambda_\delta - \lambda_\gamma$, $u : r^f + \lambda_G - \alpha - (1/2)\sigma_G^2$, $\sigma_p^2 := \sigma_G^2 + \sigma_x^2 + \sigma_y^2 + 2p\delta_\gamma\sigma_x\sigma_y - 2p_{G\delta}\sigma_G\sigma_x - 2p_{G\gamma}\sigma_G\sigma_y$ and $d\mathbf{W}_x := d\mathbf{W}_\delta$, $d\mathbf{W}_\gamma := d\mathbf{W}_\gamma$, $d\mathbf{W}_p := (1/\sigma_p)$ $(\sigma_G d\mathbf{W}_G - (\sigma_\delta/k_\delta)d\mathbf{W}_\delta - (\sigma_\gamma/k_\gamma)d\mathbf{W}_\gamma)$, the original model in the market measure becomes

$$\ln(\mathbf{S}_t) = \mathbf{x}_t + \mathbf{y}_t + \mathbf{p}_t, \tag{2.13}$$

$$d\mathbf{x}_t = -k_x x_t dt + \sigma_x d\mathbf{W}_x, \tag{2.14}$$

$$d\mathbf{y}_t = -k_y y_t dt + \sigma_x d\mathbf{W}_y, \tag{2.15}$$

$$d\mathbf{p}_t = udt + \sigma_p d\mathbf{W}_p, \tag{2.16}$$

$$Ed\mathbf{W}_x d\mathbf{W}_y = \rho_{xy}dt, Ed\mathbf{W}_x d\mathbf{W}_p = \rho_{xp}dt,$$
$$Ed\mathbf{W}_y d\mathbf{W}_p = \rho_{yp}dt, \tag{2.17}$$

where $\mathbf{x}$ is the *short-term* factor with mean-reversion speed k_x and volatility σ_x, $\mathbf{y}$ is the *medium-term* factor with mean-reversion speed k_y and volatility σ_y, $\mathbf{p}$ is the *long-term trend* factor with *growth rate u* and volatility σ_p and W_x, W_y and W_p are all Wiener processes. Note that factors $\mathbf{x}$ and $\mathbf{y}$ both have *zero* long-run means so that they will fluctuate around the trend factor $\mathbf{p}$.

In the risk-neutral measure the above system becomes

$$\ln(\mathbf{S}_t) = \mathbf{x}_t + \mathbf{y}_t + \mathbf{p}_t, \tag{2.18}$$

$$d\mathbf{x}_t = k_x\left(-x_t - \lambda_x\right)dt + \sigma_x d\mathbf{W}_x^Q, \tag{2.19}$$

$$d\mathbf{y}_t = k_y\left(-y_t - \lambda_y\right)dt + \sigma_y d\mathbf{W}_y^Q, \tag{2.20}$$

$$d\mathbf{p}_t = \left(u - \lambda_p\right)dt + \sigma_p d\mathbf{W}_p^Q \tag{2.21}$$

$$Ed\mathbf{W}_x^Q d\mathbf{W}_y^Q = \rho_{xy}dt, Ed\mathbf{W}_x^Q d\mathbf{W}_p^Q = \rho_{xp}dt,$$
$$Ed\mathbf{W}_y^Q d\mathbf{W}_p^Q = \rho_{yp}dt, \tag{2.22}$$

where $k_x\lambda_x$, $k_y \lambda_y$ and λ_p are the *risk premia* of factors $\mathbf{x}$, $\mathbf{y}$ and $\mathbf{p}$ respectively.[*]

---

[*] Note that, since these risk premia for the $\mathbf{x}$ and $\mathbf{y}$ factors are *constant*, the mean-reversion speeds of these factors are the same under the market and risk-neutral measures (see Casassus and Collin-Dufresne [2005] who assume risk premia stochastic).

We term the model (2.13) to (2.22) the *three factor (log) spot price* model. It determines spot price fluctuations in terms of three components: two mean-reverting factors representing short- and medium-term economic forces and one long-term factor that reflects the equilibrium commodity price trend and captures permanent price shocks. We note that the model belongs to the exponential affine class in the framework of Duffie *et al.* (2000).*

Solving (2.19), (2.20) and (2.21), and substituting into (2.18), together with taking logarithms of the no-arbitrage condition for the price at *t* of the futures contract with maturity *T* given by

$$F(t,T) = E_t^Q \left[\mathbf{S}_T \right] \tag{2.23}$$

in terms of the conditional expectation in the risk-neutral measure Q at t, yields $\ln F(t, T)$ in terms of the three factors at t as

$$\ln F(t,T) = \left(x_t + \lambda_x\right)e^{-k_x(T-t)} + \left(y_t + \lambda_y\right)e^{-k_y(T-t)}$$

$$+ p_t - \left(\lambda_x + \lambda_y\right) + \left(u - \lambda_p\right)(T-t)$$

$$+ \frac{1}{2} \left[\begin{array}{c} \dfrac{1-e^{-2k_x(T-t)}}{2k_x}\sigma_x^2 \\[2mm] + \dfrac{1-e^{-2k_y(T-t)}}{2k_y}\sigma_y^2 + \sigma_p^2(T-t) \\[2mm] + \dfrac{2\left(1-e^{-(k_y+k_x)(T-t)}\right)}{k_x+k_y}\rho_{xy}\sigma_x\sigma_y \\[2mm] + \dfrac{2\left(1-e^{-k_x(T-t)}\right)}{k_x}\rho_{xy}\sigma_x\sigma_p \\[2mm] + \dfrac{2\left(1-e^{-k_y(T-t)}\right)}{k_y}\rho_{yp}\sigma_y\sigma_p \end{array} \right]. \tag{2.24}$$

2.3.2 Two-Factor Convenience Yield

The price at t of the futures contract with maturity T is given in terms of the instantaneous convenience yield at t by

$$F(t,T) = S_t \exp\left(r^f(T-t) - \int_t^T \delta(t,s)ds \right), \tag{2.25}$$

where $\delta(t, s)$ is the *instantaneous convenience yield* at time t of the contract with maturity s. In our model

$$\delta(t,T) = r^f + \left(\mathbf{x}_t + \lambda_x\right)k_x e^{-k_x(T-t)} + \left(\mathbf{y}_t + \lambda_y\right)k_y e^{-k_y(T-t)}$$

$$- \left(u - \lambda_p\right) - \frac{1}{2}\left[e^{-2k_x(T-t)}\sigma_x^2 + e^{-2k_y(T-t)}\sigma_y^2 + \sigma_p^2 \right.$$

$$+ 2\rho_{xy}\sigma_x\sigma_y e^{-(k_x+k_y)(T-t)} + 2\rho_{xp}\sigma_x\sigma_p e^{-k_x(T-t)}$$

$$\left. + 2\rho_{yp}\sigma_y\sigma_p e^{-k_y(T-t)} \right]. \tag{2.26}$$

---

* See last footnote of Section 2.1.

When $T \to t$ this reduces to the (*instantaneous*) *spot convenience yield* given by

$$\delta_t = \delta(t,t) = r^f + k_x\left(\mathbf{x}_t + \lambda_x\right) + k_y\left(\mathbf{y}_t + \lambda_y\right) - u + \lambda_p$$

$$-\frac{1}{2}\left(\sigma_x^2 + \sigma_y^2 + \sigma_p^2 + 2\rho_{xy}\sigma_x\sigma_y\right) \tag{2.27}$$

$$+2\rho_{xp}\sigma_x\sigma_p + 2\rho_{yp}\sigma_y\sigma_p\big),$$

so that the spot convenience yield $\boldsymbol{\delta}$ is seen to be an affine combination of $\mathbf{x}$ and $\mathbf{y}$ factors as designed. Our calibration of the three-factor model for oil futures shows, as expected, that the $\mathbf{x}$ factor has a much higher mean-reversion speed than that of the $\mathbf{y}$ factor (see Table 2.3). As a consequence, (2.25) and (2.26) imply that the longer-term convenience yields $\boldsymbol{\delta}(t, T)$ are determined mainly by the $\mathbf{y}$ factor, while the spot convenience yield $\boldsymbol{\delta}_t$ is determined mainly by the $\mathbf{x}$ factor.

Although convenience yield is a concept of the theory of storage, in the context of the theory of normal backwardation we wish to know the shape and overall slope (i.e. *contango*, upwards slope, or *backwardation*, downwards slope) of the futures price term structure. Taking logarithms of both sides of (2.25) and differentiating the result with respect to maturity gives

$$\frac{\partial(F(t,T))/\partial T}{F(t,T)} = r^f - \delta(t,T). \tag{2.28}$$

TABLE 2.3 Parameter Estimates of Two- and Three-Factor Models

Variable	Three-Factor Model	Two-Factor Model
k_x	3.4152 (0.1147)	
k_y	0.8802 (0.0384)	1.0715 (0.0166)
u	0.0809 (0.0425)	0.0838 (0.0428)
σ_x	0.1977 (0.0078)	
σ_y	0.2817 (0.0078)	0.2866 (0.0067)
σ_p	0.1953 (0.0051)	0.1960 (0.0044)
λ_x	−0.0205 (0.0128)	
λ_y	0.1600 (0.0701)	0.0720 (0.0581)
λ_p	0.0731 (0.0425)	0.1063 (0.0428)
ρ_{xy}	−0.0794 (0.0518)	
ρ_{xp}	0.0838 (0.0431)	
ρ_{yp}	−0.0067 (0.0472)	0.0932 (0.0346)
ξ_1	0.0160 (0.0037)	0.0347 (0.0072)
ξ_2	0.0004 (0.0000)	0.0115 (0.0025)
ξ_3	0.0014 (0.0006)	0.0000 (0.0000)
ξ_4	0.0016 (0.0005)	0.0022 (0.0006)
ξ_5	0.0011 (0.0007)	0.0021 (0.0009)
ξ_6	0.0158 (0.0033)	0.0161 (0.0034)
ξ_7	0.0160 (0.0035)	0.0172 (0.0036)
Log-likelihood	**24493**	22331

Note: $\xi_1, \xi_2, \xi_3, \xi_4, \xi_5, \xi_6, \xi_7$ are, respectively, the mean absolute pricing errors of the F1, F3, F6, F9, F12, F15 and F17 contracts. The quantities in parentheses are (asymptotic) standard deviations.

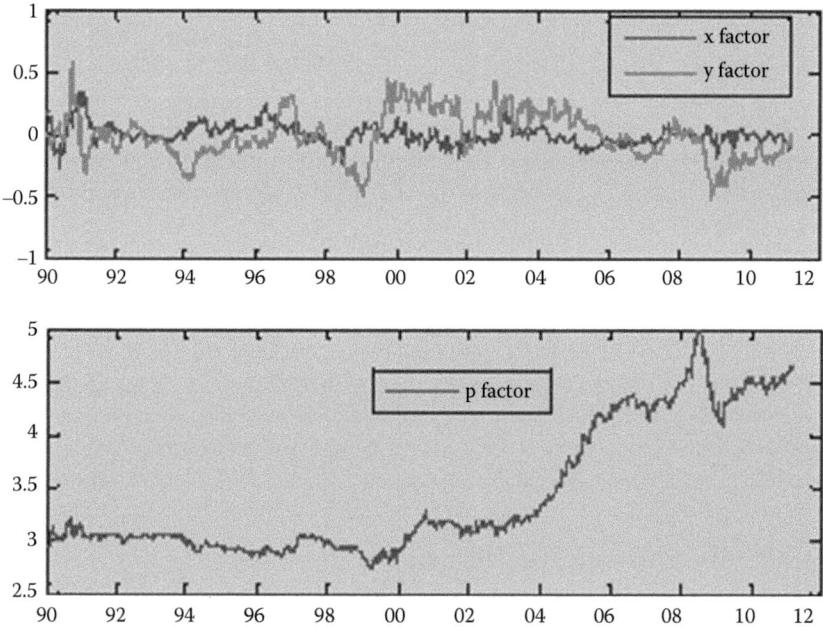

FIGURE 2.4 Estimated latent factor evolution in the three-factor model.

The convenience yield $\delta(t, T)$ therefore determines the sign of the slope of the term structure of futures prices and the **x** and **y** factors can be regarded as two components of this slope. Thus, when the instantaneous convenience yield $\delta(t, T)$ is strictly less than the instantaneous risk-free rate r^f, futures prices are in contango *locally* in maturity. However, because **x** and **y** may have different signs at a specific time t, the three-factor model is capable of reproducing the empirical near term U or humped futures curves of Figure 2.1.

2.3.3 Results

The appendix shows the state space form of the three-factor model needed for the parameter estimation filtering technique.* Since the three factors of our model are not directly observable, to calibrate it we use the *EM algorithm* procedure which alternates between the *Kalman filter* and maximum likelihood parameter estimation of the model in state space form to convergence (Schwartz 1997, Schwartz and Smith 2000, Geman and Nguyen 2005).

By F1, F3, F6, F9, F12, F15, and F17 we denote respectively the 1st, 3rd, 6th, 9th, 12th, 15th and 17th month futures contracts (in the order of their maturities) which we use in the calibration of the three-factor spot price model. Table 2.3 shows the parameter estimates. These estimated parameters are nearly all significant except for the risk premia. Both the **x** and **y** factors are significantly mean-reverting, as can be seen in Figure 2.4, which shows the estimated paths of the three factors. From the estimated parameters the short-term factor **x** has a half-life of about 2.5 months with a volatility of 20%, the medium-term factor **y** has a half-life of about 9.5 months with volatility 28% and the long-term factor **p** has a volatility of about 20%.

---

* Note that in order to make a comparison with the results of the two-factor models of Gibson and Schwartz (1990) and Schwartz and Smith (2000), we employ the same estimation method used in those papers. We also used the method proposed by Dempster and Tang (2011) to eliminate mean-reversion parameter estimation errors, but the results reported here are little changed.

2.3.4 Comparison with Two-Factor Models

To see whether or not the three-factor model is significantly better than two-factor models, we remove the **x** factor and re-estimate the resulting model with only the **y** and **p** factors.[*] Table 2.3 demonstrates that the pricing errors ($\xi_1, \ldots \xi_7$) are generally smaller than those reported for the Gibson–Schwartz model by Schwartz (1997). Comparing our two- and three-factor models, the inclusion of the **x** factor significantly improves the data fit, according to the likelihood ratio test,[†] and reduces pricing errors for short-term contracts.

2.4 Model Interpretation

In this section we study the relations between the historical paths of various economic variables and those of our latent factors (Figure 2.4), estimated in terms of the means of the sequential posterior Gaussian state distributions obtained from the Kalman filter using the optimal parameter estimates of the final iteration of the EM algorithm. Since commodities are both real and financial assets, we expect that both fundamental and financial variables will play an important role in explaining the three factors.

2.4.1 Explanatory Variable Specification

We classify our explanatory variables into four categories: (1) variables from other financial markets, such as US dollar index returns, SP500 equity index returns, etc.; (2) variables indicating the phase of the business cycle, such as the coincident business cycle index and the term spread on US interest rates; (3) variables indicating net demand for oil, such as the oil inventory level and the heating oil–crude oil spread; (4) trading variables, such as growth rate of open interest and hedging pressure for oil futures contracts.

More specifically, we utilized the following variables at weekly frequency.

2.4.1.1 Financial Variables

- **USD (y1):** Weekly returns of the US dollar index. This index measures the performance of the US dollar against a basket of currencies. It goes up when the US dollar gains strength relative to other currencies.
- **SPSOO (y2):** Weekly returns of the S&P500 equity index. Inclusion of SP500 returns controls for the possibility that investors were pursuing trading strategies in oil futures that are conditional on equity markets.
- **LIBOR (y3):** The weekly level of the three month US LIBOR rate. Casassus and Colin-Dufresne (2005) and Frankel (2008) show that interest rates tend to influence the willingness to store inventory and thus will influence the convenience yield.[‡]
- **VIX (y4):** The weekly level of the VIX index for the equity market. This index represents one measure of the market's expectation of stock market volatility over the next 30-day period. It is a weighted blend of prices for a range of options on the S&P 500 index.
- **CreditSpread (yS):** The weekly spread between Moody's BAA-rated and AAA-rated corporate bond yields, following Bailey and Chan (1993), as a proxy for the default premium.

---

[*] We have seen in Section 2.3.1 that this two-factor model is the same as the Gibson-Schwartz (1990) model.

[†] Note that this test applies to our nested case. The likelihood ratio test statistic is 4324, highly significant at the 99% confidence level for the Chi squared distribution with five degrees of freedom.

[‡] For example, a higher interest rate corresponds to a higher marginal cost of storage, a higher convenience yield and a futures term structure more likely to be in backwardation, as Keynes (1930) described. The current low interest rates have led through the opposite effect to an oil futures market in contango and a frenzy of oil storage building to exploit it physically (Bouchouev 2011).

2.4.1.2 Business Cycle Variables

- **TermSpread (y6):** The weekly spread between the 10 year and three month Treasury bond yields. The term spread between the long-term and short-term interest rates has often been found to be the most important predictor of economic recessions (see, e.g., Estrella and Mishkin [1998]).
- **CoinIndex (y7):** The *change* in weekly levels of the business cycle coincident index, as this index, obtained from the Conference Board Inc., is not stationary. However, since the coincident index is calculated at monthly frequency, we assume the weekly values of coincident index change are identical in each month.

2.4.1.3 Demand Variables

- **HOCLSpread (y8):** The weekly *log* spread between heating and crude oil prices. Since heating oil is the main product of crude oil, the (log) price spread can reflect the relative scarcity of crude oil (Casassus *et al.* 2010). Note that a high spread means that crude oil is cheap relative to heating oil and hence has low convenience yield, a phenomenon independent of heating oil, and hence spread, seasonality.
- **Inventory (y9):** US weekly crude oil inventory (excluding strategic petroleum reserves) in millions of barrels obtained from the US Energy Information Administration. Since this data is non-stationary, we follow Gorton *et al.* (2007) in first applying a Hodrick–Prescott (HP) filter to the whole time series. The detrended stationary part is used in our empirical analysis.

2.4.1.4 Trading Variables

- **OpenInterest (y10):** The weekly growth rate (log difference) of open interest for all oil contracts, obtained from the US Commodity Futures Trading Commission (CFTC).
- **HedgingPressure (y11):** The weekly hedging pressure, obtained from the US Commodity Futures Trading Commission (CFTC). This measure is calculated using the hedgers' short position less their long position normalized by total open interest.

2.4.2 SVAR Model Statement

We estimate a *structural vector auto-regression* (SVAR) model (Sims 1980, Hamilton 1994) to address the relationship between the three latent factors and the 11 explanatory variables, namely

$$BY_t = AY_{t-1} + \Sigma \varepsilon_Y, t, \tag{2.29}$$

where the vector of the variables is given by $Y_t := (y_{1,t}, \ldots, y_{11,t}, z_t')'$, with z_t a 3 vector representing the latent **x, y** and **p** factors, A and B are 14×14 matrices and $\varepsilon_{Y,t}$ is a vector of Gaussian disturbances with a spherical covariance matrix.

In structuring the SVAR matrix B we assume that all variables can influence the latent **x, y** or **p** factors, but that, consistent with our three-factor model, these factors do not (directly) influence each other. Similarly, the financial and business cycle variables do not directly influence each other, but they do affect the fundamental and trading variables. On the other hand, these latter variables do *not* influence financial and business cycle variables, i.e. the two blocks of variables are in cascade (causal Wold recursive) form. We also assume that the heating oil–crude oil spread can influence oil inventory and *vice versa*, i.e. they are determined simultaneously. Further, we assume that the fundamental demand and supply variables can influence futures trading and *vice versa*, but that open interest and hedgers' positions do not influence each other. These assumptions lead to a B matrix of (2.29), corresponding to each of the three latent factors *separately* in turn, in lower triangular form given by

$$
B = \begin{bmatrix}
1 & & & & & & & & & & & \\
0 & 1 & & & & & & & & & & \\
0 & 0 & 1 & & & & & & & & & \\
0 & 0 & 0 & 1 & & & & 0 & & & & \\
0 & 0 & 0 & 0 & 1 & & & & & & & \\
0 & 0 & 0 & 0 & 0 & 1 & & & & & & \\
0 & 0 & 0 & 0 & 0 & 0 & 1 & & & & & \\
x & x & x & x & x & x & x & 1 & & & & \\
x & x & x & x & x & x & x & x & 1 & & & \\
x & x & x & x & x & x & x & x & x & 1 & & \\
x & x & x & x & x & x & x & x & x & 0 & 1 & \\
x & x & x & x & x & x & x & x & x & x & x & 1
\end{bmatrix},
$$

with x representing non-zero elements. Note that, by using this B matrix, the full model can be estimated under our assumptions by separately estimating the resulting model for each latent factor in turn.

2.4.3 Results

To see how the exogenous variables influence our three latent variables, we first analyse the impulse response functions of the estimated SVAR model. Figures 2.5-2.7 show, for each of the **x, y** and **p** factors respectively, impulse response functions for a one standard deviation positive shock to each of the exogenous explanatory variables in turn. Note that the vertical scales in these diagrams are variable. The gray area in each represents the 95% confidence level obtained from bootstrapping.

2.4.3.1 x Factor Impulse Responses

The LIBOR rate impacts positively on the short-term convenience yield **x** factor, which is consistent with the standard argument of the theory of storage, i.e. a higher interest rate will result in a higher marginal cost of storing commodities. Thus a higher LIBOR rate should correspond to a higher convenience yield. We also see that credit spread co-moves with the **x** factor, which is consistent with Acharya *et al.* (2008) in that a higher default risk tends to encourage more commodity producers' hedging and thus a futures curve in deeper backwardation.

The log spread between heating and crude oil influences the **x** factor negatively. A high spread means a relatively low price of crude oil and less demand for it (or more crude oil inventory), corresponding to a lower convenience yield. For details of the equilibrium relationship between oil convenience yield and the heating oil-crude oil spread, see Casassus *et al.* (2010).

As shown by Hong and Yogo (2010), open interest in commodity futures forecasts commodity returns. Our result shows that a higher open interest growth rate corresponds to a futures term structure in deeper backwardation. We also see that more hedging pressure corresponds to deeper backwardation. This is consistent with Keynes (1930) and Hirshleifer (1990), i.e. the more futures contracts sold the deeper the backwardation of the futures term structure.

2.4.3.2 y Factor Impulse Responses

Similar to the situation with the **x** factor, we see that the LIBOR rate also impacts positively on the medium-term convenience yield **y** factor, which is again consistent with the theory of storage. We also see that the VIX correlates positively with the **y** factor. Since the VIX index can be seen as an indicator of the volatility of financial markets, the theory of storage says that higher volatility will lead to larger convenience yields, resulting here in a positive relationship between the VIX and the **y** factor.

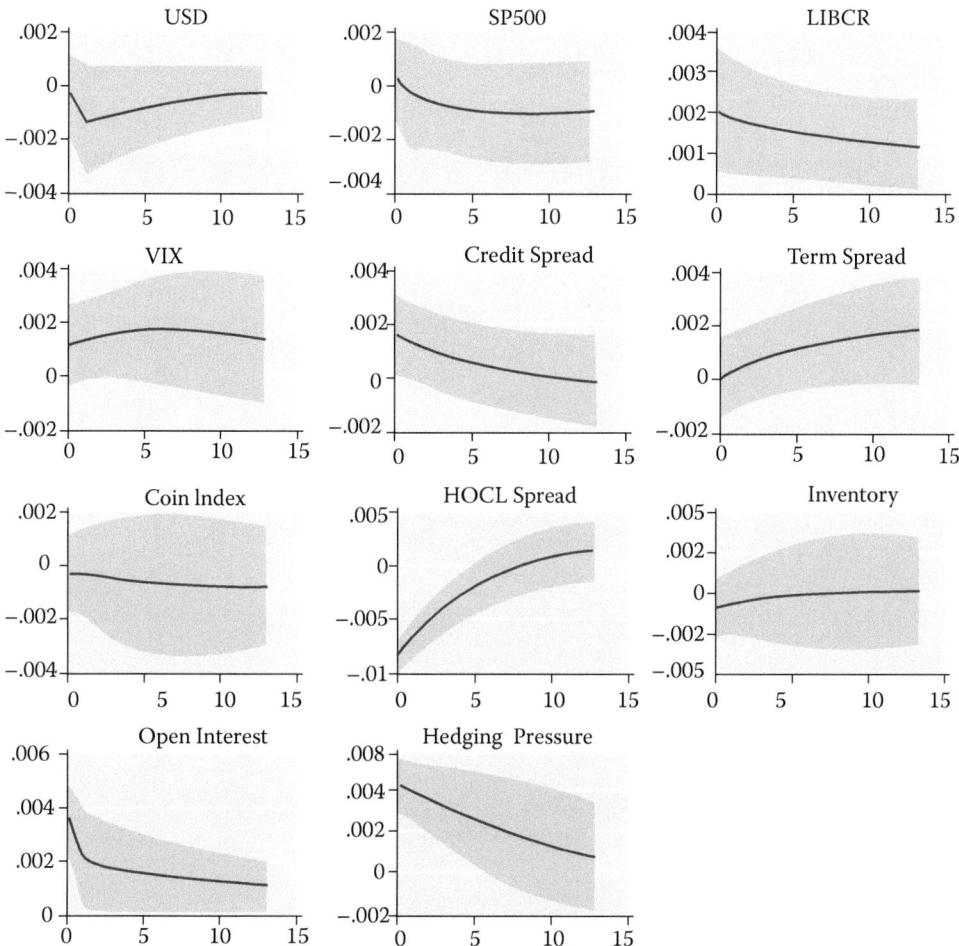

FIGURE 2.5 Impulse responses of the x factor to a one standard deviation positive shock of each of the 11 explanatory variables.

Both the term spread and the business cycle coincident index have a positive impact on the **y** factor. This is because a high term spread and coincident index both correspond to a booming state of the economy with a higher oil demand and hence a higher convenience yield. The log spread between heating and crude oil also has a negative impact on the **y** factor; the explanation for this is similar to that for the **x** factor. The inventory has a positive impact on the **y** factor, which is consistent with the theory of storage, i.e. high inventory results in higher convenience yield and deeper backwardation of the futures term structure (see (2.28)). Similar to the **x** factor, the **y** factor is also affected by the growth of open interest and the hedging pressure, but at lower impact levels.

2.4.3.3 p Factor Impulse Responses

First observe from Figure 2.7 that the impulse responses of the long-term **p** factor to explanatory variable shocks are of larger magnitude and converge faster to equilibrium than those of the **x** and **y** factors, as is consistent with its GBM dynamics.

The US dollar index co-moves negatively with the permanent shock **p** factor. This is because oil is traded in dollars, hence the depreciation of dollars should increase the price of oil due to the numeraire effect. This effect tends to be 'permanent', i.e. only affecting the long-term **p** factor in our model.

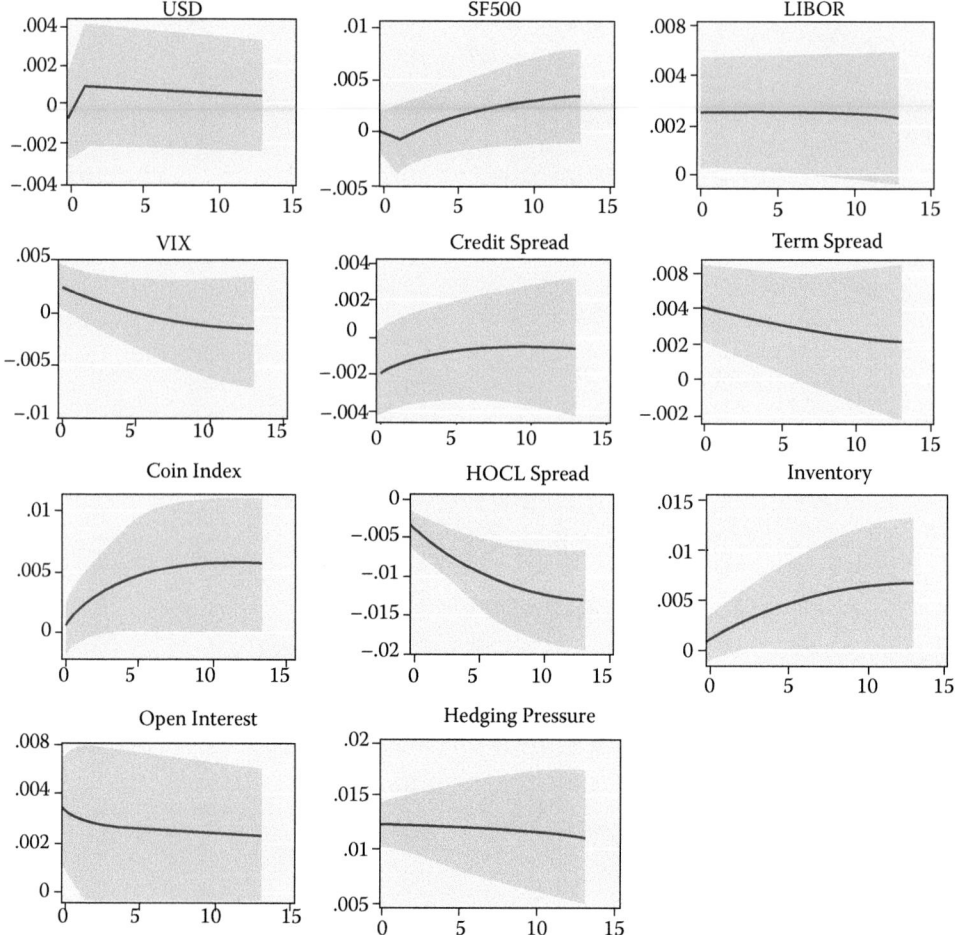

FIGURE 2.6 Impulse responses of the y factor to a one standard deviation positive shock of each of the 11 explanatory variables.

There have been several studies of the relationship between the stock market portfolio and futures prices (e.g., Dusak [1973] and Holthausen and Hughes [1978]). Using *t*-tests they found that no correlation exists between futures and market portfolio returns. However, by decomposing oil futures prices into three factors, we see that the long-term **p** factor does co-move with SP500 index returns, but the **x** and **y** factors do not.

Similar to the **x** and **y** factors, the heating oil–crude oil spread has a negative impact on the **p** factor as well.

2.4.3.4 Forecast Error Variance Decomposition

We complement the conclusions derived from impulse response analysis of our estimated SVAR model with the forecast error variance decomposition from one- and 13-step ahead rolling forecasts (see, e.g., Lütkepohl [2007]). Table 2.4 shows the variance decomposition for both one week and one quarter forecasts, with entries that are percentages of the forecast error variance of each factor accounted for by exogenous shocks to each explanatory variable, or an average of these percentages for a group of variables.

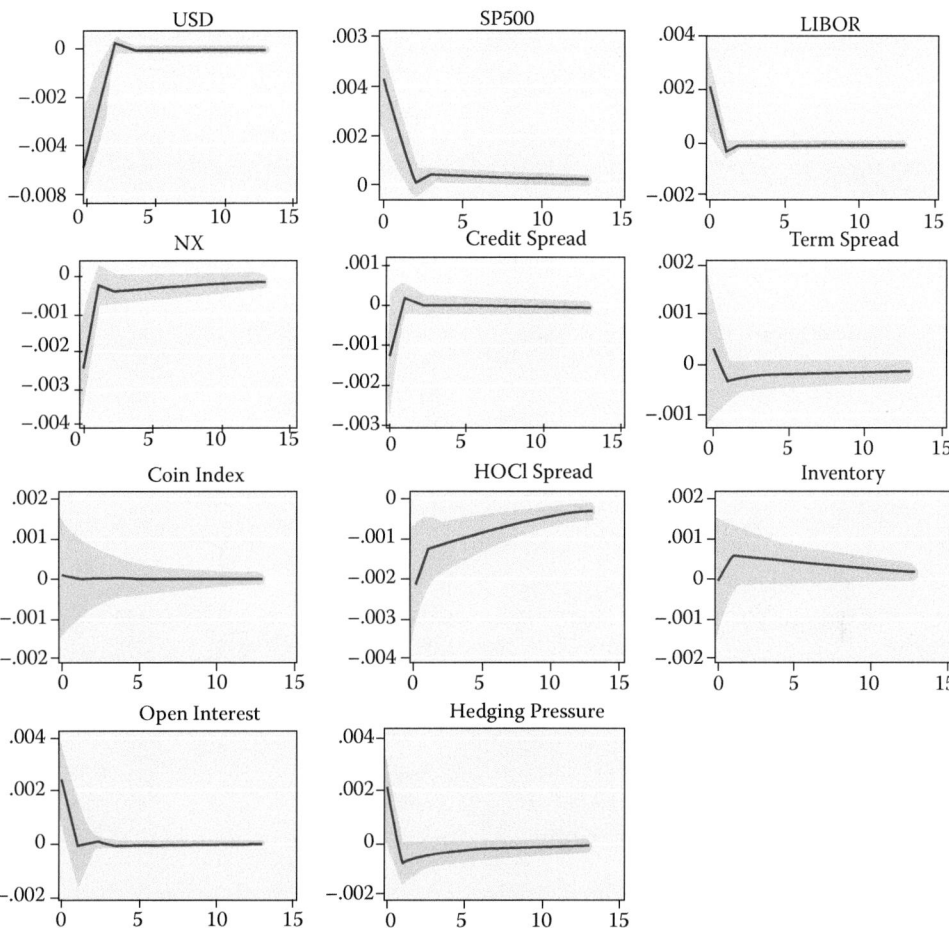

FIGURE 2.7　Impulse responses of the p factor to a one standard deviation positive shock of each of the 11 explanatory variables.

We see that the variability of the log spread between heating and crude oil strongly influences that of the **x** factor; however, this effect decreases rapidly as the forecasting horizon lengthens. Trading variables such as the growth of the open interest and the hedging pressure also play an important role. The **y** factor is influenced by the variability of the business cycle variables, term spread and coincident index. The fundamental demand and supply variables, heating-crude oil spread and the inventory also play a role, but their role is only significant at the 13-week forecasting horizon. Futures hedging pressure has a very strong influence on the variability of the medium-term **y** factor, but the long-term **p** factor is mainly influenced by the US dollar index and SP500 returns.

2.4.3.5 Summary

The financial variables mainly affect the **p** factor, however they also have a minimal influence on the **x** and **y** factors. The business cycle variables influence mainly the **y** factor; a booming state corresponding to larger **y** factors (deeper long-end backwardation). The fundamental supply and demand variables affect both **x** and **y** factors; higher net demand results in higher **x** factor (deeper short-end backwardation) and **y** factor (deeper long-end backwardation) levels. The trading variables influence both **x** and **y** factors; more intensive futures trading and stronger hedging pressure result in higher **x** and **y** factors.

TABLE 2.4 Forecast Error Variance Decomposition

Variable	x Factor		y Factor		p Factor	
	1-Week Horizon	13-Week Horizon	1-Week Horizon	13-Week Horizon	1-Week Horizon	13-Week Horizon
USD	0.02	0.2	0.03	0.06	**2.93**	**3.51**
SP500	0.00	0.25	0.00	0.35	**2.60**	**3.14**
LIBOR	0.62	0.70	0.48	0.53	0.61	0.60
VIX	0.15	0.65	0.43	0.12	0.82	0.88
Credit spread	0.38	0.17	0.26	0.08	0.23	0.23
Term spread	0.00	0.50	**1.21**	0.80	0.02	0.07
Coin index	0.01	0.14	0.01	**1.85**	0.00	0.00
HO—CO spread	**11.88**	**4.84**	1.03	**8.31**	0.64	1.59
Inventory	0.89	0.03	0.08	**2.11**	0.00	0.27
Open interest	**1.68**	0.55	0.80	0.60	0.83	0.79
Hedging pressure	**3.00**	**2.38**	**11.16**	**12.24**	0.59	0.70
Aggregate averages						
Financial	0.23	0.39	0.24	0.23	**1.44**	**1.67**
Business cycle	0.01	0.32	0.61	**1.33**	0.01	0.04
Fundamental	**6.39**	**2.44**	0.56	**5.21**	0.32	0.93
Trading	**2.34**	**1.47**	**5.98**	**6.42**	0.71	0.75

Note: This table reports the forecasting error variance decomposition for each factor over two rolling forecast horizons— one week and one quarter. Quantities in the table are percentages of the total factor variance.

2.5 Conclusion

In this paper we find that the two-factor models in the literature are not able to model the whole crude oil futures price term structure, especially at the short end. Hence we propose a three-factor model for commodity futures prices. This model is shown to be an extension of the Gibson–Schwartz (1990) (Schwartz–Smith 2000) model.

An affine combination of the **x** and **y** factors in our model represents convenience yield, while the third **p** factor models long-term trend. By regressing the three factors on several economic variables using an SVAR model, we see that the short-term **x** factor is highly correlated with demand and trading variables. The medium-term **y** factor has a relationship with the business cycle, net oil demand and trading variables. The long-term **p** factor is mainly related to financial variables. The business cycle and fundamental variables affect the movement and the shape of the oil futures price term structure; but financial and trading variables do as well. This phenomenon reflects the fact that commodities combine the characteristics of both real and financial assets.

Acknowledgements

We thank participants at the Cambridge Finance Seminar, Global Derivatives, Trading and Risk Management and the annual Cambridge-Princeton Conference for comments, Professors R. Carmona, A. Harvey, M. Pesaran and L.C.G. Rogers for helpful discussions, and three anonymous referees for criticism that forced us to materially improve the paper. Tang acknowledges financial support from the National Natural Science Foundation of China (grant No. 71171194).

References

Acharya, V., Lochstoer, L. and Ramadorai, T., Does hedging affect commodity prices? The role of producer default risk. Working paper, London Business School, 2008.

Bailey, W. and Chan, K., Macroeconomic influences and the variability of the commodity futures basis. *J. Finance*, 1993, **48**, 555–573.

Bessembinder, H., Coughenour, J., Seguin, P. and Smoller, M., Mean-reversion in equilibrium asset prices: Evidence from the futures term structure. *J. Finance*, 1995, **50**, 361–375.

Bouchouev, I., The inconvenience yield or the theory of normal contango. *Energy Risk*, September 2011.

Brennan, M., The supply of storage. *Am. Econ. Rev.*, 1958, **48**, 50–72.

Brennan, M. and Crew, N., Hedging long maturity commodity commitments with short-dated futures contracts. In *Mathematics of Derivative Securities*, edited by M.A.H. Dempster and S.R. Pliska, 1997 (Cambridge University Press: New York).

Casassus, J. and Collin-Dufresne, P., Stochastic convenience yield implied from commodity futures and interest rates. *J. Finance*, 2005, **60**, 2283–2331.

Casassus, J., Liu, P. and Tang, K., Long-term economic relationships and correlation structure in commodity markets. Working Paper, Cornell University, 2010.

Cashin, P., Liang, H. and McDermott, J., How persistent are shocks to world commodity prices. *IMF Staff Papers*, 2000, **47**, 177–217.

Cuddington, J. and Urzu, M., Trends and cycles in primary commodity prices. Working Paper, Georgetown University, Washington, DC, 1987.

Culp, C. and Miller, M., Metallgesellschaft and the economics of synthetic storage. *J. Appl. Corp. Finance*, 1995, **7**, 62–76.

Dai, Q. and Singleton, K., Specification analysis of affine term structure models. *J. Finance*, 2000, **55**, 1943–1978.

Dempster, M.A.H. and Tang, K., Estimating exponential affine models with correlated measurement errors: Applications to fixed income and commodities. *J. Bank. Finance*, 2011, **35**, 639–652.

Duffie, D., Pan, J. and Singleton, K., Transform analysis and asset pricing for affine jump-diffusions. *Econometrica*, 2000, **68**, 1343–1376.

Dusak, K., Futures trading and investor returns: An empirical investigation of commodity risk premium. *J. Polit. Econ.*, 1973, **81**, 1387–1406.

Estrella, A. and Mishkin, F., Predicting U.S. recessions: Financial variables as leading indicators. *Rev. Econ. Statist.*, 1998, **80**, 45–61.

Fama, E. and French, K., Business cycles and the behavior of metals prices. *J. Finance*, 1988, **43**, 1075–1093.

Frankel, J., An explanation for soaring commodity prices. EUVOX, 25 March 2008.

Geman, H. and Nguyen, V., Soybean inventory and forward curve dynamics. *Mgmt Sci.*, 2005, **51**, 1076–1091.

Gersovitz, M. and Paxson, C., The economies of Africa and the prices of their exports. Working Paper, Princeton Studies in International Finance No. 68, Princeton University, 1990.

Gibson, R. and Schwartz, E., Stochastic convenience yield and the pricing of oil contingent claims. *J. Finance*, 1990, **45**, 959–976.

Gorton, G., Hayashi, F. and Rouwenhorst, K., The fundamentals of commodity futures returns. Yale ICF Working Paper No. 07-08, Yale University, 2007.

Grilli, E. and Yang, M., Primary commodity prices, manufactured goods prices and the terms of trade of developing countries: What the long run shows. *World Bank Econ. Rev.*, 1988, **2**, 1–47.

Hamilton, J., *Time Series Analysis*, 1994 (Princeton University Press: Princeton NJ).

Hirshleifer, D., Hedging pressure and futures price movements in a general equilibrium model. *Econometrica*, 1990, **58**, 411–428.

Hong, H. and Yogo, M., Commodity market interest and asset return predictability. Working Paper, Princeton University, 2010.

Holthausen, D. and Hughes, J., Commodity returns and capital asset pricing. *Financ. Mgmt*, 1978, 7, 37–44.

Kaldor, N., Speculation and economic stability. *Rev. Econ. Stud.*, 1939, **7**, 1–27. Keynes, J.M., *A Treatise on Money*, 1930 (Macmillan: London).

Keynes, J.M., *A Treatise on Money: The Pure Theory of Money*, 1930 (The Economic Society).

Lütkepohl, H., *Introduction to Multiple Time Series Analysis*, 2007 (Springer: Berlin).

Mello, A. and Parsons, J., Maturity structure of a hedge matters: Lessons from the Metallgesellschaft debacles. *J. Appl. Corp. Finance*, 1995, **8**, 106–120.

Schwartz, E.S., The stochastic behavior of commodity prices: Implications for valuation and hedging. *J. Finance*, 1997, **52**, 923–973.

Schwartz, E. and Smith, J., Short term variations and long term dynamics in commodity prices. *Mgmt Sci.*, 2000, **46**, 893–911.

Sims, C., Macroeconomics and reality. *Econometrica*, 1980, **48**, 1–48.

Tang, K. and Xiong, W., Index investment and financialization of commodities. *Financial Analysts Journal*, 2012, **68**, 54–74.

Working, H., The theory of the price of storage. *Am. Econ. Rev.*, 1949, **39**, 1254–1262.

Appendix

Three-Factor Model in State Space Form

The state space form of a dynamic statistical model consists of a transition and a measurement equation. The *transition equation* describes the dynamics of the data-generating process of unobservable *state variables*. In our model this is a discrete-time version of (2.14) to (2.16). The *measurement equation* relates a multivariate time series of *observable variables,* here the future prices of different maturities, to the unobservable vector of state variables, the **x, y** and **p** factors. The measurement equation is obtained from (2.24) by adding *uncorrelated* noise to take into account *pricing errors.*[*] These errors may be caused by bid–ask spreads, non-simultaneity of the observations, etc.

In more detail, suppose the data are sampled at equally spaced times t_n, $n = 1, \ldots, N$, and that : $\Delta := t_{n+1} - t_n$, is the interval between two observations. Let $X_n := \left[x_{t_n} y_{t_n} p_{t_n} \right]'$ represent the vector of state variables at time t_n where the prime denotes transpose. Discretizing (2.14) to (2.16) we obtain the *transition equation* as

$$\mathbf{X}_{n+1} = AX_n + b + \mathbf{w}, \tag{A.1}$$

where **w** is a Gaussian random noise vector with mean 0 and covariance matrix Q and A, b and Q are given by

$$A = \begin{bmatrix} e^{-k_x \Delta} & 0 & 0 \\ 0 & e^{-k_y \Delta} & 0 \\ 0 & 0 & 1 \end{bmatrix}, \tag{A.2}$$

$$b = \begin{bmatrix} 0 & 0 & u\Delta \end{bmatrix}', \tag{A.3}$$

---

[*] But see Dempster and Tang (2011) for more general assumptions.

$$
Q = \begin{bmatrix}
\sigma_x^2 \dfrac{1-e^{(-2k_x\Delta)}}{2k_x} & \dfrac{\rho_{xy}\sigma_x\sigma_y}{k_x+k_y}\left(1-e^{-(k_x+k_y)\Delta}\right) & \dfrac{\rho_{xp}\sigma_x\sigma_p}{k_x}\left(1-e^{-k_x\Delta}\right) \\[3mm]
\dfrac{\rho_{xp}\sigma_x\sigma_p}{k_x+k_y}\left(1-e^{-(k_x+k_y)\Delta}\right) & \sigma_x^2\dfrac{1-e^{(-2k_y\Delta)}}{2k_y} & \dfrac{\rho_{yp}\sigma_y\sigma_p}{k_y}\left(1-e^{-k_y\Delta}\right) \\[3mm]
\dfrac{\rho_{xp}\sigma_x\sigma_p}{k_x}\left(1-e^{-k_x\Delta}\right) & \dfrac{\rho_{yp}\sigma_y\sigma_p}{k_y}\left(1-e^{-k_y\Delta}\right) & \sigma_p^2\Delta
\end{bmatrix}
\tag{A.4}
$$

Let $Z_n := [\ln F(t, t+\tau_1), \ldots, \ln F(t, t+\tau_M)]'$ represent log futures prices, where $\tau_1, \ldots, \tau_M$ are the *times to maturity* for these $1, \ldots, M$ futures contracts. From (2.24) the *measurement equation* becomes

$$
Z_n = C_n X_n + d_n + \varepsilon_n
\tag{A.5}
$$

where

$$
C_n = \begin{bmatrix}
e^{-k_x\tau_1} & e^{-k_y\tau_1} & 1 \\
\vdots & \vdots & \vdots \\
e^{-k_x\tau_M} & e^{-k_y\tau_M} & 1
\end{bmatrix},
\tag{A.6}
$$

$$
d_n = \begin{bmatrix}
\lambda_x\left(e^{-k_x\tau_1}-1\right)+\lambda_y\left(e^{-k_y\tau_1}-1\right)+\left(u-\lambda_p\right)\tau_1 \\
+\dfrac{1}{2}\left(\begin{array}{c} \dfrac{1-e^{-2k_x\tau_1}}{2k_x}\sigma_x^2+\dfrac{1-e^{-2k_y\tau_1}}{2k_y}\sigma_y^2+\sigma_p^2\tau_1 \\[2mm] +2\dfrac{1-e(k_x+k_y)\tau_1}{k_x+k_y}\rho_{xy}\sigma_x\sigma_y+2\dfrac{1-e^{-k_x\tau_1}}{k_x}\rho_{xp}\sigma_x\sigma_p+2\dfrac{1-e^{-k_y\tau_1}}{k_y}\rho_{yp}\sigma_y\sigma_p \end{array}\right) \\
\vdots \\
\lambda_x\left(e^{-k_x\tau_M}-1\right)+\lambda_y\left(e^{-k_y\tau_M}-1\right)+\left(u-\lambda_p\right)\tau_M \\
+\dfrac{1}{2}\left(\begin{array}{c} \dfrac{1-e^{-2k_x\tau_M}}{2k_x}\sigma_x^2+\dfrac{1-e^{-2k_y\tau_M}}{2k_y}\sigma_y^2+\sigma_p^2\tau_M \\[2mm] +2\dfrac{1-e^{-(k_x+k_y)\tau_M}}{k_x+k_y}\rho_{xy}\sigma_x\sigma_y+2\dfrac{1-e^{-k_x\tau_M}}{k_x}\rho_{xp}\sigma x\sigma_p+2\dfrac{1-e^{-k_y\tau_M}}{k_y}\rho_{yp}\sigma_y\sigma_p \end{array}\right)
\end{bmatrix},
\tag{A.7}
$$

and ε_n is an error term allowing noise in the sampling of data with covariance matrix*

$$
H = \begin{bmatrix}
\xi_1^2 & 0 & 0 \\
\vdots & \vdots & \vdots \\
0 & 0 & \xi_M^2
\end{bmatrix}.
\tag{A.8}
$$

---

* But see Dempster and Tang (2011) for more general assumptions.

3

Pricing and Hedging of Long-Term Futures and Forward Contracts with a Three-Factor Model

Kenichiro Shiraya
and
Akihiko Takahashi

3.1 Introduction

The term structure of commodity futures exhibits complex shape changes and a number of different models have been proposed for its estimation. In this paper, we propose a three-factor model to estimate the term structure of commodity futures, and then propose and verify effective hedging techniques for long-term futures and forwards estimated with the model, using as hedging instruments the short- and medium-term futures that are tradable.

Black (1976) advocated the idea of trading commodities as 'equities without dividends' and made use of geometric Brownian motion. However, given the complexity of the shapes associated with the term structure of commodities futures, the simple geometric Brownian motion model proposed by Black (1976) is not a good fit. To resolve this problem, mean reversion has been introduced. Unlike equities, when commodities prices rise, there is generally (albeit with a time lag) an increase in supply; conversely, when prices decline, supply decreases. The fact that prices are determined by the supply and

DOI: 10.1201/9781003265399-4

demand balance means that the supply side adjusts supply volumes, which has the effect of constraining the potential for commodities prices to move in a single direction. That is why it is generally considered appropriate to employ mean reversion in commodity pricing models. Much empirical research has been carried out on this subject. For example, it was verified by Bessembinder *et al.* (1995).

Nonetheless, even if mean reversion is used in a one-factor model, it is difficult to represent the complex term structure of commodities futures, leading Gibson and Schwartz (1990) to propose a model that supplements the fluctuation of spot prices with a convenience yield stochastic process, and Schwartz (1997) to propose a model that explicitly employs convenience yields and interest rates as stochastic factors.

On the other hand, different methods have been proposed that do not attempt to individually model commodities' spot prices, convenience yields or interest rates, but instead attempt direct modelling using state variables with the mean reversion of spot prices. Examples of direct modelling of spot prices include Schwartz and Smith's (2000) two-factor mean reversion model, Casassus and Collin-Dufresne's (2005) three-factor mean reversion model and Cortazar and Naranjo's (2006) *N*-factor mean reversion model.

We use three-factor Gaussian models with or without constant mean reversion. The models' parameters are estimated using a Kalman filter and have been confirmed to reproduce actual futures prices on the NYMEX WTI (light sweet crude oil) and LME Copper markets. Where our research differs from prior research is that we study cases both with and without a constant mean reversion level in the commodities price model and provide a detailed analysis not only of the model's ability to reproduce futures prices, but also its utility in hedging.

Commodities hedging is a long-debated topic. For example, Culp and Miller (1995), Mello and Parsons (1995) and many others have discussed it in terms of the Metallgesellschaft case. Culp and Miller (1995) explain that, like equities, etc., the forward prices for commodities are determined by the mechanism of 'cost of carry' and argue that long-term forward contracts can be hedged by holding short-term futures and rolling over the contract months. On the other hand, Mello and Parsons (1995) acknowledge that it is possible to use short-term futures to hedge long-term forward contracts, but criticize the hedging technique employed by Metallgesellschaft, which was to use a unit of short-term futures to hedge a unit of long-term forward contracts. They use the Gibson and Schwartz (1990) model to demonstrate that short-term prices are more sensitive to spot price changes than long-term prices and that the actual number of short-term futures required to hedge one unit of long-term forward contracts is approximately 0.3. Because of this, the trading of Metallgesellschaft, while having hedging elements, is deemed to be primarily futures speculation. Schwartz (1997) also comments on this point, using one- to three-factor models to calculate hedge positions and explaining that when one factor is used, the position is significantly less than 1, approximately 0.2–0.4, and even with two and three factors it is still, on a net basis, less than 1. Neuberger (1999) uses multiple contracts to hedge long-term exposure and shows the benefits of the simultaneous use of different hedging instruments.

Examples of research analysing not only hedge positions but also hedging errors include Brenann and Crew (1997), Korn (2005) and Buhler *et al.* (2004). Brenann and Crew (1997) attempt to use a number of different expiring futures as hedging instruments for hedges using a two-factor model, but all of the futures used as hedges expire within 6 months, and the futures to be hedged are also extremely short at no more than 2 years. Buhler *et al.* (2004) use several different models to compare and analyse performance when hedging 10-year forward contracts. However, the futures used as hedging instruments are extremely short, expiring in no more than 2 months, and the data are also only up to 1996, therefore, this analysis does not incorporate the rapid rises in commodities prices seen in recent years. Korn (2005) showed the hedging error with one- and two-factor models, but he did not show it with a three-factor model.

In this paper, we compare the hedging error using Metallgesellschaft's parallel hedging and multifactor model based hedging. More specifically, we verify the stability of hedges based on two- and three-factor models that do and do not have a constant mean reversion level, and we provide a detailed analysis of the differences in hedge effectiveness due to differences in the way in which state variables are calculated and differences in the required futures units, and hedging error rate distribution (based on simulations) due to differences in the contract months of the futures used as hedging instruments. We also use time series data to verify hedges for long-term forward contracts, for which interest rate factors have been taken into account. We find that the three-factor model without constant mean reversion level can effectively hedge long-term futures against the complex changes in term structures of recent years.

In Section 3.2, we propose a three-factor model including a two-factor model as a special case, which does not explicitly incorporate interest rates or convenience yields, and we use that model to derive an analytic solution for futures prices. Section 3.3 makes use of Kalman filters to estimate the model's parameters. Section 3.4 makes use of short- and medium-term futures to create a hedging technique for long-term futures and to analyse performance when this hedging strategy is used. Section 3.5 takes a more practical approach, analysing hedges on 'Out of Sample' and long-term forward contracts. Section 3.6 uses a simulation to analyse how the form of the distribution changes for the hedge error rate depending upon the selection of futures contract months. In the appendix, we provide the expectation and covariance of the model expressed in futures prices and notes on the numbers of units of nearer maturity futures required to hedge long-term futures.

3.2 Model

We first describe a three-factor Gaussian model used for pricing and hedging futures and forward contracts. S_t represents the spot price of a commodity at time t. The logarithm of spot price at this time is expressed by the following equation:

$$\log S_t = x_t^1, \tag{3.1}$$

where x^1 expresses a state variable corresponding to the spot price of the commodity and follows the stochastic differential equation

$$dx_t^1 = \kappa\left(x_t^2 + x_t^3 - x_t^1\right)dt + \sigma_1 dW_t^1,$$

$$dx_t^2 = -\gamma x_t^2 dt + \sigma_2 dW_t^2, \tag{3.2}$$

$$dx_t^3 = \left(\alpha - \beta x_t^3\right)dt + \sigma_3 dW_t^3,$$

where x^2 expresses a state variable corresponding to the difference between medium-term and long-term commodity futures prices, and x^3 is a state variable corresponding to the long-term portion of the term structure. $W_t^i (i = 1, 2, 3)$ have the following correlations of standard Brownian motions under an equivalent martingale measure (EMM):

$$dW_t^i \cdot dW_t^j = \rho_{ij} dt, i, j = 1, 2, 3. \tag{3.3}$$

Parameter κ expresses x^1, s speed of reversion to $x^2 + x^3$, and γ expresses x^2, s speed of attenuation. If $\gamma > 0$, then x^2 is pulled back towards 0. β expresses the speed with which x^3 reverts to α/β when $\beta \neq 0$. Therefore, intuitively, if $\kappa > \gamma > \beta > 0$, over the course of time the spot price x^1 (spot price) $\rightarrow x^2 + x^3$ (medium-term price) $\rightarrow x^3$(long-term price) with the trend expressed.

The stochastic differential equations of individual state variables can be solved analytically and expressed as follows:

$$x_t^1 = e^{-\kappa t}x_0^1 + \frac{\kappa}{\kappa - \gamma}\left(e^{-\gamma t} - e^{-\kappa t}\right)x_0^2 + \frac{\kappa}{\kappa - \beta}\left(e^{-\beta t} - e^{-\kappa t}\right)x_0^3$$

$$+ \frac{\alpha}{\beta}\left(1 - \frac{\kappa}{\kappa - \beta}e^{-\beta t} + \frac{\beta}{\kappa - \beta}e^{-"t}\right) + \sigma_1\int_0^t e^{-\kappa(t-s)}dW_s^1$$

$$+ \sigma_2\frac{\kappa}{\kappa - \gamma}\int_0^t \left(e^{-\gamma(t-s)} - e^{-\kappa(t-s)}\right)dW_s^2$$

$$+ \sigma_3\frac{\kappa}{\kappa - \beta}\int_0^t \left(e^{-\beta(t-s)} - e^{-\kappa(t-s)}\right)dW_s^3,$$

(3.4)

$$x_t^2 = e^{-\gamma t}x_0^2 + \sigma_2\int_0^t e^{-\gamma(t-s)}dW_s^2,$$

$$x_t^3 = e^{-\beta t}x_0^3 + \frac{\alpha}{\beta}\left(1 - e^{-\beta t}\right) + \sigma_3\int_0^t e^{-\beta(t-s)}dW_s^3.$$

From this the futures price is expressed as shown below.

Theorem 3.1: Using $G_T(t)$ to represent the price at time t of a futures with expiration T, under the EMM:

$$G_T(t) = E_t\left[S_T\right] = \exp\left\{\mu_{11}\left(x_t^1, x_t^2, x_t^3, T - t\right) + \frac{\Sigma_{11}(T - t)}{2}\right\}.$$

(3.5)

In this equation, E_t denotes the conditional expectation at time t. For a discussion of μ_{11} and Σ_{11} see Appendix A.

Proof: S_T has a log-normal distribution and the result can therefore be found by calculating the moment generating function of the normal distribution.

Next consider the market price of risk. $\theta(t) = (\theta_1(t), \theta_2(t), \theta_3(t))$ consists of the market price of risk for state variables x^1, x^2 and x^3. At this time, the following relationship holds true between the observed measure P and the equivalent Martingale measure Q:

$$W_t^P = W_t^Q + \int_0^t \theta(u)du.$$

(3.6)

Therefore, under measure P, the stochastic differential equations that are satisfied by individual state variables are

$$dx_t^1 = \kappa\left(x_t^2 + x_t^3 - x_t^1\right)dt + \sigma_1\theta_1(t)dt + \sigma_1 dW_t^{1,P}$$

$$dx_t^2 = -\gamma x_t^2 dt + \sigma_2\theta_2(t)dt + \sigma_2 dW_t^{2,P},$$

(3.7)

$$dx_t^3 = \left(\alpha - \beta x_t^3\right)dt + \sigma_3\theta_3(t)dt + \sigma_3 dW_t^{3,P}.$$

In particular, rewriting $\theta(t)$ with the state variables and time as $\theta(t, x^1, x^2, x^3)$:

$$\theta_1\left(t, x^1, x^2, x^3\right) = -a\left(x_t^1 - x_t^2 - x_t^3\right),$$

$$\theta_2\left(t, x^1, x^2, x^3\right) = -bx_t^2,$$

(3.8)

$$\theta_3\left(t, x^1, x^2, x^3\right) = \begin{cases} c - dx_t^3, & (\beta \neq 0), \\ c & (\beta = 0). \end{cases}$$

The stochastic differential equation (3.7) can therefore be rewritten as

$$dx_t^1 = \hat{\kappa}\left(x_t^2 + x_t^3 - x_t^1\right)dt + \sigma_1 dW_t^{1,P},$$

$$dx_t^2 = -\hat{\gamma}x_t^2 dt + \sigma_2 dW_t^{2,P}, \tag{3.9}$$

$$dx_t^3 = \left(\hat{\alpha} - \hat{\beta}x_t^3\right)dt + \sigma_3 dW_t^{3,P},$$

where

$$\hat{\kappa} = \kappa + \sigma_1 a, \quad \hat{\gamma} = \gamma + \sigma_2 b, \quad \hat{\alpha} = \alpha + \sigma_3 c, \quad \hat{\beta} = \beta + \sigma_3 d.$$

Remark 3.1: In the discussion above, when $\beta = 0$, solving for the limit will give an analytic expression. Also, when $\beta = 0$, x^3 does not have a constant mean reversion level and the model itself does not have an ultimate mean reversion level. Below, we refer to cases where $\beta \neq 0$ as the 'constant mean reversion model', and $\beta = 0$ as the 'stochastic mean reversion model'. Both types of models are essentially contained by Cortazar and Naranjo (2006) and Casassus and Collin-Dufresne (2005).

Remark 2.2: A two-factor constant mean reversion or a two-factor stochastic mean reversion model can be obtained by setting $x^2 = 0$. These models are essentially the same as in Korn (2005), who used two-factor models for an analysis of hedging. For the two-factor model in the subsequent analysis, we place the restriction $x^2 \equiv 0$ on our three-factor models.

3.3 Estimation of Parameters

This section estimates the parameters in the model. Using v_n and w_n as white noise with mean 0 and variance 1, the model described above can be expressed as the following system and observation models.
   System model:

$$x_n = F_n x_{n-1} + C_n^x + Q_n v_n,$$

$$x_n = \begin{pmatrix} x^1 \\ x^2 \\ x^3 \end{pmatrix},$$

$$F_n = \begin{pmatrix} e^{-\hat{\kappa}\Delta t} & \dfrac{\hat{\kappa}}{\hat{\kappa}-\hat{\gamma}}\left\{e^{-\hat{\gamma}\Delta t} - e^{-\hat{\kappa}\Delta t}\right\} & \dfrac{\hat{\kappa}}{\hat{\kappa}-\hat{\beta}}\left\{e^{-\hat{\beta}\Delta t} - e^{-\hat{\kappa}\Delta t}\right\} \\ 0 & e^{-\hat{\gamma}\Delta t} & 0 \\ 0 & 0 & e^{-\hat{\beta}\Delta t} \end{pmatrix},$$

$$E[Q_n] = 0, \quad G_n = \text{Cov}[Q_n] = \left(\Sigma_{ij}(\Delta t)\right),$$

$$C_n^x = \begin{pmatrix} \dfrac{\hat{\alpha}}{\hat{\beta}}\left\{1 - \dfrac{\hat{\kappa}e^{-\hat{\beta}\Delta t} - \hat{\beta}e^{-\hat{\kappa}\Delta t}}{\hat{\kappa}-\hat{\beta}}\right\} \\ 0 \\ \dfrac{\hat{\alpha}}{\hat{\beta}}\left\{1 - e^{-\hat{\beta}\Delta t}\right\} \end{pmatrix},$$

where $\Sigma_{ij}(\Delta t)$ is the covariance. For the specific formulae, see Appendix A.

Observation model:

$$y_n = H_n x_n + C_n^y + R_n w_n,$$

$$y_n = \begin{pmatrix} \log G_{T_{1_n}}(0) \\ \vdots \\ \log G_{T_{m_n}}(0) \end{pmatrix}, H_n = \begin{pmatrix} e^{-\kappa T_{1_n}} & \dfrac{\kappa}{\kappa-\gamma}\left\{e^{-\gamma T_{1_n}} - e^{-\kappa T_{1_n}}\right\} & \dfrac{\kappa}{\kappa-\beta}\left\{e^{-\beta T_{1_n}} - e^{-\kappa T_{1_n}}\right\} \\ \vdots & \vdots & \vdots \\ e^{-\kappa T_{m_n}} & \dfrac{\kappa}{\kappa-\gamma}\left\{e^{-\gamma T_{m_n}} - e^{-\kappa T_{m_n}}\right\} & \dfrac{\kappa}{\kappa-\beta}\left\{e^{-\beta T_{m_n}} - e^{-\kappa T_{m_n}}\right\} \end{pmatrix},$$

$$R_n = \begin{pmatrix} h_1^2 & & 0 \\ & \ddots & \\ 0 & & h_m^2 \end{pmatrix}, C_n^y = \begin{pmatrix} \dfrac{\Sigma_{11}(T_{1_n})}{2} + \dfrac{\alpha}{\beta}\left\{1 - e^{-\kappa T_{1_n}} - \dfrac{\kappa}{\kappa-\beta}\left\{e^{-\beta T_{1_n}} - e^{-\kappa T_{1_n}}\right\}\right\} \\ \vdots \\ \dfrac{\Sigma_{11}(T_{m_n})}{2} + \dfrac{\alpha}{\beta}\left\{1 - e^{-\kappa T_{m_n}} - \dfrac{\kappa}{\kappa-\beta}\left\{e^{-\beta T_{m_n}} - e^{-\kappa T_{m_n}}\right\}\right\} \end{pmatrix},$$

where R_n are the observational errors and $h_i, i = (1,\ldots,m)$ are the standard deviations.

In light of the computational burden, we assume that the observational errors of futures at individual maturities are independent. The parameters are estimated using the Kalman filter of this state-space representation. More specifically, the following prediction and filtering are alternately repeated and a parameter set ϑ is obtained so as to maximize the log-likelihood.

Prediction:

$$x_{n|n-1} = F_n x_{n-1|n-1} + C_n^x,$$

$$V_{n|n-1} = F_n V_{n-1|n-1} F_n^t + G_n.$$

Filtering:

$$
\begin{aligned}
d_{n|n-1} &= H_n V_{n|n-1} H_n^t + R_n, \\
K_n &= V_{n|n-1} H_n^t d_{n|n-1}^{-1}, \\
x_{n|n} &= x_{n|n-1} + K_n\left(y_n - H_n x_{n|n-1} - C_n^y\right), \\
V_{n|n} &= \left(I - K_n H_n\right) V_{n|n-1}.
\end{aligned}
$$

Log-likelihood:

$$l(\vartheta) = -\frac{1}{2}\left\{mN\log(2\pi) + \sum_{n=1}^{N}\log\left|\det\left(d_{n|n-1}\right)\right| + \sum_{n=1}^{N} u_n^t d_{n|n-1}^{-1} u_n\right\},$$

$$u_n = y_n - H_n x_{n|n-1} - C_n^y.$$

Even if optimal values are not set for the initial values of x and V, as the calculation proceeds using the Kalman filter, both approach optimal values. Therefore, the initial value problem can be avoided by discarding several steps when estimating parameters without using the likelihood calculation. Estimations of two-factor models are obtained similarly.

3.3.1 Estimation Results

The constant mean reversion model and the stochastic mean reversion model parameters were estimated using the procedure described above. The following data were used for the estimations.

3.3.1.1 NYMEX WTI (Light Sweet Crude Oil)

Data in five business day increments were used for the periods January 1997–October 2002, January 1997–October 2003 and January 1997–November 2007; futures contracts were: Front Month, 1 Dec., 2 Dec., 3 Dec., 4 Dec., 5 Dec., 6 Dec. and 7 Dec. Here, *j* Dec. indicates the *j*-th month contract expiring in December. If the Front Month is 1 Dec., it was used as the front month. Data up to 10 Dec. exists after April 2007. However, data from 8 Dec. to 10 Dec. were not used in the estimation due to their lack of reliability.

3.3.1.2 LME Copper

Data in five business day increments were used for the periods September 2002–November 2004 and September 2002–December 2007; futures contract were: Front Month, 1 Dec., 2 Dec., 3 Dec., 4 Dec., 5 Dec. and 6 Dec. If the Front Month is 1 Dec., it was used as the front month.

These are liquid and typical assets of oil and metal futures. The choice of time period is the longest period for which the data have mid-term (7 Dec. in WTI, 6 Dec. in Copper) futures. Tables 3.1-3.4 show the parameters and observational errors R_n obtained using the data described above.

Here, we note that observational errors in three-factor models are very small and that the model replicates the observed futures prices very well. For the two-factor constant mean reversion model, estimates of β using the WTI data up to 2003 and Copper data up to 2007 were $\beta \approx 0.4$, and hence, x^3 was observed to have a constant mean reversion level, but, in all other periods, both WTI and Copper had a β value of virtually 0. Therefore, x^3 does not fluctuate with a constant mean reversion level, but is rather more similar to a random walk. The parameters for Copper are significantly different between the data set up to 2004 and the data set up to 2007. When estimations are made using the data up to 2004, there is only a little more than 2 years of data used, and presumably the calculation results are biased parameters that are optimized to these 2 years. The market price of risk is expressed largely in parameters c and d for both WTI and Copper. We also observe that the standard errors of Copper's parameters are worse than those of WTIs, due, in part, to the lack of data used in the estimation. Finally, three-factor models show better fitting results than two-factor models in terms of AIC (Akaike's Information Criterion).

3.3.2 Comparison with Actual Data

This section verifies the degree of correlation between the state variables calculated with the Kalman filter using data through 2007 and settlement futures prices for NYMEX WTI and LME Copper.

As explained in Section 3.2, the state variables correspond to the term structure of futures. In this case, the state variables are assumed to have the correspondences given in Table 3.5 and the analysis seeks to determine the degree of correlation between them. Table 3.6 shows the correlations for state variables and logarithmic prices calculated from five business day increments.

Both WTI and Copper generally have high correlations, indicating that the movement of state variables roughly corresponds to the actual data. Also, the three-factor models provide higher correlations than the two-factor models.

Next, we examine whether the models can reproduce the actual term structures of futures prices. Figure 3.1 shows the term structures of two-factor and three-factor models against market prices of WTI futures on 3 November 2003, 1 November 2004, 1 November 2005, 1 November 2006 and 1 November 2007.

TABLE 3.1 Three-Factor Model (WTI)

	Constant mean reversion model						Stochastic mean reversion model					
	Nov 07	Std Err	Oct 03	Std Err	Oct 02	Std Err	Nov 07	Std Err	Oct 03	Std Err	Oct 02	Std Err
κ	1.112	0.011	1.007	0.014	1.107	0.020	1.090	0.009	1.159	0.015	1.048	0.017
γ	0.275	0.007	0.159	0.012	0.293	0.018	0.262	0.007	0.284	0.009	0.253	0.011
α	0.006	0.010	0.052	0.023	0.311	0.024	-0.009	0.001	-0.007	0.001	-0.007	0.002
β	0.004	0.001	0.021	0.008	0.110	0.007	–	–	–	–	–	–
a	0.000	–	0.000	–	0.597	1.787	0.000	–	0.000	–	0.529	1.861
b	0.000	–	0.000	–	0.000	–	0.000	–	0.000	–	0.000	–
c	0.528	0.341	0.000	–	0.070	0.219	0.482	0.266	0.062	0.318	0.063	0.347
d	0.005	0.019	0.002	0.039	0.017	0.071	–	–	–	–	–	–
$Sigma_1$	0.362	0.011	0.359	0.014	0.371	0.015	0.3630	0.010	0.387	0.015	0.363	0.013
$Sigma_2$	0.144	0.004	0.202	0.028	0.374	0.043	0.142	0.003	0.137	0.004	0.138	0.005
$Sigma_3$	0.170	0.059	0.251	0.067	0.399	0.065	0.190	0.005	0.171	0.006	0.182	0.008
$Rho_{[12]}$	0.155	0.047	0.048	0.077	-0.173	0.051	0.153	0.047	0.155	0.065	0.192	0.070
$Rho_{[23]}$	-0.601	0.205	-0.851	0.199	-0.983	0.116	-0.539	0.033	-0.599	0.052	-0.619	0.062
$Rho_{[31]}$	0.396	0.141	0.237	0.067	0.352	0.065	0.369	0.034	0.282	0.047	0.261	0.051
Front Month	0.041	0.001	0.049	0.002	0.047	0.002	0.041	0.001	0.047	0.002	0.045	0.002
1Dec	0.000	–	0.004	0.002	0.000	–	0.000	–	0.000	–	0.000	–
2Dec	0.007	0.000	0.013	0.001	0.006	0.000	0.007	0.000	0.008	0.000	0.007	0.000
3Dec	0.002	0.000	0.008	0.000	0.003	0.000	0.002	0.000	0.002	0.000	0.002	0.000
4Dec	0.004	0.000	0.001	0.000	0.004	0.000	0.004	0.000	0.005	0.000	0.005	0.000
5Dec	0.003	0.000	0.001	0.000	0.002	0.000	0.003	0.000	0.003	0.000	0.003	0.000
6Dec	0.000	–	0.001	0.000	0.002	0.000	0.000	–	0.000	–	0.000	–
7Dec	0.004	0.000	0.003	0.000	0.004	0.000	0.004	0.000	0.004	0.000	0.004	0.000
AIC	-27157.30		-16837.37		-14499.01		-27146.66		-16701.48		-14241.34	

TABLE 3.2 Three-Factor Model (Copper)

	Constant mean reversion model				Stochastic mean reversion model			
	Dec 07	Std Err	Nov 04	Std Err	Dec 07	Std Err	Nov 04	Std Err
K	0.766	0.046	0.918	0.173	0.740	0.039	0.930	0.180
γ	0.177	0.036	0.143	0.069	0.161	0.024	0.153	0.036
α	0.329	0.114	0.036	0.633	−0.059	0.010	−0.014	0.009
β	0.052	0.015	0.007	0.084	–	–	–	–
a	0.000	–	0.000	–	0.000	–	0.000	–
b	0.000	–	1.720	2.915	0.000	–	2.044	1.890
c	0.092	0.212	0.079	0.567	0.153	0.331	0.096	0.546
d	0.000	–	0.000	0.023	–	–	–	–
Sigma$_1$	0.268	0.013	0.224	0.021	0.268	0.013	0.224	0.021
Sigma$_2$	0.631	0.199	0.317	0.307	0.437	0.043	0.287	0.056
Sigma$_3$	0.658	0.201	0.319	0.350	0.440	0.039	0.287	0.046
Rho$_{\{12\}}$	0.074	0.080	0.368	0.339	0.256	0.069	0.402	0.112
Rho$_{\{23\}}$	−0.929	0.057	−0.840	0.593	−0.845	0.033	−0.801	0.072
Rho$_{\{31\}}$	0.237	0.082	0.130	0.250	0.200	0.070	0.155	0.139
Front Month	0.000	–	0.000	–	0.000	–	0.000	–
1Dec	0.007	0.000	0.006	0.001	0.007	0.000	0.006	0.001
2Dec	0.003	0.000	0.001	0.001	0.003	0.000	0.001	0.001
3Dec	0.005	0.000	0.004	0.000	0.005	0.000	0.003	0.000
4Dec	0.000	–	0.003	0.001	0.000	–	0.003	0.001
5Dec	0.010	0.001	0.002	0.002	0.010	0.001	0.002	0.002
6Dec	0.011	0.000	0.008	0.002	0.013	0.001	0.008	0.001
AIC	−9538.94		−4259.00		−9520.27		−4264.08	

Figure 3.2 shows the results for Copper on 1 December 2003, 1 December 2004, 1 December 2005, 1 December 2006 and 3 December 2007. We can see in those cases that three-factor models can replicate the actual term structures well, whereas two-factor models have some difficulty in capturing the actual term structures. In particular, the difference in fitting between the two-factor model and the three-factor model frequently occurs in 2006 and 2007 to the extent that can be observed in Figures 3.1 and 3.2.

3.4 Futures Hedging Techniques

This section describes a method for constructing a hedging strategy for one unit of a long-term futures contract and describes how the three-factor model used in this paper can be applied to this task.

The equation expressing the futures price uses state variables x^1, x^2 and x^3 so that the shape of the futures price changes according to changes in these state variables (assuming no change in the parameters). Therefore, it is possible in theory to hedge against long-term futures price fluctuations by calculating the deltas of the state variables for the long-term futures price and taking a position $\Phi = (\phi_1, \phi_2, \phi_3)^t$ in the nearer maturity futures that cancels out those deltas.

In a three-factor model, there are three factors to be hedged and therefore, futures with three different expiration maturities will be required to build the hedge portfolio. $G_{T_1}(t)$, $G_{T_2}(t)$ and $G_{T_3}(t)$ express nearer maturity futures prices of different expirations, and $G_{T_4}(t)$ the long-term futures price to be hedged. In this case, Φ is the solution to the following simultaneous equation:

$$A\Phi = b,$$

TABLE 3.3 Two-Factor Model (WTI)

	Constant mean reversion model						Stochastic mean reversion model					
	Nov 07	Std Err	Oct 03	Std Err	Oct 02	Std Err	Nov 07	Std Err	Oct 03	Std Err	Oct 02	Std Err
κ	0.218	0.004	0.042	0.003	0.511	0.007	0.334	0.006	0.379	0.008	0.395	0.007
α	−0.074	0.017	1.187	0.033	0.234	0.014	−0.009	0.001	−0.003	0.001	−0.005	0.001
β	0.012	0.001	0.390	0.010	0.084	0.002	–	–	–	–	0.000	–
a	0.000	–	0.000	–	0.000	–	0.000	–	0.000	–	0.000	–
c	0.072	0.434	2.093	1.300	0.080	0.862	0.218	0.293	0.035	0.389	0.000	–
d	0.001	0.002	1.617	0.698	0.000	0.262	–	–	–	–	–	–
Sigma$_1$	0.426	0.027	0.171	0.009	0.225	0.012	0.193	0.006	0.175	0.008	0.190	0.010
Sigma$_3$	0.520	0.036	0.865	0.129	0.222	0.076	0.180	0.005	0.160	0.005	0.167	0.006
Rho$_{[3]}$	0.941	0.034	−0.748	0.145	0.114	0.061	0.540	0.029	0.419	0.041	0.375	0.047
Front Month	0.145	0.006	0.137	0.007	0.009	0.005	0.145	0.006	0.142	0.008	0.127	0.007
1Dec	0.101	0.009	0.087	0.009	0.002	0.003	0.075	0.005	0.075	0.006	0.066	0.005
2Dec	0.029	0.002	0.016	0.001	0.000	0.001	0.016	0.001	0.016	0.001	0.015	0.001
3Dec	0.009	0.000	0.002	0.000	0.000	0.000	0.002	0.000	0.003	0.000	0.002	0.001
4Dec	0.001	0.000	0.003	0.000	0.000	0.000	0.003	0.000	0.005	0.000	0.006	0.000
5Dec	0.002	0.000	0.000	–	0.000	–	0.000	–	0.004	0.000	0.004	0.000
6Dec	0.000	–	0.004	0.000	0.000	0.000	0.005	0.000	0.000	–	0.000	–
7Dec	0.004	0.000	0.007	0.000	0.000	0.001	0.009	0.000	0.005	0.000	0.005	0.000
AIC	−22211.90		−14422.80		−12770.44		−22632.97		−14387.29		−12253.04	

TABLE 3.4 Two-Factor Model (Copper)

	Constant mean reversion model				Stochastic mean reversion model			
	Dec 07	Std Err	Nov 04	Std Err	Dec 07	Std Err	Nov 04	Std Err
κ	0.122	0.006	0.327	0.016	0.137	0.018	0.327	0.015
α	2.960	0.151	0.002	0.124	−0.032	0.014	−0.016	0.003
β	0.407	0.021	0.001	0.017	–	–	–	–
a	0.000	–	0.000	–	0.000	–	0.000	–
c	1.879	5.062	0.000	–	0.484	0.497	0.000	–
d	0.152	0.652	0.034	0.708	–	–	–	–
Sigma1	0.275	0.015	0.223	0.021	0.265	0.013	0.218	0.020
Sigma3	0.927	0.469	0.092	0.212	0.469	0.049	0.194	0.015
Rho{31}	−0.239	0.138	0.893	2.021	0.014	0.078	0.409	0.079
Front Month	0.026	0.001	0.014	0.003	0.040	0.002	0.014	0.003
1Dec	0.000	–	0.003	0.003	0.003	0.004	0.003	0.003
2Dec	0.010	0.000	0.006	0.002	0.026	0.003	0.006	0.002
3Dec	0.000	–	0.005	0.001	0.018	0.003	0.005	0.001
4Dec	0.015	0.002	0.003	0.001	0.000	–	0.003	0.001
5Dec	0.031	0.004	0.005	0.001	0.024	0.002	0.005	0.001
6Dec	0.042	0.004	0.008	0.002	0.061	0.006	0.008	0.002
AIC	−7662.46		−3927.20		−6806.08		−3922.25	

TABLE 3.5 Correspondence of State Variables

		WTI	Copper
3factor	x^1	Front Month futures Price	Front Month futures Price
	x^2	(3 DEC futures Price) − (6 DEC futures Price)*	(2 DEC futures Price) − (5 DEC futures Price)*
	x^3	6 DEC futures Price	5 DEC futures Price
2factor	x^1	2 DEC futures Price	Front Month futures Price
	x^3	6 DEC futures Price	5 DEC futures Price

* x^2 is compared with the spread between 6 Dec and 3 Dec for WTI, and the spread between 5 Dec and 2 Dec for Copper.

TABLE 3.6 Correlations

		Constant mean reversion model		Stochastic mean reversion model	
		WTI	Copper	WTI	Copper
3factor	x^1	0.926	0.923	0.927	0.924
	x^2	0.944	0.933	0.941	0.939
	x^3	0.980	0.841	0.977	0.821
2factor	x^1	0.857	0.885	0.864	0.876
	x^3	0.959	0.720	0.986	0.784

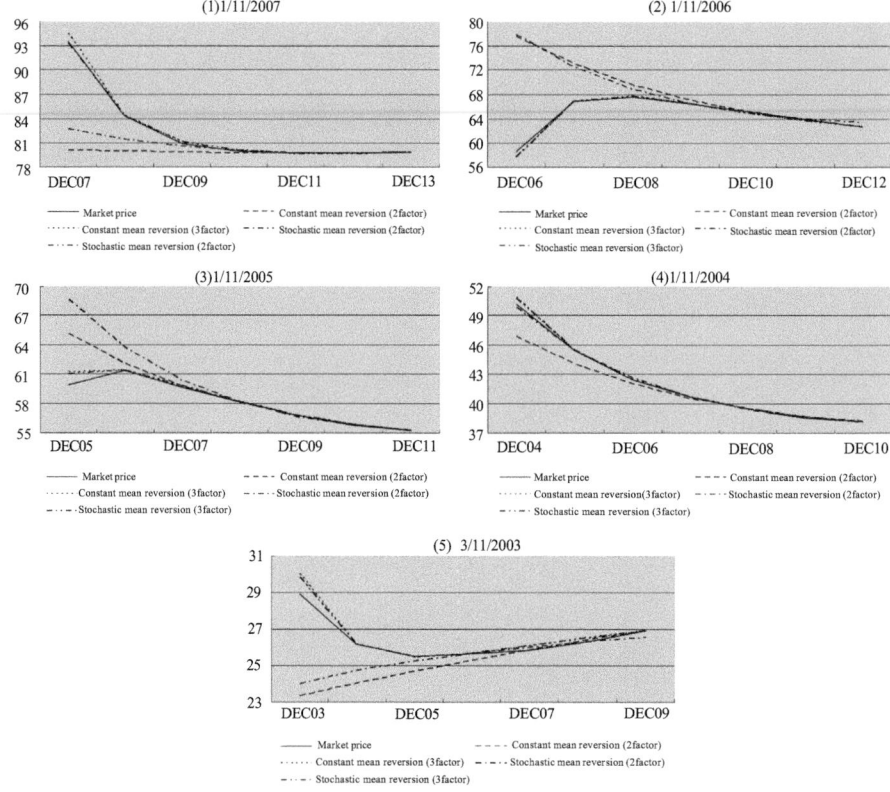

FIGURE 3.1   WTI futures term structure.

where

$$
A = \begin{pmatrix} \dfrac{\partial G_{T_1}(t)}{\partial x_1} & \dfrac{\partial G_{T_2}(t)}{\partial x_1} & \dfrac{\partial G_{T_3}(t)}{\partial x_1} \\[2ex] \dfrac{\partial G_{T_1}(t)}{\partial x_2} & \dfrac{\partial G_{T_2}(t)}{\partial x_2} & \dfrac{\partial G_{T_3}(t)}{\partial x_2} \\[2ex] \dfrac{\partial G_{T_1}(t)}{\partial x_3} & \dfrac{\partial G_{T_2}(t)}{\partial x_3} & \dfrac{\partial G_{T_3}(t)}{\partial x_3} \end{pmatrix}, \quad b = \begin{pmatrix} \dfrac{\partial G_{T_4}(t)}{\partial x_1} \\[2ex] \dfrac{\partial G_{T_4}(t)}{\partial x_2} \\[2ex] \dfrac{\partial G_{T_4}(t)}{\partial x_3} \end{pmatrix}.
$$

This paper refers to hedging using the hedging portfolio Φ as a 'delta hedge'. For a two-factor model, it is possible to construct a delta hedge in a similar way by eliminating the second factor x^2 in the corresponding three-factor model. We verify the degree of hedging error against this hedging portfolio when time series data are applied. For the purposes of this paper, the 'hedging error rate' is expressed as the final cumulative hedging error divided by the price of the instrument to be hedged at the time the hedge commences.

For comparison, we calculate the hedging error ratio for hedges such as those performed by Metallgesellschaft in which an equivalent number of nearer maturity futures is held against the futures to be hedged. This paper refers to this hedging method as the 'parallel hedge'. Metallgesellschaft hedged its long-term futures with extremely short-term futures of one to three contract months. However, given the increased liquidity of current commodities futures markets into the medium-term range, we verify the effectiveness of parallel hedges using futures of up to 6 years for WTI and up to 5 years for Copper.

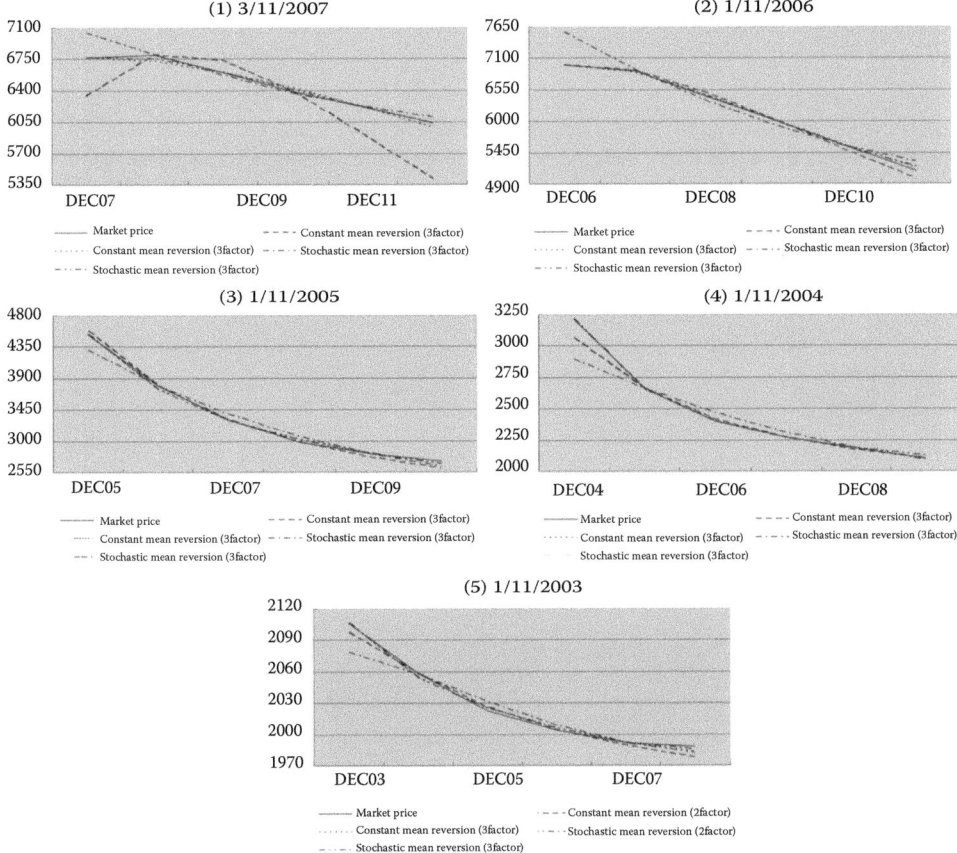

FIGURE 3.2 Copper futures term structure.

Unless specifically stated to the contrary, the discussion below refers to hedges against 10 Dec. from the front month for the WTI and 8 Dec. for Copper, for which prices are estimated by our models. For the hedging period, it is assumed that the position will be closed with an offsetting trade of a 6 Dec. futures for the WTI. In other words, a 4-year hedge is entered into that reduces the time to maturity of the instrument to be hedged from 10 to 6 years. For Copper, it is assumed that the position is closed with an offsetting trade of a 5 Dec. futures, resulting in a 3-year hedge that reduces the time to maturity of the instrument to be hedged from 8 to 5 years. For the parallel hedge, futures for the listed Decs. are used as hedge assets. For the delta hedges of three-factor models, 1, 4 and 6 Dec., and 1, 3 and 5 Dec. are used for WTI and Copper, respectively. For the delta hedges of two-factor models, 4 and 6 Dec. and 3 and 5 Dec. are used for WTI and Copper, respectively. Positions in each futures contract month are adjusted on the first business day of the month after reviewing hedging ratios each month. For both the parallel hedge and delta hedge, upon the elapse of 1 year, positions are rolled to the same contract month in the next year. (For example, if a Dec. 6 position is used to initiate a hedge on Dec. 12, after the elapse of 1 year, the Dec. 6 position used in the hedge will be rolled over to Dec. 7.) Liquidity declines the more distant the futures, but Dec. futures have comparatively high liquidity, and given the infrequency with which hedge ratios are changed and the small degree of change in the number of units required for hedging, this is considered a realistic hedge. In selecting futures contract months, this analysis uses combinations that provide the relatively small hedging error rates obtained in Section 3.6. A similar procedure is used to select futures contract months for two-factor models.

3.4.1 Hedging Error Rate of the Parallel Hedge

We first verify the hedging error rate achieved using the parallel hedge. The price of the futures contract month to be hedged is calculated based on the constant mean reversion model and data up to 2007 using parameters and state variables estimated with the Kalman filter.

Figure 3.3 shows the cumulative hedging error rates (the cumulative hedging error divided by the price of the instrument to be hedged at the time the hedge commences) using WTI and Copper time series data. In this case, for the Front Month the cumulative hedging error rate is expressed for a parallel hedge rolled over to the next-expiring contract month each month; for others, the cumulative hedging error rate is expressed for a parallel hedge with a one-year roll using Decs. for each year. The futures to be hedged are the WTI Dec. 13 and the Copper Dec. 12 and the hedge terminates at the most recently available data (2007). The horizontal axis expresses the amount of time elapsed since the commencement of the hedge; the vertical axis, the cumulative hedging error rate. The same notation is used for other graphs in this paper.

As can be observed from Figure 3.3, the error is smaller the more distant the futures used to hedge. Copper has a larger hedging error rate than WTI, indicating that the components in Copper's term structure that change in parallel are smaller than WTIs. However, even using the most distant futures with the smallest hedging error rate, the hedging error rates with a parallel hedge were still approximately 12% for WTI and approximately 32% for Copper.

3.4.2 Hedging Error Rate of the Delta Hedge

This section determines the hedging error rate for delta hedges for both the constant mean reversion model and the stochastic mean reversion model.

For each model, parameters estimated from data up to 2007 were used, and, for verification purposes, two methods were used to estimate the state variables in order to estimate long-term futures prices. In the first method, state variables were estimated using the Kalman filter ('Kalman filter state variables' hereafter); the second estimation created simultaneous linear equations for the state variables so that the futures price of the model matches the futures price of the futures contract month to be hedged, allowing state variables to be calculated by solving these equations ('simultaneous equation-based state variables' hereafter). To compare the relative precision of hedging using the model described in this paper, the verifications below give the results for the most distant futures, which was the most precise for the parallel hedge. Notation follows the practice used for parallel hedges.

Verifications were performed with different futures to be hedged, and the observed hedging error rates are summarized in Table 3.7. The parallel hedge is able to provide effective hedging when the overall term structure changes in parallel, but generates large hedging errors when there are changes in the shape of the term structure. By contrast, the delta hedge works much better than the parallel hedge (see Table 3.7). Two- and three-factor models provide relatively similar results, although three-factor models

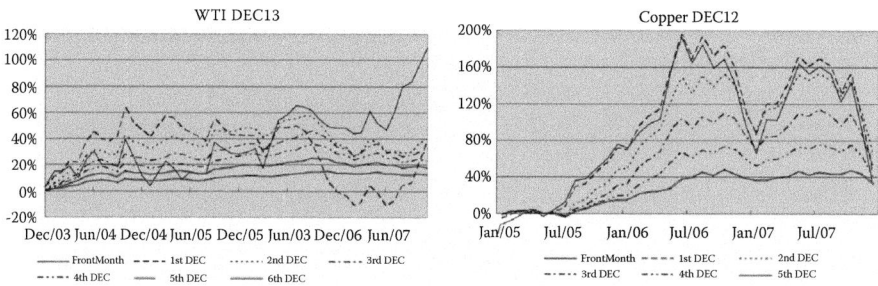

FIGURE 3.3 Cumulative hedging error rates of the parallel hedge.

TABLE 3.7 Hedging Error Rates

	Constant mean reversion model					Stochastic mean reversion model				
	Kalman Filter		Equations Based			Kalman Filter		Equations Based		
	3factor	2factor	3factor	2factor	Parallel	3factor	2factor	3factor	2factor	Parallel
	(%)	(%)	(%)	(%)	(%)	(%)	(%)	(%)	(%)	(%)
WTI DEC 07	1.0	1.3	0.7	1.3	4.0	0.7	0.8	0.4	0.4	3.8
DEC 08	1.0	0.6	0.9	0.6	8.4	0.7	1.5	0.5	0.7	8.3
DEC 09	0.8	0.8	1.0	0.8	3.4	0.5	-0.8	0.8	0.6	3.2
DEC 10	-1.5	-2.3	-2.5	-2.5	-3.5	-1.7	-0.5	-2.7	-2.5	-4.0
DEC 11	-3.1	-3.8	-3.2	-3.8	2.7	-3.1	-2.8	-3.1	-2.6	2.4
DEC 12	-3.2	-4.2	-3.7	-4.2	8.0	-3.0	0.4	-3.6	-0.9	7.8
DEC 13	-0.8	-1.1	0.0	-1.1	12.1	-0.3	0.5	0.5	3.2	12.3
Copper DEC 10	-0.4	-1.2	-0.5	-2.0	12.4	-0.1	0.0	0.0	-1.8	12.4
DEC 11	0.0	-5.2	0.2	-4.8	30.3	1.2	2.5	1.4	1.4	30.3
DEC 12	-3.1	-20.8	-6.7	-20.8	32.1	0.9	4.8	-3.2	3.7	32.0

work better for Copper. However, two-factor models have some difficulty in replicating actual term structures, as shown in Figures 3.1 and 3.2.

Comparing Kalman filter state variables and simultaneous equation-based state variables when performing a delta hedge, for Copper the estimation of Kalman filter state variables produces a large hedging error during the term of the hedge, as can be seen in Figures 3.4 and 3.5. This is presumably due to model prices not being obtained in a manner consistent with both the asset prices used in the hedge and the prices of the assets to be hedged. When using simultaneous equation-based state variables, model prices (excluding rollover timing) match the prices of the assets used in the hedge and the assets to be hedged. On the other hand, when using Kalman filter state variables, the actual prices of the assets used in the hedge differ from the model prices, resulting in a hedging error when hedging is performed. For WTI, the observational error was small for the futures contract month used in the hedge and virtually equivalent to the simultaneous equation-based state variables, indicating that there is little difference due to the method by which state variables are determined.

Because of the above result, the discussion below uses only state variables that are calculated by solving simultaneous equations for both WTI and Copper models.

3.5 Stability of the Delta Hedge

The verifications so far have estimated the parameters based on data that included the entire hedge period. However, in actual practice, the parameter estimation period and the hedge period differ. Discussions so far have also assumed that the hedges target long-term futures, but general practice is for long-term contracts to be forwards rather than futures, which requires that interest-rate factors also be taken into account.

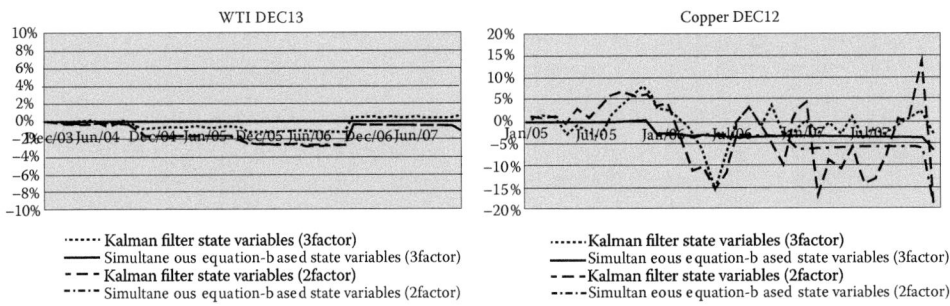

FIGURE 3.4   Cumulative hedging error rates of the delta hedge (constant mean reversion model).

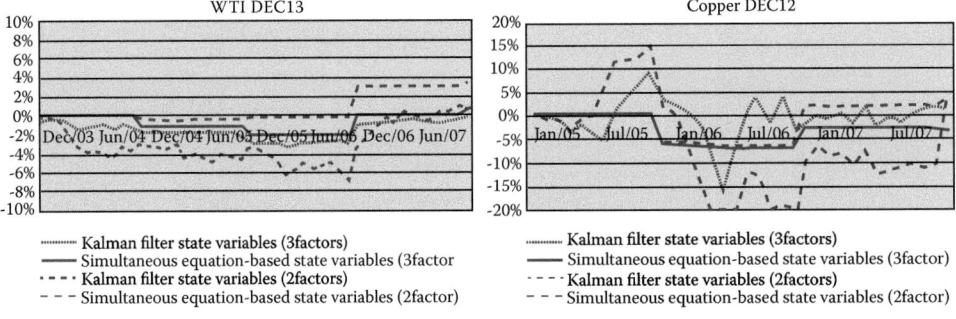

FIGURE 3.5   Cumulative hedging error rates of the delta hedge (stochastic mean reversion model).

In this section, we confirm the following settings so as to conduct the verifications in a state as close as possible to actual practice.

1. No overlap between the parameter estimation period and the hedge period.
2. Hedging against forwards.

In this paper, cases in which the entire hedge period is included in the parameter estimation period are referred to as 'In Sample', while hedges in which there are separate parameter estimation periods are referred to as 'Out of Sample'.

3.5.1 Out of Sample Hedges

To verify the effectiveness of the Out of Sample hedge, this section uses the parameters estimated in Section 3.3 with the data through 2002 or 2003 for WTI and through 2004 for Copper. The futures to be hedged is Dec. 12 (hedge period from 2002 to 2006) or Dec. 13 (hedge period from 2003 to 2007) for WTI and Dec. 12 (hedge period from 2004 to 2007) for Copper.

Table 3.8 summarizes the results of the verifications using time series data for hedging error rates when hedging long-term futures prices as estimated with the model using these parameters. For purposes of comparison, we have also listed the results for the In Sample estimations in the previous section.

For comparison between the two-factor models and the three-factor models, it is observed that the model producing the smaller absolute error In Sample also creates the smaller absolute error Out of Sample for all cases of Table 3.8.

For the three-factor stochastic mean reversion model, we confirmed that the differences in hedging error rates due to differences in the parameter estimation period were not that large for WTI. For Copper, there were differences in In Sample and Out of Sample hedging error rates, in part due to the differences in the parameters obtained for In Sample and Out of Sample. However, the error is not extreme and even though there are differences in the parameters obtained using the maximum likelihood method, hedging under the model is considered to be relatively stable. On the other hand, for the two-factor stochastic mean reversion model, there are more differences between In sample and Out of sample hedging error rates: this tendency seems stronger for Copper than for WTI.

For the constant mean reversion models, in the Out of Sample WTI, long-term price levels changed due to the sharp increases in oil prices beginning in 2003, virtually eliminating mean reversion. Nonetheless, the long-term futures prices calculated by the models revert to the mean levels observed in the data through 2002 or 2003, increasing the hedging error rates. In light of this, it is likely that the constant mean reversion model is more prone to hedging error when there are changes in mean reversion levels, etc., indicating that it is better to use the stochastic mean reversion model when prices are based on the hedge.

Due to the above results, the discussion in the next subsection uses only the three-factor stochastic mean reversion model.

TABLE 3.8 Hedging Error Rates of the Delta Hedge (Out of Sample)

		Constant mean reversion model		Stochastic mean reversion model	
		In sample (%)	Out of sample (%)	In sample (%)	Out of sample (%)
WTI	3factor (DEC 12)	−3.7	−32.1	−3.6	−3.4
	2factor (DEC 12)	−4.2	−33.5	−0.9	1.5
	3factor (DEC 13)	0.0	−7.6	0.5	0.2
	2factor (DEC 13)	−1.1	−16.5	3.2	5.3
Copper	3factor (DEC 12)	−6.7	−6.9	−3.2	−1.9
	2factor (DEC 12)	−20.8	10.0	3.7	11.0

3.5.2 Hedging Long-Term Forward Contracts

The discussions to this point have assumed that futures would be hedged, but common practice is to trade forwards for the long-term portion that is not traded on exchanges. If interest rates are deterministic or move independently from underlying assets, prices are the same for futures and forwards, but if interest rates are not deterministic, hedges must take account of their movements.

For simplicity, this discussion assumes that interest rates and underlying assets are independent, and describes hedging techniques when the instrument to be hedged is a forward and the assets used in the hedge are futures. The utility of this hedging technique is then verified using time series data.

We consider hedging long-term forwards with short-term futures in the following two steps.

1. Long-term forwards are hedged using long-term futures with the same expiration.
2. Long-term futures are hedged using the delta hedge with the nearer maturity futures as described in Section 3.4.

Because Step 2 is explained in Section 3.4, we explain the hedging technique of Step 1. The notation used is as follows: $F_T(t)$ is the price at point in time t of the forward with expiration T, $G_T(t)$ is the price at point in time t of the futures with expiration T and $P_T(t)$ is the price at point in time t of zero-coupon bond with expiration T.

In addition, $0 = t_0 < t_1 < ... < t_m = T$. The amount of change in the forward profit/loss at point in time T during the period from point in time t_i through t_{i+1} is as follows: PV at point in time t_{i+1} is expressed as $(F_T(t_{i+1}) - F_T(0))P_T(t_{i+1})$. Therefore, the amount of change in PV for the forward during the period from t_i through t_{i+1} is

$$\left(F_T\left(t_{i+1}\right) - F_T(0)\right)P_T\left(t_{i+1}\right) - \left(F_T\left(t_i\right) - F_T(0)\right)P_T\left(t_i\right). \tag{3.10}$$

In this case, (3.10) can be reformed as follows:

$$\left(F_T\left(t_{i+1}\right) - F_T(0)\right)P_T\left(t_{i+1}\right) - \left(F_T\left(t_i\right) - F_T(0)\right)P_T\left(t_i\right) = \left(F_T\left(t_{i+1}\right) - F_T\left(t_i\right)\right)P_T\left(t_i\right)$$
$$+ \left(F_T\left(t_i\right) - F_T(0)\right)\left(P_T\left(t_{i+1}\right) - P_T\left(t_i\right)\right) \tag{3.11}$$
$$+ \left(F_T\left(t_{i+1}\right) - F_T\left(t_i\right)\right)\left(P_T\left(t_{i+1}\right) - P_T\left(t_i\right)\right).$$

Thus ΔC_m, which denotes the accumulation of the last term on the right-hand evaluated at time T side of (3.11), is given by

$$\Delta C_m \approx \sum_{j=0}^{m-1} e^{\sum_{i=j+1}^{m-1} r_i\left(t_{i+1} - t_i\right)} \left(F_T\left(t_{j+1}\right) - F_T\left(t_j\right)\right)\left(P_T\left(t_{j+1}\right) - P_T\left(t_j\right)\right), \tag{3.12}$$

where the instantaneous interest rate in each period $[t_i, t_{i+1}]$ is approximated as a constant r_i. Below, we ignore ΔC_m because it expresses a negligible amount corresponding to the quadratic variation.

Assuming that the interest rate is independent of the underlying asset prices, the forward price and futures price are equivalent ($F_T(t) = G_T(t)$) and Equation (3.11) can be expressed as

$$\left(F_T\left(t_{i+1}\right) - F_T\left(t_i\right)\right)P_T\left(t_i\right) + \left(F_T\left(t_i\right) - F_T(0)\right)\left(P_T\left(t_{i+1}\right) - P_T\left(t_i\right)\right) = \left(G_T\left(t_{i+1}\right) - G_T\left(t_i\right)\right)P_T\left(t_i\right)$$
$$+ \left(G_T\left(t_i\right) - G_T(0)\right)\left(P_T\left(t_{i+1}\right) - P_T\left(t_i\right)\right). \tag{3.13}$$

The first term on the right-hand side expresses the change in the future, according to which a delta hedge is made using futures for the three nearer maturity contract months under the method described in Section 3.4. As a result, one unit of forwards can be given a proximate hedge using a portfolio comprising futures and zero-coupon bonds, as shown below.

1. A delta hedge using futures for the three nearer maturity contract months to hedge $P_T(t)$ units of futures $G_T(t)$.
2. Purchase of $G_T(t) - G_T(0)$ units of zero-coupon bond $P_T(t)$.

This hedging strategy is essentially the same as Schwartz's hedging strategy (see Schwartz 1997, pp. 963–964).

We analysed the hedging error rate when a 10-year WTI forward contract is hedged according to the above method using WTI futures and zero-coupon bonds for 4 years. Here, forward prices are assumed to be equal to theoretical futures prices calculated by our model. Note that the funds for the purchase of zero-coupon bonds and the cash flow generated by marking futures to market are invested/raised in short-term interest rates. For verification, we used In Sample parameters and, for simplification, the calculations of the zero-coupon bonds used the 8-year swap rate as spot yield; the calculations of short-term interest rates for investments and funding used the 1M LIBOR, and assume that the interest rate is independent of asset prices, thus forward price is equal to futures price. Table 3.9 shows the hedging error rates due to differences in the forwards to be hedged.

It can be seen that, in this verification, which used time series data for interest rates and futures prices, even assuming interest rates and underlying assets to be independent, the use of zero-coupon bonds and futures to hedge forwards was able to hedge virtually all of the interest rate factors generated by the difference between forwards and futures. However, if interest rates are not hedged, there are cases where large hedging errors are generated during the hedge period, as can be seen from Figure 3.6, so the idea that there does not need to be a hedge on the interest-rate portion is not supported.

TABLE 3.9 Hedging Error Rates of the Delta Hedge

	DEC07	DEC08	DEC09	DEC 10	DEC 11	DEC 12	DEC 13
Using futures and bonds	0.2%	0.3%	0.4%	−2.3%	−2.6%	−3.1%	−0.2%
Using futures	0.1%	− 1.3%	−2.4%	−9.2%	−5.8%	−10.0%	−5.6%

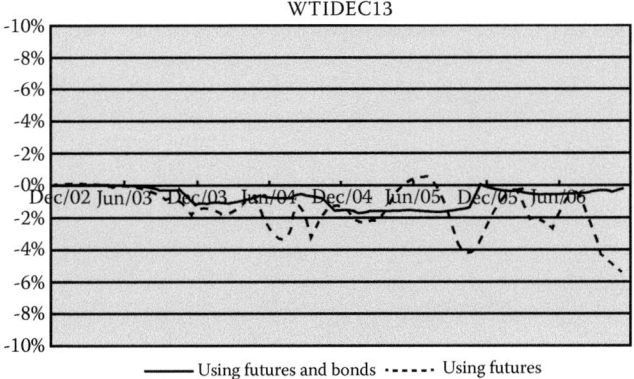

FIGURE 3.6 Cumulative hedging error rates of the delta hedge.

3.6 Measuring the Distribution of Hedging Error Rates

The analysis of hedging error rates based on time series data are limited to one of many possible paths. Thus, this section provides a simulation analysis to measure the distribution of hedging error rates resulting from fluctuations in the underlying assets. Three-factor stochastic mean reversion models are used in the simulation, where the In Sample parameters are used, and futures are hedged by futures with shorter maturities. We list below the specific procedures for the simulation.

1. Historical daily futures prices were created based on parameters and state variables estimated using the Kalman filter.
2. The error rate between the futures prices based on the model and created in Step 1 vs. the actual futures prices quoted on exchanges (for WTI, the front month and 1–6 Dec.; for Copper, the front month and 1–5 Dec.) was calculated [(Actual data–Model price)/Model price] and then the mean and covariance of the error rate were obtained. Here, it was assumed that the error follows a multidimensional normal distribution.
3. Three-dimensional normal random numbers were created and the state variables were made to fluctuate according to the model so as to create a term structure for futures.
4. Multidimensional normal random numbers according to the distribution described in Step 2 were created for the futures term structure developed in Step 3, multiplied as error and added to the original term structure (Step 3 model prices + Step 3 model prices × Random numbers following the Step 2 error distribution).
5. Simultaneous equation-based state variables were calculated on the assumption that the term structure created in Step 4 was the term structure actually observed in the market.
6. The term structure created in Step 5 was used to estimate long-term prices, hedges were taken against those prices, and the final hedging error rate measured.
7. Steps 3 to 6 were repeated for a constant number of times to find the sample mean and sample standard deviation of the hedging error rates obtained.

3.6.1 Distribution of Hedging Error Rates

We performed 5000 trials for each combination of futures contract used in the hedge according to the procedures outlined above and calculated the average and standard deviation of the hedging error rates. There was little difference in the hedging error rates due to differences in initial values, so as hedged assets we used Dec. 17 for WTI for a period of 4 years beginning 2007 and Dec. 15 for Copper for a period of 3 years beginning 2007.

Table 3.10 shows the means and standard deviations for the obtained hedging error rates. The 'contract months' column refers to which Dec. from the front month is used for the hedge. For example, 1–4–6 refers to the hedge using the 1, 4 and 6 Dec. values. Likewise, '6Y Parallel (5Y Parallel)' expresses the hedging error rate when a parallel hedge is entered into using the 6 (5) Dec. The price of the hedged assets for the parallel hedge is in the price found using 1–4–6 for WTI and 1–3–5 for Copper.

According to the results of the simulation, the appropriate selection of the futures contract months for the hedge portfolio when entering into a delta hedge has the potential for a more accurate hedge than the use of a parallel hedge. In particular, in terms of the required amounts (see Appendix B) and also the relationship between the means and standard deviations of hedging error rates, hedges for the WTI and Copper exhibited efficiency using 1, 4 and 6 Dec. and 1, 3 and 5 Dec., respectively. As a more general result, it was found that the futures used to create a hedge portfolio should, to the extent possible, have mutually disparate contract months. This is because when the futures used in the hedge are close to each other the x^1, x^2 and x^3 delta structures are similar and a greater number of futures units is required to offset the delta of the instrument hedged. The greater the number of futures units used in the hedge, the larger the error expressed as differences in the prices of the model and actual futures. Conversely, when

TABLE 3.10 Averages and Standard Deviations of Hedging Errors

Contract Months	Average (%)	Standard Deviation (%)	Contract Months	Average (%)	Standard Deviation (%)
		WTI			
1-2-3	−22.6	20.9	2-3-5	−0.3	4.5
1-2-4	−10.0	10.3	2-3-6	0.3	2.2
1-2-5	−4.3	5.0	2-4-5	−0.1	4.9
1-2-6	−1.7	2.7	2-4-6	0.4	2.2
1-3-4	−4.9	8.0	2-5-6	0.4	1.9
1-3-5	−1.5	3.9	3-4-5	−0.3	6.0
1-3-6	−0.3	2.0	3-4-6	0.3	2.5
1-4-5	−0.5	4.3	3-5-6	0.3	2.0
1-4-6	0.1	2.0	4-5-6	0.3	2.5
1-5-6	0.3	1.8		Parallel Hedge	
2-3-4	−3.0	9.3	6Y Parallel	−0.5	6.0
		Copper			
1-2-3	−14.7	11.5	2-3-4	0.3	8.5
1-2-4	−4.0	5.9	2-3-5	0.6	4.1
1-2-5	−1.1	3.7	2-4-5	−0.2	3.9
1-3-4	−0.7	6.8	3-4-5	−1.0	4.7
1-3-5	0.1	3.7		Parallel Hedge	
1-4-5	−0.2	3.7	5Y Parallel	0.4	16.3

Note: See Appendix B for more on the number of futures units required due to differences in the selection of futures months at the time the hedge is commenced.

the futures are further away from each other, they have disparate delta structures, making it more likely that hedging will not require as many units.

3.7 Conclusion

This paper has demonstrated that using a three-factor Gaussian model with appropriate estimation of parameters, it is possible to reproduce the term structures of listed commodities futures (NYMEX WTI and LME Copper) during the time period studied and that long-term futures prices can be obtained that are consistent with liquid nearer maturity contracts. It was also found that two-factor Gaussian models have some difficulty in capturing the actual term structures of futures.

Furthermore, we have proposed a hedging technique for long-term futures and forwards contracts, comparing the results from this technique with the results from the simple short-term futures-based hedging strategy used by Metallgesellschaft (parallel hedge), and have verified that our proposed strategy is stable in many different circumstances (backwardation, contango, rising prices, declining prices, etc.).

In addition, it was found that a stochastic mean reversion model offered more stable hedging than a model with a constant mean reversion level. Also, it was observed that the model producing the smaller absolute error In Sample also creates the smaller absolute error Out of Sample.

We then used simulation to measure the hedging error rates obtained due to differences in the contract months of the futures used in the hedge. It was found that the futures used to create a hedge portfolio should, to the extent possible, have mutually disparate contract months.

In summary, the three-factor model with stochastic mean reversion is useful in practice for pricing long-term futures/forward contracts and for hedging them with appropriate selected liquid instruments.

Future issues include the evaluation of option values using the model and structuring of relevant hedging techniques. Commodities generally have average-based options which makes calculation complex. It would be useful to verify the hedging techniques and their efficiency.

Acknowledgements

We thank the two anonymous referees for their efforts and comments. The views expressed in this paper are those of the authors and do not necessarily represent the views of Mizuho-DL Financial Technology Co., Ltd.

References

Bessembinder, H., Coughenour, J., Sequin, P. and Smoller, M., Mean reversion in equilibrium asset prices: evidence from the future term structure. *J. Finance*, 1995, **50**, 361–375.

Black, F., The pricing of commodity contracts. *J. Financial Econ.*, 1976, **3**, 167–179.

Brenann, M.J. and Crew, N.I., Hedging long-maturity commodity commitments with short-dated futures contracts. In *Mathematics of Derivatives Securities*, edited by M. Dempster and S. Pliska, pp. 165–190, 1997 (Cambridge University Press: Cambridge).

Bühler, W., Korn, O. and Schöbel, R., Hedging long-term forwards with short-term futures: a two-regime approach. *Rev. Deriv. Res.*, 2004, **7**, 185–212.

Casassus, J. and Collin-Dufresne, P., Stochastic convenience yield implied from commodity futures and interest rates. *J. Finance*, 2005, **60**, 2283–2331.

Cortazar, G. and Naranjo, L., An *n*-factor Gaussian model of oil futures prices. *J. Fut. Mkts*, 2006, **26**, 243–268.

Culp, C.L. and Miller, M.H., Metallgesellschaft and the economics of synthetic storage. *J. Appl. Corp. Finance*, 1995, **7**, 62–76.

Gibson, R. and Schwartz, E.S., Stochastic convenience yield and the pricing of oil contingent claims. *J. Finance*, 1990, **45**, 959–976.

Heath, D., Jarrow, R. and Morton, A., Bond pricing and the term structure of interest rates: a new methodology for contingent claims valuation. *Econometrica*, 1992, **60**, 77–105.

Korn, O., Drift matters: an analysis of commodity derivatives. *J. Fut. Mkts*, 2005, **25**, 211–241.

Mello, A.S. and Parsons, J.E., Maturity structure of a hedge matters: lessons from the Metallgesellschaft debacle. *J. Appl. Corp. Finance*, 1995, **8**, 106–121.

Miltersen, K.R. and Schwartz, E.S., Pricing of options on commodity futures with stochastic term structures of convenience yields and interest rates. *J. Financial Quant. Anal.*, 1998, **33**, 33–59.

Neuberger, A., Hedging long-term exposures with multiple short-term futures contracts?. *Rev. Financial Stud.*, 1999, **12**, 429–459.

Schwartz, E.S., The stochastic behavior of commodity prices: implications for valuation and hedging. *J. Finance*, 1997, **52**, 923–973.

Schwartz, E.S. and Smith, J.E., Short-term variations long-term dynamics in commodity prices. *Mgmt Sci.*, 2000, **46**, 893–911.

Appendix A: Expectation and Covariance Matrix of State Variables

$$E_t\left[\begin{pmatrix} x_T^1 \\ x_T^2 \\ x_T^3 \end{pmatrix}\right] = \begin{bmatrix} \mu_{11}\left(x_t^1, x_t^2, x_t^3, T-t\right) \\ \mu_{22}\left(x_t^1, x_t^2, x_t^3, T-t\right) \\ \mu_{33}\left(x_t^1, x_t^2, x_t^3, T-t\right) \end{bmatrix}$$

$$
= \begin{bmatrix}
e^{-\kappa(T-t)}x_t^1 + \dfrac{\kappa\left(e^{-\gamma(T-t)} - e^{-\kappa(T-t)}\right)}{\kappa-\gamma}x_t^2 + \dfrac{\kappa\left(e^{-\beta(T-t)} - e^{-\kappa(T-t)}\right)}{\kappa-\beta}x_t^3 + \dfrac{\alpha}{\beta}\left(1 - \dfrac{\kappa e^{-\beta(T-t)} - \beta e^{-\kappa(T-t)}}{\kappa-\beta}\right) \\[2ex]
e^{-\gamma(T-t)}x_t^2 \\[2ex]
e^{-\beta(T-t)}x_t^3 + \dfrac{\alpha}{\beta}\left(1 - e^{-\beta(T-t)}\right)
\end{bmatrix}.
$$

$$
\mathrm{Cov}_t\left[\begin{pmatrix} x_T^1 \\ x_T^2 \\ x_T^3 \end{pmatrix}\right] = \begin{bmatrix}
\Sigma_{11}(T-t) & \Sigma_{12}(T-t) & \Sigma_{13}(T-t) \\
\Sigma_{12}(T-t) & \Sigma_{22}(T-t) & \Sigma_{23}(T-t) \\
\Sigma_{13}(T-t) & \Sigma_{23}(T-t) & \Sigma_{33}(T-t)
\end{bmatrix},
$$

$$
\Sigma_{11} = \sigma_1^2 \frac{1-e^{-2\kappa(T-t)}}{2\kappa} + \sigma_2^2\left(\frac{\kappa}{\kappa-\gamma}\right)^2\left(\frac{1-e^{-2\gamma(T-t)}}{2\gamma} + \frac{1-e^{-2\kappa(T-t)}}{2\kappa} - 2\frac{1-e^{-(\kappa+\gamma)(T-t)}}{\kappa+\gamma}\right)
$$

$$
+\sigma_3^2\left(\frac{\kappa}{\kappa-\beta}\right)^2\left(\frac{1-e^{-2\beta(T-t)}}{2\beta} + \frac{1-e^{-2\kappa(T-t)}}{2\kappa} - 2\frac{1-e^{-(\kappa+\beta)(T-t)}}{\kappa+\beta}\right)
$$

$$
+2\rho_{12}\sigma_1\sigma_2\frac{\kappa}{\kappa-\gamma}\left(\frac{1-e^{-(\kappa+\gamma)(T-t)}}{\kappa+\gamma} - \frac{1-e^{-2\kappa(T-t)}}{2\kappa}\right) + 2\rho_{23}\sigma_2\sigma_3\frac{\kappa^2}{(\kappa-\gamma)(\kappa-\beta)}\left(\frac{1-e^{-(\beta+\gamma)(T-t)}}{\beta+\gamma}\right.
$$

$$
\left.-\frac{1-e^{-(\kappa+\beta)(T-t)}}{\kappa+\beta} - \frac{1-e^{-(\kappa+\gamma)(T-t)}}{\kappa+\gamma} + \frac{1-e^{-2\kappa(T-t)}}{2\kappa}\right) + 2\rho_{13}\sigma_1\sigma_3\frac{\kappa}{\kappa-\beta}\left(\frac{1-e^{-(\kappa+\beta)(T-t)}}{\kappa+\beta} - \frac{1-e^{-2\kappa(T-t)}}{2\kappa}\right),
$$

$$
\Sigma_{22} = \sigma_2^2\frac{1-e^{-2\gamma(T-t)}}{2\gamma}, \quad \Sigma_{33} = \sigma_3^2\frac{1-e^{-2\beta(T-t)}}{2\beta}, \quad \Sigma_{23} = \rho_{23}\sigma_2\sigma_3\frac{1-e^{-(\beta+\gamma)(T-t)}}{\beta+\gamma},
$$

$$
\Sigma_{12} = \rho_{12}\sigma_1\sigma_2\frac{1-e^{-(\kappa+\gamma)(T-t)}}{\kappa+\gamma} + \sigma_2^2\frac{\kappa}{\kappa-\gamma}\left(\frac{1-e^{-2\gamma(T-t)}}{2\gamma} - \frac{1-e^{-(\kappa+\gamma)(T-t)}}{\kappa+\gamma}\right)
$$

$$
+\rho_{23}\sigma_2\sigma_3\frac{\kappa}{\kappa-\beta}\left(\frac{1-e^{-(\beta+\gamma)(T-t)}}{\beta+\gamma} - \frac{1-e^{-(\kappa+\gamma)(T-t)}}{\kappa+\gamma}\right),
$$

$$
\Sigma_{13} = \rho_{13}\sigma_1\sigma_3\frac{1-e^{-(\kappa+\beta)(T-t)}}{\kappa+\beta} + \sigma_3^2\frac{\kappa}{\kappa-\beta}\left(\frac{1-e^{-2\beta(T-t)}}{2\beta} - \frac{1-e^{-(\kappa+\beta)(T-t)}}{\beta+\kappa}\right)
$$

$$
+\rho_{23}\sigma_2\sigma_3\frac{\kappa}{\kappa-\gamma}\left(\frac{1-e^{-(\beta+\gamma)(T-t)}}{\beta+\gamma} - \frac{1-e^{-(\kappa+\beta)(T-t)}}{\kappa+\beta}\right).
$$

Appendix B: The Number of Futures Units

	WTI						Copper				
	1 Dec	2 Dec	3 Dec	4 Dec	5 Dec	6 Dec	1 Dec	2 Dec	3 Dec	4 Dec	5 Dec
1-2-3	1.20	−5.65	5.48	0	0	0	2.29	−8.12	6.84	0	0
1-2-4	0.62	−2.28	0	2.69	0	0	1.06	−2.89	0	2.83	0
1-2-5	0.35	−1.19	0	0	1.87	0	0.55	−1.34	0	0	1.80
1-2-6	0.20	−0.67	0	0	0	1.49	−	−	−	−	−
1-3-4	0.22	0	−3.70	4.50	0	0	0.39	0	−3.76	4.38	0
1-3-5	0.12	0.00	−1.47	0.00	2.37	0	0.20	0	−1.35	0	2.15
1-3-6	0.07	0	−0.74	0	0	1.69	−	−	−	−	−
1-4-5	0.06	0	0	−2.96	3.93	0	0.10	0	0	−2.44	3.35
1-4-6	0.03	0	0	−1.12	0	2.11	−	−	−	−	−
1-5-6	0.01	0	0	0	−2.38	3.39	−	−	−	−	−
2-3-4	0	1.29	−5.79	5.53	0	0	0	1.64	−5.90	5.25	0
2-3-5	0	0.64	−2.25	0	2.64	0	0	0.78	−2.13	0	2.36
2-3-6	0	0.34	−1.11	0	0	1.79	−	−	−	−	−
2-4-5	0	0.23	0	−3.52	4.32	0	0	0.29	0	−2.97	3.69
2-4-6	0	0.12	0	−1.31	0	2.21	−	−	−	−	−
2-5-6	0	0.05	0	0	−2.57	3.53	−	−	−	−	−
3-4-5	0	0	1.23	−5.45	5.24	0	0	0	1.27	−4.74	4.48
3-4-6	0	0	0.58	−1.99	0	2.43	−	−	−	−	−
3-5-6	0	0	0.20	0	−3.02	3.84	−	−	−	−	−
4-5-6	0	0	0	1.04	−4.59	4.57	−	−	−	−	−

4

Planning Logistics Operations in the Oil Industry

M. A. H. Dempster,
N. Hicks Pedrón,
E. A. Medova*,
J. E. Scott and
A. Sembos

4.1 Introduction

The research presented in this paper was carried out under the European Community ESPRIT Project HChLOUSO: Hydrocarbon and Chemical Logistics under Uncertainty via Stochastic Optimization. A framework for a multiperiod supply, transformation and distribution scheduling problem under uncertainty in the product demands and spot supply costs for a consortium of oil operators was recently presented in the CORO model.[1] Our work on the implementation of alternative *stochastic programming* solutions to this problem led to the *Depot and Refinery Optimization Problem (DROP)* for strategic (tactical) level planning of the companies' activities. The DROP model is used in this paper as a basis for implementing its stochastic programming version. Earlier treatments of optimization problems in the oil industry are detailed in refs. 2–10. Various decomposition schemes for solution of the stochastic formulation of the CORO model were proposed[1] including augmented Lagrangian decomposition based on a splitting variable formulation of its deterministic equivalent.[11–14] For a background in stochastic programming the reader is referred to refs. 15–17 as well as the references in ref. 1.

The first part of this paper treats the specification, generation and solution of the deterministic DROP model in detail. We first give the mathematical formulation of the deterministic DROP model. Implementation of the model and illustrative examples are then explained.

In the second part of the paper, alternative stochastic versions of this model are studied and a number of novel related research questions and implications discussed. First the stochastic DROP model is

* *Correspondence: Dr EA Medova, Judge Institute of Management, University of Cambridge, Cambridge CB2 1AG, UK.*
  E-mail: eam28@cam.ac.uk

DOI: 10.1201/9781003265399-5

formulated and then modelling issues related to scheduling decisions and the stages of a *dynamic* stochastic programming model are treated. The industry data available for estimating the random parameters of the model with a view to their simulation, the software implementation of the stochastic model and comparisons of various solution techniques are all discussed. Finally, we draw conclusions and indicate future directions for research.

In the remainder of this introduction we briefly describe the complex practical problem addressed in the paper.

Logistics planning for a consortium of oil companies deals with supply, transformation, storage and transportation activities over a complex network structure involving (continuous flow) pipelines and other (discrete) transport means such as trucks and ships over planning horizons on different time scales. Heterogeneous yet interlinked activities across a transportation network impose a hierarchy of decisions: *strategic* decisions for the long term, *tactical* decisions for the medium term and *operational* decisions at the level of day-to-day logistics. The activities which must be undertaken during each time period are the following:

- *Supply* Each operator supplies to each depot a certain volume of crude oil within pre-agreed limits. The actual volume supplied by each operator during a specific time period is decided by the consortium and must be catered for by the operator. To ensure the operators' compliance with the system, contracts exist which specify each operator's supply obligations to the consortium.
- *Transformation* Transformation of products takes place only at refineries located at some of the depots in the network. Refining is the process of converting crude oil into usable fuel products. Part of the product volume is supplied during a time period and part of the crude oil stock existing at the end of the previous time period is processed for refining.
- *Transportation* Part of the crude oil volume supplied during a time period and part of the crude stock existing at the end of the previous time period at a depot is sent directly to the depot's refineries for transformation into usable fuels or is transported to other refinery depots. In addition, final products are transported from refineries to depots and from depots to other depots, where they are stored (stocked) temporarily. We assume the existence of a given physical transportation network and refer to a depot or refinery as a *node* of the network and to existing transportation *means* between nodes as *arcs (links)*.
- *Stocking* The volume of crude oil supplied by each operator during each time period is temporarily stocked at depots until it is transported directly to that depot's refineries or to other depots. Similarly, the usable fuel products given as output from the refineries are stocked at depots or the refineries themselves.

The individual oil companies operate according to the whole consortium's needs, sharing the raw materials once they have been supplied and sharing the refineries and the transportation means for the distribution of products. Insufficient consortium resources may cause problem infeasibility. Infeasibilities are due to lack of infrastructure resources (eg limitations, in the capacity of the network or refineries) and to imbalances in the demand requirements. The second type can be covered by *trading* in the oil commodity markets. Resort to the spot markets takes place only when the consortium's own resources do not balance product demands:

- *Spot market buying* In the case of insufficient supply or storage of crude oil at any node in any time period, the consortium may purchase crude at the spot market price.
- *Spot market selling* In the case that amounts of oil products are unneeded by the consortium at any node in any time period, it can sell them at spot market prices.

All activities involve a variety of products (*multicommodities*) and integer solutions for discrete transport means. This implies that the computational complexities of the corresponding mathematical formulations are enormous, even though this problem may be partially resolved using a *two-level hierarchical structure* of decisions: strategic (or tactical) and operational. We consider the following stages of the planning process:

- for *strategic* (tactical) logistics planning, the levels of petroleum activities for a consortium of operators are determined to minimize total cost.
- for *operational* logistics planning, given the levels of the required activities, detailed transportation schedules for each transportation means are developed for a number of *implementable* time periods containing no demand or price uncertainties.

In this paper we concentrate only on the formulation and implementation of strategic or tactical level problems. From our implementation of the deterministic CORO model[1,18] we learned that optimal decisions tend to eliminate some individual activities because of their specified costs.[19] For example, with a sufficiently low spot price specified for output product of the refineries, it was optimal simply to buy product in the spot markets. Such results demonstrate an over-simplification of the problem formulation by not properly taking into account the costs of storage facility maintenance, labor costs and many other realistic constraints. As a result the DROP model has been developed.

4.2 The Deterministic DROP Model

In our DROP model we introduce some 'operational' constraints which tend to provide more realistic solutions. This class of constraints can easily be reformulated if the consortium's overall policy changes. The DROP model has been defined so that decisions are made at the beginning of each time period and the hierarchy of decisions is given explicitly. The major features of the DROP model are:

- *Separating products into crude oil products and final products.* Unlike the CORO model the DROP model does *not* allow the supply of end-products or the spot sale of crude oil products, i.e. *pure trading* is eliminated.
- *Defining explicitly the spatial dimension of the problem.* The distribution network is specified *explicitly* in terms of a directed graph and *implicitly* by the corresponding *multicommodity flow* problem in terms of the transport means and product variables.

The *deterministic* DROP model provides optimal decisions for the magnitudes of major activities: supply, transformation, transportation, stock keeping and compensation of demand by buying or selling in the spot markets, assuming that demand at each point of the logistic network is known with *certainty* and that the demands of the various oil products are given.

4.2.1 Mathematical Formulation of the DROP Model

See the appendix for lists of sets, parameters, variables and bounds used in the following linear programme.

4.2.2 Objective Function

The consortium's objective is to minimize the overall cost g of the required activities to satisfy end-product demand over a fixed planning horizon:

$$g\left(X, Z, E, S, X^{spot}, Y^{spot}\right)$$

$$= \sum_{n \in N} \sum_{o \in O} \sum_{p \in P_i} \sum_{t \in T} c^x_{n,p,t} X_{n,o,p,t} \quad \text{SUPPLY}$$

$$+\sum_{r\in R}\sum_{f\in F_r}\sum_{t\in T}c_{r,f,t}^z Z_{r,f,t} \qquad \text{REFINING}$$

$$+\sum_{p\in P}\sum_{m\in M:p\in P_m}\sum_{(i,j)\in L_m}\sum_{t\in T}c_{i,j,p,m,t}^e E_{i,j,p,m,t} \qquad \text{TRANSPORTATION}$$

$$+\sum_{n\in N}\sum_{p\in P}\sum_{t\in T}c_{n,p,t}^S S_{n,p,t} \qquad \text{STOCK}$$

$$+\sum_{n\in N}\sum_{p\in P_1}\sum_{t\in T}c_{n,p,t}^{spot} X_{n,p,t}^{spot} \qquad \text{SPOT PURCHASE}$$

$$-\sum_{n\in N}\sum_{p\in P_2}\sum_{t\in T}p_{n,p,t}^{spot} Y_{n,p,t}^{spot} \qquad \text{SPOT SALE} \qquad (4.1)$$

4.2.3 Constraints

4.2.3.1 Operational Constraints

- *All variables* must *be non-negative, since they represent product volumes.*
- *Supply of final products is not allowed.*

$$X_{n,o,p,t}=0 \quad \forall n,o,p,t,\in N\times O\times P_2\times T \qquad (4.2)$$

$$X_{n,p,t}^{spot}=0 \quad \forall n,p,t,\in N\times P_2\times T \qquad (4.3)$$

- *Selling crude oil products is not allowed.*

$$Y_{n,p,t}^{spot}=0 \quad \forall n,p,t\in N\times P_1\times T \qquad (4.4)$$

- *Transformation of products can only take place at refineries.*

$$Z_{n,f,t}=0 \quad \forall n\in N:n\notin R,\forall f\in F,\forall t\in T \qquad (4.5)$$

- *Transformation of products can only take place at a particular refinery if the technology of interest is available at the refinery.*

$$Z_{n,f,t}=0 \quad \forall n\in N,\forall f\notin F_n,\forall t\in T \qquad (4.6)$$

- *A product can only be given as input to a refinery for transformation by a particular technology if the technology needs the product as an input.*

$$Z_{n,p,f,t}^p=0 \quad \forall n\in N,\forall p\in P_1,\forall f\notin F_p^{in},\forall t\in T \qquad (4.7)$$

- *A product will only be given as output by a transformation technology if the technology produces that product as an output.*

$$Z^p_{n,p,f,t} = 0 \quad \forall n \in N, \forall p \in P_2, \forall f \notin F^{out}_p, \forall t \in T \tag{4.8}$$

- *Transportation of products between a pair of nodes can only exist if there is a link between the nodes.*

$$E_{i,j,p,m,t} = 0 \quad \forall i,j \in N : (i,j) \notin L_m,$$
$$\forall p \in P, \forall m \in M, \forall t \in T \tag{4.9}$$

- *Transportation of a product between a pair of nodes by a particular transportation means can only exist if the product can be transported by that particular means.*

$$E_{i,j,p,m,t} = 0 \quad \forall i,j \in N, \forall p \notin P_m,$$
$$\forall m \in M, \forall t \in T \tag{4.10}$$

- *Transportation from one node to another can only be started if it is permissible for it to be started during the time period of interest.*

$$E_{i,j,p,m,t} = 0 \quad \forall i,j \in N, \forall p \in P_m,$$
$$\forall m \in M, \forall t \notin T_{i,j,p,m} \tag{4.11}$$

4.2.3.2 Product Balance Constraints

- *During each time period, the crude oil product volume set aside for transportation or refining does not exceed resources.*

$$\sum_{f \in F_n 1 : n_1 \in R} Z^p_{n_1,p,f,t} + \sum_{n_2 \in N:(n_1,n_2) \in L_m} \sum_{m \in M : p \in P_m} E_{n_1,n_2,p,m,t}$$
$$\leq S_{n_1,p,t-1} + \sum_{o \in O} X_{n_1,o,p,t} + X^{spot}_{n_1,p,t} \tag{4.12}$$
$$\forall (n_1, p, t) \in N \times P_1 \times T$$

- *During each time period, stock at the end of the previous time period is sufficient to satisfy distribution and demand requirements during the current time period.*

$$S_{n_1,p,t-1} \geq \sum_{n_2 \in N:(n_1,n_1) \in L_m} \sum_{m \in M : p \in P_m} E_{n_1,n_2,p,m,t}$$
$$+ \sum_{o \in O} d_{n_1,o,p,t} + Y^{spot}_{n_1,p,t} \tag{4.13}$$
$$\forall (n_1, p, t) \in N \times P_2 \times T$$

- *Each technology uses the correct mixture of crude oils and produces the correct proportion of final products as well.*

$$Z^p_{n,p,f,t} = g_{f,p} Z_{n,f,t} \quad \forall (n, p, f, t) \in N \times P \times F_n \times T \tag{4.14}$$

- *Product balance equation for input products, ie crude oils*

$$S_{n_1,p,t-1}$$
$$+ \sum_{o \in O} X_{n_1,o,p,t} + X^{spot}_{n_1,p,t}$$

stock of products p in node n_1 at the beginning of time period t supply of product p to node n_1 during time period t

$$+ \sum_{m \in M: p \in P_m} \sum_{n_2 \in N: (n_2, n_1) \in L_m} E_{n_2, n_1, p, m, t - \Delta_{n_2 m_1 pm}} + e_{n_1, p, t}$$

total volume of product p which arrives at node n_1 from another node during time period t volume of product p that is given as input to the refinery during time period t

$$- \sum_{f \in F_{n_1} \cap F_p^{in}: n_1 \in R} Z_{n_1, p, f, t}^P$$

$$- \sum_{m \in M: p \in P_m} \sum_{n_2 \in N: (n_1, n_2) \in L_m} E_{n_1, n_2, p, m, t}$$

total volume of product p whose transportation from node n_1 to another node starts during time period t stock of product p in node n_1 at the end of time period t

$$= S_{n_1, p, t}$$

$$\forall (n_1, p, t) \in N \times P_1 \times T$$

(4.15)

- *Product balance equation for output products, ie final products.*

$$S_{n_1, p, t-1}$$

$$+ \sum_{m \in M: p \in P_m} \sum_{n_2 \in N: (n_2, n_1) \in L_m} E_{n_2, n_1, p, m, t, -\Delta_{n_2 m_1 pm}} + e_{n_1, p, t}$$

stock of product p in node n_1 at the beginning of time period t
total volume of product p which arrives at node n_1 from another node during time period t volume of product p that is given as output by the refinery during time period t

$$+ \sum_{f \in F_{n_1} \cap F_p^{out}: n_1 \in R} Z_{n_1, p, f, t}^p$$

$$- \sum_{m \in M: p \in P_m} \sum_{n_2 \in N: (n_1, n_2) \in L_m} E_{n_1, n_2, p, m, t}$$

total volume of product p whose transportation from node n_1 to another node starts during time period t

$$- \sum_{o \in O} d_{n_1, o, p, t} - Y_{n_1, p, t}^{spot}$$

volume of product p provided to satisfy demand in node n_1 during time period t + excess product p sold in the spot market stock of product p in node n_1 at the end of time period t

$$= S_{n_1, p, t}$$

$$\forall (n_1, p, t) \in N \times P_2 \times T$$

(4.16)

4.2.3.3 Capacity Constraints

- *Pipeline capacity bounds.*

$$\sum_{t \in T_u^U} \sum_{p \in P_m : t \in T_{n1,n2p,m}} E_{n1,n2,p,m,t} \le e_{n1,n2,m,u}^c \tag{4.17}$$

$$\forall (n1,n2) \in L_m, \forall m \in M^c, \forall \in U$$

Subsets of the set of all the time periods in the planning horizon are defined for end-user purposes. These are introduced only for bounding purposes and denote an *n*-period length of time, where the *n* underlying periods might not be consecutive.[19]

- *Truck, ship or wagon capacity bounds.*

$$\sum_{p \in P_m} \sum_{(n_1,n_2) \in L_m} \sum_{t' \in T_{n_1,n_2,p,m}^t} E_{n_1,n_2 p,m,t'} \le e_{m,t}^d$$

$$\forall m \in M^d, \forall t \in T$$

Here

$$T_{n_1,n_2,p,m}^t := \begin{cases} \{t - \Delta_{n_1,n_2,p,m,...,t}\}, & \text{if } \Delta_{n_1,n_2,p,m} < t \\ \{1,...,t\}, & \text{otherwise} \end{cases} \tag{4.18}$$

- *Operator supply contract bounds.*

$$\underline{x}_{n,o,p,v} \le \sum_{t \in T_v^V} X_{n,o,p,v} \le \overline{x}_{n,o,p,v} \tag{4.19}$$

$$\forall (n,o,p,v) \in N \times O \times P \times V$$

- *Refinery capacity bounds.*

$$\underline{z}_{n,f,h} \le \sum_{t \in T_h^H} Z_{n,f,t} \le \overline{z}_{n,f,h} \quad \forall (n,f,h) \in N \times F_n \times H \tag{4.20}$$

In addition to (4.20), there is an upper bound on the 'transformation capacity' of each technology in each period of the planning horizon.

$$Z_{n,f,t} \le \overline{z}_{n,f,t}^T \quad \forall (n,f,t) \in N \times F_n \times T \tag{4.21}$$

- *Storage capacity bounds.*

$$\underline{S}_{n,pt} \le S_{n,p,t} \le \overline{s}_{n,p,t} \quad \forall (n,p,t) \in N \times P \times T \tag{4.22}$$

It also might be the case that certain product volumes need to be held in strategically important regions. This is accomplished using *entities*.

$$\underline{s_\theta} \le \sum_{n \in N_0^\Theta} \sum_{p \in P_0^\Theta} \sum_{t \in T_0^\oplus} S_{n,p,t} \le \bar{s}_\theta, \quad \forall \theta \in \Theta \tag{4.23}$$

4.2.3.4 Software Implementation

The DROP model has been written in the XPRESS-MP modelling language. [20] This simplifies the task of constructing and maintaining the model and is an aid to communication, allowing end users to verify that the model and model changes are correct.

Index sets and logical operations on binary tables are combined to specify the set of variables over which each constraint is written. The modelling software will generate a row in the problem file for each member of this set. Coefficients for the row are then calculated from the algebraic expression associated with the constraint. The modelling software also handles the creation of the right-hand side vector and simple bounds on variables. XPRESS-MP provides a variety of methods for specifying the input data. We currently store each input table in a plain text file. The matrix generated by XPRESS-MP can either be kept in memory in a sparse representation suitable for solution algorithms or written to file in MPS format.

Table 4.1 shows three problems whose data were supplied by an oil company. DROP1 is a small test problem with a transportation network consisting of only 4 nodes and a planning period of just 6 periods. DROP1 has had bounds on transportation, supply and refining capacities removed in order to achieve feasibility and for test purposes, and hence produces a zero *net* cost by applying trading profits for excess production against actual costs. DROP2 differs from DROP1 by the number of time periods (30) in the planning horizon. Also, in this problem, the bounds have not been removed but *elastic variables* ensure feasibility of the problem. DROP3 corresponds to a realistic case. The transportation network consists of 41 nodes, there are 7 products overall and there are 30 time periods in the planning horizon. All three problems have been solved using simplex algorithms from both the IBM Optimization Subroutine Library[21] and Dash's XPRESS solver [20] and also using the barrier (interior point) algorithm from Dash's solver. The machine used was an IBM RS6000/590 with 1GB RAM and 27GB disk running under AIX4.3 (Table 4.2).

To illustrate the spatiotemporal nature of the DROP model, the optimal decisions of another small example problem are analyzed below. It involves 7 nodes, two of which (2 and 4) correspond to refineries, and 5 time periods. The consortium is made of two operators, who supply two types of crude oil (products 1 and 2) to the nodes, and then share the refineries for refining the total crude oil volume into three types of final product, say motor fuel, jet fuel and fuel oil (products 3, 4 and 5). The operators also share the transportation means shown in Figure 4.1. For simplicity we assume that transportation from

TABLE 4.1   Problems

Problem	DROP1	DROP2	DROP3
Nodes	4	6	41
Products	3	3	7
Operators	2	2	1
Transport means	3	3	3
Transformation technologies	1	1	2
Planning periods	6	30	30
Rows	218	2908	30831
Columns	367	3691	110 210
Non-zeros	1096	16 227	381 036

TABLE 4.2 Solutions

Problem	Algorithm	Optimal value	Time
	OSL simplex	0.0	0.58 s
DROP1	XPRESS barrier	0.0	0. 2 s
	XPRESS simplex	0.0	0.18 s
	OSL simplex	2302.969	7.730 s
DROP2	XPRESS barrier	2302.969	3.790 s
	XPRESS simplex	2302.969	3.620 s
	OSL simplex	1167 736.907	108 min 54 s
DROP3	XPRESS barrier	1167 736.907	6 min 53 s
	XPRESS simplex	1167 736.907	24 min 21 s

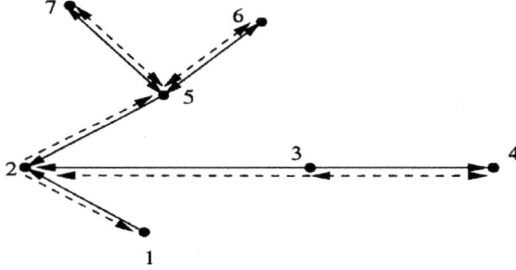

FIGURE 4.1 Existing transportation: — pipelines, --- trucks, ships or wagons.

one node to another does not take more than 1 time period, so that a product reaches its destination during the same time period as its period of departure.

The first refinery at node 2 transforms 1 unit of crude oil 1 into 1 unit of motor fuel. The second refinery at node 4 transforms 1 unit of input volume, which is a mixture of 0.6 units of crude oil 1 and 0.4 units of crude oil 2, into 0.5 units of jet fuel and 0.4 units of fuel oil.

The demand for motor fuel is equal to 5 units at all nodes from time period 4 onwards and the demand for jet fuel is equal to 5 units at nodes 2, 3 and 4 from time period 4 onwards. There is no demand for fuel oil at any node during any time period.

We analyzed two versions of the model: *with and without spot market transactions* by the operators. The optimal decisions for both operators for the model with spot market transactions are shown in Figures 4.2(a) – (e). In brief, one can see that, during time period 1, crude oil is supplied to the nodes and it is subsequently transported to refineries for transformation into final products. At the end of the period, crude oil that has not been used and final products that have been produced by the refineries are stored at depots. During periods 2, 3 and 4, all activities take place, including the sale of excess final products in the spot market. During time period 5, the final products that had been produced during the previous time period and kept in stock are given to customers to satisfy contractual demand. The excess products are sold in the spot market.

Comparison of the two versions shows the advantages of trading. Without trading there was excess fuel oil production that remained in stock until the end of the planning horizon. In the model with trading, part of the excess stock could be sold to external operators so that the consortium's cost is further minimized. In this example, the value of the objective function was 38 145 for the case without spot market transactions and 35 565 for the case with spot market transactions. The cost that would have been saved over the excess product being sold in the spot market only at the *end* of the planning horizon is 720.

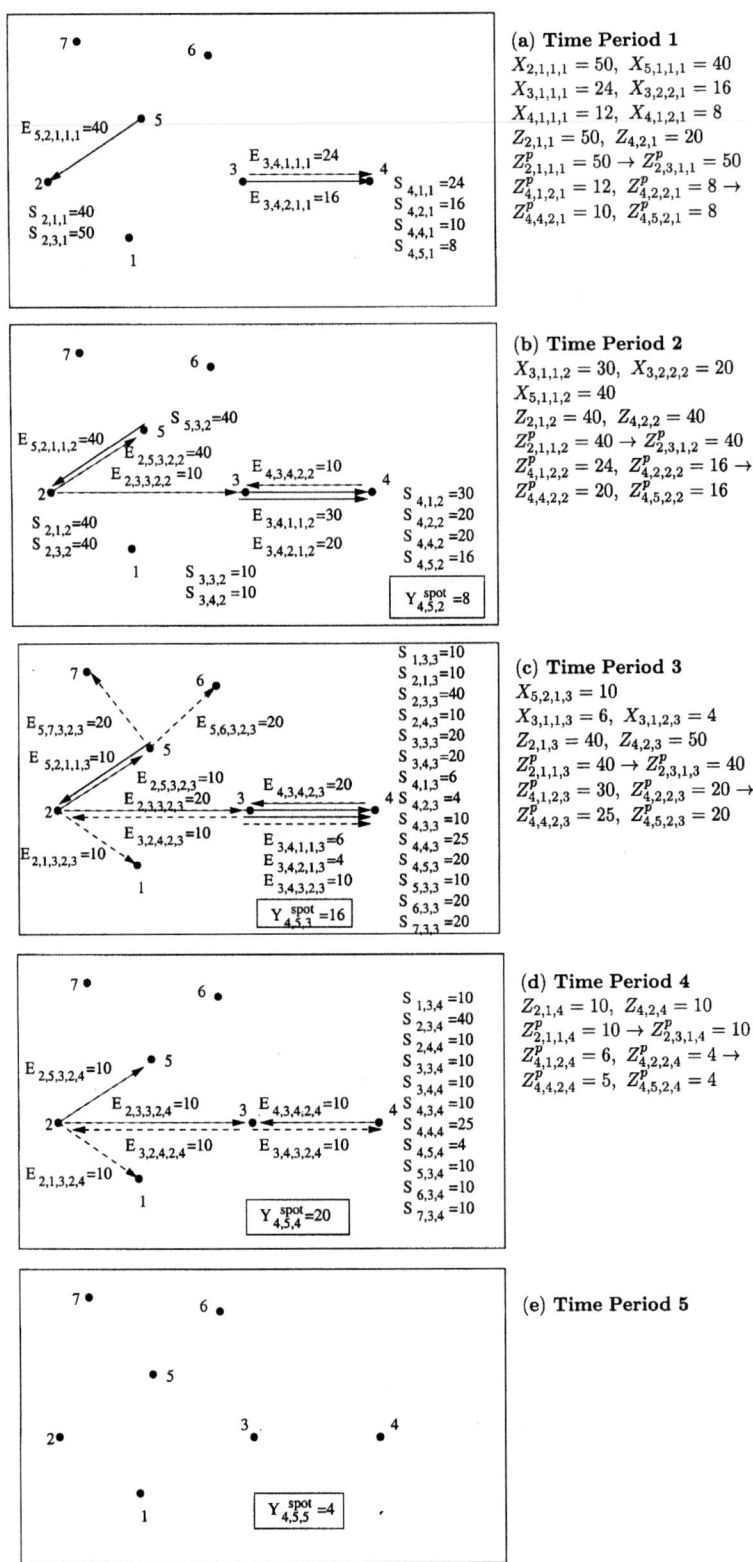

FIGURE 4.2   Optimal decisions for the model with spot market transactions.

Since future product demands and prices are uncertain in the real world, a more realistic strategic or tactical planning model in which this uncertainty is taken into account is discussed in the sequel.

4.3 Stochastic Modelling

In order to be efficient and cost-effective across the different time scales and horizons used for practical logistics planning, we assume *uncertainty* in demands and oil product prices *after* some initial time period (the first *implementable stage* of the stochastic problem). These uncertainties in the *stochastic DROPS* model are given by a number of *scenarios* over the planning horizon, whose generation will be discussed in detail. Optimal solutions of DROPS in the periods of the implementable first stage are hedged against future uncertainties and will guarantee (possibly at increased costs in the actual outcomes) consistency with operational level activities. These first stage decisions may thus be used as input to more detailed operational logistics models, for example for truck routing or ship scheduling, for optimization across the entire logistic system. In our description of the DROPS model we follow the notation developed by our group.[11,22]

4.3.1 The Stochastic DROP Model

Strategic planning for an oil consortium under uncertain demand and prices for oil products is viewed here as a *dynamic* recourse problem[15–17] in which continuous real time is divided into a number of discrete time periods of equal length called *scheduling periods* (see Figure 4.3). In accordance with ordinary usage the tth scheduling period begins at clock time $t-1$ and the first *stage* begins at clock time 0 and *decision point* T_1 while the second and successive stages begin at decision points $T_s, s = 2,\ldots,S$. The last scheduling period begins at T_{plan}, but the last stage decision point T_S may be several scheduling periods before this.

We consider a vector stochastic *data process* $\boldsymbol{\omega} := \{\boldsymbol{\omega}_t : t = 1,\ldots,T_{plan}\}$ given by

$$\boldsymbol{\omega} := \left\{ \left(\mathbf{c}_{n,p,t}^{spot} \mathbf{p}_{n,p,t}^{spot}, \mathbf{d}_{n,o,p,t} \right) : n \in N, p \in P, o \in O, \right.$$
$$\left. t = 1,\ldots,T_{plan} \right\}$$

whose realizations are *data paths* in a (Euclidean) probability space $(.\Omega; \mathcal{F})$, using boldface to denote random entities. We denote by ω_t^s a particular realization at time t of the data process *history* $\boldsymbol{\omega}^s$ up to stage s.

The vector *decision process*

$$x := \left\{ x_t := \left(\mathbf{X}_t, \mathbf{Z}_t, \mathbf{S}_t, \mathbf{X}_t^{spot}, \mathbf{Y}_t^{spot} \right) : t = 1,\ldots,T_{plan} \right\}$$

for *supply* $\mathbf{X}$, *transformation* $\mathbf{Z}$, *transportation* $\mathbf{E}$, *stock* $\mathbf{S}$, *spot* (market) *purchase* $\mathbf{X}^{spot}$ and *spot sale* $\mathbf{Y}^{spot}$ maps a data trajectory to a vector decision trajectory. (We omit here the indices of the *spatial network* structure for sets of *nodes* and *transportation means* $\{N, M\}$, *operators* $\{O\}$, *products* $\{P\}$ and *technologies* $\{F\}$ for clarity of temporal representation of the stochastic processes.)

The stages for *recourse decisions* are defined by the available information and by the requirements for planning and the latter are assumed to be made at the corresponding decision points $T_s, s = 1,\ldots,S$, immediately after the random variables are realised a (real time) instant before. The number of scheduling time periods included in each stage may thus *vary*, but *all* decisions for periods *within* a stage s are assumed to be made at T_s.

It is assumed that decisions made at any stage of recourse $s = 1,\ldots,S$ are only dependent on the information provided up to the current time stage, ie in terms of stages *nonanticipative decisions* are made. To keep track of information flow at past, present and future stages of the stochastic process, the canonical (Euclidean) probability space is equipped with a filtration $\{\mathcal{F}_s : s = 1,\ldots,T_s\}$, where $\mathcal{F}_1 \subset \mathcal{F}_2 \subset \ldots \subset \mathcal{F}_{T_s} := \mathcal{F}$, defined by the *history* ω^s of the process (ie by the data available at the time(s) t corresponding to stage s). Therefore for decisions in such time periods t

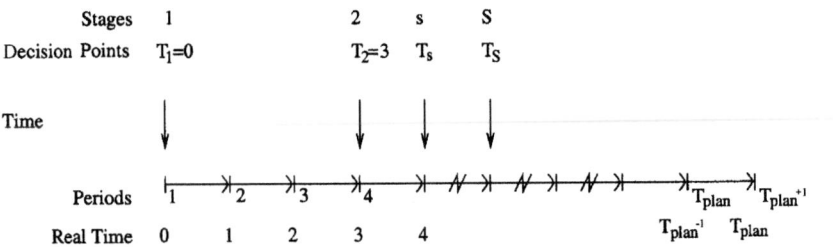

FIGURE 4.3   Alternative time measurements for dynamic stochastic programmes.

$$x_t = \mathbb{E}\{\mathbf{x}_t \mid \mathcal{F}_s\} \quad \text{a.s} \quad s = 1, \ldots, S,$$

where $\mathbb{E}\{\mathbf{x}_t \mid \mathcal{F}_s\}$ denotes conditional expectation with respect to the σ field $\mathcal{F}_s \subset \mathcal{F}$.

The stochastic DROP model is a *multistage stochastic programming* model in which the expected cost of all the underlying activities is minimized. The generic formulation is given by

$$(\text{SP})\pi_0 := \inf \mathbb{E}f(\omega, \mathbf{x}) \tag{4.24}$$

$$s.t \quad \mathbf{x}_s \in P_s(\omega^s) \quad \text{a.s} \tag{4.25}$$

$$g_s(\omega^s, x^s(\omega)) \in Q_s \quad \text{a.s} \quad s = 1, \ldots, T_s, \tag{4.26}$$

where *f* is a real-valued *objective functional* of the data paths, *g* is a *constraint function* of the data and decision trajectories, *P* is a multifunction (ie *random set* with temporal structure), $\mathbb{E}$ denotes expectation and the constraints are required to hold *almost surely* (a.s), ie with probability one.

The number of possible data paths — called *scenarios* — is assumed to be finite and the scenarios are arranged in a *scenario tree*, with *nodes* corresponding to the stages of the underlying vector stochastic data process, as depicted in Figure 4.4.

At the first or *implementable* stage, the demand forecasts and price/cost coefficients are assumed to be *given*. After the first stage, the uncertain future data is specified by a scenario tree whose paths represent the realizations of the vector stochastic process of demand and price data at the beginning of each time period along each path of the scenario tree. At each time period *t* in the *s*th stage *decisions* x_t are chosen from the *feasible set* which is defined by the constraints (4.26) as

$$x_t(\omega^s) := x_t(\omega^s, \mathbf{x}^{t-1}) \quad t = T_s, \ldots, T_{s+1} - 1 \tag{4.27}$$

and

$$x_s(\omega^s) := \left(x_{T_s}(\omega^{T_s}), x_{T_s+1}(\omega^{T_s+1}), \ldots, \right.$$
$$\left. x_{T_{s+1}-1}(\omega^{T_{s+1}-1}) \right) \quad s = 1, \ldots, S. \tag{4.28}$$

Note that (4.27) and (4.28) imply that for time periods *within* stages current decisions *can* take account of *future* information up to the time period just *before* the next stage, ie they are *anticipative within stages*. At the *stage level*, however, decisions are *nonanticipative*. When only a few (eg two) stages are employed in a multiperiod model, optimal expected costs to the planning horizon may be seriously *biased downwards* by the effects of anticipative decisions as we shall see in the following.

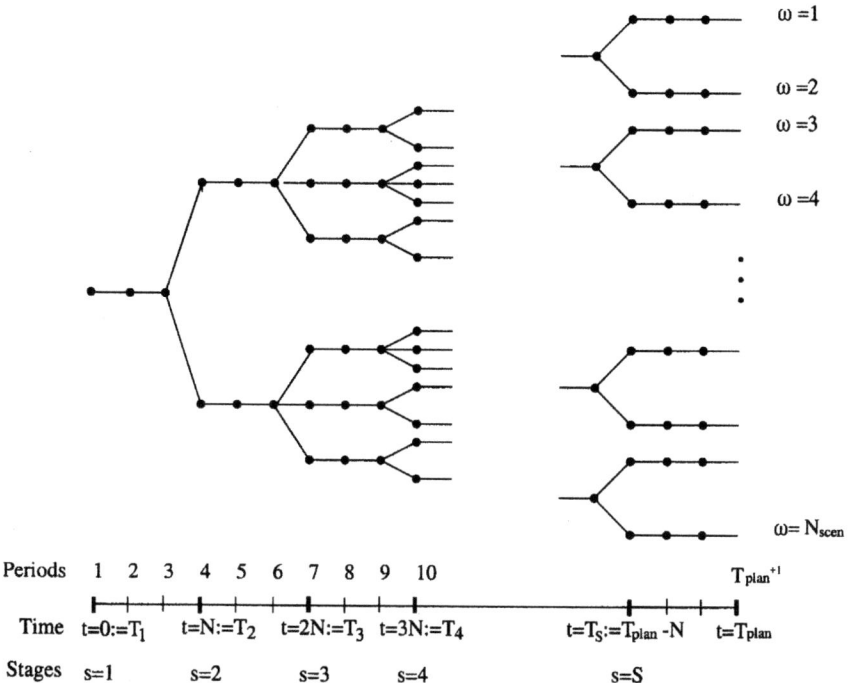

Periods 1 2 3 4 5 6 7 8 9 10 T_{plan}^{+1}

Time t=0:=T_1 t=N:=T_2 t=2N:=T_3 t=3N:=T_4 t=T_S:=T_{plan} -N   t=T_{plan}

Stages s=1   s=2   s=3   s=4   s=S

FIGURE 4.4 Scenario tree representation of the stochastic process over a discrete planning horizon with decisions (dots) at the beginning of periods $t = 1,\ldots,T_{plan}$, $s = 1,\ldots,S$ stages and N: = 3 periods in each stage.

Although the second and later stage decisions may be of 'what-if' interest for strategic logistics management at the higher levels of the consortium's company hierarchies, our focus for logistics planning is on the *implementable* first-stage scheduling decisions which are hedged against future uncertainty in demand and prices and which may be linked to deterministic *operational planning* models.

The *dynamic programming* representation of the DROP model with recourse in terms of a set of path-dependent problems takes the form

$$\pi_t\left(\omega^s\right) := \pi_s\left(\omega^s, x^{s-1}(\omega)\right)$$

$$:= \inf x_s(\omega)\left[f_s\left(\omega^s, x^{s-1}(\omega), x_s(\omega)\right)\right.$$

$$\left. +\mathbb{E}\left\{\pi_{s+1}\left(\omega^{s+1}, x^s(\omega)\right) | \mathcal{F}_s\right\}\right]$$

$$\text{s.t } x_s(\omega) \in P_s\left(\omega^s\right) \quad a.s. \tag{4.29}$$

$$g_s\left(\omega^s, x^s(\omega)\right) \in Q_s \ a.s. \tag{4.30}$$

$$x_s(\omega) = \mathbb{E}\left\{x_s(\omega) | \mathcal{F}_s\right\} \ a.s. \ s = 1,\ldots,S \tag{4.31}$$

involving the (optimal) value function sequence π_1 (the problem optimal value),$\pi_2,\ldots,\pi_S,\pi_{S+1} :\equiv 0$

The formulation of the *deterministic equivalent* (DE) of the stochastic problem (SP) in general requires the specification of a sequence of *conditional (branching) probabilities* for the generation of a scenario tree (see Figure 4.4). At the origin of the decision process ($t = 0$ in real time) the specification of the sequence of these conditional probabilities is given by

$$p(\omega_s) = p(\omega_2) p(\omega_3 \mid \omega^2) \ldots p(\omega_s \mid \omega^{s-1}) \quad s = 2, \ldots, S \tag{4.32}$$

At each stage $s = 1, \ldots, S$, the sample spaces Ω_s are here defined by $k_s = 1, \ldots, K_s$ possible states, where each *state* ω_s has probability $p(\omega_s)$ of being reached, and Ω_1 is a singleton of data corresponding to the implementable time periods. The individual *scenario probabilities* are equal to the *path* probabilities $p(\omega_s)$ of (4.32). In our implementations of the DROPS model, scenarios are generated by conditional statistical Monte Carlo sampling of an underlying continuous-state high-dimensional vector stochastic process and hence all scenarios are *equiprobable*. As with empirical density functions, however, the frequencies of specific data value intervals depend on the statistics of the data process sampled.

Computation of the conditional probabilities allows the specification of (SP) as the large-scale deterministic equivalent problem (DE), in which the decision vector is defined by $x := \left\{ x_1, x_2^1, \ldots, x_2^{k_2}, \ldots, x_{T_{plan}}^1, \ldots, x_{T_{plan}}^{K_{plan}} \right\}$. Note that $K_{T\,plan} = N_{scen}$, the *number of scenarios*. At each stage of recourse the *feasible set* at each node of the scenario tree is the intersection of the feasible sets determined for that stage by the subsequent time stages of each scenario. On the scenario tree this will correspond to all paths which descend from the chosen node of the tree. This simplifies the formulation of the problem (DE) and for the *linearly constrained* dynamic stochastic programming problem with recourse leads to a lower block-triangular constraint structure in which the stage structure remains *implicit*. For example, if time period 2 is within the implementable first stage, $k_2 := 1$, $p_2^{k_2} := 1$ and (4.35) is a single set of constraints:

$$(\text{DE}) \min_x \left\{ f_1(x_1) + \sum_{k_2=1}^{K_2} p_2^{k_2} f_2\left(x_2^{k_2}\right) + \sum_{k_3=1}^{K_3} p_3^{k_3} f_3\left(x_3^{k_3}\right) + \cdots \right.$$

$$\left. + \sum_{k_{T_{plan}}=1}^{K_{T\,plan}} p_{T_{plan}}^{k_{T\,plan}} f_{T_{plan}}\left(x_{T_{plan}}^{k_{T_{plan}}}\right) \right\} \tag{4.33}$$

s.t

$$A_1 x_1 = b_1 \tag{4.34}$$

$$B_2^{1;k_2} x_1 + A_2^{k_2} x_2^{k_2} = b_2^{k_2} \tag{4.35}$$

$$B_3^{1;k_2,k_3} x_1^{k_2} + B_3^{k_2,k_3} x_2^{k_2} + A_3^{k_2,k_3} x_3^{k_2,k_3} = b_3^{k_2,k_3}$$

$$\vdots \qquad \ddots \qquad \ddots \qquad \vdots$$

$$B_{T_{plan}}^{1;k_2,\ldots,k_{T_{plan}}} x_1 + \cdots + B_{T_{plan}}^{T_{plan}-1;k_2,\ldots,k_{T_{plan}}} x_{T_{plan}-1}^{k_2,\ldots,k_{T_{plan}}-1} \tag{4.36}$$

$$+ A_{T_{plan}}^{k_2,\ldots,k_{T_{plan}}} x_{T_{plan}}^{k_2,\ldots,k_{T_{plan}}} = b_{T_{plan}}^{k_2,\ldots,k_{T_{plan}}}$$

$$k_t = 1, \ldots, K_t$$

$$t = 2, \ldots, T_{plan} \tag{4.37}$$

$$l_1 \leq x_1 \leq u_1, \ldots, l_{T_{plan}}^{k_2,\ldots,k_{T_{plan}}} \leq x_{T_{plan}}^{k_2,\ldots,k_{T_{plan}}} \leq u_{T_{plan}}^{k_2,\ldots,k_{T_{plan}}} \tag{4.38}$$

The constraints (4.34) (4.37) can be written in standard matrix form as $Ax = b$. For *linear* problem functions the corresponding bounds (4.38) become $l \leq x \leq u$ and the corresponding *linear* objective is $c'x$, so that a *linear programme* in standard form is obtained:

$$(LP) \quad \min c'x$$

$$s.t. \quad Ax = b \tag{4.39}$$

$$l \geq x \geq u.$$

In our application the complexity of the stochastic DROP model is due to the fact that the components of the decision vector $\mathbf{x}_t := \left\{ \mathbf{X}_t, \mathbf{Z}_t, \mathbf{E}_t, \mathbf{S}_t, \mathbf{X}_t^{spot}, \mathbf{Y}_t^{spot} : t = 1, \ldots, T_{plan} \right\}$ are themselves complex vectors of decisions in space with regard to outgoing scheduling activities. The product balance constraints of the deterministic DROP model involve *linking* the decision variables for supply, transformation, transportation and stock *over time*. Therefore, decomposition techniques, which are usually applied directly to dynamic stochastic linear programming problems with Markov recourse structure, are not directly applicable without permutation of the rows and columns of A to uncover its underlying block structure.[23] It is, however, possible to trivially restate the five product balance constraints (4.12) – (4.16) of the DROP model so that all uncertainties occur in the right-hand side or objective function coefficients of the DROPS model. A detailed formulation of the stochastic DROP model is given in reference 24.

4.3.2 Industry Data Analysis and Simulation

Lack of realistic or historical data for stochastic programming applications is a common problem. A small number of 'typical' data instances have been chosen for statistical analysis here in order to formulate a model of the stochastic data process and to derive initial values for its parameters. We then generated the data process using Monte Carlo simulation which was calibrated to follow the data paths of the original scenarios provided by industry.[24-26] The *DROP model simulator* is capable of generating scenarios for any given time horizon and initial conditions and at each node of a scenario tree. It produces the required number of conditional realizations of the vector data process for specified scenario tree structures.

We have assumed that product prices and costs follow *(driftless) Brownian motion* processes. This is a common assumption for financial data processes and for simplicity the same model has been adopted here for operator demand processes.

Data generation for the DROP model is thus carried out for each node by following a system defined by the equations:

$$\mathbf{D} = \mu_D + \sigma_D dz_D$$

$$\mathbf{C} = \mu_C + \sigma_C dz_C \tag{4.40}$$

$$\mathbf{P} = \mu_P + \sigma_P dz_P,$$

Where

$D := (d_{n,o,p,t})$—demands for products of operators.

$o = 1; \ldots; O$

$C := (c_{n,p,t}^{spot})$—spot prices (costs) for crude oils

$P := (p_{n,p,t}^{spot})$—spot prices for final products

and μ denotes the conditional mean, ie *previous state*, of a process. For the test problem analysed above with 2 operators, 2 crude oils and 3 final products, this gives an 11-variable D, C, P system at each of 7 nodes to give a 77-dimensional data process.

Estimates of the parameters σ_D, σ_C and σ_P were obtained from calibrating the simulator to the historical data given by companies and the standard normal variables dz_D, dz_C and dz_P for each node n are correlated with each other in the simulator according to a correlation matrix estimated from this data.

Given the variables at time t, the *state* for the next stage at a given node is defined by

$$V(D,C,P) = \mu(D,C,P) + \Sigma(D,C,P)d\mathbf{Z},$$

where

V – column vector of variables
μ – column vector of current values μ_i, i = D, C, P
$d\mathbf{Z}$ – column vector of $d\mathbf{Z}_i$ independently distributed $N(0, 1)$
Σ – matrix $SD\,A$

with

SD – diagonal matrix with diagonal elements σ_i, i = D, C, P respectively
A – matrix such that $R = AA^T$, where R is the correlation matrix between variables.

The individual states of the multi-dimensional stochastic data process generated by the simulator must be structured according to the decision (scenario) tree of the stochastic optimization problem (see Figure 4.4). In general, any branching structure, regular or otherwise, can be defined for the scenario tree graphically or by the corresponding *nodal partition* matrix (a variant of the scenario partition matrix utilised in reference 27). This matrix identifies the tree structure for the associated stochastic program uniquely in terms of node numbers and is used by the data process generator in order to derive the states of the data process in conditional mode.

Nonanticipativity requires that all the scenarios passing through a given node in the tree share the same parameter and decision values at that time stage. All scenarios share the same parameter and decision values in the periods of the first implementable stage. In other words, first stage decisions are *hedged* against future uncertainty.

Conditional simulations are run following the nodal partition matrix order, with variables computed and initial conditions passed consistently along the tree. According to the time partition of the planning horizon, the simulator runs over the T_{plan} periods, and the final state of the data process for one node is adopted as the initial state for the next. The nodal partition matrix allows both a conditional run of the data generator — with one run for every increment in the matrix entries columnwise — and the consistent updating of the initial conditions for each period rowwise.

A stochastic LP therefore differs from a deterministic LP in that parameters, decision variables and constraints have multiple realizations for time periods beyond the first (root) period which correspond to the alternative future scenarios possible at a given time. Formulating such a model requires (a) defining its scenario structure and (b) specifying the scenario dependencies of its coefficients, variables and constraints as modifications of the deterministic model form.

The steps required for conditional scenario generation can be summarized as follows:

- Define the number of scenarios N_{scen}.
- Specify the nodal partition matrix in order to define the tree structure.
- Define the initial conditions and run the simulator to generate data paths, travelling the tree forward from the root node to the terminal node.
- Specify input variables for each conditional simulation of the data process variables.

- Complete the set of conditional data paths following the nodal partition matrix.
- Generate the model parameter (coefficient) scenarios by running the scenario generator for every trajectory of the data process.

When this set of scenarios is generated, the stochastic optimization problem must be formatted for the LP solver (eg XPRESS) to which we turn next. For a more detailed description, see reference 24.

4.3.3 Stochastic Model Implementation

For multiperiod stochastic linear programs there is a standard input format for solvers (eg MSLiP-OSL[28]) referred to as SMPS format.[29] In this format the stochastic optimization problem is identified by a *time file*, defining the dynamic structure of the problem, a *core file*, with the matrix generated along one data path, and a *stoch file*, with the stochastic information for the problem which is unique to each model coefficient scenario. Together these constitute the required triplet of *SMPS files*.

The STOCHGEN generator (Corvera-Poiré[30,31]) uses a sequence of MPS files for *nodal subproblems* to generate the SMPS files. The three main steps in the implementation of a problem using STOCHGEN are the following:

(i) Determination of the tree structure (*event (scenario) partition matrix*).
(ii) Nodal specification of the core files *(base scenario)*.
(iii) Specification of scenario (model coefficient parameter) data from the stochastic scenario generator.

Figure 4.5 shows the logical steps and the software involved in the implementation of the model. The nodal subproblem model written in XPRESS-MP sequentially generates the core file in MPS format. The data file for each LP scenario instance is retrieved sequentially by STOCHGEN following the event matrix (time stage) order resulting from using ANALYZE,[32] and is then added to the stoch file. Finally, STOCHGEN formats all information into an SMPS file which is passed to the solver, possibly after expanding to MPS format for the deterministic equivalent by a utility.

4.3.4 Analysis of Stochastic Problem Solutions

The relative value of the solution to a dynamic stochastic programming problem depends on its formulation. For example, the value of the solution may be measured in terms of the *distribution*

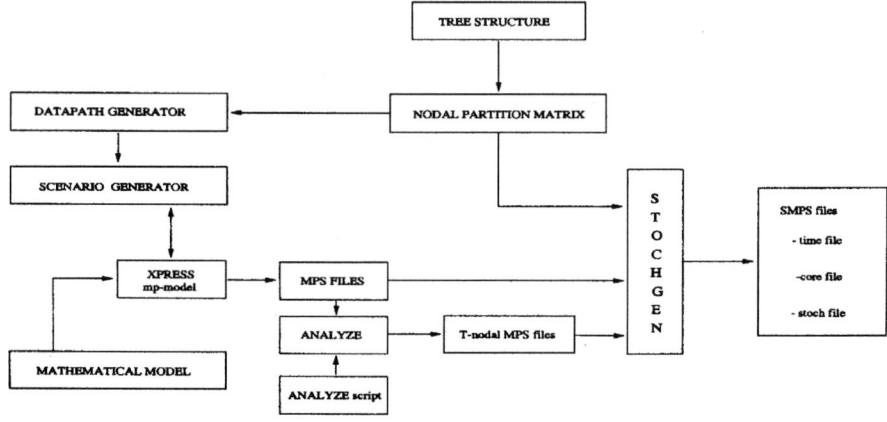

FIGURE 4.5 General specification the deterministic equivalent of a stochastic model via STOCHGEN.

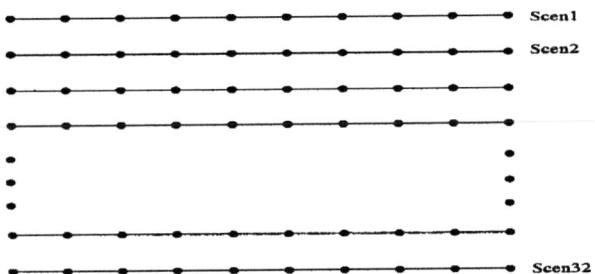

FIGURE 4.6   The distribution problem.

problem which solves the problems defined by each scenario separately and then computes the mean of their values. Here we consider three cases. The first is this distribution problem, which is closely related to engineering *scenario analysis*, the second is the *two-stage stochastic problem* and the third is the *multistage stochastic problem*. In each case we describe the level of dependence between the scenarios and examine the effect of this dependence on the value of the objective function for a small test problem. We illustrate our analysis on a problem which 'originates' from the DROP2 instance of the DROPS model discussed earlier by limiting the horizon to 10 scheduling periods. The first implementable stage (all data are deterministic) consists of the first 5 time periods and the uncertain future unfolds according to 32 scenarios for the second 5 time periods in each case.

Case 1: the distribution problem The implicit assumption associated with the distribution problem is that there is no interdependence between the scenarios, and hence each scenario can be considered as defining a separate deterministic dynamic *subproblem* to be solved individually.

Figure 4.6 illustrates the schematically 32 scenarios of the drop_ 32 test problem.

The dimensions of each scenario subproblem are:

1005	rows
1231	columns
5576	non zero elements.

The expected value of the 32 individual scenario sub-problems *after* optimization is achieved by attaching equal probability to each scenario and computing the expected optimal cost as $\sum_{i=1}^{Nscen} = p_i f_i^*$, where $p_i = \text{prob}(scenario\ i) := 1/32$ and f_i^* is the optimal objective value of the ith subproblem and hence is its *relative contribution* to the value of the expected optimal objective function of the distribution problem. The relative contribution of each scenario to the value of the objective function is given in Table 4.3 and the corresponding optimal cost of the distribution problem is equal to 272.72.

Case 2: the two-stage stochastic problem This is the case (drop_2_32) in which all data are assumed to be deterministic in the first 5 time periods, but there are 32 contingencies in time period 6. From time period 7 onwards, the data on each contingency path are assumed to be deterministic, so that the problem consists of a total of 32 alternative scenarios, beginning in time period 7, for each of which the future is known with certainty to the horizon.

Note that the branching of the 32 scenarios in the sixth time period does not now imply the absolute independence of decisions in all dynamic subproblems as in Case 1, since in the first stage common to all scenarios decisions hedge against all possible future contingencies. Figure 4.7 illustrates the overlap of the 32 scenarios in the first 5 time periods (corresponding to the first stage) and their separation from time period 6 onwards.

TABLE 4.3 Relative Contribution of Scenarios to the Optimal Value

Scenario	Distribution problem	Two-stage stochastic	Multistage stochastic
1	0	17.658623	181.807543
2	31.321884	98.602778	344.752388
3	59.672827	134.127907	302.711840
4	105.713923	165.402497	329.763348
5	450.021822	521.426496	621.144682
6	474.004827	545.114491	639.996096
7	248.888785	311.738269	384.102911
8	281.603853	344.342806	418.633773
9	520.538801	549.811937	624.638700
10	432.257438	468.014788	492.348909
11	419.067872	455.940883	532.008677
12	309.878764	339.916094	376.934610
13	276.962315	313.621423	340.920562
14	328.225666	365.430671	434.554957
15	467.013969	497.980584	610.641204
16	550.564479	597.848659	655.176105
17	421.795689	436.083906	553.390933
18	456.338699	470.838288	581.334040
19	432.410277	446.909866	531.063397
20	441.146565	455.646154	537.175416
21	97.680769	98.328548	208.456945
22	28.981334	29.629113	194.895045
23	303.198969	311.169072	458.960878
24	294.525220	308.773550	497.855173
25	0	0	0
26	25.631232	40.199764	135.938119
27	0	0	136.418608
28	118.812713	147.876116	241.084945
29	347.395162	369.510070	417.323159
30	358.077433	380.163285	442.303645
31	261.106984	278.616789	301.188785
32	184.751422	202.186116	244.275716
Value	272.72	303.22	399.12
EVPI	0	30.50	116.40

To form the deterministic equivalent, we need to attach a probability to each scenario and then aggregate them appropriately, as in (4.33)–(4.38). Its dimensions are:

18 024	rows
20 327	columns
120 772	nonzero elements.

The objective value of the deterministic equivalent corresponds to the optimal expected cost of the two-stage stochastic problem and is equal to 303.22 (with equiprobable scenarios). The relative contribution of each scenario to the value of the objective function is greater than the respective cost contribution of the same scenario to the distribution problem (see Table 4.3). This is a reflection of the scenarios' *known*

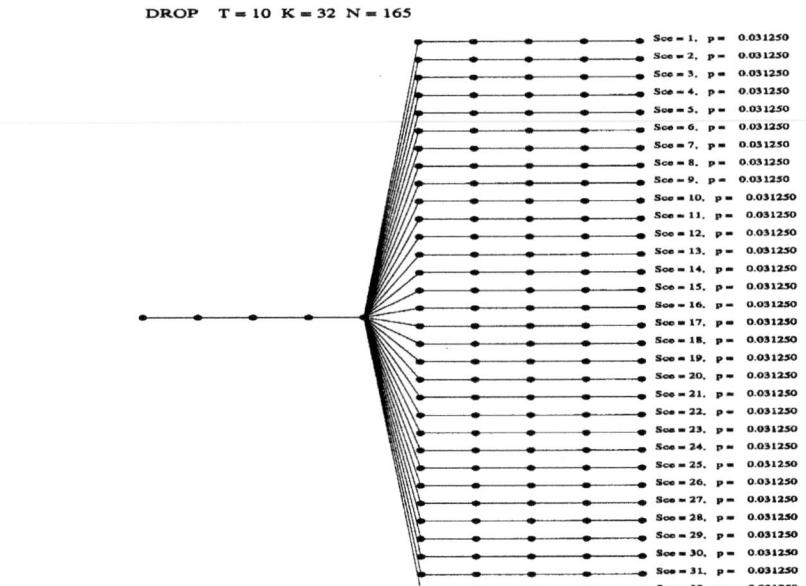

FIGURE 4.7   Tree corresponding to the two-stage stochastic problem drop_2_32.

overlap during the first stage and the need in first stage decisions to hedge against all contingencies in the second stage. As a result, the optimal expected cost of the two-stage stochastic problem is greater than the optimal expected cost of the distribution problem.

Case 3: the multistage stochastic problem This is the case (drop_ 6_32) where all data are assumed to be deterministic in the first 5 time periods, but there are two alternative possible states for the vector of stochastic parameters for each subsequent state *from time period onwards*, as shown in Figure 4.8. Thus each decision made in these non-implementable time periods faces binary uncertainty.

We form the deterministic equivalent generating the scenarios equiprobably. Its dimensions are:

9008	rows
8273	columns
79 612	nonzero elements.

The objective value of the deterministic equivalent gives us the optimal expected cost of the multistage stochastic problem, which is equal to 399.12. Note that the expected cost of the multistage stochastic problem is greater than the cost of both the distribution problem and the two-stage stochastic problem due to the greater interdependence of the scenarios. Also the two-stage problem is much larger in size, although less dense, than the multistage problem due to the presence of many more nodal subproblems, all of which are the same size.

The last line of Table 4.3 shows the *expected value of perfect information* (EVPI) for the three problems. For such problems EVPI is defined to be the difference between their optimal value and the corresponding distribution problem (with the same scenarios). Hence it represents the incremental expected value (here, the expected cost reduction) of knowing with certainty the realized future scenarios. The higher the EVPI the more future uncertainty affects the optimal value and decisions of the problem.

Figure 4.9 shows the optimistic biasing effects of ignoring uncertainty about the future in stochastic optimization problems.

The overall cost of decisions of a stochastic problem in the first implementable stage (two-stage or multistage) is always more expensive than the corresponding cost for the distribution problem. This is not

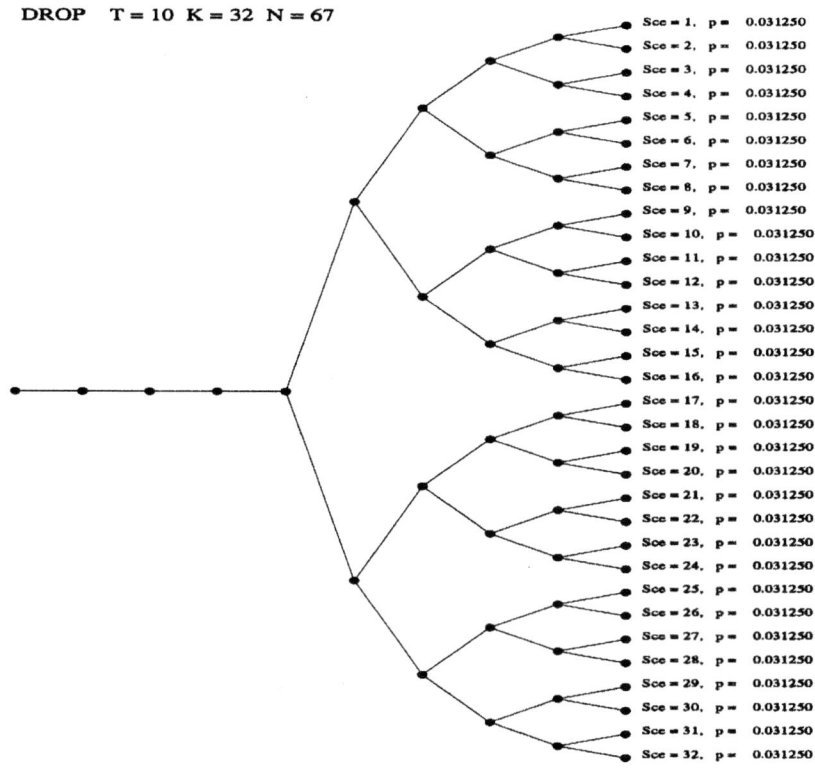

FIGURE 4.8 Tree corresponding to the multistage stochastic problem drop_6_32.

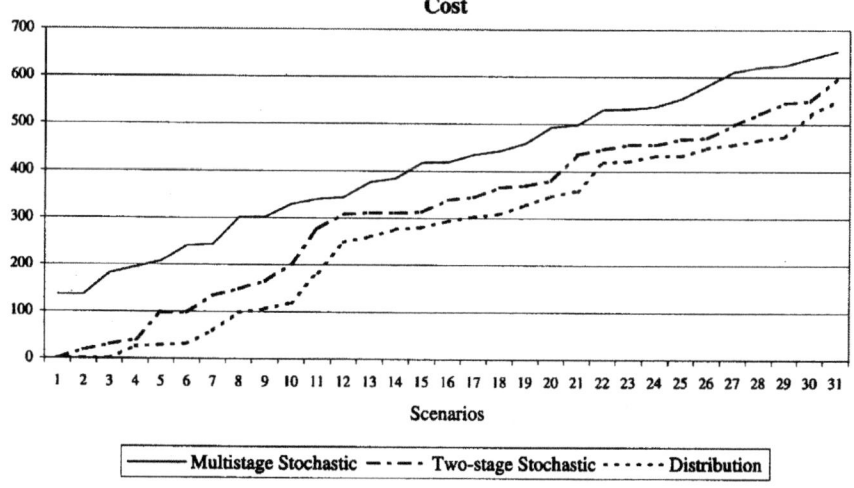

FIGURE 4.9 Relative costs of scenarios in the distribution problem, two-stage stochastic problem and multistage problem, sorted in increasing order.

necessarily true for individual decisions. Figure 4.10 illustrates some decisions given by the solution to stochastic versions with 32 scenarios of the DROP2 model as discussed earlier. It is immediately apparent that these decisions depend on the structure of the scenario tree. The need to hedge against all uncertainties in both the two-stage and the multistage stochastic cases and the overlap of the scenarios in the corresponding trees during the first five time periods (see Figures 4.7 and 4.8) enforce a unique solution for all scenarios during these periods. However, in the distribution problem, decisions are not 'averaged' since scenarios are completely independent (see Figure 4.6) and are treated as belonging to separate deterministic problems. In the distribution problem, therefore, not all scenarios lead to the same decisions in the solutions of their individual problems. That is why the volume of product 2 (in depot 3) sold in the spot market during time period 1 ranges from 0 to 0.4, and the stock of product 2 in depot 2 during time period 5 ranges from 0 to 3.5. When implementable decision values are averaged across distribution problem scenarios — a common practice in engineering scenario analysis — the result is significantly different than for either stochastic case — which are in turn significantly different from each other!

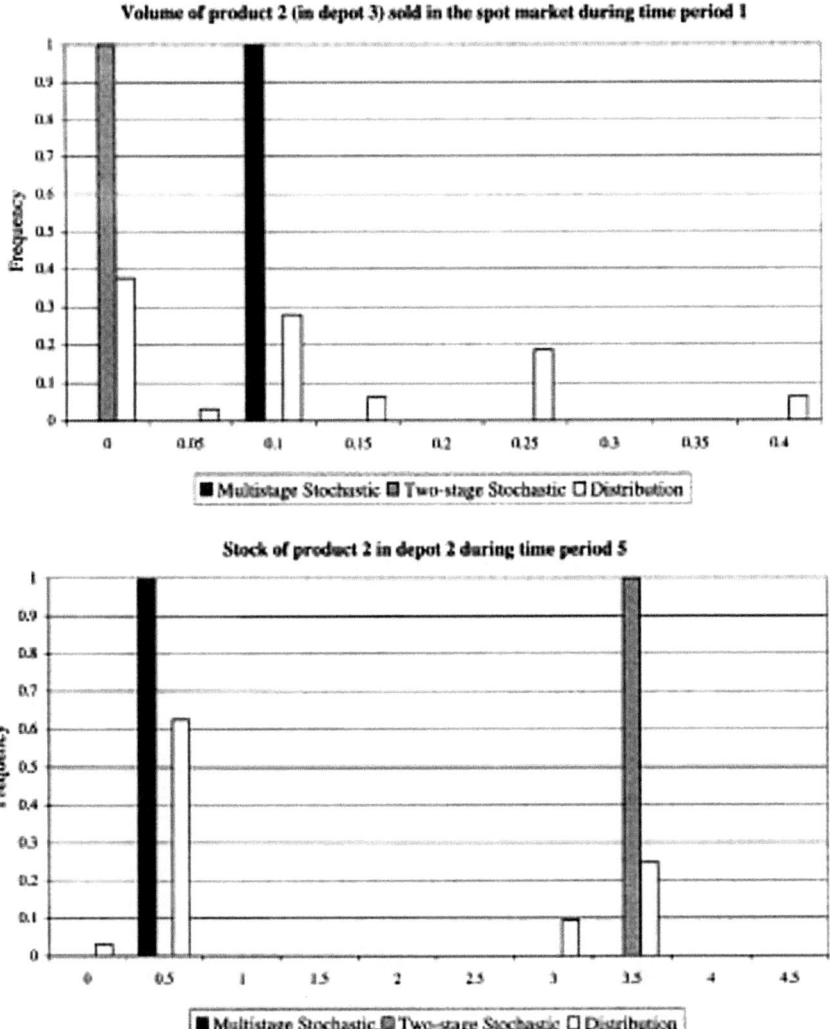

FIGURE 4.10   Comparison of various decisions for alternative stochastic formulations of the DROP model.

TABLE 4.4 Summary of Results

Problem	Rows × columns	Density	Obj. value	CPU time
drop_6_32	9627×42 637	0.02358 %	119 825.9984	61 s
drop_6_243	59 865 × 290 128	0.00369 %	123 015.9755	718 s
drop_6_1024	235 183 × 1179 159	0.00093 %	122 182.4277	3091 s

In the non-implementable periods beginning in period 6, the two-stage model makes the unrealistic assumption that decisions in these periods can be made with *perfect* knowledge of the future. On the other hand, the multistage model requires *all* decisions to be made in the face of at least two uncertain alternative futures — a minimal representation of the uncertain real world. Thus all decisions of the multistage model are more robust in the face of uncertainty than those of the two-stage model and hence lead to more realistic cost estimates.

From a purely computational view the comparison between the two-stage and the multistage stochastic problems shows significant reduction in memory requirement, as well as a more realistic increased expected cost, for the multistage problem. A summary of results for 32, 243 and 1024 scenarios is shown in Table 4.4. Note that optimal expected costs tend to rise marginally with an increasing number of scenarios, while running times of the solution algorithm rise approximately linearly. The results presented in Table 4.4 have been obtained using the XPRESS barrier solver on a Pentium II 400 MHz running under Linux 2.0.

4.4 Conclusions and Future Directions

In this paper we formulate deterministic and stochastic models of strategic planning for logistic operations in the oil industry. We implement the models using a number of existing and newly developed software tools and perform a preliminary analysis of results. We prefer the multistage stochastic formulation because of its more realistic treatment of uncertainty and its memory requirements and due to our interest in the expected value of perfect information (EVPI) techniques for scenario tree generation and the corresponding sampling algorithms.[22] EVPI gives a measure of the cost value in knowing the uncertain future – the larger the value of EVPI the more future uncertainties effect current (hedging) decisions. For stochastic logistics scheduling the solutions corresponding to the first stage are of particular interest, as they are robust decisions against future uncertainties in demand and prices and can be used with deterministic models for optimal operations scheduling, such as those being developed by our partners in the HChLOUSO project. These implementable decisions must be carefully analyzed against their costs. The development of a user-friendly interface between the problem solver and optimal decisions is a topic of future research, as are alternative solvers for very-large-scale instances of the problem treated making use of block decomposition for nested decomposition algorithms or Cholesky factorization in interior point algorithms. At present parallel schemes for the solution of such large-scale problems are under test. [23]

Acknowledgements

The continuing research reported here is supported by the Directorate General III: Industry of the European Commission (HChLOUSO, Industrial RTD Project 24897) and the UK EPSRC and ESRC Research Councils through the Centre for Financial Research. We would like to acknowledge the contributions of Amira Vejzagic-Ramhorst during the early stages of our work on the HChLOUSO project.

References

1. Escudero LF, Quintana FJ, Salmeron J (1999). Coro, a modelling and algorithmic frame-work for oil supply, transformation and distribution optimization under uncertainty. *Eur J Opl Res* **114**: 638 –656.

2. Aronofsky JS, Dutton JM and Tayyabkhan MT (1978). *Managerial Planning with Linear Programming in Process Industry Operations*. John Wiley & Sons; New York.

3. Aronofsky JS and Lee AS (1957). A linear programming model for scheduling crude oil production. *Trans AIME* **69**: 389 –403.

4. Aronofsky JS and Williams AC (1962). The use of linear programming and mathematical models in underground oil production. *Mngt Sci* **8**: 394 –407.

5. Charnes A, Cooper WW and Mellon B (1952). Blending aviation gasolines — a study in programming interdependent activities in an intergrated oil company. *Econometrica* **20**: 135 –159.

6. Charnes A, Cooper WW and Mellon B (1954). A model for programming and sensitivity analysis in an integrated oil company. *Econometrica* **22**: 193 –217.

7. Guldmann JM (1983). Supply, storage, and service reliability decisions by gas distribution utilities: a chance constrained approach. *Mngt Sci* **29**: 884 –906.

8. Hemaida RS and Kwak NK (1994). A linear goal programming model for trans-shipment problems with flexible supply and demand constraints. *J Opl Res Soc* **45**: 215 –224.

9. Klingman D *et al* (1987). An optimization based integrated short-term refined petroleum product planning system. *Mngt Sci* **33**: 813 –830.

10. Levary RR and Dean BV (1980). A natural gas flow model under uncertainty in demand. *Opns Res* **28**: 1360 –1374.

11. Dempster MAH (1988). On stochastic programming II: dynamic problems under risk. *Stochastics* **25**: 15 –42.

12. Escudero LF *et al* (1993). Production planning via scenario modelling. *Ann Oper Res* **43**: 311 –335.

13. Mulvey JM and Vladimirou H (1991). Solving multistage stochastic networks: an application of scenario aggregation. *Networks* **21**: 619 –643.

14. Escudero LF *et al* (1999). A parallel computation approach for solving multistage stochastic network problems. *Ann Oper Res* **90**: 131 –160.

15. Dempster MAH (ed) (1980). *Stochastic Programming*. Academic Press: London.

16. Ermoliev Y and Wets RJB (eds) (1988). *Numerical Techniques for Stochastic Optimization*. Springer-Verlag: Berlin.

17. Birge JR and Louveau FV (1997). *Introduction to Stochastic Programming*. Springer-Verlag: Berlin.

18. Vejzagic-Ramhorst A (1998). Optimization of stochastic networks: optimizing supply, transformation and distribution of hydrocarbon and chemical products under uncertainty. Master's Thesis, Judge Institute of Management Studies, University of Cambridge.

19. Dempster MAH *et al* (1998). Planning logistics operations in the oil industry: 1. deterministic modelling. *Research Papers in Management Studies* WP 33/98, Judge Institute of Management Studies, University of Cambridge.

20. Dash Associates Ltd (1997). *XPRESS-MP Version 10.06 User's Guide*.

21. IBM Corporation (1992). *IBM Subroutine Library (OSL)*, 4th edition.

22. Dempster MAH (1998). Sequential importance sampling algorithms for dynamic stochastic programming. Technical Report, *Research Papers in Management Studies* WP 32/98, Centre for Financial Research, Judge Institute of Management Studies, University of Cambridge.

23. Dempster MAH *et al* (2000). Solving large scale stochastic programming models for logistics planning in the oil industry. Technical Report, Centre for Financial Research, Judge Institute of Management Studies, University of Cambridge.

24. Dempster MAH *et al* (1999). Planning logistics operations in the oil industry: 2. stochastic modelling. *Research Papers in Management Studies* WP 4/99, Judge Institute of Management Studies, University of Cambridge.

25. Park SK and Miller KW (1988). *Comm ACM* **31**: 1192 –1202.

26. Press WH *et al* (1992). *Numerical Recipes in C* 2nd edn. Cambridge University Press, Cambridge.

27. Lane M and Hutchinson P (1980). A model for managing a certificate of deposit portfolio under uncertainty. In: Dempster, MAH (ed). *Stochastic Programming*. Academic Press: London, pp 473 –493.

28. Thompson RT (1997). *MSLiP-OSL User's Guide*. Judge Institute of Management Studies, University of Cambridge.

29. Birge JR *et al* (1987). A standard input format for multiperiod stochastic programs. *Math Prog Soc Committee on Algorithms Newsletter* **17**: 1 –20.

30. Corvera-Poiré X (1995). Model generation and sampling algorithms for dynamic stochastic programming. PhD Thesis, Department of Mathematics, University of Essex.

31. Corvera-Poiré X (1995). *STOCHGEN User's Manual*. Department of Mathematics, University of Essex.

32. Greenberg HJ (1994). *A Computer-Assisted Analysis System for Mathematical Programming Models and Solutions: A User's Guide for ANALYZE*. Mathematics Department, University of Colorado at Denver.

Appendix

Sets

- $N := nodes$
- $L := \{(n_1, n_2) : n_1, n_2 \epsilon N, n_1 \neq n_2\}$
- $T := \{time\ periods\}$
- $O : \{operators\}$
- $P := \{products\}, P_i := \begin{cases} \{\text{crude oils}\}, & i = 1 \\ \{\text{final products}\}, & i = 2 \end{cases}$
- $R: = \text{refineries}, R \subset N$
- $F := \{\text{transformation technologies}\}$
- $F_n : = \{\text{transformation technology available at node } n\}$
- $F_p^{in} : = \{\text{technology which uses product } p \text{ as input}\}$
- $F_p^{out} : = \{\text{technology which produces product } p \text{ as output}\}$
- $M := \{\text{transportation means}\}$
- $M^c : = \{\text{transportation means which allow continuous flows of product}\}$
- $M^d : = \{\text{transportation means which allow discrete flows of product}\}$
- $P_m : = \{p \epsilon P : p \text{ can be transported by means } m\}$
- $L_m := \{(n_1, n_2) \epsilon L \text{ means } m \text{ can be used from } n_1 \text{ to } n_2\}$
- $T_{i, j, p, m} : = \{t \epsilon T \text{ the transportation of product } p \text{ from node } i \text{ to node } j \text{ by transport means } m \text{ is allowed to start during time period } t\}$

- $U := \{\text{sets of time periods used for bounding product volumes } transported\ \}$
- $T_u^U := \{t \epsilon T : \text{time periods } t \text{ included in } u, \text{ for } u \epsilon U\}$
- $V := \{\text{sets of time periods used for bounding product volumes } supplied\}$
- $T_u^V := \{t \epsilon T : \text{time periods } t \text{ included in } \upsilon, \text{ for } \upsilon \epsilon V\}$
- $H := \{\text{sets of time periods used for bounding product volumes } refined\}$

†*Entities* are hierarchical or nested partitions of sets due to government regulations or economic considerations for supply and transportation.

- $T_h^H := \{t \in T : \text{time periods } t \text{ included in } h, \text{ for } h \in H\}$
- $\Theta :=$ entities†}

Parameters

- $c_{n,o,p,t}^x$: *cost of supplying* one unit of product p (by operator o) at node n during time period t.
- $c_{n,f,t}^z$: *cost of transforming* one unit of input product volume at node n by technology f when the transformation (ie refining) takes place *during* time period t and is output for the beginning of period t +1 at the end of period t.
- $c_{n_i,n_j,p,m,t}^e$: *cost of transporting* one unit of product p from node n_i to node n_j by means m if the transportation starts during time period t.
- $c_{n,p,t}^s$: *cost of stocking* one unit of product p at node n during period t and paid at the beginning of time period t +1.
- $c_{n,p,t}^{spot}$: *cost of* (spot) *purchasing* and supplying one unit of product p at node n during time period t paid at the beginning of period t.
- $p_{n,p,t}^{spot}$: *price charged* for (spot) *selling* one unit of product p at node n during time period t paid at the beginning of period t.
- $s_{n,p}^0$: *initial stock* volume of product p at node n at the beginning of time period 1.
- $g_{f,p}$: unit input volume used by technology f to *output* final product p.*
- $\Delta_{n_1,n_2,p,m}$: integer number of *time periods required for transportation* of product p from node n_1 to node n_2 by transport means m minus one.
- $d_{n,o,p,t}$: product p volume that must be set aside at node n at the beginning of time period t in order to guarantee satisfaction of *demand* (for operator o) during time period t.
- $e_{n,p,t}$: product p volume that arrives *in transit* at node n during time period t and whose transportation from another node started *before* the beginning of the *planning horizon*.

Variables

- $X := \{X_{n,o,p,t} : n \in N, o \in O, p \in P, t \in T\}$, where $X_{n,o,p,t}$ is the product p volume supplied by operator o to node n during time period t — (*supply*).
- $Z := \{Z_{n,f,t} : n \in N, f \in F, t \in T\}$, where $Z_{n,f,t}$ is the mixed product volume used by transformation technology f for the refining process that takes place at node (refinery) n during time period t — (*transformation*).
- $E := \{E_{i,j,p,m,t} : i, j \in N, p \in P, m \in M, t \in T\}$, where $E_{i,j,p,m,t}$ is the product p volume whose transportation from node i to node j by transportation means m starts during time period t — (*transportation*).
- $S := \{S_{n,p,t} : n \in N, p \in P, t \in T\}$, where $S_{n,p,t}$ is the product p stock volume at node n at the end of time period t (and hence at the beginning of time period t+1). This is taken to be equal to a constant vector s^0 at the beginning of time period 1 — (*stocks*).
- $X^{spot} := \{X_{n,p,t}^{spot} : n \in N, p \in P, t \in T\}$, where $X_{n,p,t}^{spot}$ is the product p volume (spot purchases) supplied from the spot market at node n during time period t — (*purchases*).
- $Y^{spot} := \{Y_{n,p,t}^{spot} : n \in N, p \in P, t \in T\}$, where $Y_{n,p,t}^{spot}$ is the product p volume sold to the spot market at node n during time period t — (*sales*).

---

* A similar scheme is used for crude oil inputs to define *input* proportions to technology f.

Bounds

- $e^c_{n_1,n_2,m,u}$: *upper bound* on the total product volume in *transit* from node n_1 to node n_2 by continuous transport means m (ie pipelines) during some time period included in the horizon segment u.

- $e^d_{m,t}$: *upper bound* on the total product volume in *transit* during time period t. This is the bound imposed by capacity on discrete transport means, such as ships, trucks and wagons.

- $\underline{x}_{n,o,p,\upsilon}$: *lower bound* on the volume of *product p supplied* by operator o to node n during the time periods included in the horizon segment υ.

- $\overline{x}_{n,o,p,\upsilon}$: *upper bound* on the volume of *product p supplied* by operator o to node n during the time periods included in the horizon segment υ.

- $\underline{z}_{n,f,h}$: *lower bound* on the product volume *refined* in refinery n by technology f during the time periods included in the horizon segment h.

- $\overline{z}_{n,f,h}$: *upper bound* on the product volume *refined* in refinery n by technology f during the time periods included in the horizon segment h.

- $\underline{s}_{n,p,t}$: *lower bound* on the product volume *stocked* at node n during time period t.

- $\overline{s}_{n,p,t}$: *upper bound* on the product volume *stocked* at node n during time period t.

- $\overline{z}^T_{n,f,t}$: *upper bound* on the product volume *transformed* by technology f at refinery n during time period t.

- $\underline{s}_\theta$: *lower bound* on the product *stock* at entity θ.

- $\overline{s}_\theta$: *upper bound* on the product *stock* at entity θ.

5

Analysing the Dynamics of the Refining Margin: Implications for Valuation and Hedging

Andrés García
Mirantes,
Javier Población and
Gregorio Serna

5.1 Introduction

The refining process converts 47% of crude oil barrels into gasoline, 24% into diesel fuel and heating oil, 13% into jet fuel oil, 4% into heavy fuel oil, 4% into liquefied petroleum gas (LPG) and 8% into other products such as asphalt.* Like crude oil, each of these products has a market price that is quoted on organized markets. Therefore, there is a relationship between refined-product prices and crude oil prices that is known as the 'refining margin'. However, most of the articles on the stochastic behaviour of commodity prices are focused either on crude oil or on the refined products derived from crude oil rather than the refining margin itself.

In this paper, we focus on the refining margin. Contrary to oil prospecting and production, refining is a margin business; a refining company buys oil and sells refined products (e.g., gasoline and heating oil) and makes a profit that is, in principle, independent of oil and product prices. In other words, the profit of a refining company (i.e. the refining margin) is related to the difference between the prices of crude oil and the refined products. However, this difference does not necessarily decrease (or increase) when the prices of crude oil and its products rise (or fall). As a result, refineries face a significant risk when,

---

* See the Oil Market Report (2006) developed by the International Energy Agency for more information about these issues.

for example, oil prices rise but product prices remain static or decline. It is clear that refining companies must somehow protect their refining margin using derivative contracts.

This risk faced by refining companies may result in unpleasant consequences that have been thoroughly studied in the literature, such as the costs of financial distress (Myers and Smith 1982, Smith and Slutz 1985), external financing costs and investment opportunities (Froot *et al.* 1993) and liquidity constraints (Holmströn and Tirole 2000, Mello and Parsons 2000). However, it is important to consider the tax incentives (see, for example, Smith and Slutz [1985] and Graham and Smith [1999]). Therefore, the refining margin deserves to be independently analysed, which is the main objective of this paper.

Previous studies, including those by Serletis (1992, 1994), Pindyck (1999), Gjolberg and Johnsen (1999), Asche *et al.* (2003) and Lanza *et al.* (2005), provide evidence of unit roots and cointegration in the prices of crude oil and refined products. In this paper, this empirical evidence is extended by showing that the refining margin is stationary and, therefore, exhibits dynamics that are different from those of crude oil or its main refined products. Furthermore, we show that crude oil, heating oil and gasoline are not only cointegrated but also exhibit common long-term dynamics, implying that the refining margin reflects only short-term effects.

The hypothesis of a common long-term trend is confirmed using different factor models to jointly explain the dynamics of commodity prices. We find that the most suitable model in terms of simplicity and fit is the one that assumes a common long-term trend for crude oil, heating oil and gasoline.

This common long-term trend model for crude oil and its main refined products is framed within the family of multi-factor models proposed by Schwartz (1997) and a related series of papers, including those by Schwartz and Smith (2000), Cortazar and Schwartz (2003) and Cortazar and Naranjo (2006). All of these multi-factor models assume that the spot price is the sum of both short- and long-term components. Long-term factors account for the long-term dynamics of commodity prices, which are assumed to follow a random walk, whereas the short-term factors account for the mean-reverting components in commodity prices. Moreover, in the cases of heating oil and gasoline, a deterministic seasonal component is added, as suggested by Sorensen (2002).*

Cortazar *et al.* (2008) have already proposed and estimated a model in which several commodities can exhibit both common and specific factors. They presented their multi-commodity model as a way to improve the estimation of futures prices for a commodity with scarce data using long-maturity futures prices that are available for another commodity. They applied their model to WTI-Brent[†] crude oil and WTI-Unleaded gasoline futures contracts but not to heating oil.

However, in contrast to the study by Cortazar *et al.* (2008), the primary objective of this paper is to model and hedge the refining margin. Specifically, we show that the refining margin exhibits dynamics that are different from those of crude oil and its main refined products because the refining margin only reflects short-term effects. Furthermore, we show that the proper way to model and hedge the refining margin is to consider the common long-term factor model because it provides a more accurate crack-spread option valuation; specifically, crack-spread option quotes imply that the refining margin is stationary, and therefore, commodities exhibit a common long-term trend. Consequently, the refining margin only reflects short-term effects; thus, a multi-commodity model with a common long-term trend is the best way to model and hedge it. In this sense, Cortazar *et al.* (2008) suggested that if the spreads between two commodities are of interest, the use of multi- commodity models should provide much more stable estimates, particularly for the commodity for which data are scarce.

The fact that crude oil, heating oil and gasoline share a common long-term trend will have straightforward implications for managing and hedging refining margin risk, specifically in terms of crack-spread

---

* In contrast to crude oil, heating oil and gasoline exhibit a strong seasonal behaviour (see, for example, Garcia *et al.* [2012]). Therefore, the model for heating oil and gasoline must account for this fact.

† WTI (West Texas Intermediate), also known as Texas light sweet, is a type of crude oil used as a benchmark in oil pricing and is the underlying commodity of New York Mercantile Exchange's oil futures contracts. Brent is another type of crude oil, which is sourced from the North Sea.

option valuation. Crack-spread options are used to protect the refining margin while simultaneously allowing market participants to take advantage of favourable changes in the spread. A crack-spread call option is a contract that gives the holder the right (but not the obligation) to buy a refined-product futures contract from the writer and sell the writer a crude oil futures contract, paying a previously agreed upon crack-spread price. A crack-spread put allows one to sell a refined product futures contract and buy a crude oil futures contract, paying a crack-spread price.

The primary organized market for crack-spread options is the NYMEX, which is now included in CME Group, where calls and puts at a one-to-one ratio between the New York Harbour unleaded gasoline or heating oil futures contract and the light sweet crude oil futures contract can be found.

Given that these crack-spread options involve several assets, the techniques used in their valuation are complex. Dempster *et al.* (2008) propose a model for valuing spread options on two commodity prices that are cointegrated and give two applications to a set of theoretical (not traded) European options on the crack-spread between heating oil and WTI crude oil and the location spread between Brent blend and WTI crude oil. Dempster *et al.* (2008) point out that the correlation between two asset returns is hard to model (Kirk 1995, Mbanefo 1997, Alexander 1999), and therefore, the crack-spread should be modelled directly. However, from our point of view, there are other reasons that must be taken into account when modelling the crack-spread. For example, as suggested by Cortazar *et al.* (2008), using two or more price series, it is possible to extract more information for estimation purposes. Therefore, a common long-term trend model for the two assets involved in the crack-spread can be an alternative to the approach suggested by Dempster *et al.* (2008).

In this paper, we present a simplified model for valuing this type of option that assumes a common long-term trend for the prices of crude oil and its main refined products (i.e. heating oil and gasoline). As opposed to Dempster *et al.* (2008), who use a theoretical (not traded) option database, we use an extensive database of American crack-spread options (heating oil vs. WTI and unleaded gasoline vs. WTI) traded at NYMEX to find that by assuming a common long-term trend for the prices of crude oil and refined products, more accurate option valuations can be obtained than when using models with more factors and parameters, including the one proposed by Dempster *et al.* (2008). This result also implies that the advantage of the model with a common long-term trend, i.e. by using two or more price series, it is possible to extract more information for estimation purposes, has more importance in terms of valuation errors than the advantage of the Dempster *et al.* (2008) model, i.e. avoiding the need to model the correlation between two asset returns.

Even more importantly, when assuming a common long-term trend, theoretical option values are lower (and closer to the real values) than when one assumes different long-term trends for each commodity. This result may be related to the fact that the option holder assumes lower risk with a common long-term trend than with different long-term trends for each commodity because, with common long-term trends, the refining margin risk reflects only short-term effects. This result can be explained as follows. Given that the refining margin is proved to be stationary, its behaviour can be better explained with a model without long-term effects because long-term effects are not stationary. By using a common long-term trend model, the (common) long-term factors disappear when computing the refining margin, which does not occur when we use the model with different long-term trends. Therefore, because the refining margin only reflects short-term effects, its volatility is bounded in time.* Consequently, as investors face lower risk, option prices must be lower.

Therefore, we can conclude that the preferred model for valuing this type of option is one that assumes a common long-term trend. Additionally, the evidence of a common long-term trend is useful not only in the valuation and hedging of commodity-contingent claims but also in defining procedures for evaluating investment projects related to natural resources, particularly when determining optimal investment rules.

---

* If the refining margin were to reflect long-term effects, which are not stationary, its volatility would tend to increase with time.

The remainder of this paper is organized as follows. Section 5.2 presents the data and some preliminary findings regarding the refining margin. We show that in contrast to crude oil and its main refined products, the refining margin is stationary, implying that it exhibits dynamics that are different from crude oil and its products. In Section 5.3, we show that crude oil, heating oil and gasoline are not only cointegrated but also exhibit common long-term trends, implying that the refining margin only reflects short-term effects. In Section 5.4, we use the results from Sections 5.2 and 5.3 to value crack-spread options, and we analyse the implications for hedging the refining margin. Specifically, Section 5.4 shows the valuation results of the crack-spread options listed on NYMEX using a model that assumes a common long-term trend for crude oil and the main refined product prices, a model that allows for a long-term trend for each commodity, a model that postulates uncorrelated sub-models for each commodity and the model by Dempster *et al.* (2008). Finally, Section 5.5 concludes with a summary and discussion.

5.2 Data and Preliminary Findings

In this section, we present some preliminary findings regarding the refining margin. Specifically, we show that the refining margin exhibits dynamics that are different from those of crude oil and its main refined products, implying that refining companies must use specific hedging products whenever they choose to cap their losses on the refining margin.

However, before presenting these preliminary findings regarding the refining margin, it is useful to briefly describe the data that will be used in this and the following sections.

5.2.1 Data Description

The data set used in this paper consists of weekly observations of WTI (light sweet) crude oil, heating oil and unleaded gasoline futures prices traded at NYMEX[*] and weekly observations of the refining margin for WTI oil on the U.S. Gulf coast.

WTI crude oil futures with maturities of 1 month to 7 years, heating oil futures with maturities from 1 to 18 months and gasoline futures with maturities of 1 to 12 months are currently being traded at NYMEX. However, in the case of gasoline, there is not enough liquidity for the futures with longer maturities. Therefore, in the cases of WTI crude oil and heating oil, our data set comprises futures prices from 1 to 18 months (664 weekly observations), from 9/9/1996 to 5/25/2009, and in the case of RBOB[†] gasoline, our data set comprises futures prices from 1 to 9 months (622 weekly observations), from 6/30/1997 to 5/25/2009.

With regard to the refining margin data, it should be noted that we can use different refining margin series depending on the type of crude oil used (WTI, Brent, Dubai, etc.) and the geographical area represented (U.S. Gulf Coast, Rotterdam, Singapore, etc.). However, to cohere with the NYMEX crude oil and refined products data described above, we will use the refining margin for WTI crude oil on the U.S. Gulf coast.[‡]

The refining margin series used in this paper was calculated (as suggested by the International Energy Agency[§]) as the gross product worth of a refinery on the U.S. Gulf Coast using the catalytic cracking method, minus the costs of crude oil and freight. The gross product worth is the weighted average value of all refined product components of a barrel of crude oil, computed by multiplying the spot price of each product by its percentage share of the yield of the barrel. Weekly observations of this refining

---

[*] Details about the contracts can be found on the NYMEX homepage.

[†] RBOB stands for 'reformulated gasoline blendstock for oxygen blending'. It is the benchmark gasoline contract on the New York Mercantile Exchange.

[‡] The results do not change if another refining margin is used. For the sake of brevity, the results for other refining margins are not presented here, although they are available from the authors upon request.

[§] Oil Market Report, Annual Statistical Supplement. International Energy Agency. http://www.omrpublic.iea.org; 2006.

TABLE 5.1 Descriptive Statistics

	WTI Crude Oil		Heating Oil		Gasoline		Ref. Margin	
	Mean	Volatility (%)	Mean	Volatility (%)	Mean	Volatility (%)	Mean	Stand. Dev.
Spot							3.58	5.21
F1	41.76	32.14	48.94	31.51	51.16	36.99		
F2	41.89	29.32	49.11	29.15	50.99	31.99		
F3	41.92	27.56	49.21	27.49	50.82	29.60		
F4	41.87	26.09	49.23	26.22	50.61	26.59		
F5	41.80	24.79	49.20	25.03	50.36	25.28		
F6	41.71	23.71	49.14	23.89	50.18	24.36		
F7	41.60	22.83	49.08	22.73	50.03	23.57		
F8	41.50	22.09	49.00	21.76	49.93	23.60		
F9	41.40	21.45	48.91	20.81	49.85	23.22		
F10	41.30	20.87	48.81	20.08				
F11	41.20	20.35	48.71	19.58				
F12	41.11	19.88	48.60	19.12				
F13	41.02	19.45	48.52	18.83				
F14	40.93	19.06	48.45	18.62				
F15	40.84	18.72	48.39	18.53				
F16	40.76	18.43	48.33	18.59				
F17	40.69	18.16	48.28	18.56				
F18	40.62	17.91	48.24	18.48				

The table shows the mean and volatility of the four commodity series prices. F1 is the futures contract closest to maturity, F2 is the contract second-closest to maturity and so on.

margin from 1/26/1997 to 5/24/2009 (representing 643 observations) were used in the tests. The main descriptive statistics of the series are summarized in Table 5.1.

5.2.2 Preliminary Findings About the Refining Margin

Previous studies, including those by Serletis (1992, 1994), Pindyck (1999), Gjolberg and Johnsen (1999), Asche *et al.* (2003) and Lanza *et al.* (2005), provide evidence of unit roots and cointegration in the prices of crude oil and refined products (i.e. heating oil and gasoline). However, to the best of our knowledge, there is no evidence of stationarity of the refining margin, which is composed not only of crude oil, heating oil and gasoline but also of other products such as liquefied petroleum gas (LPG), asphalt and kerosene.

Table 5.2 summarizes the results of the unit root tests for WTI crude oil, heating oil and gasoline prices; and the refining margin. The empirical evidence from previous studies of a unit root in crude oil, heating oil and gasoline prices is confirmed in the present work using the standard Augmented Dickey-Fuller and Philips-Perron tests. However, the refining margin does not show evidence of a unit root.[*][†] Therefore, the refining margin seems to show different dynamics than that of crude oil and its main refined products.

---

[*] To the best of the authors' knowledge, there are no spot prices for gasoline, heating oil or crude oil associated with the futures traded at NYMEX. Therefore, weekly observations of 1-month futures prices for WTI crude oil, heating oil and gasoline are used as proxies for spot prices.

[†] As discussed below, heating oil and gasoline show a seasonal effect, whereas crude oil does not (Garcia *et al.* 2012). Therefore, the refining margin must inherit a seasonal component as well. Thus, we have repeated the unit root tests after removing the seasonal component in the series, obtaining very similar results to those presented in Table 5.2.

TABLE 5.2   Unit Root Tests

Series	ADF	Phillips–Perron
WTI crude oil	−1.2409	−1.6380
	(0.6582)	(0.4625)
Heating oil	−1.8821	−1.5668
	(0.3409)	(0.4991)
Gasoline	−1.8517	−1.8534
	(0.3554)	(0.3546)
Refining margin	−5.9062***	−7.7892***
	(0.0000)	(0.0000)

The table shows the statistics of the Augmented Dickey–Fuller (ADF) and Phillips–Perron unit root tests. MacKinnon *p*-values are shown in parentheses. The results are reported with * indicating rejection of the null hypothesis of a unit root at a 10% significance level, ** indicating rejection at 5% and *** at indicating rejection at 1%.

This result has important implications in terms of hedging the position of a refining company. Refining companies face risk associated with a refining margin that is different from that of crude oil and its main refined products. Thus, refining companies must use specific hedging products that are different from those used to hedge the risk associated with crude oil and its main refined products. As discussed in the introduction, crack-spread options meet this need.

Moreover, in the next section, we show that crude oil, heating oil and gasoline prices are not only cointegrated but also exhibit common long-term dynamics.* This finding has important implications in terms of crack-spread option valuation, as will be discussed in Section 5.4.

5.3 Common Long-Term Trend Factor Models

In this section, we propose and estimate different factor models either with or without assuming common long-term trends for crude oil, heating oil and gasoline. These comparisons will demonstrate that the most suitable model in terms of both simplicity and fit is the one that assumes a common long-term trend for all three commodities. This result suggests that the three commodities are not only cointegrated, as shown in previous studies, but also share a common long-term trend, implying that the refining margin only reflects short-term effects. This finding will have straightforward implications for managing and hedging the refining margin risk, specifically in terms of crack-spread option valuation.

Given that the main objective of this paper is to characterize the dynamics of the refining margin, the possibility of modelling the refining margin directly instead of modelling each price series as a stochastic system has been considered. However, by modelling each commodity as a stochastic system, we use richer information than directly modelling the price differences because in the former case, we account for all of the information contained in the price series. In contrast, by modelling the price differences directly, we would lose some of the information contained in the original price series because one series must be subtracted from the other.†

Furthermore, by modelling each commodity as a stochastic system, we will be able to show that all three commodities share a common long-term trend, suggesting that the refining margin only reflects short-term effects.

---

* For brevity's sake, the results of the cointegration tests are not presented here, although they are available from the authors upon request.
† Specifically, by modelling each price series separately, it is possible to use futures prices with different maturities for each commodity. This is one of the main advantages of the Cortazar *et al.* (2008) multi-commodity model. However, modelling directly the price differences would force us to use futures with the same maturities for both commodities.

It seems clear that modelling each commodity separately is the way to obtain the best fit for a given data set. However, if we were to obtain a similar goodness of fit when modelling the three commodities jointly with a common long-term trend, the conclusion would be that all three commodities share a common long-term trend. It is also possible to compare the results obtained from modelling the commodities jointly with and without the assumption of a common long-term trend; if there is a common long-term trend, the results must be comparable.

This result has important implications in terms of hedging the position of a refining company. Refining companies face risk associated with a refining margin that

5.3.1 Theoretical Models

Here, we present three different specifications used to model the stochastic behaviour of the three commodities under study within the context of the two-factor model proposed by Schwartz and Smith (2000). Given the existing empirical evidence,[*] this is a reasonable approach to this kind of commodity. In this model, the log-spot price[†] (X_t) is assumed to be the sum of two stochastic factors: a short-term deviation (χ_t) and a long-term equilibrium price level (ξ_t). Thus,

$$X_t = \xi_t + \chi_t. \tag{5.1}$$

The stochastic differential equations (SDEs) for these factors are as follows:

$$\begin{cases} d\xi_t = \mu_\xi dt + \sigma_\xi dW_{\xi t}, \\ d\chi_t = -\kappa\chi_t \, dt + \sigma_\chi dW_{\chi t}, \end{cases} \tag{5.2}$$

where $dW_{\xi t}$ and $dW_{\chi t}$ can be correlated ($dW_{\xi t} \, dW_{\chi t} = \rho_{\xi\chi}dt$) and $\rho_{\xi\chi}$ represents the coefficient of correlation between long- and short-term factors.

In this model, μ_ξ and σ_ξ represent the trend and volatility, respectively, of the long-term factor, whereas κ and σ_χ represent the speed of adjustment and volatility, respectively, of the short-term factor.

This model captures the most important features of commodity prices. Specifically, as explained by Schwartz and Smith (2000), the equilibrium price level (ξ_t) is assumed to follow a geometric Brownian motion. In this motion, the drift reflects market expectations about the exhaustion of existing supplies, improvements in technology for the production and discovery of the commodity and inflation or political/regulatory effects. The short-term deviations (χ_t) are assumed to revert towards zero following an Ornstein–Uhlenbeck process. These deviations reflect short-term changes in demand, which are attenuated by market participants adjusting their inventories, accounting for the commonly observed mean-reverting feature of commodity prices.

Moreover, in the cases of heating oil and gasoline, a deterministic seasonal component is added, as suggested by Sorensen (2002).[‡] Therefore, the log spot price for heating oil and gasoline (X_t) is assumed to be the sum of two stochastic factors (χ_t and ξ_t) and a deterministic seasonal trigonometric component (α_t) (i.e. $X_t = \xi_t + \chi_t + \alpha_t$). The SDEs for ξ_t and χ_t are given by Equation (5.2) and by

$$d\alpha_t = 2\pi\varphi\alpha_t^* dt \quad \text{and} \quad d\alpha_t^* = -2\pi\varphi\alpha_t dt$$

---

[*] See, for example, Schwartz (1997).

[†] For the sake of simplicity, it is typical in the literature to formulate the model in terms of log-spot prices because log-spot prices are assumed to be normally distributed, whereas spot prices are assumed to be log-normally distributed.

[‡] Sorensen (2002) suggested introducing a deterministic seasonal component into models for agricultural commodities. Here, we use Sorensen's proposal for heating oil and gasoline, which exhibits strong seasonal behaviour (see, for example Garcia *et al.* 2012).

where α_t^* is the other seasonal factor that complements α_t, and φ is the seasonal period.

The first specification, in which each of the three commodities is modelled separately, will be the simplest one; it is represented by a six-factor model with no correlation between the factors. However, this model is not very realistic because the no-correlation assumption is clearly an undesirable property in valuing commodity-contingent claims, as discussed below.

To solve this problem, we propose a second six-factor model, a joint model for all three commodities that allows for correlation between factors. In this model, the log-spot price of crude oil (X_{1t}) is assumed to be the sum of two stochastic factors: a short-term deviation (χ_{1t}) and a long-term equilibrium price level (ξ_{1t}) (i.e. $X_{1t} = \xi_{1t} + \chi_{1t}$). However, in the cases of heating oil and gasoline, a deterministic seasonal component, α_{it}, $i = 2, 3$ is added, where subscripts 2 and 3 refer to heating oil and gasoline, respectively. Therefore, the log-spot price for heating oil and gasoline (X_{it}) will be $X_{it} = \xi_{it} + \chi_{it} + \alpha_{it}$, $i = 2, 3$. The SDEs of the factors for this joint model without a common long-term trend are

$$
\begin{aligned}
\left.\begin{aligned}
d\xi_{it} &= \mu_{\xi_i}dt + \sigma_{\xi_i}dW_{\xi_{it}}, \\
d\chi_{it} &= -\kappa_i\chi_{it}dt + \delta_{\chi_i}dW_{\chi_{it}},
\end{aligned}\right\} \quad i = 1,2,3 \\
\left.\begin{aligned}
d\alpha_{it} &= 2\pi\varphi_i\alpha_{it}^*dt \\
d\alpha_{it}^* &= -2\pi\varphi_i\alpha_{it}dt,
\end{aligned}\right\} \quad i = 2,3,
\end{aligned}
\tag{5.3}
$$

where $dW_{\xi_{1t}}$, $dW_{\xi_{2t}}$, $dW_{\xi_{3t}}$, $dW_{\chi_{1t}}$, $dW_{\chi_{2t}}$ and $dW_{\chi_{3t}}$ can show any correlation structure, resulting in 15 correlation parameters.

This second model does account for the relationships between series, but it does so in a somewhat ambiguous way. We have 15 correlations to consider, none of which are negligible. We have three correlated long-term trends, though the relationship between series does not stop there. We cannot take this correlation as the only measure because the long-term trend for crude oil is also correlated with the short-term trend for refined products. Moreover, questions regarding subjects like the general market trend cannot be answered. Therefore, in our second model, such questions are meaningless unless we assume that some combination of long-term trends is 'representative'.

This problem can be solved by means of a third specification with only one long-term trend for all three commodities. Using such a model, questions such as the one described above can be fully answered: the general trend is the common long-term trend. Furthermore, it is even possible to see the relationship between the general trend and each of the series by observing its long-run/short-run correlation coefficient. In addition, the (isolated) influence of one series on another can be directly determined by observing the short-term/short-term correlation coefficient. In this model, the log-spot price of crude oil (X_{1t}) is assumed to be the sum of two stochastic factors: a short-term deviation (χ_{1t}), which is different for each commodity; and a common long-term equilibrium price level (ξ_t), where $X_{1t} = \xi_t + \chi_{1t}$. However, in the cases of heating oil and gasoline, a deterministic seasonal component, α_{it}, $i = 2, 3$, is also added. Therefore, the log-spot price for heating oil and gasoline (X_{it}) will be $X_{it} = \xi_t + \chi_{it} + \alpha_{it}$, $i = 2, 3$. The SDEs of the factors for this joint model with a common long-term trend are

$$
\begin{aligned}
d\xi_t &= \mu_{\xi}dt + \sigma_{\xi}dW_{\xi_t}, \\
d\chi_{it} &= -\kappa_i\chi_{it}dt + \sigma_{\chi_i}dW_{\chi_{it}}, \quad i = 1,2,3, \\
\left.\begin{aligned}
d\alpha_{it} &= 2\pi\varphi\alpha_{it}^*dt, \\
d\alpha_{it}^* &= -2\pi\varphi\alpha_{it}dt,
\end{aligned}\right\} \quad i = 2,3,
\end{aligned}
\tag{5.4}
$$

where dW_{ξ_t}, $dW_{\chi_{1t}}$, $dW_{\chi_{2t}}$ and $dW_{\chi_{3t}}$ can show any correlation structures resulting in six correlation parameters.

This common long-term trend model requires two additional parameters to account for the variable quality of refining products. Therefore, even if their long-term dynamics are the same, their price levels and the effects of the long-term factor on their prices may differ. Specifically, because of differences in quality, the final price level should be different for crude oil than for refined products as well as for different refined products, if one is compared to the other. Assuming a common long-term trend for crude oil and refining products, these price level differences must be stationary; thus, they have been included in the short-term component, which is different from one commodity to the next. However, because this short-term factor has no drift and affects the log-spot price rather than the price level, it is necessary to introduce a constant (K_i) in the price level to account for this fact. Furthermore, these quality differences might lead to differences in the way that this common long-term trend affects the price dynamics of each commodity. Thus, because the long-term factor is the same for the three commodities, this factor will be multiplied by a different constant (C_i) for each commodity.

Therefore, the spot price for crude oil can be calculated as $P_{1t} = \exp(\xi_t + x_{1t})$.[*] However, in the cases of heating oil and gasoline, the spot price will be $P_{it} = K_i + \exp(C_i \cdot \xi_t + \chi_{it} + \alpha_{it})$, $i = 2, 3$.

The third model (i.e. the joint model with a common long-term trend) is preferable to the second one (i.e. the joint model without a common long-term trend); it contains fewer parameters and is therefore simpler, and it has only one long-term factor, which is an advantage in valuing long-term commodity-contingent claims.

As stated above, it is clear that the first specification, in which each commodity is modelled separately, must be the model that best fits the data. Nonetheless, if we achieve a similar goodness of fit when modelling the three commodities jointly using the third specification, the conclusion should be that all three commodities share a common long-term trend. If this is the case, the preferable specification will be the third one because it is the simplest model. Moreover, the joint model with a common long-term trend is the most appropriate model for valuing options involving several commodity prices, as we will discuss in Section 5.4.

It is also possible to test the three models proposed above with the commodities in pairs (i.e. crude oil and heating oil, crude oil and gasoline or heating oil and gasoline). These specifications in pairs will be useful for valuing the crack-spread options in Section 5.4.

Finally, it should be noted that the differences between the models presented above lie in the number of factors and the correlation assumed between them.

5.3.2 Estimation Results

Here, we present the results of the estimation of the three models with and without the assumption of a common long-term trend. The data set used in this section consists of weekly observations of WTI (light sweet) crude oil, heating oil and unleaded gasoline futures prices traded at NYMEX, as described in Section 5.2.1. When using models with more than one commodity, we have chosen to maintain a consistent time to maturity between futures contracts to avoid decompensating the short-term/long-term relations.[†] Therefore, different data sets have been used to estimate the parameters for the different models presented above.

Different data sets for gasoline and the two other commodities were used because of liquidity constraints. In the case of gasoline, as stated above, the available futures contracts are less liquid, and their maturities are shorter than those of the other two commodities' contracts. Consequently, given that in the case of heating oil there exists liquidity to support futures contracts with up to 18 months to maturity, we decided to use a data set with more futures contracts and with futures contracts that have longer maturities than gasoline (for which there is not enough liquidity for futures with maturities longer

---

[*] To avoid identification problems, in the case of crude oil, both constants are assumed to be equal to zero. Thus, the other commodity constants must be interpreted in terms of differences from crude oil.

[†] Similar results have been obtained with futures with different maturities.

than 9 months). In the case of crude oil, futures contracts with up to 7 years to maturity are available. Nevertheless, we decided to use crude oil futures contracts with the same maturity as the ones used for the other commodities to avoid decompensating the short-term/long-term effects. Schwartz (1997) realized that mean-reversion effects tend to be smaller for contracts with longer maturities. García *et al.* (2012) found evidence suggesting the same conclusion in the case of natural gas. Thus, to avoid undesirable effects, we made the time to maturity for futures contracts consistent across all commodities in models that consider more than one commodity.

Specifically, in the case of RBOB gasoline, the data set comprises contracts F1, F3, F5, F7 and F9 from 06/30/1997 to 05/25/2009, yielding 622 quotations for each contract. In this case, for instance, F1 is the contract for the month closest to maturity and F2 is the contract for the second-closest month to maturity.

There are two data sets for WTI crude oil and heating oil. The first data set comprises contracts F1, F3, F5, F7 and F9 from 06/30/1997 to 05/25/2009, yielding 622 quotations for each contract. The second data set comprises F1, F4, F7, F11, F15 and F18 from 09/09/1996 to 05/25/2009, yielding 664 quotations for each contract. The data set we employed depended on the specific case: the first data set was used when either crude oil or heating oil was used jointly with gasoline; and the second data set was used in all other situations. As explained in Schwartz (1997), because futures contracts have a fixed maturity date, the time to maturity changes as time progresses, though it remains in a narrow time interval. Therefore, as in Schwartz (1997), it is assumed that time to maturity does not change over time and is equal to, for instance, 1 month for F1 and 2 months for F2.

The models presented in Section 5.3.1 were estimated using the Kalman filter methodology, which is briefly described in Appendix A.

Table 5.3 presents the results for the first specification (i.e. the two-factor model by Schwartz and Smith [2000]) applied to each commodity (i.e. WTI crude oil, heating oil and gasoline) separately and using the different data sets described above. Table 5.4 presents the results of the second specification (i.e. the joint model without a common long-term trend). The results of the estimation of the third specification (i.e. the joint model with a common long-term trend for all three commodities) are presented in Table 5.5. Finally, Tables 5.6 and 5.7 present the results obtained when using the second and third specifications, respectively, for pairs of commodities (as will be described in Section 5.4). As stated above, in models with more than one commodity, we have chosen to use futures contracts with the same maturities for each commodity.

The first notable observation is that both the long-term trend (μ_ξ) and the speed of adjustment (κ) are positive and significantly different from zero in all cases, implying long-term growth and mean reversion in the commodity prices. This observation is consistent with the results obtained by Schwartz (1997) in the case of oil. Moreover, for each commodity, the estimated values of the long-term mean and volatility ($\mu_{\xi i}$ and $\sigma_{\xi i}$, respectively) are very similar for all specifications.

Gasoline prices exhibit a higher long-term mean and are more volatile than are the other two commodity prices; in Tables 5.3-5.7, the volatility coefficients are higher when gasoline appears in the model. This phenomenon can also be observed in Table 5.1. It is also interesting to note that in all cases, short-term volatility (σ_χ) is higher than long-term volatility (σ_ξ). This result is consistent with the results obtained by Schwartz and Smith (2000) for oil and García *et al.* (2012) for heating oil and gasoline.

In general, the market prices of risk associated with the long- and short-term factors (λ_ξ and λ_χ, respectively) are significantly different from zero, suggesting that the risk associated with these factors cannot be diversified. Additionally, the values of the market prices of risk are higher when the factor models are estimated separately (Table 5.3) than when the model is estimated jointly, with or without a common long-term trend (Tables 5.4-5.7). These results suggest that the risk associated with the long- and short-term factors is more difficult to diversify when we ignore the relationships among factors.

As expected, in the case of heating oil and gasoline, the seasonal period (φ) is roughly 1 year in all cases; this result is consistent with the findings of Garcia *et al.* (2012).

In relation to the purpose of this paper, two important results must be highlighted. First, in the specifications without a common long-term trend (Tables 5.4 and 5.6), the values of the coefficients of

TABLE 5.3 The Two-Factor Model for Each Commodity Separately

	WTI Crude Oil	WTI Crude Oil	Heating Oil	Heating Oil	Gasoline
Contracts	F1, F3, F5, F7 and F9	F1, F4, F7, F11, F15 and F18	F1, F3, F5, F7 and F9	F1, F4, F7, F11, F15 and F18	F1, F3, F5, F7 and F9
Period	06/30/1997 to 05/25/2009	09/09/1996 to 05/25/2009	06/30/1997 to 05/25/2009	09/09/1996 to 05/25/2009	06/30/1997 to 05/25/2009
Number obs	622	664	622	664	622
μ_ξ	0.1031 ** (0.0417)	0.1110*** (0.0339)	0.0998*** (0.0329)	0.1093*** (0.0322)	0.1830*** (0.0362)
κ	1.6738 *** (0.0266)	1.1599 *** (0.0110)	1.4484*** (0.0417)	1.5870*** (0.0185)	1.7403*** (0.0604)
σ_ξ	0.2118*** (0.0044)	0.1792*** (0.0036)	0.2204*** (0.0053)	0.1758*** (0.0036)	0.2245*** (0.0053)
σ_χ	0.2951 *** (0.0068)	0.3022*** (0.0067)	0.3152*** (0.0084)	0.2830*** (0.0067)	0.3919*** (0.0104)
$\rho_{\xi\chi}$	−0.0503 (0.0324)	−0.0360 (0.0306)	−0.0783* (0.0401)	0.1556*** (0.0313)	−0.2566*** (0.0411)
λ_ξ	0.2134*** (0.0418)	0.1678*** (0.0339)	0.0929*** (0.0333)	0.1341*** (0.0322)	0.2474*** (0.0373)
λ_χ	−0.1010* (0.0572)	−0.0245 (0.0568)	−0.4998*** (0.0491)	0.1991*** (0.0502)	0.2412*** (0.0638)
Φ			1.0042*** (0.0003)	0.9954*** (0.0002)	1.0002*** (0.0003)
σ_η	0.0071*** (0.0001)	0.0095*** (0.0001)	0.0115*** (0.0001)	0.0177*** (0.0002)	0.0161*** (0.0002)
Log-likelihood	23556.22	29212.38	21543.49	25668.93	19913.68
AIC	23540.22	29196.38	21525.49	25650.93	19895.68
SIC	23504.76	29160.39	21485.59	25610.44	19855.78

The table presents the results for the Schwartz and Smith (2000) two-factor model for each commodity separately (first specification). Standard errors are in parentheses. The estimated values are reported with * denoting significance at 10%, ** denoting significance at 5% and *** denoting significance at 1%.

TABLE 5.4 The Joint Model Without a Common Long-Term Trend for the Three Commodities

WTI Crude Oil (WTI), Heating Oil (HO) and Unleaded Gasoline (UG) Contracts F1, F3, F5, F7 and F9 Period 06/30/1997 to 05/25/2009 Number Obs. 622			
$\mu_{\xi WTI}$	0.1056*** (0.0257)	φ_{HO}	1.0021*** (0.0005)
$\mu_{\xi HO}$	0.1049** (0.0418)	φ_{UG}	1.0003*** (0.0003)
$\mu_{\xi UG}$	0.1601*** (0.0231)	$\rho_{\xi WTI\xi HO}$	0.7417*** (0.0032)
κ_{WTI}	2.3492*** (0.0495)	$\rho_{\xi WTI\xi UL}$	0.8620*** (0.0054)
κ_{HO}	1.4107*** (0.0789)	$\rho_{\xi WTI\chi WTI}$	0.1133*** (0.0074)
κ_{UG}	2.403*** (0.0500)	$\rho_{\xi WTI\chi HO}$	0.3901*** (0.0072)
$\sigma_{\xi WTI}$	0.2372*** (0.0018)	$\rho_{\xi WTI\chi UG}$	0.3353*** (0.0081)
$\sigma_{\xi HO}$	0.2200*** (0.0032)	$\rho_{\xi HO\chi UG}$	0.6624*** (0.0055)
$\sigma_{\xi UG}$	0.2275*** (0.0050)	$\rho_{\xi HO\chi WTI}$	0.1184*** (0.0071)
$\sigma_{\chi WTI}$	0.2754*** (0.0068)	$\rho_{\xi HO\chi HO}$	−0.1665*** (0.0144)
$\sigma_{\chi HO}$	0.3637*** (0.0142)	$\rho_{\xi HO\chi UG}$	0.1663*** (0.0107)
$\sigma_{\chi UG}$	0.3891*** (0.0100)	$\rho_{\xi UG\chi WTI}$	0.2513*** (0.0069)
$\lambda_{\xi WTI}$	0.0841*** (0.0258)	$\rho_{\xi UG\chi HO}$	0.3563*** (0.0075)
$\lambda_{\xi HO}$	0.0775* (0.0441)	$\rho_{\xi UG\chi UG}$	0.0943*** (0.0091)
$\lambda_{\xi UG}$	0.1398*** (0.0248)	$\rho_{\chi WTI\chi HO}$	0.4240*** (0.0092)
$\lambda_{\chi WTI}$	0.2070*** (0.0657)	$\rho_{\chi WTI\chi UG}$	0.4679*** (0.0099)
$\lambda_{\chi HO}$	−0.1381* (0.0774)	$\rho_{\chi HO\chi UG}$	0.4957*** (0.0187)
$\lambda_{\chi UG}$	0.0977** (0.0495)	σ_η	0.0121*** (0.0001)
Log-likelihood		63042.16	
AIC		62970.16	
SIC		62810.57	

The table presents the results obtained using the Schwartz and Smith (2000) two-factor model without a common long-term trend for the three commodities (second specification). Standard errors are in parentheses. The estimated values are reported with * denoting significance at 10%, ** denoting significance at 5% and *** denoting significance at 1%.

TABLE 5.5 The Joint Model With a Common Long-Term Trend for the Three Commodities

WTI Crude Oil (WTI), Heating Oil (HO) and Unleaded Gasoline (UG) Contracts F1, F3, F5, F7 and F9 Period 06/30/1997 to 05/25/2009 Number Obs. 622			
μ_ξ	0.1236** (0.0502)	φ_{HO}	0.9945*** (0.0007)
κ_{WTI}	1.1960*** (0.0667)	φ_{UG}	1.0055*** (0.0003)
κ_{HO}	1.2488*** (0.0666)	$\rho_{\xi\chi WTI}$	−0.2168*** (0.0102)
κ_{UG}	1.0558*** (0.0461)	$\rho_{\xi\chi HO}$	−0.2385*** (0.0072)
σ_ξ	0.2028*** (0.0043)	$\rho_{\xi\chi UG}$	−0.2583*** (0.0168)
$\sigma_{\chi WTI}$	0.3606*** (0.0054)	$\rho_{\chi WTI\chi HO}$	0.8711*** (0.0085)
$\sigma_{\chi HO}$	0.3579*** (0.0083)	$\rho_{\chi WTI\chi UG}$	0.8802*** (0.0251)
$\sigma_{\chi UG}$	0.4092*** (0.0182)	$\rho_{\chi HO\chi UG}$	0.8566*** (0.0167)
λ_ξ	0.0908* (0.0524)	κ_{HO}	1.4165** (0.5992)
$\lambda_{\chi WTI}$	0.2831** (0.1281)	κ_{UG}	1.5147*** (0.1698)
$\lambda_{\chi HO}$	−0.1011 (0.0857)	C_{HO}	0.9721*** (0.0133)
$\lambda_{\chi UG}$	0.0084 (0.1173)	C_{UG}	0.9668*** (0.0045)
σ_η	0.0169*** (0.0001)		
Log-likelihood		62125.99	
AIC		62075.99	
SIC		61965.17	

The table presents the results for the Schwartz and Smith (2000) two-factor model assuming a common long-term trend for all three commodities (third specification). Standard errors are in parentheses. The estimated values are reported with * denoting significance at 10%, ** denoting significance at 5%, and *** denoting significance at 1%.

correlation between long-term trends ($\rho_{\xi i \xi j}$) are quite large, suggesting a strong relationship between long-term trends.

Second, in the specifications with a common long-term trend (Tables 5.5 and 5.7), the values of the constants C_i are very close to one, suggesting that the common long-term trend affects all three commodities in the same way and that the price-level differences between crude oil and refined products are stationary. This fact confirms the previous finding that the refining margin is stationary and, therefore, exhibits different dynamics than crude oil or refined products. In addition, it suggests that crude oil, heating oil and gasoline prices are not only cointegrated but also share a common long-term trend.

Moreover, the values of the K_i constants in Tables 5.5 and 5.7 are positive, confirming the differences in price levels among crude oil and refined products observed in Table 5.1. Of the three commodities considered, the most expensive is gasoline, followed by heating oil and crude oil.

If we define the Schwartz Information Criterion (SIC) as $\ln(L_{ML}) - q \ln(T)$, where q is the number of estimated parameters, T is the number of observations and L_{ML} is the value of the likelihood function using the q estimated parameters, then the fit is better when the SIC is higher. The same conclusions are obtained with the Akaike Information Criterion (AIC), which is defined as $\ln(L_{ML}) - 2q$.

It is worth noting that in Table 5.3, the values of the SIC and the AIC are higher for crude oil than for the other two commodities within the same time period (06/30/1997 to 05/25/2009) and the same contracts (F1, F3, F5, F7 and F9).* The superior fit for the crude oil data can be related to the short-term imperfections in the refined product market and, particularly in the case of gasoline, the lack of liquidity in longer-maturity futures price quotes. This result is to be expected because crude oil has futures with much more liquidity at longer maturities than refined products. In the case of refined products,

---

* It should be pointed out that although in principle the values of the SIC and AIC are not directly comparable whenever the series are different, our three commodity price series have the same order of magnitude; the time period (06/30/1997 to 04/24/2006) and the futures contracts used in the estimation procedure (F1, F3, F5, F7 and F9) are the same.

TABLE 5.6 The Joint Model Without a Common Long-Term Trend for Pairs of Commodities

	WTI Crude Oil (i) and Heating Oil (j)	WTI Crude Oil (I) and Gasoline (j)	Heating Oil (i) and Gasoline (j)
Contracts	F1, F4, F7, F11, F15 and F18	F1, F3, F5, F7 and F9	F1, F3, F5, F7 and F9
Period	09/09/1996 to 05/25/2009	06/30/1997 to 05/25/2009	06/30/1997 to 05/25/2009
Number obs.	664	622	622
$\mu_{\xi j}$	0.1048*** (0.0378)	0.1159*** (0.0427)	0.1184*** (0.0297)
$\mu_{\xi j}$	0.1248* (0.0372)	0.1479*** (0.0468)	0.1330*** (0.0173)
κ_i	1.1549*** (0.0163)	1.5571*** (0.0323)	2.0123*** (0.0547)
κ_j	1.5461*** (0.0147)	2.0262*** (0.0758)	2.2203*** (0.0573)
$\sigma_{\xi j}$	0.1824*** (0.0038)	0.2434*** (0.0056)	0.2077*** (0.0044)
$\sigma_{\xi j}$	0.1741*** (0.0037)	0.2740*** (0.0094)	0.2374*** (0.0056)
$\sigma_{\chi i}$	0.3091*** (0.0075)	0.2577*** (0.0049)	0.2672*** (0.0064)
$\sigma_{\chi j}$	0.3051****** (0.0086)	0.4047*** (0.0117)	0.3809*** (0.0088)
$\lambda_{\xi j}$	0.0901** (0.0377)	0.2931*** (0.0427)	−0.0194 (0.0305)
$\lambda_{\xi j}$	0.0827** (0.0370)	0.1551*** (0.0505)	0.2115*** (0.0179)
$\lambda_{\chi i}$	0.1716** (0.0633)	0.2491*** (0.0524)	−0.0534 (0.0533)
$\lambda_{\chi j}$	0.4984** (0.0697)	0.4998*** (0.0768)	0.4999*** (0.0001)
φ_i			1.0035*** (0.0003)
φ_j	0.9955*** (0.0002)	1.0033*** (0.0003)	0.9994*** (0.0005)
$\rho_{\xi j \xi j}$	0.8774*** (0.0091)	0.9424*** (0.0049)	0.6803*** (0.0220)
$\rho_{\xi j \chi i}$	−0.0679** (0.0335)	−0.3623*** (0.0298)	0.1512*** (0.0362)
$\rho_{\xi j \chi j}$	0.0859** (0.0349)	−0.1849*** (0.0437)	0.2752*** (0.0279)
$\rho_{\xi j \chi i}$	0.2304*** (0.0314)	−0.2486*** (0.0507)	0.5084*** (0.0213)
$\rho_{\xi j \chi j}$	0.1356*** (0.0355)	−0.1036 (0.0675)	−0.2050*** (0.0338)
$\rho_{\chi i \chi j}$	0.7125*** (0.0192)	0.8328*** (0.0073)	0.0458 (0.0345)
σ_η	0.0142*** (0.0001)	0.0125*** (0.0001)	0.0142*** (0.0001)
Log-likelihood	54209.19	42004.52	40908.09
AIC	54169.19	41964.52	40868.09
SIC	54079.22	41875.86	40779.43

The table presents the results for the Schwartz and Smith (2000) two-factor model without a common long-term trend for pairs of commodities. Standard errors are in parentheses. The estimated values are reported with * denoting significance at 10%, ** denoting significance at 5%, and *** denoting significance at 1%.

particularly gasoline, maturities above several months are not as liquid as would be required to calibrate such models.

The relative fit of the models to our three commodity price series can be assessed by evaluating their predictive ability. The in-sample predictive ability of the Schwartz and Smith (2000) two-factor model is presented in Table 5.8 for the three specifications considered in Section 5.3.1. We found that the differences among the three models, in terms of bias and the root mean squared error, are small. Moreover, in Table 5.8, it can be observed that the root mean squared error values for crude oil are generally lower than the corresponding values obtained for the other two commodities, confirming our previous prediction.

The out-of-sample predictive ability of the Schwartz and Smith (2000) two-factor model for the three specifications is presented in Table 5.9. The results were obtained by valuing the contracts corresponding to the period 06/16/2003 to 05/25/2009 with the parameters obtained estimating the models with data from 06/30/1997 to 06/09/2003 (311 observations in each period). As expected, out-of-sample pricing errors are slightly higher than the corresponding in-sample values. As before, the differences between the three models are small.

TABLE 5.7 The Joint Model With a Common Long-Term Trend for Pairs of Commodities

	WTI Crude Oil (i) and Heating Oil (j)	WTI Crude Oil (i) and Gasoline (j)	Heating Oil (i) and GasOline (j)
Contracts	F1, F4, F7, F11, F15 and F18	F1, F3, F5, F7 and F9	F1, F3, F5, F7 and F9
Period	09/09/1996 to 05/25/2009	06/30/1997 to 05/25/2009	06/30/1997 to 05/25/2009
Number obs.	664	622	622
μ_ξ	0.1053** (0.0425)	0.1238** (0.0534)	0.1321*** (0.0331)
κi	1.1689*** (0.0142)	1.1804*** (0.0338)	1.3344*** (0.0433)
κj	1.3696*** (0.0164)	1.1047*** (0.0287)	1.0773*** (0.0281)
σ_ξ	0.2019*** (0.0187)	0.2393*** (0.0069)	0.2114*** (0.0052)
$\sigma_{\chi i}$	0.7989*** (0.0126)	0.3377*** (0.0093)	0.3443*** (0.0105)
$\sigma_{\chi j}$	0.6080*** (0.0001)	0.3462*** (0.0092)	0.3683*** (0.0110)
λ_ξ	0.0718** (0.0416)	0.0571* (0.0328)	0.0439* (0.0249)
$\lambda_{\chi i}$	0.4589*** (0.0707)	−0.1255 (0.0997)	−0.0111 (0.0617)
$\lambda_{\chi j}$	0.2463*** (0.0810)	−0.7679*** (0.0734)	−0.4750*** (0.0495)
φ_i			1.0016*** (0.0003)
φ_j	0.9944*** (0.0002)	1.0036*** (0.0002)	1.0011*** (0.0006)
$\rho_{\xi\chi i}$	−0.3323* (0.1904)	−0.3753*** (0.0347)	−0.2651*** (0.0414)
$\rho_{\xi\chi j}$	−0.2322 (0.1861)	−0.2047*** (0.0415)	−0.2340*** (0.0456)
$\rho_{\chi i\chi i}$	0.9532*** (0.0001)	0.8137*** (0.0184)	0.8242*** (0.0167)
K_i			
K_j	0.5594*** (0.1312)	−1.6122*** (0.2819)	1.6669*** (0.2270)
C_i			
C_j	1.0025*** (0.0034)	0.8862*** (0.0061)	0.8932*** (0.0049)
σ_η	0.0161*** (0.0001)	0.0165*** (0.0001)	0.0174*** (0.0001)
Log-likelihood	53061.99	41007.69	40517.87
AIC	53029.99	40975.69	40483.87
SIC	52958.02	40904.76	40408.51

The table presents the results for the Schwartz and Smith (2000) two-factor model assuming a common long-term trend for pairs of commodities. Standard errors are in parentheses. The estimated values are reported with * denoting significance at 10%, ** denoting significance at 5%, and *** denoting significance at 1%.

As a result, given that the predictive ability of the two models is very similar, we can conclude that all three commodities exhibit the same long-term trend. Consequently, because the joint model that assumes common long-term trends is the simplest one, we do not need a second long-term factor when jointly modelling the three commodities.

It is also worth noting that although previous studies have already found that the prices of crude oil and its main refined products are cointegrated, this conclusion is extended in the present work because we find that they also exhibit a common long-term trend. Moreover, to the best of our knowledge, this is the first time that a factor model with a common long-term trend for the prices of crude oil and its main refined product has been proposed and estimated. Additionally, the results of the estimation of this common long-term trend model suggest that the price-level differences between crude oil and refined products are stationary, confirming that the refining margin is stationary and therefore, exhibits different dynamics from those of crude oil and refined products.

Finally, the fact that crude oil and refined product prices share a common long-term trend has important implications for managing and hedging the refining margin risk. With a single long-term trend, the refining margin reflects only short-term effects. However, if we assume different long-term trends for each commodity, the refining margin reflects long- and short-term effects, implying higher volatility and therefore, a higher value of derivatives for the refining margin.

TABLE 5.8 In-Sample Predictive Ability

Panel A. Model for the Commodities Separately								
WTI			Heating Oil			Gasoline		
Contract	ME	RMSE	Contract	ME	RMSE	Contract	ME	RMSE
F1	−0.0005	0.0455	F1	0.0154	0.047	F1	−0.0053	0.0509
F3	0.0013	0.0394	F3	0.0159	0.0438	F3	−0.004	0.0436
F5	−0.0004	0.0353	F5	0.0144	0.0398	F5	−0.0039	0.0381
F7	−0.0009	0.0323	F7	0.0126	0.0344	F7	−0.0038	0.0345
F9	0.0006	0.0309	F9	0.0119	0.0327	F9	−0.0023	0.0337

Panel B. Joint Model Without a Common Long-Term Trend								
WTI			Heating oil			Gasoline		
Contract	ME	RMSE	Contract	Mean	RE	Contract	ME	RMSE
F1	−0.0053	0.0463	F1	0.0021	0.0452	F1	−0.0068	0.0476
F3	−0.0028	0.0386	F3	0.0016	0.0389	F3	−0.005	0.0396
F5	−0.0031	0.0325	F5	0.0013	0.0341	F5	−0.0033	0.0329
F7	−0.0031	0.0284	F7	−0.0006	0.028	F7	−0.0033	0.0285
F9	−0.0016	0.0266	F9	−0.001	0.0263	F9	−0.0028	0.0274

Panel C. Joint Model With a Common Long-Term Trend								
WTI			Heating Oil			Gasoline		
Contract	ME	RMSE	Contract	ME	RMSE	Contract	ME	RMSE
F1	−0.0103	0.049	F1	−0.0121	0.0551	F1	−0.0033	0.0467
F3	−0.0069	0.039	F3	−0.0089	0.0442	F3	−0.0033	0.0381
F5	−0.0082	0.033	F5	−0.0068	0.0391	F5	−0.0031	0.0336
F7	−0.0105	0.03	F7	−0.0077	0.0337	F7	−0.0043	0.0315
F9	−0.0125	0.0292	F9	−0.0083	0.0309	F9	−0.005	0.0321

The table presents the mean error (ME), calculated as real minus predicted values, and the root mean squared error (RMSE), to analyse the in-sample predictive power ability of the Schwartz and Smith (2000) two-factor model for the three commodities separately (panel A), and for all three commodities both without a common long-term trend (panel B) and with a common long-term trend (panel C). The time period is 06/30/1997 to 05/25/2009 (622 weekly observations for each commodity).

In the next section, we will use these results to value the main products used to protect the refining margin, i.e. the crack-spread options listed by NYMEX.

5.4 Crack-Spread Option Valuation

It is well known that commodity markets have grown quickly during recent years. Four types of investors can be distinguished in commodity markets: investors seeking a portfolio diversification tool, particularly pension funds; investors seeking a source of alpha, particularly hedge funds; European banks, which use commodity derivatives to structure products that they then retail to their customers; and energy companies protecting their profits.

Energy companies include refining companies that are willing to hedge their refining margin with specific hedging products. Crack-spread options are the most popular products for this purpose. As stated above, a crack-spread call option is a contract that gives the holder the right (but not the obligation) to buy a refined product futures contract from the writer and sell the writer a crude oil futures contract, paying a previously agreed-upon crack-spread price. A crack-spread put allows one to sell a refined product futures contract and buy a crude oil futures contract, paying a crack-spread price.

TABLE 5.9   Out-of-Sample Predictive Ability

Panel A. Model for the commodities separately								
WTI			Heating Oil			Gasoline		
Contract	ME	RMSE	Contract	ME	RMSE	Contract	ME	RMSE
F1	−0.003	0.0456	F1	−0.0062	0.045	F1	−0.0045	0.0556
F3	0.0015	0.0402	F3	0.0036	0.0438	F3	−0.0044	0.0472
F5	0.0004	0.0379	F5	0.0043	0.0401	F5	−0.0044	0.0429
F7	0.0001	0.0363	F7	0.0021	0.0354	F7	−0.003	0.0422
F9	0.0016	0.0354	F9	−0.0018	0.0414	F9	0.0001	0.0535
Panel B. Joint Model Without a Common Long-Term Trend								
WTI			Heating Oil			Gasoline		
Contract	ME	RMSE	Contract	Mean	RE	Contract	ME	RMSE
F1	−0.0085	0.0471	F1	0.0002	0.0448	F1	−0.0073	0.058
F3	−0.0029	0.0404	F3	0.004	0.0435	F3	−0.0049	0.0477
F5	−0.003	0.0379	F5	0.0024	0.0396	F5	−0.0044	0.0438
F7	−0.0026	0.036	F7	0.001	0.0355	F7	−0.003	0.0423
F9	−0.0004	0.0347	F9	−0.0001	0.0448	F9	−0.0002	0.0571
Panel C. Joint Model With a Common Long-Term Trend								
WTI			Heating Oil			Gasoline		
Contract	ME	RMSE	Contract	ME	RMSE	Contract	ME	RMSE
F1	−0.0192	0.0528	F1	−0.0365	0.0622	F1	0.021	0.0572
F3	−0.0073	0.0411	F3	−0.0103	0.049	F3	0.0016	0.0477
F5	−0.0049	0.0388	F5	0.0054	0.0424	F5	−0.0166	0.0516
F7	−0.0047	0.0383	F7	0.0171	0.0399	F7	−0.0313	0.058
F9	−0.005	0.0388	F9	0.0254	0.0471	F9	−0.0422	0.0657

The table presents the mean error (ME), calculated as real minus predicted values, and the root mean squared error (RMSE) to analyse the out-of-sample predictive power ability of the Schwartz and Smith (2000) two-factor model for the three commodities separately (panel A) and for all three commodities jointly without a common long-term trend (panel B) and with a common long-term trend (panel C). The results are obtained valuing the contracts corresponding to the period 06/16/2003 to 05/25/2009 using the parameters obtained by estimating the models with data from 06/30/1997 to 06/09/2003 (311 observations in each period).

In valuing this type of product, it is essential to consider our previous results. In particular, it is crucial to realize that the refining margin dynamics are completely different from those of crude oil or refined products. As shown above, the price dynamics of crude oil and refined products are integrated, whereas the refining margin dynamics are stationary. Even more importantly, crude oil and refined products are not only cointegrated but also exhibit a common long-term trend; as we will observe, this finding is of critical importance in managing and hedging refining margin risk, primarily in the valuation of crack-spread options. Specifically, the values of crack-spread options must be lower if we assume a common long-term trend for crude oil and refined products than if the refining margin is integrated because in the case of a common long-term trend, the refining margin risk reflects only short-term effects. Therefore, there is less volatility and the writer of the crack-spread option faces lower risk than in cases where there are different long-term trends for each commodity. Consequently, we conclude that we must consider a refining margin model that only includes short-term effects (i.e. a model in which crude oil and its main refining products share the long-term factors).

In this section, we compare the pricing performance of the model with a common long-term trend, the model that allows for a long-term trend for each commodity, the model that postulates uncorrelated sub-models for each commodity, and the approach by Dempster *et al.* (2008) for directly modelling the

TABLE 5.10   Descriptive Statistics for Crack-Spread Options

	% Sample	Mean Number of Observations Series	Mean K ($)	Mean Price ($)
Panel A: Heating oil vs. WTI (169 series, 6663 observations)				
Put options	74.66	42.86	8.24	1.0111
Call options	25.44	29.37	8.67	1.3044
Panel B: Gasoline vs. WTI (188 series, 20,243 observations)				
Put options	67.55	143.66	8.50	1.1647
Call options	32.45	32.75	6.03	2.0986

The table shows the main descriptive stats of the crack-spread options. The time period is November 2001 to May 2009 for gasoline vs. WTI crack-spread options, and there are exercise prices from 3 to 17 dollars. For heating oil vs. WTI crack-spread options, there are data corresponding to contracts maturing from August 2000 to November 2009, with exercise prices from 3 to 14 dollars.

crack-spread.* We show that the most suitable method of valuing crack-spread options is to assume a common long-term trend for WTI crude oil and gasoline and for WTI and heating oil.

5.4.1 Data

The data set used in the estimation procedure consists of two sets of daily observations of crack-spread put and call options quoted at the NYMEX. The first set consists of heating oil vs. WTI data, and the second one consists of gasoline vs. WTI data.

In the NYMEX, there are only quotations for these two types of crack-spread options (i.e. heating oil vs. WTI and gasoline vs. WTI), and contracts mature each month for the following 18 months at different strike prices. Specifically, the strike prices are the one at-the-money strike price, five additional strikes both above and below the established at-the-money strike price in $0.25 (25¢) increments, three additional out-of-the-money strike prices above and below those strikes at $1.00 intervals and two more strikes added above and below at $2.00 intervals. Options traded at NYMEX are American-style, and thus, the holder can exercise his or her right at any time.

The market for this type of option is much less liquid than the market for futures. Due to the scarcity of data, we have chosen to use daily rather than weekly data. As described below, this has forced us to make same minor changes to the model. Specifically, for the case of gasoline vs. WTI crack-spread options, we have daily data from November 2001 to May 2009 (20,243 observations), with exercise prices ranging from $3 to $17. In the case of the heating oil vs. WTI crack-spread options, we have daily data corresponding to contracts maturing from August 2000 to November 2009 (6663 observations), with exercise prices from $3 to $14. A brief summary of these options is given in Table 5.10.

5.4.2 Option Valuation Methodology

In this study, we have a set of crack-spread American options that we wish to replicate. There is no close analytic expression of their price, so a simulation must be used. In addition, our models are either three-dimensional (i.e. the common trended model) or four-dimensional (i.e. when we consider one trend for each commodity), severely narrowing the methods we can use.

There are several methods for valuing American options. Each method involves the discretization of the state space, such that at each point in time and for every value of the factors, the option holder must decide whether to exercise his right; therefore, the decision can only be based on future dynamics. However, when the number of factors increases, one encounters the 'curse of dimensionality' (a concept

---

* A short summary of the Dempster *et al.* (2008) model is presented in Appendix B.

conceived by Bellman) because full discretization is almost infeasible due to the computational complexity inherent in there being more than three factors.

To minimize this problem, one of three main approaches can be selected (Bally *et al.* 2005). One approach is to perform state aggregation to develop a 'synthetic indicator' such that the holder of the option can decide whether to exercise it using only this indicator rather than three or more factors (Barraquand and Martineau 1995). The second approach computes conditional expectations via Malliavin Calculus.[*] The third one uses base functions (Tsitsiklis and VanRoy 1999, Longstaff and Schwartz 2001). Among all these approaches, the one by Longstaff and Schwartz (2001), called Least Squares Monte Carlo, has become one of the most popular models in the literature for valuing American options and will be the approach used in this paper. However, the point here is not the specific method used to value the crack-spread options but rather whether the valuation results are better when a common long-term trend is assumed than when different long-term trends are assumed.

The parameters and dynamics needed for option valuation, for the joint models with and without a common long-term trend and for the uncorrelated model, were estimated in Section 5.3 using weekly data. Specifically, given that crack-spread options involve two commodities, we used the results in Tables 5.6 and 5.7.[†] However, changing the model to daily dynamics is not a problem in continuous time. We simply re-sampled the model using an aliasing algorithm to obtain information on daily states. The RMSE is slightly higher, though the difference is small.

5.4.3 Commodity Price Dynamics

For the purpose of option valuation, we need a full description of the model. In matrix form, the state dynamics can be described as follows:

$$dZ_t = \left(\mu + AZ_t\right)dt + dW_t. \tag{5.5}$$

To clarify matters, let U_t be a unit of Brownian motion (i.e. $dU_t dU_t^T = Idt$) and rewrite (5.5) as

$$dZ_t = \left(\mu + AZ_t\right)dt + R\,dU_t. \tag{5.6}$$

The models used for crack-spread option valuation will be the same as those proposed in Section 5.3 for the commodities in pairs: the model for two commodities with no correlation (i.e. the first specification), the joint model without a common long-term trend (i.e. the second specification), the joint model with a common long-term trend (i.e. the third specification), together with the model by Dempster *et al.* (2008).

In the case where two models are combined, we have two equations in matrix form:

$$dZ_{it} = \left(\mu_i + A_i Z_{it}\right)dt + R_i\,dW_{it}\,(i = 1,2). \tag{5.7}$$

Therefore, if we assume no correlation, the global model is

$$\begin{pmatrix} dZ_{1t} \\ dZ_{2t} \end{pmatrix} = \left[\begin{pmatrix} \mu_1 \\ \mu_2 \end{pmatrix} + \begin{pmatrix} A_1 & 0 \\ 0 & A_2 \end{pmatrix}\begin{pmatrix} Z_{1t} \\ Z_{2t} \end{pmatrix}\right]dt + \begin{pmatrix} R_1 & 0 \\ 0 & R_2 \end{pmatrix}\begin{pmatrix} dU_{1t} \\ dU_{2t} \end{pmatrix}. \tag{5.8}$$

---

[*] See Fournié *et al.* (1993) for numerical applications and Bouchard and Touzi (2004) for general theory and variance reduction.

[†] As stated in Appendix B, the model by Dempster *et al.* (2008) is estimated using the same procedure (the Kaman filter) and the same data sets as the rest of the models presented in Section 5.3.

which is equivalent to estimating both models separately. If we allow a free correlation structure, the model is

$$
\begin{pmatrix} dZ_{1t} \\ dZ_{2t} \end{pmatrix} = \left[\begin{pmatrix} \mu_1 \\ \mu_2 \end{pmatrix} + \begin{pmatrix} A_1 & 0 \\ 0 & A_2 \end{pmatrix} \begin{pmatrix} Z_{1t} \\ Z_{2t} \end{pmatrix} \right] dt + R^* \begin{pmatrix} dU_{1t} \\ dU_{2t} \end{pmatrix},
\tag{5.9}
$$

where R^* is a free triangular matrix.

For parameter estimation purposes (see Appendix A), we use Kalman filter equations to estimate $Z_t|_{t-1} = E[Z_t/Z_1,\ldots,Z_{t-1}]$, and as an intermediate result, $Z_{t-1}|_{t-1} = E[Z_{t-1}/Z_1,\ldots,Z_{t-1}]$. This process (estimating using current or even future information) is termed 'aliasing' in the Kalman filter literature, and the series $Z_{t|t}$ will be used as initial states for option valuation.

5.4.4 Results

Table 5.11 presents several metrics to analyse the in-sample predictive ability of the models for heating oil vs. WTI and gasoline vs. WTI crack-spread options. The models considered are the (three-factor) joint model with a common long-term trend, the (four-factor) joint model without a common long-term trend, the (two-factor) model for both commodities separately and the model by Dempster *et al.* (2008).

The statistics presented in Table 5.11 are the mean bias (real minus predicted), the root mean squared error, the bias standard deviation and the mean absolute error. The values shown in the table are the mean of the different means or standard deviations for each option series.

It can be observed that we achieve the best results with the common long-term trend model (i.e. the three-factor model). The common long-term trend model achieves even better results than the one by Dempster *et al.* (2008). The uncorrelated model achieves the worst results, thus, confirming our hypothesis regarding common trending. It is worth noting that, as expected, the theoretical option values are lower (and closer to the real values) when we assume a common long-term trend than when we allow for different long-term trends. This result may be related to the fact that there is less volatility and that

TABLE 5.11 Crack-Spread Option Valuation Results. Error Descriptive Statistics

Statistic	Three-Factor Model (Joint Model with Common Trend)	Four-Factor Model (Joint Model Without Common Trend)	Two-Factor Model (Uncorrelated Model)	Two-Factor Model For The Spread
Panel A: Heating oil vs. WTI crack-spread options				
Mean bias (real – predicted)	−1.4268	−3.1037	−3.0287	−2.6674
RMSE	1.1859	3.4725	3.2371	3.1717
Bias standard deviation	0.3982	0.4376	0.4381	0.3352
Mean absolute error	1.7378	3.3738	3.1429	3.0988
Panel B: Gasoline vs. WTI crack-spread options				
Mean bias (real – predicted)	−0.7465	−1.3125	−1.3605	−1.6759
RMSE	1.3637	1.6608	1.6884	2.3714
Bias standard deviation	0.5190	0.5566	0.5847	1.0432
Mean absolute error	1.2173	1.4912	1.5148	1.9204

The table presents several metrics to analyse the in-sample predictive power ability of the models under study: the joint model with a common trend, the joint model without a common trend, the model tracking commodities separately and the two-factor model for the crack-spread by Dempster *et al.* (2008). The time period is November 2001 to May 2009 for gasoline vs. WTI crack-spread options (20,243 daily observations) and August 2000 to November 2009 for heating oil vs. WTI crack-spread options (6663 daily observations). For each series, we have used the corresponding statistic. These results correspond to the mean of these multiple means or standard deviations.

the writer of the option assumes lower risk when there is a common long-term trend for all three commodities. The fact that all three commodities share a common long-term trend implies that the refining margin risk reflects only short-term effects; therefore, there is less volatility, and the writer of the option faces lower risk than in the case when there are different long-term trends for each commodity. With different long-term trends for each commodity, the refining margin reflects long- and short-term effects, implying higher volatility; therefore, crack-spread option values are higher.

The better performance of the common long-term trend model compared to the model by Dempster *et al.* (2008) also has important implications. As explained in the Section 5.1, the approach by Dempster *et al.* (2008) is advantageous because by directly modelling the spread, we can avoid modelling the correlation between two asset returns, which is difficult (Kirk 1995, Mbanefo 1997, Alexander 1999). On the other hand, the common long-term trend model proposed in this paper is advantageous because, using two or more price series, it is possible to extract more information for estimation purposes (Cortazar *et al.* 2008). *A priori,* it is difficult to know which of the two approaches gives better valuation results. In this paper, we show that at least with our database, the common long-term trend model gives better valuation results in terms of option pricing errors.

Finally, we can conclude that because the valuation errors obtained using the joint model with a common long-term trend are smaller than those obtained using the joint model without a common long-term trend or the uncorrelated model, the preferred model is the one that assumes a common long-term trend, which confirms the Section 5.3 findings. This result also implies that the advantage of the model that has a long-term trend in common (i.e. the possibility of extracting more information for estimation purposes using two or more price series) is more important with respect to valuation errors than the advantage of the Dempster *et al.* (2008) model (i.e. avoiding the need to model the correlation between two asset returns).

This result also provides additional evidence suggesting that the refining margin exhibits dynamics that are different from those of crude oil and refined products, which has crucial implications not only in terms of crack-spread option valuation but also in terms of managing and hedging the refining margin risk.

5.5 Conclusion

In this paper, we show that, in contrast to crude oil and its main refined products (i.e. gasoline and heating oil), the refining margin is stationary. Moreover, we found evidence suggesting that these three commodities are not only cointegrated, as shown in previous papers, but also share common long-term dynamics, suggesting that the refining margin only reflects short-term effects.

We confirmed the hypothesis of a common long-term trend for crude oil, heating oil and gasoline by proposing different factor models to jointly estimate the dynamics of these three commodities. These factor models were framed within the multi-factor model family proposed by Schwartz (1997) and a related series of papers by Schwartz and Smith (2000), Cortazar and Schwartz (2003), Cortazar and Naranjo (2006) and Cortazar *et al.* (2008), among others.

Furthermore, we found that among different factor models with and without common long-term trends, the most suitable model in terms of simplicity and fit is the one that assumes a common long-term trend for all three commodities.

The fact the these three commodities share a common long-term trend has important implications in terms of managing and hedging the risk faced by a refining companies; specifically, it has important implications in the valuation of crack-spread options, which are used to protect the refining margin. Given that these crack-spread options involve several assets, the techniques used for their valuation are complex. In this paper, we present a simplified valuation model for this type of option, which assumes a common long-term trend for crude oil, heating oil and gasoline.

We found that by assuming a common long-term trend for the prices of crude oil and refined products, option valuation was more accurate than using models with more factors and parameters, including the model by Dempster *et al.* (2008), which directly models the spread. Therefore, we conclude that the preferred model for valuing this kind of option is one that assumes a common long-term trend; such a model not only yields better results in terms of pricing errors but is also simpler and easier to implement than standard techniques. Moreover, these results point out that using two or more price series, it is possible to extract more information for estimation purposes, which has more importance in terms of valuation errors than avoiding the need to model the correlation between two asset returns.

In summary, the primary finding of this paper is that the refining margin exhibits dynamics that are different from those of crude oil and its main refined products because the refining margin risk only reflects short-term effects. This finding has important implications for crack-spread option valuation, and more generally, it also has many other implications for the process of managing and hedging the risks faced by refining companies.

References

Alexander, C., Correlation and cointegration in energy markets. In *Managing Energy Price Risk*, 2nd ed., edited by V. Kamisnky, pp. 573–606, 1999 (Risk Publications: London).

Asche, F., Gjolberg, O. and Völker, T., Price relationships in the petroleum market: an analysis of crude oil and refined product prices. *Energy Econ.*, 2003, **25**, 289–301.

Bally, V., Caramellino, L. and Zanette, A., Pricing American options by Monte Carlo methods using a Malliavin Calculus approach. *Monte Carlo Meth. Applic.*, 2005, **11**, 97–133.

Barraquand, J. and Martineau, D., Numerical valuation of high dimensional multivariate American securities. *J. Financ. Quantit. Anal.*, 1995, **30**(3), 383–405.

Bouchard, B. and Touzi, N., On the Malliavin approach to Monte Carlo approximation of conditional expectations. *Finance Stochast.*, 2004, **8**(1), 45–71.

Cortazar, G., Milla, C. and Severino, F., A multicommodity model for futures prices: using futures prices of one commodity to estimate the stochastic process of another. *J. Fut. Mkts.*, 2008, **28**(6), 537–560.

Cortazar, G. and Naranjo, L., An N-factor Gaussian model of oil futures prices. *J. Fut. Mkts.*, 2006, **26**(3), 209–313.

Cortazar, G. and Schwartz, E.A., Implementing a stochastic model for oil futures prices. *Energy Econ.*, 2003, **25**, 215–218.

Dempster, M.E., Medova, E. and Tang, K., Long term spread option valuation and hedging. *J. Bank. Finance*, 2008, **32**, 2530–2540.

Fournié, E., Lasry, J.M., Lebouchoux, J., Lions, P.L. and Touzi, N., Applications of Malliavin calculus to Monte Carlo methods in finance. *Finance Stochast.*, 1993, **3**, 391–412.

Froot, K.A., Schwarfstein, D.S. and Stein, J.C., Risk management: coordinating corporate investment and financial policies. *J. Finance*, 1993, **48**, 1629–1658.

García, A., Población, J. and Serna, G., The stochastic seasonal behavior of natural gas prices. *Eur. Financ. Mgmt.*, 2012, **18**, 410–443.

Gjolberg, O. and Johnsen, T., Risk management in the oil industry: can information on long-run equilibrium prices be utilized? *Energy Econ.*, 1999, **21**(6), 517–527.

Graham, J.R. and Smith, C.W., Tax incentives to hedge. *J. Finance*, 1999, **54**, 2241–2262.

Holmström, B. and Tirole, J., Liquidity and risk management. *J. Money, Credit Bank,* 2000, **32**, 295–319.

International Energy Agency. Oil Market Report. Annual Statistical Supplement, 2006. Available online at: http://www.omrpublic.iea.org

Kirk, E., Correlation in the energy markets. In: *Management Energy Price Risk*, pp. 177–186, 1995 (Risk Publications: London).

Lanza, A., Manera, M. and Giovannini, M., Modeling and forecasting cointegrated relationships among heavy oil and product prices. *Energy Econ.*, 2005, **27**, 831–848.

Longstaff, F. and Schwartz, E., Valuing American options by simulations: a simple least squares approach. *Rev. Financ. Stud.*, 2001, **14**(1), 113–147.

Mayers, D., Smith, J. and Clifford, W., On the corporate demand for insurance. *J. Business*, 1982, **55**, 281–296.

Mbanefo, A., Co-movement term structure and the valuation of energy spread options. In *Mathematics of Derivative Securities*, edited by M.A.H. Dempster and R.R. Pliska, pp. 99–102, 1997 (Cambridge University Press: Cambridge).

Mello, A.S. and Parsons, J.E., Hedging and liquidity. *Rev. Financ. Stud.*, 2000, **13**, 127–153.

Pindyck, R.S., The long-run evolution of energy prices. *Energy J.*, 1999, **20**(2), 1–27.

Schwartz, E.S., The stochastic behavior of commodity prices: implication for valuation and hedging. *J. Finance*, 1997, **52**, 923–973.

Schwartz, E.S. and Smith, J.E., Short-term variations and long-term dynamics in commodity prices. *Mgmt Sci.*, 2000, **46**(7), 893–911.

Serletis, A., Unit root behavior in energy futures prices. Energy J., 1992, **13**, 119–128.

Serletis, A., A cointegration analysis of petroleum futures prices. *Energy Econ.*, 1994, **16**(2), 93–97.

Smith, C.W. and Slutz, R.M., The determinants of firms hedging polities. *J. Financ. Quantit. Anal.*, 1985, **20**, 391–405.

Sorensen, C., Modeling seasonality in agricultural commodity futures. *J. Fut. Mkts*, 2002, **22**, 393–426.

Tsitsiklis, J.N. and VanRoy, B., Optimal stopping for Markov processes: Hilbert space theory, approximation algorithm and an application to pricing high-dimensional financial derivatives. *IEEE Trans. Autom. Control*, 1999, **44**, 1840–1851.

Appendix A: Estimation Methodology

The Kalman filter technique is a recursive methodology that estimates the unobservable time series, the state variables or the factors (Z_t) based on an observable time series (Y_t) that depends on these state variables. The *measurement equation* accounts for the relationship between the observable time series and the state variables:

$$Yt = dt + MtZt + \eta t, t = 1, \ldots Nt, \tag{A.1}$$

where Y_t, $d_t \in R^n$, $M_t \in R^{nxm}$, $Z_t \in R^h$, h is the number of state variables, or factors, in the model, and $\eta_t \in R^n$ is a vector of serially uncorrelated Gaussian disturbances with zero mean and covariance matrix H_t.

In the estimation procedure, a discrete time version of this equation is necessary; in the case of the joint model with a common long-term trend for the three commodities, this equation is given by the following expressions:

$$Y_t = \begin{pmatrix} \ln F_{T1}^1 \\ \vdots \\ \ln F_{Tn}^1 \\ \ln F_{T1}^2 \\ \vdots \\ \ln F_{Tn}^2 \\ \ln F_{T1}^3 \\ \vdots \\ \ln F_{Tn}^3 \end{pmatrix}, dt = \begin{pmatrix} A^1(T_1) \\ \vdots \\ A^1(T_n) \\ A^2(T_1) \\ \vdots \\ A^2(T_n) \\ A^3(T_1) \\ \vdots \\ A^3(T_n) \end{pmatrix}$$

$$M_t = \begin{pmatrix} 1 & e^{-k_1 T_1} & 0 & 0 \\ \vdots & \vdots & \vdots & \vdots \\ 1 & e^{-k_1 T_n} & 0 & 0 \\ 1 & 0 & e^{-k_2 T_1} & 0 \\ \vdots & \vdots & \vdots & \vdots \\ 1 & 0 & e^{-k_2 T_n} & 0 \\ 1 & 0 & 0 & e^{-k_3 T_1} \\ \vdots & \vdots & \vdots & \vdots \\ 1 & 0 & 0 & e^{-k_3 T_n} \end{pmatrix},$$

and F_{T1}^i is the price of a futures contract for the commodity 'i' ($i = 1, 2, 3$) with maturity at time '$T_1 + t$' traded at time t. In principle, it would be possible to use a different number of futures contracts for each commodity; however, in this work, we consider it more suitable to use the same number ('n') of futures contracts for all commodities.

The *transition equation* accounts for the evolution of the state variables:

$$Z_t = c_t + T_t Z_{t-1} + \psi_t, \quad t = 1, \ldots N_t, \tag{A.2}$$

where $c_t \in R^h$, $T_t \in R^{h \times h}$ and $\psi_t \in R^h$ is a vector of serially uncorrelated Gaussian disturbances with zero mean and covariance matrix Q_t.

In the case of the joint model with a common long-term trend for the three commodities, the discrete time version of this equation, which is needed in the estimation procedure, is given by the following expressions:

$$Z_t = \begin{pmatrix} \xi_{1t} \\ \chi_{1t} \\ \chi_{2t} \\ \chi_{3t} \end{pmatrix}, c_t = \begin{pmatrix} \mu_{\xi 1} \Delta t \\ 0 \\ 0 \\ 0 \end{pmatrix}$$

$$T_t = \begin{pmatrix} 1 & 0 & 0 & 0 \\ 0 & e^{-k_1 \Delta t} & 0 & 0 \\ 0 & 0 & e^{-k_2 \Delta t} & 0 \\ 0 & 0 & 0 & e^{-k_3 \Delta t} \end{pmatrix}$$

and

$$Var(\psi_t) = \begin{pmatrix} \sigma_{\xi_1}^2 \Delta t & \sigma_{\xi_1}\sigma_{\chi_1}\rho_{\xi_1\chi_1}\left(1-e^{-k_1\Delta t}\right)/k_1 & \sigma_{\xi_1}\sigma_{\xi_2}\rho_{\xi_1\xi_2}\Delta t & \sigma_{\xi_1}\sigma_{\chi_2}\rho_{\xi_1\chi_2}\left(1-e^{-k_1\Delta t}\right)/k_2 \\ \sigma_{\xi_1}\sigma_{\chi_1}\rho_{\xi_1\chi_1}\left(1-e^{-k_1\Delta t}\right)/k_1 & \sigma_{\chi_1}^2\left(1-e^{-2k_1\Delta t}\right)/(2k_1) & \sigma_{\chi_1}\sigma_{\chi_2}\rho_{\chi_1\chi_2}\left(1-e^{-(k_1\Delta t+k_2\Delta t)}\right)/(k_1+k_2) & \sigma_{\chi_1}\sigma_{\chi_3}\rho_{\chi_1\chi_3}\left(1-e^{-(k_1\Delta t+k_3\Delta t)}\right)/(k_1+k_3) \\ \sigma_{\xi_1}\sigma_{\chi_2}\rho_{\xi_1\chi_2}\left(1-e^{-k_1\Delta t}\right)/k_2 & \sigma_{\chi_1}\sigma_{\chi_2}\rho_{\chi_1\chi_2}\left(1-e^{-(k_1\Delta t+k_2\Delta t)}\right)/(k_1+k_2) & \sigma_{\chi_2}^2\left(1-e^{-2k_2\Delta t}\right)/(2k_2) & \sigma_{\chi_2}\sigma_{\chi_3}\rho_{\chi_2\chi_3}\left(1-e^{-(k_2\Delta t+k_3\Delta t)}\right)/(k_2+k_3) \\ \sigma_{\xi_1}\sigma_{\chi_3}\rho_{\xi_1\chi_3}\left(1-e^{-k_3\Delta t}\right)/k_3 & \sigma_{\chi_1}\sigma_{\chi_3}\rho_{\chi_1\chi_3}\left(1-e^{-(k_1\Delta t+k_3\Delta t)}\right)/(k_1+k_3) & \sigma_{\chi_2}\sigma_{\chi_3}\rho_{\chi_2\chi_3}\left(1-e^{-(k_2\Delta t+k_3\Delta t)}\right)/(k_2+k_3) & \sigma_{\chi_3}^2\left(1-e^{-2k_3\Delta t}\right)/(2k_3) \end{pmatrix} \lim_{x \to \infty}$$

Here, $Y_t|_{t-1}$ is the conditional expectation of Y_t, and Ξ_t is the covariance matrix of Y_t conditional on all the information available at time $t - 1$. After omitting unessential constants, the log-likelihood function can be expressed as

$$l = -\sum_t \ln|\Xi_t| - \sum_t \left(Y_t - Y_{t|t-1}\right)'\Xi_t^{-1}\left(Y_t - Y_{t|t-1}\right). \tag{A.3}$$

Appendix B: Modelling the Crack-Spread Directly

Dempster *et al.* (2008) propose a model for valuing spread options on two commodity prices that are cointegrated. Let x_t denote the (latent) mean-reverting spot spread and y_t the mean-reverting process, with a zero long-run mean, representing the deviation from the long-term equilibrium of spot spreads. The expressions for x_t and y_t are*

$$dx_t = k\left(\theta + \phi(t) + y_y - x_t\right)dt + \sigma dW,$$
$$dy_t = -\kappa_2 y_t dt + \sigma_2 dW_2. \tag{B.1}$$

In this model, x and y revert towards θ and zero, with mean reversion speeds κ and κ_2, respectively. dW and dW_2 are two standard Brownian motions with correlation ρ. $\varphi(t)$ accounts for the seasonality effects in the spread, as in Dempster *et al.* (2008) and Sorensen (2002).

The model is estimated according to the Kalman filter methodology using two crack-spread series: the difference between heating oil and WTI crude oil prices and the difference between unleaded gasoline (RBOB) and WTI crude oil prices. In the case of heating oil vs. WTI crude oil, the data set used in the estimation comprises contracts F1, F4, F7, F11, F15 and F18, from 09/09/1996 to 05/25/2009 (664 weekly observations each contract). In the case of RBOB gasoline vs. WTI crude oil, the data set comprises contracts F1, F3, F5, F7 and F9, from 06/30/1997 to 05/25/2009 (622 weekly observations).

---

* To be consistent with our previous notation, the model by Dempster *et al.* (2008) is formulated here under the real measure.

6

Long-Term Spread Option Valuation and Hedging

M.A.H. Dempster,
Elena Medova
and Ke Tang

6.1 Introduction

Commodity spreads are important for both investors and manufacturers. For example, the price spread between heating oil and crude oil (*crack* spread) represents the value of production (including profit) for a refinery firm. If an oil refinery in Singapore can deliver its oil both to the US and the UK, then it possesses a real option of diversion which directly relates to the spread of WTI and Brent crude oil prices. There are four commonly used spreads: spreads between prices of the same commodity at two different locations (*location* spreads) or times (*calendar* spreads), between the prices of inputs and outputs (*production* spreads) or between the prices of different grades of the same commodity (*quality* spreads).*

A *spread option* is an option written on the difference (*spread*) of two underlying asset prices S_1 and S_2, respectively. We consider European options with *payoff* the greater or lesser of $S_2(T)-S_1(T)-K$ and 0 at maturity T for *strike price K* and focus on spreads in the commodity (especially energy) markets (for both spot and futures). In pricing spread options it is natural to model the spread by modelling each asset price separately. Margrabe (1978) was the first to treat spread options and gave an analytical solution for strike price zero (the *exchange option*). Closed form valuation of a spread option is not available if the two underlying prices follow geometric Brownian motions (see Eydeland and Geman, 1998).

---

* For more details on these concepts see Geman (2005a).

DOI: 10.1201/9781003265399-7

Hence various numerical techniques have been proposed to price spread options, such as for example the Dempster and Hong (2000) fast Fourier transformation approach. Carmona and Durrleman (2003) offer a good review of spread option pricing.

Many researchers have modelled the spread using two underlying commodity spot prices (the two price method) in the unique *risk neutral measure* as*

$$
\begin{aligned}
d\mathbf{S}_1 &= \left(r - \delta_1\right)S_1\,dt + \sigma_{1,1}\,S_1\,d\mathbf{W}_{1,1}, \\
d\delta_1 &= k_1\left(\theta_1 - \delta_1\right)dt + \sigma_{1,2}\,d\mathbf{W}_{1,2}, \\
d\mathbf{S}_2 &= \left(r - \delta_2\right)S_2\,dt + \sigma_{2,1}S_2\,d\mathbf{W}_{2,1}, \\
d\delta_2 &= k_2\left(\theta_2 - \delta_2\right)dt + \sigma_{2,2}\,d\mathbf{W}_{2,2},
\end{aligned}
\tag{6.1}
$$

where $\mathbf{S}_1$ and $\mathbf{S}_2$ are the spot prices of the commodities and δ_1 and δ_2 are their convenience yields, and $\mathbf{W}_{1,1}$, $\mathbf{W}_{1,2}$, $\mathbf{W}_{2,1}$ and $\mathbf{W}_{2,2}$ are four correlated Wiener processes. This is the classical Gibson and Schwartz (1990) model for each commodity price in a complete market.[†] The return correlation $\rho_{12} := E[d\mathbf{W}_{1,1}\,d\mathbf{W}_{2,1}]/dt$ plays a substantial role in valuing a spread option; trading a spread option is equivalent to trading the correlation between the two asset returns. However, Kirk (1995), Mbanefo (1997) and Alexander (1999) have suggested that return correlation is very volatile in energy markets. Thus assuming a *constant* correlation in (6.1) is inappropriate.

But there is another longer term relationship between two asset prices, termed *cointegration*, which has been little studied by asset pricing researchers. If a cointegration relationship exists between two asset prices the spread should be modelled *directly* over the long term horizon. Soronow and Morgan (2002) proposed a one factor mean reverting process to model the location spread directly, but do not explain under what conditions this is valid nor derive any results.[‡] See also Geman (2005a) where diffusion models for various types of spread option are discussed.

In this paper, we use two factors to model the spot spread process and fit the futures spread term structure. Our main contributions are threefold. First, we give the first statement of the economic rationale for mean reversion of the spread process and support it statistically using standard cointegration tests on data. Second, the paper contains the first test of mean-reversion of *latent* spot spreads in both the risk neutral and market measures. Third, we give the first latent multi-factor model of the spread term structure which is calibrated using standard state-space techniques, i.e. Kalman filtering.

The paper is organized as follows. Section 6.2 gives a brief review of price cointegration together with the principal statistical tests for cointegration and the mean reversion of spreads. Section 6.3 proposes the two factor model for the underlying spot spread process and shows how to calibrate it. Section 6.4 presents option pricing and hedging formulae for options on spot and futures spreads. Sections 6.5 and 6.6 provide two examples in energy markets which illustrate the theoretical work and Section 6.7 concludes.

---

[*] Boldface is used throughout to denote random entities – here conditional on S1 and S2 having realized values S1 and S2 at time t which is suppressed for simplicity of notation.

[†] We adopt this model for three reasons. (1) It fits futures contract prices much better than the one factor mean-revertinglog price model as shown by Schwartz (1997). (2) In examining the historical commodity prices used here, we show that WTI and Brent crude oil and heating oil prices are not mean-reverting. This has been found by many others, e.g. Girma and Paulson (1999) and Geman (2005b). (We also show that the spread is mean-reverting.) Since the Gibson–Schwartz model has a GBM backbone, we believe it matches historical commodity prices better. (3) Schwartz (1997) shows that futures volatility in the one factor model will decay to close to zero after ten years, but using the Enron dataset he shows that the volatility for market futures with maturities longer than 2 years fluctuates around 12%. However the two factor Gibson–Schwartz model can match the volatility term structure quite well.

[‡] We are grateful to an anonymous referee for this reference.

6.2 Cointegrated Prices and Mean Reversion of the Spread

A *spread* process is determined by the dynamic relationship between two underlying asset prices and the *correlation* of the corresponding returns time series is commonly understood and widely used. *Cointegration* is a method for treating the long run dynamic equilibrium relationships between two asset *prices* generated by market forces and behavioural rules. Engle and Granger (1987) formalized the idea of integrated variables sharing an equilibrium relation which turns out to be either stationary or to have a lower degree of integration than the original series. They used the term cointegration to signify co-movements among trending variables which could be exploited to test for the existence of equilibrium relationships within the framework of fully dynamic markets.

In general, the return correlation is important for short term price relationships and the price cointegration for their long run counterparts. If two asset prices are cointegrated (6.1) is only useful for *short term* valuation even when the correlation between their returns is known exactly. Since we wish to model long term spread we shall investigate the cointegration (long term equilibrium) relationship between asset prices. First we briefly explain the economic reasons why such a long run equilibrium exists between prices of the same commodity at two different locations, prices of inputs and outputs and prices of different grades of the same commodity.[*]

The *law of one price* (or *purchasing power parity*) implies that cointegration exists for prices of the same commodity at different locations. Due to market frictions (trading costs, shipping costs, etc.) the same good may have different prices but the mispricing cannot go beyond a threshold without allowing market arbitrages (Samuelson, 1964). Input (raw material) and output (product) prices should also be cointegrated because they directly determine supply and demand for manufacturing firms. There also exists an equilibrium involving a threshold between the prices of a *commodity* of different grades since they are substitutes for each other. Thus the spread between two spot commodity prices reflects the profits of producing (production spread), shipping (location spread) or switching (quality spread). If such long-term equilibria hold for these three pairs of prices, cointegration relationships should be detected in the empirical data.

In empirical analysis economists usually use equations (6.2) and (6.3) to describe the cointegration relationship:

$$\mathbf{S}_{1t} = c_t + d\mathbf{S}_{2t} + \varepsilon_t, \tag{6.2}$$

$$\varepsilon_t - \varepsilon_{t-1} = \omega\varepsilon_{t-1} + \mathbf{u}_t, \tag{6.3}$$

where $\mathbf{S}_1$ and $\mathbf{S}_2$ are the two asset prices and $\mathbf{u}$ is a Gaussian disturbance. Engle and Granger (1987) demonstrate that the error term ε_t in (6.2) must be *mean reverting* (6.3) if cointegration exists. Thus the *Engle–Granger* two step test for cointegration directly tests whether ω is a significantly negative number using an augmented Dicky and Fuller (1979) test. Note that (6.2) can be seen as the dynamic equilibrium of an economic system. When the trending prices $\mathbf{S}_1$ and $\mathbf{S}_2$ deviate from the long run equilibrium relationship they will revert back to it in the future.

For both location and quality spreads $\mathbf{S}_1$ and $\mathbf{S}_2$ should ideally follow the *same* trend, i.e. d should be equal to 1.[†] Since gasoline and heating oil are cointegrated substitutes, the d value could be 1 for both the heating oil/crude oil spread and the heating oil/gasoline spread (Girma and Paulson, 1999). For our spreads of interest – production and location – d is treated here as 1.

---

[*] Calendar spreads can be modelled using the models for individual commodities such as the models proposed by Schwartz (1997) and Schwartz and Smith (2000). In this paper two different commodities are considered.

[†] However for production spreads such as the spark spread (the spread between the electricity price and the gas price) d may not be exactly 1. Usually 3/4 of a gas contract is equivalent to 1 electricity contract so that investors trade a 1 electricity/3/4 gas spread which represents the profit of electricity plants (Carmona and Durrleman, 2003).

Letting $\mathbf{x}_t$ denote the spread between two cointegrated spot prices $\mathbf{S}_1$ and $\mathbf{S}_2$ it follows from (6.2) and (6.3) in this case that

$$\mathbf{x}_t - x_{t-1} = c_t - c_{t-1} - \omega\left(c_{t-1} - x_{t-1}\right) + \mathbf{u}_t, \tag{6.4}$$

i.e. the spread of the two underlying assets is *mean reverting*. No matter what the nature of the underlying $\mathbf{S}_1$ and $\mathbf{S}_2$ processes,* the spread between them can behave quite differently from their individual behaviour. This suggests modelling the spread *directly* over a long run horizon because the cointegration relationship has a substantial influence in the long run. Such an approach gives at least three advantages over alternatives since it: (1) avoids modelling the correlation between the two asset returns; (2) catches the long run equilibrium relationship between the two asset prices; (3) yields an analytical solution for spread options (cf. Geman, 2005a).

6.2.1 Cointegration Tests

The Engle–Granger two step test is the most commonly used test for the cointegration of two time series. We first need to test whether each series generates a *unit root* time series. If two asset price processes are unit root but the spread process is not, there exists a cointegration relationship between the prices and the spread will not deviate outside economically determined bounds. The augmented *Dickey–Fuller* (ADF) test may be used to check for unit roots in the two asset price time series. The ADF test statistic uses an ordinary least squares (OLS) auto regression

$$\mathbf{S}_t - S_{t-1} = \pi_0 - \pi_1 S_{t-1} + \sum_{i=1}^{p} \pi_{i+1}\left(S_{t-i} - S_{t-i-1}\right) + \eta_t \tag{6.5}$$

to test for unit roots, where $\mathbf{S}_t$ is the asset price at time t, π_i, $i = 0, \ldots, p$, are constants and η_t is a Gaussian disturbance. If the coefficient estimate of π_1 is negative and exceeds the critical value in Fuller (1976) then the null-hypothesis that the series has no unit root is rejected. We can use an extension of (6.3) corresponding to (6.2) to test the cointegration relationship:

$$S_{1t} = c_t + dS_{2t} + \varepsilon_t,$$

$$\varepsilon_t - \varepsilon_{t-1} = \chi_0 + \chi_1 \varepsilon_{t-1} + \sum_{i=1}^{p} \chi_{i+1}\left(\varepsilon_{t-i} - \varepsilon_{t-i-1}\right) + u_t. \tag{6.6}$$

When estimate of χ_1 is significantly negative the hypothesis that cointegration exists between the two underlying asset price processes is accepted (Hamilton, 1994).

Another way to test for cointegration of two time series is the Johansen method, where the cointegration relationship is specified in the framework of an error-correcting *vector auto-regression* (VAR) process (see Johansen, 1991 for details). For the two examples of this paper we test the cointegration relationship using both the Engle–Granger and Johansen tests, which agree in both cases.

6.2.2 Risk Neutral and Market Measure Spread Mean Reversion Tests

Testing the mean reversion of the spot spread is a key step in testing cointegration of the two underlying price processes. However usually only futures spreads are observed in the market and spot spreads are

---

* Especially for commodities where many issues have to be considered, such as jumps, seasonality, etc. Hence no commonly acceptable model exists for all commodities.

latent (i.e. unobservable). The futures spread with fixed maturity date is a martingale in the risk neutral measure and is thus not mean reverting. If we assume a constant risk premium the futures spread with fixed maturity date is also not mean reverting in the market measure. It is therefore difficult to test empirically the mean reversion of the spot spread using futures markets data without options data. We nevertheless propose novel methods to test the mean reversion of the spot spread in both the risk neutral and market measures which do not require derivative prices.

We can use *ex ante* market data analysis to test whether risk neutral investors *expect* the future spread to revert in the *risk neutral measure*. This approach uses relations between spread *levels* and the *spread term structure slope* defined as the price change between the maturities of two futures spreads. A negative relationship between the spot spread level (or short term futures spread level) and the futures spread term structure slope shows that risk neutral investors expect mean reversion in the spot spread. Indeed, since each futures price equals the trading date expectation of the delivery date spot price in the risk neutral measure the current term structure of the futures spread reveals where risk neutral investors expect the spot spread to be at delivery.[*] To discover this negative relationship in the risk neutral measure we estimate the time series model

$$\chi_L - \chi_S = \zeta + \gamma \chi_S + \varepsilon, \tag{6.7}$$

where χ_L and χ_S are, respectively, long-end and short-end spread levels in the futures spread term structure and ε is a noise term. If the estimate of γ is significantly negative there is evidence that the spot spread is *ex ante* mean reverting in the risk neutral measure.

To detect mean reversion of the spot spread in the *market measure* we construct a time series using *historical* futures data which preserves the mean reversion of the latent spot spreads. In Section 6.3 we will see that futures spreads with a *constant* time to maturity preserve the mean reversion of the spot spread price.[†] Thus to test mean reversion empirically we use an ADF test based on

$$F(t + \Delta t, t + \Delta t + \tau) - F(t, t + \tau)$$

$$= \chi_0 + \chi_1 \cdot F(t, t + \tau) + \sum_{i=0}^{p} \chi_{i+2}[F(t - i\Delta t, t - i\Delta t + \tau) \tag{6.8}$$

$$-F(t - (i+1)\Delta t, t - (i+1)\Delta t + \tau)] + \varepsilon_{t+\Delta t},$$

where $F(t, t + \tau)$ is the *futures spread* of maturity $t + \tau$ observed at t, Δt is the sampling time interval and ε_t is a random disturbance. Note that $F(t, t + \tau)$ and $F(t + \Delta t, t + \Delta t + \tau)$ are spreads relating to different futures contracts. If the estimate of χ_1 is significantly negative then the spot spread is deemed to be mean reverting.

6.3 Modelling the Spread Process

We now present the two factor spot spread models in both the risk neutral and market measures.

---

[*] Bessembinder *et al.* (1995) attempt to discover *ex ante* mean reversion in commodity spot *prices* by this technique.

[†] Since usually the spot prices are not directly observable, we use futures with short maturity to represent the spot prices (cf. Clewlow and Strickland, 1999). Therefore, in performing an Engle–Granger two step test for mean reversion we use futures spread time series with short and constant time to maturity because they can represent both the latent spot spread and preserve its mean reversion.

6.3.1 Spread Process in the Risk Neutral Measure

In the risk neutral measure the underlying *spot spread* process x and the *long run factor* y satisfy

$$dx_t = k(\theta + \varphi(t) + y_t - x_t)dt + \sigma dW,$$

$$dy_t = -k_2 y_t dt + \sigma_2 dW_2, \tag{6.9}$$

$$EdWdW_2 = \rho dt,$$

where x and y are two *latent* mean reverting factors with, respectively, *long run means* θ and 0 and *mean reversion speeds* k and k_2 and ϕ is a *seasonality* function specifying the seasonality in the spread.

The **x** factor represents the mean reverting spot spread and the **y** factor is also a mean reverting process with zero long run mean representing deviation from the long term equilibrium of the spot spreads. We can interpret the dynamics of the spot spread x_t as reverting to a *stochastic* long run mean $\theta + y_t$, which itself reverts (eventually) to θ. The seasonality function ϕ induced by the seasonality of the individual commodity prices is specified (cf. Durbin and Koopman, 1997; Richter and Sorensen, 2002) as

$$\varphi(t) = \sum_{i=1}^{K} \left[\alpha_i \cos(2\pi i t) + \beta_i \sin(2\pi i t) \right], \tag{6.10}$$

where α_i and β_i are constants.

Many tests for cointegration assume the long run relationship between two commodity prices is constant during the period of study. Modelling the spread by one mean reverting **x** factor is consistent with such a relationship. However in reality the long run relationship between the underlying commodities can change due to inflation, economic crises, changes in consumers' behaviour, etc. Gregory and Hansen (1996) have attempted to identify structural breaks of the cointegration relationship. We assume the long run relationship changes continuously and adopt the second **y** factor to reflect these changes. Our model nests the one factor model by setting the **y** factor to be identically zero. In Sections 6.5 and 6.6 we will compare the ability of the one and two factor models to fit observed futures spreads.

6.3.2 Spread Process in the Market Measure

To calibrate our model to market data we need a version in which risk is priced. We can incorporate risk premium processes for *x* and *y* in the drifts of our risk neutral models to return to the market measure. Previous studies assume *constant* risk premia when modelling *Ornstein–Uhlenbeck* (OU) processes (see, e.g. Hull and White, 1990; Schwartz, 1997; Geman and Nguyen, 2005) and similarly we assume that the two factor model in the market measure satisfies

$$dx_t = \left[k(\theta + \varphi(t) + y_t - x_t) + \lambda \right] dt + \sigma dW$$

$$dy_t = \left(-k_2 y_t + \lambda_2 \right) \sigma_2 dW_2 \tag{6.11}$$

$$EdWdW_2 = \rho dt,$$

where λ and λ_2 are the *risk premia* of the **x** and **y** processes, respectively.

With starting time *v* and starting position x_v, x_s and y_s at time *s* can be expressed as

$$x_S = \chi_v e^{-k(s-v)} + \left(\theta + \frac{\lambda}{k} \right) \left[1 - e^{-k(s-v)} \right] + \frac{y_v k}{k - k_2} \left[e^{-k_2(s-v)} - e^{-k(s-v)} \right]$$

$$+ \frac{\lambda_2}{k_2} \left(1 - e^{-k\Delta t} \right) - \frac{\lambda_2 k}{k_2(k - k_2)} \left(e^{-k_2 \Delta t} - e^{-k\Delta t} \right) + G(v, s) \tag{6.12}$$

$$+ \frac{K\sigma_2}{k - k_2} \int_v^s \left[e^{-k_2(s-u)} - e^{-k(s-u)} \right] dW_2(u) + \int_v^s e^{-k(s-t)} \sigma dW(t),$$

$$\mathbf{y}_s = y_v e^{-k_2(s-v)} + \int_V^s e^{-k_2(s-u)} \sigma_2 d\mathbf{W}_2(u),$$

(6.13)

where $G(v, s)$ denotes the *seasonality effect* given by

$$G(v,s) = \int_v^s k e^{-k(s-r)} \varphi(r) dr.$$

(6.14)

Then the standard deviation of $\mathbf{x}_s$ becomes

$$b_s = \sqrt{A_1 + A_2 + 2\rho A_3},$$

(6.15)

where

$$A_1 := \frac{\sigma^2}{2k}\left[1 - e^{-2k(s-v)}\right],$$

$$A_2 := \left(\frac{1}{2k_2}\left[1 - e^{-2k_2(s-v)}\right] + \frac{1}{2k}\left[1 - e^{-2k_2(s-v)}\right]\right.$$

$$\left. - \frac{2}{(k+k_2)}\left[1 - e^{-(k_2+k)(s-v)}\right]\right) \cdot \frac{k^2 \sigma_2^2}{(k-k_2)^2},$$

$$A_3 := \frac{k\sigma_2 \sigma}{k-k_2}\left(\frac{1}{k+k_2}\left[1 - e^{-(k+k_2)(s-v)}\right] - \frac{1}{2k}\left[1 - e^{-2k_2(s-v)}\right]\right).$$

$$\text{As } s \to \infty, b_s \to \sqrt{\frac{\sigma^2}{2k} + \frac{\sigma_2^2}{2k_2(1 + k_2/k)} + \frac{\rho\sigma\sigma_2}{2(k+k_2)}}, \text{ a constant.}$$

On the other hand it is easy to show in the two price extended Gibson and Schwartz (1990) model (6.1) that both $\mathbf{S}_1$ and $\mathbf{S}_2$ are non-stationary (i.e. *not* mean reverting). One easy way to show this is to write (6.1) in the vector format, as for $i = 1, 2$,

$$\begin{bmatrix} d \ln \mathbf{S}_i \\ d\delta_i \end{bmatrix} = \begin{bmatrix} r - \frac{1}{2}\sigma_{i,1}^2 \\ k_i\theta_i \end{bmatrix} dt + \psi \begin{bmatrix} \ln \mathbf{S}_i \\ \delta_i \end{bmatrix} dt + \begin{bmatrix} \sigma_{i,1} dW_{i,1} \\ \sigma_{i,2} dW_{i,2} \end{bmatrix}, \text{ where } \psi : \begin{bmatrix} 0 & -1 \\ 0 & -k_i \end{bmatrix}.$$

As shown in Arnold (1974), only if the real parts of all the eigenvalues of ψ are strictly negative is $\ln \mathbf{S}_i$ stationary. However inspection of ψ shows that one eigenvalue is zero, so that $\ln \mathbf{S}_i$ is not stationary.

Moreover, the variances of both $\mathbf{S}_1$ and $\mathbf{S}_2$ increase to infinity with time. Thus the variance of the spread will also blow up asymptotically. This is not consistent with the behaviour of spreads between cointegrated commodity prices in historical market data (Villar and Joutz, 2006).

6.3.3 Futures Pricing

Define $F(t, T, x_t)$ as the *futures spread* (the spread of two futures prices) of maturity T observed in the market at time t when the *spot spread* is x_t. In the risk neutral measure the spot spread process $\mathbf{x}$ must satisfy the *no arbitrage condition*

$$E\left[X_T \mid X_t\right] = F(t, T, X_t),$$

(6.16)

i.e. in the absence of arbitrage the conditional expectation in the risk neutral measure of the spot spread at T with respect to the realized spot spread x_t at t is the futures spread observed at time $t < T$. This must hold because it is costless to enter a futures spread (long one future and short the other).

Thus, for the two factor model

$$F(t,T;x_t) = x_t e^{-k(T-t)} +, \left[1 - e^{-k(T-t)}\right]$$
$$+ \frac{y_t k}{k - k_2}\left[e^{-k_2(T-t)} - e^{-k(T-t)}\right] + G(t,T). \tag{6.17}$$

From Ito's lemma it follows that the *risk neutral* futures spread $F(t, T)$ process with fixed maturity date T satisfies

$$dF(t,T) = e^{-k(T-t)}\sigma d\mathbf{W} + k\phi\sigma_2 d\mathbf{W}_2, \tag{6.18}$$

where $\phi := \left[e^{-k_2(T-t)} - e^{-k(T-t)}\right]/\left(k - k_2\right)$. In the market measure (6.18) becomes

$$dF(t,T) = \left(\lambda e^{-k(T-t)} + k\lambda_2\phi\right)dt + e^{-k(T-t)}\sigma d\mathbf{W} + k\phi\sigma_2 d\mathbf{W}_2. \tag{6.19}$$

The futures spread with *fixed* maturity date T following (6.19) is *not* mean reverting.

However, defining $\tau := T - t$ as the constant time to maturity, the futures spread (6.17) can be rewritten as

$$\mathbf{F}(t,t+\tau) = x_t e^{-k\tau} + e^{-k(t+\tau)}\int_t^{t+\tau}(\theta + \varphi(t))e^{ku}k\,du + \phi y_t. \tag{6.20}$$

Differentiating (6.20) with respect to t the process of futures spreads with a constant time to maturity τ satisfies

$$d\mathbf{F}(t,t+\tau) = e^{-k\tau}d\mathbf{x_t} + k\phi d\mathbf{y_t} - k^2 e^{-k(t+\tau)}[$$
$$\times \int_t^{t+\tau}(\theta + \varphi(t))e^{ku}du\Bigg)\Bigg]dt + ke^{-k(t+\tau)}[\theta + \varphi(t)] \tag{6.21}$$
$$\times\left[e^{k(t+\tau)} - e^{kt}\right]dt.$$

Substituting for the integral from (6.20) and using (6.11)–(6.13) we obtain

$$d\mathbf{F}(t,t+\tau) = k\left[\theta + \varphi(t) + y_t e^{-k\tau} + \frac{\lambda e^{-k\tau}}{k} + \phi\lambda_2\right.$$
$$-F(t,t+\tau)]dt + e^{-k\tau}\sigma d\mathbf{W} + k\phi\sigma_2 d\mathbf{W}_2. \tag{6.22}$$

Note that if we set $y_t := 0$, $\lambda_2 := 0$, $\sigma_2 := 0$ in (6.22), we obtain the analogous process for the one factor model. From (6.22) a futures spread with constant time to maturity is mean reverting with the *same* mean-reversion speed as the spot process in both the two factor model and its one factor restriction (with $y \equiv 0$). Note that when $\tau \to 0$ (6.22) converges to (6.11).

6.3.4 Calibration

A difficulty with the calibration of the two factor model is that the factors (or state variables) are not directly observable, i.e. they are latent. An approach to maximum likelihood estimation of the model is to pose the model in *state space* form and use the *Kalman filter* to estimate the latent variables. Harvey (1989) and Hamilton (1994) give good descriptions of estimation, testing and model selection for state space models.

The state space form consists of transition and measurement equations. The *transition equation* describes the dynamics of the underlying unobservable data-generating process. The *measurement equation* relates a multivariate vector of observable variables, in our case future prices of different maturities, to the unobservable state vector of *state variables*, in our case the **x** and **y** factors. The measurement equation is given by (6.17) with the addition of uncorrelated disturbances to take account of pricing errors. These errors can be caused by bid-ask spreads, non-simultaneity of the observations, etc.

More precisely, suppose the data are sampled at equally spaced times t_n, $n = 1, \ldots, N$, with $\Delta t := t_{n+1} - t_n$ for the interval between two observations. Let $X_n := \left[x_{t_n} y_{r_{tu}} \right]^{\mathrm{T}}$ represent the vector of state variables at time t_n. We obtain the transition equation from the discretization of (6.11) in the form

$$\mathbf{X}_{n+1} = A\mathbf{X}_n + b_n + \mathrm{w}, \tag{6.23}$$

where **w** is a serially independent multivariate normal innovation with mean 0 and covariance matrix Q, and A, b_n and Q are given by

$$A := \begin{bmatrix} e^{-k\Delta t} & \dfrac{k}{k-k_2}\left(e^{-k_2\Delta t} - e^{-k\Delta t} \right) \\ 0 & e^{-k_2\Delta t} \end{bmatrix},$$

$$b_n := \begin{bmatrix} \dfrac{\lambda_2}{k_2}\left(1 - e^{-k\Delta t}\right) - \dfrac{\lambda_2 k}{k_2\left(k-k_2\right)}\left(e^{-k_2\Delta t} - e^{-k\Delta t} \right) \\ +\left(\theta + \dfrac{\lambda}{k} \right)\left(1 - e^{-k\Delta t}\right) + G\left(t_n, t_n + \Delta t\right) \\ \dfrac{\lambda_2}{k_2}\left(1 - e^{-k_2\Delta t}\right) \end{bmatrix},$$

$$Q := \begin{bmatrix} V_x & V_{xy} \\ V_{xy} & V_y \end{bmatrix}$$

with

$$V_x := \frac{1 - e^{-2k\Delta t}}{2k}\sigma^2 + \left\{ \frac{1}{2k_2}\left[1 - e^{-2k_2\Delta t}\right] + \frac{1}{2k}\left[1 - e^{-2k\Delta t}\right] \right.$$
$$\left. -\frac{2}{k+k_2}[1 - e^{-(k+k_2)\Delta t}] \right\} \frac{k^2\sigma_2^2}{\left(k-k_2\right)^2}$$
$$+ \frac{2\rho k\sigma_2\sigma}{k-k_2}\left[\frac{1}{k+k_2}\left(1 - e^{-(k+k_2)\Delta t}\right) - \frac{1}{2k}\left(1 - e^{-2k\Delta t}\right) \right],$$

$$V_y := \sigma_2^2\left(\frac{1 - e^{-2k_2\Delta t}}{2k_2} \right.$$

$$V_{xy} := \frac{k\sigma_2}{k-k_2}\left[\frac{1 - e^{-2k_2\Delta t}}{2k_2} - \frac{1 - e^{-(k+k_2)\Delta t}}{k+k_2} \right] + \frac{\rho\left(1 - e^{-(k+k_2).\Delta t}\right)}{k+k_2}.$$

In preliminary study we found that the estimated correlation $\hat{\rho}$ between the long and short end fluctuations was insignificant. This makes economic sense in that long-end movements are slow and driven by fundamentals, while short-term movements are random, fast and driven by market trading activities. In the model calibration we thus assume the correlation ρ to be zero.[*]

Let

$$Z_n := \left[\ln\left(F\left(T_n, t_n + \tau_1\right)\right) \ldots \ln\left(F\left(t_n, t_n + \tau_M\right)\right) \right]^{\mathrm{T}},$$

where $\tau_1, \ldots, \tau_M$ are *times to maturity* for $1, \ldots, M$ futures contracts. Thus the measurement equation takes the form,

$$\mathbf{Z}_n = C_n X_n + d_n + \varepsilon_n, \tag{6.24}$$

where

$$C_n := \begin{bmatrix} e^{-k\tau_1} & \dfrac{k}{k - k_2}\left(e^{-k_2\tau_1} - e^{-k\tau_1}\right) \\ \vdots & \vdots \\ e^{-k\tau_M} & \dfrac{k}{k = k_2}\left(e^{-k_2\tau_M} - e^{-k\tau_M}\right) \end{bmatrix},$$

$$d_n := \begin{bmatrix} \theta\left(1 - e^{-k\tau_1}\right) + G\left(t_n, t_n + \tau_1\right) \\ \vdots \\ \theta\left(1 - e^{-k\tau_M}\right) + G\left(t_n, t_n + \tau_m\right) \end{bmatrix}$$

and ε_n is a serially independent multivariate normal innovation with mean 0 and covariance matrix $\xi^2 I_M$. To reduce the parameters to be estimated we assume the variance of the futures spread pricing errors of all maturities to be the same. This assumption is based on the practical requirement that our model should price the futures of different maturities equally well. We also assume that the futures spread pricing errors are independent across different maturities and their common variance is denoted by ξ^2.

6.4 Spread Option Pricing and Hedging

If the underlying asset price follows a Gaussian process the European *call* price[†] with maturity T on this asset can be calculated as

$$c = B\frac{b_s}{\sqrt{2\pi}} \exp\left[-\frac{\left(a_s - K\right)^2}{2b_s^2} \right] + B\left(a_s - K\right)\Phi\left(\frac{a_s - K}{b_s}\right), \tag{6.25}$$

where B is the price of a discount bond, a_s and b_s are, respectively, the mean and standard deviation of the underlying at maturity, and K is the strike price of the option and Φ denotes the cumulative distribution function of the normal distribution. Its *delta* is given by

$$\Delta_c = \frac{\partial C}{\partial a_s} = B\Phi\left(\frac{a_s - K}{b_s}\right). \tag{6.26}$$

---

[*] Assuming that innovations in the long and short run are uncorrelated has been used to analyse the long-run and short-run components of stock prices (e.g. Fama and French, 1988).

[†] We show here only call price, the put price can be obtained by using put-call parity.

Since a spread can be seen as simultaneously long one asset and short the other the delta hedge yields an *equal volume hedge*, i.e. long and short the same value of commodity futures contracts. The spread distribution is Gaussian in our model so that (6.25) and (6.26), respectively, can be used to price and hedge spread options. Procedures to hedge long term options with short term futures are well established, see e.g. Brennan and Crewe (1997), Neuberger (1998) and Hilliard (1999).

If we price and hedge options on the *spot spread*, then $a_s := F(t, T)$ which can be calculated from (6.17) and b_s is given by (6.15). If we price the options on the *futures spread*, a_s is the market observed futures spread and

$$b_s := \sqrt{A_1^F + A_2^F + 2\rho A_3^F} \,, \tag{6.27}$$

where

$$A_1^F := \frac{\sigma^2}{2k}\left[e^{-2k(T-R)} - e^{-2k(T-t)}\right],$$

$$A_2^F := \frac{k^2\sigma_2^2}{\left(k-k_2\right)^2}\left\{\frac{1}{2k_2}\left[e^{-2k_2(T-R)} - e^{-2k_2(T-t)}\right] + \frac{1}{2k}\left[e^{-2k(T-R)} - e^{-2k(T-t)}\right]\right.$$

$$\left. -\frac{2}{\left(k+k_2\right)}\left[e^{-(k_2+k)(T-R)} - e^{-(k_2+k)(T-t)}\right]\right\},$$

$$A_3^F := \frac{k\sigma\sigma_2}{k-k_2}\left\{\frac{1}{k+k_2}\left[e^{-(k+k_2)(T-R)} - e^{-(k+k_2)(T-t)}\right]\right.$$

$$\left. -\frac{1}{2k}\left[e^{-2k(T-R)} - e^{-2k(T-t)}\right]\right\},$$

when the option maturity is R, the futures maturity is $T \geq R$ and the current time is t.

Since this paper focuses on 'long term' option valuation and hedging, we shall now discuss the option maturities appropriate for using our valuation models. These depend on the mean-reversion speeds k and k_2. The average *decay half-lives* $\ln 2/k$ and $\ln 2/k_2$ of the mean reverting spread process can be used to represent its mean-reversion strength. If the spread option maturity is longer than the average half decay time, our methodology is appropriate to value the option. We expect the traditional two price spread option model (6.1) to overvalue these longer term options because of the variance blow-up phenomenon discussed previously. Mbanefo (1997) noted that long-term (longer than 90 days) crack spread options will be overvalued if mean-reversion of the spreads is not considered. Before turning to examples we remark that jumps and stochastic volatility are not important for determining *long term* theoretical or empirical option prices (Bates, 1996; Pan, 2002) so that the present parsimonious model is appropriate for this purpose.

6.5 Crack Spread: Heating Oil/WTI Crude Oil

The *crack spread* between the prices of heating oil and WTI crude oil represents the gross revenue from refining heating oil from crude oil. In Section 6.2 we saw that a deviation from the (equilibrium) input and output price relationship can exist for short periods of time, but a prolonged large deviation could lead to the production of more end products until the output and input prices are brought nearer their long-term equilibrium.

6.5.1 Data

The data for modelling the crack spread consist of NYMEX daily futures prices of WTI crude oil (CL) and heating oil (HO) from January 1984 to January 2005. The times to maturity of these futures range from 1 month to more than 2 years. In order to test for unit roots, a data point is collected each month by taking the price of the futures contract with one month (fixed) time to maturity. For example, if the trading day is 20 February 2000 then the futures contract taken for the time series is the 20 March 2000 maturity future. We also create a long-end crack spread with time to maturity 1 year. The methodology is exactly the same as with the 1 month time series, but due to data unavailability we only use data from January 1989 to January 2005 to construct the long-end crack spread.

To calibrate the two factor model we calculate monthly the futures spreads for five futures contracts from January 1989 to January 2005. The time step Δt is thus chosen to be 1 month and the futures contracts chosen have 1, 3, 6, 9 and 12 month times to maturity.

6.5.2 Unit Root and Cointegration Tests

Figure 6.1 shows the 1 month futures prices of crude oil and heating oil. First, we conduct the ADF test on the individual heating and crude oil prices.

The first step of the Engle–Granger two step test (Table 6.1) does not reject the hypothesis that both crude oil and heating oil are unit-root time series, which is consistent with the empirical findings in Girma and Paulson (1999) and Alexander (1999). This also agrees with the results in Schwartz (1997) that the Gibson–Schwartz two factor model fits the oil data much better than the one factor mean reverting log price model (the first model in Schwartz, 1997) because the individual commodity prices are unlikely to be mean reverting (see also Routledge *et al.*, 2000).*

Next we effect the second step in the Engle–Granger test, i.e. test the spread time series by estimating (6.8). We find a very strong mean reverting speed significant at the 1% level, which suggests that cointegration does exist in the data. In other words, the mean-reversion of the crack spread does not appear to be caused by the separate mean-reversion of the heating and crude oil prices but by the long run equilibrium (cointegration) between them. We also employ the Johansen cointegration test, which again shows a significant cointegration relationship.

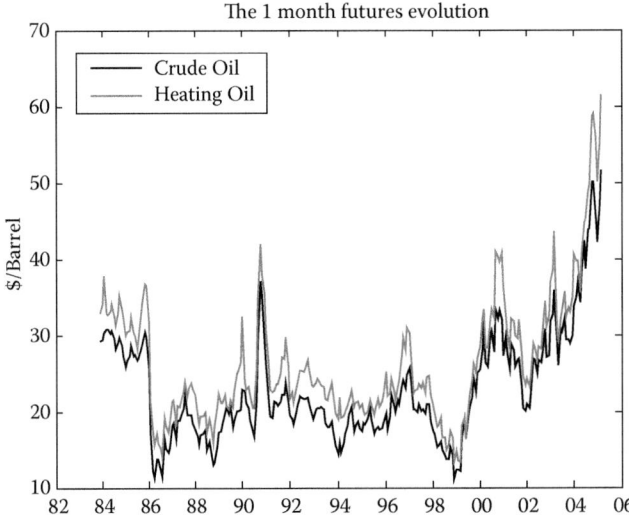

FIGURE 6.1 The 1 month futures prices of crude oil and heating oil.

---

* This is why we chose the Gibson–Schwartz model (6.1) as our benchmark.

TABLE 6.1 Cointegration Test for Crude Oil and Heating Oil

	No. of observations	Lag (*p*)	π_1 or χ_1	*t* value
Engle–Granger two step tests				
Crude oil	257	6	0.020	1.03
Heating oil	257	6	0.019	1.13
Crack spread	257	6	−0.31	−3.97*
Johansen test				
Likelihood ratio trace statistic: 46.81 (significant at 99%)[a]				

[a]*Note:* that the null hypothesis is no cointegration, which is rejected at the 99% confidence level.
* Significant at the 1% level.

TABLE 6.2 Regression (6.7) Parameter Estimates for the Crack Spread

	S	γ
Value	2.16	−0.55
t-Stat.	15.69*	−16.39*
No. of observations	154	
R^2	57.37%	

* Significant at the 1% level.

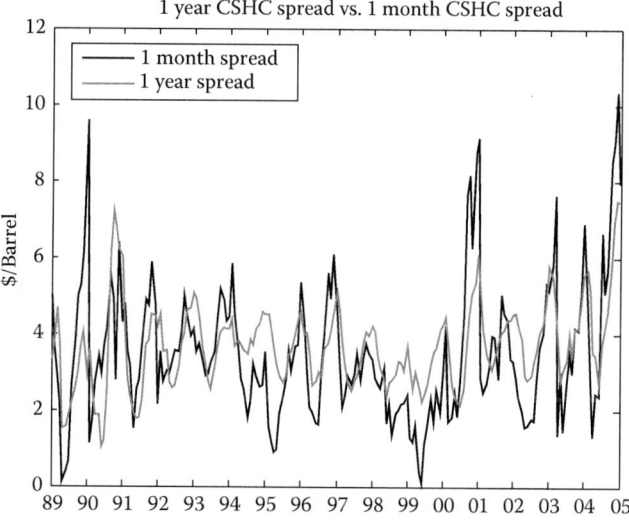

FIGURE 6.2 The 1 month and 1 year crack spreads.

To test the mean-reversion in the risk neutral measure, we estimate the regression (6.7) using the 1 year crack spread as the long-end futures spread and the 1 month crack spread as the short-end futures spread. The results are given in Table 6.2 for the 1 year and 1 month crack spreads depicted in Figure 6.2.

The estimate of γ is significantly less than zero. Thus both the market and risk neutral tests give evidence that the spot crack spread is mean reverting.

6.5.3 Model Calibration

We assume that the seasonality of the crack spread follows an *annual* pattern and that ϕ can be specified as

$$\phi(t) = \alpha \cos(2\pi t) + \beta \sin(2\pi t).$$

Our simple method of dealing with seasonality aims to keep the model parsimonious. Thus the $G(v,s)$ function in (6.12) becomes

$$G(v,s) = \frac{\alpha k^2}{k^2 + 4\pi^2} \left[\cos(2\pi s) - \cos(2\pi v)e^{-k(s-v)} + \frac{2\pi}{k}(\sin(2\pi s) \right.$$

$$\left. -\sin(2\pi v)e^{-k(s-v)}) \right] + \frac{\beta k^2}{k^2 + 4\pi^2} \left[\sin(2\pi s) - \sin(2\pi v)e^{-k(s-v)} \right.$$

$$\left. -\frac{2\pi}{k}\left(\cos(2\pi s) - \cos(2\pi v)e^{-k(s-v)}\right) \right].$$

Our calibration results are shown in Table 6.3. The market prices of risk λ and λ_2 are insignificantly different from zero in the two factor model,[*] but estimates of all the other parameters (α, β, σ, σ_2, k, k_2 and θ) are significant. The estimated seasonality function ϕ is shown in Figure 6.3, where we can see that in the winter time the spread is more valuable than in summer. Since crude oil does not exhibit seasonality, the seasonality of the crack spread is caused by the seasonality of heating oil.[†] The *mean error* (ME) and the *root mean squared error* (RMSE) are plotted in Figure 6.4. The large pricing errors from 1990 to 1991 are due to the Gulf War.

The asymptotic estimate of the standard deviation of the spread is $2.42. Also, assuming zero correlation between the two factors, we can examine the ratio of the short-end and long-end variance – A1:A2 in (6.15) – as time goes to infinity. It is about 2.5:1 in this example. To see whether the **y** factor provides a statistical improvement, we can compare the differences in log-likelihood scores with and without the **y** factor. The relevant statistic for this comparison is the chi-squared *likelihood ratio test* statistic (Hamilton, 1994) with three degrees of freedom and the 99th percentile of this distribution is 11.34. We thus ran the calibration with only the **x** factor and obtained its likelihood. Given that the likelihood ratio statistic is 282, the improvements provided by the **y** factor are highly significant. The correlation between the filtered long and short factors is 0.0056, which is consistent with the assumption that the correlation between these two factors is zero.

6.5.4 Futures Spread Option Valuation

The average half-life is about 7 months for the **x** factor so that using the methodology of this paper is appropriate to valuing an option longer than 7 months.

TABLE 6.3   Parameter Estimates for the Two Factor Model

	k	θ	λ	σ	k_2	λ_2	σ_2	α	β	ξ
Value	3.39	3.58	−1.33	5.32	0.66	0.25	1.63	1.61	−1.47	0.38
Std. error	0.13	0.05	1.44	0.08	0.27	0.41	0.13	0.08	0.18	0.06
Log likelihood	−771									

Note: We assume the correlation ρ between the two factors is zero.

---

[*] Probably due to alternating periods of spread backwardation and contango implying corresponding periods of positive and negative market prices of risk due to shifts in input and output supply/demand balances (Geman, 2005a).
[†] Heating oil is known to be more expensive in winter than in summer due to high demand in the winter.

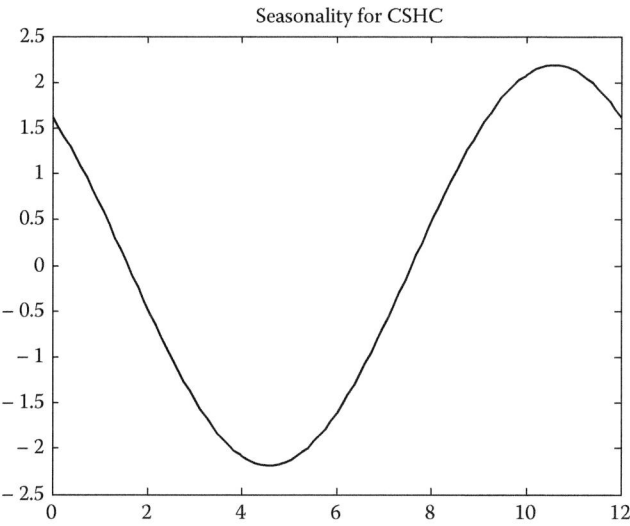

FIGURE 6.3   The estimated seasonality function φ for the crack spread.

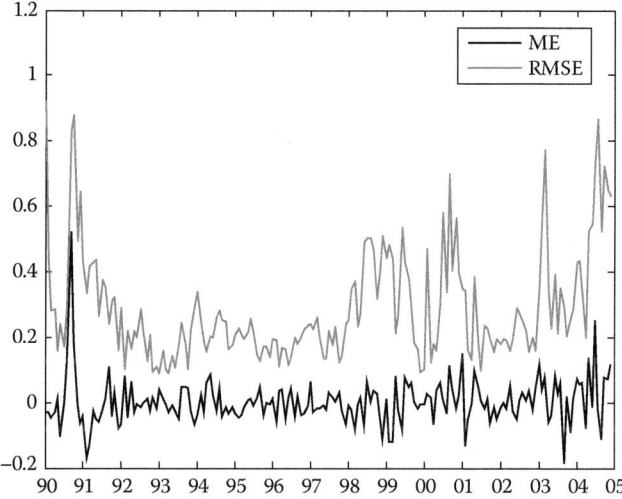

FIGURE 6.4   The mean pricing error (ME) and root mean squared error (RMSE) for the crack spread.

For comparison it is necessary to calculate option values from a model which ignores the cointegration effect. Thus, we must simulate both the crude and heating oil futures prices and then calculate the spread option value* (a two price method). We utilize the Gibson and Schwartz (1990) model for each commodity. Appendix A gives a detailed justification of this method.

On 3 January 2005 the HO06N (heating oil future with maturity July 2006) contract had a value 44.2 ($/Barrel) and implied volatility 29.9%; on the same day the CL06N (crude oil future with maturity July 2006) traded at 39.78 ($/Barrel) and implied volatility 28.3%. Using the parameters in Tables 6.3–6.5 give, respectively, the European option values and deltas on the *futures* crack spread with maturity July 2006 on this date. As a comparison, we also list the option values and deltas for the one factor model

---

* Since there is no analytical solution when the strike is not zero a convenient way to calculate the option value is by Monte Carlo simulation.

TABLE 6.4 Comparison of Crack Spread Option Values

Strike ($)	2	3	4	5	6
One factor spread model	2.41	1.61	0.98	0.52	0.24
Two factor spread model	2.48	1.72	1.10	0.64	0.33
Two price model	3.87	3.32	2.83	2.39	2.03

TABLE 6.5 Comparison of Crack Spread Option Deltas

Strike ($)	2		3		4		5		6	
Underlying	HO	CL	HO	CL	HO	CL	HO	CL	HO	CL
One factor spread model	0.84	−0.84	0.72	−0.72	0.55	−0.55	0.37	−0.37	0.21	−0.21
Two factor spread model	0.81	−0.81	0.69	−0.69	0.54	−0.54	0.39	−0.39	0.24	−0.24
Two price model	0.64	−0.56	0.57	−0.53	0.55	−0.43	0.43	−0.42	0.41	−0.36

HO, heating oil; CL, crude oil.

(with $y := 0$). Since as previously noted the correlation between energy price returns is difficult to obtain andvery volatile, when simulating the two prices we here assume $\hat{\rho}_{12}$ to be the constant 20 year correlation between the crude and heating oil 1 month futures prices (0.89 in our calibration).

From Table 6.4 we can see that the option value from the one factor model is typically smaller than that from the two factor model and the latter is much smaller than that from the two price model. Since this latter model does not consider mean-reversion (the cointegration) of the spread, its spread distribution at maturity is wider than that of a cointegrated model and thus yields a larger option value. Put simply, a non-cointegrated model ignores the long run equilibrium between crude and heating oil prices and thus *overprices* the option. In Table 6.5, both the one factor and two factor models yield an equal volume hedge but the two price model does not. As is well known, the less disperse the underlying terminal distribution, the more sensitive are the option deltas to the strike prices.* Thus the one-factor model yields the most sensitive deltas and the two price model has the least sensitive deltas amongst the three models.

6.6 Location Spread: Brent/WTI Crude Oil

We define the *location spread* as the price of WTI crude oil (CL) minus the price of the Brent blend crude oil (ITCO). WTI is delivered in the USA and Brent in the UK.

6.6.1 Data

NYMEX daily futures prices of WTI crude oil were described in the previous example. The daily Brent futures prices are from January 1993 to January 2005. The time to maturity of the Brent futures contracts range from 1 month to about 3 years. As in the previous example, monthly data is used to test for the unit root in Brent oil prices. We also create a monthly long end location spread with time to maturity of 1 year ranging from January 1993 to January 2005. In order to calibrate our model, we calculate monthly the futures spread with five maturities from January 1993 to January 2005. The five contracts involved are the 1, 3, 6, 9 and 12 month futures.

---

* The *sensitivity* is defined as the ratio of the change of the deltas to the change of the strike prices.

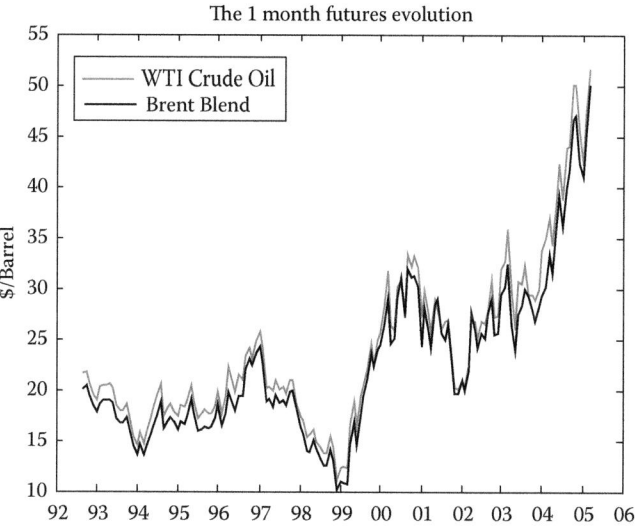

FIGURE 6.5 The 1 month futures prices of WTI and Brent crude oil.

TABLE 6.6 Cointegration Test for Brent and Wti Crude Oil

	No. of observations	Lag (p)	π_1 or χ_1	t-Stat $_1$
Engle–Granger two step tests				
Brent blend	152	6	0.04	2.48
WTI	257	6	0.02	1.03
Location spread	152	6	−0.28	−3.45*
Johansen test				
Likelihood ratio trace statistic: 26.74 (significant at 99%)[a]				

[a]*Note:* that the null hypothesis is no cointegration, which is rejected at the 99% confidence level.
*Significant at the 1% level.

6.6.2 Unit Root and Cointegration Tests

As from the previous example we know that the WTI crude oil price follows a unit root process, in this example we need only conduct the ADF test on Brent crude oil prices. Figure 6.5 shows the 1 month futures prices of WTI crude oil and Brent blend.

Similar to WTI crude oil, the Brent blend price is also a unit root process, but the corresponding location spread appears to be a mean reverting process. This again suggests the existence of a long run equilibrium in the data. The Johansen test also confirms the existence of the cointegration relationship (see Table 6.6). The estimate of γ in (6.7) is strongly negative, so that the market appears to expect the spot spread to be mean reverting in the risk neutral measure (Table 6.7). The 1 year and 1 month spread evolution is depicted in Figure 6.6.

Hence both the market and risk neutral analyses support the mean reversion of the spot spread.

6.6.3 Model Calibration

We did not find evidence of seasonality in the location spread. Thus, we set $\varphi(t) :\equiv 0$. Table 6.8 lists the calibration results for our model. Figure 6.7 shows the pricing errors of the model. We see that the

TABLE 6.7 Regression (6.7) Parameter Estimates for the Location Spread

	ζ	γ
Value	0.92	−0.70
t-Stat.	11.57*	−15.54*
No. of observations	152	
R^2	63.65%	

*Significant at the 1% level.

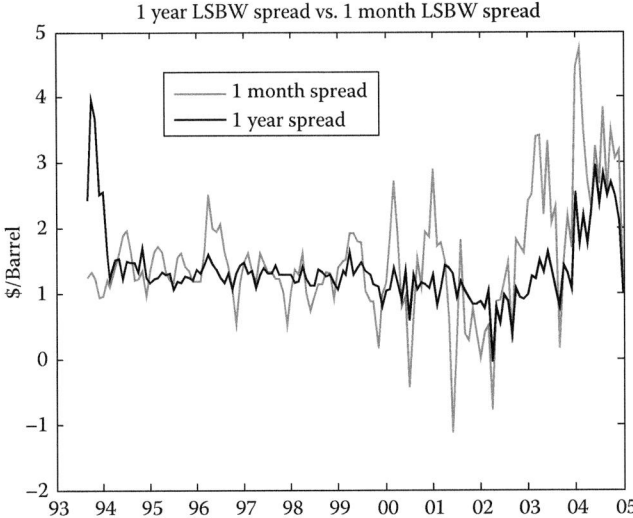

FIGURE 6.6 The 1 month and 1 year location spreads.

asymptotic standard deviation of the spread is estimated to be \$2.98. The ratio of short-end to the long-end variance (A_1:A_2 in (6.9)) is 1:5, i.e. the long end (second factor) movement of the spread accounts for much more variance than that of the short end (first factor). From the likelihood ratio test the two factor model is significantly better than the one factor model in explaining the observed LSBW spread data.*

6.6.4 Futures Spread Option Valuation

The average half-life is about 2.5 years, thus the methods in this paper should be used to price an option of maturity longer than 2.5 years.

On 1 December 2003 the ITCO06Z (Brent blend crude oil future with maturity December 2006) contract had a value 24.62 (\$/Barrel) and implied volatility 19.2%; on the same day the CL06Z (WTI crude oil future with maturity December 2006) had a value 25.69 (\$/Barrel) and implied volatility 19.2%. Tables 6.9 and 6.10, respectively, show the European option values and deltas on the futures spread with maturity December 2006 using the one and two factor model and the two price model. Note that $\hat{\rho}_{12}$ is estimated to be 0.94 for the 1 month WTI and Brent oil data. The correlation between the two factors is 0.04 which is again consistent with our zero correlation assumption.

---

* The likelihood ratio statistic is 628.

TABLE 6.8 Parameter Estimates of the Two Factor Model

	k	θ	λ	σ	k_2	λ_2	σ_2	ξ
Value	5.04	0.00	0.50	3.93	0.14	0.21	1.45	0.12
Std. error	0.28	0.00	1.18	0.28	0.02	0.28	0.10	0.03
Log likelihood	−23.8							

Note: We assume the correlation ρ between the two factors is zero.

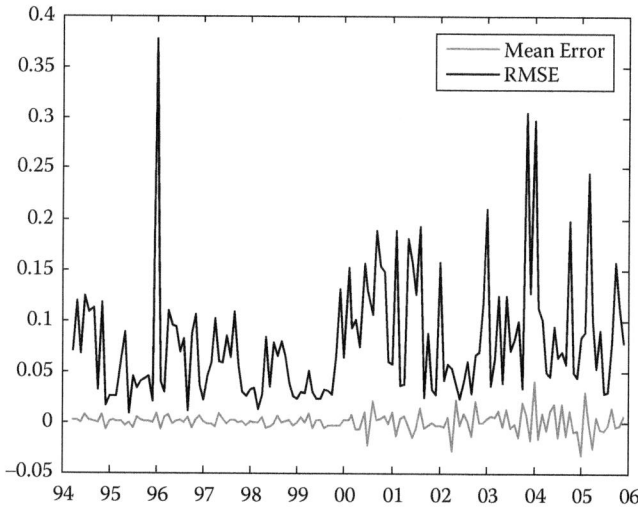

FIGURE 6.7 The mean pricing error (ME) and root mean squared error (RMSE) for the location spread.

TABLE 6.9 Comparison of Location Spread Option Values

Strike ($)	−1	0	1	2	3
One factor spread model	1.96	1.19	0.60	0.24	0.08
Two factor spread model	2.11	1.43	0.89	0.49	0.24
Two price model	2.27	1.61	1.10	0.71	0.45

TABLE 6.10 Comparison of Location Spread Option Deltas

Strike ($)	−1		0		1		2		3	
Underlying	CL	ITCO	CL	ITCO	CL	ITCO	CL	ITCO	CL	ITCO
One factor spread model	0.81	−0.81	0.68	−0.68	0.47	−0.47	0.26	−0.26	0.10	−0.10
Two factor spread model	0.74	−0.74	0.62	−0.62	0.47	−0.47	0.32	−0.32	0.19	−0.19
Two price model	0.69	−0.63	0.55	−0.53	0.48	−0.43	0.42	−0.34	0.34	−0.30

CL, WTI crude oil; ITCO, Brent crude oil.

Table 6.9 shows that, as in the previous example, the option value of the one factor model is typically smaller than that of the two factor model and the latter is much smaller than the two price model.

As before, by ignoring cointegration the two price model tends to over value the long term option. We obtain a pattern of deltas similar to the previous example (see Table 6.5) and the explanation for this is the same.

6.7 Conclusion

In this paper we have developed spread option pricing models for the situation when the two underlying prices of the spread are cointegrated. Since the cointegration relationship is important for the long run dynamical relationship between the two prices, contingent claim evaluation based on spreads should take account of this relationship for long maturities. We model the spread process *directly* using a two factor model, i.e. we model directly the dynamic deviation from the long run equilibrium which cannot be specified correctly by modelling the two underlying assets separately. We also propose two methods (risk neutral and market) for testing data for mean-reversion of the spread process. The corresponding spread option can be priced and hedged analytically. In order to illustrate the theory we study two examples – of crack and location spreads, respectively. Both spread processes are found to be mean reverting. From likelihood ratio tests the second **y** factor is found to be important in explaining the crack and location spread data. The option values from our model are quite different from those of standard models, but they are consistent with the practical observations of Mbanefo (1997).

Acknowledgements

We are grateful to HelyetteGeman and an anonymous referee for comments which materially improved the paper. This chapter is reprinted with permission from the *Journal of Banking and Finance*, 2011, **35**, 639–652.

References

Alexander, C., 1999. Correlation and cointegration in energy markets. In: Kaminsky, V. (Ed.), *Managing Energy Price Risk*, second ed. Risk Publications, London, pp. 573–606.

Arnold, L., 1974. *Stochastic Differential Equations: Theory and Applications*. John Wiley & Sons, New York.

Bates, D., 1996. Jump and stochastic volatility: Exchange rate processes implicit in deutsche mark options. *Review of Financial Studies* 9, 69–107.

Bessembinder, H., Coughenour, J., Seguin, P., Smoller, M., 1995. Mean-reversion in equilibrium asset prices: Evidence from the futures term structure. *Journal of Finance* 50, 361–375.

Black, F., 1976. The pricing of commodity contracts. *Journal of Financial Economics* 3, 167–180.

Brennan, M., Crewe, N., 1997. Hedging long maturity commodity commitments with short-dated futures contracts. In: Dempster, M.A.H., Pliska, S.R. (Eds.), *Mathematics of Derivative Securities*. Cambridge University Press, New York, pp. 165–189.

Carmona, R., Durrleman, V., 2003. Pricing and hedging spread options. *SIAM Review* 45, 627–685.

Clewlow, L., Strickland, C., 1999. Valuing Energy Options in a One Factor Model Fitted to Forward Prices.Working Paper, University of Technology, Sydney, Australia.

Dempster, M., Hong, S., 2000. Pricing spread options with the fast Fourier transform. In: Geman, H., Madan, D., Pliska, S.R., Vorst, T. (Eds.), *Mathematical Finance*, Bachelier Congress. Springer-Verlag, Berlin, pp. 203–220.

Dicky, D., Fuller, W., 1979. Distribution of the estimates for autoregressive time series with a unit root. *Journal of the American Statistical Association* 74, 427–429.

Durbin, J., Koopman, S., 1997. Monte Carlo maximum likelihood estimation of non-Gaussian state space models. *Biometrika* 84, 669–684.

Engle, R., Granger, C., 1987. Cointegration and error correction: Representation, estimation and testing. *Econometrica* 55, 251–276.

Eydeland, A., Geman, H., 1998. Pricing power derivatives. RISK, January.

Fama, E., French, K., 1988. Permanent and temporary components of stock prices. *Journal of Political Economy* 96, 246–273.

Fuller, W., 1976. *Introduction to Statistical Time Series*. John Wiley & Sons, New York.

Geman, H., 2005a. *Commodities and Commodities Derivatives: Modelling and Pricing for Agriculturals Metals and Energy*. John Wiley & Sons, Chichester.

Geman, H., 2005b. Energy commodity prices: Is mean reversion dead? *Journal of Alternative Investments* 8, 34–47.

Geman, H., Nguyen, V., 2005. Soybean inventories and forward curves dynamics. *Management Science* 51, 1076–1091.

Gibson, R., Schwartz, E., 1990. Stochastic convenience yield and the pricing of oil contingent claims. *Journal of Finance* 45, 959–976.

Girma, P., Paulson, A., 1999. Risk arbitrage opportunities in petroleum futures spreads. *Journal of Futures Markets* 19, 931–955.

Gregory, A., Hansen, B., 1996. Residual-based tests for cointegration in models with regime shifts. *Journal of Econometrics* 70, 99–126.

Hamilton, J., 1994. *Times Series Analysis*. Princeton University Press.

Harvey, A., 1989. *Forecasting Structural Time Series Models and the Kalman Filter*. Cambridge University Press, Cambridge.

Hilliard, J., 1999. Analytics underlying the metallgesellschaft hedge: Short term futures in a multiperiod environment. *Review of Finance and Accounting* 12, 195–219.

Hilliard, J., Reis, J., 1998. Valuation of commodity futures and options under stochastic convenience yields, interest rates, and jump diffusions in the spot. *Journal of Financial and Quantitative Analysis* 33, 61–86.

Hull, J., White, A., 1990. Pricing interest-rate derivative securities. *Review of Financial Studies* 3, 573–592.

Johansen, S., 1991. Estimation and hypothesis testing of cointegration vectors in Gaussian vector autoregressive models. *Econometrica* 59, 1551–1580.

Kirk, E., 1995. Correlation in the energy markets. In: *Managing Energy Price Risk*. Risk Publications, London, pp. 71–78.

Margrabe, W., 1978. The value of an option to exchange one asset for another. *Journal of Finance* 33, 177–186.

Mbanefo, A., 1997. Co-movement term structure and the valuation of energy spread options. In: Dempster, M.A.H., Pliska, S.R. (Eds.), *Mathematics of Derivative Securities*. Cambridge University Press, Cambridge, pp. 99–102.

Neuberger, A., 1998. Hedging long-term exposures with multiple short-term futures contracts. *Review of Financial Studies* 12, 429–459.

Pan, J., 2002. The jump-risk premia implicit in options: Evidence from an integrated time-series study. *Journal of Financial Economics* 63, 3–50.

Richter, M., Sorensen, C., 2002. Stochastic Volatility and Seasonality in Commodity Futures and Options: The Case of Soybeans. Working Paper, Copenhagen Business School.

Routledge, B., Seppi, D., Spatt, C., 2000. Equilibrium forward curves for commodities. *Journal of Finance* 55, 1297–1338.

Samuelson, P., 1964. Theoretical notes on trade problems. *Review of Economics and Statistics* 46, 145–154.

Schwartz, E., 1997. The stochastic behaviour of commodity prices: Implications for valuation and hedging. *Journal of Finance* 52, 923–973.

Schwartz, E., Smith, J., 2000. Short-term variations and long-term dynamics in commodity prices. *Management Science* 46, 893–911.

Soronow, D., Morgan, C., 2002. Modelling Locational Spreads in Natural Gas Markets. Working Paper, Financial Engineering Associates. http://www.fea.com/resources/pdf/a_modeling_locational .pdf.

Villar, J., Joutz, F., 2006. The Relationship Between Crude Oil and Natural Gas Prices. Working Paper, Department of Economics, The George Washington University.

Appendix

Two Price Method of Simulating Spreads

In the risk neutral measure, Hilliard and Reis (1998) show that the futures price $H_i(t, T)$, $i = 1, 2$, in the Gibson–Schwartz model (6.1) follows

$$\frac{dH_i(t,T)}{H_i(t,T)} = \sigma_{i,1}dW_{i,1} - \frac{1 - e^{-k_t(T-t)}}{k_i}\sigma_{i,2}dW_{i,2}, \tag{A.1}$$

where T is the maturity date of the futures and $\rho_{12}\, dt = E[dW_{i,1}\, d\text{-}W_{i,2}]$. Thus by integrating (1.1) the spot prices $S_{i,\,T}$ at maturity T are given by

$$S_{i,T} = H_i(T,T) = \exp\left(-\frac{1}{2}v_i^2 + v_i \cdot \varepsilon_i\right), \tag{A.2}$$

where

$$\begin{aligned}
v_i^2 &:= \int_t^T \left[\sigma_{i,1}^2 + \sigma_{i,2}^2\left(\frac{1 - e^{-k_i(T-S)}}{k_i}\right)^2 - 2\rho_{12}\sigma_{i,1}\sigma_{i,2}\frac{1 - e^{-k_i(T-S)}}{k_i}\right]ds \\
&= \sigma_{i,2}^2(T - t) - \frac{2\sigma_{i,1}\sigma_{i,2}\rho_{12}}{k_i}\left(T - t - \frac{1 - e^{-k_1(T-t)}}{k_i}\right) \\
&\quad + \frac{\sigma_{i,2}^2}{k_i^2}\left(T - t - \frac{2 - 2e^{-k_1(T-t)}}{k_i} + \frac{1 - e^{-k_1(T-t)}}{2k_i}\right)
\end{aligned} \tag{A.3}$$

and ε_i is a normal random variable with mean zero and standard deviation one.

Defining the *average* volatility per unit time $\sigma_{av} := \sqrt{v^2 / (T - t)}$ in terms of the cumulative volatility of each, (A.2) shows that the simulation of each spot price S_T is exactly the same as simulating a Black (1976) driftless GBM model with volatility σ_{av}. Thus the options on S_T can also be priced by the Black (1976) formula with σ_{av} as the input volatility. Moreover, σ_{av} can be *observed* directly from the market as the implied volatility of an option on a future with the same maturity as the futures contract. However in order to simulate the spread, we also need to know the estimated correlation $\hat{\rho}_{12}$ between ε_1 and ε_{12} which, as discussed in the introduction, is notoriously volatile and difficult to obtain. In Sections 6.5 and 6.6 of this paper we assume $\hat{\rho}_{12}$ to be a *constant* represented by the historical correlation between the one monthfutures prices.

2

Other
Commodities

7

A Rough Multi-Factor Model of Electricity Spot Prices

Mikkel Bennedsen

7.1 Introduction

This paper proposes a new mathematical model of electricity spot prices which parsimoniously accounts for the most important statistical properties of such prices. These *stylized facts* are empirical features consistently observed in the time series of electricity prices and at least four stylized facts are by now well-documented in the literature: (i) seasonality which is attributable to month-to-month variation in prices, mainly due to the seasons, and day-to-day variations, mainly due to weekly dependence of demand; (ii) spikes which are large up- or downwards movements followed by rapid reversion to previous levels often attributed to the non-storable nature of electricity combined with the fact that the energy grid needs to be in complete balance at all times. Because of very inelastic demand the price can surge (or drop) suddenly if the supply is limited (or in surplus) e.g. due to plant outages or unforeseen weather conditions. (iii) Extreme variability of prices as compared to traditional financial time series such as storable commodities or stocks; and (iv) mean-reversion of prices which has been found by several authors such as Weron and Przybylowicz (2000), Kaminski (2004) and Pilipovic (2007). The wish

DOI: 10.1201/9781003265399-9

to incorporate these stylized facts into a mathematical model of electricity spot prices has motivated a large literature starting with the seminal work of Schwartz (1997) who introduced a model of commodity prices based on the Ornstein-Uhlenbeck (OU) process. Since Schwartz (1997), this model has been extended in several ways and applied to electricity price series, e.g. the jump-diffusion model of Cartea and Figueroa (2005) and the threshold model of Geman and Roncoroni (2006). Benth et al. (2007) generalized the Ornstein-Uhlenbeck framework to the so-called *multi-factor model*, where the (log-)price of electricity is modelled as

$$S_t = \Lambda_t + X_t + Y_t, \quad t \geq 0$$

where Λ is a seasonal term, X is the *base signal* accounting for smaller day-to-day variations in the price, and Y is a *spike signal* accounting for large up- or down-movements and the ensuing rapid reversion to previous levels. A typical approach is to model Λ by a deterministic sinusoidal function, X as a Gaussian OU process and Y as a non-Gaussian (i.e. Lévy driven) OU process (e.g. Meyer-Brandis and Tankov, 2008; Benth et al., 2012). This model has been particularly useful as it elegantly accounts for the stylized facts (i)-(iv).

Meanwhile, a possible fifth stylized fact has been found in a different strand of literature, namely that the time series of electricity spot prices exhibit (v) rough characteristics (often termed *antipersistence* in the literature). Specifically, a number of studies show that the *roughness index* of electricity spot price time series is negative, meaning that the paths of the spot prices are very *rough* as compared to a model based on the Brownian motion such as the Brownian-driven OU process. This was for instance found for the Nordic market, Nord Pool, in Simonsen (2002) and Erzgräber et al. (2008), for the Spanish market in Norouzzadeh et al. (2007), and the for Czech market in Kristoufek and Lunackova (2013). Although roughness/antipersistence of the price series on several energy markets is by now well established, very few authors have utilized their findings to actually employ and estimate a rough model for the spot prices. A notable exception is Rypdal and Løvsletten (2013) who compared a rough model based on a multifractal random walk with a (non-rough) model of Ornstein-Uhlenbeck type and concluded that the latter best described the Nord Pool spot prices from May 1992 to August 2011.

The main contribution of the present paper is to add to the above literature and merge the multi-factor approach with the rough approach. In short, we propose a rough multi-factor model of electricity spot prices, i.e. a multi-factor model where the base signal, X, is modelled by a rough process. This will allow us to account for *all* the stylized facts (i)-(v), contrary to the above mentioned previous studies employing either a multi-factor model (accounting for stylized facts (i)-(iv)) or a rough/antipersistent model (accounting for stylized fact (v) only). What is more, our setup accomodates *testing*, in a rigorous statistical sense, for the presence of roughness in the price series; this is a novel contribution in the literature, as previous studies have focused on point estimates of the roughness index in lieu of confidence intervals and hypothesis testing. In fact, as we shall see, our approach provides a direct statistical hypothesis test of whether the base signal is best described by a Brownian-driven OU process (the null) or of a rough process (the alternative). Additionally, the multi-factor setup allows us to account for spikes *before* estimating and making inference on the roughness index; this is crucial because spikes can be detrimental to the estimation as we illustrate in the Appendix.

As an application of the model we consider a simple forecasting exercise, showing that including a rough component provides superior short-run forecasts of the base signal, as compared to three benchmark models. The exercise indicates that agents in the energy markets might improve their forecasts of prices by including a rough component in their models.

The rest of the paper is structured as follows. In Section 7.2, we take a closer look at the six data sets we consider. Section 7.3 details the concept of roughness as it relates to time series of electricity prices. Section 7.4 introduces the modelling framework we propose; this section also contains an estimation procedure for the model. Section 7.5 is concerned with the implications of our model as it relates to arbitrage and to forecasting. In Section 7.6, we present our empirical work where we find evidence of roughness in five out of the six markets considered. Section 7.7 discusses the findings and concludes. The Appendix contains an in-depth look at the estimation procedure by detailing the individual steps using the German EEX data set. Using simulations, the Appendix also briefly investigates possible sources of biases when estimating the roughness index of electricity prices and illustrates how the multi-factor framework helps in alleviating these.

7.2 The Data

Electricity markets are organized as day-ahead markets where participants submit their bid/ask price for a given amount of electricity in a given hour of the day. The exchange then matches supply and demand, which yields 24 spot prices, one for each hour, for the following day.[*] In the following we define the 'spot price' as the *peak load price* which is computed as the mean of the spot prices from 9 a.m. to 8 p.m., the hours where load (demand) is highest. We consider data from six different European exchanges, as shown in Table 7.1 and plotted in Figure 7.1. All data series are excluding weekends as these tend not to add a lot of information not contained in the Friday price (Meyer-Brandis and Tankov, 2008). It is well known that electricity prices are subject to strong seasonal effects and before working with the data we therefore pre-process it by de-seasonalizing as explained in Section 7.4.1 below. Some empirical characteristics of these de-seasonalized prices and their increments are seen in Table 7.2. Looking at the table we see that the (log as well as raw) increments exhibit significant departures from normality. Indeed, the series display some skewness and a large amount of (excess) kurtosis ranging from 9.5 to 650 in the raw increment series and 6.6 to 116 in the log-increments. Naturally then, when conducting a formal test for normality (Jarque-Bera), we reject the null of Gaussianity of the increments in all time series considered.

The question of whether the *levels* of electricity prices constitute a stationary time series has been much debated in the literature. In Table 7.3 we therefore run two tests for a unit root in the data; the Phillips-Perron (PP) test and the augmented Dickey-Fuller (ADF) test with automatic lag selection.[†] Both tests reject the null of a unit root in the raw prices series as well as the log-prices at all reasonable significance levels. That is, we find evidence of stationarity in the de-seasonalized prices and we therefore proceed to model all price series as stationary in the following.

7.3 Roughness in Electricity Prices

In this section, we explain how roughness relates to one-dimensional time series and hence to electricity prices. Let X be a zero-mean, real-valued time series with stationary increments, and define its (second order) *variogram*:

TABLE 7.1 Data Summary

Abbr.	Region	Start	End	Observations
EEX	Germany	Jan. 1, 2008	Jan. 23, 2015	1844
NP	Nordic	Jan. 1, 2008	Jan. 23, 2015	1844
PN	France	Jan. 1, 2008	Jan. 23, 2015	1844
GME	Italy	Jan. 1, 2008	Dec. 19, 2014	1819
UKAPX	United Kingdom	Jan. 1, 2008	Jan. 23, 2015	1844
NLAPX	Netherlands	Jan. 1, 2008	Jan. 23, 2015	1844

---

[*] Unlike the five other markets, which records spot prices every hour, the UK market which we consider in this paper has prices for every 30 minutes, i.e. 48 prices per day.

[†] As implemented in the MFE Toolbox of Kevin Sheppard, http://www.kevinsheppard.com/MFE_Toolbox.

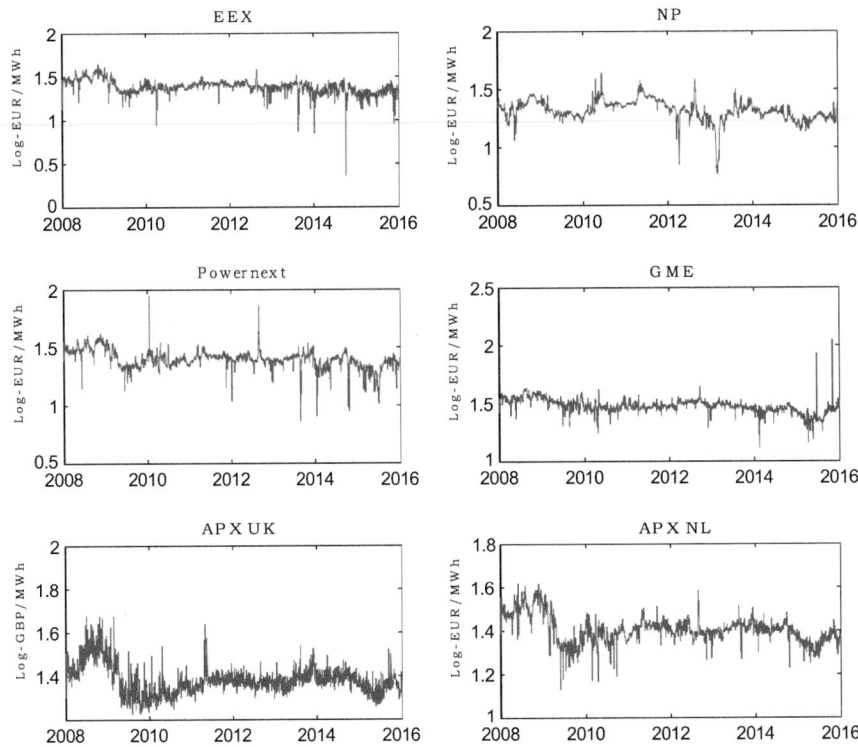

FIGURE 7.1 Illustration of the six data sets: log-price evolution in log-Euros per Mwh (UK APX is in log-British Pounds).

$$\gamma_2(h) := \frac{1}{2}\mathbb{E}\left[\left|X_{t+h} - X_t\right|^2\right], \quad h \in \mathbb{R}$$

The concept of roughness refers to a particular behavior of the variogram for $h \approx 0$. That is, if it holds that

$$\gamma_2(h) = c_2\,|h|^{2\alpha+1} + o\left(|h|^{2\alpha+1+\beta}\right), \quad h \downarrow 0, \tag{7.1}$$

for a contant $c_2 > 0$ and some $\alpha \in (-1/2, 0)$, then we say that X is *rough*. In the following, we will therefore call α the *roughness index* or *roughness parameter*. The reason for this terminology is that when $\alpha < 0$, X will have sample paths which are rougher than the samples paths of the Brownian motion (the Brownian motion has $\alpha = 0$). What is meant by this is that, under quite general assumptions, a process X with roughness parameter α will be Hölder continuous of order γ for all $\gamma \in (0, \alpha + 1/2)$ (see e.g. Proposition 2.1. in Bennedsen, 2016, for a precise statement of this result). It should be noted, that one usually allows for α to take values in $(-1/2, 1/2)$; the values where $\alpha > 0$ will correspond to processes with trajectories *smoother* than those of the Brownian motion. However, as we will see, the electricity data considered in this paper will always give rise to estimates of α which are negative and our focus will therefore be on the rough case, i.e. on $\alpha < 0$. A last thing to note is the well known fact that processes with roughness index $\alpha \neq 0$ are necessarily *not semimartingales*; see Proposition 2.2. in Bennedsen (2016) for a proof and Section 7.5 below for a discussion of the implications of using a rough statistical model for electricity prices.

A typical example of a Gaussian process with non-trivial roughness index is the fractional Brownian motion (fBm) with Hurst parameter $H \in (0, 1)$, corresponding to $\alpha = H - \frac{1}{2}$. Here $\alpha = 0$ (so $H = \frac{1}{2}$)

TABLE 7.2 Descriptive Statistics of Electricity Price Increments

| | Panel A: Raw increments | | | | | | Panel B: Log-increments | | | | | |
	EEX	NP	Powernext	GME	APX UK	APX NL	EEX	NP	Powernext	GME	APX UK	APX NL
nObs	1848	1844	1844	1819	1844	1844	1848	1844	1844	1819	1844	1844
Mean	0.028	0.011	0.023	0.024	0.013	0.015	0.001	0.000	0.000	0.000	0.000	0.000
Min	−61.273	−60.748	−1064.780	−2195.649	−115.513	−50.791	−2.191	−0.573	−2.669	−3.518	−0.860	−0.708
Max	99.152	97.258	1080.133	2182.405	150.305	41.812	2.053	0.884	2.830	3.345	1.292	0.917
SD	9.710	5.790	39.397	79.447	14.430	7.830	0.203	0.099	0.194	0.215	0.184	0.127
Skew	0.558	2.569	0.725	−0.193	0.218	0.018	0.052	0.747	0.349	0.060	0.101	0.421
Kurt	12.549	78.412	616.925	653.490	21.017	9.474	23.412	15.590	57.139	115.594	6.610	10.128
JB	7114	438737	28943260	32052661	24942	3219	32066	12343	225113	960318	1004	3956
P-value	0.001	0.001	0.001	0.001	0.001	0.001	0.001	0.001	0.001	0.001	0.001	0.001

Descriptive statistics for the increments of the de-seasonalized (see Section 7.4.1) electricity prices for the various power exchanges. Weekends are excluded. Panel A (left) is for the raw increments, while Panel B (right) is for the log-increments (log-returns). "nObs" denotes the number of observations, while "JB" is the Jarque-Bera test statistic.

TABLE 7.3 Testing for Unit Roots in Electricity Price Series

	Panel A: *Raw prices*						Panel B: *Log-prices*					
	EEX	NP	Powernext	GME	APX UK	APX NL	EEX	NP	Powernext	GME	APX UK	APX NL
PP	−14.486	−9.594	−32.437	−40.137	−17.399	−11.086	−18.391	−7.448	−16.002	−23.681	−15.453	−12.147
P-value	0.001	0.001	0.001	0.001	0.001	0.001	0.001	0.001	0.001	0.001	0.001	0.001
ADF	−4.136	−4.481	−10.276	−19.057	−2.696	−3.365	−5.832	−4.830	−6.206	−3.990	−3.046	−3.320
P-value	0.000	0.000	0.000	0.000	0.007	0.000	0.000	0.000	0.000	0.000	0.003	0.000

Unit root tests of the de-seasonalized (see Section 7.4.1) electricity prices for the various power exchanges. Weekends are excluded. Panel A (left) is for the raw prices, while Panel B (right) is for the log-prices. PP stands for the Phillips-Perron test for a unit root, while ADF is the augmented Dickey-Fuller test with automatic lag selection.

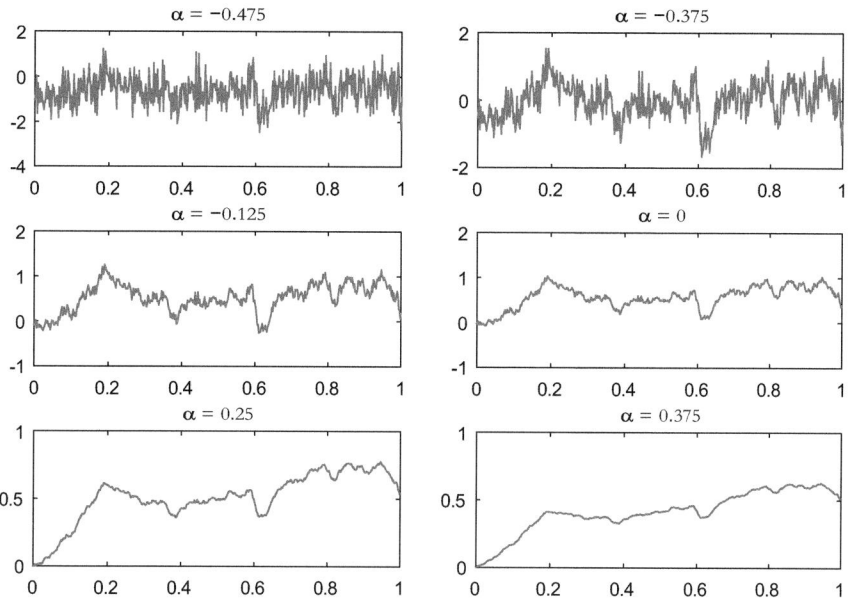

FIGURE 7.2 Simulated paths of an fBm for varying values of the roughness index α. Recall that the Hurst index is $H = \alpha + \frac{1}{2}$, so that $\alpha = 0$ corresponds to the process being a standard Brownian motion. The same random numbers were used in the six simulation, hence the only difference is the roughness of the simulated paths as dictated by the value of α.

implies that the fBm is a standard Brownian motion. To illustrate the concept of roughness visually, several simulated fBm paths for different values of α are shown in Figure 7.2. The differing roughness of the paths, when α changes, is evident: we see how a low α corresponds to very rough paths while positive values results in increasingly smooth paths. Relating this back to time series of electricity prices, other authors have often found $\alpha < 0$ for these data sets, i.e. that the paths of the spot prices are rougher than what a Brownian motion-based model, such as the Brownian-driven OU model, would suggest. The next section introduces a model for electricity prices which is compatible with this observation.

7.4 A Rough Multi-Factor Modelling Framework

Consider the *multi-factor decomposition* of the electricity spot price

$$S_t = \Lambda_t + X_t + Y_t, \quad t \geq 0, \tag{7.2}$$

where $\Lambda = \{\Lambda_t\}_{t \geq 0}$, $X = \{X_t\}_{t \geq 0}$, and $Y = \{Y_t\}_{t \geq 0}$ are the seasonal component, the base signal and the spike signal respectively. This decomposition is the basis of the multi-factor model and has been extensively studied and succesfully employed to electricity prices in several studies, such as Benth et al. (2007), Meyer-Brandis and Tankov (2008) and Hayfavi and Talasli (2014). The decomposition (7.2) can be seen as both an arithmetic and geometric model of the spot price by letting $S_t = P_t$ or $S_t = \log P_t$ accordingly, where $P = \{P_t\}_{t \geq 0}$ denotes the spot price of electricity. Note, that only the arithmetic model will allow for the possibility of negative spot prices, which has recently been observed in some markets. However, as the data we work with is the peak load price, which is an average of the prices in 12 consecutive hours, and as negative prices do not tend to persist for many hours at the time, we do not observe any negative (or zero) prices in any of the time series under consideration. Hence, we proceed to work with the geometric model, $S_t = \log P_t$, in the following. We want to note and stress that we also performed the analysis using the arithmetic model (i.e. using non-log prices), which did not yield qualitative differences and actually resulted in even rougher paths (i.e. a more negative roughness index α) than what we will see in the following. That is, the conclusions drawn in this paper concerning roughness of electricity prices hold a fortiori when considering an arithmetic model. As mentioned above, in the multi-factor framework the processes X and Y in (7.2) are usually modelled as (possibly non-Gaussian) OU processes but here we take a more general approach when modelling X, as explained in Section 7.4.2 below.

7.4.1 Modelling the Seasonal Component

Following Meyer-Brandis and Tankov (2008) we will take the seasonal component, Λ, to be a deterministic sinusoidal function plus an intercept and a trend. In full generality one could consider a more general stochastic process as a model for the seasonal component, which would, most likely, be more realistic. However, since our focus is on the rough nature of the prices, we will not pursue this here. We thus model the seasonal component as

$$\Lambda_t = c_1 + c_2 t + a_1 \sin\left(\frac{2\pi t}{5}\right) + a_2 \cos\left(\frac{2\pi t}{5}\right) + a_3 \sin\left(\frac{2\pi t}{260}\right) + a_4 \cos\left(\frac{2\pi t}{260}\right), \quad t \geq 0, \tag{7.3}$$

where $c_1, c_2, a_i \in \mathbb{R}$, $i = 1, \ldots, 4$, are unknown coefficients to be estimated. The first two terms in (7.3) model the overall level and trend, while the following two takes care of the weekly seasonal component and the last two are for the yearly seasonal pattern (recall that weekends have been removed).

Inspecting the price series in Figure 7.1 and given that we reject the presence of a unit root in the series, there seems to be no evidence of any trend in the data; indeed, in Section 7.6 we will obtain estimates of c_2 which are practically zero. One could therefore argue that the linear trend term in the seasonal function is superfluous, but since the specification (7.3), including the linear trend component, is standard in the literature, we will keep it in the sequel.

7.4.2 Modelling the Base Component

As explained in the introduction, several earlier studies have found evidence of rough characteristics of the prices and we therefore propose to model the base component X as a rough process. A mathematical process which is particularly suited for our needs is the *volatility modulated Brownian semistationary* (*BSS*) *process* (Barndorff-Nielsen and Schmiegel, 2007, 2009),

$$X_t = \int_{-\infty}^{t} (t-s)^{\alpha} e^{-\lambda(t-s)} \sigma_s dW_s, \quad t \geq 0, \tag{7.4}$$

where $\lambda > 0$, $\alpha \in \left(-\frac{1}{2}, \frac{1}{2}\right)$ and $\sigma = \{\sigma_t\}_{t \geq 0}$ is a stochastic volatility process, possibly correlated with the standard Brownian motion W. As shown in Proposition 2.1. of Bennedsen et al. (2015) the roughness index of X, as defined in Section 7.3, is equal to α, and the estimators we consider later in this paper will therefore be valid for this model. We only consider X_t for $t \geq 0$; the integration from minus infinity in (7.4) is purely for modelling reasons since it causes X to be a (strictly) stationary process as long as σ is stationary as well, which we assume henceforth. As discussed in Section 7.2 and evidenced in Table 7.2, stationarity is a desirable feature for our purposes, as the (deseasonalized) electricity prices we work with are found to be stationary. A few additional comments on the *BSS* process X are in order. First off, (7.4) extends the popular OU model, which is the go-to process in the multi-factor literature. To see this, consider (7.4) with $\alpha = 0$; now X indeed reduces to the (volatility modulated) OU process. This will provide the basis of our statisitical test for roughness later: we will test the null hypothesis $H_0 : \alpha = 0$ against the alternative $H_a : \alpha < 0$, which gives rise to a statistically rigorous way to decide if the base signal is best described by a rough $\mathcal{BSS}$ process or whether one can do away with an OU process as is the standard choice in the multi-factor literature. In other words, this methodology provides a direct test comparing the OU model with the rough $\mathcal{BSS}$ model.

Secondly, (7.4) implies that the autocorrelations of the process decay asymptotically with exponential rate λ. The parameter λ is given in inverse units of time (days) and λ^{-1} is termed the *characteristic time scale* of the process (e.g., Rypdal and Løvsletten, 2013). An interesting question is to assess whether the electricity price data actually display a characteristic time scale; following the suggestion of an anonymous referee, we explore this question further in Section 7.4.2.1 below. In particular, we propose a simulation-based test for whether or not a characteristic time scale exists in the data.

Lastly, the framework in equation (7.4) allows for non-Gaussianity through volatility modulation; in fact, in the multi-factor literature, the base signal is often modelled as a Gaussian OU process, despite studies such as Benth et al. (2012) finding evidence of non-Gaussianity. Indeed, in Section 7.6.3 we also find evidence of non-Gaussianity. The Section 7.4.2.2 below details how the $\mathcal{BSS}$ framework can accomodate this through the stochastic volatility component σ.

7.4.2.1 Models with and without Characterisic Time Scale in Their Autocorrelations

In the literature on statistical properties of electricity prices, there are opposing views on which mathematical frameworks are adequate for modelling the data. On the one hand, authors such as Weron and Przybylowicz (2000) and Simonsen (2002) argue for a rough ("antipersistent") model without characteristic time scale in its autocorrelation function, such as the fractional Brownian motion. On the other hand, the OU-based models (e.g., Schwartz, 1997; Cartea and Figueroa, 2005; Meyer-Brandis and Tankov, 2008; Benth et al., 2012) implicitly assumes the presence of a characteristic time scale; however, very few studies have presented explicit evidence that such a time scale should actually be included in the model. A notable exception is Rypdal and Løvsletten (2013), where the authors find evidence that the Nordic energy market do display a characteristic time scale; however, Rypdal and Løvsletten (2013) are unable to conclude whether this finding is simply caused by seasonal effects in the data. The multi-factor setup allows us to bypass such doubts; since we filter out the seasonality (and spikes) in a preliminary step, any characteristic time scale found in the data will be in the base signal, i.e. it will be an inherent feature of the underlying stochastic process driving the prices.

One of the advantages of the $\mathcal{BSS}$ framework is that it conveniently and parsimoniously allows for *both* roughness (antipersistence) *and* a characteristic time scale in the autocorrelations of the model. The former is captured by the parameter α and the latter by λ. This is a flexible and powerful approach to decoupling the small-scale (roughness) behavior of the process from the long-term (memory) properties; see Bennedsen et al. (2016) for a further discussion of this decoupling and how it relates to financial time series. Loosely speaking, in the context of our proposed $\mathcal{BSS}$ model (7.4), $\lambda = 0$ corresponds to no characteristic time scale in the model, while $\lambda > 0$ corresponds to a characteristic time scale of λ^{-1} days.*

---

* Strictly speaking, the $\mathcal{BSS}$ process X in (7.4.3) does not exist for $\lambda = 0$, as the integrand should be square integrable for the integral to be defined in the Itô sense. However, if one considers the *modified* $\mathcal{BSS}$ process

$$X_t^\# := \int_{-\infty}^{t} \left((t-s)^\alpha e^{-\lambda(t-s)} - (-s)_+^\alpha e^{-\lambda(-s)} \right) \sigma_s dW_s, \quad t \geq 0$$

where $y+ := \max(0, y)$ denotes the positive part of $y \in \mathbb{R}$, this process does exist for $\lambda = 0$. Indeed, this process reduces (up to a constant factor) to a *Type I fractional Brownian motion* (Marinucci and Robinson, 1999) when $\lambda = 0$. Note further, that the *increments* of processes X and $X^\#$ have similar behavior; in particular their variograms (B.1) coincide. When $\lambda > 0$ it is also the case that $\{X_t^\#\}_{t\geq 0} = \{X_t - X_0\}_{t\geq 0}$. Although the modified $\mathcal{BSS}$ process $X^\#$ is non-stationary, and therefore not suited for modelling of the base signal, cf. the discussion in Section 7.2, it does highlight the close connection between the fractional Brownian motion and the $\mathcal{BSS}$ process when $\lambda \approx 0$.

As argued above, the $\mathcal{BSS}$ model allows us to test for the presence of roughness (antipersistence) by testing the null hypothesis $H_0 : \alpha = 0$ and we will do so in the empirical section below. However, given the discussion of this section, another interesting question is to investigate whether the electricity prices display a characteristic time scale in their autocorrelations as well, such as was found in Rypdal and Løvsletten (2013), or whether a rough model without such a characteristic scale is more adequate.

We would thus like to test the null hypothesis $\lambda = 0$, but as this value is not permitted in the model (7.4), we consider an alternative way to test whether the data display a characteristic time scale.* Specifically, we propose a Monte Carlo approach where we assess the distribution of our proposed estimator of λ, when the underlying process does *not* possess a characteristic time scale in its autocorrelations. We then compare the estimated value of λ, obtained from electricity data, to the Monte Carlo distribution of $\check{\lambda}$ obtained under the null of no characteristic time scale. In this way, the test will gauge whether an estimate of λ obtained from the data really corresponds to a characteristic scale in the autocorrelations, or whether the observed scale is "spurious", i.e. whether it might as well, by chance, have arisen from a process without any characteristic scale. To be precise, our proposed test proceeds as follows.

We simulate under the null of the process having no characteristic time scale (i.e. "$\lambda = 0$"). Since the fBm is the "canonical" rough model without such a time scale (it is *self-similar*), we choose this model as our data generating process under the null. Then we estimate λ using a method-of-moments procedure, which we explain in Section 7.4.4.2 below. By performing these Monte Carlo simulations a large number of times, we can infer the distribution of the estimator of λ under the null that the process has no characteristic autocorrelation scale (here taking to mean that X is an fBm). Then, when we estimate λ from electricity price series we can perform inference by comparing the estimated value to the (Monte Carlo) distribution of $\hat{\lambda}$ under the null. For instance, if the estimate obtained from the data is larger than 95% of the Monte Carlo estimates from the simulations, we reject the null at a 5% significance level, and conclude that the data indeed appear to possess a characteristic time scale in its autocorrelations. Similarly, a P-value can be calculated from the relative position of the estimated value of λ from the data, compared to the Monte Carlo estimates; for instance, if the estimated value from the data is greater than 98% of the Monte Carlo estimates, but smaller than the remaining 2%, the P-value of the null will be 2%.

7.4.2.2 Stochastic Volatility: Marginal Distribution of the Base Component

The stochastic volatility process allows for flexible modelling of the marginal distributon of the $\mathcal{BSS}$ process X. When the SV process is constant, $\sigma(t) = 1$ for all t say, the resulting process X is Gaussian. When σ is stochastic, however, we get, conditionally on σ,

$$X_t | \sigma \sim N\left(0, \xi_t^2\right), \quad t \geq 0, \tag{7.5}$$

where

$$\xi_t^2 := \int_0^\infty x^{2\alpha} e^{-2\lambda x} \sigma_{t-x}^2 \, dx, \quad t \geq 0.$$

This shows that the distribution of X_t for $t \geq 0$ is a mean-variance mixture and the form of σ will be determining the marginal distribution of the $\mathcal{BSS}$ process. Specifying σ^2 as a *Lévy semistationary process (LSS)* process, i.e.

$$\sigma_t^2 = \int_{-\infty}^t k(t-s)v_s dZ_s, \quad t \geq 0,$$

---

* Although $\lambda = 0$ *is* permitted for the modified $\mathcal{BSS}$ process X^*, this value for λ is still on the boundary of the parameter space, making testing non-standard. We therefore proceed with the simulation-based test as described in the following.

where k is a kernel function, v is volatility of volatility and Z is a subordinator (a non-decreasing Lévy process), provides a particularly powerful and convenient modelling framework.

As the marginal distribution of electricity spot price series has been found (e.g. Barndorff-Nielsen et al., 2013; Veraart and Veraart, 2014, and also Section 7.6.3) to be well described by the Normal Inverse Gaussian (NIG) distribution, it is an attractive feature of the $\mathcal{BSS}$ framework that it accomodates this distribution for X_t. Indeed, as shown in Barndorff-Nielsen et al. (2013), when X is given as in (7.4) and when letting $k(x) = \dfrac{x^{-2\alpha-1}e^{-2\nu x}}{\Gamma(2\alpha+1)\Gamma(-2\alpha)}, v_t := 1$ for all t and thus

$$\sigma_t^2 = \frac{1}{\Gamma(2\alpha+1)\Gamma(-2\alpha)}\int_{-\infty}^{t}(t-s)^{-2\alpha-1}e^{-2\lambda(t-s)}dZ_s, \quad t \geq 0, \tag{7.6}$$

with Γ being the gamma function, then, by the stochastic Fubini theorem,

$$\xi_t^2 = \int_{-\infty}^{t}e^{-2\lambda(t-s)}dZ_s, \quad t \geq 0. \tag{7.7}$$

In other words, $\xi^2 = \left\{\xi_t^2\right\}_{t\in\mathbb{R}}$ is a Lévy driven OU process with mean reversion parameter 2λ. As argued in Barndorff-Nielsen et al. (2013), (7.7) implies the exisitence of a Lévy process Z, such that ξ_t^2 has the Inverse Gaussian distribution; the upshot is that X, by (7.5), is NIG distributed (Barndorff-Nielsen and Halgreen, 1977). In Section 7.6.3 we illustrate how the NIG distribution fits the empirical distribution of the base signal much better than the Gaussian distribution does, strengthening our view that stochastic volatility is needed in the model. Note also, that the parameters in (7.6) are the same as those of the $\mathcal{BSS}$ process (7.4). In other words, when modelling the SV component as in (7.6), and thus obtaining the NIG distribution as the marginal distribution of X, we need only estimate the parameters $\theta = (\alpha, \lambda)^T$ of the $\mathcal{BSS}$ process to also estimate the parameters from the SV process σ.

7.4.2.3 An Alternative Rough Component

A perhaps more well-known choice for a stationary rough component is the fBm-driven OU process, i.e.

$$X_t^H = \int_{-\infty}^{t}e^{-\lambda(t-s)}dB_s^H, \quad t \geq 0 \tag{7.8}$$

where $\lambda > 0$ and B^H is a fractional Brownian motion with Hurst index $H \in (0, 1)$. Recall, that in terms of the roughness index we have $\alpha = H - \dfrac{1}{2}$, so that $H = 1/2$ corresponds to B^H being a standard Brownian motion, while $H < 1/2$ indicates paths rougher than the Bm. This fBm-OU framework was indeed what was employed in Rypdal and Løvsletten (2013). However, we wish to employ a more flexible rough model with a more tractable autocorrelation structure than (7.8) – which is not available in closed form – and we therefore prefer to use the $\mathcal{BSS}$ process (7.4), which ACF is in closed form and given below in Equation 7.10. In Section 7.6.4 we will compare the two models when investigating if a rough process can be utilized for increased forecast performance of prices, see also Section 7.5.2.

7.4.3 Modelling the Spike Component

When examining paths of electricity prices such as the ones seen in Figure 7.1, it seems reasonable to model the very extreme moves by a *spike process*, i.e. a process which suddenly jumps a large (positive or negative) value and rapidly reverts back to the previous level. The reasons for such moves are numerous and were briefly discussed in the introduction. A standard and very general approach to model the spikes in the multi-factor framework is using a non-Gaussian OU model for Y, such as was done in e.g. Meyer-Brandis and Tankov (2008) and Benth et al. (2012). That is, we take

$$Y_t = \int_{-\infty}^{t} e^{-\lambda_Y(t-s)} dL_s, \quad t \geq 0,$$

where $\lambda_Y > 0$ controls the speed of mean reversion and L is a pure jump Lévy process, e.g. a compound Poisson process. Typically, λ_Y will be quite large, implying fast decay of the spikes. Again, the integration from minus infinity is to ensure stationarity of Y. To focus on the rough nature of the base signal we will be wholly agnostic about the distribution and general form of the spike component, Y. Examining the dynamics and distributional properties of the spike signal Y is an interesting subject in itself; we leave such investigations in the rough framework to future studies.

For simplicity, we assume that X and Y are independent; this will imply that the de-seasonalized prices, $S - \Lambda = X + Y$, to be (strictly) stationary.

7.4.4 Estimation of the Rough Multi-Factor Model

The estimation procedure we propose and will use in the empirical study in Section 7.6 is straightforward given the above discussion. In a first step, we wish to estimate the seasonal component (7.3). However, the estimation can be very sensitive to outliers (spikes), and we therefore construct a smoothed prices series on which we estimate (7.3) by OLS. Specifically, the time series we use for this OLS regression is obtained by a block moving average of the log-prices, where the window length is 15 observations on each side of an individual observation. After having estimated the seasonal parameters we subtract the seasonal signal, (7.3), from the (original, non-smoothed) log- prices, leaving us with the time series $\{S_t - \Lambda_t\}_{t\geq 0} = \{X_t + Y_t\}_{t\geq 0}$. The next step is to separate the base signal X from the spike signal Y. We do this using the *hard thresholding* algorithm of Meyer-Brandis and Tankov (2008) – an example of such filtering is given in Section A.2 where we study the EEX data set in detail. Finally, after subtracting the jump process we are left with the base signal $\{X_t\}_{t\geq 0} = \{S_t - \Lambda_t - Y_t\}_{t\geq 0}$, which we proceed to model as the $\mathcal{BSS}$ process (7.4).

This procedure of estimating the different components separately is of course not optimal; at a minimum one loses efficiency and estimates can plausibly become biased as well. An alternative approach is to try to estimate the model in one go; this was recently done using particle Markov chain Monte Carlo methods in Brix et al. (2015). However, these methods are not straighforwardly applicable to the rough model of this paper; further, in simulations (not reported here for brevity) we found good performance of the piece-wise estimation procedure. For these reasons, we prefer the simple approach of estimating the signals separately and will do so in the following.

7.4.4.1 Estimating the Roughness Index of Time Series of Electricity Prices

Many estimators of the roughness index exist and the choice about which one to use is up for debate. Although most studies consider only one particular estimator, we recommend the use several of them to check if they agree. For this reason, B contains a brief review of a number of different estimators of α and in the empirical section below we will apply them all to the data. We do this for two reasons; firstly, we use them as a robustness check to see whether the estimators agree. Secondly, presenting the estimations using all of the well-known (and some not so well-known) estimators should make it more comparable to earlier studies.

For our purposes, we face two major obstacles when estimating the roughness index of electricity prices; (i) jumps (spikes) in the price process and (ii) relatively low frequency (daily) data. In Appendices C.1 and C.2 we will look closer at how the estimators behave under (i) and (ii), respectively. We find that the former causes a downward bias in the estimation of the roughness index, while the latter is innocuous in our setup. The downward bias introduced by (i), i.e. the spikes, is the reason for us filtering out the spike signal Y, before estimating α. This has not been done in previous studies, which leads us to conjecture that earlier studies might have found values of α which are more negative than they are in reality, cf. Section 7.6 and in particular Table 7.6.

Let us briefly give some details on one of these estimators of the roughness index, namely the *change-of-frequency* (COF) estimator of Corcuera et al. (2013). The reasons for going into detail with this particular estimator, is that it comes with a central limit theorem (CLT), so that inference can be conducted easily.[*] Suppose we have $n \in \mathbb{N}$ equidistant observations of a process X observed in a fixed time interval $[0, 1]$, say. The COF statistic compares the *power variations* of the (second order) differences of the observations at two different frequencies:

$$\text{COF}_{p,n} = \frac{\sum_{k=5}^{n} \left| X_{k/n} - 2X_{(k-2)/n} + X_{(k-4)/n} \right|^p}{\sum_{k=3}^{n} \left| X_{k/n} - 2X_{(k-1)/n} + X_{(k-2)/n} \right|^p},$$

where $p > 0$ is a power parameter to be chosen by the researcher. Corcuera et al. (2013) show that when X is a rough process with fractal index $\alpha \in (-1/2, 1/2)$ in the sense of Section 7.3, then as $n \to \infty$

$$\text{COF}_{p,n} \to 2^{(2\alpha+1)p/2} \quad \text{in probability.} \tag{7.9}$$

This asymptotic regime, where the time period is fixed but the number of observations goes to infinity (so that the time between observations goes to zero), is often called *in-fill asymptotics*. Using (7.9), a straight forward estimator of α is

$$\hat{\alpha}_p := \frac{\log_2\left(\text{COF}_{p,n} \right)}{p} - \frac{1}{2}, \quad p > 0.$$

which we will simply refer to as COF(p) in the following. Both $p = 1$ and $p = 2$ can be found recommended in the literature; in what follows we will use both values. Note, however, that a feasible CLT is only available when $p = 2$, which will make this the most important value for us in the empirical section. We refer to Corcuera et al. (2013) for the details on the CLT.

7.4.4.2 Estimating the Mean Reversion Parameter λ

As λ controls the mean reversion (at least on longer scales), we seek to estimate this parameter using the dependence structure of the base signal X. We do this using our estimate of α and the autocorrelation function (ACF) of the $\mathcal{BSS}$ process (7.4), which can be shown to be equal to (e.g. Barndorff-Nielsen, 2012)

$$\rho(h) = Corr\left(X_{t+h}, X_t \right) = \frac{2^{-\alpha+\frac{1}{2}}}{\Gamma(\alpha+1/2)} (\lambda h)^{\alpha+\frac{1}{2}} K_{\alpha+\frac{1}{2}}(\lambda h), \quad h \in \mathbb{R}, \tag{7.10}$$

where Γ is the gamma function and $K_\nu(x)$ is the modified Bessel function of the third kind with parameter ν, evaluated at x, see, e.g., Abramowitz and Stegun (1972). To estimate λ, we employ a method-of-moments procedure by fitting (7.10), via non-linear least squares, to the empirically observed autocorrelations of the electricity (base signal) prices. In simulations, we found this two-step method of first estimating α and then plugging this estimate $\hat{\alpha}$ into (7.10) to estimate λ via method-of-moments, to work very well in a setup similar to our empirical applications below (simulation results not reported here, but available upon request).

---

[*] For a simulation-based, and therefore more computationally heavy, approach to inference on the fractal index see Bennedsen (2016).

7.5 Implications of a Rough Model of Electricity Prices

Rough models of financial prices are non-standard and will as such have different implications than a usual model based on a non-rough process. We briefly consider some of these implications here.

7.5.1 Arbitrage

It is well known that models with roughness index $\alpha \neq 0$, such as the fBm-OU model (7.8) with $H \neq 1/2$ and the rough $\mathcal{BSS}$ process in (7.4) are, in general, *not* semimartingales. Classically, using such a process as the basis of the price of a financial asset would constitute an arbitrage opportunity (or, more precisely, a free lunch with vanishing risk), as was shown in Delbaen and Schachermayer (1994), see also Rogers (1997). One could fear that introducing a rough model for the price of electricity would induce the same adverse effects. Fortunately, there are several reasons why this need not be. Most importantly, the electricity spot market is wholly different from the usual financial markets in that one can not buy electricity and hold it in a portfolio for a later date. For this reason, the usual 'buy-and-hold' arbitrage arguments break down and we can not rely on the classical no-arbitrage results such as the one in Delbaen and Schachermayer (1994). As an additional assurance, Pakkanen (2011) showed that $\mathcal{BSS}$ processes such as the one in (7.4) will not be the source of arbitrage when transaction costs are considered.

7.5.2 Forecasting

Since roughness refers to the small scale behavior of the time series under consideration, it is plausible that including a rough term will provide superior forecasts of prices *at least in the short run*. Since our three components in (7.2) are assumed independent, we will focus on forecasting the base signal only; the (deterministic) seasonal component is then easily adjusted for, as are the spike signal Y, given a model for this. However, as we explained in Section 7.4.3 we do not go into details with the spike part in this work. Hence, we will focus on predicting X in the following. The two forecasting methods we outline here will be empirically tested in Section 6.4 and visually illustrated for the EEX data set in A.4.

7.5.2.1 Forecasting the $\mathcal{BSS}$ Process

Ideally, for a forecast horizon $h > 0$ (days, say) we would like to compute

$$\hat{X}_{t+h} = \mathbb{E}\left[X_{t+h} \,\middle|\, \mathcal{F}_t\right], \tag{7.11}$$

where $\mathcal{F}_t = \sigma\{X_s, s \in [0,t]\}$ is the filtration generated by the base signal X. Unfortunately, for the $\mathcal{BSS}$ process, the expression for (7.11) is an open problem. Instead, we will use a linear combination of past observed values to predict the future, i.e. we use the *best linear predictor*. The best linear predictor is given in terms of the autocorrelation function of the $\mathcal{BSS}$ process, which we recall is given by Equation (7.10). The best linear predictor of X_{t+h} given $X_1, X_2, \ldots, X_t$ is (e.g. Grimmett and Stirzaker, 2001, Section 7.9.2.)

$$\breve{X}_{t+h} = \sum_{i=1}^{t} a_i X_i, \quad h \geq 1,$$

where $a_1, a_2, \ldots, a_t \in \mathbb{R}$ solves

$$\sum_{i=1}^{t} a_i \rho(|i-j|) = \rho(h+j), \quad 1 \leq j \leq t.$$

To use this method in practice, we estimate the parameters of the $\mathcal{BSS}$ model, plug them into the functional form of the ACF in (7.10), and use this in the above to arrive at the best linear predictor.

7.5.2.2 Forecasting the fBm-OU Process

In the case of forecasting, there are advantages in choosing the fBm-based OU process in Equation (7.8) as a model of the base signal; the fBm has been extensively studied and in particular, we have a closed-form expression for the conditional expectation (7.11). To be precise, Nuzman and Poor (2000) show that for $H < \dfrac{1}{2}$

$$\mathbb{E}\left[B_{t+h}^{H} \mid \mathcal{F}_{t}^{H}\right] = \frac{\cos(H\pi)}{\pi} h^{H+1/2} \int_{-\infty}^{t} \frac{B_{s}^{H}}{(t-s+h)(t-s)^{H+1/2}} ds,$$

where $\mathcal{F}_{t}^{H} = \sigma\left\{B_{s}^{H}, s \leq t\right\}$ is the filtration generated by the fBm B^{H} during its entire history (i.e. from the infinite past until time t). Now, as argued in Gatheral et al. (2014), for reasonably small time scales h and for small values of the mean reversion parameter λ, we have the approximation

$$\mathbb{E}\left[X_{t+h}^{H} \mid \mathcal{F}_{t}^{H}\right] \approx \frac{\cos(H\pi)}{\pi} h^{H+1/2} \int_{-\infty}^{t} \frac{X_{s}^{H}}{(t-s+h)(t-s)^{H+1/2}} ds, \tag{7.12}$$

where X^{H} is the fBm-OU process in (7.8). We approximate this expression using the observed values of X^{H} together with a straightforward Riemann sum:

$$\hat{X}_{t+h}^{H} = \frac{\cos(H\pi)}{\pi} h^{H+1/2} \sum_{i=1}^{\lfloor t/\delta \rfloor} \frac{X_{i\delta}^{H}}{(t-(i-r)\delta+h)(t-(i-r)\delta)^{H+1/2}} \delta, \quad r \in (0,1]$$

where $\delta > 0$ is the observation frequency (i.e. $\delta = 1$ day in our case) and $H = \alpha + \dfrac{1}{2}$ is the Hurst exponent. Note, that we face a rather arbitrary choice of $r \in (0, 1]$ when evaluating the integral at the points $s = (i-r)\delta$ for $i = 1, \ldots, \lfloor t/\delta \rfloor$; in Section 7.6.4 we will simply choose r such that the forecast performance is optimized for the shortest forecast horizon $h = 1$ day.

7.6 Empirical Analysis

In this section we apply the estimation procedure explained in Section 7.4.4 to the data in Table 7.1; in A we will illustrate the approach by considering the German EEX data in detail. The estimation results for all six data sets are seen in Tables 7.4-7.6. Table 7.4 shows the estimates from the seasonal function in Equation (7.3): we see no noticable trend (c_2), while there is a level effect (c_1) and some seasonality in all data series (a_1, a_2, a_3, a_4).

7.6.1 Roughness of Electricity Price Series

Table 7.5 contains the most important information for our purposes. From the COF(2) estimator we have estimates of the roughness index α, which are negative across the board, indicating that the time series indeed do display a degree of roughness. Similarly, when conducting the formal hypothesis test of $H_0 : \alpha = 0$ against the alternative $H_a : \alpha < 0$, we reject H_0 for all markets under consideration, except Nord Pool, indicating that in these energy markets the rough $\mathcal{BSS}$ process is to be preferred over the OU model as a model of the base signal. That is, we conclude that five out of the six markets studied here display rough characteristics in their base signals.

To further check this fact, we employ nine different estimators of α, which are discussed in B; the results are seen in Table 7.6 which corroborates the findings of this section. Indeed, *all* estimates of α

TABLE 7.4   Estimation of Seasonality Parameters

	c_1	c_2	a_1	a_2	a_3	a_4
EEX	4.303	−0.000	0.030	−0.006	−0.079	0.071
	(0.011)	(0.000)	(0.007)	(0.007)	(0.007)	(0.007)
NP	3.938	−0.000	0.017	0.000	0.010	0.101
	(0.013)	(0.000)	(0.009)	(0.009)	(0.009)	(0.009)
Powernext	4.333	−0.000	0.024	−0.008	−0.060	0.118
	(0.011)	(0.000)	(0.007)	(0.007)	(0.008)	(0.007)
GME	4.666	−0.000	0.011	−0.008	−0.064	0.041
	(0.009)	(0.000)	(0.006)	(0.006)	(0.006)	(0.006)
APX UK	4.112	−0.000	0.005	0.001	−0.031	0.019
	(0.012)	(0.000)	(0.009)	(0.009)	(0.009)	(0.009)
APX NL	4.270	−0.000	0.016	−0.011	−0.046	0.065
	(0.011)	(0.000)	(0.007)	(0.007)	(0.008)	(0.007)

Estimates of the parameters of the seasonal function (7.2) with standard errors in parentheses.

are negative. Note that if one does not filter out the spikes (Panel A), the α estimates from Nord Pool using either of the GEN(1), GEN(2) and DFA estimators would – incorrectly – lead one to find $\alpha < 0$ in this market. Further, this is even the case after filtering out the spikes (Panel B) for the estimators ABS and AGG. Recall, that the DFA estimator has traditionally been popular in the literature. Therefore, it is evident that it is important to first filter out the spikes and, further, that which estimator one uses really does matter as ABS and AGG in general give more extreme results than do the other estimators.

In Figure 7.3 we plot the estimates of α for all markets using the COF(2) estimator, as well as their 95% confidence bands. This is done to both the time series including spikes (blue squares) and the base signal (red crosses). A few conclusions emerge from this. First off, the estimates from two of the six markets (Powernext and GME) become noticeably less negative after filtering out the spikes, strengthening the claim that spikes might cause downward bias in the roughness estimates (see also C.1). Even more salient is the observation that the standard deviation of the estimator is generally huge when one does not filter out spikes before estimation, leading to too wide confidence intervals, and thus to (possibly) wrong inference. For instance, $\alpha = 0$ is contained in the confidence intervals for Nord Pool, Powernext, GME, and almost for EEX as well, when the analysis is done on the data including spikes. Conversely, when the analysis is done on the base signal, only the confidence bands from the Nord Pool market contain zero.

The above shows that accounting for spikes prior to estimation of α is crucial and the rough multifactor framework is thus ideally suited for modelling electricity prices, since it allows us to do this in a straightforward manner. We note again that most authors in the rough/antipersistent literature have generally *not* done any attempt to account for spikes, thus leading us to conjecture that earlier estimates in the literature might have been *too negative*.

7.6.2 Characteristic Time Scale of Electricity Price Series

Turning again to Table 7.5, we also see estimates of the mean reversion parameter λ from both the $\mathcal{BSS}$ model and the OU model. The value of $1/\hat{\lambda}$ is also reported (in days); this is the "characteristic time scale" of the process, i.e. a measure of how fast the autocorrelations decay (exponentially) to zero. In the OU model, $1/\lambda$ corresponds to the *half-life* of shocks to the process, i.e. the expected waiting time until the effect of a shock to the process is halved. For the $\mathcal{BSS}$ process with ACF (7.9) this interpretation is only valid approximately in finite samples, i.e. in the long run. For the base signal, we see that the characteristic time scale for the $\mathcal{BSS}$ model varies from approximately 20 to 90 days. These numbers are

TABLE 7.5 Estimating and Testing for Roughness

	Panel A: *Log-prices*					
	EEX	NP	Powernext	GME	APX UK	APX NL
$\hat{\alpha}$	−0.204	−0.045	−0.300	−0.405	−0.306	−0.161
std($\hat{\alpha}$)	0.100	0.069	0.168	0.230	0.049	0.061
Test-stat	−2.044	−0.646	−1.792	−1.763	−6.283	−2.668
P-value	0.020	0.259	0.037	0.039	0.000	0.004
Reject at 5 perc	1	0	1	1	1	1
$\hat{\lambda}_{BSS}$	0.120	0.043	0.034	0.019	0.021	0.052
$1/\lambda_{BSS}$ (days)	8.320	23.332	29.595	52.749	48.493	19.224
P-value	0	0	0.0002	0.0578	0.0116	0
$\hat{\lambda}_{OU}$	0.296	0.056	0.218	0.463	0.192	0.131
$1/\hat{\lambda}_{OU}$ (days)	3.376	17.768	4.581	2.159	5.216	7.636
P-value	0	0	0	0	0	0

	Panel B: *Filtered log-prices (base signal)*					
	EEX	NP	Powernext	GME	APX UK	APX NL
$\hat{\alpha}$	−0.211	−0.059	−0.198	−0.246	−0.308	−0.191
std($\hat{\alpha}$)	0.037	0.046	0.045	0.044	0.040	0.040
Test-stat	−5.683	−1.281	−4.410	−5.597	−7.664	−4.825
P-value	0.000	0.100	0.000	0.000	0.000	0.000
Reject at 5 perc	1	0	1	1	1	1
$\hat{\lambda}_{BSS}$	0.048	0.026	0.033	0.031	0.011	0.019
$1/\lambda_{BSS}$ (days)	20.960	38.723	30.553	32.067	88.295	51.584
P-value	0	0	0	0.0002	0.0832	`
$\hat{\lambda}_{OU}$	0.149	0.037	0.105	0.144	0.144	0.074
$1/\hat{\lambda}_{OU}$ (days)	6.722	27.085	9.549	6.958	6.957	13.517
P-value	0	0	0	0	0	0

Estimating the roughness index, α, and testing $H_0 : \alpha = 0$ for various data sets. Panel A is using the standard approach of directly estimating the roughness index of the de-seasonalized prices. In contrast, Panel B is after filtering the seasonal and spike signals from the data, i.e. the estimations are done on the base signal. $\hat{\alpha}$ is estimated using the COF (2) estimator. The test statistic and P-value refer to the hypothesis test $H_0 : \alpha = 0$ against the alternative $H_a : \alpha < 0$. In the bottom half of the table the estimates of λ are reported from both the $\mathcal{BSS}$ model and the OU model. Here, the P-values refer to the test for a characteristic autocorrelation time scale, as explained in Section 7.4.2.1. For this test, 5 000 Monte Carlo replications of an fBm with Hurst index $H = \hat{\alpha} + 1/2$ where used to approximate the distribution of $\hat{\lambda}$ under the null.

significantly smaller for the OU model; the reason for this is, that the roughness in the $\mathcal{BSS}$ model helps in fitting the autocorrelations in the short run: a negative value of α implies a sharp decrease in autocorrelation for the very short lags. This allows the λ parameter in the $\mathcal{BSS}$ model to better fit the ACF at the longer lags as compared to the OU model. In other words, not accounting for the roughness in electricity prices results in severe under estimation of the characteristic time scale, leading the researcher to conclude that shocks to the process die out more quickly than what is the case in actuality.

Table 7.5 also contains P-values for the test of the presence of a characteristic time scale, detailed in Section 7.4.2.1. Looking at the base signal in Panel B, the P-values are all very small. In fact, for the OU model, the P-values are all equal to zero. All this tells us, however, is that a model without *both* roughness

TABLE 7.6   Estimating the Roughness Index of All Data Sets Using Various Estimators

	Panel A: *Log-prices*					
	EEX	NP	Powernext	GME	APX UK	APX NL
COF (2)	−0.204	−0.045	−0.300	−0.405	−0.306	−0.161
COF (1)	−0.198	−0.022	−0.197	−0.260	−0.306	−0.184
Variogram	−0.270	−0.088	−0.294	−0.421	−0.342	−0.224
Madogram	−0.243	−0.052	−0.211	−0.290	−0.343	−0.201
GEN (2)	−0.340	−0.135	−0.286	−0.464	−0.365	−0.329
GEN (1)	−0.301	−0.112	−0.231	−0.341	−0.382	−0.296
ABS	−0.345	−0.278	−0.293	−0.350	−0.399	−0.273
DFA	−0.375	−0.169	−0.320	−0.454	−0.402	−0.335
AGG	−0.403	−0.350	−0.370	−0.431	−0.431	−0.307

	Panel B: *Filtered log-prices (base signal)*					
	EEX	NP	Powernext	GME	APX UK	APX NL
COF (2)	−0.211	−0.059	−0.198	−0.246	−0.308	−0.191
COF (1)	−0.230	−0.033	−0.194	−0.210	−0.314	−0.218
Variogram	−0.220	−0.073	−0.178	−0.253	−0.314	−0.198
Madogram	−0.232	−0.052	−0.175	−0.218	−0.314	−0.190
GEN (2)	−0.248	−0.082	−0.170	−0.267	−0.328	−0.222
GEN (1)	−0.242	−0.082	−0.176	−0.252	−0.349	−0.229
ABS	−0.323	−0.235	−0.294	−0.346	−0.446	−0.297
DFA	−0.309	−0.092	−0.243	−0.300	−0.369	−0.291
AGG	−0.334	−0.291	−0.341	−0.365	−0.460	−0.316

Estimating the roughness index α for various data sets and various estimators. The estimators are reviewed in Appendix B. In Panel A estimation is done on the de-seasonalized log-prices. In contrast, Panel B (right) is after filtering the seasonal and spike signals from the data, i.e. the estimations are done on the base signal.

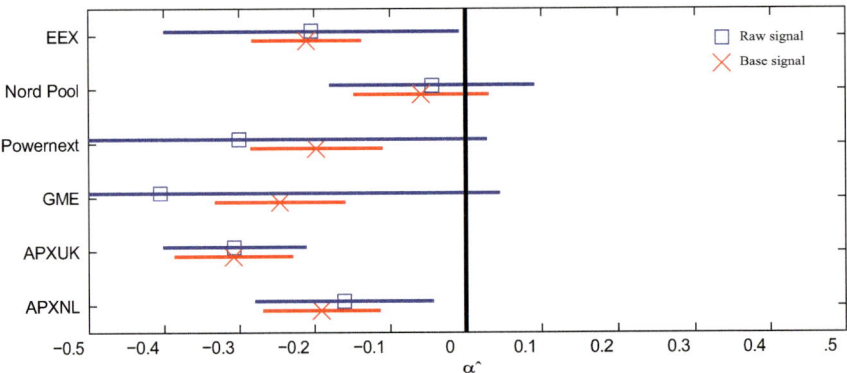

FIGURE 7.3   Estimation of the roughness index of the spot prices from various energy exchanges. Estimation is done on the raw log-prices (blue squares) and on the base signal, i.e. after filtering out the seasonal component and the spike signal (red crosses). The solid lines correspond to 95% confidence intervals.

and a characteristic time scale in the autocorrelations is a poor model for the base signal (in essence, the test is whether a Brownian motion would be an adequate model). More interestingly, the P-values are also very low when we consider the $\mathcal{BSS}$ model; only for one market (APX UK) is the P-value above 5%. This shows, that even after including a rough/antipersistent component, it is still important to also allow for a characteristic time scale in the model.

7.6.3 Marginal Distribution of the Base Signal

As explained in Section 7.4.2.2, the $\mathcal{BSS}$ framework accomodates the NIG distribution as its marginals; we now take a closer look at how this distribution fits the empirical distribution of the base signal. Figure 7.4 shows that the NIG distribution indeed fits very well the empirically observed marginal distribution of the base signal, a fact that was also found in e.g. Barndorff-Nielsen et al. (2013) and Veraart and Veraart (2014). The plots in the left column of Figure 7.4 simply plots the empirical density of the base signal against a fitted Gaussian and a fitted NIG density:* in most cases the NIG distribution fits the data much better than do the Gaussian distribution, especially around the center of the distribution. To

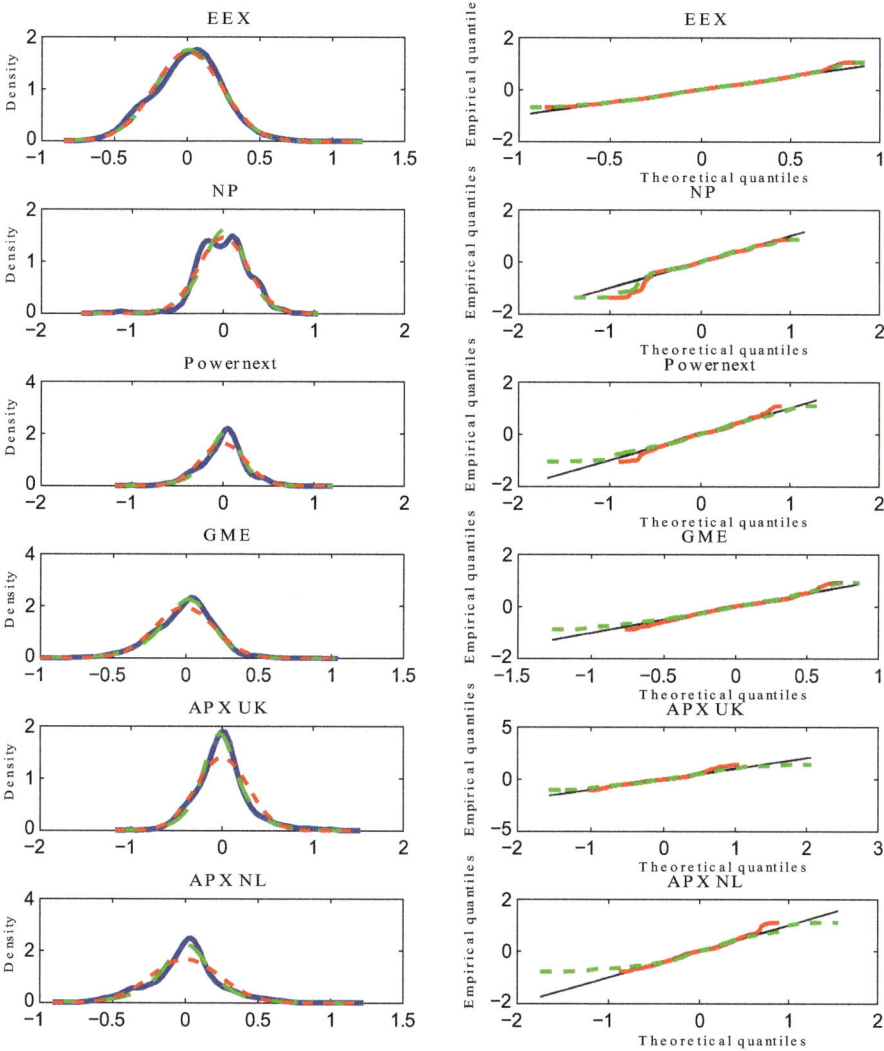

FIGURE 7.4 Left plots: Empirical marginal distribution of the base signal (blue solid line) superimposed with a fitted NIG density (green dashed line) and the corresponding fitted Gaussian density (red dashed line). Right plots: QQ-plots; the solid red curve denotes the Gaussian distribution and the green dashed line denotes the NIG distribution. If the data fits one of these distributions, the corresponding curve will lie on the black diagonal line.

---

* The NIG distribution were fitted using a maximum likelihood procedure.

examine the fit in the tails better, we present QQ-plots in the right column of Figure 7.4: here we again see that, for most markets, the NIG distribution is better at capturing the tail behavior of the data, indicating fat tails of the marginal density of the time series under consideration. This is evidence of the inadequacy of the Gaussian OU process as a model of the base signal and thus indicates the presence of stochastic volatility. In summary, in view of the arguments in Section 7.4.2.2, the $\mathcal{BSS}$ framework is able to accurately capture the empirically observed marginal density of the extracted base signal of electricity prices.

7.6.4 Forecasting

In Table 7.7 we present the results of a forecasting exercise. We took the extracted base signal from each of the six exchanges and applied the forecasting approaches of Section 7.5.2 using different values of forecast horizon h. We then compared the forecasts from the $\mathcal{BSS}$ model with those from three benchmark models, the fBm model, the OU model, and a random walk (RW) model. The RW model is simply a "no-change" forecast, while the (discretized) OU model can be written as an AR(1) model and therefore easily forecast using standard methods. Table 7.7 shows root mean squared forecast errors (RMSFE) of the three approaches FBM, OU, and RW as compared to the $\mathcal{BSS}$ approach. That is, the numbers reported are

$$\text{Relative RMSFE}(x) = \frac{\text{RMSFE}(\mathcal{BSS})}{\text{RMSFE}(x)}, \quad x = \text{FBM, OU, RW}.$$

Thus, values less than one indicate that the $\mathcal{BSS}$-based forecasts fare better than the alternative model. Some conclusions emerge: firstly, the fBm-based forecasts perform very poorly in general; this could be

TABLE 7.7 Forecasting the Base Signal

		RMSE relative to $\mathcal{BSS}$					
		$h = 1$	$h = 2$	$h = 4$	$h = 8$	$h = 12$	$h = 16$
EEX	RW	0.928	0.909	0.880	0.835	0.796	0.781
	OU	0.968	0.980	0.993	1.005	1.004	0.988
	FBM	0.821	0.846	0.861	0.856	0.842	0.844
NP	RW	1.007	0.990	0.978	0.945	0.914	0.882
	OU	1.015	1.005	1.006	1.001	1.000	0.996
	FBM	0.694	0.833	0.913	0.952	0.952	0.944
Powernext	RW	0.958	0.948	0.919	0.858	0.813	0.781
	OU	0.988	1.001	1.012	1.025	1.034	1.032
	FBM	0.866	0.888	0.890	0.867	0.845	0.830
GME	RW	0.921	0.904	0.900	0.885	0.884	0.873
	OU	0.958	0.954	0.943	0.910	0.901	0.907
	FBM	0.760	0.810	0.858	0.892	0.913	0.918
APX UK	RW	0.870	0.842	0.833	0.807	0.813	0.796
	OU	0.910	0.905	0.909	0.871	0.842	0.819
	FBM	0.680	0.747	0.814	0.858	0.887	0.895
APX NL	RW	0.942	0.924	0.902	0.874	0.853	0.835
	OU	0.964	0.963	0.967	0.973	0.973	0.964
	FBM	0.943	0.938	0.934	0.924	0.912	0.904

Forecasting the base signal using the four models, $\mathcal{BSS}$, FBM, OU and RW. The forecast horizon is h days and the numbers in the tables are root mean squared (RMSE) forecast errors relative to the $\mathcal{BSS}$ model. Values less than one favors the $\mathcal{BSS}$ model, while values greater than one favors the alternative method.

an indication that the model is not describing the data well, but more likely it is because the approximation in (7.2) is poor for our purpose. Note, that the autocorrelation of the OU-fBm model (7.7) is not available in closed form, hence making prediction using the best linear predictor, as was done for the $\mathcal{BSS}$ process, difficult, at least without resorting to approximations or numerical calculations of the autocorrrelation function. This makes rough models with tractable autocorrelation structures, such as the $\mathcal{BSS}$ process, preferable from a forecasting standpoint. Secondly, the $\mathcal{BSS}$ forecast method generally fares better than the (non-rough) OU method, when the market in question display roughness, i.e. for all markets except Nord Pool (where we could not reject $H_0 : \alpha = 0$, see Table 7.5). Lastly, the RW model performs poorly, especially for longer horizons, indicating that including some form of mean reversion in the model is indeed important for forecasting.

7.7 Conclusion

This article has conducted an in-depth investigation of the apparent roughness of time series of electricity spot prices. Similar to other authors we find evidence of rough (i.e. "antipersistent") characteristics in electricity spot prices: in particular, in five out of the six markets studied, we reject the null hypothesis of a zero roughness index in favor of the alternative hypothesis of a negative index. Only in the Nord Pool market could we not reject the null $\alpha = 0$. Contrary to previous studies, we sought to avoid possible sources of biases in the estimators of the roughness index by filtering the prices for seasonality and spikes *before* estimating the degree of roughness in the data. We considered two pieces of evidence, which indicate that the $\mathcal{BSS}$ model is indeed to be preferred over the OU model for most electricity price series. Firstly, we simply find statisically significant values of the roughness index and that these are negative. Secondly, the results of our forecasting exercise showed that there are gains to be had in this area by applying the rough $\mathcal{BSS}$ model; a very interesting avenue of future research would be to include a rough component in a more realistic forecasting setup, such as the one considered in Weron (2014).

Although the OU model was found to be inadequate, the results of this paper show that it is important for a mathematical model of (the base signal of) electricity prices to contain a characteristic time scale in the autocorrelation function implied by the model. Indeed, when we performed the statistical tests for such a time scale, the P-values were all very low (only one market had a P-value above 5%, namely the APX UK market with a P-value of 8.3%). In particular, this finding rules out *self-similar* models, such as the fractional Brownian motion, as a model of the base signal. It also extends a finding in Rypdal and Løvsletten (2013), where the authors found a characteristic time scale in the Nord Pool market, but were unable to conclude whether this was simply due to seasonal effects in the data. In contrast, the multi-factor approach taken here, where we in an initial step filter out seasonality and spikes, allows us to conclude that even after controlling for jumps and seasonal effects, a characteristic time scale still exists in electricity price series. What is more, the $\mathcal{BSS}$ process seems to offer a particular convenient and tractable model which features both roughness and a characteristic time scale in its autocorrelations.

The analysis in the present work was done on log-prices, but we stress that the approach would apply equally well to raw prices. In fact, we conducted the same analysis on these prices (not given here for brevity) and the results on roughness hold a fortiori: we find even more extreme (negative) values of the roughness index α in this case.

Finally, in the model presented here, we all but ignored three very important components of electricity prices, namely the seasonality process, the spike process, and the stochastic volatility process. Further research into a full-fledged rough model and the implications for spikes, stochastic volatility, and stochastic seasonality is left for future research.

Acknowledgements

I would like to thank the editor and two anonymous referees for valuable comments, which resulted in considerable improvements of the paper. I would also like to thank Professor Asger Lunde for comments on an earlier draft of this paper as well as for many insightful discussions concerning statistical modelling of energy prices and Solveig Sørensen for competent and meticulous proof reading of the manuscript. The research has been supported by CREATES (DNRF78), funded by the Danish National Research Foundation, by Aarhus University Research Foundation (project "Stochastic and Econometric Analysis of Commodity Markets"), and by Aage and Ylva Nimbs Foundation.

References

Abramowitz, M. and I. A. Stegun (1972). *Handbook of mathematical functions with formulas, graphs, and mathematical tables* (10th ed.), Volume 55. United States Department of Commerce.

Barndorff-Nielsen, O. E. (2012). Notes on the gamma kernel. *Thiele Research Reports* (03).

Barndorff-Nielsen, O. E., F. E. Benth, and A. E. D. Veraart (2013). Modelling energy spot prices by volatility modulated Lévy-driven Volterra processes. *Bernoulli 19* (3), 803–845.

Barndorff-Nielsen, O. E., J. M. Corcuera, and M. Podolskij (2011). Multipower variation for Brownian semistationary processes. *Bernoulli 17* (4), 1159–1194.

Barndorff-Nielsen, O. E. and C. Halgreen (1977). Infinite divisibility of the hyperbolic and generalized inverse Gaussian distributions. *Z. Wahrscheinlichkeitstheorie verw. Gebiete 38*, 309–311.

Barndorff-Nielsen, O. E. and J. Schmiegel (2007). Ambit processes: with applications to turbulence and tumour growth. In *Stochastic analysis and applications*, Volume 2 of *Abel Symp.*, pp. 93–124. Berlin: Springer.

Barndorff-Nielsen, O. E. and J. Schmiegel (2009). Brownian semistationary processes and volatility/ intermittency. In *Advanced financial modelling*, Volume 8 of *Radon Ser. Comput. Appl. Math.*, pp. 1–25. Berlin: Walter de Gruyter.

Bennedsen, M. (2016). Semiparametric inference on the fractal index of Gaussian and conditionally Gaussian time series data. Working paper available at arXiv: 1608.01895.

Bennedsen, M., A. Lunde, and M. S. Pakkanen (2015). Hybrid scheme for Brownian semistationary processes. *Working paper available at arXiv:1507.03004.*

Bennedsen, M., A. Lunde, and M. S. Pakkanen (2016). Decoupling the short- and long-term behavior of stochastic volatility. Working paper.

Benth, F. E., J. Kallsen, and T. Meyer-Brandis (2007). A non-Gaussian Ornstein–Uhlenbeck process for electricity spot price modeling and derivatives pricing. *Applied Mathematical Finance 14* (2).

Benth, F. E., R. Kiesel, and A. Nazarova (2012). A critical empirical study of three electricity spot price models. *Energy Economics 34* (5), 1589–1616.

Brix, A. F., A. Lunde, and W. Wei (2015). A generalized Schwartz model for energy spot prices - estimation using a particle MCMC method. CREATES Working paper 2015-46.

Cartea, A. and M. G. Figueroa (2005). Pricing in electricity markets: a mean reverting jump diffusion model with seasonality. *Applied Mathematical Finance 12* (4), 313–335.

Corcuera, J. M., E. Hedevang, M. S. Pakkanen, and M. Podolskij (2013). Asymptotic theory for Brownian semistationary processes with application to turbulence. *Stochastic Process. Appl. 123* (7), 2552–2574.

Delbaen, F. and W. Schachermayer (1994). A general version of the fundamental theorem of asset pricing. *Mathematische Annalen 300* (1), 463–520.

Erzgräber, H., F. Strozzi, J. Zaldívar, H. Touchette, E. Gutiérrez, and D. K. Arrowsmith (2008). Time series analysis and long range correlations of Nordic spot electricity market data. *Physica A 287*, 6567–6574.

Gatheral, J., T. Jaisson, and M. Rosenbaum (2014). Volatility is rough. *Working paper.*

Geman, H. and A. Roncoroni (2006). Understanding the fine structure of electricity prices. *Journal of Business 79* (3).

Gneiting, T., H. Sevcikova, and D. B. Percival (2012). Estimators of Fractal Dimension: Assessing the Roughness of Time Series and Spatial Data. *Statistical Science 27* (2), 247–277.

Grimmett, G. and D. Stirzaker (2001). *Probability and Random Processes* (3rd ed.). Oxford University Press.

Hayfavi, A. and I. Talasli (2014). Stochastic multifactor modeling of spot electricity prices. *Journal of Computational and Applied Mathematics 259* (B), 434–442.

Kaminski, V. (2004). *Managing Energy Price Risk: Third Edition*. Risk Books.

Kristoufek, L. and P. Lunackova (2013). Long-term memory in electricity prices: Czech market evidence. *Czech Journal of Economics and Finance 63* (5), 407–424.

Marinucci, D. and P. M. Robinson (1999). Alternative forms of fractional Brownian motion. *Journal of Statistical Planning and Inference*.

Matteo, T. D. (2007). Multi-scaling in finance. *Quantitative Finance 7* (1), 21–36.

Meyer-Brandis, T. and P. Tankov (2008). Multi-factor jump-diffusion models of electricity prices. *International Journal of Theoretical and Applied Finance 11* (5), 503–528.

Norouzzadeh, P., W. Dullaert, and B. Rahmani (2007). Anti-correlation and multifractal features of Spain electricity spot market. *Physica A 380*, 333–342.

Nuzman, C. J. and V. H. Poor (2000). Linear estimation of self-similar processes via Lamberti's transformation. *Journal of Applied Probability 37* (2), 429–452.

Pakkanen, M. S. (2011). Brownian semistationary processes and conditional full support. *International Journal of Theoretical and Applied Finance 14* (4), 579–586.

Peng, C.-K., S. V. Buldyrev, S. Havlin, M. Simons, H. E. Stanley, and A. L. Goldberger (1994). Mosaic organization of DNA nucleotides. *Physical Review E 49* (2), 1685–1689.

Pilipovic, D. (2007). *Energy Risk: Valuing and Managing Energy Derivatives*. McGraw-Hill.

Rogers, L. C. G. (1997). Arbitrage with fractional Brownian motion. *Mathematical Finance 7*, 95–105.

Rypdal, M. and O. Løvsletten (2013). Modeling electricity spot prices using mean-reverting multifractal processes. *Physica A 392* (1), 194–207.

Schwartz, E. (1997). The stochastic behavior of commodity prices: Implications for valuation and hedging. *The Journal of Finance 52* (3), 923–973.

Simonsen, I. (2002). Measuring anti-correlations in the Nordic electricity spot market by wavelets. *Physica A 233*, 597–606.

Taqqu, M. S., V. Teverovsky, and W. Willinger (1995). Estimators for long-range dependence: An empirical study. *Fractals 3*, 785–798.

Veraart, A. E. D. and L. A. M. Veraart (2014). Modelling electricity day-ahead prices by multivariate Lévy semistationary processes. In F. E. Benth, V. A. Kholodnyi, and P. Laurence (Eds.), *Quantitative Energy Finance*, pp. 157–188. New York: Springer.

Weron, R. (2014). Electricity price forecasting: A review of the state-of-the-art with a look into the future. *International Journal of Forecasting 30*, 1030–1081.

Weron, R. and B. Przybylowicz (2000). Hurst analysis of electricity price dynamics. *Physica A 286*, 462–468.

Appendix A: An Empirical Case Study: The German EEX Market

We analyze the German EEX spot market from January 1, 2008 to January 23, 2015, excluding weekends using the estimation approach outlined in the article. We work on the peak load price, i.e. the average price over Hour 9 to Hour 20 each day, yielding 1844 daily observations. The raw price (top) and log-price (bottom) are seen in Figure 7.5.

A.1 De-seasonalizing the Data

We do OLS using the seasonal function (7.2) on the log-prices as explained in Section 7.4.1. The resulting fit is seen in Figure 7.6 (top) where we also show the residuals $\{\log P_t - \Lambda_t\}_{t \geq 0} = \{X_t + Y_t\}_{t \geq 0}$ (bottom).

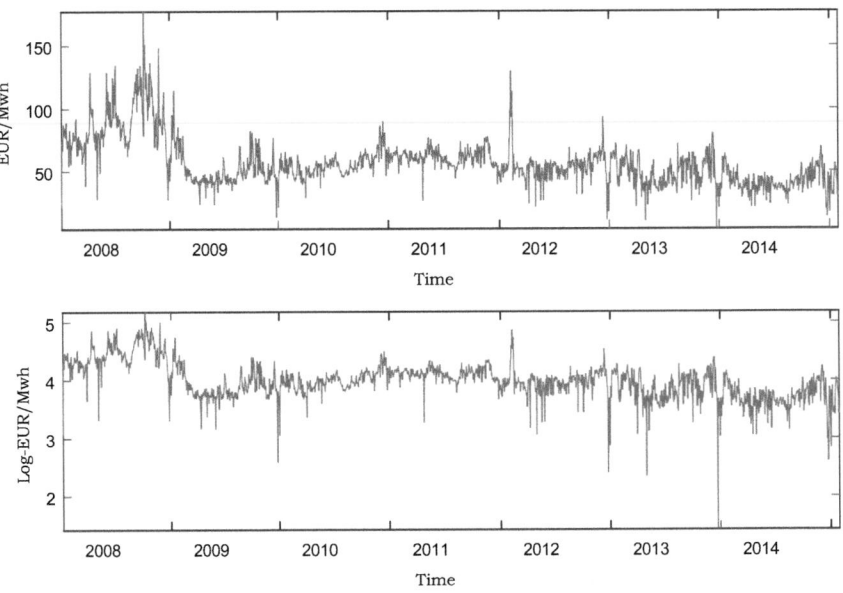

FIGURE 7.5   Raw price (top) and log-price (bottom) of EEX peak load prices from January 1, 2008 to January 23, 2015.

A.2 Filtering the Jump Process

We use the hard thresholding method described in Meyer-Brandis and Tankov (2008) and filter out the path of the spike process Y. One could estimate the mean reversion parameters and use them in the thresholding algorithm, but both Meyer-Brandis and Tankov (2008) and Benth et al. (2012) find that tuning them by eye works better. We follow their recommendations and choose $\lambda_1 = \dfrac{1}{115}$ and $\lambda_2 = \dfrac{1}{4}$. There is an additional tuning parameter, ϵ, controlling how many jumps to remove in an initial step.

Specifically, we remove $\epsilon = 0.045 = 4.5\%$ of the largest (in absolute value) increments of the process before we calculate the (target) standard deviation of the resulting process. Then we insert back the ϵ values and then run the thresholding algorithm, removing spikes until the standard deviation of the base signal equals the target standard deviation. The result is seen in Figure 7.7.

A.3 Analysis of the Residual Base Signal

The filtered base signal from the hard thresholding algorithm is now assumed to be $\{X_t\}_{t\geq 0} = \{\log P_t - \Lambda_t - Y_t\}_{t\geq 0}$, where Y is the filtered spike signal; see Figure 7.8. We will model X as the $\mathcal{BSS}$ process (4.3), which will be fit to the empirical base signal — specifically, we will estimate α using the COF (2) estimator and then λ from the empirical autocorrelation (7.9). We then test the hypothesis $H_0 : \alpha = 0$ to see if we can do away with an OU model or if there is still the need for a roughness parameter in the base signal; the result of this test was seen in Table 7.5, Panel B, where the null hypothesis is rejected in favor of $H_a : \alpha < 0$.

We also estimate the mean-reversion parameter of an OU process, λ_{OU}, which is likewise fitted to the OU ACF, $\rho_{OU}(h) = e^{-\lambda_{OU} h}$. For the estimations involving the autocorrelation function we use $\left\lfloor \sqrt{N} \right\rfloor + 1 = 43$ lags, where $N = 1844$ is the total number of observations of the base signal X. As discussed above, the rough nature of the $\mathcal{BSS}$ process causes the estimate of λ to be lower than in the OU model ($\hat{\lambda} = 0.048$ compared to $\hat{\lambda}_{OU} = 0.149$). This is evidence that the α parameter helps in the rapid

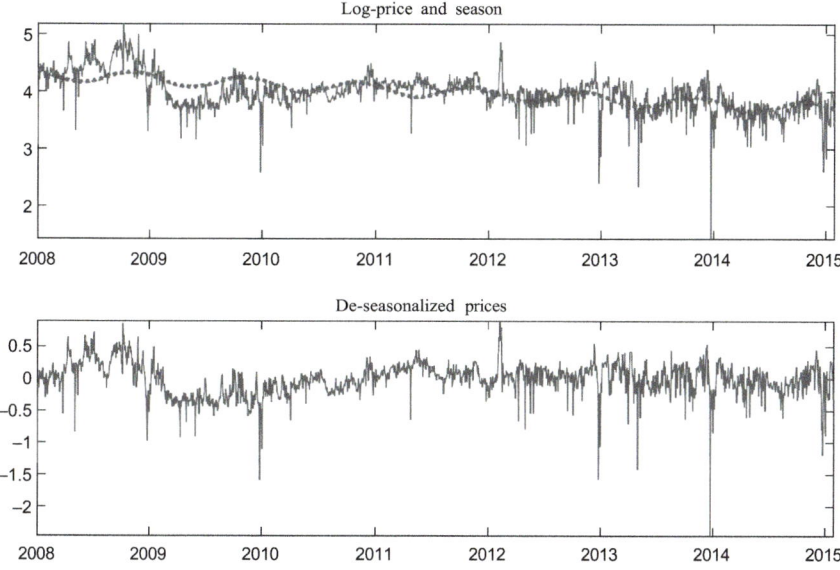

FIGURE 7.6   Log-prices and fitted seasonality function (top) and residuals from the non-linear least squares regression (bottom).

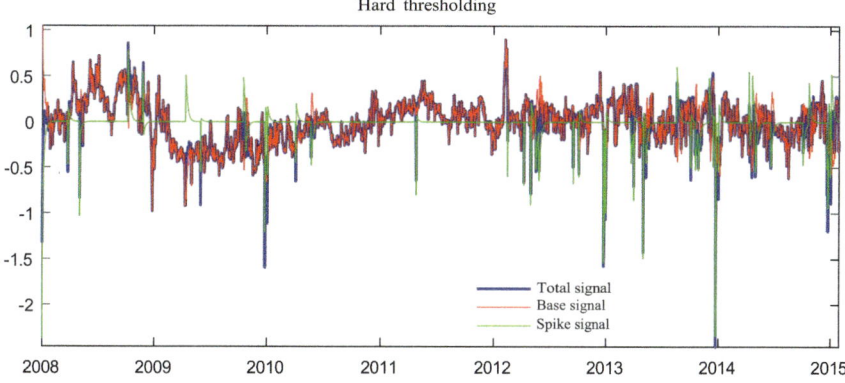

FIGURE 7.7   Result of hard thresholding where we see the initial signal (blue), the filtered base signal (red), and the filtered spike signal (green).

mean reversion around the very short lags, allowing λ to take care of the longer lags and hence fit the ACF better at these lags.

A.4 Forecasting the Base Signal

We here take a slightly closer look at the forecasting exercise of Section 7.6.4. Figure 7.9 plots the cumulative squared forecast errors (CSFE) for the three benchmark models as compared to the $\mathcal{BSS}$ forecasts, when applied to the EEX data set. That is, we plot

$$\text{Excess CSFE}(x) = \text{CSFE}(x) - \text{CSFE}(\mathcal{BSS}), \quad x = \text{FBM,OU,RW},$$

so that postive values are in favor of the $\mathcal{BSS}$ model. We use the first 400 observation for estimation and then forecast h periods into the future. After this we re-estimate the model the next day, forecast h

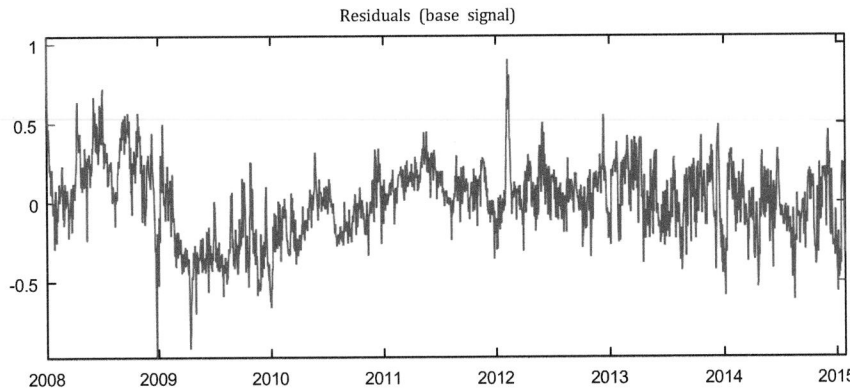

FIGURE 7.8   Residuals of log-prices after filtering for seasonality and spikes. This signal will be modelled as a $\mathcal{BSS}$ process.

periods into the future and repeat. The figure shows how the $\mathcal{BSS}$ model fares best, but as the horizon h increases, the performance of the OU model gets closer to that of the $\mathcal{BSS}$. Both the fBm and RW models perform very poorly in general for the reasons discussed in Section 6.4.

Appendix B: More Estimators of the Roughness Index

In this section, we will briefly review some of the estimators of the roughness index α and explain how they fit in our setup; we refer to Gneiting et al. (2012) a more detailed exposition. For our purposes, we face two major obstacles when estimating the roughness index of electricity prices; (i) jumps (spikes) in the price process and (ii) relatively low frequency (daily) data. In Appendices C.1 and C.2 we will look closer at how the estimators behave under (i) and (ii), and we find that the former causes a downward bias in the estimation of the roughness index, while the latter is innocuous in our setup. The downward bias introduces by (i), i.e. the spikes, is the reason for us filtering out the spike signal Y, before estimating α. This has not been done in previous studies, which leads us to conjecture that earlier studies might have found values of α which are more negative than they are in reality, cf. Section 7.6 and in particular Table 7.6.

The most straightforward estimation of the roughness index is to use relations similar to Equation (7.1). For a stationary Gaussian process X we actually have the more general (see Gneiting et al., 2012) result concerning the variogram of order $p > 0$,

$$\gamma_p(h) = \frac{1}{2}\mathbb{E}\left[\left|X_{t+h} - X_t\right|^p\right] = c_p\,|h|^{(2\alpha+1)p/2} + o\left(|h|^{(2\alpha+1+\beta)p/2}\right), \quad \text{as } h \downarrow 0, \tag{B.1}$$

where again α is the roughness index of the process, $\beta \geq 0$ and $c_p > 0$.* Now, for a $p > 0$, an estimator of $a = (2\alpha + 1)p/2$ is obtained by regressing $\log|h|$ on an estimate of the log-variogram, i.e.

$$\log\hat{\gamma}_p(h) = C + a\log|h| + \varepsilon_h, \quad h = 1, 2, \ldots, m, \tag{B.2}$$

where C is a constant, m is a "bandwidth" parameter, and E_h an error term. We then estimate $\hat{\alpha}(p) = \hat{a}/p - \frac{1}{2}$, where $\hat{a}$ is obtained from the OLS regression in (B.2) with

---

* Note, that even though this definition is stated for Gaussian processes, it equally well applies to volatility modulated $\mathcal{BSS}$ (and OU) processes as shown in Proposition 2.1. of Bennedsen et al. (2015).

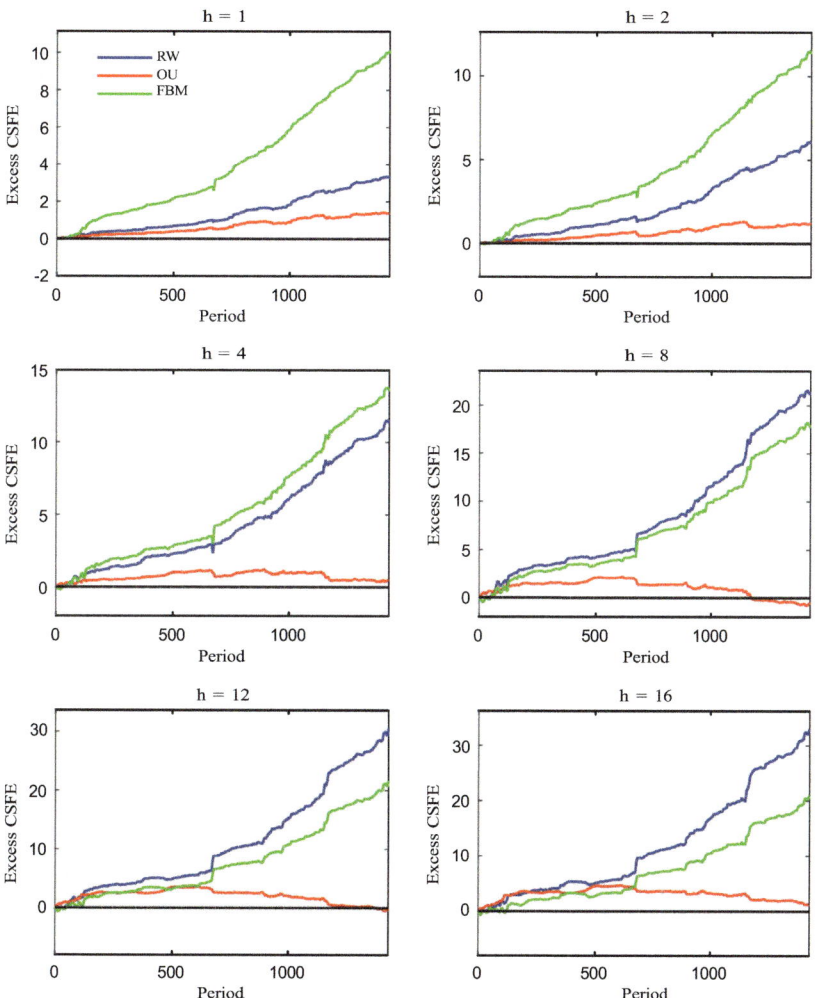

FIGURE 7.9 Cumulated squared forecast errors for h periods ahead forecast using the three benchmark models, fBm, OU and a random walk (RW) as compared to the $\mathcal{BSS}$ model.

$$\hat{\gamma}_p(h) = \frac{1}{N-h} \sum_{i=1}^{N-h} |X_{i+h} - X_i|^p.$$

Intuitively, this estimator will work, since the second part of (B.1) will be asymptotically negligible as $h \to 0$.* For this reason, it is prudent to use only very few lags of h in this regression, i.e. to set $m = 2$ or $m = 3$. In our application we set $m = 3$. For the power parameter p, a standard choice is $p = 2$ so that $\gamma_p(h) = \gamma_2(h)$ is the (second order) variogram of the process. However, $p = 1$ – which means that $\gamma_p(h) = \gamma_1(h)$ is the so-called *madogram* – has been found to be more robust to outliers (see C.1 and Gneiting et al., 2012; Bennedsen, 2016). In the following we will employ this method using both $p = 1$ and $p = 2$, and will term them "madogram" and "variogram" respectively.

---

* The properties of this estimator has been widely studied. See, e.g., Bennedsen (2016) Proposition 3.1. for a proof of consistency of the estimator.

Besides the straightforward method discussed above, there exist a wealth of more sophisticated estimators of the roughness index α (often in the form of an estimator of the Hurst index $H = \alpha + \frac{1}{2}$), see Taqqu et al. (1995) for a partial list along with a Monte Carlo study of their finite sample performances. We focus here on the one-dimensional, i.e. time series, case. Popular estimators used in the literature include Detrended Fluctuation Analysis (DFA) of Peng et al. (1994) and estimation of the generalized Hurst exponent (GEN), see e.g. Matteo (2007). Here we also consider the absolute value method (ABS) and the aggregated variance method (AGG), see Taqqu et al. (1995), as well as the Change-of-Frequency (COF) estimator of Barndorff-Nielsen et al. (2011). Both the GEN and the COF estimators rely on choosing a power parameter p, as was the case of the madogram/variogram in the regression (B.2). Thus, in the following we will write $GEN(p)$ and $COF(p)$ for the respective estimator using the power parameter p, typically $p = 1$ or $p = 2$. We note, that the $COF(2)$ estimator comes with a CLT (Barndorff-Nielsen et al., 2011), which makes this estimator particularly attractive for our purpose, since it will allow us to do inference on the roughness parameter and test whether it is statistically different from zero (recalling that zero corresponds to the $\mathcal{BSS}$ process X being actually a OU process). Of the above estimators, DFA has arguably been the most popular one in the electricity price literature as it controls for the possible non-stationarity induced by seasonal effects.

These estimators are applied to the data of Section 6 in Table 7.6.

Appendix C: Estimating the Degree of Roughness in Practice

C.1 Roughness Estimation in the Presence of Spikes

To gauge the effect that possible jumps in the price process have on the estimators of the roughness index, we here conduct a small Monte Carlo study. Table 7.8 shows the bias of the various estimators, when outliers are introduced in a rough process. We learn a few things from Table 7.8. Firstly, all estimators are virtually unbiased when no outliers are present. Secondly, when introducing outliers, all estimators become biased *downwards*; that is, in the presence of spikes, the estimate one obtains of the roughness index will indicate rougher paths than otherwise. Indeed, with only one outlier inserted, the estimators based on the power $p = 2$, i.e. the variogram, COF(2) and GEN(2), become noticably biased. In contrast, the estimators based on $p = 1$, i.e. COF(1), GEN(1), ABS and the Madogram are significantly less biased, even for a moderate amount of spikes in the process. This robustness when $p = 1$ supports a similar findings in Gneiting et al. (2012) and Bennedsen (2016). However, when spikes

TABLE 7.8   Estimating the Roughness Index in the Presence of Outliers

	Number of outliers									
	0	1	2	3	4	5	6	7	8	9
Madogr.	−0.001	−0.020	−0.035	−0.049	−0.061	−0.072	−0.082	−0.091	−0.099	−0.106
Variogr.	−0.001	−0.125	−0.165	−0.186	−0.199	−0.207	−0.213	−0.217	−0.221	−0.223
COF(1)	−0.008	−0.030	−0.048	−0.064	−0.078	−0.090	−0.100	−0.109	−0.118	−0.125
COF(2)	−0.005	−0.137	−0.175	−0.195	−0.206	−0.214	−0.219	−0.223	−0.226	−0.228
GEN(1)	−0.002	−0.018	−0.045	−0.069	−0.091	−0.087	−0.107	−0.109	−0.114	−0.124
GEN(2)	−0.004	−0.128	−0.195	−0.235	−0.266	−0.250	−0.264	−0.257	−0.249	−0.256
ABS	−0.015	0.005	0.005	0.005	−0.036	−0.075	−0.075	−0.087	−0.008	−0.030
DFA	0.013	−0.068	−0.101	−0.121	−0.136	−0.140	−0.147	−0.150	−0.154	−0.158
AGG	−0.009	−0.057	−0.057	−0.057	−0.110	−0.190	−0.190	−0.200	−0.078	−0.099

Bias of the roughness index estimators when outliers are placed in the data series. The simulated process, X, is a $\mathcal{BSS}$ process with $\alpha = −0.25$ and $\lambda = 1$ on $[0, 1]$ with $N = 500$ observations. See Bennedsen et al. (2015) for a method to simulate the $\mathcal{BSS}$ process. Outliers of size $\pm 5std(X_t)$ are then introduced equidistantly into the process. 5 000 Monte Carlo replications.

become more frequent, say more than 5 spikes (constituting 5% of the observations) even these estimators – except maybe ABS – also become biased. Given that most electricity price series arguably contain somewhat frequent spikes, this is a possible source of significant downwards bias. Indeed, this effect might explain the very low estimates of the roughness index found by previous authors, e.g. Simonsen (2002), Norouzzadeh et al. (2007) and Erzgräber et al. (2008).

C.2 Roughness Estimation using Low-Frequency Data

Besides spikes, another problem arises when we try to estimate the roughness index of (peak load) electricity prices. Electricity prices are recorded daily and are as such inherently sampled at a relatively low frequency. This is worrisome because the very definition of roughness is a local one; the roughness index is describing the very fine structure of the path of the stochastic process. This is also evident in the discussion in Section 7.3, where we saw that the definition of the roughness index is given in terms of in-fill asymptotics. In the same vein, the estimators such as the ones through the variogram in Equation (B.1) rely on the distance between observations, h, being very small. The question becomes whether it is possible at all to estimate the roughness index of a stochastic process observed so infrequently. Fortunately, the answer is in the affirmative. To see this, we conduct a Monte Carlo study where we simulate $N = 500$ observations of a rough process on $[0, T]$ and estimate its roughness index α. We do this with varying values of T and thus of the sampling frequency $dt = T/N$. Choosing $T = N$ and thinking of T as days, this mimics daily data.

The results of the Monte Carlo study are seen in Table 7.9. Again a few conclusions emerge. As expected, the bias increases when we sample less frequently; further, this is a *downwards* bias, i.e. the paths of the price process are estimated to be *rougher* than what the underlying process truly is. This behavior is amplified when we increase the value of the mean reversion parameter λ, as also seen in the table. The reason for this is that strong mean reversion can have the same effect as a negative roughness index when the data is observed infrequently. Indeed, a negative roughness index causes the increments of the process to be negatively correlated and mean reversion mimics this. The upshot is that when we sample infrequently (e.g. $dt = 1$), low roughness indices can simply be a product of strong mean reversion. As in the case of spikes, investigated in the previous section, this is a possible further source of (downward) bias in previous studies. Fortunately, when estimating the mean reversion parameter λ in the data, we often find it on the order $\tilde{\lambda} \sim 001 - 010$; see e.g. Benth et al. (2012) and also Table 7.5 in Section 7.6 where the range $0.011 - 0.048$ is found for the data in Table 7.1. Looking again at Table 7.9 we

TABLE 7.9 Estimating the Roughness Index with Low Frequency Data

	$\lambda = 0.1$			$\lambda = 0.5$			$\lambda = 1$		
$dt =$	0.1	0.5	1	0.1	0.5	1	0.1	0.5	1
Madogram	−0.001	−0.004	−0.010	−0.004	−0.034	−0.073	−0.011	−0.072	−0.137
Variogram	−0.001	−0.005	−0.010	−0.004	−0.034	−0.073	−0.011	−0.073	−0.137
COF(1)	−0.006	−0.007	−0.009	−0.006	−0.019	−0.050	−0.009	−0.050	−0.116
COF(2)	−0.004	−0.004	−0.005	−0.003	−0.017	−0.047	−0.006	−0.048	−0.113
GEN(1)	−0.003	−0.018	−0.039	−0.017	−0.097	−0.153	−0.040	−0.153	−0.203
GEN(2)	−0.005	−0.019	−0.040	−0.018	−0.098	−0.155	−0.041	−0.154	−0.204
ABS	−0.019	−0.073	−0.133	−0.075	−0.218	−0.252	−0.131	−0.253	−0.259
DFA	0.013	−0.003	−0.031	−0.002	−0.100	−0.152	−0.031	−0.152	−0.186
AGG	−0.013	−0.068	−0.129	−0.070	−0.214	−0.248	−0.127	−0.249	−0.256

Bias of the roughness index estimators when sampling infrequently and for different values of the mean-reversion parameter λ. The simulated process, X, is a BSS process with $\alpha = -0.25$ and $\lambda \in \{0.1, 0.5, 1, 2\}$ on $[0, T]$ with $N = 500$ observations and for T $\in \{50, 250, 500\}$. This corresponds to sampling the process at $dt \in \{0.1, 0.5, 1\}$ (days). 5 000 Monte Carlo replications.

notice that several of the estimators, such as *COF*(1) and *COF*(2), are not very biased in this situation, even for daily data ($dt = 1$). This makes us confident that it is, after all, feasible to estimate the roughness index, even using infrequently sampled data. This also holds for the DFA estimator, but not for the ABS and AGG estimators.

C.3 Lessons from the Simulation Experiment

The two previous sections above show that one should be careful when estimating the roughness index of electricity prices. In particular, spikes in the time series will cause the estimators to be downward biased. The same is true for data sampled with low frequency, especially when the mean reversion parameter, λ, is large in magnitude. We conclude that it is important to filter the prices for spikes before estimating the value of the roughness index. Further, it is preferable to use an estimator that is not too biased in the low frequency environment. Given the above and the fact that there is a CLT attached to it, we recommend using the COF(2) estimator. However, in practice it is prudent to use several estimators, including ones that are robust to outliers, such as the ones using $p = 1$; we recommend employing several estimators and only trusting the results when most of them agree.

8

Investing in the Wine Market: A Country-Level Threshold Cointegration Approach

Lucia Baldi,
Massimo Peri and
Daniela Vandone

8.1 Introduction

In recent years, investments in commodities, via commodity futures and commodity index funds, have grown rapidly. The appeal of investing in commodities is generally attributed to the low correlation with traditional stocks, which allows for portfolio diversification benefits, that is reduction of risk for any given level of expected return (Erb and Harvey 2006, Gorton and Rouwenhorst 2006, Sanning et al. 2007, Geman and Kharoubi 2008, Masset and Henderson 2008, Buyuksahin et al. 2010, Chong and Miffre 2010, Masset and Weisskopf 2010), and the evidence that profitable trading strategies can be constructed. Indeed, a number of studies find that commodity returns can be predicted by a number of variables (Hong and Yogo 2011), and profitable momentum and term structure strategies can be implemented (Schwarz and Szakmary 1994, Crain and Lee 1996, Brooks et al. 2001, Yang et al. 2001, Erb and Harvey 2006, Gordon et al. 2007, Miffre and Rallis 2007, Fuertes et al. 2010, Stoll and Whaley 2010, Szakmary et al. 2010). The majority of these studies were conducted in the framework of cointegration analysis and threshold cointegration analysis; from a methodological standpoint, this allows the analysis of the interdependence between asset classes, their relationships, and the speed of adjustment to the long-run equilibrium.

In our paper we focus on wine, a commodity of growing importance not only in European countries (often referred to as 'Old World Countries'), that have historically dominated this market, but also in the so-called 'New World Countries', e.g. China, South Africa, Chile, that have registered astonishing rates of growth, at times with extraordinary speed, entering not only the lower-quality segment, but also the medium–high segment, once the exclusive domain of traditional long-established producers (Aylward 2003, Aylward and Turpin 2003).

DOI: 10.1201/9781003265399-10

The aim of this study is to analyse the long-run relationship between wine share price indexes and general stock market indexes, examining their different speeds of adjustment to the long-run equilibrium. The goal is to provide traders with signals of information inefficiency that could be exploited to make profitable investment strategies.

Indeed, according to Fama (1965), in an efficient market all available information is instantaneously and completely reflected in stock prices. Thus, it is not possible to forecast future price evolution on the basis of previous stock price variations because this information is already integrated in the present price. Conversely, if there is no efficiency in the market, then active traders can anticipate price movements over the short run and make profitable investments from buying and selling stocks (Siourounis 2002). Specifically, if two economic series, such as financial indexes, are cointegrated, this implies that movements of one series can be used to predict fluctuations of the other: in other words, traders can anticipate the evolution of the dynamics of a stock market index by knowing that of another stock market index: the key question we pose is whether investors can exploit the dynamics of stock markets to predict wine indexes returns and thus make profitable investments by buying and selling stocks.

Following the seminal paper of Balke and Fomby (1997), we use threshold cointegration rather than cointegration since this econometric technique allows for non-linear adjustments in the long-run equilibrium (Hokkio and Rush 1989, Anderson 1997, Perez-Quiros and Timmermann 2000, Maasoumi and Racine 2002). Indeed, since the efficiency hypothesis assumes no transaction costs, free and symmetric information, as well as rational investors, studying stock prices and efficiency using linear cointegration techniques corresponds to the assumption of symmetric, linear and continuous stock price adjustment dynamics. These set of assumptions seem to be very constraining, since markets present friction, such as transaction costs, noise traders and imitative behaviour, and this can imply price adjustment dynamics towards fundamental values that are discontinuous and non-linear (Enders and Siklos 2001, Shively 2003, Prat and Jawadi 2009).

The economic rationale for considering the possibility of a non-linear rather than a linear type of adjustment to the long-run equilibrium is that the first allows prices to adjust in a different way to large or small deviations from the long-run equilibrium level. This implies that the dynamic behaviour of the rate of return differs according to the size of the deviation. In fact, this methodology captures those adjustments that are active only when deviations from the equilibrium exceed a threshold, which is often represented by transaction costs (Sercu et al.1995, Jawadi and Koubaa 2004, Aslandis and Kouretas 2005). In fact, traders may not act immediately as prices move, due to the possibility of 'mis-price deepening' (Shleifer 2000), they only act when the expected profits exceed the costs. In this sense, the threshold cointegration constitutes an adequate specification, since the error correction mechanism is active only above a certain size of the variation compared with the equilibrium.

In other words, in the linear cointegration approach the adjustment parameters are assumed to be constant within the period analysed, while in the threshold cointegration approach the error correction terms (ECTs) are inactive when the value is inside a given range, but are active above a certain threshold. When the deviation from equilibrium is above the critical threshold, agents enter the market to implement profitable arbitrage activities, moving the system back to the equilibrium (McMillan 2003, 2005).

This study makes use of the Mediobanca Global Wine Industry Share Price Index and the composite stock market indexes for a selection of countries belonging to the Old World Countries and the New World Countries—Australia, Chile, China, France and the United States—where wine shares are listed. The Mediobanca Wine Index encompass companies operating in the wine industry listed on stock markets worldwide.

The originality of this work is twofold. Firstly, the threshold cointegration methodology is applied to the wine sector; to the authors' knowledge, no previous study has investigated the long-term dynamics between the share price of wine companies and whole stock market indexes. However, as underlined above, such an analysis is relevant since it enables a better understanding of financial investment opportunities due to wine share price adjustment dynamics and stock market inefficiencies. Secondly, the focus of the analysis is not on fine wine, as in most of the current literature, but on non-fine wine (which will be referred to as normal wine in the remaining part of the article). The interest in normal wine stocks as an asset class that traders can use for investment purposes is due to the growing size of the wine market in countries not historically suited to the production of wine (such as China, Australia and Chile, where large companies have recently become established in the market), their increasing stock price volatility and better performance. This paper presents the first use of the Mediobanca Global Wine Industry Share Price Index in an academic context, presenting this databank to the wider research community through one of the uses that can be made of this resource.

The paper is organized as follows. Section 8.2 describes the theoretical framework. Section 8.3 presents the dataset used for the purpose of the study and a brief analysis of index performance. Section 8.4 proposes the econometric methodology and Section 8.5 develops the empirical results. Section 8.6 includes the discussion and final remarks.

8.2 Wine: The Theoretical Framework

The economic literature on commodities presents a wide range of interesting studies that have analysed the potential of wine as a financial investment (Fogarty (2006) presents a detailed overview of the subject). Most investigated has been the estimation of the rate of return of wines, comparing this value with the return from other common assets, both in the mean value and in the conditional volatility or covariance (see, among others, Krasker (1979), Weil (1993), Burton and Jacobsen (2001) and Sanning *et al.* (2007)).

This literature has been interested only in fine and rare wines, and, using auction data of specific wines or composite indexes (e.g. *LiveEx)*, its results have encompassed a broad mix of empirical findings that has provided financial guidance in areas such as prices (Jaeger 1981, Weil 1993, Fogarty 2006), buyer's premium (De Vittorio and Ginsburgh 1996), speculative bubbles (Jovanovic 2007), excellence of vintage and respective ranking (Jones and Storchmann 2001, Masset and Henderson 2008), fluctuations in inventories (Bukenya and Labys 2007) and portfolio diversification strategies. Specifically, as far as diversification is concerned, Sanning *et al.* (2007) use the Fama–French three-factor model and the Capital Asset Pricing Model to directly assess the risk–return profile of wines as compared with equities and find that investment-grade wines benefit from low exposure to market risk factors, thus offering a valuable dimension for portfolio diversification. Masset and Henderson (2008) find that investing in the wine market might achieve an attractive performance in terms of both average returns and volatility since wine returns are only slightly correlated with other assets and, as such, they can be used to reduce the risk of an equity portfolio. Moreover, Fogarty (2006) finds that the performance of Australian wines is comparable to that of Australian equities, and Burton and Jacobsen (2001) demonstrate that the returns of a wine portfolio consisting only of wines from the 1982 vintage compare favourably with that of the Dow Jones.

Among these studies, French fine wines are the most commonly analysed, together with Australian high-quality production.

Despite the extensive empirical literature on the wine market, to the best of our knowledge there are no studies that have analysed, from a financial viewpoint, non-fine wine. However, in recent years the importance of normal wine has grown rapidly, specifically in New World Countries. Indeed, while historically the wine market has been dominated by European countries (often referred to as Old World countries), since the beginning of the 1990s new producing countries have found their way into the market, showing strong competitive potential due to their innovative strategies in production and trade (Campbell and Guibert 2006). Europe (in particular, France, Italy, Germany and Spain) still occupies a leading position on the world wine market, accounting globally for 49% of growing areas and 63% of wine production (data from the FAO databank for the year 2007). However, wine is also currently produced in Argentina (accounting for 9% of world production), the USA (8%), China (5%), Australia (4%), South Africa (4%) and Chile (3%). Contrary to many traditional winegrowing countries, where production has dropped by 25% compared with volumes in the 1990s, 'New World' countries have registered astonishing rates of growth, at times with extraordinary speed (as in the case of China), entering not only the lower quality segment, but also the medium–high segment, once the exclusive domain of traditional long-established producers (Aylward 2003, Aylward and Turpin 2003).

8.3 The Mediobanca Index: Evidence on Stock Performance

8.3.1 The Data

The analysis presented in this paper makes use of data on the wine share price index and the composite stock market index of the stock exchange for five countries belonging both to Old and New World Countries: Australia, Chile, China, France and the US. The dataset covers the period starting January 1, 2001 up to the end of February 2009. All series are expressed in euro and appear in the econometric model in logarithmic form. The wine series is the Mediobanca Global Wine Industry Share Price Index from Mediobanca,[*] which covers companies operating in the wine industry, listed on regulated stock markets and quoted for at least six months. Prices are computed daily and represent a financial benchmark of wine, measuring and monitoring the dynamic of risk and return of wine stocks. The index is calculated for each of those countries whose stock had traded at least three titles that meet some specific selection criteria.[†]

Data on the composite stock market series are daily prices supplied by Datastream, and represent the performance of the whole stock market for a given country. Specifically, the data used is for the Australian S&P/ASX300 index, the Chilean IPSA index, the Chinese SSE index, the French CAC40 index and the US S&P500 index.[‡]

All indexes are 'capitalization-weighted', that is the components are weighted according to the total market value of their outstanding shares.

Both series are 'price' indexes, expressing the dynamics of stock prices alone, without the component of income represented by the distribution of dividends. 'Total return' indexes, which also include dividends, are available for all series, but the pure price index is preferred. The rationale behind this choice is that the dividend policy adopted by each company is not relevant in the analysis presented here, as, in general, dividends do not reveal the level of volatility that would be necessary to influence the null hypothesis of 'no cointegration' among a set of share price indices (Dwyer and Wallace 1992, Subramianan 2009).

8.3.2 Performance

Figure 8.1 shows the cumulative stock return for wine indexes over the period from 2001 to 2009. At the end of this eight-year period France, the US and Australia had achieved almost similar total cumulated returns (respectively 37%, 37% and 40%), although with very different dynamics. In particular, the French wine index suffered a decline between the beginning of 2001 and mid-2002, followed by a stable rise and then steeper growth from the beginning of 2005 up to the beginning of 2008. During this last period the French wine index almost tripled. Since mid-2008, however, the index has decreased in line with other financial assets and stock markets as a result of the global economic crisis.

---

[*] http://www.mbres.it/

[†] The Mediobanca index includes wine companies selected according to the following characteristics: companies listed on regulated markets; series of quotes of at least six months; at least 50% of revenues must come from initiating wine; commitment as direct management in the production cycle. The panel index is comprised of 42 stocks and has an aggregate market capitalization of €14.3 bn.

[‡] The S&P/ASX300 index is the stock market index of Australian stocks listed in the Australian Securities Exchange; the IPSA is a stock market index composed of the 40 stocks with the highest average annual trading volume in the Santiago Stock Exchange; the SSE is the composite index from the Shanghai Stock Exchange; CAC40 is the benchmark French stock market index and includes the 40 most significant stocks in terms of liquidity; and, finally, the S&P500 includes the prices of 500 large-cap common stocks actively traded on the two largest American stock markets, NYSE and NASDAQ.

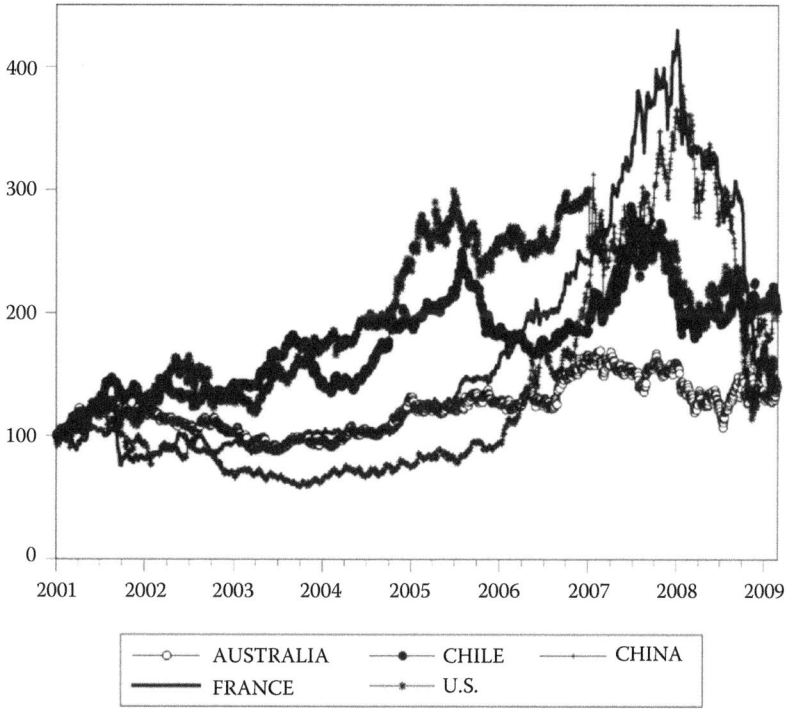

FIGURE 8.1 Cumulative stock return for wine indexes for the five countries considered. **Source: Our calculations on Mediobanca data.**

In Australia, after an initial bearish trend, the wine index showed a stable pattern over the period, while in the US the wine index rose between 2003 and 2006 and later declined, although to a lesser extent than its French counterpart.

Chile and China reached a higher level of cumulative returns over the 2001–2009 period (respectively 101% and 94%), but again with different patterns. In particular, the value range of the Chilean index went from a minimum of 99.50 points to a maximum of 290 points, while in China the index declined significantly until the end of 2005 and then climbed sharply at the beginning of 2008. The strong performance of the index for China continued into 2009.

Estimates of the average daily return and volatility for the wine and composite index were also calculated and are summarized in Table 8.1. For France, the US, Chile and China the average daily return of wine indexes is higher than that of composite indexes (respectively CAC40, S&P500, IPSA, and SSE) over the 2001–2009 period. More specifically, while the average return of wine indexes is always positive, the average return of composite indexes is negative for all countries except Chile. The most significant difference is that from the US, where the average daily return of the wine index is 95% while the S&P500 average daily return is –28%. Looking at yearly intervals, it becomes apparent that only rarely do wine indexes yield negative returns, almost always outperforming the composite indexes.

Australia is the only exception. There, apart from the 2001–2002 interval, the wine index average return is lower than that of the S&P300. The Australian wine sector has achieved many successes in recent decades through government measures promoting exports and low taxes. Moreover, the Australian wine industry is characterized by high levels of concentration (four companies accounting for over 75% of production), providing economies of scale in producing value-for-money wines. However, after a planting boom during the mid to late nineties, from 2001 onwards this country has gone through a very difficult period mainly due to strong supply pressure. As a result, a number of listed

TABLE 8.1   Daily Wine and Composite Index Returns and Volatility

	Average daily return (%)		Cond. volatility[a] (%)	
	Wine	Comp.	Wine	Comp.
Australia				
2001–2002	13.81	−1.24	1.02	1.08
2003–2004	−2.49	19.42	0.90	0.78
2005–2006	0.49	23.18	1.21	0.90
2007–2008[b]	−12.40	−5.05	1.81	2.69
All period	22.40	30.96	1.37	1.27
Chile				
2001–2002	29.02	−10.40	0.90	1.28
2003–2004	11.33	36.38	0.90	1.13
2005–2006	0.90	32.09	0.80	0.95
2007–2008[b]	17.19	3.71	1.40	1.73
All period	75.54	32.84	1.05	1.28
China				
2001–2002	−4.19	−14.98	4.54	1.70
2003–2004	2.50	−9.74	1.32	1.38
2005–2006	58.92	7.79	1.67	1.44
2007–2008[b]	21.80	77.40	2.46	2.69
All period	41.14	−28.39	2.03	1.93
France				
2001–2002	−7.09	−25.73	1.28	1.95
2003–2004	5.02	11.06	0.91	1.02
2005–2006	40.07	21.63	0.90	0.80
2007–2008[b]	22.83	−11.89	1.45	3.07
All period	67.45	−26.53	1.15	1.78
U.S.				
2001–2002	28.15	−14.58	1.68	1.69
2003–2004	31.91	4.95	1.15	1.13
2005–2006	9.74	12.90	1.37	0.81
2007–2008[b]	−28.03	−13.58	2.51	1.90
All period	95.05	−28.52	1.87	1.36

Source: Own calculations with Mediobanca and Datastream data.

[a] All the volatilities were computed via the GARCH(1,1) model except for the USA composite index within the 2003–2004 range where the GARCH(2,1) model was used.

[b] The 2007–2008 range also includes 42 observations for 2009.

companies have announced profit downgrades (including Southcorp company, Australia's largest wine producer), perhaps contributing to making wine stocks less of a profitable investment.

Elsewhere, the risk features of both indexes are more heterogeneous. Chile and France show a lower wine index volatility, whereas for Australia, the US and China the volatility of the wine index is slightly higher than that of the composite. It is interesting to note that during the financial crisis in the period 2007–2008, all countries except the US show a conditional volatility of the wine index that is lower than that of the composite index.

Overall, what emerges from the analysis of the returns is that, in general, wine indexes performed well during the last decade and investors could have earned greater returns by investing in these indexes rather than solely investing in the domestic composite indexes, although in some cases they would have been exposed to greater volatility.

The superior performance of the wine index is clearly visible from the graphs shown in Figure 8.2, which represent the cumulative abnormal return of the wine indexes, that is the market-adjusted abnormal return.

The abnormal return is estimated by subtracting the composite stock market index from the return of the wine index. The resulting evidence is quite interesting. Apart from Australia, whose wine industry suffered a crippling financial crisis during the mid-2000s (see above), the cumulative abnormal returns remained significantly positive for almost all years and in all countries.

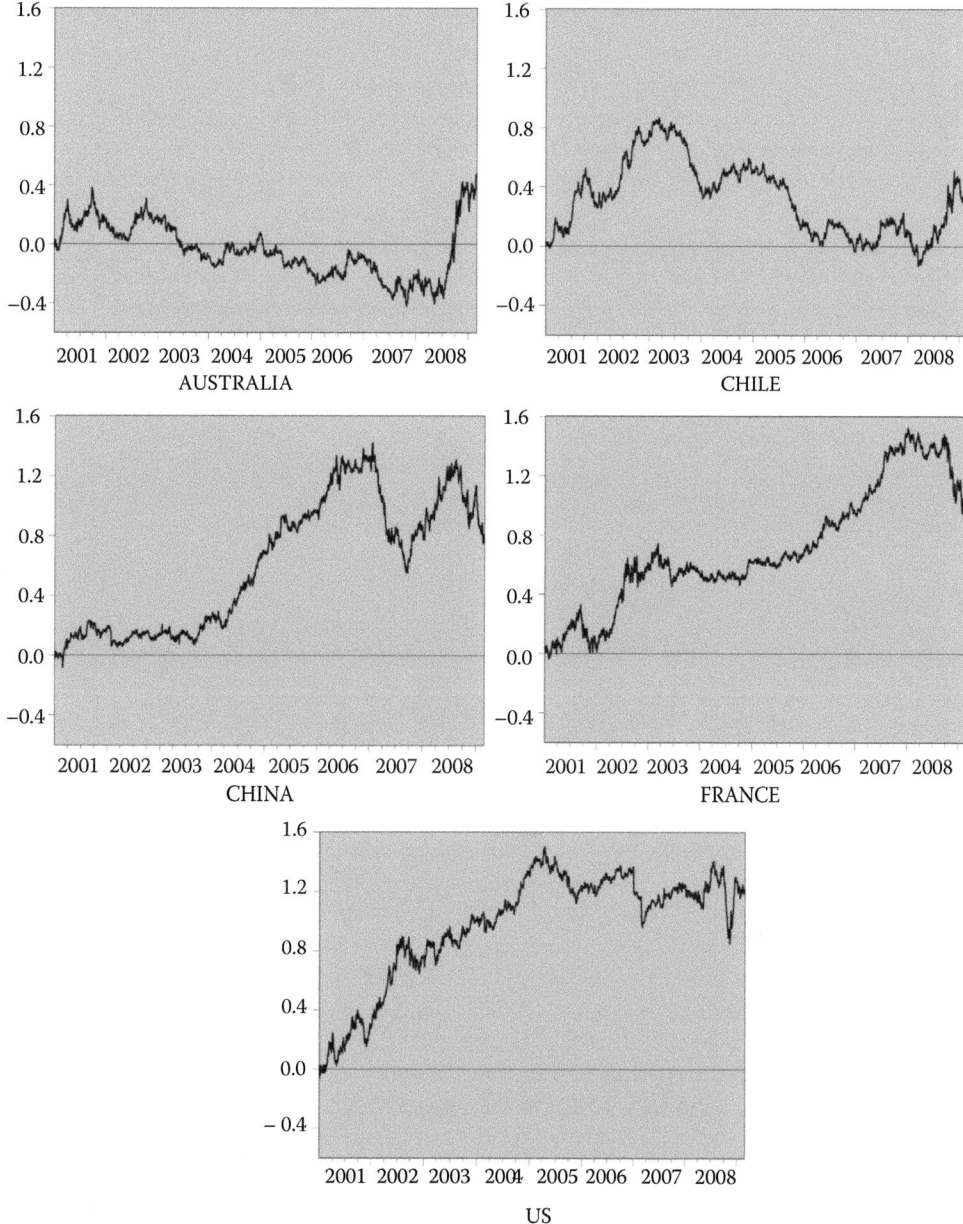

FIGURE 8.2 Abnormal return in each of the five countries considered. **Source: Our calculations on Mediobanca data.**

From Figure 8.2 it can be observed that the US registered the best performance of the countries under analysis. This is not surprising given that the American wine industry leads the group of New World wine producers and is the world's fourth largest wine producer, however remaining a net importer of wine (Canning and Perez 2008, Goodhue *et al.* 2008). The growth of the sector started several decades ago and continues to this day, not having had any apparent slowdown recorded since its early days. As in other New World wine industries, firm concentration is particularly high, even if the industry has been evolving recently with the proliferation of new wineries (Insel 2008). In addition, the sector benefits from a concentrated and efficient distribution system for products. These factors together help explain the good performance of US wine stocks in the financial markets.

Elsewhere, the abnormal return for France is high, but lower than that of the US. The country, traditionally one of the largest world producers and consumers of wine, has a fragmented industry subject to too many controls (Terblanche *et al.* 2008). Since 2001, French wines have been going through a slowdown period, losing market share both in domestic and in export markets, a situation exacerbated by negative currency effects arising from a strong euro (Hussain *et al.* 2008).

On the other hand, the Chilean wine sector has recorded a notable production and export record during the last decade and many wine firms have developed the competence to be present in an increasingly large number of countries (Giuliani and Bell 2005, Gwynne 2008). Concha Y Toro, the market leader, has become the most successful Chilean exporting company, transforming its business radically and becoming the world's first winery to list on the New York Stock Exchange. It should not be surprising, therefore, that the Chilean wine sector's performance in terms of return should have been positive.

Finally, the Chinese wine industry, although still oriented towards the domestic market (which is still relatively small and is dominated by the medium- and low-quality segment), has made very rapid progress in recent years (Jenster and Cheng 2008, Mitry *et al.* 2009), with the performance of wine in the stock market having always been satisfactory, surpassing that of the Shanghai Stock Exchange.

As a concluding remark, it can therefore be argued that, in general, wine indexes have performed well compared with composite indexes during the last decade, and that investors could have earned greater returns by investing in this market.

8.4 Econometric Methodology

Evidence of cointegration (Granger 1981) among several indexes of stock prices suggests that series have the tendency to move together in the long run even if they experience short-term deviations from their common equilibrium path (Masih and Masih 1997, Patra and Poshakwale 2008). These traditional models assume that the adjustment process to maintain *equilibrium* occurs at every time period. However, many situations, and in particular stabilizations of commodity prices, are often characterized by discrete interventions. In recent years, two main classes of models have been proposed in the literature to characterize this kind of non-linear adjustment process. One class considers Markov-switching Vector Error Correction models, assigning probabilities to the occurrence of different regimes (Hamilton 1989, Krolzig 1997). The second class is based on Tong and Lim's (1980) approach using a Self-exciting Autoregressive Model where the regimes that have occurred in the past and the present are known with certainty, 'certainty' being established using statistical techniques. In this framework, Balke and Fomby (1997) introduced the concept of 'threshold cointegration', a feasible estimation methodology that allows the adjustment process to move differently in separate regimes. They hypothesized that this movement towards a long-run equilibrium may not occur in every time period, but only when the deviation from equilibrium exceeded a critical threshold. Following Balke and Fomby (1997), in this paper we apply a threshold vector error correction model (TVECM), with a threshold effect based on an error correction term. In the case of two regimes, Balke and Fomby (1997) present a TVECM of order $L + 1$ that takes the form

$$\Delta x_t = \begin{cases} A_1' X_{t-1}(\beta) + u_t, & \text{if } w_{t-1}(\beta) \leq \gamma \text{ regime1} \\ A_2' X_{t-1}(\beta) + u_t, & \text{if } w_{t-1}(\beta) > \gamma \text{ regime2} \end{cases} \tag{8.1}$$

where

$$X_{t-1}(\beta) = \left\{ \begin{array}{c} 1 \\ w_{t-1}(\beta) \\ \Delta x_{t-1} \\ \Delta x_{t-2} \\ \vdots \\ \Delta x_{t-1} \end{array} \right\} \tag{8.2}$$

and x_t is a p-dimensional time series I(1) cointegrated with one ($p \times 1$) cointegrating vector β, $w_t(\beta)$ is the ECT, and u_t is the error term assumed to be an iid Gaussian sequence with a finite covariance matrix. A_1 and A_2 are matrices of coefficients describing the dynamics in each regime, while γ is the threshold parameter. Values of w_{t-1} below or above the threshold γ allow the coefficients to switch between regimes 1 and 2; in particular, the estimated coefficients of w_{t-1} of each regime denote the different adjustment speeds of the series towards equilibrium.

Hansen and Seo (2002) provided an estimation method for TVECM via maximum likelihood, which involves a joint grid search over the threshold parameter and cointegrating vector. In order to test for threshold cointegration, Tsay (1989, 1998) proposed non-parametric non-linearity tests, while Andrews (1993), Hansen (1996), Balke and Fomby (1997) and Lo and Zivot (2001) presented different methods of estimation based on the Lagrange Multiplier (LM) statistics. More recently, Hansen and Seo (2002) developed two SupLM (Supremum Lagrange Multiplier) tests for a given or estimated β using a parametric bootstrap method to calculate asymptotic critical values with the respective p-values. The first test is denoted as

$$\sup LM^0 = \sup_{\gamma_L \le \gamma \le \gamma_U} LM(\beta_0, \gamma)$$

and would be used when the true cointegrating vector β is known *a priori.* The second test is used when the true cointegrating vector $\tilde{\beta}$ is unknown and the test statistic in this more general case corresponds to

$$\sup LM = \sup_{\gamma_L \le \gamma \le \gamma_U} LM(\tilde{\beta}, \gamma)$$

where $\tilde{\beta}$ is the null estimate of the cointegrating vector. In these tests, the search region $[\gamma_L, \gamma_U]$ is set so that γ_L is the π_0 percentile of $\tilde{w}_{t-1}$, where $\tilde{w}_{t-1} = w_{t-1}(\tilde{\beta})$, and γ_U is the $(1 - \pi_0)$ percentile.[*]

8.5 Empirical Results

To implement the asymmetric cointegration approach, we carried out several steps. First, we tested the degree of integration of the variables via the Augmented-Dickey Fuller test (ADF) and the Philips–Perron test (PP). Subsequently, cointegration (Johansen 1988, Johansen and Juselius 1990) and Granger causality (Granger 1969) between the price pairs (wine and composite index) were tested for each of the countries analysed. The following step entailed a test for the presence of threshold cointegration. Finally, TVECM was run using the Hansen and Seo (2002) procedure.

Table 8.2 shows the results of the ADF and PP tests, where Δ in front of variable names indicates the differentiated series. It emerges that all the series are I(1) with and without trend.

Since the price series have a unit root, the presence of cointegration between the series can be tested following the Johansen approach using the Trace and Maximum-Eigenvalue tests. Both tests were conducted including an intercept in the cointegrating equations and estimating the model with a linear

---

[*] Andrews (1993) argued that setting π_0 between 0.05 and 0.15 is generally a good choice.

TABLE 8.2   Test for Unit Root and Stationarity[a]

		No trend		With trend	
		ADF	PP	ADF	PP
Australia	Wine	−1.904	−1.817	−2.442	−2.331
	Δ Wine	−45.599	−45.665	−45.588	−45.653
	Composite	−0.980	−0.952	0.191	0.294
	Δ Composite	−45.995	−46.007	−46.032	−46.048
Chile	Wine	−2.312	−2.304	−2.160	−2.432
	Δ Wine	−42.741	−43.048	−42.767	−43.063
	Composite	−0.770	−0.895	−1.508	−1.713
	Δ Composite	−38.502	−38.390	−38.493	−38.381
China	Wine	−0.561	−0.561	−1.502	−1.503
	Δ Wine	−44.661	−44.638	−44.659	−44.636
	Composite	−1.098	−1.130	−1.205	−1.230
	Δ Composite	−46.186	−46.211	−46.229	−46.248
France	Wine	−0.803	−0.866	0.887	0.371
	Δ Wine	−43.223	−43.858	−43.248	−43.870
	Composite	−1.360	−1.110	−1.329	−1.060
	Δ Composite	−48.517	−48.975	−48.508	−48.967
U.S.	Wine	−1.869	−1.855	−0.604	−0.522
	Δ Wine	−46.838	−46.846	−46.922	−46.940
	Composite	−1.306	−0.877	−1.734	−1.298
	Δ Composite	−48.517	−48.975	−48.508	−48.967

[a] 1% critical value: ADF and PP −3.430; ADF and PP with trend −3.960.

trend. The results shown in Table 8.3 indicate the presence of a linear cointegration relationship only in France. In the other countries the results indicate the absence of a cointegration vector at the 0.05 critical value, leading to the conclusion that the Australian, Chilean, Chinese and American wine share price indexes and composite stock market indexes have an unlikely long-term linear relationship. Turning to market information efficiency, cointegration estimates may lead to the conclusion that, generally, traders cannot use movements of one series to predict fluctuations of the other.

However, as already pointed out, traders act on the market only when expected profits exceed costs, such as transaction fees and other market friction, that is only when the deviation from the long-run equilibrium is over a critical threshold. Consequently, we perform a threshold cointegration analysis. The presence of a threshold was estimated via the application of the Hansen and Seo (2002) SupLM test (when β is estimated) using a parametric bootstrap method with 3000 replications. The results of the tests are reported in Table 8.4. The residual bootstrap value of the SupLM test provides evidence of the presence of threshold cointegration for all the cases studied except for Australia.

The robustness of the threshold results is also supported by the rejection of the null of the equality of ECT coefficients between the two regimes detected. Apart from the Australian case of no threshold cointegration, the *p*-value of the Wald test is significant for all countries.

The threshold value identified in the French series detects the presence of two regimes with different adjustment speeds in the long-run equilibrium. Table 8.5 reports the estimated coefficients for the TVECMs and the related graphs that exhibit the error correction effect, i.e. the estimated regression functions of the wine and composite index as a function of ECT, holding the other variables constant. The first regime, defined as the *usual regime,* includes the majority (94%) of the observations, while the second, defined as *unusual,* contains the remaining 6% of observations. As can be seen from the figure,

TABLE 8.3 Cointegration Test between the Wine and Composite Indexes

Series	Hypothesized No. of CE(s)	Trace test	0.05 Critical value	Max-Eigen test	0.05 Critical value
Australia	None	6.882	15.410	6.270	14.070
	At most 1	0.612	3.760	0.612	3.760
Chile	None	12.227	15.410	10.363	14.070
	At most 1	1.864	3.760	1.864	3.760
China	None	8.124	15.410	5.730	14.070
	At most 1	2.394	3.760	2.394	3.760
France	None	23.735	15.410	22.640	14.070
	At most 1	1.097	3.760	1.097	3.760
U.S.	None	6.591	15.410	6.290	14.070
	At most 1	0.301	3.760	0.301	3.760

Lag(s) interval: lag=1 for Australia and China, lag=2 for France, United States and Chile (selected by the Akaike Information Criterion in VAR). Trend assumption: linear deterministic trend.

TABLE 8.4 Threshold Cointegration Test

	Australia	Chile	China	France	USA
Test statistic value (SupLM)	15.177	25.282	20.650	23.178	24.500
Residual bootstrap value	0.133	0.049	0.030	0.066	0.040
Threshold value		3.480	−1.348	−9.477	0.284
Estimate of the cointegration vector		0.381	1.562	3.624	1.212
Wald test for equality of ECM coefficient		5.416	10.736	8.231	7.242
p-Value		0.067	0.005	0.016	0.027

TABLE 8.5 Threshold VECMs between the French Wine Index and the CAC40 Composite Index

	Usual regime (94% of obs.)		Unusual regime (6% of obs.)	
	Wine	Comp.	Wine	Comp.
W_{t-1}	−0.001	0.000	−0.008	0.057
	−1.830	*0.696*	*−0.520*	*2.480*
Intercept	−0.009	0.005	−0.074	0.533
	−1.730	*0.639*	*−0.538*	*2.466*
Δ Wine $_{t-1}$	−0.009	−0.013	0.037	−0.143
	−0.315	*−0.376*	*0.330*	*−0.757*
Δ Wine $_{t-2}$	0.000	0.014	0.273	0.166
	−0.012	*0.347*	*2.070*	*0.768*
Δ Composite $_{t-1}$	0.080	−0.076	0.160	0.143
	3.541	*−2.504*	*2.665*	*1.455*
Δ Composite $_{t-2}$	0.037	−0.016	0.020	−0.090
	1.795	*−0.466*	*0.263*	*−0.946*

Note: t-Student obtained by Eicker-White standard errors and reported in italics.

in the usual regime the ECT coefficients are quite close to zero, indicating that the variables are close to a random walk.

In the unusual regime the speed of the adjustment coefficient of the composite index (CAC40) is significant and higher with respect to the wine index. In particular, when the gap between the two price series exceeds a critical threshold ($\gamma > -9.477$) the speed of the domestic stock market index's response in restoring the long-run equilibrium is seven time faster than the wine share price index. Therefore, assuming that the long-run relationship is governed by the composite index, the different speeds of adjustment could be used by investors to achieve profitable gains. Hence, when the price gap is over a critical threshold, informed investors operating in the wine sector exploiting market inefficiency can make gainful investments just by looking at the price adjustment dynamics of the domestic stock market and forecasting fluctuations in wine stocks.[*]

Considerations similar to those relating to the French case could be formulated for the United States, whose TVECM results are reports in Table 8.6. Here the usual regime includes 61% of observations and, as for France, it is close to a random walk. Only when the price gap is over a critical threshold does the adjustment coefficient become active in restoring the long-run equilibrium. In contrast to the findings for France, however, in the United States the speed of adjustment of the composite index to the long-run equilibrium is slower, but the adjustment process involves substantially more observations (39%). On the other hand, the composite index is three times faster than the wine share price index's response to the disequilibrium. Hence, also in the United States there is a boundary of profitable arbitrage in managing wine share price indexes. In particular, in the United States the differences in the adjustment speed are smaller, but more frequent than in France. Consequently, agents have more opportunities to make profitable investments, but with a shorter operating time span.

The findings for France and the United States differ from those related to the two developing counries analysed here, whose TVECM results appear in Tables 8.7 and 8.8. In Chile, the unusual regime

TABLE 8.6 Threshold VECMs between the North America Wine Index and the S&P500 Composite Index

	Usual regime (61% of obs.)		Unusual regime (39% of obs.)	
	Wine	Comp.	Wine	Comp.
W_{t-1}	−0.001	0.001	0.006	0.018
	−0.503	*1.039*	*0.790*	*2.811*
Intercept	0.001	0.000	−0.004	−0.008
	1.821	*−0.635*	*−1.176*	*−3.059*
Δ Wine $_{t-1}$	−0.005	0.000	0.015	0.005
	−0.113	*−0.008*	*0.308*	*0.111*
Δ Wine $_{t-2}$	0.019	−0.004	0.033	0.006
	0.530	*−0.108*	*0.687*	*0.159*
Δ Composite $_{t-1}$	−0.063	−0.103	−0.082	−0.214
	−1.804	*−2.824*	*−1.235*	*−3.886*
Δ Composite $_{t-2}$	0.029	−0.026	−0.250	−0.115
	0.665	*−0.656*	*−2.912*	*−1.320*

Note: t-Student obtained by Eicker-White standard errors and reported in italics.

---

[*] For the French case, all these results may appear ambiguous. In fact, the French series results are cointegrated but also threshold cointegrated. Moreover, we obtained a high level of threshold values and a small percentage of observations in the unusual regime. These results may lead to the conclusion that, in this case, the model just identifies outliers rather than genuine non-linearity behaviour. However, we obtained good statistic results for the significance of the threshold, allowing rejection of the null of the equality of the ECT coefficients.

TABLE 8.7 Threshold VECMs between the Chile Wine Index and the IPSA Composite Index

	Usual regime (90% of obs.)		Unusual regime (10% of obs.)	
	Wine	Comp.	Wine	Comp.
W_{t-1}	−0.002	0.003	−0.056	−0.064
	−1.151	*1.511*	*−1.723*	*−1.933*
Intercept	0.007	−0.010	0.193	0.225
	1.253	*−1.501*	*1.705*	*1.941*
Δ Wine $_{t-1}$	0.028	0.013	0.156	−0.096
	0.858	*0.330*	*1.642*	*−1.196*
Δ Wine $_{t-2}$	0.069	0.100	0.059	0.087
	2.170	*2.378*	*0.561*	*1.258*
Δ Composite $_{t-1}$	0.064	0.187	0.193	0.224
	2.942	*5.319*	*1.695*	*3.390*
Δ Composite $_{t-2}$	0.017	−0.080	−0.300	−0.144
	0.755	*−2.018*	*−3.003*	*−1.593*

Note: *t*-Student obtained by Eicker-White standard errors and reported in italics.

TABLE 8.8 Threshold VECMs between the China Wine Index and the SSE Composite Index

	Usual regime (68% of obs.)		Unusual regime (32% of obs.)	
	Wine	Comp.	Wine	Comp.
W_{t-1}	0.002	0.002	0.017	0.016
	1.305	*1.258*	*3.490*	*3.270*
Δ Intercept	0.004	0.004	0.018	0.018
	1.319	*1.055*	*3.509*	*3.539*
Δ Wine $_{t-1}$	−0.048	−0.009	0.198	0.084
	−1.189	*−0.200*	*3.226*	*1.460*
Δ Composite $_{t-1}$	0.042	0.025	−0.169	−0.100
	0.909	*0.562*	*−2.997*	*−1.812*

Note: *t*-Student obtained by Eicker-White standard errors and reported in italics.

starts as a critical threshold of 3.48, but this includes only 10% of observations, while in China, where the threshold is −1.35, the usual regime includes 39% of observations. In both countries the ECTs of the usual regime exhibit small levels of significance and minimal dynamics, whilst becoming significant for both indexes in the unusual regime.

In the case of Chile, the coefficient of the speed of adjustment of the composite index is slightly larger than for the wine coefficient, providing evidence of a reduced profitable space for investment.

In China, where most of the shares are held by Chinese retail investors, the results need to be interpreted with caution, due to the low degree of market openness to foreign investment.[*] Nevertheless, the

---

[*] Financial indicators provided by the Institutional Profiles Database (CEPII 2009) show that, in the period considered, the Chinese market and, to a lesser extent, the Chilean market, exhibited a low level of financial openness compared with France, the United States and Australia.

TVECMs give similar and significant ECT coefficients in the unusual regime for both series, showing no space for arbitrage, hence no profitable investments.

8.6 Conclusion

This paper makes use of the Mediobanca Global Wine Industry Share Price Index with the aim of considering the wine market as a possible financial investment. The analyses include five wine-producing countries, Australia, Chile, China, France and the U.S., between January 1, 2001 to the end of February 2009.

A first-step investigation of the Mediobanca Index returns and abnormal returns shows that, apart from Australia, whose wine industry suffered a financial crisis during the mid-2000s, in all countries considered the wine indexes outperformed the composite indexes, revealing investment in wine stocks as a profitable investment *per se.*

We then focus our analyses on the long-run relationship between wine share price indexes and general stock market indexes in the same producing countries, examining their speed of adjustment to the long-run equilibrium. This is done using non-linear cointegration to capture price adjustments, which are activated when deviations from the equilibrium values exceed some threshold.

The results confirm the existence of threshold cointegration between wine and composite indexes for the period under study for all countries except Australia. In particular, in more mature markets (i.e. France and the US), when the gap between the wine index and the composite index exceeds a critical threshold, the speed of adjustment of the wine index is slower than that of the composite index. This means that wine price deviations from equilibrium last a longer time and informed traders can anticipate wine price movements over the short run and make profitable investments as a result of the weak-form efficiency and different speeds of adjustment.

In less mature markets (i.e. Chile and China), the wine index and the composite index are still non-linearly cointegrated, but there is no marked difference in the speed of adjustment between composite and wine prices. In those countries, results from threshold cointegration analysis lead to the conclusion that these markets are still not sufficiently effective to allow certain types of trading strategies.

Although the results may need to be interpreted with caution, the evidence from Chile and China is likely to be the consequence of the different economic situations that characterize the different countries in the analysis, which include a different level of development of financial markets, and a different level of market openness, as outlined earlier. Erb *et al.* (1997) and Garten (1997) have already noted that, in general, emerging markets are complex and a proper understanding of the dynamics they experience hinges upon many factors. In particular, in an emerging context, financial markets may be 'thin': their size may be comparatively small in terms of market capitalization, number of listed companies and trading volumes. Furthermore, they may be characterized by a different level of free-market capitalism and democracy, as well as other specific factors that shape the economic and political context that sets them apart within the countries under analysis.

References

Anderson, H.M., Transaction costs and nonlinear adjustment toward equilibrium in the US Treasury Bill markets. *Oxford Bull. Econ. Statist.*, 1997, **59**, 465–484.

Andrews, D.W.K., Testing for parameter instability and structural change with unknown change point. *Econometrica*, 1993, **61**, 821–856.

Aslanidis, N. and Kouretas, G.P., Testing for two-regime threshold cointegration in the parallel and official markets for foreign currency in Greece. *Econ. Model.*, 2005, **22**(4), 665–682.

Aylward, D.K., A documentary of innovation support among New World wine industries. *J. Wine Res.*, 2003, **14**(1), 31–43.

Aylward, D.K. and Turpin, T., New wine in old bottles: A case study of innovation territories in 'New World' wine production. *Int. J. Innov. Mgmt.*, 2003, **7**(4), 501–525.

Balke, N.S. and Fomby, T.B., Threshold cointegration. *Int. Econ. Rev.*, 1997, **38**, 627–646.

Brooks, C., Rew, A.G. and Ritson, S., A trading strategy based on the lead–lag relationship between the spot index and futures contract for the FTSE 100. *Int. J. Forecast.*, 2001, **17**, 31–44.

Bukenya, J. and Labys, W.C., Do fluctuations in wine stocks affect wine prices? Working Paper No. 37317, American Association of Wine Economists, 2007.

Burton, J. and Jacobsen, J.P., The rate of return on wine investment. *Econ. Inquiry*, 2001, **39**, 337–350.

Buyuksahin, B., Haigh, M.S. and Robe, M.A., Commodities and equities: Ever a "Market of one"?. *J. Altern. Invest.*, 2010, **12**, 76–95.

Campbell, G. and Guibert, N., Old World strategies against New World competition in a globalising wine industry. *Br. Food J.*, 2006, **108**(4), 233–242.

Canning, P. and Perez, A., Economic geography of the U.S. wine industry. Working Paper, No. 22, American Association of Wine Economists, 2008.

CEPII, Institutional Profiles Database, 2009. Available online at: http://www.cepii.fr

Chong, J. and Miffre, J., Conditional correlation and volatility in commodity futures and traditional asset markets. *J. Altern. Invest.*, 2010, **12**, 61–75.

Crain, S.J. and Lee, J.H., Volatility in wheat spot and futures markets, 1950–1993: Government farm programs, seasonality, and causality. *J. Finance*, 1996, **51**, 325–344.

Dawson, P.J., Sanjuan, A.I. and White, B., Structural breaks and the relationship between barley and wheat futures prices on the London International Financial Futures Exchange. *Rev. Agric Econ.*, 2010, **28**, 585–594.

De Vittorio, A. and Ginsburgh, V., Pricing red wines of Medoc vintages from 1949 to 1989 at Christie's auctions. *J. Soc. Statist. Paris*, 1996, **137**, 19–49.

Dwyer, G. and Wallace, M., Cointegration and market efficiency. *J. Int. Money Finance*, 1992, **11**, 318–327.

Enders, W. and Siklo, P.L., Cointegration and threshold adjustment. *J. Bus. Econ. Statist.*, 2001, **19**, 166–176.

Erb, C.B. and Harvey, C.R., The strategic and tactical value of commodity futures. *Financ. Anal. J.*, 2006, **62**, 69–97.

Erb, C.B., Harvey, C.R. and Viskanta, T.E., Demographics and international investments. *Financ. Anal. J.*, 1997, **53**(4), 14–28.

Fama, E.F., The behaviour of stock market prices. *J. Bus.*, 1965, **38**, 31–105.

Fogarty, J.J., The return to Australian fine wine. *Eur. Rev. Agric. Econ.*, 2006, **33**, 542–561.

Fuertes, A.M., Miffre, J. and Rallis, G., Tactical allocation in commodity futures markets: Combining momentum and term structure signals. *J. Bank. Finance*, 2010, **34**, 2530–2548.

Garten, J., Troubles ahead in emerging markets. *Harv. Bus. Rev.*, 1997, **75**, 38–50.

German, H. and Kharoubi, C., WTI crude oil futures in portfolio diversification: The time-to-maturity effect. *J. Bank. Finance*, 2008, **32**, 2553–2559.

Giuliani, E. and Bell, M., The micro-determinants of meso-level learning and innovation: Evidence from a Chilean wine cluster. *Res. Policy*, 2005, **34**(1), 47–68.

Goodhue, R.E., Green, R.D., Heien, D.M. and Martin, P.L., California wine industry evolving to compete in 21st century. *Calif. Agric.*, 2008, **62**(1), 12–18.

Gorton, G.B., Hayashi, F. and Rouwenhorst, G.K., The fundamentals of commodity futures returns. Working paper, National Bureau of Economic Research, 2007.

Gorton, G.B. and Rouwenhorst, G.K., Facts and fantasies about commodity futures. *Financ. Anal. J.*, 2006, **62**, 47–68.

Granger, C.W.J., Investigating causal relations by econometric models and cross-spectral methods. *Econometrica*, 1969, **37**, 424–438.

Granger, C.W.J., Some properties of time series data and their use in econometric model specification. *J. Econometr.*, 1981, **16**, 121–130.

Gwynne, R., UK retail concentration, Chilean wine producers and value chains. *Geograph. J.*, 2008, **174**(2), 97–108.

Hamilton, J.D., A new approach to the economic analysis of nonstationary time series and the business cycle. *Econometrica*, 1989, **57**(2), 357–384.

Hansen, B.E., Inference when a nuisance parameter is not identified under the null hypothesis. *Econometrica*, 1996, **64**, 413–430.

Hansen, B.E. and Seo, B., Testing for two-regime threshold cointegration in vector error-correction models. *J. Econometr.*, 2002, **110**, 293–318.

Hokkio, C. and Rush, M., Market efficiency and cointegration: An application to Sterling and Deutschmark exchange rates. *J. Int. Money Finance*, 1989, **8**, 75–88.

Hong, H. and Yogo, M., What does futures market interest tell us about the macroeconomy and asset prices? Working paper, Princeton University, 2011.

Hussain, M., Cholette, S. and Castaldi, R., An analysis of globalization forces in the wine industry: Implications and recommendations for wineries. *J. Glob. Mktg*, 2008, **21**(1), 33–47.

Insel, B., The U.S. wine industry. *Business Econ.*, 2008, January.

Jaeger, E., To save or savor: The rate of return to storing wine: Comment. *J. Polit. Econ.*, 1981, **89**, 584–592.

Jawadi, F. and Koubaa, Y., Threshold cointegration between stock returns: an application of STECM Models. Working paper series, University of Paris X-Nanterre, 2004.

Jenster, P. and Cheng, Y., Dragon wine: Developments in the Chinese wine industry. *Int. J. Wine Bus. Res.*, 2008, **20**(3), 244–259.

Johansen, S., Statistical analysis of cointegrating vectors. *J. Econ. Dynam. Control*, 1988, **12**, 231–254.

Johansen, S. and Juselius, K., Maximum likelihood estimation and inference on cointegration – with application to the demand for money. *Oxford Bull. Econ. Statist.*, 1990, **52**, 169–210.

Jones, G.V. and Storchmann, K.H., Wine market prices and investment under uncertainty: An econometric model for Bordeaux Crus Classés. *Agric. Econ.*, 2001, **26**(2), 115–133.

Jovanovic, B., Bubbles in prices of exhaustible resources. Working Paper No. 13320, National Bureau of Economic Research, Cambridge, MA, 2007.

Krasker, W.S., The rate of return to storing wine. *J. Polit. Econ.*, 1979, **87**, 1363–1367.

Krolzig, H.M., Markov-switching Vector Autoregressions. Modelling, Statistical Inference and Applications to Business Cycle Analysis, 1997 (Springer: Berlin).

Lo, M. and Zivot, E., Threshold cointegration and nonlinear adjustment to the Law of One Price. *Macroecon. Dynam.*, 2001, **5**, 533–576.

Maasoumi, E. and Racine, J., Entropy and predictability of stock market returns. *J. Econometr.*, 2002, **107**, 291–312.

Masih, A. and Masih, R., Dynamic linkages and the propagation mechanism driving major international stock markets: An analysis of pre- and post-crash eras. *Q. Rev. Econ. Finance*, 1997, **37**, 859–885.

Masset, P. and Henderson, C., Wine as an alternative asset class. Price evolution over the period 1996–2007, market segmentation and portfolio diversification. Paper presented at the Second Annual Conference, American Association of Wine Economists (AAWE), Portland, Oregon, 2008.

Masset, P. and Weisskopf, J.P., Raise your glass: Wine investments and the financial crises. SSRN Research Papers, 2010.

McMillan, D.G., Non-linear predictability of UK stock market returns. *Oxford Bull. Econ. Statist.*, 2003, **65**, 557–573.

McMillan, D.G., Non-linear dynamics in international stock market returns. *Rev. Financ. Econ.*, 2005, **14**, 81–91.

Miffre, J. and Rallis, G., Momentum strategies in commodity futures markets. *J. Bank. Finance*, 2007, **31**, 1863–1886.

Mitry, D.J., Smith, D.E. and Jenster, P.V., China's role in global competition in the wine industry: A new contestant and future trends. *Int. J. Wine Res.*, 2009, **1**, 19–25.

Patra, T. and Poshakwale, S.S., Long-run and short-run relationship between the main stock indexes: Evidence from the Athens stock exchange. *Appl. Financ. Econ.*, 2008, **18**, 1401–1410.

Perez-Quiros, G. and Timmermann, A., Firm size and cyclical variations in stock returns. *J. Finance*, 2000, **55**, 1229–1262.

Prat, G. and Jawadi, F., Nonlinear stock price adjustment in the G7 countries. Working paper, University of Paris Ouest La Defense Nanterre, 2009.

Sanning, L.W., Shaffer, S. and Sharratt, J.M., Alternative investments: The case of wine. Working Paper No. 37322, American Association of Wine Economists, 2007.

Schroeder, T.C. and Goodwin, B.K., Price discovery and cointegration for live hogs. *J. Fut. Mkts.*, 1991, **11**, 685–696.

Schwarz, T.V. and Szakmary, A.C., Price discovery in petroleum markets: Arbitrage, cointegration, and the time interval of analysis. *J. Fut. Mkts.*, 1994, **14**, 147–167.

Sercu, P., Uppal, R. and Van Hulle, C., The exchange rate in the presence of transaction costs: Implications for tests of purchasing power parity. *J. Finance*, 1995, **50**(4), 1309–1319.

Shively, P.A., The nonlinear dynamics of stock prices. *Q. Rev. Econ. Finance*, 2003, **43**, 505–517.

Shleifer, A., Inefficient Markets. An Introduction to Behavioural Finance, 2000 (Clarendon Lectures in Economics) (Oxford University Press: Oxford).

Siourounis, G.D., Modelling volatility and test for efficiency in emerging capital markets: The case of the Athens stock exchange. *Appl. Financ. Econ.*, 2002, **1**, 47–55.

Stoll, H.R. and Whaley, R.E., Commodity index investing and commodity futures prices. *J. Appl. Finance*, 2010, **20**(1), 7–46.

Subramanian, U., Cointegration of stock markets in East Asia in the boom and early crisis period. Working Paper Series, Addis Ababa University, 2009.

Szakmary, A.C., Shen, Q. and Sharma, S.C., Trend-following trading strategies in commodity futures: A re-examination. *J. Bank. Finance*, 2010, **34**, 409–426.

Terblanche, N.S., Simon, E. and Taddei, J.-C., The need for a marketing reform: The wines of the Loire region. *J. Int. Food Agribus. Mktg*, 2008, **20**(4), 113–138.

Tong, H. and Lim, K.S., Threshold autoregression, limit cycles and cyclical data. *J. R. Statist. Soc., Ser. B: Methodological*, 1980, **42**, 245–292.

Tsay, R.S., Testing and modelling threshold autoregressive process. *J. Am. Statist. Assoc.*, 1989, **84**, 231–240.

Tsay, R.S., Testing and modelling multivariate threshold models. *J. Am. Statist. Assoc.*, 1998, **93**, 1188–1202.

Weil, R.L., Do not invest in wine, at least in the U.S. unless you plan to drink it, and maybe not even then. Paper presented at the 2nd International Conference of the Vineyard Data Quantification Society, Verona, Italy, 1993.

Yang, J., Bessler, D.A. and Leatham, D.J., Asset storability and price discovery in commodity futures markets: A new look. *J. Fut. Mkts.*, 2001, **21**, 279–300.

9

Cross-Market Soybean Futures Price Discovery: Does the Dalian Commodity Exchange Affect the Chicago Board of Trade?

Liyan Han,
Rong Liang
and Ke Tang

9.1 Introduction

Soybeans, a source of high-quality vegetable protein, are widely used as food, fodder and industrial raw materials in the fields of health care and biodiesel. However, due to the limited amount of arable land and production technology, together with China's lack of transgenic varieties, China's soybean production has been restricted. However, China's demand for soybeans has been increasing, leading to an excess of soybean imports over its production since 2001. In 2008–2009, China's soybean consumption was 51.21 million tons, while its production was only 15.55 million tons; the remaining amount was imported. Figure 9.1 demonstrates trends in China's soybean production, consumption and import behaviour from 1994 to 2010.

Soybean futures, as one of the most actively traded species of agricultural futures, play a key role in the price discovery and hedging of soybeans and related products. Because soybean futures are simultaneously traded in many markets globally, the price formation in one market is not only affected by its domestic market information, but also by the related international market information. The information of one market can also be transferred to other markets. However, because different markets have

million metric tons

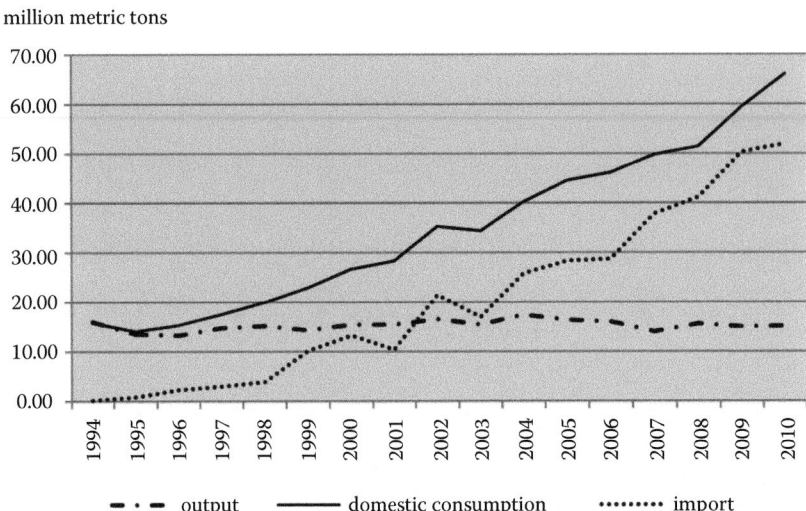

FIGURE 9.1   The time series of China's soybean production, imports and consumption. The sample period is 1994 to 2010.

different information processing capabilities, transaction costs, regulations and liquidity, even if they are information-linked, their relative contribution to the price discovery is not the same.

Currently, the world's major soybean futures markets include the Chicago Board of Trade (CBOT) of the United States, China's Dalian Commodity Exchange (DCE) and Japan's Tokyo Grain Exchange (TGE). The CBOT holds an important position in the soybean futures trading around the world. Over the past 10 years, the average trading volume of the CBOT soybean futures has been close to 20 million lots per year. Being highly liquid, the CBOT soybean market provides a convenient channel for investing and hedging, thus attracting not only the soybean-related production and circulation enterprises in and out of the United States, but also the international and domestic investment and speculative funds. Currently, there are two soybean futures contracts traded in the CBOT: one is a large-scale contract with a trading unit of 5000 bushels, and the other is a mini-contract with a trading unit of 1000 bushels. The DCE in China is another important soybean futures market in the world, at least in terms of trading volume. In 2008, the trading volume of soybean futures in the DCE was ranked second globally, surpassed only by that in the CBOT. The first trade of soybean futures contracts in the DCE can be traced back to 1993. The underlying product of that contract is soybeans, and the delivery goods can be both genetically modified soybeans and non-genetically modified soybeans. In 2001, the national agriculture policy and related management measures for genetically modified products were introduced, in line with which the DCE soybean contract was split into No. 1 yellow soybean contracts and No. 2 yellow soybean contracts in 2002. The No. 1 yellow soybean contract only allows the delivery of non-genetically modified soybeans and started trading on March 15, 2002. This contract represents the overall condition of China's domestic soybeans and is currently the world's largest-traded non-genetically modified soybean futures. The No. 2 yellow soybean contract started trading on December 22, 2004, and the delivery goods are genetically modified soybeans. However, its turnover is much smaller compared with the No. 1 contract, and there is thus a lack of liquidity. Table 9.1 provides the details of the No. 1 yellow soybean contracts of the DCE and the soybean futures of the CBOT, with which we can easily compare the two contracts.

China, as the major soybean import country in the world, may have played a key role in the global price discovery of soybean, which is the main issue that we investigate in this paper. By examining the relationship between the Chinese and the global soybean futures market, we can better understand China's position in the pricing of soybean futures, which can provide guidance for production, trade and policy making decisions.

TABLE 9.1 Details of the DCE No.1 Soybean Futures Contract and the CBOT Soybean Futures Contract

	Dalian Commodity Exchange	Chicago Board of Trade
Product	No.1 soybeans	Soybeans
Trading unit	10MT/contract	5000 bushels (~136 metric tons)
Price quote	CNY/MT	Cents per bushel
Tick size	1 CNY/MT	1/4 of one cent per bushel
Daily price limit	4% of last settlement price (temporarily 5%)	$0.70 per bushel expandable to $1.05 and then to $1.60 when the market closes at limit bid or limit offer
Contract months	Jan, Mar, May, July, Sep, Nov	Jan, Mar, May, July, Aug, Sep, Nov
Trading hours	9:00–11:30 a.m., 1:30–3:00p.m. Beijing Time, Monday–Friday	9:30 a.m.–1:15 p.m. Central Time, Monday–Friday
Last trading day	10th trading day of the delivery month	The business day prior to the 15th calendar day of the contract month
Last delivery day	7th day after the last trading day of the delivery month	Second business day following the last trading day of the delivery month
Deliverable grades	Quality standards of No.1 soybeans	#2 Yellow at contract price, #1 Yellow at a 6 cent/bushel premium,
		#3 Yellow at a 6 cent/bushel discount
Delivery location	The warehouses appointed by the DCE	The warehouses appointed by the CBOT

Our paper focuses on the impact of the DCE on the CBOT in the price discovery process of soybean futures. We first explore the relationship between the DCE and the CBOT futures trading and non-trading hour returns with a Structural Vector Auto-Regression (SVAR) model, which can capture the lead–lag relationship between the two markets. The results show that the impacts of both the DCE on the CBOT and the CBOT on the DCE are significant and of a similar magnitude. A Vector Error Correction (VEC) model on the closing price of the soybean futures of the two markets further confirms this conclusion, which suggests that the closing prices of the DCE and the CBOT are cointegrated and that both the CBOT and the DCE adjust according to this long-term relationship. These results signify that information flows from the DCE to the CBOT, and *vice versa*. This finding differs from the results of many existing studies, which mainly show that the CBOT is the dominant market and that the DCE is a satellite market. For example, previous research on this subject, such as Fung *et al.* (2003), find that the DCE is in a subordinate position in the price discovery process, i.e. the CBOT influences the DCE, but not *vice versa*. In this paper, using more *recent* data from 2002 to 2011, we find a different result, i.e. that the DCE significantly influences the CBOT in the soybean futures price discovery.

The research of cross-market price discovery was first proposed by Garbade and Silber (1979). Based on a study of the New York Stock Exchange and regional exchanges, the authors propose the concept of the 'dominant market' and the 'satellite market'. They posit that the regional exchanges are the satellite markets and that the price changes of the satellite markets lag behind those of the dominant markets. In other words, new information is first reflected in the price of the dominant market and, then, in the price of the satellite markets. Harris *et al.* (1995) examine the transactions of the IBM stock on the New York Stock Exchange, Pacific Stock Exchange and Midwest Stock Exchange with an error correction model, and the result shows that there is a long-term cointegration relationship among all three markets. Kleidon and Werner (1996) study the intraday patterns of volatility, volume and spreads of the stocks that are cross-listed in the UK and New York Stock Exchange, and they find that the order flow for the cross-listed securities are not fully integrated. Eun and Sabherwal (2003) study the Canadian stocks that are traded both on the Toronto Stock Exchange and the U.S. exchange. The authors find that prices are cointegrated in these two markets and that price adjustments exist in both markets. The role that the U.S. exchange plays in the price discovery of these shares is weaker overall than that in the Toronto Stock Exchange, but for some stocks, the U.S. exchange has a larger contribution. Grammig *et al.* (2005)

investigate the price discovery of three stocks of German companies that are simultaneously listed on the New York Stock Exchange and Frankfurt Stock Exchange. They conclude that most price discovery first occurs in the Frankfurt market, and the New York Stock Exchange only plays a minor role in the price discovery of these stocks.

Regarding commodity futures, Booth and Ciner (1997) study the price series of the commodity futures from 1993 to 1995 in the Japan TGE and the U.S. CBOT. With a dynamic VAR model, they examine the relationship between the futures market prices of the two countries and the spillover effects of price volatility. The results show that the CBOT corn futures are dominant in the transmission of price information, but the TGE can also affect the prices of the CBOT commodity futures. Tse and Booth (1997) investigate the information transmission of the New York and London oil futures market with a Vector Error Correction (VEC) model and find that the New York oil futures market takes a dominant position in the information transmission process.

Our paper proceeds as follows. Section 9.2 presents the data. Section 9.3 employs an SVAR model to test the information transfer between the DCE and CBOT soybean markets. As a robustness check, Section 9.4 utilizes a VEC model to investigate the price adjustment among two exchanges. Section 9.5 concludes the paper.

9.2 The Data

The selected data of the DCE No. 1 soybean futures prices and of the CBOT soybean futures prices are dated from March 2002, when the No. 1 yellow soybeans were first launched in China, to September 2011. We choose the No. 1 soybean futures from the DCE because the No. 1 yellow soybean contract has a much larger trading volume and, hence, more liquidity than the No. 2 contract.

9.2.1 The Construction of Continuous Price Series

Because futures contracts of different months are traded simultaneously, it is of practical importance to choose the one that best represents the commodity prices. In previous studies, there are generally two approaches. One is to select the price of the front-month futures contract and, when the front-month contract enters the delivery month, to select the next front-month contract as the representative. Another approach focuses on the trading volume and selects the contract with the largest trading volume as the representative (Booth and Ciner 1997).

For the CBOT soybean futures, the front-month contract is generally the most actively traded; therefore, prices (returns) of the front-month contract are used for the CBOT. In China, however, the front-month soybean futures contract is not the most actively traded. It is the contracts of January, May and September that are generally more actively traded due to the seasonal soybean production and circulation pattern. Nonetheless, the second approach means that the price series may switch between the front contract, the second front contract and possibly the third front contract, which could result in a mismatch issue with the CBOT price series. Therefore, we still draw on the first approach for the DCE and select the prices (returns) of the front-month contract. We obtain the daily DCE and CBOT futures prices from the Wind database and Datastream, respectively.

Because holidays in the United States and China are not exactly the same and futures trading stops during the holidays, the trading dates of soybean futures in the DCE and those in the CBOT do not overlap entirely. Following Tse *et al.* (1996) and Xu and Fung (2005), we remove the days for which transactions only exist in one market and construct the price (return) series with a complete match between these two markets. The prices of the DCE futures are adjusted to U.S. dollars using the daily exchange rate from the Wind database. Figure 9.2 depicts the sequence of the normalized closing prices of the DCE soybean futures contract and the CBOT soybean futures contract. Note that the DCE price series is converted into US dollar denomination with the exchange rate from the Wind database. The two price series show a similar trend.

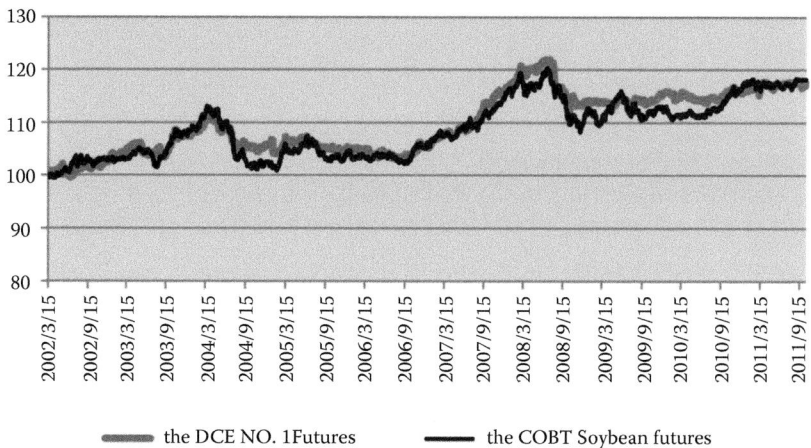

FIGURE 9.2 The time series of the price of the DCE No.1 soybean futures and the CBOT soybean futures. The price series of DCE soybean futures is converted into US dollars with the corresponding exchange rate. The futures prices are normalized to be 100 with March 15, 2002 as the base period. The sample starts on March 15 2002, and ends on September 2, 2011.

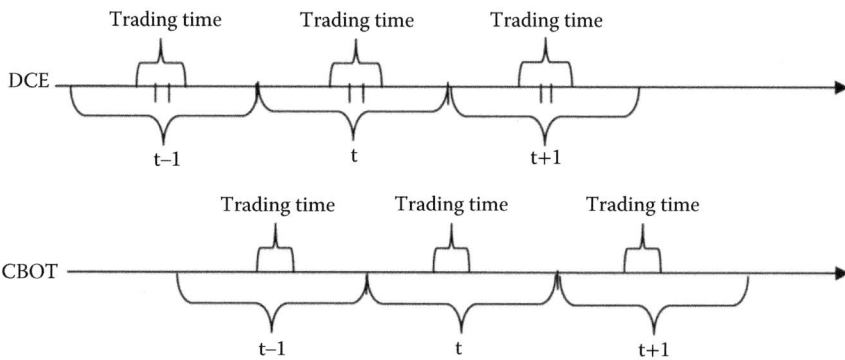

FIGURE 9.3 The time difference of the DCE and the CBOT exchanges. It shows the relative length of the trading time and non-trading time, as well as the correspondence of the DCE and the CBOT trading and non-trading time due to the time difference.

9.2.2 Calculation of Futures Returns

Before we present the calculation of different futures returns, it should be noted that the trading times of the two markets do *not* overlap. Due to the time difference, the calendar time of the Chinese market is ahead of the U.S. market, and when the DCE is trading, the CBOT is closed, and *vice versa*. Specifically, under the standard time, there is a 14-hour time difference between the DCE and the CBOT, while under the daylight saving time, the time difference is 13 hours. Under the standard time, the DCE opens at 9:00 Beijing time (Chicago time previous day 19:00) and closes at 15:00 Beijing time (Chicago time 1:00); the CBOT opens at 23:30 Beijing time (Chicago time 9:30) and closes at 3:10 Beijing time (Chicago time 13:10). During the daylight saving time period, the case is similar and there is still no overlap of the trading time. Figure 9.3 shows the time line of the non-trading and trading hours for both the CBOT and DCE markets. To examine the price response to information, two classes of returns should be considered: trading hour returns and non-trading hour returns. For example, the information generated in the trading hours of the Chinese market may affect the non-trading hour returns of the U.S. market, i.e.

the U.S. market's opening price at date t may reflect digested information generated from the Chinese market on date t.

The trading and non-trading hour returns at date t for the DCE and the CBOT are defined as follows:

DCE soybean trading hour returns: $CN\_INR_t = CN\_CLOSE\_t - CN\_OPEN_t$,
DCE soybean non-trading hour returns: $CN\_OUTR_t = CN\_OPEN - CN\_CLOSE_{t-1}$,
CBOT soybean trading hour returns: $US\_INR_t = US\_CLOSE_t - US\_OPEN_t$,
CBOT soybean non-trading hour returns: $US\_OUTR_t = US\_OPEN_t - US\_CLOSE_{t-1}$,

where CN_INR_t is the trading hour returns of the DCE soybean futures at date t, CN_OUTR_t is the non-trading hour returns of the DCE soybean futures at date t, CN_CLOSE_t is the log of the closing price of the DCE soybean futures at date t, and CN_OPEN_t is the log of the opening price of the DCE soybean futures at date t. Similarly, the trading and non-trading hour returns for the CBOT soybean futures are US_INR_t and US_OUTR_t, and US_CLOSE_t and US_OPEN_t correspond to the log of the closing price and opening price, respectively, of the CBOT soybean futures at date t. Panel A of Table 9.2 gives the respective summary statistics of the different returns for DCE and CBOT soybean futures.

From panel A, we can see that the DCE and the CBOT returns, on average, are zero, negatively skewed and relatively thin-tailed. Before estimating the SVAR model, we need to first test the stationarity of the return series. To ensure a sound conclusion, ADF, PP and KPSS tests are used for the unit root test. Note that we exclude both trend and intercept in the test since the plot (not given here) indicates that the return series seem to fluctuate around zero. Panel B of Table 9.2 gives the test result. All three tests

TABLE 9.2 Summary Statistics and the Unit Root Test of the Return of the DCE No.1 Soybean Futures and the Cbot Soybean Futures. Panel A Gives the Summary Statistics, Where the First Two Columns Present the Statistics of the Trading and Non-Trading Hour Returns of the DCE No. 1 Soybean Futures While the Last Two Columns Give Statistics of the Trading and Non-Trading Hour Returns of the CBOT Futures. Panel B Presents the Unit Root Test Result of the ADF, PP and KPSS Tests. Values in Parentheses are the Corresponding *P*-Values. The Sample Period is March 15, 2002, to September 2, 2011

Test statistic	CN_INR	CN_OUTR	US_INR	US_OUTR
		Panel A: Summary statistics		
Mean	0.00024	0.00021	6.08E−05	0.00042
	(0.2581)	(0.399)	(0.8234)	(0.1195)
Median	0	4.46E−05	0.00039	0.00050
Max	0.08744	0.08662	0.07633	0.0734
Min	−0.07599	−0.12867	−0.074	−0.23307
Std. dev.	0.0104	0.0119	0.0131	0.0131
Skewness	−0.115	−0.307	−0.285	−2.586
Kurtosis	12.53	15.49	5.21	49.88
Jarque–Bera	8819.6	15,178.7	504.2	215,954
	(0.000)	(0.000)	(0.000)	(0.000)
Observations	2330	2330	2330	2330
		Panel B: Unit root test		
ADF	−47.71	−48.79	−49.11	−48.93
	(0.0001)	(0.0001)	(0.0001)	(0.0001)
PP	−4.72	−48.80	−49.16	−48.98
	(0.0001)	(0.0001)	(0.0001)	(0.0001)
KPSS	0.245	0.124	0.146	0.097

Note: For the KPSS test, all the series are tested including the intercept; critical values at the 10%, 5%, and 1% significance levels are 0.347, 0.463, and 0.7439, respectively.

indicate that the four return series are all stationary. In this case, we need not consider the possible cointegration problem in the Structural VAR model.

9.3 The SVAR Model

This section investigates the relationship between the DCE and the CBOT in soybean futures price discovery. In previous research on cross-market price discovery with non-overlapping data, the primary method has been the VAR model with some restrictions on the coefficients or the VEC model. However, the SVAR model can better capture the contemporaneous relationship among the related variables. Hence, we first use the SVAR model to explore the relationship between DCE and CBOT soybean futures. We now report the model setup and estimation result.

9.3.1 Model Setup

The SVAR model can capture the interaction between variables of the same period. A SVAR model can be expressed as[*]

$$\mathbf{B}\mathbf{y}_t = \Gamma_0 + \Gamma_1 \mathbf{y}_{t-1} + \Gamma_2 \mathbf{y}_{t-2} + \ldots + \Gamma_p \mathbf{y}_{t-p} + \mu_t,$$
$$t = 1, 2, \ldots, T,$$

(9.1)

where matrix $\mathbf{B}$ captures the relationship between variables of the same period. μ_t is i.i.d. with a mean of zero and is known as the structural disturbance vector. If the matrix $\mathbf{B}$ is invertible, the model can be simplified as follows:

$$\mathbf{y}_t = \mathbf{B}^{-1}\Gamma_0 + \mathbf{B}^{-1}\Gamma_1 \mathbf{y}_{t-1} + \mathbf{B}^{-1}\Gamma_2 \mathbf{y}_{t-2} + \ldots + \mathbf{B}^{-1}\Gamma_p \mathbf{y}_{t-p} + \mathbf{B}^{-1}u_t, t = 1, 2, \ldots, T.$$

(9.2)

After imposing some constraints on the matrix $\mathbf{B}$, we can estimate the SVAR model and obtain the identifiable structural impact.

For the SVAR model of the four return series of the DCE soybean futures and CBOT soybean futures, we define $\mathbf{y}_t$ as follows:

$$y_t := \begin{pmatrix} CN\_OUTR_t \\ CN\_INR_t \\ US\_OUTR_t \\ US\_INR_t \end{pmatrix}.$$

Since this SVAR model contains four vectors, we need six constraints to identify it. When making constraints on matrix $\mathbf{B}$, we follow the rule that returns appearing earlier can influence the returns appearing later. As shown in Figure 9.3, due to the time difference, the trading time of the DCE at date t corresponds to the non-trading time of the CBOT at date t. Because the duration of the non-trading time is longer than the duration of the trading time, the non-trading hour's returns of one market can be affected by the other market's trading hour returns of the same date, while the trading hour returns of one market are not subject to the influence of the other market's non-trading hour returns of the same date. For example, as illustrated in Figure 9.3, when the DCE closes at time t, the CBOT is not yet open; therefore, the opening price of the CBOT at day t can be affected by the DCE's trading time information at day t. Because the CBOT's non-trading hour returns are calculated as the CBOT's open price at time t divided by the previous day's closing price, it therefore may also be affected by the DCE's trading hour

---

[*] Note that bold letters denote a matrix or vector in this paper.

returns at day t. The relationships between the returns of other pairs of the DCE and the CBOT returns can be analysed in a similar way. With Equation (9.1) and the definition of vector $\mathbf{y}_t$, we thus impose the following restrictions according to the order of occurrence for the DCE and CBOT trading hour and non-trading hour returns.

First, neither the CBOT trading and non-trading hour returns nor the DCE trading hour returns can affect the DCE non-trading hour returns of the same day. Thus, $b_{12} = b_{13} = b_{14} = 0$.

Second, only DCE non-trading hour returns of the same date can affect the DCE trading hour returns. Note that the CBOT trading hour and non-trading hour returns have no effect on the same day's DCE trading hour returns because these two returns are not even known when the DCE closes at date t. Thus, $b_{23} = b_{24} = 0$.

Third, the DCE trading and non-trading hour returns of the same date can affect the CBOT non-trading hour return, while the CBOT trading hour returns of the same date have no effect on the CBOT non-trading hour returns. Thus, $b_{34} = 0$.

Fourth, the DCE trading and non-trading hour returns and the CBOT non-trading hour returns of the same date can all affect the CBOT trading hour returns.

In summary, $\mathbf{B}$ in the SVAR model can be expressed as

$$
\mathbf{B} = \begin{pmatrix} 1 & 0 & 0 & 0 \\ -b_{21} & 1 & 0 & 0 \\ -b_{31} & -b_{32} & 1 & 0 \\ -b_{41} & -b_{42} & -b_{43} & 1 \end{pmatrix}.
$$

Note that there is a minus sign in front of the coefficients. This is for the convenience of interpretation, which will soon become clear with the following explanation. After estimating the matrix $\mathbf{B}$, we can write the model in the structural form and obtain a sense of the impact of the DCE on the CBOT.

To make the interpretation easier, we can rearrange Equation (9.1) into the following form:

$$
\mathbf{y}_t = \Gamma_0 + (\mathbf{I} - \mathbf{B})\mathbf{y}_t + \Gamma_1 \mathbf{y}_{t-1} + \Gamma_2 \mathbf{y}_{t-2} + \ldots + \Gamma_p \mathbf{y}_{t-p} + \boldsymbol{\mu}_t,
$$

$$
t = 1, 2, \ldots, T
$$

(9.3)

where

$$
\mathbf{I} - \mathbf{B} = \begin{pmatrix} 0 & 0 & 0 & 0 \\ b_{21} & 0 & 0 & 0 \\ b_{31} & b_{32} & 0 & 0 \\ b_{41} & b_{42} & b_{43} & 0 \end{pmatrix},
$$

which shows the contemporaneous relationship of $\mathbf{y}_t$. This also explains why we set a minus sign in front of the coefficients of matrix $\mathbf{B}$. Collecting the relevant terms, we can write the model more explicitly as follows:

$$
CN\_OUTR_t = a_1 + \sum_{i=1}^{n} c_{1i} CN\_OUTR_{t-i}
$$

$$
+ \sum_{i=1}^{n} d_{1i} CN\_INR_{t-i}
$$

$$
+ \sum_{i=1}^{n} f_{1i} US\_OUTR_{t-i}
$$

$$
+ \sum_{i=1}^{n} g_{1i} US\_INR_{t-i} + h_1 Mon_t + \varepsilon_{1t},
$$

(9.4)

$$CN\_INR_t = a_2 + \sum_{i=0}^{n} c_{2i} CN\_OUTR_{t-i}$$

$$+ \sum_{i=1}^{n} d_{2i} CN\_INR_{t-i} + \sum_{i=1}^{n} f_{2i} US\_OUTR_{t-i} \tag{9.5}$$

$$+ \sum_{i=1}^{n} g_{2i} US\_INR_{t-i} + h_2 \, Mon_t + \varepsilon_{2t},$$

$$US\_OUTR_t = a_3 + \sum_{i=0}^{n} c_{3i} CN\_OUTR_{t-i}$$

$$+ \sum_{i=0}^{n} d_{3i} CN\_INR_{t-i}$$

$$+ \sum_{i=1}^{n} f_{3i} US\_OUTR_{t-i} \tag{9.6}$$

$$+ \sum_{i=1}^{n} g_{3i} US\_INR_{t-i} + h_3 \, Mon_t + \varepsilon_{3t},$$

$$US\_INR_t = a_4 + \sum_{i=0}^{n} c_{4i} CN\_OUTR_{t-i}$$

$$+ \sum_{i=0}^{n} d_{4i} CN\_INR_{t-i} + \sum_{i=0}^{n} f_{4i} US\_OUTR_{t-i} \tag{9.7}$$

$$+ \sum_{i=1}^{n} g_{4i} US\_INR_{t-i} + h_4 \, Mon_t + \varepsilon_{4t}.$$

In the above equations, when the subscript i is zero, the contemporaneous value at date t of that explanatory variable can affect the left-hand side variable. For example, in (9.5), the summation symbol of CN_OUTR_t has a subscript of 0, indicating that the non-trading hour returns of the DCE soybean futures can affect its trading hour returns at the same date. The 0 subscripts in the other equations are interpreted similarly. Actually, it is easy to see that $c_{20} = b_{21}$, $c_{30} = b_{31}$, $d_{30} = b_{32}$, $c_{40} = b_{41}$, = and $f_{40} = b_{43}$. Also note that we include dummy variable Mon in the SVAR model, which equals 1 when the day is Monday, and 0 otherwise. This is to capture the well-documented weekend effect, that is, Monday returns tend to be significantly negative.

In addition, as Figure 9.3 shows, although the CBOT market may not affect the DCE market of the same calendar date, the lag one period effect may still be significant. In other words, the information generated in the CBOT at date $t-1$ may affect the DCE at date t. This potential impact can be verified by examining the coefficients of Γ_1 (see Equation (9.3)).* For this model, Γ_1 is in the following form:

$$\Gamma_1 = \begin{pmatrix} \Gamma_{11} & \Gamma_{12} & \Gamma_{13} & \Gamma_{14} \\ \Gamma_{21} & \Gamma_{22} & \Gamma_{23} & \Gamma_{24} \\ \Gamma_{31} & \Gamma_{32} & \Gamma_{33} & \Gamma_{34} \\ \Gamma_{41} & \Gamma_{42} & \Gamma_{43} & \Gamma_{44} \end{pmatrix},$$

---

* In the empirical analysis, we allow a lag of order 2 in the SVAR; hence, $\Gamma 2$ also influences the DCE returns. However, because its magnitude is quite small from our empirical results, we omit it when presenting the value of $\Gamma 2$.

where Γ_{11} represents the influence of CN_OUTR_{t-1} on CN_OUTR_t, and Γ_{12} represents the influence of CN_INR_{t-1} on CN_OUTR_t. The other parameters can be interpreted similarly. The coefficients that we focus on are Γ_{13}, Γ_{14}, Γ_{23} and Γ_{24}. Here, Γ_{13} and Γ_{14} represent the effect of the t–1 non-trading and trading hour returns of the CBOT on the date t non-trading hour returns of the DCE, while Γ_{23} and Γ_{24} represent the effect of the t–1 non-trading and trading hour returns of the CBOT on the date t trading hour returns of the DCE. With a detailed analysis of $\mathbf{\Gamma_1}$ and $\mathbf{B}$, we can better assess the information transfer between CBOT and DCE. Also note that Γ_{ij} actually equals c_{ij} in Equations (9.4)–(9.7).

9.3.2 Estimation Results

The BIC statistics indicate that the appropriate lag order of the SVAR model is 2. The detailed estimation results for the b_{ij} values are shown in Table 9.3. As seen from the table, except for the fact that b_{41} and b_{42} are not significant, all of the coefficients are significant at the 1% significance level. Because we are interested in the impact of the two exchanges, here we focus on the analyses of b_{31}, b_{32}, b_{41} and b_{42}. Among these coefficients, b_{31} and b_{32} represent the impact of the non-trading and trading hour returns of the DCE soybean futures on the non-trading hour returns of the CBOT soybean futures, and b_{41} and b_{42} represent the impact of the DCE soybean futures on the trading hour returns of the CBOT soybean futures. It can be seen that both the non-trading and trading hour returns shock of the DCE soybean futures has a significant *positive* impact on the CBOT soybean futures non-trading hour returns, while neither the non-trading nor trading hour returns of the DCE soybean futures have a significant influence on the CBOT soybean futures' trading hour returns. The reason why this happens may be that the time lag between the DCE returns and the CBOT soybean futures' trading hour return is relatively long, and the information of the DCE may have already been reflected in the opening price of the CBOT soybean futures.

Table 9.4 gives the estimated results for $\mathbf{\Gamma_1}$. As we can see, except for Γ_{24}, the estimated results for Γ_{13}, Γ_{14} and Γ_{23} are all significant at the 5% significance level. Both the non-trading and trading hour returns of the lag one date CBOT soybean futures have a significant positive impact on the DCE soybean futures' non-trading hour returns, and the non-trading hour returns of the lag one date CBOT soybean

TABLE 9.3 Estimated Results of Matrix **B**, Which Represents the Relationship between Variables of the Same Period in the SVAR Model. The Variables in the SVAR Model are in the order of *CN_OUTR*, *CN_INR*, *US_OUTR* and *US_INR*. The Values in Brackets are the Corresponding *T*-Statistics. *** and ** Denote the 1% and 5% Significance Level, Respectively. The Sample Period is March 15, 2002 to September 2, 2011

	Coefficient
b_{21}	−0.358***
	[−19.88]
b_{31}	0.292***
	[11.35]
b_{41}	0.0004
	[0.015]
b_{32}	0.245***
	[8.95]
b_{42}	0.034
	[1.18]
b_{43}	−0.068***
	[−3.21]

TABLE 9.4 Estimated Results of Matrix $\boldsymbol{\Gamma}_1$, Which represents The Lag One Period Effect of the Independent Variables on the Dependent Variables in the SVAR Model. The Variables in the SVAR Model are in the order of *CN_OUTR, CN_INR, US_OUTR* and *US_INR*. The Values in Brackets are the Corresponding T-Statistics. *** and ** Denote the 1% and 5% Significance Level, Respectively. The Sample Period is March 15, 2002 to September 2, 2011

Dependent variable	Independent variable			
	CN_OUTR_{t-1}	CN_INR_{t-1}	US_OUTR_{t-1}	US_INR_{t-1}
CN_OUTR	−0.086***	−0.173***	0.179***	0.296***
	[−3.73]	[−7.16]	[9.89]	[17.02]
CN_INR_t	−0.028	−0.042	0.025**	0.080
	[0.11]	[0.86]	[−2.29]	[−1.58]
US_OUTR_t	0.045	0.027	−0.059	−0.032**
	[0.75]	[−0.65]	[−0.77]	[2.35]
US_INR_t	0.012	0.016	−0.027	−0.017
	[0.39]	[0.61]	[−1.27]	[−1.01]

futures also have a significant positive impact on the DCE soybean futures' trading hour returns. This indicates that the CBOT prices influence the DCE prices, which is consistent with Fung *et al.* (2003).

To conclude, the DCE and the CBOT influence each other in both directions. It needs to be emphasized that the DCE has a significant effect on the CBOT price discovery. With China's large-scale and still increasing import demand, this result is reasonable.

9.3.3 Impulse Response Analysis

In the SVAR model, by orthogonalizing the impulse response function, we can investigate the impact of each variable on the other variables separately. The responses of the CBOT non-trading and trading hour returns to one shock of the DCE trading and non-trading hour returns are illustrated in Figure 9.4. It can be seen that the response of the CBOT non-trading hour returns to one unit of positive shock of the DCE non-trading hour returns in the first period is positive, but the response quickly dampens down in the second period. The pattern and the magnitude of the response to one unit of positive shock of the DCE trading hour returns are quite similar.

As shown in Figure 9.4, the response of the CBOT trading hour returns to one unit of positive shock of the DCE non-trading hour returns in the first period is only marginal, and soon dampens down in the following period. It is worth noting that the magnitude of the response here is much smaller than the response of the CBOT non-trading hour returns to the DCE shock. However, this is anticipated if we consider the time difference between the DCE returns and the CBOT trading hour returns.

To provide a more comprehensive picture, we now examine how the DCE responds to the CBOT non-trading and trading hour returns, which are illustrated in Figure 9.5. Consistent with our estimates of the coefficients of matrix $\boldsymbol{\Gamma}_1$, the impulse response analysis in the upper panels of Figure 9.5 also indicates that the CBOT non-trading and trading hour returns shock will positively affect the DCE non-trading hour returns in the next period. The magnitude of the response of the DCE non-trading hour returns to the CBOT shock is similar to, albeit a little larger than, that of the CBOT non-trading hour returns to the DCE shock shown in Figure 9.4.

The lower panels of Figure 9.5 show that the DCE trading hour returns respond negatively to the CBOT shock in the previous period. This finding is consistent with the estimation result of $\boldsymbol{\Gamma}_1$ in Table 9.4. A reasonable explanation is that the opening price of the DCE soybean futures overreacts to the information from the CBOT, and the closing price of the DCE soybean futures reverts to the normal level, leading to a negative response of the DCE trading hour returns to the CBOT return shock.

Response to Cholesky One S.D. Innovations ±2 S.E.

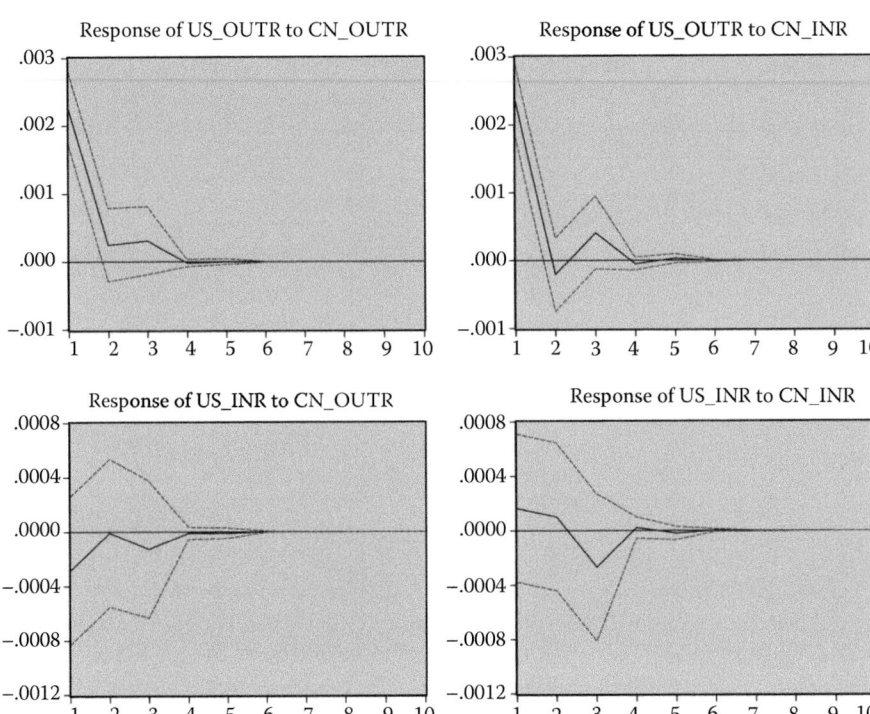

FIGURE 9.4 The responses of the CBOT non-trading and trading hour returns to one unit of positive shock of the DCE non-trading and trading hour returns. The sample period is March 15, 2002 to September 2, 2011.

To sum up, the impulse response analysis indicates that the impact of the DCE market on the CBOT is significantly positive, pointing to the reality that the China soybean futures market is playing an important role in global soybean price discovery.

9.3.4 Forecasted Error Variance Decomposition

In line with the cross-market price discovery study of Grammig *et al.* (2005), we also explore the 'information share' of each variable in the price discovery process, which is defined as the proportion of the innovation variance in the variable *i* that can be attributed to innovations in the variable *j*. The information shares are obtained by forecasted error variance decomposition (FEVD).

When there is contemporaneous correlation among the innovations, identifying an independent information share for each variable without further restrictions is not possible. However, the upper and lower bounds can be provided using the Cholesky factorization for the covariance matrix of the error term. The Cholesky factorization will give the upper bound of the share for the variable ordered first and a lower bound of the share for the variable ordered last.

However, since there are four variables in the SVAR model, it is impossible to give the upper bound and the lower bound of each variable at the same time while maintaining that the total information shares sum to 100. Thus, we consider two sets of ordering; one is as the order of the SVAR model, *D_OUTR D_ INR US_OUTR US_INR*, which approximates the upper bound for the DCE market, and the other ordering is *US_OUTR US_INR D_OUTR D_INR*, which approximates the upper bound for the CBOT market.

Table 9.5 presents the results of the FEVD for the DCE non-trading and trading hour returns. As shown in the table, the non-trading and trading hour returns of the CBOT together contribute approximately 13%–16% to the forecasting of the DCE non-trading hour returns. Their contribution to the DCE

Response to Cholesky One S.D. Innovations ±2 S.E.

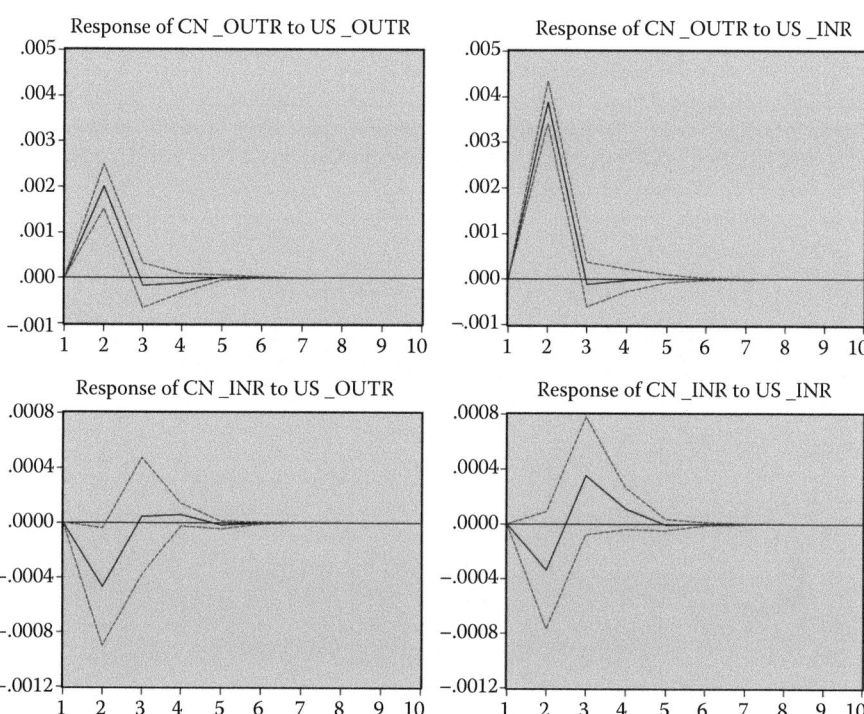

FIGURE 9.5 The responses of the DCE non-trading and trading hour returns of the soybean futures to one unit of positive shock of the CBOT non-trading and trading hour returns. The sample period is March 15, 2002 to September 2, 2011.

trading hour returns is much weaker, which is consistent with the result of the impulse response analysis. It is also worth noting that the fraction of the forecasting error of the DCE non-trading hour returns attributable to the returns of the CBOT significantly differs from zero in the second period, which is consistent with the time difference between the two markets. The variance decomposition for the CBOT non-trading and trading hour returns is shown in Table 9.6. Approximately 0.2%–6% of the CBOT non-trading hour returns forecast error is attributable to the DCE trading and non-trading hour returns.

Combining the results from Tables 9.5 and 9.6, the DCE and the CBOT both have a significant effect on each other, but the magnitude of the impact of the CBOT on the DCE is larger. It seems that the information transmission in the two markets is bidirectional, but the CBOT is still in the dominant position.

In the following section, as a robustness check, we use the VEC model to test the price interdependence between these two exchanges.

9.4 The VEC Model

In this section, we first present the model setup, we then show the estimation results, and, finally, we present an analysis of the results. Note that the VEC model emphasizes the long-term cointegration relationship, and hence we omit the short-term time difference between the two markets in the VEC model.

9.4.1 Model Setup

A cointegrated VAR system can be written in the following form:

$$\Delta \mathbf{y}_t = \zeta_1 \Delta \mathbf{y}_{t-1} + \zeta_2 \Delta \mathbf{y}_{t-2} + \ldots + \zeta_{p-1} \Delta \mathbf{y}_{t-p+1} + \alpha + \mathbf{K} \mathbf{A}' \mathbf{y}_{t-1} + \varepsilon_t, \tag{9.8}$$

TABLE 9.5 Variance Decomposition of the Non-Trading and Trading Hour Return of the DCE No. 1 Soybean Futures. Panel A is the Variance Decomposition for the DCE Non-Trading Hour Return, While Panel B is for the DCE Trading Hour Return. In Each Panel, The First Column is the Period; The Second to the Fifth Columns Correspond to the Percentage of the Forecasted Error that can be Attributed to *CN_OUTR, CN_INR, US_OUTR* and *US_INR*, Respectively, with the Cholesky ordering of *CN_OUTR CN_INR US_OUTR US_INR*; The Sixth to the Ninth Columns Correspond to the Percentage of the Forecasted Error that can be Attributed to *CN_OUTR, CN_INR, US_OUTR* and *US_INR*, Respectively, with the Cholesky ordering of *US_OUTR US_INR CN_OUTR CN_INR*. The Sample Period is March 15, 2002 to September 2, 2011

Period	CN_OUTR	CN_INR	US_OUTR	US_INR	CN_OUTR	CN_INR	US_OUTR	US_INR
				Panel A: Variance decomposition of *CN_OUTR*				
1	100	0	0	0	97.02	0.00	2.97	0.01
2	85.52	1	2.84	10.64	83.00	1.87	4.67	10.45
3	85.07	1.5	2.84	10.59	82.56	2.39	4.65	10.40
4	85.05	1.51	2.85	10.59	82.54	2.40	4.66	10.40
5	85.05	1.51	2.85	10.59	82.54	2.40	4.66	10.40
6	85.05	1.51	2.85	10.59	82.54	2.40	4.66	10.40
7	85.05	1.51	2.85	10.59	82.54	2.40	4.66	10.40
8	85.05	1.51	2.85	10.59	82.54	2.40	4.66	10.40
9	85.05	1.51	2.85	10.59	82.54	2.40	4.66	10.40
10	85.05	1.51	2.85	10.59	82.54	2.40	4.66	10.40
				Panel B: Variance decomposition of *CN_INR*				
1	14.5	85.5	0	0	16.32	82.61	1.01	0.07
2	14.47	85.22	0.2	0.11	16.26	82.36	1.20	0.17
3	14.52	85.05	0.21	0.22	16.30	82.19	1.22	0.29
4	14.52	85.04	0.21	0.23	16.30	82.18	1.22	0.30
5	14.52	85.04	0.21	0.23	16.30	82.18	1.22	0.30
6	14.52	85.04	0.21	0.23	16.30	82.18	1.22	0.30
7	14.52	85.04	0.21	0.23	16.30	82.18	1.22	0.30
8	14.52	85.04	0.21	0.23	16.30	82.18	1.22	0.30
9	14.52	85.04	0.21	0.23	16.30	82.18	1.22	0.30
10	14.52	85.04	0.21	0.23	16.30	82.18	1.22	0.30
Cholesky ordering:					Cholesky ordering:			
CN_OUTR CN_INR US_OUTR US_INR					*US_OUTR US_INR CN_OUTR CN_INR*			

where $\mathbf{A}'\mathbf{y}_{t-1}$ is the long-term cointegration relationship between the series, while $\mathbf{K}$ reflects the short-term reaction to the deviation from this long-term relationship. The economic implication of a VEC model is that if there is a long-term equilibrium relationship between the relevant variables, the short-term variation of these variables can be viewed as a partial adjustment to the long-term relationship.

We employ a VEC model for the closing log price of the CBOT and the DCE to explore the information transmission mechanism between the two markets. The VEC model for the closing price of the CBOT and the DCE is as follows:

$$\Delta CN\_CLOSE_t = \sum_{i=1}^{p} \zeta_{1i} \Delta CN\_CLOSE_{t-i}$$

$$+ \sum_{i=1}^{q} \gamma_{1i} \Delta US\_CLOSE_{t-i} + \alpha_1 \qquad (9.9)$$

$$+ k_1 \left(CN\_CLOSE_{t-1} + a \right.$$

$$\left. \times US\_CLOSE_{t-1} \right) + h_1 \, Mon_t + \varepsilon_{1t},$$

TABLE 9.6 Variance Decomposition of the Non-Trading and Trading Hour Return of the CBOT Soybean Futures. Panel A is the Variance Decomposition for the CBOT Non-Trading Hour Return, While Panel B is for the CBOT Trading Hour Return. In Each Panel, The First Column is the Period; The Second to the Fifth Columns Correspond to the Percentage of the Forecasted Error that can be Attributed to CN_OUTR, CN_INR, US_OUTR and US_INR, Respectively, with the Cholesky Ordering of CN_OUTR CN_INR US_OUTR US_INR; The Sixth to the Ninth Columns Correspond to the Percentage of the Forecasted Error that can be Attributed to CN_OUTR, CN_INR, US_OUTR and US_INR, Respectively, with the Cholesky ordering of US_OUTR US_INR CN_OUTR CN_INR. The Sample Period is March 15, 2002 to September 2, 2011

Period	CN_OUTR	CN_INR	US_OUTR	US_INR	CN_OUTR	CN_INR	US_OUTR	US_INR
			Panel A: Variance decomposition of US_OUTR					
1	2.97	3.23	93.80	0.00	0.00	0.00	100.00	0.00
2	3.00	3.24	93.53	0.24	0.05	0.02	99.70	0.23
3	3.03	3.32	92.96	0.69	0.10	0.09	99.12	0.69
4	3.03	3.32	92.96	0.69	0.10	0.09	99.12	0.69
5	3.03	3.32	92.95	0.69	0.10	0.09	99.11	0.69
6	3.03	3.32	92.95	0.69	0.10	0.09	99.11	0.69
7	3.03	3.32	92.95	0.69	0.10	0.09	99.11	0.69
8	3.03	3.32	92.95	0.69	0.10	0.09	99.11	0.69
9	3.03	3.32	92.95	0.69	0.10	0.09	99.11	0.69
10	3.03	3.32	92.95	0.69	0.10	0.09	99.11	0.69
			Panel B: Variance decomposition of US_INR					
1	0.05	0.02	0.44	99.50	0	0	0.43	99.57
2	0.05	0.02	0.50	99.43	0.001	0.016	0.48	99.50
3	0.06	0.06	0.60	99.27	0.003	0.034	0.62	99.35
4	0.06	0.06	0.60	99.27	0.003	0.034	0.62	99.35
5	0.06	0.06	0.60	99.27	0.003	0.034	0.62	99.35
6	0.06	0.06	0.60	99.27	0.003	0.034	0.62	99.35
7	0.06	0.06	0.60	99.27	0.003	0.034	0.62	99.35
8	0.06	0.06	0.60	99.27	0.003	0.034	0.62	99.35
9	0.06	0.06	0.60	99.27	0.003	0.034	0.62	99.35
10	0.06	0.06	0.60	99.27	0.003	0.034	0.62	99.35
Cholesky ordering:					Cholesky ordering:			
CN_OUTR CN_INR US_OUTR US_INR					US_OUTR US_INR CN_OUTR CN_INR			

$$\Delta US\_CLOSE_t = \sum_{i=1}^{p} \zeta_{2i}\Delta CN\_CLOSE_{t-i}$$

$$+\sum_{i=1}^{q} \gamma_{2i}\Delta US\_CLOSE_{t-i} + \alpha_2 \quad (9.10)$$

$$+k_2\big(CN\_CLOSE_{t-1} + a$$

$$\times US\_CLOSE_{t-1}\big) + h_2\,Mon_t + \varepsilon_{2t},$$

where CN_CLOSE_t denotes the log of the closing price of the DCE soybean futures at day t, which is adjusted to U.S. dollars with the daily exchange rate from the Wind database, and US_CLOSE_t denotes the log of the closing price of the CBOT soybean futures at day t. $\Delta CN\_CLOSE_t := CN\_CLOSE_t - CN\_CLOSE_{t-p}$, and the other first differences of the price series are defined similarly. In this model, $CN\_CLOSE_t + a \times US\_CLOSE_t$ reflects the long-term relationship between the closing prices of the two markets, and $k_i (i = 1, 2)$ reflects how each market adjusts to the long-term relationship. With this model, we can further test the relationship between the two markets. If k_1 is significant while k_2 is not, then the DCE reacts to the long-term relationship between the two markets while the CBOT does not, which means that the information is transmitted from CBOT to the DCE. Other cases can be interpreted similarly. Note that, as in the SVAR model, we also include dummy variable Mon to account for the widely discussed weekend effect.

9.4.2 Estimation Results

We first conduct the unit root test for the price series with three tests, the ADF, PP and KPSS tests. In this way, we can ensure a sounder conclusion since the ADF and PP tests have low test power. Since the variables are in logarithms, a time trend, which implies an ever-increasing (or decreasing) rate of change in the first difference of the variable, is unlikely. Thus, we exclude the time trend but include an intercept in the unit root test. Panel A of Table 9.7 gives the test result. For both the series, the ADF and PP tests cannot reject the null hypothesis, indicating the existence of a unit root. The KPSS test gives results consistent with the ADF and PP tests, which reject the null hypothesis of no unit root. Thus all three test statistics indicate the existence of a unit root. Panel A of Table 9.7 also gives the unit root test result for the first difference of the price series. Note that neither the time trend nor an intercept is included in the test for the difference of the price series. The test results indicate the stationary of the two variables.

Since the unit root test indicates that both of the closing price series have a unit root, we conduct the Johansen cointegration test and the result is shown in panel B of Table 9.7. Both the trace statistic and max-eigen statistic suggest that there is one cointegrating equation at the 5% significance level, which means that the logarithm prices of soybean futures in the DCE and the CBOT markets have similar stochastic trends and could form a long-run equilibrium relationship.

Table 9.8 shows the estimated result of the VEC model. As we can see from panel A of Table 9.8, the parameter of the cointegration equation is significant, and the sign of the estimated coefficient indicates that there is a positive long-term relationship between the closing price of the CBOT and the DCE. Panel B of Table 9.8 shows the estimated coefficient for the error correction term. The result suggests that both the DCE and the CBOT closing prices adjust with the long-term cointegration relationship, and the short-term adjustment is significant for both markets. Specifically, the significantly negative k_1 implies that a larger CN_CLOSE_{t-1} or a smaller US_CLOSE_{t-1} relative to the long-run equilibrium will result

TABLE 9.7   Test Results of the Unit Root Test and the Cointegration Test. Panel A Gives the ADF, PP and KPSS Tests of the CBOT and the DCE Soybean Futures Price, as well as Their First Difference. The Value in Parentheses is the Corresponding P-Value. Panel B Gives the Cointegration Test. Here ** Denotes Significant at the 5% Significance Level. The Sample Period is from March 15, 2002 to September 2, 2011 and the Series are the Log of the Original Price Series

Test statistic	CN_CLOSE	US_CLOSE	ΔCN_CLOSE	ΔUS_CLOSE
ADF	−1.53	−1.25	−32.93	−48.46
	(0.5187)	(0.6556)	(0)	(0.0001)
PP	−1.55	−1.32	−51.95	−48.53
	(0.5063)	(0.6243)	(0.0001)	(0.0001)
KPSS	4.448	4.117	0.101	0.051

Panel B: Cointegration test.				
Hypothesized No. of CE(s)	Trace statistic	Max-eigen statistic	5% Critical value	
			Trace statistic	Max-eigen statistic
None*	23.80**	19.55**	20.26	15.89
At most 1	4.25	4.25	9.16	9.16

Note: For the KPSS test, all the series are tested including the intercept; critical values at the 10%, 5%, and 1% significance level are 0.347, 0.463, and 0.7439, respectively.

TABLE 9.8 Estimated Results of the VEC Model. Panel A Gives the Cointegration Relationship between the Closing Price of the CBOT and the DCE Soybean Futures Price, While Panel B Gives the Estimated Coefficients in the Error Correction Equation. Here, *** Denotes the 1% Significance Level. The Sample Period is from March 15, 2002 to September 2, 2011 and the Series are the Log of the Original Price Series

CN_CLOSE_{t-1}	ΔUS_CLOSE_{t-1}	Constant
Panel A: The cointegration equation.		
1	-0.926^{***}	0.139
	$[-16.53]$	

	Panel B: The error correction equation.	
Independent variable	Dependent variable	
	ΔCN_CLOSE_t	ΔUS_CLOSE_t
Cointegration equation	-0.011^{***}	0.008^{*}
	$[-3.68]$	$[1.95]$
ΔCN_CLOSE_{t-1}	-0.130^{***}	0.007
	$[-6.23]$	$[0.23]$
ΔCN_CLOSE_{t-2}	-0.054^{***}	0.017
	$[-2.70]$	$[0.56]$
ΔUS_CLOSE_{t-1}	0.210^{***}	-0.001
	$[14.12]$	$[-0.02]$
ΔUS_CLOSE_{t-2}	0.030^{**}	0.008
	$[2.02]$	$[0.38]$
Constant	-0.001	0.001
	$[-0.48]$	$[0.14]$
Mon	0.003^{***}	-0.001
	$[4.13]$	$[-0.03]$

in a negative adjustment on *CN_CLOSE_t*, i.e. *CN_CLOSE_t* tends to revert to the long-run cointegration relationship. Similarly, the significantly positive k_2 implies that a larger *US_CLOSE_{t-1}* or a smaller *CN_CLOSE_{t-1}* relative to the long-run equilibrium will result in a negative adjustment on *US_CLOSE_t*, i.e. *US_CLOSE_t* tends to revert to the long-run cointegration relationship. We can thus indicate that the price from either the DCE or the CBOT tends to adjust to the price of the other exchange, i.e. the prices on both exchanges tend to influence each other.

Nonetheless, the error correction term in the equation with ΔCN_CLOSE_t being the dependent variable is greater in magnitude than that of the equation with ΔUS_CLOSE_t as the dependent variable. This indicates that when the long-term equilibrium relationship is perturbed, it is the DCE price that makes the greater adjustment in order to re-establish the equilibrium. In other words, the CBOT price leads the DCE price in price discovery (Tse 1999).

9.4.3 Impulse Response and FEVD

We also carry out an impulse response analysis and forecast the error variance decomposition for the VEC model. The result for the impulse response analysis is shown in Figure 9.6. As demonstrated in the figure, the CBOT closing price responds positively to an innovation in the DCE closing price, and the response is persistent. The DCE closing price also has a significant positive response to an innovation in the CBOT closing price, but the pattern of the response is different from that of the CBOT to the DCE shock. The response of the CBOT to the DCE shock immediately reaches a high level and then continues

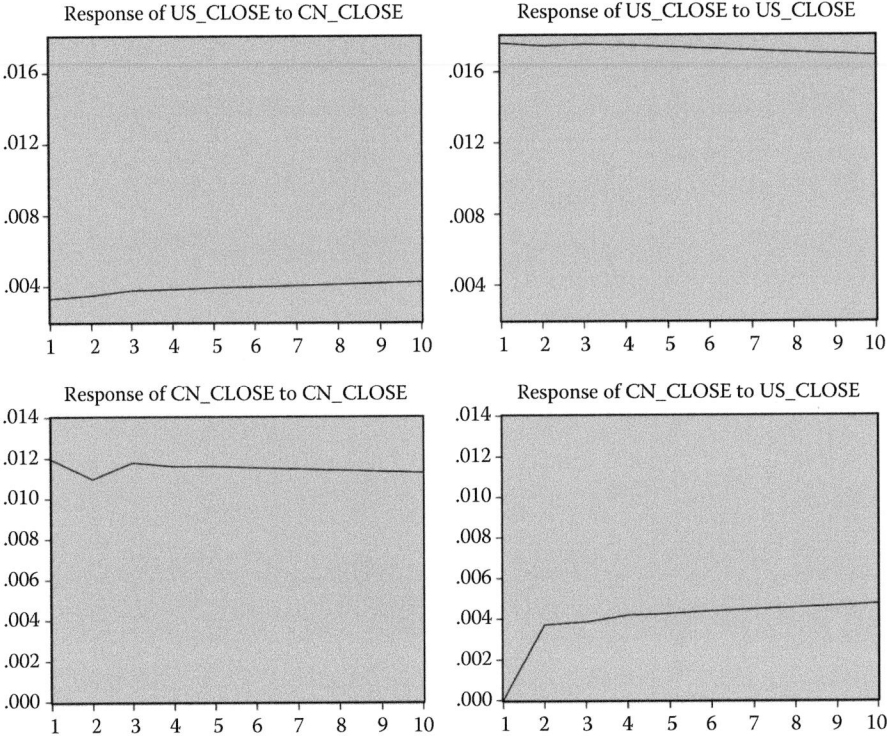

Response to Cholesky One S.D. Innovations

FIGURE 9.6 The impulse response analysis in the VEC model for the closing price of the DCE and the CBOT soybean futures. The sample period is March 15, 2002 to September 2, 2011, and the series are the log of the original price series.

to increase gradually; however, the response of the DCE to the CBOT is almost 0 in the first period and then jumps to a high level in the second period, which persists for a long time. Nonetheless, the magnitude of the two responses to the other market is similar.

Using the method of Hasbrouck (1995), we analyse the information share attributable to each market with the forecasted error variance decomposition. An upper bound for the DCE's information share can be obtained by placing the DCE's price first in the Cholesky decomposition, while a lower bound may be obtained by placing its price last. The same reasoning applies to the CBOT. The results are presented in Table 9.9 with the ordering of *CN_CLOSE US_CLOSE* on the left and the ordering of *US_CLOSE CN_CLOSE* on the right. The result indicates that about 0.1%–5% of the forecasted error from the CBOT can be attributed to the DCE, while about 5%–25% of the forecasted error from the DCE can be attributed to the CBOT. It is clear from the result that although information flows bidirectionally in the two markets, the CBOT still leads the DCE.

9.5 Conclusion

In this paper, we examine the role that the DCE plays in the price discovery of soybean futures. We employ the SVAR and VEC models to investigate the information transfer between soybean futures prices on the DCE and the CBOT. In addition to the well-documented fact that the CBOT significantly affects the DCE, our analysis also emphasizes that the DCE simultaneously has a significant

TABLE 9.9 Forecasted Error Variance Decomposition of the DCE and CBOT Closing Price of the Soybean Futures in the VEC Model. Panel A is the Variance Decomposition for the DCE Closing Price, While Panel B is for the Cbot Closing Price. In Each Panel, the First Column is the Forecast Period, The Second and the Third Columns Correspond to the Percentage of the Forecasted Error that can be Attributed to *CN_CLOSE* and *US_CLOSE* With the Cholesky Ordering of *CN_CLOSE US_CLOSE*, While The Fourth and Fifth Columns Correspond to the Percentage of the Forecasted Error that can be Attributed to *CN_CLOSE* and *US_CLOSE* with the Cholesky Ordering of *US_CLOSE CN_CLOSE*. The Sample Period is March 15, 2002 to September 2, 2011 and the Series are the Log of the Original Price Series

Period	CN_CLOSE	US_CLOSE	CN_CLOSE	US_CLOSE
Panel A: Variance decomposition of *CN_CLOSE*				
1	100.00	0.00	96.57	3.43
2	95.02	4.98	86.58	13.42
3	93.35	6.65	83.09	16.91
4	92.09	7.91	80.78	19.22
5	91.26	8.74	79.31	20.69
6	90.59	9.41	78.18	21.82
7	90.04	9.96	77.27	22.73
8	89.55	10.45	76.50	23.50
9	89.11	10.89	75.81	24.19
10	88.69	11.31	75.18	24.82
Panel B: Variance decomposition of *US_CLOSE*				
1	3.43	96.57	0.00	100.00
2	3.62	96.38	0.01	99.99
3	3.89	96.11	0.03	99.97
4	4.06	95.94	0.04	99.96
5	4.21	95.79	0.06	99.94
6	4.34	95.66	0.07	99.93
7	4.47	95.53	0.09	99.91
8	4.59	95.41	0.11	99.89
9	4.71	95.29	0.13	99.87
10	4.82	95.18	0.15	99.85
	Cholesky ordering: CN_CLOSE US_CLOSE		Cholesky ordering: US_CLOSE CN_CLOSE	

impact on the CBOT. More importantly, the magnitude of the impacts is similar in both directions from the results of the forecasted error variance decomposition (information share) in the SVAR and VEC models. This result logically leads us to the conclusion that the DCE is already playing an important role in the global price discovery of soybean futures, and the information revealed in its trading can affect global market pricing. Because China is now the largest soybean importer in the world and the trading volume of soybean futures on the DCE is second only to the CBOT, it is reasonable that the Chinese market is playing a prominent role in the global price discovery of soybean futures.

Acknowledgements

We thank Shang Liu for his excellent research assistance with preliminary data analysis. Han acknowledges the support from the National Natural Sciences Foundation of China (NSFC), project No. 70831001. Tang acknowledges financial support from the National Natural Science Foundation of China (grant No. 71171194).

References

Booth, G.G. and Ciner, C., International transmission of information in corn futures markets. *J. Multinatl Financ. Mgmt*, 1997, **7**, 175–187.

Eun, C.S. and Sabherwal, S., Cross-border listings and price discovery: evidence from U.S.-listed Canadian stocks. *J. Finance*, 2003, **58**, 549–575.

Fung, H.G., Leung, W.K. and Xu, X.E., Information flows between the U.S. and China commodity futures trading. *Rev. Quantit. Finance Account*, 2003, **21**, 267–285.

Garbade, K.D. and Silber, W.L., Dominant and satellite markets: a study of dually-traded securities. *Rev. Econ. Statist.*, 1979, **61**, 455–460.

Grammig, J., Melvin, M. and Schlag, C., Internationally cross-listed stock prices during overlapping trading hours: price discovery and exchange rate effects. *J. Empir. Finance*, 2005, **12**, 139–164.

Harris, F.H., McInish, T.H., Shoesmith, G.L. and Wood, R.A., Cointegration, error correction, and price discovery on informationally linked security markets. *J. Financ. Quantit. Anal.*, 1995, **30**, 563–579.

Hasbrouck, J., One security, many markets: determining the contributions to price discovery. *J. Finance*, 1995, **50**, 1175–1199.

Kleidon, A.W. and Werner, I.M., U.K. and U.S. trading of British cross-listed stocks: an intraday analysis of market integration. *Rev. Financ. Stud.*, 1996, **9**, 619–664.

Tse, Y., Price discovery and volatility spillovers in the DJIA index and futures markets. *J. Fut. Mkts*, 1999, **19**(8), 911–930.

Tse, Y. and Booth, G.G., Information shares in international oil futures markets. *Int. Rev. Econ. Finance*, 1997, **6**, 49–56.

Tse, Y., Lee, T.-H. and Booth, G.G., The international transmission of information in Eurodollar futures markets: a continuously trading market hypothesis. *J. Int. Money Finance*, 1996, **15**, 447–465.

Xu, X.E. and Fung, H.G., J. Int. Financ. Mkts, Inst. *Money*, 2005, **15**, 107–124.

10

The Structure of Gold and Silver Spread Returns

10.1 Introduction

The price dynamics of precious metals, generally, and gold and silver, in particular, has long been a matter of popular concern and fascination. Recently, a number of authors have successfully modelled the stochastic nature of precious metal returns (e.g., Urich (2000), Casassus and Collin-Dufresne (2005), Cochran *et al.* (2012) and Baur (2013)), while new tools from econometrics have demonstrated important insights into the long-term price relationships between bivariate combinations of precious metal prices (e.g., Escribano and Granger (1998) and Ciner (2001)).

The objective of this study is to report the long-run price dynamics of the bivariate relationship between gold and silver. First, we investigate the spread, measured as the price difference between gold and silver trading as a futures contract. This is novel since previous analyses tend to focus on the price of individual series. Then, we extend the work of Ciner (2001) and Figuerola-Ferretti and Gonzalo (2010), who apply the cointegration techniques of Johansen (1991) and Escribano and Granger (1998) to identify the presence of a long-term equilibrium in the gold and silver markets, to the investigation of the presence of long-term dependence—or long memory—effects in the time-series of the gold- silver spreads.

The investigation and subsequent identification of complex dynamics, including long-memory processes, in various financial time series has provoked debate in the empirical finance literature (e.g., Mandelbrot (2001) and Bianchi and Pianese (2007)), especially in terms of how the presence of these processes affect asset market efficiency (e.g., Eom *et al.* (2008) and McCauley *et al.* (2008)) and asset pricing (e.g., Ellis and Hudson (2007) and Takami *et al.* (2008)). Furthermore, a key contribution of this paper is that we are able to assess the economic implications of the identified long-memory process by applying a series of trading rules to the gold-silver spread time-series. Trading rules may be used to test the efficiency of markets (e.g., Brock *et al.* (1992)), since no trading rule should consistently be able to outperform a simple buy-hold strategy in the long term.

Long-memory processes are typically associated with the hyperbolic decay of the autocorrelation function and are easily measured using the classical rescaled adjusted range (RAR) technique of Hurst (1951), although other approaches, such as detrended fluctuation analysis (e.g., Wan *et al.*'s (2011a,b)

DOI: 10.1201/9781003265399-12

analysis of the gold and oil markets respectively) may be used.* One key advantage of the RAR approach is that, given a daily time series of length N, the RAR may be estimated over a rolling sample (n) to produce a series of (daily) statistics (of length $N - n + 1$). In our case, we set $n = 22$ and 66, representing 1 month and 3 months, respectively, and utilise a time series of Hurst statistics of 1682 daily observations. The RAR approach also has the benefit of providing an insight into the direction of the equilibrium reverting process since it allows for differentiation between processes that revert to their long-term mean after an information shock, such as from news announcements (Christie-David *et al.* 2000), and those that progressively move away from the long-term mean after each new shock.

One feature common to most financial time series is the presence of short-memory effects (e.g., Lo (1991)), typically observed as short-term autocorrelation, which may be associated with lingering liquidity effects in a financial market. To overcome this feature of the gold–silver spread returns, we filter the series using AutoRegressive Moving Average (ARMA) techniques prior to estimating the RAR. This approach has been used previously by Szilagyi and Batten (2007) and Batten and Hamada (2009) to prefilter residuals from foreign exchange and electricity returns before testing for fractality using the rescaled range approach. The ARMA filtering approach therefore accommodates any short-term autocorrelated innovations that may be present in the return process.

We make two main contributions in this paper. First, the gold–silver spread returns reveal time-varying non-linearities, which can be identified as fractality using commonly applied Hurst tests. The use of the local Hurst coefficient, estimated over a rolling sample period (or window) of 22-day and 66-day periods, highlights the time-varying nature of this phenomenon. Time dependence in the Hurst coefficient has been identified by a number of empirical studies investigating the dynamics of financial prices, including Carbone *et al.* (2004), Batten *et al.* (2008) and Grech and Pamela (2008), amongst others, and so we add to this literature. Second, we test the performance of simple trading rules based upon the Hurst coefficient and, importantly, find that these rules outperform both a simple moving-average strategy (which is typically applied to trending series by traders) and a buy-hold strategy. This finding adds to a developing literature that utilises the Hurst coefficient as an investment and trading tool (e.g., Clark (2005)).

These findings are also consistent with other researchers that identify time-varying long-term dependence in other financial markets (Lo 1991, Batten *et al.* 2005, Cajueiro and Tabak 2007, 2008, Du and Ning 2005, Grech and Pamula 2008), other forms of non-linearities or complex structure in financial returns (Jiang and Zhou 2008, Takami *et al.* 2008), or demonstrated the economic gains from trading strategies designed to exploit stock market inefficiencies—specifically trend following trading systems (Brock *et al.* 1992, Bessembinder and Chan 1998, Eom *et al.* 2008), or the possible effects of thin and nonsynchronous trading, typical of emerging markets (Bley 2011).

Next we briefly provide information on the gold and silver futures data used for the study. Then the results from price analysis using rescaled range analysis are reported and the trading strategy applied. The final section allows for some concluding remarks.

10.2 Data

Lucey and Tulley (2006) provide a detailed account of studies investigating trading in the international gold and silver markets. Gold is an important reserve asset and in recent times has become an important part of the monetary regime in emerging economies (see Taguchi 2011). We investigate the price of two contracts trading on the New York Mercantile Exchange (NYMEX): the deliverable 100 troy ounce nominal COMEX gold and the deliverable 5000 troy ounce nominal COMEX silver contracts. In an economic sense these futures contract are fully arbitrageable against gold and silver trading in a variety of other worldwide cash and futures markets. Open-outcry trading commences at 08:20 h/08:25h and

---

* Owczarczuk (2012) provides an excellent discussion of this technique. The links between oil and other commodity markets is discussed in Mohanty *et al.* (2010) and Ewing and Malik (2013).

ends at 13:30h/13:25 h (gold/silver). Trading is also available simultaneously (termed side-by-side trading) on the GLOBEX electronic trading system available on the Chicago Mercantile Exchange (CME). Our data comprises 1746 daily observations of the near month COMEX gold and silver contract at the start of trading from January 1999 to December 2005. This number is reduced to 1682 once the first 64 observations are used to begin the rolling Hurst estimations.

We first estimate the interday returns (ΔP_t) for the price spread (P_t) between gold (G_t) and silver (S_t), where $P_t = G_t - S_t$. Allow $\Delta P_t = \log(P_t) - \log(P_{t-1})$ where the interval $t-1 \to t$, is 1 day. Individual asset returns (for gold and silver) are also measured as $\Delta G_t = \log(G_t) - \log(G_{t-1})$ and $\Delta S_t = \log(S_t) - \log(S_{t-1})$ where the interval $t-1 \to t$, is 1 day.

Note that to the extent the returns of the underlying assets (ΔG_t and ΔS_t) are themselves random processes, then $\Delta P_t = \varepsilon_t$, with ε_t being a random variable, which is expected to have a mean of zero. ΔP_t should also be both mean stationary and uncorrelated over various time increments, which is a requirement for efficient markets in the sense of Fama (1998), although, as is well known, these features are rarely present in financial markets (McCauley *et al.* 2008).

The descriptive statistics of our data are reported in Table 10.1. Over the sample period the spread varied from a maximum of US$474.31 to a minimum of US$248.63. The spread mean was US$329.32. For the return series (Figures 10.1(d) and 10.1(e)) the mean was slightly positive in all cases (ΔP_t, ΔG_t, ΔS_t equalling 0.00013, 0.00014 and 0.00013, respectively) with variation on a similar scale as shown in

Figures 10.1(d) and 10.1(e). Silver was more volatile over the entire sample period than the spread, or gold, measured by the standard deviation (ΔP_t, ΔG_t, ΔS_t equalling 0.0044, 0.0044 and 0.0062, respectively) and the coefficient of variation (ΔP_t, ΔG_t, ΔS_t equalling 3059, 3060 and 4733, respectively). However, this was not the case when the coefficient of variation (CV) was estimated over a 22-day (equal to one calendar month) rolling window (i.e. $CV_n = \mu_n/\sigma_n$, where $n = 22$). The CV_{22} for gold and silver are illustrated in Figures 10.1(a) and 10.1(b). It is clear from these figures that silver appears more stable, with less significant peaks and troughs in the CV_{22} of returns than gold. Such differences would likely impact upon the dynamics of the long-term equilibrium between the two metals.

To demonstrate the time-varying nature of the relationship between gold and silver we also estimate and then plot in Figure 10.1(c) the rolling 22-day correlation between gold and silver returns ($GS\rho_{22}$). Over the entire sample period the correlation is high and positive ($GS\rho_{1746} = 0.686$, $p = 0.000$). However,

TABLE 10.1 Descriptive Statistics of the Gold–Silver Spread
(Daily Spreads From January 1999 to December 2005)

	Gold–silver spread	Gold–silver spread return
	$P_t = G_t - S_t$	$\Delta P_t = \log(P_t) - \log(P_{t-1})$
Mean	329.32	0.00013
Median	309.44	0.00015
Maximum	474.31	0.04464
Minimum	248.63	−0.02873
Standard deviation	62.55	0.00440
Skewness	0.505	0.902
Kurtosis	1 801	16.411
Jarque-Bera	172.34	12 839.13
Probability	0.000	0.00000
Observations	1682	1682

The original sample contains 1746 observations. The later 22-day (66-day) estimation of the Hurst coefficient is based on the returns from $R_0 \to R_{t-22}$ (and $R_0 \to R_{t-66}$) observations. To allow convergence to a stable Hurst coefficient we simply report the subsequent 1682 observations.

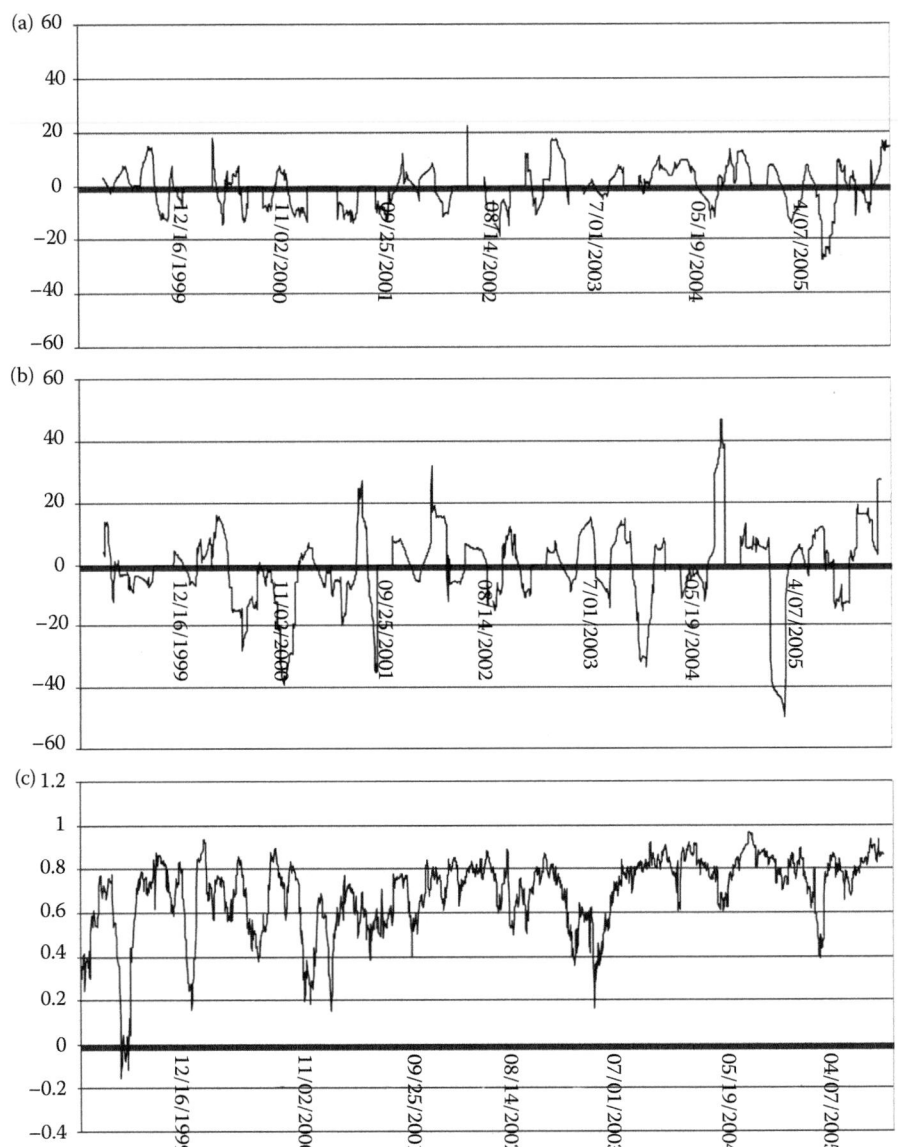

FIGURE 10.1 (a) The coefficient of variation of daily silver returns estimated on a 22-day rolling window. (b) The coefficient of variation of daily gold returns estimated on a 22-day rolling window. (c) The correlation between daily gold and silver returns estimated on a 22-day rolling window.

estimating $GS\rho_{22}$ the range varies from a maximum of 0.9675 to a minimum of −0.1524. Nonetheless, since a positive correlation is maintained over the sample period, trading strategies based on mean reversion of the spread to its average may in fact provide profitable opportunities for market participants. This finding is consistent with widely held views in commodity markets that despite the fundamental differences between the two markets, gold and silver prices tend to move together (Lucey and Tulley 2006), thereby offering the possibility for various trading and portfolio strategies that exploit mean reversion in the spread returns. This possibility is considered further when the rescaled range statistic is estimated in the next section and when trading strategies based upon the local Hurst exponent are implemented in Section 10.4.

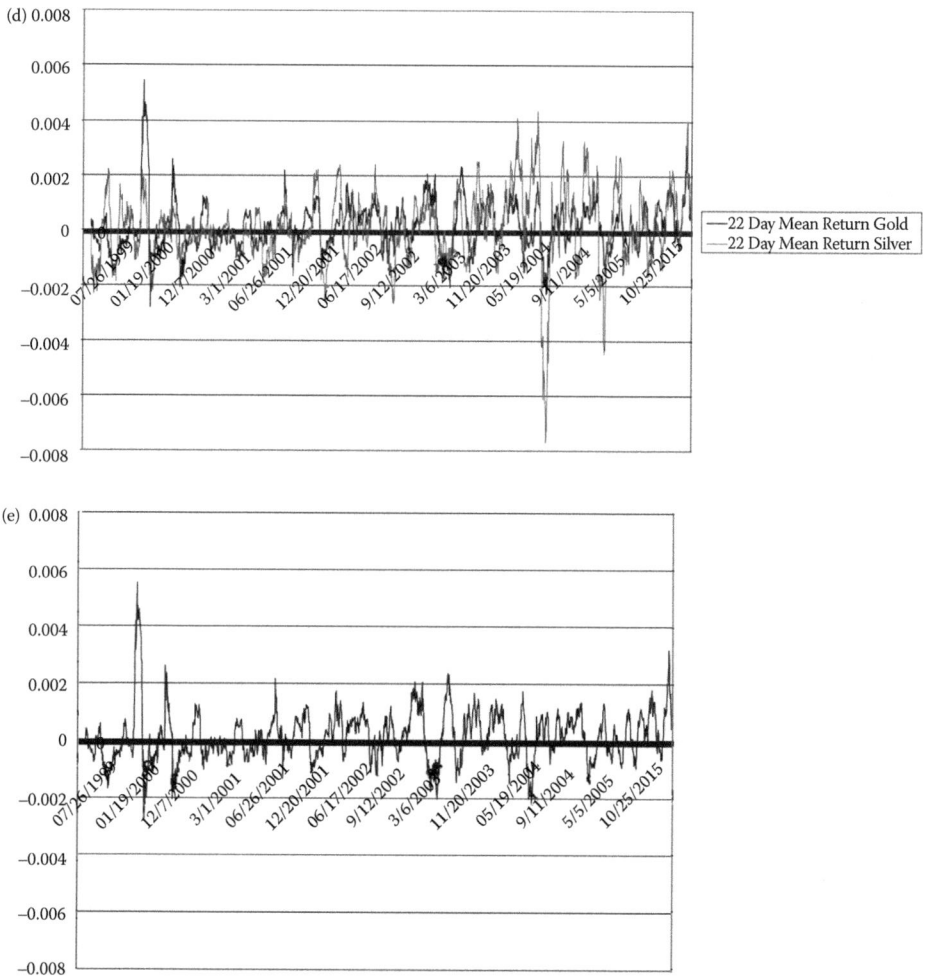

FIGURE 10.1 (CONTINUED) (d) The average 22-day daily gold and silver returns estimated on a rolling window. (e) The average 22-day gold–silver spread returns estimated on a rolling window.

10.3 Rescaled Adjusted Range Analysis

The presence of long-term dependence in the spread returns (ΔP_t) between gold and silver may be measured using statistical techniques based on the Hurst (1951) rescaled range analysis, after accommodating for any short-term autocorrelated innovations in the return process (Batten *et al.* 2008). The filtering process is readily accomplished via ARMA models and forms the basis for RAR (Szilagyi and Batten 2007, Batten and Hamada 2009). Of specific interest is the residual ψ_t after applying various filters (AR(0) → ARMA(2,1)) to P_t. Consider an ARMA(2,1) model of the form

$$\Delta P_t = \alpha_0 + \beta_1 \Delta P_{t-1} + \beta_2 \Delta P_{t-2} + X_1 \lambda \psi_{t-1} + \psi_t, \tag{10.1}$$

which systematic analysis is found to provide the best fit to the data with $\beta_1 = 0.4802$ ($p = 0.06$), $\beta_2 = 0.1109$ ($p = 0.000$) and $X_1 = 0.5780$ ($p = 0.024$).

For each ψ_t over the subsample n, the classical rescaled adjusted range $(R/\sigma)_n$ of Hurst (1951) and Mandelbrot and Wallis (1969) is calculated as

$$\left(\frac{R}{\sigma}\right)n = \frac{1}{\sigma_n}\left[\max_{1\le k\le n}\sum_{j=1}^{k}\left(\psi_j - \mu_n^\psi\right) - \min_{1\le k\le n}\sum_{j=1}^{k}\left(\psi_j - \mu_n^\psi\right)\right], \tag{10.2}$$

where μ_n is the mean and σ_n is the standard deviation of ψ_t over an overlapping sample of length n

$$\sigma_n = \left[1/n\sum_{j=1}^{n}\left(\psi_j - \mu_n^\psi\right)^2\right]^{0.5}. \tag{10.3}$$

In order to capture the time-varying nature of the dependence in ψ_t this study employs a local measure of the Hurst exponent (h) as used by Szilagyi and Batten (2007) and Batten and Hamada (2009). The local version (h) of the Hurst exponent is then estimated for ($N- n + 1$) times overlapping subseries of these various length n:

$$h_n = \frac{\log(R/\sigma)_n}{\log_n}, \tag{10.4}$$

where n is set to either 22 days or 66 days, which is equivalent to a standard one- and three-month period. This procedure in effect creates a time-series of exponent values,[*] the change in whose value can be measured over time. The averages of the local Hurst (h) for the entire sample period are summarised in Table 10.2. The top row in this table records the filtering technique applied to ψ_t. These four techniques range from AR(0)—no filtering—to ARMA(2, 1) as per Equation (10.1).

Recall from Hurst (1951) that, under the null hypothesis of no long-term dependence, the value of $h_n = 0.5$ (a white noise process). For time-series exhibiting positive long-term dependence, the observed value of the exponent is $h_n > 0.5$. Time-series containing negative dependence are mean-reverting and alternatively characterised by $h_n < 0.5$ (termed pink noise by Mulligan (2004)). Note that while the original sample contains 1746 observations, estimation of the 22-day (66-day) local Hurst coefficient is based on the returns from $R_0 \rightarrow R_{t-22}$ (and $R_0 \rightarrow R_{t-66}$) observations. To allow convergence to a stable Hurst coefficient we ignore the first 64 estimations from the original sample.

From Table 10.1, the mean for h_{22} varies from 0.7070 for AR(0) to 0.7170 for ARMA(2,1). The scores for h_{66} are lower, suggesting the series is becoming more random as the sample length n in Equation (10.4) increases, and vary from 0.6487 for AR(0) to 0.6496 for ARMA(2,1). Note also that these different filters have little effect on the size of the h-statistic as evidenced by the overlapping of the confidence intervals. This is due to the fact that the long memory, by definition, relates to the lingering effects of hyperbolic decay in the autocorrelation of the return series. One economic explanation for this property is due to

TABLE 10.2   Local Hurst Exponents Estimated Using 22- and 66-Day Rolling Windows

Overall sample	22-Day rolling window (1 month)				66-Day rolling window (3 months)			
	AR(0)	AR(1)	AR(2)	ARMA(2,1)	AR(0)	AR(1)	AR(2)	ARMA(2,1)
Mean	0.7071	0.7370	0.7189	0.7170	0.6487	0.6667	0.6577	0.6496
Standard deviation	0.1314	0.1328	0.1312	0.1283	0.0737	0.0741	0.0740	0.0725
95% Confidence interval	0.7007–0.7133	0.7239–0.7366	0.7123–0.7521	0.7046–0.7168	0.6451–6523	0.6631–6703	0.6541–6613	0.6460–6532

---

[*] The filtering using the ARMA(2,1) affects the residuals of Equation (10.1). It is the residual series ψ_t that is used to estimate the $(R/\sigma)n$ of Hurst (1951) using $(R/\sigma)n = 1/\sigma n$. The transformation shown in Equation (10.4) is then applied to estimate the local Hurst (h) over a 22-day and 66-day rolling window.

liquidity effects in the market. Even allowing for a 95% confidence interval, these local Hurst coefficients are consistent with positive long-term dependence. For positively dependent processes, another price movement further away from the mean (or long-term equilibrium) will follow the earlier movement away from equilibrium. Thus, they are trend-reinforcing processes (Mulligan 2004), such that the spread should tend to become larger, or smaller, depending on whether the previous change in price was positive, or negative.

Nonetheless, a plot of h_{22} and h_{66} over the entire sample period (shown in Figures 10.2(a) and 10.2(b)) shows considerable variation in the statistics and rare episodes when the statistic was below 0.5000. In the case of h_{22} the minimum value was 0.3296 and the maximum was 1.1667, while for h_{66} the minimum value was 0.4071 and the maximum was 0.9399. Values below 0.5000 are consistent with negative dependence where a price movement towards the equilibrium should follow a movement away from equilibrium. Note that this high level of the Hurst exponent (greater than 1) appears to be a statistical artefact due to extreme volatility in the daily gold–silver spread price which moved from USD$250 to US$320 in the period from 9/7/1999 to 11/3/1999 (nearly a 30% move). Removing the spreads from the period 10/1/1999 to 10/18/1999 eliminates this anomaly. Alternately, if the estimation sample length is

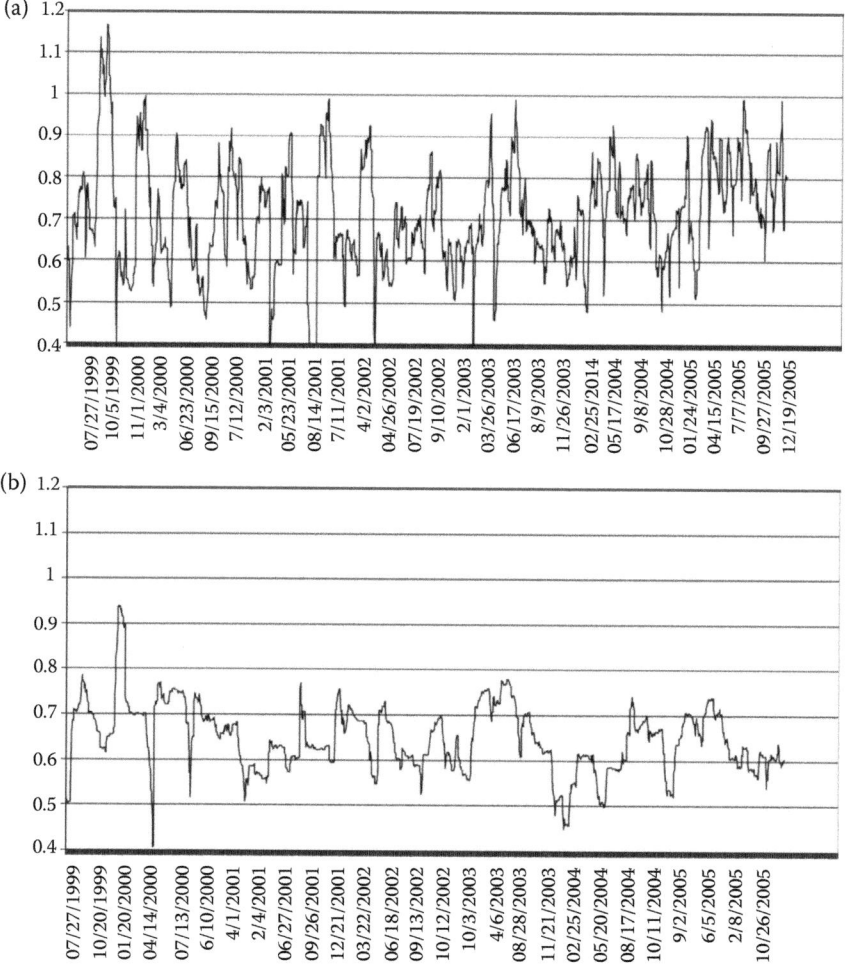

FIGURE 10.2 (a) The local Hurst estimated using an ARMA(2,1) filter on the spread return between daily gold and silver (estimated on a 22-day rolling window). (b) The local Hurst estimated using an ARMA(2,1) filter on the spread return between daily gold and silver (estimated on a 66-day rolling window).

increased to 66, then the anomaly dissipates. Removing these extreme observations does affect the estimated average Hurst value. For example, in Table 10.2 the ARMA(2,1) 22-day Hurst is now 0.7048 versus the reported number of 0.7170 and, for the ARMA(2,1), the 66-day Hurst is 0.6475 versus the reported number of 0.6496, which lie just within the confidence interval.

10.4 Trading Implications

To investigate the trading implications we first divide the unfiltered return series (ΔP_t) into varying holding periods (HP) from 1 day to 22 days from the start day return at t_0 (i.e. R_0). These holding period returns are reported in Table 10.3 and reflect the reward to the investor for a buy-hold investment strategy for these specified number of days. Note that kurtosis and skewness decline with holding period length, while the mean return increases as days are added to the holding period. However, the return over the holding period on an average daily basis is not linear. This is clearer in Figure 10.3, which plots the average daily return over the varying holding period lengths (reported in the third column of Table 10.3). Figure 10.3 clearly shows that there is an advantage to holding a portfolio for the next 2 days instead of just one day, but this advantage declines quickly for the next 6 days, after which it increases again steadily for the next period. The optimal holding period (that is the holding period with the largest average daily return) is a period of 15 days from today.

A number of trading strategies were then applied to the return series. The first of these is one commonly applied to trend-following systems: a moving average. Brock *et al.* (1992), in a leading paper, demonstrated the success of moving average trading systems over others. While moving averages of varying length could be considered, for the sake of brevity we investigate a moving average of length 22, which matches the other estimations for the local Hurst coefficient.

TABLE 10.3   Returns Over Different Holding Periods (1 Day to 22 Days)

Return (R) window	Mean	Average daily	Standard deviation	Kurtosis	Skewness
R_0	0.0001443	0.000144349	0.004417358	0.8478223	12.916320
$R_0 \rightarrow R_1$	0.0002910	0.000145520	0.005920985	1.0070222	8.732371
$R_0 \rightarrow R_2$	0.0004361	0.000145371	0.007283346	1.0934505	9.641540
$R_0 \rightarrow R_3$	0.0005775	0.000144373	0.008408646	0.9301216	6.773518
$R_0 \rightarrow R_4$	0.0007179	0.000143574	0.009547577	1.0231929	6.256877
$R_0 \rightarrow R_5$	0.0008565	0.000142749	0.010560701	1.0968331	7.211523
$R_0 \rightarrow R_6$	0.0009934	0.000141918	0.011531883	1.1706315	8.118051
$R_0 \rightarrow R_7$	0.0011374	0.000142175	0.012337697	1.1952264	7.778539
$R_0 \rightarrow R_8$	0.0012822	0.000142462	0.013090957	1.1863138	7.273252
$R_0 \rightarrow R_9$	0.0014258	0.000142576	0.013770619	1.2489172	7.404415
$R_0 \rightarrow R_{10}$	0.0015711	0.000142826	0.014435979	1.2597065	6.978512
$R_0 \rightarrow R_{11}$	0.0017207	0.000143395	0.015047962	1.2411078	6.672399
$R_0 \rightarrow R_{12}$	0.0018772	0.000144402	0.015588680	1.2269747	6.462010
$R_0 \rightarrow R_{13}$	0.0020384	0.000145601	0.016069123	1.2125824	6.145907
$R_0 \rightarrow R_{14}$	0.0021926	0.000146175	0.016539234	1.2111407	5.949431
$R_0 \rightarrow R_{15}$	0.0023446	0.000146540	0.016999905	1.1954588	5.660918
$R_0 \rightarrow R_{16}$	0.0024945	0.000146736	0.017448983	1.1795708	5.321579
$R_0 \rightarrow R_{17}$	0.0026419	0.000146773	0.017885264	1.1508532	4.945140
$R_0 \rightarrow R_{18}$	0.0027895	0.000146814	0.018343986	1.0978180	4.601016
$R_0 \rightarrow R_{19}$	0.0029360	0.000146802	0.018774185	1.0558345	4.244551
$R_0 \rightarrow R_{20}$	0.0030790	0.000146621	0.019170964	1.0150762	3.913951
$R_0 \rightarrow R_{21}$	0.0032216	0.000146438	0.019545214	0.9705729	3.697957
$R_0 \rightarrow R_{22}$	0.0033633	0.000146233	0.019907824	0.9214373	3.398959

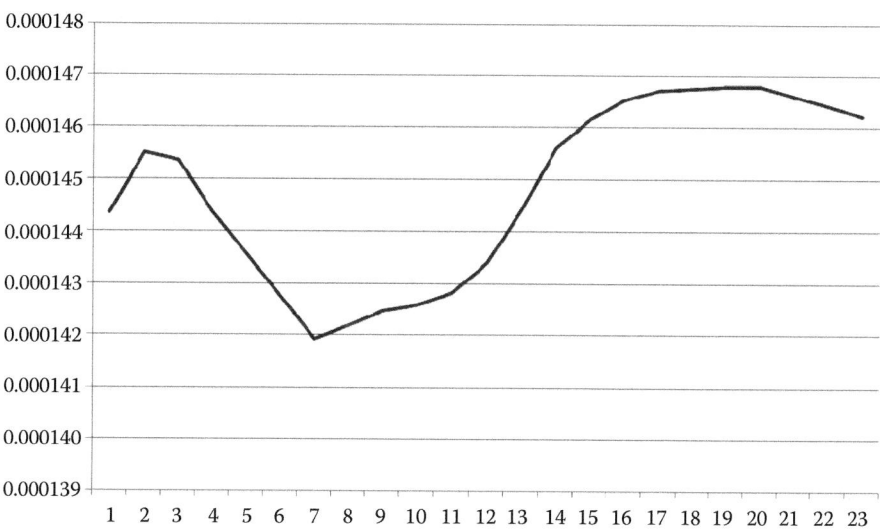

FIGURE 10.3 Plot of average daily gold–silver spread returns for holding periods from one day to one month.

The first rule that is applied is to buy when the 22-day moving average price is greater than the price today (i.e. $P_t > $ MA22 in Table 10.4). If this rule is triggered, the subsequent returns to the next day (HP = $R_0 \rightarrow R_1$), the next 5 days (HP = $R_0 \rightarrow R_5$), the next 10 days (HP = $R_0 \rightarrow R_{10}$) and the next 20 days (HP = $R_0 \rightarrow R_{20}$) are reported in the first four rows of Table 10.4. The frequency of this rule is reported as $N = 0$ (reject) and $N = 1$ (accept). In all cases there were more accepted applications of the rule than rejected. However, while mean returns from the accepted application of the rule ($N = 1$) steadily increased from 0.00133 for the next day to 0.00252 for the next 20 days, an F-test of difference in the means shows that only the next one-day and 5-day holding periods were statistically different from the alternate (rejected) portfolio returns ($N = 0$). Importantly, the returns for the return on the average buy-hold portfolio held for 20 days (reported in Table 10.3 as 0.003079) is higher than the return for an accepted application of the rule for 20 days (Table 10.4, $N = 1$, mean 0.00252). Thus, while the moving average rule offers potential short-term gains, these appear to diminish as the holding period is lengthened. This short-term gain is consistent with trading models that exploit the autoregressive features of the series. These were previously identified in the ARMA model (Equation (10.1)), which identifies positive one- and two-day autocorrelation in the return series.

The next set of rules that are applied utilise the time-series of local Hurst coefficients, whose averages are reported in Table 10.2. For the sake of brevity, only the unfiltered local Hurst (AR0) estimated over 22 days is reported, although similar findings apply for the Hurst (AR0) 66-day estimations. There are two conditions considered to provide the perspective when the return time series is mean reverting (H22 < 0.4; 27 instances when the rule is accepted), or trend-reinforcing (H22 > 0.6; 1342 instances when the rule is accepted). The same holding period lengths for the moving average rule are applied (next day, next 5 days, next 10 days and next 20 days). Apart from one exception (for H22 < 0.4, for the next day holding period), which is the likely consequence of the positive autoregressive structure mentioned earlier, the F-test of the difference in the other mean returns shows significant differences in returns over the alternate portfolio. Also, these returns are significantly different to the returns from the average buy- hold portfolios of varying holding periods reported in Table 10.3. For example, from Table 10.4, for the Hurst rule of H22 > 0.6, for the 20-day holding period, the mean return is 0.00461 compared with the 20-day buy-hold of 0.003079 (Table 10.2). Note that to effect a profitable trade the investor is required to sell when H22 < 0.4 (a bet on mean reversion), and buy when H22 > 0.6 (a bet on trend-reinforcement).

TABLE 10.4   Analysis of Variance (Anova) of Differences in Mean Returns Based on Various Trading Rules Over Different Holding Periods

Trading rule	Mean		Standard deviation		F-Test	p-Value	Adjusted R^2 (%)
	$N = 0$	$N = 1$	$N = 0$	$N = 1$			
$P_t > MA22$ $HP = R_0 \rightarrow R_1$	812 / 889	−0.00085 / 0.00133	0.004834 / 0.00665		58.59	0.000	3.28
$P_t > MA22$ $HP = R_0 \rightarrow R_5$	810 / 887	−0.00001 / 0.00164	0.00846 / 0.01221		10.18	0.001	0.54
$P_t > MA22$ $HP = R_0 \rightarrow R_{10}$	808 / 884	0.00108 / 0.00206	0.01265 / 0.01603		1.96	0.162	0.06
$P_t > MA22$ $HP = R_0 \rightarrow R_{20}$	808 / 874	0.00387 / 0.00252	0.01979 / 0.01876		2.08	0.150	0.06
$P_p\ H22 < 0.4$ $HP = R_0 \rightarrow R_1$	1,674 / 27	0.00031 / −0.00073	0.00591 / 0.00829		0.81	0.368	0.00
$P_p\ H22 < 0.4$ $HP = R_0 \rightarrow R_5$	1,670 / 27	0.00098 / −0.00681	0.01061 / 0.00775		14.40	0.000	0.78
$P_p\ H22 < 0.4$ $HP = R_0 \rightarrow R_{10}$	1,665 / 27	0.00181 / −0.01184	0.01449 / 0.00836		23.81	0.000	1.33
$P_p\ H22 < 0.4$ $HP = R_0 \rightarrow R_{20}$	1,655 / 27	0.00345 / −0.01425	0.01925 / 0.01072		22.71	0.000	1.28
$P_p\ H22 > 0.6$ $HP = R_0 \rightarrow R_1$	359 / 1342	−0.00021 / 0.00042	0.00621 / 0.00588		3.14	0.076	0.13
$P_p\ H22 > 0.6$ $HP = R_0 \rightarrow R_5$	359 / 1338	−0.00071 / 0.00127	0.01166 / 0.01028		9.96	0.002	0.53
$P_p\ H22 > 0.6$ $HP = R_0 \rightarrow R_{10}$	359 / 1333	−0.00128 / 0.00237	0.01442 / 0.01445		18.05	0.000	1.00
$P_p\ H22 > 0.6$ $HP = R_0 \rightarrow R_{20}$	359 / 1323	−0.00214 / 0.00461	0.01727 / 0.01953		35.36	0.000	2.00

The original sample contains 1746 observations. The 22-day Hurst coefficient is based on the previous 22 observations, which allows 1746–22 Hurst (H) estimations. To allow convergence to a stable Hurst coefficient we ignore the next 22 observations. The holding period (HP) return (R) varies from the return to the next day (HP = $R_0 \rightarrow R_1$) to the next 20 days (HP = $R_0 \rightarrow R_{20}$). Therefore, the sum of $N = 0$ and $N = 1$ varies from 1701 (for a 1-day holding period) to 1652 (for a 20-day holding period). The trading rules are buy when the price at $t = 0$ > previous 22-day Moving Average ($P_t > MA22$), sell the price at $t = 0$ when the previous 22-day Hurst coefficient is < 0.4 (P_p, H22 < 0.4), and buy the price at $t = 0$ when the previous 22-day Hurst coefficient is > 0.6 (P_p, H22 > 0.6).

10.5 Conclusions

This study investigates the relationship between gold and silver trading as a futures contract on COMEX from January 1999 to December 2005. During this period the correlation between gold and silver returns was positive and high, even though the relationship itself was unstable. We apply techniques from fractal geometry after accommodating underlying autoregressive behaviour to investigate the long-term dynamics of the spread between these two contracts. Using a local Hurst exponent we find episodes of both positive and negative dependence, although the positive dependent relationship appears to be dominant. This last finding is suggestive of a time-varying fractal structure in the spread returns. Positive dependence (consistent with a Hurst coefficient > 0.5) in the gold–silver spread returns suggests the series will not immediately revert to its average or long-term mean, thereby offering traders limited profit opportunities. To test this proposition we test a number of simple trading rules based upon the Hurst coefficient.

We find that trading rules requiring the investor to sell when the local Hurst coefficient, estimated over a 22-day window, is less than 0.4 (a bet on mean reversion), and buying when the local Hurst coefficient, estimated over a 22-day window, is greater than 0.6 (a bet on trend-reinforcement) outperforms a simple buy-hold and moving-average strategy. This result adds to a growing body of work that finds significant arbitrage possibilities remaining in financial markets despite the advent of improved pricing and information technology (Avellaneda and Lee 2010).

References

Avellaneda, M. and Lee, J.-H., Statistical arbitrage in the US equities market. *Quant. Finance*, 2010, **10**(7), 761–782.

Batten, J.A., Ellis, C. and Fetherston, T., Statistical illusions and long-term return anomalies on the Nikkei. *Chaos, Solitons & Fractals*, 2005, **23**, 1125–1136.

Batten, J.A., Ellis, C. and Fetherston, T., Sample period selection and long-term dependence: New evidence from the Dow Jones. *Chaos, Solitons & Fractals*, 2008, **36**(5), 1126–1140.

Batten, J.A. and Hamada, M., The compass rose pattern in electricity prices. *Chaos: Interdisc. J. Nonlin. Sci.*, 2009, **19**, 043106.

Baur, D.G., The autumn effect of gold. *Res. Int. Bus. Finance*, 2013, **27**(1), 1–11.

Bessembinder, H. and Chan, K., Market efficiency and the returns to technical analysis. *Financ. Mgmt*, 1998, **27**(2), 5–17.

Bianchi, S. and Pianese, A., Modelling stock price movements: Multifractality or multifractionality? *Quant. Finance*, 2007, **7**(3), 301–319.

Bley, J., Are GCC stock markets predictable? *Emerg. Mkts Rev.*, 2011, **12**(3), 217–237.

Bradley, T.E. and Malik, F., Volatility transmission between gold and oil futures under structural breaks. *Int. Rev. Econ. Finance*, 2013, **25**, 113–121.

Brock, W., Lakonishok, J. and LeBaron, B., Simple technical trading rules and the stochastic properties of stock returns. *J. Finance*, 1992, **47**(5), 1731–1764.

Cajueiro, D.O. and Tabak, B.M., Long-range dependence and market structure. *Chaos, Solitons & Fractals*, 2007, **31**(4), 995–1000.

Cajueiro, D.O. and Tabak, B.M., Testing for long-range dependence in world stock markets. *Chaos, Solitons & Fractals*, 2008, **37**(3), 918–927.

Carbone, A., Castelli, G. and Stanley, H.E., Time-dependent Hurst exponent in financial time series. *Physica A*, 2004, **344**(1/2), 267–271.

Casassus, J. and Collin-Dufresne, P., Stochastic convenience yield implied from commodity futures and interest rates. *J. Finance*, 2005, **60**, 2283–2331.

Christie-David, R., Chaudhry, M. and Koch, T.W., Do macroeconomics news releases affect gold and silver prices? *J. Econ. Bus.*, 2000, **52**, 405–421.

Ciner, C., On the long run relationship between gold and silver prices. *Glob. Financ. J.*, 2001, **12**, 299–303.

Clark, A., The use of Hurst and effective return in investing. *Quant. Finance*, 2005, **5**(1), 1–8.

Cochran, S.J., Mansur, I. and Odusami, B., Volatility persistence in metal returns: A FIGARCH approach. *J. Econ. Bus.*, 2012, **64**(4), 287–305.

Du, G. and Ning, X., Multifractal properties of Chinese stock market in Shanghai. *Physica A*, 2005, **387**(1), 261–269.

Ellis, C. and Hudson, C., Scale-adjusted volatility and the Dow Jones index. *Physica A*, 2007, **378**(2), 374–384.

Eom, C., Choi, S., Oh, G. and Jung, W.S., Hurst exponent and prediction based on weak-form efficient market hypothesis of stock markets. *Physica A*, 2008, **387**(18), 4630–4636.

Escribano, A. and Granger, C.W.J., Investigating the relationship between gold and silver prices. *J. Forecast.*, 1998, **17**, 81–107.

Fama, E.F., Market efficiency, long term returns, and behavioural finance. *J. Financ. Econ.*, 1998, **49**, 283–306.

Figuerola-Ferretti, I. and Gonzalo, J., Price discovery and hedging properties of gold and silver markets. Working Paper presented at the 2010 European Financial Management Conference, Aarhus. Paper dated January 19, 2010.

Grech, D. and Pamula, G., The local Hurst exponent of the financial time series in the vicinity of crashes on the Polish stock exchange market. *Physica A*, 2008, **387**(16/17), 4299–4308.

Hurst, H.E., Long-term storage capacity of reservoirs. *Trans. Am. Soc. Civ. Engrs*, 1951, **116**, 770–799.

Jiang, Z.Q. and Zhou, W.X., Multifractality in stock indexes: Fact or Fiction? *Physica A*, 2008, **387**(14), 3605–3614.

Johansen, S., Cointegration and hypothesis testing of cointegration vectors in Gaussian vector autoregressive models. *Econometrica*, 1991, **59**(6), 1551–1580.

Lo, A.W., Long-term memory in stock market prices. *Econometrica*, 1991, **59**(5), 1279–1313.

Lucey, B. and Tully, E., Seasonality, risk and return in daily COMEX gold and silver data 1982–2002. *Appl. Financ. Econ.*, 2006, **16**, 319–333.

Mandelbrot, B.B., Stochastic volatility, power laws and long memory. *Quant. Finance*, 2001, **1**(6), 558–559.

Mandelbrot, B.B. and Wallis, J.R., Robustness of the rescaled range R/S in the measurement of noncyclic long-run statistical dependence. *Water Resour. Res.*, 1969, **5**(5), 967–988.

Matos, J., Gama, S., Ruskin, M.A., Sharkasi, H. and Crane, M., Time and scale Hurst exponent analysis for financial markets. *Physica A*, 2008, **387**(15), 3910–3915.

McCauley, J.L., Bassler, K.E. and Gunaratne, G.H., Martingales, nonstationary increments, and the efficient market hypothesis. *Physica A*, 2008, **387**(15), 3916–3920.

Mohanty, S., Nandha, M. and Bota, G., Oil shocks and stock returns: The case of the Central and Eastern European (CEE) oil and gas sectors. *Emerg. Mkts Rev.*, 2010, **11**(4), 358–372.

Mulligan, R., Fractal analysis of highly volatile markets: An application to technology equities. *Q. Rev. Econ. Finance*, 2004, **44**, 155–179.

Owczarczuk, M., Long memory in patterns of mobile phone usage. *Physica A*, 2012, **391**(4), 1428–1433.

Szilagyi, P.G. and Batten, J.A., Covered interest parity arbitrage and temporal long-term dependence between the US dollar and the Yen. *Physica A*, 2007, **376**, 409–421.

Taguchi, H., Monetary autonomy in emerging market economies: The role of foreign reserves. *Emerg. Mkts Rev.*, 2011, **12**(4), 371–388.

Takami, Y., Tabak, B. and Miranda, M., Interest rate option pricing and volatility forecasting: An application to Brazil. *Chaos, Solitons & Fractals*, 2008, **38**(3), 755–763.

Urich, T.J., Modes of fluctuation in metal futures prices. *J. Fut. Mkts*, 2000, **20**, 219–241.

Wang, Y., Wei, Y. and Wu, C., Analysis of the efficiency and multifractality of gold markets based on multifractal detrended fluctuation analysis. *Physica A*, 2011a, **390**(5), 817–827.

Wang, Y., Wei, Y. and Wu, C., Detrended fluctuation analysis on spot and futures markets of West Texas Intermediate crude oil. *Physica A*, 2011b, **390**(5), 864–875.

11

Gold and the U.S. Dollar: Tales from the Turmoil

Paolo Zagaglia and
Massimiliano Marzo

'(Gold's) quantity cannot increase at the same rate as you can print money, which will eventually weaken the US dollar', Faber said on Thursday in a live interview.

Marc Faber on CNBC, March 4, 2010

'There is still a link on a day-to-day and weekly basis between gold and the dollar, so if the euro weakens gold will come under pressure', said Standard Bank analyst Walter de Wet. 'We might test down towards 1.10$ if the euro goes to 1.30$'.

CNBC Commentary, March 4, 2010

11.1 Introduction

Various accounts of the recent financial turmoil have stressed the propensity of investors to turn away from risky securities into 'safe' assets. This flight to safety has also taken the form of a renowned interest in gold as an asset class. In fact, it is typically argued that the price of this precious metal is uncorrelated with both stock and bond prices during episodes of market crash (see, e.g. Baur and Lucey (2010) and Baur and McDermott (2010)). Gold is also identified as a 'hedge' against fluctuations in the U.S. dollar (USD) on average (Baur and Lucey 2010).* The properties of safe asset and hedging capabilities suggest that the dollar price of gold should increase when the bilateral exchange of the U.S. dollar against other currencies depreciates.

The international role of gold dates back to the early years of the nineteenth century when several countries adopted the 'gold standard'. In this arrangement, the value of a currency was backed by government holdings of gold. The gold standard collapsed in 1971 with the decision of President Richard Nixon to end the dollar convertibility into gold.

---

* Several contributions indicate that gold provides stability to industry portfolios. For instance, Davidson *et al.* (2003) show that standard international asset pricing models prescribe a systematic exposure to gold.

DOI: 10.1201/9781003265399-13

Cappie *et al.* (2005) point out that two reasons are typically suggested for the use of gold as a hedging instrument or safe asset against exchange rate risk. First, a number of financial products are available that track the price of gold, despite the fact that they do not involve the property of the physical commodity. For instance, there are commodity Exchange Traded Funds that are linked to gold. Second, gold is often pointed at as a protection against currency fluctuations worldwide, not only for the U.S. dollar (see also Sjaastad and Scacciavillani (1996)). The results of Cappie *et al.* (2005) show that the hedging power of gold for the U.S. dollar has varied widely since 1971. In particular, they argue that the degree of protection offered by the dollar depends on largely unpredictable events.*

In this paper, we study how the relation between gold prices and the U.S. dollar has been affected by the turmoil that erupted in financial markets in 2007. In other words, we provide evidence on whether gold has behaved as a safe asset against currency fluctuations since 2007.

The tests for gold as a safe asset or as a hedging instrument consider the patterns of conditional correlation between gold returns and the returns of alternative assets, and how these change through time. A negative correlation is indicative of hedging capabilities. We consider a more general approach, and shed light on properties of the relation between gold and the U.S. dollar that have not been considered previously. We provide a new definition based on the evolution of the pattern of 'contagion', on one hand, and 'comovements', on the other, during the turmoil period. This allows us to reinterpret some previous results available in the literature under a novel perspective, using methods that shed new light on old issues.

The idea of contagion is widely used in empirical studies of financial market crises. The aim is to investigate how exogenous changes in volatility are transmitted between markets. We believe that this is an issue of major importance for understanding both the sources and the channels through which risk propagates. Gauging how volatility 'spills over' across assets requires making structural assumptions about the relations between two assets.† The validity of these assumptions can then be verified both through standard statistical tests, as well as by considering whether their model implications are reasonable.

Our structural approach differs noticeably from the 'reduced-form' evidence on the hypothesis of hedging tool or safe asset for gold. An example of an application for this method can be found, amongst others, in Cappie *et al.* (2005). This key study uses information from the statistical coefficient of univariate generalized autoregressive conditional-heteroskedasticity (GARCH) models to come to conclusions on the hedging properties of gold for the U.S. dollar. In our paper, we dive into the question of how large the volatility spillovers are that arise from changes to gold returns onto the USD. Our method allows us to study the role of specific channels of transmission of instability across asset markets.

Since our sample data includes the recent financial turmoil, we also consider the joint tail behaviour of both gold and the USD returns. To put it differently, we investigate how the relation between gold and the USD is affected by extreme events within an episode of financial-market turbulence.

To deal with these two main questions, this paper employs two different econometric frameworks to evaluate the change in the comovements between gold and the U.S. dollar. First, we provide evidence on how the impact of volatility shocks and spillovers has changed during the turmoil. We estimate the standard bivariate GARCH models proposed by Engle and Kroner (1995). In order to uncover the role played by market linkages in the propagation of volatility shocks, we consider the extension to the structural BEKK model discussed by Spargoli and Zagaglia (2008). This amounts to using the information from time-varying conditional heteroskedasticity to identify causal relations between the volatility movements in the price of gold and the U.S. dollar (Rigobon 2003).

---

* We should stress that the price of gold is also affected by some of the driving factors of the dollar exchange rate, such as changes in the inflation rate (see, e.g. McCown and Zimmerman (2006)) and the release of macroeconomic news (Christie-David *et al.* (2000)).

† By 'structural' we refer to the hypothesis on the distinction between the causes of purely exogenous changes of variables and the endogenous impact of these shocks.

In the following step, we study the evolution of comovements between extreme prices. We use the measure of contagion proposed by (Cappiello *et al.*, 2005) to investigate whether the probability of observing closer comovements has increased since August 2007. The framework of Cappiello *et al.* (2005) is based on the computation of the probability of a variable falling below a threshold conditional on the same pattern for the other variable. Thresholds are obtained through quantile estimation. In this statistical model, a high conditional probability of comovement implies a strong codependence between the variables.

Our results indicate that the outbreak of the turmoil in August 2007 has not increased the uncertainty of either gold, or the USD. Rather, the triggering event for market volatility was the bankruptcy of Lehman Brothers in September 15, 2008. The volatility links between gold and the foreign exchange market exacerbate the transmission of shocks and the resulting uncertainty in price movements.

We uncover two features of gold that are disregarded in the literature. We show that exogenous increases in market uncertainty have tended to produce reactions of gold prices that are more stable than those of the U.S. dollar during the turmoil period. Hence, gold prices are affected by market uncertainty to a smaller extent than the USD exchange rate. For a given level of correlation between these two assets, this suggests that the volatility shock originating from gold prices and transmitted to the USD has a more limited size. Overall, this is a source of value added for gold in currency portfolios.

Furthermore, we uncover a feature of gold that is typically disregarded. This consists of its ability to generate comovements with the U.S. dollar exchange rate that are stable through time. Strikingly, such stability has survived the extreme events that have taken place during the recent phases of market disruption. Also, it characterizes both tail episodes of abrupt market swings, as well as conditions of milder changes. In other words, even though our empirical methods focus on a 'crisis within a crisis', we confirm the previous findings suggesting that gold can be considered a safe asset.

The results presented here have important implications for risk management. In this area, the use of the concept of volatility shocks is both appealing from a methodological point of view, and for the purpose of practical applications. Understanding the determinants of conditional volatility is relevant for modelling the risk factors and how these manifest themselves. For instance, in the context of risk budgeting for currency portfolios, we should expect gold to provide a contribution to portfolio risk—following an increase in market-wide uncertainty—smaller than that generated by the USD. There are also implications in terms of portfolio management strategies. Including holdings of gold and selected USD-bilateral rates removes a source of exchange-rate risk, namely the risk of transmission of volatility shocks. In particular, we suggest that gold shields a portfolio from the volatility spillovers potentially arising from movements in the U.S. dollar. This feature does not depend on the severity of market fluctuations.

The paper is organized as follows. Section 11.2 presents the dataset and discusses some properties of the series. Section 11.3 elaborates a structural BEKK model to provide an interpretation of the effects of volatility changes. In Section 11.4 we focus on the relationship at the tails of the distributions, and present evidence of comovement patterns of extreme price changes based on quantile regressions. Section 11.5 concludes the paper.

11.2 The Dataset

In this paper we use daily data for spot contracts of exchange rates between the U.S. dollar and the euro and the U.S. dollar and the British pound. Our dataset also includes prices for spot contracts of gold negotiated at the Chicago Board of Trade, expressed in U.S. dollars for 100 ounces.* Both the exchange rate and gold price data are extracted from Bloomberg. The dataset spans from October 13, 2004 to

---

* In order to deal with the problematic issue of asynchronous trading, we use both gold prices and exchange rates sampled at 1.30 p.m. ET. This time identifies the closing of the open outcry session for gold derivatives on weekdays.

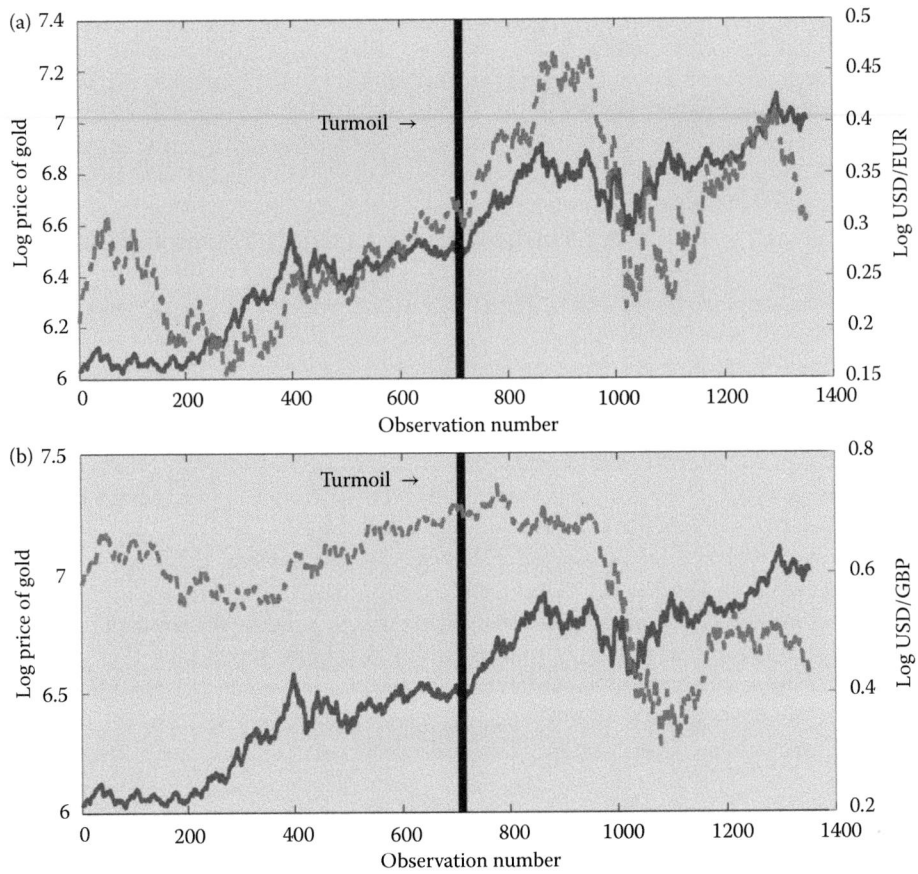

FIGURE 11.1   Data series.

March 5, 2010.* We consider August 9, 2007 as the starting date for the outbreak of the turmoil in financial markets worldwide. During that day, BNP Paribas froze the redemption of three investment funds, and the resulting panic forced the European Central Bank to start extraordinary measures for the supply of liquidity in the euro interbank market.

Figure 11.1 plots the data series (in logarithms), and Table 11.1 reports some descriptive statistics. During the turmoil, the U.S. dollar depreciates, on average, with respect to both the euro and the British pound. The average return of gold increases as well. The returns on both exchange rates and gold are more volatile during the turmoil. Interestingly, the increase in volatility for the exchange rates is far higher than for gold. The changes in the kurtosis coefficients confirm that the USD exchange rate is subject to events more extreme than gold.

It is also instructive to consider the two scatter plots of the (logs) bilateral exchange rates versus the price of gold. Figure 11.2 shows the plots, emphasizing the periods before and during the turmoil. The USD/EUR exhibits a positive correlation with the price of gold both before and during the turmoil. This can be interpreted as evidence suggesting that investors have used gold as a hedge for the USD/EUR exchange rate. For the USD/GBP exchange rate though, the turmoil appears to be characterized by a

---

* We choose October 13, 2004 as the starting observation in our dataset because we intend to use a roughly equal amount of time-series information to characterize the periods before and during the turmoil experience. In other words, this is only meant to avoid the possible distortions that may arise if the sample is skewed towards tranquil periods for financial markets.

TABLE 11.1 Descriptive Statistics of the Asset Returns (In Per cent)

	Before the turmoil				During the turmoil			
	Mean	St. dev.	Skewness	Kurtosis	Mean	St. dev.	Skewness	Kurtosis
Gold	0.0687	1.1287	−0.8199	7.6902	0.0761	1.6003	0.1552	5.4622
USD/EUR	0.0165	0.4834	−0.0014	3.6719	−0.0019	0.8002	0.3979	6.3870
USD/GBP	0.0178	0.4816	0.0819	3.5273	−0.0420	0.8942	−0.3278	7.5810

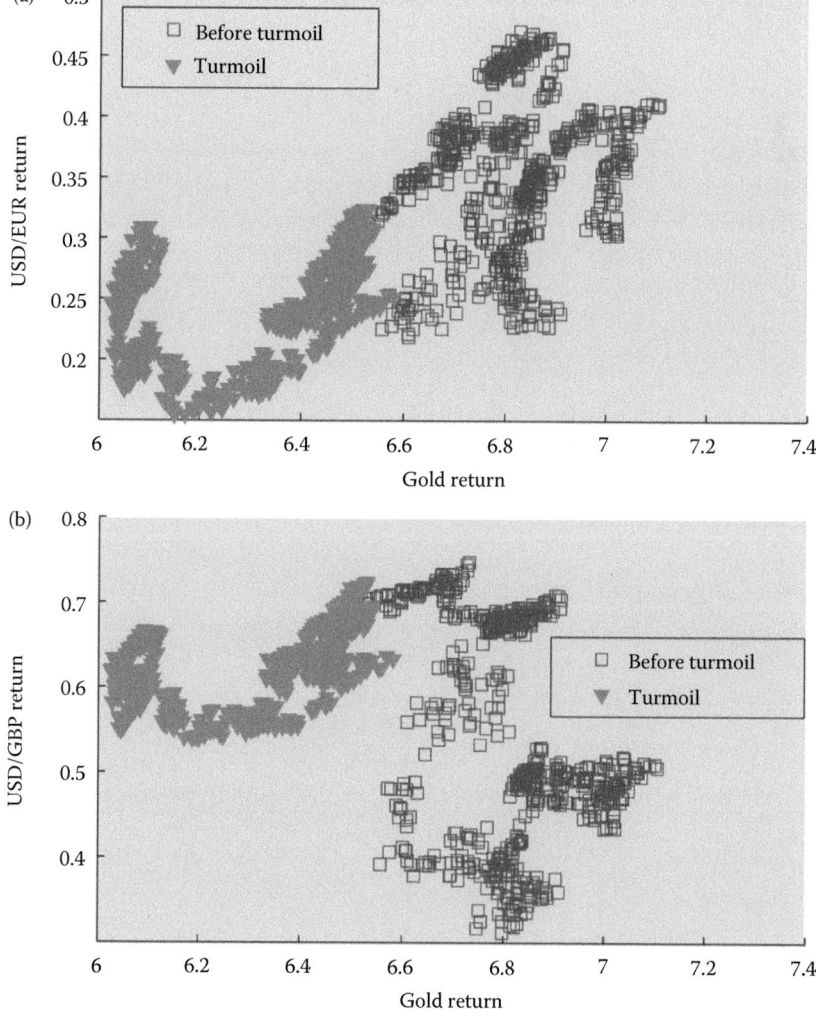

FIGURE 11.2 Scatter plots.

break in the correlation with the price of gold. The turmoil has apparently induced investors to hold gold for hedging purposes also against a fall of the USD/GBP exchange rate.

11.3 Some Structural Evidence on Volatility Spillovers

In order to understand how the hedging power of gold changes during the turmoil, we investigate how the pattern of transmission of shocks with the U.S. dollar evolves. For that purpose, we estimate the structural BEKK model discussed by Spargoli and Zagaglia (2008).

We assume that the joint evolution of a bilateral USD exchange rate return and gold returns can be summarized by a structural vector-autoregressive (VAR) model,

$$Ax_t = \Psi + \Phi(L)x_t + \eta_t, \tag{11.1}$$

where x_t is the vector of the vector of returns, ψ is a vector of constants, A is a matrix of structural parameters, and $\eta_t \ N(0, h_t)$ is a vector of structural shocks. The structural innovations exhibit conditional heteroskedasticity. We use the BEKK-GARCH model of Engle and Kroner (1995),

$$h_t = CC' + Gh_{t-1}G' + T\eta_{t-1}\eta'_{t-1}T'. \tag{11.2}$$

In model (11.1)–(11.2), the regressors are not exogenous because their source of variation is represented by the dependent variable in the same equation through another equation in the system. In order to achieve identification of the relations modelled in the VAR we rely on heteroskedasticity. This idea was originally introduced by Wright (1928) and recently developed by Rigobon (2003). The heteroskedasticity approach to identification amounts to using the information from time-varying volatility as a source of exogenous variation in the endogenous variables. To see this, let us consider the reduced-form VAR model

$$x_t = c + F(L)x_t + v_t, \tag{11.3}$$

where $c = A^{-1}\Psi$, $F(L) = A^{-1}\Phi(L)$ and $v_t = A^{-1}\eta_t$ are the reduced-form innovations, whose variance–covariance matrix is a combination of the variance–covariance matrix of the structural-form innovations, that is

$$H_t = Bh_tB', \tag{11.4}$$

$$H_t = BCC'B' + BGh_{t-1}G'B' + BT\eta_{t-1}\eta'_{t-1}T'B'. \tag{11.5}$$

In this formulation the variance–covariance matrix of the reduced-form innovations is a function of the structural innovations, which the econometrician does not know. However, we can use the equality to show that

$$\eta\eta' = Av_tv'_tA', \tag{11.6}$$

and represent it in terms of the reduced-form innovations as

$$H_t = BCC'B' + BGAH_{t-1}A'G'B' + BTAv_{t-1}v'_{t-1}A'T'B'. \tag{11.7}$$

This reduced form is then used for the estimation.

After estimating the model, we compute impulse-response functions. In structural GARCH models, these functions show the impact that a shock produces on the conditional second moments of the variables in the system. However, differently from the impulse-response functions for a standard VAR, the impulse responses of a structural GARCH depend both on the magnitude of the shock and on the period during which the shock itself takes place. This is due to the fact that the residuals enter the model in quadratic form. Hence, differently from the case of linear models, the magnitude of the effects of a shock is not proportional to the size of the shock itself. This allows us to compute a distribution of impulse responses following each shock. For this purpose, we use the concept of Volatility Impulse Response Functions (VIRF) proposed by Hafner and Herwartz (2006). The impulse-response function for a vech-GARCH model can be written as

$$V_t(\xi_0) = E\left(\text{vech}(H_t)\big|\xi_0, I_{t-1}\right) - E\left(\text{vech}(H_t)\big|I_t\right). \tag{11.8}$$

The response at time t of the variances and covariances following a shock η_t in $t = 0$, denoted $V_t(\xi_0)$, is equal to the difference, conditioned on the information set I_{t-1} at time $t-1$ and on the shock η_0, of the variance (or covariance) at t from its expected value conditional on the information set of the previous period.

We used standard likelihood methods to estimate two structural BEKK(p, q) models.[*] The variables modelled are the returns of each bilateral exchange rate and the return on gold. Each model has order $p=1$ and $q=1$.[†] The parameter estimates are listed in Table 11.2. The standard errors are computed using the delta method as in Spargoli and Zagaglia (2008). It should be stressed that the estimated parameters for the BEKK are statistically significant at standard confidence levels.[‡]

Figure 11.3 reports the conditional variances of both the structural BEKK model, as well as the variances from a reduced-form model that disregards the issue of identification of causal relations. Two

TABLE 11.2   Parameter Estimates of the Structural BEKK

Parameter	USD/EUR–gold		USD/GBP–gold	
	Point estimate	t-Stat.	Point estimate	t-Stat.
$c_{1,1}$	0.2908	3.6784	0.2908	2.7094
$c_{1,2}$	1.0144	12.8313	0.3608	1.5084
$c_{2,2}$	2.1310	26.9553	2.6808	3.5785
$a_{1,2}$	−4.2570	−53.8473	2.8468	2.0312
$a_{2,1}$	2.8960	36.6318	1.8961	7.6318
$g_{1,1}$	0.8438	10.6733	−0.7600	9.0153
$g_{1,2}$	−0.2194	−2.7752	0.7600	2.4137
$g_{2,1}$	−0.1291	−1.6330	0.7600	6.2190
$g_{2,2}$	0.7600	9.6133	−0.7600	4.6201
$t_{1,1}$	0.4715	5.9641	0.6100	3.4309
$t_{1,2}$	0.2745	3.4722	0.7745	6.0121
$t_{2,1}$	0.2399	3.0345	0.2935	4.0922
$t_{2,2}$	0.3840	4.8573	0.5945	4.8022

[*] We maximize the likelihood functions by simulated annealing.
[†] We select the model that delivers the largest value for the likelihood.
[‡] Since we use a derivative-free optimization method for the likelihood, we compute standard errors through the 'delta' method.

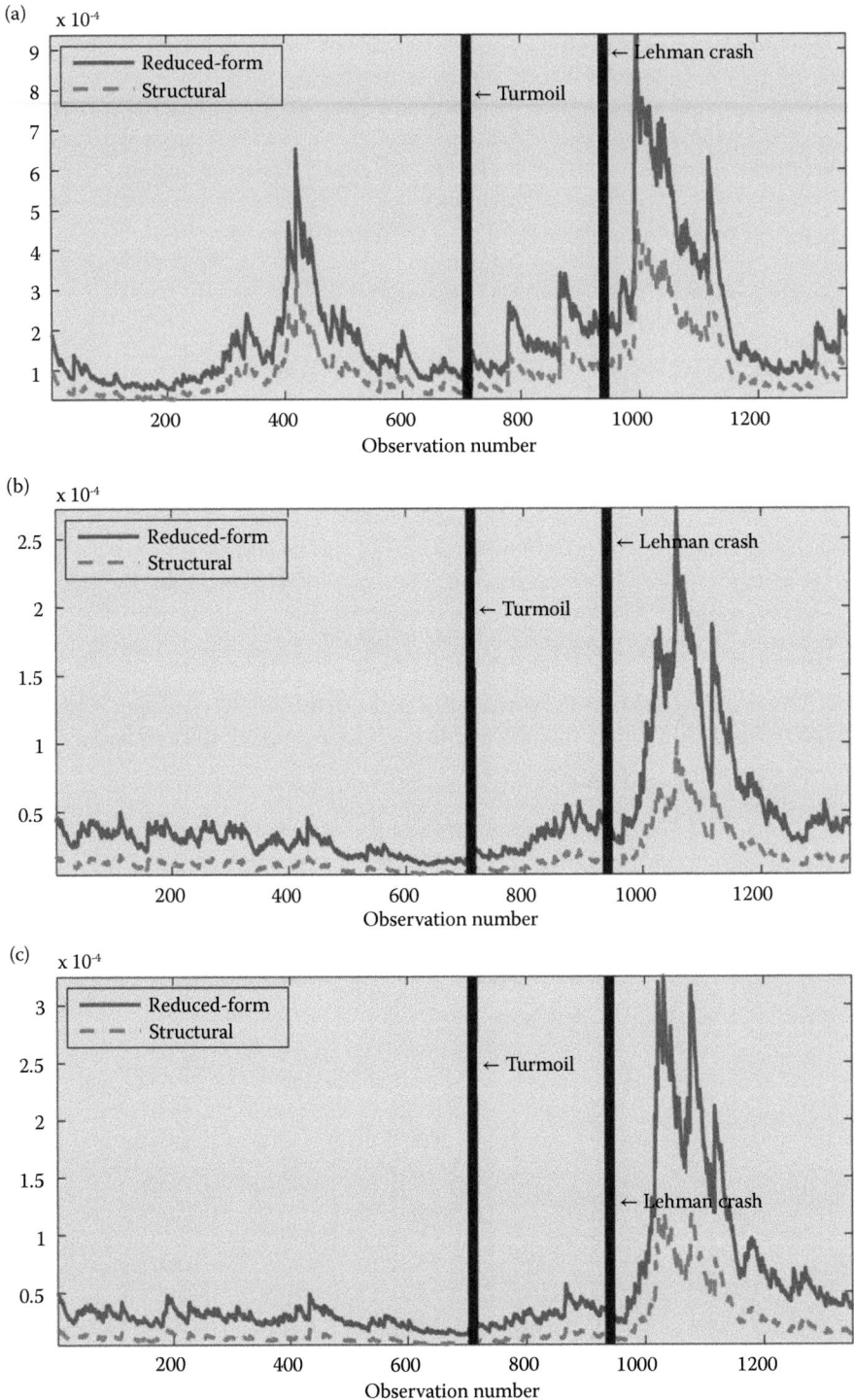

FIGURE 11.3   Estimated reduced-form and structural variances.

observations emerge. The first is that the outbreak of the turmoil alone did not enhance the pattern of uncertainty characterizing the returns of either gold or the bilateral exchange rates. Rather, there is a surge in volatility in correspondence with the bankruptcy of Lehman Brothers of September 15, 2008. The second point is that the reduced-form estimates lay on top of the structural variances. This suggests that the links between gold and the foreign exchange markets exacerbate the transmission of shocks and, thus, the uncertainty in the price movements.

Figure 11.4 plots the structural and reduced-form estimates of the correlations. The turmoil has no clear-cut effect on the dynamics of the correlations. On the other hand, following the Lehman crash, there is an increase in the average reduced-form correlations, in comparison with the structural correlations, that lasts until the beginning of 2009. This indicates that the linkages between markets enhance the tendency of the assets to move together.

In order to gain some understanding of the transmission of shocks, Figure 11.5 reports the means of the distributions of volatility-impulse responses following a one standard deviation shock. The returns of gold are stable, in the sense that the marginal change in their uncertainty is not substantially affected by the turmoil. The reaction of the variance of the bilateral exchange rates is, instead, magnified. Figure 11.5

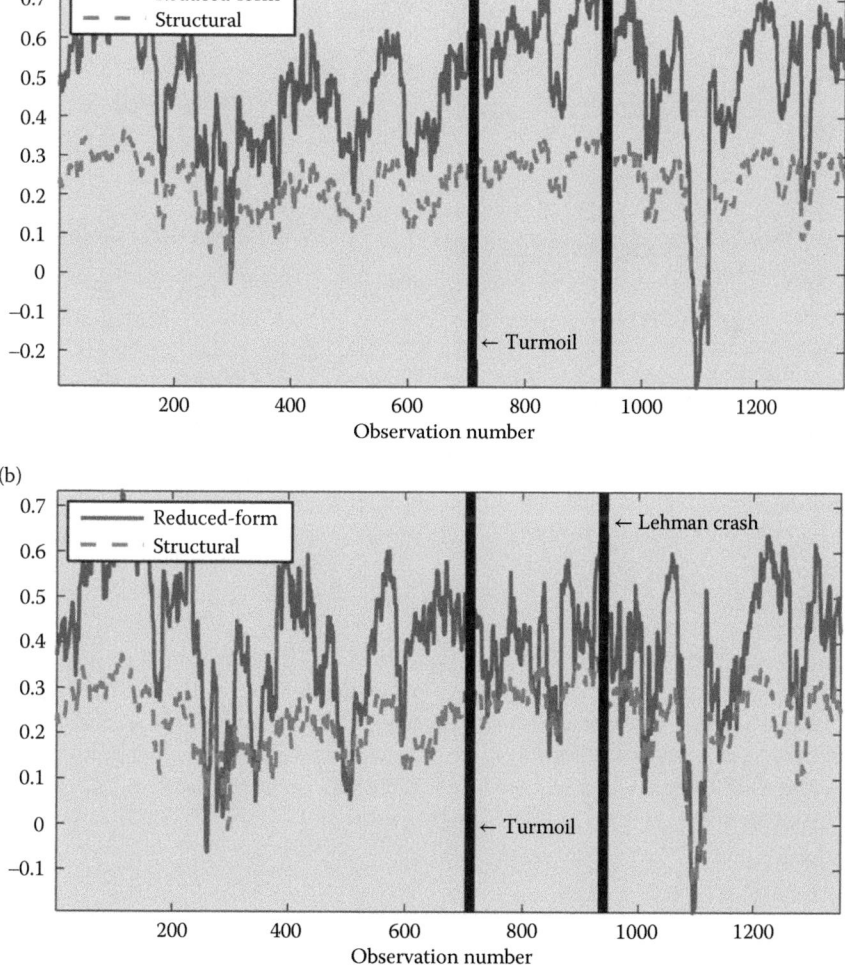

FIGURE 11.4 Estimated reduced-form and structural correlations.

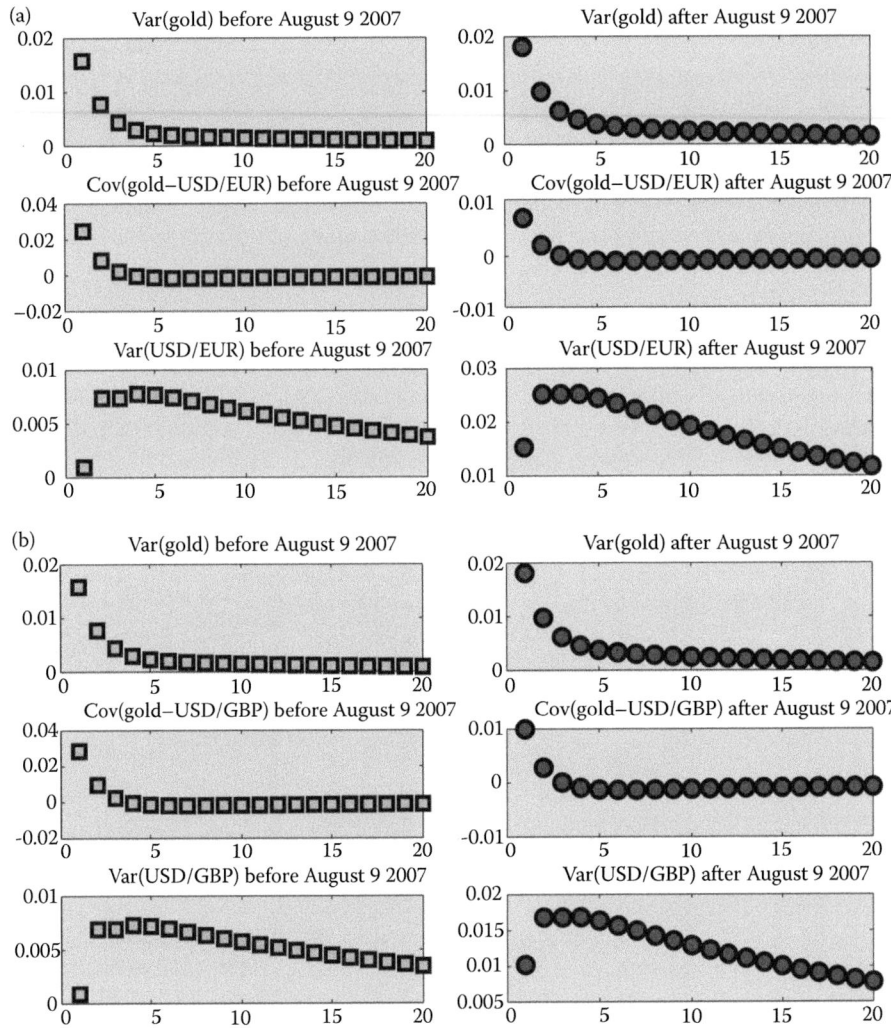

FIGURE 11.5 Mean of the distribution of volatility-impulse responses before and after August 9, 2007.

also shows that the price of gold and the bilateral exchange rates tend to decouple as the response of the covariance on impact falls for the turmoil period.

There are two issues with this type of introductory evidence. We are looking at the mean of the distribution of volatility-impulse responses. There may be relevant information that is carried over in the relation between gold and foreign exchange rates at the tail of the distribution. The second aspect concerns the lack of a criterion of statistical significance of the figures discussed earlier. Both these issues are addressed in the next section, where we consider a framework for testing how the tail relation between gold and foreign exchange rates has changed after the turmoil.

11.4 A Look at the Tails

Standard tests for comovements rely on the estimation of correlations between asset returns. These tests are, however, typically significant both to the presence of heteroskedasticity, and to departures from normality in the empirical distributions of two returns. The comovement box of Cappiello *et al.* (2005) relies on semiparametric methods to provide a robust method for analysing comovements.

Let $\{r_{i,t}\}_{t=1}^{T}$ and $\{r_{j,t}\}_{t=1}^{T}$ denote the time series of returns on two different assets, namely a USD exchange rate and gold. Define by $q_{\theta,i}^{r_i}$ the θ-quantile of the conditional distribution of $r_{i,t}$ at time t. $F_t(r_i, r_j)$ denotes the conditional cumulative joint distribution of the two asset returns. Finally,

$$F_{it}^{-}\left(r_i\middle|r_j\right) := \text{prob}\left(r_{i,t} \le r_i\middle|_i\, r_{j,t} \le r_j\right), \tag{11.9}$$

$$F_{it}^{+}\left(r_i\middle|r_j\right) := \text{prob}\left(r_{i,t} \ge r_i\middle|r_{j,t} \ge r_j\right), \tag{11.10}$$

The conditional probability,

$$p_t(\theta) := \begin{cases} F_{it}^{-}\left(q_{\theta,t}^{r_i}\middle|q_{\theta,t}^{r_j}\right) & \text{if } \theta \le 0.5, \\ F_{it}^{+}\left(q_{\theta,t}^{r_i}\middle|q_{\theta,t}^{r_j}\right) & \text{if } \theta > 0.5, \end{cases} \tag{11.11}$$

can be used to represent the characteristics of $F_t(r_i, r_j)$. In fact, $p_t(\theta)$ measures the probability that the returns at maturity i are below its θ-quantile, conditional on the same event occurring at maturity j.

The information about $p_t(\theta)$ is summarized in the so-called 'comovement box'. This is a square with unit size where $p_t(\theta)$ is plotted against θ. Since the shape of $p_t(\theta)$ depends on the joint distribution of the two time series, it can be derived only by numerical simulation.*

The framework of Cappiello *et al.* (2005) can also be used to test whether the dependence between two markets has changed over time. Given a cut-off date of a specific event, we can estimate the conditional probability of comovements in two different periods, and plot the estimated probabilities in a graph. Differences in the intensity of comovements can then be detected. This idea can be formalized in a simple way. Denote by $p^A(\theta) := A^{-1}\sum_{t<\tau} p_t(\theta)$ and $p^B(\theta) := B^{-1}\sum_{t<\tau} p_t(\theta)$ the average conditional probabilities before and after a certain event occurs at a threshold τ, with A and B the number of corresponding observations. Let $\Delta(\underline{\theta}, \overline{\theta})$ denote the area between $p^A(\theta)$ and $p^B(\theta)$. A measure of 'contagion' or 'spillovers' between the two markets can be obtained by noting that contagion increases if

$$\Delta(\underline{\theta}, \overline{\theta}) = \int_{\underline{\theta}}^{\overline{\theta}}\left[p^B(\theta) - p^A(\theta)\right]d\theta > 0. \tag{11.12}$$

We stress that, unlike the standard measures of correlation, $\Delta(\underline{\theta}, \overline{\theta})$ allows us to study changes in codependence over specific quantiles of the distribution.

Several steps are followed to construct the comovement box and test for differences in conditional probabilities. First, we estimate univariate time-varying quantiles using the Conditional Autoregressive Value at Risk (CAViaR) model proposed by Engle and Manganelli (2004). For each series and each quantile, we create an indicator variable that takes the value one if the return is lower than this quantile, and zero otherwise. Then we regress the θ-quantile indicator variable on market j on the θ-quantile indicator on market i. The estimated regression coefficients provide a measure of the conditional probabilities of comovements, and of their changes across regimes.

The time-varying quantiles of the returns are estimated using the CAViaR model of Engle and Manganelli (2004). The quantiles of the returns r_t are assumed to follow the autoregressive model

$$q_t(\beta_\theta) = \beta_{\theta,0} + \sum_{i=1}^{q}\beta_{\theta,i}q_{t-i} + \sum_{i=1}^{p}l\left(\beta_{\theta,j}, r_{t-j}, \Omega_t\right), \tag{11.13}$$

---

* The interested reader can refer to Cappiello *et al.* (2005, p. 9) for a thorough discussion on the comovement box.

where Ω_t denotes the information set at time t. The autoregressive terms of the quantiles are meant to capture the clustering of volatility that is typical of financial variables. Including a predetermined information set allows us instead to consider the interaction between the quantiles and the conditions of the market. Following Cappiello *et al.* (2005), we estimate the time-varying quantiles using the following specification of the CAViaR:

$$q_t(\beta_\theta) = \beta_{\theta,0} + \beta_{\theta,1}d_t + \beta_{\theta,2}r_{t-1} + \beta_{\theta,3}q_{t-1}(\beta_\theta)$$

$$-\beta_{\theta,2}\beta_{\theta,3}r_{t-2} + \beta_{\theta,4}|r_{r-1}|. \tag{11.14}$$

The dummy variable d_t ensures that the periods of high and low volatility have the same proportion of quantile exceedances.

In order to investigate the specification of the CAViaR model, we compute the DQ test of Engle and Manganelli (2004). The null hypothesis of the DQ test consists of the lack of autocorrelation in the exceedances of the quantiles. Figure 11.6 reports the p-values for 99 conditional quantiles, together with the p-values for unconditional quantiles. The specification with unconditional quantiles is rejected over the entire domain.

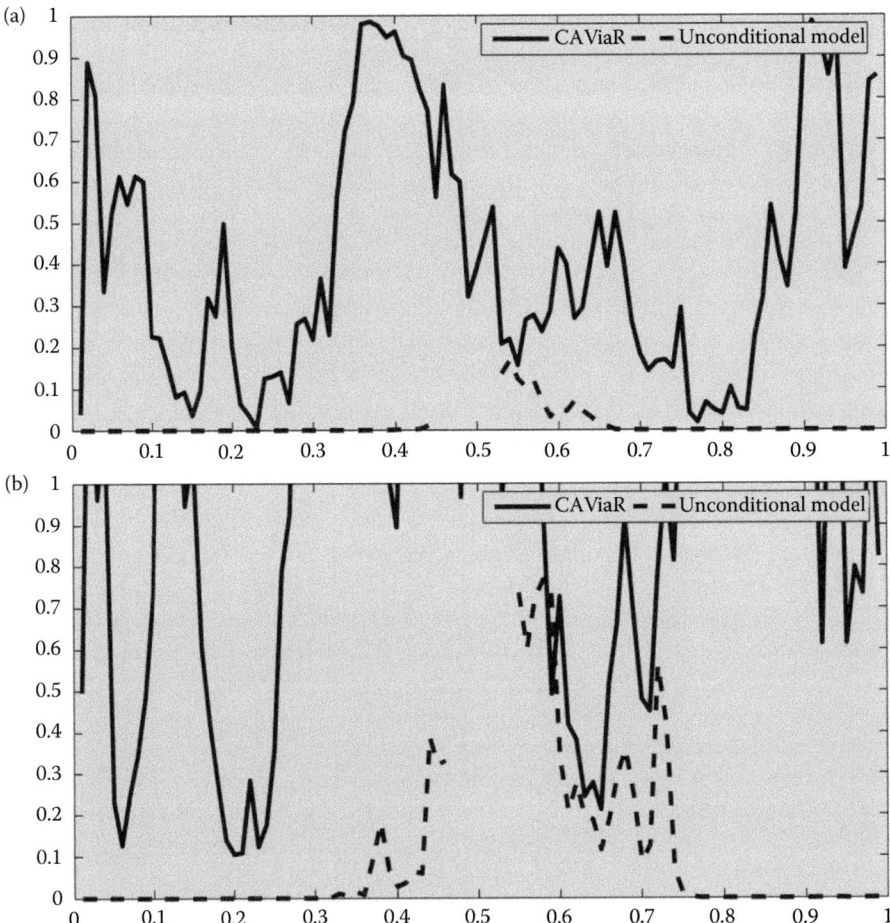

FIGURE 11.6 p-Values of the dynamic quantile test.

Figure 11.7 plots the estimates of the conditional probabilities of comovements before and during the turmoil period. We also report confidence bands of plus/minus twice the standard errors around the estimates of the probabilities of comovements. Two observations emerge from Figure 11.7. The first is that extreme events due to the turmoil provide relevant information on the relation between the USD exchange rate and gold prices. This can be seen by comparing the changes in the positions of the tails of the distributions before and during the turmoil. In other words, the relation of comovements between

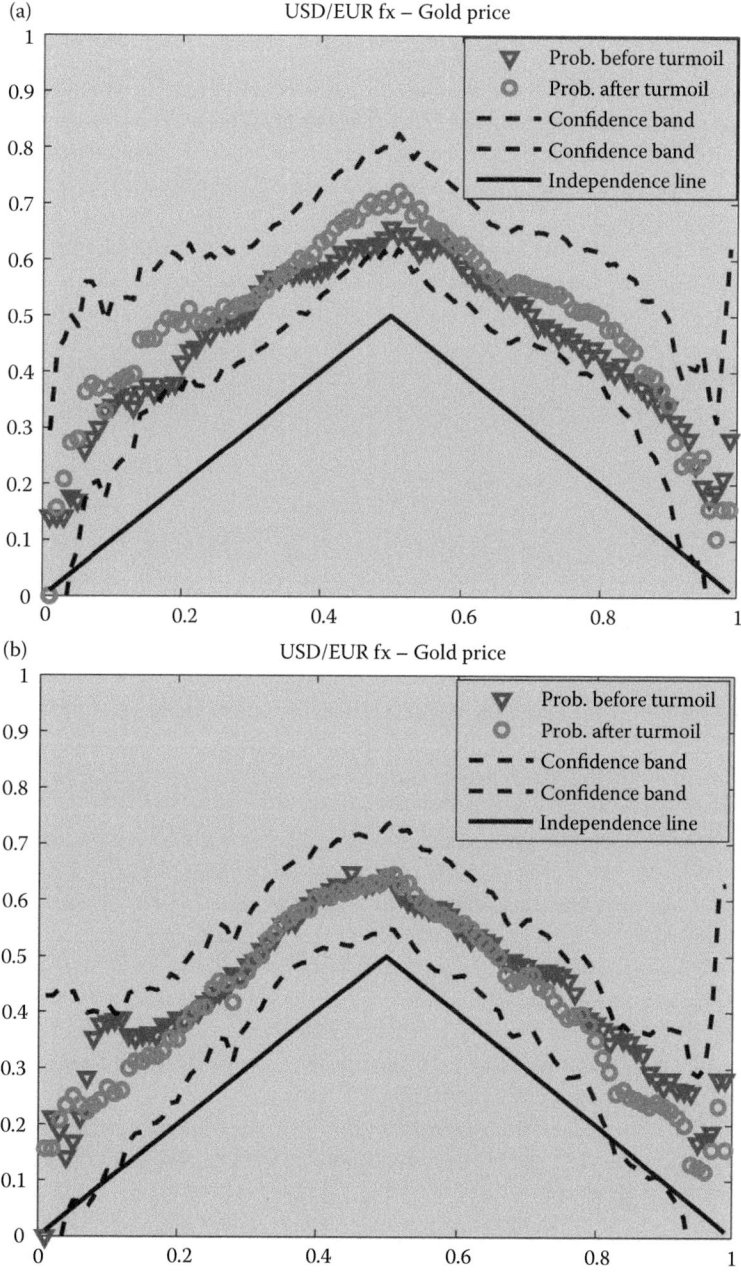

FIGURE 11.7 Estimated tail codependence.

TABLE 11.3 Test of Difference in Tail Coincidences
between Periods Before and During the Turmoil

	Lower tail: $\theta \leq 0.5$		Higher tail: $\theta \geq 0.5$	
	$\hat{\delta}(0, 0.5)$		$\hat{\delta}(0.5,1)$	
	Stat.	s.e.	Stat.	s.e.
USD/EUR–gold	1.7555	2.1640	0.4084	2.5282
USD/GBP–gold	2.2261	2.7650	1.0981	2.3332

the USD exchange rates and gold changes across time, especially when extreme events materialize in financial markets. The second point of interest is that the probability of comovement induced by the turmoil is not statistically different from the comovement probabilities characterizing the pre-turmoil period. In more intuitive terms, this suggests that the transmission mechanism of shocks between bilateral exchange rates and the price of gold is not affected by the turmoil.

Table 11.3 reports the results of the quantile tests for contagion for two parts of the distributions. All the test statistics have a positive sign, indicating an increase in comovement after the turmoil, on average. However, the null hypothesis of conditional comovement between the USD exchange rates and gold is rejected strongly in all cases. These findings uncover an important source of stability generated by gold. In this paper we stress the ability of this asset to shield portfolios from the effects of spillovers due to swings in the U.S. dollar, independently from how extreme the market conditions are.

11.5 Conclusion

The relation between gold prices and the U.S. dollar exchange rate has been the subject to intense scrutiny. Using recently developed econometrics methods, we reconsider the case for gold as a safe asset against fluctuations in the USD. Following the account of numerous commentators, we investigate whether the beneficial properties of holding gold have survived the recent turbulence in financial markets that started in August 2007.

Using methods from the literature on financial crises, we study the impact of the turmoil on two sources of these properties. We consider the evolution of the pattern of volatility spillovers between gold prices and the dollar. We investigate whether the turmoil has caused an increase in contagion, defined as the probability of extreme tail events in both gold prices and the dollar.

On general grounds, our results confirm the widely held belief that gold can be considered a financial asset. Several qualifications apply though. We find that the defining moment in the transmission of volatility shocks between gold and the USD coincides with the Lehman Brothers bankruptcy. Even if we consider the tails of the bivariate distributions, we find that holding gold generates patterns of comovements with the dollar that have survived the recent episodes of market turbulence. Most of all, we show that exogenous volatility shocks tend to generate reactions of gold prices that are more stable than those of the U.S. dollar. In other words, gold prices react to market uncertainty to a smaller extent than the USD. Since a smaller reaction of volatility is then transmitted to the USD, this pins down a key source of stability for gold holdings in currency portfolios.

To our knowledge, this is the first contribution that discusses the structural features of volatility transmission in the gold market. The relevant question concerns the nature of the determinants of the changes in contagion. Thus, the analysis of this paper can be extended along different and, yet, promising avenues for future research. For instance, it would be important to understand how deep the relation between gold prices and the U.S. dollar is around the business cycle. We can also consider the issue of whether the properties of gold discussed earlier arise also in relation to other asset classes. We might expect U.S. stocks to represent a relevant candidate.

The available literature has not yet quantified the advantages of gold holdings in a systematic framework of portfolio allocation when extreme events take place. In this sense, it would be interesting to compute the implications for standard risk measures, such as the Value-at-Risk or the expected tail loss. Other types of multivariate models of tail behaviour could be used. In a related work-in-progress, we consider bivariate copulas to model the joint determinants of co-exceedances. Finally, the recent experience of financial market turbulence suggests that structural breaks in the conditional mean, variance and correlations are important factors that should be accounted for in the relation between assets. This hints at the advantages of using models with time varying parameters for understanding the portfolio implications of changing relations between gold prices and the U.S. dollar.

Acknowledgements

We are very grateful to two anonymous referees for their helpful comments and suggestions on a previous draft. This research was carried out by the authors for purely academic purposes and without financial support from any private institution.

References

Baur, D.G. and Lucey, B.M., Is gold a hedge or a safe haven? An analysis of stocks, bonds and gold. *Financ. Rev.*, 2010, **5**(3), 217–229.

Baur, D.G. and McDermott, T.K., Is gold a safe haven? International evidence. *J. Bank. Finance*, 2010, **34**(8), 1886–1898.

Cappie, F., Mills, T.C. and Wood, G., Gold as a hedge against the dollar. *J. Int. Financ. Mkts, Inst. Money*, 2005, **15**, 343–352.

Cappiello, L., Gerard, B. and Manganelli, S., Measuring comovements by regression quantiles. ECB Working Paper 501, 2005.

Christie-David, R., Chaudhry, M. and Koch, T.W., Do macroeconomics news releases affect gold and silver prices. *J. Econ. Bus.*, 2000, **52**(5), 405–421.

Davidson, S., Faff, R. and Hillier, D., Gold factor exposures in international asset pricing. *J. Int. Financ. Mkts*, Inst. Money, 2003, **13**, 271–289.

Engle, R.F. and Kroner, F.K., Multivariate simultaneous generalized ARCH. *Econometr. Theory*, 1995, **11**, 122–150.

Engle, R. and Manganelli, S., CAViaR: Conditional autoregressive value at risk by regression quantiles. *J. Bus. Econ. Statist.*, 2004, **22**, 367–381.

Forbes, K.J. and Rigobon, R., No contagion, only interdependence: Measuring stock market comovements. *J. Finance*, 2002, **57**, 2223–2261.

Hafner, C.M. and Herwartz, H., Volatility impulse responses for multivariate GARCH models: An exchange rate illustration. *J. Int. Money Finance*, 2006, **25**(5), 719–740.

McCown, J.R. and Zimmerman, J.R., Is gold a zero-beta asset? Analysis of the investment potential of precious metals. Working Paper Series, Oklahoma City University, 2006.

Rigobon, R., Identification through heteroskedasticity. *Rev. Econ. Statist.*, 2003, **85**(4), 777–792.

Rigobon, R. and Sack, B., Measuring the reaction of monetary policy to the stock market. *Q. J. Econ.*, 2003, **118**(2), 639–669.

Sjaastad, L.A. and Scacciavillani, F., The price of gold and the exchange rate. *J. Int. Money Finance*, 1996, **15**(6), 879–897.

Spargoli, F. and Zagaglia, P., The comovements along the forward curve of natural gas futures: A structural view. *J. Energy Mkts*, 2008, **1**(3), 23–47.

Wright, P., *The Tariff on Animal and Vegetable Oils*, 1928 (Macmillan: New York).

12

Application of a TGARCH-Wavelet Neural Network to Arbitrage Trading in the Metal Futures Market in China

Lei Cui, Ke Huang,
and H.J. Cai

12.1 Introduction

Worldwide, metal futures markets play an indispensable role in both individual and corporate investment. During its liquidity boom, the metal futures market in China has been rife with fraud and price manipulation because of the inefficient market environment.* This feature makes statistical arbitrage an appropriate quantitative trading strategy in this market. Generally, a statistical arbitrage consists of three main parts (Steve *et al.* 2004): obtaining arbitrage pairs, establishing the spread series and setting trading thresholds and stopping limits.

The GARCH-based financial time series model, which was especially designed to address volatility clustering, has proven to be successful at modelling many aspects of financial time series (Bollerslev *et al.* 1992, Weber and Prokopczuk 2011). Many studies have applied the GARCH model to determine ideal trading pairs and to predict the spread standard deviation in statistical arbitrage. However, when applying the GARCH model to predict the spread standard deviation, Christian and Zakoian (2010)

---

* Detailed information about Chinese futures markets can be found at 'http://www.world-exchanges.org/news-views/views/chinese-futures-markets-coming-booming-decade'.

DOI: 10.1201/9781003265399-14

found that positive and negative information have a different influence on prediction, which is consistent with the asymmetric behaviour commonly exhibited by the stock market (Wang and Wang 2011). Thus, asymmetric GARCH models have been utilized to analyse the volatility of financial time series. He (2008) and Lin *et al.* (2010) applied the threshold GARCH (TGARCH) model to study asymmetric effects in the Chinese metal futures market. Their works, however, primarily study daily trading data to determine long-term equilibrium asymmetric effects that exist in the market. In our study, attention is paid to short-term asymmetric effects, which are more meaningful for high-frequency trading models.

For high-frequency statistical arbitrage, for which the holding period is just a few minutes, setting the proper trading threshold is very important. Therefore, the traditional method to set the trading threshold, which utilizes the historical optimal value and is described by Gatev *et al.* (1999), becomes unreasonable. Following the method established by Kantz (1994), we study the characteristics of the optimal trading threshold series which maximize profits and note its chaotic properties, which leads us to consider using learning algorithms, especially neural networks, to predict arbitrage trading thresholds. Neural networks have been widely applied to predict price trends in equity markets using daily prices (Creamer 2012). However, McMillan and Speight (2006) argue that for the strong microstructure relationship between informed and noise traders in futures markets, it is risky to apply neural networks alone in trading models. One of the most effective methods to solve this problem is to utilize the wavelet transform to alter the noise during the signal process (Zhang *et al.* 2001, Jaisimha *et al.* 2010, Chang *et al.* 2011). Thus, Li and Kuo (2008) suggest that wavelet neural networks (WNNs) are a promising prediction model for establishing more effective trading models.

Studies by Locke and Venkatesh (1997) argue that transaction costs are an essential factor for investments in futures market. Costs influence the market liquidity (Brennan and Subrahmanyam 1996) and bid-ask spread (Glosten and Harris 1988). For quantitative trading methods such as statistical arbitrage, transaction costs, especially the commission level, can even determine the profitability of arbitrage (Chung 1991). Thus, the impact of changes in the commission level on the cumulative yields and total commission in our model will be discussed in detail.

Previous studies concerning quantitative arbitrage models in the Chinese futures market focused on the cointegration relationships among different types of contracts. For instance, Fung *et al.* (2010) established a trading strategy based on the cointegration relationship between aluminium and copper futures price series in the Shanghai Futures Exchange (SHFE). However, such methods fail to provide accurate trading recommendations. Thus, our study combines the advanced theories of non-linear dynamics and artificial neural networks with a traditional statistical arbitrage model to create a new arbitrage strategy for metal futures markets. There are two primary contributions of our study. First, based on the sketch model proposed by Huang *et al.* (2012), we apply threshold GARCH (TGARCH) models to capture short-term price cointegration and extend the model to the metal futures market in China. We also further study adjustments of the model to fit different market situations based on the TGARCH model. Second, to the best of our knowledge, our model is the first to predict trading thresholds in the Chinese metal futures market using WNNs. With deliberately designed backtesting programs that take the margin rate, margin call, slippage and commission levels into account, we use high-frequency data to demonstrate the robustness of the WNN model compared with typical predication methods, such as the backpropagation neural network (BP) and historical optimal value (HO) models, and to test the profitability of statistical arbitrage using TGARCH-WNN under different commission levels.

The remainder of the paper is organized as follows: first, an outline of the TGARCH-WNN statistical arbitrage model, which combines cointegration with the TGARCH and WNN models, is given. Then, the high-frequency data and contract pairs used in this paper are described in the sample selection section. The final part of the paper is the empirical study section, which contains four subsections. In the first subsection, discussions about the application of the TGARCH model to eliminate heteroscedasticity and an analysis of asymmetric effects under distinctive market situations are presented. In the second subsection, the Lyapunov exponent is utilized to demonstrate the chaotic properties of the trading threshold series. In the third subsection, a comparison of the prediction accuracy and market performance of the WNN, BP and HO models is made. Finally, the impact of commissions is discussed in detail.

12.2 Model Development: TGARCH-WNN Statistical Arbitrage

There are two parts to the TGARCH-WNN model. In the first part, TGARCH is applied to find the cointegration relationship between contracts and to predict the standard deviation of the price spread. In the second part, WNNs are used to predict the trading thresholds.

12.2.1 Mean Spread with TGARCH

Without loss of generality, it is assumed that there are only two contracts X and Y in the arbitrage pair and that the price series of X and Y are x and y, respectively.[*]

- y serves as the dependent variable and x serves as the independent variable. If y is cointegrated with x, then x and y are a pair of potential trading candidates for statistical arbitrage. Moreover, x and y will be regressed using TGARCH (1, 1) to avoid the influence of heteroscedasticity in the time series. The threshold GARCH model is applied here because positive news and negative news have distinct impacts on variance prediction. The regression functions are as follows:

mean function:

$$y_t = \beta_0 + \beta_1 x_t + u_t \tag{12.1}$$

variance function:

$$\sigma_t^2 = \omega + \alpha u_{t-1}^2 + \gamma u_{t-1}^2 d_{t-1} + \beta \sigma_{t-1}^2, \tag{12.2}$$

where u_t are the disturbance terms in the mean function, σ_t is the conditional standard deviation series and d_t is determined by u_t. When $u_t < 0$, $d_t = 1$ and when $u_t \geq 0$, $d_t = 0$. In the mean function (12.1), positive news ($u_t \geq 0$) and negative news ($u_t < 0$) have different influences on the conditional variance: the influence level of positive news is α; the influence level of negative news is $\alpha + \gamma$. If $\gamma > 0$, there is a leverage effect, which means the asymmetric effect will increase the volatility of variance; if $\gamma < 0$, the asymmetric effect will decrease the volatility of variance.

- From the above model, we can obtain residual series:

$$u_t = y_t - \beta_0 - \beta_1 x_t. \tag{12.3}$$

There is evidence that the Phillips–Perron (PP) test has more power than the augmented the Dickey-Fuller test to examine stationarity.[†] Thus, the PP test is applied to check whether the residual series of (12.3) is stationary. If it is stationary, it can be concluded that there is a cointegration relationship between $\{y_t\}$ and $\{x_t\}$, which means that this contract pair can be used for statistical arbitrage. Let $\{1, -\beta_1\}$ be the integrated vector and $\{u_t\}$ be the spread series which is renamed as $\{spread_t\}$.

---

[*] When there are more than two contracts $\{y_t, x_{1t}, x_{2t}, \ldots, x_{nt}\}$, we define their mean function as $y_t = \beta_0 + \beta_1 x_{1t} + \ldots + \beta_n x_{nt} + u_t$.
[†] Concrete evidence can be found in study by Davidson and Mackinnon (1993).

- If the {$spread_t$} passes the PP test, {$spread_t$} is decentralized such that the equilibrium value of the series is zero. Such a residual series is denoted as mean spread: $mspread_t = spread_t - mean(spread_t)$.

Before moving forward to WNN, the general idea of statistical arbitrage is presented here. As Figure 12.1 shows, in the process of arbitrage, the $mspread_t$ series of the contract pair (denoted as e_t) is monitored over time. Let $k_1\sigma_t$ and $k_2\sigma_t$ be the values of the trading thresholds (the dashed lines in Figure 12.1), where σ_t is the standard deviation of $mspread_t$ at time t, and can be predicted from function 12.2. Trading is triggered when $|e_t|$ deviates from the equilibrium by more than $|k_1\sigma|$ or $|k_2\sigma|$, namely $e_t > k_1\sigma$ or $e_t < -k_2\sigma$. Let x and y be the contracts defined in Equation 12.1. The detailed trading rules are:

When $e_t > k_1\sigma$, buy β contracts of X and sell one contract of Y. Close the positions when e_t decreases below zero.

When $e_t < -k_2\sigma$, sell β contracts of X and buy one contract of Y. Close the positions when e_t increases above zero.

However there are times when mspread deviates far away from the average. This situation is very dangerous, so stop losses (the solid lines is Figure 12.1) should be set to avoid large money losses. Let the upper and lower stop losses be s_1 and s_2, then positions should also be closed when $e_t > s_1$ or $e_t < -s_2$. The distribution of mspread is assumed to be $N(0, \sigma_t)$ (Noortwijk *et al.* 2007). Thus let $s_1 = n\sigma_t$ and $s_2 = -n\sigma_t$ where $n\sigma_t$ is the significance value of a small probability event.

12.2.2 Wavelet Theory and Chaotic Properties of Time Series

Now, the only question left is to predict the optimal {$k1, k2$}. For simplicity, we call the {$k1, k2$} series the trading threshold series in the following section. This paper applies WNN to predict the appropriate trading threshold. To introduce WNN, two points must be addressed first: what is wavelet theory and why it is sensible to apply WNN to predict trading thresholds?

Wavelet theory is a new method to transform signals into simpler and lower dimensional formations. Unlike the traditional Fourier transform, which loses time information when transforming a time series, the wavelet transform accurately retains time information. Moreover, the wavelet transform is useful for eliminating noise when predicting the trading threshold value.

It is well known that neural networks are an ideal method to make predictions when time series have chaotic properties. Therefore, the chaotic properties of trading threshold series should be confirmed before using WNN for prediction.

The Lyapunov exponent is applied to prove that the trading threshold series has chaotic properties. If the Lyapunov exponents of the series have stable positive values as the embedding dimension increases,

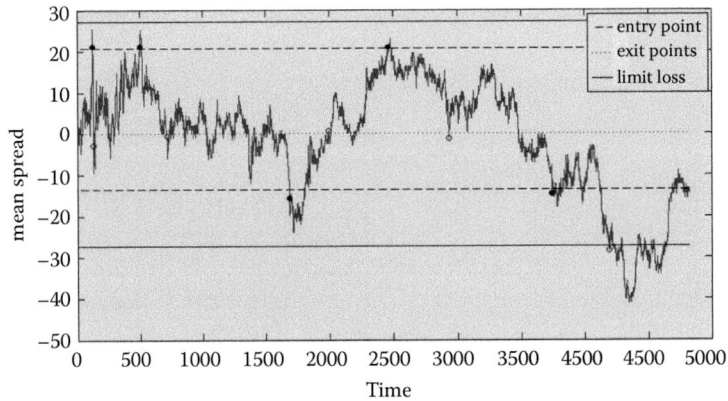

FIGURE 12.1 Arbitrage process of a contract pair.

which is also defined as having positive largest Lyapunov exponents, then the series has chaotic properties. The detailed process can be found in the study by Wolf *et al.* (1985) and the concise procedure is presented as follows:

- The k series is denoted by $\{k(t), t = 1, 2, 3,\ldots, M\}$.[*] To get the average period, the k series is transformed by Fast Fourier Transform (FFT) and get: $F(u) = \sum_{i=1}^{M} k(i)e^{-j(u-1)f_i}$ where j is an imaginary number. In the transformation, the frequency function is: $f_i = 2\pi(i-1)/M$.

- In order to find the optimal point k^4 of the phase space, we minimize the Euclidean distance $L(t')$ between $K(t')$ and $K(t)^\dagger$ such that $\|t - t'\| > T_m$, where t' denotes the base point and satisfies $t'_n = t'_{n-1} + q, n = 1, 2, \ldots, M - (m-1)\tau, t = 1, 2, \ldots, t' - 1, t' + 1, \ldots$

- Increase the embedding dimension m until Lyapunov exponents[‡] become stationary which is defined as Largest Lyapunov exponent. If Largest Lyapunov exponent converges to a positive number, then $\{k(t), t = 1, 2, 3,\ldots\}$ has the properties of chaos.

Here two improvements have been done to meet the special character of arbitrage spread series. Firstly, historical trading thresholds $\mathbf{k}(t)$ which maximize profits in each period make up the time series. Secondly, when calculating Lyapunov exponents, the traditional way to find the average period is unconvincing with regard to the special time series here. Instead, frequency-weighted average period is calculated by (12.4) (Ellen *et al.* 1998).

$$T = \frac{\sum_{i=2}^{M} \dfrac{F(i)}{f_i}}{\sum_{i=2}^{M} f_i} \tag{12.4}$$

12.2.3 Wavelet Neural Networks

In this paper, a tight WNN is applied to make a one-step-ahead forecast. This type of neural network is based on the typical BP neural network and applies the wavelet basis function as the hidden nodes transfer function. Apart from wavelet nodes, some traditional multilayers of BP neural networks are added in the hidden layers to increase the convergence rate. In the network, signals spread forward, whereas error series spread backward, as shown in Figure 12.2.

In Figure 12.2, $K_1, K_2,\ldots, K_n$ are WNN's input parameters.[§] $Y_1, Y_2,\ldots,Y_m$ are WNN's predictions, ω_{ij} and ω_{jk} are, respectively, wavelet nodes' input and output weight.

When the input signals are k_i ($k_i \in K_i$, $i = 1, 2,\ldots, n$), the wavelet functions are:

$$f(j): f_j\left[\frac{\sum_{i=1}^{n} \omega_{ij}x_i - \gamma_j}{\alpha_j}\right] j = 1, 2, \quad ,l. \tag{12.5}$$

In the function (12.5) above, $f(j)$ is the output of j's hidden node; f_j is a wavelet basis function.

---

[*] There is an upper k and lower k for each period and the k series here means either the upper or lower k series.

[†] $K(t)$ is the reconstructed phase space: $K(t) = \{k(t), k(t + \tau),\ldots, k(t + (m-1)\tau)\}$, where m is the embedding dimension and τ is the delay time which equals the time step q.

[‡] The Lyapunov exponent can be written as $LE(m) = \dfrac{1}{m}\sum_{i=1}^{m}\dfrac{1}{k}\log_2\dfrac{L(t_i)}{L(t_{i-1})}$.

[§] In our model input parameters are $\{k(t), t = 1, 2, 3, \ldots\}$.

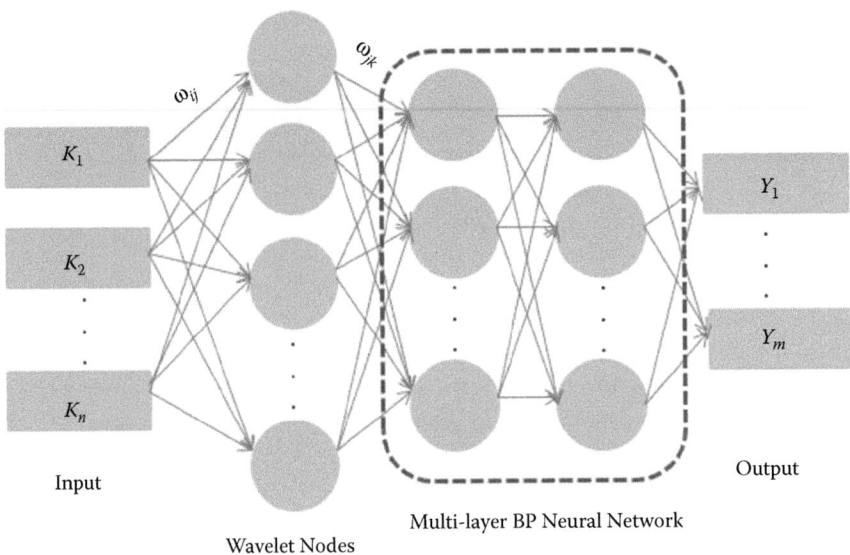

FIGURE 12.2   Architecture of WNN.

To decide the threshold in the WNN, the Mexican Hat Function is applied as mother function (Zhou and Hojjat 2003): $y(x) = \frac{2}{\sqrt{3}} \pi^{-\frac{1}{4}} \left(1 - x^2\right) e^{-\frac{x^2}{2}}$.

The scales and shifts for each wavelet node are optimized by including them as parameters within the usual training algorithm as in a multilayer BP neural network. Thereby the optimal set of scales and shifts appropriate for the prediction task is determined.

Utilizing the theory above, the main process of WNN prediction is as follows:

- Initializing the WNN: Assign initial value to parameters in the network structure including learning rate, epoch size, shifts and scales.
- Data preprocessing: Determine the size of testing and training sets. In the model, optimal $\{k(t),\ t = 1, 2, 3 \ldots\}$ series is the same series mentioned in Section 12.2.2. Denote the value of the k series as k_{it}, where $i = 1, 2$; $t = 1, 2 \ldots, m$ ($i = 1$ means upper threshold and $i = 2$ means lower threshold). Thus, k series of m periods are used for training are $\{k_{i1}, k_{i2}, k_{i3}, \ldots, k_{im}\}$. If n sample points are used to predict one point, the series should be reconstructed, as shown in Table 12.1.
- Training WNN: The input the series shown in Table 12.1 input is into the WNN to get the prediction vector $\left\{k'_{in+1}, k'_{in+2}, k'_{in+3}, \ldots, k'_{im}\right\}$ Compare $\left\{k'_{in+1}, k'_{in+2}, k'_{in+3}, \ldots, k'_{im}\right\}$ with $\{k_{i1}, k_{i2}, k_{i3}, \ldots, k_{im}\}$ and calculate the value of target function $E(\theta)$:

$$E(\theta) = \frac{1}{2} \sum_{j=n+1}^{m} \left(k'_{ij} - k_{ij}\right)^2 \quad i = 1, 2. \tag{12.6}$$

TABLE 12.1   Input and Target of WNN

Input	Target Value
$k_{i1}, k_{i2}, k_{i3}, \ldots, k_{in}$	$k_{i(n+1)}$
$k_{i2}, k_{i3}, k_{i4}, \ldots, k_{i(n+1)}$	$k_{i(n+2)}$
$k_{i(m-n)}, k_{i(m-n+1)}, k_{i(m-n+2)}, \ldots, k_{i(m-1)}$	k_{im}

- According to the value of $E(\theta)$, revise the parameters in the WNN until $E(\theta)$ is smaller than the adaptive tolerance.
- Predict the k value and apply the well-trained WNN to predict k's trading threshold value.

12.3 Sample Selection

The empirical study in this paper is focused on the commodities in the SHFE, which is an exchange in China that primarily trades metals. Taking liquidity and trading volume into account, only primary contracts are considered. A typical calendar spread arbitrage pair, zinc, is selected to conduct empirical testing for two practical reasons.* First, zinc's relatively high trading volume ensures that there are enough market orders for us to set up positions when the trading signal is triggered. Second, unlike deformed steel bars and copper, for which the commission level is based on a percentage of the transaction volume, the commission level of zinc is based on the number of lots in the transaction.† When the commission level is based on the number of lots, it is easier to calculate parameters such as the lower bound of the trading threshold and the total commission.‡

Generally, two contracts are chosen to construct zinc calendar spread arbitrage: the primary contract, which has the largest trading volume, and the secondary contract, which has the second largest trading volume. Because the trading of the primary contract is more active than that of the secondary contract, it is reasonable to assume that the price of the secondary contract is influenced by the price of the primary contract. Thus, the price of the primary contract serves as the independent variable, whereas the price of the secondary contract serves as the dependent variable.

Data of different frequencies are applied in each part of the empirical studies to fit some special needs. When analysing the cointegration relationships of the contract pairs using the TGARCH model, it is essential to capture the characteristics of short-term equilibrium and avoid the noise in the price series. Thus, closing price series sampled at one-minute intervals are used here. It is important to apply high-frequency data to test the performance of quantitative trading models, such as the statistical arbitrage model in this paper. Thus, tick-by-tick intraday data from SHFE are utilized in the trading simulation. We trade and quote level two data, which means that the sample series in the trading simulation are recreated at one-second intervals. Each point in the series is the volume-weighted average price (bid price, bid volume, ask price and ask volume).

To make the empirical studies more realistic, four points should be emphasized here. First, according to the standard of the SHFE, the contract size of zinc is 5 tons per lot and the margin rate is assumed to be 10%. Second, the trading hours each day are separated into three parts, period 1 (9:00–10:15), period 2 (10:20–11:30) and period 3 (13:30–15:00). A pair of trading thresholds is calculated for each period in a trading day. For example, there were 20 trading days in May 2010; thus, there are 60 points for each k series. Third, the contract pair can be utilized to perform statistical arbitrage only when the mspread series is stationary; thus, the mspread series generated using two-day data before the trading date is used to perform the PP test. For example, if we decide to do the arbitrage on 2011/11/21, then we utilize the data from 2011/11/19 to 2011/11/20 to conduct the PP test. If the series is proven to be stable, then we conclude that the cointegration relationship will still hold on 2011/11/21. Fourth, to demonstrate the effectiveness of TGARCH-WNN, the model should be examined under distinct market circumstances. Therefore, three typical trading periods are selected to conduct the empirical study. The characteristics of these periods are described in Table 12.2 and Figure 12.3.

The samples selected here are all cointegrated pairs of stationary price series that passed the PP test. Like other statistical arbitrages, the trading frequency of the TGARCH-WNN arbitrage model is high

---

* The TGARCH-WNN model can be applied to all primary contracts in the SHFE.

† In the SHFE, the commission level of deformed steel bars is three ten thousandths of the turnover, whereas the commission level of copper is one ten-thousandth of the turnover.

‡ A detailed discussion of this issue is in Section 12.4.4.

TABLE 12.2   Market Conditions of Three Typical Trading Periods

Trading Period	Contract Pair	Market Conditions
2010/07/20–2010/07/23	zn09&zn10	Both contracts are falling continuously
2010/04/27–2010/04/30	zn06&zn07	Both contracts are rising continuously
2011/11/21–2011/11/24	zn01&zn02	Both contract prices are stable

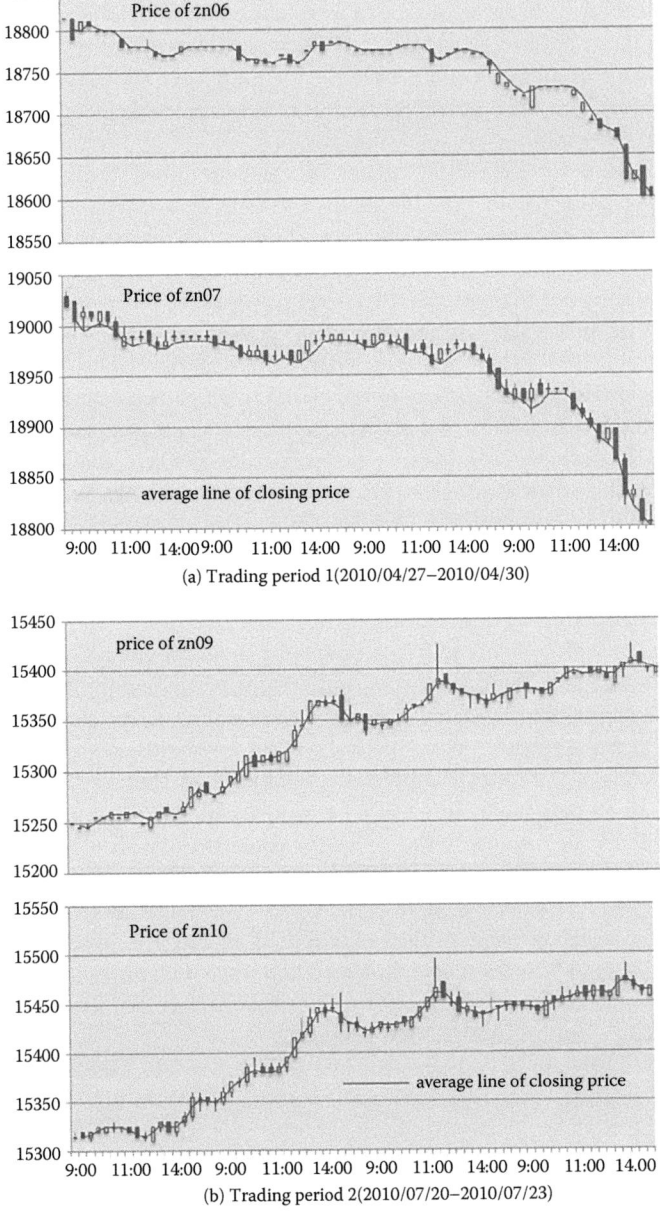

FIGURE 12.3   Sample price movement of zinc calendar spread arbitrage.

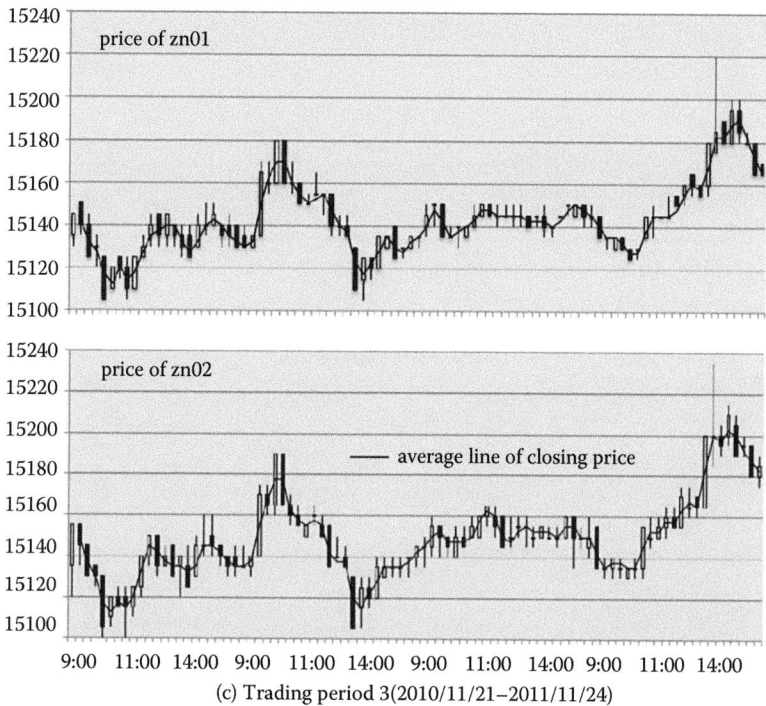

(c) Trading period 3(2010/11/21–2011/11/24)

FIGURE 12.3 (CONTINUED)

and all positions are closed within the day; thus, long-term market situations have little influence on the arbitrage. That is the reason why only short-term market trends are considered here. Therefore, each market situation contains four days of tick-by-tick data.

12.4 Empirical Study

The empirical study is separated into four parts. In the first part, a discussion of the application of the TGARCH model to eliminate heteroscedasticity and an analysis of asymmetric effects under distinct market situations are presented. In the second part, the Lyapunov exponent is utilized to prove the chaotic properties of the k series. In the third part, a comparison of the prediction accuracy and market performance of the WNN, BP and HO models is made. Finally, the impact of commission level is discussed in detail.

12.4.1 Application of the TGARCH Model

As mentioned in Section 12.2.1, to eliminate heteroscedasticity in the regression and to take asymmetric effects into consideration when predicting the *mspread* series' standard deviation,* the TGARCH regression model is applied. Here statistical tests are conducted to demonstrate the effectiveness of TGARCH at removing heteroscedasticity and present detailed characteristics of asymmetric effects in different market situations. Three typical market situations are considered.† Consistent with the points mentioned in Section 12.3, two-day minute data are selected to conduct statistical tests.

---

* As mentioned in Equation (12.2).
† As shown in Table 12.2 and Figure 12.3.

TABLE 12.3 ARCH-LM Test (lags:2) Result Comparison

Sample information		Test result of TGARCH		Test result of OLS	
Contract pair	Trading time	F-statistics	Prob. F(2450)	F-statistics	Prop. F(2450)
zn06&zn07	**2010/04/26–2010/04/27**	0.103336	0.9018	20.73649	0.0000
zn09&zn10	**2010/07/21–2010/07/22**	0.148928	0.8617	19.62015	0.0000
zn01&zn02	**2011/11/21–2011/11/22**	0.242065	0.7851	100.4303	0.0000

TABLE 12.4 Regression Results

Parameters[a]	Coefficient	p-value	Coefficient	p-value	Coefficient	p-value
	2010/04/26–2010/04/27		**2010/07/21–2010/07/22**		**2011/11/21–2011/11/22**	
Mean function:						
x_t	1.027	0.0000	1.028	0.000	1.173	0.0000
Constant	−305.148	0.0000	−378.483	0.0000	−2621.516	0.0000
TGARCH residuals:						
u_{t-1}^2	−0.052	0.0124	0.501	0.0000	0.210	0.0025
$u_{t-1}^2 d_{t-1}$	0.109	0.0000	−0.506	0.0000	−0.057	0.4759
σ_{t-1}^2	0.518	0.0000	−0.011	0.0000	0.629	0.0000
Constant	4.060	0.0000	26.558	0.0000	6.540	0.0212

[a] The parameters in the table are according to function (12.1) and function (12.2).

Table 12.3 presents the ARCH-LM test results of the TGARCH regression and OLS regression of contract pairs in different market situations. The null hypothesis of the ARCH-LM test is that there is no heteroscedasticity in the regression. From Table 12.3, we can see that the p values of the OLS regression are zero, which means that heteroscedasticity is apparent in all market situations. In contrast, the tests results of the TGARCH regression have significantly positive *p* values, which means TGARCH regression eliminates heteroscedasticity effectively.

It is important to understand the properties of asymmetric effects under different market circumstances when predicting the standard deviation of price spread. Thus, we examine the significance of the $u_{t-1}^2 d_{t-1}$ term in the function (12.2) and the sign its coefficient.

From Table 12.4, the *p* value of $u_{t-1}^2 d_{t-1}$'s coefficient in the periods 2010/04/26–2010/04/27 and 2010/07/21–2010/07/22 is less than 0.05; thus there are asymmetric effects when the price of both contracts increases or decreases unidirectionally. However, the coefficient of $u_{t-1}^2 d_{t-1}$ in the two periods has different signs. In the period 2010/04/26–2010/04/27, when the contract prices demonstrate a downward trend, $u_{t-1}^2 d_{t-1}$'s coefficient has a positive value which means that the asymmetric effect increases the variance of the price spread, whereas in the period 2010/07/21–2010/07/22, when the contract prices increase unidirectionally, the asymmetric effect decreases the variance of the price spread because of the negative value of $u_{t-1}^2 d_{t-1}$'s coefficient. This result is reasonable because the unidirectional change of the contract price implies that the powers of positive news and negative news are different. When price increases unidirectionally, positive news dominates the market; when price decreases unidirectionally, negative news dominates the market. Thus, the influence of the news on the variance of the price spread is distinct under these circumstances.

However, in the period 2011/11/21–2011/11/22, when the zinc price is stable, the *p* value of $u_{t-1}^2 d_{t-1}$'s coefficient is much greater than 0.05 which means the asymmetric effect is not apparent in this period. The reason is because positive news and negative news have the same power in the market. Then, Equation (12.2) can be simplified to

$$\sigma_t^2 = \omega + \alpha u_{t-1}^2 + \beta \sigma_{t-1}^2, \tag{12.7}$$

The result of the asymmetric effect analysis above is critical, because it helps to adjust the trading model with the TGARCH-WNN trading model to fit different market situations: when the contract price is changing unidirectionally, (12.2) is applied to predict *mspread*'s standard deviation, whereas when the contract price is stable, (12.7) is utilized to forecast *mspread*'s standard deviation.

12.4.2 Testing Chaotic Properties Using Lyapunov Exponents

As mentioned in Section 12.2.2, to demonstrate that it is rational to predict optimal trading threshold series using WNN, the chaotic properties of the sample series are tested by calculating the Lyapunov exponents of each series. Because the chaotic property is a long-term property of the trading threshold series, instead of applying samples from three typical situations, two-month data that contain different market circumstances are used to establish *k* series and calculate the Lyapunov exponents.[*] Specifically, we apply tick-by-tick data from March 1, 2010, to April 30, 2010, to generate the optimal trading threshold series in the test.

Figure 12.4 and Table 12.5 show the details of the test. From Figure 12.4 and Table 12.5, the testing series has stable positive largest Lyapunov exponents, which demonstrates the properties of chaos in

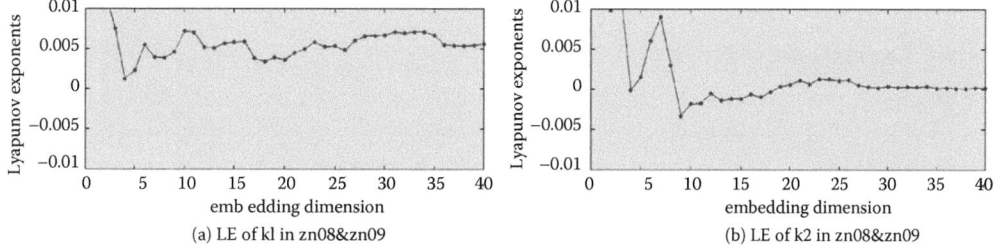

(a) LE of k1 in zn08&zn09   (b) LE of k2 in zn08&zn09

FIGURE 12.4 Lyapunov exponents.

TABLE 12.5 Lyapunov Exponents of Two Contract Pairs

Embedding Dimension[a]	LE	Embedding Dimension	LE
k1 of LE:			
32	0.0071	36	0.0054
33	0.0067	37	0.0054
34	0.0055	38	0.0055
35	0.0055	39	0.0057
k2 of LE:			
32	0.00029	36	0.00016
33	0.00044	37	0.00012
34	0.00018	38	0.00023
35	0.00022	39	0.00020

[a] Based on our data test, only when embedding dimension increases to above 30 do Lyapunov exponents tend towards stability. We thus display the beginning embedding dimension from above 30.

---

[*] As mentioned in Section 12.3, only contracts with the largest trading volumes are used to conduct arbitrage. Because the primary contracts of zinc futures shift monthly, the contracts used to calculate the *k* series are different each month. In addition, the short-run market conditions alter several times within a month according to our backtesting. Thus, based on the above two points, it is reasoned that two months of data contain enough different market conditions.

such periods. As mentioned above, two months of data contain sufficient different market conditions. Thus, we can reason that the three typical market conditions in Section 12.3 will also have chaotic properties, which is the premise for the following forecast of the optimal trading threshold series.

12.4.3 Robustness Check of the TGARCH-WNN, Backpropagation Neural Network and Historical Optimal Methods

In traditional statistical arbitrage models, trading thresholds are trading signals that maximize the cumulative yield historically in the latest trading term. This method is called the historical optimal (HO) method. The backpropagation (BP) neural network, as a major artificial intelligence method, has been widely used for financial time series prediction. Thus, to demonstrate the robustness of the WNN model, comparisons among one-step-ahead forecasts specified by WNN, BP and HO in terms of prediction accuracy and return-rate performance under different market situations are conducted in backtesting. To construct optimal architecture networks for the WNN and the BP neural network, trading threshold series generated from four-months data (from December 1, 2009, to March 31, 2010) are utilized for training and trading threshold series obtained from the three periods described in Section 12.3 are used for out-of-sample performance evaluation. The optimal parameters of WNN are presented in Table 12.6. For the BP neural network, the same statistical tests are conducted; the result is almost the same as for BP. The only difference is that the BP network contains two hidden layers, with 13 and 11 nodes, respectively.

12.4.3.1 Comparison of Prediction Accuracy

The prediction series generated by the three models are compared with the target series, which maximizes the cumulative yield in each period. The mean absolute positive error* is utilized as the standard of prediction accuracy. Table 12.7 shows the results of the comparison.

From Table 12.7, the predictions from WNN have smaller absolute positive errors than those from BP and HO in all three periods. To see how accurate the prediction of the trading thresholds is for the three models, some typical prediction points are shown in Figure 12.5.[†] It is obvious that the predictions of the trading thresholds in the HO model deviate seriously from those of the other two models. Thus, only the accurate predictions of the BP and WNN models are discussed in detail here. From Figure 12.5, although sometimes the prediction of BP is closer to the optimal value than is the prediction from WNN, the BP prediction is seriously affected by the noise in the mspread series. Thus, its prediction value series is more volatile, whereas the prediction value series from WNN is more stable and less affected by noise in mspread series because of the wavelet transform (Table 12.8).

TABLE 12.6   Architecture Network of WNN

Input Nodes	Output Nodes	Node Parameters		Momentum Study Rate	Iteration
		Weight Study Rate	Parameter Study Rate		
6	1	0.01	0.001	0.7	5000
Hidden layer		Hidden layer			
		Wavelet	BP neural network		
Number of layer		1	2		
Number of hidden nodes		8	8–10		

---

* As mentioned in function (12.6).

† For a clearer comparison, only the data from 2010/7/20 to 2010/7/23 are shown in the figure.

TABLE 12.7 Accuracy Test Comparison of Three Models

Testing Period	K	WNN (%)	BP Neural Network (%)	Historical Optimal (%)
2010/04/27–2010/04/30	k1	19.72	24.71	30.14
	k2	16.98	20.37	23.49
2010/07/20–2010/07/23	k1	16.68	18.12	22.71
	k2	17.49	18.37	22.47
2011/11/21–2011/11/24	k1	17.90	21.56	23.46
	k2	17.46	22.24	18.16

However, the mean absolute positive error does not necessarily reflect the information value of the forecast because it is a measure of fit rather than a measure of profitable signalling. Therefore, we continue to test the market performance of these models.

12.4.3.2 Comparison of Rate of Return under Different Circumstances

Robustness checks of the rate of return of WNN, BP and HO are performed under different market conditions. The commission level applied here is the average commission level asked by futures brokers, which is 1.25 times that specified by SHFE, 10 yuan per lot. Three essential assumptions should be emphasized here. First, to make the test more realistic, all positions are closed within the trading day to avoid margin calls. Second, all the money earned is put into the arbitrage process again. Third, the slippage loss per trade is 2 points.

From Table 12.3 and Figure 12.6, the TGARCH-WNN model performs much better than the other two trading models in the three typical periods in two aspects: it gains the highest profit and it has the highest winning rate.

Apart from the comparison result, other important phenomena should be emphasized. First, comparing the cumulative yield column and trading frequency column for the three periods, which are shown in Table 12.3, all trading models gain much more money and have a higher trading frequency in period 2011/11/21–2011/11/24 than in the periods 2010/04/27–2010/04/30 and 2010/07/20–2010/07/23. This means that although TGARCH-WNN has positive return in all three periods, the time when both contract prices are stable is the most ideal situation to conduct statistical arbitrage using the TGARCH-WNN model. Second, special attention should be paid to the testing result for the period 2010/04/27–2010/04/30. From the first picture in Figure 12.6, there is a sudden decrease in the cumulative yield. This is because there is a large jump in the price at approximately 10:00 a.m. on April 30. The abrupt change of price makes the system lose approximately 30 000 yuan in a single trade. Thus, applying statistical arbitrage alone can be quite risky, and some risk management strategies should be combined with the arbitrage system for real trades.

12.4.3.3 The Impact of Commission

In this section, an essential factor in statistical arbitrage, the commission level, is discussed. Because the trading frequency of the arbitrage is high, the level of commission has a great influence on the profit earned. To make a profit, the yield of the arbitrage must be greater than the commission charged per trade. Thus, when calculating the trading threshold, the unit commission level determines the lower bound of the trading threshold, which in turn affects the result of the WNN. However, the exact impact of the commission level is still obscure. To give a clear view of the impact, the zinc calendar spread arbitrage's performance under different commission levels is tested in the typical situations mentioned in Section 12.3. In the test, the commission level starts from the minimum level specified by SHFE (8 yuan/lot) and increases by 1 yuan each time to 14 yuan per lot. In the backtesting, special attention is paid to the changes in cumulative yield, number of trades, trading frequency and total commission, which are shown in Table 12.9. The backtesting results demonstrate some important phenomena and some essential phenomena deserve our special attention.

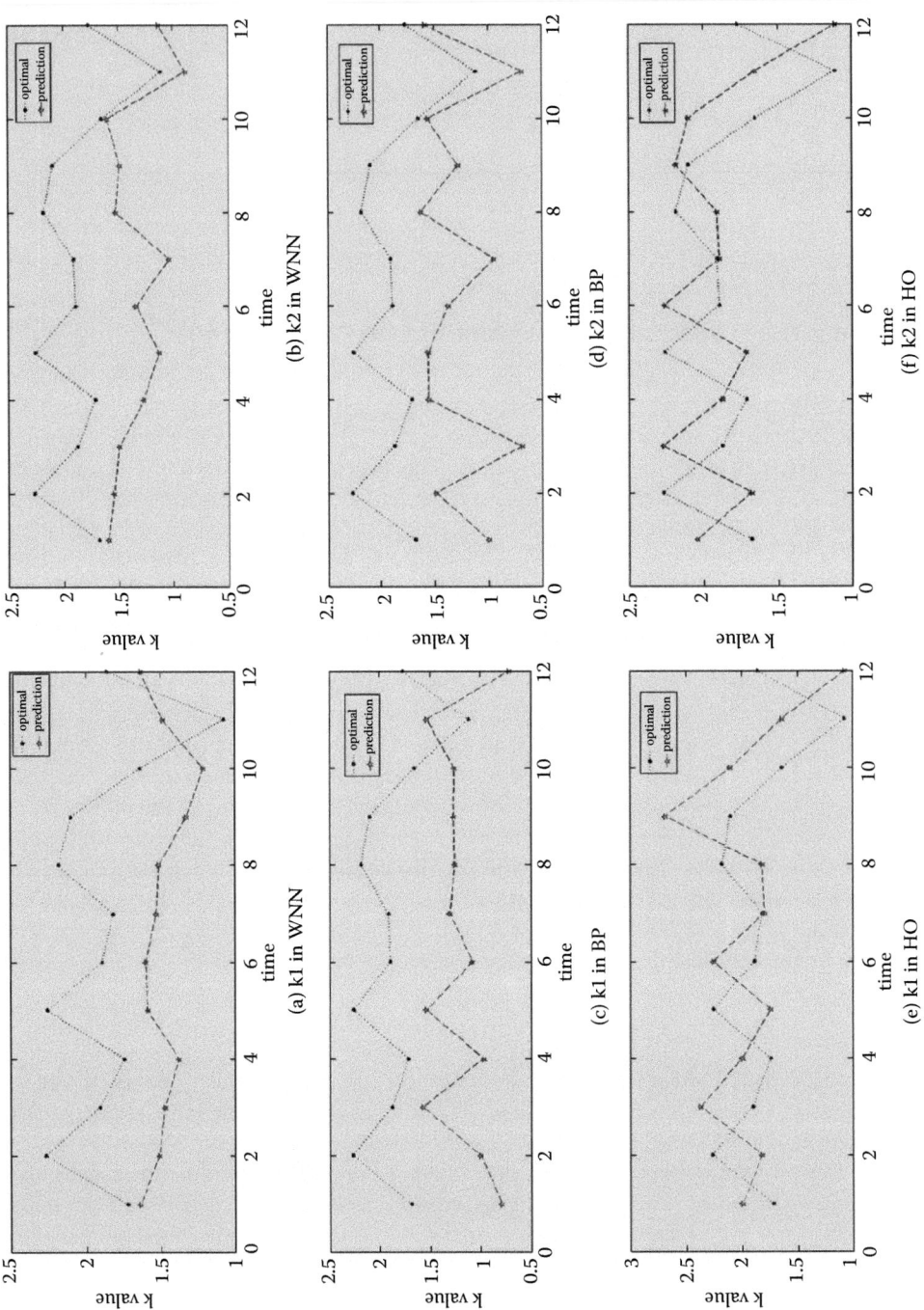

FIGURE 12.5   Prediction accuracy result of contract pair zn09 & zn10.

TABLE 12.8 Backtesting Results for Three Trading Models

Trading Models	Number of Trades	Trading Frequency (per min)	Winning Rate[a] (%)	Cumulative Yield[b] (%)
2010/04/27–2010/04/30:				
WNN	485	0.539	89.42	6.14
BP	546	0.607	88.64	5.35
HO	419	0.466	86.81	4.86
2010/07/20–2010/07/23:				
WNN	789	0.877	90.11	54.580
BP	675	0.750	87.70	50.400
HO	617	0.686	86.22	48.155
2011/11/21–2011/11/24:				
WNN	1259	1.410	90.39	105.330
BP	1299	1.440	90.07	89.005
HO	1004	1.12	87.75	78.279

[a] *Winning rate = $(m/n) \times 100$ where m is number of trade that yields positive profit (offset − transaction fee > 0) and n is total number of trade.*

[b] The initial margin in the test is 100000 yuan.

First, from Figure 12.7, the impact of the commission change is distinct for the three periods. In the periods 2010/4/27–2010/4/30 and 2010/7/20–2010/7/23, the cumulative yield decreases linearly by approximately 50% each time when the commission level increases by one yuan and drops below zero when the commission is approximately 13 yuan per lot. In the period 2011/11/21–2011/11/24, the cumulative yield decreases exponentially with a declining decreasing rate. In general, however, statistical arbitrage with TGARCH-WNN is very sensitive to changes in the commission level.

Second, from discussion with futures brokers, we learned that when the trading frequency is approximately 0.5 times per minute, they generally charge less than 1.5 times the base commission level (12 yuan per lot for zinc), and the lowest commission level charged can even be 1.125 times the base commission level (9 yuan per lot for zinc). Example commissions charged by mainstream futures brokers are shown in Table 12.10. For such commission levels, statistical arbitrage with TGARCH-WNN is still profitable.

Third, from Figure 12.8, the trading frequency of in the three periods decreases linearly to approximately the same level. A mathematical reason may be provided for this phenomenon as follows:

Without loss of generality, let j be the commission level and k_{jt} be the $k1$ value under commission level j in period t. (For $k2$, the notation is the same.) Let σ_t be the conditional standard deviation of mspread at time t. Then the trading threshold and conditional standard deviation at t are $k_{jt}\sigma_t$ and $n\sigma_t$, respectively.* According to the trading strategy mentioned in Section 12.2.1, when $mspread_t$ equals to the upper trading threshold at time t, namely $y_t - \beta_0 - \beta_1 x_t - \mu = k_{jt}\sigma_t$, we should buy one unit of contract Y and sell β_1 units of contract X. The position should be closed when $mspread_t$ equals 0 at time t'. According to functions 12.1 and 12.2, the revenue of this trade is $y_t - y_{t'} + \beta(x_{t'} - x_t) = k_{jt}\sigma_t$. However, this trade also generates a commission fee of $(1 + \beta_1)j$. Thus, to make a positive profit, $k_{jt}\sigma_t$ should be greater than $(1 + \beta_1)j$. Furthermore, it is explicitly required that the trading threshold value should be less than the stop-loss value, as discussed in Section 12.2.1. Therefore, the feasible zone of k_{jt} is $U_t = \left\{ k_{jt} \mid \dfrac{(1+\beta_1)j}{\sigma_t} < k_{jt} < n \right\}$.

For commission levels i and j, if $j > i$, then $\sigma_t k_{jt} \geq \sigma_t k_{it}$ and it is harder for the mspread series to hit $\sigma_t k_j$ than to hit $\sigma_t k_i$ to set the positions. Thus, the trading frequency decreases when the commission level increases. To see why the trading frequency decreases to same level, it should be noticed that the

---

* We obtain this estimated value from the assumption that mspread is normally distributed with $(0, \sigma_t)$.

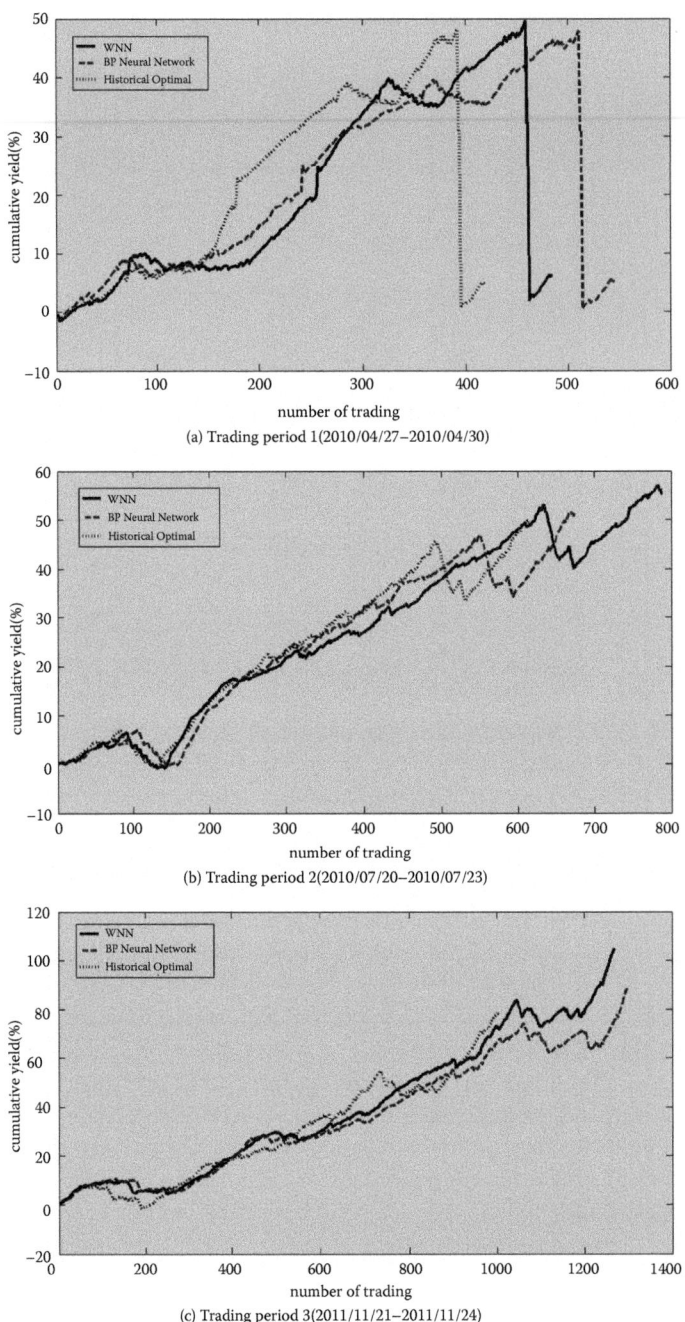

(a) Trading period 1(2010/04/27–2010/04/30)

(b) Trading period 2(2010/07/20–2010/07/23)

(c) Trading period 3(2011/11/21–2011/11/24)

FIGURE 12.6 Cumulative yields of different prediction methods under three typical market situations.

augmentation of the commission level narrows the feasible zones U_1, U_2 and U_3 to approximately the same level. Thus, k_{j1}, k_{j2} and k_{j3} are likely to obtain approximately the same value. Moreover, the distributions of the same contract pair's mspread in different periods are similar. Thus, the mspread series will have nearly the same probability of hitting the trading threshold lines, which implies that the is identical, as shown in Figure 12.8.

TABLE 12.9 Back Testing Result of Different Commission Levels in Three Typical Periods

Commission Level (Yuan Per Lot)	Cumulative Yield (%)	Number of Trade	Trading Frequency (%)	Total Commission (Yuan)
2010/04/27–2010/04/30:				
8	28.236	440	0.489	47032
9	14.560	379	0.421	43983
10	6.410	393	0.437	46890
11	−2.857	371	0.412	47201
12	−3.292	376	0.417	51396
13	−10.570	310	0.344	43160
14	−20.842	281	0.312	40320
2010/07/20–2010/07/23:				
8	81.758	721	0.801	108846
9	61.593	576	0.639	80316
10	40.380	550	0.613	78460
11	26.604	421	0.468	63723
12	8.782	416	0.462	63084
13	−1.810	351	0.390	54080
14	−9.961	268	0.297	42868
2011/11/21–2011/11/24:				
8	232.295	1168	1.298	221440
9	123.742	954	1.06	160704
10	66.445	738	0.820	115640
11	28.380	618	0.687	96910
12	5.586	505	0.561	76032
13	−3.267	384	0.427	57746
14	−2.987	273	0.303	44156

Fourthly, it seems contradictory to perceive that the total commission generated by arbitrage decreases as the commission level increases, as shown in the total commission columns of Table 12.9. However, apart from the commission level, there are two other factors that also affect the total commission: the number of trades and the commission per trade. Their changes account for this phenomenon. To understand the phenomenon mathematically, under commission level j, let the total commission be T_j, the number of trades be N_j, the commission per trade be C_j, the real-time equity be E_j, the real-time initial margin be M_j, the real-time price be P_j and the number of contracts purchased and sold be l_j, respectively. It is assumed that all the money earned is put into the arbitrage process again; thus, $E_j = M_j$ whenever positions are set. As mentioned in Section 12.3, the margin rate is 10%; then the expression of the total commission is as follows:*

$$T_j = N_j \times C_j = N_j \times \left(j l_j \right) = N_j \times \left(j \frac{M_j}{0.1 P_j} \right) \tag{12.8}$$

For commission level i and j, if $i < j$, then the arbitrage trade can earn more money with the lower commission level i each time. As time goes by, $M_i > M_j$, which in turn leads to $C_i > C_j$. It can also be observed from the number of trade columns in Table 12.9 that $N_i > N_j$. Thus, according to (12.8), higher C and N values will most likely lead to higher total commission T.

---

* For simplicity, the strategy of how many contracts to buy and sell specifically is ignored here.

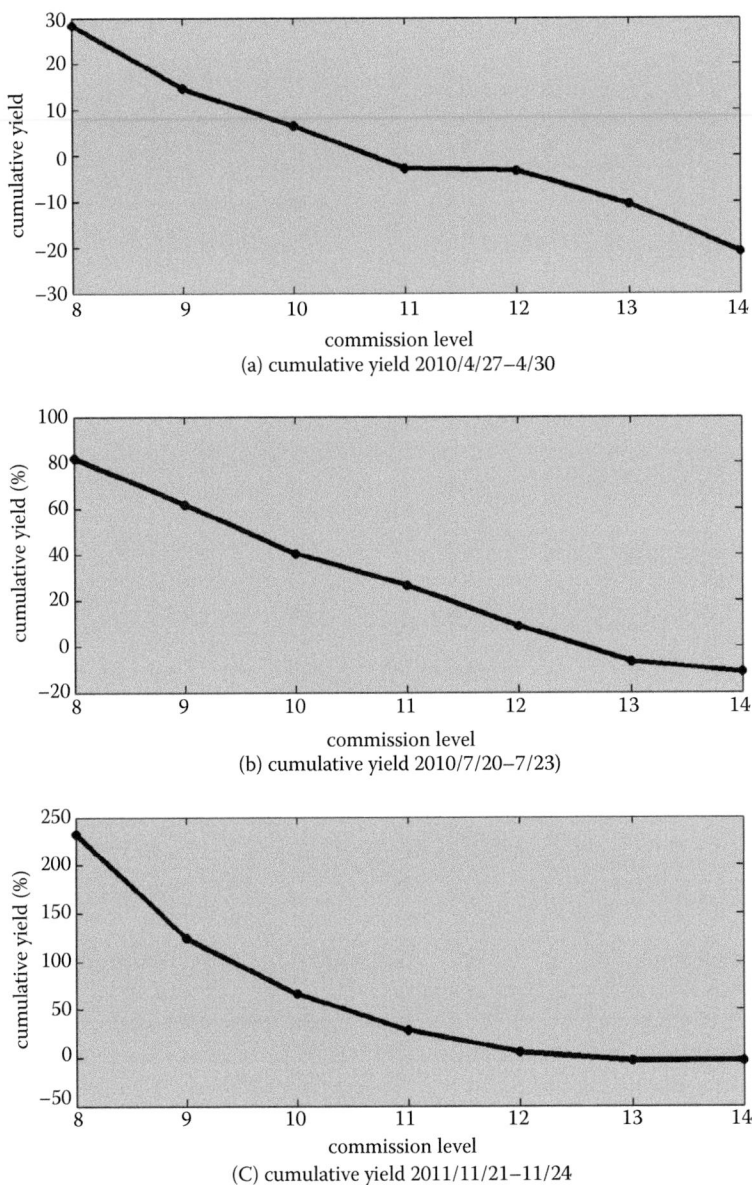

FIGURE 12.7   Cumulative yield under different commission levels in typical periods.

TABLE 12.10   Commission Level Charged by Futures Brokers

Futures Brokers	Commission Charge for Zinc (Yuan Per Lot)
China International Futures Co., Ltd.	10
Wonder Futures	12
Baocheng Futures Co., Ltd.	10
Changjiang Futures	9
Mailyard Futures	12

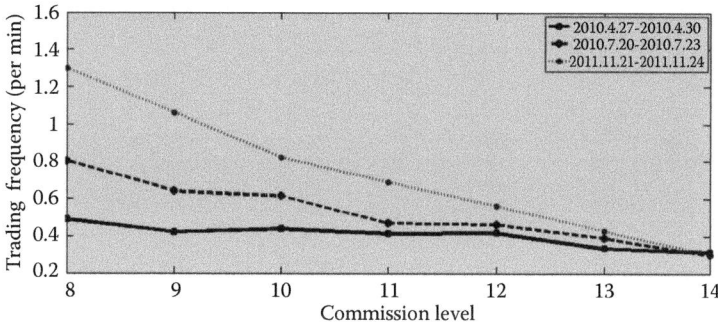

FIGURE 12.8 Trading frequency of different periods with distinctive commission level.

12.5 Conclusion

In this paper, a new statistical arbitrage model based on TGARCH-WNN is described. The empirical study focuses on the Chinese metal futures market. It is shown that the regression of futures price series is negatively impacted by heteroscedasticity. While analysing asymmetric effects under different market situations, we find that asymmetric effects are apparent when the price is trending, whereas it is faint when the contract price is stable. We apply different algorithms according to the distinct characteristics of each typical period. The robustness check in the empirical study demonstrates that when using the wavelet transform to eliminate noise in the series, the predictions made by WNN are not only more accurate and stable than those from BP and HO but also generate more profits when applied in the arbitrage process. From the commission impact test, it is shown that the TGARCH-WNN statistical arbitrage model is very sensitive to changes in the commission level. However, it is still profitable for the average commission level charged by futures brokers.

Furthermore, with highly profitable opportunities for statistical arbitrage, a statistical arbitrage model using WNN could have many potential applications and is of great commercial value. Both investors and futures companies can take advantage of this model to gain stable and considerable profit.

Finally, it is extremely significant to note that although the empirical study demonstrates that statistical arbitrage using TGARCH-WNN is very profitable, it is also quite risky. The trading model alone cannot address sudden price changes very well, which can thus cause severe losses. Therefore, to put this model into practice, a risk management system should be constructed along with the trading model.

Acknowledgements

This research was supported by Natural Science Foundation of China (11171247). We would like to thank the editor and Elizabeth Dawes, as well as the anonymous referees for their insightful comments. We would like to acknowledge the help received from the Shanghai Futures Exchange and CSMAR Solution in providing the data. The authors thanks Rui Cao and Xiaoyong Cui from Central University of Finance and Economics, Zhuo Huang from CCER in Peking University, Yulei Luo from Hong Kong University and Wei Zou from IAS in Wuhan University for many helpful comments and suggestions.

References

Bollerslev, T., Chou, R. and Kroner, K., ARCH modelling in finance. *J. Econom.*, 1992, **52**, 5–59.

Brennan, M.J. and Subrahmanyam, A., Market microstructure and asset pricing: On the compensation for illiquidity in stock returns. *J. Financ. Econ.*, 1996, **41**, 441–464.

Chang, P.C., Liao, T.W., Lin, J.J. and Fan, C.Y., A dynamic threshold decision system for stock trading signal detection. *Appl. Soft Comput.*, 2011, **11**, 3998–4010.

Christian, F. and Zakoian, J.M., GARCH Models: Structure, Statistical Inference and Financial Applications, 2010 (John Wiley & Sons: Hoboken).

Chung, Y.P., A transaction data test of stock index futures market efficiency and index arbitrage profitability. *J. Finance*, 1991, **46**, 1791–1809.

Creamer, G., Model calibration and automated trading agent for Euro futures. *Quant. Finance*, 2012, **12**, 531–545.

Davidson, R. and Mackinnon, J.G., *Estimation and Inference in Econometrics*, 1993 (Oxford University Press: New York).

Ellen, E.R., Norman, A.A. and Jonathan, D.B., Simplified frequency content estimates of earthquake ground motions. *J. Geotech. Geoenviron. Eng.*, 1998, **124**, 150–159.

Fung, H.G., Liu, Q.F. and Tse, Y., The information flow and market efficiency between the US and Chinese aluminum and copper futures market. *J. Futures Market*, 2010, **30**, 1192–1209.

Gatev, E., Goetzmann, W.N. and Rouwenhorst, K.G., Pairs Trading: Performance of a relative-value arbitrage rule. Working Paper, 1999.

Glosten, L.R. and Harris, L.E., Estimating the components of the bid-ask spread. *J. Financ. Econ.*, 1988, **21**, 123–142.

He, Q., Value at risk for repo interest rate based on TGARCH. *Appl. J. Financ. Econ*, Proceedings of 2008 IEEE International Conference on Service Operations and Logistics, and Informatics. IEEE/SOLI 2008, **1**, 425–428.

Huang, K., Cui, L. and Cai, H.J., A trading model in Chinese commodity futures market using GARCH-wavelet neural networks. In *Proceedings – 2012* 3rd International Conference on E-Business and E-Government, pp. 4925–4929, 2012.

Jaisimha, M., Imon, P. and Oleg, S., Wavelet decomposition for intra-day volume dynamics. *Quant. Finance*, 2010, **10**, 917–930.

Kantz, H., A robust method to estimate the maximal Lyapunov exponent of a time series. *Phys. Lett. A*, 1994, **185**, 77–87.

Li, S.T. and Kuo, S.C., Knowledge discovery in financial investment for forecasting and trading strategy through wavelet-based SOM networks. *Expert Syst. Appl.*, 2008, **34**, 935–951.

Lin, N., Lu, H. and Zheng, Q., An empirical study on the existence of bubble in Chinese stock market: Based on TGARCH model. In *Proceedings – 2010* 2nd IEEE International Conference on Information and Financial, *Engineering*, pp. 87–90, 2010.

Locke, P.R. and Venkatesh, P.C., Futures market transaction costs. *J. Futures Markets*, 1997, **17**, 229–245.

McMillan, D.G. and Speight, E.H., Nonlinear dynamics and competing behavioral interpretations: Evidence from intra-day FTSE-100 index and futures data. *J. Futures Market*, 2006, **26**, 343–368.

Noortwijk, J.M., Weide, J.A.M., Kallen, M.J. and Pandey, M.D., Gamma processes and peaks-over-threshold distributions for time-dependent reliability. *Reliab. Eng. Syst. Saf.*, 2007, **92**, 1651–1658.

Steve, H., Robert, J., Melvyn, T. and Mitch, W., Testing market efficiency using statistical arbitrage with applications to momentum and value models. *J. Financ. Econ.*, 2004, **73**, 525–565.

Wang, P. and Wang, P., Asymmetry in return reversals or asymmetry in volatilities?—New evidence from new markets. *Quant. Finance*, 2011, **11**, 271–285.

Weber, M. and Prokopczuk, M., American option valuation: Implied calibration of Garch pricing models. *J. Futures Markets*, 2011, **31**, 971–994.

Wolf, A., Jack, B.S., Harry, L.S. and John, A.V., Determing Lyapunov exponents from a time series. *Physica*, 1985, **16D**, 285–317.

Zhang, B.L., Coggins, R., Jabri, M.A., Dersch, D. and Flower, B., Multiresolution forecasting for futures trading using wavelet decompositions. *IEEE Trans. Neural Networks*, 2001, **12**, 765–775.

Zhou, Z.Q. and Hojjat, A., Time-frequency signal analysis of earthquake records using Mexican hat wavelets. *Gen. Introductory Civil Eng. Constr.*, 2003, **18**, 379–389.

13

Multivariate Continuous-Time Modeling of Wind Indexes and Hedging of Wind Risk

Fred E. Benth,
Troels S.
Christensen and
Victor Rohde*

13.1 Introduction*

In the power market, producers in general face market risk in the sense of uncertainty of the prices at which they can sell their generated power. The intermittent nature of renewable energy sources such as wind and photo voltaic power production adds yet another layer of risk, known as volumetric risk in the sense that the produced amount of electricity is uncertain due to the dependence on weather. Globally, so-called power purchase agreements and subsidies from governments have minimized the market risk for renewable power producers. In contrast, the volumetric risk has only recently been addressed in Germany—and only for wind power producers (WPPs)—by the introduction of the exchange-traded wind power futures (WPF) contracts. The underlying of a WPF contract is an index between zero and one representing the overall utilization of the installed German wind power production. By taking an

---

* Corresponding author. Email: victor@math.au.dk

DOI: 10.1201/9781003265399-15

appropriate position in WPF contracts, the lost income of the WPPs implied by low wind scenarios is (partially) offset by the position in WPF contracts, hence minimizing the volumetric risk. Due to the prioritization ofthe cheapest power producers in Germany, the opposite part of the WPF market is typically conventional power producers (CPPs) such as gas-fired power plants. By taking an appropriate position in exchange-traded WPF contracts, CPPs can hedge their exposure to the cheap electricity generated by WPPs.

In consequence of the recent introduction of the WPF market, the related literature is limited. Gersema and Wozabal (2017) consider the WPF market in great detail, and propose an equilibrium pricing model. They find that the willingness to engage in the WPF market is greater for the WPPs compared to CPPs. In other words, the hedging benefits of the WPF contracts are greater for WPPs than CPPs. This is supported by the results in Christensen and Pircalabu (2018) who employ an ARMA-GARCH copula framework to the joint modelling of one site-specific wind index and the underlying WPF index. In Benth and Pircalabu (2018) modeling of the underlying WPF index is considered and closed-form formulas for the WPF price and the price of European options written on the WPF index are derived. A common feature of the mentioned articles is that they do notconsider the simultaneous modeling of more than two wind indexes. Considering a portfolio of wind generation sites, where the production of electricity at each site is represented by the wind index at each wind site, a simultaneous model for the wind indexes at several wind sites is motivated for e.g. risk management of this portfolio. To the best of our knowledge, such a simultaneous model is unexplored in the literature.

One way of constructing such a simultaneous model is by using Lévy processes. Continuous-time modeling using univariate Ornstein-Uhlenbeck (OU) processes driven by non-decreasing Lévy processes, like the compound Poisson process with exponential jumps, have been studied extensively, and used to model, for example, stochastic volatility of financial assets, wind, electicity prices, and temperature (see Barndorff–Nielsen and Shephard 2001, Benth and Benth 2007, Benth *et al.* 2007, Benth and Pircalabu 2018). A detailed treatment of Lévy processes can be found in Sato (1999). The multivariate modeling of more than two stochastic processes using multidimensional non-Gaussian Lévy processes is, however, more limited. Here we mention the work of Leoni and Schoutens (2008) and Semeraro (2008) that introduce the multivariate construction by subordination of Brownian motions, and the work of Ballotta and Bonfiglioli (2016) using linear transformations of Lévy processes.

Our contribution to the literature is twofold. Firstly, we propose a joint model for the simultaneous behaviour of wind indexes that allows for a parsimonious representation of the correlation structure. This model can be seen as the multivariate version of the model presented in Benth and Pircalabu (2018). As a consequence of the scarce literature on such multivariate models, we propose an alternative model for comparison reasons. Secondly, we suggest the idea of so-called tailor-made WPF contracts to eliminate the volumetric risk of WPPs completely. Employing our proposed joint model of wind indexes, we investigate the hedging benefits of exchange-traded WPF contracts for an owner of a portfolio of tailor-made WPF contracts, and comment on the risk premium of tailor-made WPF contracts. We show that this construction is beneficial for both parties of the tailor-made WPF contracts.

The rest of the paper is organized as follows. Section 13.2 presents the data of wind indexes we later analyze in greater detail and also serves as motivation for the proposed model. In section 13.3 we present models for the joint behaviour of wind indexes and corresponding estimation procedures. In section 13.4 we present the estimation results of the two models. Section 13.5 discusses the hedging of wind power production using WPF contracts implied by the proposed models. Lastly, section 13.6 concludes.

13.2 Data Presentation

The empirical observation period spans from 1 July 2016 to 30 June 2019, which corresponds to 1095 daily observations for each considered wind index. The data set consists of

(1) A daily wind index at three wind sites provided byCentrica Energy Trading. The wind index at wind site i is calculated by

$$\frac{Q_i(t)}{h(t)C_i},$$

where $h(t)$ is the number of hours in day t, $Q_i(t)$ is the power production at day t at site i, and C_i is the installed capacity at site i. Figure 13.1 shows the approximate geographical locations for the three wind sites.*

(2) A daily German wind index provided by Nasdaq, representing the German utilization of wind power plants. The acronym used for it is NAREX WIDE (NAsdaq Renewable indeX WInd DE (Germany)) and is used as the underlying for WPF contracts traded on Nasdaq. We will simply denote it as the German wind index in the remaining part of the paper.

The wind indexes are bounded between zero and one. Figure 13.2 shows all four wind indexes, and the corresponding autocorrelation function for each wind index. In all four cases, the wind index displays a pronounced yearly cycle consistent with the observations made in Benth and Pircalabu (2018) and Christensen and Pircalabu (2018). Since the German wind index is by construction made up of all wind power production in Germany, the behaviour of the German wind index is less extreme compared to the individual wind indexes. To concretize, a value of zero for the German wind index is not observed

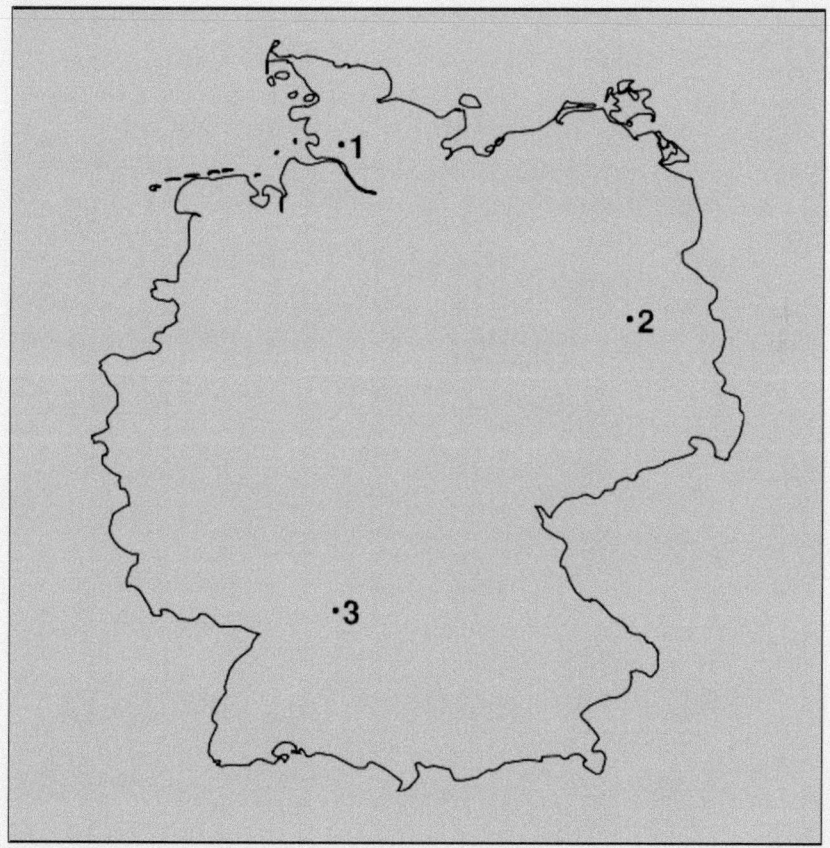

FIGURE 13.1   Locations of wind sites with site ID in Germany.

---

* The locations are approximate due to confidentially issues.

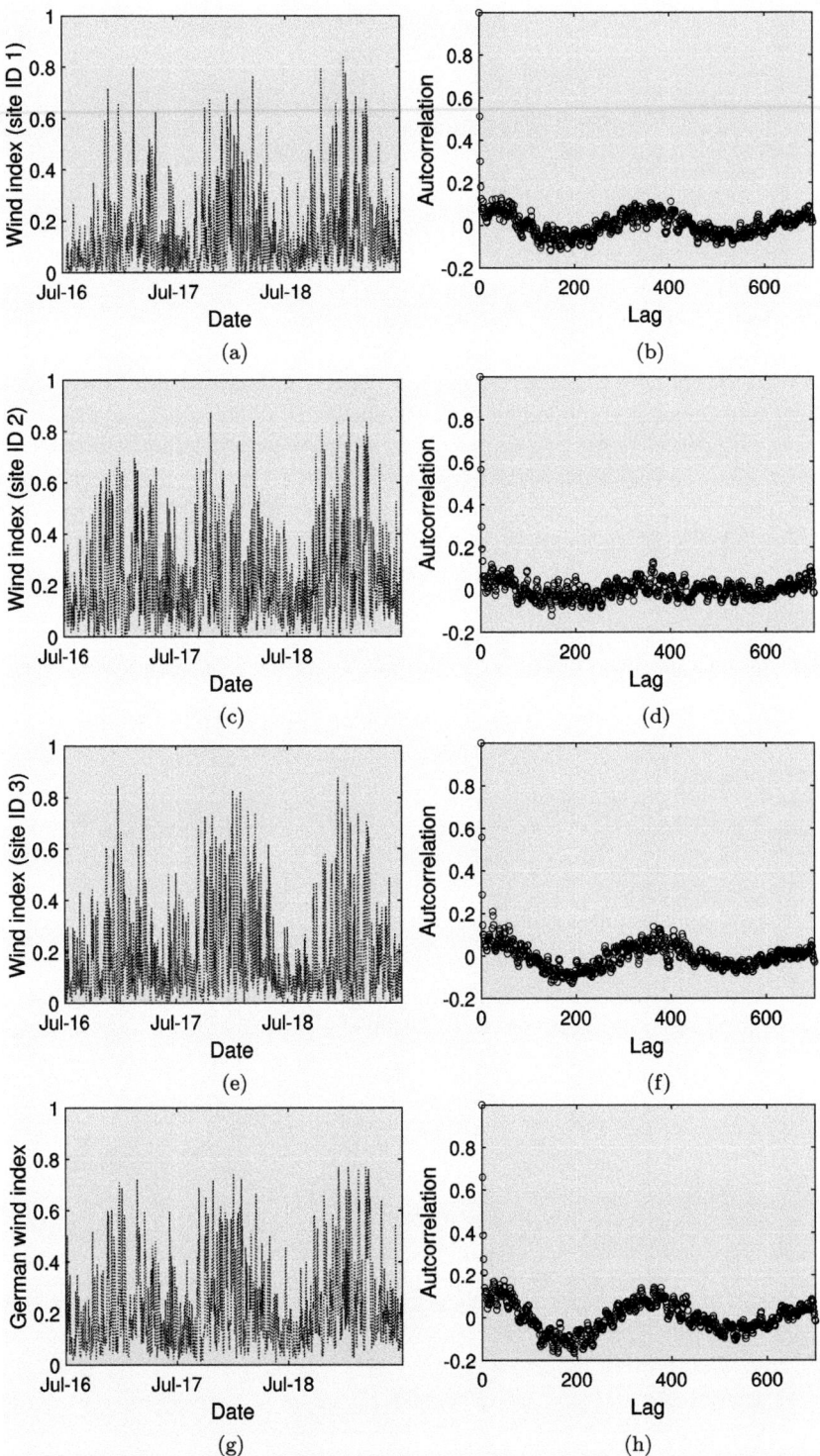

FIGURE 13.2   All four wind indexes with corresponding empirical auto-correlation function. (a) Wind index at site 1. (b) Autocorr. of site 1. (c) Wind index at site 2. (d) Autocorr. of site 2. (e) Wind index at site 3. (f) Autocorr. of site 3. (g) German wind index and (h) Autocorr. of German.

in our observation period, whereas it is for all three wind sites. Also the maximum attained value for each site wind index is higher than the maximum value of the German index; however, it does not reach one in any of the cases.

13.3 Model Description

13.3.1 General Model Considerations

Let n denote the number of wind sites and $P_i(t)$ the ith wind index. We assume that the ith wind index can be described by

$$P_i(t) = 1 - e^{-S_i(t)X_i(t)}, \quad i = 1,...,n, \tag{13.1}$$

where $S_i(t): \mathbb{R} \to \mathbb{R}^+$ is a deterministic function intended to filter out potential seasonal effects, and $X_i(t)$ is a mean-reverting stochastic process satisfying $X_i(t) \geq 0$ for all t. The intention of $X(t) = (X_1(t), ..., X_n(t))^\top$ is to capture the short-term uncertainty and the dependence between the n wind indexes. By this specification we are ensured that $P_i(t) \in [0, 1)$.

The proposed model in equation (13.1) distinguishes itself from the specification in Benth and Pircalabu (2018), where the natural extension of their univariate setup to the present multivariate setup would be

$$P_i(t) = S_i(t)e^{-X_i(t)}. \tag{13.2}$$

with appropriate choices of $X_i(t)$ and $S_i(t)$. We do, however, prefer equation (13.1) over equation (13.2) for the following reasons. Firstly, due to our specification with regard to the deterministic part $S_i(t)$ of the model, we do not face any potential model inconsistencies as is the case of equation (13.2). We refer the interested reader to Benth and Pircalabu (2018) for more information and discussion. Secondly, as discussed in section 13.2, the wind index at a given site can attain a value of zero, whereas, on the other hand, we have not observed full utilization of the capacity at a single wind site. Since Benth and Pircalabu (2018) consider the German wind index separately, which is never zero or one due to the construction of it, equation (13.2) is applicable without any modifications. Lastly, equation (13.1) implies that increased values of $S_i(t)$ and $X_i(t)$ translate to an increased value of $P_i(t)$, which ismore intuitively appealing. In equation (13.2), increased values of $S_i(t)$ will still translate to increased values of $P_i(t)$, but here increased values of $X_i(t)$ translates to decreased values of $P_i(t)$.

Moving on to the seasonal components of the model, we include the following yearly seasonality motivated by the observations made in Figure 13.2,

$$S_i(t) = a_i + b_i \sin(2\pi t / 365) + c_i \cos(2\pi t / 365),$$

$$i = 1,...,n,$$

where $a_i, b_i, c_i \in \mathbb{R}$. With N being the number of observations, the coefficients are determined by

$$\min_{a_i, b_i, c_i} \sum_{t=1}^{N} \left[-\log\left(1 - P_i(t)\right) - S_i(t) \right]^2, \quad i = 1,...,n.$$

Having obtained the estimated seasonal functions, the observed values of $X_i(t)$ implied by the estimated seasonal function $\hat{S}_i(t)$ can then be calculated as

$$X_i(t) = \frac{-\log\left(1 - P_i(t)\right)}{\hat{S}_i(t)}. \tag{13.3}$$

We will in the following discuss two approaches for modeling $X_i(t)$.

13.3.2 A Gamma Model

In this section a multivariate model for $n-1$ wind indexes and the German wind index is discussed, which we refer to as the gamma model in the sequel. In section 13.4 we will considerthe case $n = 4$. We start by introducing the noise process. In particular, we say a Lévy process L is a compound Poisson process with exponential jumps and parameters $\alpha > 0$ and $\beta > 0$ if

$$L(t) = \sum_{i=1}^{N(t)} J_i$$

Where $(N(t))_{t\in\mathbb{R}}$ is a Poisson process with intensity α and J_i, $i \in$ N, are independent exponentially distributed random variables with parameter β. We say a random variable has an exponential distribution with parameter β if it has density $x \mapsto 1_{[0,\infty)}(x)\beta e^{-\beta x}$.

The gamma model assumes X is a multidimensional Lévy-driven Ornstein-Uhlenbeck (OU) process,

$$dX(t) = -\Lambda X(t)dt + \Sigma_L\, dL(t). \tag{13.4}$$

Here, L is an n-dimensional Lévy process where the i'th entry is an independent compound Poisson process with exponential jumps, variance equal to one, and parameters α_i and β_i for $i = 1,...,n$. Furthermore, Λ is a diagonal matrix, diag $(\lambda_1, ..., \lambda_n)$ with $\lambda_i > 0$ for $i = 1, ..., n$. We assume Σ_L is given by

$$\Sigma_L = \begin{pmatrix} \sigma_{1,1} & 0 & \cdots & 0 & \sigma_{1,n} \\ 0 & \sigma_{2,2} & \cdots & 0 & \sigma_{2,n} \\ \vdots & \vdots & \ddots & 0 & \vdots \\ 0 & 0 & 0 & \sigma_{n-1,n-1} & \sigma_{n-1,n} \\ 0 & 0 & 0 & 0 & \sigma_{n,n} \end{pmatrix} \tag{13.5}$$

and that all entries of Σ_L are non-negative. Due to the form of Σ_L, each individual wind index has an idiosyncratic risk associated to it through one of the first $n-1$ compound Poisson process $L_1, ...,L_{n-1}$. Furthermore, all sites and the German index share a systematic risk through the n'th compound Poisson process L_n. A similar construction is also considered in Ballotta and Bonfiglioli (2016) where a multivariate model is proposed for modeling financial products written on more than one underlying asset.

13.3.2.1 Distribution of $P_n(t)$ in the Gamma Model

The process associated with the German wind index, X_n, is an OU process driven by one compound Poisson process with exponential jumps, and it therefore has a gamma distribution as its stationary distribution. The processes $X_1, ...,X_{n-1}$ on the other hand are sums of two independent gamma distributions. The density of a sum of two independent gamma distributions does not, in general, have a closed form, and thus, there does not exist simple expressions for the densities of the individual site index similar to the one for the German index stated in proposition 13.3.1.

PROPOSITION 13.3.1 *The stationary distribution of $P_n(t)$ in the gamma model has density*

$$f_{P_n(t)}(x) = \frac{(-\log(1-x))^{\alpha_n-1}(1-x)^{\beta_n/S_n(t)-1}}{S_n(t)^{\alpha_n}} \quad x \in (0,1) \tag{13.6}$$

Proof This is a direct consequence of $X_n(t)$ being gamma distributed with shape α_n and rate β_n (see for example Barndorff–Nielsen and Shephard 2001). ∎

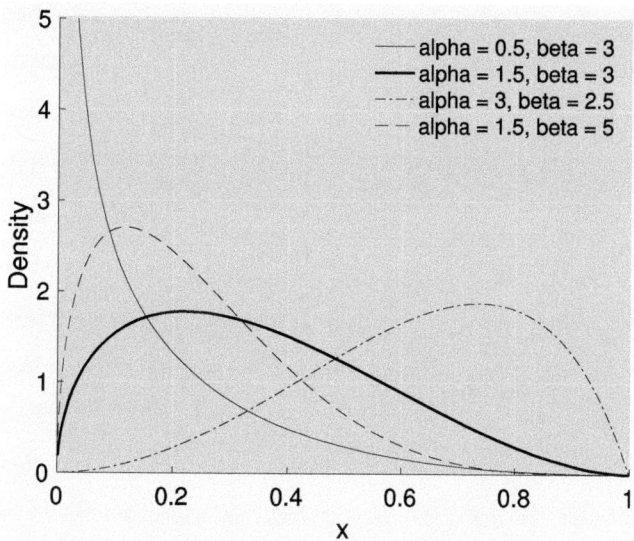

FIGURE 13.3 Different variations of the density in (13.6).

In Figure 13.3 the density of P_n is depicted for different α_n and β_n (with $S_n(t) = 1$). We see that the densities implied by the gamma model are rather flexible and able to cover both low and high utilization scenarios.

13.3.2.2 Covariance between Wind Indexes in the Gamma Model

We now give (semi-)analytical expressions of the covariances implied by the gamma model, which will be useful for fast calculation of the minimum variance hedges discussed in section 13.5.

Before we state the result, let us introduce some notation to help making a concise statement. To this end define the $n \times (n + 1)$ matrix $\tilde{\Sigma}_L$ by

$$\tilde{\Sigma}_L = \begin{pmatrix} \sigma_{1,1} & 0 & \cdots & 0 & 0 & \sigma_{1,n} \\ 0 & \sigma_{2,2} & \cdots & 0 & 0 & \sigma_{2,n} \\ \vdots & \vdots & \ddots & 0 & 0 & \vdots \\ 0 & 0 & 0 & \sigma_{n-1,n-1} & 0 & \sigma_{n-1,n} \\ 0 & 0 & 0 & 0 & 0 & \sigma_{n,n} \end{pmatrix}$$

where $\sigma_{i,j}$ is the (i,j)'th entry of Σ_L. Let $\tilde{\sigma}_{i,j}$ denote the (i,j)'th entry of $\tilde{\Sigma}_L$. Furthermore, define $\tilde{\alpha}, \tilde{\beta} \in \mathbb{R}^{n+1}$ by

$$\tilde{\alpha} = \left(\alpha_1, \ldots, \alpha_{n-1}, 0, \alpha_n \right)^\top \quad \text{and}$$

$$\tilde{\beta} = \left(\beta_1, \ldots, \beta_{n-1}, 1, \beta_n \right)^\top,$$

and denote the i'th entry of $\tilde{\alpha}$ and $\tilde{\beta}$ by $\tilde{\alpha}_i$ and $\tilde{\beta}_i$. We now give the expressions of the covariances of the gamma model. The proof is relegated to the Appendix.

PROPOSITION 13.3.2 *Consider $s \leq t$ and define*

$$f_{i,j}(u) = S_i(t)\tilde{\sigma}_{i,n+1}e^{-\lambda_i(t-s+u)} + S_j(s)\tilde{\sigma}_{j,n+1}e^{-\lambda_j u},$$

$$i, j = 1, \ldots, n.$$

Then

$$\text{cov}\big(P_i(t), P_j(s)\big)$$

$$= \left(\frac{\tilde{\beta}_i}{\tilde{\beta}_i + \tilde{\sigma}_{i,i}S_i(t)}\right)^{\tilde{\alpha}_i/\lambda_i} \left(\frac{\tilde{\beta}_{n+1} + \tilde{\sigma}_{i,n+1}S_i(t)e^{-\lambda_i(t-s)}}{\tilde{\beta}_{n+1} + \tilde{\sigma}_{i,n+1}S_i(t)}\right)^{\tilde{\alpha}_{n+1}/\lambda_i}$$

$$\times \left(\frac{\tilde{\beta}_j}{\tilde{\beta}_j + \tilde{\sigma}_{j,j}S_j(s)}\right)^{\tilde{\alpha}_j/\lambda_j} \left[\left(1 + \frac{f_{i,j}(0)}{\tilde{\beta}_{n+1}}\right)^{\tilde{\alpha}_{n+1}f_{i,j}(0)/f'_{i,j}(0)}\right.$$

$$\times \exp\left\{\tilde{\alpha}_{n+1}\int_0^\infty \frac{\mathrm{d}}{\mathrm{d}u}\left(\frac{f_{i,j}(u)}{\frac{\mathrm{d}}{\mathrm{d}u}f_{i,j}(u)}\right)\log\left(1 + \frac{f_{i,j}(u)}{\tilde{\beta}_{n+1}}\right)\mathrm{d}u\right\} \tag{13.7}$$

$$- \left(\frac{\tilde{\beta}_{n+1}}{\tilde{\beta}_{n+1} + \tilde{\sigma}_{i,n+1}S_4(t)e^{-\lambda_i(t-s)}}\right)^{\tilde{\alpha}_{n+1}/\lambda_i}$$

$$\times \left.\left(\frac{\tilde{\beta}_{n+1}}{\tilde{\beta}_{n+1} + \tilde{\sigma}_{j,n+1}S_j(s)}\right)^{\tilde{\alpha}_{n+1}/\lambda_j}\right]$$

for $i, j = 1, \ldots, n$, $i \neq j$, and

$$\text{cov}\big(P_i(t), P_i(s)\big)$$

$$= \left(\frac{\tilde{\beta}_i + \tilde{\sigma}_{i,i}S_i(t)e^{-\lambda_i(t-s)}}{\tilde{\beta}_i + \tilde{\sigma}_{i,i}S_i(t)}\right)^{\tilde{\alpha}_i/\lambda_i} \left(\frac{\tilde{\beta}_{n+1} + \tilde{\sigma}_{i,n+1}S_i(t)e^{-\lambda_i(t-s)}}{\tilde{\beta}_{n+1} + \tilde{\sigma}_{i,n+1}S_i(t)}\right)^{\tilde{\alpha}_{n+1}/\lambda_i}$$

$$\times \left[\left(\frac{\tilde{\beta}_i}{\tilde{\beta}_i + \left(S_i(t)e^{-\lambda_i(t-s)} + S_i(s)\right)\tilde{\sigma}_{i,i}}\right)^{\tilde{\alpha}_i/\lambda_i}\right.$$

$$\times \left(\frac{\tilde{\beta}_{n+1}}{\tilde{\beta}_{n+1} + \left(S_i(t)e^{-\lambda_i(t-s)} + S_i(s)\right)\tilde{\sigma}_{i,n+1}}\right)^{\tilde{\alpha}_{n+1}/\lambda_i} \tag{13.8}$$

$$- \left(\frac{\tilde{\beta}_i^2}{\left(\tilde{\beta}_i + \tilde{\sigma}_{i,i}S_i(t)e^{-\lambda_i(t-s)}\right)\left(\tilde{\beta}_i + \tilde{\sigma}_{i,i}S_i(s)\right)}\right)^{\tilde{\alpha}_i/\lambda_i}$$

$$\times \left.\left(\frac{\tilde{\beta}_{n+1}^2}{\left(\tilde{\beta}_{n+1} + \tilde{\sigma}_{i,n+1}S_i(t)e^{-\lambda_i(t-s)}\right)\left(\tilde{\beta}_{n+1} + \tilde{\sigma}_{i,n+1}S_i(s)\right)}\right)^{\tilde{\alpha}_{n+1}/\lambda_i}\right]$$

for $i = 1, \ldots, n$.

The integral in equation (13.7) is the only non-analytical part of the expression, but we argue in remark A.3 that this integral is small and can be coarsely approximated without significant effect. This allows us to maintain a fast computational speed when calculating the covariances. To further this point, we find a compute time, when evaluating (13.7) or (13.7) numerically 1,000 times, of around 0.03 seconds on a standard laptop implemented in MATLAB®R2018b.

13.3.2.3 Identification of Parameters in the Gamma Model

Let A_{var} be the $n \times n$ matrix given by

$$\left(\Lambda_{var}\right)_{i,j} = \frac{1}{\lambda_i + \lambda_j}.$$

Furthermore, denote by $\circ$ the Hadamard product. Then the following result will be used to estimate the parameters of the gamma model. Again, we relegate the proof of proposition 13.3.3 to the Appendix.

PROPOSITION 13.3.3 *The mean of X is*

$$\mathbb{E}[X(0)] = \Lambda^{-1}\Sigma_L\beta / 2 \tag{13.9}$$

and the auto-covariance of X is

$$\text{cov}(X(0), X(t)) = \left(\Lambda_{var} \circ \left(\Sigma_L\Sigma_L^\top\right)\right)e^{-\Lambda t} \tag{13.10}$$

for $t \geq 0$.

The parameters of the gamma model will be estimated in three steps. First, the mean-reversion matrix Λ will be fitted to the empirical auto-correlation function based on the first 25 lags. From (13.10), it follows that the model auto-correlation function of X_i is $t \mapsto e^{-\lambda_i t}$. To find $\hat{\lambda}_i$, the estimate of λ_i, we therefore minimize

$$\sum_{t=1}^{25}\left(\hat{\rho}_i(t) - e^{-\hat{\lambda}_i t}\right)^2$$

such that $\hat{\lambda}_i > 0$ *for* $i = 1, \ldots, n$, where $\hat{\rho}_i(t)$ is the empirical auto-correlation function of X_i.

Next, $\hat{\Sigma}_L$ is chosen such that the model matches the empirical covariances. In particular, we choose $\hat{\Sigma}_L$ to minimize

$$\left\|\hat{\Sigma}_X - \Lambda_{var} \circ \left(\hat{\Sigma}_L\hat{\Sigma}_L^\top\right)\right\|^2$$

where $\hat{\Sigma}_X$ is the sample covariance of X, $\|\cdot\|$ is the Frobenius norm and the minimization is done over matrices $\hat{\Sigma}_L$ with non-negative entries of the form in (13.5).

Finally, we discuss how the parameters α and β are estimated. First, we choose $\hat{\beta} = \left(\hat{\beta}_1, \ldots, \hat{\beta}_n\right)$ to minimize

$$\left\|\hat{\mu}_X - \Lambda^{-1}\hat{\Sigma}_L\hat{\beta} / 2\right\|^2$$

such that $\hat{\beta}_i > 0$, where $\hat{\mu}_X$ is the empirical mean of X.

It is not too difficult to show that $\text{var}\left(L_i(1)\right) = 2\alpha_i / \beta_i^2$ and, since the compound Poisson processes are assumed to have unit variance, it therefore follows that

$$1 = \text{var}\left(L_i(1)\right) = \frac{2\alpha_i}{\beta_i^2}.$$

Consequently, we take $\hat{\alpha}_i = \hat{\beta}_i^2 / 2$.

13.3.3 A Lognormal Model

In this section we present a lognormal model relying on different assumptions than the gamma model. We assume that $G(t) := \log X(t)$ can be modelled as a multidimensional Gaussian Ornstein-Uhlenbeck process,

$$dG(t) = -\Upsilon(G(t) - \Theta)dt + \Sigma dB(t), \tag{13.11}$$

where $(B(t))_{t \in \mathbb{R}}$ is an n-dimensional Brownian motion, $\Upsilon \in \mathbb{R}^{n \times n}$ is a diagonal matrix, $\Sigma \in \mathbb{R}^{n \times n}$ is a lower triangular matrix, and $\Theta \in \mathbb{R}^n$.

It is well-known (see e.g. Meucci 2009) that the stationary distribution of $G(t)$, when the diagonal elements of Υ all are positive, is normal with mean Θ. The autocovariance of $G(t)$ is given in, for example, Meucci (2009), and it is the same as for the gamma model found in Prop. 13.3.3. In particular, we find

$$\Sigma_G(t) := \mathrm{cov}(G(0), G(t)) = \left(\Upsilon_{\mathrm{var}} \circ \left(\Sigma \Sigma^\top \right) \right) e^{-\Upsilon t}, \quad t \geq 0. \tag{13.12}$$

Here, Υ_{var} is the $n \times n$ matrix given by

$$\left(\Upsilon_{var} \right)_{i,j} = \frac{1}{v_i + v_j},$$

where v_i is the i'th entry of Υ, $i = 1, ..., n$.

Consequently, the stationary distribution of $X(t)$ is multivariate lognormal with expected value of $X_i(0)$ being

$$\mathbb{E}\left[X_i(0) \right] = \exp\left(\Theta_i + \frac{\Sigma_G(0)_{ii}}{2} \right),$$

while the autocovariance is

$$\mathrm{cov}\left(X_i(0), X_j(t) \right) = \mathbb{E}\left[X_i(0) \right] \mathbb{E}\left[X_j(0) \right] \left(e^{\Sigma_G(t)_{ij}} - 1 \right) \tag{13.13}$$

for $i, j = 1, ..., n$ (see e.g. Halliwell 2015 for more information on the multivariate lognormal distribution). This implies that the autocorrelation of $X(t)$ is

$$\mathrm{corr}\left(X_i(0), X_j(t) \right) = \frac{\exp\left(\Sigma_G(0)_{ij} e^{-tv_j} \right) - 1}{\sqrt{\left(e^{\Sigma_G(0)_{ii}} - 1 \right)\left(e^{\Sigma_G(0)_{jj}} - 1 \right)}}, \tag{13.14}$$

for $i, j = 1, ..., n$.

13.3.3.1 Distribution of $P_i(t)$ in the Lognormal Model

Having the results for $X(t)$ from the previous section in mind, the stationary distribution of $P_i(t)$ is given in proposition 13.3.4.

PROPOSITION 13.3.4 *The stationary distribution of $P_i(t)$ is characterized by the density*

$$f_{P_i(t)}(x) = \frac{-1}{(1-x)\log(1-x)\sqrt{\Sigma_G(0)_{i,i}}} \phi$$

$$\times \left(\frac{\log\left(-\frac{\log(1-x)}{S_i(t)} \right) - \Theta_i}{\sqrt{\Sigma_G(0)_{i,i}}} \right), \tag{13.15}$$

where $\Sigma_G(0)_{i,i}$ is the i'th element of the diagonal of $\Sigma_G(0)$, Θ_i is the i'th element of Θ, and $\phi(\cdot)$ is the density of the standard normal distribution.

To investigate the density of $P_i(t)$ in more detail, consider for a moment a more generic version of equation (13.15), given by

$$f(x|\mu,\sigma)$$

$$= \frac{-1}{(1-x)\log(1-x)\sigma}\phi\left(\frac{\log(-\log(1-x))-\mu}{\sigma}\right). \tag{13.16}$$

As can be seen in Figure 13.4, showing examples of the density given different values of μ and σ in equation (13.16), the distribution is rather flexible and capable of attaining quite different forms.

13.3.3.2 Covariance between Wind Indexes in the Lognormal Model

Deriving the covariances between wind indexes in the lognormal model is closely related to the derivation of the Laplace transform of the lognormal distribution. To the best of our knowledge, no closed-form has been derived for the Laplace transform of the lognormal distribution, but there exist approximations, see e.g. Asmussen *et al.* (2016). With regard to this paper, we refer the interested reader to Asmussen *et al.* (2016) and the references herein for further information, and employ numerical integration by exploiting our knowledge of the distribution of $G(t)$ to determine the covariances between the wind indexes.

13.3.3.3 Identification of Parameters in the Lognormal Model

To identify the parameters of the model, we employ the method of moments as in the gamma model case. We first identify $\Sigma_G(0)$ by exploiting equation (13.13),

$$\Sigma_G(0)_{ij} = \log\left(\frac{\hat{\Sigma}_{X,ij}}{\hat{\mu}_i\hat{\mu}_j}+1\right), \tag{13.17}$$

with $\hat{\mu}_i$ being the empirical mean of $X_i(t)$, and $\hat{\Sigma}_{X,ij}$ is the (i,j)th entry of the empirical covariance between $X_i(0)$ and $X_j(0)$.

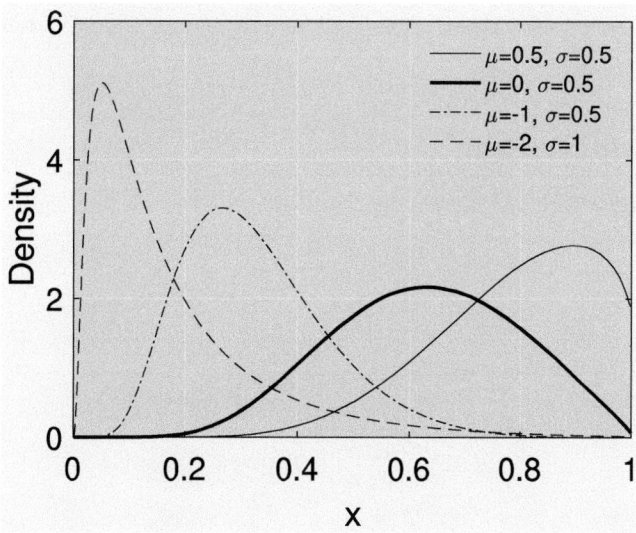

FIGURE 13.4 Different variations of the density in equation (13.16).

Having obtained an estimate of $\Sigma_G(0)$ and remembering the model implied autocorrelation in equation (13.14), we identify v_i by minimizing

$$\sum_{t=1}^{25}\left(\hat{\rho}_i(t)-\frac{\exp\left(\hat{\Sigma}_G(0)_{ii}e^{-tv_i}\right)-1}{\sqrt{\left(e^{\hat{\Sigma}_G(0)ii}-1\right)\left(e^{\hat{\Sigma}_G(0)ii}-1\right)}}\right)^2,$$

where $\hat{\rho}_i(t)$ is the empirical autocorrelation function of $X_i(0)$ and $X_i(t)$. Here, as in the gamma model, we use the first 25 lags of the empirical auto-covariance function to estimate λ_i. With $\breve{\Upsilon}$, consisting of the estimated v_i for $i = 1, ..., n$ in the diagonal, at hand, we identify $\Sigma\Sigma^\top$ by

$$\Sigma\Sigma^\top = \hat{\Sigma}_G(0) \oslash \hat{\Upsilon}_{\text{var}},$$

where $\oslash$ is the Hadamard division defined for two matrices A and B by $A \oslash B = A_{ij} / B_{ij}$. Lastly, we determine Θ by

$$\Theta_i = \log\left(\mu_i\right) - \frac{\hat{\Sigma}_G(0)_{ii}}{2}, \quad i = 1, ..., n. \tag{13.18}$$

13.3.4 Comparison of the Gamma and Lognormal Model

The covariances between indexes in the gamma model can be calculated fast using Proposition 13.3.2 to find the optimal hedging strategy (see section 13.5). The noise in the gamma model also has a compelling interpretation, where an idiosyncratic risk is associated to each site index and a systematic risk is associated to all site indexes and the German index. On the other hand, the lognormal model gives rise to closed-form expressions of the densities of all indexes as opposed to only the German index in the gamma model. The lognormal model is simple in the sense that the underlying process is a Gaussian driven OU process. This makes it possible to do numerical analysis based on Gaussian theory.

Both the gamma and lognormal model have straightforward and fast estimation procedures, making them easy to implement. Furthermore, as we will see in section 13.4, both models capture the autocorrelation of X_i, the cross-autocorrelations between X_i and X_j, and the stationary distribution of X_i well.

13.4 Estimation Results

In this section we summarize and discuss the estimation results. As a starting point, we consider the fitted seasonal functions. In Table 13.1 we report the fitted parameters for all four wind indexes.

TABLE 13.1 Fitted Seasonal Parameters for the Four Wind Indexes

	$\hat{a}_i$	$\hat{b}_i$	$\hat{c}_i$
Site 1	0.1721	−0.0491	−0.0804
Site 2	0.2848	−0.0405	−0.0956
Site 3	0.2294	−0.0322	−0.1226
German	0.2732	−0.0298	−0.1285

13.4.1 Gamma Model

Figure 13.5 shows the theoretical autocorrelation implied by the estimated gamma model compared to the empirical autocorrelation. The fit to the empirical autocorrelation is convincing, and it is worth noticing that the cross-autocorrelations match well even though the model has only been estimated to the marginal autocorrelation functions.

In Figure 13.6 the histogram of X_i and the model density based on a simulation are shown. We see that the distribution of the wind sites implied by the gamma model have a convincing fit to the empirical distributions. The German wind index takes values close to zero more frequently than predicted by the gamma model, but besides this, the distribution of the gamma model gives a satisfying fit to the empirical distribution.

We report the estimated parameters in Table 13.2. The parameters α_4 and β_4 are considerably larger than α_i and β_i for $i= 1, 2, 3$. This implies that the systematic risk factor L_4 jumps much more frequently than L_i, $i = 1, 2, 3$, but that the jumps of L_4 are relatively small compared to the jumps of L_i, $i = 1, 2, 3$. This aligns well with the intuition that the systematic risk is associated to the wind utilization of the whole of Germany.

To further assess the model, we report in Table 13.3 the mean, variance, skewness, and kurtosis of the gamma model along with the empirical and lognormal equivalents for the German wind index.* The first two moment of the gamma model agrees with the empirical version as expected from the estimation procedure, where we match the gamma model to the empirical mean and variance. Further, the empirical skewness and kurtosis are captured within a reasonable precision by the gamma model.

13.4.2 Lognormal Model

Figure 13.7 shows the theoretical autocorrelation implied by the estimated lognormal model compared to the empirical autocorrelation. As in the gamma model case, the lognormal model captures the autocorrelation and cross-autocorrelation well, in particular taking into account that only the autocorrelation is used to estimate the parameters affecting both the autocorrelations and the cross-autocorrelations.

Figure 13.8 shows histograms of the marginal distributions and the fitted lognormal densities. The lognormal model provides overall a decent fit, but seems to capture the distribution of the German wind index better than the site indexes. The estimated Θ and Υ for the lognormal model is reported in Table 13.4 and the estimated Σ is

$$\hat{\Sigma} = \begin{bmatrix} 1.0987 & 0 & 0 & 0 \\ 0.6763 & 0.5886 & 0 & 0 \\ 0.4902 & 0.3505 & 0.8376 & 0 \\ 0.6381 & 0.2539 & 0.2162 & 0.3035 \end{bmatrix}.$$

Although the speed of mean reversion parameters $\hat{\upsilon}_i$ differ in the lognormal model compared to the gamma model, the same pattern is observed, with the German wind index being the most persistent.

Returning to Table 13.3, the lognormal model matches the empirical mean and variance as a results of the estimation procedure, but it does not capture the higher order standardized moments. This indicates that the lognormal model does not capture the whole distribution of the data as well as the gamma model.

---

* Since the same quantities for the site wind indexes are not relevant in the remaining part of the paper, we have chosen to omit them.

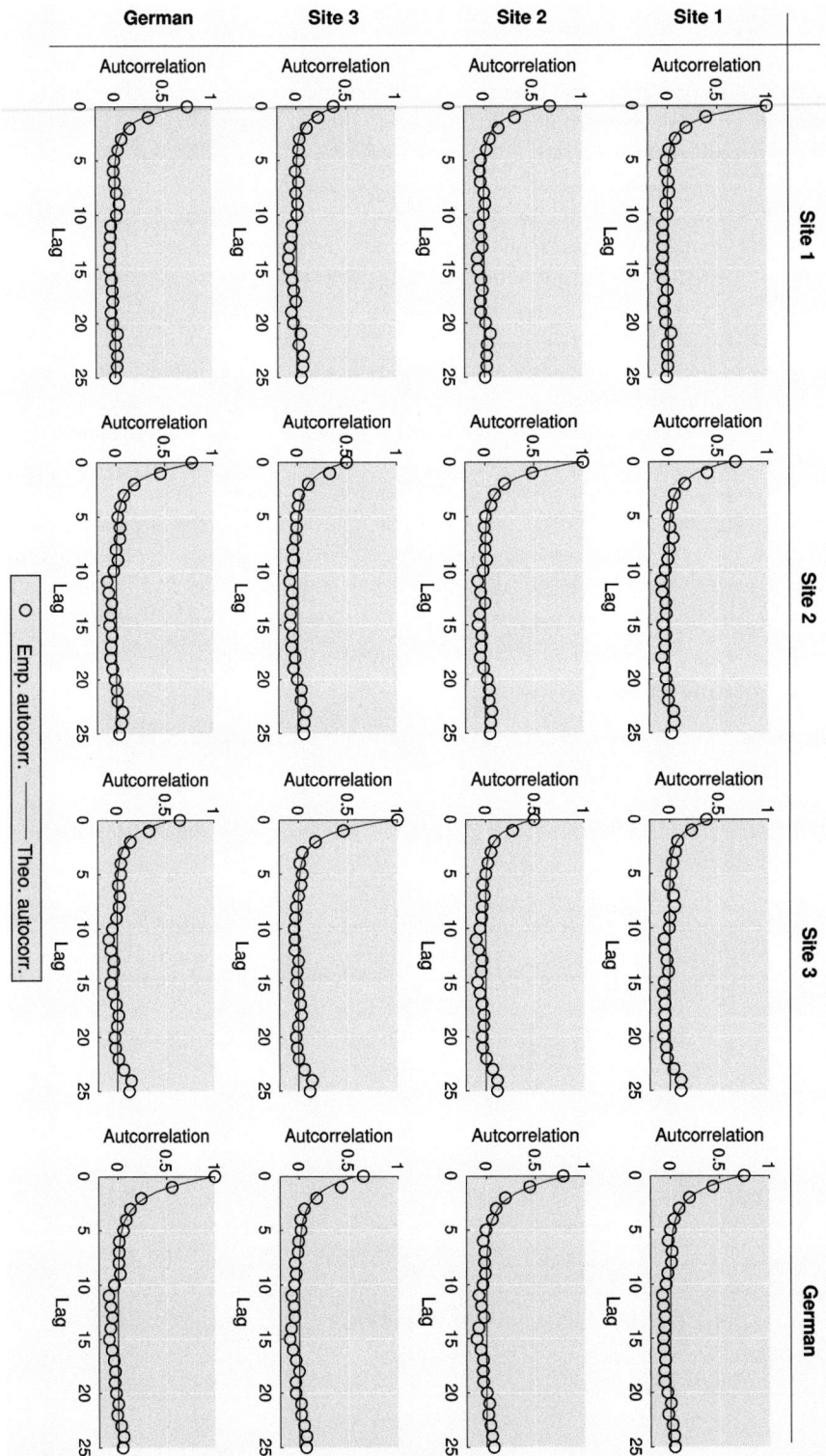

FIGURE 13.5 Empirical autocorrelation and theoretical autocorrelation implied by the fitted gamma model. The (i, j)'th panel shows cor$(Xi(0), Xj(t))$ for $t = 0, 1, ..., 25$.

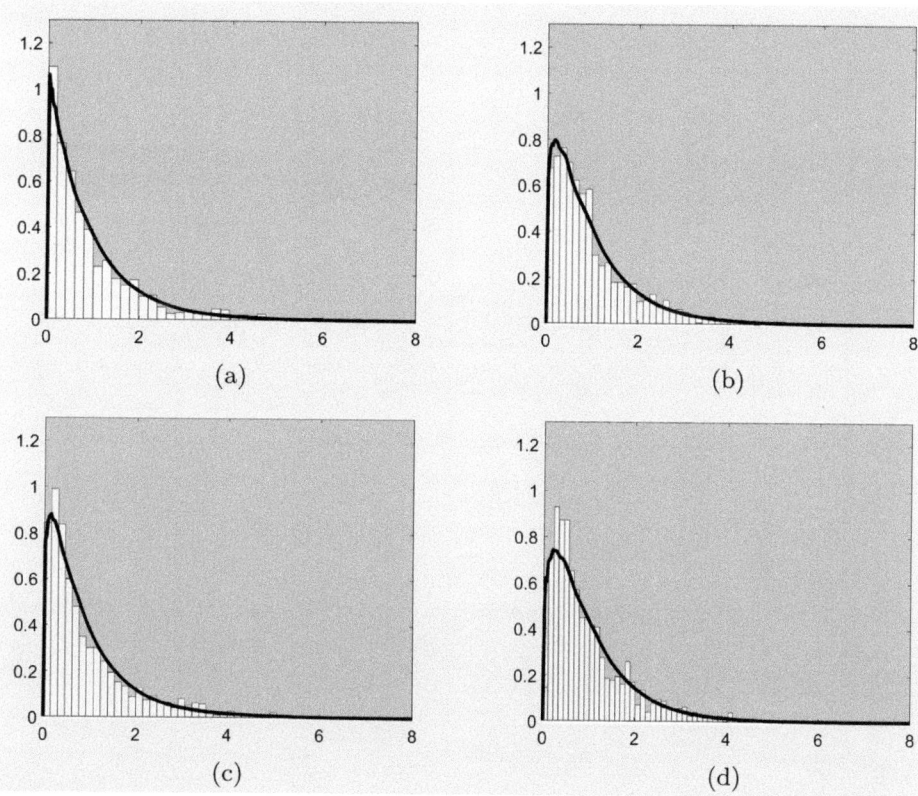

FIGURE 13.6 Histograms of $Xi(t)$ with the fitted densities of the gamma model. (a) Site 1. (b) Site 2. (c) Site 3 and (d) German.

TABLE 13.2 Estimated Parameters in the Gamma Model

	$\hat{\alpha}_i$	$\hat{\beta}_i$	$\hat{\lambda}_i$	$\hat{\sigma}_{i,i}$	$\hat{\sigma}_{i,4}$
Site 1	0.0271	0.2328	0.8977	1.0305	1.1593
Site 2	0.0538	0.3282	0.7589	0.6101	0.9792
Site 3	0.1383	0.5260	0.8513	1.1674	0.8247
German	0.8960	1.3387	0.6539	(−)	0.9781

TABLE 13.3 Mean, Variance, Skewness, and Kurtosis of the German Wind Index in the Gamma and Lognormal Model together with the Empirical Values for the German Wind Index

	Mean	Variance	Skewness	Kurtosis
Gamma	1.00	0.73	1.71	7.38
Lognormal	1.00	0.73	3.17	24.98
Empiricial	1.00	0.73	1.67	6.27

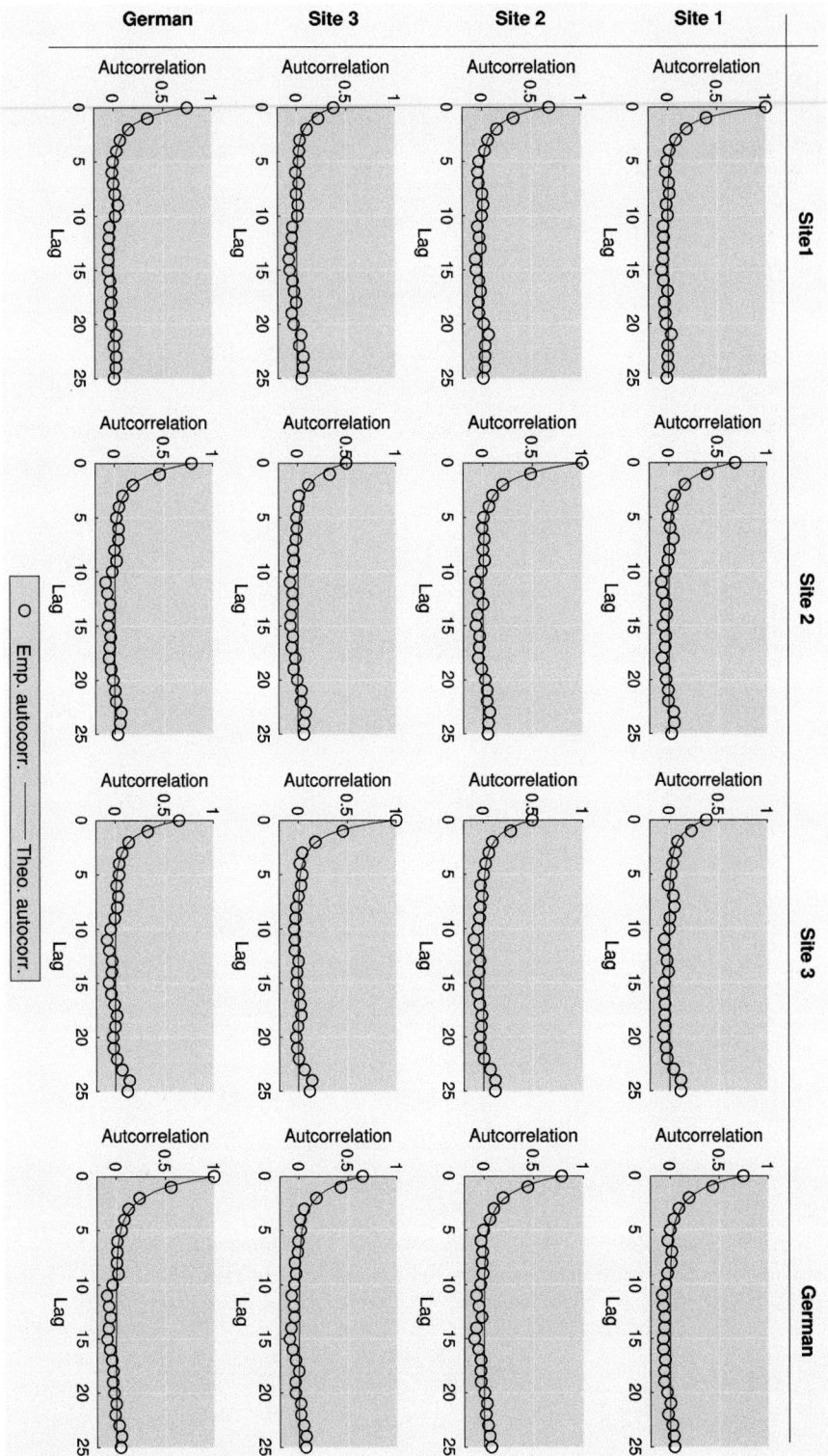

FIGURE 13.7 Empirical autocorrelation and theoretical autocorrelation implied by the fitted lognormal model. The (i, j)'th panel shows $\mathrm{cor}(Xi(0), Xj(t))$ for $t = 0, 1, ..., 25$.

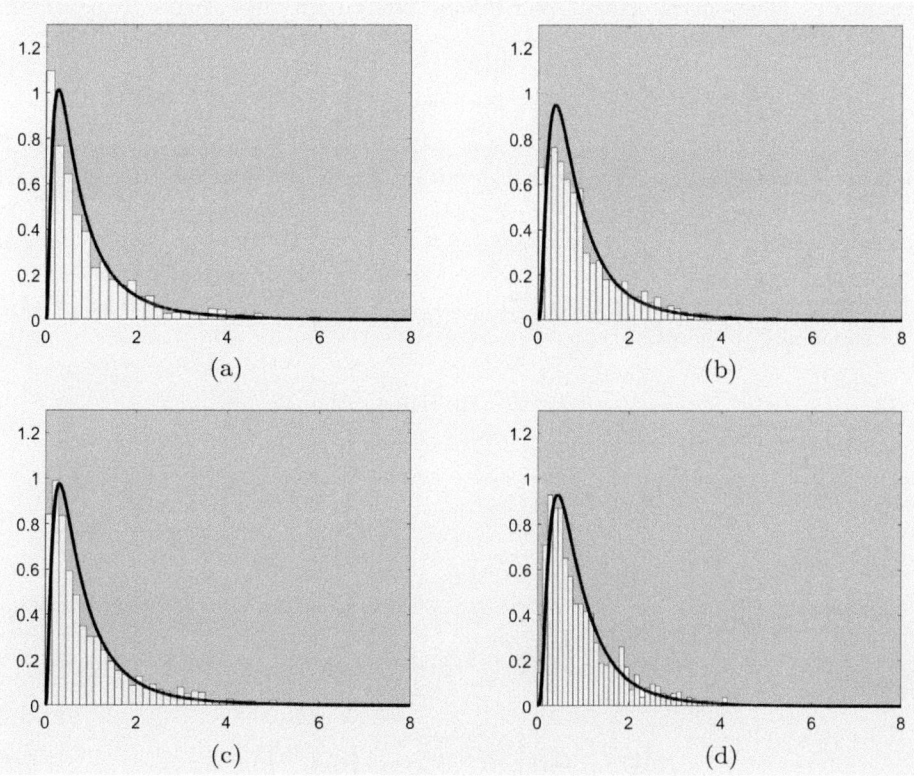

FIGURE 13.8 Histograms of $Xi(t)$ with fitted lognormal densities. (a) Site 1. (b) Site 2. (c) Site 3 and (d) German.

TABLE 13.4 Estimated
Parameters in the Lognormal Model

	$\hat{\Theta}_i$	$\hat{\upsilon}_i$
Site 1	−0.4282	0.7080
Site 2	−0.3215	0.6341
Site 3	−0.3803	0.6837
German	−0.2711	0.5607

13.5 Hedging Wind Power Production

In the following we denote the German wind index at day t by $P_n(t)$. An exchange-traded WPF contract is written on the underlying daily wind index, $P_n(t)$. The payoff of a long position in such a contract is

$$H\left(\bar{P}_n(S,T) - P_n\left(t_0,S,T\right)\right)X, \tag{13.19}$$

where H is the number of hours during the delivery period $[S, T]$, $P_n(t_0, S, T)$ is the WPF price agreed on at time $t_0 < S < T$, X is a specified tick size, and

$$\bar{P}_n(S,T) = \frac{1}{T - S + 1}\sum_{t=S}^{T}P_n(t).$$

From equation (13.19) it is apparent that a short position results in a positive payoff in low wind scenarios according to the short position equivalent to equation (13.19),

$$H\left(P_n\left(t_0, S, T\right) - \bar{P}_n(S, T)\right)X.$$

That is, if the realization of $\bar{P}_n(S, T)$ is lower than $P_n(t_0, S, T)$. This is favourable for a WPP, since this payoff will offset the loss in income from the long position in wind power production.

To be more specific, let C_i denote the capacity of WPP i, and let $P_i(t)$ denote the daily wind index/utilization of WPP i such that $C_i P_i(t)$ is the actual production of power. Further assume that the WPP receives a fixed price Q_i per produced MWh. The long position in wind power production for WPP i doing the period $[S, T]$ is therefore

$$\bar{P}_i(S, T) C_i H Q_i,\tag{13.20}$$

where

$$\bar{P}_i(S, T) = \frac{1}{T - S + 1} \sum_{t=S}^{T} P_i(t).\tag{13.21}$$

Assume that the WPP takes a position $\gamma_i \in \mathbb{Z}$ in WPF contracts with delivery period being $[S, T]$. The payoff from taking this position and the long position in wind power production results in a portfolio with payoff

$$H\bar{P}_i(S, T) C_i Q_i + \gamma_i H\left(\bar{P}_i(S, T) - P_n\left(t_0, S, T\right)\right)X.\tag{13.22}$$

From equation (13.22) it is clear that perfectly hedging the volumetric risk would mean to choose γ_i such that $H\bar{P}_i(S, T) C_i Q_i = -\gamma_i H\bar{P}_n(S, T)X$, resulting in the deterministic payoff $\gamma_i H P_n(t_0, S, T)X$. However, the problem for the WPP is that the stochastic terms $\bar{P}_i(S, T)$ and $\bar{P}_n(S, T)$ are not perfectly dependent, and hence obtaining the deterministic payoff $\gamma_i H P_n(t_0, S, T)X$ is not possible. In fact, as shown in Christensen and Pircalabu (2018), it is far from optimal using the exchange-traded WPF contracts for hedging purposes for a single WPP, depending on the dependence structure between the site-specific wind index and the underlying index of the WPF contract.

13.5.1 Perfect Hedging of Volumetric Risk Using Tailor-Made Wind Power Futures

Tailor-made over-the-counter WPF contracts is a way of perfectly hedging the volumetric risk. Instead of going short the exchange-traded WPF contract, the WPP could instead go short an over-the-counter WPF contract with the underlying being $P_i(t)$ instead of $P_n(t)$. In the following we therefore consider the situation of an energy management company (EMC) acting as counterparty of these tailor-made WPF contracts from $n-1$ different WPPs in Germany. Let $H\left(\bar{P}_i(S, T) - P_i\left(t_0, S, T\right)\right)C_i Q_i$ be the payoff of a long position in a tailor-made WPF contract for WPP i. Thus, from the point of view of the EMC, the payoff of acting as counterparty for $n-1$ different WPPs and taking a position $\gamma \in \mathbb{Z}$ in the exchange-traded WPF contract is

$$R_C(\gamma) = \sum_{i=1}^{n-1} H\left(\bar{P}_i(S, T) - P_i\left(t_0, S, T\right)\right)C_i Q_i$$

$$+ \gamma H\left(\bar{P}_n(S, T) - P_n\left(t_0, S, T\right)\right)X,\tag{13.23}$$

while the payoff from the point of view of the ith WPP is

$$R_{WPP,i} = H\bar{P}_i(S,T)C_iQ_i + H\left(P_i(t_0,S,T) - \bar{P}_i(S,T)\right)C_iQ_i$$
$$= HP_i(t_0,S,T)C_iQ_i$$

We argue that this construction can be beneficial for both the individual WPPs and the EMC: Firstly, the individual WPPs obtain a perfect hedge of their volumetric risk, and secondly, with an appropriate number of WPPs and distribution of the WPPs geographically, the portfolio of tailor-made WPF contracts approximately replicates the exchange-traded WPF contract. The EMC will thereby be able to hedge its volumetric risk by taking an appropriate position in the exchange-traded WPF, as in equation (13.23). A premium has to be paid from the individual WPP to the EMC in order to transfer the WPP's volumetric risk to the EMC. We return to this discussion in section 13.5.2.3. However, the motivation for the EMC to engage in such tailor-made WPF contracts lies in this premium and the size of it compared to the premium in the exchange-traded WPF market. We mention in passing that to offset the potential residual volumetric risk of the portfolio described by equation (13.23), additional instruments could be added to the portfolio. This is outside the scope of present paper, and we leave this as future work.

13.5.2 Minimum Variance Hedge of a Tailor-Made WPF Contracts Portfolio

In this section we discuss a minimum variance hedge of a portfolio consisting of tailor-made WPF contracts for the EMC. That is from equation (13.23) we define the objective to

$$\min_{\gamma} \text{var}\left(R_C(\gamma)\right).$$

The variance is

$$\text{var}\left(R_C(\gamma)\right)$$

$$= \text{var}\left[\sum_{i=1}^{n-1} H\left(\frac{1}{T-S+1}\sum_{t=S}^{T}P_i(t) - P_i(t_0,S,T)\right)C_iQ_i\right.$$

$$\left. + \gamma H\left(\frac{1}{T-S+1}\sum_{t=S}^{T}P_n(t) - P_n(t_0,S,T)\right)X\right]$$

$$= \sum_{i=1}^{n-1}\sum_{j=1}^{n-1}\left(\frac{H}{T-S+1}\right)^2 C_iQ_iC_jQ_j\sum_{t=S}^{T}\sum_{s=S}^{T}\text{cov}\left(P_i(t),P_j(s)\right) \qquad (13.24)$$

$$+ \left(\gamma\frac{H}{T-S+1}X\right)^2\sum_{t=S}^{T}\sum_{s=S}^{T}\text{cov}\left(P_n(t),P_n(s)\right)$$

$$+ 2\sum_{i=1}^{n-1}\left(\frac{H}{T-S+1}\right)^2\gamma XC_iQ_i\sum_{t=S}^{T}\sum_{s=S}^{T}\text{cov}\left(P_n(t),P_i(s)\right).$$

It follows from equation (13.24) that the optimal position of WPF contracts is

$$\gamma = -\frac{\sum_{i=1}^{n-1}C_iQ_i\sum_{t=S}^{T}\sum_{s=S}^{T}\text{cov}\left(P_n(t),P_i(s)\right)}{X\sum_{t=S}^{T}\sum_{s=S}^{T}\text{cov}\left(P_n(t),P_n(s)\right)}. \qquad (13.25)$$

Besides the fact that the dependencies between the stochastic variables impact the size of γ, the size of each wind site measured by C_i and the price paid for each MWh to each wind site measured by Q_i both translate linearly to the size of γ. Therefore, the larger the wind site or the higher the price paid for each MWh, the larger γ will be in absolute terms (all other things being equal).

13.5.2.1 In-Sample Hedging Effectiveness

In the following we consider the case of an EMC that needs to hedge its portfolio of tailor-made WPF from one year ahead to two years ahead. The considered wind sites are the ones depicted in Figure 13.1. We assume that the contract specifications for each site is as shown in Table 13.5. Further, we assume that $X = 100$ EUR. The estimated parameters of the gamma and lognormal model are the ones reported in section 13.4.

In Table 13.6 we present the hedging results for the gamma and lognormal model. We include the case with all three sites and the German WPF in the portfolio, and then three cases where we only include one of the wind sites and the German WPF. In each case, we report the model-implied optimal position of exchange-traded WPF contracts, $\check{\gamma}$, and the variance reduction (in percentage) implied by the model calculated by $\left[\text{var}\left(R_C(0) \right) - \text{var}\left(R_C(\check{\gamma}) \right) \right] / \text{var}\left(R_C(0) \right)$. It is apparent that the portfolio with all three sites outperforms the three other cases, confirming the diversification approach of the EMC discussed in section 13.5.1.

The fact that the difference regarding $\check{\gamma}$ is small indicate that both models could be used interchangeably to determine an appropriate hedge, though the difference in variance reduction will mislead in a risk management context. In other words, if the wind indexes are driven by the gamma (lognormal) model, and one uses the lognormal (gamma) model to determine hedges, the variance reduction implied by the used model is wrong, while the hedging quantity is relatively close to the optimal hedge.

TABLE 13.5 Fictional Contract Specifications for the Sites in Figure 13.1

Site ID	Capacity in MW, C_i	Price in EUR/MWh, Q_i
1	100	30
2	100	30
3	100	30

TABLE 13.6 Optimal Hedging Quantity γ Implied by the Gamma and Lognormal Model for Different Portfolios Consisting of Different Wind Sites, and the Corresponding Variance of the Portfolio Excluding the Exchange–Traded WPF Contract, and the Variance of the Portfolio When the Optimal Hedge is Employed. Additionally, We Show in All Cases the Associated Variance Reduction in Percentage

Case	Sites in portfolio	$\hat{\gamma}$	var(RC (0))	var(RC ($\hat{\gamma}$))	Variance reduction (%)
			Gamma		
1	1,2,3	−63.60	$8.12 \cdot 10^{11}$	$1.31 \cdot 10^{11}$	83.83
2	1	−18.87	$8.55 \cdot 10^{10}$	$2.56 \cdot 10^{10}$	70.12
3	2	−26.79	$1.55 \cdot 10^{11}$	$3.41 \cdot 10^{10}$	77.96
4	3	−17.94	$1.25 \cdot 10^{11}$	$7.07 \cdot 10^{10}$	43.42
			Lognormal		
1	1,2,3	−64.00	$7.06 \cdot 10^{11}$	$1.21 \cdot 10^{11}$	82.89
2	1	−18.97	$8.12 \cdot 10^{10}$	$2.98 \cdot 10^{10}$	63.33
3	2	−25.10	$1.34 \cdot 10^{11}$	$4.42 \cdot 10^{10}$	67.04
4	3	−19.92	$1.13 \cdot 10^{11}$	$5.66 \cdot 10^{10}$	50.05

Comparing equations (13.22) to (13.23), the cases 2, 3, and 4 represent the variance reduction implied by the model if the individual wind sites were to hedge their power production themselves by using the exchange-traded WPF contract. From a social welfare point of view, the sum of variances of case 2, 3, and 4 is approximately 8% larger for both models compared to the variance of case 1. So not only does the model suggest that tailor-made WPF contracts constitute an obvious way of mitigating uncertainty for wind power producers, but also as a way of optimizing the integration of wind power penetration in the electricity grid from a total variance perspective.

13.5.2.2 Out-of-Sample Hedging Effectiveness

In this section we consider the same portfolio of wind sites as in the previous case study, and the specifications of the sites are therefore specified in Table 13.5. However, here we assess the model on out-of-sample observations. We assume that an EMC has bought tailor-made WPF contracts at the three sites for the period from 2 July 2018 to 30 June 2019, corresponding to 364 days or 52 weeks. We employ a weekly minimum variance hedging strategy, meaning that the EMC has a naked position in a portfolio of tailor-made WPF contracts for the entire period with the exception of the front week. To concretize, the first position taken in exchange-traded WPF contracts is the contract with a weekly delivery period from 2 July 2018 to 8 July 2018. The position is taken based on a model that is estimated by using two years of observations ending the last trading day before the delivery period of the weekly exchange-traded WPF contract. With the delivery period starting the 2 July 2018, the last trading day turns out to be 29 June 2018. Then we step one week ahead and determine the appropriate hedge for the week starting 9 July 2018 and ending 15 July 2018, but again only by employing two years of in-sample observations to estimate the model (the estimation period again ends on the last day where one can trade the weekly exchange-traded WPF contract). In this way we end up with 52 hedging quantities, where each quantity is calculated using different estimated parameters of the model due to the moving two-year observation period.

A comment on the model specifications is in place. The stationarity of the models in section 13.3 might seem unreasonable in the present context, given the short period of time between an estimation date and the corresponding start date of delivery of the exchange-traded WPF contract. However, we also implemented the models that take the conditional distribution into account, resulting in similar results. For the sake of keeping the presentation as clear as possible, we have therefore only chosen to present the stationary versions of the models.

The resulting optimal hedge quantities are depicted in Figure 13.9, indicating a seasonal pattern with more exchange-traded WPF contracts needed during spring compared to autumn. Considering equation (13.25), this is the result of the fact that the difference between the sum of the autocovariances of the German wind index and the sum of the autocovariances between the German wind index and the site indexes increases. To assess the hedging effectiveness, we calculate the corresponding implied weekly payoff, $R_C(\check{\gamma})$, for each weekly hedge quantity, $\check{\gamma}$. Since we have a variance minimizing perspective, we force a simplistic view on $P_i(t_0, S, T)$ for all wind indexes. Specifically, we assume that for each i, $P_i(t_0, S, T)$ for all weeks during the out-of-sample period from 2 July 2018 to 30 June 2019 is the mean of $P_i(t)$ over the first estimation period spanning 1 July 2016 to 29 June 2018,

Figure 13.10 shows a histogram of the payoffs of the portfolio of tailor-made WPF contracts and the exchange-traded WPF contract acting as hedging instrument. Compared to Figure 13.10(a), the variances in Figure 13.10(b,c) are clearly reduced. In fact, the variance reduction in percentage of using the exchange-traded WPF contracts as hedging instrument is 93.64% for the gamma model and 93.62% for the lognormal model.

13.5.2.3 Risk Premium of Wind Power Futures

Since tailor-made WPF contracts are, by construction, traded over-the-counter, it is worth to discuss the risk premium of such contracts. As a reference point, we consider the risk premium of the exchange-traded WPF contracts. We define the risk premium as the model implied WPF contract price under

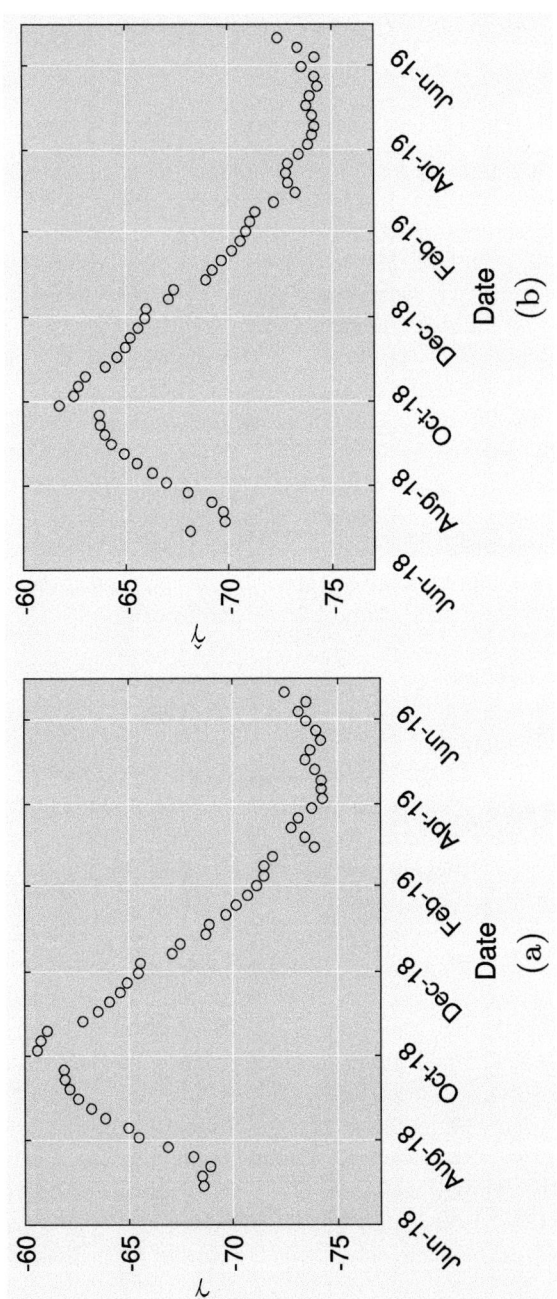

FIGURE 13.9 Variance minimizing hedge quantity, $\hat{\gamma}$, implied by (a) the gamma model and (b) the lognormal model for the 52 weeks covering the period from 2 July 2018 to 30 June 2019. (a) Gamma model and (b) Lognormal model.

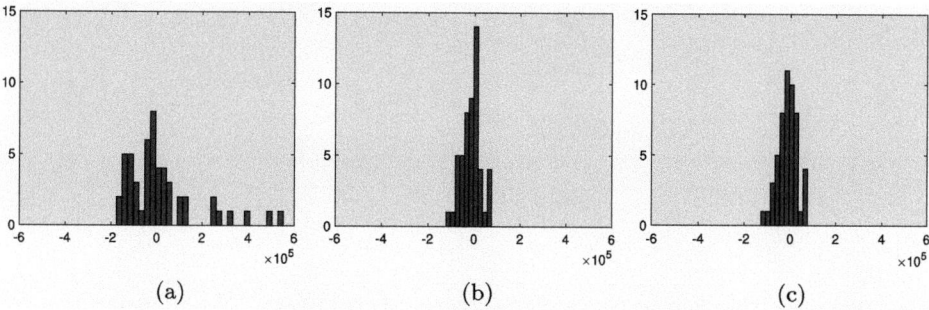

FIGURE 13.10 Histograms of (a) the payoff for EMC by not hedging the portfolio of tailor-made WPF contracts with exchange-traded WPF contracts, and (b)–(c) using the gamma and lognormal model to find the position in exchange-traded WPF contracts used as a hedging instrument for the portfolio of tailor-made WPF contracts. (a) No hedge. (b) With hedge, gamma model and (c) With hedge, lognormal model.

the physical measure subtracted from the observed exchange-traded WPF contract price. The model implied price is defined by $\mathbb{E}\left[\bar{P}_n(S,T)\right]$, meaning that the risk premium $RP(t_0, S, T)$ is

$$RP\left(t_0, S, T\right) = \bar{P}_n\left(t_0, S, T\right) - \mathbb{E}\left[\bar{P}_n(S, T)\right] \tag{13.26}$$

on day t_0 for the delivery period $[S, T]$. The observed quoted exchange-traded WPF prices are obtained from NASDAQ OMX. As concluded in section 13.5.2.2, the stationarity of the models does not imply different results compared to the conditional versions of the models for such long time periods, so to ease the presentation, we only consider the unconditional expected value here.*

We limit ourselves to yearly and quarterly exchange-traded WPF contracts for two reasons. First, it is unlikely that the tailor-made WPF contracts in general will be specified for a short delivery such as a week as a result of such non-standardized instrument. Secondly, as concluded in Benth and Pircalabu (2018), fundamentals impact the information premium of exchange-traded WPF contracts with a short delivery period (e.g. a week) and a short period of time to delivery, which we would like to avoid. Thirdly, to assess the seasonal differences we also consider quarterly contracts.

For the period from 1 July 2016 to 30 June 2019, we show $\mathbb{E}\left[\bar{P}_n(S,T)\right]$ and $\bar{P}_n\left(t_0, S, T\right)$ in Figure 13.11 for the frontyear (that is, for a given date, the front year denotes the nextyear). The quoted prices are fairly constant throughout the entire period, which could be a consequence of illiquidity of exchange-traded WPF contracts.

The risk premium is 0.011 for the gamma model and −0.013 for the lognormal model on average. Since we are considering a yearly WPF contract we can ignore the seasonality and use the empirical mean to assess the risk premium. The empirical risk premium is 0.011, agreeing with the gamma model. This is likely a consequence of the gamma model having a better fit to the distribution of the German wind index as discussed in section 13.4 (see also Table 13.3).

Figure 13.12(a) shows the model implied and quoted prices for the front quarter, and Figure 13.12(b) shows the corresponding risk premium. The mean of the risk premium in this case is −0.014 for the gamma model and −0.016 for the log-normal model. The seasonal variation in the prices peaks for contracts with delivery during Q4 and Q1, simply since more wind is present during these quarters. This is also reflected in the model-implied prices. The peaks in the risk premium are observed for contracts with delivery during Q3 and Q24. One explanation of this could be non-aligned incentives to engage in

---

* Despite the fact that $RP(t_0, S, T)$ still depends on t_0 through $\bar{P}_n(t_0, S, T)$, the assumption of stationarity is to some degree confirmed by the constant pattern of $\bar{P}_n(t_0, S, T)$ observed in Figures 13.11 and 13.12(a).

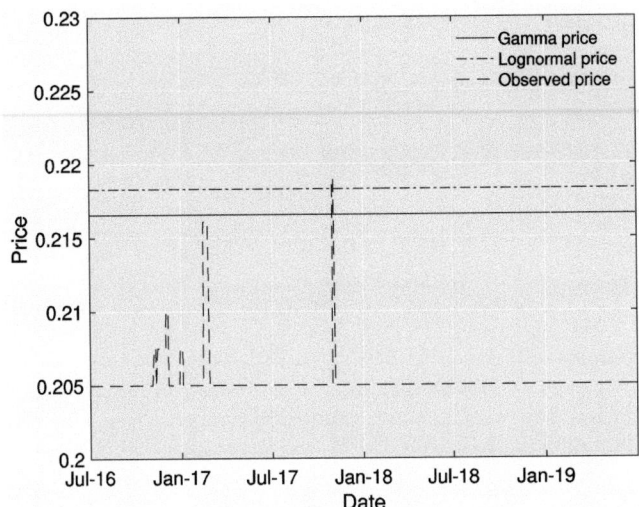

FIGURE 13.11 The model implied and quoted price for the front year for the period from 1 July 2016 to 30 June 2019. Notice that the date refers to the observation date; i.e. the date where the contract is quoted.

the WPF market throughout the year for the buying and selling side. Christensen *et al.* (2019) shows that the hedging benefits are greater for CPPs during Q3 and Q4 compared to Q1 and Q2; hence, during Q3 and Q4, CPPs are more interested in WPF contracts and thus willing to pay more. In particular, this is observed for the contract with delivery during Q3 in 2017, where the risk premium even turns slightly positive for both models. We note that the observed spike for this contract, where the value of the risk premium becomes almost 0.03, corresponds to the trading activity on a single day.

A negative risk premium is in line with the findings in Benth and Pircalabu (2018) and Gersema and Wozabal (2017). One might argue that this is expected from a hedging benefit perspective, since the hedging benefits in general are greater for the selling side than the buying side (see Gersema and Wozabal 2017, Christensen and Pircalabu 2018, Christensen *et al.* 2019). Continuing this argument, the risk premium is likely to be even more negative in the tailor-made WPF market as a result of the perfect hedge implied by the tailor-made WPF contracts for WPPs. However, from the perspective of the individual WPP, this extra risk premium associated with the tailor-made WPF contract compared to the exchange-traded WPF contract has to be weighted against the deterministic payoff implied by the tailor-made WPF contract. Opposite, from the perspective of the EMC, this negative risk premium constitutes the motivation for engaging in the tailor-made WPF market and thereby taking over the volumetric risk.

13.6 Conclusion

In this paper, we propose and compare two multivariate continuous-time models, the gamma and lognormal model, for the joint behaviour of wind indexes. We discuss the properties of the models, and propose estimation procedures. Empirically, we apply the models to a joint model for the wind indexes at three different wind sites in Germany, and the German wind index that represents the overall utilization of wind power production in Germany. We find that both models are able to capture the autocorrelation structure well. However, the gamma model captures the skewness and kurtosis of the German wind index better than the lognormal model.

The models are applied to a variance-minimizing hedging strategy of a portfolio consisting of long positions in so-called tailor-made wind power futures contracts at the three wind sites, and a short

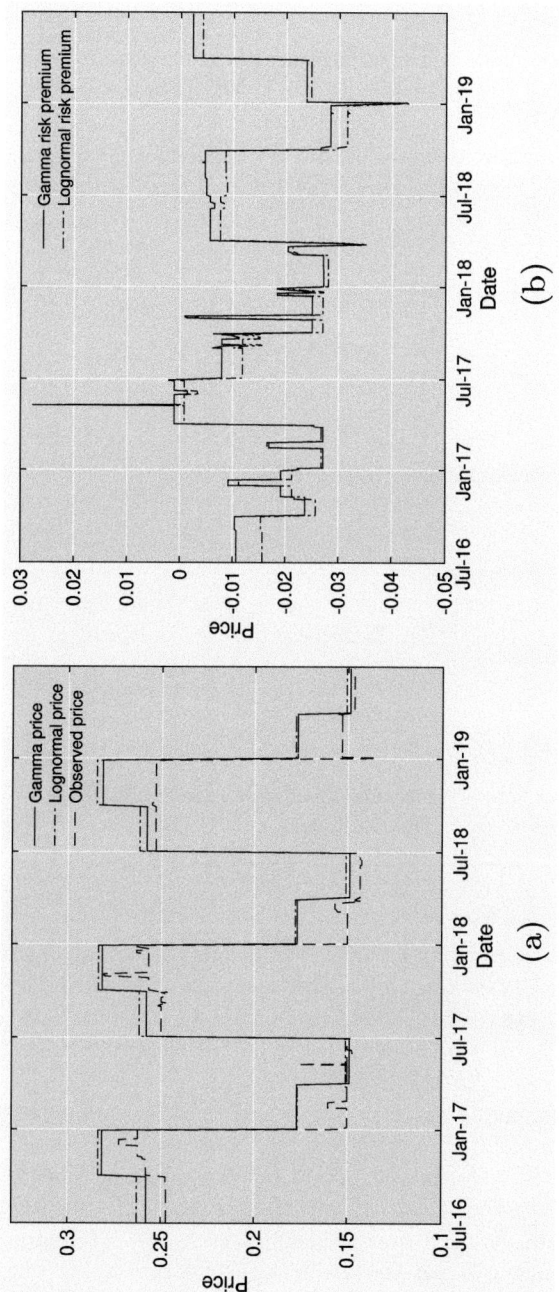

FIGURE 13.12 (a) the model implied and quoted price for the front quarter, and (b) the risk premium for the front quarter. The observations period is from 1 July 2016 to 30 June 2019. Notice that the date refers to the observation date; i.e. the date where the contract is quoted. (a) Model and quoted price and (b) Risk premium.

position in the exchange-traded wind power futures contract. The hedging effectiveness is assessed in both an in-sample and out-of-sample context. Both models indicate that a significant variance reduction can be obtained by hedging the portfolio with the exchange-traded wind power futures contracts in-sample as well as out-of-sample. Further, the hedging benefits are greater for the portfolio of tailor-made wind power futures compared to hedging each individual wind site with exchange-traded wind power futures contracts.

The risk premium of the exchange-traded wind power futures contracts is examined, where we find that the gamma model implies a more reliable estimate of the risk premium. A negative risk premium is observed in line with other findings in the literature for both yearly and quarterly contracts. Even though the tailor-made wind power futures contracts give each wind power producer a perfect volumetric hedge for her wind power production, we argue that it is likely that the risk premium for a tailor-made wind power futures contract is even more negative compared to the exchange-traded contract.

Acknowledgment

We are grateful for the careful reading and comments from two referees.

Disclosure Statement

No potential conflict of interest was reported by the authors.

Funding

Troels Sønderby Christensen is supported by the Innovation Fund Denmark under Grant 5189-00117B. Victor Rohde is supported by the Danish Council for Independent Research under Grant DFF-4002-00003.

References

Asmussen, S., Jensen, J.L. and Rojas-Nandayapa, L., On the Laplace transform of the lognormal distribution. *Methodol. Comput. Appl. Probab.*, 2016, **18**, 441–458.

Ballotta, L. and Bonfiglioli, E., Multivariate asset models using Lévy processes and applications. *Eur. J. Finance*, 2016, **22**(13), 1320–1350.

Barndorff–Nielsen, O.E. and Shephard, N., Non-Gaussian Ornstein-Uhlenbeck-based models and some of their uses in financial economics. J. R. Stat. Soc. Ser. B (Stat. Methodol.), 2001, **63**(2), 167–241.

Benth, F.E. and Benth, J. Š., The volatility of temperature and pricing of weather derivatives. *Quant. Finance*, 2007, 7(5), 553–561.

Benth, F.E. and Pircalabu, A., A non-Gaussian Ornstein-Uhlenbeck model for pricing wind power futures. *Appl. Math. Finance*, 2018, **25**, 36–65.

Benth, F.E. and Rohde, V., On non-negative modeling with CARMA processes. *J. Math. Anal. Appl.*, 2019, **476**(1), 196–214.

Benth, F.E., Kallsen, J. and Meyer-Brandis, T., A non-Gaussian Ornstein-Uhlenbeck process for electricity spot price modeling and derivatives pricing. *Appl. Math. Finance*, 2007, **14**(2), 153–169.

Christensen, T.S. and Pircalabu, A., On the spatial hedging effectiveness of German wind power futures for wind power generators. *J. Energy Markets*, 2018, **11**, 71–96.

Christensen, T.S., Pircalabu, A. and Høg, E., A seasonal copula mixture for hedging the clean spark spread with wind power futures. *Energy Econom.*, 2019, **78**, 64–80.

Gersema, G. and Wozabal, D., An equilibrium pricing model for wind power futures. *Energy Econom.*, 2017, **65**, 64–74.

Halliwell, L.J., *The Lognormal Random Multivariate*, 2015 (Casualty Actuarial Society E-Forum).

Leoni, P. and Schoutens, W., Multivariate smiling. *Wilmott Mag.*, 2008, March.

Meucci, A., Review of statistical arbitrage, cointegration, and multivariate Ornstein-Uhlenbeck, 2009. Available at SSRN: https://ssrn.com/abstract 1404905,

Rajput, B.S. and Rosinski, J., Spectral representations of infinitely divisible processes. *Probab. Theory. Relat. Fields.*, 1989, **82**(3), 451–487.

Sato, K.I., *Lévy Processes and Infinitely Divisible Distributions*, 1999 (Cambridge University Press).

Semeraro, P., A multivariate variance gamma model for financial application. *Inter. J. Theor. Appl. Finance*, 2008, **11**(1), 1–18.

Appendix: Theoretical Results for the Gamma Model

This appendix is dedicated to proving propositions 13.3.2 and 13.3.3. We start by proving proposition 13.3.3, which the lemma below is a first step towards. We will use some standard results about continuous-time moving averages, all of which can be found in Rajput and Rosinski (1989).

The following Lemma is well-known, but we give a proof for the sake of completeness.

LEMMA A.1 *Let $t \geq 0$ and consider the two one-dimensional processes*

$$Y_1(t) = \int_{-\infty}^{t} f_1(t-u)\mathrm{d}Z(u) \quad \text{and} \quad Y_2(t) = \int_{-\infty}^{t} f_2(t-u)\mathrm{d}Z(u) \tag{A.1}$$

for functions f_1 and f_2 in $L^1(\mathbb{R}) \cap L^2(\mathbb{R})$, and where Z is a one-dimensional Lévy process with second moment. Then

$$\mathbb{E}\big[Y_1(0)\big] = \int_0^\infty f_1(u)\mathrm{d}u\,\mathbb{E}[Z(1)]$$

and

$$\mathbb{E}\Big[\big(Y_1(0) - \mathbb{E}\big[Y_1(0)\big]\big)\big(Y_2(t) - \mathbb{E}\big[Y_2(t)\big]\big)\Big]$$

$$= \int_0^\infty f_1(u)f_2(t+u)\mathrm{d}u\,\mathrm{var}(Z(1)).$$

Proof Let $\psi_{Y_1(0),Y_2(t)}$ be the cumulant generating function of $(Y_1(0), Y_2(t))$ and ψ_Z be the cumulant generating function of Z. Then

$$\psi_{Y_1(0),Y_2(t)}(x)$$

$$= \log\mathbb{E}\Big[\exp\big\{x_1 Y_1(0) + x_2 Y_2(t)\big\}\Big]$$

$$= \int_0^t \psi_Z\big(x_2 f_2(u)\big)\mathrm{d}u + \int_0^\infty \psi_Z\big(x_1 f_1(u) + x_2 f_2(t+u)\big)\mathrm{d}u.$$

It follows that for $n_1, n_2 \in \mathbb{N}_0$ with $n_1 + n_2 \leq 2$,

$$\frac{\mathrm{d}^{n_1+n_2}}{\mathrm{d}x_1^{n_1}\mathrm{d}x_2^{n_2}}\psi Y_1,Y_2(x)$$

$$= \int_0^t f_2^{n_2}(u)\psi_Z^{(n_2)}\big(x_2 f_2(u)\big)\mathrm{d}u$$

$$+ \int_0^\infty f_1^{n_1}(u)f_2^{n_2}(u)\psi_Z^{(n_1+n_2)}\big(x_1 f_1(u) + x_2 f_2(u)\big)\mathrm{d}u.$$

where $\psi_Z^{(n_1+n_2)}$ denotes the $n_1 + n_2$ times derivative of ψ_Z. We conclude that

$$\mathbb{E}\big[Y_1(0)\big] = \frac{\mathrm{d}}{\mathrm{d}x_1}\psi_{Y_1(0),Y_2(t)}(0) = \int_0^\infty f_1(u)\mathrm{d}u\,\mathbb{E}[Z(1)].$$

Assume now, without loss of generality, $\mathbb{E}[Z(1)] = 0$. Then

$$\mathbb{E}\Big[\big(Y_1(0) - \mathbb{E}\big[Y_1(0)\big]\big)\big(Y_2(t) - \mathbb{E}\big[Y_2(t)\big]\big)\Big]$$

$$= \frac{\mathrm{d}^2}{\mathrm{d}x_1\,\mathrm{d}x_2}\psi_{Y_1(0),Y_2(t)}(0)$$

$$= \int_0^\infty f_1(u)f_2(t+u)\mathrm{d}u\,\mathrm{var}[Z(1)] \qquad \blacksquare$$

Proof of proposition 13.3.3 Let $\sigma_{i,k}$ denote the (i,k)'th entry of Σ_L. Then, using Lemma A.1,

$$\mathbb{E}\big[X_i(t)\big] = \sum_{k=1}^n \mathbb{E}\bigg[\int_{-\infty}^t e^{-\lambda_i(t-u)}\sigma_{i,k}\mathrm{d}L_k(u)\bigg]$$

$$= \sum_{k=1}^n \frac{1}{\lambda_i}\sigma_{i,k}\beta_k/2$$

$$= \big(\Lambda^{-1}\Sigma_L\beta/2\big)_i.$$

This gives (13.9). Assume now, without loss of generality, $\mathbb{E}[L(1)] = 0$. Then, using Lemma A.1 again,

$$\mathbb{E}\big[X_i(0)X_j(t)\big] = \mathbb{E}\Bigg[\bigg(\sum_{k=1}^n \int_{-\infty}^0 e^{-\lambda_i(-u)}\sigma_{i,k}\mathrm{d}L_k(u)\bigg)$$

$$\times \bigg(\sum_{k=1}^n \int_{-\infty}^t e^{-\lambda_j(t-u)}\sigma_{j,k}\mathrm{d}L_k(u)\bigg)\Bigg]$$

$$= \sum_{k=1}^n \sigma_{i,k}\sigma_{j,k}\int_0^\infty e^{-\lambda_j(t+u)}e^{-\lambda_i u}\mathrm{d}u$$

$$= \frac{e^{-\lambda_j t}}{\lambda_i + \lambda_j}\sum_{k=1}^n \sigma_{i,k}\sigma_{j,k}$$

$$= \Big(\big(\Lambda_{\mathrm{var}}\circ\Sigma_L\Sigma_L^\top\big)e^{-\Lambda t}\Big)_{i,j}$$

from which (13.10) follows. $\blacksquare$

We now turn to prove proposition 13.3.2. Initially, we give the next result which is a special case of Benth and Rohde (2019, Theorem 4.8), but again, we give a proof for the sake of completeness.

PROPOSITION A.2 *Let L be a compound Poisson process with intensity $\alpha > 0$ and exponential jumps with parameter $\beta > 0$. Consider $t \in \mathbb{R}$, $\lambda, \mu > 0$ and $x_1, x_2 \in \mathbb{R}$ with $x_1 + x_2 < \beta$. Furthermore, assume $x_1 x_2 \geq 0$ and $x_1 \neq 10$, and let $f(t) = x_1 e^{-\lambda t} + x_2 e^{-\mu t}$. Then*

$$\log \mathbb{E}\left[\exp\left\{\int_{-\infty}^{t} f(t-u)\mathrm{d}L(u)\right\}\right]$$

$$= \alpha \frac{f(0)}{f'(0)} \log\left(1 - \frac{f(0)}{\beta}\right) \tag{A2}$$

$$+ \alpha \int_{0}^{\infty} \left(\frac{f(u)}{f'(u)}\right)' \log\left(1 - \frac{f(u)}{\beta}\right) \mathrm{d}u,$$

where

$$\left|\left(\frac{f(u)}{f'(u)}\right)'\right| \leq \frac{(\lambda - \mu)^2}{2\lambda\mu} \tag{A3}$$

for all $u \geq 0$.

Proof Initially, note that f/f' is bounded and

$$\left|\left(\frac{f(u)}{f'(u)}\right)'\right| = \frac{\left|f'(u)^2 - f(u)f''(u)\right|}{f'(u)^2}$$

$$= \frac{x_1 x_2 (\lambda - \mu)^2 e^{-(\lambda+\mu)u}}{x_1^2 \lambda^2 e^{-2\lambda u} + x_2^2 \mu^2 e^{-2\mu u} + 2x_1 x_2 \lambda\mu e^{-(\lambda+\mu)u}}$$

$$\leq \frac{(\lambda - \mu)^2}{2\lambda\mu}.$$

This gives the bound on $(f/f')'$. Additionally, we find that

$$\left|\left(\frac{f(u)}{f'(u)}\right)'\right| = \frac{x_1 x_2 (\lambda - \mu)^2 e^{-(\lambda+\mu)u}}{x_1^2 \lambda^2 e^{-2\lambda u} + x_2^2 \mu^2 e^{-2\mu u} + 2x_1 x_2 \lambda\mu e^{-(\lambda+\mu)u}}$$

$$= \frac{x_1 x_2 (\lambda - \mu)^2}{x_1^2 \lambda^2 e^{-(\lambda-\mu)u} + x_2^2 \mu^2 e^{-(\mu-\lambda)u} + 2x_1 x_2 \lambda\mu}.$$

We conclude that $(f(u)/f'(u))' = O(e^{-|\lambda-\mu|u})$ as $u \to \infty$. Thus, all integrals below are convergent and the integration by parts is justified. Next let

$$\psi(u) = \log \mathbb{E}[\exp(uL(1))] = \alpha \frac{u}{\beta - u}$$

be the cumulant-generating function of $L(1)$ and let $\phi(u) = -\alpha \log(1 - u/\beta)$ be the cumulant-generating function of a gamma distribution with shape α and rate β (see for example Benth and Pircalabu 2018). Note that $\psi(u) = u\phi'(u)$. Then, using integration by parts,

$$\log \mathbb{E}\left[\exp\left\{\int_{-\infty}^{t} f(t-u)\mathrm{d}L(u)\right\}\right]$$

$$= \int_{0}^{\infty} \psi(f(u))\mathrm{d}u$$

$$= \int_{0}^{\infty} \frac{f(u)}{f'(u)}(\phi(f(u)))'\mathrm{d}u$$

$$= -\frac{f(0)}{f'(0)}\phi(f(0)) - \int_{0}^{\infty}\left(\frac{f(u)}{f'(u)}\right)'\phi(f(u))\mathrm{d}u. \qquad \blacksquare$$

REMARK A.3 Considering the proof of proposition A.2 there are two approaches to calculate

$$\log \mathbb{E}\left[\exp\left\{\int_{-\infty}^{t} f(t-u)\mathrm{d}L(u)\right\}\right]. \tag{A4}$$

Either by calculating

$$\int_{0}^{\infty} \psi(f(u))\mathrm{d}u \tag{A5}$$

or

$$-\frac{f(0)}{f'(0)}\phi(f(0)) - \int_{0}^{\infty}\left(\frac{f(u)}{f'(u)}\right)'\phi(f(u))\mathrm{d}u. \tag{A6}$$

Here, ψ and ϕ are the cumulant-generating function of $L(1)$ and of a gamma distribution with shape α and rate β as defined in the proof of proposition A.2. By (A3), the integral in (A6) will be small whenever $(\lambda - \mu)^2/(2\lambda\mu)$ is small. In the application we consider we are concerned with the case where $\lambda = \hat{\lambda}_i$ and $\mu = \hat{\lambda}_j$ for some $i, j = 1, 2, 3, 4$, where $\hat{\lambda}_i$ and $\hat{\lambda}_j$ are given in Table 13.2. We have

$$\max_{i,j} \frac{\left(\hat{\lambda}_i - \hat{\lambda}_j\right)^2}{2\hat{\lambda}_i\hat{\lambda}_j} = 0.0506,$$

and therefore, indeed, that $(\lambda - \mu)^2/(2\lambda\mu)$ is small in the case relevant to us. The integral in (A6) has ϕ in the kernel whereas (A5) has ψ, making a direct comparison more difficult. We do, however, have

$$\phi(u) = \alpha u + O\left(u^2\right) \text{ and } \psi(u) = \alpha u + O\left(u^2\right) \text{ as } u \to 0$$

(by a Taylor approximation argument), indicating that ϕ and ψ are of comparable size, at least for small values. Furthermore, by numerical comparison, we have found them to be of similar size. We conclude that the kernel of (A6) is expected to be considerably smaller than the kernel of (A5). We therefore prefer to do the calculation in (A6) instead of (A5) since we can do a much more coarse approximation for a desired precision of an approximation of (A4).

PROPOSITION A.4 *Let L be a compound Poisson process with intensity $\alpha > 0$ and exponential jumps with parameter $\beta > 0$. Consider $s < t$, $\lambda > 0$ and $x < \beta$. Then*

$$\mathbb{E}\left[\exp\left\{x \int_s^t e^{-\lambda(t-u)} dL(u)\right\}\right] = \left(\frac{\beta - xe^{-\lambda(t-s)}}{\beta - x}\right)^{\alpha/\lambda}$$

and

$$\mathbb{E}\left[\exp\left\{x \int_{-\infty}^t e^{-\lambda(t-u)} dL(u)\right\}\right] = \left(\frac{\beta}{\beta - x}\right)^{\alpha/\lambda} \tag{A7}$$

Proof Let

$$\psi(t) = \log \mathbb{E}[\exp(tL(1))] = \alpha \frac{t}{\beta - t}$$

be the cumulant-generating function of L. Then

$$\log \mathbb{E}\left[\exp\left\{\int_s^t f(t-u) dL(u)\right\}\right]$$

$$= \int_0^{t-s} \psi\left(e^{-\lambda u}\right) du$$

$$= \frac{\alpha}{\lambda}\left(\log\left(\beta - xe^{-\lambda(t-s)}\right) - \log(\beta - x)\right).$$

A similar calculation gives (A7). $\blacksquare$

Proof of theorem 3.2 For notional convenience, let

$$\tilde{L}(t) = \left(L_1(t), \ldots, L_{n-1}(t), 0, L_n(t)\right)^\top \in \mathbb{R}^{n+1}.$$

First consider (13.7) and assume $i \neq j$. We have

$$X_i(t) = \int_{-\infty}^t e^{-\lambda_i(t-u)} \tilde{\sigma}_{i,i} d\tilde{L}_i(u) + \int_s^t e^{-\lambda_i(t-u)} \tilde{\sigma}_{i,n+1} d\tilde{L}_{n+1}(u)$$

$$+ \int_{-\infty}^s e^{-\lambda_i(t-u)} \tilde{\sigma}_{i,n+1} d\tilde{L}_{n+1}(u)$$

and

$$X_j(s) = \int_{-\infty}^s e^{-\lambda_j(s-u)} \tilde{\sigma}_{j,j} d\tilde{L}_j(u) + \int_{-\infty}^s e^{-\lambda_j(s-u)} \tilde{\sigma}_{j,n+1} d\tilde{L}_{n+1}(u).$$

Next, note that $\text{cov}(UV, W) = \text{cov}(V, UW) = \mathbb{E}[U]\text{cov}(V, W)$ for a random variable U independent of the random variables V and W.

Applying this, and the above, we conclude that

$$\text{cov}\Big(P_i(t), P_j(s)\Big)$$

$$= \text{cov}\Big(e^{-S_i(t)X_i(t)}, e^{-S_j(s)X_j(s)}\Big)$$

$$= \mathbb{E}\left[\exp\left\{-S_i(t)\int_{-\infty}^{t} e^{-\lambda_i(t-u)}\tilde{\sigma}_{i,i}\mathrm{d}\tilde{L}_i(u)\right\}\right]$$

$$\times \mathbb{E}\left[\exp\left\{-S_i(t)\int_{s}^{t} e^{-\lambda_i(t-u)}\tilde{\sigma}_{i,n+1}\mathrm{d}\tilde{L}_{n+1}(u)\right\}\right] \qquad\text{(A8)}$$

$$\times \mathbb{E}\left[\exp\left\{-S_j(s)\int_{-\infty}^{s} e^{-\lambda_j(t-u)}\tilde{\sigma}_{j,j}\mathrm{d}\tilde{L}_j(u)\right\}\right]$$

$$\times \text{cov}\Bigg(\exp\left\{-S_i(t)\int_{-\infty}^{s} e^{-\lambda_i(t-u)}\tilde{\sigma}_{i,n+1}\mathrm{d}\tilde{L}_i(u)\right\},$$

$$\exp\left\{-S_j(s)\int_{-\infty}^{s} e^{-\lambda_j(s-u)}\sigma_{j,n+1}\mathrm{d}\tilde{L}_{n+1}(u)\right\}\Bigg)$$

Expressions of the three expectations in equation (A8) are given in proposition A.4. Furthermore,

$$\text{cov}\Bigg(\exp\left\{-S_i(t)\int_{-\infty}^{s} e^{-\lambda_i(t-u)}\tilde{\sigma}_{i,n+1}\mathrm{d}\tilde{L}_i(u)\right\},$$

$$\exp\left\{-S_j(s)\int_{-\infty}^{s} e^{-\lambda_j(s-u)}\tilde{\sigma}_{j,n+1}\mathrm{d}\tilde{L}_{n+1}(u)\right\}\Bigg)$$

$$= \mathbb{E}\left[\exp\left\{-\int_{-\infty}^{s} f_{i,j}(s-u)\mathrm{d}\tilde{L}_{n+1}(u)\right\}\right]$$

$$- \mathbb{E}\left[\exp\left\{-S_i(t)\int_{-\infty}^{s} e^{-\lambda_i(t-u)}\tilde{\sigma}_{i,n+1}\mathrm{d}\tilde{L}_{n+1}(u)\right\}\right]$$

$$\times \mathbb{E}\left[\exp\left\{-S_j(s)\int_{-\infty}^{s} e^{-\lambda_j(s-u)}\tilde{\sigma}_{j,n+1}\mathrm{d}\tilde{L}_{n+1}(u)\right\}\right]$$

for which expressions are given in propositions A.2 and A.4.

Next, consider (13.8). We write

$$X_i(t) = \int_{s}^{t} e^{-\lambda_i(t-u)}\Big(\tilde{\sigma}_{i,i}\mathrm{d}\tilde{L}_i(u) + \tilde{\sigma}_{i,n+1}\mathrm{d}\tilde{L}_{n+1}(u)\Big)$$

$$+ \int_{-\infty}^{s} e^{-\lambda_i(t-u)}\Big(\tilde{\sigma}_{i,i}\mathrm{d}\tilde{L}_i(u) + \tilde{\sigma}_{i,n+1}\mathrm{d}\tilde{L}_{n+1}(u)\Big)$$

and

$$X_i(s) = \int_{-\infty}^{s} e^{-\lambda_i(s-u)}\Big(\tilde{\sigma}_{i,i}\mathrm{d}\tilde{L}_i(u) + \tilde{\sigma}_{i,n+1}\mathrm{d}\tilde{L}_{n+1}(u)\Big).$$

Consequently,

$$
\mathrm{cov}\big(P_i(t),P_i(s)\big)
$$

$$
= \mathrm{cov}\Big(e^{-S_i(t)X_i(t)},e^{-S_i(s)X_i(s)}\Big)
$$

$$
= \mathbb{E}\!\left[\exp\!\left\{-S_i(t)\int_s^t e^{-\lambda_i(t-u)}\Big(\tilde{\sigma}_{i,i}\mathrm{d}\tilde{L}_i(u)+\tilde{\sigma}_{i,n+1}\mathrm{d}\tilde{L}_{n+1}(u)\Big)\right\}\right]
$$

$$
\times \mathrm{cov}\bigg(\exp\!\left\{-S_i(t)\int_{-\infty}^s e^{-\lambda_i(t-u)}\Big(\tilde{\sigma}_{i,i}\mathrm{d}\tilde{L}_i(u)+\tilde{\sigma}_{i,n+1}\mathrm{d}\tilde{L}_{n+1}(u)\Big)\right\},
$$

$$
\exp\!\left\{-S_i(s)\int_{-\infty}^s e^{-\lambda_i(s-u)}\Big(\tilde{\sigma}_{i,i}\mathrm{d}\tilde{L}_j(u)+\tilde{\sigma}_{i,n+1}\mathrm{d}\tilde{L}_{n+1}(u)\Big)\right\}\bigg)
$$

Again, expressions for the expectation in (A8) can be found using proposition A.4. Finally,[****]

$$
\mathrm{cov}\bigg(\exp\!\left\{-S_i(t)\int_{-\infty}^s e^{-\lambda_i(t-u)}\Big(\tilde{\sigma}_{i,i}\mathrm{d}\tilde{L}_i(u)+\tilde{\sigma}_{i,n+1}\mathrm{d}\tilde{L}_{n+1}(u)\Big)\right\},
$$

$$
\exp\!\left\{-S_i(s)\int_{-\infty}^s e^{-\lambda_i(s-u)}\Big(\tilde{\sigma}_{i,i}\mathrm{d}\tilde{L}_i(u)+\tilde{\sigma}_{i,n+1}\mathrm{d}\tilde{L}_{n+1}(u)\Big)\right\}\bigg)
$$

$$
= \mathbb{E}\!\left[\exp\!\left\{-\int_{-\infty}^s \tilde{\sigma}_{i,i}\Big(S_i(t)e^{-\lambda_i(t-s)}+S_i(s)\Big)e^{-\lambda_i(s-u)}\mathrm{d}\tilde{L}_i(u)\right\}\right]
$$

$$
\times \mathbb{E}\!\left[\exp\!\left\{-\int_{-\infty}^s \tilde{\sigma}_{i,n+1}\Big(S_i(t)e^{-\lambda_i(t-s)}+S_i(s)\Big)e^{-\lambda_i(s-u)}\mathrm{d}\tilde{L}_{n+1}(u)\right\}\right]
$$

$$
- \mathbb{E}\!\left[\exp\!\left\{-\int_{-\infty}^s \tilde{\sigma}_{i,i}S_i(t)e^{-\lambda_i(t-u)}\mathrm{d}\tilde{L}_i(u)\right\}\right]
$$

$$
\times \mathbb{E}\!\left[\exp\!\left\{-\int_{-\infty}^s \tilde{\sigma}_{i,n+1}S_i(t)e^{-\lambda_i(t-u)}\mathrm{d}\tilde{L}_{n+1}(u)\right\}\right]
$$

$$
\times \mathbb{E}\!\left[\exp\!\left\{-\int_{-\infty}^s \tilde{\sigma}_{i,i}S_i(s)e^{-\lambda_i(s-u)}\mathrm{d}\tilde{L}_i(u)\right\}\right]
$$

$$
\times \mathbb{E}\!\left[\exp\!\left\{-\int_{-\infty}^s \tilde{\sigma}_{i,n+1}S_i(s)e^{-\lambda_i(s-u)}\mathrm{d}\tilde{L}_{n+1}(u)\right\}\right]
$$

whose expressions are given in proposition A.4. ∎

3

Commodity Prices and Financial Markets

14

Short-Horizon Return Predictability and Oil Prices

Jaime Casassus and
Freddy Higuera

14.1 Introduction

The predictability of stock returns is a controversial topic. Until recently, the prevailing view was that returns could be predicted by the business cycle at long horizons (Cochrane 2005) and that this evidence was significantly stronger than for short horizons. But in recent years, several studies have questioned the existence of such return predictability. For example, Boudoukh *et al.* (2008) show that the dominant findings in this literature are solely the consequence of the high persistence of the predictors. Also recently, Welch and Goyal (2008), who compare the out-of-sample predictive performance of a large number of popular predictors with the prevailing average excess stock return, find that none of these variables predicts stock returns at short horizons better than the historical average return. The current challenge is to propose a variable that predicts equity returns at short horizons and is robust against the new tests suggested in the literature.*

    This paper shows that unexpected changes in oil prices are a significant predictor of excess stock market returns at short horizons. We measure unexpected oil price shocks by short-term futures returns on crude oil contracts. Our sample period is 1983Q2–2009Q4 and is restricted by the existence of crude oil futures prices. Based on in-sample tests and on the macroeconomic literature (i.e. Hamilton and Herrera (2004)), we use four lags of this variable for the predictive regressions. Our predictive variable has deep

---

* In the words of Welch and Goyal (2008) the challenge is still open: ". . . our article suggests only that the profession has yet to find some variable that has meaningful and robust empirical equity premium forecasting power . . .".

DOI: 10.1201/9781003265399-17

macroeconomic roots and allows us to connect the short-horizon predictability of equity returns with the business cycle. Indeed, the existence of negative Granger-causality of oil price shocks on both equity returns and production growth in a trivariate VAR confirms that oil price shocks are a leading variable and are countercyclical. This evidence suggests that increases in oil prices precede recessions and declines in excess stock returns.

We find that, at horizons of one to three quarters, the oil price shocks exhibit predictive performance which is both statistically and economically significant. Obtaining significant results in a sample period after the oil crisis in the 1970s is not an easy task since most variables lose their forecasting power within this period (Welch and Goyal 2008). In terms of global performance, our variable is better than all other variables considered (consumption–wealth ratio, price–dividend ratio, output gap and risk-free rate) with an $\overline{R}^2$ of 6%. This meaningful in-sample result was also detected in out-of-sample tests. Our variable exhibited the best out-of-sample R^2, which was close to 1.2% at one quarter. For longer time horizons, however, no variable showed a significant predictive performance.

Furthermore, oil price shocks have other virtues such as no persistence (they do not produce the pattern reported by Boudoukh *et al.* (2008)), they are directly observable by futures changes (unlike variables such as consumption–wealth ratio and product gap, which must be estimated), they have no correlation with the predictive regression's disturbances (do not generate the bias analysed by Stambaugh (1999)) and they are a high-frequency variable available at no cost. All of these characteristics are valued in the practice of portfolio management. To our knowledge, these results position oil price shocks as the best short-term forecasting variable today.

Our study is motivated by two lines of research: the relationship between GDP and oil prices and the linkage between equity returns and oil price shocks. Because stock return predictability has been detected at business-cycle frequencies, the first strand of literature justifies the relationship between oil prices and the macroeconomy. Intuitively, there are good reasons for believing that oil price leads the business cycle, as nine of ten recessions in the United States since World War II have been preceded by an increase in oil prices (Hamilton 2008). This has not gone unnoticed by economists and has generated a substantial amount of research, particularly given the fact that oil expenditure represents only 4% of GDP. This literature motivates the main question we address in this paper: Given that oil price shocks precede changes in GDP, do they also have some predictive power for stock market returns? The second line of related research, although less voluminous than the first, provides empirical evidence that past oil shocks have an impact on future equity returns. The following section studies these topics in more detail.

Our paper is also related to recent studies of short-term predictability. For example, Ang and Bekaert (2007) report evidence of predictability, at horizons up to one year, using both the dividend–price ratio and the interest rate. Campbell and Thompson (2008), using a much longer sample, find that the out-of-sample forecasting power of several variables improves significantly when certain restrictions are imposed, and that trading on these predictors can lead to significant welfare benefits when compared with trading on the historical average return. More recently, Cooper and Priestley (2009) propose the output gap as a new forecasting variable for stock market returns. This variable is robust against the tests of Boudoukh *et al.* (2008) and Welch and Goyal (2008). Finally, Bakshi *et al.* (2011) find that the Baltic Dry Index, a shipping activity variable that is tied to the business cycle, has predictive ability for a range of stock markets.

While our purpose is to study the effect of oil on the stock market, there is an emerging literature on the financialization of commodities that could extend the short-term predictability of oil to other commodities. Indeed, the emergence of commodities as an asset class has increased the speculative trading in futures markets, especially for those commodities belonging to futures indices, such as the Goldman Sachs Commodity Index and the S&P Commodity Index. Tang and Xiong (2012) find that these indexed commodities have become increasingly correlated with oil since 2004 which makes us believe that these assets would also have some predictive power on stock returns.* Unfortunately, the sample period start-

---

* See Masters (2008), Buyuksahin and Robe (2010), Etula (2010), Acharya *et al.* (2011), Basak and Pavlova (2012) and Singleton (2012) for more studies concerning the financialization of commodities and the effect of speculators on commodity prices.

ing from 2004 is not long enough to do a formal study of predictability, so we leave the answer to this question for future research.

The rest of the paper is structured as follows. The next section reviews the relation between oil prices and both the business cycle and excess market returns, and also presents our variable. Section 14.3 contains the results of in-sample predictability at a quarterly horizon. Section 14.4 reports the out-of-sample tests. Section 14.5 discusses the evidence of predictability at longer horizons. Section 14.6 analyses the impact of predictability in the cross section of expected returns. Section 14.7 concludes.

14.2 Oil Price, the Business Cycle and Stock Returns

This section presents the mechanisms that relate the oil prices to the business cycle and excess stock market returns. It also presents our variable.

14.2.1 Oil Price and the Macroeconomy

The study of the relationship between oil prices and the macroeconomy begins with the seminal work of Hamilton (1983). He uses Sims's (1980) six-variable VAR and bivariate VARs with quarterly data for the 1948–1980 period to show that oil price shocks strongly Granger-caused the U.S. GNP growth rate and the unemployment rate. He finds that an increase in the oil price is followed by four successive quarters of lower GNP growth rates. As shown by Ferderer (1996), the common transmission channels of oil price shocks to the real economy are: inflation, terms of trade (Huntington 2007), the capital utilization rate (Finn 2000) and increasing returns to scale in production (Aguiar-Conraria and Wen 2007).

Subsequent studies (Mork 1989, Lee *et al.* 1995, Hooker 1996, Hamilton 2008) note a weakening of this relationship when data from 1980 onwards are included (which coincides with OPEC's loss of control of the oil market).* This turned attention towards nonlinear relationships between the variables (Mork 1989, Ferderer 1996, Hamilton 1996, 2003). This evidence required new explanations to understand the asymmetric impact of oil on the economy. The most commonly cited asymmetric mechanisms are: monetary policy (Ferderer 1996, Bernanke *et al.* 1997, Balke *et al.* 2002, Hamilton and Herrera 2004, Leduc and Sill 2004), imperfect intersectoral mobility of factors (Lilien 1982, Hamilton 1988, Davis and Haltiwanger 2001, Lee and Ni 2002), investment irreversibility (Bernanke 1983), wage rigidities (Lee *et al.* 1995), and interest rates (Balke *et al.* 2002). Also, Carruth *et al.* (1998) document a strong and significant relationship between the U.S. rate of unemployment and oil prices.

Recently, Kliesen (2008) added to the standard regression the variable CFNAI (Chicago Fed National Activity Index), which is the first principal component of 85 monthly indicators of real economic activity, and finds that oil has a significant impact on the U.S. macroeconomic performance. In addition, Cologni and Manera (2009) find a negative influence of oil shocks on GDP growth, although they reject the hypothesis that real GDP growth has no effect on oil prices. Kilian (2008) finds empirical evidence that exogenous oil supply shocks have caused significant impacts on the GDP of G-7 countries.† Cologni and Manera (2008) observe that oil price shocks affect only the GDP in Italy and in the U.S., albeit temporarily. For almost all of the countries in their sample, oil shocks affect inflation and nominal exchange rates. Gronwald (2008) concludes that only oil shocks that exceed a certain threshold affect the real sector of the economy, while 'normal' positive shocks generate significant nominal impacts. On the contrary, Barsky and Kilian (2004) find inaccuracies in the explanations that support most theories and conclude that 'disturbances in the oil market are likely to matter less for U.S. macroeconomic performance than has commonly been thought'.

The above empirical evidence, although not free of debate, largely supports the existence of a significant relationship between oil price shocks and the business cycle in both economic and statistical terms.

---

* For example, Hamilton (2008) finds that while in the 1949–1980 period an increase of 10% in the oil price predicts that the GDP growth would be 2.9% slower four quarters later, this figure for the 1949–2005 period is only 0.7%.

† This somehow contradicts Kilian (2009), who finds that the impacts of oil demand shocks are more significant than supply shocks.

14.2.2 Oil Price and the Financial Market

Contrary to what has occurred with the relationship between oil shocks and the macroeconomy, the linkage between oil and the financial markets has received little attention. Moreover, this reduced literature has not been conclusive about this relationship. For a deeper analysis, Table 14.1 presents detailed information on the empirical studies reviewed in this section. The table also shows the multiple variables proxying for oil shocks that have been used in the literature.

At a country level, Jones and Kaul (1996) find that oil price shocks produce significant changes in stock returns. The relation can be explained by changes in cash flows and discount rates for the United States and Canada, however they find an overreaction in the United Kingdom and Japan. Driesprong *et al.* (2008) find that oil price shocks have significant predictive power in developed economies, but not for emerging countries. They find an initial underestimation by agents of the impacts of the shock that is slowly corrected later. Park and Ratti (2008) find evidence that oil price shocks have a significant negative impact on real returns of several net importing countries, unlike what occurs in Norway, a net exporter, where the impact is positive.

Following Kilian (2009), Kilian and Park (2009) break down oil shocks into three classes: supply, aggregate demand and specific demand (or precautionary demand) shocks. According to their results, these shocks explain 6%, 5% and 11% of the long-run variation of real stock returns, respectively. They do not find a significant response of stock returns to oil supply shocks, however they do find a positive response to global demand shocks and a negative response to precautionary demand shocks. Sector-level evidence suggests that the mechanism that transmits oil shocks to stock returns is through the demand for industrial products, and not, as widely believed, through the production costs of the firms. Apergis and Miller (2009) criticize the methodology used by Kilian (2009) and Kilian and Park (2009) due to the use of both stationary and non-stationary variables in their VAR specification. They differentiate the $I(1)$ variables and carry out the same breakdown as Kilian (2009), but using only $I(0)$ variables. Instead of including the equity returns in the same VAR, as in Kilian and Park (2009), Apergis and Miller (2009) use a second VAR with the three types of shocks and the stock market returns of each country. Using a sample composed of the G-7 Group and Australia, they conclude that the effects of oil shocks, although statistically significant, produce a minor impact on stock returns.

Some papers report a significant intratemporal correlation between oil shocks and portfolio returns. Nandha and Faff (2008) calculate a two-factor model that includes unexpected returns on the world market and the oil market. They find a significant negative impact of oil that is largely symmetrical on all sectors except for mining and oil/gas. On the other hand, Bachmeier (2008) regresses portfolio returns on shocks that are contemporaneous with the oil price. He finds that oil shocks have a significant negative impact on returns.

Finally, Huang *et al.* (1996) find that daily oil futures returns have no correlation with stock returns, except for returns on oil companies. Ciner (2001) tests for nonlinear Granger-causality of oil futures returns on stock market returns and finds a significant relationship. Using monthly data, Sadorsky (1999) finds that, in the U.S. market, the price of oil affects the financial market, but that the effect in the other direction is insignificant. He also finds that oil price shocks have an asymmetric effect on industrial production and real stock returns, with positive shocks having a greater impact than negative ones.

On the theoretical side, Wei (2003) builds a general equilibrium model to estimate the impact of an oil price shock on the value of a firm that faces investment irreversibility. His model predicts that an oil shock will have only a small impact. Consequently, he is unable to explain the massive decline in the stock market in 1974 after the oil shock of 1973.

14.2.3 Stock Returns and the Business Cycle

This section shows that stock market returns vary considerably with the business cycle. This suggest that a variable that anticipates the cycle, such as oil price shocks, may also do a good job at predicting stock returns.

TABLE 14.1 Literature on Financial Markets and Oil Prices. The Table Provides Information on Studies on Financial Markets and Oil Prices. The Variables Used Are: Δf, the Log Returns on Oil Futures Prices; Δs, the Log Returns on Nominal Spot Prices; Δs_r, the Log Returns on Real Spot Prices; Sop, the Oil Prices Scaled by Volatility; Unexpected Changes in Oil Prices (Lee *et al.* 1995); *Nopi*, Net Oil Price Increases (Hamilton 1996, 2003); *Vol*, Rolling Volatility of Oil Price Changes; $x^+ = \max(0, x)$; $x^- \min(0, x)$.

Paper	Subject	Frequency	Data Stock Market	Data Oil Series	Data Oil Variable
Huang *et al.* (1996)	Joint dynamics	Daily	S&P 500, industrial portfolios, oil companies	Heating and crude oil futures	Δf
Jones and Kaul (1996)	Market efficiency	Quarterly	Real country indexes	PPI oil and related products	Δs
Sadorsky (1999)	Joint dynamics	Monthly	S&P 500/CPI	PPI fuels/CPI	$\Delta s, \Delta s_r^+, \Delta s_r^-, Sop, Sop^+, Sop^-$
Ciner (2001)	Joint dynamics	Daily	S&P 500	Heating and crude oil futures	Δf
Driesprong *et al.* (2008)	Predictability	Monthly	Country indexes, sector indexes	Brent, WTI, Dubai, Arab Light, Brent futures, oil futures	$\Delta s, \Delta f$
Park and Ratti (2008)	Joint dynamics	Monthly	Real country indexes	Brent/PPI all commodities	$\Delta s_r, Sop, Nopi, \Delta s_r^+, \Delta s_r^-, Sop^+, Sop^-, Vol$
Kilian and Park (2009)	Oil demand and supply shocks	Monthly	Real CRSP value weighted, industry portfolios	EIA refiner acquisition cost/CPI	Δs_r
Apergis and Miller (2009)	Oil demand and supply shocks	Monthly	Real country indexes	EIA refiner acquisition cost/CPI	Δs_r

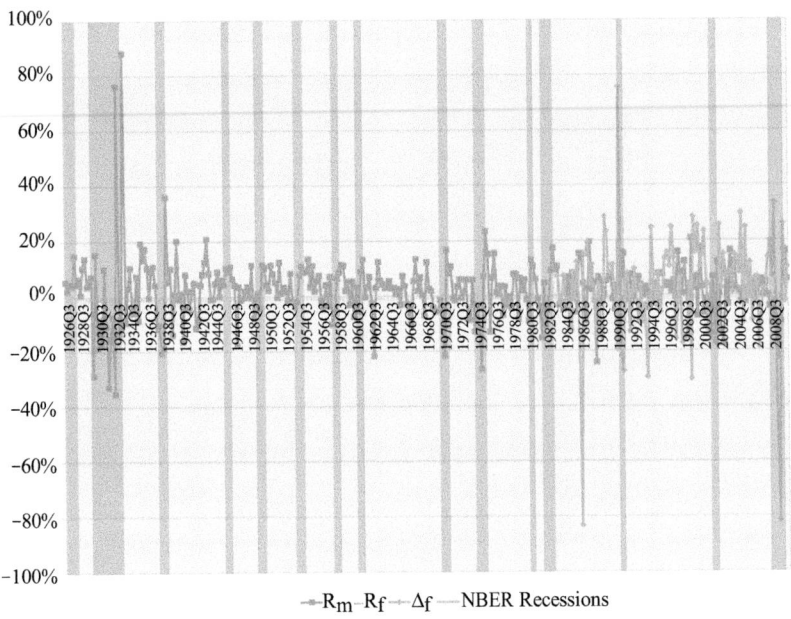

$-\!\!-R_m\!-\!R_f\!\!-\!\!-\!\Delta_f\!-\!\!-$ NBER Recessions

FIGURE 14.1   Excess stock market returns, oil shocks and the business cycle, 1926Q3–2009Q4.

TABLE 14.2   Stock Market Returns and the Business Cycle. Conditional Excess Stock Returns for the Sample Period 1926Q3 to 2009Q4. Classification of States Based on NBER Business Cycle Dates

State	Frequency	Average Excess Return	Standard Deviation of Excess Returns
Expansion	252	2.9	9.1
Peak	15	−5.5	8.1
Recession	52	−3.0	17.3
Trough	15	12.3	10.0
Total	334	2.0	11.3

Figure 14.1 shows the excess market returns from 1926Q3 to 2009Q4 and the shaded areas represent the NBER (National Bureau of Economic Research) recession periods.* The stock returns tend to be negative and grow during recessions, reaching peaks towards the end of each one. In fact, in our sample period the maximum return was reached by the end of the Great Depression of the 1930s. Nevertheless, this commonly cited contracyclical character of stock returns is evident only at the end of recessions; at the beginning of and during recessions, these returns are highly procyclical. For example, the minimum in the sample period was also reached during the Great Depression. In addition, expansions are characterized on average by positive returns, although they are less volatile than those generated during recessions.

NBER's business cycle dates enable the sample to be divided into four stages: expansion, peak, recession and trough. Table 14.2 shows the first two conditional sample moments for the excess stock returns. As can be seen in the table, the most frequent stage of the economy is expansion, and during this phase of the cycle the average return is positive (2.9%) and greater than its historical average (2.0%). On the

---

* For longer sample periods we use the stock returns and the risk-free rate available from Ken French's web page, which contains data from September 1926. Stock returns are from the value-weighted CRSP index and the risk-free rate is from the 3-month Treasury bills from Ibbotson Associates.

contrary, during the recession stage the average excess return on the market has a similar magnitude as it does during the expansion stage, but with the opposite sign (−3.0%) and almost twice the volatility. Finally, during the peaks (troughs), expectations about the state of the economy are negative (positive) and therefore excess stock returns are highly negative (positive) in these stages.

14.2.4 Measuring Oil Price Shocks

Considering the evidence mentioned above, it seems natural to propose oil price shocks as a leading variable with the potential to forecast stock market returns. However, to maximize its predictive power, it is essential to consider only unanticipated changes in the oil price. Although oil spot returns have been a widely used variable in the literature (see Table 14.1), we do not consider them, because they contain some components that are clearly anticipated by market participants, such as the interest rate and the convenience yield.* One way to address this issue is to estimate unexpected oil changes from a model for the spot price dynamics, but this procedure still has disadvantages in that it depends on the model specification and that the information set used by the econometrician to estimate the conditional mean may not coincide with that of the market. Unexpected oil price changes can only be captured with an objective and precise estimate of the expected spot price in the future.

We measure unexpected oil price shocks by short-term futures returns on crude oil. Using Fama and French's (1987) methodology and cointegration tests, Switzer and El-Khoury (2007) show that oil futures prices have significant predictive power for future spot prices. Moreover, Ma (1989) and Kumar (1992) confirm that futures prices, in addition to being unbiased predictors of spot prices, exceed the predictive capacity of a random walk and a wide variety of models. This evidence suggests that unexpected changes in oil prices are correctly captured by our proposed variable.† Therefore, we assume that quarterly unexpected oil shocks are proxied by oil futures returns, i.e.

$$\Delta f(t) \equiv f^1(t) - f^4(t-3) \approx s(t) - \mathbb{E}_{t-3}[s(t)], \tag{14.1}$$

where $s(t)$ is the log oil spot price and $f_\tau(t)$ is the log oil futures price of a contract that matures in τ months.

Data on oil futures prices are from NYMEX, which began trading these contracts in March 1983. Therefore, our sample period goes from 1983Q2 to 2009Q4.‡ Figure 14.1 confirms that recessions are preceded by positive oil shocks, while during recessions these shocks are rapidly reversed. The great variability of the oil shocks is evidence of their predictive potential. Another visual characteristic is the low persistence of the series, which prevents it from being subject to the critiques of existing predictive variables. The optimal number of lags of our variable to be considered in this study was determined using the Akaike Information Criterion (AIC) in Ordinary Least Squares (OLS) regressions of the excess stock market return on lagged oil shocks. In this study we consider four lags of our variable, because this number minimizes the AIC (see Table 14.3). Interestingly, the number of lags coincides with that obtained in the macroeconomic literature (Hamilton 2003, 2008).

---

* The convenience yield of crude oil is a benefit for immediate ownership of physical units of the commodity attributed to the benefit of protecting regular production from temporary shortages of oil stocks (see, for example, Casassus and Collin-Dufresne (2005)).

† A direct way to test that futures prices are unbiased estimators of future spot prices is to verify the existence of a risk premium associated with this contract. In our sample, the *t*-stat of the null hypothesis that the average logarithmic futures return is zero is 1.13. Although this informal test only detects a constant risk premium, the results obtained by Switzer and El-Khoury (2007) suggest that, in oil futures, this finding is not incompatible with the existence of predictive power for spot prices.

‡ We use quarterly data to better capture the aggregate impact of oil shocks and make our results more compatible with the macroeconomic evidence. Moreover, this allows us to include in the analysis the consumption–wealth ratio, a variable which is only available at a quarterly frequency (Lettau and Ludvigson 2001a).

TABLE 14.3   Optimal Lags of Oil Price Shocks. OLS Regressions of
Excess Stock Returns on Lags of Oil Price Shocks, 1983Q2–2009Q4.
All Estimations Use the Full Sample and Include a Constant

Lags of oil price shocks	AIC
0	−4.884
1	−4.891
2	−4.867
3	−4.846
4	−4.893
5	−4.870
6	−4.844
7	−4.824
8	−4.803

14.2.5 Oil Price, the Business Cycle and Stock Returns

To show evidence of the relationship among the oil price, the business cycle and excess stock returns, we study the joint dynamics of these variables using a vector auto-regression analysis with four lags. Our proxy for the stock market is the value-weighted CRSP index, from which we obtain the quarterly returns on the market portfolio (R_m). We proxy the risk-free rate (R_f) with the 3-month constant-maturity treasury yields from the Federal Reserve Board of Governors. Following Cooper and Priestley (2009), we use the total Industrial Production Index (IP) from the Fed as a measure of output and a proxy for the business cycle. Table 14.4 shows the maximum likelihood estimates for the VAR(4) model. As is common in macroeconomic series, industrial production growth rate ($\Delta\%IP$) is the easiest series to predict (its R^2 is 51%) and its own lags have useful information for forecasting its future values. On the other hand, the predictive power of the excess stock return ($R_m − R_f$) on %IP is a clear signal that the financial market correctly anticipates future economic growth. Moreover, in our sample the Granger-causality of oil shocks (Δf) on economic growth is also verified, evidence that is in line with the macroeconomic studies mentioned above.

The table also shows that, for the excess stock returns, the lack of significance of its own lags suggests the stock market is efficient and that the initial underreaction and later correction documented by Driesprong *et al.* (2008) is not seen at quarterly frequency.[*] There is also evidence of inverse causality with the industrial production growth rate, which is probably the consequence of an adjustment process of previous expectations about actual economic growth. Furthermore, and as expected, oil shocks demonstrate a significant predictive power for excess stock returns, which will be explored in greater depth in the following sections.

Finally, the results reveal that oil shocks cannot be predicted with any of the lagged variables, which is evidence that our measure for oil shocks effectively captures unanticipated changes in this variable. This result implies that our variable is exogenous to the U.S. economy; however, it may not be so to global aggregate demand. Therefore, the critique of Kilian (2009) regarding the endogeneity of oil price shocks is still valid for our variable.

The following sections analyse and test the predictive power of oil price shocks for stock returns.

---

[*] If the adjustment process of returns after an oil shock is slow, this should be evidenced to some degree in the excess stock return, either positively if the adjustment is gradual or negatively if it is excessive and requires future reversals.

TABLE 14.4   VAR Estimation Results, 1983Q2–2009Q4. Maximum Likelihood Estimates of the VAR(4) Model for the Rate of Growth of Industrial Production ($\Delta\%IP(t)$), Excess Stock Market Returns ($R_m(t) - R_f(t)$) and Log Returns on Crude Oil Futures ($\Delta f(t)$). Asymptotic t-Stat in Parentheses. Granger Causality Was Tested Using the Asymptotic Wald Test. The *All* Row at the Bottom of the Table Refers to All Coefficients Except the Constant

	$\Delta\%IP(t)$	$R_m(t) - R_f(t)$	$\Delta f(t)$
Constant	0.000	0.011	0.020
	(0.45)	(1.20)	(0.88)
$\Delta\%IP(t-1)$	0.276	0.956	−0.205
	(3.04)	(1.25)	(−0.11)
$\Delta\%IP(t-2)$	−0.027	−0.852	0.404
	(−0.26)	(−0.99)	(0.19)
$\Delta\%IP(t-3)$	−0.113	−1.215	0.460
	(−1.10)	(−1.40)	(0.22)
$\Delta\%IP(t-4)$	0.347	2.327	−1.194
	(3.70)	(2.93)	(−0.62)
$R_m(t-1) - R_f(t-1)$	0.061	0.038	0.269
	(5.59)	(0.41)	(1.19)
$R_m(t-2) - R_f(t-2)$	0.030	−0.069	0.191
	(2.52)	(−0.68)	(0.77)
$R_m(t-3) - R_f(t-3)$	0.005	−0.038	0.268
	(0.42)	(−0.37)	(1.07)
$R_m(t-4) - R_f(t-4)$	0.026	0.019	−0.087
	(2.12)	(0.19)	(−0.35)
$\Delta f(t-1)$	0.010	0.065	0.035
	(2.09)	(1.59)	(0.35)
$\Delta f(t-2)$	−0.006	−0.044	−0.183
	(−1.22)	(−1.04)	(−1.79)
$\Delta f(t-3)$	−0.012	−0.004	0.027
	(−2.35)	(−0.10)	(0.26)
$\Delta f(t-4)$	−0.005	−0.119	−0.057
	(−1.00)	(−2.74)	(−0.54)
R^2	0.51	0.18	0.06
$\overline{R^2}$	0.44	0.07	−0.06
p-Value Granger causality test			
All	0.00	0.03	0.85
$\Delta\%IP$	0.00	0.04	0.98
$R_m - R_f$	0.00	0.94	0.53
Δf	0.01	0.04	0.49

14.3 Short-Horizon Predictability of Stock Returns

This section focuses on the in-sample predictability of stock returns at a quarterly horizon. The predictive performance of our variable is evaluated and compared with the performance of the following variables: the risk-free rate (Campbell 1987), the log dividend–price ratio (Fama and French 1988), the consumption–wealth ratio (Lettau and Ludvigson 2001a) and the output gap (Cooper and Priestley

2009). The log dividend–price ratio ($d - p$) was calculated from the value-weighted CRSP index using the methodology described by Ang and Bekaert (2007). The consumption–wealth ratio (*cay*) and its individual components are from Martin Lettau's web page and sampled at a quarterly frequency. The output gap (*gap*) is constructed using the total Industrial Production Index.* The output *gap* is estimated with data from 1948Q1 to replicate the series from Cooper and Priestley (2009). The variables *cay*(*t*) and *gap*(*t*) are assumed to be known at the start of time $t + 1$, and therefore can be used to forecast excess stock returns. We omit any complication due to the look-ahead bias in these variables and the normal delay in the publication and subsequent revisions of these and other macroeconomic series.

Table 14.5 presents the main statistics of the predictive variables and Figure 14.2 provides graphical evidence. The upper panel of Table 14.5 shows that our variable (Δf) and $d - p$ exhibit more volatility than the variables R_f, *cay* and *gap*. Unlike the other predictors, Δf shows very low persistence; in fact, its first-order serial correlation is as low as that of $R_m - R_f$. The lower panel of the table shows that the intratemporal correlation of our variable with the excess stock return is very low, rejecting Δf as a possible pricing factor. Both the low persistence of our variable and its low correlation with excess stock returns discard the idea that the forecastability of stock returns from oil returns might be due to a common factor driving both variables. Another important characteristic of our variable is its low correlation with other predictive variables, which suggests that Δf contains business cycle information not captured by the existing predictors. On the contrary, the existing variables show high levels of correlation (in absolute terms) among themselves, revealing the presence of redundant information.

Next we turn our attention to evaluating the predictive performance of our variable and the other variables considered here. We begin with the evidence of in-sample predictability at a quarterly horizon, for which we estimate the following regression:

$$R_m(t) - R_f(t) = \alpha + \beta' X(t-1) + \varepsilon(t), \tag{14.2}$$

where $X(t - 1)$ is a vector of known predictors at $t - 1$ and β its associated coefficient vector. It should be emphasized that $X(t - 1)$ can include one variable, several variables or several lags of the same variable.

TABLE 14.5 Statistics for 1983Q2–2009Q4. Autocorrelation is the First-Order Serial Correlation

	$R_m - R_f$	Δf	R_f	$d - p$	*cay*	*gap*
Average	0.016	0.021	0.012	−3.781	0.005	−0.014
Standard deviation	0.087	0.195	0.006	0.386	0.019	0.063
Autocorrelation	0.031	0.031	0.972	0.974	0.910	0.977
			Correlation matrix			
	$R_m - R_f$	Δf	R_f	$d - p$	*cay*	*gap*
$R_m - R_f$	1.000					
Δf	−0.039	1.000				
R_f	0.006	−0.035	1.000			
$d - p$	0.139	−0.101	0.524	1.000		
cay	−0.108	−0.100	0.348	0.472	1.000	
gap	−0.121	0.111	−0.210	−0.847	−0.601	1.000

* Cooper and Priestley (2009) define four methods for calculating the output gap, but the main one is the quadratic version of *gap* (based on its greater correlation with procyclical variables), which we also use here. The output gap is calculated from the following regression: $ip(t) = a + bt + ct^2 + gap(t)$, where $ip(t)$ is the log industrial production, t is a time trend and $gap(t)$ is the error term.

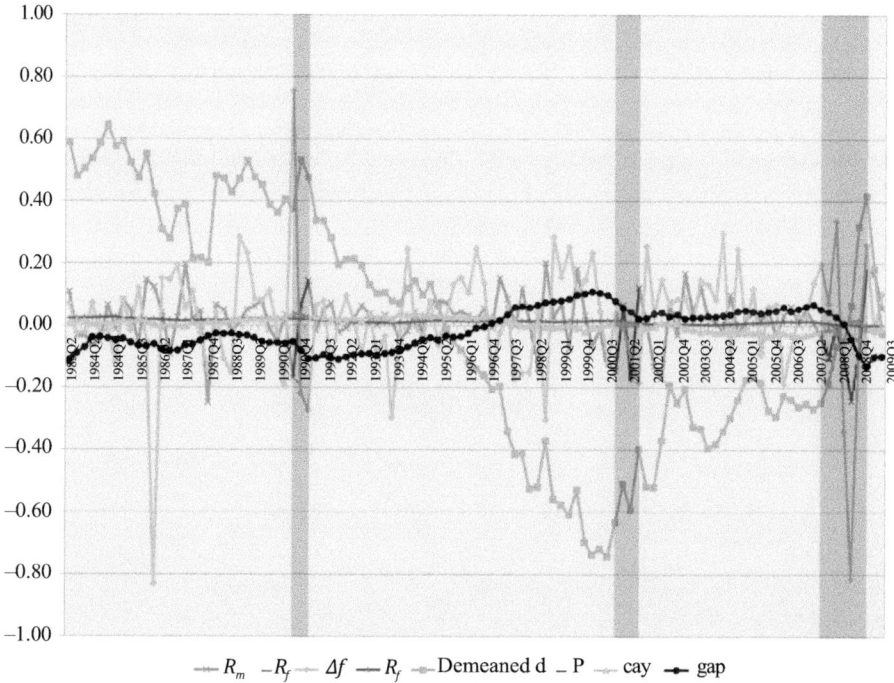

FIGURE 14.2 Excess stock market returns and predictor variables, 1983Q2–2009Q4.

Table 14.6 shows the OLS results of Equation (14.2). We report asymptotic t-stats and Wald tests that correct for serial correlation and heteroscedasticity using Newey and West (1987).* The columns present the results for each predictor variable. The results indicate that our variable has the best in-sample predictive performance with an $\overline{R^2}$ of 6%. The Wald test corresponds to the null hypothesis that all the coefficients in Equation (14.2) are zero, except the constant. This statistic is highly significant for the oil price shocks (p-value of 2%). The annual cumulative impact of our variable is –0.106 and is calculated from the sum of the coefficients corresponding to the four lags of Δf. This impact is also economically significant. To see this, consider a one-time increase in Δf of one standard deviation (19.5% in our sample, see Table 14.5). This change leads to a 2.1% decrease in expected quarterly excess returns on the value-weighted CRSP index (–0.106 × 19.5% = –2.1%), which is equivalent to 29.6% (2.1%/7.0% = 29.6%) of the historical average annual excess return.

The dynamics of the distributed lags can be explained with aggregate demand shocks (Kilian and Park 2009). A positive shock to the global demand for industrial commodities produces both a direct positive impact and an indirect negative one in the financial market. The direct impact manifests as an increase in both the oil price and economic growth with a consequent positive stock return. Increased economic growth pushes the oil price even higher, and thus indirectly affects negatively the future expected economic growth and expected stock returns. The final impact will depend on the relative magnitudes of both impacts.[†] The direct impact is initially stronger, which explains the positive significant effect of the first lag of Δf. Later, the indirect negative impact of the second and third lags begins to gather some

---

* Following Newey and West (1994), we chose lag length equal to $floor[4 \cdot (T/100)^{2/9}]$, where $floor[x]$ denotes the integer part of x.

[†] Note from our VAR analysis in the previous section that the signs of the four lags of oil shocks are the same for both industrial production and excess stock returns (see Table 14.4). This result gives support to our aggregate demand based explanation.

TABLE 14.6 Predictive Regressions for Excess Stock Returns, 1983Q2–2009Q4. OLS Regressions for Excess Stock Returns on the Predictor Variables in the First Row. All Tests are Based on Covariance Matrices of Coefficients Corrected for Heteroscedasticity and Serial Correlation Using Newey and West (1987). Lag Length in the Newey–West Estimator Is *floor*[4. $(T/100)^{2/9}$], Where *Floor*[*X*] Denotes the Integer Part of *X* (Newey and West 1994). Asymptotic *t*-Stat in Parentheses. At the Bottom of the Table, the *P*-Value Is for the Asymptotic Wald Test and *All* Refers to All Coefficients Except the Constant

	Δf	R_f	$d - p$	cay	gap
Constant	0.019	0.013	0.154	0.014	0.015
	(2.59)	(0.55)	(2.04)	(1.52)	(1.66)
$\Delta f(t-1)$	0.068				
	(1.95)				
$\Delta f(t-2)$	−0.035				
	(−0.53)				
$\Delta f(t-3)$	−0.024				
	(−0.48)				
$\Delta f(t-4)$	−0.115				
	(−2.68)				
$R_f(t)$		0.403			
		(0.25)			
$d(t-1) - p(t-1)$			0.036		
			(1.75)		
$cay(t-1)$				0.729	
				(2.09)	
$gap(t-1)$					−0.219
					(−1.58)
R^2	0.10	0.00	0.02	0.02	0.02
$\overline{R^2}$	0.06	−0.01	0.01	0.01	0.02
p-Value Wald *All*	0.02	0.80	0.08	0.04	0.11

strength, although not of a sufficient magnitude to cancel out the initial positive impact (see the sign and low significance of the following two lags). One year after the unexpected aggregate demand shock, the indirect effect becomes dominant; in other words, the high price of oil causes a deceleration in the economy. This is manifested by the negative and significant coefficient of $\Delta f(t-4)$ which is also responsible for the cumulative negative impact reported.

The third column of Table 14.6 shows the results for the interest rate as a predictor. This variable has the worst predictive performance in our sample, with a $\overline{R^2}$ of −0.01. Contrary to the findings of previous studies (Campbell 1987), the coefficient that accompanies this variable is positive, although not significant. This poor performance is associated with the low volatility of this variable in our sample. Its standard deviation is 0.006 (see Table 14.5), a value well below the 0.032 reported by Ang and Bekaert (2007) for 1935Q2–2001Q4. The last three columns contain the results for the $d - p$, *cay* and *gap* variables. All of these have intuitive signs, although their coefficients are lower in absolute value than those reported in previous studies (see Ang and Bekaert (2007), Lettau and Ludvigson (2001a) and Cooper and Priestley (2009), respectively).[*] Moreover, the R^2 statistic for these variables (all about 2%) suggest that they have similarly poor predictive power. Among these variables, *cay* has a better performance, being the only one significant at the 5% level.

---

[*] According to Stambaugh (1999), the estimation of the β coefficient in a predictive regression of $d - p$ is biased. This bias further reinforces our conclusion regarding this variable's lack of forecasting power.

TABLE 14.7 Predictive Regressions: Additional Controls, 1983Q2–2009Q4. OLS Regressions for Excess Stock Returns on the Predictor Variables in the First Row. All Tests Are Based on Covariance Matrices of Coefficients Corrected for Heteroscedasticity and Serial Correlation Using Newey and West (1987). Lag Length in the Newey–West Estimator Is *Floor*[4. $(T/100)^{2/9}$], Where *floor* [x] Denotes the Integer Part of X (Newey and West 1994). Asymptotic *t*-Stat in Parentheses. At the Bottom of the Table, the *p*-Value Is for the Asymptotic Wald Test, *All* Refers to All Coefficients Except the Constant and Δf Refers to the Coefficients Associated with the Four Lags of That Variable

	Δf & R_f	Δf & $d - p$	Δf & cay	Δf & gap	Δf & All
Constant	0.012	0.140	0.015	0.017	0.269
	(0.57)	(2.02)	(1.81)	(2.01)	(1.03)
$\Delta f (t-1)$	0.067	0.075	0.075	0.076	0.078
	(1.96)	(2.13)	(2.07)	(2.20)	(1.87)
$\Delta f (t-2)$	−0.036	−0.026	−0.029	−0.025	−0.028
	(−0.53)	(−0.41)	(−0.47)	(−0.38)	(−0.47)
$\Delta f (t-3)$	−0.026	−0.017	−0.023	−0.013	−0.026
	(−0.49)	(−0.33)	(−0.45)	(−0.26)	(−0.53)
$\Delta f (t-4)$	−0.116	−0.108	−0.114	−0.105	−0.117
	(−2.72)	(−2.50)	(−2.67)	(−2.33)	(−2.54)
$Rf(t)$	0.650				−1.708
	(0.45)				(−0.64)
$d(t-1) - p(t-1)$		0.032			0.061
		(1.68)			(1.01)
$cay(t-1)$			0.784		0.985
			(2.00)		(1.78)
$gap(t-1)$				−0.172	0.303
				(−1.50)	(0.93)
R^2	0.10	0.11	0.12	0.11	0.14
$\overline{R^2}$	0.05	0.07	0.08	0.06	0.06
p-Value Wald *All*	0.02	0.00	0.00	0.00	0.00
p-Value Wald Δf	0.01	0.01	0.01	0.02	0.02

The previous results suggest that the oil shocks have a significant in-sample forecasting power for short horizons. We now see whether our variable is robust to the inclusion of the other predictors considered here. To evaluate this, we estimate the following extended predictive regression:

$$R_m(t) - R_f(t) = \alpha + \sum_{j=1}^{4} \beta_j \Delta f(t-j) + \theta'Z(t-1) + \in (t), \qquad (14.3)$$

where $Z(t-1)$ is a vector of predictor variables and θ its associated coefficients vector. Lack of robustness in the predictive power of our variable should be reflected in changes in sign and/or loss of significance in the coefficients that accompany the lags of oil shocks.

The results of the estimation of Equation (14.3) are presented in Table 14.7. The last row contains the *p*-value of an asymptotic Wald test for the combined null hypothesis that all the coefficients associated with the lags of our variable are zero. The columns show the results of including each of the other variables in the predictive regression of Δf, while the inclusion of all of them is considered in the last column. First, given the low correlation of our variable with the others, the forecasting power of our variable remains intact. According to the Wald test for the coefficients of Δf, in all of the estimates these coefficients keep their joint significance. In addition, their signs and individual significance remain the same, and they are roughly the same size. Second, consistent with previous results, the greatest increase in predictive power is reached when our variable is used in combination with *cay*, obtaining an $\overline{R^2}$ of 8%. Third, contrary to what occurs with the oil shock coefficients, the results of the last column provide evidence of great instability in the predictive power of the other variables. None of these are significant. The coefficients of R_f and *gap* experience changes in sign and those of $d - p$ and *cay* vary dramatically in size. Of course, this evidence is consistent with the high correlation between these variables reported in Table 14.5.

In summary, this section demonstrates that our variable has significant and robust short-horizon in-sample forecasting power for stock returns. Out-of-sample predictability, also at a quarterly horizon, is considered in the next section.

14.4 Out-of-Sample Predictability of Stock Returns

In-sample predictive performance is essential for establishing the existence of predictability. However, in order for a predictor to be used by an investor, it must also demonstrate a good out-of-sample performance. That is, a predictive variable must be able to forecast excess returns reasonably well with information available at the time of the forecast, which is not guaranteed by the in-sample tests of Equations (14.2) and (14.3), as the coefficients are estimated using the full sample.

Welch and Goyal (2008) conclude that it is very difficult to find variables with short-horizon out-of-sample forecasting power that outperform the average excess return in a recent sample period. In fact, when considering the sub-period from 1965 to 2005, they only find variables that outperform the prevailing historical average return at a five-year horizon. Although the out-of-sample predictive performance can be increased by imposing certain restrictions, as shown by Campbell and Thompson (2008), here we choose to keep the simplicity and linearity of the predictive model. We also test the other variables' out-of-sample performance despite their poor in-sample predictive power.

In order to contribute to this discussion, we compare forecasts from nested linear models to determine whether each variable has predictive content for stock returns. The prevailing historical average of excess stock returns is used as a benchmark. Therefore, we define the following benchmark and competing models:

$$\text{benchmark: } R_m(t+1) - R_f(t+1) = \alpha_1 + u_1(t+1), \tag{14.4}$$

$$\text{competing: } R_m(t+1) - R_f(t+1) = \alpha_2 + \beta' X(t) + u_2(t+1), \tag{14.5}$$

where the coefficients α_1, α_2 and β are estimated recursively. The sample size T is divided into in-sample and out-of-sample portions. Q is defined as the minimum number of observations used for estimating the coefficients and $P \equiv T - Q$ denotes the maximum number of one-step-ahead predictions. Thus, forecasts of $R_m(t+1) - R_f(t+1)$, $t = Q, \ldots, T-1$, are generated recursively using the two linear models in Equations (14.4) and (14.5), where all coefficients are re-estimated with new observations as forecasting moves forward through time.

Our assessment of out-of-sample predictability involves three metrics that are presented with more detail in Appendix A. The first approach is the forecast encompassing test of Clark and McCracken (2001) that verifies whether the competing forecasts incorporate any useful information absent in the benchmark forecasts. Their statistic ENC-NEW tests the null hypothesis that the benchmark model encompasses the competing model. This test suffers from a look-ahead bias, since ENC-NEW depends on a parameter that is estimated with the full sample. The second metric we use is the MSE-F test developed by McCracken (2007) that compares the predictive accuracy between nested models measured as the mean squared forecasting errors (MSE). Although this test does not have the best small-sample properties, it enables testing of out-of-sample predictive power under the same conditions that an investor faces in reality. The final measure of out-of-sample forecasting performance is the out-of-sample R^2,

R_{OS}^2, proposed by Campbell and Thompson (2008). Following Welch and Goyal (2008), we use bootstrap in order to address size distortions in the t-stat for long horizons (Ang and Bekaert 2007) and to make the correct inference from nested out-of-sample predictability tests (Clark and McCracken 2005). We also consider covariance matrices of coefficients corrected for heteroscedasticity and autocorrelation that arise from the use of distributed lags and overlap of returns (Newey and West 1987).

In general, when performing out-of-sample tests in a small sample, there is a trade-off with the number of in-sample observations that is hard to resolve. On the one hand, the objective is to use a relatively large in-sample proportion of the sample (Q/T), so that the out-of-sample forecasts are done with estimates which are as similar as possible to those obtained with the full sample. But at the same time, as suggested by the results of Inoue and Kilian (2005), the out-of-sample proportion (P/T) must be large enough to prevent significant differences in power between the in-sample and out-of-sample tests. Thus, to achieve a reasonable level of power without producing excessive forecasting errors at the beginning of the out-of-sample sub-period, the optimal choice should be around $\pi = P/Q = 1$. However, to make our test more rigorous, we choose 1997Q4, which is when the Asian crisis hit the U.S. economy, as the starting point for the out-of-sample sub-period. That is, given that our adjusted (for lags) sample spans the period 1984Q2–2009Q4, our choice implies the following sample portions: $Q = 54$, $P = 49$ and $\pi = 0.91$.

The results of the out-of-sample tests are presented in Table 14.8. All of the tests coincide in that our variable is the only one with out-of-sample forecasting power for the excess stock returns at a 10% significance level.*

TABLE 14.8   Out-of-Sample Predictability Tests, 1983Q2–2009Q4. Out-of-Sample Tests of Stock Return Predictability. Each Column Reports the Results Using the Predictor from the First Row. Out-of-Sample Period Is from 1997Q4 to 2009Q4. ENC-NEW, MSE-F and R_{os}^2 Statistics Are Described in Equations (A.6), (A.7) and (A.8). Asymptotic Critical Values for the ENC-NEW Test Are from Table 1 of Clark and Mccracken (2001) Using $\pi = 1.0$. Asymptotic Critical Values for the MSE-F Test Are from Table 4 of Mccracken (2007) Using $\pi = 1.0$. Bootstrapped p-Values and Critical Values Are Based on the Methodology of Clark and Mccracken (2005)

	Δf	R_f	$d-p$	cay	gap
ENC-NEW					
Sample value	2.218	−0.555	0.202	1.630	0.063
0.10 Asymptotic critical value	2.169	0.984	0.984	0.984	0.984
0.05 Asymptotic critical value	3.007	1.584	1.584	1.584	1.584
Bootstrapped p-value	0.077	0.791	0.278	0.103	0.505
0.10 Bootstrapped critical value	1.939	1.325	1.325	1.653	2.163
0.05 Bootstrapped critical value	2.970	2.033	2.276	2.559	3.294
MSE-F					
Sample value	0.603	−1.287	−0.034	0.261	−0.750
0.10 Asymptotic critical value	0.545	0.751	0.751	0.751	0.751
0.05 Asymptotic critical value	1.809	1.548	1.548	1.548	1.548
Bootstrapped p-value	0.081	0.655	0.172	0.197	0.463
0.10 Bootstrapped critical value	0.443	0.789	0.431	1.335	1.816
0.05 Bootstrapped critical value	1.420	1.860	1.268	2.237	3.006
R_{os}^2					
Sample value	0.012	−0.027	−0.001	0.005	−0.016
Bootstrapped p-value	0.081	0.655	0.172	0.197	0.463
0.10 Bootstrapped critical value	0.009	0.016	0.009	0.027	0.036
0.05 Bootstrapped critical value	0.028	0.037	0.025	0.044	0.058

---

* Our results could be distorted by a possible look-ahead bias in the *cay* and *gap* variables, since for the sake of simplicity, both were calculated using the full sample of observations. Strictly speaking, in an out-of-sample test, forecasting of the excess stock returns implies the estimation of the coefficients with data only up to the prevailing quarter.

Although the *cay* variable is marginally significant according to the ENC-NEW test (its bootstrapped *p*-value is 10.3%), as explained above, the only test that measures forecasting power under the effective conditions faced by a potential investor is the MSE-F test. Our variable (Δf) has the highest and most significant R^2_{os} among all of the variables considered; however, the size of this statistic is only 1.2%. The table also shows that, given the wide differences between the bootstrapped and asymptotic critical values, controlling for considerations of small sample and differences in the relative out-of-sample portion (i.e. π) is essential for a reliable inference, especially when highly persistent predictor variables are used. Also, as a consequence of the close relationship between the MSE-F and the R^2_{os} statistics shown in Appendix A, the inference using the bootstrap method produces the same results for both tests.

Finally, the low R^2_{os} for all the predictors is evidence that out-of-sample forecasting of stock returns has become an increasingly difficult challenge in recent times (one of the main points emphasized by Welch and Goyal (2008)). The forecasting ability of a predictor variable depends exclusively on its capacity to successfully summarize the conditioning information used by the market participants, which has become increasingly complex.

14.5 Long-Horizon Predictability of Stock Returns

This section examines the in-sample predictability of stock returns at longer horizons. The evidence presented below is based on the following long-horizon regression:

$$R_m^h(t+h) - R_f^h(t+h) = \alpha_h + \beta_h' X(t) + \varepsilon^h(t+h), \tag{14.6}$$

where $R_i^h(t+h) = \prod_{j=1}^{h}\left(1 + R_i^1(t+j)\right) - 1$ is the *h*-period return for asset $i = m, f$ and $R_i^1(t+j)$ is the respective one-period return from time $t+j-1$ to $t+j$.

The evidence of long-run predictability has been the subject of a great deal of criticism. According to Boudoukh *et al.* (2008) long-horizon regressions are misleading for persistent predictors.[*] That is, long-horizon regressions associated with the variables R_f, $d - p$, *cay* and *gap* cannot show anything different to what was shown in Section 14.3 due to their high persistence (see Table 14.5). On the other hand, given the almost null persistence of our variable, the Δf shocks are absolutely short-lived. Therefore, at the most our variable could have forecasting power for stock returns over a one-year horizon.

Following Kilian (1999), we adapt the bootstrap algorithm described in Appendix A to support the inference from the long-horizon regressions presented here. In addition, to evaluate the impact of the findings of Boudoukh *et al.* (2008) regarding persistent predictors, we report both expected regression coefficient and the R^2 statistic at the *h*th horizon conditional on their one-period counterpart, under the null of non-predictability. The expected coefficients are given by the following equations:

$$\mathbb{E}\left[\hat{\beta}_h \mid \hat{\beta}_1 = \hat{\beta}_1^*\right] = \left(1 + \frac{\rho\left(1-\rho^{h-1}\right)}{1-\rho}\right)\hat{\beta}_1^*, \tag{14.7}$$

$$\mathbb{E}\left[R_h^2 \mid R_1^2 = R_1^{2^*}\right] = \frac{\left(1 + \left\{\left[\rho\left(1-\rho^{h-1}\right)\right]/(1-\rho)\right\}\right)^2}{h} R_1^{2^*}, \tag{14.8}$$

---

[*] Other problems documented for long-horizon regressions are: (1) serial correlation in residuals induced by the overlap of observations; (2) inefficient use of the data that provide spurious forecasts about the dynamics of variables, especially for non-exogenous predictors (Campbell 1991); and (3) by aggregating returns, long-horizon regressions invalidate the inference from standard asymptotic methods (Valkanov 2003).

TABLE 14.9 Long-Horizon Predictability, 1983Q2–2009Q4. OLS Regressions of $R_m^h(t+h) - R_f^h(t+h)$ on the Predictor Variables Using the Same Number of Observations. All Tests are Based on Covariance Matrices of Coefficients Corrected for Heteroscedasticity and Serial Correlation (Newey and West 1987). Lag Length in the Newey–West Estimator is $Floor[4 \cdot (T/100)^{2/9}] + (h-1)$, Where $Floor[x]$ Denotes the Integer Part of x. Bootstrapped p-Values for t-Stats in Parentheses. $\mathbb{E}\left[\hat{\beta}_h \middle| \hat{\beta}_1 = \hat{\beta}_1^*\right]$ and $\mathbb{E}\left[R_h^2 \middle| R_1^2 = R_1^{2^*}\right]$ Are Described in Equations (14.7) and (14.8), Respectively. p-Value Wald *All* Is the Bootstrapped p-Value for the Asymptotic Wald Test That All Coefficients Except the Constant Are Zero

Forecast horizon in quarters (h)	$h=3$	$h=4$	$h=8$	$h=12$	$h=16$	$h=20$
$\Delta f(t)$	0.153	0.019	−0.019	0.105	0.112	−0.143
	(0.01)	(0.44)	(0.45)	(0.33)	(0.35)	(0.28)
$\Delta f(t-1)$	−0.051	−0.057	−0.057	0.029	−0.147	−0.323
	(0.27)	(0.31)	(0.39)	(0.47)	(0.30)	(0.18)
$\Delta f(t-2)$	−0.081	−0.142	−0.128	−0.105	−0.051	−0.380
	(0.24)	(0.18)	(0.28)	(0.35)	(0.42)	(0.14)
$\Delta f(t-3)$	−0.208	−0.146	−0.154	−0.119	−0.138	−0.432
	(0.05)	(0.14)	(0.21)	(0.31)	(0.30)	(0.08)
R^2	0.12	0.05	0.02	0.01	0.01	0.04
$\bar{R}^2$	0.07	0.00	−0.03	−0.04	−0.04	0.00
p-Value Wald *All*	0.02	0.58	0.82	0.75	0.71	0.45
$R_f(t+1)$	0.682	0.926	1.129	−0.084	−3.804	5.304
	(0.39)	(0.38)	(0.38)	(0.57)	(0.47)	(0.34)
$\mathbb{E}\left[\hat{\beta}_h \middle| \hat{\beta}_1 = \hat{\beta}_1^*\right]$	−0.125	−0.165	−0.317	−0.456	−0.585	−0.702
R^2	0.00	0.00	0.00	0.00	0.00	0.00
$\mathbb{E}\left[R_h^2 \middle| R_1^2 = R_1^{2^*}\right]$	0.00	0.00	0.00	0.00	0.00	0.00
$\bar{R}^2$	−0.01	−0.01	−0.01	−0.01	−0.01	−0.01
p-Value Wald *All*	0.87	0.87	0.90	0.99	0.78	0.85
$d(t) - p(t)$	0.083	0.111	0.220	0.324	0.440	0.672
	(0.13)	(0.14)	(0.18)	(0.18)	(0.18)	(0.14)
$\mathbb{E}\left[\hat{\beta}_h \middle| \hat{\beta}_1 = \hat{\beta}_1^*\right]$	0.078	0.102	0.190	0.266	0.331	0.388
R^2	0.06	0.08	0.13	0.14	0.17	0.25
$\mathbb{E}\left[R_h^2 \middle| R_1^2 = R_1^{2^*}\right]$	0.05	0.06	0.10	0.14	0.16	0.17
$\bar{R}^2$	0.04	0.06	0.12	0.13	0.16	0.24
p-Value Wald *All*	0.28	0.29	0.36	0.36	0.36	0.28
$cay(t)$	2.501	3.436	7.493	12.588	18.904	25.065
	(0.26)	(0.28)	(0.30)	(0.36)	(0.25)	(0.10)
$\mathbb{E}\left[\hat{\beta}_h \middle| \hat{\beta}_1 = \hat{\beta}_1^*\right]$	2.867	3.662	6.212	7.986	9.222	10.083
R^2	0.07	0.10	0.21	0.30	0.44	0.49
$\mathbb{E}\left[R_h^2 \middle| R_1^2 = R_1^{2^*}\right]$	0.09	0.11	0.16	0.17	0.17	0.17
$\bar{R}^2$	0.06	0.09	0.20	0.30	0.43	0.48
p-Value Wald *All*	0.27	0.30	0.32	0.39	0.26	0.11

(*Continued*)

TABLE 14.9 (CONTINUED)   Long-Horizon Predictability, 1983Q2–2009Q4. OLS Regressions of $R_m^h(t+h) - R_f^h(t+h)$ on the Predictor Variables Using the Same Number of Observations. All Tests are Based on Covariance Matrices of Coefficients Corrected for Heteroscedasticity and Serial Correlation (Newey and West 1987). Lag Length in the Newey–West Estimator is $Floor[4 \cdot (T/100)^{2/9}] + (h-1)$,

Where $Floor[x]$ Denotes the Integer Part of x. Bootstrapped p-Values for t-Stats in Parentheses.

$\mathbb{E}\left[\hat{\beta}_h \middle| \hat{\beta}_1 = \hat{\beta}_1^*\right]$ and $\mathbb{E}\left[R_h^2 \middle| R_1^2 = R_1^{2*}\right]$ Are Described in Equations (14.7) and (14.8), Respectively. p-Value Wald *All* Is the Bootstrapped p-Value for the Asymptotic Wald Test That All Coefficients Except the Constant Are Zero

Forecast horizon in quarters (h)	h=3	h=4	h=8	h=12	h=16	h=20
$gap(t)$	−0.444	−0.578	−1.548	−2.635	−4.161	−5.938
	(0.60)	(0.62)	(0.54)	(0.45)	(0.33)	(0.25)
$\mathbb{E}\left[\hat{\beta}_h \middle| \hat{\beta}_1 = \hat{\beta}_1^*\right]$	−0.345	−0.454	−0.862	−1.229	−1.558	−1.853
R^2	0.04	0.05	0.14	0.21	0.34	0.43
$\mathbb{E}\left[R_h^2 \middle| R_1^2 = R_1^{2*}\right]$	0.02	0.03	0.05	0.07	0.08	0.09
$\bar{R}^2$	0.02	0.03	0.13	0.20	0.33	0.43
p-Value Wald *All*	0.64	0.66	0.57	0.47	0.34	0.26

where $\hat{\beta}_1^*$ is the actual estimate of the regression coefficient in Equation (14.6) for $h = 1$, ρ is the first-order serial correlation coefficient of the predictor variable and R_1^{2*} is the actual estimate of the R^2 statistic for $h = 1$, also from Equation (14.6).

Our analysis covers return horizons up to five years ($h = 20$). For each horizon we consider the same number of observations and the same sample period 1984Q1–2004Q4 (i.e. 84 observations for each horizon). Table 14.9 presents the OLS estimates of Equation (14.6) for $h = 3, 4, 8, 12, 16, 20$. We report boot-strapped p-values for t-stats and Wald tests that correct for serial correlation and heteroscedasticity using Newey and West (1987). Because of overlapping observations, we increase the lag length for the Newey–West estimator by $h - 1$. The table shows that our variable has significant forecasting power for stock returns up to an horizon of three quarters. For $h = 3$ the signs of the coefficients associated with Δf lags are the same as those for $h = 1$, the $\bar{R}^2$ is 7% and the bootstrapped p-value for the test that all coefficients for oil lags are zero is 0.02. At longer horizons, as expected, there is no evidence of predictability with our variable.

Regarding the other variables, Table 14.9 shows that, at the 10% significance level, none of the variables demonstrates forecasting power for stock returns at long horizons. The *cay* variable is marginally significant at a five-year horizon; however, given the absence of predictive ability at other horizons, it is very likely that this result can be explained by the look-ahead bias. The variable R_f has so little forecasting power that despite its high persistence, it does not exhibit the pattern predicted by Boudoukh *et al.* (2008). Instead, its regression coefficient changes sign through horizons and its $\bar{R}^2$ is always negative. Finally, the other persistent predictor variables ($d - p$, *cay* and *gap*) effectively follow the pattern predicted by Boudoukh *et al.* (2008). That is, both regression coefficients and R^2 are always (in absolute value) growing with the horizon. In fact, both the regression coefficients and the R^2 values look pretty much like those predicted by Equations (14.7) and (14.8). These results reinforce the scepticism with which the evidence of predictability with highly persistent variables should always be viewed.

14.6 Implications for the Cross Section of Expected Returns

In this section we study the cross sectional implications of our variable. Despite the clear relationship between predictability, time-varying risk premium and conditional asset pricing models, there is an important methodological asymmetry between studies on predictability and those on conditional asset pricing. Papers on predictability have only sought to prove that the equity risk premium is variable and do not consider cross sectional tests. On the other hand, studies on conditional asset pricing only present evidence of in-sample predictability and generally lack the empirical rigor of the predictability studies.

In our opinion, if a variable exhibits significant forecasting power for stock market returns, it should also be tested as a conditioning variable to explain the cross section of expected stock returns. For this reason and for greater robustness, we offer empirical evidence in the cross-section of expected returns through CAPM and CCAPM models conditioned on our variable.[*] In addition, the empirical performance of our conditional model is compared with the performance of the same models conditioned on the other predictor variables considered in the previous sections, as well as the following unconditional models: CAPM (Sharpe 1964), CCAPM (Breeden and Litzenberger 1978) and the three-factor FF model (Fama and French 1993).

The standard equilibrium condition for any asset pricing model is given by

$$1 = \mathbb{E}\Big[M(t+1) \cdot \big(1 + R_i(t+1)\big) \big| \Omega(t) \Big], \tag{14.9}$$

where $M(t+1)$ is the stochastic discount factor or pricing kernel and $\Omega(t)$ is the agent's information set at time t. We assume that the pricing kernel is exponentially affine on the pricing factor:

$$M(t+1) = \exp\Big(b^0(t) + b^1(t) F(t+1) \Big), \tag{14.10}$$

where $b^i(t) = b_0^i + b_1^i z_1(t) + \cdots + b_L^i z_L(t), i = 0,1$, and $F(t+1)$ is the risk factor. We use the aggregate market return, $R_m(t+1)$, as the risk factor for the CAPM models, and the real per-capita consumption growth rate, $\Delta c(t+1)$, for the CCAPM models. We allow for more than one conditioning variable, in particular we use $L = 4$ for our models, whereas $L = 1$ for conditional models on other variables. The exponentially affine form in Equation (14.10), as opposed to the standard linear assumption, is to keep the pricing kernel positive and to avoid the large negative values implied in linear models (Nagel and Singleton 2011).

As noted by Lewellen *et al.* (2010), conditional asset pricing models present serious problems in pricing a risk-free portfolio. Although they report high R^2 in the cross-section, this is typically achieved at the expense of estimated intercepts that are substantially greater than their theoretical values (i.e. the risk-free rate). To address this, we follow Nagel and Singleton (2011) and include the risk-free portfolio as an extra asset in our study.

We consider the orthogonal relationship between the risk-free rate and the pricing kernel for the following extra moment condition:[†]

$$0 = \mathbb{E}\left[M(t+1) - \frac{1}{1 + R_f(t+1)} \right]. \tag{14.11}$$

---

[*] Multiple conditioning variables have been used in the literature. For example, Lettau and Ludvigson (2001b) propose CAPM and CCAPM models conditioned on the *cay* variable. Lustig and Van Nieuwerburgh (2005) derive a conditional CCAPM on the housing collateral ratio (ratio of housing wealth to human wealth). Santos and Veronesi (2006) present a conditional CAPM on the labour income to consumption ratio (fraction of consumption funded by labour income). In these studies the evidence of predictability is only an intermediate step that enables the authors to fulfil the first requisite needed for the existence of a conditional asset pricing model, or in other words, that the equity risk premium is variable. Then, given that the predictability condition is satisfied, there should be a conditional model that correctly values the cross section of returns using the proposed predictor.

[†] The grounds for this moment condition is that the risk-free rate is known at time t, therefore $1/\big[1 + R_f(t+1)\big] = \mathbb{E}\big[M(t+1) \big| \Omega(t) \big]$. By taking the unconditional mean on both sides yields equation (14.11).

In addition, we use the beta representation of Equation (14.9) to produce the unconditional expressions for the expected risk-free return and the expected return of asset i given by

$$\mathbb{E}\left[R_f(t+1)\right] = \frac{1}{\mathbb{E}[M(t+1)]} - 1, \tag{14.12}$$

$$\mathbb{E}\left[R_i(t+1)\right] = \mathbb{E}\left[R_f(t+1)\right] - \mathbb{E}\left[1 + R_f(t+1)\right]$$
$$\times \operatorname{Cov}\left[M(t+1), R_i(t+1)t\right], \tag{14.13}$$

where in Equation (14.12) we are assuming that the risk-free asset is unconditionally orthogonal to $M(t+1)$ (i.e. the risk-free asset is a zero-beta asset). Equations (14.12) and (14.13) can be used to assess the fit of an estimated model to the cross section of average returns.

For the cross sectional tests we use the standard 25 portfolios of Fama and French (1993) ordered by size and book-to-market, in addition to the risk-free portfolio, therefore $N = 26$. These series as well as the SMB (small minus big) and HML (high minus low) factors from the FF three-factor model are available from Ken French's web page. The real per-capita consumption series was constructed using data from the Bureau of Economic Analysis. Specifically, we construct our quarterly series from nominal consumption of non-durables and services, seasonally adjusted, per capita (NIPA Table 7.1). Real consumption was calculated by deflating the nominal series by the PCE (personal consumption expenditures) price index, 2005 = 100 (NIPA Table 2.3.4).

Tables 14.10–14.12 present the results of the estimation by the Generalized Method of Moments (GMM) for the moment conditions in Equations (14.9) and (14.11).[*] We report the asymptotic t-stats, and the Wald and J_T tests based on the covariance matrices of pricing errors corrected for heteroscedasticity and serial correlation using the Newey–West estimator. We use the root of mean square errors (RMSE) to measure the fit of an estimated model to the cross-section of average returns. Figures 14.3–14.15 plot the fitted expected returns for the 26 portfolios against their realized average returns.[†]

Table 14.10 and Figures 14.3–14.5 show the results for the unconditional CAPM, CCAPM and FF three-factor models. The J_T tests do not reject any conditional or unconditional model, so our inference is based solely on the t-stats and Wald tests. The latter tests the hypothesis that all coefficients except the constant are zero. According to the Wald test, no model is significant at the 5% level; nevertheless, the FF three-factor model presents the best cross sectional adjustment ($RMSE = 0.55\%$).

Table 14.11 and Figures 14.6–14.10 present the results for the conditional CAPM models. A Wald test for the null hypothesis that the conditional CCAPM model does not improve the adjustment relative to the unconditional CCAPM model is included in the p-value Wald CAPM row (i.e. it tests whether the additional coefficients in the conditional CCAPM model are zero). The table shows that the CAPM conditional on Δf exhibits by far the best forecasting power for the cross-section of expected returns ($RMSE = 0.34\%$); moreover, in accordance with both Wald tests, it is the only significant one at the 5% level. The effect of the risk factor $R_m(t)$ is significant only through its interaction with $\Delta f(t-4)$. This implies that a positive oil price shock accompanied by a subsequent decline in market return ($\Delta f(t-4) \cdot R_m(t) < 0$) causes a drop in future portfolio returns. This effect strongly suggests that our variable also has forecasting power for returns on more disaggregated stock portfolios. Finally, CAPM models

---

[*] We use the identity weighting matrix for all estimates, based on the following reasons. First, we do not have theoretical arguments for giving more or less importance to a particular portfolio. Second, the number of moment conditions ($N = 26$) is large relative to our sample size ($T = 103$), so this choice avoids dealing with estimates that depend on unstable and near singular error covariance matrices.

[†] The 25 portfolios, sorted by size and book-to-market ratio, are labelled with two digits. The first digit refers to the size quintile (1 indicating the smallest firms, 5 the largest), and the second digit refers to book-to-market quintile (1 indicating the portfolio with the lowest book-to-market ratio, 5 with the highest).

TABLE 14.10 Unconditional Asset Pricing Models, 1983Q2–2009Q4. Gmm Estimates of Pricing Kernel Coefficients for the Unconditional Models. The Models Are Estimated Using Returns on the Fama and French (1993) Portfolios and a Risk-Free Portfolio ($N = 26$). The Identity-Weighting Matrix Is Used in All Estimates. All Tests Are Based on Covariance Matrices of Errors Corrected for Heteroscedasticity and Serial Correlation (Newey and West 1987). Lag Length in the Newey–West Estimator Is $floor[4 \cdot (T/100)^{2/9}]$, Where $floor[x]$ Denotes the Integer Part of x. Asymptotic t-Stat in Parentheses. At the Bottom of the Table, We Report the P-Value for the J_t Test of the Null That All Pricing Errors Are Zero. p-Value Wald *All* Is the p-Value for the Asymptotic Wald Test That All Coefficients Except the Constant Are Zero. Rmse Is the Root of Mean Square Errors and Measures the Fit of the Estimated Model to the Cross Section of Average Returns

	CAPM	CCAPM	FF Three-Factor
Constant	−0.027	0.506	0.022
	(−0.66)	(1.31)	(0.60)
$R_m(t)$	−0.139		
	(−0.11)		
$\Delta c(t)$		−121.603	
		(−1.21)	
$R_m(t) - R_f(t)$			−2.058
			(−1.25)
SMB(t)			0.259
			(0.09)
HML(t)			−3.986
			(−1.80)
p-Value J_T	0.44	0.43	0.32
p-Value Wald *All*	0.91	0.23	0.28
RMSE (%)	0.78	0.74	0.55

conditioned on $d - p$ and *gap* slightly outperform the FF three-factor model, but none of these models is statistically significant.

Table 14.12 and Figures 14.11–14.15 display the outcomes for conditional CCAPM models. Based on these outcomes, we note that the CCAPM conditional on Δf is the only one that outperforms the FF three-factor model; however, it is not significant at the 5% level. Also, the CCAPM model conditioned on R_f is statistically significant but is outperformed by several conditional and unconditional models ($RMSE = 0.60\%$).

In summary, the conditional CAPM on our variable has significant predictive power for the cross-section of expected stock returns and has a better fit than all unconditional and conditional models considered here.

14.7 Conclusions

Although the predictability of stock market excess returns has been associated with the business cycle, the evidence that supports this relationship is far from being conclusive. Moreover, when the sample is extended to include the period of the sub-prime crisis, none of the popular predictors exhibit forecasting power for market excess returns. In this paper, we show that such a relationship exists.

We find that unexpected oil price changes, a non-persistent variable with deep macroeconomic roots, have significant forecasting power for stock returns at short horizons. Our variable, proxied by futures returns on crude oil, shows statistically and economically significant predictive power for stock returns at horizons from one to three quarters. Its predictive power outperforms those of the risk-free rate, the dividend–price ratio, the consumption–wealth ratio and the output gap, with quarterly $\bar{R}^2$ between 6% and 7%. This result is robust against the inclusion of other variables and out-of-sample tests. However,

TABLE 14.11 Conditional CCAPM Models, 1983Q2–2009Q4. GMM Estimates of Pricing Kernel Coefficients for the Conditional CCAPM Models on Variables Are in the First Row. The Models Are Estimated Using Returns on Fama and French's (1993) 25 Portfolios and the Risk-Free Asset ($N = 26$). The Identity-Weighting Matrix Is Used in all Estimates. All Tests Are Based on Covariance Matrices of Errors Corrected for Heteroscedasticity and Serial Correlation. Lag Length in the Newey–West Estimator Is $Floor[4 \cdot (T/100)^{2/9}]$, Where $Floor[x]$ Denotes the Integer Part of X. Asymptotic T-Stat in Parentheses. At the Bottom of Table, We Report the p-Value for the J_t Test of the Null That All Pricing Errors Are Zero. The p-Values Presented Are for the Asymptotic Wald Tests. *All* Means All Coefficients Except the Constant. *CCAPM* Means the Coefficients That Are Not in the Unconditional CAPM Model. RMSE Is the Root of Mean Square Errors and Measures the Adjustment of an Estimated Model to the Cross Section of Average Returns

	Δf	R_f	$d-p$	cay	gap
Constant	−1.262	0.266	1.962	−0.413	0.007
	(−2.61)	(0.41)	(0.38)	(−0.75)	(0.09)
$\Delta f(t-1)$	−2.744				
	(−2.18)				
$\Delta f(t-2)$	−3.239				
	(−2.68)				
$\Delta f(t-3)$	−1.387				
	(−0.46)				
$\Delta f(t-4)$	6.932				
	(3.07)				
$R_f(t)$		−39.419			
		(−0.66)			
$d(t-1)-p(t-1)$			0.520		
			(0.38)		
$cay(t-1)$				42.908	
				(1.13)	
$gap(t-1)$					0.283
					(0.03)
$R_m(t)$	1.157	−7.826	−59.513	1.602	−1.634
	(0.59)	(−1.42)	(−1.96)	(0.44)	(−0.96)
$\Delta f(t-1) \cdot R_m(t)$	−20.573				
	(−1.21)				
$\Delta f(t-2) \cdot R_m(t)$	−17.374				
	(−1.63)				
$\Delta f(t-3) \cdot R_m(t)$	11.929				
	(0.54)				
$\Delta f(t-4) \cdot R_m(t)$	−63.751				
	(−2.92)				
$Rf(t) \cdot R_m(t)$		785.090			
		(1.42)			
$(d(t-1)-p(t-1)) \cdot R_m(t)$			−15.186		
			(−1.90)		
$cay(t-1) \cdot R_m(t)$				−251.228	
				(−0.95)	
$gap(t-1) \cdot R_m(t)$					89.443
					(1.79)
p-Value J_T	0.07	0.32	0.33	0.33	0.35
p-Value Wald *All*	0.00	0.46	0.19	0.45	0.25
p-Value Wald *CAPM*	0.00	0.33	0.15	0.52	0.16
RMSE (%)	0.34	0.67	0.53	0.72	0.53

TABLE 14.12   Conditional CCAPM Models, 1983Q2–2009Q4. GMM Estimates of Pricing Kernel Coefficients for the Conditional CCAPM Models on Variables Are in the First Row. The Models Are Estimated Using Returns on Fama and Frenchs (1993) 25 Portfolios and the Risk-Free Asset ($N = 26$). The Identity-Weighting Matrix Is Used in all Estimates. All Tests Are Based on Covariance Matrices of Errors Corrected for Heteroscedasticity and Serial Correlation. Lag Length in the Newey–West Estimator Is $Floor[4 . (T/100)^{2/9}]$, Where $Floor[X]$ Denotes the Integer Part of x. Asymptotic t-Stat in Parentheses. At the Bottom of Table, We Report the p-Value for the J_t test of the Null That All Pricing Errors Are Zero. The p-Values Presented Are for the Asymptotic Wald Tests. *All* Means all Coefficients Except the Constant. *CCAPM* Means the Coefficients That Are Not in the Unconditional CCAPM Model. RMSE Is the Root of Mean Square Errors and Measures the Adjustment of an Estimated Model to the Cross-Section of Average Returns

	Δf	R_f	$d - p$	cay	gap
Constant	0.811	1.033	1.877	0.375	0.485
	(2.25)	(1.48)	(0.44)	(0.86)	(1.50)
$\Delta f(t-1)$	−0.483				
	(−0.38)				
$\Delta f(t-2)$	−1.858				
	(−2.15)				
$\Delta f(t-3)$	0.384				
	(0.29)				
$\Delta f(t-4)$	−0.407				
	(−0.17)				
$R_f(t)$		−186.224			
		(−2.18)			
$d(t-1) - p(t-1)$			0.364		
			(0.32)		
$cay(t-1)$				19.948	
				(0.76)	
$gap(t-1)$					−0.253
					(−0.04)
$\Delta c(t)$	−308.738	−320.549	−390.098	−160.464	−144.812
	(−2.77)	(−1.83)	(−0.41)	(−1.57)	(−1.51)
$\Delta f(t-1) \cdot \Delta c(t)$	116.051				
	(0.35)				
$\Delta f(t-2) \cdot \Delta c(t)$	−134.842				
	(−0.59)				
$\Delta f(t-3) \cdot \Delta c(t)$	343.002				
	(1.09)				
$\Delta f(t-4) \cdot \Delta c(t)$	651.304				
	(1.46)				
$R_f(t) \cdot \Delta c(t)$		34600.413			
		(2.88)			
$(d(t-1) - p(t-1)) \cdot \Delta c(t)$			−68.620		
			(−0.27)		
$cay(t-1) \cdot \Delta c(t)$				1269.334	
				(0.34)	
$gap(t-1) \cdot \Delta c(t)$					1302.922
					(1.05)
p-Value J_T	0.08	0.34	0.33	0.32	0.34
p-Value Wald *All*	0.10	0.03	0.44	0.33	0.47
p-Value Wald *CCAPM*	0.24	0.01	0.93	0.74	0.54
RMSE (%)	0.48	0.60	0.73	0.69	0.68

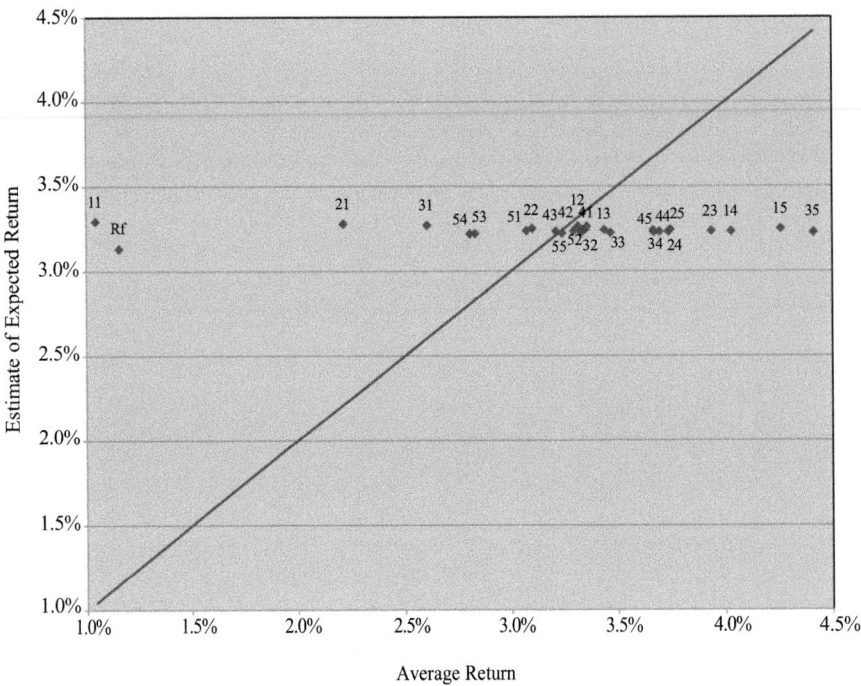

FIGURE 14.3   Realized vs. expected returns by CAPM.

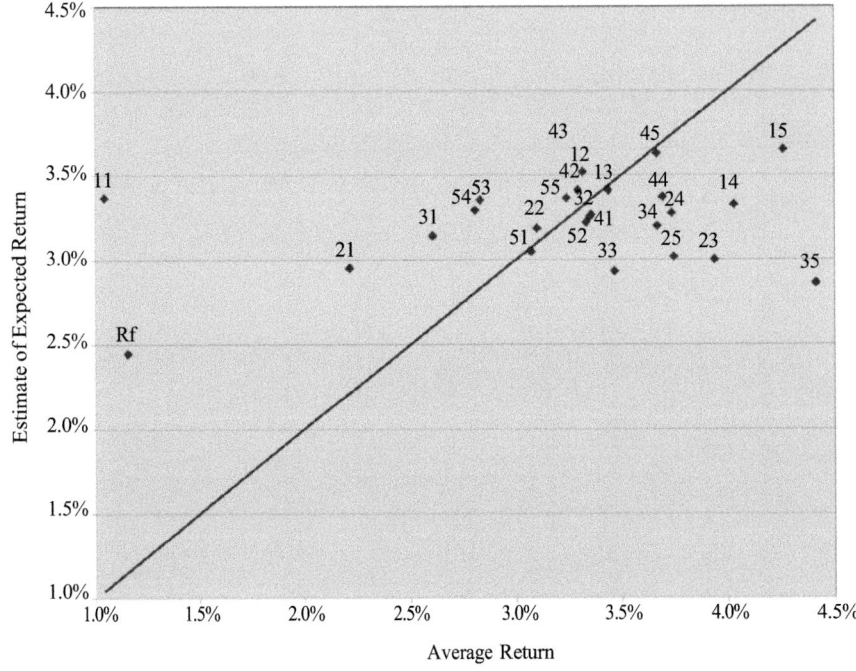

FIGURE 14.4   Realized vs. expected returns by CCAPM.

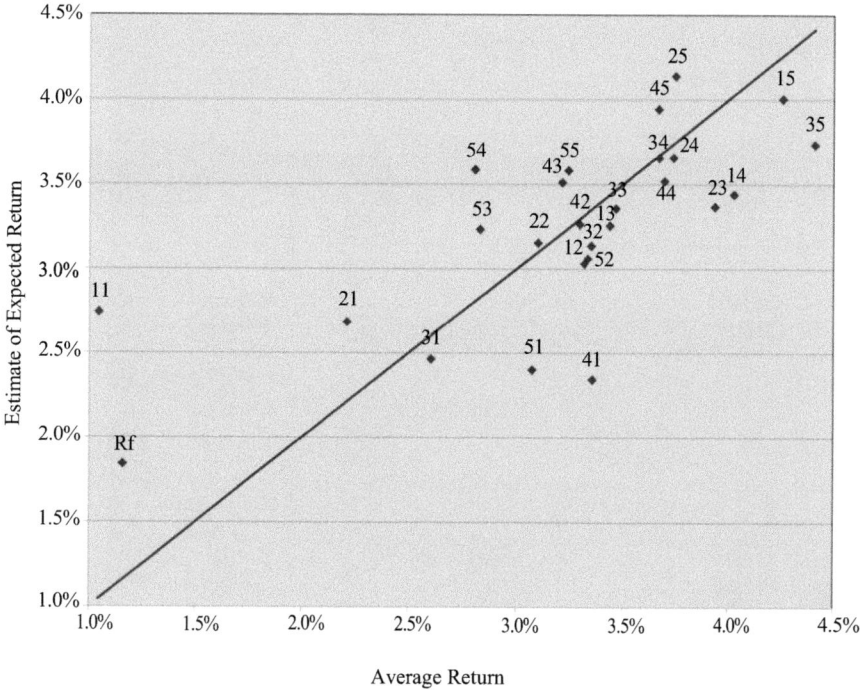

FIGURE 14.5   Realized vs. expected returns by FF three-factor.

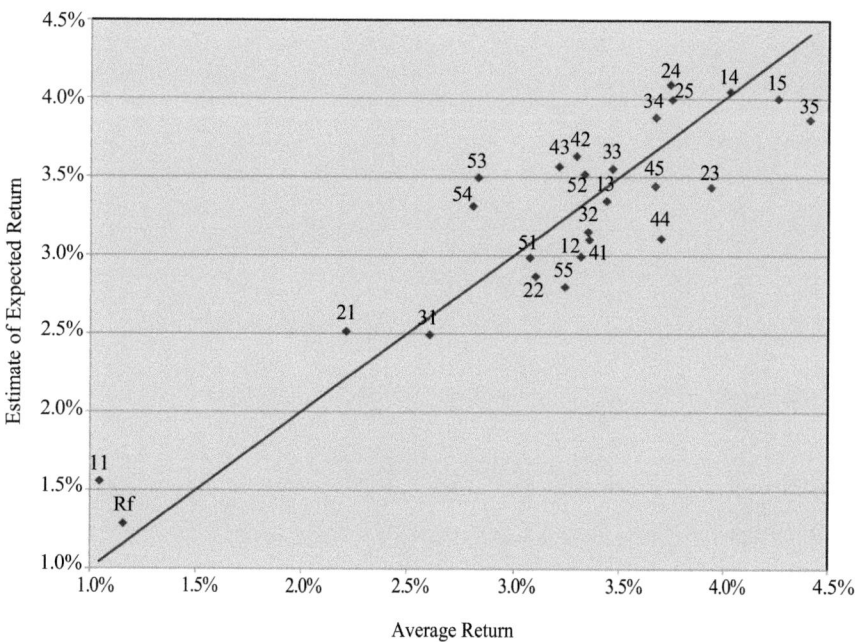

FIGURE 14.6   Realized vs. expected returns by CAPM conditioned on Δ*f*.

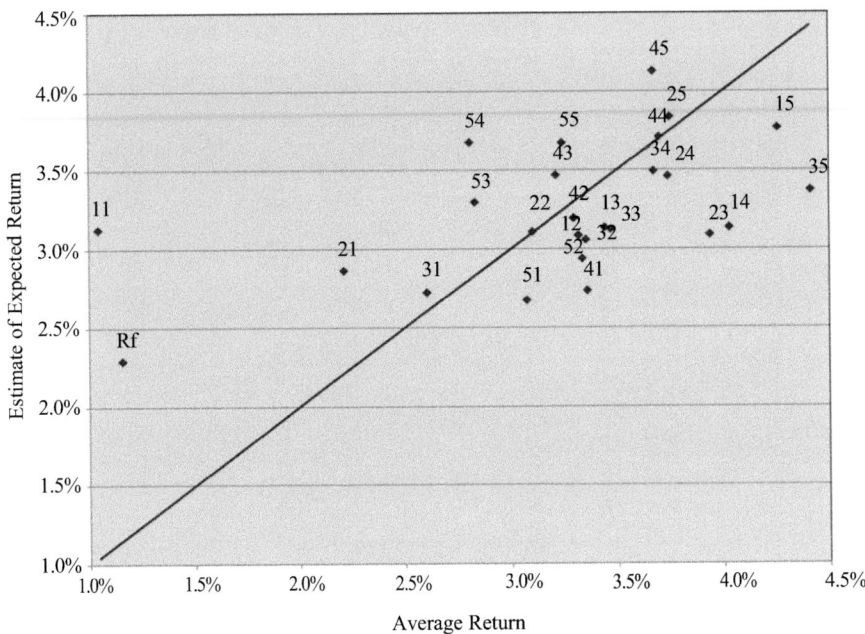

FIGURE 14.7   Realized vs. expected returns by CAPM conditioned on R_f.

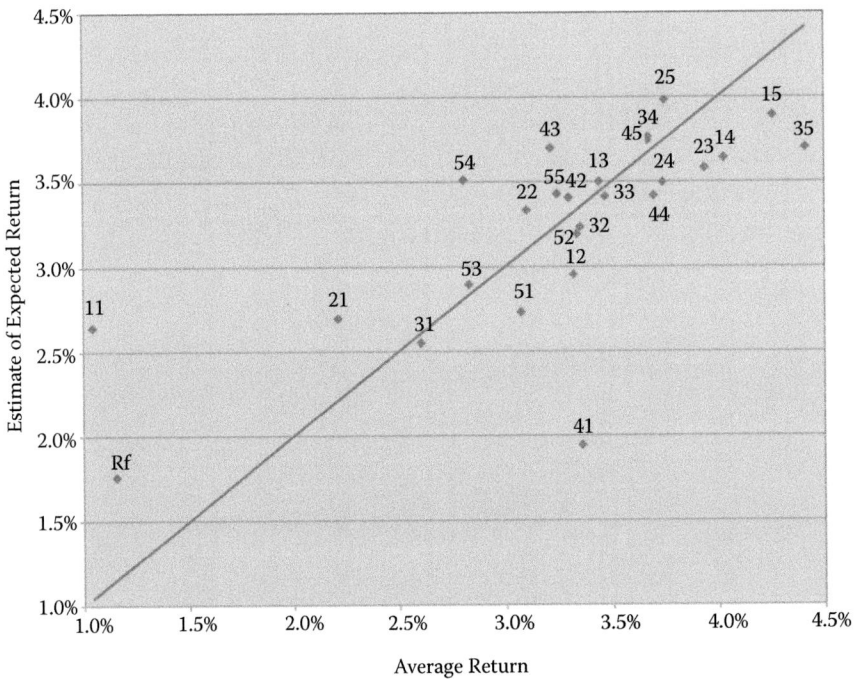

FIGURE 14.8   Realized vs. expected returns by CAPM conditioned on $d - p$.

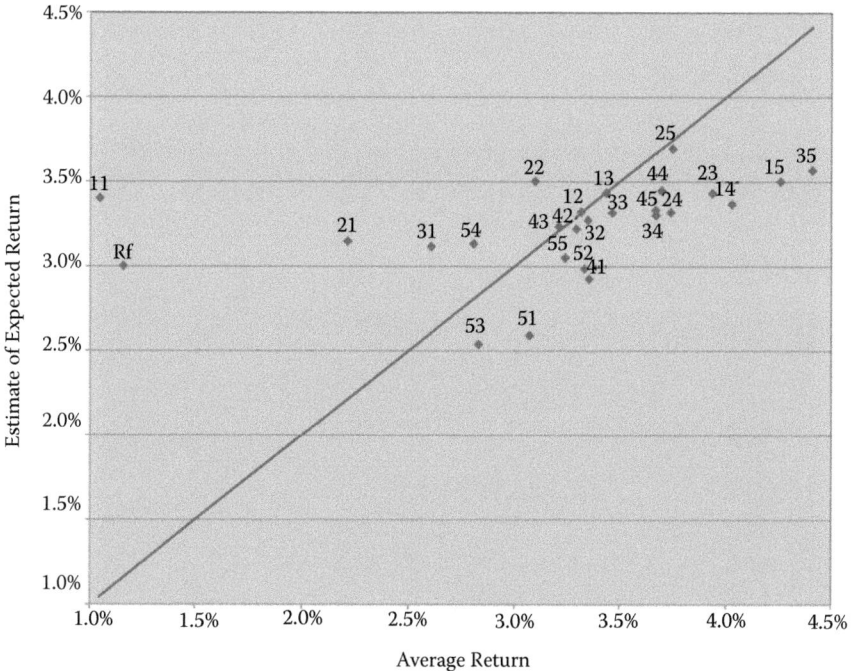

FIGURE 14.9   Realized vs. expected returns by CAPM conditioned on *cay*.

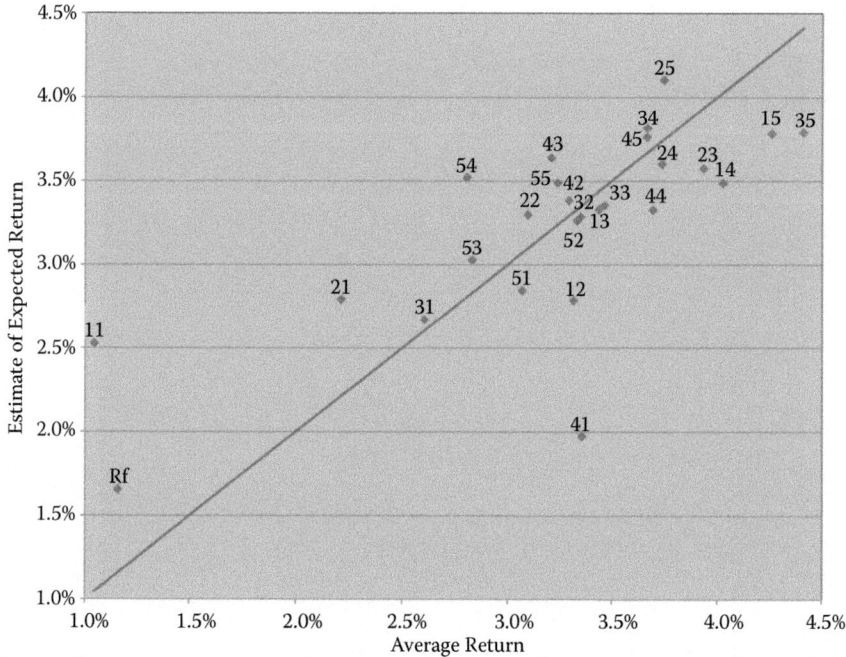

FIGURE 14.10   Realized vs. expected returns by CAPM conditioned on *gap*.

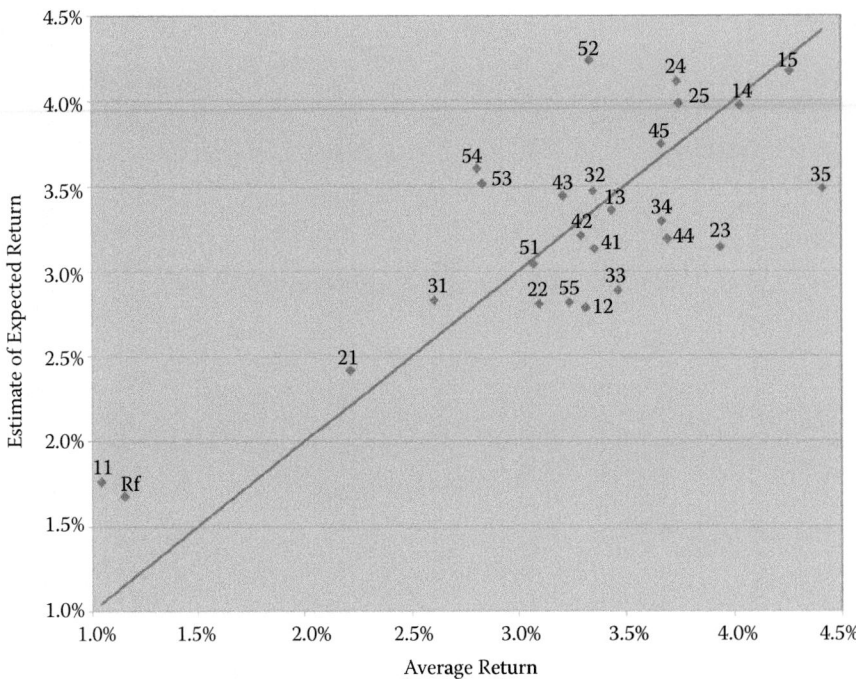

FIGURE 14.11   Realized vs. expected returns by CCAPM conditioned on Δ*f*.

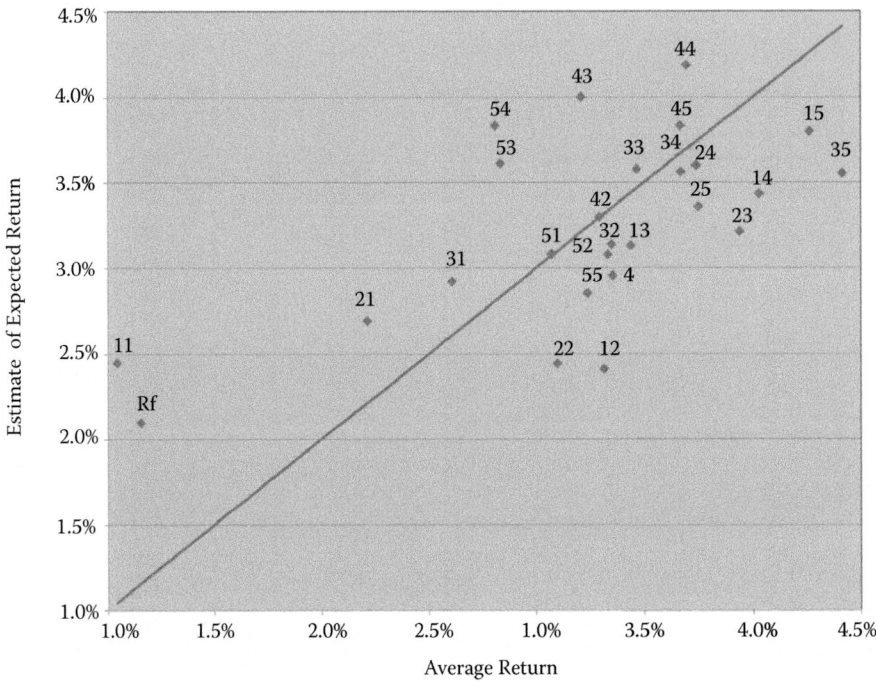

FIGURE 14.12   Realized vs. expected returns by CCAPM conditioned on R_f.

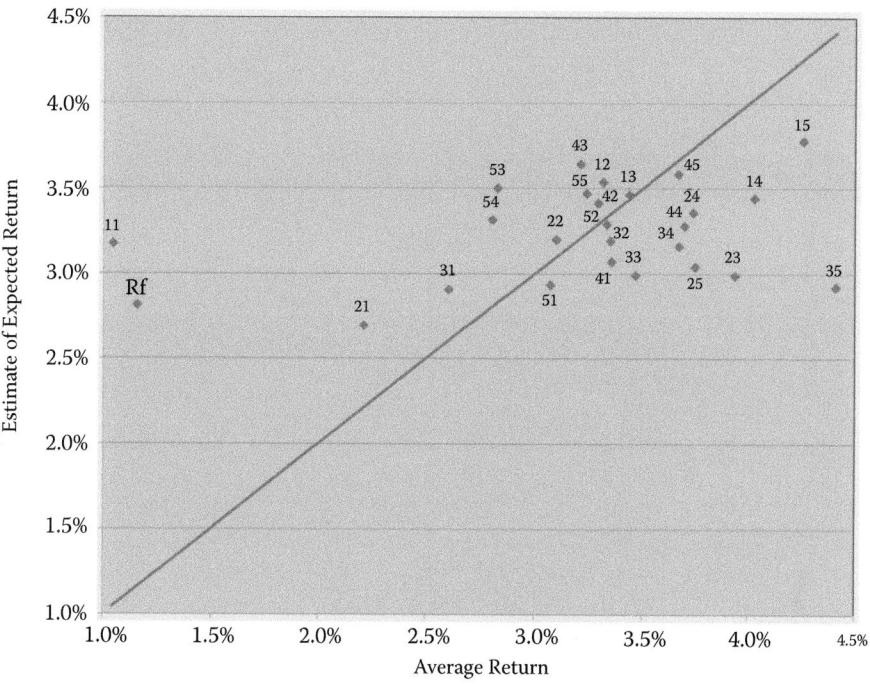

FIGURE 14.13 Realized vs. expected returns by CCAPM conditioned on $d - p$.

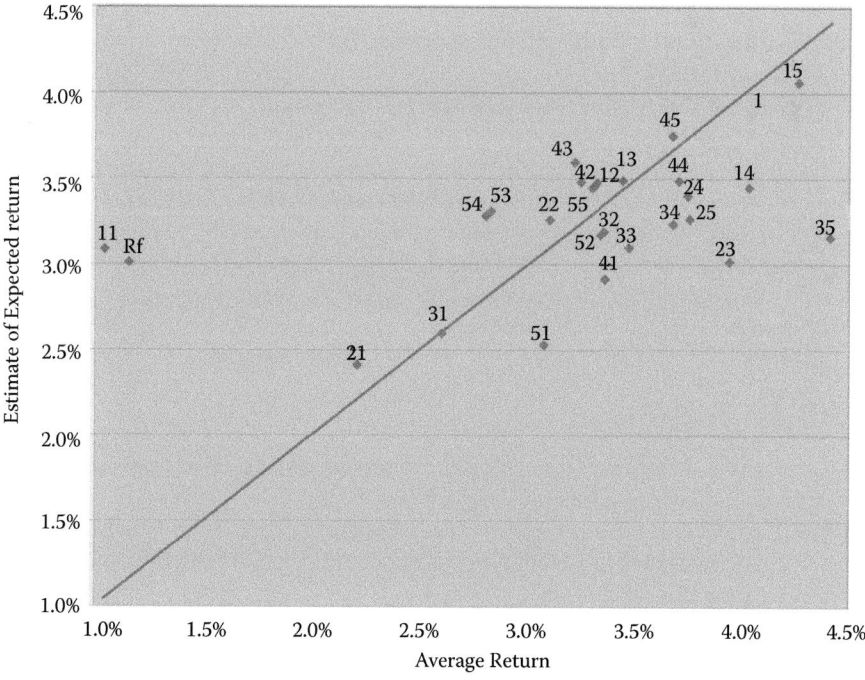

FIGURE 14.14 Realized vs. expected returns by CCAPM conditioned on *cay*.

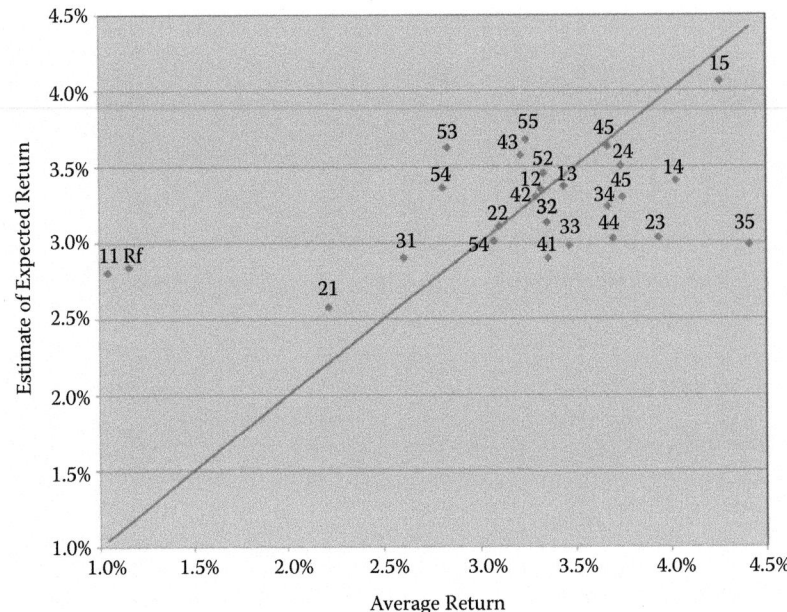

FIGURE 14.15   Realized vs. expected returns by CCAPM conditioned on *gap*.

at longer horizons, none of the variables displays significant forecasting ability. Our results also validate the recent findings of Boudoukh *et al.* (2008) that unstable results in previous studies in this literature are due to the high persistence of the predictors used.

Our variable also shows a significant forecasting power for the cross section of expected returns. We build a conditional CAPM model on oil price shocks, which shows high statistical significance and better adjustment than all conditional and unconditional models considered, including the Fama and French (1993) three-factor model.

From a practical perspective, unlike variables based on macroeconomic series, such as the consumption–wealth ratio and the output gap, our variable can be directly observed and is available on a daily basis at no cost. These characteristics make use of our variable by potential investors highly feasible.

Finally, an open question motivated by the emerging literature on the financialization of commodities is how well is the forecasting ability of other commodities on stock returns. The correlation of indexed commodities and oil has increased dramatically since 2004 because of the speculative trading in futures markets, suggesting that these commodities would also have some predictive power. We leave this topic for future research.

Acknowledgements

We would like to thank Augusto Castillo, Gonzalo Cortazar, Rodrigo Fuentes, Borja Larrain, Chris Telmer, Jose Tessada, Eduardo Walker and seminar participants at UC. Any errors or omissions are the responsibility of the authors. Casassus acknowledges financial support from FONDECYT (grant No. 1110841) and from Grupo Security through FinanceUC. Higuera acknowledges financial support from CONICYT and FinanceUC.

References

Acharya, V.V., Lochstoer, L.A. and Ramadorai, T., Limits to arbitrage and hedging: Evidence from commodity markets. NBER Working Paper No. 16875, 2011.

Aguiar-Conraria, L. and Wen, Y., Understanding the large negative impact of oil shocks. *J. Money, Credit Bank.*, 2007, **39**(4), 925–944.

Ang, A. and Bekaert, G., Stock return predictability: Is it there? *Rev. Financ. Stud.*, 2007, **20**(3), 651–707.

Apergis, N. and Miller, S.M., Do structural oil-market shocks affect stock prices? *Energy Econ.*, 2009, **31**(4), 569–575.

Bachmeier, L., Monetary policy and the transmission of oil shocks. *J. Macroecon.*, 2008, **30**(4), 1738–1755.

Bakshi, G., Panayotov, G. and Skoulakis, G., The Baltic Dry Index as a predictor of global stock returns, commodity returns, and global economic activity. Working Paper, University of Maryland, 2011.

Balke, N.S., Brown, S.P.A. and Yucel, M.K., Oil price shocks and the U.S. economy: Where does the asymmetry originate? *Energy J.*, 2002, **23**(3), 27–52.

Barsky, R.B. and Kilian, L., Oil and the macroeconomy since the 1970s. *J. Econ. Perspect.*, 2004, **18**(4), 115–134.

Basak, S. and Pavlova, A., A model of financialization of commodities. Working paper, London Business School, 2012.

Berkowitz, J. and Kilian, L., Recent developments in bootstrapping time series. *Econometr. Rev.*, 2000, **19**(1), 1–48.

Bernanke, B.S., Irreversibility, uncertainty, and cyclical investment. *Q. J. Econ.*, 1983, **98**(1), 85–106.

Bernanke, B.S., Gertler, M. and Watson, M.W., Systematic monetary policy and the effects of oil price shocks. *Brook. Pap. Econ. Activ.*, 1997, **1997**(1), 91–157.

Boudoukh, J., Richardson, M. and Whitelaw, R.F., The myth of long-horizon predictability. *Rev. Financ. Stud.*, 2008, **21**(4), 1577–1605.

Breeden, D.T. and Litzenberger, R.H., Prices of state-contingent claims implicit in option prices. *J. Business*, 1978, **51**(4), 621–651.

Buyuksahin, B. and Robe, M.A., Speculators, commodities and cross-market linkages. Working paper, American University, 2010.

Campbell, J.Y., Stock returns and the term structure. *J. Financ. Econ.*, 1987, **18**(2), 373–399.

Campbell, J.Y., A variance decomposition for stock returns. *Econ. J.*, 1991, **101**(405), 157–179.

Campbell, J.Y. and Thompson, S.B., Predicting excess stock returns out of sample: Can anything beat the historical average? *Rev. Financ. Stud.*, 2008, **21**(4), 1509–1531.

Carruth, A.A., Hooker, M.A. and Oswald, A.J., Unemployment equilibria and input prices: Theory and evidence from the United States. *Rev. Econ. Statist.*, 1998, **80**(4), 621–628.

Casassus, J. and Collin-Dufresne, P., Stochastic convenience yield implied from commodity futures and interest rates. *J. Finance*, 2005, **60**(5), 2283–2332.

Ciner, C., Energy shocks and financial markets: Nonlinear linkages. *Stud. Nonlin. Dynam. Econometr.*, 2001, **5**(3), 203–212.

Clark, T.E. and McCracken, M.W., Tests of equal forecast accuracy and encompassing for nested models. *J. Econometr.*, 2001, **105**, 85–110.

Clark, T.E. and McCracken, M.W., Evaluating direct multistep forecasts. *Econometr. Rev.*, 2005, **24**(4), 369–404.

Cochrane, J.H., *Asset Pricing*, 2005 (Princeton University Press: Princeton).

Cologni, A. and Manera, M., Oil prices, inflation and interest rates in a structural cointegrated VAR model for the G-7 countries. *Energy Econ.*, 2008, **30**(3), 856–888.

Cologni, A. and Manera, M., The asymmetric effects of oil shocks on output growth: A Markov-switching analysis for the G-7 countries. *Econ. Model.*, 2009, **26**(1), 1–29.

Cooper, I. and Priestley, R., Time-varying risk premiums and the output gap. *Rev. Financ. Stud.*, 2009, **22**(7), 2801–2833.

Davis, S.J. and Haltiwanger, J., Sectoral job creation and destruction responses to oil price changes. *J. Monet. Econ.*, 2001, **48**(3), 465–512.

Diebold, F.X. and Mariano, R.S., Comparing predictive accuracy. *J. Business Econ. Statist.*, 1995, **13**(3), 253–263.

Driesprong, G., Jacobsen, B. and Maat, B., Striking oil: Another puzzle? *J. Financ. Econ.*, 2008, **89**(2), 307–327.

Etula, E., Broker{dealer risk appetite and commodity returns. Working paper, Federal Reserve Bank of New York, 2010.

Fama, E.F. and French, K.R., Commodity futures prices: Some evidence on forecast power, premiums, and the theory of storage. *J. Business*, 1987, **60**(1), 55–73.

Fama, E.F. and French, K.R., Dividend yields and expected stock returns. *J. Financ. Econ.*, 1988, **22**(1), 3–25.

Fama, E.F. and French, K.R., Common risk factors in the returns on stocks and bonds. *J. Financ. Econ.*, 1993, **33**(1), 3–56.

Ferderer, J.P., Oil price volatility and the macroeconomy. *J. Macroecon.*, 1996, **18**(1), 1–26.

Finn, M.G., Perfect competition and the effects of energy price increases on economic activity. *J. Money, Credit Bank.*, 2000, **32**(3), 400–416.

Gronwald, M., Large oil shocks and the US economy: Infrequent incidents with large effects. *Energy J.*, 2008, **29**(1), 151–171.

Hamilton, J.D., Oil and the macroeconomy since World War II. *J. Polit. Econ.*, 1983, **91**(2), 228–248.

Hamilton, J.D., A neoclassical model of unemployment and the business cycle. *J. Polit. Econ.*, 1988, **96**(3), 593–617.

Hamilton, J.D., This is what happened to the oil price–macroeconomy relationship. *J. Monet. Econ.*, 1996, **38**(2), 215–220.

Hamilton, J.D., What is an oil shock?. *J. Econometr.*, 2003, **113**(2), 363–398.

Hamilton, J.D., Oil and the macroeconomy. In *The New Palgrave Dictionary of Economics*, 2nd ed., edited by S.N. Durlauf and L.E. Blume, 2008 (Palgrave Macmillan: Basingstoke). Available online at: http://www.dictionaryofeconomics.com/article?id=pde2008_E000233

Hamilton, J.D. and Herrera, A.M., Comment: Oil shocks and aggregate macroeconomic behavior: The role of monetary policy. *J. Money, Credit Bank.*, 2004, **36**(2), 265–286.

Harvey, D.I., Leybourne, S.J. and Newbold, P., Tests for forecast encompassing. *J. Business Econ. Statist.*, 1998, **16**(2), 254–259.

Hooker, M.A., What happened to the oil price–macroeconomy relationship? *J. Monet. Econ.*, 1996, **38**(2), 195–213.

Huang, R.D., Masulis, R.W. and Stoll, H.R., Energy shocks and financial markets. *J. Fut. Mkts*, 1996, **16**(1), 1–27.

Huntington, H.G., Oil shocks and real U.S. income. *Energy J.*, 2007, **28**(4), 31–46.

Inoue, A. and Kilian, L., In-sample or out-of-sample tests of predictability: Which one should we use? *Econometr. Rev.*, 2005, **23**(4), 371–402.

Jones, C. and Kaul, G., Oil and the stock markets. *J. Finance*, 1996, **51**(2), 463–491.

Kilian, L., Small-sample confidence intervals for impulse response functions. *Rev. Econ. Statist.*, 1998, **80**(2), 218–230.

Kilian, L., Exchange rates and monetary fundamentals: What do we learn from long-horizon regressions? *J. Appl. Econometr.*, 1999, **14**(5), 491–510.

Kilian, L., A comparison of the effects of exogenous oil supply shocks on output and inflation in the G7 countries. *J. Eur. Econ. Assoc.*, 2008, **6**(1), 78–121.

Kilian, L., Not All oil price shocks are alike: Disentangling demand and supply shocks in the crude oil market. *Am. Econ. Rev.*, 2009, **99**(3), 1053–1069.

Kilian, L. and Park, C., The impact of oil price shocks on the U.S. stock market. *Int. Econ. Rev.*, 2009, **50**(4), 1267–1287.

Kliesen, K.L., Oil and the U.S. macroeconomy: An update and a simple forecasting exercise. *Fed. Reserve Bank St. Louis Rev.*, 2008, **90**(5), 505–516.

Kumar, M.S., The forecasting accuracy of crude oil futures prices. *Staff Papers – International Monetary Fund*, 1992, **39**(2), 432–461.

Leduc, S. and Sill, K., A quantitative analysis of oil-price shocks, systematic monetary policy, and economic downturns. *J. Monet. Econ.*, 2004, **51**(4), 781–808.

Lee, K. and Ni, S., On the dynamic effects of oil price shocks: A study using industry level data. *J. Monet. Econ.*, 2002, **49**(4), 823–852.

Lee, K., Ni, S. and Ratti, R.A., Oil shocks and the macroeconomy: The role of price variability. *Energy J.*, 1995, **16**(4), 39–56.

Lettau, M. and Ludvigson, S.C., Consumption, aggregate wealth, and expected stock returns. *J. Finance*, 2001a, **56**(3), 815–849.

Lettau, M. and Ludvigson, S.C., Resurrecting the (C)CAPM: A cross sectional test when risk premia are time-varying. *J. Polit. Econ.*, 2001b, **109**(6), 1238–1287.

Lewellen, J., Nagel, S. and Shanken, J., A skeptical appraisal of asset pricing tests. *J. Financ. Econ.*, 2010, **96**(2), 175–194.

Lilien, D.M., Sectoral shifts and cyclical unemployment. *J. Polit. Econ.*, 1982, **90**(4), 777–793.

Lustig, H.N. and Van Nieuwerburgh, S.G., Housing collateral, consumption insurance, and risk premia: An empirical perspective. *J. Finance*, 2005, **60**(3), 1167–1219.

Ma, C.W., Forecasting efficiency of energy futures prices. *J. Fut. Mkts*, 1989, **9**(5), 393–419.

Masters, M.W., Testimony before Committee on Homeland Security and Governmental Affairs of the United States Senate, May 2008.

McCracken, M.W., Asymptotics for out of sample tests of Granger causality. *J. Econometr.*, 2007, **140**, 719–752.

Mork, K.A., Oil and the macroeconomy when prices go up and down: An extension of Hamilton's results. *J. Polit. Econ.*, 1989, **97**(3), 740–744.

Nagel, S. and Singleton, K.J., Estimation and evaluation of conditional asset pricing models. *J. Finance*, 2011, **66**(3), 873–909.

Nandha, M. and Faff, R., Does oil move equity prices? A global view. *Energy Econ.*, 2008, **30**(3), 986–997.

Newey, W.K. and West, K.D., A simple, positive semi-definite, heteroskedasticity and autocorrelation consistent covariance matrix. *Econometrica*, 1987, **55**(3), 703–708.

Newey, W.K. and West, K.D., Automatic lag selection in covariance matrix estimation. *Rev. Econ. Stud.*, 1994, **61**(4), 631–653.

Park, J. and Ratti, R.A., Oil price shocks and stock markets in the U.S. and 13 European countries. *Energy Econ.*, 2008, **30**(5), 2587–2608.

Sadorsky, P., Oil price shocks and stock market activity. *Energy Econ.*, 1999, **21**(5), 449–469.

Santos, T. and Veronesi, P., Labor income and predictable stock returns. *Rev. Financ. Stud.*, 2006, **19**(1), 1–44.

Sharpe, W.F., Capital asset prices: A theory of market equilibrium under conditions of risk. *J. Finance*, 1964, **19**(3), 425–442.

Sims, C.A., Macroeconomics and reality. *Econometrica*, 1980, **48**(1), 1–48.

Singleton, K.J., Investor ows and the 2008 boom/bust in oil prices. Working paper, Stanford University, 2012.

Stambaugh, R.F., Predictive regressions. *J. Financ. Econ.*, 1999, **54**(3), 375–421.

Stine, R.A., Estimating properties of autoregressive forecasts. *J. Am. Statist. Assoc.*, 1987, **82**(400), 1072–1078.

Switzer, L.N. and El-Khoury, M., Extreme volatility, speculative efficiency, and the hedging effectiveness of the oil futures markets. *J. Fut. Mkts*, 2007, **27**(1), 61–84.

Tang, K. and Xiong, W., Index investment and the financialization of commodities. *Financ. Anal. J.*, 2012, **68**(6), 54–74.

Valkanov, R., Long-horizon regressions: Theoretical results and applications. *J. Financ. Econ.*, 2003, **68**(2), 201–232.

Wei, C., Energy, the stock market, and the Putty-Clay Investment model. *Am. Econ. Rev.*, 2003, **93**(1), 311–323.

Welch, I. and Goyal, A., A comprehensive look at the empirical performance of equity premium prediction. *Rev. Financ. Stud.*, 2008, **21**(4), 1455–1508.

Appendix: Tests for Out-of-Sample Predictability

This appendix presents the three metrics we use to test the out-of-sample performance of the predictors. First, we estimate the forecast errors for the benchmark and competing models in Equations (14.4) and (14.5) as

$$\text{benchmark: } \hat{u}_1(t+1) = \left[R_m(t+1) - R_f(t+1) \right] - \hat{\alpha}_1(t), \tag{A.1}$$

$$\text{competing: } \hat{u}_2(t+1) = \left[R_m(t+1) - R_f(t+1) \right]$$

$$-\hat{\alpha}_2(t) - \hat{\beta}(t)' X(t), \tag{A.2}$$

for $t = Q, \ldots, T - 1$, and the coefficients $\hat{\alpha}_2(t)$, $\hat{\alpha}_2(t)$ and $\hat{\beta}(t)$ are estimated with data through periods 1, $\ldots, t$. Then, one-step-ahead forecasts from the competing model can be compared with forecasts from the benchmark model (that is, a restricted version of the competing model) by using statistics based on the time series $\hat{u}_1(t+1)$ and $\hat{u}_2(t+1)$.

The first test we use for out-of-sample predictability is the forecast encompassing test of Clark and McCracken (2001). To clarify how it works, we follow Harvey *et al.* (1998) and specify a regression of the excess stock return on a weighted average of forecasted values from the benchmark and competing models:

$$R_m(t+1) - R_f(t+1)$$
$$= (1-\lambda)\left[\alpha_1\right] + \lambda\left[\alpha_2 + \beta' X(t)\right] + v(t+1), \tag{A.3}$$

where $0 \leq \lambda \leq 1$ and $v(t+1)$ is an error term. Substituting both forecasts from Equations (14.4) and (14.5) yields

$$u_1(t+1) = \lambda\left[u_1(t+1) - u_2(t+1)\right] + v(t+1), \tag{A.4}$$

Then, as λ is also the coefficient of the regression model in equation (A.4):

$$\lambda = \frac{\text{Cov}\left[u_1(t+1), u_1(t+1) - u_2(t+1)\right]}{\text{Var}\left[u_1(t+1) - u_2(t+1)\right]}. \tag{A.5}$$

Thus, the combined forecast will have a smaller expected squared error than the benchmark model forecast unless the covariance between $u_1(t+1)$ and $u_1(t+1) - u_2(t+1)$ is zero (i.e. $\lambda = 0$). This way, Clark and McCracken (2001) tests the null hypothesis that $\lambda \leq 0$ and is given by

$$\text{ENC} - \text{NEW} = \frac{P \sum_{t=Q}^{T-1} \left(\hat{u}_1(t+1)^2 - \hat{u}_1(t+1)\hat{u}_2(t+1) \right)}{\sum_{t=Q}^{T-1} \hat{u}_2(t+1)^2} \tag{A.6}$$

Under the null hypothesis that the benchmark model encompasses the competing model, the covariance between series $u_1(t+1)$ and $u_1(t+1) - u_2(t+1)$ will be less than or equal to zero. Under the alternative that the competing model contains added information, the covariance should be positive. Hence, the encompassing test presented above is one-sided. Clark and McCracken (2001) demonstrate that the limiting distribution of ENC-NEW is not normal when the forecasts are nested under the null, but they provide asymptotic critical values for this statistic.

The second test used is the one developed by McCracken (2007). This test, unlike the one proposed by Diebold and Mariano (1995) in the context of non-nested models, allows for comparison of predictive accuracy between nested models. In particular, we use it to test for equality of the mean squared forecasting errors (MSE) from the benchmark and competing models, which is given by

$$
MSE - F = \frac{P \sum_{t=Q}^{T-1} \left(\hat{u}_1(t+1)^2 - \hat{u}_2(t+1)^2 \right)}{\sum_{t=Q}^{T-1} \hat{u}_2(t+1)^2},
$$

$$
= P \left[\frac{MSE_1 - MSE_2}{MSE_2} \right]
$$

(A.7)

where $MSE\, j = \sum_{t=Q}^{T-1} \hat{u}_j(t+1)^2 / P, j = 1,2$. Based upon the value of this statistic the null of equal MSE is either rejected or not rejected. McCracken (2007) shows that when the two models are nested the alternative is one-sided, rather than two-sided. Moreover, since the asymptotic distribution of MSE-F under the null is non-standard, tables of asymptotically valid critical values are provided by McCracken (2007).

Clark and McCracken (2001) use simulations to examine the small-sample properties of the ENC-NEW and MSE-F tests. They report that although both tests have good sample size properties, the ENC-NEW test is clearly the more powerful out-of-sample test of predictive ability. While this evidence indicates that the inference from the ENC-NEW test is more reliable, Welch and Goyal (2008) highlight an important problem of encompassing tests in general. The ENC-NEW test uses the entire out-of-sample test to estimate the parameter λ, but an investor trying to use a combined forecast to predict $R_m(t+1) - R_f(t+1)$ will only have the information available up to t to calculate the combination coefficient λ.

The final measure of out-of-sample forecasting performance is the out-of-sample R^2, R_{OS}^2 This statistic is the analogue to the in-sample R^2 and in terms of our notation is computed as

$$
R_{OS}^2 = 1 - \frac{\sum_{t=Q}^{T-1} \hat{u}_2(t+1)^2}{\sum_{t=Q}^{T-1} \hat{u}_1(t+1)^2} = \frac{MSE_1 - MSE_2}{MSE_1}
$$

$$
= \frac{MSE_1 - F}{P} \left(\frac{MSE_2}{MSE_1} \right).
$$

(A.8)

As can be seen from Equation (A.8), if R_{OS}^2 is positive then the competing model has a lower MSE than the benchmark model. Also, as shown in the last equality, R_{OS}^2 is not a statistic that provides new information with respect to the other tests, since it is merely a scaled-up version of the MSE-F statistic.[*] That is, predictor variables with greater MSE-F will also exhibit greater R_{OS}^2. Then, this could also be considered a test of equal MSE, assuming that it has an asymptotic distribution.[†]

---

[*] The MSE-F, as derived originally by McCracken (2007), is designed to be used with any loss function. Here we use only one particular case.

[†] Just as the in-sample R^2 has an adjusted counterpart for degrees of freedom, $\bar{R}^2$, Welch and Goyal (2008) use a version of the R_{OS}^2 adjusted for degrees of freedom. However, since the forecasting errors are not part of the OLS estimation, and therefore there is no loss of degrees of freedom in its calculation, we consider it inappropriate to adjust this statistic.

As mentioned above, in the context of one-step-ahead forecasts, Clark and McCracken (2001) and McCracken (2007) provide asymptotic critical values for the ENC-NEW and MSE-F statistics, respectively. These critical values depend on two parameters: $\pi = P/Q$ and $K_2 - 1$, the number of variables included in $X(t)$. Because the tables with the critical values for these non-standard tests do not contain the particular value of π chosen by us, we follow Clark and McCracken (2005) and obtain these values with an inference technique based on bootstrapping. In addition, based on the bootstrapped time series, we obtain the empirical distribution of the R_{OS}^2 statistic and critical values for the tests. In particular, we use a parametric bootstrap Berkowitz and Kilian (2000) and our algorithm has five steps, which we briefly describe below: *

1. We estimate a bivariate VAR for the excess stock return, $R_m(t) - R_f(t)$, and the variable $X(t)$ under the null hypothesis of non-predictability. The model is estimated with OLS and using the full sample. The excess return is modelled according to Equation (14.4) and for the variable $X(t)$ the optimal number of lags of $R_m(t) - R_f(t)$ and $X(t)$ were chosen with the AIC criterion.
2. The coefficients of the VAR were adjusted for the small-sample bias using the procedure of Kilian (1998) with 10 000 bootstrap draws.
3. We bootstrapped 999 time series for the excess stock return and the variable $X(t)$ by drawing from the rescaled sample residuals with replacement (Berkowitz and Kilian 2000), using the adjusted VAR coefficients and initial observations selected by sampling from actual data (Stine 1987).
4. Each artificial bivariate time series is used to estimate the benchmark and competing models (Equations (14.4) and (14.5)) in a recursive way. Forecasting errors are calculated according to Equations (A.1) and (A.2) and using the sample portions described above. The ENC-NEW, MSE-F and R_{OS}^2 statistics were calculated based on these estimated forecasting errors with Equations (A.6), (A.7) and (A.8).
5. For each statistic, critical values are simply computed as percentiles of the corresponding empirical distribution. The p-values are calculated using the standard method.

---

* For more details on this methodology, see Clark and McCracken (2005).

15

Time-Frequency Analysis of Crude Oil and S&P 500 Futures Contracts

Joseph McCarthy
and Alexei G. Orlov

15.1 Introduction

We use frequency-domain techniques to study the effects of oil price movements on U.S. stock returns. We also seek to help explain the dynamics of the stock market by examining levels and returns in the crude oil market. Specifically, we study the linkages between volatility in the futures price of crude oil and the futures price of S&P 500 contracts, as well as the relationship between the volume of crude oil futures and the volume of the S&P 500 futures contracts. We also investigate the lead–lag relationship between crude oil and stock prices, stock returns and trading volumes.

Conventional wisdom (as delineated by, for example, Sauter and Awerbuch (2003)) suggests that changes in oil prices should be negatively related to macroeconomic and financial indicators. Higher oil prices may depress economic activity, thereby reducing personal income and wealth. Incomes may become lower due to higher unemployment and lower wages, while wealth may be negatively affected through reduced stock values of companies held by individuals.

There are several channels through which oil prices can affect financial volatility. First, capital market theory predicts that higher future costs associated with increased oil prices would translate into higher risk of holding the affected assets. Since the increased risk is bound to manifest in higher volatility, one would expect higher oil prices to be reflected in higher financial volatility and instability. Second, if there is a (negative) relationship between oil prices and the stock market in levels, this relationship is likely to be carried on to the second moments. It should be noted that volatility of oil prices *per se* would be indicative of a riskier economic environment, thereby reinforcing the predicted positive link between volatilities of oil prices and financial indicators.

DOI: 10.1201/9781003265399-18

Finally, one can contend that oil price movements and financial performance find their highest correlation at a lag or a lead: for example, the expected future higher price of oil may negatively affect the current value of assets and positively affect current financial volatility. We use cross-spectral and wavelet analyses to answer empirically the following questions stemming from the theoretical premises outlined above.

> H1: Is there a negative (or positive) contemporaneous correspondence between oil and stocks, either in levels or returns?
>
> H2: Is there a lead–lag relationship among trading volumes and prices in the oil and stock markets? If so, what is the lag or lead time that produces the strongest correspondence?
>
> H3: Is there a negative (or positive) correspondence between the (levels of) oil prices and stock returns at a lag or a lead? If so, what is the lag or lead time that produces the strongest correspondence?

One of the first comprehensive studies on the relationship between oil prices and macroeconomic activity was Hamilton (1983), who found that higher oil prices led to slower economic growth in the U.S. for the period 1948–1980. Hamilton's conclusion was not entirely surprising, as his time span included two major oil shocks. Subsequent studies confirmed the negative relationship between oil prices and U.S. economic activity for more recent periods, as well as for other oil-consuming countries (e.g. Hooker (1996), Rotemberg and Woodford (1996), Hamilton (2000) and Yang *et al.* (2002)). These studies report a fairly substantial decrease in economic activity due to higher oil prices: a 10% oil price increase results in a decline in GDP growth of 0.6 to 2.5% (Sauter and Awerbuch 2003).[*]

In contrast to the established negative relationship between oil prices and the economy in the 1970s and 1980s, more recently lower prices of oil and natural gas did not appear to help propel the U.S. and other economies out of recession. Our ability to unequivocally declare the breakdown of negative correlations between oil prices and economic and financial performance is, of course, impaired by the complications of the ongoing financial crisis with its various causes and effects. It is thus imperative to revisit the empirical analysis of oil prices using more extensive data sets as well as frequency domain techniques that can help refine the conventional time-domain analyses.

It should be pointed out that prior to the mid-1980s, most oil price changes were price increases, whereas since the late 1980s we started to observe both positive and negative changes in oil prices (Sauter and Awerbuch 2003). This increased oil price volatility should prompt a re-examination of some of the earlier studies, as their conclusions may not pertain to the most recent patterns in the data. In fact, Mork (1989), Mork *et al.* (1994), Lee *et al.* (1995) and Lee and Zeng (2011), among others, report the asymmetric effects of oil prices for U.S. and other industrialized economies: the impact on economic activity of oil price increases can be quite negative, while the impact of price decreases was insignificant.[†]

It should be made clear that the focus of our paper is rather narrow: we study the relationship between oil and security prices, as well as the relationship among trading volumes and prices. We do not purport to explain the state of the macroeconomy (i.e. unemployment, GDP, etc.). Instead, we use movements in S&P 500 futures contracts to proxy for financial volatility and investigate the link between said volatility and the oil market.

The subject of this paper is important in yet another respect. The debate surrounding fossil-based technologies *vis-a-vis* green energy is ongoing and impacts a large amount of investment capital. Although the primary focus of this debate is on climate change and the sustainability of natural resources, the effect of old versus new technologies on the state of national economies also plays a non-trivial role in this debate. It is thus interesting to determine if oil price fluctuations have disruptive effects on equity

---

[*] Some researchers (e.g. Ciner (2001)) find an even stronger relationship between oil prices and the stock market in the 1990s relative to the 1970s or 1980s.

[†] Ferderer (1996) suggests that the asymmetric effects of oil prices are due to (i) sectorial shifts of specialized labour and capital, and (ii) the uncertainty that may induce companies to postpone their investment. The latter issue of the impact of oil price uncertainty on investment was further investigated by Elder and Serletis (2010).

markets. If so, there is an additional argument to be made in favour of fostering the development and refinement of non-fossil energy sources. If, on the other hand, there are no such effects found in the most recent data, one cannot comfortably offer the financial stability argument when advocating the switch to alternative energy sources.

The objective of this paper is to use frequency domain techniques, such as wavelets and cross-spectra, to examine the association between the daily prices of crude oil futures and daily S&P 500 futures closing prices over the past several decades. We believe that the wavelet and cross-spectral analyses employed in this paper offer insights regarding the relationship between oil prices and financial indicators that are not apparent from a conventional time-domain framework. We also believe that our research introduces a useful way of looking at an important question: Are the recent fluctuations in the oil markets associated with greater comovements of oil prices and financial indicators? Our research allows one to judge to what extent the relationship between oil and the stock market has changed over time and if there is a lag–lead effect of (expected) oil price changes on the financial stability of the U.S. economy.*

Our extensive data sample includes a wide range of price movements, both with respect to levels, as well as rates and direction of change. We find a positive correspondence between oil prices and S&P 500 in levels, and our findings cast doubt on the purported negative relationship between oil and the U.S. stock market. Also, our analysis suggests that oil prices lead oil volume, while S&P 500 trading volume leads S&P 500 prices. These findings persist over a large number of time scales and across a wide range of Fourier frequencies.

15.2 Literature Review

Most empirical studies that focus on the role of oil prices in the economy, starting with Hamilton's (1983) seminal paper and continuing with Hooker (1996), Rotemberg and Woodford (1996), Hamilton (2000) and others, report significant negative effects of oil prices on macroeconomic activity. One approach to measure the effect of oil prices on financial indicators is through estimating a beta for oil. Recall that an asset's beta measures the covariance between that asset's value and the value of a diversified portfolio or a market index. Thus, an estimated negative beta for oil would indicate a negative covariance risk and, therefore, a negative relationship between oil prices and the stock market. Awerbuch (1993), Awerbuch and Deehan (1995) and Bolinger *et al.* (2002) use this approach and report a negative beta for oil and natural gas. Such a result is indicative of a possible 'double whammy effect' (Sauter and Awerbuch 2003) of the higher oil prices: consumers feel the pinch at the pump and when they pay their utility bills, and they also see the decline in their wealth due to the lower value of their assets.

Many of the oil price increases, particularly after 1986, have been followed by large decreases (Hamilton 1996). Thus oil price volatility is likely to have played an increasingly important role in the last two decades. Lee *et al.* (1995) emphasize that oil price volatility may be an important variable in accounting for economic fluctuations. Including both the oil price volatility measure and the magnitude of an oil price shock (relative to the trend) in their analysis, Lee *et al.* (1995) conclude that an oil price shock has a smaller impact on the macroeconomy in a high volatility environment, i.e. when such a shock is less of a surprise. Ferderer (1996) reports that output is negatively affected by both levels and volatility of oil prices, with volatility having a more pronounced effect, and points to a possible lag between oil price changes and macroeconomic activity of about a year. A similar—and more assertive—conclusion is reached by Hooker (1996): oil price levels do not have predictive power over output, while volatility of oil prices does.

The possibility of a nonlinear relationship between oil prices and the stock markets has been advocated by Ciner (2001) and Reboredo (2010), among others. In particular—and most relevant to our

---

* Only recently have researchers started to rigorously investigate the possibility that oil prices may have predictive power in forecasting stock markets (i.e. Driesprong *et al.* (2008)).

study—Ciner finds a strong bidirectional causality between crude oil futures returns and the S&P 500 index returns. The interdependence of the stock and oil futures prices has recently been confirmed by Vo (2011). Thus, there appears to be a nonlinear feedback effect between oil and the stock market. The semiparametric and non-parametric methods used in this paper do not confine us to a set of linear relationships and are flexible enough to find non-linear links between oil prices and stock returns.

Although macroeconomic activity is closely related to financial indicators and, in particular, stock market performance, it must be acknowledged that the value of the stock market is not always synonymous with the current state of the economy. Huang *et al.* (1996) and Jones and Kaul (1996) were among the first comprehensive studies that examined the interplay between oil prices and changes in stock market fluctuations. Jones and Kaul (1996) report that oil prices help predict stock market returns in the U.S., Canada and Japan, and that the effect of oil prices on stock markets is negative both contemporaneously and at a lag. The results suggest that oil shocks affect the stock markets through their effect on current and future real cash flows, which is consistent with the cash-flow dividend valuation model (Campbell 1991).

Huang *et al.* (1996) study the link between daily oil futures returns and daily U.S. stock returns using vector autoregressions (VARs). Unlike Jones and Kaul (1996), they did not find any evidence that oil futures returns have an impact on stock market indices such as S&P 500. However, the findings of Jones and Kaul are confirmed by Sadorsky (1999), who applies VARs to monthly data and reports a negative and significant impact of an oil shock on stock returns. Higher oil prices lead to higher production costs, thereby putting downward pressure on the value of assets. Sadorsky also emphasizes the importance (and asymmetry) of oil price volatility shocks in accounting for fluctuations in real stock returns.

Sawyer and Nandha (2006) consider the question of whether oil prices are a global factor in asset returns. They conclude that the effect of oil on stock prices is both disaggregated and heterogeneous. Indeed, Abdelaziz *et al.* (2008) examine the impact of oil prices on the stock markets of four Middle-East countries and find that oil prices have a strong and positive impact on these markets. Driesprong *et al.* (2008) find that higher oil prices result in lower stock returns in the following month. This negative relationship between oil prices and stock returns at a lag is significant both statistically and economically for many countries and for the world market index. The authors interpret their finding using the 'underreaction hypothesis': it takes time for investors to fully react to changes in the oil prices and thus stock markets do not fully reflect these changes until later.

Miller and Ratti (2009) report a negative long-run relationship between oil prices and stock markets for the 1970s and 1990s for several OECD countries, but virtually no relationship for the 1980s and 2000s. Agusman and Deriantino (2008) found that oil price changes did not have a significant impact on industry stock returns in Indonesia, a member of OPEC. In a recent U.S. study, Gogineni (2009) examined the impact of rising oil prices on the stock returns of a wide variety of industries. The author found diverse and heterogeneous results across industries with some showing a quite favourable impact from rising oil prices, whereas other industries, such as airlines, had a negative impact. It therefore appears that the extant literature has not yet reached a consensus with respect to the empirical relationship between oil prices and broad financial indicators, and the research conducted in our paper provides a useful approach to refining previous results.

The application of wavelets in the field of finance is relatively new. However, given the versatility of wavelets, increased applications in finance can be expected. As Raihan *et al.* (2005) have noted, wavelets are particularly helpful in analysing non-stationary price series. Wavelet decomposition has also been applied to irregular time series to better analyse intraday trading volume (Manchaldore *et al.* 2010).

A recent study of commodity futures by Elder and Jin (2009) used wavelet analysis to ascertain the degree of fractional integration in various commodities. Applying the wavelet transform to different commodity time series allows the authors to obtain the sample variance of the time series from the wavelet detail coefficients at each scale. Wavelet transforms completely preserve all of the energy of a time series and the contribution of energy at each time scale, allowing an analysis of both high-frequency and low-frequency variation (Walker 1999). Scale-by-scale analysis also allows for a more detailed examination of the relationship between futures prices of the S&P 500 and futures prices of oil. An example of wavelet scale-by-scale analysis applied to the crude oil market is the paper by Jammazi and Aloui (2010), where the authors use multi-resolution analysis to highlight periods of extreme oil price volatility.

15.3 Data

The data for this study are the daily futures closing prices for oil and S&P 500 over the 28-year period from March 30, 1983, through March 4, 2011, as reported by Price-data.com. This continuous price data set is based upon the futures price of the near month contract, as this contract generally has the most open interest and thus the greatest liquidity. Prior to expiration of the futures contract, the position is rolled over into the next near month contract. Rolling over the contract prior to expiration allows Price-data.com to maintain a continuous time series of futures prices. In practice, the front month contract is rolled to the next month when its open interest is greater than the contract about to expire. The time period of study includes a wide variety of market conditions, such as the market crash of October 1987, the tech bubble of 2000, and the recent stock market volatility through the beginning of 2011. The time period of study also includes the period of extreme volatility in the oil market preceding the First Gulf War in spring 1991 as well as the very volatile year of 2008. In total, we have 7045 data points in our study. Figure 15.1 presents our data in levels (first two graphs), as well as log-differences (last two graphs). Panel (a) of Figure 15.1 depicts oil futures closing prices, while panel (b) presents S&P 500 futures prices. Panel (c) of Figure 15.1 shows returns on oil futures, and, finally, panel (d) plots returns on S&P 500 contracts. A review of the descriptive statistics in Table 15.1 shows a high degree of skewness and kurtosis.

Calculation of the traditional Pearson correlation coefficient is presented in Table 15.2 and corroborates the results of both wavelet analysis and Fourier analysis. The correlation matrix in Table 15.2 shows that, over our 28-year period of study, there is a pronounced positive correlation between crude oil futures and S&P 500 futures prices. In Table 15.3 we show the cross-correlations over 22 trading days (one calendar month of data). Again, we note the pronounced and persistently positive association between crude oil and the S&P 500. Finally, in Figure 15.2 we plot the cross-correlation function for +/−22 lags (number of trading days in a month) and show very strong statistical significance at all lags as all of the plotted values are far beyond the confidence limits for 99.9%.

15.4 Methodology

15.4.1 Cross-Spectral Analysis

We conduct a cross-spectral analysis of the daily futures price of crude oil and the futures price of the S&P 500 contract between March 1983 and March 2011. We use frequency-domain techniques to determine the relative importance of cycles of different frequencies in accounting for the comovement among prices and returns on oil futures and the S&P 500. Further, we study how cospectra (which are the real component of the cross-spectra) change during two periods—before and after 1996 (i.e. the first and second halves of our sample). The goal of spectral analysis is to determine how important cycles of different frequencies are in accounting for volatility and comovement of the time series (Hamilton 1994). According to Granger (1966), one of the advantages of spectral methods is that they do not require specification of a model and so the results are not based on any rigid modelling assumptions.

To perform cross-spectral analysis, we first use the finite Fourier transform to decompose the data into a sum of sine and cosine waves of different amplitudes and wavelengths to obtain periodograms. According to the spectral representation theorem (Hamilton 1994), any covariance-stationary process x_t can be expressed as the finite Fourier transform decomposition of x_t:

$$x_t = \bar{x} + \sum_{k=1}^{m} \left[a_k \cos(\omega_k t) + b_k \sin(\omega_k t) \right], \tag{15.1}$$

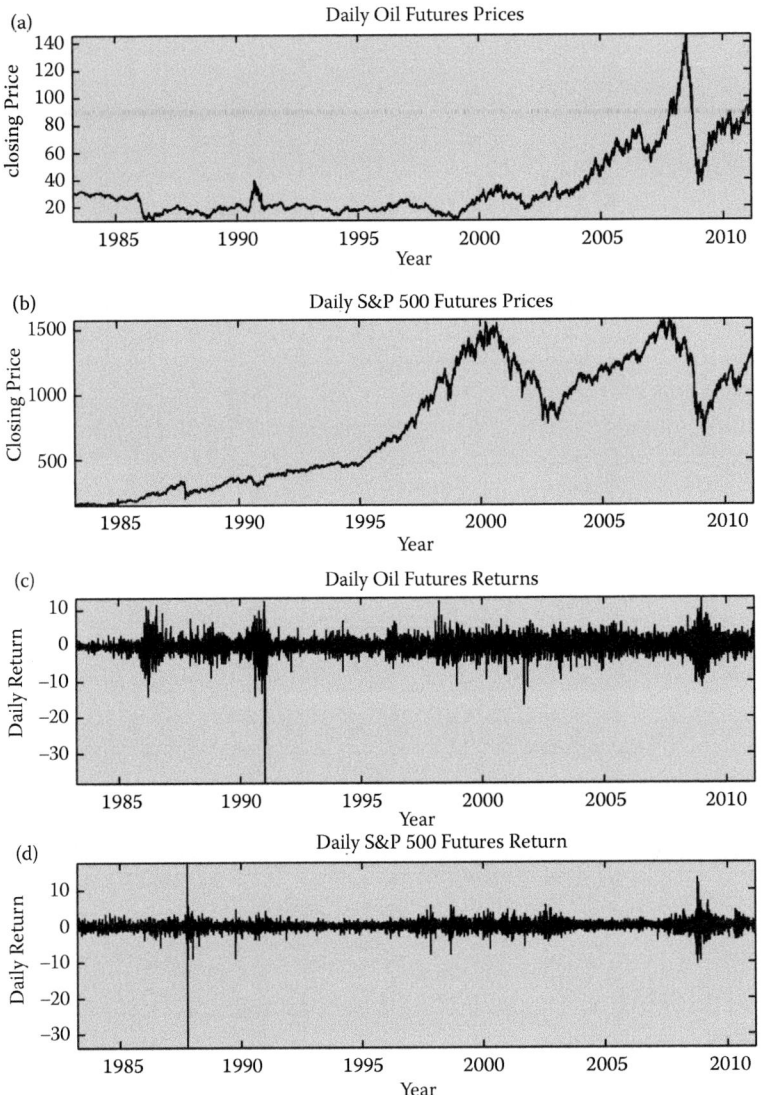

FIGURE 15.1   (a, b) Daily futures closing prices and (c, d) daily futures returns.

where t is the time subscript ($t = 1, 2, \ldots, n$), n is the number of observations in the time series, $\bar{x}$ is the mean value of x, m is the number of frequencies in the Fourier decomposition,* a_k are the cosine coefficients, b_k are the sine coefficients, and ω_k are the Fourier frequencies ($\omega_k = 2\pi k/n$). Such an approach allows one to describe the value of x_t as a sum of periodic functions of different amplitudes and wavelengths. Each time series is thus decomposed into a number of orthogonal components associated with various frequencies.† We then calculate the amplitude cross-periodograms for each pair of time series as

$$J_k^{xy} = \frac{n}{2}\left(a_k^x a_k^y + b_k^x b_k^y\right) + i\frac{n}{2}\left(a_k^x b_k^y - b_k^x a_k^y\right), \tag{15.2}$$

---

* $m = n/2$ if n is even, and $m = (n-1)/2$ if n is odd.
† Since the value of cos (ωt) repeats itself every $2\pi/\omega$ periods, a frequency ω corresponds to a period of $2\pi/\omega$.

TABLE 15.1 Descriptive Statistics: S&P 500 and Oil Futures Daily Prices and Per Cent Daily Returns

Variable	Oil Prices	Oil Returns	S&P 500 Prices	S&P 500 Returns
Mean	34.1679	0.0178	766.5020	0.0305
Median	25.0050	0.0444	770.0000	0.0621
Maximum	145.8600	13.3403	1576.2000	17.7493
Minimum	10.4200	−38.4071	148.8500	−33.7004
Std. dev.	23.8764	2.1757	443.4322	1.2654
Skewness	1.7548	−0.9477	0.1137	−2.4779
Kurtosis	5.7047	20.6091	1.4951	87.4236
Jarque–Bera	5762	92063	680	2099081
Probability	0.000	0.000	0.000	0.000
Observations	7044	7044	7044	7044

where i represents the imaginary unit $\sqrt{-1}$. A cross-periodogram is a sample analogue of the population cross-spectrum and it shows the contribution of the kth harmonic to the total covariance between two data series.

The cross-periodogram is, admittedly, a volatile and inconsistent estimator of the cross-spectrum. In addition, it does not become more accurate with an increase in sample size. To overcome this deficiency, cospectral density estimates are produced by smoothing the real part of the cross-periodogram ordinates.[*] The idea behind such non-parametric (or kernel) estimation is to use frequencies $\{\omega_k, \omega_{k\pm1}, \omega_{k\pm2}, \dots, \omega_{k\pm h}\}$ in estimating the cospectrum at ω_k. So the bandwidth parameter h and the relative weights (that must sum to unity) given to each frequency fully characterize the kernel. Because the kernel estimate is an average over a number of frequencies, and because estimates of the cospectrum at ω_k and ω_l are approximately independent for large n and $k \neq l$ (Hamilton 1994), kernel estimates are less volatile and provide more consistent estimates of the cross-spectrum than the cross-periodogram. We use a triangular weight function (with $h = 21$) in the moving average applied to the cross-periodogram to form smoothed cospectral density estimates.[†]

Finally, we compare the cospectra for all pairs of data series before and after 1996 (which is the midpoint of our sample). The cross-spectrum $s_{xy}(\omega)$ integrates to the unconditional covariance, and the quadrature spectrum $q_{xy}(\omega)$ (the imaginary part of the cross-spectrum) integrates to zero since $q_{xy}(-\omega) = -q_{xy}(\omega)$ (Hamilton 1994). Therefore, the area under the cospectrum (or the real component of the cross-spectrum) is equal to the covariance between x and y. Note that the cospectrum may be positive over some frequencies and negative over others.

15.4.2 Wavelets

Wavelets are small waves. By design, the distance from the peak of the wave to average sea-level is exactly offset by the distance from average sea-level to the trough of the wave. Mathematically, this admissibility condition is written as

$$\int_{-\infty}^{+\infty} \Psi(t)\mathrm{d}t = 0. \tag{15.3}$$

---

[*] There is an obvious trade-off: smoothing reduces the variance of the estimator but introduces a bias.
[†] The main conclusions of this paper are immune to using other relative weights (or kernels) applied to the periodogram ordinates to form the cospectral density estimates.

TABLE 15.2 Correlation Matrix

	Price$_{Oil}$	Price$_{S\&P}$	Volume$_{Oil}$	Volume$_{S\&P}$	Return$_{Oil}$	Return$_{S\&P}$	ΔVolume$_{Oil}$	ΔVolume$_{S\&P}$	Return$^2_{Oil}$	Return$^2_{S\&P}$
Price$_{Oil}$	1	0.564***	0.740***	−0.330***	0.025**	−0.018	0.011	0.012	0.004	0.011
	[0]	[0.000]	[0.000]	[0.000]	[0.038]	[0.131]	[0.380]	[0.322]	[0.743]	[0.370]
Price$_{S\&P}$	0.564***	1	0.562***	0.099***	0.022***	−0.002	0.003	0.016	0.010	−0.005
	[0.000]	[0]	[0.000]	[0.000]	[0.070]	[0.902]	[0.798]	[0.190]	[0.391]	[0.691]
Volume$_{Oil}$	0.740***	0.562***	1	−0.191***	0.008	−0.030***	0.009	0.009	0.059***	0.028**
	[0.000]	[0.000]	[0]	[0.000]	[0.490]	[0.012]	[0.434]	[0.446]	[0.000]	[0.020]
Volume$_{S\&P}$	−0.330***	0.099***	−0.191***	1	−0.019	−0.064***	0.002	0.068***	0.026**	0.076***
	[0.000]	[0.000]	[0.000]	[0]	[0.120]	[0.000]	[0.842]	[0.000]	[0.030]	[0.000]
Return$_{Oil}$	0.025**	0.022**	0.008	−0.019	1	0.048***	0.008	−0.019	−0.210***	−0.022**
	[0.038]	[0.070]	[0.490]	[0.120]	[0]	[0.000]	[0.503]	[0.117]	[0.000]	[0.065]
Return$_{S\&P}$	−0.018	−0.002	−0.030***	−0.064***	0.048***	1	0.003	−0.041***	0.029**	−0.262***
	[0.131]	[0.902]	[0.012]	[0.000]	[0.000]	[0]	[0.822]	[0.001]	[0.016]	[0.000]
ΔVolume$_{Oil}$	0.011	0.003	0.009	0.002	0.008	0.003	1	0.008	0.012	−0.003
	[0.380]	[0.798]	[0.434]	[0.842]	[0.503]	[0.822]	[0]	[0.480]	[0.300]	[0.823]
ΔVolume$_{S\&P}$	0.012	0.016	0.009	0.068***	−0.019	−0.041***	0.008	1	0.007	0.006
	[0.322]	[0.190]	[0.446]	[0.000]	[0.117]	[0.001]	[0.480]	[0]	[0.582]	[0.639]
Return$^2_{Oil}$	0.004	0.010	0.059***	0.026+*	−0.210***	0.029**	0.012	0.007	1	0.037***
	[0.743]	[0.391]	[0.000]	[0.030]	[0.000]	[0.016]	[0.300]	[0.582]	[0]	[0.002]
Return$^2_{S\&P}$	0.011	−0.005	0.028**	0.076***	−0.022**	−0.262***	−0.003	0.006	0.037***	1
	[0.370]	[0.691]	[0.020]	[0.000]	[0.065]	[0.000]	[0.823]	[0.639]	[0.002]	[0]

P-Values in brackets. *, **, and *** denote 10%, 5% and 1% significance levels, respectively.

TABLE 15.3   Cross-Correlation Matrix

	Lag:	Lead:
i	Oil,S&P 500$(-i)$	Oil,S&P 500$(+i)$
0	0.563	0.563
1	0.564	0.562
2	0.564	0.561
3	0.564	0.560
4	0.565	0.559
5	0.565	0.559
6	0.565	0.558
7	0.566	0.557
8	0.566	0.556
9	0.566	0.555
10	0.566	0.554
11	0.567	0.553
12	0.567	0.552
13	0.567	0.551
14	0.567	0.551
15	0.567	0.550
16	0.568	0.549
17	0.568	0.548
18	0.568	0.547
19	0.568	0.546
20	0.568	0.546
21	0.569	0.545
22	0.569	0.544

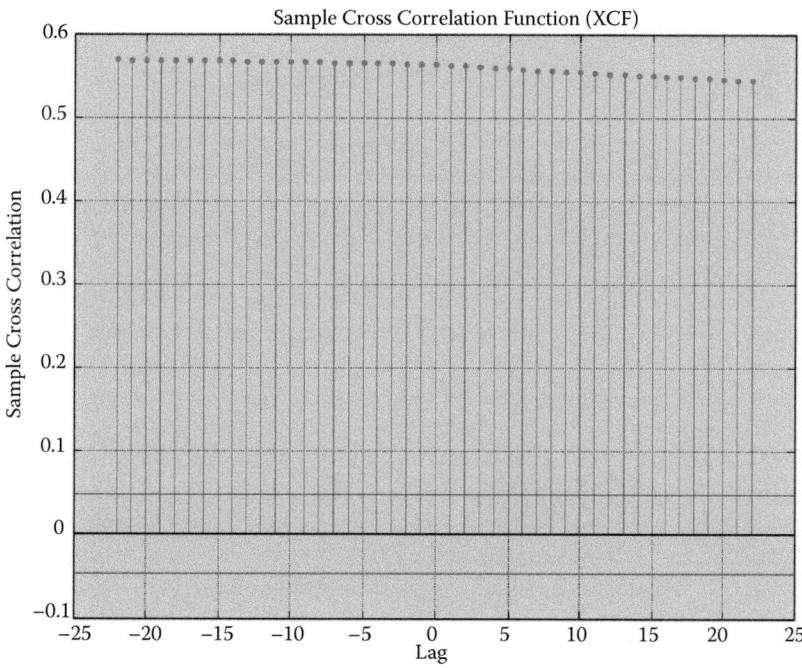

FIGURE 15.2   Cross-correlations. The variance of the cross-correlation coefficient under the null hypothesis of zero correlation is approximately $1/n$, where n is the length of the series. Thus the blue boundary lines in the graph denote 99.9% confidence limits: $4\left(1/\sqrt{n}\right)=0.047653$.

An additional condition is that the energy of the series being studied must be preserved. This second condition is written as (Gençay *et al.* 2002)

$$\int_{-\infty}^{+\infty} |\Psi(t)|^2 \, dt = 1. \tag{15.4}$$

In the field of signal analysis, the French mathematician Joseph Fourier discovered that complex signals could be broken down into a sum of sine waves of differing frequencies. However, as a time series is split into an increasing number of frequencies, it becomes near impossible to tell at what point in time a sudden spike in a given frequency may have occurred.

In the development of wavelet analysis, the French seismologist Jean Morlet was particularly interested in maintaining the time point location of unusual breaks in a signal. His solution was that instead of splitting a signal into a number of different frequencies, he would keep the signal intact and change the length of the analysing wavelet (often by powers of 2, i.e. dyadic) (Burke-Hubbard 1998). Changing the length of the wavelet at each level of resolution gives rise to the phraseology: multi-resolution analysis (MRA) (Mallat 1999). Wavelets of relatively short length are particularly good at capturing the high-frequency components of a time series, whereas longer length wavelets capture the low-frequency longer enduring components of a time series. As one proceeds from one level of resolution to the next, a graphical summary of the pronounced information effects is often presented as a wavelet power spectrum that is made up of the squared absolute value of the wavelet coefficients at each level of resolution.

The Morlet wavelet is given by

$$\Psi(t) = \frac{1}{\sqrt{2\pi}} e^{-i\omega t} e^{-t^2/2}, \tag{15.5}$$

where $i = \sqrt{-1}$, ω is the central frequency (most often set to $\omega = 6$ or $\omega = 2\pi$), and t is time. In order to implement the Morlet wavelet, one convolves a time series $x(t)$ with the wavelet. The continuous wavelet transform by length (translation) and scale (dilation) is then obtained (Gençay *et al.* 2002):

$$W(s,u) = x(t) * \Psi_{u,s}(t), \tag{15.6}$$

where '*' indicates convolution, u is length, s is scale, and $\Psi_{u,s}(t) = (1/\sqrt{s})\Psi[(t-u)/s]$. We apply the wavelet transform to the time series for daily oil futures and S&P 500 futures. For our study, we are interested in the association both within and across the data series over a wide range of scales.

A wavelet transform may be either discrete or continuous. The discrete wavelet transform (DWT) gives the most compact representation of a signal, but because its wavelet spectrum in turn contains discrete blocks, it is difficult to examine the evolution of changing frequencies over time (Wang and Sassen 2008). Although the Morlet wavelet is a widely used analysing wavelet, other possibilities might include the 'Paul' or the 'Derivative of Gaussian' wavelet. The Derivative of Gaussian includes only real values and thus would not provide important information with respect to different data series being in or out of phase with each other. A more serious alternative to the Morlet wavelet would be the Paul wavelet. Based upon an important comparison paper by De Moortel *et al.* (2004), the Morlet wavelet has favourable characteristics with respect to capturing and tracking different frequency oscillations, while the Paul wavelet has an advantage with respect to time localization. Given that the time location of our daily data series is already known and that we are particularly concerned with tracking changes in frequency oscillations (especially large changes), we believe that the Morlet wavelet is more appropriate for our work. Addison (2005) also considered several different wavelet transforms and selected the Morlet wavelet given its versatility and widespread acceptance in many different disciplines.

15.4.3 Cross Wavelets

Following Torrence and Compo (1998), if time series X and Y have theoretical Fourier power spectra P_k^x and P_k^y, then the cross-wavelet distribution is given as

$$\frac{\left|W_n^x(s)W_n^y(s)\right|}{\sigma_x\sigma_y} \Rightarrow Z_v(p)\frac{\sqrt{P_k^x P_k^y}}{v},$$

where W_n^x and W_n^y are the n wavelet coefficients of X and Y, respectively, at scale s, σ_x and σ_y are the respective standard deviations, and P is the power spectra of the $k = 0, \ldots, n/2$ frequency of X and Y.

For an autoregressive process of lag(1), α is the autocorrelation coefficient and

$$P_k = \frac{1-\alpha^2}{1+\alpha^2 - 2\alpha\cos[2\pi(k/N)]}.$$

For $v = 1$ (real wavelets), $Z(95\%) = 2.182$, and for $= 2$ (complex wavelets), $Z(95\%) = 3.999$.

15.4.4 Cross Wavelet Phase Angles

A distinct advantage of working with a complex wavelet such as the Morlet wavelet transform is that you receive both real and imaginary results. While the real values are important for calculating cross correlations across different time scales, imaginary values provide important information with respect to whether or not two time series are in phase with each other and, if they are not, measurement of the circular angular difference between the two series can be used to ascertain which time series leads the other, and with wavelets we can determine this by both scale and time location.

From Zar (1999) and Grinsted *et al.* (2004), the mean of a circular set of angles for (a_i, $i = 1, \ldots, n$) is defined as

$$a_m = \arg(X,Y), \quad \text{with } X = \sum_{i=1}^{n}\cos\left(a_i\right) \text{ and } Y = \sum_{i=1}^{n}\sin\left(a_i\right),$$

and the circular standard deviation is

$$s = \sqrt{-2\ln\frac{R}{n}},$$

where

$$R = \sqrt{X^2 + Y^2}$$

15.5 Empirical Results

Our paper analyses daily futures data from March 30, 1983, to March 4, 2011. We use frequency-domain techniques to determine the relative importance of cycles of different frequencies in accounting for the comovement among prices and volume on crude oil and the S&P 500 futures contracts. Further, we study if—and how—cospectra change from the first to the second half of the sample (after 1996) as oil price volatility had become more pronounced.

Wavelet transforms are adept at capturing short-term high-frequency fluctuations as well as pronounced long-term trends. Our analysis across time scales offers a comprehensive view of multiple interactions in the futures contracts of oil and the S&P 500 futures contracts. By working with a complex wavelet, such as the Morlet wavelet, we also analyse the degree to which the time series are in phase with each other.

15.5.1 Cospectral Densities

All of our time series are non-stationary in levels, but stationary in log-differences, as evidenced by unit root tests.* Therefore, legitimate cross-spectral results can be obtained only when the methodology is applied to log-differences (or returns). We apply the cross-spectral methodology summarized in the previous section to three pairs of time series: cospectral densities of (1) oil price returns and oil volume changes, (2) oil price returns and S&P 500 returns, and (3) oil volume changes and S&P 500 returns. Figures 15.3 and 15.4 plot the cospectral densities of log-differences of our time series against frequency.

Figure 15.3 presents cospectral densities of oil price returns and volume (solid line), oil price returns and S&P 500 returns (dotted line), and changes in oil volume and S&P 500 returns (dashed line). It is easy to see that the correlation between log differences of oil prices and the stock market is close to zero at virtually all frequencies. This is an interesting result in light of the capital market and cash-flow theories summarized in Sections 15.1 and 15.2. Correlation for the other two pairs of series—oil prices and volume and oil volume and S&P 500—is higher in absolute value, but positive and negative association is distributed fairly evenly across most frequencies. We observe a slightly stronger relationship between

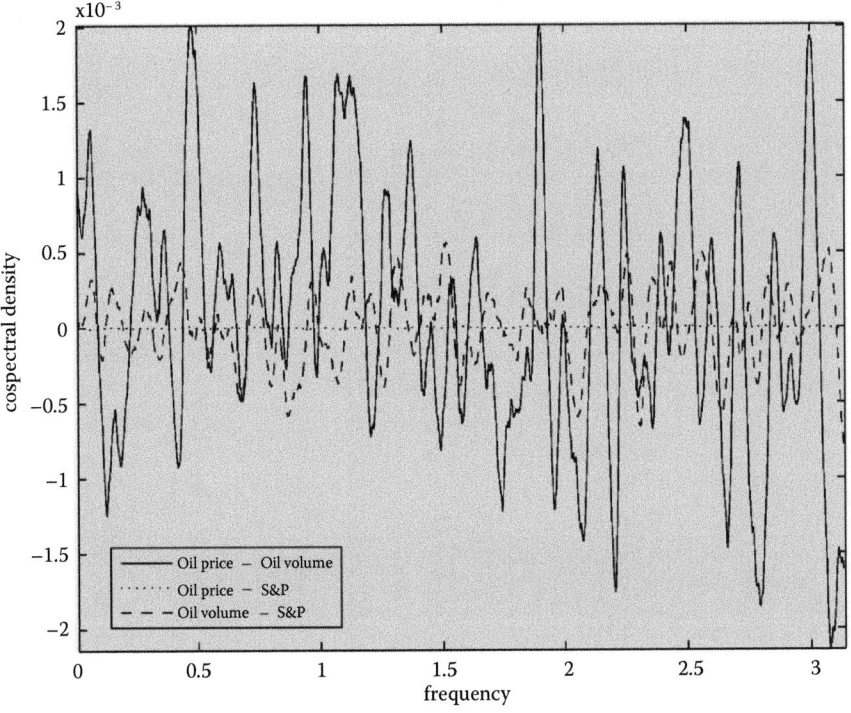

FIGURE 15.3   Cospectral densities of oil price returns and oil volume changes, oil price and S&P 500 returns, oil volume changes and S&P 500 returns, 1983–2011.

---

* Augmented Dickey–Fuller and Phillips–Perron tests were performed for oil futures contract closing prices and S&P 500 futures contract closing prices. The hypothesis of a unit root in levels could not be rejected even at the 10% significance level; log-differencing the data, on the other hand, yields stationary time series at the 1% significance level.

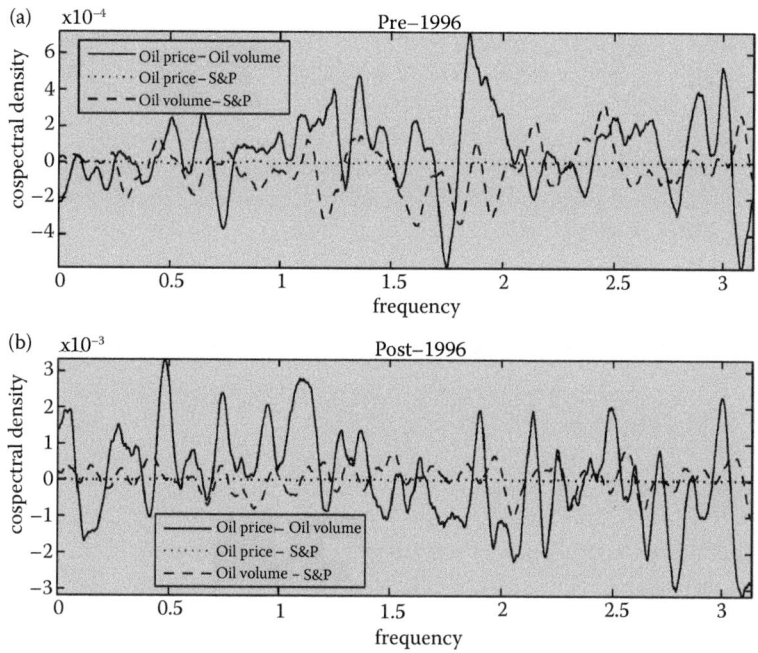

FIGURE 15.4 Cospectral densities of oil price returns and oil volume changes, oil price and S&P 500 returns, oil volume changes and S&P 500 returns, 1983–1996 (a) and 1997–2011 (b).

oil volume and S&P 500 over higher frequencies, namely for $\omega \geq 1.5$ (corresponding to the periodicity of four trading days or less).

Next, we are interested in analysing how the correlations among the three time series (oil price returns, oil volume changes and S&P 500 returns) changed over time and, more specifically, after the 1980s when oil price volatility became more pronounced. To this end, we compare the cospectral densities for two subsamples: from March 1983 to December 1996, and from January 1997 to March 2011. Panels (a) and (b) of Figure 15.4 report our results. The most important difference between the two subsamples is that covolatility along most of the frequency components for the oil price–oil volume pair and the oil volume–S&P pair is almost one order of magnitude higher during the second half of the sample. This is consistent with the higher overall volatility in the oil and stock markets after the 1980s. Panel (b) suggests that positive covolatility between oil price and oil volume in the post-1996 subsample is concentrated in the relatively low frequencies of $\omega \leq 1.2$, which corresponds to cycles with a duration of more than five working days.

There are no other striking differences between the two subsamples. As with the entire sample, we observe both positive and negative covariances between log-differences of oil prices and S&P 500 contract returns. Both panels of Figure 15.4 suggest that the correlations between oil contract returns and S&P returns are virtually zero at all frequencies. Thus, the fact that we observed more negative changes in the post-1980s period relative to the 1980s did not affect the lack of correlation between log differences of oil prices and the stock returns.[*]

---

[*] As with the analysis of the entire sample, we find that, quite predictably, oil futures volume is more closely associated with oil prices than with S&P 500 returns.

15.5.2 Wavelets

Figures 15.5 through 15.11 report the results of the wavelet analysis. Figure 15.5 depicts the continuous wavelet transform for the S&P 500. We note a pronounced effect between the scales at 64 and 128 over the observation period from 4500 to 5000. This indicates a lasting impact over three to six months (there are approximately 22 trading days per month). This impact occurred during calendar years 2000 and 2001 when the tech bubble rose to its highest peak and subsequently burst. The contour lines mark the 95% statistical significance, with the lighter shaded areas indicating the strongest impact.

Figure 15.6 plots the continuous wavelet transform for the oil futures prices. We note that the two most significant effects occur between scales 128 and 256 (6 to 12 months) over the observation periods 2000 and 6250. These periods correspond to the first time that the U.S. was at war with Iraq and is indicative of the disruption to the oil markets at that time, as well as the huge speculative oil bubble buildup and collapse in 2008. Again, the contour lines mark the 95% statistical significance, with the lighter shaded areas indicating the strongest impact.

Figure 15.7 plots a three-dimensional view of the wavelet power spectrum (squared wavelet detail coefficients). The wavelet power spectrum illustrates the pronounced effect at observation period 2000 corresponding to the first U.S. war with Iraq and its impact on the oil market, and an even more pronounced impact at approximately observation number 6250 corresponding to the dramatic increases in oil prices in summer 2008, followed by the dramatic decline in what many considered to be a huge speculative bubble. In Figure 15.8, the wavelet power spectrum for S&P 500 shows the most pronounced effect occurring at approximately observation period 4500, which corresponds to the tech bubble and its subsequent collapse at the turn of the century. Also notice the strongest scale effect at 1024 (trading days) corresponding to the 4 year presidential impact on stock prices.

In Figure 15.9, the dark bold contour lines mark the 95% significance threshold level where we see the strongest wavelet correlation between the futures price of oil and the futures trading volume of oil. The cross-wavelet correlation is strongest at 128 trading days (6 months) and at 256 trading days (1 year).

Figure 15.9 also shows the statistically significant phase angles plotted by directional arrows. An arrow pointing to the right indicates that both time series are exactly in phase with each other, whereas

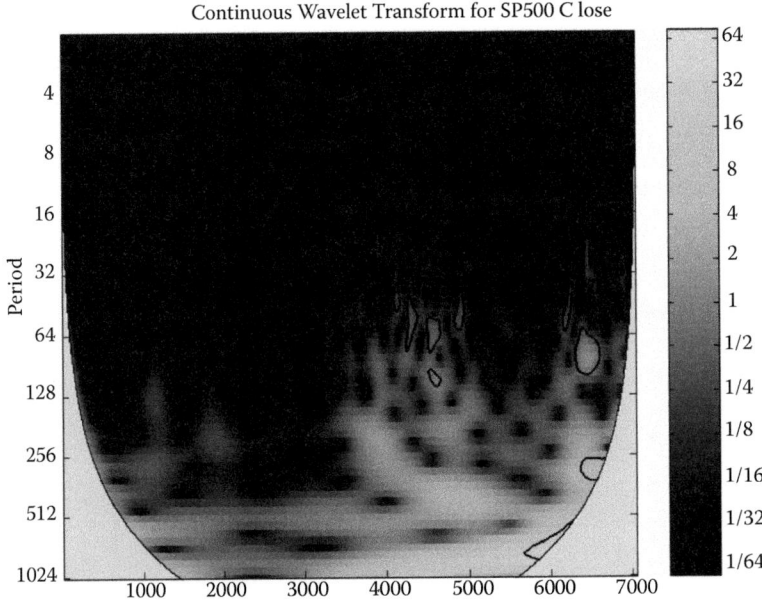

FIGURE 15.5   Continuous wavelet transform—S&P 500 futures closing prices.

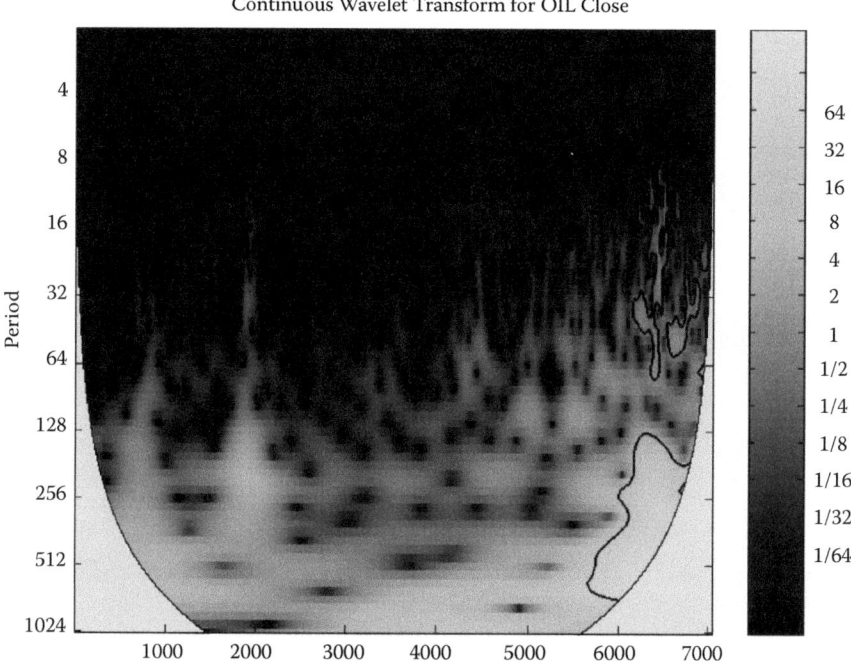

FIGURE 15.6 Continuous wavelet transform—Oil futures closing prices.

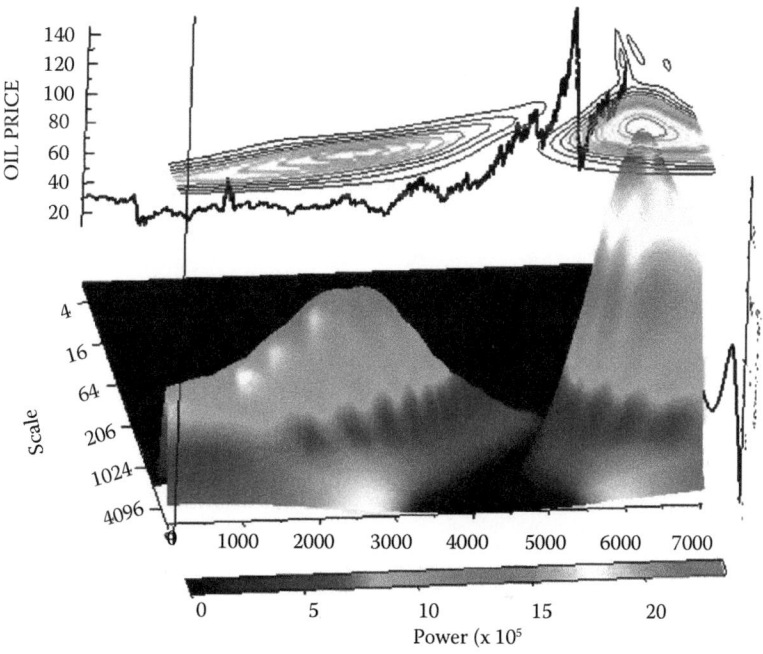

FIGURE 15.7 Three-dimensional view of the wavelet power spectrum—Oil futures.

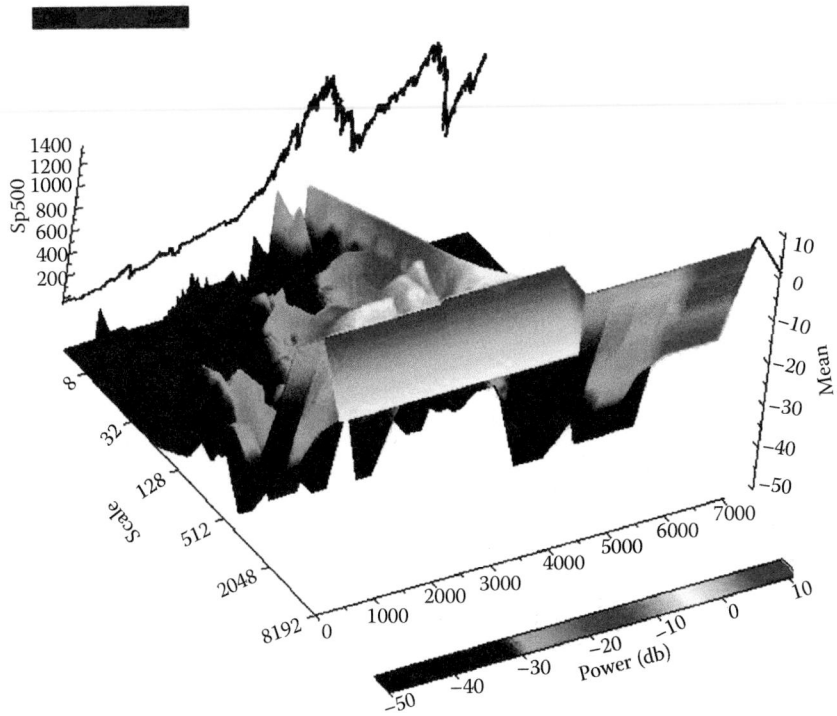

FIGURE 15.8   Three-dimensional view of the wavelet power spectrum—S&P 500 futures.

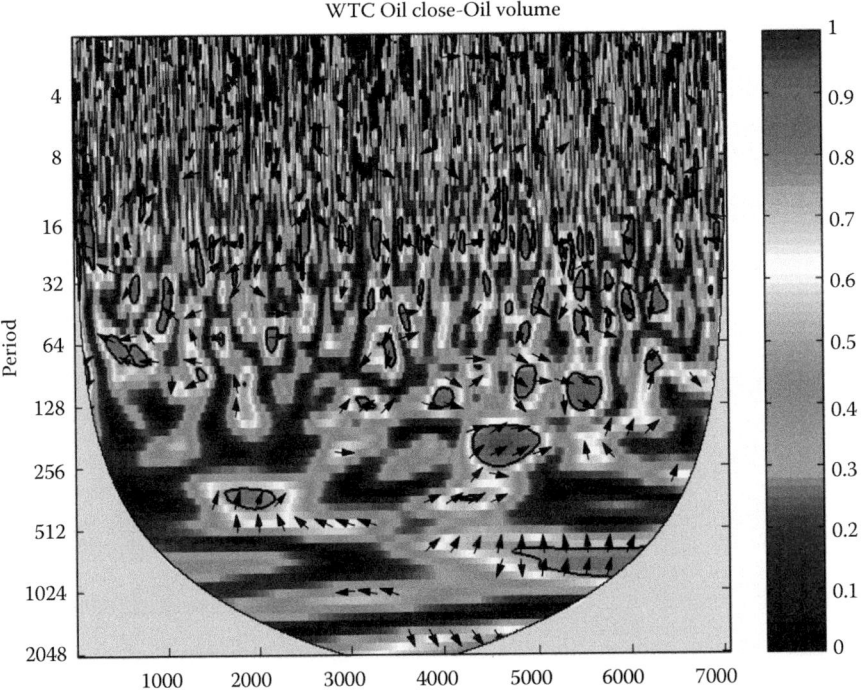

FIGURE 15.9   Cross-correlations: Oil prices and oil volume.

an arrow pointing to the left indicates that the two time series are completely out of phase with each other. An arrow pointing straight up indicates that the first data series (oil futures price) leads the second data series (oil futures volume) and an arrow pointing straight down indicates the second series leads the first. As we can see from Figure 15.9, the arrows predominantly point up, indicating that oil prices lead oil volume, or they point to the right, indicating that the two data series are in phase with each other. Figure 15.9 suggests that oil prices lead oil volume for about 256 days (one year) in the early 2000s.

In Figure 15.10, we observe the wavelet cross-correlations between oil prices and S&P 500 prices. The strongest field of correlation is found in the earlier part of our study (approximately the first time the U.S. went to war in Iraq). The directional phase angle arrows predominately point up and to the left in the earlier part of our study, indicating a mix between oil prices leading S&P 500 futures prices and S&P 500 futures prices being out of phase with oil futures prices. Towards the end of the study, the phase arrows point to the right, indicating that the two series move together. Note that the movement together occurs over a wide range of scales.

Finally, in Figure 15.11, the predominate cross-correlations occur over a 128 trading day period (6 months). Note that the statistically significant directional phase angle arrows mostly point down, indicating that increased trading volume leads to higher S&P futures prices by approximately 6 months. Arrows that are pointing to the left in this figure indicate that volume and prices are out of phase.

15.6 Conclusions

Our results indicate that, over the 28-year period of study, there is a persistent positive association between crude oil futures prices and S&P 500 futures prices. This finding is observed over a large number of time scales and over a very wide range of prices. With respect to returns, cross-spectral analysis indicates that there is virtually no contemporaneous relationship between oil futures returns and S&P 500 returns at most frequencies. The positive correspondence between oil futures prices and S&P 500 futures prices is contrary to the purported negative relationship between oil prices and the U.S. stock market reported by Sauter and Awerbuch (2003) and earlier by Jones and Kaul (1996).

Our seemingly counterintuitive findings may be due to the likely presence of nonlinearities in the relationship between the oil market and the macroeconomy suggested by the extant literature (e.g.

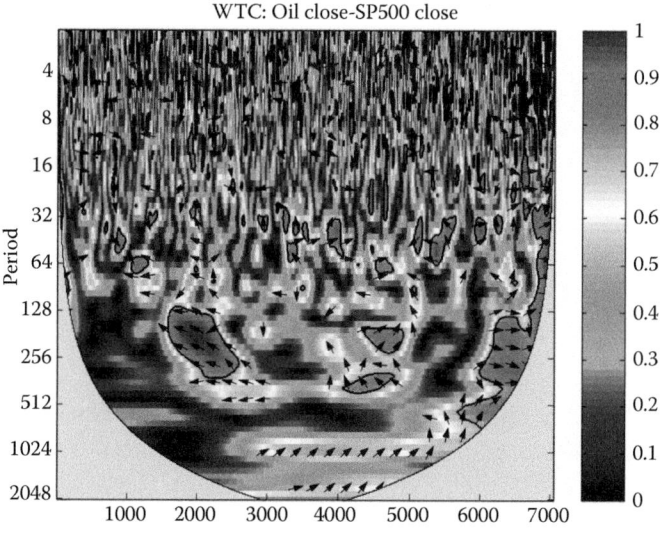

FIGURE 15.10 Cross-correlations: Oil price and S&P 500 close.

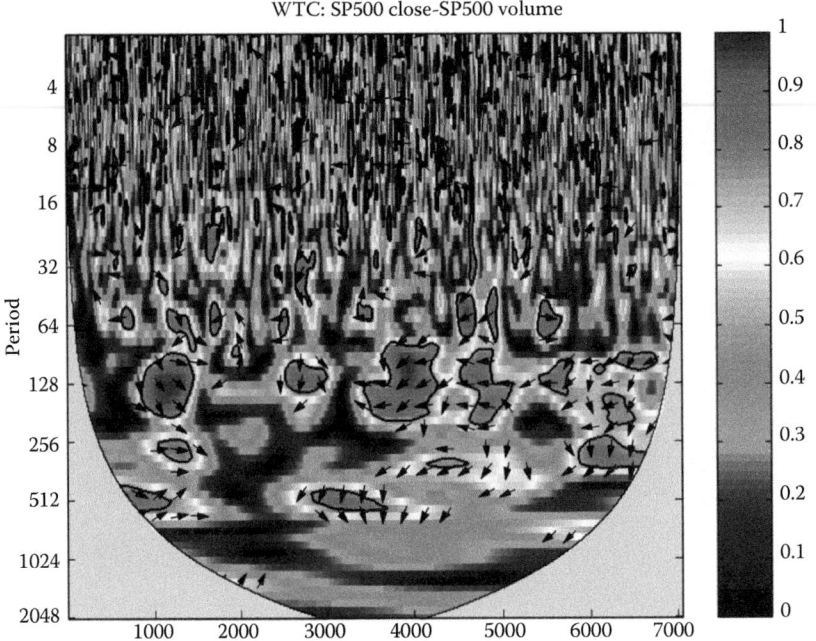

FIGURE 15.11   Cross-correlations: S&P 500 and S&P 500 volume.

Lee *et al.* (1995) and Hamilton (1996)). Tang and Xiong (2010), who find that commodity prices indeed became increasingly more correlated with equity markets after 2000, point to the rise of investments in commodity futures in the mid-2000s, which suggests that commodity prices can be affected by financial factors in addition to changes in supply and demand. Another possible explanation lies in the bi-directional causality between oil futures returns and the S&P 500 returns emphasized by Ciner (2001). The cross-spectral and wavelet analyses employed in our paper allow for nonlinear feedback effects between oil prices and financial markets.

We also analyse the relationship between futures prices and trading volume. The directional phase angle plots reported in the empirical section suggest that oil prices lead oil volume and S&P 500 trading volume leads S&P 500 futures prices. In particular, oil prices lead oil volume for about 256 days (one year) in the early 2000s. Generally speaking, our results can be viewed as reflective of the investors' bounded rationality and the gradual information diffusion hypothesis (Hong and Stein 1999). Our empirical analysis indicates that investors and policy makers should pay close attention to trading volume when analysing stock market movements.

Conventional wisdom and past research (see, for example, Sauter and Awerbuch (2003) as well as works cited therein) suggest that changes in oil prices should be negatively related to macroeconomic and financial indicators. Although it is tempting to assume that higher oil prices bode ill for companies' stock valuations, our paper shows that this assumption is spurious. Contrary to the conventional wisdom, our findings point to the positive association between oil and S&P 500 futures for a large number of time scales and across a wide range of Fourier frequencies. This finding of a positive association between oil prices and stock markets is logical from an economic standpoint. When the economy is recovering from a recession and economic conditions start to improve, stock prices generally go up. During an expansion, wealth is positively affected through higher net worth of the companies and the corresponding higher value of personal assets. Consumers' incomes also become higher due to better and more abundant employment opportunities, as well as due to higher wages during an expansion. Higher demand for consumer and industrial goods due to increased public wealth and incomes, coupled with better balance sheets of companies, lead to a greater demand for energy, which in turn drives up prices of oil. Thus, higher economic activity is associated with higher oil prices. Conversely, when the economy begins to falter, stock market valuations of firms become lower. These negative movements in the stock markets decrease public wealth and companies' net worth, thereby further depressing economic activity. Lower levels of economic activity, consumers' wealth and companies' net worth, as well as the ensuing weaker demand for energy, put downward pressure on oil prices and volume. The main findings of our paper are, therefore, consistent with economic theory.

This paper finds no evidence of disruptive effects of oil price movements on the financial markets. Although there may be many reasons to switch from fossil fuels such as oil to alternative forms of renewable energy (e.g. balance of payments problem, rising concentrations of CO_2, dependence on unstable regimes), our results indicate that an unfavourable impact on U.S. equity markets from rising oil prices is not supported by empirical research.

The results reported in this paper have important practical implications for investors: it is possible to improve the accuracy of the forecast of future returns by studying oil futures prices and futures volumes. Unlike most of the literature that examines the relationship between oil prices and the stock market, we use daily data on oil futures and S&P 500 futures, which allows us to conduct a more detailed analysis of the said relationship. Careful empirical investigations of oil–stock market relationships can have important implications for portfolio selection (e.g. Arouri and Nguyen (2010)). Finally, the frequency-domain techniques used in this paper help circumvent the issue of possible endogeneity of oil prices with respect to the U.S. economy that often plagues the literature's attempts to discern the impact of oil price changes on the U.S. stock market (Kilian 2009, Kilian and Park 2009).

Acknowledgements

We would like to thank two anonymous referees for invaluable comments. A grant for futures data is gratefully acknowledged; information is available at http://www.price-data.com/.

Software Acknowledgements

1. IDL (Interactive Data Language), http://www.ittvis.com/ProductServices/IDL.aspx
2. Grinsted, A., Moore, J.C. and Jevrejeva, S., http://www.pol.ac.uk/home/research/waveletcoherence/
3. Torrence, C. and Compo, G., http://atoc.colorado.edu/research/wavelets/
4. S W A N (Software for Waveform Analysis), CNRS/LPCE, Orleans, France, Lagoutte, D., Brochot, J.Y. and Latremoliere, P., ftp://lpce.cnrs-orleans.fr/projects/swan/pub/swan–2.42/doc/swan–at.pdf

References

Abdelaziz, M., Chortareas, G. and Cipollini, A., Stock prices, exchange rates, and oil: Evidences from Middle East oil-exporting countries. In *Proceedings of the Middle East Economic Association*, New Orleans, LA, January 4–7, 2008.

Addison, P.S., Wavelet transforms and the ECG: A review. *Physiol. Measure.*, 2005, **26**(5), R155–R199.

Agusman, A. and Deriantino, E., Oil price and industry stock returns: Evidence from Indonesia. In *21st Australasian Finance and Banking Conference*, August 25, 2008.

Arouri, M.E.H. and Nguyen, D.K., Oil prices, stock markets and portfolio investment: Evidence from sector analysis in Europe over the last decade. *Energy Policy*, 2010, **38**(8), 4528–4539.

Awerbuch, S., The surprising role of risk and discount rates in utility integrated-resource planning. *Electric. J.*, 1993, **6**(3), 20–33.

Awerbuch, S. and Deehan, W., Do consumers discount the future correctly? A market-based valuation of fuel switching. *Energy Policy*, 1995, **23**(1), 57–69.

Bolinger, M., Wiser, R. and Golove, W., Quantifying the value that wind power provides as a hedge against volatile natural gas prices. Lawrence Berkeley National Laboratory Working Paper, 2002.

Burke-Hubbard, B., The World According to Wavelets: The Story of a Mathematical Technique in the Making, 1998 (AK Peters Ltd).

Campbell, J.Y., A variance decomposition for stock returns. *Econ. J.*, 1991, **101**, 157–179.

Ciner, C., Energy shocks and financial markets: Nonlinear linkages. *Stud. Nonlin. Dynam. Econometr.*, 2001, **5**(3), 203–212.

De Moortel, I., Munday, S.A. and Hood, A.W., Wavelet analysis: The effect of varying basic wavelet parameters. *Solar Phys.*, 2004, **222**(2), 203–228.

Driesprong, G., Jacobsen, B. and Maat, B., Striking oil: Another puzzle? *J. Financ. Econ.*, 2008, **89**(2), 307–327.

Elder, J. and Jin, H.J., Fractional intergration in commodity futures returns. *Financ. Rev.*, 2009, **44**(4), 583–602.

Elder, J. and Serletis, A., Oil price uncertainty. *J. Money, Credit Bank.*, 2010, **42**(6), 1137–1159.

Ferderer, J.P., Oil price volatility and the macroeconomy. *J. Macroecon.*, 1996, **18**(1), 1–26.

Gençay, R., Selcuk, F. and Whitcher, B., An Introduction to Wavelets and Other Filtering Methods in Finance and Economics, 2002 (Academic Press: New York).

Gogineni, S., Oil and the stock market: An industry level analysis. Working Paper Series, University of Oklahoma, Division of Finance, 2009.

Granger, C.W.J., The typical spectral shape of an economic variable. *Econometrica*, 1966, **34**(1), 150–161.

Grinsted, A., Jevrejeva, S. and Moore, J., Application of the cross wavelet transform and wavelet coherence to geophysical time series. *Nonlin. Process. Geophys.*, 2004, **11**, 561–566.

Hamilton, J.D., Oil and the macroeconomy since World War II. *J. Polit. Econ.*, 1983, **21**(2), 228–248.

Hamilton, J.D., *Time Series Analysis*, 1994 (Princeton University Press: Princeton).

Hamilton, J.D., This is what happened to the oil price–macroeconomy relationship. *J. Monet. Econ.*, 1996, **38**(2), 215–220.

Hamilton, J.D., What is an oil shock? NBER Working Paper 7755, 2000.

Hong, H. and Stein, J.C., A unified theory of underreaction, momentum trading and overreaction in asset markets. *J. Finance*, 1999, **54**(6), 2143–2184.

Hooker, M.A., What happened to the oil price–macroeconomy relationship? *J. Monet. Econ.*, 1996, **38**(2), 195–213.

Huang, R.D., Masulis, R.W. and Stoll, N.R., Energy shocks and financial markets. *J. Fut. Mkts*, 1996, **16**(1), 1–27.

Jammazi, R. and Aloui, C., Wavelet decomposition and regime shifts: Assessing the effects of crude oil shocks on stock market returns. *Energy Policy*, 2010, **38**(3), 1415–1435.

Jones, C.M. and Kaul, G., Oil and the stock markets. *J. Finance*, 1996, **51**(2), 463–91.

Kilian, L., Not all oil price shocks are alike: Disentangling demand and supply shocks in the crude oil market. *Am. Econ. Rev.*, 2009, **99**(3), 1053–1069.

Kilian, L. and Park, C., The impact of oil price shocks on the U.S. stock market. *Int. Econ. Rev.*, 2009, **50**(4), 1267–1287.

Lee, C.-C. and Zeng, J.-H., The impact of oil price shocks on stock market activities: Asymmetric effect with quantile regression. *Math. Comput. Simul.*, 2011, **81**(9), 1910–1920.

Lee, K., Ni, S. and Ratti, R.A., Oil shocks and the macroeconomy: The role of price variability. *Energy J.*, 1995, **16**(4), 39–56.

Mallat, S., *A Wavelet Tour of Signal Processing*, 2nd ed., 1999 (Academic Press: New York).

Manchaldore, J., Palit, I. and Soloviev, O., Wavelet decomposition for intra-day volume dynamics. *Quantit. Finance*, 2010, **10**(8), 917–930.

Miller, J.I. and Ratti, R.A., Crude oil and stock markets: Stability, instability, and bubbles. *Energy Econ.*, 2009, **31**(4), 559–568.

Mork, K.A., Oil and the macroeconomy, when prices go up and down: An extension of Hamilton's results. *J. Polit. Econ.*, 1989, **97**(3), 740–744.

Mork, K.A., Olsen, O. and Mysen, H.T., Macroeconomic responses to oil price increases and decreases in seven OECD countries. *Energy J.*, 1994, **15**(4), 19–35.

Raihan, S.Md., Wen, Y. and Zeng, B., Wavelet: A new tool for business cycle analysis. Federal Reserve Bank of St. Louis Working Paper 2005-050, 2005.

Reboredo, J.C., Nonlinear effects of oil shocks on stock returns: A Markov-switching approach. *Appl. Econ.*, 2010, **42**(29), 3735–3744.

Rotemberg, J.J. and Woodford, M., Imperfect competition and the effects of energy price increases on economic activity. *J. Money, Credit, Bank.*, 1996, **28**(4), 550–577.

Sadorsky, P., Oil price shocks and stock market activity. *Energy Econ.*, 1999, **21**, 449–469.

Sauter, R. and Awerbuch, S., Oil price volatility and economic activity: A survey and literature review. IEA Working Paper, 2003.

Sawyer, K.R. and Nandha, M., How oil moves stock prices. Working Paper Series, University of Melbourne, Department of Finance, 2006.

Tang, K. and Xiong, W., Investment and nancialization of commodities. NBER Working Paper No. 16385, 2010.

Torrence, C. and Compo, G.P., A practical guide to wavelet analysis. *Bull. Am. Meteorol. Soc.*, 1998, **79**, 61–78.

Vo, M., Oil and stock market volatility: A multivariate stochastic volatility perspective. *Energy Econ.*, 2011, **33**(3), 956–965.

Walker, J.S., A Primer on Wavelets and their Scientific Applications, 1999 (CRC Press: Boca Raton).

Wang, L. and Sassen, K., Wavelet analysis of cirrus multiscale structures from lidar backscattering: A cirrus uncinus complex case study. *J. Appl. Meteorol. Climatol.*, 2008, **47**(10), 2645–2658.

Yang, C.W., Hwang, M.J. and Huang, B.N., An analysis of factors affecting price volatility of the US oil market. *Energy Econ.*, 2002, **24**, 107–119.

Zar, J.H., *Biostatistical Analysis*, 1999 (Prentice Hall: Engelwood Cliffs, NJ).

16

Sectoral Stock Return Sensitivity to Oil Price Changes: A Double-Threshold FIGARCH Model

**Elyas Elyasiani,
Iqbal Mansur and
Babatunde Odusami**

16.1 Introduction

Despite the substantial adverse effects of oil price shocks on the U.S. economy since the OPEC oil embargo of 1973, research on the association between oil price changes and the U.S. stock market has been relatively limited.* In particular, no study has been conducted on the threshold effect of the oil price changes on the equity return distribution at the sector level. Identification of threshold values for oil price effects is important because basic models overlooking this non-linearity distort the effects of oil shocks on both the stock market and the real economy. Moreover, the findings in the existing literature are conflicting in terms of significance, magnitude and the direction of the oil shock effects. The

---

* The impact of oil price shocks on the U.S. and the global economy is considerable. It is estimated that the oil price shocks of 2004–2008 cost the U.S. economy nearly $1.9 trillion (http://www.fueleconomy.gov/FEG/oildep.shtml). Similarly, the International Energy Agency (IEA) estimates that for every $1 increase in oil prices, the world GDP is adversely affected by $25 billion (http:// www.iea.org). Along these lines, Greene and Tishchishyna (2000) report that from 1970 to 2000, oil price movements resulted in a $7 trillion loss in terms of costs to the U.S. Yang *et al.* (2002) also argue that higher oil prices engender recessions and unemployment in oil-consuming nations, as they curtail economic activity (Awerbuch and Sauter 2005). These findings have major implications for the formulation of the monetary, fiscal and energy policies of oil exporting/importing countries. Moreover, dependence on imported oil has major implications in terms of political leverage and political subordination, examples of which were clearly manifested in the aftermath of the 1973 oil embargo and the 1979 sharp increase in oil prices.

DOI: 10.1201/9781003265399-19

pioneering work of Chen *et al.* (1986) suggests that the risk associated with oil price changes was not priced in the U.S. stock market even during the 1968–1977 period when OPEC played a critical role in setting oil prices. However, Jones and Kaul (1996) demonstrate that changes in real oil returns did have a detrimental effect on output and real stock returns in the U.S., Canada, Japan and the U.K. Similarly, Huang *et al.* (1996) find that futures returns on heating and crude oil did lead individual oil company and petroleum industry stock returns during the 1980s, but failed to exert an impact on other industry stock indices or the S&P 500 index. Oil futures volatility was also found to lead the petroleum stock index volatility, but not other industry or market indices.

Sadorsky (1999) employs a Generalized Autoregressive Conditionally Heteroskedastic (GARCH) model to investigate the interaction of oil price shocks and the stock market. He also uses the vector auto regression (VAR) technique to examine the interaction between return on oil and industrial production, interest rate and real stock returns. His results provide evidence of unidirectional and asymmetric shock transmission from oil to stocks. Moreover, positive oil price shocks and oil price volatility shocks explain a larger portion of the forecast error variance in real stock returns than do their negative counterparts, at least for some sub-periods.

Faff and Brailsford (1999) investigate the sensitivity of the Australian industry equity returns to oil price changes over the period July 1983 to March 1996. Their results suggest that oil is an important factor in describing the return generating process of Australian industries. However, the direction of the effect is industry-specific; oil, gas and diversified resources industries show positive sensitivity, while paper and packaging, and transport industries demonstrate negative sensitivity to oil shocks.* Building on Sadorsky (1999), Huang *et al.* (2005) use a multivariate threshold model to investigate the impact of real oil return and its volatility on industrial production, interest rate and real stock returns for the U.S., Canada and Japan. Using monthly data from 1970 to 2002, they uncover the existence of differential thresholds for oil return and oil return volatility in their impacts on each of these three economies. The threshold values are found to be dependent on the extent to which an economy is reliant on imported oil. Their findings suggest that when oil return, or its volatility, are below their corresponding threshold levels, they have limited impacts on the macroeconomic variables considered, while if they are above the threshold levels, changes in oil returns demonstrate greater explanatory power than those in oil return volatility.

Given the vital importance of oil to the U.S. economy, the relationship between sectoral stock returns and oil futures returns, and sensitivity of the stock return mean and volatility to oil shocks require further scrutiny. The main goal of this study is to provide a detailed analysis of the association between the equity return distributions of different sectors in the U.S. economy and changes in the oil returns. To this end, we take the following steps: (i) we model the sectoral stock return effects of oil return changes within a generalized Double-Threshold Fractionally Integrated GARCH (DT-FIGARCH) framework, (ii) we determine whether there is a threshold level of oil returns for each of the sectors considered, beyond which the oil return effects on equity return levels change in character and magnitude; (iii) we investigate whether a threshold effect is evident in the conditional volatility of sector returns; and (iv) we examine the effect of oil shocks on the return of each sector and its volatility within this generalized framework.

The contributions of this study are as follows: (i) this is the first study to apply a double-threshold FIGARCH methodology in order to allow for long memory and to assess the threshold effects of oil movement on both the first and the second moments of the return distribution for a large number of sectors; (ii) we test for the existence of, and identify the values of the threshold oil returns for different sectors and establish the asymmetry of the responses of sector returns, and their volatilities, to oil return changes when the latter is above/below the threshold; (iii) we provide strong evidence of statistically significant long-memory processes in the conditional variances of sector returns and (iv) we establish that the longevity of shock memory in the conditional volatility of returns is sector-specific and responsive to threshold regime shifts.

---

* The magnitude of the distortion due to model misspecification can be considerable (see footnote Section 16.2.3).

Our findings have several policy implications. The primary advantage of modelling conditional variance using a FIGARCH model is that it not only captures the short-run dynamics (as in the standard GARCH model), but also the long-run persistence that decays at a slower hyperbolic pace. Therefore, this long-memory model is capable of carrying information over a year or longer and delineating its pattern of effect. The implication of the long memory is that policy makers should not solely consider the effects of the latest turmoil or extreme events in the formulation of overall or industry-specific policies; they should also account for the impacts of the events in the distant past in this regard. Additionally, if volatility decay exhibits a mean-reverting behaviour, policy makers may be able to take appropriate actions to smooth the effect of a shock, shorten its duration and accelerate the mean-reversion process. Examples of such policy actions include the release of oil from the U.S. Strategic Petroleum Reserve (SPR) in the aftermath of Hurricane Katrina, the 2008 suspension of oil purchases for SPR in order to ease the upward pressure on oil prices, tax credits used to encourage investment in renewable energy and tax surcharges introduced to reduce oil consumption. Correct understanding and appropriate modelling of stock return volatility is also important because it can aid investors in their quest for optimal portfolio selection and risk management.

Analysis of stock return behaviour at the sector level is appealing to stock investors because while globalization has reduced the benefits of international risk diversification, sectors across national markets continue to demonstrate independent movements and, hence, sector-based investment portfolios can be helpful in controlling risk. In this context, Moskowitz and Grinblatt (1999) find that investment strategies based on 'industry-momentum' produce holding-period returns that are substantially larger than 'individual stock' momentum-based strategies. Moreover, findings about the sector-specific risk and return effects of oil price changes have important implications for planning decisions by producers in these sectors and government assessment of cost/benefits of regulatory policies designed to promote the use of alternative energy sources. Similarly, understanding sector-based volatility is important for determining the sectoral cost of capital, determination of sectoral shock persistence and asset allocation decisions, because volatility changes are likely to alter these decisions and outcomes.

Extant studies mostly focus on the first moment of the return distribution, overlooking the second moment or the volatility equation. Given that financial time series are known to exhibit GARCH properties, the assumption of a fixed-variance error term in the mean equation results in erroneous inferences. Even studies that do account for GARCH properties generally adopt a single-regime scenario, overlooking the presence of non-linearities such as threshold and asymmetry effects. Given that the degree of volatility persistence is likely to be sensitive to whether the return on oil (R_{oil}) takes values above or below the threshold, these thresholds need to be recognized, identified and accounted for. Sadorsky (1999) arbitrarily chooses the zero change level as the threshold for all macro-variables he studies. In practice, however, idiosyncratic characteristics of economic sectors are likely to result in non-zero and dissimilar threshold values for different sectors. We remedy this shortcoming by adopting a Tsay (1998) type methodology to estimate sector-specific threshold values for each period studied. We specify volatility as a fractionally integrated (FIGARCH (p, δ, q)) model, developed by Baillie *et al.* (1996). This model adequately captures the decay mechanism of the autocorrelation function of the conditional volatility process, allowing one to properly analyse the long-term persistence of shocks to the volatility process. The remainder of the paper is organized as follows. Section 16.2 provides the data definitions and methodology. Empirical results are presented in Section 16.3 while the conclusions are given in Section 16.4.

16.2 Data and Methodology

16.2.1 Data Description

Ten sectors, constructed based on two-digit SIC codes, are analysed in this study and roughly sorted into four major types:.

   a. Oil substitutes; Coal Mining (SIC 12), Electric–Gas Services (SIC 49);
   b. Oil-related; Oil and Gas Extraction (SIC 13), Petroleum Refining (SIC 29);

c. Oil-consuming; Building Construction (SIC15), Chemicals & Allied Products (SIC 28), Rubber & Plastic (SIC30), Transport Equipment (SIC 37) and Air Transportation (SIC 45);

d. Financial; Depository Institutions (SIC 60).

Sector return data are collected from the Centre for Research in Security Prices (CRSP) database. The Fama–French factors used in the model include the market risk premium (RM), the size factor (SMB) and the value factor (HML) and are obtained from the Kenneth R. French website (www.mba.tuck.dartmouth.edu). SMB is the return on a portfolio long on small firms and short on large firms (small minus large firm returns). HML is the return on a portfolio long on stocks with high book-to-market ratio and short on stocks with low book-to-market ratio (high minus low market-to-book firm returns).

The focus of this study is the *ex ante* effect of oil price movements on industry returns. Hence, returns on crude oil futures are favoured over other oil time series as an indicator of oil return changes. Moreover, Sadorsky (2001) shows that spot prices of oil are more susceptible to transitory random noise than oil futures prices.* Following Guo and Kliesen (2005), the return on one-month crude oil futures traded on the New York Mercantile Exchange (NYMEX) is used as the oil return variable. The daily return on oil (R_{oil}) is calculated as $\log(p_t/p_{t-1})$ where p is the one-month crude oil futures price over the period of April 4, 1983, to December 29, 2006.

We distinguish between periods of tranquil and highly volatile oil market conditions because market reactions to shocks and, hence, threshold levels are likely to differ under these circumstances. The oil market volatility as measured by the standard deviation (SD) of oil prices was relatively high in the 1980s, low in the 1990s and very high in the 2000s. Specifically, the 1983–1991 sub-period exhibits a moderate level of volatility (SD of 6.31%) followed by relative calm in the 1991–1998 sub-period (SD of 2.71%) and extreme volatility in the 1998–2006 sub-period (SD of 15.49%). The corresponding oil price ranges (averages) for these three periods were $10.42–40.42 (22.55), $10.72–26.62 (19.15) and $17.45–77.03 (39.81), respectively.[†]

The oil price data show that, since 1984, the oil market has seen two distinct structural shifts, one in 1990 and the other in 1998 (Figure 16.1). We employ the Zivot and Andrews (1992) (ZA) test to determine the exact location of the structural changes in this market. Using the full sample period, the ZA-test procedure identifies January 24, 1991, as the first break point. The test is repeated for the remaining observations and the second break point is found to be on December 10, 1998.[‡§] Estimation is carried out

---

* A host of oil price series are used in the literature: refiners' acquisition cost for crude oil (Mork 1989), PPI for oil (Jones and Kaul 1996, Sadorsky 1999), net oil price increase dummy (Lee and Ni 2002), crude petroleum price (Davis and Haltiwanger 2001, Hamilton 2003), PPI of oil relative to inflation (Hooker 2002), average of world spot crude oil prices deflated by county-specific PPI (Huang *et al.* 2005) and log difference of multiple oil series such as Arab Light, West Texas, Dubai, Brent, Brent Futures and NYMEX Futures (Driesprong *et al.* 2008). Due to space considerations a detailed discussion of this issue is not presented here. See Elyasiani *et al.* (2011) for more on the preference for oil futures returns over oil spot returns.

† We report a similar set of statistics for oil futures return in Table 16.1.

‡ The advantage of the ZA procedure is that, unlike Perron (1989), no *a-priori* knowledge of the break points in the trend function is necessary. Instead, a data-dependent algorithm is used to determine the structural breaks. Thus, the algorithm produces unconditional unit-root test results. The ZA-test statistics for unit-root tests are –28.72 and –24.74 for the first and second structural break points, respectively. The critical value at the 1% level is 5.57.

§ An unparalleled level of economic and political activity affected oil prices both in January 1991 and December 1998. A major factor that contributed to the global uncertainty of oil price movements was the first Gulf War (U.S. and allied countries' air strikes on Iraq beginning on January 16, Iraqi SCUD missile attacks on Israel on January 18 and 22, and the World's largest oil spill caused by the Iraqi forces in Kuwait on January 23). Following the Gulf War, crude oil prices entered a period of steady decline until 1994. The price cycle then turned up. The U.S. economy was strong and the Asian Pacific region was booming. Asian consumption accounted for all but 300,000 barrels per day of that gain and contributed to a price recovery that extended into 1997 (www.wtrg.com). The rapid growth in Asian economies came to a halt in 1998, leading to a lower consumption of oil. Higher OPEC production, coupled with lower consumption led to a sharp decline in oil prices, which continued through December 1998, recovered in early 1999 and continued to rise throughout 2000 (www.wtrg.com).

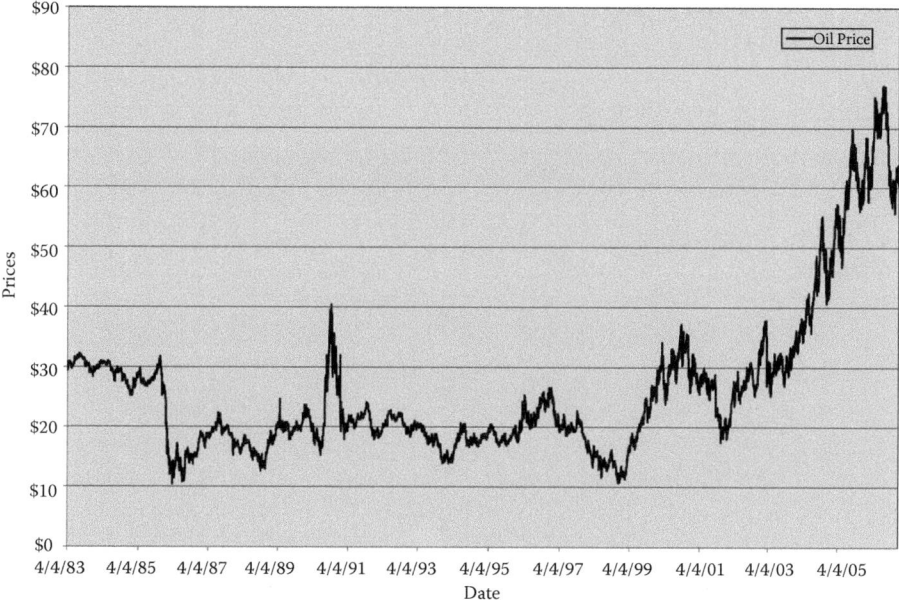

FIGURE 16.1 NYMEX One Month Oil Futures Prices—April 4, 1983, to December 29, 2006.

for these three distinct sub-periods separately. However, to save space, only the results of the latter two periods (tranquil and highly volatile, respectively) are presented. This allows us to contrast the threshold effects of oil return under these dissimilar environments. We will refer to the 1991–1998 and 1998–2006 as the first (low volatility) and second (high volatility) periods, respectively. The 1983–1991 period results are rather similar in scope to those obtained for the 1998–2006 period (available upon request).

Table 16.1 presents the summary statistics for the sample period (1/25/1991 to 12/29/06).* Panel A (panel B), reports the results for the 1991–1998 (1998–2006) sub-period. In the low volatility sub-period, the Depository Institutions sector had the highest return while Petroleum Refinery (oil-related sector) had the lowest return. In the same sub-period, the Coal Mining sector had the highest unconditional risk (SD 1.83%) of daily equity return, while Electric–Gas services had the lowest (SD 0.46%). 'For the high volatility sub-period (1998–2006), the data show that the Oil and Gas Extraction sector (an oil supplier industry) had the highest average daily return while the Air Transportation and Transport Equipment sectors (an oil user industry) had the lowest average daily return. The highest level of unconditional risk (SD 1.63%) is seen for the Building sector while the lowest (SD 0.52%) is for the Depository Institutions sector. On a risk-per-unit of return basis (coefficient of variation), the Petroleum Refinery sector had the highest risk in the low volatility sub-period, while Air Transportation had the highest in the high volatility sub-period. Depository Institutions had the lowest risk-per-unit of return in both sub-periods.

During the low volatility period, the skewness coefficients of daily returns for the Coal, Building, Petroleum Refinery, Plastic and Rubber and Air Transportation sectors were positive while those of Oil–Gas Extraction, Transport Equipment, Electric–Gas Services and Depository Institutions sectors were negative. In the high volatility sub-period, skewness is negative for all sector returns, except Coal and Building Construction. The skewness and kurtosis values and the Jarque-Bera (1987) (J–B) statistics are highly significant for all sectors in both sub-periods, indicating that the unconditional distributions of returns are non-normal. The Ljung–Box (1978) test statistics for the return series also reject the null

---

* Summary statistics of the data for the 1983 to 1991 sub-period were also produced but are not reported. The Jarque–Bera and the Ljung–Box statistics in all sub-periods favour the application of a GARCH-type model.

TABLE 16.1 Summary Statistics

Series	Mean (%)	Std (%)	Min (%)	Max (%)	Skew	Kurtosis	J-B (p-value)	Q(24)	ADF(4)	ADF(4,t)	PP(0)	PP(4)	CV
						Panel A: January 25, 1991, to December 10, 1998							
Coal	0.147	1.830	−7.923	9.748	0.436	1.668	0.000	165.221	−22.651	−22.672	−58.482	−60.036	12.419
Oil–Gas Extraction	0.146	0.809	−5.734	4.750	−0.602	5.287	0.000	294.346	−15.814	−16.018	−34.905	−35.388	5.525
Building	0.173	1.374	−5.821	10.238	0.612	3.634	0.000	107.227	−18.236	−18.282	−44.409	−44.625	7.921
Chemical	0.127	0.837	−6.680	3.815	−1.217	7.705	0.000	332.129	−15.740	−15.814	−34.773	−35.293	6.602
Petroleum Refinery	0.072	0.860	−4.679	8.017	0.535	6.090	0.000	29.487	−18.909	−18.988	−44.253	−44.301	11.869
Plastic and Rubber	0.119	0.776	−4.020	6.352	0.325	4.963	0.000	83.406	−16.961	−17.181	−45.661	−45.764	6.520
Transport Equipment	0.119	0.736	−5.556	6.982	−0.295	8.496	0.000	100.207	−16.905	−17.011	−40.463	−40.833	6.202
Air Transportation	0.117	1.213	−6.558	6.295	0.154	2.548	0.000	50.600	−17.399	−17.425	−44.191	−44.269	10.345
Electric-Gas Services	0.081	0.459	−3.383	2.179	−0.433	3.293	0.000	47.679	−18.205	−18.201	−40.839	−41.000	5.693
Depository Inst.	0.176	0.513	−3.969	4.144	−0.654	9.386	0.000	752.827	−14.437	−15.045	−34.425	−35.242	2.907
ROIL	−0.035	1.921	−10.168	14.231	0.217	4.875	0.000	41.784	−20.766	−20.786	−43.649	−43.651	−54.759
RM	0.054	0.762	−6.620	4.810	−0.697	8.820	0.000	45.893	−20.535	−20.536	−41.509	−41.473	14.168
SMB	−0.008	0.469	−2.450	2.200	−0.136	1.417	0.000	81.534	−18.007	−18.227	−41.515	−41.609	−57.521
HML	0.016	0.423	−2.400	2.840	0.290	4.159	0.000	117.862	−19.288	−19.318	−36.308	−36.310	27.041
						Panel B: December 11, 1998, to December 29, 2006							
Coal	0.116	1.522	−7.954	8.183	0.008	1.589	0.000	27.635	−19.532	−19.546	−42.971	−42.944	13.081
Oil–Gas Extraction	0.141	1.280	−5.242	5.187	−0.172	0.992	0.000	111.165	−18.598	−18.601	−36.707	−36.593	9.051
Building	0.134	1.628	−6.935	24.585	1.856	26.081	0.000	34.104	−19.573	−19.593	−42.386	−42.408	12.129
Chemical	0.118	1.196	−9.169	7.639	−0.208	4.260	0.000	247.060	−16.945	−16.995	−35.791	−36.044	10.120
Petroleum Refinery	0.121	1.175	−4.819	4.419	−0.099	0.678	0.000	32.314	−19.228	−19.237	−43.210	−43.198	9.713
Plastic and Rubber	0.087	0.931	−5.068	4.448	−0.057	1.804	0.000	85.329	−17.612	−17.609	−41.740	−42.007	10.648
Transport Equipment	0.069	0.950	−6.481	4.795	−0.241	2.719	0.000	85.432	−18.018	−18.013	−40.277	−40.576	13.750
Air Transportation	0.069	1.462	−19.275	6.985	−0.914	16.666	0.000	57.972	−19.269	−19.264	−40.955	−41.133	21.210
Electric-Gas Services	0.073	0.766	−6.278	3.544	−0.514	4.417	0.000	48.780	−19.545	−19.555	−41.916	−42.038	10.433
Depository Inst.	0.077	0.476	−2.461	2.550	−0.057	1.668	0.000	48.344	−18.313	−18.309	−42.904	−42.988	6.178
ROIL	0.086	2.346	−16.545	8.113	−0.588	2.954	0.000	23.235	−21.115	−21.136	−45.014	−45.030	27.408
RM	0.016	1.131	−6.650	5.310	0.092	2.367	0.000	44.497	−21.422	−21.431	−44.499	−44.505	71.546
SMB	0.025	0.630	−4.570	2.900	−0.578	3.697	0.000	88.119	−17.877	−17.884	−40.998	−41.152	25.300
HML	0.030	0.676	−4.930	3.900	−0.102	4.920	0.000	102.635	−20.294	−20.307	−40.005	−39.975	22.409

hypothesis of no autocorrelation in both sub-periods, except for Coal Mining, Petroleum Refinery and Building sectors in the high volatility sub-period. Evidence of non-normality and the presence of serial correlation in the return distribution strongly support the application of a GARCH-type methodology.[*][†]

16.2.2 Methodology and Models

Our threshold framework of the oil effect on sector returns combines the FIGARCH (1, δ, 1) volatility specification of Baillie *et al.* (1996) with a four-factor generalization of Tsay (1998) to form a double-threshold FIGARCH (1, δ, 1) model (DT-FIGARCH):

$$R_{it} = \sum_{j=1}^{J=2} I_t^j \left(b_0^{(j)} + b_1^{(j)} RM_t + b_2^{(j)} SMB_t + b_3^{(j)} HML_t \right.$$

$$\left. + b_4^{(j)} R_{\text{oil},t-d} + \varepsilon_t \right), \varepsilon_t \sim N\left(0, \sigma_t\right) \Big| \Phi_t,$$

$$\sigma_t^2 = \sum_{j=1}^{J=2} I_t^j \left(\frac{\omega^{(j)}}{1 - \beta^{(j)}(L)} \right. \tag{16.1}$$

$$\left. + \left[1 - \frac{\left(1 - \phi^{(j)}(L)\right)\left(1 - (L)\right)^{\delta^{(j)}}}{1 - \beta^{(j)}(L)} \right] \varepsilon_t^2 \right),$$

$j = 1$ when $R_{\text{oil},t-d} < r$, and

$j = 2$ when $R_{\text{oil},t-d} \geq r$.

The critical values for ADF(4), ADF(4,t), PP(0) and PP(4) are 3.45, 3.96, 3.43 and 3.43, respectively at the 1% level. Q(24) is the Ljung-Box test for the 24th order serial correlation in return. The critical value for Q(24) at the 1% and 5% levels are 42.98 and 36.42, respectively. J-B is the Jarque-Bera joint normality test statistics.

In this model, $R_{i,t}$ is the excess return on sector i ($i = 1, 2, \ldots, 10$), the market risk premium (RM), size (SMB) and book-to-market (HML) are the Fama–French factors, R_{oil} is the return on oil as defined previously, I_t^j is an indicator function for regime j ($j = 1$ and 2) and ε_t is the normal random error with mean of zero and variance σ^2, conditional on the information set Φ_t available at time t. In the volatility equation, $\omega^{(j)}$ is a constant, $\varphi^{(j)} \equiv [1 - \alpha^{(j)}(L) - \beta^{(j)}(L)](1 - L)^{-1}$ is the moving average (MA) component of the short-term volatility dynamics where $\alpha^{(j)}$ and $\beta^{(j)}$ are the ARCH and GARCH parameters, respectively, L denotes the lag operator, and, finally, δ is the fractional order of differencing, $0 < \delta < 1$, which accounts for long-term persistence in the FIGARCH specification. We describe the components of the model below.

---

[*] Sadorsky (1999) finds evidence that transmission of shocks from oil to stock returns is unidirectional. Hence, endogeneity of the latter can be assumed away.

[†] Augmented Dickey–Fuller (Dickey and Fuller 1979, 1981) and the Phillips–Perron (Perron 1988, Phillips and Perron 1988) tests indicate that all returns are stationary. For the explanatory variables, the significance of the values pertaining to skewness (except for *HML*), kurtosis, J–B, Q(24), and all stationarity tests are similar to those of the sector returns. We also estimated the Threshold-GARCH (1, 1) specification for all sectors over the three sub-periods. Results show that, for an overwhelming number of cases, either shock persistence is very high, or returns follow an IGARCH process, highlighting the permanence of the shock effects on conditional variance. To study the possibility of permanence of shock effects on volatility, the more flexible FIGARCH model is appropriately suited.

16.2.2.1 The Return and Conditional Volatility Generating Processes

To estimate the industry returns, in excess of the risk-free rate, we propose a four-factor threshold model that adds return on oil (R_{oil}) to the market (RM), size (SMB) and book-to-market (HML) factors used by Fama and French (1992), to account for the oil price effects on sector return.[*] The mean equation specification of our model follows the generalization of Tsay (1998), which allows for exogenous threshold variables ($Z_{t-d} = R_{oil,t-d}$) and the simultaneous estimation of the delay parameter d and threshold variable r.

Li and Li (1996) and Brooks (2001) are among the few studies employing a Threshold GARCH-type model to address asymmetries in equity and currency returns, respectively. These studies use a Self-Exciting Threshold Autoregressive (SETAR)-GARCH specification to capture the asymmetries in equities return levels and variances.[†] Several studies have also found the prevalence of IGARCH properties in high-frequency financial data, implying indefinite persistence of the innovations in the conditional variance process. Poterba and Summers (1986) have also documented that the extent to which conditional volatility affects stock prices through a time-varying risk premium depends on the degree of shock persistence; transitory shocks to conditional volatility have smaller impacts on prices than long-term shocks. Thus, a precise depiction of the degree of volatility persistence is crucial to asset valuation. Indeed, neither the stationary GARCH nor the IGARCH specification has desirable features from an asset pricing point of view. Baillie *et al.* (1996) show that IGARCH properties observed in asset pricing data may be the trace of a mean-reverting long-memory FIGARCH data-generating process and propose the latter model as a way to allow for an intermediate degree of volatility persistence.[‡]

The FIGARCH ($1, \delta, 1$) model nests the GARCH ($1, 1$) and the IGARCH ($1, 1$) processes. When $\delta = 0$, FIGARCH becomes a GARCH process, with an exponential decay of the conditional variance. When $\delta = 1$, FIGARCH becomes IGARCH, where shocks to the conditional variance persist indefinitely. The FIGARCH ($1, \delta, 1$) process is strictly stationary and ergodic if $0 \leq \delta \leq 1$ (Baillie *et al.* 1996). Indeed, since the conditional variance under the FIGARCH specification follows a hyperbolic rate of decay, it allows us to observe the true extent of long memory in the conditional variance of returns. Our model falls into the class of threshold models transformed by Tsay (1989, 1998) and Li and Li (1996).[§] It combines two piecewise-linear regimes to form a non-linear specification capable of modelling a wide array of empirical properties in asset returns, such as jumps, mean reversion and asymmetric and cyclical movements. The DT-FIGARCH model is a two-regime specification, each of which is piecewise linear with respect to oil returns $R_{oil,\,t-d}$. When the oil return is below (above) the threshold value (r), the trajectory of sector returns follows the first (second) regime. Thus, in regime 1 (regime 2) the coefficients of the model account for oil return effects that are below (above) the threshold value.

---

[*] To determine the appropriateness of the four-factor model used here versus a two-factor process (RM and R_{oil}), we also estimated a model excluding the other two factors. The log likelihood ratio test rejects the two-factor model for all sectors at 1%, clearly establishing the superiority of our four-factor specification.

[†] Unlike the threshold approach used here, where the regimes are determined by a structural indicator (R_{oil}), in a Self-Exciting Threshold Autoregressive model, the behaviour of a return series (R_t) changes once the series itself enters a different regime. The switch from one regime to the next is dependent on the past values of the series itself.

[‡] Misspecification of volatility persistence has serious consequence on asset valuation. To assess this effect, we ran GARCH-M and FIGARCH-M models of the above specification. Using the coefficients of the volatility risk premium obtained from GARCH-M and FIGARCH-M models, a 10 basis points increase in conditional volatility leads to a $7.13 billion and $3.14 billion decline in the total market capitalization of the Oil and Gas industries, respectively. Detailed results can be obtained from the authors.

[§] The advantages of the Tsay (1998) threshold method are threefold. First, this method is not an exclusively self-exciting threshold autoregressive type model, where the threshold value is determined by its own returns, it rather allows for exogenous variables to be considered for threshold. Second, it allows for the estimation of the delay parameter (d) and the threshold value (r) simultaneously, using the Akaike Information Criteria, instead of scatter plots of predictive residuals used by Tsay (1989) that require subjective assessment. Third, the test procedure to detect change points is relatively simple and has an asymptotic Chi-squared distribution.

16.2.2.2 The Estimated Model

The mean and volatility equations described earlier are presented in more familiar terms below where all variables are as defined previously. The coefficient estimates for the model are obtained using the quasi-maximum likelihood method.*

$$R_{i,t} \begin{cases} \begin{Bmatrix} \left(b_0^1 + b_1^1 RM_t + b_2^1 SMB_t \right. \\ \left. + b_3^1 HML_t + b_4^1 R_{\text{oil},t-d} + \varepsilon_t \right), \end{Bmatrix} & \text{if } R_{\text{oil},t-d} < r, \\[2em] \begin{Bmatrix} \left(b_0^2 + b_1^2 RM_t + b_2^2 SMB_t \right. \\ \left. + b_3^2 HML_t + b_4^2 R_{\text{oil},t-d} + \varepsilon_t \right), \end{Bmatrix} & \text{if } R_{\text{oil},t-d} \geq r, \end{cases}$$

$$\sigma_t^2 = \begin{cases} \begin{Bmatrix} \omega_0^1 + \beta_0^1 \sigma_{t-1}^2 + \left[1 - \beta_0^1(L) \right. \\ \left. - \left(1 - \phi_0^1(L)\right)(1 - (L))^{\delta^1} \right] \varepsilon_t^2, \end{Bmatrix} & \text{if } R_{\text{oil},t-d} < r, \\[2em] \begin{Bmatrix} \omega_0^2 + \beta_0^2 \sigma_{t-1}^2 + \left[1 - \beta_0^2(L) \right. \\ \left. - \left(1 - \phi_0^2(L)\right)(1 - (L))^{\delta^2} \right] \varepsilon_t^2, \end{Bmatrix} & \text{if } R_{\text{oil},t-d} \geq r. \end{cases}$$

Oil price effects may impact industry earnings via direct or indirect channels of influences. Specifically, oil price effects on industry returns may manifest themselves through the sensitivity of industry returns to oil prices as well as indirectly via the sensitivity of other risk factors to oil price movements. If oil price effects are subject to a threshold at the industry level, then it is likely that oil price changes above (below) the threshold trigger differential readjustments in how oil price and other risk factors are priced into industry returns. In the model described above, when the threshold variable (R_{oil}) exceeds the threshold value (r), it triggers a shift in the slopes of all the model factors with a delay window of (d) periods.[†] We allow for shifts in all of the slopes for two reasons. First, this approach allows us to maintain generality and to test whether such shifts do occur, rather than assuming they do or they do not. Second, Fama and French (1997) argue that bad news about future cash flows tends to raise the industry's risk loading on *SMB* and *HML* factors. Accordingly, in our model, the increase in oil prices, if large enough, would tend to limit the cash flows of the oil-consuming sectors and to shift up the coefficients on the *SMB* and *HML* factors. Similarly, industries with large positive *HML* slopes (i.e. high BV/MV) are likely to have experienced surprise negative conditions (or to be relatively more distressed) and tend to sell low (Fama and French 1997). Along the same lines, the *HML* slopes would tend to shift higher (lower) with higher oil prices for the oil-consuming (substitute) sectors. In brief, the Fama–French argument implies that when oil return increases are above a positive threshold ($j = 2$), the coefficients $\left(b_2^2, b_3^2\right)$, which measure the sectoral effects of *SMB* and *HML* for the oil-consuming sectors, will be larger in magnitude than their counterparts $\left(b_2^1, b_3^1\right)$ when oil return increases are below the threshold ($j = 1$) because they have to

---

* Appendix A presents a discussion of the simultaneous estimation of the threshold coefficient (r) and the delay parameter (d) for each sector.

† The issue of how many thresholds should be estimated and the economic interpretation of the threshold values are still unresolved in the extant literature. Meyer and von Carmon-Taubadel (2004) provide an extensive review of the asymmetric price transmission literature and conclude that the magnitude interpreted as the minimum incentives required by the economic agent to elicit price or return of threshold values have never been interpreted in an economic sense. The estimated values are generally adjustments.

spend more on their oil consumption than in the former regime. To test this view, we allow for a shift in the slope of these factors.*

16.2.3 Testable Hypotheses

Two main sets of hypotheses are tested. First, the appropriateness of the FIGARCH specification, and second the asymmetry of the effects of the model factors (R_{oil}, RM, SMB, HML) under the two regimes ($J = 1, 2$). Correct specification of the stock return patterns has major implications on the empirical findings and related policy conclusions. For example, since the systematic risk measure (beta) of a sector is used to estimate the cost of capital for that sector, its accurate measurement is essential for deriving a correct estimate of the cost of capital and making optimal investment decisions. In particular, if projects on alternative energy sources are assessed based on incorrect cost of capital estimates, the choices made will be unreliable. Similarly, accurate measurement of sectoral sensitivities to oil price changes can serve as a basis for policies concerning energy source diversity (e.g. development of renewable energy such as fossil fuel, atomic energy, etc.) and can help determine optimal hedging positions.

Finding of dissimilar (asymmetric) effects under the two regimes is also important because it may necessitate differential regulatory policies by the government, and dissimilar investment and managerial decisions under these regimes. In particular, if the effects of oil price changes on the stock returns are symmetric, the impacts of oil price increases and oil price decreases (bad news and good news) would be similar in magnitude and counterbalancing. However, if oil price increases harm a sector more than oil price declines improve its conditions, firms in the sector would need to take steps to hedge the shocks, or to be prepared to absorb them. Investors would also have to take this pattern of behaviour into account when investing in the sector and the government may consider subsidies to moderate the adverse effects from the shocks.

The hypotheses concerning the functional form are as follows.

H_{11}: The proper specification for sector returns is GARCH (1, 1) (the effect of a shock decays at an exponential rate): $H_0 : \delta^1 = \delta^2 = 0$

H_{12}: The proper specification for sector returns is IGARCH (shocks persist indefinitely): $H_0 : \delta^1 = \delta^2 = 1$

H_{13}: The proper specification for sector returns is the Fama–French model: $H_0 : b_4^1 = b_4^2 = \beta_0^1 = \beta_0^2 = \phi_0^1 = \phi_0^2 = \delta^1 = \delta^2 = 0$

Asymmetry hypotheses are given below. In each of the first four cases, symmetry indicates that sensitivity of the sectoral return to the market, SMB, HML, or R_{oil} is the same, regardless of whether the changes in these variables occur when oil returns are above or below the threshold (r). The last hypothesis, H_{25}, investigates whether the volatility decay process is symmetric.

H_{21}: Market risk premia (slopes) are symmetric for oil return shocks above and below the threshold: $H_0 : b_1^1 = b_1^2$

H_{22}: SMB risk premia (slopes) are symmetric for oil return shocks above and below the threshold: $H_0 : b_2^1 = b_2^2$

H_{23}: HML risk premia (slopes) are symmetric for shocks above and below the threshold $H_0 : b_3^1 = b_3^2$

H_{24}: Oil return risk premia (slopes) are symmetric for oil return shocks above and below the threshold: $H_0 : b_4^1 = b_4^2$

H_{25}: Volatility decay process is symmetric for oil price shocks above and below the threshold: $H_0 : \delta^1 = \delta^2$

---

* An alternative would be to introduce a shift only in the slope of the oil return variable (R_{oil}). This would likely produce a bigger shift in the slope of R_{oil}. Currently, the effects are divided among all four factors. Hence, our specification may underestimate the shift in the slope of R_{oil}.

16.3 Empirical Findings⋆

16.3.1 Tests of Model Specification

It is crucial to determine the precise measure of volatility persistence (memory) because of its implications for asset valuation. A FIGARCH specification is employed here because it nests the GARCH and IGARCH models and it also allows us to observe the true extent of long memory in the conditional variance of returns. Another major issue is whether oil shock effects show similar decay patterns when oil returns are below and above the threshold because differential decay patterns and different decay speeds call for different policy actions by the oil producers, oil consumers, and government regulators.

Results for model specification tests are reported in Table 16.2. Based on the statistics reported in this table, the basic GARCH, IGARCH and the Fama–French specifications (H_{11}, H_{12}, H_{13}) are rejected for almost all sectors in both the volatile and calm oil price periods (Petroleum Refinery is the exception in the calm period). The implication is that these specifications are indeed too restrictive to capture the behavioural patterns of sectoral stock returns and their application may produce distorted results.[†] The rejection of both GARCH and IGARCH specifications also indicates that oil price shock effects do not decline exponentially (geometrically), nor persist indefinitely. Instead, these effects decay at a slower hyperbolic pace, reflecting a long-memory pattern. In addition, the rejection of the Fama–French model indicates that the menu of factors chosen by these authors is too restrictive, and other factors, at least the oil returns (R_{oil}), are likely to play a role in describing asset returns. Overall, these findings indicate that predictions and policy recommendations based on the functional forms rejected by the data are misleading and can fall short of or run counter to their intended objectives.

16.3.2 Tests of Asymmetry

An interesting question is whether the systematic risks associated with the Fama–French and oil return factors included in the model change when oil return shocks become large enough to exceed the corresponding thresholds. The answer to this question would help us reach a more accurate assessment of the overall sectoral risk and has implications for investors and firm managers. Hypotheses H_{21}, H_{22}, H_{23} and H_{24} investigate these symmetry issues, namely whether the slopes of the model factors are identical when oil returns fall short and exceed the threshold values. A second question is whether oil shock effects show similar decay patterns when oil returns are below and above the threshold, because differential decay patterns and different decay speeds call for different policy actions by the oil producers, oil consumers and government regulators. Hypothesis H_{25} ($H_0 : \delta^1 = \delta^2$) tests whether the speed of memory decay is identical for smaller oil returns falling below the threshold and the larger ones positioned above it. The results of symmetry tests can provide an answer to these questions.

The critical values of χ^2(2d.f) = 5.9, 4.6, χ^2(1d.f) = 3.8, 2.7 and χ^2(8d.f) = 15.5, 13.3 at the 5% and 10% levels, respectively.

According to the test results in Table 16.2, in the volatile period (1998–2006), the symmetry of the response to market (RM) and HML factors is unanimously rejected for all 10 sectors, and for all, except

---

⋆ Diagnostic statistics for each model including the Log Likelihood value, the LjungBox portmanteau test for up to 20th order serial correlation in the standardized residuals $Q(20)$ and squared standardized residuals $Q^2(20)$ are presented at the bottom of Tables 16.3(A) and 3(B). The $Q^2(20)$ statistics suggest that the FIGARCH model adequately captures the autocorrelations in the conditional variance of sector returns.

† The magnitude of the distortion due to model misspecification can be considerable. We contrasted the oil return effects on the Oil–Gas Extraction sector derived from the single-regime GARCH and the model used here for 1998–2006. The differences are statistically significant and sizeable. Under GARCH, the ROIL coefficient was 0.21 ($t = 31.8$) while under FIGARCH the ROIL coefficient was 0.19 ($t = 70.1$) above the threshold, and 0.21 ($t = 79.35$) below it, indicating an overestimation of two basis points under GARCH. With a market capitalization of Oil–Gas Extraction of $793 008 million in 2006 for firms with SIC 13, a two basis point overestimation translates into approximately ($793 008 million×two basis points) or $158.60 million.

TABLE 16.2 Hypotheses Concerning Functional Form and Symmetry

Hypotheses Concerning Functional Form:

H_{11}: The proper specification is GARCH(1,1): $\delta^1 = \delta^2 = 0$
H_{12}: The proper specification is GARCH: $\delta^1 = \delta^2 = 1$
H_{13}: The proper specification is Fama–French: $b_4^1 = b_4^2 = \beta_0^1 = \beta_0^2 = \Phi_0^1 = \Phi_0^2 = \delta^1 = \delta^2 = 0$

Symmetry Hypotheses:

H_{21}: Market risk premiums are symmetric: $b_1^1 = b_1^2$
H_{22}: SMB premiums are symmetric: $b_2^1 = b_2^2$
H_{23}: HML premiums are symmetric: $b_3^1 = b_3^2$
H_{24}: Oil price effects are symmetric: $b_4^1 = b_4^2$
H_{25}: Volatility decay process is symmetric: $\delta^1 = \delta^2$

	Coal (SIC12)	Electric-Gas Services (SIC49)	Oil-Gas Extraction (SIC13)	Petroleum Refinery (SIC29)	Building (SIC15)	Air Transportation (SIC45)	Chemical (SIC28)	Plastic and Rubber (SIC30)	Transport Equipment (SIC37)	Depository Inst. (SIC60)
						January 25, 1991, to December 10, 1998, Sub-Period				
H_0: Model Specification										
H11: 2d.f.	879.47	1377.08	306.22	3.23E-12	901.31	2286.05	248.94	154.63	431.51	3991.12
H12: 2d.f.	1919.5	2674.51	1521.22	3486.72	4582.92	57648.78	2604.19	2107.99	2757.24	1618.66
H13: 8d.f.	7.76E+04	2.37E+05	3.40E+04	4.26E+04	2.73E+05	1.75E+07	6.62E+04	4.26E+04	2.73E+05	1.97E+05
H_0: Symmetry										
H21: 1d.f.	2.15	12.04	0.03	18.15	0.06	3.46	535.44	5.55	23.76	89.44
H22: 1d.f.	18.3	126.99	179.08	2.18	45.28	72.58	501.15	6.3	85.34	10.16
H23: 1d.f.	1.26E-03	80.4	30.52	1.45	2.17	23.38	104.6	27.07	44.63	0.02
H24: 1d.f.	1.94	3.6	0.4	0.3	0.65	0.73	5.59E-05	1.23	2.79	1.91
H25: 1d.f.	189.41	298.88	2.41	2.88E-12	24.62	78.76	108.2	32.36	289.87	411.92
						December 11, 1998, to December 29, 2006, Sub-Period				
H_0: Model Specification										
H11: 2d.f.	1153.09	1500.75	506.87	119.68	17965.15	302.58	1143.79	79.69	1476.86	351.96
H12: 2d.f.	1458.12	8871.43	3118.18	1871.46	8.76E+06	2622.21	1497.71	2473.21	1881.27	1562.25
H13: 8d.f.	6.76E+05	1.44E+05	1.92E+05	7.05E+04	2.05E+04	2.94E+03	5.04E+05	3.23E+05	2.51E+05	6.97E+04
H_0: Symmetry										
H21: 1d.f.	104.47	593.95	20.97	9.81	79.86	1586.94	621.56	192	119.28	196.57
H22: 1d.f.	59.29	28.32	0.13	41.95	17.02	5.29	16.08	45.71	98.87	13.8
H23: 1d.f.	6.73	180.98	8.99	40.86	472.5	205.58	40.59	4.95	646.46	751.66
H24: 1d.f.	0.8	3.15	5.55	1.2	4.74	0.04	6.73	17.4	8.49	0.61
H25: 1d.f.	66.76	105.15	146.71	3.54	104.73	79	67.33	0.19	456.62	8.85

for the Oil–Gas Extraction sector, for the *SMB* factor. Symmetry results are less uniform in the low volatility period (1991–1998). Nevertheless, symmetry is always rejected in more than two-thirds of the cases for each of the three Fama–French factors. These findings indicate that the systemic risks (the slopes) of the three Fama–French factors do indeed change when the oil return passes the threshold level during both the calm and volatile periods, although the shift (regime change) occurs in more sectors when oil prices are more volatile and larger overall. We also note that the direction of the shift is different for different sectors. In particular, in both periods, the market risk for the oil-substitute sector, Coal, falls while that for the oil-consuming sector, Air Transportation, increases, when oil return exceeds the threshold, indicating that the former benefits while the latter suffers (becomes riskier) when oil returns exceed the threshold. This pattern is observed over a broader spectrum of industries in the less volatile period.

These findings provide only mixed support for the Fama–French (1997) argument that bad news about future cash flows tends to raise the industry's risk loading on *SMB* and *HML*. The bad news here for the oil-consuming sector is the increase in the oil return above the threshold level. For the Air Transportation and Transport Equipment sectors, the increase in oil return does, as expected, shift up the *SMB* and *HML* risk in the calm period, but it fails to do so in the volatile period. Similarly, for the oil-substitute sector, Coal, the results are mixed; an increase in oil return does reduce the *SBM* and *HML* risk in the volatile period, but not in the calm period. This indicates that only when markets are volatile do oil price changes translate into better conditions for the oil-substitute sector. These findings provide considerable evidence against the extant studies that assume identical coefficients for the risk factors in their models at all levels of oil return change as a part of their maintained hypotheses.

It is notable that the shift in the slope (systematic risk) of the fourth factor, the oil return (R_{oil}), is less frequent in both sub-periods, and more so in the calm periods, compared with the other three factors. This is perhaps an indication that oil return, although a significant factor, plays a less important role in determining industry returns than the traditional Fama–French factors. This issue will be further discussed in Section 16.3.3.3 when analysing the economic significance of the model factors. A stronger explanation may be that the effects of increases in oil return are mostly picked up by changes in the oil return level (R_{oil}) itself, rather than by a change in its coefficient. In addition, it is likely that the Fama–French factors, e.g. the market return, also proxy for oil price changes. It follows that if the interdependence of the oil and market returns is stronger when oil returns are higher (above threshold), market return may pick up some of the asymmetry in response to the oil return and weaken the asymmetry test for the latter variable.

Based on the slope shifts in the four factors, two conclusions can be drawn. First, the effect of oil price changes includes both direct and indirect components. Thus, the overall effect should be assessed by adding up the effects due to the slope shifts of all four factors in the model, including oil return and Fama–French factors. If the indirect effects are not accounted for in government policy decisions and managerial planning by firms, the resulting cost/benefits analysis of oil price changes will be distorted and, hence, the adopted policy decisions will be suboptimal. Second, findings based on simpler models are suspect as they fail to account for the asymmetry of sectoral responses to changes in the model factors when oil returns are below and above the threshold.

The hypothesis of identical speeds of decay in volatility when oil returns are below and above the threshold (H_{25}) is rejected strongly for all sectors, except for Oil–Gas Extraction and Petroleum Refinery sectors in the calm period, and the Plastic and Rubber sector in the volatile period. The results of this hypothesis indicate that the volatility decay process is indeed dissimilar for oil return changes below and above the threshold and, hence, these dissimilarities have to be accounted for. The estimated values of the decay parameter–, reported in Tables 16.3(A) and 16.3(B), are found to be, with very few exceptions, smaller for oil returns positioned above the threshold in both periods, indicating a faster decay (less persistence) when return increases are larger. This may happen because unusually large oil price increases are more likely to be considered transitory and, hence, their effects die down more quickly. The dissimilarity of the volatility decay process is also overlooked in the literature.

TABLE 16.3A   Double-Threshold FIGARCH(1, δ,1) Results January 25 , 1991 to December 10,1998 Sub–Period

1	2	3	4Electric–	5	6	7	8	9	10	11	12
Coefficient	Variable	Coal (SIC12)	Gas Services (SIC49)	Oil–Gas Extraction (SIC13)	Petroleum Refinery (SIC29)	Building (SIC15)	Air Transportation (SIC45)	Chemical (SIC28)	Plastic and Rubber (SIC30)	Transport Equipment (SIC37)	Depository Inst. (SIC60)
b_0^2	CONSTANT	135E-03	2.34E-04	1.45E-03	-1.40E-03	1.90E-03	1.46E-04	5.72E-04	1.93E-04	1.39E-04	1.63E-03
		2.82	10.593	14.833	-22.80	7.872	0.797	23.628	3.487	5.744	33.396
b_0^1	CONSTANT	-1.21E-04	5.26E-05	4.41E-04	-8.95E-04	1.79E-04	1.85E-04	5.16E-04	8.29E-04	7.00E-04	1.05E-03
b_1^2	RM	-1.074	1.442	15.252	-15.259	2.739	4.993	18.924	16.288	11.668	53.695
		0.785	0.597	0.839	0.685	1.1	1.192	1.002	0.714	0.95	0.712
b_1^1	RM	10.975	204.311	96.893	104.036	20.01	40.255	489.536	84.894	218.994	126.141
		0.893	0.576	0.837	0.733	1.086	1.135	0.892	0.688	0.895	0.658
b_2^2	SMB	47.076	110.593	197.539	57.111	260.329	145.161	208.285	90.632	85.772	496.254
		0.126	0.31	0.509	0.332	1.344	1.199	0.921	0.685	0.728	0.544
b_2^1	SMB	1.213	93.814	30.217	20.989	24.021	23.627	224.24	57.693	191.481	31.813
		0.586	0.222	0.768	0.266	0.946	0.75	0.725	0.644	0.601	0.488
b_3^2	HML	19.76	31.25	80.58	20.888	48.38	53.141	93.666	59.628	45.459	161.814
		0.411	0.328	0.256	0.536	0.557	0.634	-0.015	0.22	0.475	0.485
b_3^1	HML	3.189	63.872	10.412	29.796	5.144	11.339	-5.024	18.581	78.817	32.605
		0.406	0.418	0.395	0.471	0.717	0.356	0.035	0.308	0.364	0.483
b_4^2	R_{oil}	11.804	48.895	72.9	23.922	63.413	25.782	9.026	25.277	23.497	190.624
		-0.034	0.003	0.071	0.04	0.043	-0.05	-0.007	0.005	-0.005	0.009
b_4^1	R_{oil}	-0.995	1.961	5.735	5.908	1.222	-1.747	-6.807	0.797	-2.968	0.876
		0.015	-0.01	0.063	0.035	0.015	-0.025	-0.007	-0.004	0.008	-0.005
ω_0^2	CON. VOL.	1.803	-1.463	56.361	4.795	3.802	-6.489	-0.488	-0.774	1.07	-3.424
		-1.86E-13	-7.27E-12	-2.43E-14	1.84E-13	0.00E+00	0.00E+00	0.00E+00	-1.13E-14	-1.77E-14	-8.91E-08

(Continued)

TABLE 16.3A (CONTINUED) Double–Threshold FIGARCH(1, δ,1) Results January 25 , 1991 to December 10,1998 Sub–Period

Coefficient	Variable	Coal (SIC12)	Electric–Gas Services (SIC49)	Oil–Gas Extraction (SIC13)	Petroleum Refinery (SIC29)	Building (SIC15)	Air Transportation (SIC45)	Chemical (SIC28)	Plastic and Rubber (SIC30)	Transport Equipment (SIC37)	Depository Inst. (SIC60)
	CON. VOL.	-2.53E-08	-7.40E-05	-4.12E-07	0.023	9.40E-10	8.96E-06	-2.42E-07	-4.67E-07	-0.015	-0.226
ω_0^1		1.14E-06	9.96E-14	6.20E-07	8.73E-15	0.00E+00	0.00E+00	1.55E-07	-6.22E-13	4.52E-13	4.87E-16
		0.734	3.11E-07	64.374	9.50E-07	-1.72E-07	0.00E+00	0.262	-0.001	5.38E-05	1.81E-08
β_0^2	GARCH	0.581	0.636	0.231	0.944	0.949	1.332	0.144	0.375	0.418	0.364
		15.895	37.784	6.521	267.238	23.642	129.905	5.937	10.624	15.993	17.536
β_0^1	GARCH	0.849	0.848	0.507	0.561	0.777	0.753	0.575	0.271	0.77	0.97
		64.787	28.632	22.981	27.865	68.77	117.436	12.712	9.486	23.999	112.454
Φ_0^2	PHI	0.748	0.484	-3.86E-08	0.887	1.045	1.285	0.107	0.3979	0.4453	0.216
		23.288	25.983	0	180.745	36.131	135.535	4.476	11.676	17.598	9.34
Φ_0^1	PHI	0.458	0.131	0.232	0.736	0.613	0.624	-5.00E-09	5.13E-07	2.18E-08	0.112
		20.065	5.606	9.255	43.126	40.318	77.311	-1.01E-07	1.82E-05	5.02E-07	6.046
δ^2	DECAY	0.019	0.198	0.267	0.129	0.123	0.078	0.159	0.1015	0.0839	0.351
		0.608	11.635	8.304	9.845	3.229	7.629	9.654	4.203	4.801	20.073
δ^1	DECAY	0.582	0.832	0.323	0	0.341	0.223	0.67	0.279	0.999	0.951
		28.482	30.032	16.302	0	26.991	29.682	13.751	12.136	19.967	50.299
LL Value		5335.09	8770.41	7502.96	7068.489	6170.7	6457.57	8198.49	7377.32	7982.39	8785.82
Q(20)		174.09	48.56	234.58	52.03	63.89	28.62	212.26	89.76	64.39	368.33
$Q^2(20)$		16.09	11.16	7.05	33.72	13.72	10.5	9.22	8.31	16.64	11.82
Threshold (%)		1.51	-0.687	0.977	-0.477	1.621	1.158	-1.447	-0.309	-0.805	1.588
Delay (days)		4	3	1	7	6	7	3	2	4	7

t–values are below the coefficients. Significant *t*–values at the 5% level are highlighted in bold. The critical *t*–values are 1.64, 1.96, 2.57 and Q(20) are 28.41, 31.41 and 37.57 at the 10%, 5% and 1% levels, respectively.

16.3.3 Sector Analysis*

The estimation results for the 1991–1998 (calm) and 1998–2006 (volatile) periods are reported in Tables 16.3(A) and 16.3(B), respectively.[†] In the 1991–1998 period, the threshold values (r) are positive for sectors such as Coal, Oil–Gas Services, Building, Air Transport and Depository Institutions, indicating that, for these sectors, regime shifts or changes in sensitivities occur when oil returns are rising. Among these industries, the Oil–Gas Extraction industry exhibits the lowest threshold level (0.977%) and Building and Depository Institutions have the largest threshold values (1.621%, 1.588%). These figures indicate that although it does take a considerable change in the return on oil to trigger a regime shift for either of these industries, the requirement is much smaller for the former than for the latter industries. It is notable that the remaining industries have negative thresholds, indicating that regime shifts or changes in factor sensitivities occur when oil return is falling. The magnitudes again vary across sectors.

16.3.3.1 Analysis of Oil Effects (R_{oil} Coefficient) Below and Above the Threshold

Assessing the effect of R_{oil} on sector return is of particular interest because of the major macroeconomic implications of oil price changes, especially given the spectacular rise in oil prices and the tremendous oil price volatility during the recent decades. The figures in Tables 16.3(A) and 16.3(B) delineate the relationship between sectoral returns and R_{oil}. The direct marginal effect of oil price increases on sector return is determined by the magnitude of the R_{oil} coefficient b_4^2 when R_{oil} equals or exceeds the threshold (r), and by b_4^1 when it falls short of it. We analyse the results according to the four groups of industries considered and do so selectively within each group to save space.

16.3.3.1.1 Oil-Substitute Sectors

The results for the Coal sector, an oil-substitute, are presented in column 3 of Tables 16.3(A) and 16.3(B). The threshold and the coefficients for this industry are $r = 1.51\%$, $b_4^1 = 0.015$ and $b_4^2 = -0.034$ (insignificant) in the calm period, and $r = 1.13\%$, $b_4^1 = 0.083$ and $b_4^2 = 0.103$ in the volatile period. These figures indicate that, in the volatile period, the increase in oil return needed for a regime change is smaller ($r = 1.13\%$ instead of 1.51%) and the oil effect on the Coal sector is much stronger, compared with the calm period, both when R_{oil} is below the threshold ($b_4^1 = 0.083$ versus $b_4^2 = 0.015$) and above it ($b_4^2 = 0.103$ versus b_4^2 insignificant). Specifically, in the more volatile period, the increase in sector return due to a one percentage point change in the oil return is 8.3 basis points (bp) compared with 1.5 bp during the calm period, when the increase in R_{oil} falls short of the threshold, and 10.3 bp compared with zero when R_{oil} exceeds the threshold.

The Coal sector does show a strong substitution effect during the period of oil price volatility as higher oil returns lead to greater coal sector returns. The coefficient of R_{oil} in this period is larger and more stringently significant compared with the calm oil price period, reflecting the greater sensitivity of market participants under volatile conditions when oil price changes are more difficult to predict. This effect is slightly stronger when oil return surpasses the threshold. During the calm period, substitution is weaker, and when oil return surpasses the threshold, it loses significance. The lower strength of the relationship between oil return and risk premia when the return exceeds the threshold, in both periods, may indicate that when the oil return rises beyond a certain level, oil users are motivated to economize further on the

---

* As explained in the appendix, threshold values can be positive or negative and can be larger or smaller during periods of high oil price volatility compared to calm periods.

† The model was also estimated for the 1983–1991 period. Results are rather similar to the 1998–2006 period. A number of alternative specifications of Threshold-GARCH models were estimated as well. These include GARCH (1, 1), GARCH (1, 1)-M, and GARCH (1, 1) with market volume in the variance equation. Regardless of the GARCH specification used, the persistence of shocks (the sum of the ARCH and GARCH parameters), is 1.0 or close to 1.0 for most sectors, rendering the inferences unreliable. We also estimated FIGARCH (1, δ, 0) models. For space limitation, only the results of FIGARCH (1, δ, 1) are presented here. Other results can be obtained from the authors.

TABLE 16.3B Double-Threshold FIGARCH(1, δ,1) Results December 11, 1998, to December 29, 2006, Sub-Period

1	2	3	4	5	6	7	8	9	10	11	12
Coefficient	Variable	Coal (SIC12)	Electric–Gas Services (SIC49)	Oil–Gas Extraction (SIC13)	Petroleum Refinery (SIC29)	Building (SIC15)	Air Transportation (SIC45)	Chemical (SIC28)	Plastic and Rubber (SIC30)	Transport Equipment (SIC37)	Depository Inst. (SIC60)
b_0^2	CONSTANT	9.01E-04	3.31E-04	6.27E-04	5.62E-04	8.59E-04	6.98E-05	5.50E-04	2.37E-04	2.16E-04	2.66E-04
		6.406	8.997	7.742	4.263	19.700	0.434	12.951	4.040	3.646	19.163
b_0^1	CONSTANT	-1.65E-04	4.93E-05	2.34E-04	2.88E-04	3.05E-04	-2.47E-04	8.19E-04	2.57E-04	-1.03E-04	6.80E-04
		-3.78	1.515	3.673	4.081	3.953	-4.997	11.652	5.049	-2.258	25.16
b_1^2	RM	0.738	0.649	0.89	0.899	0.955	1.342	0.782	0.722	0.798	0.445
		76.256	305.544	103.527	58.976	425.55	217.662	245.428	126.086	133.642	377.24
b_1^1	RM	0.858	0.743	0.789	0.849	1.045	1.02	0.931	0.624	0.872	0.409
		130.617	232.156	215.166	166.525	106.947	196.423	183.095	151.569	266.207	178.731
b_2^2	SMB	0.449	0.193	0.53	0.204	0.615	0.61	0.762	0.369	0.366	0.241
		19.654	31.219	28.803	10.254	109.342	22.947	95.699	38.383	40.171	86.004
b_2^1	SMB	0.636	0.241	0.526	0.35	0.667	0.543	0.815	0.448	0.475	0.253
		76.882	36.434	45.741	33.788	59.247	42.92	77.527	68.659	78.083	152.787
b_3^2	HML	0.773	0.693	0.973	0.992	0.351	0.955	0.132	0.425	0.436	0.361
		17.437	204.501	89.478	55.392	78.13	45.652	29.197	46.365	47.101	177.98
b_3^1	HML	0.889	0.777	0.858	0.864	0.808	0.636	0.209	0.448	0.708	0.274
		156.494	148.447	92.209	97.616	39.325	82.526	18.53	82.031	130.007	111.954
b_4^2	R_{oil}	0.103	0.002	0.198	0.106	-0.075	-0.088	-0.014	-0.029	-0.001	-0.015
		4.626	0.453	70.192	4.346	-2.721	-4.452	-5.068	-3.561	-0.168	-8.742
b_4^1	R_{oil}	0.083	0.011	0.216	0.133	-0.014	-0.092	0.003	6.67E-03	-0.017	-0.013
		25.888	4.253	79.35	47.225	-2.807	-35.693	0.517	2.541	-9.661	-6.926
ω_0^2	CON. VOL.	-3.10E-13	0.00E+00	-8.14E-12	-9.58E-13	-8.32E-11	7.55E-06	5.67E-14	8.70E-07	-4.93E-15	-4.12E-14
		-1.48E-07	-7.43E-06	-4.53E-06	-2.64E-07	-1.67E-04	1.43	2.66E-05	0.957	-1.07E-04	-7.00E-07

(Continued)

TABLE 16.3B (CONTINUED) Double-Threshold FIGARCH(1, δ,1) Results December 11, 1998, to December 29, 2006, Sub-Period

1	2	3	4	5	6	7	8	9	10	11	12
Coefficient	Variable	Coal (SIC12)	Electric–Gas Services (SIC49)	Oil–Gas Extraction (SIC13)	Petroleum Refinery (SIC29)	Building (SIC15)	Air Transportation (SIC45)	Chemical (SIC28)	Plastic and Rubber (SIC30)	Transport Equipment (SIC37)	Depository Inst. (SIC60)
ω_0^1	CON. VOL.	-2.67E-07	0.00E+00	1.21E-06	1.82E-13	2.79E-11	-4.39E-15	-9.45E-14	-2.06E-14	-1.61E-07	-2.59E-14
		-0.149	2.04E-08	0.895	1.46E-07	0	0	-0.029	0	-24.578	-1.96E-07
β_0^2	GARCH	0.658	0.991	0.511	0.166	0.887	2.61E-01	0.769	0.525	0.387	0.707
		26.553	42.91	18.339	4.541	396.468	5.555	46.169	17.32	15.126	33.214
β_0^1	GARCH	0.811	0.612	0.7	0.498	0.943	0.51	0.773	0.408	0.897	0.394
		56.277	43.45	35.248	21.334	135.009	22.53	32.974	16.715	42.431	12.327
Φ_0^2	PHI	0.446	1.039	0.464	0	0.833	0.232	0.526	0.448	0.28	0.513
		16.891	66.233	16.231	0	334.108	5.095	25.081	14.488	10.519	21.791
Φ_0^1	PHI	0.411	0.299	0.463	0.369	0.995	0.275	0.289	0.387	0.103	0.15
		19.758	19.274	20.473	15.262	212.56	10.675	10.92	15.81	4.517	4.648
δ^2	DECAY	0.333	0.057	0.148	0.149	0.069	0.024	0.382	0.161	0.177	0.283
		15.214	2.437	5.959	4.513	27.078	0.87	21.933	6.103	9.095	13.391
δ^1	DECAY	0.57	0.385	0.283	0.218	0.013	0.304	0.641	0.148	0.919	0.383
		30.283	31.115	13.597	10.388	4.342	17.388	24.871	7.506	35.492	13.953
	LL Value	6095.84	8281.94	6568.97	6840.2	6119.08	6518.31	7582.38	7290.29	7825.51	9344.29
	Q(20)	15.67	31.57	52.49	33.7	63.89	21.708	198.93	37.196	75.69	275.48
	Q²(20)	12.68	26.63	11.07	19.4	13.72	17.507	17.05	14.685	8.77	8.28
	Threshold (%)	1.13	0.951	0.621	2.104	1.756	2.311	-0.541	0.795	0.499	-0.562
	Delay (days)	7	1	1	6	5	7	6	7	2	4

t–values are below the coefficients. Significant *t*–values at the 5% level are highlighted in bold. The critical *t*–values are 1.64, 1.96, 2.57 and Q(20) are 28.41, 31.41 and 37.57 at the 10%, 5% and 1% levels, respectively.

use of energy and to switch to alternative energy sources such as solar power and wind. Moreover, protective hedges against oil price increases are more likely to be triggered under higher oil prices.

Electric–Gas Services may also be considered an oil substitute sector, although it is an oil consumer as well. For this industry, the threshold and the coefficients are $r = -0.68\%$, $b_4^1 = -0.010$ (insignificant), $b_4^2 = 0.003$, and $r = 0.95\%$, $b_4^1 = 0.011$, $b_4^2 = 0.002$ (insignificant) (column 4). The effect of oil return on the Electric–Gas Services sector is positive when statistically significant, as expected from an oil-substitute sector. Significant cases occur when oil returns rise above the threshold in the clam period and below it in the volatile period. Excessively high oil prices in the volatile period may have discouraged demand for this industry along with that for oil, in favour of other energy sources, and may have also made the passing on of the additional costs to the consumers difficult, offsetting any gains due to the substitution effect.

16.3.3.1.2 Oil-Related Sectors

The results for the oil-related industries, namely the Oil–Gas Extraction and Petroleum Refinery sectors, are reported in columns 5 and 6, respectively. The threshold and the oil return coefficients for the Oil–Gas Extraction sector in the calm and volatile periods are, respectively, $r = 0.98\%$, $b_4^1 = 0.06$, $b_4^2 = 0.07$ and $r = 0.62\%$, $b_4^1 = 0.216$, $b_4^2 = 0.198$. According to these figures, the effect of oil return increases on this sector is always positive, as expected from an oil-related industry. As the oil return increases, this industry experiences higher demand and higher prices, resulting in increased profitability, that in turn results in higher stock returns. In addition, as observed for the Coal industry, the threshold for the volatile period is smaller, indicating that it takes less of an increase in oil return to trigger a regime change. This may be because changes in the calm atmosphere of the earlier sub-period are more likely to be considered temporary, requiring a bigger oil price increase for a regime shift, or perhaps because, under volatile conditions, market participants are more accepting of a regime change. Another possibility is that the prevailing expectation formation mechanism was one of mean reversion such that market players expected a reversal in price trends, unless a considerably large increase occurred, convincing the market participants that it was time for a regime shift. As for a comparison between the two periods, the marginal effect of oil return on the Oil–Gas Extraction sector return is much higher in the volatile period than in the calm period (21.6 bp to 19.8 bp, compared with 6 bp to 7 bp in the clam period). For the Petroleum Refinery sector (column 6), the threshold and the oil return coefficients for the calm and volatile periods are, respectively, $r = -0.48\%$, $b_4^1 = 0.035$, $b_4^2 = 0.040$ and $r = 2.10\%$, $b_4^1 = 0.133$, $b_4^2 = 0.106$. This indicates that when oil return rises, the sector will benefit in all cases, but much more so during the volatile period, although these effects require a larger oil price increase to exceed the threshold. These findings are by and large consistent with expectations for an oil-related industry.

16.3.3.1.3 Oil-Consuming Sectors

This group of industries includes Building Construction, Air Transportation, Chemical, Plastic and Rubber and Transport Equipment (columns 7–11). The expectation may be that oil price increases will harm the stock returns of the firms in the oil-consuming industries. However, market conditions and market power may alter these effects. For example, if an industry has sufficient market power to pass on the additional costs due to oil price increases, or even more than the additional costs incurred, to the consumers, it may benefit from higher oil prices or, at least, it may remain unscathed. Similarly, if an industry is hedged against oil price increases, it need not be affected negatively. In a hedged scenario, the pattern of the effect will depend on the conditions under which the hedge becomes effective, how close it is to a complete hedge, and what the partial hedge provides.*

---

* Southwest Airlines' experience exemplifies how hedging can be beneficial or detrimental to the profitability of a firm. Southwest hedged against higher fuel prices and purchased long-term contracts to buy most of its fuel at $51 a barrel through 2009. The value of these hedges soared as oil prices continued to increase throughout 2006 and 2007. However, in 2008, oil prices dramatically decreased to $40 a barrel. Consequently, Southwest lost $56.0 million, or 8 cents per share in the fourth quarter of 2008 (www.forbes.com).

Moreover, while the direct effect from oil price increases on a given industry may be negative, the industry may be subject to indirect positive effects, counterbalancing the direct effects. For example, if oil price increases lead to portfolio reshuffling by investors in favour of a particular oil-using industry, this industry's stock return may advance with higher oil prices, at least partially offsetting the direct impact.* Similarly, if sales contracts are indexed to oil prices, or if higher oil prices lead to supply side inflation and raise the industry's product prices, the sector may become neutral to oil price changes or may even benefit from them. The nature of the expectation formation mechanism also plays a role in determining the effect. In brief, since these industries are quite varied in terms of size and character, they are expected to react differently to macroeconomic changes. Our results manifest these dissimilarities in terms of the delay period, size and sign of the threshold and the changes in the magnitude of the oil return coefficient when oil returns fall below or exceed the threshold values. Based on the symmetry tests reported in Table 16.2, some sectors do show asymmetric responses to oil return changes in the calm period of 1991–1998 with the number increasing considerably in the more volatile 1998–2006 period. We examine only a sample of the oil-consuming industries below.

The results for the Building Construction industry are reported in column 7 of Tables 16.3(A) and 16.3(B). The threshold value and the coefficients of the oil return effects for the calm and volatile periods are $r = 1.62\%$, $b_4^1 = 0.015$, $b_4^2 = 0.043$ (insignificant) and $r = 1.75\%$, $b_4^1 = -0.014$, $b_4^2 = -0.075$, respectively. The threshold for this industry is rather large in both periods. Moreover, the response of the industry to oil return shocks is symmetric in the first but asymmetric in the second period (Table 16.2). These figures indicate that the industry either benefits from or remains unaffected by oil return increases when oil price markets are relatively steady (calm period), while it suffers when oil prices are highly volatile, especially when oil price increases exceed the threshold. The positive sign of the return on the oil coefficient during the first period may occur because when oil prices are steady for a length of time, building constructors may be better able to pass on the higher cost of fuel to the home buyers. The builders may have also effectively hedged the increase in oil prices.

The situation is quite dissimilar in the 1998–2006 period, when oil prices are higher and highly volatile. In this period, the industry suffers from the oil price increases and its losses only multiply in magnitude from –0.014 (1.4 bp) to –0.075 (7.5 bp) once they pass the threshold. Specifically, during this period, any increase in the price of oil, regardless of its being below or above the threshold, contributes to lower profitability of the Building Construction sector, which in turn leads to a decline in sector return. This scenario is compatible with the commonly expected oil price effects on the oil-consuming sectors. The strong negative effect in this period, especially for oil price increases above the threshold, may be because these increases were generally large and, hence, could not be fully passed on to the consumers. Moreover, home prices had been rising throughout the 1990s and early 2000 and were expected, at least by some observers, to decline in the coming periods. This may have made real estate buyers reluctant to absorb the price increments due to higher oil prices. Overall, given that oil price increases were high and volatile and they occurred when home prices had been rising for a good length of time, oil price changes were heavily reflected in profitability and stock returns of the building industry. These price increases may have also been larger than expected and not hedged by the industry.

Another example of a major oil-consuming industry is Air Transportation (column 8), for which the threshold and the oil return coefficients for the two periods are, respectively, $r = 1.15\%$, $b_4^1 = -0.025$,

---

* Higher oil prices provide more investment opportunities for various sectors. For example, in the oil exploration area, higher prices permit exploration of oil in remote areas/deep sea that were not cost efficient before. Higher oil prices also lead to product innovations and development of substitutes for oil-using products. The auto industry is a perfect example of this phenomenon. As oil prices increase, consumers flock to electric and plug-in-hybrid vehicles. Currently, the auto industry views hybrid cars as the next big profit center. The U.S. is the largest hybrid car market in the world, with sales accounting for 60%–70% of global hybrid sales. According to J.D. Power and Associates, hybrid-electric vehicle sales volumes in the country are expected to grow 268% between 2005 and 2012. Presently, there are only 12 hybrid models available in the U.S., which would increase to 52 by 2012 (http://www.dailymarkets.com/stock/2011/05/24/analyst-interviews-auto-industry-outlook-and-review-4/).

$b_4^2 = -0.050$ and $r = 2.31\%$, $b_4^1 = -0.092$, $b_4^2 = -0.088$. The effect of oil on this industry is strong and clear-cut. A few points are notable. First, threshold values are positive and large, indicating that this industry requires a substantial increase in oil prices for a regime change. Second, oil shock effects are universally negative, indicating a loss due to oil price increases, regardless of their being below or above the threshold, or during the calm or volatile period. This result makes sense because fuel is a major cost item for the airline industry. Third, the magnitude of the oil effect is much greater in the volatile period than in the calm period, perhaps because the average price is much higher and the industry may be unable to pass on further increases in the cost of fuel to its clients. Fourth, when the price increases exceed the threshold, the effect on the industry doubles in the low volatility period, although it remains rather unchanged in the high volatility period because the effect is already very high. In other words, given the strong sensitivity of this sector to oil prices in this latter period, the industry shows symmetric effects or equal vulnerability to oil return changes when these returns are below and above the threshold.

For the Chemical, Plastic and Transportation Equipment sectors (columns 9–11) the effects are mostly insignificant when oil returns fall below the threshold, but mostly significant when they exceed the threshold. The direction of the effect for the significant cases is negative for Chemical and Plastic but positive for the Transport Equipment sector. Since these sectors are not oil-intensive, these findings are reasonable. These results may have also received a contribution from the market demand and supply conditions and the relative market power of each industry, which may or may not allow the increased costs of fuel to be passed on to the consumers of these industries' products. Asymmetry of oil shock effects holds for both periods (Table 16.2).

16.3.3.1.4 Depository Institution (DI) Sector

The threshold and the oil return coefficients for the DI sector for the two periods are, respectively, $r = 1.58\%$ $b_4^1 = -0.005$, $b_4^2 = 0.009$ (insignificant) and $r = -0.56\%$ $b_4^1 = -0.013$ and $b_4^2 = -0.015$ (column 12). Unlike the sectors discussed earlier, DIs do not produce or consume significant amounts of oil as a part of their core business. Their profitability is derived from their exposure to loans, deposits and investments in businesses and households affected by oil price changes. Hence, changes in the composition of the loans, investments and deposits and changes in the spreads and the default rates of the loans, due to oil price shocks, largely determine the sensitivity of the DI's premium to the oil return. Thus, the DI sector may benefit or suffer from oil price inflation depending on the mix of the borrowers and depositors.

Our findings indicate that, in both periods, there is an inverse relationship between changes in oil return and those of DI stock return, although in one case the effect is insignificant (calm period when oil returns exceed the threshold). Given that there are many more oil-consuming than oil-producing sectors among the DI customers, these findings are reasonable because increased fuel costs increase the probability of customers' loan default. In effect, DIs mirror the behaviour of their oil-consuming clients. Particular banks with large deposits from oil-producing countries or sectors may potentially benefit from positive effects, but they do not seem to dominate the sample.

16.3.3.2 Summary of Sector Returns

To summarize the effect of oil return changes on different industries, it can be stated that, during the less volatile period in the oil market, R_{oil} played a less important role in determining sector returns. During this period (1991–1998), three sectors, Oil–Gas Extraction, Petroleum Refinery and Air Transportation, show significant R_{oil} effects in both regimes ($j = 1, 2$), while the Plastic and Rubber sector displays no effect in either regime and the rest show mixed results (Table 16.3A). For three sectors, Coal, Building and DIs, the R_{oil} effect is significant only when oil returns are below the threshold ($j = 1$), and for another three sectors, Chemical, Transportation Equipment and Electric-Gas Services, the effect is prevalent only when the oil return exceeds it ($j = 2$). In contrast, for the more volatile oil price period (1998–2006), the R_{oil} sensitivity results, reported in Table 16.3(B), are highly robust (uniform); R_{oil} sensitivity is significant for seven out of the 10 sectors considered in both regimes ($j = 1, 2$). Of the other three sections,

Transport Equipment, and Electric–Gas Services sectors show R_{oil} sensitivity only in regime $j = 1$, and the Chemical sector shows R_{oil} sensitivity only in regime $j = 2$. The results of the symmetry of R_{oil} sensitivity during the volatile oil price period also suggest a marked difference in the R_{oil} effect when oil returns are below and above the threshold ($j = 1, 2$). During this period, the R_{oil} sensitivity between $j = 1$ and $j = 2$ is statistically different for six out of the 10 sectors: Oil–Gas, Building, Chemical, Plastic and Rubber, Transport Equipment and Electric–Gas Services sectors. On the other hand, in the calm period, only two sectors, Transport Equipment and Electric–Gas Services, show asymmetric effects.

Moreover, our findings show that there is salient information value for asset pricing in oil return shocks that is not accounted for in existing models. Two points are notable. First, we find that oil return shocks exceeding the threshold motivate slope shifts for all of the risk factors, not just the oil return factor, for most sectors, and more so for the period of oil price volatility. These shifts are assumed away in single-regime models as the oil price sensitivity of each factor in the model is taken to be identical for all shocks. Overlooking these slope shifts may lead to considerable underestimation (overestimation) of the oil shock impact and erroneous policy recommendations or consumer budget allocations. Second, the oil return is found to be an important asset pricing factor in its own right, as discussed below, which is unaccounted for within the framework of the Fama–French three-factor model. The omission of this factor from the model is likely to produce misleading results and incorrect inferences.

16.3.3.3 Economic Significance

To examine the economic significance of the factors in the model, the effect of one standard deviation change in each factor on the return of each sector is calculated by multiplying the coefficient estimate for the factor by its respective standard deviation, reported in Table 16.1. The results are displayed in Tables 16.4(A) and 16.4(B). According to these results, although the economic significance of the oil return variable falls short of those for the Fama–French factors in both calm and volatile oil price periods, there are some cases in which the economic effect of this variable is large enough to exceed that of the growth (*HML*) and/or size (*SMB*) variables. The economic magnitudes of the oil return effects are found to be much larger in the more recent volatile oil price period (e.g. SIC 12 and SIC 45) and these effects more frequently exceed those of the Fama–French growth and size variables (e.g. SIC 13 and SIC 29).

16.3.4 The FIGARCH Volatility Equation

A main contribution of this study is to model the persistence of shocks to the volatility of sector returns using the FIGARCH specification. We study the nature of shock persistence due to oil return changes to determine whether the effects are subject to decay and, if so, whether they follow an exponential (geometric) or a slower hyperbolic rate of decline. The parameters in the FIGARCH (1, δ, 1) volatility equation, $\omega_0^j, \beta_0^j, \phi_0^j$ and δ^j ($j = 1, 2$), are presented in Tables 16.3(A) and 16.3(B). The lack of significance of the constant term, the time-invariant component of the volatility equation $\left(\omega_0^j\right)$, in both sub-periods, indicates that sectoral return volatility is entirely driven by past and current shocks; it does not contain a homoskedastic component.* This stands in sharp contrast to most extant studies that assume return volatility is invariant.

The parameter β_0^j for the autoregressive term σ_{t-1}^2 in the volatility equation is the indicator of short-term volatility decay; the greater the value of β_0^j, the greater the shock persistence (the slower the decay) will be. Values of this parameter less than unity indicate the absence of explosive volatility behaviour, as

---

* This finding is similar to those obtained by Baillie *et al.* (2002) in modelling U.S. inflation.

TABLE 16.4A Economic Significance of Double-Threshold FIGARCH(1, δ, 1) Results January 25, 1991, to December 10, 1998, Sub-Period

1 Coefficient	2 Variable	3 Coal (SIC12)	4 Electric-Gas Services (SIC49)	5 Oil-Gas Extraction (SIC13)	6 Petroleum Refinery (SIC29)	7 Building (SIC15)	8 Air Transportation (SIC45)	9 Chemical (SIC28)	10 Plastic and Rubber (SIC30)	11 Transport Equipment (SIC37)	12 Depository Inst. (SIC60)
b_1^2	RM	0.754	0.573	0.805	0.658	1.056	1.144	0.962	0.685	0.912	0.684
b_1^1	RM	0.857	0.553	0.804	0.704	1.043	1.090	0.856	0.660	0.859	0.632
b_2^2	SMB	0.330	0.175	0.287	0.187	0.757	0.675	0.519	0.386	0.410	0.306
b_2^1	SMB	0.209	0.125	0.432	0.150	0.533	0.422	0.408	0.363	0.338	0.275
b_3^2	HML	0.209	0.167	0.130	0.272	0.283	0.322	-0.008	0.112	0.241	0.246
b_3^1	HML	0.206	0.212	0.201	0.239	0.364	0.181	0.018	0.156	0.185	0.245
b_4^2	R_{oil}	0.035	0.007	0.166	0.094	0.101	-0.117	-0.016	0.012	-0.012	0.021
b_4^1	R_{oil}		-0.023	0.148	0.082		-0.059	-0.016	0.012	0.019	

TABLE 16.4B Economic Significance for Double-Threshold FIGARCH(1, δ, 1) Results December 11, 1998, to December 29, 2006, Sub-Period

1 Coefficient	2 Variable	3 Coal (SIC12)	4 Electric-Gas Services (SIC49)	5 Oil-Gas Extraction (SIC13)	6 Petroleum Refinery (SIC29)	7 Building (SIC15)	8 Air Transportation (SIC45)	9 Chemical (SIC28)	10 Plastic and Rubber (SIC30)	11 Transport Equipment (SIC37)	12 Depository Inst. (SIC60)
b_1^2	RM	0.708	0.623	0.854	0.863	0.917	1.288	0.751	0.693	0.766	0.427
b_1^1	RM	0.824	0.713	0.757	0.815	1.003	0.979	0.894	0.599	0.837	0.393
b_2^2	SMB	0.253	0.109	0.298	0.115	0.346	0.343	0.429	0.208	0.206	0.136
b_2^1	SMB	0.358	0.136	0.296	0.197	0.376	0.306	0.459	0.252	0.267	0.142
b_3^2	HML	0.393	0.352	0.494	0.504	0.178	0.485	0.067	0.216	0.221	0.183
b_3^1	HML	0.452	0.395	0.436	0.439	0.410	0.323	0.106	0.228	0.360	0.139
b_4^2	R_{oil}	0.241		0.464	0.248	−0.176	−0.206	−0.033	−0.068	−0.040	−0.035
b_4^1	R_{oil}	0.195	0.026	0.506	0.312	−0.033	−0.216		0.016		−0.030

well as the failure of the constant variance models in accurately describing the sectoral stock indexes.[*†] Oil shocks affect both the real and financial sectors of the economy, altering the way investors perceive risk and form expectations in response to the arrival of oil price news. The coefficient ϕ_0^j, expected to lie between one and zero ($1 \geq \phi_0^j \geq 0$), describes the moving-average (MA) component of the short-term volatility dynamics and provides an approximation to the level of short-term shock persistence as its value affects the degree of dependence of the current volatility on lagged squared errors. Although within a FIGARCH model $\phi_0^j = 1$ suggests an IGARCH process, Baillie *et al.* (1996) point out that the interpretation of the decay process is substantially different for these two models. Specifically, while IGARCH imposes complete persistence in its impulse response weights, FIGARCH implies a long-memory behaviour (a slow rate of decay) as determined by the magnitude of δ^j, producing a mean-reverting pattern. In the low volatility period, all ϕ_0^j values, with the exception of the Building and Air Transportation sectors ($j = 2$, $\phi_0^j = 1.045$ and 1.285, respectively), are less than unity, satisfying the regularity conditions. In four cases (three in regime 1 and one in regime 2), ϕ_0^j is insignificant, indicating that the memory does not include a moving-average component.[‡] The results are rather similar between low and high volatility periods.

The finding of regime shifts in the volatility equation at threshold oil return, evidenced by the dissimilarity of the volatility equation in the two regimes, denotes the existence of oil-driven structural changes in the parameters of the volatility equation, rendering the models overlooking this feature suspect. It follows that the short-term volatility structure of sector returns is sensitive to changes in the oil industry, and the estimates of oil shock effects based on simpler models are likely to be distorted. Previous research (e.g. Hamilton and Susmel (1994) and Gray (1996)) has documented that overlooking structural changes in the GARCH parameters creates spuriously high persistence in conditional volatility, as well as poor forecasting performance. There are also economic justifications for allowing GARCH parameters to vary across regimes as some industries, e.g. the financial sector, are able to adjust to shocks quickly, while others, e.g. Coal Mining, take much longer to absorb them because of the nature of their production processes. Under a single-regime specification, the parameters of the conditional covariance matrix are assumed to stay constant over time. The threshold approach explicitly recognizes the possibility of shifts in these parameters.

Parameter δ^j is a measure of long-term memory decay. As it captures the long-range dependence in volatility, larger values of this parameter indicate a slower pace of decay, or a higher level of persistence, in the long run. In particular, when this parameter takes a value of unity, the model is described as an IGARCH process according to which shocks persist permanently, while a zero value of this parameter reduces the model to a GARCH process with a more rapid exponential pace of decay. The presence of long memory (hyperbolic decay) in the conditional variance of sector returns is strongly supported by the magnitude and significance of the fractional difference coefficients δ^1 and δ^2. The decay parameter δ^j is significant and in an acceptable range for all sectors in both periods with one exception (Petroleum Refinery in the first period in regime $j = 1$). Thus, the FIGARCH coefficients simultaneously present the long-term dynamics captured by significant decay parameters, and the short-term dynamics represented by significant autoregressive and moving-average parameters (β and ϕ). Large and significant values for $0 < \delta < 1$ imply strong support for the hyperbolic decay and condition of high persistence,

---

[*] Only the GARCH parameter for the Air Transportation sector exceeds the unity value in regime $j = 2$ during the calm period (β_0^2), suggesting that its conditional volatility is non-stationary in the short run. However, this path is truncated at periodic intervals by the switch to the stationary regime as exhibited by the β_0^1 value being less than unity in regime $j = 1$. A chi-square test of equality of the GARCH parameters (H_0: $\beta_0^1 = \beta_0^2$) is rejected at the 1% level.

[†] Other specifications, including GARCH (1, 1) and FIGARCH (1, δ, 0), were also tested. The former model fails to converge in a number of cases. The latter produces results similar to those obtained above.

[‡] In such a situation, a FIGARCH (1, δ, 0) process may better model the return volatility. Even though ϕ_0^j is found to be insignificant, the fractional difference parameter, δ, still properly captures the long-memory decay process.

as opposed to the conventional exponential decay associated with the GARCH (1, 1), or indefinite persistence with the IGARCH (1, 1) processes. Additional support for hyperbolic decay is provided by the rejection of hypotheses H_{11} (GARCH model: $\delta^1 = \delta^2 = 0$) and H_{12} (IGARCH model: $\delta^1 = \delta^2 = 1$), the results for which are reported in Table 16.2. These results are consistent with, and reinforce, the finding of long memory for squared index returns and individual stock returns of Lobato and Savin (1998).

Further analysis of the long-range volatility dependence parameters δ^1 and δ^2 reveals that the pattern of long-term decay in response to volatility shocks is regime-dependent because the coefficient δ^1 pertaining to oil return changes below the threshold ($j = 1$) is larger in magnitude than its counterpart δ^2 when oil returns exceed the threshold ($j = 2$), both in the first and second periods. This indicates that when oil returns are below the threshold the decay is slower and shock effects are more persistent, regardless of the degree of oil price volatility. The dissimilarity of the decay process under the two regimes is further strengthened by the rejection of the asymmetry tests (H_{25}: $\delta^1 = \delta^2$) discussed earlier. The findings of regime shifts and the dissimilarity of parameter values under calm and volatile oil price periods also support the view that oil price shocks influence investor perception of risk in the short run as well as bringing about structural changes in risk determination processes and thereby exerting profound influences on the way investors perceive risk and form expectations in the long run.

16.4 Conclusions

This study is the first to provide a detailed assessment of the effects of oil return changes on the equity return distributions of 10 major industrial and financial sectors in the U.S. within the framework of a Double-Threshold FIGARCH (1, δ, 1) model. This framework allows us to delineate the magnitude and persistence of oil shock effects at the sector level, which can help in the assessment of sectoral diversification benefits by investors, hedging decisions by firm managers and the formulation of regulatory actions by government. The issue of sectoral effects deserves special attention, in particular because market and country factors have become more and more correlated over time, due to technological advancement, deregulation and globalization, and thus less helpful in diversifying portfolio risk, while sector-specific risk continues to play an important role in diversification, investment and risk management decisions.

This study addresses the long-memory property in the return volatility of various sectors in the U.S. in response to oil shocks within a generalized asset pricing model including the Fama–French factors together with return on oil as explanatory variables. We find strong support for the proposition that the return on oil (R_{oil}) improves upon the basic Fama–French factor model in describing asset returns. Unlike extant studies, we introduce and identify an oil return threshold for each of the sectors considered, beyond which the impacts of the explanatory variables in the model change in magnitude and, sometimes, even in terms of the direction of the effect. We also find that GARCH, IGARCH and Fama–French models are inadequate for describing sectoral returns and accounting for long-term persistence and asymmetric responses to oil shocks.

The threshold effects of oil return changes are sector-specific, as well as period-specific. In particular, during the period of volatile oil markets with an upward trend in oil prices, the threshold values are predominantly positive, whereas during the less volatile and downward trend in the price of oil the threshold values have mixed signs. Moreover, we find that during the less volatile period in the oil market, oil prices played a less significant role in determining sector returns in comparison with the period when oil prices were rising and more volatile. Oil prices also played a less important role when they were located below the threshold rather than above it.

Hypotheses regarding the symmetry of coefficients for above and below threshold oil return levels are largely rejected for the factors included as explanatory variables in sector return models. For the mean equation coefficients, threshold asymmetry is evident for all risk factors in the model, market, size, value and R_{oil}, although less frequently significant for the latter variable. These findings support the need for separating volatile and less volatile periods and the superiority of a threshold specification for modelling sectoral returns over single-regime specifications. We also extend the literature by analysing the conditional volatility of returns. The results reveal the concurrent prevalence of long-term volatility dynamics (δ) and short-term volatility dynamics (β). The finding of highly significant δ values between $0 < \delta < 1$ supports the stationarity of the volatility process and establishes the validity of the hyperbolic decay (high shock persistence), as opposed to the conventionally used GARCH (1, 1) with exponential (rapid) decay, and the IGARCH (1, 1) process indicating indefinite persistence. This suggests that the appropriate methodology to model sector returns is FIGARCH (1, δ, 1). Finally, the results show that the regime shifts and long-term structural changes caused by oil price shocks have a profound influence on the way investors perceive long-term risk and form their expectations.

Acknowledgements

This paper was presented at the 7th Annual International Conference on Business at the Athens Institute for Education and Research, Athens, Greece. The authors would like to thank the participants at the conference for comments. All errors are ours.

References

Awerbuch, S. and Sauter, R., Exploiting the oil–GDP effect to support renewable deployment. SPRU Working Paper No. 129, October 2004, 2005.

Baillie, R.T., Bollerslev, T. and Mikkelsen, H.O., Fractionally integrated generalized autoregressive conditional heteroskedasticity. *J. Econometr.*, 1996, **74**, 3–30.

Baillie, R.T., Han, Y.W. and Kwon, T.-G., Further long memory properties of inflationary shocks. *Southern Econ. J.*, 2002, **68**(3), 496–510.

Brooks, C., A double-threshold GARCH model for the French franc/Deutschmark exchange rate. *J. Forecast.*, 2001, **20**, 135–143.

Chen, N., Roll, R. and Ross, S.A., Economic forces and the stock market. *J. Business*, 1986, **59**, 383–403.

Choi, K. and Hammoudeh, H., Volatility behavior of oil, industrial, commodity and stock markets in a regime-switching environment. *Energy Policy*, 2010, **38**, 4388–4399.

Davis, S.J. and Haltiwanger, J., Sectoral job creation and destruction response to oil price changes. *J. Monet. Econ.*, 2001, **48**, 465–512.

Dickey, D. and Fuller, W., Distribution of the estimators for autoregressive time series with a unit root. *J. Am. Statist. Assoc.*, 1979, **74**, 427–431.

Dickey, D. and Fuller, W., Likelihood ratio statistics for autoregressive time series with a unit root. *Econometrica*, 1981, **49**, 1057–1072.

Driesprong, G., Jacobson, B. and Benjamin, M., Striking oil: Another puzzle? *J. Financ. Econ.*, 2008, **89**, 307–327.

Elyasiani, E., Mansur, I. and Odusami, B., Oil price shocks and industry stock return. *Energy Econ.*, 2011, **33**(5), 966–974.

Faff, R. and Brailsford, T., Oil price and the Australian stock market. *J. Energy Finance Devel.*, 1999, **4**, 69–87.

Fama, E. and French, K., The cross-section of expected stock returns. *J. Finance*, 1992, **47**, 427–465.

Fama, E. and French, K., Industry costs of equity. *J. Financ. Econ.*, 1997, **43**, 153–193.

Gray, S.F., Modeling the conditional distribution of interest rates as a regime switching process. *J. Financ. Econ.*, 1996, **42**, 27–62.

Greene, D.L. and Tishchishyna, N.I., *Costs of Oil Dependence: A 2000 Update*, 2000 (Oak Ridge National Laboratory: Oak Ridge, Tenn.).

Guo, H. and Kliesen, K.L., Oil price volatility and U.S. macroeconomic activity. *Fed. Res. Bank St. Louis Rev.*, 2005, **87**(6), 669–683.

Hamilton, J.D., What is an oil shock? *J. Econometr.*, 2003, **113**, 363–398.

Hamilton, J.D., Oil and the macroeconomy. In *The New Palgrave Dictionary of Economics*, 2nd ed., edited by S.N. Durlauf and L.E. Blume, 2008 (Houndmills: U.K./ Palgrave Macmillan: New York).

Hamilton, J.D. and Susmel, R., Autoregressive conditional heteroskedasticity and change in regime. *J. Econometr.*, 1994, **64**, 307–333.

Hammoudeh, S., Yuan, Y., Chiang, T. and Nandha, M., Symmetric and asymmetric US sector return volatilities in presence of oil, financial and economic risks. *Energy Policy*, 2010, **38**, 3922–3932.

Hooker, M.A., Are oil shocks inflationary? Asymmetric and nonlinear specifications versus changes in regime. *J. Money, Credit Bank.*, 2002, **34**, 540–561.

Huang, B., Hwang, M.J. and Peng, H., The asymmetry of the impact of oil price shocks on economic activities: An application of the multivariate threshold model. *Energy Econ.*, 2005, **27**, 455–476.

Huang, R.D., Masulis, R.W. and Stoll, H.R., Energy shocks and financial markets. *J. Fut. Mkt*, 1996, **16**, 1–27.

Jarque, C.M. and Bera, A.K., A test for normality of observations and regression residuals. *Int. Statist. Rev.*, 1987, **55**, 163–172.

Jones, C.M. and Kaul, G., Oil and the stock markets. *J. Finance*, 1996, **51**, 463–491.

Keane, M.P. and Prasad, E.S., The employment and wage effects of oil price changes: A sectoral analysis. *Rev. Econ. Statist.*, 1996, **78**, 389–400.

Kilian, L., Exogenous oil supply shocks: How big are they and how much do they matter for the U.S. economy? *Rev. Econ. Statist.*, 2008a, **90**, 216–240.

Kilian, L., The economic effects of energy price shocks. *J. Econ. Liter.*, 2008b, **46**, 871–1009.

Kilian, L., Not all oil price shocks are alike: Disentangling demand and supply shocks in the crude oil market. *Am. Econ. Rev.*, 2009, **99**, 1053–1069.

Lee, K.S. and Ni, S., On the dynamic effects of oil price shocks: A study using industry level data. *J. Monet. Econ.*, 2002, **49**, 823–852.

Li, C.W. and Li, W.K., On a double threshold autoregressive heteroscedastic time series model. *J. Appl. Econometr.*, 1996, **11**, 253–274.

Ljung, G.M. and Box, G.E.P., On a measure of a lack of fit in time series models. *Biometrika*, 1978, **65**, 297–303.

Lobato, I. and Savin, N.E., Real and spurious long-memory properties of stock market data. *J. Business Econ. Statist.*, 1998, **16**, 261–283.

Meyer, J. and von Cramon-Taubadel, S., Asymmetric price transmission: A survey. *J. Agric. Econ.*, 2004, **55**, 581–611.

Mork, K.A., Oil and the macroeconomy when prices go up and down: An extension of Hamilton's results. *J. Polit. Econ.*, 1989, **97**, 740–744.

Moskowitz, T. and Grinblatt, M., Do industries explain momentum? *J. Finance*, 1999, **54**, 1249–1290.

Perron, P., Trends and random walks in macroeconomic time series: Further evidence from a new approach. *J. Econ. Dynam. Control*, 1988, **12**, 297–332.

Perron, P., The great crash, the oil price shock, and the unit root hypothesis. *Econometrica*, 1989, **57**, 1361–1401.

Phillips, P.C.B. and Perron, P., Testing for a unit root in time series regression. *Biometrika*, 1988, **75**, 335–346.

Poterba, J. and Summers, L., The persistence of volatility and stock market fluctuations. *Am. Econ. Rev.*, 1986, **76**, 1142–1151.

Sadorsky, P., Oil price shocks and stock market activity. *Energy Econ.*, 1999, **21**, 449–469.

Sadorsky, P., Risk factors in stock returns of Canadian oil and gas companies. *Energy Econ.*, 2001, **23**, 17–28.

Tsay, R.E., Testing and modeling threshold autoregressive process. *J. Am. Statist. Assoc.*, 1989, **84**, 231–240.

Tsay, R.E., Testing and modeling multivariate threshold models. *J. Am. Statist. Assoc.*, 1998, **93**, 1188–1202.

Yang, C.W., Hwang, M.J. and Huang, B.N., An analysis of factors affecting price volatility of the US oil market. *Energy Econ.*, 2002, **24**, 107–119.

Zivot, E. and Andrews, D.W.K., Further evidence on the great crash, the oil-price shock, and the unit-root hypothesis. *J. Business Econ. Statist.*, 1992, **10**, 251–270.

Appendix: Threshold and Delay Parameter Values

In a threshold model, a smaller threshold value for a sector means that a comparatively smaller increase in oil return is sufficient to trigger a regime shift for that sector. Threshold values can be positive or negative. In the latter case a regime change occurs when the oil return is falling. This is possible because the slowing of oil price declines can alter the expectations about future changes in oil prices, or even the expectation generation mechanism itself. For example, if expectations are extrapolative and subject to a momentum effect, slowing of oil price reductions may lead to expectations of further slowing and even a reversal in the oil price trend towards a continuum of price increases. In this scenario, a negative threshold is reasonable because a slowing in the oil price declines can be sufficient to alarm some investors about future oil price increases, prompting a selloff (or additional purchases) of the stocks in the sectors being harmed (bolstered) by the change in the oil price patterns.

Threshold values may be larger or smaller during highly volatile periods (1998–2006) compared with calm periods (1991–1998). If high oil price volatility leads economic agents to consider the oil price changes as mostly transitory, it will take more of a change in oil prices to trigger regime shifts and the effect will also decay more quickly. In this case the threshold value (r) in a volatile period may be larger than that for a calm period. Conversely, if several months of steady oil price levels in a calm period create a perception that steady prices are the norm and any changes are temporary, then the threshold may be larger during the calm period. Smaller threshold values may occur when expectations about a price increase are firmer (more certain), price changes are widely considered to be permanent, and/or agents are less tolerant of these changes. In general, a relatively smaller or larger threshold may be manifested during volatile or calm periods, depending on the prevailing expectation formation mechanism and characteristics of the industry. Delays in regime shifts (d) may also be shorter or longer depending on the same factors, as well as the speed of information transmission in the financial markets, and the composition of market players between individual and institutional, and between domestic and foreign investors. Change in the factor slopes (regime shifts) may be upward or downward. That is, the slopes b_i^2, in effect for oil return variations above the threshold, can be smaller or larger than b_i^1, for changes below the threshold. When b_i^1 is smaller in value than b_i^2, sectoral sensitivity to the factors increases when we move above the threshold. On the contrary, if b_i^1 exceeds b_i^2, once we move above the threshold, there will be a downward movement in factor sensitivities. Fama and French (1997) have provided some theoretical explanations for these slope shifts.

17

Long–Short Versus Long-Only Commodity Funds

John M. Mulvey

17.1 Introduction

Commodity investments have gained considerable interest over the past decade. For traditionally diversified investors, an allocation to a fund that invests exclusively in commodity markets offers not only a hedge against inflation, but also effective diversification due to its low correlation with traditional asset classes. Over the long run, commodity investment funds show equity-like returns, but are accompanied by lower volatility and shortfall risk.

Fundamental flaws in traditional portfolio models became apparent during the severe 2008/09-crash period. Among other issues, many investors had assumed correlations in return among asset categories would approximate historical values. Under this assumption, the investor would be adequately and safely diversified to 'protect' capital. Unfortunately, most asset categories suffered together and even sophisticated investors, such as university endowment funds, lost substantial value—close to 30% for even the elite universities.

Absolute performance hedge funds failed to protect investor capital. In Figure 17.1, we see that the correlation matrix among hedge fund categories became almost completely covered with ones. Facts of particular note are: (1) most hedge-fund categories lost money during this period; and (2) the Goldman Sachs Commodity Index (GSCI) suffered a precipitous drop of 51.4% over the specified period, including a maximum 75+% drawdown value. In Figure 17.1, two categories—dedicated short bias and managed futures—stand out with very low correlation to the equity markets represented by the FTSE 500.

17.2 Sources of Alpha in Commodity Markets

The commodity segment of the managed futures domain can provide exceptional diversification from equities and fixed income. Commodity futures markets are among the oldest exchanges in the world, such as the Dojima rice futures market, which began in 1710 in Osaka, Japan. In recent years, investors have turned to owning commodities and other real assets to protect themselves against long-term risks.

As the world population exceeds seven billion, the demand for basic commodities bumps against supply constraints for land, energy and agricultural products, possibly resulting in pricing disruptions. Even safe drinking water is becoming a scarce commodity in many parts of the world.

DOI: 10.1201/9781003265399-20

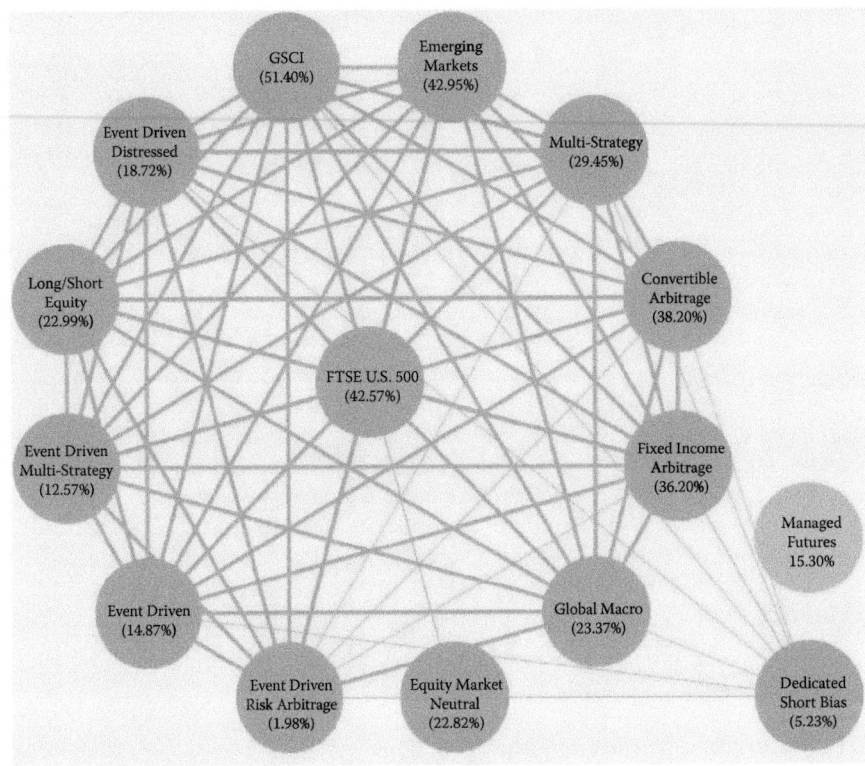

FIGURE 17.1 Correlations among major hedge fund categories—Feb. 2008 to Feb. 2009 (heavy line = correlation > 0.5; light line = correlation between 0.2 and 0.5; no line = correlation < 0.2).

A related risk is inflation. Many countries are experiencing extraordinarily low nominal interest rates and massive deficit spending plans in order to overcome the fallout from the 2008/09 crash. The current level of negative real interest rates in a number of countries likely will contribute to future inflation.

Political risks, such as disruptions caused by oil embargos, wars and terrorist attacks, present another concern. Owning raw materials can be profitable during turbulent periods caused by political factors. Further, it is likely that equities will drop very quickly when a political crisis erupts. The 1973 oil embargo, for example, precipitated a substantial increase in energy prices, accompanied with higher inflation.

Last, there is a small, but still significant risk due to weather and catastrophic shocks such as crop freezes, hurricanes and tsunamis. Many commodity prices spike when these events occur.

Studies have shown the presence of trends and regime changes in commodity prices (Erb and Harvey 2006, Miffre and Rallis 2007, Shen *et al.* 2007). These patterns are due to the diffusion of information, inventory conditions, weather and political risks. Since most commodities are employed for consumption, either final or intermediate, consumers and producers render hedging decisions on an ongoing basis as a function of their core businesses. Many commodity-trading strategies are based on sustained price swings—either positive or negative—as a function of regime changes.

A second source of alpha involves the shape of the futures curve. In most commodities, the price of a futures contract is not determined by arbitrage. Rather, supply, demand and inventory considerations are paramount. Thus, for example, backwardation occurs when inventories are low and short-term spikes in demand occur. Tactics based on the shape of the futures curve can lead to positive performance (Brennan *et al.* 1997, Gorton *et al.* 2008). A cause of the positive expected return for the traditional future curve strategy—selling in contango and buying in backwardation—goes back to the early work of Keynes around 1930. Therein, the general concept was that producers would be the predominant hedgers, thus resulting in backwardation as the normal course of events. In this situation, taking the other side of the bet—purchasing—would provide a positive expected performance, and empirical evidence supports this claim. Today, however, consumers such as the airlines and even pension plans who are attempting to hedge against inflation risks, have become the majority of hedgers, which results in a greater degree of contango. In this environment, there is a positive expected return for selling when the futures curve has severe contango (again empirical evidence supports this claim). Also see Bouchouev (2012) and Dempster *et al.* (2012) for further discussion.

There is controversy as to whether alpha is present in commodity strategies. Bhardwaj *et al.* 2008 indicates that commodity hedge funds rarely earn much more than the risk-free rate. This study was completed before the 2008/09 crash, however, wherein managed futures funds outperformed other categories by a wide margin (Figure 17.1). Lintner (1983) provides an early report showing the superior performance of selected commodity funds for improving performance relative to traditional assets.

17.3 Temporal Performance Measures

For long-term investors such as pension plans and university endowments, the most critical elements for their continued success are: (1) the growth of capital/wealth over time and (2) the investor's ability to support funding needs such as paying legal liabilities and other cash outflows. Traditional risk measures such as Sharpe ratios do not provide much information about the long-term financial health of the investor under a particular asset allocation mix. Instead, a multi-period portfolio model can be applied to determine more realistic future stochastic outcomes (e.g. Mulvey *et al.* (2008) and Dempster and Medova (2011)). In a multi-period context, it can be readily shown that long-term investors must prepare themselves to avoid a substantial loss of capital during periods of turbulence such as the 2008–09-crash period.

Given adequate historical evidence, we can construct temporal risk measures. One of the simplest is *maximum drawdown*. This value provides a limited starting point for analysing risks. Of course, max-drawdown, like most risk measures, is to some degree a backward-looking metric. Still, if we have observed a specified historical drawdown, we may well see a similar level of drawdown over future periods.

A related measure is the Ulcer index, which evaluates not only the amount of maximum drawdown, but also its duration. First, we define the historical wealth path: w_t: for $t = \{1, 2, \dots, T\}$, where the number of periods equals T. Next, the high water mark and drawdown are defined, respectively, as follows at each time period t: h_t: = maximum $\{w_{t1}\}$ for $t1 = \{1, 2, \dots, t\}$ and $d_t = (h_t - w_t)$. Note that d_t is greater than or equal to zero.

Let u be the average drawdown over the entire historical period:

$$u = \sum_{t=1}^{T} d_t / T$$

Then the *Ulcer index* is the standard deviation of drawdown values:

$$Ulcer = \sum \left(\sum_{t=1}^{T} (d_t - u)^2 / T \right)^{1/2}$$

A related measure is the *Calmar index*—the average over fixed time periods, for instance 36 months, of return/the max-drawdown over this specified look-back period.

For highly structured decisions such as blackjack, the investor can measure the long-run growth of capital by reference to the Kelly investment strategy—by maximizing expected log (capital)—and its siblings. Unfortunately, the Kelly strategy can suffer relative long and sustained drawdown periods. Mulvey *et al.* (2011) combine the Kelly objective function with explicit drawdown constraints. A feature of multiperiod models involves excess return over traditional static portfolio models. Herein, the level of excess return depends upon the volatility of the securities, diversification benefits and total transaction costs. In the case of commodity futures markets, volatility can be large, while transaction costs can be low— the ideal environment for achieving rebalancing gains. One of the simplest ways to achieve rebalancing gains is to apply a fixed-mix or even an equal-weighted portfolio of securities (see, for example, Dempster *et al.* (2011)). We will be evaluating the advantages of an equal-weighted portfolio of commodity tactics.

17.4 A Relative-Value Commodity Approach

Passive indexing strategies have become well established over the past 30 years. Index strategies are designed to match a well-defined market segment with a low cost (and possibly tax efficient) approach to investing. Passive indexing strategies can be pertinent for large institutional investors, due to their low cost and fees as well as their transparency and scalability.

Figures 17.2 and 17.3 show the wealth path for the S&P 500 equity index alongside two popular commodity indices—the Goldman Sachs Commodity Index (GSCI) and the Dow Jones/UBS index (DJUBS). These two commodity indices encompass long-only investments in the most actively traded commodities and serve as benchmarks for institutional investors.

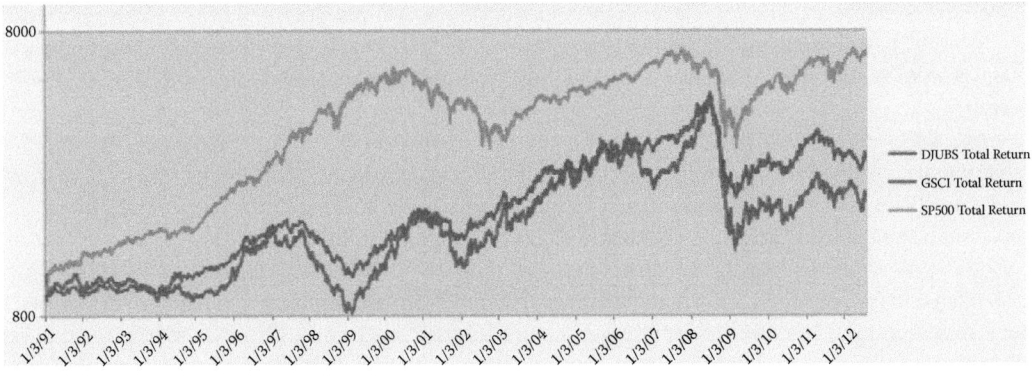

FIGURE 17.2   Time series of two commodity indices and S&P 500 (1991–2012).

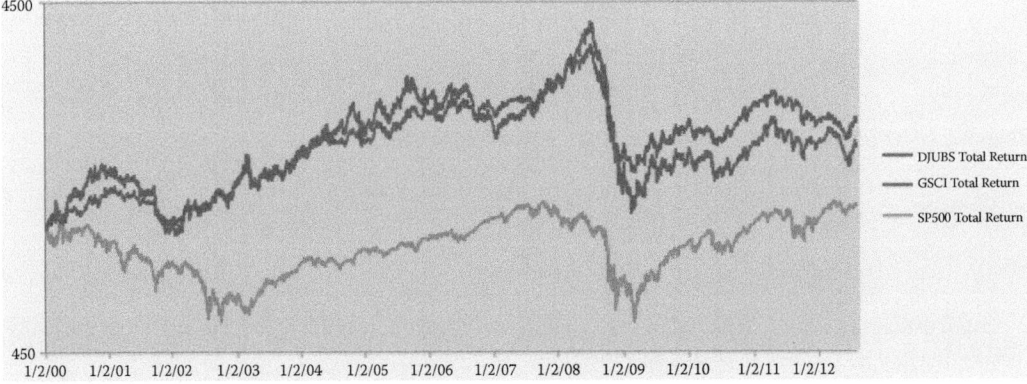

FIGURE 17.3   Time series of two commodity indices and S&P 500 (2000–2012).

The price patterns of the GSCI and DJUBS and equity markets such as the S&P 500 are roughly similar at first glance. The S&P 500 outperformed over the entire October 1991 to June 2012 period in terms of returns. Over the past 12 years, however, the results have flipped, with commodity indices beating the equity markets in terms of total return (Figure 17.3). It is clear that selecting a relevant time period is critical to relative comparisons.

In Figure 17.4, we see that the investible version of the GSCI (symbol GSG) achieved much lower performance than the spot index over the past 12 years. Underperformance is largely due to the presence of contango; commodities in contango produce lower returns due to the negative roll return. Several prominent commodities, including oil, have gone from backwardation to contango over the past decade as demand for these commodities has increased relative to their supply.

As mentioned, a critical issue for long-term investors involves drawdown risks. Unfortunately, both the GSCI and DJUBS experienced large drawdown values in 2008 equal to 70+%—at the exact time that equity markets crashed—and high Ulcer index values since drawdown continues.

To address drawdown and negative roll returns of long-only strategies, we propose a commodity portfolio via six established tactics: long and short momentum, long and short futures curve, trend following and breakout. Each of the first four tactics is based on a relative ranking of the commodities under study.

Briefly, we employ the six tactics to capture alpha embedded in commodity markets, while carefully balancing the long and short positions in the portfolio—in order to minimize drawdowns and produce positive returns with excellent diversification characteristics as compared with traditional assets (Mulvey 2012). Table 17.1 shows the performance of each of the individual six tactics over 1991 to 2012. Four of the tactics have roughly similar Sharpe ratios; however, adding the 'inferior' short tactics to a portfolio improves upon the overall performance over the full 1991 to 2012 time period.

A significant feature of futures (and forward) markets is the opportunity to be implemented as an overlay strategy without direct capital allocation. Herein, the investor employs their usual allocation to traditional assets such as stocks and bonds, e.g. 60/40. These assets provide margin requirements for the commodity futures contracts. Thus, the return of the commodity fund is additive for the investor. Especially important, therefore, is the diversification between traditional assets and the overlay strategies (discussed below).

First, we show the advantage of combining long and short directed funds. Figure 17.5 graphs the performance of the long and short momentum tactics. The long-momentum tactic outperforms short

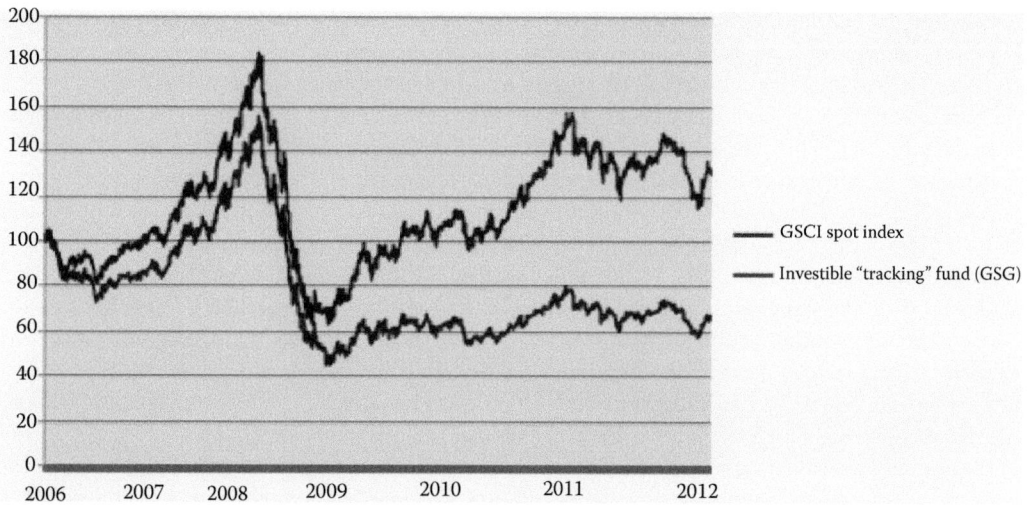

FIGURE 17.4 Time series of GSCI index and investible 'tracking' fund (GSG)—source Bloomberg.

TABLE 17.1   Performance of Six Individual Tactics Over 1991–2012

Oct. 1991–Jun. 2012	Momentum Long	Momentum Short	Futures Curve Long	Futures Curve Short	Trend Following	Breakout
Geo. Return	8.76%	−0.25%	7.87%	4.64%	6.24%	6.51%
Volatility	15.21%	15.90%	15.42%	14.71%	9.21%	9.08%
Sharpe ratio**	0.38	−0.20	0.32	0.11	0.35	0.39
Drawdown	49.10%	52.16%	52.16%	48.64%	15.85%	15.89%
Ret/Drawdown	0.18	0.00	0.15	0.10	0.39	0.41
Ulcer index	15.19%	27.97%	15.18%	20.55%	5.80%	5.14%
UPI	0.58	−0.01	0.52	0.23	1.08	1.27

** 3% risk free rate.

Note: Data is back-tested only. All investors should be aware that future results may not be the same as historical performance.

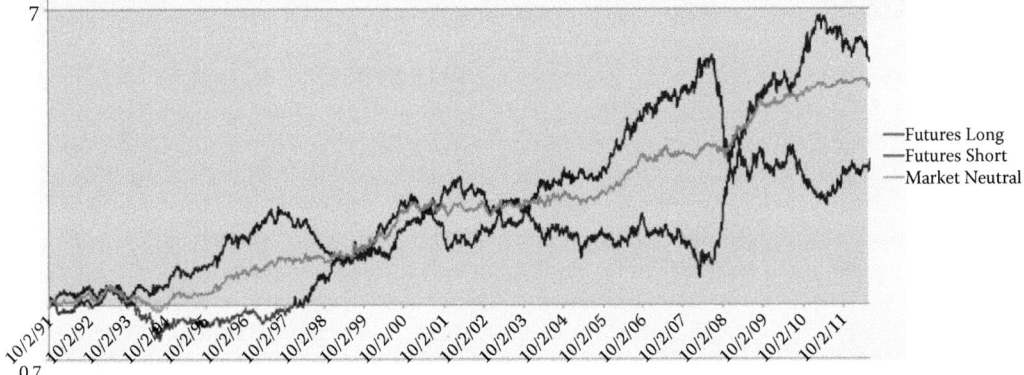

FIGURE 17.5   Futures curve tactics (long and short) and market-neutral combination—1991 to 2012. (Results are from back test experiments. Readers should be aware of the limitations of results based solely on back tests performance.)

momentum over most of the entire span. However, during both crash periods—2001/02 and 2008/09—the short-momentum tactic outperformed its long-only counterpart by a wide margin. The two tactics combine 50/50 to provide a more stable return pattern—higher return per Ulcer Index. Remember that the cost of maintaining the low returning short-momentum tactic is modest due to the overlay structure; the short tactics provide low-cost 'insurance'. The low Sharpe ratio of the short-momentum tactic understates the importance of this tactic for sound, long-term asset allocation.

A similar characteristic occurs with the long- and short-futures curve tactics. Again, the long-futures tactic has the better long-term performance as compared with the short-futures tactic, but does suffer from sharp drawdowns. Accordingly, the market-neutral 50/50 version has superior return/risk characteristics from the standpoint of return per drawdown or return per Ulcer Index. Table 17.2 lists the performance of a market neutral version (equal weighted) of the long and short versions of both the momentum and futures curve. The Ulcer index for the equal-weighted version is far superior to any of the four individual tactics—dropping below 4% versus 15%–27% for the tactics. Thus, the market-neutral portfolio has a high Ulcer performance index (UPI)—return per Ulcer.

We next turn to evaluating a portfolio of the six mentioned tactics. Figure 17.6 and Table 17.3 depict the performance of an equal-weighted portfolio. In addition, we show that performance can be improved by applying regime detection for determining the tilting of long and short positions (Mulvey *et al.* 2011). We found that three regimes are most appropriate: (1) growth (66%), (2) transition (16%) and (3) contraction (18%). We tilt to the long (or short) side during regime 1 (or 3); the level of tilt is approximate 20%. The

TABLE 17.2 Market-Neutral Combination of Momentum and Futures-Curve Tactics Versus SP 500

Oct. 1991– Jun. 2012	Momentum Long	Momentum Short	Futures Curve Long	Futures Curve Short	Market Neutral	Momentum Regime	S&P 500
Geo. Return	8.76%	−0.25%	7.87%	4.64%	6.24%	7.86%	5.81%
Volatility	15.21%	15.90%	15.42%	14.71%	6.15%	6.23%	18.78%
Sharpe ratio**	0.38	−0.20	0.32	0.11	0.53	0.78	0.15
Drawdown	49.10%	52.16%	52.16%	48.64%	9.99%	7.75%	56.78%
Ret/Drawdown	0.18	0.00	0.15	0.10	0.62	1.01	0.10
Ulcer Index	15.19%	27.97%	15.18%	20.55%	3.73%	2.87%	19.85%
UPI	0.58	−0.01	0.52	0.23	1.67	2.74	0.29

** 3% risk free rate.

Note: Data is back-tested only. All investors should be aware that future results may not be the same as historical performance.

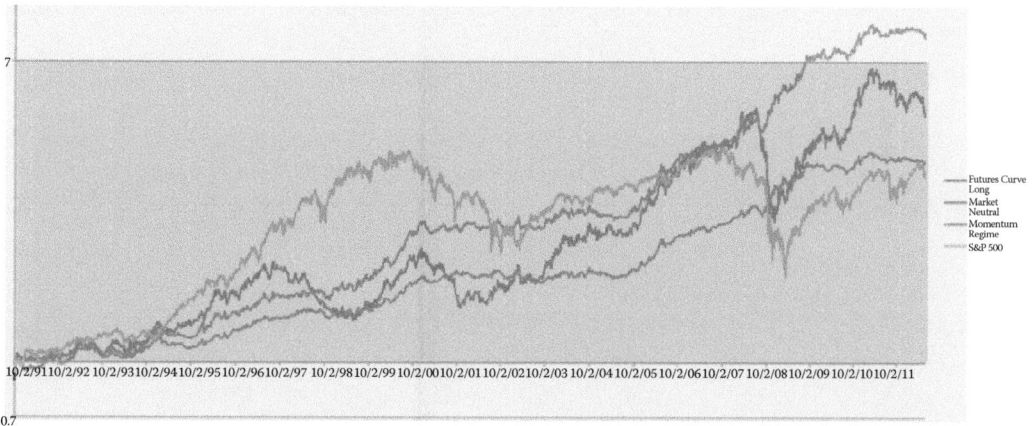

FIGURE 17.6 Wealth paths for long futures-curve tactic, equal-weighted combination and regime detection.

TABLE 17.3 Performance of a Portfolio of the Six Tactics

Oct. 1991–Jun. 2012	Equal Weight	Regime Detection
Geo. Return	6.36%	10.67%
Volatility	6.24%	8.60%
Sharpe ratio**	0.54	0.89
Drawdown	8.73%	10.36%
Ret/Drawdown	0.73	1.03
Ulcer index	3.35%	3.89%
UPI	1.90	2.74

** 3% risk free rate.

Note: Data is back-tested only. All investors should be aware that future results may not be the same as historical performance.

portfolio of six tactics improves performance over the previous four tactics since there are additional sources of uncertainties—leading to high rebalancing gains (e.g. Luenberger (1998), Dempster *et al.* (2003) and Mulvey *et al.* (2007)).

Last, we form a simplified combination of the commodity strategies—equal-weighted and regime-weighted—along with the S&P 500 equity index (Figure 17.7 and Table 17.4). Regime detection is based

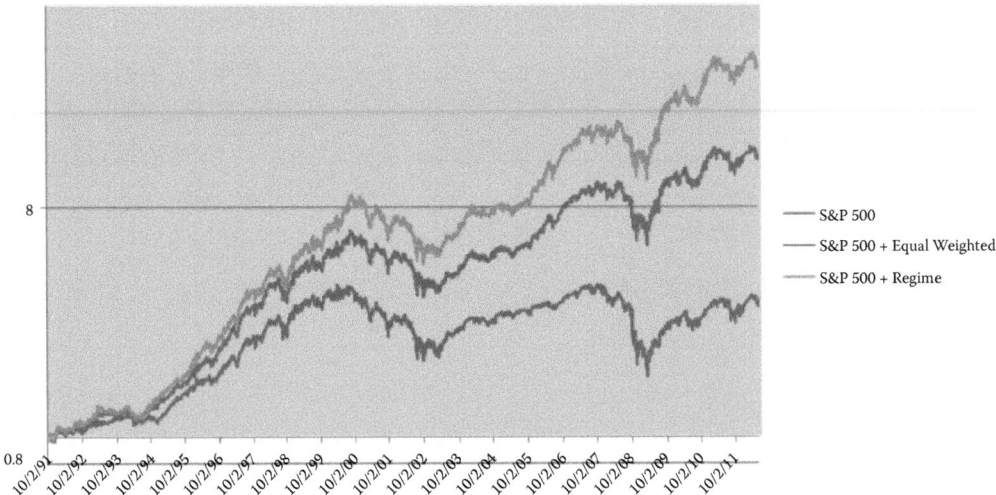

FIGURE 17.7 Combining the commodity portfolios (equal-weighted and regime-weighted) as overlays with equities.

TABLE 17.4 Performance of Combined Equity and Commodity Fund

Oct. 1991–Jun. 2012	S&P 500	S&P 500 + Equal Weighted	S&P 500 + Regime
Geo. Return	5.81%	12.66%	17.16%
Volatility	18.78%	19.23%	20.37%
Sharpe ratio**	0.15	0.50	0.69
Drawdown	56.78%	45.06%	46.26%
Ret/Drawdown	0.10	0.28	0.37
Ulcer index	19.85%	14.31%	13.68%
UPI	0.29	0.67	1.04

** 3% risk free rate.

Note: Data is back-tested only. All investors should be aware that future results may not be the same as historical performance.

on a hidden Markov model, with several decades of out-of-sample performance (Fraser 2008, Mulvey *et al.* 2011, Prajogo 2011). The addition of the commodity portfolio as an overlay strategy greatly improves total return while modestly reducing drawdown values. The overlay strategies are additive to portfolio performance. Clearly, there are more desirable ways to combine commodities with traditional assets, but even this stylized version demonstrates the advantages of adding a long–short commodity fund to an equity portfolio.

17.5 Conclusions

This paper discusses the advantages of long–short commodity funds as meaningful diversifiers within a portfolio of traditional assets. The managed futures category of hedge funds, including commodity funds, performed particularly well during the 2008/09 crash periods. We have seen similar results in previous crash periods including the Asian currency crisis in 1997–98 and the Russian debt debacle and LTCM in 1998–99.

Positive performance during crashes can be attributed to: (1) the ready ability to go long or short; (2) deep liquidity and low transaction costs allowing for dynamic asset allocation and (3) the opportunity to take advantage of volatility via rebalancing gains and regime changes. Each element provides a small advantage. When combined, a portfolio of commodity tactics can substantially improve overall investment performance, especially when traditional assets are doing poorly.

The advantages of long–short commodity funds are described with attention to short positions during sharp downturns. We are currently working with FTSE to develop a dynamic index based on these principles (Mulvey 2012).

References

Bhardwaj, G., Gorton, G.B. and Rouwenhorst, K.G., Fooling some of the people all of the time: the inefficient performance and persistence of commodity trading advisors. Working Paper 8(21), Yale ICF, 2008.

Bodie, Z. and Rosansky, V.I., Risk and return in commodity futures. *Financ. Anal. J.*, 1980, **36**, 27–39.

Bouchouev, I., The inconvenience yield or the theory of normal contango. Quant. Finance, 2012, this issue.

Brennan, D., Williams, J. and Wright, B.D., Convenience yield without the convenience: a spatial-temporal interpretation of storage under backwardation. *Econ. J.*, 1997, **107**, 1009–1022.

Dempster, M.A.H., Evstigneev, I.V. and Schenk-Hoppe, K.R., Exponential growth of fixed-mix strategies in stationary asset markets. *Finance Stochast.*, 2003, **7**, 263–276.

Dempster, M.A.H., Evstigneev, I.V. and Schenk-Hoppe, K.R., Growing wealth with fixed mix strategies. In *The Kelly Capital Growth Investment Criterion*, edited by L. MacLean, E. Thorp and W. Ziemba, pp. 427–458, 2011 (World Scientific: Singapore).

Dempster, M.A.H. and Medova, E., Asset-liability management for individual investors. *Br. Actuar. J.*, 2011, **16**(2), 405–464.

Dempster, M.A.H., Medova, E. and Tang, K., Determinants of oil futures prices and convenience yields. *Quant. Finance*, 2012, this issue.

Erb, C.B. and Harvey, C.R., The strategic and tactical value of commodity futures. *Financ. Anal. J.*, 2006, **62**(2), 69–97.

Fraser, A.M., Hidden Markov models and dynamical systems. Society for Industrial and Applied Mathematics, 2008.

Gorton, G.B., Hayashi, F. and Rouwenhorst, K.G., The fundamentals of commodity futures returns. Working Paper, Yale University, 2008.

Gorton, G.B. and Rouwenhorst, K.G., Facts and fantasies about commodity futures. *Financ. Anal. J.*, 2006, **62**(2), 47–68.

Greer, R.J., The nature of commodity index returns. *J. Altern. Invest.* 2000, **3**, 45–52.

Jeanneret, P., Monnin, P. and Scholz, S., Protection potential of commodity hedge funds. *J. Altern. Invest.*, 2011, **13**(3), 43–52.

Lintner, J., The potential role of managed commodity-financial futures accounts (and/or funds) in portfolio of stocks and bonds. Working Paper, Harvard University, 1983.

Luenberger, D., *Investment Science*, 1998 (Oxford University Press: Oxford).

Miffre, J. and Rallis, G., Momentum in commodity futures markets. *J. Bank. Finance*, 2007, **31**(6), 1863–1886.

Mulvey, J.M., The role of managed futures and commodity funds: protecting wealth during turbulent periods. *J. Indexes*, 2012, **15**(4), 38–45.

Mulvey, J.M., Bilgili, M. and Vural, T., A dynamic portfolio of investment strategies: applying capital growth with drawdown penalties. In *The Kelly Capital Growth Investment Criterion*, edited by L. MacLean, E. Thorp and W. Ziemba, pp. 735–752, 2011 (World Scientific: Singapore).

Mulvey, J.M., Simsek, K., Zhang, Z., Fabozzi, F. and Pauling, W., Assisting defined-benefit pension plans. *Oper. Res.*, 2008, **56**(5), 1066–1078.

Mulvey, J.M., Ural, C. and Zhang, Z., Improving performance for long-term investors: wide diversification, leverage and overlay strategies. *Quant. Finance*, 2007, **7**(2), 1–13.

Prajogo, A., Analyzing patterns in the equity market: ETF investor sentiment and corporate cash holdings. Ph.D. Dissertation, Princeton University, 2011.

Shen, Q., Szakmary, A. and Sharma, S., An examination of momentum strategies in commodity futures markets. *J. Fut. Mkts*, 2007, **26**(3), 227–256.

18

The Dynamics of Commodity Prices

Chris Brooks and
Marcel Prokopczuk

18.1 Introduction

Following the seminal work of Samuelson (1965), it is now widely accepted that commodity prices fluctuate randomly. Understanding the nature of this stochastic behavior is of crucial importance for decision makers engaging in commodity markets. Yet traditionally, with the probable exception of gold, commodities have not been in the focus of investors. However, interest has grown enormously over the last decade for a variety of reasons. First, following the relatively poor performances of stocks and Treasuries, investors have sought previously unexplored asset classes as potential new sources of returns. Second, the low correlation of commodity returns with equities and their ability to provide a hedge against inflation make them useful additions to portfolios. The liberalization of numerous markets has also increased corporates' requirements for hedging. This increased investment and hedging interest has led to a fast growth of the commodity derivatives market.

The aim of this paper is to study the stochastic behavior of commodity prices, both from an individual perspective but also concerning the cross-market linkages between commodities, and between commodities and the equity market. As such, this is the first piece of research to comprehensively apply a range of stochastic volatility models to commodities from several market segments. A good understanding of commodity prices' behavior and their interdependences as well as their relation to the equity market is important for investors, producers, consumers, and also policymakers.

We study the following issues for six major commodity markets. First, we investigate whether volatility does indeed behave stochastically and we estimate several models for the returns and volatility processes. Second, we examine the volatility of volatility, its persistence, and whether changes in prices and volatility are correlated. We additionally allow for jumps in both prices and volatility. We also investigate the linkages between the commodity markets considered. We determine whether volatilities are correlated

DOI: 10.1201/9781003265399-21

across markets and whether the prices or the volatilities of different commodities jump at the same time. Moreover, we analyse the same questions for the linkages between each commodity and the equity market.

Our main findings are as follows. First, we find that, within the stochastic volatility framework, the models that allow for jumps provide a considerably better fit to the data than those that do not, although there is little to choose between the models allowing for jumps in returns only and those allowing for jumps in both returns and volatility. Second, we observe alternate signs in the relationships between returns and volatilities for different commodities – negative for crude oil and equities, close to zero for gasoline and wheat, and positive for gold, silver, and soybeans. We attribute these differences to variations in the relative balances of speculators and hedgers across the markets. We also find evidence of considerable differences in both the intensity and frequency of jumps, although all commodities are found to exhibit more frequent jumps than the S&P 500. We conclude that commodities have very different stochastic properties, and therefore that it is suboptimal to consider them as a single, unified asset class. To analyse the economic implications of the differences across commodities and between model specifications, as exemplars we employ applications focused on options pricing and hedging.

The stochastic behavior of prices, equity markets, and especially equity index markets, has received a great deal of attention. The non-normality of equity returns has been documented extensively. Motivated by the poor performance of the Black–Scholes–Merton options pricing formula that produced the well-known smile phenomenon, researchers have extended the simple Brownian motion in various directions. Merton (1976) was probably the first to suggest adding a discontinuous jump component to the continuous Brownian price process, and Ball and Torous (1985) empirically confirmed the merits of this approach. Scott (1987) and Heston (1993) suggested modeling volatility stochastically, and the latter study is able to derive a semi-closed-form solution for European option prices in the environment where volatility is stochastic. The two ideas were combined by Bates (1996) and Bakshi *et al.* (1997), who proposed employing stochastic volatility and price jumps to improve the description of the asset's price process with the aim of enhancing the accuracy of options pricing. Finally, Duffie *et al.* (2000) developed a general affine framework for asset prices, and suggested a model with stochastic volatility, price jumps, and, additionally, jumps in volatility. Andersen *et al.* (2002), Chernov *et al.* (2003), Eraker *et al.* (2003), and Eraker (2004) studied various versions of stochastic volatility models, with the latter two concluding that jumps in prices and volatility improve the description of S&P 500 price dynamics as well as the pricing of options written on the index. Asgharian and Bengtsson (2006) used this class of models to study the cross-market dependencies of various equity index markets.

Compared with this array of research on equity markets, state-of-the-art empirical studies on commodity markets are sparse. Of the few examples there are of such studies, Brennan and Schwartz (1985), Gibson and Schwartz (1990), Schwartz (1997), and Schwartz and Smith (2000) study the ability of continuous-time Gaussian factor models with constant volatility to describe the stochastic behavior of the futures curve, mostly considering the crude oil market. Also studying the crude oil market only, Larsson and Nossman (2011) examine the empirical performance of several stochastic volatility models with jumps. Sorensen (2002) and Manoliu and Tompaidis (2002) apply the model of Schwartz and Smith (2000) to the agricultural and natural gas futures markets, respectively. Casassus and Collin-Dufresne (2005) mainly study the nature of risk premia in four different commodity markets; they also allow for the possibility of discontinuous price jumps, which Aiube *et al.* (2008) conclude are important in the crude oil market. A three-factor model incorporating prices, interest rates and the convenience yield is constructed by Liu and Tang (2011) to capture the stochastic relationship between the convenience yield level and its volatility for industrial commodities. Tang (2012) develops another three-factor model with a stochastic long-run mean. The possibility of stochastic volatility in commodities is considered by Geman and Nguyen (2005) for the soybean market, and by Trolle and Schwartz (2009) for crude oil.*

---

* There is an enormous number of papers that model time-varying volatilities and correlations within a discrete-time GARCH-type framework. In the commodities area, many of these are on the oil market and focus on the objective of determining effective hedge ratios. A full survey of such work is beyond the scope of this paper, but relevant studies include Serletis (1994), Ng and Pirrong (1996), Haigh and Holt (2002), Pindyck (2004), Sadorsky (2006), Alizadeh *et al.* (2008), and Wang *et al.* (2008).

Papers considering multiple commodity markets and their dependencies rely mainly on simple return analyses. For example, Erb and Harvey (2006) and Gorton and Rouwenhorst (2006) study the benefits of investing in commodity markets; Kat and Oomen (2007a,b) conduct statistical analyses of commodity returns. In a paper related to ours, Du *et al.* (2011) estimate a bivariate stochastic volatility model for crude oil and two agricultural markets. However, the focus of their analysis is on pure volatility spillovers and their model does not allow for jumps in prices nor in volatility.

The remainder of the chapter is structured as follows. Section 18.2 describes the models estimated and the procedures employed to obtain the parameters, while Section 18.3 presents and discusses the data and the empirical results. Section 18.4 analyses the economic implications of employing different models in terms of options valuation and hedging errors. Finally, Section 18.5 concludes, while further details of the Markov chain Monte Carlo procedure are presented in an appendix.

18.2 Models and Estimation

18.2.1 Spot Price Models

The first model specification employed includes stochastic volatility only, and is therefore denoted SV. The log spot price $Y_t = \log S_t$ is assumed to follow the dynamics

$$dY_t = \mu(\tau)dt + \sqrt{V_t}\,dW_t^Y, \tag{18.1}$$

where W_t^Y is a standard Brownian motion.* To capture potential seasonal effects in the price dynamics, we assume a trigonometric function for the drift component in (18.1)

$$\mu(\tau) = \overline{\mu} + \eta\sin(2\pi(\tau + \zeta)), \tag{18.2}$$

where $\tau \in [0, 1]$ denotes the time fraction of the year elapsed. The average drift rate is denoted by $\overline{\mu}$. The parameter $\eta > 0$ controls the amplitude of the function and therefore captures the strength of the seasonal effect, whereas ζ governs the periodicity of the process' drift capturing the form of the seasonality.† For markets that do not show any seasonal behavior we set $\eta = 0$, yielding a constant drift. For the volatility $\sqrt{V_t}$, we consider the square-root process

$$dV_t = \kappa\big(\theta - V_t\big)dt + \sigma\sqrt{V_t}\,dW_t^V, \tag{18.3}$$

where W_t^V is a second Brownian motion with $dW_t^Y dW_t^V = \rho dt$. The variance process is mean-reverting towards the long-run mean θ with speed K. The parameter σ captures the volatility of volatility (vol-of-vol) and the kurtosis of returns increases for higher values of σ. The correlation between returns and volatility captures the skewness of the returns distribution. Positive values imply a skew to the right, and negative values a skew to the left.

This specification guarantees the positiveness of price and volatility at all times. Except for the seasonal adjustment, it is identical to the model proposed by Heston (1993).

Empirical evidence suggests that a pure continuous specification of the spot price cannot capture all salient features observed in financial data, and in particular the possibility of rapid price movements,

---

* Alternatively, one could specify the spot price dynamics as a mean-reverting process. We have done this, however the empirical results were inferior to the non-stationary Brownian motion.

† We follow Sorensen (2002) and Richter and Sorensen (2002) and model the seasonal component using a simple sine function. An alternative approach would be to introduce monthly dummy variables in the drift as suggested by Borovkova and Geman (2007). However, as this would increase the number of parameters substantially, we prefer the more parsimonious approach based on sine functions.

i.e. jumps. To allow for jumps in the price dynamics, we follow Bates (1996) and add a Poisson process N_t^Y to specification (18.1), yielding the SVJ model

$$dY_t = \mu(\tau)dt + \sqrt{V_t}dW_t^Y + Z_t dN_t^Y. \tag{18.4}$$

The intensity λ_Y of N_t^Y is assumed to be constant and the jump sizes are assumed to be generated by a normal distribution, i.e. $Z_t \sim N(\mu_Y, \sigma_Y^2)$. Assuming that jumps occur infrequently but are relatively large (which is the natural perception of jumps as opposed to the diffusion components in prices), the jump component will mainly affect the tails of the return distribution. The specifications for the variance process and the seasonality adjustment remain unchanged.

Bakshi *et al.* (1997) and Bates (2000) find that although the inclusion of jumps in returns helps to describe the behavior of equity prices and the pricing of options, the model is still severely misspecified. Jumps in returns are transient, however, and hence a more persistent component is needed. Therefore, Duffie *et al.* (2000) introduce a model specification allowing for jumps in both returns and volatility. The return process remains identical to (18.4), but the variance process is amended by the jump process N_t^V, i.e.

$$dV_t = \kappa(\theta - V_t)dt + \sigma\sqrt{V_t}dW_t^V + C_t dN_t^V. \tag{18.5}$$

We assume that prices and volatility jump simultaneously, i.e. $N_t^V = N_t^Y$, which is commonly denoted as a SVCJ model. This assumption is motivated by the idea that periods of stress when prices jump are often accompanied by high levels of uncertainty, resulting in a jump of volatility. However, the jump sizes are not equal. Following Duffie *et al.* (2000), we assume the variance jump size to be exponentially distributed, i.e. $C_t \sim \exp(\mu_V)$. To allow for dependence between the jump sizes in returns and volatility, the jumps in returns are conditionally normally distributed with $Z_t | C_t \sim N(\mu_Y + \xi C_t, \sigma_Y^2)$. Due to the inclusion of the jump component in the variance process, the long-run mean changes to $\theta + \mu_V \lambda / \kappa$.

18.2.2 Estimation Approach

In this section, we briefly outline the Markov chain Monte Carlo (MCMC) estimation approach we use to estimate all models. MCMC belongs to the class of Bayesian simulation-based estimation techniques. The main advantage of the MCMC methodology is the fact that it allows us to estimate the unknown model parameters and the unobservable state variables, i.e. the volatility, the jump times and the jump sizes, simultaneously in an efficient way. Jacquier *et al.* (1994) show how MCMC methods can be used to estimate the parameters and latent volatility process of a stochastic volatility model. Johannes *et al.* (1999) extend this approach for jumps in returns and, finally, Eraker *et al.* (2003) estimate the model including jumps in returns and volatility via MCMC methods.[*]

In order to be able to estimate the models, it is necessary to express them in discretized form. Using a simple Euler discretization, the SVCJ model is given as[†]

$$Y_t = Y_{t-\Delta t} + \mu(\tau)\Delta t + \sqrt{V_{t-\Delta t}}\,\varepsilon_t^Y + Z_t J_t \tag{18.6}$$

and

$$V_t = V_{t-\Delta t} + \kappa(\theta - V_{t-\Delta t})\Delta t + \sigma\sqrt{V_{t-\Delta t}}\,\varepsilon_t^V + C_t J_t. \tag{18.7}$$

---

[*] For an excellent introduction to MCMC estimation techniques, see Johannes and Polson (2006).

[†] As we work with daily data, the discretization bias is negligible.

The innovations ε_t^Y and ε_t^V are normal random variables, i.e. $\varepsilon_t^Y \sim N(0, \Delta t)$, and $\varepsilon_t^V \sim N(0, \Delta t)$ with correlation ρ. The jump times J_t take the value one if a jump occurs and zero if not, i.e. $J_t \sim \text{Ber}(\lambda \Delta t)$. The discretized versions of the SVJ and SV models are obtained by dropping the respective jump components. In the following, we set $\Delta t = 1$, i.e. one day.

The general idea of the MCMC methodology is to break down the high-dimensional posterior distribution into its low-dimensional complete conditionals of parameters and latent factors which can be efficiently sampled from. The posterior distribution $p(\Theta, V, J, Z, C \mid Y)$ provides sample information regarding the unknown quantities given the observed quantities (prices). By Bayes rule we have

$$
\begin{aligned}
p(\Theta, V, J, Z, C | Y) &\propto p(Y | V, J, Z, C, \Theta) \\
&\quad p(V, J, Z, C | \Theta) p(\Theta),
\end{aligned}
\tag{18.8}
$$

where Y is the vector of observed log prices, V, J, Z and C contain the time series of volatility, jump times and jump sizes, respectively, Θ is the vector of model parameters, $p(Y \mid V, J, Z, C, \Theta)$ is usually called the likelihood, $p(V, J, Z, C \mid \Theta)$ provides the distribution of the latent state variables, and $p(\Theta)$ the prior, reflecting the researcher's beliefs regarding the unknown parameters. To keep the influence of the prior small, we specify extremely uninformative priors.

As $p(\Theta, V, J, Z, C \mid Y)$ is high dimensional, it is not possible to directly sample from it. Therefore, it is necessary to simplify the problem by breaking down the posterior distribution into its complete conditional distributions which fully characterize the joint posterior. Whenever possible, we use conjugate priors which allow us to directly sample from the conditional. If this is not possible, we rely on a Metropolis algorithm. For more details on the precise specifications, see the appendix.

The output of the simulation procedure is a set of G draws $\{\Theta^{(g)}, V^{(g)}, J^{(g)}, Z^{(g)}, C^{(g)}\}_{g=1:G}$ that forms a Markov chain and converges to $p(\Theta, V, J, Z, C \mid Y)$. Estimates of the parameters, the volatility paths, and the jump sizes are obtained by simply taking the mean of the posterior distribution. For the jump times, one additional step is required. As each draw of J is a set of Bernoulli random variables, i.e. taking on the value one or zero, the mean over all draws will provide a time series of jump probabilities. To obtain estimates for the jump times, we follow Johannes *et al.* (1999) and identify the jump times by choosing a threshold probability, i.e. we estimate the jump times $\hat{J}$ as

$$
\hat{J}_t = \begin{cases} 1, & \text{if } p(J_t = 1) > \alpha, \\ 0, & \text{if } p(J_t = 0) \leq \alpha. \end{cases}
\tag{18.9}
$$

The threshold α is chosen such that the number of jumps identified corresponds to the estimate of the jump intensity λ.

18.3 Data and Empirical Results

18.3.1 Data

In this paper, we employ daily spot price data for a variety of commodities traded in the US. The series are chosen to reflect the relative importance of those markets, and also to ensure the availability of as long a span of high quality data as possible. We consider two energy commodities, namely crude oil (CL) and gasoline (HU). From the metal markets, we include gold (GC) and silver (SI) in the study. Finally, we consider soybeans (S) and wheat (W) as representatives of the agricultural commodities market. All commodity data are obtained from the Commodity Research Bureau.* Additionally, we employ

---

* See www.crbtrader.com.

TABLE 18.1   Descriptive Statistics

	CL	HU	GC	SI	S	W	SP
Mean	0.0186	0.0189	0.0204	0.0165	0.0076	0.0008	0.0311
Std. dev.	2.6975	2.7803	1.0076	1.7621	1.5226	2.1192	1.1920
Skewness	−0.6996	−0.3748	0.1052	−1.1182	−0.6291	−0.3507	−1.3842
Kurtosis	18.0155	10.0331	11.5751	17.7078	8.1172	9.1778	32.8066
Min	−40.0011	−31.4158	−7.2327	−23.6716	−13.1820	−20.4226	−22.8997
Max	21.2765	23.0015	10.2073	13.4045	7.8667	13.1596	10.9572

Note: This table reports descriptive statistics for the daily log returns scaled by 100, i.e. Daily percentage returns. CL stands for crude oil, HU for gasoline, GC for gold, SI for silver, S for soybeans, W for wheat and SP for the S&P 500. The sample period is January 2, 1985, to March 31, 2010.

S&P 500 index data (obtained from Bloomberg) to enable us to put the results into the perspective of the existing literature on equity dynamics and to investigate the relationship of commodity and equity markets in Subsection 18.3.4. The data period covered is more than 25 years, spanning January 1985 to March 2010, and yielding 6290 observations per commodity and for the S&P 500.

Table 18.1 provides descriptive summary statistics and Figure 18.1 shows time series of the six commodity markets considered. Several points are worth noting. Compared with the S&P 500, the mean return is smaller for all commodities. The lowest average return is observed for the wheat market, which is barely positive, and the highest for the gold market. The standard deviation is higher than the S&P 500 for all but the gold market. The highest levels of volatility are observed for the energy commodities, with annualized values of 42.82% and 44.13%, which is more than twice the volatility of 16.00% and 18.92% observed in the gold and the equity markets. The kurtosis is, however, the highest for the S&P 500, while the lowest values are observed in the agricultural markets. However, these values are still higher than can be explained by a simple normal distribution. The smallest and largest returns are observed for the energy markets, both twice as large as for the S&P 500, which is to some extent surprising as the kurtosis levels are substantially lower. The gold and soybeans markets exhibit the smallest range between the minimum and maximum values[*]

18.3.2 Univariate Analysis

Table 18.2 reports the estimation results for the SV model, i.e. the stochastic volatility model without jumps. All but the parameters affecting the mean equation and some of the correlations are statistically significant. The estimates for the speed of mean reversion, κ, are of similar size in all commodity markets and are also comparable to those of the S&P 500. Only the estimate for the silver market is somewhat higher, indicating a slightly less persistent volatility process. The long-term variance levels, θ, of all but the gold market are significantly higher than the corresponding equity estimates. The S&P 500 annualized long-term volatility is 17.85% $\left(=\sqrt{\theta \cdot 252}\right)$, which is very close to the unconditional volatility, and also of similar size as in other studies.[†] The highest levels of annualized long-term volatility are observed for the crude oil and gasoline markets, at 36.00% and 37.61% respectively. The difference across the commodity markets is considerable. The vol-of-vol parameters, σ, are highly significant, which confirms the stochastic nature of volatility in the commodity markets considered. Again, with the exception of the

---

[*] It is worth noting that the time series of gasoline prices exhibits a significant price spike, i.e. an upward jump with a subsequent downward jump, in the year 2005. This feature, more often observed in markets with no or limited storage possibilities (e.g. electricity or natural gas), cannot be captured very well by the models employed. We refer to Geman and Roncoroni (2006) and Nomikos and Andriosopoulos (2012) for modelling approaches including spikes. In this study, we have decided to stay within the popular class of affine jump-diffusion stochastic volatility models such that the powerful transform analysis developed by Duffie *et al.* (2000) remains applicable.

[†] For example, Eraker *et al.* (2003) report 15.10% for the period 1985 to 1999.

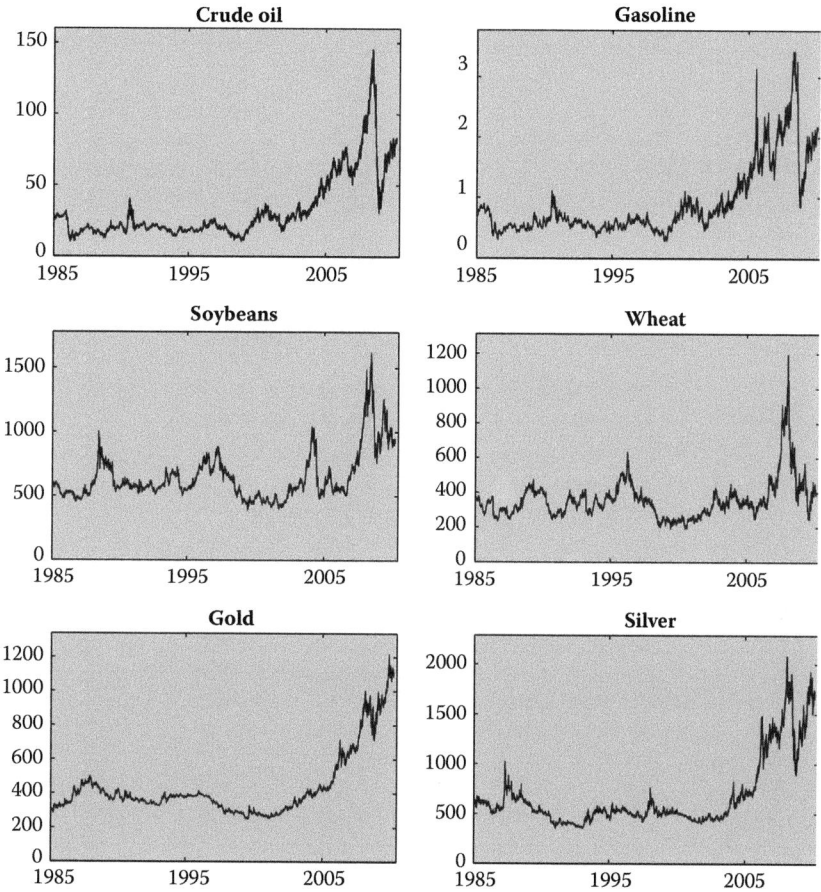

FIGURE 18.1   Price series. This figure shows the historical price series for the six commodities considered. All figures are in US dollars.

gold market, higher values than for the equity market are observed, implying heavier tails and greater deviance from normality. Moreover, there are substantial differences across markets—for example, the crude oil series exhibits a vol-of-vol that is twice as big as for soybeans.

Figure 18.2 plots the fitted volatility process in annualized percentages obtained from the SV model for the six commodities. It is evident that the crude oil and gasoline series are the most variable overall, while gold in particular is the least. The oil-based products showed two particular spikes in volatility during 1986 and 1990, both of which can be tied to economic events that occurred at those times. In March 1986, Saudi Arabia changed its policy and significantly expanded production to increase market share; other OPEC members followed, leading to price drops in both crude oil and gasoline. August 1990 marked the start of the first gulf war when Iraq invaded Kuwait. Furthermore, we can observe the effect of Hurricane Katrina hitting the US gulf coast in August 2005, leading to a short-term supply bottleneck in gasoline (although it had little noticeable effect on crude oil). Wheat was the only commodity in the sample that showed an upward trend in volatility from the late 1990s onwards. Moreover, it is also of interest to note that the Lehman default of September 2008 increased levels of volatility and uncertainty across all markets.

The estimates for the correlation of the underlying and the variance process, ρ, are most interesting, as we can observe different signs for different markets. The crude oil market shows a negative correlation, the gasoline and wheat markets (almost) zero correlations, and the gold, silver, and soybean markets significant positive correlations. For equity markets, ρ is usually found to be negative, which

TABLE 18.2   Parameter Estimates: SV

	μ	κ	θ	σ
CL	0.0575 (0.0230)	0.0204 (0.0033)	5.1432 (0.4032)	0.3676 (0.0247)
HU	0.0333 (0.0265)	0.0182 (0.0031)	5.6137 (0.4452)	0.3356 (0.0238)
GC	0.0161 (0.0082)	0.0140 (0.0025)	0.9697 (0.1243)	0.1358 (0.0085)
SI	0.0330 (0.0155)	0.0291 (0.0041)	2.7776 (0.2155)	0.3082 (0.0185)
S	0.0335 (0.0149)	0.0157 (0.0029)	2.2171 (0.2275)	0.1917 (0.0121)
W	0.0060 (0.0206)	0.0197 (0.0033)	3.8276 (0.3040)	0.2740 (0.0202)
SP	0.0334 (0.0101)	0.0177 (0.0025)	1.2261 (0.1206)	0.1597 (0.0089)

	ρ	η	ζ
CL	−0.1266 (0.0512)		
HU	−0.0039 (0.0533)	0.0702 (0.0218)	0.0528 (0.0793)
GC	0.3826 (0.0481)		
SI	0.3138 (0.0459)		
S	0.3206 (0.0512)	0.0460 (0.0118)	0.1363 (0.0692)
W	0.0223 (0.0542)	0.0946 (0.0136)	0.4186 (0.0425)
SP	−0.5678 (0.0387)		

Note: This table reports the means and standard deviations (in parentheses) of the posterior distributions for each parameter of the SV model. CL stands for crude oil, HU for gasoline, GC for gold, SI for silver, S for soybeans, W for wheat and SP for the S&P 500. The sample period is January 2, 1985, to March 31, 2010.

is confirmed by the estimate of −0.57. The negativity of ρ in equity markets is usually explained by the leverage effect, as discussed by Black (1976). However, this argument does not apply to commodity markets. A possible explanation for the different correlations might be the fraction of hedgers and speculators in the markets. Assuming that hedging activities reduce volatility, whereas speculation activity increases volatility, this line of argument would indicate that, in the crude oil market, the fraction of hedgers over speculators increases with rising prices. In the metal and the soybean markets, the opposite would hold true—that is, more hedging takes place for lower prices, while speculation dominates in times of higher than average prices.[*]

Table 18.3 provides the parameter estimates for the SVJ model, i.e. the model including Poisson price jumps. The rates of mean reversion, κ, substantially decrease in all cases and are now lower for every commodity compared with the S&P 500, implying a higher persistence of the volatility process. The long-term volatility levels (of the diffusion component) decrease, which is a technical effect, as part of the price variation is now captured by the jump component. Interestingly, however, the change for the S&P 500 is minimal compared with the commodity markets. The vol-of-vol parameters, σ, also decrease for the same reason, in that part of the excess kurtosis is now captured by the jump component. Again, the changes in the commodity markets are much bigger than for the equity case, indicating that the jump component plays an even bigger role in these markets. The correlation estimates become more pronounced compared with the SV model, i.e. the negative values decrease, whereas the positive values increase.

The jump intensity estimates, λ, lie between 0.022 for wheat and 0.088 for silver, providing evidence of significant differences between markets. In the oil, gasoline and wheat markets, jumps occur about six times per year. In the soybeans and gold market, we annually observe 12 to 13 jumps on average, whereas the silver price jumps about 22 times per year. These numbers are significantly larger than those

---

[*] We note that these hypotheses cannot be substantiated without considering the relative balance between hedgers and speculators in the respective markets. However, a detailed analysis of this aspect is beyond the scope of this paper.

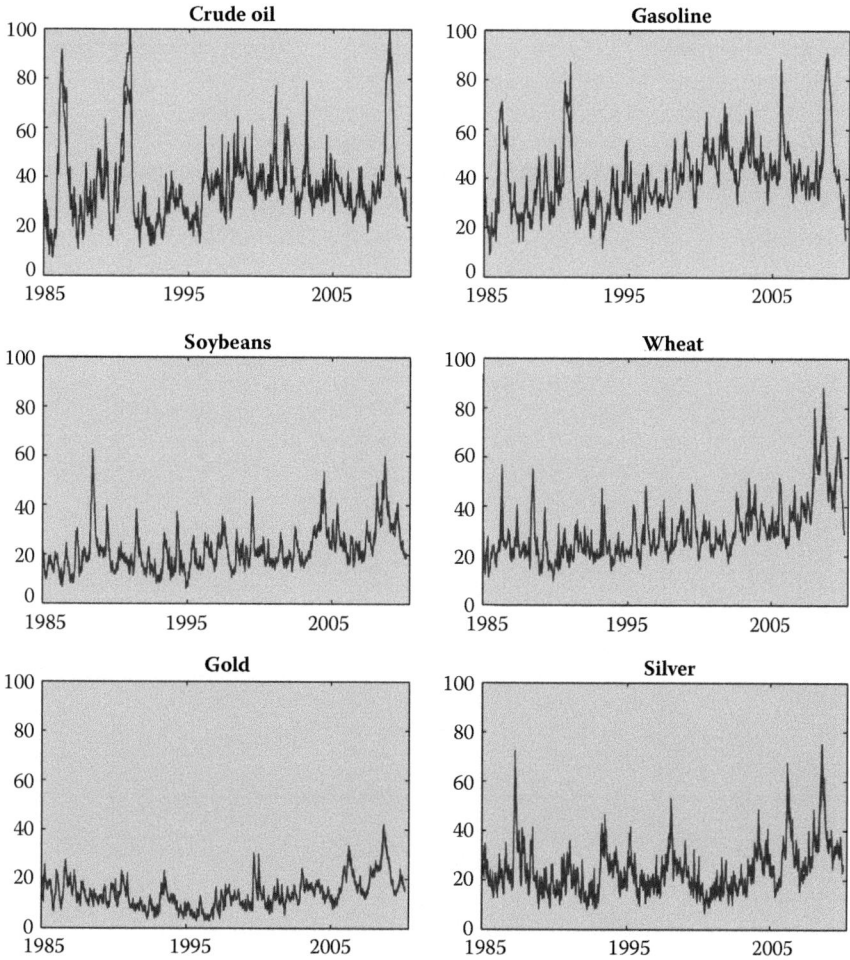

FIGURE 18.2   Estimated volatility. This figure shows the estimated volatility processes for the six considered commodities under the SV model (Annualized in per cent).

observed for equity indices. In our sample period, we estimate the S&P 500 to jump only 1.8 times per year (Eraker *et al.* (2003) estimated 1.5 times per year for their sample period). The mean jump sizes, μj, are negative for all markets, although only significant in some instances. Taking into account the standard deviation of jump sizes, σj, one can observe a huge variation. A two-sigma interval covers almost as much probability mass on the positive line as on the negative. The most notable cases are the two energy markets. A plus two-sigma event in these two markets corresponds to a +14% and +12% (daily!) price jump, respectively. Analogously, a minus two-sigma event corresponds to price drops of –16% and –14%. The results for the other commodity markets are qualitatively similar, but less extreme. For the equity market, on the other hand, the mean jump size is significantly negative, and 75% of the probability mass lies on the negative line.

Lastly, Table 18.4 reports the estimates for the SVCJ models, i.e. the models including contemporaneous jumps in prices and volatility. The speed of mean reversion, κ, reverts back to values comparable to or even higher than in the SV model, indicating a lower persistence when including jumps. This makes intuitive sense as the inclusion of jumps induces the need for higher mean-reversion speeds directly after jumps (as only positive jumps are allowed). The long-term volatility levels further decrease, as part of the variation is now captured by the volatility jump component. We still observe significant

TABLE 18.3 Parameter Estimates: SVJ

	μ	κ	θ	σ	ρ
CL	0.0654 (0.0221)	0.0096 (0.0019)	3.9773 (0.4431)	0.2106 (0.0159)	−0.2894 (0.0648)
HU	0.0413 (0.0274)	0.0100 (0.0027)	4.5790 (0.6264)	0.2182 (0.0222)	−0.0202 (0.0744)
GC	0.0159 (0.0081)	0.0096 (0.0019)	0.7863 (0.1201)	0.1001 (0.0059)	0.4660 (0.0591)
SI	0.0512 (0.0163)	0.0125 (0.0025)	2.0745 (0.2420)	0.1649 (0.0175)	0.5507 (0.0653)
S	0.0555 (0.0160)	0.0105 (0.0021)	1.9251 (0.2390)	0.1439 (0.0105)	0.3477 (0.0620)
W	0.0091 (0.0204)	0.0092 (0.0021)	3.2832 (0.3668)	0.1682 (0.0165)	−0.0443 (0.0727)
SP	0.0380 (0.0100)	0.0132 (0.0024)	1.2217 (0.1401)	0.1381 (0.0094)	−0.6417 (0.0369)

	λ	μJ	σJ	η	ζ
CL	0.0249 (0.0047)	−1.4218 (0.7941)	7.7162 (0.7108)		
HU	0.0241 (0.0071)	−1.1414 (0.7866)	6.5266 (0.8605)	0.0909 (0.0197)	0.0369 (0.0568)
GC	0.0531 (0.0117)	−0.0492 (0.1471)	1.8276 (0.1794)		
SI	0.0879 (0.0163)	−0.4627 (0.1901)	2.7129 (0.2476)		
S	0.0471 (0.0107)	−0.8150 (0.2658)	2.2938 (0.2473)	0.0360 (0.0177)	0.1464 (0.1169)
W	0.0225 (0.0064)	−0.7974 (0.6487)	4.7285 (0.5420)	0.0944 (0.0116)	0.4215 (0.0363)
SP	0.0071 (0.0023)	−2.5866 (0.8356)	2.8435 (0.5423)		

Note: This table reports the means and standard deviations (in parentheses) of the posterior distributions for each parameter of the SVJ model. CL stands for crude oil, HU for gasoline, GC for gold, SI for silver, S for soybeans, W for wheat and SP for the S&P 500. The sample period is January 2, 1985, to March 31, 2010.

TABLE 18.4 Parameter Estimates: SVCJ

	μ	κ	θ	σ	ρ	λ
CL	0.0622 (0.0227)	0.0173 (0.0041)	2.4719 (0.4271)	0.2085 (0.0219)	−0.1623 (0.0623)	0.0221 (0.0043)
HU	0.0441 (0.0275)	0.0200 (0.0041)	3.0628 (0.6079)	0.2151 (0.0343)	−0.0348 (0.0721)	0.0182 (0.0063)
GC	0.0124 (0.0082)	0.0185 (0.0032)	0.4098 (0.0700)	0.0961 (0.0080)	0.3391 (0.0553)	0.0172 (0.0047)
SI	0.0283 (0.0159)	0.0374 (0.0076)	1.2995 (0.2321)	0.2110 (0.0213)	0.2583 (0.0594)	0.0305 (0.0098)
S	0.0338 (0.0157)	0.0262 (0.0046)	1.0821 (0.1656)	0.1447 (0.0158)	0.2960 (0.0657)	0.0174 (0.0056)
W	0.0164 (0.0207)	0.0240 (0.0045)	1.7065 (0.2534)	0.1373 (0.0249)	−0.0146 (0.0791)	0.0177 (0.0038)
SP	0.0421 (0.0100)	0.0225 (0.0041)	0.7874 (0.0787)	0.1264 (0.0098)	−0.5831 (0.0395)	0.0041 (0.0012)

	μJ	σJ	ξ	μV	η	ζ
CL	−1.5291 (1.1932)	7.9167 (0.7397)	−0.0170 (0.4167)	2.3176 (0.9789)		
HU	−2.0728 (1.7065)	6.7416 (0.9186)	0.1425 (0.2745)	4.0436 (1.3707)	0.0851 (0.0204)	0.0277 (0.0684)
GC	−0.3294 (0.5197)	2.7819 (0.3301)	0.5546 (0.6923)	0.5108 (0.1470)		
SI	−0.6968 (0.6375)	3.9967 (0.4951)	−0.0099 (0.3746)	1.5656 (0.6734)		
S	−0.7981 (0.8480)	2.8019 (0.3667)	−0.0612 (0.2793)	1.7129 (0.6099)	0.0456 (0.0124)	0.1455 (0.0715)
W	−1.6252 (0.9708)	4.9274 (0.5244)	0.0478 (0.2229)	3.0470 (0.7534)	0.0902 (0.0109)	0.4144 (0.0393)
SP	−4.4478 (0.9411)	2.2486 (0.5251)	0.0519 (0.2012)	2.7594 (0.7914)		

Note: This table reports the means and standard deviations (in parentheses) of the posterior distributions for each parameter of the SVCJ model. CL stands for crude oil, HU for gasoline, GC for gold, SI for silver, S for soybeans, W for wheat and SP for the S&P 500. The sample period is January 2, 1985, to March 31, 2010.

TABLE 18.5   Model Comparison

	SV	SVJ	SVCJ
CL	27,934	26,756	26,779
HU	29,169	28,444	28,534
GC	16,096	14,634	14,871
SI	23,683	21,458	21,776
S	21,569	20,657	20,992
W	25,395	24,609	24,725
SP	17,437	16,969	16,819

Note: This table reports the DIC scores for the model specifications examined: SV, SVJ, and SVCJ. CL stands for crude oil, HU for gasoline, GC for gold, SI for silver, S for soybeans, W for wheat and SP for the S&P 500. The sample period is January 2, 1985, to March 31, 2010.

differences across markets, ranging from 10% for the gold market to 27.5% for the gasoline market. The equity market lies, at 14%, closer to the lower border of this interval. The vol-of-vol parameter, σ, slightly decreases for most markets; the silver market is a notable exception, where σ increases from 0.16 to 0.21. The correlation of the diffusion components mostly decreases in absolute terms; as prices and volatility are always jumping contemporaneously, part of the correlation is now captured by the jump components. Interestingly, the jump intensities decrease significantly in some instances, such as the gold, silver, and soybean markets, whereas it remains almost unchanged in the crude oil market. Changes in the average price jump and its volatility are mostly relatively small. This is opposed to the corresponding result for the S&P 500, where the mean jump size almost doubles. The jump dependence parameter ξ is not estimated very precisely, a phenomenon that has already been observed by Eraker *et al.* (2003). As the values are close to zero, only modest dependence between the jump sizes is observed. The average volatility jump size lies between 0.51 for the gold, and 4.04 for the gasoline market.[*] Although model choice is not the main focus of this paper, in order to compare the fit of the three different models to the data, we make use of the Deviance Information Criterion (DIC) proposed by Spiegelhalter *et al.* (2002) and in particular applied to stochastic volatility models by Berg *et al.* (2004). The DIC can be regarded as a generalization of the Akaike Information Criterion (AIC), and trades off adequacy and complexity in a similar fashion.[†] As for the AIC, smaller DIC values indicate a better model fit.

The DIC scores are reported in Table 18.5. First, we can observe that the fit of the SV model is by far the worst for every market, providing evidence of the benefits of including a price jump component. Comparing the DIC scores of the SVJ and the SVCJ models is less conclusive as most of the scores are close to each other for a given commodity. However, the SVJ values are always lower for all commodity markets, whereas the SVCJ obtains the lowest score for the S&P 500.

A second way to compare the fit of the models to the data is by simulating price paths and comparing the average time series moments of the simulated data with those obtained from the real data. Table 18.6 reports the results of this exercise. Comparing these numbers with the descriptive statistics reported in Table 18.1, we can observe that the standard deviation is matched relatively well by the SV and SVJ

---

[*] As an additional analysis, we have split the entire data set into two equal subsets and re-estimated the models for each of these. From these estimations we could observe that the most interesting parameters, such as the vol-of-vol parameters σ, the correlation ρ and the jump intensity λ do not change substantially across sub-period. Naturally, there were also some changes of the estimated parameters across the subsamples. However, it is noteworthy that these mainly affect the drift parameters, which is not surprising, and also to some extent the parameters governing the jump sizes. The latter observation needs to be interpreted with caution, as the decrease of the sample size makes the estimation of the jump component, the occurrence of which is rare by definition, quite difficult. In order to keep the presentation manageable, we do not present these results but they are available from the authors upon request.

[†] AIC equals DIC in the special case of flat priors.

TABLE 18.6   Simulated Moments

	CL	HU	GC	SI	S	W	SP
Panel A: SV							
Mean	0.0525	0.0239	0.0200	0.0330	0.0035	0.0091	0.0347
Std. dev.	2.2220	2.3116	0.9772	1.6444	1.4570	1.9325	1.0698
Skewness	−0.0299	0.0081	0.0215	0.0477	−0.0206	−0.0122	−0.0653
Kurtosis	4.0779	3.8389	3.9488	4.0809	3.8000	3.8225	3.8885
Panel B: SVJ							
Mean	0.0264	0.0173	0.0126	0.0100	0.0183	0.0104	0.0204
Std. dev.	2.3261	2.3329	0.9557	1.6184	1.4453	1.9166	1.1265
Skewness	−0.3910	−0.2842	0.0292	−0.1908	−0.2374	−0.1430	−0.4611
Kurtosis	11.6217	7.6318	5.9566	5.6459	4.7935	6.1708	6.2511
Panel C: SVCJ							
Mean	0.0094	0.0058	0.0025	0.0006	0.0062	0.0050	0.0131
Std. dev.	1.7049	1.8143	0.8876	1.1933	1.0588	1.5885	0.8706
Skewness	−0.5723	−0.3398	−0.1958	−0.2931	−0.1866	−0.3641	−0.3637
Kurtosis	16.8886	8.6686	10.9397	9.2814	6.5732	8.1980	17.0958

Note: This table reports the average first four central moments of simulated returns using the estimated parameters. CL stands for crude oil, HU for gasoline, GC for gold, SI for silver, S for soybeans, W for wheat and SP for the S& P 500.

models. The SV model is, however, not able to match the kurtosis levels found in the data. The SVJ model does a much better job and also matches the skewness closer than the SV model. In terms of skewness and kurtosis, the SVCJ model is by far the best, in many cases close to the observed levels. However, the SVCJ model does not fit the levels of standard deviations as well as the SVJ model. Overall, as the DIC values for the SVJ and SVCJ models are always close together and the analysis of simulated moments does not provide a clear conclusion as to which of the two models to prefer, we will, in the spirit of robustness, use both in the subsequent analysis. It is clear from the above analysis, however, that the jumps in returns play an important role.

18.3.3 Cross-Commodity Market Analysis

The MCMC estimation approach employed not only provides us with parameter estimates but also with estimates for the latent state variables, i.e. volatility, jump times and jump sizes. We can use this information to analyse the cross-market dependence structure of these state variables.[*] In a first step, we calculate the correlations of the differences in volatility, i.e. $\sqrt{V_t} - \sqrt{V_{t-1}}$, to analyse the extent to which the volatilities in the various markets move together. Table 18.7 reports these correlations, as well as return correlations to enable us to put the volatility results into perspective.

As expected, return correlations are highest between related commodities belonging to the same market segment. However, this correlation is still far from perfect. In particular, the moderate degree of correlation between crude oil and gasoline of 0.49 is interesting. The strongest correlation is observed for gold and silver with a value of 0.68. Return correlations between commodities of different segments are weak or close to zero, never exceeding 0.11.

---

[*] Alternatively, one could try to estimate a multivariate version of the model employed in order to analyse these dependencies. This approach is, however, computationally infeasible as it would involve a 14×14 covariance matrix when considering all seven markets together.

TABLE 18.7 Correlations of Returns and Volatilities

	CL	HU	GC	SI	S	W	SP
			Panel A: Returns				
CL	1.0000						
HU	0.4894	1.0000					
GC	0.0539	0.0104	1.0000				
SI	0.0404	0.0060	0.6849	1.0000			
S	0.1094	0.1098	0.0006	−0.0106	1.0000		
W	0.0969	0.0755	0.0030	−0.0157	0.3941	1.0000	
SP	0.0298	0.0459	−0.0418	−0.0630	0.0792	0.0700	1.0000
			Panel B: SVJ volatilities				
CL	1.0000						
HU	0.3084	1.0000					
GC	−0.0084	0.0410	1.0000				
SI	−0.0257	0.0090	0.6375	1.0000			
S	−0.0574	−0.0157	−0.0028	0.0124	1.0000		
W	0.0368	0.1163	0.0340	0.0440	0.1146	1.0000	
SP	0.0398	0.0433	0.0470	0.0429	−0.0151	0.0418	1.0000
			Panel C: SVCJ volatilities				
CL	1.0000						
HU	0.3051	1.0000					
GC	0.0392	0.0244	1.0000				
SI	0.0118	0.0274	0.4175	1.0000			
S	−0.0058	0.0041	0.0067	0.0282	1.0000		
W	−0.0003	0.0265	0.0081	0.0314	0.1045	1.0000	
SP	0.0289	0.0592	0.0403	0.0369	0.0299	0.0358	1.0000

Note: This table reports return and volatility correlations. Panel A reports the correlations of returns. Panel B displays the correlations of changes in volatility calculated from the estimated volatility process of the SVJ model. Panel C displays the correlations of changes in volatility calculated from the estimated volatility process of the SVCJ model. CL stands for crude oil, HU for gasoline, GC for gold, SI for silver, S for soybeans, W for wheat, and SP for the S&P 500. The sample period is January 2, 1985, to March 31, 2010.

Looking at the volatility correlations, one can identify a similar pattern. There are, however, some differences. For example, the volatility correlation between soybeans and wheat is substantially smaller than the corresponding return correlation, indicating that the prices move more closely together than the volatilities in these markets. Another interesting point is the correlation of crude oil with the agricultural commodities. The returns are mildly correlated, indicating some dependence across markets; the volatilities' correlations are, on the other hand, close to zero. Comparing the results for the SVJ and the SVCJ model, it is interesting to observe that, for some instances, the estimated correlation is almost identical (e.g. CL–HU or S–W), whereas in other instances, the result changes substantially (e.g. GC–SI). This might be a consequence of the fact that the number of jumps identified in the gold and silver markets changes substantially between the two model variants (this can be seen from the changes in the estimated value of λ).

Next, we analyse the simultaneous jump probabilities for each pair of commodities. To do this, we simply count the numbers of simultaneous jumps and divide it by the sample size T, i.e. we calculate the following quantity: $\sum_{t=1}^{T} \hat{J}_t^i \hat{J}_t^j / T$. To put these numbers into perspective, we also compute the

TABLE 18.8 Simultaneous Jump Probabilities (%)

	CL	HU	GC	SI	S	W	SP
			Panel A: SVJ				
CL	–	0.06	0.12	0.21	0.12	0.06	0.02
HU	0.40	–	0.12	0.21	0.11	0.05	0.02
GC	0.17	0.14	–	0.45	0.24	0.11	0.04
SI	0.17	0.25	2.23	–	0.43	0.20	0.07
S	0.08	0.13	0.32	0.62	–	0.11	0.04
W	0.08	0.05	0.22	0.29	0.46	–	0.02
SP	0.10	0.02	0.13	0.10	0.05	0.02	–
			Panel B: SVCJ				
CL	–	0.04	0.04	0.08	0.03	0.04	0.01
HU	0.35	–	0.03	0.06	0.03	0.03	0.01
GC	0.08	0.05	–	0.06	0.03	0.03	0.01
SI	0.10	0.06	0.65	–	0.06	0.06	0.02
S	0.03	0.06	0.06	0.08	–	0.03	0.01
W	0.03	0.05	0.02	0.06	0.21	–	0.01
SP	0.03	0.03	0.05	0.06	0.06	–	0.03

Note: This table reports the simultaneous jump probabilities of returns in the lower triangular. The upper triangular reports probabilities assuming independence. CL stands for crude oil, HU for gasoline, GC for gold, SI for silver, S for soybeans, W for wheat, and SP for the S&P 500. The sample period is January 2, 1985, to March 31, 2010.

probability of a simultaneous jump assuming independence which is given by the product of the two jump intensities. Table 18.8 provides the results.

We can observe that the simultaneous jump probabilities of commodities belonging to the same segment (i.e. the pairs CL–HU, GC–SI, and S–W) are about 4–10 times higher than one would expect under independence. For all other pairs, the probabilities are very similar, indicating that jumps of non-related commodities are independent of each other. Compared with the results for the dependence in volatility, there are some interesting differences to be observed. The simultaneous jump probability between soybeans and wheat is halved from 0.46% to 0.21% when including jumps in volatility. This is in contrast to the volatility correlation, which remains almost equal. Furthermore, one can observe a substantial decrease of simultaneous jump probabilities for most cases, whereas the correlations of volatilities remain rather constant, or even increase, when including volatility jumps.

18.3.4 Commodity and Equity Markets

We now investigate the dependencies between the individual commodity markets and the equity market, i.e. the S&P 500. It is well known that return correlations between commodities and equities are quite low, which is the reason why commodities are often considered to be a diversifier for a traditional portfolio of stocks and bonds. From Table 18.7, it can be seen that the same holds true for volatility, i.e. equity and commodity volatilities are almost uncorrelated. Consequently, commodities not only serve as a return diversifier, but as a volatility diversifier at the same time.

Table 18.8 also contains the daily simultaneous jump probabilities in equity and commodity markets as well as the corresponding probabilities assuming independence. For the SVJ model, the greatest difference is observed for the gold (0.13% vs. 0.04%) and oil (0.10% vs. 0.02%) markets, indicating that jumps in these two commodities are related to jumps in the equity market. When considering the results for the SVCJ model, one can see that the joint jump probabilities are all very small. However, compared with

TABLE 18.9 Conditional Jump Sizes in Equity and Commodity Markets

SVJ	CL	HU	GC	SI	S	W
SP/xx	−3.60	−4.43	−4.95	−5.70	−6.51	−8.32
xx/SP	3.87	14.84	0.34	−1.11	−2.03	−2.82
SVCJ						
SP/xx	−4.39	−5.41	−5.45	−4.87	−5.05	−5.06
xx/SP	9.11	7.01	0.02	−0.54	−2.47	−5.07

Note: This table reports the average jump sizes of the S&P 500 conditional on the occurrence of a jump in a commodity market (reported in line 'SP/xx') and the average jump size in the commodity market conditional on the occurrence of a jump in the S&P 500 (reported in line 'xx/SP'). 'xx' stands for the respective commodity reported in the columns. CL stands for crude oil, HU for gasoline, GC for gold, SI for silver, S for soybeans, W for wheat, and SP for the S&P 500. The sample period is January 2, 1985, to March 31, 2010.

the previous results for the SVJ model, the differences from the case assuming independence become more pronounced, indicating that the likelihood of a simultaneous jump in volatility is higher than a simultaneous jump in prices. This is especially true for the soybeans market.

When a simultaneous jump occurs, it might be that the prices in both markets jump in the same, or alternatively, in opposite directions. Moreover, the jumps might be more or less pronounced than average jumps. To investigate this issue, we compute the average jump sizes for each commodity market conditional on the occurrence of a jump in the equity market and *vice versa*. Table 18.9 reports the results. Interestingly, the average jump size in the equity market is much more negative than the unconditional average jump size of −2.59 (SVJ model). Surprisingly, this effect is strongest for the soybean and wheat markets, i.e. a jump in the agricultural markets induces an average jump size that is two to three times bigger than normal. A consideration of the average jump sizes in the commodity markets given that the equity market jumps reveals that the energy commodities tend to jump upward if the equity market jumps downwards. This is different to their unconditional average jump size which is negative. The other commodity jumps react less to jumps in the equity market; the results for the SVCJ model are qualitatively similar but with some differences in terms of strength of the effects.

18.4 Economic Implications

18.4.1 Options Valuation

To analyse how the differences of the commodity dynamics impact upon economic quantities of interest we consider the pricing and hedging of options. One of the advantages of the continuous-time models employed in this study is that semi-closed-form options pricing formulas exist. These formulas are based on Fourier analysis, which was initially proposed by Heston (1993). Duffie *et al.* (2000) provide general solutions for all affine models and for the SVCJ model in particular.

In the following, we use the model parameters estimated in the previous section to analyse the implications for the value of European call options. As we have estimated all parameters from historical price data under the physical measure, we need to make an assumption regarding the market price of volatility and jump risk. Broadie *et al.* (2007) review the most recent literature on the sign and size of these risk premia for the S&P 500, concluding that the evidence is inconclusive; most studies report insignificant risk premia from a statistical and/or economic point of view. Due to these findings, and for simplicity, in the following we assume zero volatility and jump risk premia.*

---

* The results would, of course, change if this assumption were not valid, especially if the nature of the risk premia differs across commodities. This is a non-trivial question that we leave for future research.

TABLE 18.10   Option Prices

	Absolute			Relative (%)		
	ITM	ATM	OTM	ITM	ATM	OTM
			Panel A: SV			
CL	7.01	4.07	2.11	172	100	52
HU	7.14	4.26	2.32	168	100	55
GC	5.23	1.77	0.38	295	100	22
SI	6.01	3.00	1.31	200	100	44
S	5.80	2.69	1.03	215	100	38
W	6.51	3.53	1.67	185	100	47
SP	5.49	1.99	0.35	276	100	17
			Panel B: SVJ			
CL	7.15	4.20	2.21	170	100	53
HU	7.17	4.28	2.33	167	100	54
GC	5.24	1.77	0.37	295	100	21
SI	6.02	2.98	1.25	202	100	42
S	5.81	2.68	0.99	217	100	37
W	6.51	3.51	1.64	185	100	47
SP	5.54	2.06	0.38	269	100	18
			Panel C: SVCJ			
CL	6.65	3.56	1.66	187	100	47
HU	6.79	3.80	1.89	179	100	50
GC	5.12	1.38	0.20	372	100	14
SI	5.82	2.60	0.94	223	100	36
S	5.53	2.19	0.63	253	100	29
W	6.07	2.87	1.09	211	100	38
SP	5.38	1.71	0.19	314	100	11

Note: This table reports the option prices for a horizon of one month. The price of the underlying is normalized to 100. The strike price is set to 95 (ITM), 100 (ATM), and 105 (OTM). The left part (absolute) reports the options price, the right part (relative) reports the options price relative to the ATM option price. CL stands for crude oil, HU for gasoline, GC for gold, SI for silver, S for soybeans, W for wheat and SP for the S&P 500.

To make the results comparable, we assume that each underlying currently trades at $S = 100$. The current variance level is assumed to be equal to its long-run average, and the risk-free rate is assumed to be zero. Using the estimated parameter values, we calculate the values of at-the-money (ATM; $X = S$), in-the-money (ITM; $X = 0.95S$), and out-of-the-money (OTM; $X = 1.05S$) call options where X denotes the strike price. Table 18.10 reports the results of this computation. The left columns report the value of the option, and the right columns express the option's value relative to the value of the corresponding ATM option.

One can observe large differences across assets. Naturally, options written on the underlyings with the highest volatility levels are most expensive. The value of the gold and S&P 500 OTM options is already close to zero (although they are only 5% out-of-the-money), whereas the values for crude oil and gasoline remain at higher levels. The differences in jump intensities, jump amplitudes and the correlations of returns with volatilities lead to some interesting differences in the relative values of ITM and OTM options across commodities.* For example, comparing the soybeans and wheat markets, one can

---

* When considering implied volatilities instead of relative values, these differences would manifest themselves in the shape of the volatility smile/skew.

observe that the relative value of ITM options is higher for soybeans, whereas the relative value of OTM options is higher for wheat. Comparing the prices for the SV and SVJ with the SVCJ model, one can see that prices are lower for the latter, i.e. neglecting the possibility of volatility jumps leads to increased option prices.

To further compare the consequences of the different behavior of the commodities considered for options valuation, we analyse the values of the most popular exotic contracts, i.e. barrier options. To save space, we focus our attention on at-the-money down-and-out call and put options and vary the value of the barrier between 90% and 100% of the current spot level. To take the different levels of volatility in the markets into account, we normalize the value of the barrier option by the value of a corresponding plain-vanilla option. Consequently, all resulting values must be between zero and one.

Figure 18.3 displays the results of these computations. One can observe that, especially for the down-and-out put options, the value is quite distinct across commodities, even though we have already adjusted for the absolute volatility level by normalizing with the values of plain-vanilla options. This is due to the fact that the probability of hitting the barrier is more sensitive to the volatility and jump level than the simple option value. Inspecting the figures more closely, one can observe differences between model specifications. For example, the relative value of the down-and-out put option on wheat for a barrier of 90% is around 0.4 for the SV and SVJ models but increases to 0.5 in the SVCJ model, whereas the value of the corresponding silver option remains at 0.6 for all three models. From this result, one can deduce that the inclusion of jumps in the volatility dynamics is more relevant for wheat options. A similar observation can be made for soybeans and the energy commodities.

18.4.2 Options Hedging

Lastly, we investigate the consequences of the different stochastic behavior of commodities and different models in terms of the delta hedging of options. Consider, for example, the influence of the correlation parameter. Depending on the options position, one is long or short the underlying and long or short volatility. For example, a short put position translates into a long position in the underlying and a short position in volatility. Thus, depending on the sign of the correlation, one is either naturally diversified or not. By implementing a standard delta hedge, one neglects this relationship, which will have consequences on the hedging outcome. Similarly, the jump frequency as well as the mean and standard deviation of the jump sizes will impact upon the hedging success.

We set up our simulation analysis as follows. We consider a short position in an ATM put option with a maturity of one month. The hedging horizon τ is one week, the current variance level is assumed to be equal to the long-run mean, and interest rates are assumed to be zero. To investigate the potential hedging error, we assume that the true price dynamics are given by either the SV, the SVJ, or the SVCJ model and we calculate the corresponding options value.* We then consider a standard delta hedging strategy, i.e. we calculate the option's delta using Black's formula to set up the hedging portfolio. Next, we simulate the price and volatility dynamics until T and calculate the new option value to obtain the pay-off of the hedging portfolio. This procedure is repeated 10,000 times to obtain a distribution of hedging errors.

Table 18.11 reports the mean, standard deviation, skewness, and kurtosis of the hedging errors. The mean hedging error is negative in all cases and larger (in absolute terms) for the energy commodities, which is as expected due to the higher volatility levels. The same observation applies to the standard deviation of the errors. When considering the skewness and kurtosis, one can observe some interesting differences across commodities and models. Considering the SV model first, we see that all distributions are skewed to the left, i.e. big negative outcomes are more likely than big positive outcomes. Interestingly, the skews for the agricultural commodities and the S&P 500 are bigger than for crude oil. Moreover, looking at the kurtosis of the hedging errors, it becomes clear that crude oil exhibits the thinnest tails, i.e. extreme negative (and positive) outcomes are less likely than for the other markets.

---

* As in the previous section, we assume a risk premium of zero.

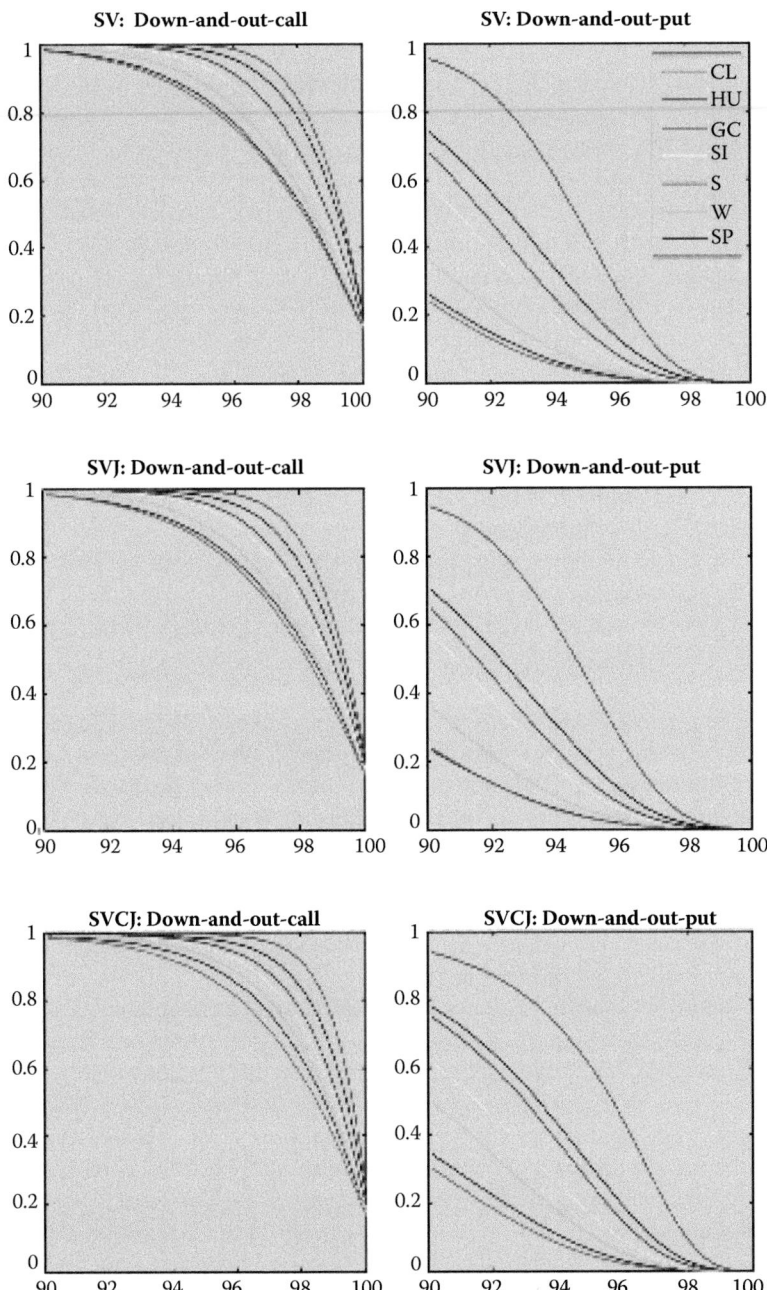

FIGURE 18.3 Down-and-out options. This figure displays the value of Down-and-out barrier options as a fraction of the corresponding plain vanilla options value. The values on the *x*-axis correspond to the Knock-out barrier. CL stands for crude oil, HU for gasoline, GC or gold, SI for silver, S for soybeans, W for wheat and SP for the S&P 500.

The picture completely changes when introducing jumps in the price and variance processes. Whereas the mean and volatility of the hedging errors remain at similar levels, or even decrease, the skewness, and, most importantly, the kurtosis, drastically increase. This holds true for all markets, but

TABLE 18.11 Hedging Error

	CL	HU	GC	SI	S	W	SP
			Panel A: SV				
Mean	−0.4441	−0.4864	−0.1964	−0.3395	−0.3018	−0.3955	−0.2319
Std	0.8305	0.8618	0.3510	0.6464	0.5272	0.7072	0.3781
Skew	−1.0725	−1.3783	−1.4260	−1.3480	−1.4964	−1.4902	−1.4691
Kurt	2.5727	4.1403	4.2224	4.1102	4.4456	4.6943	4.1992
			Panel B: SVJ				
Mean	−0.4896	−0.4745	−0.1967	−0.3341	−0.3028	−0.3990	−0.2309
Std	0.8772	0.7944	0.3237	0.5446	0.4823	0.6497	0.3659
Skew	−3.4585	−2.5196	−2.1526	−2.4347	−1.8658	−2.5869	−2.0724
Kurt	22.9281	12.2694	8.5793	10.5993	5.7959	13.1202	8.3335
			Panel C: SVCJ				
Mean	−0.4037	−0.4062	−0.1444	−0.2634	−0.2192	−0.2997	−0.1873
Std	0.8425	0.7275	0.2879	0.4967	0.3785	0.5319	0.3034
Skew	−4.7612	−3.4210	−3.5175	−2.9353	−2.1156	−3.7515	−2.4437
Kurt	38.8046	24.8022	23.9426	16.6159	8.5904	25.8874	15.2599

Note: This table reports the mean, standard deviation (std), skewness (skew), and kurtosis (kurt) of the hedging error when delta-hedging an atm option using black's formula. The option has a maturity of one month, and the hedging horizon is one week. CL stands for crude oil, HU for gasoline, GC for gold, SI for silver, S for soybeans, W for wheat, and SP for the S&P 500.

with different rates. Under the SVCJ model, the kurtosis of crude oil rockets from 2.6 to more than 38, which is now more than twice the value observed for the S&P 500. By contrast, the rise in kurtosis for the soybeans case is least dramatic with an increase from 4.4 to 8.6. Overall, one can make out substantial differences across commodities and models when considering the third and fourth moments of the hedging errors, which translate into a significant degree of model risk when delta hedging in these markets.

18.5 Conclusions

This paper has examined the stochastic behavior of the prices and volatilities of a sample of six of the most important commodity markets. Using a Bayesian Markov chain Monte Carlo estimation approach, three separate stochastic volatility-type models are estimated and compared for each commodity price series, and for the S&P 500 by way of comparison. Fairly intuitively, correlations between the returns are high for pairs of commodities from the same sub-class but almost zero across sub-classes. The same pattern holds for the relationships between the simultaneous jump probabilities: within market segments, jumps often occur in tandem, whereas they are essentially independent across segments.

We are able to demonstrate that not only are return correlations between commodities and the stock index low, as is well documented, but the correlations between commodity and stock volatilities are also low. This is an important result since it shows that commodities may be an even more useful portfolio constituent than previously thought as they can act as a volatility diversifier as well, which is important for options portfolios or any portfolio of securities with embedded contingent claims. The paper examines whether jumps occur simultaneously across commodities and equities; there is some evidence that jumps in the two asset classes do occur together, notably for gold and oil, although the results vary somewhat between models.

Finally, we test the economic impact of employing one stochastic volatility model rather than another by considering the pricing of European calls and barrier options written on each commodity, and by evaluating the effectiveness of delta hedging. Again, we find substantial differences both across commodities and between models, indicating the heterogeneous nature of this asset class and the importance of judiciously choosing the most appropriate specification for each individual series.

Acknowledgements

We thank two anonymous referees of Quantitative Finance and discussants and participants at the German Finance Association Meeting (DGF), Regensburg, 2011, the Meeting of the Swiss Society for Financial Research (SGF), Zurich, 2012 and the European Meeting of the Financial Management Association (FMA), Istanbul, 2012, for helpful comments.

References

Aiube, F., Baidya, T. and Tito, E., Analysis of commodity prices with the particle filter. *Energy Econ.*, 2008, **30**, 597–605.

Alizadeh, A., Nomikos, N. and Pouliasis, P., A Markov regime switching approach for hedging energy commodities. *J. Bank. Finance*, 2008, **32**, 1970–1983.

Andersen, T., Benzoni, L. and Lund, J., An empirical investigation of continuous-time equity return models. *J. Finance*, 2002, **62**, 1239–1284.

Asgharian, H. and Bengtsson, C., Jump spillover in international equity markets. *J. Financ. Econometr.*, 2006, **4**, 167–203.

Bakshi, G., Cao, G. and Chen, C., Empirical performance of alternative option pricing models. *J. Finance*, 1997, **52**, 2003–2049.

Ball, C. and Torous, W., On jumps in common stock prices and their impact on call option pricing. *J. Finance*, 1985, **40**, 155–173.

Bates, D., Jumps and stochastic volatility: exchange rate processes implicit in Deutschemark options. *Rev. Financ. Stud.*, 1996, **9**, 69–108.

Bates, D., Post-'87 crash fears in S&P 500 futures options. *J. Econometr.*, 2000, **94**, 181–238.

Berg, A., Meyer, R. and Yu, J., Deviance information criterion for comparing stochastic volatility models. *J. Business Econ. Statist.*, 2004, **22**, 107–120.

Black, F., Studies of stock price volatility changes. In Proceedings from the American Statistical Association, Business and Economic Statistics Section, pp. 177–181, 1976 (American Statistical Association).

Borovkova, S. and Geman, H., Seasonal and stochastic effects in commodity forward curves. *Rev. Deriv. Res.*, 2007, **9**(2), 167–186.

Brennan, M. and Schwartz, E., Evaluating natural resource investments. J. Business, 1985, **58**, 135–157.

Broadie, M., Chernov, M. and Johannes, M., Model specification and risk premia: evidence from futures options. *J. Finance*, 2007, **62**, 1453–1490.

Casassus, J. and Collin-Dufresne, P., Stochastic convenience yield implied from commodity futures and interest rates. *J. Finance*, 2005, **60**, 2283–2331.

Chernov, M., Ghysels, E., Gallant, A. and Tauchen, G., Alternative models for stock price dynamics. *J. Econometr.*, 2003, **116**, 225–257.

Du, X., Yu, C. and Hayes, D., Speculation and volatility spillover in the crude oil and agricultural commodity markets: a Bayesian analysis. *Energy Econ.*, 2011, **33**, 497–503.

Duffie, D., Pan, J. and Singleton, K., Transform analysis and asset pricing for affine jump-diffusions. *Econometrica*, 2000, **68**(6), 1343–1376.

Eraker, B., Do stock prices and volatility jump? Reconciling evidence from spot and option prices. *J. Finance*, 2004, **59**, 1367–1403.

Eraker, B., Johannes, M. and Polson, N., The impact of jumps in volatility and returns. *J. Finance*, 2003, **58**, 1269–1300.

Erb, C. and Harvey, C., The strategic and tactical value of commodity futures. *Financ. Anal. J.*, 2006, **62**, 69–97.

Geman, H. and Nguyen, V.-N., Soybean inventory and forward curve dynamics. *Mgmt Sci.*, 2005, **51**, 1076–1091.

Geman, H. and Roncoroni, A., Understanding the fine structure of electricity prices. *J. Business*, 2006, **79**, 1225–1261.

Gibson, R. and Schwartz, E., Stochastic convenience yield and the pricing of oil contingent claims. *J. Finance*, 1990, **45**, 959–976.

Gorton, G. and Rouwenhorst, G., Facts and fantasies about commodity futures. *Financ. Anal. J.*, 2006, **62**(2), 47–68.

Haigh, M. and Holt, M., Crack spread hedging: accounting for time-varying volatility spillovers in the energy futures markets. *J. Appl. Econometr.*, 2002, **17**(3), 269–289.

Heston, S., A closed-form solution for options with stochastic volatility with applications to bond and currency options. *Rev. Financ. Stud.*, 1993, **6**, 327–343.

Jacquier, E., Polson, N. and Rossi, P., Bayesian analysis of stochastic volatility models. *J. Business Econ. Statist.*, 1994, **12**, 371–389.

Johannes, M., Kumar, R. and Polson, N., State dependent jump models: How do US equity indices jump. Working Paper, 1999.

Johannes, M. and Polson, N., MCMC methods for financial econometrics. In Handbook of Financial Econometrics, edited by Y. Ait-Sahalia and L. Hansen, 2006 (Elsevier: Amsterdam).

Kat, H. and Oomen, R., What every investor should know about commodities, Part I: univariate return analysis. *J. Invest. Mgmt*, 2007a, **5**, 4–28.

Kat, H. and Oomen, R., What every investor should know about commodities, Part II: multivariate return analysis. *J. Invest. Mgmt*, 2007b, **5**, 40–64.

Koop, G., Bayesian Econometrics, 2003 (Wiley: New York).

Larsson, K. and Nossman, M., Jumps and stochastic volatility in oil prices: time series evidence. *Energy Econ.*, 2011, **33**, 504–514.

Liu, P. and Tang, K., The stochastic behavior of commodity prices with heteroskedasticity in the convenience yield. *J. Empir. Finance*, 2011, **18**, 211–224.

Manoliu, M. and Tompaidis, S., Energy futures prices: term structure models with Kalman filter estimation. *Appl. Math. Finance*, 2002, **9**, 21–43.

Merton, R., Option pricing when underlying stock returns are discontinuous. *J. Financ. Econ.*, 1976, **3**, 125–144.

Ng, V. and Pirrong, C., Price dynamics in refined petroleum spot and futures markets. *J. Empir. Finance*, 1996, **2**, 359–388.

Nomikos, N. and Andriosopoulos, K., Modelling energy spot prices: empirical evidence from nymex. Energy Econ., 2012, **34**, 1153–1169.

Pindyck, R., Volatility and commodity price dynamics. *J. Fut. Mkts*, 2004, **24**, 1029–1047.

Richter, M. and Sorensen, C., Stochastic volatility and seasonality in commodity futures and options: the case of soybeans. Working Paper, 2002.

Sadorsky, P., Modeling and forecasting petroleum futures volatility. *Energy Econ.*, 2006, **28**, 467–488.

Samuelson, P., Proof that properly anticipated prices fluctuate randomly. *Ind. Mgmt Rev.*, 1965, **6**, 41–49.

Schwartz, E., The stochastic behavior of commodity prices: implications for valuation and hedging. J. Finance, 1997, **52**, 923–973.

Schwartz, E. and Smith, J., Short-term variations and long-term dynamics in commodity prices. *Mgmt Sci.*, 2000, **46**, 893–911.

Scott, L., Option pricing when the variance changes randomly: theory, estimation, and an application. *J. Financ. Quantit. Anal.*, 1987, **22**, 419–438.

Serletis, A., A cointegration analysis of petroleum futures prices. *Energy Econ.*, 1994, **16**, 93–97.

Sorensen, C., Modeling seasonality in agricultural commodity futures. *J. Fut. Mkts*, 2002, **22**(5), 393–426.

Spiegelhalter, D., Best, N., Carlin, B. and van der Linder, A., Bayesian measures of model complexity and fitting. *J. R. Statist. Soc. B*, 2002, **64**, 583–639.

Tang, K., Time-varying long run mean of commodity prices and the modelling of futures term structure. *Quant. Finance*, 2012, **12**, 781–790.

Trolle, A. and Schwartz, E., Unspanned stochastic volatility and the pricing of commodity derivatives. *Rev. Financ. Stud.*, 2009, **22**, 4423–4461.

Wang, T., Wu, J. and Yang, J., Realized volatility and correlation in energy futures markets. *J. Fut. Mkts*, 2008, **28**, 993–1011.

Appendix: MCMC Estimation Details

This appendix provides more detailed information on the MCMC estimation procedure. For a general introduction to MCMC techniques, see, for example, Koop (2003). The Gibbs sampling technique allows one to draw each parameter and state variable of the joint posterior sequentially. For many parameters, conjugate priors can be used to derive the conditional posterior distribution, which is a standard distribution and therefore easily sampled from. Details are given below.

μ we use a standard normal distribution as the prior

κ, θ we use a truncated normal distribution bounded at zero and hyperparameters of 0 and 1 as priors for each parameter

σ, ρ as the posteriors for σ and ρ are not known, we follow the re-parameterization suggested by Jacquier *et al.* (1994) and use the inverse Gamma distribution with hyperparameters 1 and 200 as priors

η we use an exponential distribution with a hyperparameter of 0.05 as a prior, yielding a truncated normal distribution as a posterior

ζ the posterior distribution for ζ is non-standard, and we apply an independence Metropolis algorithm, drawing from the uniform distribution on the unit interval

λ we use a Beta distribution with hyperparameters 2 and 40

μ_V we use a Gamma distribution with hyperparameters 1 and 1

μ_J we use a Normal distribution with hyperparameters 0 and 100 as priors

σ_J^2 we use an Inverse Gamma distribution with hyperparameters 5 and 20 as priors

ξ we use a standard normal distribution as a prior

The posteriors of J_t, Z_t and C_t are non-standard but are provided in the appendix of Eraker *et al.* (2003). The conditional posterior distribution of the variance path V is also non-standard and given by

$$p(V|Y,J,Z,C,\Theta) \propto \prod_{t=1}^{T} p\left(V_t | V_{t-1}, V_{t+1}, Y, J, Z, C, \Theta\right), \tag{A.1}$$

where the posterior function for each V_t is given as

$$p\left(V_t | V_{t-1}, V_{t+1}, Y, J, Z, C, \Theta\right) \propto \frac{1}{V_t} e^{-(1/2)\left(\omega_1 + \omega_2 + \omega_3\right)}, \tag{A.2}$$

with

$$\omega_1 = \frac{\left(Y_{t+1} - \mu(\tau) - J_{t+1} Z_{t+1}\right)^2}{V_t}, \tag{A.3}$$

$$\omega_2 = \frac{\left(V_t - \theta\,\kappa - (1-\kappa)V_{t-1} - \sigma\rho\left(Y_t - \mu(\tau) - J_t Z_t\right) - J_t C_t\right)^2}{\sigma^2 V_{t-1}\left(1-\rho^2\right)}, \tag{A.4}$$

$$\omega_3 = \frac{\left(V_{t+1} - \theta\kappa - (1-\kappa)V_t - \sigma\rho\left(Y_{t+1} - \mu(\tau) - J_{t+1}Z_{t+1}\right) - J_{t+1}C_{t+1}\right)^2}{\sigma^2 V_t\left(1-\rho^2\right)}. \tag{A.5}$$

We use a random walk Metropolis algorithm. The volatility of the error term in this procedure is calibrated such that the acceptance probability is within the range 30%–50%. See Koop (2003) for details of calibrating the Random Walk Metropolis algorithm. Each parameter (and state variable) is sampled 100,000 times (i.e. $G = 100,000$) and we discard the first 30,000 'burn-in' draws as is standard practice with MCMC estimations.

19

Short-Term and Long-Term Dependencies of the S&P 500 Index and Commodity Prices

Michael Graham,
Jarno Kiviaho and
Jussi Nikkinen

19.1 Introduction

The volatility in global markets emanating from the recent financial crisis has presented real and momentous challenges for portfolio management as cross-market dependencies have increased.[*] In the last two decades alone, financial markets have endured a catastrophic collapse in confidence resulting from the Mexican, Asian and Russian crises. The most recent financial crisis, which started in late 2007, has been different from these earlier crises given its global and pervasive impact across different markets. Many studies have shown that the various asset markets have all been significantly affected by the recent events in unforeseen ways that are not mirrored by the experiences of past crises.

In this paper, we investigate whether commodities could potentially provide some diversification benefits for the stock market investor during periods of financial turmoil. We do this by examining the short-term and long-term comovement of the returns series for S&P 500 and the S&P GSCI® commodity index using wavelet coherency analysis. We also examine dependencies between the S&P 500 and 10 sub-indexes of the S&P GSCI® commodity index.[†]

---

[*] Financial crisis interferes with the resource allocation process that moves funds to agents with the most productive investment opportunities in the financial system.

[†] This paper deals with dependence in time of crisis. Other studies have concentrated on extreme comovement in international financial markets. See Longin and Solnik (2001), Ang and Chen (2002), Patten (2004), Garcia and Tsafack (2009), Adel and Salma (2012) and references therein for a discussion of this literature.

DOI: 10.1201/9781003265399-22

The seminal portfolio theory of Nobel Laureate Professor Harry Markowitz (1952) implicitly posits that uniquely including assets with less than perfect correlations in a portfolio reduces its inherent diversifiable risk. Commodities as an asset class has shaken off its reputation as a comparatively unknown asset class to become a new unique asset class (see Gorton and Rouwenhorst (2006a,b) for more on this), with the potential to reduce systematic risk in a portfolio. Currently, many academics (see, e.g. Billingsley and Chance (1996), Jensen *et al.* (2002) and Gorton and Rouwenhorst (2006a,b)) and market participants consider commodities to be an important element of being diversified[*] given that they show low correlations (Gorton and Rouwenhorst 2006a,b, Ibbotson Associates 2006) with traditional assets (e.g. equities and bonds) that are robust to extreme events. This is because factors that influence commodity prices (e.g. weather and the geopolitical situation, supply constraints in physical production and event risk) differ from those that affect the value of traditional assets (Geman 2005). Significantly, assets tied to the S&P GSCI index as of 31 December 2010 amounted to a staggering $US100 billion.[†] Following this intense interest in commodities, the financial press have also keenly reported on this market and published many articles discussing the role of commodities in diversified portfolios.[‡]

The recent global financial crisis has raised concerns, particularly among some market observers, that the diversification argument in favour of commodities investment is weak. For example, in the fourth quarter of 2008, financial markets experienced simultaneous price declines in every financial asset, with the exception of the U.S. Treasuries. Tang and Xiong (2010) and Silvennoinen and Thorp (2010) suggest that the growing prominence of commodities index investors increases comovement between commodity markets and traditional asset markets. Kyle and Xiong (2001) also posit that portfolio rebalancing by commodity index investors can spill over price volatility from other markets to commodity markets, potentially exposing commodities to time-scale dependencies with traditional asset markets. Given the importance and practical implications of asset dependencies in asset allocation decisions and risk management, we use an enhanced methodology, the three-dimensional analysis of wavelet coherency, to investigate the comovement of the returns series for S&P 500 and the S&P GSCI® commodity index for a 10-year period encompassing the height of the current global financial crisis. Low comovement during the crisis would imply that investing in commodities as part of a diversified equity portfolio can reduce portfolio volatility during periods of financial crisis.

Our empirical investigations add to the body of knowledge on stock market and commodities market dependencies and offer portfolio managers novel evidence of the observed stock and commodity market dependencies. The current literature intimates that the commodity market and equity market dependencies vary over time (Chance 1994). However, studies examining the usefulness of commodities as part of a portfolio comprising traditional financial assets typically make no distinction between short-term and long-term investors in their analyses (see, e.g. Jensen *et al.* (2000, 2002), Gorton and Rouwenhorst (2006b), Ibbotson Associates (2006) and Yamori (2011)). Thus the observed time variation of asset dependencies lacks an explicit interpretation from the point of view of investors in different time horizons. This difference is important because short-term and long-term investors would be interested in different fluctuations in the comovements of the commodities and equity markets (Candelon *et al.*

---

[*] The value of commodities is best achieved as part of a diversified portfolio and not as a stand-alone investment. Historically, commodities have shown inferior returns and greater volatility when compared with other assets (Erb and Harvey 2006). However, when combined with conventional investment assets (e.g. stocks and bonds), they can enhance portfolio performance. See also, for instance, the article of Mark Huamani on investing in commodities and diversification: http://www.jpmorgan.com/tss/General/Investing_in_Commodities/1159337364479 (accessed 1 September 2011).

[†] http://www.bloomberg.com/news/2011-01-07/s-p-gsci-says-commodity-assets-against-index-rose-as-high-as-100-billion. html (accessed 1 September 2011).

[‡] See, for instance, http://www.ft.com/intl/cms/s/1/23fadee4-4803-11db-a42e-0000779e2340.html#axzz1YDqkOknx (an article by John Authers, *Financial Times*, September 22, 2006) (accessed 8 November 2010).

2008).* We investigate this difference and explore the potential time-varying behaviour of stock and commodities market dependencies in a unified empirical framework.

The wavelet coherency methodology employed in this study offers a refinement in terms of analysis as both time and frequency domains are simultaneously taken into consideration to measure market dependencies in an integrated framework. Accordingly, both transient and long-term associations between stock and commodity markets can be analysed concurrently. The output of this empirical tool enables researchers and analysts to zero in on regions where equity market and commodity market dependencies are higher for both short-term and long-term investors. In such regions, the benefits of portfolio diversification, in terms of risk management, are lower.

The evidence presented implies that combining an equity portfolio with commodities can provide diversification benefits in the short term. The dependencies between the S&P 500 and S&P GSCI® commodity indexes as well as between S&P 500 and 10 sub-sets of the S&P GSCI® commodity index all show weak fluctuations at the highest frequencies. Moreover, this result is shown to be robust during the current financial crisis. In the long term, however, our results show that there are relatively reduced gains from combining an equity portfolio with commodities as the onset of the current financial crisis ushered in a period of increased dependencies between stock index returns and commodity index returns. These results are, in general, confirmed when we examine comovements between the equity portfolio, S&P 500, and 10 sub-indexes of the commodity index, and the S&P GSCI® commodity index, but with notable exceptions. For example, the evidence presented also shows enhanced diversification benefits by combining equity (S&P 500 total return) and the S&P GSCI® Precious Metal total return in the long term as well as in the short term.

The remainder of the paper is organized as follows. Section 19.2 briefly reviews the relevant literature on commodity and portfolio diversification. Section 19.3 presents the wavelet methodology used to analyse comovements of stock and commodity market returns. The data employed in the study and our empirical findings are discussed in Sections 19.4 and 19.5, respectively, and Section 19.6 concludes.

19.2 Brief Literature Review

The identification of dependencies between financial assets is a key ingredient in risk assessment and portfolio management. Modern portfolio theory demonstrates that the value of diversification is bounded by the systematic risk of the market as a whole. Gorton and Rouwenhorst (2006a, p. 20), however, point out that commodities have *some power at diversifying the systematic component of risk*. Additionally, this asset class is strongly correlated to the inflation rate and changes in the rate of inflation, which provides a hedge against rising inflation (see, for example, Anson (1999)). It is therefore not surprising that vast amounts of money have been invested in commodity linked indexes by hedge funds and pension funds (Akey 2005).

Numerous academic studies have investigated the benefits of commodities for portfolio diversification and as a source of investment returns. For instance, Lummer and Siegel (1993), Ankrim and Hensel (1993) and Kaplan and Lummer (1998) find that allocating resources to this asset class offers excellent diversification benefits for equity and bonds in a mean–variance static asset allocation framework. Fortenbery and Hauser (1990) and Conover *et al.* (2010) also note the risk reduction benefit of including commodities in a portfolio of traditional assets without sacrificing returns. Furthermore, Satyanarayan and Varangis (1996), Abanomey and Mathur (1999) and Garreth and Taylor (2001) find enhancements in portfolio returns for a given level of risk (an upward shift in the efficient frontier) if an allocation of commodities is included in an international stock portfolio. Similarly, Greer (1994) finds that the effect of equal weights invested in S&P 500 and commodities produces superior portfolio returns with a lower standard deviation relative to investing only in the S&P 500. Anson (1999), Jensen *et al.* (2000, 2002),

---

* Utilizing wavelet coherency methodology is not novel, however. The methodology has been applied to analyse financial time-series (see, e.g. Rua and Nunes (2009), Nikkinen *et al.* (2011), Graham and Nikkinen 2011) and Graham *et al.* (2012)).

Ibbotson Associates (2006), Gorton and Rouwenhorst (2006b), Idzortek (2007) and Nguyen and Sercu (2010) all note the diversification advantages, in terms of portfolio performance, of commodity indexes. Additionally Gorton *et al.* (2005) and Belousova and Dorfleitner (2012) generalize the U.S-based conclusion to Japanese and European investors.

Erb and Harvey (2006), exploring the tactical and strategic opportunities that the inclusion of commodities in traditional portfolios present, however, cast doubt on the role of commodities in long-term asset allocation. Furthermore, Cheung and Miu (2010) show that the diversification benefit of including commodities in a portfolio of traditional assets is not convincing. Similarly, Daskalaki and Skiadopoulos (2011) find that optimal portfolios that include only the traditional assets show superior performance, thus challenging earlier findings advocating the benefits of commodities in investors' portfolios. Given the mixed results, an examination of equity and commodity market comovement utilizing an integrated framework offers a fresh understanding of the nature of the dependency between the two asset classes and its implications for asset allocation.

19.3 Measuring Stock and Commodities Market Dependencies Using Wavelet Squared Coherency

Wavelet coherency analysis is used to examine dependencies between the stock market and the commodity market returns. We briefly review the concept in this section.

Wavelet analysis is a mathematical tool for extracting scale/frequency information from data (see, e.g. Daubechies (1992) or Mallat (1998) for a thorough exposition). In contrast to the other popular frequency analysis method, e.g. Fourier analysis, wavelets preserve the time-localized information of the data and are therefore ideal for analysing non-stationary time-series. Wavelet transforms can be categorized into the discrete wavelet transform and its continuous version. The output from the continuous transform is more redundant than its discrete counterpart and thus more readily applicable for feature extraction purposes. Our presentation closely follows Torrence and Compo (1998) and Grinsted *et al.* (2004).

Inputs to wavelet analysis are a time-series x_n, $n = 0, \ldots, N$, with equal time steps δt, and a wavelet function $\psi_0(\eta)$, where η is a dimensionless time parameter. The wavelet function must fulfil the conditions of having a zero mean and being localized both in frequency and time. We utilize the popular wavelet function for feature extraction purposes, the Morlet wavelet, as the basis function for wavelet transform. This function enables the identification and isolation of signals, as it affords a balance between localization of time and frequency. In its simple form, the Morlet wavelet is represented as

$$\psi_n(\eta) = \pi^{1/4} e^{i\omega_0 \eta} e^{-0.5\eta^2}, \tag{19.1}$$

with the dimensionless frequency parameter $\omega_0 = 6$ providing a good balance between localization of time and frequency. The wavelet transform has the ability to estimate the spectral characteristics of signals as a function of time and can provide both the time-varying power spectrum and the phase spectrum needed for the calculation of coherence. The continuous wavelet transform, $W_n(s)$, of a discrete sequence x_n at time n and scale s with the wavelet $\psi_0(\eta)$ of choice is defined as the convolution

$$W_n(s) = \frac{1}{\sqrt{s}} \sum_{n'=1}^{N} x_{n'} \psi_0^* \left[\frac{(n'-n)\delta t}{s} \right], \tag{19.2}$$

where * denotes the complex conjugate. The resulting (complex) wavelet coefficients can then be used to describe certain features of the time-series. By taking the absolute value of the coefficients, the results can be interpreted as the amplitude or *power* of the series: $|W_n(s)|$. The wavelet power spectrum or variance can be obtained by squaring the power: $|W_n(s)|^2$. Since we are interested in the comovements of two time-series (say x_n and y_n), we calculate the cross-wavelet transform: $W_n^{XY}(s) = W_n^X(s)W_n^Y(s)^*$, where W_n^X and W_n^Y are wavelet transforms of x_n and y_n, respectively. The cross-wavelet power $\left|W_n^{XY}(s)\right|$ can be interpreted as depicting the local covariance between the two time-series at each scale and frequency. Thus it indicates areas in the scale–time space where the two time-series share high common power.

Finally, to obtain a practical measure of the dependencies of two time-series, we calculate the wavelet squared coherency. This is done by squaring the cross-wavelet power and dividing the result by the product of the two wavelet spectra:

$$R_n^2(s) = \frac{\left|S\left(s^{-1}W_n^{XY}(s)\right)\right|^2}{S\left(s^{-1}\left|W_n^X(s)\right|^2\right) \cdot S\left(s^{-1}\left|W_n^Y(s)\right|^2\right)} \tag{19.3}$$

where S is a smoothing operator. Since the resulting squared coherency is normalized to take values between 0 and 1, it is reminiscent of the classic correlation coefficient.

To assess the statistical significance level of the wavelet coherence, we create a background spectrum by simulating a large number of white noise time-series pairs. The mean background power spectrum is assumed to represent the mean power spectrum of the actual time-series. High power in the actual spectrum compared with the background spectrum can then be assumed to be *"a true feature with a certain percent confidence"* (Torrence and Compo 1998). We use the 95% confidence level as in the original paper.

19.4 Data

19.4.1 Commodity and Stock Indexes

The commodity returns series utilized in this study is the Standard and Poor Goldman Sachs Commodity Index (S&P GSCI®) of commodity sector returns.* S&P GSCI® is a tradable index and a benchmark for investment in the commodity market. It measures the performance in the physical commodities market. The construction of the S&P GSCI® is on a world production-weighted basis, where the quantity of each commodity in the index is determined by the average quantity of production in the last five years of available data and encompasses the principal physical non-financial commodities that are the subject of active, liquid futures markets.† The liquidity constraint is aimed at promoting cost-effective implementation and true investability.

The S&P GSCI® index is largely diversified across a broad range of commodity sectors, which enables a high level of diversification, both across sub-sectors and within each sub-sector. The diversification also reduces the impact of idiosyncratic events, which significantly impact individual commodity markets, but are muffled when aggregated to the level of S&P GSCI®. The index is seen as a good indicator of general price movements and inflation in the world economy by market participants and researchers. The S&P GSCI total return reflects the performance of total return investment in commodities and

---

* We acknowledge the existence of alternatives, e.g. public futures funds, commodity trading advisors, and private commodity pool operators, through which investors can invest in commodities (Edwards and Liew 1999). However, using the S&P GSCI® is advantageous because it allows for a clean examination without the contamination of superior/inferior trading techniques of other active public futures funds. Our choice of index also avoids biases associated with databases that report futures returns. Akey (2005) provides an excellent review of available passive commodity indexes.

† See the following link for further details on the components, weights, and construction of the S&P GSCI® index: http://www.standardandpoors.com/servlet/BlobServer?blobheadername3=MDT-Type&blobcol=urldata&blobtable=Mungo Blobs&blobheadervalue2=inline%3B+filename%3DSP_GSCI_FAQ_Web.pdf&blobheadername2=Content-Disposition &blobheadervalue1=application%2Fpdf&blobkey=id&blobheadername1=content-type&blobwhere=1243748755917&bl obheadervalue3=UTF-8.

is computed as the product of three variables: (i) the value of S&P GSCI on the preceding S&P GSCI Business Day, (ii) one plus the sum of the Contract Daily Return and the Treasury Bill Return on the S&P GSCI Business Day on which the calculation is made, and (iii) one plus the Treasury Bill Return for each non-S&P GSCI Business Day since the preceding S&P GSCI Business Day.[*]

As indicated above, the computation of the index recognizes that idiosyncratic events may significantly impact on individual commodity sectors. Jensen *et al.* (2000) also show that commodity prices are sensitive to economic conditions and are likely to vary across commodities. Consequently, we do not rely exclusively on the S&P GSCI® composite index and investigate the performance of 10 sub-indexes as well: S&P GSCI® Energy, S&P GSCI® Light-Energy, S&P GSCI® Non-Energy, S&P GSCI® Reduced Energy, S&P GSCI® Agriculture, S&P GSCI® Livestock, S&P GSCI® Petroleum, S&P GSCI® Industrial Metal, S&P GSCI® Precious Metals and S&P GSCI® Softs. The sub-indexes of the S&P GSCI® are constructed using the same rules regarding world production weights, the methodology for rolling and other functional characteristics as disclosed in the S&P GSCI® manual.

The stock market returns data used in this paper are drawn from the S&P 500 index which includes the 500 leading companies in leading industries of the U.S. economy. The index only includes large companies with market capitalization in excess of US$3.5 billion and covers approximately 75% of U.S. equities. Given that it includes a significant segment of the total value of the equity market, investors, academics and market participants alike consider the index to be an ideal proxy for the total market return. The components of the S&P 500 index are classified according to the Global Industry Classification Standard (GICS). All data used in this study is sourced from Datastream.

19.4.2 Data Description

Table 19.1 presents descriptive statistics for the S&P 500 index and each S&P GSCI® index used in this paper. Between January 1999 and December 2009, the S&P 500 posted a maximum (minimum) return of 0.102 (–0.164). The commodities total return index, S&P GSCI®, shows a maximum (minimum) return of 0.123 (–0.131) for the same period. The mean return of 0.00141 produced by the equity index for the sample period is superior to that obtained by the S&P GSCI®, which achieved a mean return of 0.00084. As Table 19.1 shows, two S&P GSCI® sub-indexes, energy and petroleum, outperformed the S&P 500 index. The sub-indexes for non-energy, agriculture, livestock and softs posted negative mean returns for the same period. The return distribution for all market indexes seems to be non-normal and shows positive skewness. The S&P GSCI® appears to be more volatile than the S&P 500.

Table 19.2 presents the pairwise correlation of returns for all pairs of market indexes used in the study. The S&P 500 is positively correlated to the S&P GSCI®, with a correlation coefficient of 0.165, as well as to all the sub-indexes for the commodity index. All the correlations between the S&P 500 and the S&P GSCI commodity indexes (the main and sub-indexes) are below 0.3, indicating low comovements of asset returns. Table 19.2 also sheds light on the variable correlation of returns between that of the main commodity index, S&P GSCI® and its sub-indexes. Some sub-indexes, e.g. energy, petroleum, reduced energy and light energy, exhibit high correlations (above 0.9) with the main index. Others have correlation coefficients below 0.5 (industrial metals, precious metals, agriculture, livestock, softs) with the main

---

[*] To reflect the performance of a total return investment in commodities, four separate but related indexes have been developed based on the S&P GSCI: (1) the S&P GSCI Spot Index, which is based on the price levels of the contracts included in the S&P GSCI; (2) the S&P GSCI Excess Return Index (S&P GSCI ER), which incorporates the returns of the S&P GSCI Spot Index as well as the discount or premium obtained by 'rolling' hypothetical positions in such contracts forward as they approach delivery; (3) the S&P GSCI Total Return Index (S&P GSCI TR), which incorporates the returns of the S&P GSCI ER and interest earned on hypothetical fully collateralized contract positions on the commodities included in the S&P GSCI; and (4) the S&P GSCI Futures Price Index (S&P GSFPI), which is intended to serve as a benchmark for the fair value of the futures contracts on the S&P GSCI traded on the Chicago Mercantile Exchange (CME).

TABLE 19.1 Descriptive Statistics for the S&P 500 Stock Index and the S&P GSCI Commodity Index Returns from January 1999 through December 2009 for a Total of 767 Weekly Observations

Market	Mean	Median	Std Dev	Skewness	Kurtosis	Min	Max
S&P 500	0.00141	0.00326	0.0246	4.327	−0.582	−0.164	0.102
Commodity	0.00084	0.00137	0.0313	1.698	−0.433	−0.131	0.123
Energy	0.00145	0.00364	0.0439	1.076	−0.339	−0.186	0.173
Petroleum	0.00213	0.00447	0.0450	1.751	−0.402	−0.213	0.207
Non-energy	−0.00004	0.00064	0.0183	3.414	−0.497	−0.094	0.073
Reduced energy	0.00059	0.00127	0.0263	2.344	−0.541	−0.119	0.102
Light energy	0.00036	0.00135	0.0221	2.969	−0.614	−0.106	0.092
Industrial metal	0.00089	0.00094	0.0287	3.662	−0.047	−0.133	0.147
Precious metals	0.00141	0.00077	0.0239	3.926	−0.031	−0.127	0.119
Agriculture	−0.00072	−0.00006	0.0264	2.056	−0.127	−0.131	0.115
Livestock	−0.00070	−0.00053	0.0195	2.241	−0.610	−0.119	0.057
Softs	−0.00029	0.00012	0.0277	1.370	−0.060	−0.124	0.112

index. S&P GSCI® livestock is shown to have the lowest correlation with the main index. From Table 19.2 we can also note that S&P GSCI® livestock exhibits the lowest correlations for all pairs of sub-indexes.

19.5 Empirical Analysis of Comovements

In this section we present evidence on the wavelet squared coherency for the S&P 500 and the S&P GSCI® commodity index as well as that between S&P 500 and 10 sub-indexes of the S&P GSCI® commodity index. As indicated above, the wavelet squared coherency analysis involves two time-series given as input in a time–frequency space. Consequently, the wavelet squared coherency is presented using contour plots as it involves three dimensions, frequency, time and wavelet squared coherency power (similar to Rua and Nunes (2009)).

Frequency and time are represented on the vertical and horizontal axes, respectively. Frequency is converted into time units (years) to simplify interpretation and ranges from the highest frequency of one week (top of the plot) to the lowest frequency of four years (bottom of the plot). The wavelet squared coherency is depicted by a grey scale in the figures. The scale for the wavelet coherency is interpreted in terms of the darkness of the grey colour, where an increasing value of the wavelet squared coherency corresponds to a deepening darkness of grey and imitates the height in a surface plot. Subsequently, through a visual assessment of the graphs, we can identify both the time intervals (on the horizontal axis) and frequency bands (on the vertical axis) where the two markets (commodity and stock) move together.

The frequency scale allows us to differentiate the effects of comovements between the short term and the long term. We consider short term to relate to frequencies ranging from two weeks to six months. Long term can be thought of as investing in frequencies ranging from two to three years. In this unified framework, a dark grey area at the bottom (top) of the figures would indicate strong comovements at low (high) frequencies, whereas a dark grey area on the left-hand (right-hand) side signifies strong comovements at the start (end) of the sample period.

The thick black continuous line in the figures sequesters regions where the wavelet squared coherency is statistically significant at the 5% level, i.e. the wavelet squared coherency is statistically significant within such a delimited time–frequency area. The significance level was simulated as explained in Section 19.3 using a Monte Carlo method of 10,000 sets of two white noise time-series with the same length as the series under analysis.

Figure 19.1 depicts the standard wavelet squared coherency between the S&P 500 and the S&P GSCI® commodity index and reveals several interesting features. First, we find no evidence of comovements

TABLE 19.2 Pearson's Correlation Coefficients for the S&P 500 Index and the S&P GSCI Commodity Index Returns. *, **, *** Denote Statistical Significance at the 5%, 1% and 0.1% Level, Respectively

	S&P 500	Commodity	Energy	Petroleum	Non-energy	Reduced energy	Light energy	Industrial metals	Precious metals	Agriculture	Livestock
Commodity	0.165***										
Energy	0.133***	0.973***									
Petroleum	0.137***	0.933***	0.955***								
Non-energy	0.232***	0.529***	0.351***	0.345***							
Reduced energy	0.184***	0.981***	0.922***	0.885***	0.645***						
Light energy	0.209***	0.927***	0.828***	0.797***	0.788***	0.978***					
Industrial metals	0.264***	0.410***	0.304***	0.308***	0.632***	0.470***	0.547***				
Precious metals	0.057	0.309***	0.237***	0.244***	0.448***	0.349***	0.400***	0.348***			
Agriculture	0.136***	0.409***	0.255***	0.245***	0.861***	0.517***	0.648***	0.258***	0.256***		
Livestock	0.114***	0.193***	0.124***	0.119***	0.332***	0.233***	0.278***	0.094***	0.046	0.090**	
Softs	0.141***	0.328***	0.232***	0.238***	0.549***	0.389***	0.462***	0.261***	0.241***	0.588***	0.015

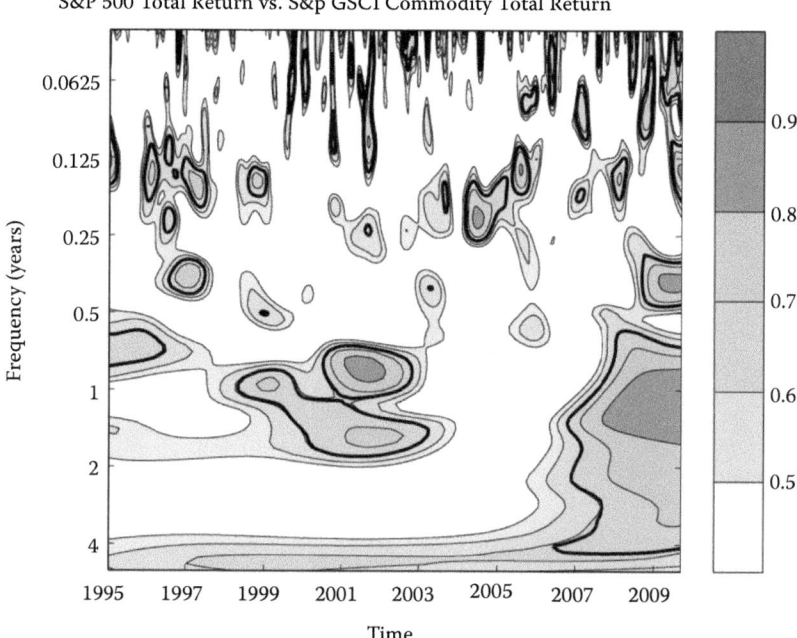

S&P 500 Total Return vs. S&p GSCI Commodity Total Return

FIGURE 19.1   Comovement of the S&P 500 stock index and the S&P GSCI commodity index. This figure presents the wavelet squared coherency between the S&P 500 index returns and the S&P GSCI commodity index returns. Time and frequency are represented on the horizontal and vertical axes, respectively. Frequency is converted into years. The wavelet squared coherency is represented by the grey scale where increasing darkness of the grey scale corresponds to an increasing value and mimics the height in a surface plot. The heavy black line isolates the statistically significant area at the 5% significance level.

of the S&P 500 total return and the S&P GSCI® commodity index total return in the higher frequencies, indicating diversification benefits in the short term. The onset of the financial crisis post-2007 does not alter this short-term diversification benefit. Secondly, the comovement of returns between the two indexes at low frequencies before 2007 is very weak. The conclusion that may be drawn here is that long-term investors derived diversification benefits by combining the two assets in a portfolio during this time period. There is, however, some notable evidence of increasing comovements at low, rising to medium, frequencies towards the end of the series (after 2007). This perceptible rise in market dependencies between the two return index series coincides with the onset of the most recent financial crisis. As has been noted in the literature, in periods of deleveraging, correlations between uncorrelated assets can increase to +1. This appears to be the case with the long-term fluctuations of the two financial assets return series. Thus the evidence implies that including commodities in an equity portfolio may not provide significant diversification benefits for long-term investors during a period of financial crisis.

Given the possible variation in sensitivity of commodity prices to economic conditions, we further extend our empirical investigations by examining the comovement between the S&P 500 total return and the sub-indexes of the S&P GSCI® commodity total returns. Figure 19.2 reports the results of the examination of the comovements between the S&P 500 total return and three sub-indexes of the S&P GSCI® commodity index total return: S&P GSCI® Energy Total Return, S&P GSCI® Light Energy Total Return and S&P GSCI® Red Energy Total Return. The results mirror the finding for the comovements between the S&P 500 total return and the S&P GSCI® commodity index total return discussed above. Once again, a clear distinction can be made between the short-term and long-term diversification benefits of combining the two assets. We also find weak dependencies between the S&P 500 total return and

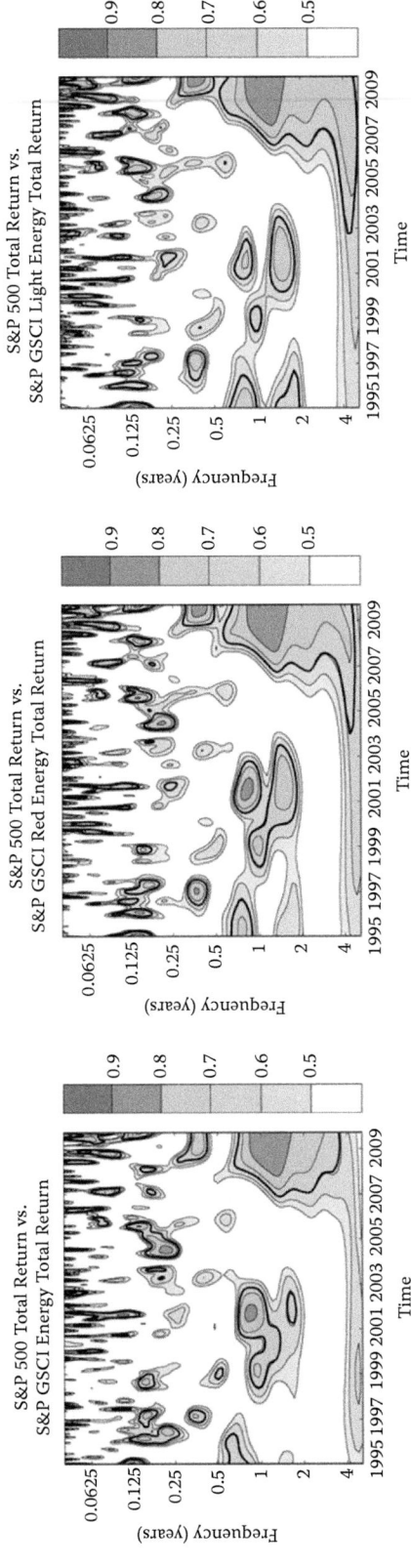

FIGURE 19.2 Comovement of the S&P 500 index and the S&P GSCI Energy sub-indexes. This figure presents the wavelet squared coherency between the S&P 500 index returns and three S&P GSCI Energy indexes: the S&P GSCI Energy Total returns, the S&P GSCI® Red Energy Total Return and the S&P GSCI® Light Energy Total Return. Time and frequency are represented on the horizontal and vertical axes, respectively. Frequency is converted into years. The wavelet squared coherency is represented by the grey scale where increasing darkness of the grey scale corresponds to an increasing value and mimics the height in a surface plot. The heavy black line isolates the statistically significant area at the 5% significance level.

the S&P GSCI® Energy total return, the S&P 500 total return and the S&P 500 GSCI® Red Energy total return and the S&P 500 total return and the S&P 500 GSCI® Light Energy total return at higher frequencies for the whole data period. Analogous to the results discussed above, there is also evidence of rising comovement between the S&P 500 and the S&P GSCI® sub-indexes after 2007. Thus, the short-term diversification benefits across time of combining equity and commodities extend to these sub-indexes of S&P 500 GSCI®.

Figure 19.3 shows the result of examining the comovements between the S&P 500 total return and the S&P GSCI® non-Energy total return. We find no consistent evidence of comovement between the two indexes in the short term. Parallel to the documented relationship between the S&P 500 and S&P GSCI® commodity index above, we find evidence of increasing levels of long-term comovements from low, rising to intermediate, frequencies towards the end of the series.

The comovement of returns between the S&P 500 total return and the S&P GSCI® industrial metals total return and that of the S&P 500 and S&P GSCI® precious metals total return reveals some important details (Figure 19.4). The short-term and long-term evidence regarding the comovement between the S&P GSCI® precious metals total return is very weak at best. Thus this segment of the S&P GSCI® index provides significant diversification benefits for both long-term and short-term investors. The evidence only suggests strong short-term diversification benefits when S&P 500 and S&P GSCI® industrial metals are considered.

Examining the comovement of returns between the S&P 500 total return and the S&P GSCI® agriculture sub-indexes, we find weak comovement between the S&P 500 total and the S&P GSCI® Agriculture

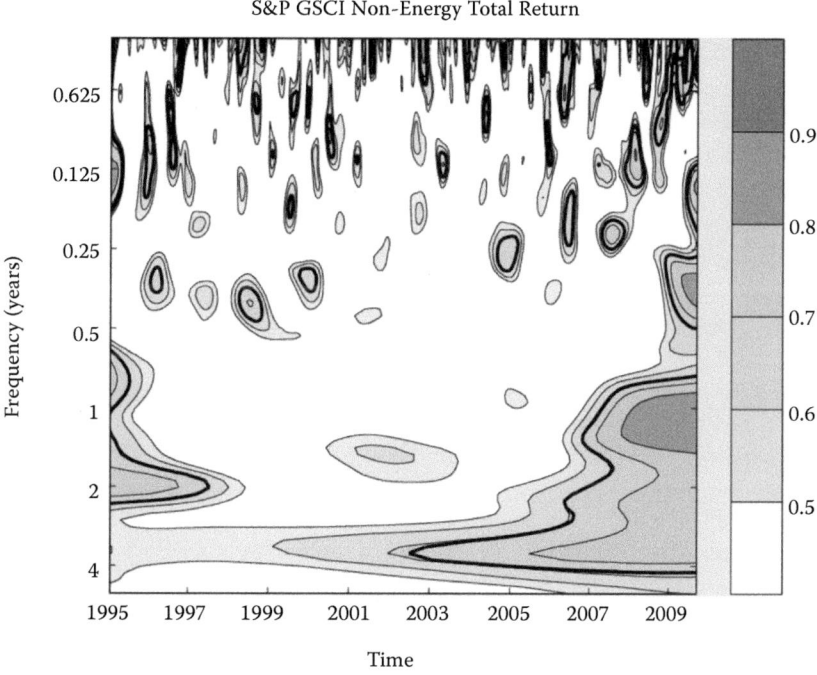

FIGURE 19.3 Comovement of the S&P 500 index and the S&P GSCI non-Energy sub-index. This figure presents the wavelet squared coherency between the S&P 500 index returns and the S&P GSCI non-Energy index. Time and frequency are represented on the horizontal and vertical axes, respectively. Frequency is converted into years. The wavelet squared coherency is represented by the grey scale where increasing darkness of the grey scale corresponds to an increasing value and mimics the height in a surface plot. The heavy black line isolates the statistically significant area at the 5% significance level.

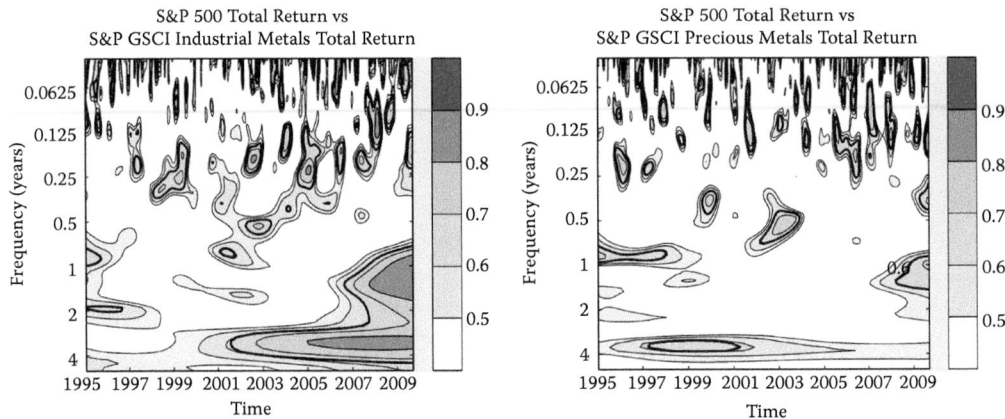

FIGURE 19.4 Comovement of the S&P 500 index and the S&P GSCI Metals sub-indexes. This figure presents the wavelet squared coherency between the S&P 500 index returns and the S&P GSCI Industrial Metals index and the S&P 500 index returns and the S&P GSCI Precious Metals index. Time and frequency are represented on the horizontal and vertical axes, respectively. Frequency is converted into years. The wavelet squared coherency is represented by the grey scale where increasing darkness of the grey scale corresponds to an increasing value and mimics the height in a surface plot. The heavy black line isolates the statistically significant area at the 5% significance level.

total return, both in the short and long term (see Figure 19.5). The evidence, however, indicates relatively stronger comovement between the two indexes towards the end of the series. This result is replicated for the comovement of returns between the S&P 500 total return and the S&P GSCI® Livestock total return. We also find that combining an equity portfolio and S&P GSCI® Softs provides important diversification benefits for both the short-term and long-term investors (see Figure 19.5). The onset of the financial crisis does not seem to impact on these benefits.

19.6 Conclusion

The diversification properties of commodities have long been understood by market participants. The recent financial crisis has, however, raised apprehension that the diversification line of reasoning in favour of commodities investment is not robust. This brings to the forefront the question of the independence of commodity index returns. This study therefore investigates the comovements of returns of the S&P 500 and the S&P GSCI Commodity Index using an enhanced methodology, the three-dimensional analysis of wavelet coherency, during the period January 1999 to December 2009. The sample periods enable us to compare the pre-financial crisis era comovements with that of the onset of the crisis. The wavelet methodology represents a refinement and allows the examination of the time-and-frequency varying comovements of stock markets within a unified framework and enables a distinction between short-term and long-term benefits of diversification.

Generally, we find weak evidence of comovement between the S&P 500 total return and the S&P GSCI® Commodity Index total return for our sample period. This implies that commodities still have diversification value when combined with equities during the recent financial crisis. The argument for diversification benefits is even stronger when we segment our analyses and examine the gains from the point of view of short-term and long-term investors. From the point view of short-term investors, the onset of the current financial crisis has not dampened the diversification value of commodities. The pre-financial crisis period and the onset of the financial crisis at the end of 2007 show a similar lack of correlation at high frequencies. From the point of view of the long-term investor, however, we note a rising comovement of returns for the two indexes after 2007, which coincides with the onset of the

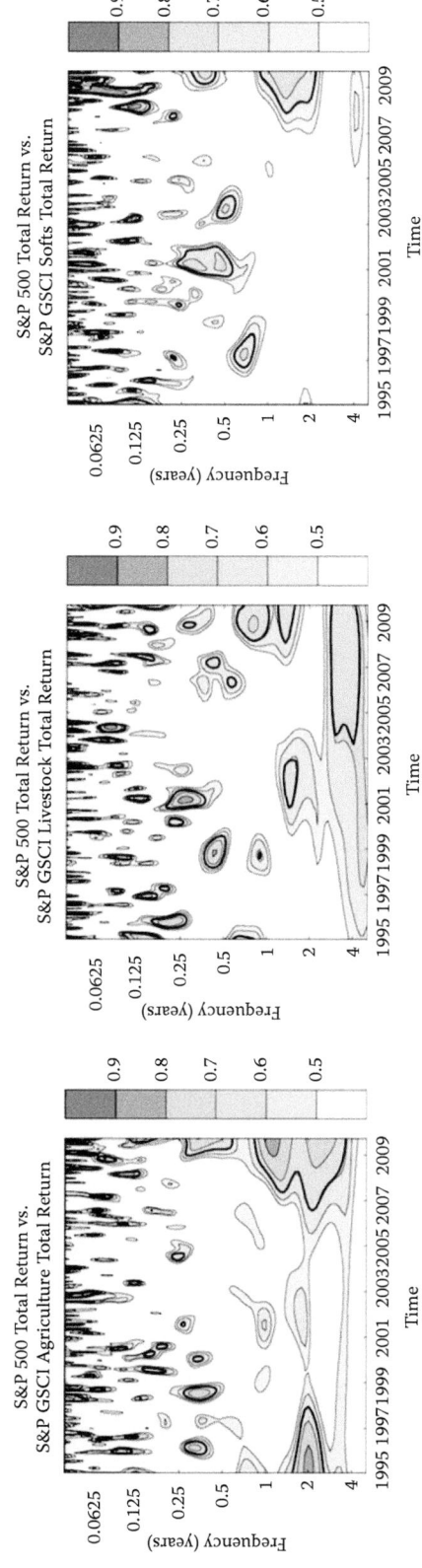

FIGURE 19.5 Comovement of the S&P 500 index and three S&P GSCI sub-indexes. This figure presents the wavelet squared coherency between the S&P 500 index returns and three S&P GSCI sub-indexes: the S&P GSCI Agricultural Total Return, the S&P GSCI Livestock Total Return and the S&P GSCI Softs Total Return. Time and frequency are represented on the horizontal and vertical axes, respectively. Frequency is converted into years. The wavelet squared coherency is represented by the grey scale where increasing darkness of the grey scale corresponds to an increasing value and mimics the height in a surface plot. The heavy black line isolates the statistically significant area at the 5% significance level.

financial crisis. It should be noted, though, that the comovement is not strong. Thus, a case for limited long-term diversification gains could be made during the financial crisis.

Earlier studies alluded to the possible variation in the sensitivity of commodity prices to economic conditions across commodities. As a consequence, we further investigate the comovement of the S&P 500 and 10 sub-indexes of the S&P GSCI index. This empirical exercise generally confirms our earlier findings. For the short-term investor, the current financial crisis shows no measurable impacts on the diversification benefits of combining equity and commodities. From the point of view of the long-term investor, typified by fluctuations at low frequencies, the evidence suggests an increase in comovement at the onset of the financial crisis. The evidence, however, suggests weak correlations. Thus there may be limited benefits even in time of crisis. There are some important distinctions that can be made for combining equity and the various segmented commodity indexes. For instance, the diversification benefits of combining equity (S&P 500 total return) and the S&P GSCI® Precious Metal total return is robust for both long-term and short-term investors across time.

Future research may construct different portfolios (with and without commodities) based on wavelet analysis (high-versus low-dependence portfolios) and compare their realized risk-adjusted performance.

Acknowledgements

We would like to thank the anonymous reviewers and the editor of the journal for their insightful comments. The financial support of the Academy of Finland (project 132913) is also gratefully acknowledged.

References

Abanomey, W.S. and Mathur, I., The hedging benefits of commodity futures in international portfolio construction. *J. Altern. Invest.*, 1999, **2**, 51–62.

Adel, B. and Salma, J., The Greek financial crisis, extreme comovement, and contagion effects in the EMU: a copula approach. *Int. J. Account. Financ. Rep.*, 2012, **2**, 289–307.

Akey, R.P., Commodity: a case for active investment. *J. Altern. Invest.*, 2005, **8**, 8–29.

Aloui, R., Ben Aïssa, M.S. and Nguyen, D.K., Global financial crisis, extreme interdependences, and contagion effects: the role of economic structure? *J. Bank.* Finance, 2011, **35**, 130–141.

Ang, A. and Chen, J., Asymmetric correlations of equity portfolios. *J. Financ. Econ.*, 2002, **63**, 443–494.

Ankrim, E.M. and Hensel, C.R., Commodities in asset allocation: a real asset alternative to real estate? *Financ. Anal. J.*, 1993, **49**, 20–29.

Anson, M., Maximizing utility with commodity futures diversification. *J. Portfol. Mgmt*, 1999, **25**, 86–94.

Bartram, S.M. and Bodnar, G.M., No place to hide: the global crisis in equity markets in 2008/2009. *J. Int. Money Finance*, 2009, **28**, 1246–1292.

Belousova, J. and Dorfleitner, G., On the diversification benefits of commodities from the perspective of euro investors. *J. Banking & Finan.*, 2012, **36**, 2455–2472.

Billingsley, R.S. and Chance, D.M., Benefits and limitations of diversification among commodity trading advisors. *J. Portfol. Mgmt*, 1996, **23**, 65–80.

Candelon, B., Piplack, J. and Straetmans, S., On measuring synchronization of bulls and bears: the case of East Asia. *J. Bank. Finance*, 2008, **32**, 1022–1035.

Chance, D.M., Managed futures and their role in investment portfolios. The Research Foundation of the Institute of Chartered Financial Analysts. Charlottesville, VA, 1994.

Cheung, C.S. and Miu, P., Diversification benefits of commodity futures. *J. Int. Financ. Mkts, Inst. & Money*, 2010, **20**, 451–474.

Claessens, S., Kose, M.A. and Terrones, M.E., The global financial crisis: how similar? How different? How costly? *J. Asian Econ.*, 2010, **21**, 247–264.

Conover, C.M., Jensen, G.R., Johnson, R.R. and Mercer, J.M., Is now the time to add commodities to your portfolio? *J. Invest.*, 2010, **19**, 10–19.

Daskalaki, C. and Skiadopoulos, G.S., Should investors include commodities in their portfolios after all? New evidence. *J. Banking & Finan.*, 2011, **35**, 2606–2626.

Daubechies, I., Ten lectures on wavelets (CBMS-NSF Regional Conference Series in Applied Mathematics). Society for Industrial & Applied Mathematics, 1992.

Dwyer, G.P. and Tkac, P., The financial crisis of 2008 in fixed income markets. *J. Int. Money Finance*, 2009, **28**, 1293–1316.

Edwards, F.R. and Liew, J., Managed commodity funds. *J. Futures Markets*, 1999, **19**, 377–411.

Erb, C.B. and Harvey, C.R., The tactical and strategic value of commodity futures. *Financ. Anal. J.*, 2006, **62**, 69–97.

Fortenbery, T.R. and Hauser, R.J., Investment potential of agricultural futures contracts. *Am. J. Agric. Econ.*, 1990, **72**, 721–726.

Garcia, R. and Tsafack, G., Dependence structure and extreme comovement in international equity and bond markets. *J. Bank. Finance*, 2009, **8**, 1954–1970.

Garrett, I. and Taylor, N., Intraday and interday basis dynamics: evidence from the FTSE100 Index Futures Market. *Stud. Nonlin. Dynam. Econometr.*, 2001, **5**, 133–152.

Geman, H., *Commodities and Commodity Derivatives: Modeling and Pricing for Agriculturals, Metals and Energy*, 2005 (Wiley: Chichester).

Gorton, G., Hayashi, F. and Rouwenhorst, K.G., Commodity futures: a Japanese perspective. Yale ICF Working Paper No. 05–27, 2005.

Gorton, G. and Rouwenhorst, K.G., Facts and fantasies about commodity futures. *Financ. Anal. J.*, 2006a, **62**, 47–68.

Gorton, G. and Rouwenhorst, K.G., Are commodity futures too risky for your portfolio? Hogwash! Knowledge Wharton, 2006b.

Graham, M. and Nikkinen, J., Co-movement of the Finnish and international stock markets: a wavelet analysis. *Eur. J. Finance*, 2011, **17**, 409–425.

Graham, M., Kiviaho, J. and Nikkinen, J., Integration of 22 emerging stock markets: a three dimensional analysis. *Glob. Finance J.*, 2012, **23**, 34–47.

Greer, R.J., Methods for institutional investment in commodity futures. *J. Deriv.*, 1994, **2**, 28–36.

Grinsted, A., Moore, J.C. and Jevrejeva, S., Application of the cross wavelet transform and wavelet coherence to geophysical time series. *Nonlin. Process. Geophys.*, 2004, **11**(5/6), 561–566.

Ibbotson Associates, *Strategic Asset Allocation and Commodities*, 2006 (Ibbotson Associates: Chicago, IL).

Idzorek, T.M., Commodities and strategic asset allocation. In *Intelligent Commodity* Investing, edited by H. Till and J. Eagleeye, 2007 (Risk Books: London).

Jensen, G., Johnson, R.R. and Mercer, J.M., Efficient use of commodity futures in diversified portfolios. *J. Fut. Mkts*, 2000, **5**, 489–506.

Jensen, G., Johnson, R.R. and Mercer, J.M., Tactical asset allocation and commodity futures. *J. Portfol. Mgmt*, 2002, Summer, 100–110.

Kaplan, P.D. and Lummer, S.L., Update: GCSI collateralized futures as a hedging and diversification tool for institutional portfolios. *J. Invest.*, 1998, Winter, 11–17.

Kenourgios, D. and Padhi, P., Emerging markets and financial crises: regional, global or isolated shocks? *J. Multinatl Financ. Mgmt*, 2012, **22**, 24–38.

Kyle, A.S. and Xiong, W., Contagion as a wealth effect. *J. Finance*, 2001, **56**, 1401–1440.

Longin, F. and Solnik, B., Extreme correlation of international equity markets. *J. Finance*, 2001, **56**, 649–676.

Lummer, S.L. and Siegel, L.B., GSCI collateralized futures: a hedging and diversification tool for institutional portfolios. *J. Invest.*, 1993, **2**, 75–82.

Mallat, S., *A Wavelet Tour of Signal Processing*, 1998 (Academic Press: San Diego, CA).

Markowitz, H.M., Portfolio selection. *J. Finance*, 1952, **7**, 77–91.

Merath, J., Commodities are part of an optimized portfolio. Global Trends, Credit Suisse, 2006.

Nguyen, V.T.T. and Sercu, P.M.F.A., Tactical asset allocation with commodity futures: implications of business cycle and monetary policy (May 26, 2010). In *Paris December 2010 Finance Meeting EUROFIDAI – AFFI*, 2010. Available online at SSRN: http://ssrn.com/abstract=1695889.

Nikkinen, J., Pynnönen, S., Ranta, M. and Vähämaa, S., Crossdynamics of exchange rate expectations: a wavelet analysis. *Int. J. Finance Econ.*, 2011, **16**, 205–217.

Patten, A., On the out of sample importance of skewness and asymmetric dependence for asset allocation. *J. Financ. Econometr.*, 2004, **2**, 130–168.

Rua, A. and Nunes, L.C., International comovement of stock returns: a wavelet analysis. *J. Empir. Finance*, 2009, **11**, 632–639.

Satyanarayan, S. and Varangis, P., Diversification benefits of commodity assets in global portfolios. *J. Invest.*, 1996, **5**, 69–78.

Silvennoinen, A. and Thorp, S., Financialization, crisis and commodity correlation dynamics. Working Paper, Queensland University of Technology, 2010.

Tang, K. and Xiong, W., Index investing and the financialization of commodities. Working Paper, Princeton University, 2010.

Torrence, C. and Compo, G.P., A practical guide to wavelet analysis. *Bull. Am. Meteorol. Soc.*, 1998, **79**, 61–78.

Yamori, N., Co-movement between commodity market and equity market: does commodity market change? *Mod. Econ.*, 2011, **2**, 335–339.

20

Commodity Markets through the Business Cycle

Julien Chevallier,
Mathieu Gatumel
and Florian Ielpo

20.1 Introduction

What is the relationship between the global macroeconomic environment and commodity markets (e.g. agricultural, metal and energy products)? From January 1987 to December 2010, the correlation between commodities and more traditional assets has been historically low: for example, the correlation between the WTI and the S&P 500 is equal to 3% on average. The correlation between gold and US government bonds can be rounded to 0%. Within commodities, the correlations are rather low as well in the long run: for example, coffee and sugar are only related by a correlation coefficient of 8%. One of the largest correlations can be found between the WTI and gold—22%—which is much lower than the typical correlation found between the S&P 500 and the Russell 3000 indices (equal to 85%).

In contrast with this view, it appears that since 2008, these correlation levels have been changing dramatically. In the long run, the correlation between the WTI and the Russell 3000 index was close to 5%. However, from 2008 to 2010, this correlation reached 28%—and the same comment arises for the S&P 500 index. Within commodities, a similar assessment can be made. Instead of the 8% coefficient found previously, the correlation between coffee and sugar is now closer to 35%.

What happened to the attractive 'uncorrelation' of commodity markets with other asset classes (Miffre and Rallis 2007, Geman and Ohana 2008, Daskalaki and Skiadopoulos 2011)?

One possible answer to this surge in correlation could be to assume some sort of stronger integration of these assets into global financial markets: the growing inclusion of commodity futures in strategies built by hedge funds, or in the balanced mandates managed by asset managers, may impact the

DOI: 10.1201/9781003265399-23

behaviour of these futures with respect to other markets. How can we measure this integration phenomenon? How integrated now are commodity markets with equity and bond markets?

This paper aims at gathering different pieces of evidence to contribute to a better understanding of these key questions. Indeed, previous literature has already documented that some cross-market link- ages do exist between commodity markets and the global macroeconomic environment (Caballero *et al.* 2008, Chng 2009, Tang and Xiong 2010, Chan *et al.* 2011, Dionne *et al.* 2011, Chevallier 2012). Besides, commodity price modelling has emerged as a vivid area of research in the financial literature (Miltersen 2003, Tang 2012).

First, if an increasing integration between commodity and more traditional asset markets has been taking place over the past 10 years, this must be measurable by the reaction of these markets to economic news. Thus, we develop a commodity by commodity analysis of the impact of news. This analysis has been performed recently by Elder *et al.* (2012) for metal futures only. Our results reveal an interesting pattern: the response of commodity prices to economic surprises is strong during global downturns, and weak during expansion periods. In addition, by slicing the information available into two-year subsets, the results show a very limited support to the 'financial integration' theory. Our results suggest that commodity markets have been over-reacting to economic news during the 2008–2009 crisis, similarly to what happened in 2001. During 2009–2010, our estimations reveal indeed a decreased sensitivity of commodity markets to the economic news flow.

Second, instead of evaluating how commodity markets are affected by economic surprises, we seek to discuss the relationship between the business cycle in various economic zones and the performance of the different commodity sectors. Namely, we investigate whether commodity markets are related to a given economic zone or to the worldwide business cycle—mainly through the global demand for raw materials and energy. To do so, we consider how commodity prices evolve through the business cycle based on the class of Markov regime-switching models. This approach has been applied successfully to commodity markets in previous literature (Alizadeh *et al.* 2008, Andersen 2010). We apply this methodology for the USA, the Eurozone and China. Our results cast some light on the strong relationship that appears between commodity markets and the underlying business cycle. More especially, we are able to detect an increased sensitivity to economic activity in China.

Moreover, we provide a more detailed analysis of the US business cycle by decomposing its phases based on the joint behaviour of production growth, inflation, unemployment and retail sales behaviour. The US cycle is broken down into five different types of regimes: strong expansion, medium expansion, stalling, slowdown and strong crisis. For each of these regimes, we characterize the performance of commodities by assessing which phase coincides with rising or falling commodity prices. Besides, we identify during which periods commodity markets are most likely to grow or fall.

The remainder of the paper is structured as follows. Section 20.2 details the incorporation of news in commodity markets. Section 20.3 relates commodity markets to the business cycles. Section 20.4 briefly concludes.

20.2 The Reaction of Commodity Markets to Economic News

The investigation of the reaction of commodity markets to economic news is based on the intuition that liquid financial markets integrate efficiently new information.

20.2.1 Measuring the Impact of Price Discovery on Asset Prices

One way to determine what is priced on a given market is to measure the impact of news on asset prices. Note that $R_{i,t}$ is the actual value for the release figure i at time t and $F_{i,t}$ is the market consensus for this figure.* The 'surprise' component in this economic news $S_{i,t}$ is given by:

---

* The series $F_{i,t}$ is taken from the Bloomberg survey data-set. It represents the median value of a survey of economists. Given that Bloomberg is one of the most widespread information services used by practitioners and market participants, it seems reasonable to use this data-set.

$$S_{i,t} = R_{i,t} - F_{i,t}. \tag{20.1}$$

Given that various news may have different scales, it is very common to scale $S_{i,t}$ by its full sample standard deviation. By scaling the surprises by $\sigma_i = \sqrt{V(S_{i,t})}$, we make the various announcements comparable and it will thus be possible to rank the main movers on commodity markets. Denote $s_{i,t}$ the scaled variable:

$$S_{i,t} = \frac{R_{i,t} - F_{i,t}}{\sigma_i}. \tag{20.2}$$

For a given financial asset, the return on investment on day t is denoted r_t. This sampling frequency allows us to focus on the impact of surprises on the fundamental value of the asset studied.* Formally, the news analysis is performed by running the following sets of regressions:

$$r_t = \phi_0 + \phi_1 r_{t-1} + \sum_{i=1}^{n} \mathbb{I}_i \beta_i \times s_{i,t} + \sigma_t \varepsilon_t \tag{20.3}$$

$$\log \sigma_t^2 = \omega + \alpha \varepsilon_{t-1} + \theta |\varepsilon_{t-1}| + \log \sigma_{t-1}^2. \tag{20.4}$$

with $\mathbb{I}_i$ a dummy variable equal to 1 whenever the news i is released at time t, and 0 otherwise n is the total number of news. The analysis focuses mainly on the sensitivity of r_t to $s_{i,t}$, that is to say β_i. Note that:

- Whenever β_i is statistically different from 0, it seems reasonable to assume that this asset's price incorporates the economic information associated with the surprise component $s_{i,t}$.
- The sign of β_i is essential to the market interpretation of the data.
- Equations (20.3) and (20.4) incorporate two types of robustness checks: an autoregressive component through $\phi_0 + \phi_1 r_{t-1}$, and a time-varying volatility component through the dynamics of the exponential GARCH (EGARCH) model.

Overall, this methodology corresponds to standard practice in order to gauge the impact of news on financial returns. We propose to apply it to the analysis of commodity markets.

20.2.2 Database of News

In this section, we present the Bloomberg data-set of 16 series of surprises stemming from three economic zones: the USA, the EMU and China. For the news associated to Europe and the USA, the data-set starts in 1999 and ends in 2011. For the news associated to China, the data-set starts in 2007 and ends in 2011.

The news cover different aspects of the business cycle: real activity with macroeconomic data (such as industrial production) or surveys (such as the ISM or the German IFO), inflation dynamics with the consumer price index (CPI), monetary variables with the Fed Target Rate and the ECB's minimum refinancing rate.†

A detailed list of these news is provided in Table 20.1, along with the correlation between the US GDP and each of these news. The interested reader can find additional background information in the Appendix.

---

* Although an approach based on a higher frequency can also be interesting to commodity investors, a price variation that would not last at least one day can hardly be regarded as a fundamental change in the value of the investigated asset.

† Note that the Chinese decision rate is not modified through a prescheduled decision meeting, as in the case of the Fed and the ECB. For this reason—and due also to the lack of consistency that it would introduce in the analysis—we have discarded the announcements by the Public Bank of China (PBOC).

TABLE 20.1 Database of Economic News from Bloomberg

	Type of data	Geographical zone	Correlation with US GDP (%)
Non-Farm Payroll	Employment	USA	77
ISM	Economic survey	USA	84
Jobless Claims	Employment	USA	−68
US CPI MoM	Inflation	USA	22
US Retail Sales	Economic activity	USA	27
Fed Target Rate	Central bank rate	USA	3
US GDP	Economic activity	USA	100
ZEW Eco. Sent.	Economic survey	Germany	20
IFO Expectations	Economic survey	Germany	75
EMU CPI	Inflation	EMU	26
EMU GDP	Economic activity	EMU	80
FR Bus. Conf.	Economic survey	France	70
ECB Ref. Rate	Central bank rate	EMU	−13
China CPI YoY	Inflation	China	45
China Ind. Prod.	Economic activity	China	73
China PMI	Economic survey	China	41

Except for the weekly Jobless Claims—that is the weekly number of new jobless applications in the USA—and the ECB decision rate, the majority of economic news is found to have a positive correlation with the US business cycle (as measured by GDP growth).

The weakest positive correlation is obtained for the German ZEW: this survey gathers the opinion of market participants regarding the current state of the German economy. Another German survey—the IFO survey—is computed in a very different way, by collecting the opinion of purchasing managers. In this respect, the IFO is closer in its spirit to the famous US ISM survey, that is why we find a strong correlation with the US GDP in the latter case.

Before turning to the analysis of commodities, we illustrate this methodology by showing the estimation results obtained when using traditional assets.

20.2.3 An Example: S&P 500, US 10-Year and USD

Before dealing with the impact of news on commodity markets, we start first with a review of the impact of news on standard assets: the S&P 500 index, the US 10-year bond rate and the trade weighted value of the US dollar. This preliminary step can be seen as a benchmark of comparison with commodities. Table 20.2 contains the estimation analysis obtained with these three assets.

Out of the seven US economic indicators, the 10-year US bond rate responds to six of them, i.e. to all figures except the GDP itself. The US dollar index also responds positively to the ISM, the Non-Farm Payrolls and the Fed decision rate. The S&P 500 responds positively to the retail sales, and negatively to the Fed action. When strictly focusing on the US economy, the equity index is clearly less sensitive to economic indicators than the 10-year bond rate and the US dollar. Besides, equity markets are also prone to react to economic news coming from other countries: the S&P 500 reacts positively to the ECB decision rate, and negatively to European news such as the ZEW survey (unlike the US dollar which reacts negatively to good news coming from the ZEW survey). Finally, the US dollar is found to react negatively to a positive surprise in the French business confidence index, and positively to an unexpected increase in the CPI.

Next, we investigate in more detail the behaviour of commodity markets.

TABLE 20.2 Impact of News on the S&P 500 Index, the US 10-Year Rate and the Trade Weighted Value of the US Dollar

	S&P 500	10Y US	US Dollar
Intercept	−7.15** (−4.68)	−6.35** (−4.21)	−1.25 (−0.82)
AR	0.02 (1.23)	−0.02 (−1.05)	0 (0.07)
Non-Farm Payroll	−0.09 (−0.99)	*0.75** (6.31)*	*0.19** (5.07)*
ISM	0.1 (1.01)	*0.55** (4.62)*	*0.13** (3.29)*
Jobless Claims	−0.03 (−0.34)	*−0.33** (−2.82)*	−0.04 (−1.04)
US CPI MoM	−0.11 (−1.19)	*0.4** (3.3)*	0.04 (0.91)
US Retail Sales	*0.29** (2.59)*	*0.7** (4.92)*	0.07 (1.42)
Fed Target Rate	*−0.25* (−1.86)*	*0.4** (2.29)*	*0.19** (3.4)*
US GDP	0.06 (0.33)	−0.21 (−0.96)	−0.02 (−0.26)
ZEW Economic Sentiment	*−0.29** (−2.46)*	−0.04 (−0.24)	*0.08* (1.76)*
IFO Expectations	−0.03 (−0.22)	0.16 (0.96)	−0.04 (−0.85)
EMU CPI Flash Estimate	−0.11 (−0.9)	0 (0.01)	0.03 (0.73)
EMU GDP	0.13 (0.71)	−0.08 (−0.32)	0.03 (0.41)
France Business Confidence	0.05 (0.45)	0.21 (1.47)	*−0.09* (−1.84)*
ECB Refinancing Rate	*0.37** (3.32)*	−0.02 (−0.16)	0.03 (0.63)
China CPI YoY	0.02 (0.14)	−0.05 (−0.33)	*0.08* (1.7)*
China Industrial Production	−0.01 (− 0.11)	0.03 (0.19)	−0.02 (−0.51)
China PMI	0.08 (0.51)	0.05 (0.27)	0.02 (0.4)

Notes: T-stats are between brackets. Italic figures indicate statistically significant figures at a 10% risk level.

20.2.4 Commodity by Commodity Analysis

In this section, we report the results obtained when estimating Equations (20.3) and (20.4) commodity by commodity.

20.2.4.1 Database for Commodity Prices

Table 20.3 gives the characteristics of each commodity futures price series included in the news analysis. Closest-to-maturity futures employed, in the sense that rollover returns are taken into account from the expiring contract to the one with the second-to-maturity contract to avoid the so-called 'Samuelson effect' (with a rollover date 15 days before maturity). The data comes from Bloomberg with a monthly frequency.[*]

20.2.4.2 Precious Metals

Tables 20.4 and 20.5 present the results obtained with a subset of precious metals, i.e. gold, silver and platinum. When considering the full sample, we obtain the results reproduced in Table 20.4. When considering the potential reaction of precious metals depending on the phase of business cycle—i.e. 'expansion' or 'recession'—as indicated by the NBER Business Cycle Dating Committee,[†] we obtain the results reproduced in Table 20.5. In each table, the Goldman Sachs Commodity Index (GSCI) for precious metals is used as a benchmark.[‡]

---

[*] This comment applies in the remainder of the paper.

[†] The NBER methodology of dating business cycle is authoritative and well known, and widely used by financial practitioners as well.

[‡] The GSCI is a broadly diversified index designed for investors seeking to store value in commodity markets.

TABLE 20.3 Database of Commodity Futures Used for the News Analysis

Commodity	Exchange	Unit
Gold	COMEX Gold Futures price	US Dollar per troy ounce
Silver	COMEX Silver Futures price	US Dollar per troy ounce
Platinum	COMEX Platinum Futures price	US Dollar per troy ounce
Aluminium	LME Aluminium Futures price	US Dollar per tonne
Copper	COMEX Copper Futures price	US Dollar per pound
Nickel	LME Nickel Futures price	US Dollar per tonne
Zinc	LME Zinc Futures price	US Dollar per tonne
Lead	LME Lead Futures price	US Dollar per tonne
WTI	NYMEX WTI Light Sweet Crude Oil Futures price	US Dollar per barrel
Brent	ICE Brent Crude Futures price	US Dollar per barrel
Gasoil	RBOB Gasoline Futures price	US Dollar per gallon
Natural Gas	Henry Hub Natural Gas Futures price	US Dollar per mmBtu
Heating Oil	NYMEX Heating Oil Futures price	US Dollar per gallon
Corn	CBOT Corn Futures price	US Dollar per bushel
Wheat	CBOT Wheat Futures price	US Dollar per bushel
Coffee	ICE Coffee C Futures price	US Dollar per pound
Sugar	NYMEX Sugar No.11 Futures price	US Dollar per pound
Cocoa	ICE Cocoa Futures price	US Dollar per tonne
Cotton	ICE Cotton No.2 Futures price	US Dollar per pound
Soybean	CBOT Soybean Futures price	US Dollar per short ton
Rice	CME Rough Rice Futures price	US Dollar per hundred weight

Source: Bloomberg.

TABLE 20.4 Estimation of the Impact of Economic News over Selected Precious Metals

	Gold	Silver	Platinum	GSCI Prec. Metals
Intercept	−1.89 (−1.25)	−1.73 (−1.15)	4.11** (2.73)	1.52 (0.99)
AR	0.03** (2.16)	0.04 (1.59)	0.03 (1.2)	0.04** (2.19)
Non-Farm Payroll	−0.17** (−2.16)	−0.18 (−1.32)	−0.18* (−1.7)	−0.16* (−1.92)
ISM	*−0.03 (−0.37)*	0.05 (0.39)	0.07 (0.7)	*−0.03 (−0.32)*
Jobless Claims	−0.05 (−0.59)	−0.27* (−1.91)	−0.1 (−0.96)	−0.05 (−0.57)
US CPI MoM	0.06 (0.71)	0.11 (0.75)	0.11 (1.06)	0.03 (0.38)
US Retail Sales	−0.03 (−0.31)	−0.04 (−0.23)	−0.02 (−0.13)	−0.1 (−1.01)
Fed Target Rate	−0.04 (−0.33)	−0.23 (−1.12)	−0.09 (−0.58)	−0.02 (−0.19)
US GDP	−0.01 (−0.04)	0.08 (0.29)	*0.38** (1.98)*	−0.03 (−0.21)
ZEW Eco. Sent.	0.07 (0.76)	0.03 (0.16)	*−0.21 (−1.6)*	0.09 (0.86)
IFO Expectations	0.13 (1.22)	0.18 (0.92)	0.2 (1.41)	*0.08 (0.72)*
EMU CPI	*−0.05 (−0.46)*	−0.2 (−1.14)	*−0.18 (−1.37)*	*−0.09 (−0.85)*
EMU GDP	−0.08 (−0.49)	−0.11 (−0.37)	−0.04 (−0.19)	−0.08 (−0.49)
FR Bus. Conf.	*0.12 (1.32)*	*0.02 (0.15)*	0.1 (0.8)	*0.13 (1.25)*
ECB Ref. Rate	−0.06 (−0.67)	0.14 (0.87)	0.12 (0.99)	−0.07 (−0.7)
China CPI YoY	−0.03 (−0.33)	0.04 (0.21)	−0.06 (−0.51)	−0.1 (−0.98)
China Ind. Prod.	0 (0.03)	0.01 (0.08)	0.26** (2.18)	0.05 (0.54)
China PMI	*−0.43** (−3.58)*	*−0.48** (−2.23)*	−0.23 (−1.45)	*−0.5** (−3.92)*

Notes: T-stats are between brackets. Italic figures indicate statistically significant figures at a 10% risk level.

TABLE 20.5   Estimation of the Impact of Economic News over Selected Precious Metals Depending on the NBER Phase of the Business Cycle

		Gold	Silver	Platinum	GSCI Prec. Metals
	Intercept	−2.15 (−1.41)	−1.66 (−1.08)	4.38** (2.88)	1.29 (0.84)
	AR	0.03** (2.2)	0.05* (1.64)	0.03 (1.23)	0.04** (2.2)
REC	Non-Farm Payroll	0.03 (0.17)	−0.32 (−0.86)	−0.42 (−1.51)	0.07 (0.33)
	ISM	*−0.36** (−2.04)*	−0.49 (−1.54)	−0.34 (−1.41)	*−0.41** (−2.19)*
	Jobless Claims	−0.06 (−0.45)	−0.24 (−1.02)	−0.18 (−1.01)	−0.01 (−0.04)
	US CPI MoM	−0.19 (−1.29)	−0.36 (−1.37)	−0.24 (−1.23)	−0.22 (−1.41)
	US Retail Sales	−0.02 (−0.15)	0.02 (0.08)	0.17 (1.03)	−0.1 (−0.78)
	Fed Target Rate	0.01 (0.05)	−0.17 (−0.66)	−0.12 (−0.6)	0.02 (0.1)
	US GDP	0.1 (0.24)	0.22 (0.3)	*1.98** (3.54)*	−0.05 (−0.11)
	ZEW Eco. Sent.	0.29 (1.44)	−0.26 (−0.73)	*−0.92** (−3.41)*	0.12 (0.58)
	IFO Expectations	−0.1 (−0.55)	−0.41 (−1.21)	0.15 (0.61)	*−0.36* (−1.81)*
	EMU CPI	*−0.39** (−1.99)*	−0.49 (−1.41)	*−0.71** (−2.75)*	*−0.36* (−1.78)*
	EMU GDP	−0.62 (−1.18)	−0.8 (−0.85)	−0.73 (−1.05)	−0.48 (−0.88)
	FR Bus. Conf.	*0.6** (3.72)*	*0.74** (2.58)*	0.3 (1.38)	*0.53** (3.11)*
	ECB Ref. Rate	−0.06 (−0.55)	0.09 (0.44)	0.15 (0.99)	−0.05 (−0.4)
	China CPI YoY	0.36 (1.14)	0.45 (0.79)	−0.01 (−0.03)	−0.28 (−0.84)
	China Ind. Prod.	−0.11(−0.51)	0.25 (0.62)	0.09 (0.29)	0.04 (0.19)
	China PMI	*−0.51** (−3.88)*	*−0.66** (−2.82)*	−0.25 (−1.44)	*−0.59** (−4.26)*
EXP	Non-Farm Payroll	*−0.21** (−2.49)*	−0.17 (−1.15)	−0.14 (−1.31)	*−0.2** (−2.29)*
	ISM	0.06 (0.67)	0.18 (1.19)	0.17 (1.51)	0.07 (0.77)
	Jobless Claims	−0.02 (−0.19)	−0.24 (−1.41)	−0.04 (−0.32)	−0.05 (−0.49)
	US CPI MoM	*0.16* (1.74)*	*0.31* (1.83)*	*0.26** (2.07)*	0.13 (1.37)
	US Retail Sales	−0.04 (−0.31)	−0.1 (−0.4)	−0.25 (−1.3)	−0.09 (−0.61)
	Fed Target Rate	−0.03 (−0.17)	−0.17 (−0.52)	0.05 (0.23)	0 (0.01)
	US GDP	0 (0)	0.08 (0.28)	0.18 (0.89)	−0.02 (−0.12)
	ZEW Eco. Sent.	0.01 (0.09)	0.11 (0.57)	0 (−0.02)	0.07 (0.6)
	IFO Expectations	*0.22* (1.68)*	*0.42* (1.81)*	0.23 (1.3)	*0.28** (2.01)*
	EMU CPI	0.1 (0.88)	−0.05 (−0.25)	0.03 (0.17)	0.03 (0.27)
	EMU GDP	−0.02 (−0.14)	−0.04 (−0.12)	0.03 (0.13)	−0.04 (−0.22)
	FR Bus. Conf.	−0.12 (−1.03)	−0.34 (−1.6)	0 (0.01)	−0.07 (−0.61)
	ECB Ref. Rate	−0.1 (−0.66)	0.16 (0.59)	0.06 (0.32)	−0.14 (−0.91)
	China CPI YoY	−0.07 (−0.75)	−0.02 (−0.12)	−0.1 (−0.74)	−0.09 (−0.83)
	China Ind. Prod.	0.02 (0.22)	−0.04 (−0.2)	*0.29** (2.24)*	0.05 (0.48)
	China PMI	0.38 (1.11)	*1.03* (1.66)*	0.4 (0.87)	0.34 (0.92)

Notes: REC stands for 'Recession' and EXP for 'Expansion' according to the NBER Business Cycle Dating Committee. T-stats are between brackets. Italic figures indicate statistically significant figures at a 10% risk level.

The results may be summarized as follows:*

1. *Gold*: From a long-term perspective, gold reacts negatively to positive surprises in Non-Farm Payrolls and the Chinese PMI. In this respect, sustained and accelerating growth regimes are thus likely to hinder the rise of gold prices. We do not find evidence of a reaction of gold prices to surprises concerning inflation news. When breaking down the phases of the business cycle

---

* To conserve space, we skip the results of the EGARCH model, where all parameters are highly significant. They can be obtained upon request to the authors.

(Table 20.5), we obtain a different picture. During recessions, gold reacts negatively to the US ISM, the EMU CPI and the Chinese PMI. Two out of the three figures having a statistical impact on gold are economic surveys. These surveys are to be monitored cautiously to anticipate periods of rising gold prices. During expansion periods, gold displays a negative sign with the Non-Farm Payrolls, but a positive sign with the US CPI and the German IFO. Hence, during expansion, positive surprises concerning inflation—i.e. when the inflation index rises more than expected—should trigger a surge in gold prices. Overall, we identify that gold has a very different reaction to economic news depending on the underlying economic regime.

2. *Silver*: From a long-term perspective, silver reacts negatively to positive surprises in the Jobless Claims and the Chinese PMI. The picture is more complex than for gold: the negative sign with the Chinese PMI underlines the 'safe haven' role of silver during recessions, whereas its negative reaction to an unexpected increase in the number of unemployed people in the USA tells us a different story. This effect comes from the key fact that many precious metals are used both as safe havens (Baur and McDermott 2010, Beaudry *et al.* 2011) and for industrial purposes. During recession periods, silver reacts positively to negative surprises in the Chinese PMI, and negatively to positive surprises in the French business confidence indicator. During expansion periods, silver reacts positively to positive surprises in the US CPI, the Chinese PMI and the IFO Expectations. It has a weaker sensitivity to the Jobless Claims.

3. *Platinum*: During the full sample (Table 20.4), platinum shows a negative sensitivity to Non-Farm Payroll, the German ZEW and the EMU CPI, and a positive reaction to the US GDP and the Chinese industrial production. The most striking fact for platinum is that it features a stronger relationship to economic news than gold and silver: while gold only reacts to two news, platinum reacts to five of the news considered. We also clearly see the mix between the safe haven characteristic of the average precious metal, and the apparent use of such metal for industrial purposes. When considering only recession periods, the strongest sensitivity of platinum is obtained for the US GDP: an unexpected decrease in the US GDP would lead to a strong decline in the prices of platinum. Two minor effects come from the sign of the German ZEW and the EMU CPI, in a contracyclical way. Expansion periods are characterized by a rise in the platinum price, whenever the US CPI and the Chinese industrial production surprises are on the upside. Similarly to the GSCI benchmark for precious metals, platinum can be used for industrial purposes. Hence, it reacts positively to positive news about economic growth, and it is traditionally used as a hedge against inflation.

20.2.4.3 Industrial Metals

In the case of industrial metals, we use the following subset of markets: Aluminium, Copper, Nickel, Zinc and Lead. As for precious metals, we report the estimation results obtained during the full sample (Table 20.6), and during expansion vs. recession periods (Table 20.7).

Our results can be summarized as follows:

1. *Aluminium:* Over the 1999–2011 period, the price of aluminium has been rising with Non-Farm Payrolls, the US ISM and ECB rates. On the contrary, it reacted negatively to increases in the number of unemployed people in the US, as measured by the Jobless Claims. Hence, in the long run, the aluminium price seems to increase when employment and production perspectives are improving, and to decrease when these conditions deteriorate. In Table 20.7, we observe that recession periods lead to an heightened sensitivity of aluminium to the world business cycle: it decreases with declining payrolls, ECB rates but increases with German leadings. On the contrary, during expansion periods, the aluminium price seems to react weakly to economic news (only to the US ISM).

2. *Copper*: The case of copper is very similar to aluminium, with a large discrepancy in terms of market reaction to economic news over recessions and expansions. During the full sample, the

TABLE 20.6 Estimation of the Impact of Economic News over Selected Industrial Metals

	Aluminium	Copper	Nickel	Zinc	Lead	GSCI Ind. Metals
Intercept	−3.04** (−1.98)	−4.7** (−3.07)	0.71 (0.46)	−2.61* (−1.71)	5.39** (3.52)	−4.53** (−2.95)
AR	0 (0.21)	0.03 (1.13)	0.02 (0.64)	0.01 (0.5)	0.03 (0.86)	0.02 (1.07)
Non-Farm Payroll	*0.22** (2.32)*	0.15 (1.19)	0.16 (0.97)	0.2 (1.52)	0.09 (0.59)	*0.19* (1.8)*
ISM	*0.16* (1.65)*	*0.37** (2.96)*	0.24 (1.39)	0.18 (1.37)	0.21 (1.48)	*0.21** (1.99)*
Jobless Claims	*−0.18* (−1.89)*	*−0.28** (−2.26)*	*−0.32* (−1.92)*	−0.21 (−1.59)	−0.2 (−1.38)	*−0.24** (−2.33)*
US CPI MoM	−0.11 (−1.17)	0.04 (0.29)	−0.28 (−1.63)	−0.12 (−0.86)	0.03 (0.2)	−0.06 (−0.61)
US Retail Sales	0.16 (1.4)	−0.04 (−0.29)	*−0.53** (−2.63)*	0.02 (0.12)	−0.22 (−1.24)	0 (−0.03)
Fed Target Rate	0.05 (0.33)	0.03 (0.16)	0.05 (0.2)	0.05 (0.26)	0.27 (1.28)	0.08 (0.51)
US GDP	0.12 (0.69)	0.37 (1.61)	0.02 (0.06)	*0.5** (2.01)*	−0.05 (−0.17)	0.24 (1.25)
ZEW Eco. Sent.	−0.12 (−0.99)	*−0.27* (−1.73)*	−0.21 (−1)	0.16 (0.97)	−0.24 (−1.3)	−0.2 (−1.56)
IFO Expectations	−0.18 (−1.36)	−0.26 (−1.54)	−0.27 (−1.14)	0 (0)	−0.14 (−0.69)	−0.2 (−1.41)
EMU CPI	−0.06 (−0.53)	*−0.26* (−1.66)*	*−0.41* (−1.93)*	−0.11 (−0.67)	−0.16 (−0.9)	−0.14 (−1.06)
EMU GDP	−0.27 (−1.39)	−0.4 (−1.6)	*−0.78** (−2.27)*	−0.35 (−1.3)	−0.46 (−1.58)	*−0.39* (−1.84)*
FR Bus. Conf.	0.08 (0.66)	0.21 (1.38)	0.19 (0.9)	−0.09 (−0.56)	0.26 (1.49)	0.2 (1.54)
ECB Ref. Rate	*0.27* (2.42)*	*0.53** (3.66)*	0.32 (1.63)	0.09 (0.59)	−0.08 (−0.45)	*0.35** (2.88)*
China CPI YoY	0.07 (0.57)	0.03 (0.18)	−0.09 (−0.42)	−0.01 (−0.08)	0.09 (0.53)	0.06 (0.47)
China Ind. Prod.	0.06 (0.54)	−0.15 (−1.06)	−0.09 (−0.47)	−0.17 (−1.11)	*−0.41** (−2.48)*	−0.08 (−0.63)
China PMI	0.05 (0.31)	0.23 (1.18)	0.05 (0.2)	0.28 (1.34)	*0.49** (2.19)*	0.26 (1.6)

Notes: T-stats are between brackets. Italic figures indicate statistically significant figures at a 10% risk level.

price of copper increases with the ISM, the US unemployment and the ECB main refinancing rate, and decreases with the German leadings and the EMU CPI. During recession periods, the copper price drops with decreases in US leadings, US unemployment and US GDP, and with lowered ECB rate and Chinese PMI (surprisingly to the downside). The negative EMU GDP and German IFO lead to increases in the price of copper. This result may be explained by the fact that we focus on the NBER recession dates, whereas Europe has effectively entered the financial crisis during late 2008—i.e. almost one year after the USA. During expansion periods, the copper price is found to rise only with positive ISM surprises.

3. *Nickel*: During the full sample, the price of nickel displays a negative sensitivity to surprises in the US Retail Sales, Jobless Claims, the EMU CPI and the EMU GDP. This surprising pattern features clearly a counter-cyclical reaction to economic news. When considering recession periods, we uncover again a negative sensitivity of the nickel price to the German IFO, the US Retail Sales, Jobless Claims and the EMU GDP. This commodity records nonetheless a positive reaction to increases in the ECB main refinancing rate, and in the French business confidence survey. During expansion periods, the price of nickel reacts positively to unexpected increases in the US ISM, and negatively to unexpected increases in the US CPI and the EMU GDP. This finding underlines the potential complexity of the relationship between industrial metals and the economic newsflow.

4. *Zinc*: During the full sample, the price of zinc only reacts positively to surprises in the US GDP. During recession periods, this sensitivity to the US GDP appears again, even with an increased importance. Besides, the price of zinc displays strongly procyclical reactions to the ISM, the US unemployment, the US GDP, the ECB rate and the Chinese PMI. On the contrary, during expansion periods, the price of zinc exhibits two kinds of reactions: one positive to the US ISM, and one negative to the ECB main refinancing rate. Hence, zinc prices are more likely to rise when US surveys bring positive surprises, and when the ECB is in the process of lowering its main refinancing rate.

TABLE 20.7 Estimation of the Impact of Economic News over Selected Industrial Metals Depending on the NBER Phase of the Business Cycle

		Aluminium	Copper	Nickel	Zinc	Lead	GSCI Ind. Metals
	Intercept	−3.06** (−1.98)	−4.47** (−2.9)	0.45 (0.29)	−2.78* (−1.8)	5.06** (3.27)	−4.67** (−3.03)
	AR	0.01 (0.3)	0.03 (1.23)	0.02 (0.64)	0.01 (0.47)	0.03 (0.93)	0.02 (1.17)
REC	Non-Farm Payroll	*0.9** (3.54)*	*0.78** (2.37)*	0.47 (1.05)	*0.76** (2.15)*	0.22 (0.57)	*0.76** (2.75)*
	ISM	0.12 (0.54)	0.19 (0.69)	−0.48 (−1.24)	−0.37 (−1.22)	−0.02 (−0.05)	−0.02 (−0.09)
	Jobless Claims	−0.23 (−1.39)	*−0.49** (−2.32)*	*−0.89** (−3.1)*	*−0.41* (−1.79)*	−0.32 (−1.31)	*−0.42** (−2.39)*
	US CPI MoM	−0.21 (−1.16)	0.24 (1.04)	0.11 (0.35)	−0.12 (−0.47)	0.17 (0.61)	0.02 (0.08)
	US Retail Sales	0.22 (1.43)	−0.11 (−0.54)	*−0.95** (−3.57)*	0.04 (0.19)	−0.27 (−1.19)	−0.02 (−0.15)
	Fed Target Rate	0.05 (0.3)	*0 (0.01)*	0.31 (0.97)	0.06 (0.23)	0.37 (1.34)	0.1 (0.52)
	US GDP	0.63 (1.23)	*1.46** (2.19)*	0.36 (0.4)	*1.38* (1.92)*	*1.46* (1.87)*	*1.1* (1.95)*
	ZEW Eco. Sent.	*−0.46* (−1.87)*	*−1.06** (−3.31)*	−0.64 (−1.47)	0.51 (1.48)	*−0.78** (−2.09)*	*−0.85** (−3.16)*
	IFO Expectations	*−0.62** (−2.7)*	*−1.12** (−3.78)*	*−0.92** (−2.28)*	−0.39 (−1.22)	−0.47 (−1.35)	*−0.82** (−3.28)*
	EMU CPI	−0.14 (−0.59)	−0.29 (−0.93)	−0.63 (−1.51)	−0.24 (−0.73)	−0.57 (−1.59)	−0.14 (−0.54)
	EMU GDP	−0.72 (−1.12)	−0.7 (−0.85)	*−1.95* (−1.72)*	*−2.39** (−2.67)*	*−3.28** (−3.38)*	−1.05 (−1.5)
	FR Bus. Conf.	0.25 (1.27)	*0.6** (2.36)*	*0.59* (1.68)*	−0.19 (−0.7)	*0.94** (3.17)*	*0.41* (1.89)*
	ECB Ref. Rate	*0.37** (2.59)*	*0.81** (4.42)*	*0.48* (1.92)*	*0.42* (2.11)*	−0.17 (−0.79)	*0.54** (3.53)*
	China CPI YoY	0.18 (0.47)	0.09 (0.19)	−0.48 (−0.7)	*1.02* (1.88)*	0.55 (0.94)	0.17 (0.4)
	China Ind. Prod.	0.28 (1.02)	−0.29 (−0.82)	0.25 (0.5)	−0.63 (−1.62)	−0.67 (−1.59)	−0.04 (−0.14)
	China PMI	0.16 (0.98)	*0.41** (1.98)*	0.24 (0.83)	*0.39* (1.74)*	*0.61** (2.52)*	*0.29* (1.67)*
EXP	Non-Farm Payroll	0.11 (1.06)	0.04 (0.31)	0.1 (0.56)	0.1 (0.7)	0.06 (0.37)	0.09 (0.81)
	ISM	*0.19* (1.81)*	*0.44** (3.19)*	*0.43* (2.28)*	*0.34* (2.26)*	*0.28* (1.77)*	*0.28** (2.38)*
	Jobless Claims	−0.17 (−1.47)	−0.2 (−1.35)	−0.03 (−0.17)	−0.14 (−0.83)	−0.13 (−0.72)	−0.16 (−1.25)
	US CPI MoM	−0.07 (−0.62)	−0.05 (−0.33)	*−0.46** (−2.28)*	−0.1 (−0.64)	−0.03 (−0.17)	−0.1 (−0.8)
	US Retail Sales	0.1 (0.54)	0.05 (0.22)	0.03 (0.09)	−0.03 (−0.11)	−0.14 (−0.52)	0.03 (0.15)
	Fed Target Rate	0.06 (0.26)	−0.01 (−0.03)	−0.49 (−1.26)	0.05 (0.15)	0.07 (0.2)	0 (0.01)
	US GDP	0.05 (0.26)	0.22 (0.89)	−0.02 (−0.07)	0.37 (1.39)	−0.23 (−0.8)	0.12 (0.58)
	ZEW Eco. Sent.	−0.02 (−0.12)	−0.02 (−0.13)	−0.06 (−0.23)	0.06 (0.32)	−0.06 (−0.29)	0 (−0.02)
	IFO Expectations	0.03 (0.19)	0.14 (0.69)	0.04 (0.14)	0.2 (0.91)	−0.01 (−0.02)	0.09 (0.53)
	EMU CPI	−0.06 (−0.41)	−0.27 (−1.52)	−0.32 (−1.33)	−0.07 (−0.37)	−0.03 (−0.17)	−0.13 (−0.86)
	EMU GDP	−0.22 (−1.1)	−0.38 (−1.43)	*−0.66* (−1.84)*	−0.15 (−0.53)	−0.18 (−0.6)	−0.33 (−1.47)
	FR Bus. Conf.	0 (0.01)	0.03 (0.15)	−0.01 (−0.03)	−0.03 (−0.15)	−0.08 (−0.35)	0.11 (0.68)
	ECB Ref. Rate	0.1 (0.56)	0.07 (0.27)	0.12 (0.37)	*−0.42* (−1.64)*	0.05 (0.17)	0.03 (0.16)
	China CPI YoY	0.05 (0.39)	0.02 (0.11)	−0.02 (−0.11)	−0.11 (−0.66)	0.04 (0.23)	0.05 (0.35)
	China Ind. Prod.	0.02 (0.15)	−0.12 (−0.81)	−0.15 (−0.72)	−0.08 (−0.47)	*−0.36** (−2)*	−0.08 (−0.64)
	China PMI	−0.64 (−1.49)	−0.75 (−1.37)	−0.35 (−0.46)	−0.07 (−0.11)	0.02 (0.03)	0.35 (0.75)

Notes: REC stands for 'Recession' and EXP for 'Expansion' according to the NBER Business Cycle Dating Committee. T-stats are between brackets. Italic figures indicate statistically significant figures at a 10% risk level.

5. *Lead:* During the full sample, the lead price reacts positively to positive surprises in the Chinese PMI, and to negative surprises in the Chinese industrial production. When looking at the NBER Business Cycle phases, we find that this negative sensitivity to industrial production is only valid during expansion periods. This result may be related to the aggressive monetary policy conducted by the Public Bank of China over the post-2008 period. In addition, during expansionary phases, we find a positive sensitivity of the lead price to the US ISM. During recession periods, this commodity displays a positive sensitivity to the Chinese PMI—as in the full sample case—and to the US GDP, and a negative reaction to the EMU GDP.

To sum up, we find common patterns across industrial metals: most of them display a procyclical behaviour during recessions, a weak reaction to news during expansion periods and a significant sensitivity to Chinese news during recessions.

20.2.4.4 Energy

Tables 20.8 and 20.9 present the estimation results of Equations (20.3) and (20.4) in the case of energy commodities.

The main findings may be summarized as follows:

1. *WTI and Brent:* crude oil prices show a rather weak long-term relationship to economic news. During the full sample, the only market mover on these markets is the US Jobless Claims, with a negative sign.* During recession periods, crude oil prices seem to be negatively related to US Jobless Claims, US Retail Sales, the German IFO, the German ZEW and the EMU CPI. In the meantime, these commodity prices also exhibit a positive reaction to increases in the ECB Refinancing Rate. This latter result illustrates that a rising ECB target rate is interpreted as a good news on crude oil markets, as it reveals that the world economy is roaring. During expansion periods, we cannot identify any significant market mover on these markets.

2. *Gasoil:* The full sample analysis reveals two market movers for the gasoil price: the US Retail Sales and the EMU CPI with negative signs. Hence, the gasoil market features an interesting

TABLE 20.8 Estimation of the Impact of Economic News over Selected Energy Markets

	WTI	Brent	Gasoil	Natural gas	Heating oil	GSCI energy
Intercept	−0.95 (−0.61)	−4.46** (−2.92)	0.25 (0.16)	−1.27 (−0.82)	−3.4** (−2.21)	−2.1 (−1.37)
AR	0.05 (1.39)	0.05 (1.62)	0.05 (1.5)	0.01 (0.2)	0.05 (1.43)	0.03 (1.15)
Non-Farm Payroll	0.15 (0.92)	0.09 (0.57)	0.19 (1.26)	0.31 (1.23)	0.16 (1.02)	0.12 (0.79)
ISM	0.2 (1.22)	0.04 (0.23)	0.09 (0.61)	0.29 (1.14)	0.13 (0.8)	0.09 (0.6)
Jobless Claims	−0.58** (−3.47)	−0.34** (−2.17)	0 (0.02)	0.16 (0.63)	−0.33** (−2.03)	−0.25* (−1.68)
US CPI MoM	−0.1 (−0.6)	−0.07 (−0.41)	−0.05 (−0.3)	0.02 (0.08)	0.01 (0.07)	−0.04 (−0.27)
US Retail Sales	−0.27 (−1.36)	−0.29 (−1.54)	−0.43** (−2.41)	−0.13 (−0.43)	−0.22 (−1.15)	−0.25 (−1.4)
Fed Target Rate	0 (0.01)	0.23 (1.02)	0.26 (1.22)	−0.21 (−0.57)	0.1 (0.42)	0.08 (0.36)
US GDP	0.21 (0.69)	0.1 (0.35)	0.04 (0.14)	0.24 (0.52)	0.19 (0.64)	0.22 (0.79)
ZEW Eco. Sent.	−0.19 (−0.93)	−0.13 (−0.65)	−0.08 (−0.41)	−0.39 (−1.23)	−0.27 (−1.31)	−0.21 (−1.15)
IFO Expectations	−0.06 (−0.24)	−0.07 (−0.3)	−0.25 (−1.23)	0.18 (0.51)	0.04 (0.18)	−0.03 (−0.15)
EMU CPI	−0.31 (−1.48)	−0.19 (−0.98)	−0.39** (−2.07)	0.25 (0.79)	−0.29 (−1.43)	−0.21 (−1.11)
EMU GDP	0.05 (0.14)	0.09 (0.27)	0.07 (0.23)	0.5 (0.96)	0.05 (0.15)	0.1 (0.32)
FR Bus. Conf.	0.16 (0.77)	0.22 (1.14)	0.15 (0.8)	0.44 (1.42)	0.21 (1.04)	0.19 (1.06)
ECB Ref. Rate	0.28 (1.43)	*0.3* (1.64)	−0.06 (−0.33)	0.18 (0.61)	0.18 (0.96)	*0.29* (1.66)
China CPI YoY	0.03 (0.13)	0.01 (0.07)	0.06 (0.36)	0.15 (0.48)	0.19 (1)	0.06 (0.31)
China Ind. Prod.	−0.09 (−0.49)	−0.1 (−0.53)	−0.1 (−0.6)	0 (−0.02)	−0.01 (−0.07)	−0.13 (-0.79)
China PMI	0.25 (0.97)	0.07 (0.27)	0.02 (0.07)	0.29 (0.74)	0.12 (0.46)	0.14 (0.6)

Notes: T-stats are between brackets. Italic figures indicate statistically significant figures at a 10% risk level.

---

* Note that the Brent price also displays a positive sensitivity to the ECB Refinancing Rate.

TABLE 20.9 Estimation of the Impact of Economic News over Selected Energy Markets Depending on the NBER Phase of the Business Cycle

		WTI	Brent	Gasoil	Natural Gas	Heating Oil	GSCI Energy
	Intercept	−0.65 (−0.42)	−4.05** (−2.64)	0.62 (0.4)	−1.25 (−0.81)	−3.3** (−2.13)	−1.78 (−1.16)
	AR	0.05 (1.43)	0.05* (1.66)	0.05 (1.45)	0.02 (0.3)	0.05 (1.49)	0.04 (1.19)
REC	Non-Farm Payroll	0.63 (1.44)	0.67 (1.59)	*0.82** (2.07)*	1.03 (1.52)	*0.77* (1.78)*	*0.8** (2.03)*
	ISM	0.5 (1.3)	−0.02 (−0.04)	*−0.59* (−1.72)*	−0.65 (−1.12)	−0.36 (−0.98)	−0.22 (−0.65)
	Jobless Claims	−1.31** (−4.63)	−0.71** (−2.62)	−0.05 (−0.21)	0.26 (0.61)	*−0.61** (−2.23)*	*−0.58** (−2.3)*
	US CPI MoM	0.01 (0.04)	−0.09 (−0.31)	0.25 (0.89)	−0.26 (−0.53)	−0.04 (−0.15)	0 (0.02)
	US Retail Sales	−0.48* (−1.82)	−0.51** (−2.01)	−0.59** (−2.49)	−0.11 (−0.26)	−0.38 (−1.49)	*−0.41* (−1.75)*
	Fed Target Rate	−0.01 (−0.04)	0.26 (0.88)	0.23 (0.82)	−0.4 (−0.84)	0.04 (0.13)	−0.01 (−0.03)
	US GDP	1.31 (1.46)	*1.6* (1.87)*	0.23 (0.28)	0.3 (0.22)	0.82 (0.93)	1.16 (1.45)
	ZEW Eco. Sent.	−1.21** (−2.81)	−0.93** (−2.25)	−0.21 (−0.54)	*−1.44** (−2.18)*	−1.35** (−3.22)	−1.23** (−3.19)
	IFO Expectations	−1.02** (−2.56)	−1.01** (−2.64)	−0.87** (−2.44)	−0.51 (−0.84)	−0.51 (−1.32)	−0.88** (−2.46)
	EMU CPI	−0.73* (−1.77)	−0.94** (−2.37)	−0.7* (−1.88)	*1.11* (1.75)*	−0.94** (−2.34)	−0.73* (−1.96)
	EMU GDP	0.56 (0.5)	0.16 (0.15)	−0.16 (−0.16)	2.44 (1.42)	0.12 (0.11)	0.71 (0.71)
	FR Bus. Conf.	*0.91** (2.66)*	*0.96** (2.93)*	0.37 (1.22)	0.71 (1.36)	*0.73** (2.18)*	*0.81** (2.64)*
	ECB Ref. Rate	*0.55** (2.24)*	*0.54** (2.28)*	*−0.02 (−0.1)*	0.25 (0.66)	*0.29 (1.21)*	*0.48** (2.2)*
	China CPI YoY	−0.34 (−0.51)	0.26 (0.4)	*1.33** (2.19)*	0.12 (0.12)	0.6 (0.91)	−0.08 (−0.13)
	China Ind. Prod.	−0.61 (−1.27)	−0.61 (−1.34)	*−1.08** (−2.49)*	−0.41 (−0.56)	−0.53 (−1.13)	−0.64 (−1.48)
	China PMI	0.4 (1.42)	0.27 (1.02)	0.23 (0.93)	0.4 (0.93)	0.31 (1.15)	0.32 (1.28)
EXP	Non-Farm Payroll	0.08 (0.46)	0 (−0.03)	0.07 (0.44)	0.17 (0.64)	0.06 (0.33)	0 (0.01)
	ISM	0.16 (0.87)	0.07 (0.4)	0.27 (1.6)	*0.52* (1.84)*	0.27 (1.48)	0.18 (1.11)
	Jobless Claims	−0.21 (−1.03)	−0.17 (−0.87)	0.02 (0.12)	0.08 (0.24)	−0.18 (−0.91)	−0.09 (−0.48)
	US CPI MoM	−0.15 (−0.74)	−0.05 (−0.26)	−0.15 (−0.84)	0.15 (0.5)	0.05 (0.24)	−0.05 (−0.29)
	US Retail Sales	0.02 (0.07)	0.01 (0.02)	−0.24 (−0.86)	−0.13 (−0.27)	0.01 (0.02)	−0.01 (−0.05)
	Fed Target Rate	−0.02 (−0.06)	0.18 (0.48)	0.23 (0.65)	0.2 (0.34)	0.19 (0.5)	0.18 (0.51)
	US GDP	0.08 (0.24)	−0.08 (−0.27)	0 (0.02)	0.21 (0.43)	0.12 (0.39)	0.1 (0.34)
	ZEW Eco. Sent.	0.12 (0.5)	0.12 (0.54)	−0.02 (−0.09)	−0.09 (−0.24)	0.06 (0.28)	0.09 (0.44)
	IFO Expectations	0.39 (1.38)	0.36 (1.36)	0.04 (0.17)	0.49 (1.15)	0.29 (1.05)	0.36 (1.43)
	EMU CPI	−0.15 (−0.64)	0.04 (0.2)	−0.3 (−1.4)	−0.08 (−0.21)	−0.08 (−0.33)	−0.04 (−0.17)
	EMU GDP	0 (−0.01)	0.08 (0.23)	0.09 (0.29)	0.31 (0.57)	0.04 (0.13)	0.04 (0.12)
	FR Bus. Conf.	−0.21 (−0.86)	−0.13 (−0.56)	0.05 (0.24)	0.32 (0.83)	−0.06 (−0.23)	−0.1 (−0.47)
	ECB Ref. Rate	−0.13 (−0.39)	−0.08 (−0.27)	−0.11 (−0.38)	0.04 (0.08)	0.01 (0.04)	−0.03 (−0.12)
	China CPI YoY	0.06 (0.29)	−0.02 (−0.1)	−0.06 (−0.3)	0.12 (0.38)	0.14 (0.7)	0.06 (0.33)
	China Ind. Prod.	−0.01 (−0.03)	0 (−0.01)	0.09 (0.48)	0.07 (0.22)	0.08 (0.4)	−0.05 (−0.26)
	China PMI	0.41 (0.55)	−0.44 (−0.62)	*−1.2* (−1.81)*	−1.04 (−0.92)	−0.47 (−0.65)	−0.33 (−0.5)

Notes: REC stands for 'Recession' and EXP for 'Expansion' according to the NBER Business Cycle Dating Committee. T-stats are between brackets. Italic figures indicate statistically significant figures at a 10% risk level.

counter-cyclical behaviour. During recession periods, we identify various market movers: the price of gasoil displays a negative reaction to the US Retail Sales, the ISM, German surveys, the EMU CPI and the Chinese industrial production. Besides, gasoil shows a positive reaction to the Chinese CPI and the US Non-Farm Payroll. During expansionary phases, we cannot relate the gasoil price to economic news at statistically significant levels. This latter result is conform to what has been found for crude oil prices.

3. *Natural Gas:* The full sample analysis reveals no reaction of the natural gas market to economic news. Expansion periods are characterized by one market mover: the US ISM with a procyclical pattern. During recession periods, the natural gas price shows a negative reaction to the German ZEW, and a positive reaction to the EMU CPI. Overall, natural gas seems to be the energy market with the weakest link to the economic newsflow, even when accounting for periods of recession.

4. *Heating Oil:* In line with the WTI and the Brent prices, the heating oil price reacts only to the ISM—positively—over the long run. During expansion periods, there is no reaction to economic news at statistically significant levels. During recession periods, the results are similar to crude oil prices. In addition, we uncover a positive sensitivity to the US Non-Farm Payroll.

To sum up, energy markets seem to share a common pattern, i.e. a weak reaction to economic news during periods of economic expansion. This limited link during expansion periods might reflect the fact that energy commodities are influenced by local factors on the physical side of the market, as in the case of the natural gas.* During recession periods, the US unemployment seems to play a leading role in the evolution of these markets.

20.2.4.5 Agricultural Commodities

In this section, we detail the results obtained with agricultural commodities in Tables 20.10–20.13.
  The main empirical findings can be summarized as follows:

1. *Corn:* During the full sample, corn prices exhibit a negative relationship with the Fed Target Rate, the ZEW Economic Sentiment index and the Chinese PMI. According to this long-term perspective, corn prices seem to be contra-cyclical—fearing for instance a tighter monetary policy from the Fed. These negative sensitivities are also valid during periods of recession. However, we also find a positive reaction to the US Non-Farm Payroll and the US CPI. To sum up, corn prices are found to increase when the Fed lowers its target rate, and when the US payroll and inflation figures are rising. During expansion periods, this commodity market does not seem to react statistically significantly to economic news.

TABLE 20.10   Estimation of the Impact of Economic News over Selected Agricultural Markets

	Corn	Wheat	Coffee	Sugar	Cocoa
Intercept	3.79** (2.48)	0.22 (0.14)	−0.91 (−0.59)	−0.13 (−0.08)	−0.6 (−0.39)
AR	0.03 (0.98)	0.02 (0.5)	0.01 (0.16)	0.01 (0.17)	0.01 (0.33)
Non-Farm Payroll	0.15 (1.12)	0.22 (1.53)	0 (0.02)	−0.04 (−0.24)	0.03 (0.24)
ISM	−0.13 (−0.96)	0.2 (1.35)	−0.1 (−0.56)	0.16 (1.01)	0.05 (0.37)
Jobless Claims	0.09 (0.71)	0.16 (1.11)	−0.15 (−0.84)	−0.1 (−0.62)	0.02 (0.12)
US CPI MoM	0.18 (1.34)	0.23 (1.56)	0.27 (1.47)	0.16 (0.96)	0.09 (0.63)
US Retail Sales	0.12 (0.73)	0.03 (0.19)	*0.54** (2.51)*	0.24 (1.23)	−0.04 (−0.25)
Fed Target Rate	*−0.38* (−1.95)*	−0.32 (−1.54)	−0.25 (−0.96)	−0.13 (−0.57)	0 (−0.01)
US GDP	0.19 (0.78)	0.18 (0.67)	0.3 (0.89)	0.13 (0.44)	−0.21 (−0.81)
ZEW Eco. Sent.	*−0.41** (−2.46)*	−0.29 (−1.62)	*−0.44* (−1.94)*	−0.32 (−1.57)	0.04 (0.23)
IFO Expectations	−0.14 (−0.75)	−0.04 (−0.18)	0.13 (0.51)	0.18 (0.81)	0.16 (0.81)
EMU CPI	−0.21 (−1.27)	*−0.4** (−2.22)*	*−0.48** (−2.16)*	*−0.52** (−2.57)*	*−0.32* (−1.77)*
EMU GDP	0.14 (0.53)	0.11 (0.38)	0.24 (0.65)	0.29 (0.88)	−0.17 (−0.59)
FR Bus. Conf.	0.11 (0.71)	*0.31* (1.73)*	0.22 (1.01)	*0.32* (1.64)*	*0.31* (1.79)*
ECB Ref. Rate	0.19 (1.19)	0.26 (1.54)	0.34 (1.6)	0.3 (1.59)	−0.15 (−0.92)
China CPI YoY	−0.22 (−1.35)	−0.24 (−1.36)	0.23 (1.09)	*0.59** (3.03)*	−0.28 (−1.6)
China Ind. Prod.	−0.05 (−0.35)	−0.04 (−0.25)	0.21 (1.02)	*0.33* (1.8)*	0.02 (0.11)
China PMI	*−0.39* (−1.86)*	*−0.39* (−1.73)*	0.26 (0.94)	0.41 (1.63)	−0.06 (−0.25)

Notes: T−stats are between brackets. Italic figures indicate statistically significant figures at a 10% risk level.

---

* We wish to thank a referee for highlighting this point.

TABLE 20.11   Estimation of the Impact of Economic News over Selected Agricultural Markets

	Cotton	Soybean	Rice	GSCI Agri.
Intercept	2.35 (1.53)	−0.87 (−0.57)	4.62** (3.01)	2.46 (1.61)
AR	0 (0.11)	0.02 (0.72)	0.02 (0.79)	−0.01 (−0.27)
Non-Farm Payroll	0.18 (1.28)	0.16 (1.39)	−0.01 (−0.05)	0.14 (1.56)
ISM	0.11 (0.77)	0.1 (0.86)	0.11 (0.88)	0.08 (0.84)
Jobless Claims	−0.12 (−0.89)	0.12 (1.07)	0 (−0.01)	0.05 (0.52)
US CPI MoM	0.04 (0.32)	*0.22* (*1.91*)	0.19 (1.43)	*0.18** (2)*
US Retail Sales	0.22 (1.3)	0.01 (0.11)	0.08 (0.51)	0.09 (0.79)
Fed Target Rate	0.04 (0.19)	−0.12 (−0.72)	0.06 (0.29)	−0.23* (−1.75)
US GDP	0.18 (0.7)	0.11 (0.5)	0.24 (1.01)	0.15 (0.91)
ZEW Eco. Sent.	−0.21 (−1.22)	−0.32** (−2.18)	0.05 (0.33)	−0.31** (−2.76)
IFO Expectations	0 (0.01)	−0.09 (−0.54)	−0.1 (−0.59)	−0.06 (−0.5)
EMU CPI	*−0.37** (−2.15)*	−0.38** (−2.62)	−0.01 (−0.08)	−0.36** (−3.2)
EMU GDP	−0.41 (−1.46)	−0.05 (−0.2)	0.18 (0.68)	0.05 (0.26)
FR Bus. Conf.	*0.33** (1.98)*	0.13 (0.92)	0.25 (1.6)	*0.19* (1.69)*
ECB Ref. Rate	0.19 (1.16)	0.03 (0.19)	0.12 (0.8)	*0.2* (1.88)*
China CPI YoY	*0.36** (2.18)*	−0.12 (−0.87)	0.08 (0.5)	−0.08 (−0.74)
China Ind. Prod.	0.03 (0.19)	0.19 (1.46)	−0.05 (−0.32)	0.03 (0.24)
China PMI	0.04 (0.17)	−0.12 (−0.67)	−0.03 (−0.17)	−0.19 (−1.36)

Notes: T-stats are between brackets. Italic figures indicate statistically significant figures at a 10% risk level.

2. *Wheat:* The price of wheat behaves similarly to corn, with two minor differences: (i) during recession periods, the wheat price does not react to the US payroll figures, and (ii) during expansion periods, this commodity price exhibits some sensitivity to the German IFO, i.e. to European economic conditions.

3. *Coffee:* The price of coffee is found to react positively to the US Retail Sales, and negatively to the German ZEW and the EMU CPI. By disentangling recession from expansion periods, we get the insight that the reaction to US Retail Sales and the EMU CPI occurs during recessions. On the contrary, the negative reaction to surprises in the German ZEW is a feature of expansion periods. Moreover, during recession periods, we uncover that the price of coffee exhibits a positive sensitivity to the ECB main refinancing rate.

4. *Sugar:* During the full sample, we find a positive reaction of the sugar price to the Chinese PMI and CPI, and to the French business confidence index. Conversely, our results highlight a negative reaction to the EMU CPI. During periods of recession, the sugar price shows a statistically significant and positive reaction to the US Non-Farm Payroll and to the French business confidence index. During expansion periods, this commodity price reacts negatively to positive surprises in the EMU CPI, and to negative surprises in Chinese CPI.

5. *Cocoa:* In the long term, the price of cocoa has a positive reaction to the French business confidence index, and a negative reaction to the EMU CPI. During recession periods, the cocoa price reacts positively to the US Non-Farm Payroll, to the French business confidence index and to the Chinese industrial production. However, we also record a negative reaction when it comes to positive surprises in the EMU CPI. During expansion periods, we cannot detect any statistically significant reaction of the cocoa price to economic news.

6. *Cotton:* The full sample analysis reveals that the cotton price reacts negatively to surprises in the EMU CPI, and positively to surprises in the French business confidence index and the Chinese CPI. During recession periods, the price of cotton shows a negative reaction to the German IFO and the EMU CPI. Besides, we uncover a positive reaction to the French business confidence

TABLE 20.12   Estimation of the Impact of Economic News over Selected Agricultural Markets Depending on the NBER Phase of the Business Cycle

		Corn	Wheat	Coffee	Sugar	Cocoa
	Intercept	4.01** (2.61)	0.62 (0.4)	−0.87 (−0.56)	−0.2 (−0.13)	−0.24 (−0.16)
	AR	0.03 (0.88)	0.01 (0.35)	0.01 (0.16)	0.01 (0.15)	0.01 (0.33)
REC	Non-Farm Payroll	*0.66* (1.88)*	0.37 (0.97)	0.31 (0.64)	*0.72* (1.67)*	*0.7* (1.85)*
	ISM	−0.03 (−0.09)	0.25 (0.74)	−0.32 (−0.79)	−0.16 (−0.43)	−0.08 (−0.24)
	Jobless Claims	−0.02 (−0.08)	0.28 (1.12)	0.03 (0.1)	0 (0.01)	−0.08 (−0.31)
	US CPI MoM	*0.5** (1.97)*	*0.51* (1.86)*	0.29 (0.85)	0.1 (0.32)	0.22 (0.8)
	US Retail Sales	0.13 (0.61)	0.14 (0.6)	*0.88** (3.1)*	0.25 (0.96)	0.08 (0.33)
	Fed Target Rate	*−0.53** (−2.11)*	*−0.57** (−2.1)*	−0.39 (−1.16)	−0.2 (−0.65)	0.22 (0.8)
	US GDP	0.79 (1.1)	0.67 (0.86)	0.42 (0.44)	0.02 (0.02)	0.1 (0.13)
	ZEW Eco. Sent.	*−0.87** (−2.52)*	*−0.73* (−1.95)*	−0.39 (−0.84)	−0.35 (−0.83)	−0.24 (−0.63)
	IFO Expectations	*−1.1** (−3.44)*	*−1.19** (−3.42)*	−0.14 (−0.32)	−0.36 (−0.92)	−0.16 (−0.47)
	EMU CPI	−0.53 (−1.59)	*−0.9** (−2.5)*	*−0.95** (−2.12)*	−0.59 (−1.46)	*−0.82** (−2.29)*
	EMU GDP	0.91 (1.02)	0.7 (0.72)	0.4 (0.33)	−0.05 (−0.05)	−1.39 (−1.44)
	FR Bus. Conf.	0.37 (1.33)	0.38 (1.28)	0.26 (0.7)	*0.66** (1.98)*	*0.68** (2.31)*
	ECB Ref. Rate	0.23 (1.17)	0.42* (1.96)	*0.55** (2.05)*	0.36 (1.51)	−0.34 (−1.62)
	China CPI YoY	−0.24 (−0.45)	−0.02 (−0.03)	0.59 (0.8)	−0.32 (−0.48)	−0.94 (−1.61)
	China Ind. Prod.	−0.62 (−1.61)	−0.67 (−1.59)	0.12 (0.23)	0.31 (0.66)	*0.83** (1.98)*
	China PMI	*−0.45** (−2.01)*	*−0.5** (−2.04)*	0.25 (0.82)	0.38 (1.39)	0 (0)
EXP	Non-Farm Payroll	0.07 (0.47)	0.19 (1.26)	−0.05 (−0.26)	−0.17 (−0.97)	−0.08 (−0.51)
	ISM	−0.14 (−0.92)	0.19 (1.18)	−0.04 (−0.22)	0.25 (1.41)	0.11 (0.66)
	Jobless Claims	0.16 (0.99)	0.11 (0.6)	−0.25 (−1.15)	−0.15 (−0.77)	0.09 (0.51)
	US CPI MoM	0.06 (0.36)	0.13 (0.74)	0.27 (1.23)	0.18 (0.91)	0.01 (0.04)
	US Retail Sales	0.1 (0.41)	−0.1 (−0.39)	0.08 (0.25)	0.24 (0.79)	−0.2 (−0.76)
	Fed Target Rate	−0.25 (−0.81)	−0.03 (−0.09)	−0.05 (−0.12)	0.02 (0.06)	−0.37 (−1.11)
	US GDP	0.1 (0.38)	0.09 (0.32)	0.28 (0.78)	0.15 (0.46)	−0.25 (−0.87)
	ZEW Eco. Sent.	−0.27 (−1.44)	−0.17 (−0.83)	*−0.46* (−1.8)*	−0.32 (−1.38)	0.13 (0.65)
	IFO Expectations	0.33 (1.46)	*0.54** (2.19)*	0.26 (0.84)	0.43 (1.57)	0.3 (1.25)
	EMU CPI	−0.08 (−0.43)	−0.21 (−1.01)	−0.32 (−1.24)	*−0.49** (−2.09)*	−0.14 (−0.7)
	EMU GDP	0.07 (0.24)	0.05 (0.17)	0.22 (0.58)	0.33 (0.95)	−0.05 (−0.16)
	FR Bus. Conf.	0.02 (0.12)	0.32 (1.45)	0.2 (0.74)	0.16 (0.66)	0.13 (0.63)
	ECB Ref. Rate	0.1 (0.41)	−0.04 (−0.15)	−0.06 (−0.16)	0.15 (0.47)	0.16 (0.57)
	China CPI YoY	−0.21 (−1.29)	−0.26 (−1.45)	0.19 (0.85)	*0.68** (3.32)*	−0.21 (−1.17)
	China Ind. Prod.	0.05 (0.31)	0.07 (0.42)	0.22 (1)	0.33 (1.63)	−0.14 (−0.77)
	China PMI	0.43 (0.73)	0.64 (1)	0.53 (0.66)	0.61 (0.84)	−0.07 (−0.11)

Notes: REC stands for 'Recession' and EXP for 'Expansion' according to the NBER Business Cycle Dating Committee. T-stats are between brackets. Italic figures indicate statistically significant figures at a 10% risk level.

index and to the ECB Refinancing Rate. During expansionary phases, the price of cotton exhibits a positive reaction to surprises in the IFO index and in the Chinese CPI index and the Chinese CPI. During recession periods, the price of cotton shows a negative reaction to the German IFO and the EMU CPI. Besides, we uncover a positive reaction to the French business confidence index and to the ECB Refinancing Rate. During expansionary phases, the price of cotton exhibits a positive reaction to surprises in the IFO index and in the Chinese CPI.

7. *Soybean:* During the full sample, we record a positive sensitivity of the soybean price to surprises in the US CPI. In addition, we detect a negative reaction to surprises in the German ZEW and the

TABLE 20.13 Estimation of the Impact of Economic News over Selected Agricultural Markets Depending over the NBER Cycles

		Cotton	Soybean	Rice	GSCI Agri.
	Intercept	2.34 (1.52)	−0.3 (−0.19)	4.65** (3.02)	2.99* (1.94)
	AR	0 (0.08)	0.01 (0.56)	0.02 (0.88)	−0.01 (−0.39)
REC	Non-Farm Payroll	0.45 (1.23)	0.31 (1.02)	−0.25 (−0.72)	*0.45* (1.88)*
	ISM	−0.17 (−0.54)	−0.14 (−0.53)	0.42 (1.41)	0 (0.01)
	Jobless Claims	−0.14 (−0.59)	0 (−0.01)	0.27 (1.21)	0.07 (0.43)
	US CPI MoM	−0.06 (−0.23)	0.27 (1.23)	0.26 (1.05)	*0.37** (2.18)*
	US Retail Sales	0.18 (0.82)	0.11 (0.59)	0.21 (1.01)	0.15 (1.06)
	Fed Target Rate	0.11 (0.44)	−0.27 (−1.23)	−0.08 (−0.32)	*−0.4** (−2.36)*
	US GDP	0.7 (0.95)	−0.04 (−0.07)	0.87 (1.25)	0.53 (1.08)
	ZEW Eco. Sent.	−0.48 (−1.34)	*−0.76** (−2.52)*	−0.52 (−1.56)	*−0.7** (−2.96)*
	IFO Expectations	*−0.83** (−2.5)*	*−0.7** (−2.53)*	*−0.59* (−1.91)*	*−0.73** (−3.35)*
	EMU CPI	*−1.34** (−3.9)*	*−0.86** (−2.99)*	0.19 (0.59)	*−0.8** (−3.55)*
	EMU GDP	−0.79 (−0.86)	−0.39 (−0.5)	0.24 (0.27)	0.24 (0.39)
	FR Bus. Conf.	*0.76** (2.68)*	0.38 (1.61)	*0.75** (2.81)*	*0.4** (2.12)*
	ECB Ref. Rate	*0.43* (2.13)*	0 (−0.01)	−0.03 (−0.18)	*0.29* (2.14)*
	China CPI YoY	−0.03 (−0.06)	−0.36 (−0.77)	*0.93* (1.78)*	−0.16 (−0.43)
	China Ind. Prod.	0.38 (0.96)	−0.49 (−1.47)	−0.31 (−0.82)	−0.37 (−1.4)
	China PMI	0.07 (0.3)	0 (−0.02)	−0.18 (−0.83)	−0.22 (−1.46)
EXP	Non-Farm Payroll	0.12 (0.84)	0.13 (1.06)	0.04 (0.27)	0.09 (0.91)
	ISM	0.19 (1.25)	0.17 (1.29)	0.03 (0.22)	0.11 (1.05)
	Jobless Claims	−0.11 (−0.68)	0.2 (1.39)	−0.12 (−0.78)	0.04 (0.4)
	US CPI MoM	0.08 (0.48)	0.21 (1.51)	0.18 (1.15)	0.11 (1.05)
	US Retail Sales	0.27 (1.08)	−0.1 (−0.48)	−0.08 (−0.35)	−0.01 (−0.04)
	Fed Target Rate	−0.04 (−0.13)	0.08 (0.31)	0.23 (0.78)	−0.06 (−0.28)
	US GDP	0.11 (0.42)	0.12 (0.54)	0.16 (0.63)	0.1 (0.56)
	ZEW Eco. Sent.	−0.13 (−0.65)	−0.19 (−1.17)	0.22 (1.21)	−0.21 (−1.6)
	IFO Expectations	*0.39* (1.69)*	0.2 (1.05)	0.11 (0.52)	*0.25* (1.65)*
	EMU CPI	−0.03 (−0.13)	−0.22 (−1.3)	−0.08 (−0.43)	−0.2 (−1.51)
	EMU GDP	−0.37 (−1.26)	−0.01 (−0.05)	0.17 (0.62)	0.03 (0.15)
	FR Bus. Conf.	0.13 (0.61)	0.02 (0.13)	0.02 (0.09)	0.1 (0.75)
	ECB Ref. Rate	−0.27 (−1.03)	0.08 (0.37)	0.3 (1.19)	0.03 (0.19)
	China CPI YoY	*0.4** (2.31)*	−0.11 (−0.77)	−0.02 (−0.1)	−0.07 (−0.65)
	China Ind. Prod.	−0.04 (−0.25)	*0.31** (2.2)*	0 (0.02)	0.09 (0.84)
	China PMI	0.56 (0.92)	−0.47 (−0.91)	0.48 (0.83)	0.36 (0.88)

Notes: REC stands for 'Recession' and EXP for 'Expansion' according to the NBER Business Cycle Dating Committee. T-stats are between brackets. Italic figures indicate statistically significant figures at a 10% risk level.

EMU CPI. During recession periods, the price of soybean exhibits a negative relationship with surprises in the German ZEW, the German IFO and the EMU CPI. During expansion periods, we find one market mover for this commodity price at statistically significant levels, i.e. the Chinese industrial production (with a positive sign).

8. *Rice:* During the full sample and expansionary phases, we are unable to detect any significant influence on the price of rice. During recession periods, we identify one negative relationship to the German IFO, and two positive reactions to the French business confidence index and the Chinese CPI.

Overall, agricultural products are characterized by a rather complicated pattern, as we find little similarities within that particular class of commodities. It seems that agricultural markets are more sensitive to local factors (e.g. geography and climate) than other commodities.* Finally, it is noteworthy to remark that we find almost no market mover for this type of commodity during expansionary phases (as in the case of energy markets).

During 1999–2011, can we identify an increasing sensitivity of commodity markets to the business cycle? As discussed in the next section, a positive reply would provide some empirical support to the view that commodities are increasingly correlated with financial markets.

20.2.5 Rolling Analysis

In what follows, we perform a rolling analysis, i.e. we run regressions similar to the previous section by using a rolling three-year sample. We are therefore able to compute the percentage of news having a statistical impact on commodity markets with a clearer view of how it evolves through time.

In addition, for each of the estimates obtained, we compute the average absolute of β as detailed in Equation (20.3). This provides us with a sensitivity index of the reaction of commodities to economic news.

Formally, we compute the number of news whose β has a t-statistic greater than 1.64 (in absolute value). This corresponds to the quantile of a scaled Gaussian distribution at the 10% risk level. The sensitivity index for a given market—consistently with Equation (20.3)—is given by:

$$B = \frac{1}{I}\sum_{i=1}^{I}\left|\hat{\beta}_i\right|, \tag{20.5}$$

with B the sensitivity index, $\hat{\beta}_i$ the estimated sensitivities to the individual news and I the total number of news.

Figure 20.1 shows the evolution of the percentage of market movers for the GSCI sub-indices. Figure 20.2 contains the evolution of the sensitivity scores. Two main stylized facts can be listed:

1. Commodity markets exhibit a time-varying sensitivity to business cycle indicators. In Figures 20.1 and 20.2, the percentage of news to which commodity markets react and the cumulated sum of the absolute β vary strongly over time. For industrial metals and agricultural products, this time-varying pattern follows closely the phases of the US business cycle, according to the NBER. For energy markets, our results show a strong reaction to economic news during 2008–2009, especially what concerns the sensitivity score. These markets display a high sensitivity to Chinese news in 2008, and then to US news in 2009. The sensitivity to Chinese news is also observable in 2009 for industrial metals and agricultural products. Finally, precious metals are characterized by an increased sensitivity to news during 2004–2006. This period corresponds to rising inflation and tighter monetary policy. Under such circumstances, commodity markets typically display an increased sensitivity to economic signals. Precious metals also show an increased sensitivity to European news after the 2008–2009 crisis, i.e. their price tends to rise with the level of risk aversion.

2. The reaction of commodity markets to news depends on recession/expansion phases, but the picture looks globally more complex. In Figures 20.1 and 20.2, we notice that precious metals react to European news despite the end of the US economic crisis. The European sovereign crisis could therefore have led to a rising level of risk aversion that triggered a surge in the price of precious metals (through a flight-to-quality event). A similar comment can be made during the 2004–2005 US slowdown in leading indicators: agricultural prices have shown an increased sensitivity to economic news, both in terms of number and average market beta.

Having detailed the reaction of each type of commodity to the economic newsflow, we investigate more closely in the next section how commodity prices vary along the business cycle in various geographical zones.

---

* We wish to thank a referee for stressing this fact.

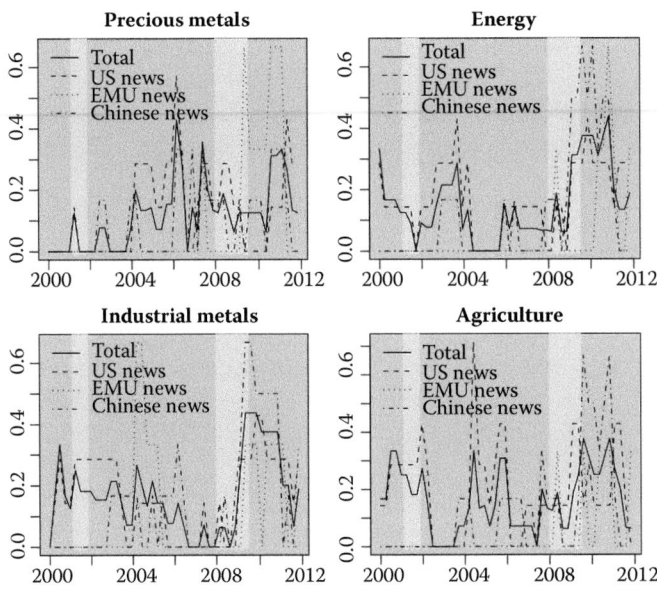

FIGURE 20.1 Percentage of news with a statistical impact over different sectors of the commodity market.

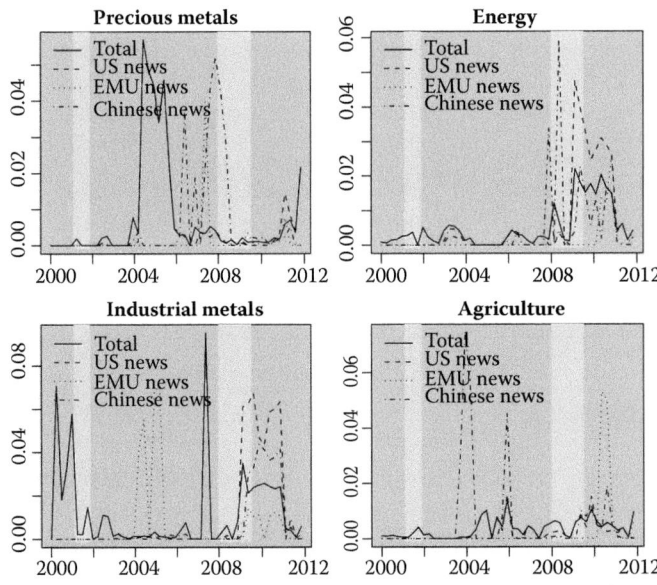

FIGURE 20.2 Rolling sensitivity index of news with a statistical impact over different sectors of the commodity market.

20.3 Economic Regimes and Commodity Markets as an Asset Class

This section is devoted to the understanding of the stylized behaviour of commodity prices depending on various phases of the business cycle. First, we present the structure of the Markov regime-switching model. Second, we investigate to which business cycle commodity markets seem to be mostly related. Third, we determine during which phase the commodity prices are more likely to rise or fall.

20.3.1 Measuring the Business Cycle

We recall here how to measure business cycles according to Hamilton (1989)'s methodology. Let X_t be a vector of variables representing the position of a given economy in the business cycle at time t. This economy is composed of N states, which are defined by a different expectation and covariance of X_t such that for the state $i \in [1:N]$:

$$X_t \sim N(\mu_i, \Sigma_i), \tag{20.6}$$

with μ_i, the vector of expectations for the elements of X_t, and Σ_i, the covariance matrix. When X_t is conditionally Gaussian, its unconditional distribution is a mixture of Gaussian distributions, that can encompass non-normality. The probability at each time t for the economy to be assigned to state i is time-varying. However, the transition matrix—i.e. the matrix containing the unconditional probabilities to move from state 1 to state 2—are fixed. Let us denote P this transition matrix. The parameters driving the model are the first two moments for each state, the probabilities involved in the transition matrix and the initial probabilities at time $t = 0$.

One of the main challenges for this model lies in the estimation of the parameters,[*] which are estimated by maximizing the log-likelihood function for a given time series. This log-likelihood function is given by:

$$L(\theta) = \sum_{t=1}^{T} \log f\left(X_t | \mathcal{F}_{t-1}, \theta\right), \tag{20.7}$$

with $f(.)$ the conditional density of X_t, T the total number of observations and θ a vector containing the parameters to be estimated. $\mathcal{F}_{t-1}$ represents the information available at time t:

$$\mathcal{F}_{t-1} = \sigma\{X_{t-1}, X_{t-2}, \ldots, X_1\}. \tag{20.8}$$

For a given θ, the conditional density at time t is:

$$f\left(X_t | \mathcal{F}_{t-1}, \theta\right) = \sum_{i=1}^{N} f\left(X_t | \mathcal{F}_{t-1}, \theta, S_t = i\right) P\left(S_t = i | \mathcal{F}_{t-1}\right), \tag{20.9}$$

with S_t the underlying state variable at time t. Intuitively, S_t is defined as the latent state of the economy, which is unknown to the econometrician but which can be inferred from the data by using a recursive filter in the spirit of the Kalman filter. Indeed, the central feature of the model is that the transition between regimes is governed by a Markov chain that assigns the probability of falling into the ith regime, according to the unobserved state variable.

$P\left(S_t = i | \mathcal{F}_{t-1}\right)$ is the probability to be assigned the state i at time t, conditionally upon the information available at time $t - 1$. The expression of $f\left(X_t | \mathcal{F}_{t-1}, \theta, S_t = i\right)$ is given by:

$$f\left(X_t | \mathcal{F}_{t-1}, \theta, S_t = i\right)$$
$$= \frac{1}{(2\pi)^{\frac{k}{2}} |\Sigma_i|} \exp\left\{-\frac{1}{2}(X_t - \mu_i)^T \Sigma_i^{-1} (X_t - \mu_i)\right\}, \tag{20.10}$$

---

[*] See Hamilton (1996) for a detailed presentation.

with k the size of the vector X_t (i.e. the number of variables used to measure the phase of the business cycle). The parameters of the Markov regime-switching model are estimated by selecting θ such that:

$$\theta^* : \max_\theta L(\theta) \qquad (20.11)$$

As mentioned above, the estimation process requires the use of a recursive filter—known as the Expectation Minimization (EM) algorithm—detailed in Hamilton (1989)'s original article. As is standard practice in the literature, we use the EM filter to obtain explicitly the parameters of the Markov regime-switching model for a given θ.

20.3.2 To Which Business Cycle Are the Commodity Markets Related?

To our best knowledge, previous literature identifies the USA as the leading business cycle for commodity markets (Frankel and Hardouvelis 1985, Barnhart 1989). This view is justified by the fact that the US GDP is closely related to the worldwide business cycle, especially given the openness of this economy. Besides, most commodity prices are labelled in US dollar, and the economic and monetary regimes under which the USA evolve can arguably be of a primary importance for commodity investors.

This hypothesis needs to be verified empirically. That is why we propose to conduct a formal statistical analysis of the performance of commodity markets through seven different domestic regimes in the USA, Europe, Germany, Canada, Australia, Brazil and China. These business cycles have been selected on the ground of the potential interactions between these economies and commodity markets. For instance, Australia and Canada are well-known commodity producers, while China, Europe and the USA are mostly on the demand side of commodities. Brazil lies in between the two categories, as both a producer and a consumer. Taken together, these case studies will help us understanding whether there is any regional-specific behaviour at stake in the price development of commodity markets, given that our sample includes various regions covering North and South America, Europe, Asia and Oceania.

During 1993–2011, we have gathered for each of these countries seasonally adjusted measures of industrial production (IP) and CPI. We choose to relate the business cycle to these two series for several reasons. First, the industrial production is a well-known cyclical measure of the business cycle. Whenever production capacities are tense (idle), we can infer that the economy is in a boom (bust). Second, standard approaches exist to compute IP and CPI figures, which eases the comparability across countries. Third, IP and CPI are available with a monthly frequency, whereas the GDP is only accessible for quarterly data.

Hence, this higher number of observations will translate into more precise estimates. Taken together, those three reasons explain our choices when it comes to the data selection for the macroeconomic variables of interest.[*] This data-set is composed of monthly series.

We estimate the Markov regime-switching models as detailed in Section 20.3.1. We set the number of regimes $N = 2$: Regime 1 is characterized as 'expansion' (i.e. production and prices are rising together), while regime 2 is referred to as 'recession' (i.e. prices record a slightly negative variation and production goes through a sharp drop). Estimation results are presented in Table 20.14 and in Figure 20.3.

According to Table 20.14, during expansion periods, the US industrial production is expected to grow by 3.94%/month, while during recession periods it decreases by 7.27%/month. Similar comments arise for the different countries investigated here.[†] Another common feature of these different cycles consists in the duration of each regime: expansion periods typically last for a longer period of time than recessions. For instance, in Germany, the economy is expected to spend 18.2% of the sample in recession, with the remaining data points falling into expansionary phases.

---

[*] Note that for Europe the sample starts in 1996.

[†] In Brazil, the recession was characterized by a surge in inflation during the so-called 'Tequila crisis' in the mid-1990s.

TABLE 20.14 Descriptive Statistics of the Domestic Cycles Estimated with the Markov Regime-Switching Model

			USA	Europe	Germany	Australia	Canada	China	Brazil
Expansion	Ind. Prod.	μ	3.949	3.101	3.696	4.512	2.977	15.291	4.179
		σ	0.661	0.91	1.012	0.806	1.116	1.636	1.922
	Inflation	μ	2.624	2.127	1.785	3.756	2.242	10.506	5.881
		σ	0.278	0.297	0.343	0.476	0.41	1.819	0.747
	Stat.	Pr. of expansion	0.984	0.981	0.969	0.936	0.963	0.978	0.988
		Freq. (months)	192	163	148	134	196	138	171
		Freq. (%)	0.828	0.901	0.818	0.59	0.856	0.603	0.747
Recession	Ind. Prod.	μ	−7.265	−12.28	−4.522	0.153	−4.561	12.153	−0.495
		σ	1.806	3.675	4.864	0.856	1.895	1.364	4.418
	Inflation	μ	1.827	0.234	0.351	1.78	0.104	0.557	41.933
		σ	0.994	0.371	0.347	0.358	0.617	0.53	13.333
	Stat.	Pr. of recession	0.92	0.849	0.878	0.918	0.821	0.967	0.947
		Freq. (months)	40	18	33	93	33	91	58
		Freq. (%)	0.172	0.099	0.182	0.41	0.144	0.397	0.253

Figure 20.3 shows the associated smoothed transition probabilities, along with the NBER business cycle reference dates (published by the NBER's Business Cycle Dating Committee*) in grey. Recessions start at the peak of a business cycle and end at the trough.

Estimates for the USA correspond most of the time to the NBER phases of the business cycle. In the rest of the world, the domestic cycles exhibit strong specificities. For instance, in 2008, the Eurozone has entered a recessionary regime later than the USA, and also exited later. In 2011, Germany has been impacted by a slowdown, without being followed by any other country.

Next, for each of these domestic cycles, we attempt to relate the performance of the S&P 500 (as a proxy of equity markets) compared to the GSCI and its sub-indices (Agriculture, Energy, Industrial Metals and Precious Metals). By doing so, we aim at evaluating how commodity markets behave along the business cycle, by explicitly taking into account country- and market-specific effects.

We are interested in the forecasting horizon that leads to the greatest discrimination between asset performances during expansionary and recessionary regimes. Thus, we need to find h such that excess returns—once adjusted from volatility—are as different as possible.

Let $SR_h^{(i)}(m)$ be the Sharpe ratio obtained during state m for the asset i when considering that commodity markets are ahead of the business cycle by h months. We select h such that:

$$\max_h \sum_{j=1}^{N} \sum_{k=1, k \neq j, k>j}^{N} \left(SR_h^{(i)}(m_j) - SR_h^{(i)}(m_k) \right)^2, \quad (20.12)$$

with m_j, the jth, state of the economy, and N, the total number of states. With $N = 2$, we have:

$$\max_h \left(SR_h^{(i)}(m_1) - SR_h^{(i)}(m_2) \right)^2. \quad (20.13)$$

This maximum spread between the Sharpe ratios can be used to measure the influence of a given phase of the business cycle on a given market. The results obtained are presented during the full sample (1993–2011) in Table 20.15, and during the subsample period of 'commodity boom' (2004–2011) in Table 20.16.

---

* See more on the NBER Business Cycle Expansions and Contractions at http://www.nber.org/cycles.html.

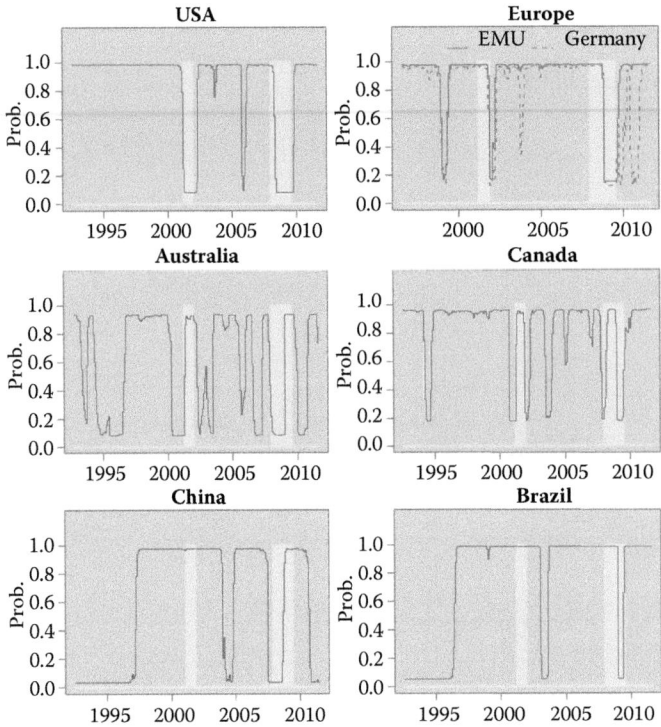

FIGURE 20.3   Business conditions during 1993–2011 estimated with the Markov regime-switching model.

The following comments arise:

- During the full sample, the GSCI shows a strong connection to the business cycle located in the USA, Europe, Germany and Canada. The maximum score ranges from 0.23 in the USA to 0.42 in Europe. Therefore, we uncover that the US business cycle does not display the strongest relationship with commodity markets. This surprising finding is consistent across the GSCI sub-indices and during the sub-sample as well.

- The agricultural and energy sectors have the highest correlation with the business cycle measures. For instance, the maximum score is equal to 0.51 concerning the relationship between the GSCI energy-index and the European business cycle.

- During the full sample, industrial metals are not found to be the most cyclical sector. However, during 2004–2011, industrial metals are characterized by the strongest connection with the business cycle measures (with a maximum score equal to 0.66 in China). This result illustrates the influence of the sustained Chinese growth on the demand for raw materials in the industrial sector.

- In the long run, we find a limited evidence of any connection between commodity markets and emerging countries such as China and Brazil, or with Australia. The bulk of the long-term relationship between macroeconomic conditions and commodity markets seems to occur in Europe and North America. For agricultural products, however, notice that the maximum score (equal to 0.19) is reached in China.

- During 2004–2011, the influence of Brazil and China has been growing: for industrial metals, the maximum score is obtained with the Chinese business cycle and a one-month forward looking horizon. Similarly, the influence of the Brazilian business cycle is as strong as the USA's for the energy and industrial metals sectors (with a maximum score close to 0.4).

TABLE 20.15   Worldwide Business Cycles and Commodities Performances: 1993–2011

	AU	BR	CA	CH	EU	GE	US
GSCI Agriculture							
Recession	0.28%	−0.18%	−1.53%	1.22%	−4.20%	−3.10%	−2.77%
Expansion	−0.17%	0.11%	0.27%	−0.76%	0.10%	0.40%	0.49%
Lead	1	5	7	1	8	7	1
Score max.	0.04	0.03	0.14	0.19	0.25	0.25	0.24
GSCI Energy							
Recession	1.15%	−1.85%	−3.56%	0.50%	−10.17%	−5.82%	−3.63%
Expansion	−0.19%	1.15%	1.08%	0.22%	1.72%	1.96%	1.08%
Lead	1	2	5	8	3	3	1
Score max.	0.07	0.16	0.2	0.03	0.51	0.37	0.22
GSCI Industrial Metals							
Recession	−0.22%	−0.94%	−2.68%	0.52%	−6.10%	−2.76%	−2.80%
Expansion	1.07%	1.10%	1.15%	0.59%	1.40%	1.42%	1.23%
Lead	1	3	5	1	6	5	6
Score max.	0.11	0.17	0.29	0	0.51	0.3	0.3
GSCI Precious Metals							
Recession	1.14%	0.98%	1.68%	1.04%	0.41%	1.74%	0.64%
Expansion	0.39%	0.64%	0.56%	0.46%	0.82%	0.63%	0.69%
Lead	8	1	1	6	5	2	8
Score max.	0.08	0.06	0.14	0.1	0.07	0.06	0.03
GSCI							
Recession	0.57%	−1.04%	−2.99%	0.70%	−5.89%	−3.02%	−2.16%
Expansion	0.55%	1.18%	1.19%	0.38%	1.45%	1.55%	1.10%
Lead	1	3	5	8	3	3	5
Score max.	0.01	0.18	0.27	0.06	0.42	0.31	0.23
SP500							
Recession	−0.14%	−0.39%	−1.48%	0.75%	−4.74%	−1.68%	−2.71%
Expansion	0.95%	0.79%	0.87%	0.34%	0.98%	0.87%	1.11%
Lead	1	6	3	5	5	5	6
Score max.	0.13	0.13	0.23	0.09	0.45	0.23	0.38

- Most of the commodities are found to be procyclical: during expansion periods, commodity prices record a positive momentum. On the contrary, recession periods are characterized by retreating prices. This behaviour is very similar to the S&P 500.
- Precious metals stand out as the only exception. First, they display on average the weakest relationship to any business cycle. Second, they record either negative or close-to-zero performances during periods of growth, and positive returns during periods of recession. This finding is consistent with the role of gold as a 'safe haven' where to store value during recessions (Baur and McDermott 2010, Beaudry *et al.* 2011), and as a hedge against inflation during expansion regimes.

TABLE 20.16 Worldwide Business Cycles and Commodities Performances: 2004–2011

	AU	BR	CA	CH	EU	GE	US
			GSCI	Agriculture			
Recession	−1.61%	−3.83%	−4.67%	−5.02%	−5.44%	−6.62%	−2.76%
Expansion	1.38%	0.90%	1.11%	0.80%	1.06%	1.05%	1.06%
Lead	8	8	4	1	3	3	2
Score max.	0.22	0.29	0.37	0.33	0.37	0.48	0.25
			GSCI	Energy			
	AU	BR	CA	CH	EU	GE	US
Recession	−2.54%	−8.20%	−6.42%	−10.54%	−9.47%	−13.94%	−5.78%
Expansion	1.99%	1.42%	1.24%	1.34%	1.20%	1.66%	1.53%
Lead	8	8	3	1	1	3	2
Score max.	0.24	0.39	0.32	0.41	0.41	0.65	0.31
			GSCI	Industrial Metals			
	AU	BR	CA	CH	EU	GE	US
Recession	−1.42%	−5.10%	−4.49%	−8.90%	−7.87%	−9.01%	−4.83%
Expansion	3.37%	2.61%	2.22%	2.44%	2.64%	2.43%	2.95%
Lead	8	8	8	1	3	4	5
Score max.	0.36	0.47	0.4	0.66	0.64	0.61	0.5
			GSCI	Precious Metals			
	AU	BR	CA	CH	EU	GE	US
Recession	1.97%	2.06%	1.88%	2.91%	0.97%	1.67%	1.70%
Expansion	1.12%	1.33%	1.36%	1.22%	1.53%	1.42%	1.42%
Lead	1	1	1	8	8	1	7
Score max.	0.11	0	0.02	0.09	0.07	0.04	0.06
			GSCI				
	AU	BR	CA	CH	EU	GE	US
Recession	−1.43%	−5.47%	−4.58%	−8.92%	−5.87%	−10.38%	−3.40%
Expansion	2.88%	2.08%	2.04%	2.13%	1.98%	2.32%	2.11%
Lead	8	8	3	1	3	3	2
Score max.	0.33	0.42	0.37	0.53	0.4	0.67	0.34
			SP500				
	AU	BR	CA	CH	EU	GE	US
Recession	−0.85%	−2.10%	−2.78%	−5.96%	−4.95%	−5.06%	−3.51%
Expansion	1.28%	0.88%	0.99%	1.16%	1.25%	1.05%	1.44%
Lead	6	8	5	1	3	5	4
Score max.	0.28	0.27	0.38	0.66	0.54	0.51	0.51

- Finally, when the S&P 500 seems to be anticipating the business cycle by five to six months, commodities present a larger variety of forecasting horizons (ranging from 1 to 8). During the full sample, these horizons are closer to 3–5 when focusing only on the highest maximum score. Hence, it seems that commodity markets contain less anticipation from market agents than equity markets.

Having detailed the relationship between commodity prices and the business cycle in emerging and developed countries, we investigate the following question: during which phase of the business cycle are commodity prices more likely to rise or fall?

20.3.3 Commodity Performances Depending on the Nature of Each Economic Regime

In this section, we use a large data-set of US economic time series during 1984–2011. We select the following key economic indicators: the industrial production, the consumer price index, consumer good inventories, durable good inventories, the unemployment rate and the Fed Target Rate. Hence, we capture various characteristics of the US business cycle, going from inventory to inflation cycles. The dataset is composed of monthly series, and comes from Bloomberg.

Concerning the number of regimes necessary to model the US economy, previous literature focuses on three states (Sichel 1994, Boldin 1996, Clements and Krolzig 2003). As pointed by Hamilton (1996), the assumption that the process describing the data presents a given number N of regimes cannot be tested by using the usual likelihood ratio test, as specific regularity conditions are not fulfilled.[*]

To test explicitly that a Markov regime-switching model with two states is superior to another model with three states for instance, we use Vuong (1989)'s test based on the distributional goodness-of-fit that the model is able to provide. This approach has been applied recently to US equity and credit markets by Ielpo (2012). Formally, to compare a Markov regime-switching model with N_i states to another model with N_j states, we need to compute the following test statistics:

$$t_{N_t, N_J} = \frac{1}{T} \sum_{t=1}^{T} \left(\log f_{\hat{\theta}N_t} \left(X_t | \mathcal{F}_{t-1} \right) - \log f_{\hat{\theta}N_j} \left(X_t | \mathcal{F}_{t-1} \right) \right),$$

(20.14)

Under the null hypothesis that the Markov regime-switching model with N_i states provides an equivalent fit to the model with N_j states, this statistic is distributed as:

$$\frac{t_{N_t, N_J}}{\hat{\sigma}_T} \sqrt{T} \sim N(0,1),$$

(20.15)

with σ_T the standard deviation associated to the statistic t_{N_t, N_J}. The alternative hypothesis is that the test statistic is different from 0, and that the forecasting ability of both models is not the same. The standard deviation $\hat{\sigma}_T$ is estimated by using the Newey–West long-run variance (HAC) estimator.

The best model is selected so that it is preferred to models with a lower number of states, but equivalent to models with a higher number of states. The resulting model specification should be parsimonious to avoid the problem of overfitting (Bradley and Jansen 2004), while being as consistent as possible with the joint distribution of returns.

When performing Vuong (1989)'s test, we find that $N = 5$ as shown in Table 20.17. We obtain two expansion regimes, two recession regimes and one 'stalling regime'. These regimes are displayed in Figure 20.4, with descriptive statistics given in Table 20.18. The following comments arise:

- Regimes 1 and 5 are *expansionary* regimes:
  i. Regime 5 is characterized by a 3.6%/month increase in industrial production, rising inflation, receding unemployment, but still—on average—decreasing Fed's decision rates. Inventories are weakly evolving.
  ii. Regime 1 is very similar to Regime 5. As we observe strongly building inventories, and an increasing Fed Target Rate, we label this regime as 'strong growth'.

---

[*] Nonetheless, Hamilton (1996) presents a variety of tests to determine whether an additional state is required to model the dynamics of the data.

TABLE 20.17   Vuong (1989)'s Test to Select the Appropriate Number of Regimes

	1 state	2 states	3 states	4 states	5 states	6 states
1 state		−7.4	−10.05	−11.4	−13.31	−12.7
2 states			−6.48	−10.62	−13.49	−13.3
3 states				−5.01	−10.22	−8.47
4 states					−5.16	−4.86
5 states						−1.72
6 states						

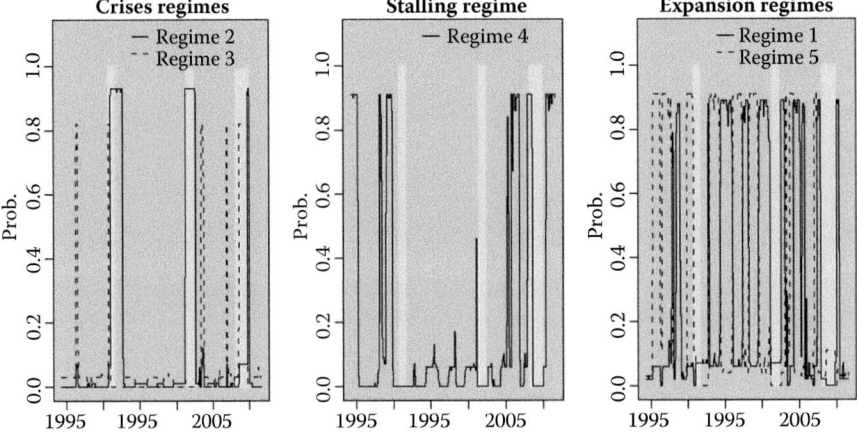

FIGURE 20.4   Business conditions during 1984–2011 estimated with the Markov regime-switching model.

TABLE 20.18   Descriptive Statistics of the Estimated Economic Regimes

		Strong exp.	Slowdown	Strong Crisis	Stalling	Medium Expans.
		Reg. 1	Reg. 2	Reg. 3	Reg. 4	Reg. 5
Indus. Prod.	μ	0.041	0	−0.076	0.021	0.036
	σ	0.009	0.017	0.024	0.011	0.007
CPI	μ	0.029	0.025	0.007	0.036	0.031
	σ	0.003	0.004	0.017	0.006	0.004
Good Inv.	μ	0.055	−0.065	−0.048	0.095	0.005
	σ	0.008	0.016	0.027	0.011	0.008
Dur. Good Inv.	μ	0.048	−0.079	−0.041	0.099	0.006
	σ	0.009	0.016	0.023	0.011	0.01
Unempl. rate	μ	−0.382	1.172	2.276	−0.307	−0.411
	σ	0.18	0.213	0.527	0.265	0.199
Fed TR	μ	0.971	−2.629	−1.379	−0.348	−0.399
	σ	0.407	0.565	0.513	1.042	0.358
	Freq. (months)	92	40	27	67	102
	Freq. (%)	0.28	0.122	0.082	0.204	0.311

- Regime 2 and 3 are *recessionary* regimes:
  i. Regime 2 is characterized by a strong decline in industrial production, with unemployment rate rising rapidly and lowered Fed decision rates. By looking at Figure 20.4, we understand that this regime corresponds to the burst of an economic crisis.
  ii. Regime 3 is a typical slowdown period: industrial production decreases—but not as much as for Regime 2—and the Fed decreases even more aggressively its decision rate, so as to limit as much as possible the building of unemployment. When examining the transition matrix presented in Table 20.19, the persistence of this regime is weaker than the persistence of Regime 2.
- Regime 4 is a 'stalling' regime, i.e. between expansion and recession. Industrial production grows by 2%/month, whereas good and durable good inventories are building up sharply, underlining the weak underlying consumers' demand.

Table 20.20 presents the optimal forecasting horizon that should be retained to link the five regimes to the performances of commodities. We use the same metric as presented in Equation (20.13) to which we refer as the 'maximum score'. Based on these results, we can describe the stylized performances of commodity sectors, depending on the phase of the business cycle.

Table 20.21 presents the performance by regime for each of the assets considered. The main findings can be summarized as follows:

- When entering the crisis, all commodities deliver strongly negative returns, except precious metals (with an increase by 1.29%/month on average). Hence, we verify the role of 'safe haven' for precious metals when entering a recession. Other commodity markets logically underperform

TABLE 20.19 Estimated Transition Matrix between Regimes

	Reg. 1 (%)	Reg. 2 (%)	Reg. 3 (%)	Reg. 4 (%)	Reg. 5 (%)
Reg. 1	89	1	0	6	4
Reg. 2	7	93	0	0	0
Reg. 3	0	7	82	0	11
Reg. 4	2	0	3	91	3
Reg. 5	6	0	3	0	91

TABLE 20.20 Optimal Forecasting Horizon for the Various Asset Classes

	GSCI Agri.	GSCI Energy	GSCI Ind. Metals	GSCI Prec. Metals	GSCI Total	SP500	10Y US	US Dollar
Leading +0	0.16	0.25	0.2	0.07	0.24	0.12	0.09	0.14
Leading +1	0.14	0.27	0.2	0.11	0.29	0.21	0.05	0.08
Leading +2	0.26	0.27	0.24	0.11	0.3	0.24	0.09	0.06
Leading +3	0.25	0.26	0.22	0.13	0.29	0.24	0.08	0.1
Leading +4	*0.31*	*0.3*	0.32	0.14	*0.32*	0.25	0.08	0.06
Leading +5	0.22	0.26	*0.35*	*0.18*	0.3	0.29	0.16	0.07
Leading +6	0.19	0.21	0.31	0.11	0.26	0.31	0.17	0.1
Leading +7	0.17	0.18	0.29	0.21	0.26	0.36	0.22	0.1
Leading +8	0.22	0.19	0.36	0.15	0.28	*0.41*	*0.3*	*0.13*
Leading +9	0.19	0.13	0.37	0.14	0.24	0.4	0.24	0.11
Leading +10	0.11	0.05	0.34	0.08	0.18	0.39	0.18	0.15
Leading +11	0.17	0.12	0.34	0.18	0.22	0.36	0.33	0.18
Leading +12	0.27	0.16	0.27	0.2	0.27	0.37	0.31	0.27

TABLE 20.21   Performance of Various Indices Depending on Each of the Economic Regimes

	Strong exp. State 1 (%)	Slowdown State 2 (%)	Strong Crisis State 3 (%)	Stalling State 4 (%)	Medium Expans. State 5 (%)
GSCI Agri.	0.47	−0.72	−2.87	1.27	0.27
GSCI Energy	2.63	0.32	−4.14	1.60	−0.18
GSCI Ind. Metals	2.11	0.21	−2.12	3.44	−0.45
GSCI Prec. Metals	0.55	−0.51	1.29	1.25	0.05
GSCI Total	1.76	−0.34	−2.63	1.36	−0.37
SP500	1.37	−0.80	−2.43	0.84	1.20
10Y US	1.36	−0.91	−2.58	−0.03	−1.66
US Dollar	−0.31	0.11	−0.29	0.24	−0.30

compared to the S&P 500. For instance, the energy sector records a drop by −4.14%/month vs. −2.43%/month for equities.

- During the prolonged slowdown period, central banks are stimulating the economy. As a consequence, we find that energy and industrial metals commodities are improving on average by, respectively, 0.32%/month and 0.21%/month. Other cyclical commodities exhibit decreasing prices.

- During the medium expansion regime, agricultural prices are rising by 0.27%/month. On the contrary, energy and industrial metals prices are decreasing by, respectively, −0.18%/month and −0.45%/month. More importantly, the S&P 500 outperforms all commodities. Investors would therefore not increase their exposure to commodity markets during periods of medium growth.

- During strong expansions, energy and industrial metals prices show positive returns, and even outperform equity returns. This strong expansionary phase is typically a period of 'commodity boom' for investors.

- During the stalling regime, all commodity prices are rising and outperform the returns on the S&P 500. With roaring inflation, even the precious metals deliver positive returns, consistent with the idea that these metals are also used as a hedge against inflation. Industrial metals are found to outperform all commodity indices.

By decomposing the business cycle into five regimes, we have successfully shown in this section that (long) investors should invest into commodities especially during the stalling regime, i.e. at the end of the economic cycle (before the recession). According to our estimates, the stalling regime corresponds to 20% of the data sample.

20.4 Conclusion

The contributions of the paper are twofold: (i) evaluating in a regression framework with EGARCH effects the influence of economic news on each type of commodity (agricultural products, energy markets, industrial and precious metals), and (ii) studying the evolution of commodity prices along the business cycle in different geographic regions based on the class of Markov regime-switching models.

From an investment point of view, the main results may be summarized as follows. First, investors should not add commodities to their portfolio on the ground of a low correlation of the commodity with standard financial assets (such as bonds and equities). Indeed, we uncover a strong correlation of commodity markets with risky assets during economic downturns. Second, the economic influences on commodity prices are rather complex: the patterns detected for each type of commodity sub-index vary greatly between the geographical zones in Europe, the US, China, etc. Third, there is a cyclical rotation amongst commodity markets: during periods of strong growth, investors should overweigh industrial metals and energy vs. agricultural products and industrial metals. Finally, 'stalling' economic regimes correspond to periods during which commodities outperform the S&P 500.

References

Alizadeh, A.H., Nomikos, N.K. and Pouliasis, P.K., A Markov regime switching approach for hedging energy commodities. *J. Bank. Financ.*, 2008, **32**(9), 1970–1983.

Andersen, L., Markov models for commodity futures: Theory and practice. *Quant. Finance*, 2010, **10**(8), 831–854.

Barnhart, S.W., The effects of macroeconomic announcements on commodity prices. *Am. J. Agr. Econ.*, 1989, **17**(2), 389–403.

Baur, D.G. and McDermott, T.K., Is gold a safe haven? International evidence. *J. Bank. Financ.*, 2010, **34**, 1886–1898.

Beaudry, P., Collard, F. and Portier, F., Gold rush fever in business cycles. *J. Monetary Econ.*, 2011, **58**, 84–97.

Boldin, M.D., A check on the robustness of Hamilton's Markov Switching model approach to the economic analysis of the business cycle. *Stud. Nonlinear Dyn. E.*, 1996, **1**, 35–46.

Bradley, M.D. and Jansen, D.W., Forecasting with a nonlinear dynamic model of stock returns and industrial production. *Int. J. Forecasting*, 2004, **20**(2), 321–342.

Caballero, R.J., Farhi, E. and Gourinchas, P.O., Financial crash, commodity prices, and global imbalances. *Brookings Pap. Eco. Act.*, 2008, Fall **2008**, 1–54.

Chan, K.F., Treepongkaruna, S., Brooks, R. and Gray, S., Asset market linkages: Evidence from financial, commodity and real estate assets. *J. Bank. Financ.*, 2011, **35**(6), 1415–1426.

Chevallier, J., Global imbalances, cross-market linkages, and the financial crisis: A multivariate Markov-switching analysis. *Econ. Modell.*, 2012, **29**(3), 943–973.

Chng, M.T., Economic linkages across commodity futures: Hedging and trading implications. *J. Bank. Financ.*, 2009, **33**(5), 958–970.

Clements, M.P. and Krolzig, H.M., Business cycle asymmetries: Characterization and testing based on Markov-switching autoregressions. *J. Bus. Econ. Stat.*, 2003, **21**(1), 196–211.

Daskalaki, C. and Skiadopoulos, G., Should investors include commodities in their portfolios after all? New evidence. *J. Bank. Financ.*, 2011, **35**(10), 2606–2626.

Dionne, G., Gauthier, G., Hammami, K., Maurice, M. and Simonato, J.G., A reduced form model of default spreads with Markov switching macroeconomic factors. *J. Bank. Financ.*, 2011, **35**, 1984–2000.

Elder, J., Miao, H. and Ramchander, S., Impact of macroeconomic news on metal futures. *J. Bank. Financ.*, 2012, **36**(1), 51–65.

Frankel, J.A. and Hardouvelis, G.A., Commodity prices, money surprises and fed credibility. *J. Money Credit Bank.*, 1985, **17**, 425–438.

Geman, H. and Ohana, S., Time-consistency in managing a commodity portfolio: A dynamic risk measure approach. *J. Bank. Financ.*, 2008, **32**(10), 1991–2005.

Hamilton, J., A new approach to the economic analysis of nonstationary time series and the business cycle. *Econometrica*, 1989, **57**(2), 357–384.

Hamilton, J., *Time Series Analysis*, 2nd ed., 1996 (Princeton University Press: Princeton, NJ).

Ielpo, F., Equity, credit and the business cycle. *Appl. Financ. Econ.*, 2012, **22**(12), 939–954.

Miffre, J. and Rallis, G., Momentum strategies in commodity futures markets. *J. Bank. Financ.*, 2007, **31**(6), 1863–1886.

Miltersen, K.L., Commodity price modelling that matches current observables: A new approach. *Quant. Finance*, 2003, **3**(1), 51–58.

Sichel, D., Inventories and the three phases of the business cycle. *J. Bus. Eco. Stat.*, 1994, **12**, 269–277.

Tang, K., Time-varying long-run mean of commodity prices and the modeling of futures term structures. *Quant. Finance*, 2012, **12**, 781–790.

Tang, K. and Xiong, W., Index investing and the financialization of commodities. NBER Working Paper #16385, Cambridge, MA, 2010.

Vuong, Q.H., Likelihood ratio tests for model selection and non-nested hypotheses. *Econometrica*, 1989, **57**(2), 307–333.

Appendix: Database of News

The Appendix provides background information on the Bloomberg data-set of news used in the paper.

- *Non-Farm Payroll*: number of jobs added or lost in the US economy over the last month, not including jobs relating to the farming industry. It includes goods-producing, construction and manufacturing companies.
- *ISM*: survey from the Institute for Supply Management based on comments from purchasing managers in the manufacturing sector. It provides the earliest clues of how the US economy has fared during the previous four weeks.
- *Jobless Claims*: the weekly number of new jobless applications in the US.
- *US CPI MoM*: Consumer Price Index Month on Month.
- *US Retail Sales*: Retail and Food Services Sales in million US Dollars.
- *Fed Target Rate*: interest rate at which depository institutions actively trade balances held at the Federal Reserve.
- *US GDP*: US Gross Domestic Product.
- *ZEW Eco. Sent.*: survey from the Mannheim Centre for European Economic Research (ZEW). Expectations from 350 financial experts for the Euro-zone, Japan, Great Britain and the USA.
- *IFO Expectations*: survey from the Munich Society for the Promotion of Economic Research (IFO). 7000 monthly survey responses from firms in manufacturing, construction, wholesaling and retailing. Firms are asked to give their assessments of the current business situation and their expectations for the next six months.
- *EMU CPI*: Euro Area Consumer Price Index from Eurostat.
- *EMU GDP*: Euro Area Gross Domestic Product from Eurostat.
- *FR Bus. Conf.*: France Business Confidence index from INSEE.
- *ECB Ref. Rate*: minimum refinancing rate from the European Central Bank.
- *China CPI YoY*: China Consumer Price Index Year on Year
- *China Ind. Prod.*: China Industrial Production index.
- *China PMI*: China Purchasing Managers' Index.

Overall, the news cover different aspects of the business cycle:

- **real activity** with macroeconomic data (such as industrial production) or surveys,
- **inflation** dynamics with the CPI,
- **monetary** variables with the Fed Target Rate and the ECB's minimum refinancing rate.

21

A Hybrid Commodity and Interest Rate Market Model

Kay F. Pilz and
Erik Schlögl

21.1 Introduction

Traditionally, market risks in different asset classes, such as FX, fixed income or commodities, have been modelled separately, although modelling approaches incorporating several sources of market risk are becoming increasingly popular. The aim of this paper is to contribute to the more integrated approach by applying the multi-currency LIBOR Market Model (LMM) as presented by Schlögl (2002b) jointly to interest rates and commodities. The domestic fixed income market will be interpreted in the usual way for the LMM, i.e. with a given bond market paying in a certain currency (say USD), whereas the foreign market will be associated with a commodity market (e.g. crude oil), with the physical commodity as its currency, and the 'convenience yield' as its rate of interest. The contribution of this paper goes beyond a simple 're-interpretation' of the multi-currency LMM in that issues arising in the application of the model to actual commodity market data are specifically addressed. Firstly, liquid market prices are only available for options on commodity futures, rather than forwards, thus the difference between forward and futures prices must be explicitly taken into account in the calibration. Secondly, we construct a procedure to achieve a consistent fit of the model to market data for interest rate options, commodity options and historically estimated correlations between interest rates and commodity prices. This is achieved in a two-stage process, calibrating first to the interest rate and commodity markets separately, followed by an orthonormal transformation of the commodity volatility vectors to 'rotate' the commodity volatilities relative to the interest rate volatilities in such a manner as to achieve the desired correlations between the two markets. Thirdly, we illustrate the use of the model on actual market data and demonstrate a way of efficiently calculating commodity spread options.

DOI: 10.1201/9781003265399-24

For the market data illustration, we chose WTI Crude Oil as the commodity and US dollars as the 'domestic' currency. Thus the 'exchange rate' is the WTI Crude Oil price, i.e. the WTI Crude Oil futures quotes—when converted to forward prices using an appropriate convexity correction—can be interpreted as forward exchange rates between the US dollar economy and an economy where value is measured in terms of units of WTI Crude Oil (and where convenience yields play the role of interest rates). In this example, the model is calibrated to USD at-the-money caplets and swaptions, as well as WTI Crude Oil futures and ATM options on futures.

The model presented here builds on the prior literature on the LMM, starting from the seminal work of Miltersen *et al.* (1997), Brace *et al.* (1997) and Jamshidian (1997). The model construction follows Musiela and Rutkowski (1997) and the calibration method builds on the approach first proposed by Pedersen (1998) for the basic LMM.

The earliest work combining commodity and interest rate risk (based on dynamics of the continuously compounded short rate, without reference to a full model calibrated to an initial term structure) goes back to Schwartz (1982). Subsequently, models for commodity price dynamics incorporating stochastic convenience yields have been constructed by a number of authors, for example Gibson and Schwartz (1990), Cortazar and Schwartz (1994), Schwartz (1997), Miltersen and Schwartz (1998), and Miltersen (2003). These models are typically constructed on the basis of a Heath *et al.* (1992) term structure model with generalized (possibly multi-factor) Ornstein/Uhlenbeck dynamics, i.e. continuously compounded convenience yields (and possibly interest rates rates) are normally distributed. While in theory these models allow some freedom to calibrate to market data, the published work does not contain much evidence as to whether an effective calibration to available commodity and interest rate options—even at the money—can be achieved. This type of fit to observed market prices, in particular to at-the-money options, is a strength of the LIBOR Market Models that we exploit in the present paper.

The paper is organized as follows. The basic notation, the results of the single- and multi-currency LMM and their interpretation in the context of commodities are presented in Section 21.2. In Section 21.3 the calibration of the commodity part of the Commodity LMM to plain vanilla options is discussed. In all major commodity markets, only options with the same maturity as the underlying forward are traded liquidly. Hence, no volatility term structure for a particular forward commodity price can be extracted only from the options written on this forward. To overcome this problem, option prices for all maturities are used to calibrate simultaneously a volatility term structure, which is desired to be as time-homogeneous and smooth as possible. In keeping with the assumptions of the LMM, we thus focus on the volatility term structure only, i.e. with time dependence but no strike dependence. In Section 21.4 the relationship between futures and forwards in the model is presented, which permits calibration of the model to futures as well as forwards. The calibration of the interest rate part of the hybrid Commodity LMM will not be discussed in detail in this paper, because this problem has already been addressed by many authors and most methods should be compatible with our model. However, in Section 21.5 we discuss how both separately calibrated parts—the interest rate and the commodity part—of the model can be merged in order to have one underlying d-dimensional Brownian motion for the joint model and still match the market prices used for calibration of the particular parts. Section 21.6 illustrates the application of the Commodity LMM to real market data. Finally, Section 21.7 addresses the pricing of commodity spread options in the model.

21.2 The Commodity LIBOR Market Model

21.2.1 The Interest Rate Market

The construction of the LMM for the domestic interest rate market follows the presentation of Musiela and Rutkowski (1997). Assume a given probability space $\left(\Omega, \{\mathcal{F}_t\}_{t\in[0,T^*]}, \mathbb{P} \right)$, where the underlying filtration $\{\mathbb{F}_t\}_{t\in[0,T^*]}$ coincides with the P-augmentation of the natural filtration of a d-dimensional standard

Brownian motion W. Let $T^\star$ be a fixed time horizon, then a family of bond prices is any family of strictly positive real-valued adapted processes $B(t, T)$ for $t \in [0, T]$, with $B(T, T) = 1$ for every $T \in [0, T^\star]$. The bond price $B(t, T)$ for $T \in [t, T^\star]$ is the amount that has to be invested at time t to receive one unit of the domestic currency at time T.

Corresponding to assumptions (BP.1) and (BP.2) of Musiela and Rutkowski (1997), we have that the bond price $B(t, T)$ for $t \leq T \leq T^\star$ is a strictly positive special semi-martingale (see Musiela and Rutkowski (1997, p. 263) for the definition) and that the forward process,

$$F_B\left(t, T, T^\star\right) := \frac{B(t, T)}{B\left(t, T^\star\right)}, \tag{21.1}$$

follows a martingale under the $T^\star$-forward measure $\mathbb{P}$. Equivalently, the bond price process satisfies

$$B(t, T) = \mathbb{E}_{\mathbb{P}}\left[\left. \frac{B\left(t, T^\star\right)}{B\left(T, T^\star\right)} \right| \mathcal{F}_t\right], \tag{21.2}$$

for all $t \in [0, T]$.

We present some conclusions from these assumptions, which will be of use later and can be found in Musiela and Rutkowski (1997). Firstly, there exists a $\mathbb{R}^d$-valued process $\gamma(t, T, T^\star)$, such that the forward process has the representation

$$dF_B\left(t, T, T^*\right) = F_B\left(t, T, T^*\right)\gamma\left(t, T, T^*\right) \cdot dW_{T^*}(t). \tag{21.3}$$

By Itô's formula and Girsanov's theorem it follows that, for any given $S, T \in [0, T^\star]$, the dynamics of $F_B(t, S, T)$ for $t \in [0, S \wedge T]$ can be written as

$$dF_B(t, S, T) = F_B(t, S, T)\gamma(t, S, T) \cdot dW_T(t), \tag{21.4}$$

with

$$\gamma(t, S, T) = \gamma\left(t, S, T^*\right) - \gamma\left(t, T, T^*\right)$$

$$dW_T(t) = dW_{T^*}(t) - \gamma\left(t, T, T^*\right)dt. \tag{21.5}$$

Hence, $F_B(t, S, T)$ is an exponential (local) martingale under the $\mathbb{P}_T$ measure, and since both $\mathbb{P}$ and $\mathbb{P}_T$ are absolutely continuous, the Radon–Nikodým derivative is given by the Doléans exponential

$$\frac{d\mathbb{P}_T}{d\mathbb{P}} = \varepsilon_T\left(\int_0^{\cdot} \gamma\left(u, T, T^*\right) \cdot dW_{T^*}(u)\right), \quad \mathbb{P}_T\text{-a.s.}$$

The measure $\mathbb{P}_T$ is usually denoted as the T-forward measure, and $\mathbb{P}_{T^*} = \mathbb{P}$ additionally as the terminal measure.

For practical purposes it is often convenient to use a discrete-tenor version of the LMM.[*] Therefore, we assume that the time horizon $T^\star = N\delta$ is a multiple of a fixed period δ. Then, the LIBOR forward rate

---

[*] Where an extension to continuous tenor is required, it is typically more practical to start with a discrete-tenor version of the LMM and extend it using 'interpolation by daycount fractions' as in Schlögl (2002a) in order to avoid the infinite-dimensional state variables resulting from the continuous-tenor extensions proposed by Brace *et al.* (1997) and Musiela and Rutkowski (1997).

$L(t, T)$ as seen at time t, for an investment of one currency unit at time T with pay-off $B(t, T)/B(t, T + \delta)$ at $T + \delta$, can be written as

$$dL(t,T) = L(t,T)\lambda(t,T) \cdot dW_T(t), \tag{21.6}$$

for $t \in [0, T]$. The volatility function of LIBOR relates to the volatility function of the forward process by

$$\gamma(t,T,T+\delta) = \frac{\delta L(t,T)}{1+\delta L(t,T)}\lambda(t,T). \tag{21.7}$$

To shorten notation, define $T_i = i\delta$ (for $0 \leq i \leq N$).

21.2.2 The Commodity Market

The approach incorporating a commodity as followed in this section parallels the multi-currency LMM introduced by Schlögl (2002b). The commodity market can naturally be considered as a 'foreign interest rate market', where the currency is the physical commodity itself, e.g. as in our example in Section 21.6, the value of any asset is measured in units (barrels) of crude oil instead of dollars. The bond prices $C(t, T)$ quote (as seen at time t) the amount of the commodity that has to be invested at time t to physically receive one unit of the commodity at time T. Corresponding LIBORs can be interpreted in the same way, i.e. the yield of $C(t, T)$ is the convenience yield (adjusted for storage costs, if applicable). Although these rates have a natural interpretation, they are usually not traded (liquidly) for most of the commodities. We will discuss below how this model can still be calibrated to the instruments commonly traded in the commodities market.

Some assumptions are required with regard to the commodity market. Firstly, all assumptions made in constructing the LMM on the domestic interest rate market in the previous section are assumed to be true for the commodity market as well. This particularly includes the process $C(t, T)$ for $0 \leq t \leq T \leq T^*$ to be adapted to the filtration $\{\mathcal{F}_t\}_{t \in [0,T^*]}$. We will denote the corresponding T_i-forward measures, Brownian motions and volatility functions by $\tilde{\mathbb{P}}_{T_i}$, $\tilde{W}_{T_i}$ and $\tilde{\gamma}(t,T_i,T_{i+1})$, respectively.

Secondly, we postulate the existence of a spot exchange rate process $X(t)$, which is a positive special semi-martingale under $\mathbb{P}_{T^*}$. $X(t)$ is the spot price of the commodity at time t.

A commodity bond $C(t, T)$ converted by the spot exchange rate is then a traded asset in the domestic market (denominated in domestic currency), and hence its forward value,

$$X(t,T_i) = \frac{C(t,T_i)X(t)}{B(t,T_i)} \quad (0 \leq i \leq N), \tag{21.8}$$

is a martingale under $\mathbb{P}_{T_i}$ and is called the T_i-forward exchange rate, and in the present context this is the forward price of the commodity. If the dynamics of the forward exchanges rates are written in terms of

$$dX(t,T_i) = X(t,T_i)\sigma_X(t,T_i) \cdot dW_{T_i}(t) \quad (0 \leq i \leq N), \tag{21.9}$$

in which the volatility functions are not necessarily deterministic, then it is shown by Schlögl (2002b) that, under no-arbitrage restrictions, the volatility functions must satisfy the relation

$$\sigma_X(t,T_{i-1}) = \tilde{\gamma}(t,T_{i-1},T_i) - \gamma(t,T_{i-1},T_i) + \sigma_X(t,T_i) \tag{21.10}$$

$$(1 \leq i \leq N).$$

This generally leaves two ways to link the interest rate and commodity markets in order to have an arbitrage-free multi-currency LMM. The first would be to calibrate the single-currency LMM models separately, i.e. determine the volatility functions $\gamma(t, T_i, T_{i+1})$ and $\tilde{\gamma}(t, T_i, T_{i+1})$ (or $\lambda(t, T_i)$ and $\tilde{\lambda}(t, T_i)$ equivalently) for $i = 1, \ldots, N - 1$, and then calibrate the forward exchange rate volatility $\sigma_X(\cdot, T_i)$ for one arbitrary forward time T_i. Using (21.10), all other forward exchange rate volatilities $\sigma_X(\cdot, T_j)$ for $1 \le j \ne i \le N$ can be derived. This approach seems especially appropriate when linking two (real) interest rate markets (like USD and AUD), because LIBORs (or similar) and currency forwards are liquidly traded for all major currencies.

In order to link an interest rate market with a commodity market, the second approach seems to be more appropriate. The fixed income market volatilities $\gamma(t, T_i, T_{i+1})$ (or, equivalently, $\lambda(t, T_i)$) are calibrated as for a single-currency LMM—in our example in Section 21.6, this will be the USD market calibrated to at-the-money caplets and swaptions. Then, for all forward times $T_0, \ldots, T_N$, volatility functions $\sigma_X(\cdot, T_i)$ are calibrated to the commodity options market (in Section 21.6, we specifically consider options on WTI Crude Oil futures), i.e. the forward commodity prices are assumed to follow log-normal dynamics under the appropriate probability measure, rather than the simple-compounded convenience yields.[*] The volatility functions $\tilde{\gamma}(t, T_i, T_{i+1})$ and $\tilde{\lambda}(t, T_i)$ can now be derived from relations (21.10) and (21.7), respectively.

Remark 1: If a deterministic volatility function σ_X is chosen for the commodity forward process, it will be shown in Section 21.4 that the corresponding futures process is not log-normally distributed, apart from the futures price at maturity. Nevertheless, the difference between forward and futures prices can be expressed in terms of volatilities for the commodity forwards and forward interest rates.

21.3 Calibration with Time-Dependent Volatilities

Since the Commodity LMM is based on commodity forwards, we have to calibrate to forward implied volatilities or plain vanilla option prices written on forwards. However, commodities futures rather than forwards are most liquidly traded (consider, for example, the WTI Crude Oil futures in the market data example in Section 21.6) and thus forward prices have to be deduced from the futures. As we are specifically concerned with integrating commodity and interest rate risk, it is not adequate to equate forward prices with futures prices, as is still common among practitioners. The following Section 21.4 describes how to approximate the difference between futures and forwards as well as the implied forward volatilities, in order to apply the calibration methods proposed in the present section.

We first state some notational conventions and modify slightly the notation in Section 21.2, adapted from the prior literature. Since in our setup the forward exchange rate $X(t, T)$ is a commodity forward, we will write $F(t, T)$ instead, where t is termed the process time (aka calendar time) and T the forward time. Accordingly, the volatility of the forwards will be denoted by $\sigma(t, T)$, instead of $\sigma_X(t, T)$, and is the instantaneous volatility at time t of the forward with forward time T.

We assume we have forward processes $F(\cdot, T_1), \ldots, F(\cdot, T_N)$ with expiries $T_1, \ldots, T_N$ and we further think of T_0 as 'now'. Times to maturity for an arbitrary calendar time $t \ge 0$ are given by $x_i = T_i - t$ for $i = 0, 1, \ldots, N$. Finally, market prices for call options on $F(\cdot, T_i)$ with pay-off $[F(T_i, T_i) - K]^+$ are assumed to be available and denoted by C_i^{mkt} for $i = 1, \ldots, N$.

For calibration we further need to determine vectors of calendar times $t_c = (0, t_1, \ldots, t_{n_c})$ and times to maturity $x_f = (x_0, x_1, \ldots, x_{n_f})$, which define a grid for $V = (v_{i,j})_{1 \le i \le n_c, 1 \le j \le n_f}$, the matrix of piecewise constant instantaneous volatilities. The entry $v_{i,j}$ represents the volatility corresponding to forward $F(t, T)$ with $t_{i-1} \le t < t_i$ and $x_{j-1} \le T - t < x_j$. The number of forward times n_f in the volatility matrix need

---

[*] Assuming the convenience yields to be log-normal also does not appear reasonable in light of the fact that, empirically, convenience yields can become negative.

not coincide with the number of traded forwards N, and especially in regions of large forward times a rougher spacing can be chosen for x_f, since volatilities tend to flatten out with increasing forward time. Let $v(t, T)$ denote the (annualized) implied volatility as seen at time t for forward time T, then the relation between time-dependent instantaneous and implied volatilities is given by

$$\frac{1}{T-t}\int_t^T \sigma^2(s,T)\mathrm{d}s = v^2(t,T). \tag{21.11}$$

Identity (21.11) is utilized to compute a (model) option price C_i^{mod} (for $i = 1, \ldots, N$) from the volatility matrix V with Black's formula (Black, 1976). Note that the assumption of deterministic volatilities for commodity forwards made in the previous section permits us to apply Black's formula in a consistent manner. In Appendix A we show how to compute the integral when volatilities are piecewise constant. Now, the first calibration criterion measures the *quality of fit* by considering the squared differences between market and model call prices. In order to weight the various fit criteria we assign different weights η to each of them,

$$q = \eta_q \sum_{j=1}^N \left(C_j^{\mathrm{mkt}} - C_j^{\mathrm{mod}}\right)^2. \tag{21.12}$$

As already mentioned in the introduction, in commodity markets, typically only options with maturities coinciding with the maturity of the underlying forward exist. In order to construct a full volatility term structure for a particular forward, it is necessary to use information gained from other options on forwards expiring before the particular one. Following Pedersen (1998), to this end several heuristic concepts can be employed, which more or less rely on market practice and experience. We will refer to these as *smoothness criteria*. The first is time-homogeneity, which assumes that the (unknown) volatility at future calendar time $t = T_2 - T_1$ of a forward $F(\cdot, T_2)$ having then time to maturity T_1 and time to maturity T_2 now, will be 'similar' to the (known) volatility the forward $F(\cdot, T_1)$ with time to maturity T_1 has now. In other words, under strict time-homogeneity, $\sigma(t, T)$ can be expressed as a function of time to maturity $T - t$ only. This is usually too restrictive to be able to fit the model to the market in general, so instead of enforcing strict time-homogeneity, we penalize departures from time-homogeneity in the first smoothness criterion. In terms of the volatility matrix V, this means that volatilities with different calendar times but the same time to maturity should not differ too much, so the penalty function is

$$s_1 = \eta_1 \sum_{j=1}^{n_f} \sum_{i=1}^{n_c-1} \left(v_{i+1,j} - v_{i,j}\right)^2. \tag{21.13}$$

Another typical characteristic is the so-called Samuelson effect, which states that forward volatilities tend to decrease with increasing time to maturity. In order to be consistent with this behaviour, the volatility matrix has to be monotonically decreasing in forward time, and violations of this property are penalized by

$$s_2 = \eta_2 \sum_{i=1}^{n_c} \sum_{j=1}^{n_f-1} \left(\max\left\{v_{i,j+1} - v_{i,j}, 0\right\}\right)^2. \tag{21.14}$$

Finally, we enable a criterion that imposes the volatility term structure to be smooth in time to maturity for each fixed calendar time. Assigning a large weight to this criterion would force the volatility to be flat in the forward time direction, which is usually not desirable, however with a small weight this criterion contributes to a smoother volatility surface,

$$s_3 = \eta_3 \sum_{i=1}^{n_c} \sum_{j=1}^{n_f-1} \left(v_{i,j+1} - v_{i,j} \right)^2. \tag{21.15}$$

To minimize the objective value $q + s_1 + s_2 + s_3$, a least-squares optimization method is used, one that allows for non-linear dependencies between model parameters and objective function values.

Remark 2: The quality-of-fit criterion q and the smoothness criteria s_1, s_2, s_3 are constructed such that the absolute squared differences between model and market option prices and between neighbouring volatilities, respectively, are minimized. Alternatively, relative deviations could be used for minimization, but we did not experience substantial differences in the quality of fit or the smoothness of volatilities when doing so.

Remark 3: Regarding the relationship between calendar times and forward times of the volatility matrix and the times to maturity of the available options on forwards, two important points should be noted. Firstly, to be able to price options on all of the forwards, the maturity of the longest available forward has to be smaller than or equal to the latest calendar time and the longest time to maturity, $T_N \leq \min\{t_{n_c}, x_{n_f}\}$. Here, we assume that all options mature at the same time as their underlying forwards. Secondly, in the context of the integral in (21.11), volatilities $\sigma(t, T)$ at any calendar time t with forward time $T > T_N$, or, equivalently, $t + x > T_N$, have no impact on the price of any plain vanilla option used for calibration. In terms of the piecewise constant volatility matrix, this means that volatilities v_{ij} with $t_{i-1} + x_{j-1} > T_N$ have no contribution to the quality-of-fit criterion and are therefore determined only by smoothness criteria.

We want to conclude this section by briefly describing the concept of volatility factor decomposition and factor reduction, which can either be included in the calibration process, or applied to the final volatility matrix after calibration. As will be demonstrated in the following Sections 21.4 and 21.5, volatility factor decomposition has to be included in the calibration process when calibrating to futures and options on futures. The method has to be applied separately for every calendar time, which is therefore fixed to arbitrary t_i in the following, and let v_i denote the ith row (corresponding to calendar time t_i) of V, written as a column vector. Together with the exogenously given correlation matrix C, which is assumed to be constant over calendar time, the covariance matrix is calculated by

$$\Sigma = \left(v_i v_i^\top \right) \odot C, \tag{21.16}$$

where $\top$ means transposition and $\odot$ multiplication by components (Hadamard product). Σ is symmetric and positive definite, hence can be decomposed into $\Sigma = RD^{1/2}(RD^{1/2})^\top$. The columns of $R = \left(r_{j,k} \right)_{1 \leq j,k \leq n_f}$ consists of orthonormal eigenvectors and the diagonal matrix $D = \left(\xi_{j,k} \right)_{1 \leq j,k \leq n_f}$ of the eigenvalues of Σ. This representation allows us to reduce the number of stochastic factors, because usually the first d eigenvalues (when ordered decreasingly in D and the columns of R accordingly) account for more than around 95% of the overall variance, with d substantially smaller than n_f. Hence, R and D can be reduced to matrices $R \in \mathbb{R}^{n_f \times d}$ and $D \in \mathbb{R}^{d \times d}$ by retaining only the first d columns in R and the upper $d \times d$ submatrix in D, respectively. Instead of v_i we now use the matrix $U = \left(u_{j,k} \right)_{1 \leq j \leq n_f, 1 \leq k \leq d}$ (the calendar time index i is omitted for notational convenience) with entries $u_{j,k} = r_{j,k}\sqrt{\xi_k}$, which relates to v_i by

$$v_{i,j}^2 = \sum_{k=1}^{d} u_{j,k}^2 = \sum_{k=1}^{d} r_{j,k}^2 \xi_k \quad (j = 1, \ldots, n_f). \tag{21.17}$$

The computation of the variance in (21.11), as described in Appendix A, allows also for representations of the volatilities in decomposed form.

Remark 4: The correlation matrix C in (21.16) is the correlation matrix of the forward returns, not of the futures returns.* However, for the convexity adjustment suggested in Section 21.4, both correlation matrices coincide.

Remark 5: When used as in (21.16), the calculation of a correlation matrix from a historical time series should be consistent with the forward time concept of the volatility matrix. That means that if the volatility matrix is given for calendar times t and absolute forward times T, the correlations should be calculated from instruments with the same absolute maturity times. Correspondingly, if the volatility matrix is given for calendar times t and times to maturity x, the correlations should be calculated from instruments with the same times to maturity. Typically, futures are quoted in the market for absolute maturity times and forwards (if they are quoted at all) have constant times to maturity. Although, in practice, both methods typically do not exhibit substantial differences, an interpolation could be performed before estimating the correlations in order to switch from absolute maturity times to times to maturity or *vice versa*. For a discussion of different correlation concepts and their effect on calibration and pricing in the context of the LMM, see Choy *et al.* (2004).

21.4 Futures/Forward Relation and Convexity Correction

The calibration method in Section 21.3 is applicable only when forwards and options on forwards are available. This section presents the approximate conversion of futures prices to forward prices for all relevant data for calibration, in order to apply the methods of the previous section when only futures and options on futures are available (such as in the case of WTI Crude Oil as considered in the market data example in Section 21.6).

We introduce the notation $G(t, T)$ for a futures price at time t with maturity T, and, as before, $F(t, T)$ will be the corresponding forward price. From no-arbitrage theory we know that $F(T, T) = G(T, T)$ and that prices of plain vanilla options on forwards and futures must coincide, whenever the maturities of the option, forward and futures are the same. This allows us to use the call prices of options on futures for calibration of forwards, and we only have to ensure that the (virtual) forwards have the same maturities as the futures. Due to Equation (21.9) the forward $F(\cdot, T_i)$ is an exponential martingale under the T_i-forward measure, and since it has deterministic volatility, it is log-normally distributed under $\mathbb{P}_{T_i}$. A change of measure from the T_i-forward to the spot risk-neutral measure relates the Brownian motions by

$$dW_{T_i}(t) = \eta(t, T_i)dt + dW_{\mathbb{Q}}(t) \quad (1 \leq i \leq N),$$

and $\eta(\cdot, T_i)$ relates to the volatility of the process $F_B(t, T_i, T_i + \delta)$ by

$$\eta(t, T_{i+1}) - \eta(t, T_i) = \gamma(t, T_i, T_{i+1})$$

$$= \frac{\delta L(t, T_i)}{1 + \delta L(t, T_i)} \lambda(t, T_i) \quad (1 \leq i \leq N - 1), \tag{21.18}$$

with $\eta(t, T_1) = 0$ if we are using 'interpolation by daycount fractions' to extend the model to continuous tenor as in Schlögl (2002a). Hence, for $1 \leq i \leq N$,

---

* Strictly speaking, it is the matrix of quadratic covariation of the forward processes.

$$F(T_i, T_i)$$

$$= F(t, T_i) \exp \left\{ \int_t^{T_i} \sigma(u, T_i) \cdot dW_{T_i}(u) - \frac{1}{2} \int_t^{T_i} \left\| \sigma(u, T_i) \right\|^2 du \right\}$$

$$= F(t, T_i) \exp \left\{ \int_t^{T_i} \sigma(u, T_i) \cdot dW_{\mathbb{Q}}(u) - \frac{1}{2} \int_t^{T_i} \left\| \sigma(u, T_i) \right\|^2 du \right\}$$

$$\times \exp \left\{ \int_t^{T_i} \sigma(u, T_i)^{\mathsf{T}} \eta(u, T_i) du \right\},$$

where $\| \cdot \|$ denotes the inner product norm. Furthermore, futures follow the general relation

$$G(t, T) = \mathbb{E}_{\mathbb{Q}} \left[S(T) | \mathcal{F}_t \right] \tag{21.19}$$

(see, e.g. Cox *et al.* (1981) and Miltersen and Schwartz (1998)), where $S(t)$ is the spot price that satisfies by no-arbitrage constraints $S(t) = F(t, t) = G(t, t)$ for all t. Putting these relations together, the futures in (21.19) can be expressed as

$$G(t, T_i)$$

$$= \mathbb{E}_{\mathbb{Q}} \left[F(T_i, T_i) | \mathcal{F}_t \right]$$

$$= F(t, T_i) \mathbb{E}_{\mathbb{Q}} \left[\exp \left\{ \int_t^{T_i} \sigma(u, T_i) \cdot dW_{\mathbb{Q}}(u) - \frac{1}{2} \int_t^{T_i} \left\| \sigma(u, T_i) \right\|^2 du \right\} \right. \tag{21.20}$$

$$\left. \times \exp \left\{ \int_t^{T_i} \sigma(u, T_i)^{\mathsf{T}} \eta(u, T_i) du \right\} \middle| \mathcal{F}_t \right],$$

and is obviously not log-normally distributed, since $\eta(\cdot, T_i)$ depends on the forward interest rates. The difficulty here is to calculate an expectation of a random process value under a measure for which the process is not a martingale. Similar techniques as utilized for convexity correction in interest rate theory can be applied in order to obtain an expression under the expectation operator that is a log-normally distributed random variable with respect to the $\mathbb{Q}$-measure.[*]

A simple and widely used way is to make the second term in the expectation deterministic by 'freezing' the level dependence of $\eta(\cdot, T_i)$ with respect to the currently observed forward curve, i.e. defining $\bar{\eta}(\cdot, T_i)$ by

$$\bar{\eta}(t, T_{i+1}) - \bar{\eta}(t, T_i) = \frac{\delta L(0, T_i)}{1 + \delta L(0, T_i)} \lambda(t, T_i). \tag{21.21}$$

Then all volatility terms in (21.20) are deterministic and we have

$$G(t, T_i) = F(t, T_i) \exp \left\{ \int_t^{T_i} \sigma(u, T_i)^{\mathsf{T}} \bar{\eta}(u, T_i) du \right\} \quad (1 \le i \le N). \tag{21.22}$$

In Appendix A we show how to compute the integral in (21.22) when the volatility functions $\sigma(t, T)$ and $\bar{\eta}(t, T)$ are piecewise constant.

---

[*] See, for example, Pelsser (2000, chapter 11) for details on convexity correction.

Remark 6: Based on the convexity correction (21.22), the identity of the correlation matrices for forward and futures returns follows immediately.

21.5 Merging Interest Rate and Commodity Calibrations

So far the interest rate market and the commodity market have been considered separately, with the exception of the convexity correction in (21.22) that involved the forward interest rate volatility. In this section we focus on linking both volatility matrices for building a joint Commodity LMM. The linkage is controlled by the correlation matrix between the interest rate and commodity forwards. The whole correlation matrix, i.e. for any pair of forwards or forward rates within the set of commodity forwards and forward interest rates, is assumed to be constant over time. Further, we assume that the calendar time discretization, for which the matrices of the piecewise constant volatility for commodities and interest rates have been calibrated, coincide. The forward time discretization may differ.

The method to be described has to be applied separately for every calendar time, which is therefore fixed to some t_i in the following. The same techniques as applied for factor reduction in the context of Section 21.3 will be employed here. We add to the notation of the aggregate volatility matrix V from Section 21.3 the subscript 'C' for 'commodity' in order to distinguish it from its interest rate equivalent, which is denoted by 'I'. Further, we decompose the volatility matrices V_C (obtained from the calibration in Section 21.3) and V_I (obtained from the LMM calibration, for example by the method described by Pedersen (1998)) to matrices

$$U_C = \left(u_{j,k}^C \right)_{1 \le j \le n_f, 1 \le k \le d_C}, \quad U_I = \left(u_{j,k}^I \right)_{1 \le j \le m_f, 1 \le k \le d_I},$$

where m_f is the number of interest forward rates and d_C and d_I are the number of factors with significant eigenvalues for commodity forwards and interest forward rates, respectively. The columns of each matrix correspond to the stochastic factors and the rows to different forward times. Moreover, each column vector is orthogonal to the other column vectors within the same matrix.

We start to link both calibrations by choosing a parameter $d \ge d_C, d_I$ that determines the number of relevant factors for both commodities and interest rates. d will also be the dimension of the vector of Brownian motions in the joint Commodity LMM. If d is chosen to be greater than d_C or d_I, respectively, the corresponding matrices $U_C \in \mathbb{R}^{n_f \times d_C}$ and $U_I \in \mathbb{R}^{m_f \times d_I}$ have to be enlarged to matrices $U_C \in \mathbb{R}^{n_f \times d}$ and $U_I \in \mathbb{R}^{m_f \times d}$, simply by adding zero columns at the end. The forward interest rates now depend only on factors that relate to non-zero columns, but it is different for commodity forwards. Our fitting procedure, presented below, subsequently modifies U_C and therefore determines which factors will have an impact solely on the commodity forwards, which will have no impact on the commodity forwards and which will contribute to both commodity forwards and forward interest rates, and therefore to the cross-correlation.

The separate calibrations for commodities and interest rates are merged by matching the model intrinsic cross-covariance matrix $U_C U_I^\mathsf{T}$ with the cross-covariance matrix calculated from the exogenously given cross-correlation matrix C_{CI},

$$\Sigma_{CI}^{\text{target}} = \sqrt{\text{diag}\left\{ v_{Ci} v_{Ci}^\mathsf{T} \right\} \text{diag}\left\{ v_{Ii} v_{Ii}^\mathsf{T} \right\}^\mathsf{T}} \odot C_{CI},$$

where diag (applied to a matrix) returns the diagonal of the matrix as column vector and the square root has to be applied component-wise. As in Section 21.3, v_{Ci} and v_{Ii} denote the ith row (corresponding to calendar time t_i) of V_C and V_I, respectively, written as column vectors. In order to achieve $U_C U_I^\mathsf{T} \approx \Sigma_{CI}^{\text{target}}$ we exploit the property of multivariate normal distributed random variables to be invariant to orthonormal rotations. That means we have to find a matrix Q satisfying $QQ^\mathsf{T} = I_d$ (where I_d denotes the $d \times d$-identity matrix) that minimizes

$$r_1 = \zeta_1 \left\| \Sigma_{\text{CI}}^{\text{target}} - U_\text{C} Q U_\text{I}^\mathsf{T} \right\|,$$

with respect to some matrix norm, e.g. the Frobenius norm. The weight factor ζ_1 may be necessary when further constraints have to be controlled by the loss function as well, as will be discussed below. Alternatively, it is also possible to define the cross-correlation matrix to be the target matrix and minimize

$$r_1 = \zeta_1 \left\| C_{\text{CI}} - \left(U_\text{C} Q U_\text{I}^\mathsf{T} \right) \odot \left(\text{diag}\left\{ v_{\text{C}i} v_{\text{C}i}^\mathsf{T} \right\} \text{diag}\left\{ v_{\text{I}i} v_{\text{I}i}^\mathsf{T} \right\}^\mathsf{T} \right)^{-1/2} \right\|,$$

subject to the same orthonormality constraint for Q. The exponent $-1/2$ has to be applied component-wise.

As with calibration, the optimization for obtaining a Q can be performed by any non-linear optimization procedure that allows for non-linear constraints, or by non-linear least-squares algorithms such as Levenberg/Marquardt, in which the distance of QQ^T to the identity matrix I_d is again controlled by a matrix norm,

$$r_2 = \zeta_2 \left\| QQ^\mathsf{T} - I_d \right\|,$$

ensuring that Q is as close as possible to a valid orthonormal rotation preserving the original interest rate and commodity calibrations.

Remark 7: The cross-correlations between commodity forwards and interest rate forwards are much lower than the correlations within the asset classes, and estimation from historical data appears to be much more volatile for the cross-correlations than for the correlations within the asset classes. For example, the structure of the cross-correlation matrix between WTI Crude Oil forwards and USD interest rate forwards in Figure 21.7 of Section 21.6 can hardly be explained by obvious rationales. Therefore, in practice, one might wish to specify a flat cross-correlation founded on particular market views. This can be realized in a straightforward manner in our approach, but the example in Section 21.6 below shows that it is also possible to adequately match a more complicated cross-correlation structure. In the end, there is a trade-off between the quality of the cross-correlation fit and the number of stochastic factors in the model. In view of computational efficiency and model parsimony a limited number of factors is desirable and perhaps more important than exactly fitting a given cross-correlation structure.

Remark 8: In our approach the basis transformation Q applies to the commodity volatility matrix U_C. Alternatively, one could choose the interest rate volatility matrix U_I for transformation. This would not change criteria r_1 and r_2, because Q can also be interpreted as Q^T with inverse Q, and the transformed model covariance matrix would read as $U_\text{C}(U_\text{I}Q)^\mathsf{T}$.

The determination of the basis transformation matrices concludes the calibration of the model, if the market instruments are forwards and options on forwards. However, if the calibration is carried out for futures and options on futures, the model option prices would change with any non-trivial basis transformation, since the forward prices are modified by means of the convexity correction (21.22).

This implies that the whole calibration process has to be iterated by, first, refitting the volatility matrix V_C, where in each step of the optimization the corresponding basis transformation matrix Q has to be multiplied by the volatility decomposition U_C (for each calendar time) before the model call prices and the value of the loss function are computed. Secondly, for the refitted V_C and U_C, new basis transformations have to be determined in order to still match the cross-correlations.

Both steps, the fitting of V_C and Q, have to be iterated until a sufficient smoothness and quality of fit subject to the market option prices and the exogenously given cross-correlation is reached. For the real data example of the following section, this occurs already after three iterations in the case with six stochastic factors.

For clarity we summarize the steps of the whole calibration process to futures and options on futures.

(I) **Preliminary calculations applied to the LMM calibration outcome.**
   1. Computations for each calendar time t_i ($1 \leq i \leq n_I$).
       a. Computation of the covariance matrix Σ_{Ii} as in (21.16).
       b. Decomposition of Σ_i into U_{Ii} using PCA in the way described at the end of Section 21.3.
(II) **Iteration until a sufficient quality of fit based on the criteria on matching market option prices, smoothness and cross-correlation is reached.**
   1. Calibration of V_C.
          Minimization of the penalty function given in (d):
       a. Computations for each calendar time t_i ($1 \leq i \leq n_C$).
            i.   Computation of the covariance matrix Σ_{Ci} by (21.16).
            ii. Decomposition of Σ_{Ci} into U_{Ci} using PCA as described at the end of Section 21.3.
            iii. Multiplication by the basis transform Q_i resulting from the previous iteration, $U_{Ci}Q_i$ (Q_i is the identity matrix in the very first iteration).
       b. Computation of forward prices from market futures prices using (21.22).
       c. Computation of model prices for options on forwards using (21.17) and the Black formula.
       d. Calculation of the loss value $q + s_1 + s_2 + s_3$ as defined in Section 21.3.
   2. Fit of the basis transformations.

Fitting $\{Q_i\}_{1 \leq i \leq n_c}$ subject to the penalty function $r_1 + r_2$ as described in this section.

Remark 9: The number of factors d_C can affect the number of iterations required for the calibration process to converge satisfactorily. Equations (21.21) and (21.22) demonstrate that if the cross-correlation is zero, futures and forward prices coincide. On the other hand, the first fit of the volatility matrices U_C is made without any consideration of cross-correlation. Hence (unintentionally) assigning in the computation of U_C the most contributing eigenvalues and eigenvectors of V_C to those stochastic factors of the d-dimensional Brownian motion, which also represent strong V_I contributions, would generate a rather high cross-correlation in the first calibration step at the end of Step II.1 above. In this case the forward curve would tend to depart from the futures curve. However, if the exogenous cross-correlation has a low level, the basis transformation would modify the volatility matrix U_C in a way that not only reduces the high model cross-correlation generated in the first calibration step, but also returns the forward curve in the direction of the futures curve by reducing the magnitude of the convexity correction. An iteration of the procedure (as described above) will typically (in all real data scenarios that we have considered) force the objective variables V_C and Q to a balanced state with adequate calibration results. As demonstrated in the following section, the convergence is much slower if d_C is large relative to d_I, since a large d_C allows for more de-correlation than a small value.

Finally, in the last iteration of the procedure described above we end up with decomposed volatility matrices U_C and transformation matrices Q for each calendar time, and assembling these resulting n_c-many matrices U_CQ and U_I into three-dimensional arrays $\Lambda_C \in \mathbb{R}^{n_c \times n_f \times d}$ and $\Lambda_I \in \mathbb{R}^{n_c \times m_f \times d}$ the calibration of the hybrid Commodity LMM is finished.* Note that Λ_I basically consists of piecewise constant forward interest rate volatilities $\lambda(t, T)$ as occurring in (21.6), obtained from the separate LMM

---

* Since we had to equalize the calendar time discretization for commodities and interest rates in order to be able to fit a transformation matrix Q, the first dimension of Λ_I is of size n_c. The choice to adapt the forward interest rate discretization of calendar time to the commodity discretization is arbitrary and any other discretization could be used.

calibration, at most modified by some interpolation on these calibrated volatilities in order to obtain Λ_I. Λ_C is the result of the calibration method as described in the previous sections, together with the transformation presented in this section in order to match cross-correlations.

This allows us to write the dynamics of the hybrid Commodity LMM as follows. Let W be a d-dimensional Brownian motion and denote by $\lambda_{i,j}^I$ and $\lambda_{i,j}^C$ the d-dimensional vectors in Λ_I and Λ_C of volatilities for calendar times $t \in [t_{i-1}, t_i)$ and times to maturity $x \in [x_{j-1}, x_j)$. Then, the dynamics of the forward interest rates $L(t, T)$ as given in (21.6) can be written as

$$dL(t,T) = L(t,T)\lambda_{i,j}^I \cdot dW_T(t),$$

and the dynamics of the commodity forwards $F(t, T)$ as given in (21.9) as

$$dF(t,T) = F(t,T)\lambda_{i,j}^C \cdot dW_T(t)$$

for all maturity times T satisfying $T - t \in [x_{j-1}, x_j)$ and all calendar times $t_{i-1} \le t < t_i$ (for some $1 \le i \le n_c$ and $1 \le j \le m_f$ or $1 \le j \le n_f$, respectively).

21.6 Real Data Example

We demonstrate the performance and applicability of the Commodity LIBOR Market Model by calibrating it to real data.[*] We have chosen May 5, 2008, for calibration and WTI Crude Oil as commodity, hence the US Dollar (USD) forward rate as interest rate. As can be seen in Figure 21.1, the nearest WTI Crude Oil future price of 119.97 USD was not too far from its peak in July 2008. The futures curve is in 'backwardation' and covers a rather large range of about 10 USD within the first five years of maturity. The 3-month forward rates show a less extreme pattern than the commodity futures. An application to 2009 data produced results comparable to those presented in this example regarding the fit to commodity and fixed income market data, but at the cost of a slightly rougher forward interest rate volatility surface.

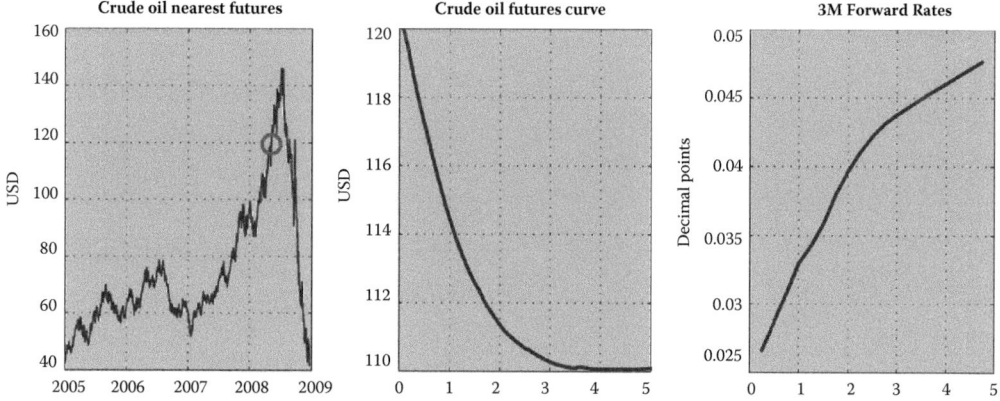

FIGURE 21.1 The commodity and interest market for calibration date May 5, 2008. Left: The WTI Crude Oil nearest futures between 2005 and the end of 2008. The circle indicates the calibration date. Middle: The futures curve as seen at calibration date with maturities up to five years. Right: The 3-month USD forward rates for reset dates (expiries) between 3 months and 4 years and 9 months.

---

[*] The data were taken from the SuperDerivatives platforms SD-IR and SD-CM.

The calibration of the (classical) LMM was done as proposed by Pedersen (1998). Figure 21.2 shows the resulting volatility surface and the correlation matrix as used for calibration, which was estimated historically from the time series of forward rates covering the 3 months before May 5, 2008. Caplets, caps and swaptions were used for calibration and the fit to market prices is quite good, as Figure 21.3 demonstrates. Note that we have to employ the forward interest rate curve up to 6 years forward in time, in order to calculate the convexity correction for a commodity volatility surface with 3 years calendar and 3 years forward time. The calibration of the WTI Crude Oil futures was achieved by the method described in Section 21.3. The market instruments are futures and options on futures traded on the New York Mercantile Exchange. Figure 21.4 shows the calibrated volatility surface and the historically estimated correlation matrix using 3 months of futures prices before the calibration date. Calendar and forward times go out to 3 years, and although on the exchange futures with expiries in every month are traded, we chose the calendar and forward time vectors to be unequally spaced (while still calibrating

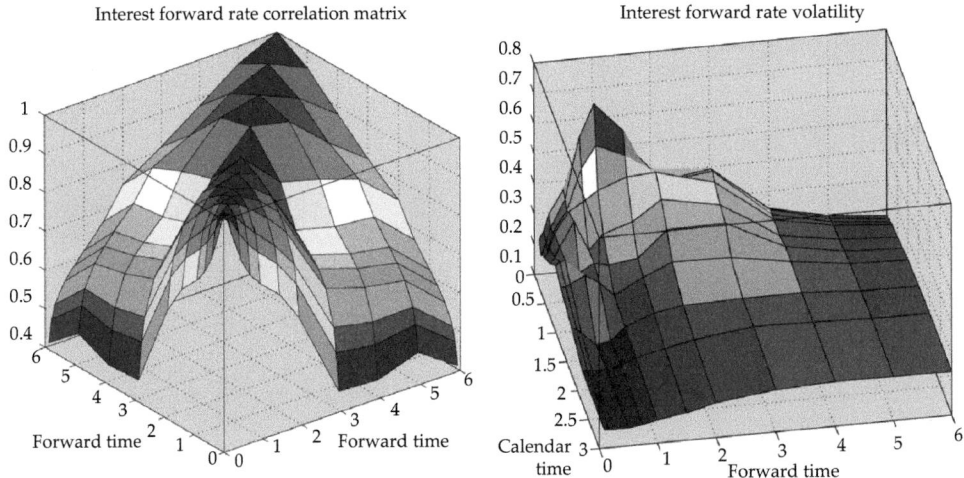

FIGURE 21.2   Left: The calibrated volatility matrix. Right: The historically estimated correlation matrix as used for calibration.

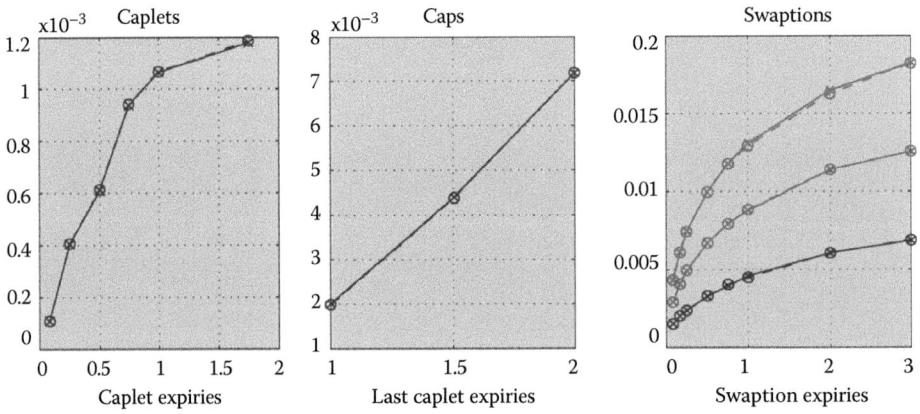

FIGURE 21.3   Market prices versus model prices. The market prices are represented by crosses connected by solid lines, the model prices by circles connected by dashed lines. Left: Caplets. Middle: Caps. Right: Swaptions with 1 year, 2 year and 3 year tenors (from bottom to top).

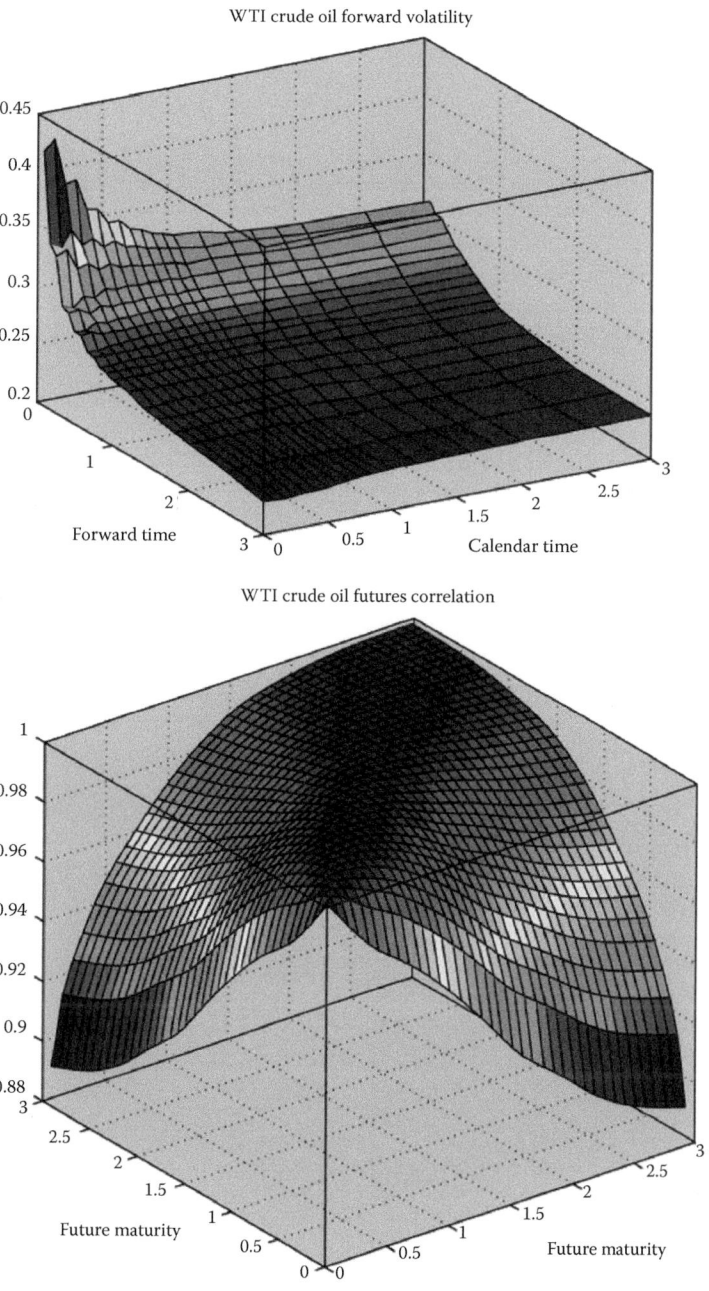

FIGURE 21.4 Left: The calibrated commodity volatility surface. Right: The historically estimated correlation matrix as used for calibration.

to all traded instruments), with 1 month difference up to 1 year, 2.4 months difference between 1 and 2 years and 6 months difference between 2 and 3 years. This speeds up the calibration without losing much structure in the volatility surface, since the market views futures with long maturity as having almost flat volatility. For weighting in the calibration objective function we have chosen $\eta_q = 1$ (fit to call prices), $\eta_1 = 0.1$ (time homogeneity, i.e. smoothness in the calendar time direction), $\eta_3 = 0.01$ (smoothness in the forward time direction) and $\eta_2 = 0.1$ (decreasing monotonicity in the forward time direction). As initial guess we have interpolated the implied volatilities to the forward time grid for the first calendar time t_1 and then used constant extrapolation to all other calendar times.

The model fit to the commodity call prices is very good, as illustrated in the right graph of Figure 21.5. The left graph shows the difference between futures and forwards as calculated by the convexity correction.

In Figure 21.6, particular slices of the forward volatility structure are examined. The left graph shows the term structure of volatilities for certain calendar times, whereas the right graph illustrates the evolution of volatilities over the lifetime of different forwards.

Finally, we link both separately calibrated volatility matrices to one set of stochastic factors. Table 21.1 shows how much of the overall variance, i.e. of the sum of variances over all factors, can be explained by the leading factors, when the factors are sorted according to decreasing contribution to total variance of the commodity forwards. The first two factors already account for more than 99% of the overall variance, hence a reasonable choice would be $d_C = 2$. Since Pedersen's calibration method for the LMM part of the model already includes a spectral decomposition of the interest forward rates covariance matrix, it is not meaningful to do it again here, because the number of factors with a reasonably large contribution should be the same as the number of factors chosen for the LMM calibration. If it is necessary to interpolate the forward interest rate volatility matrix in order to match the calendar times of the commodity volatility matrix, the forward rate covariance matrix will change and, hence, eigenvalue decompositions of the calendar time adjusted covariance matrices yield different results than eigenvalue decompositions of the original covariance matrices as used for calibration. However, these differences should not be substantial as long as the calibrated volatility matrix is sufficiently smooth in calendar time. For the LMM calibration we have chosen the number of factors to be $d_I = 4$, which again covers about 99% of the overall variance.

We now have to determine the parameters for the final cross-correlation fit. First, we choose the total number of factors in the joint model to be $d = 6$ and allow the commodity volatility matrix to disperse over all six factors, whereas the interest rate volatilities should remain on the first four factors, since we apply the basis transform to the commodity volatilities (see also Remark 8). Figure 21.7 demonstrates the result of the transformation. The left graph shows two surfaces, the

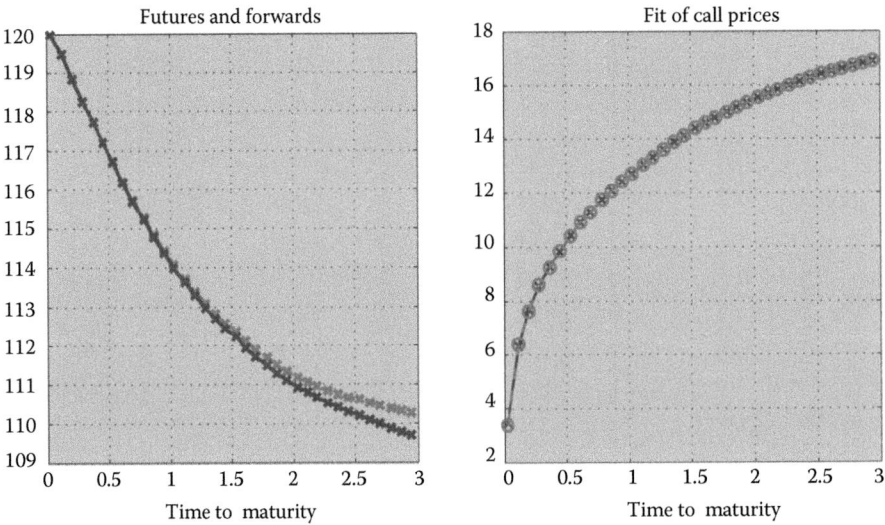

FIGURE 21.5 Left: The differences between futures (dashed red line) and forwards (solid blue line) as calculated from the convexity correction. Right: Fit of model prices (red circles connected by dashed line) to market prices (blue crosses connected by solid line).

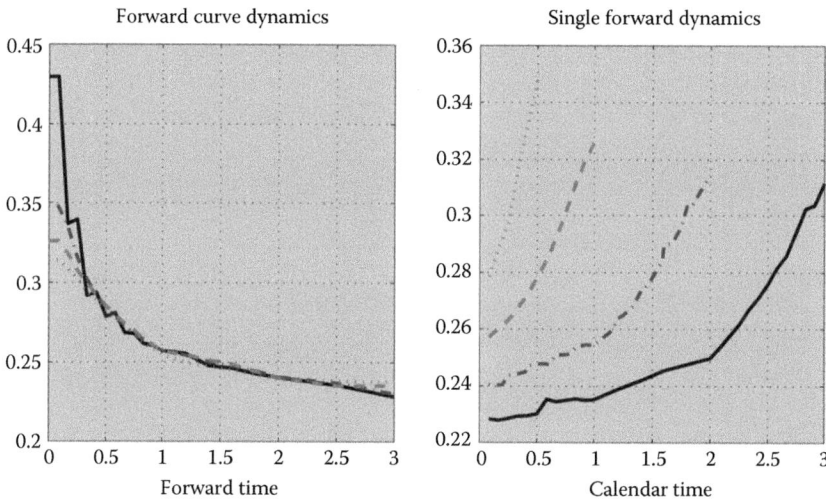

FIGURE 21.6 Left: Forward volatility structures at different calendar times (solid black = 0 (calibration date), dotted-dashed blue = 6 month, dashed red = 1 year, dotted green = 2 years). Right: Evolution of forward volatilities over the lifetime (solid black = forward with 3 years maturity, dashed-dotted blue = 2 years, dashed red = 1 year, dotted green = 6 month).

TABLE 21.1 Results of the Eigenvalue Decomposition of the Commodity Forward Covariance Matrix for the First Calendar Time t_1. The Maximum Number of Factors Coincides with the Number of Forward Times Greater Than zero. The Second Row Shows, for the ith Factor, the Percentage of Overall Variance That Can Be Generated by the First i Factors

	No. of factors							
	1	2	3	4	5	6	...	19
Percentage of overall variance	98.138	99.835	99.957	99.979	99.988	99.992	...	100

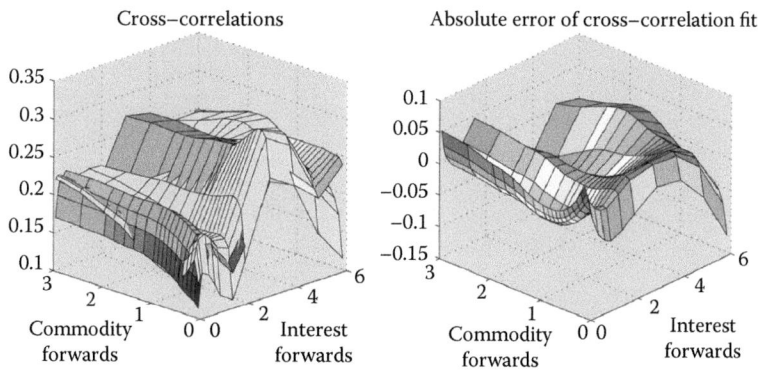

FIGURE 21.7 The cross-correlation fit with $d_C = d = 6$. Left: The target cross-correlation matrix (coloured) as estimated from historical futures returns and the cross-correlation matrix for the calendar time that fitted worst (grey). Right: Differences between target and fitted cross-correlations.

exogenously given target cross-correlation matrix (as calculated from the historical time series covering 3 months before the calibration time) and the transformed correlation matrix for the particular calendar time that fitted worst with respect to the Frobenius norm. The right graph illustrates the absolute differences between the target cross-correlation and the fitted cross-correlation matrix of the left graph. For optimization we applied a non-linear Levenberg/Marquardt algorithm with scale parameters $\zeta_1 = 1$ (quality of fit) and $\zeta_2 = 10$ (orthonormality of the transformation matrix). Without the application of a basis transformation, i.e. by optimizing only subject to the quality of fit and smoothness criteria, the cross-correlation ranges between 0.5 and 1 for all commodity and interest rate forwards over all calendar times.

By construction, the aggregated commodity volatility matrix, as shown in the left graph of Figure 21.4, does not change. How the decomposed factors change is illustrated in Figure 21.8. The left graph shows only the first four factors of the original calibration without considering cross-correlations, since the others are very close to zero. The right graph demonstrates how the basis transformation spreads the contribution of the relevant factors of the original calibration to all available factors in order to match the cross-correlation coefficients.

The quality of the cross-correlation fit can be substantially improved by increasing d_C and hence d. Allowing for six more factors, i.e. setting $d_C = d = 12$, results in cross-correlation fits as shown in Figure 21.9.

The choice of $d_C - d_I = 8$ independent factors for the commodity forwards allows the fitted basis transformation matrices in the very first iteration to have a strong de-correlation effect. This is especially true compared with the high cross-correlation (between 0.5 and 1.0) obtained from fitting V_C without any consideration of cross-correlation in step II.1 of the very first iteration. Table 21.2 demonstrates that, for $d_C = 12$, more iterations are required to find a balance for the fits of V_C and the cross-correlation matrix.

Figure 21.10 shows the forward prices and the fit to the market call prices for both calibrations. Whereas the forwards almost coincide, the fit to market call prices seems to be worse in the case of $d_C = 12$, especially for options with long maturities. The left graph of Figure 21.11, which is a detail of the right graph of Figure 21.10, confirms this and explains the larger loss value ℓ_1. The smoothness of the

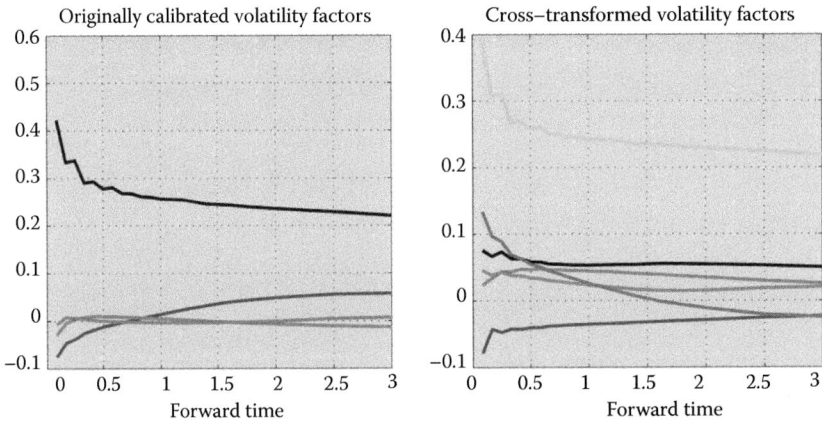

FIGURE 21.8 The factorized commodity volatilities of the first calendar time t_1. Left: The first four factors from the initial calibration, without considering cross relations. The fifth and sixth factors are not shown since they are almost zero. Right: All six factors after applying the basis transformation in order to match the cross-correlations. The fifth and sixth factors are represented by those two lines that have the largest value at forward time closest to zero.

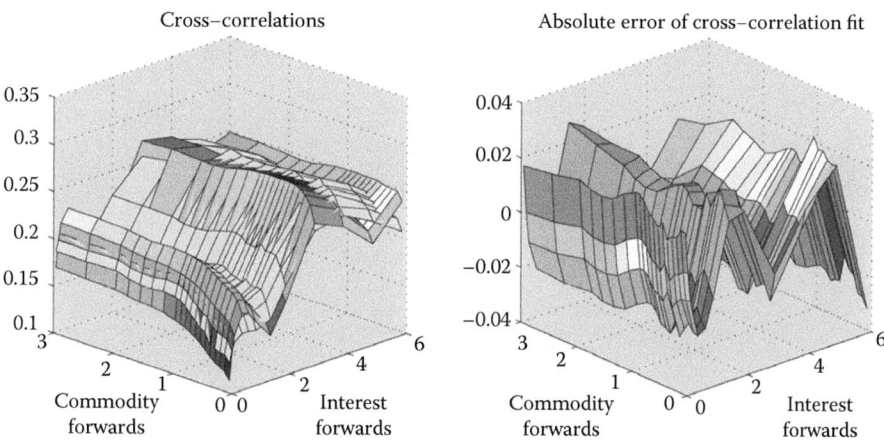

FIGURE 21.9 The cross-correlation fit with $d_C = d = 12$. Left: The target cross-correlation matrix (coloured) as estimated from historical futures returns and the cross-correlation matrix for the calendar time that fitted worst (grey). Right: Differences between target and fitted cross-correlations.

TABLE 21.2 Calibration with $d_C = 6$ and Three Iterations Compared with $d_C = 12$ and 12 Iterations. The value ℓ_1 Denotes the Final Loss Function Value of the V_C Optimization in Step II.1. The Value ℓ_{2i} Denotes the Frobenius Norm of the Difference between the Model and Target Cross-Correlation Matrix for Calendar Time t_i as Induced by the Fitted Q_i in Step II.2

		Iteration					
		1	2	3	4	5	6
$d_C = 6$	ℓ_1	0.0261	0.0265	0.0263	–	–	–
	$\min\{\ell_{2i}\}$	0.2373	0.2409	0.2414	–	–	–
	$\max\{\ell_{2i}\}$	0.3544	0.3922	0.3972	–	–	–
$d_C = 12$	ℓ_1	0.0261	0.9852	0.9550	0.7856	0.7828	0.6816
	$\min\{\ell_{2i}\}$	0.2004	0.2005	0.2005	0.2000	0.2000	0.2000
	$\max\{\ell_{2i}\}$	0.2551	0.2636	0.2543	0.2879	0.2868	0.2539

	Iteration					
	7	8	9	10	11	12
ℓ_1	0.6367	0.6301	0.6327	0.5090	0.4146	0.3992
$\min\{\ell_{2i}\}$	0.2001	0.2000	0.2000	0.2001	0.2012	0.2012
$\max\{\ell_{2i}\}$	0.2538	0.2537	0.2537	0.2538	0.2539	0.2537

volatility matrix V_C is lower for $d_C = 12$, as is demonstrated by the right graph of Figure 21.11. This further contributes to a higher ℓ_1.

21.7 Pricing Spread Options

This section examines an approximation approach in order to price spread options on forwards and futures following Kirk (1995). Spread options can be used to hedge the risk of roll over costs, i.e. the price difference between two futures or forwards with different maturities. Moreover, spread options depend on the correlation and the difference between the volatilities of the involved futures or forwards, and

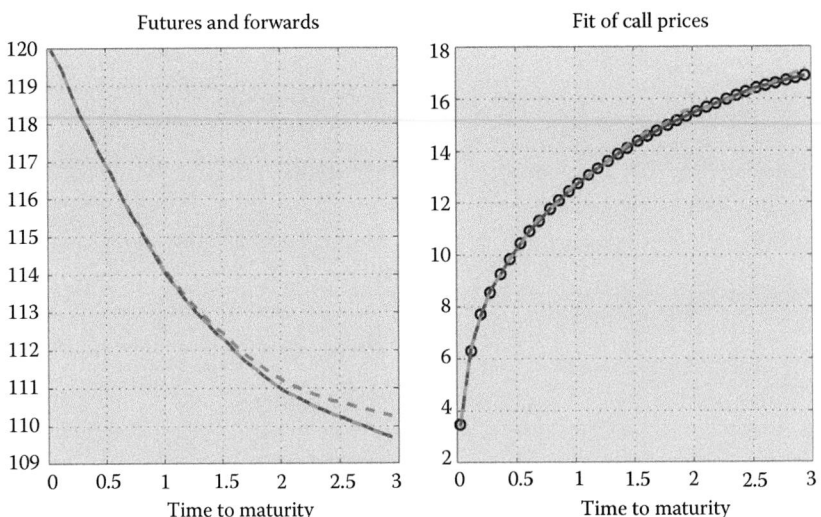

FIGURE 21.10 Comparison of the calibration results for $d_C = 6$ with three iterations and $d_C = 12$ with 12 iterations. Left: The market future prices (red dashed line) and the forward prices as calibrated with $d_C = 6$ (blue solid line) and $d_C = 12$ (green dotted-dashed line). Right: The market call prices (black circles) and the model call prices for $d_C = 6$ (solid blue line) and $d_C = 12$ (green dotted-dashed line).

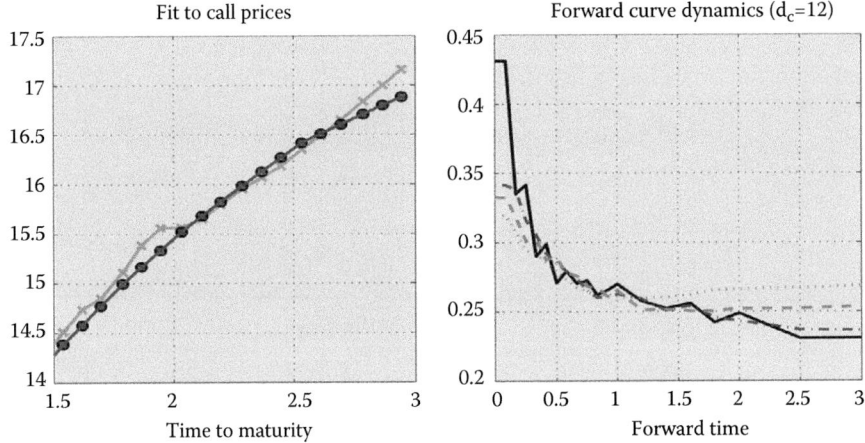

FIGURE 21.11 Comparison of the calibration results for $d_C = 6$ with three iterations and $d_C = 12$ with 12 iterations. Left: A detail of the right graph of Figure 21.10. The market call prices (black circles) and the model call prices for $d_C = 6$ (solid blue line) and $d_C = 12$ (green dotted-dashed line). Right: Forward volatilities as calibrated with $d_C = 12$ for different calendar times (solid black = 0 (calibration date), dotted-dashed blue = 6 months, dashed red = 1 year, dotted green = 2 years). See the left graph of Figure 21.6 for the corresponding volatilities calibrated with parameter $d_C = 6$.

hence can be employed to reduce the risk generated by these parameters. We will first consider forwards and then sketch an analogous derivation for futures.

Let T_i, T_j denote two forward maturities with $T_i < T_j$ and we set the option expiration time to T_i. According to market practice, the spread is defined by $S_{i,j}(t) = F(t, T_j) - F(t, T_i)$ and the spread call option value is given by

$$C_{\text{Spread}}^{\text{Fwd}}\left(0,T_i,T_j,K\right)$$

$$= \mathbb{E}_{\mathbb{Q}}\left[D\left(T_i\right)\left(S_{i,j}\left(T_i\right)-K\right)^+\Big|\mathcal{F}_0\right] \tag{21.23}$$

$$= B\left(0,T_i\right)\mathbb{E}_{T_i}\left[\left(F\left(T_i,T_j\right)-F\left(T_i,T_i\right)-K\right)^+\Big|\mathcal{F}_0\right].$$

From Section 21.2 we know that

$$dF\left(t,T_i\right)=F\left(t,T_i\right)\sigma\left(t,T_i\right)\cdot dW_{T_i}(t) \tag{21.24}$$

$$dF\left(t,T_j\right)=F\left(t,T_j\right)\sigma\left(t,T_j\right)^{\mathsf{T}}\left(\sum_{\ell=i}^{j-1}\frac{\delta L\left(t,T_\ell\right)}{1+\delta L\left(t,T_\ell\right)}\lambda\left(t,T_\ell\right)\right)dt$$

$$+F\left(t,T_j\right)\sigma\left(t,T_j\right)\cdot dW_{T_i}(t). \tag{21.25}$$

By 'freezing' the level dependence of $L(\cdot, T_\ell)$ with respect to the currently observed forward curve as described in Section 21.4, both processes become geometric Brownian motions,

$$dF\left(t,T_i\right)=F\left(t,T_i\right)\sigma\left(t,T_i\right)\cdot dW_{T_i}(t), \tag{21.26}$$

$$dF\left(t,T_j\right)\approx F\left(t,T_j\right)\sigma\left(t,T_j\right)^{\mathsf{T}}\Gamma_{i,j-1}(t)dt$$

$$+F\left(t,T_j\right)\sigma\left(t,T_j\right)\cdot dW_{T_i}(t), \tag{21.27}$$

with

$$\Gamma_{i,j-1}(t)=\sum_{\ell=i}^{j-1}\frac{\delta L\left(0,T_\ell\right)}{1+\delta L\left(0,T_\ell\right)}\lambda\left(t,T_\ell\right). \tag{21.28}$$

The idea of Kirk (1995) is to assume that, instead of $F(t, T_i)$ in (21.26), $F(t, T_i) + K$ is a geometric Brownian motion with adjusted volatility,

$$d\left(F\left(t,T_i\right)+K\right)=\left(F\left(t,T_i\right)+K\right)\frac{F\left(0,T_i\right)}{F\left(0,T_i\right)+K}\sigma\left(t,T_i\right)\cdot dW_{T_i}(t). \tag{21.29}$$

This allows us to apply Margrabe's approach Margrabe (1978) for options to exchange one asset for another by rewriting the pay-off in (21.23) such that

$$C_{\text{Spread}}^{\text{Fwd}}\left(0,T_i,T_j,K\right)$$

$$= B\left(0,T_i\right)\mathbb{E}_{T_i}\left[\left(F\left(T_i,T_i\right)+K\right)\left(\frac{F\left(T_i,T_j\right)}{F\left(T_i,T_i\right)+K}-1\right)^+\Big|\mathcal{F}_0\right]. \tag{21.30}$$

The solution of (21.29) is

$$\frac{F(t,T_i)+K}{F(0,T_i)+K}$$

$$= \exp\left\{-\frac{1}{2}\left(\frac{F(0,T_i)}{F(0,T_i)+K}\right)^2 \int_0^t \sigma(s,T_i)^\mathsf{T}\sigma(s,T_i)\mathrm{d}s\right.$$

$$\left. +\left(\frac{F(0,T_i)}{F(0,T_i)+K}\right)\int_0^t \sigma(s,T_i)\cdot\mathrm{d}W_{T_i}(s)\right\},$$

where the right-hand side is an exponential martingale that defines a Radon–Nikodým derivative $\mathrm{d}\hat{\mathbb{P}}/\mathrm{d}\mathbb{P}$. Hence, the call price (21.30) can be written as

$$C_\text{Spread}^\text{Fwd}\left(0,T_i,T_j,K\right)$$

$$= B(0,T_i)\left(F(0,T_i)+K\right)\mathbb{E}_{\hat{\mathbb{P}}}\left[\left(Y(T_i)-1\right)^+\bigg|\mathcal{F}_0\right], \tag{21.31}$$

where the dynamics of $Y(t) = F(t, T_j)/(F(t, T_i) + K)$,

$$\mathrm{d}Y(t) = Y(t)\left[\sigma(t,T_j)^\mathsf{T}\Gamma_{i,j-1}(t)\mathrm{d}t\right.$$

$$\left. +\left(\sigma(t,T_j)-\frac{F(0,T_i)}{F(0,T_i)+K}\sigma(t,T_i)\right)\cdot\mathrm{d}W_{\hat{\mathbb{P}}}(t)\right],$$

is derived from (21.27) and (21.29) using Itô's product rule and

$$\mathrm{d}W_{\hat{\mathbb{P}}}(t) = \mathrm{d}W_{T_i}(t)-\frac{F(0,T_i)}{F(0,T_i)+K}\sigma(t,T_i)\mathrm{d}t.$$

Standard techniques applied to (21.31) yield

$$C_\text{Spread}^\text{Fwd}\left(0,T_i,T_j,K\right)$$

$$= B(0,T_i)\left[F(0,T_j)e^{\int_0^{T_i}\sigma(s,T_j)^\mathsf{T}\Gamma_{i,j-1}(s)\mathrm{d}s}N(d_+)\right. \tag{21.32}$$

$$\left. -\left(F(0,T_i)+K\right)N(d_-)\right],$$

where $N(\cdot)$ is the cumulative standard normal distribution function and

$$d_\pm = \frac{1}{\zeta}\left[\ln\left(\frac{F(0,T_j)}{F(0,T_i)+K}\right)+\int_0^{T_i}\sigma(s,T_j)^\mathsf{T}\Gamma_{i,j-1}(s)\mathrm{d}s\pm\frac{1}{2}\zeta^2\right],$$

$$\zeta^2 = \int_0^{T_i}\left\|\sigma(s,T_j)-\frac{F(0,T_i)}{F(0,T_i)+K}\sigma(s,T_i)\right\|^2\mathrm{d}s.$$

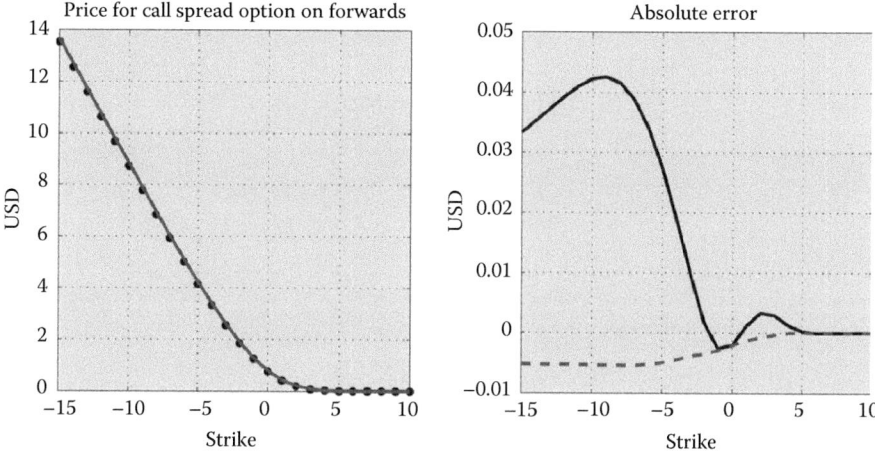

FIGURE 21.12 Call spread options on forwards for the market as calibrated in Section 21.6. The maturities for the commodity forwards are 1 year and 1 year plus 3 months. Left: The black dots show Monte Carlo prices with simulated commodity forwards and simulated interest forward rates (i.e. no 'freezing'). The blue solid line indicates the prices calculated using (21.32). Right: The black solid line shows the absolute difference between prices calculated by the closed-form formula in (21.32) and the Monte Carlo prices, i.e. the difference between the curves in the left graph. The dashed blue line shows the difference between the closed-form formula prices and a Monte Carlo simulated price using (21.27) and (21.29) but without 'freezing' in (21.28).

Figure 21.12 demonstrates the applicability of this approximation approach for the market scenario as calibrated in Section 21.6. The commodity forward maturities are chosen to be 1 year for T_i and 1 year plus 3 months for T_j. Since the forward interest rate maturities do not exactly match with the commodity forward maturities, we have interpolated the commodity forward prices in order to obtain values for $F(\cdot, T_i)$ and $F(\cdot, T_j)$. For an arbitrage-free interpolation in the maturities of the forward interest rates we refer to Schlögl (2002a). The resulting prices for call spread options on these forwards calculated by (21.32) and reference prices are shown for strikes between −15 and 10 in the left graph of Figure 21.12. Reference prices are computed by Monte Carlo simulation using Equations (21.24) and (21.25), where both commodity forwards as well as the interest forward rates $L(t, T_\ell)$ in (21.24) were simulated by the same set of 500,000 Brownian motion paths for all strikes.

The differences between the prices calculated by (21.32) and the Monte Carlo prices is shown by the upper curve of the right graph of Figure 21.12 and result from two approximations in the derivation of the formula. First, from 'freezing' the forward interest rates in (21.27) in order to make $\Gamma_{i,j-1}(t)$ deterministic and, secondly, from inserting the strike K in the dynamics of $F(t, T_i)$ as done in (21.29). The lower curve in the right graph shows the error resulting from the second approximation only. It is the difference between the prices calculated from (21.32) and Monte Carlo simulated prices from the dynamics given by (21.25) (i.e. without 'freezing' of the forward interest rates) and by (21.29).

We conclude this section by sketching the derivation of the analogous spread call options formula for futures instead of forwards. From the forward dynamics (21.26) and relation (21.22) we obtain for futures the general dynamics

$$dG(t,T_k) = -G(t,T_k)\sigma(t,T_k)^{\mathsf{T}}\,\bar{\eta}(t,T_k)dt$$
$$+ G(t,T_k)\sigma(t,T_k)\cdot dW_{T_k}(t) \quad (1 \le k \le N). \tag{21.33}$$

Applying Kirk's idea gives for the first futures with maturity T_i,

$$d\big(G(t,T_i)+K\big)$$

$$= -\big(G(t,T_i)+K\big)\frac{G(0,T_i)}{G(0,T_i)+K}\sigma(t,T_i)^{\mathsf{T}}\,\bar{\eta}(t,T_i)\mathrm{d}t \tag{21.34}$$

$$+\big(G(t,T_i)+K\big)\frac{G(0,T_i)}{G(0,T_i)+K}\sigma(t,T_i)\cdot\mathrm{d}W_{T_i}(t),$$

which is the analogue to (21.29). For the second futures with maturity T_j we employ (21.5) in order to obtain the dynamics under $\mathrm{d}W_{T_i}$,

$$dG(t,T_j)=G(t,T_j)\sigma(t,T_j)^{\mathsf{T}}\big[\Gamma_{i,j-1}(t)-\bar{\eta}(t,T_j)\big]\mathrm{d}t$$

$$+G(t,T_j)\sigma(t,T_j)\cdot\mathrm{d}W_{T_i}(t).$$

Using (21.21) we know that

$$\Gamma_{i,j-1}(t)=\sum_{\ell=i}^{j-1}\frac{\delta L(0,T_\ell)}{1+\delta L(0,\ell)}\lambda(t,T_\ell)$$

$$=\sum_{\ell=i}^{j-1}\bar{\eta}(t,T_{\ell+1})-\bar{\eta}(t,T_\ell),$$

and the dynamics of the second futures is given by

$$dG(t,T_j)=-G(t,T_j)\sigma(t,T_j)^{\mathsf{T}}\,\bar{\eta}(t,T_i)\mathrm{d}t$$

$$+G(t,T_j)\sigma(t,T_j)\cdot\mathrm{d}W_{T_i}(t),$$

which is the analogue to (21.27). Solving (21.34) yields the relation

$$\frac{G(t,T_i)+K}{G(0,T_i)+K}\exp\left\{\frac{G(0,T_i)}{G(0,T_i)+K}\int_0^t\sigma(s,T_i)^{\mathsf{T}}\,\bar{\eta}(s,T_i)\mathrm{d}s\right\}$$

$$=\exp\left\{-\frac{1}{2}\left(\frac{G(0,T_i)}{G(0,T_i)+K}\right)^2\int_0^t\sigma(s,T_i)^{\mathsf{T}}\sigma(s,T_i)\mathrm{d}s\right.$$

$$\left.+\frac{G(0,T_i)}{G(0,T_i)+K}\int_0^t\sigma(s,T_i)\cdot\mathrm{d}W_{T_i}(s)\right\},$$

in which the right-hand side is an exponential martingale and defines a Radon–Nikodým derivative $\mathrm{d}\hat{\mathbb{P}}/\mathrm{d}_{T_i}$. This allows us to write the spread call option prices as

$$C_{\text{Spread}}^{\text{Fut}}\left(0, T_i, T_j, K\right)$$

$$= B(0, T_i) \mathbb{E}_{T_i}\left[\left(G(T_i, T_j) - G(T_i, T_i) - K\right)^+\bigg|\mathcal{F}_0\right]$$

$$= B(0, T_i) \mathbb{E}_{T_i}\left[\left(G(T_i, T_i) + K\right)\left(\frac{G(T_i, T_j)}{G(T_i, T_i) + K} - 1\right)^+\bigg|\mathcal{F}_0\right]$$

$$= B(0, T_i)\left(G(0, T_i) + K\right)$$

$$\times \exp\left\{-\frac{G(0, T_i)}{G(0, T_i) + K}\int_0^{T_i}\sigma(s, T_i)^\top \overline{\eta}(s, T_i)ds\right\}$$

$$\times \mathbb{E}_{\widehat{\mathbb{P}}}\left[\left(Y(T_i) - 1\right)^+\bigg|\mathcal{F}_0\right],$$

now with $Y(t) = G(t, T_j)/(G(t, T_i) + K)$. The same standard techniques as for spread options on forwards yield

$$C_{\text{Spread}}^{\text{Fut}}\left(0, T_i, T_j, K\right)$$

$$= B(0, T_i)\left[G(0, T_j)\exp\left\{\int_0^{T_i}\sigma(s, T_j)^\top \overline{\eta}(s, T_i)ds\right\}N(d_+)\right.$$

$$-\left(G(0, T_i) + K\right)\exp\left\{-\frac{G(0, T_i)}{G(0, T_i) + K}\right.$$

$$\times \left.\left.\int_0^{T_i}\sigma(s, T_i)^\top \overline{\eta}(s, T_i)ds\right\}N(d_-)\right],$$

with

$$d_\pm = \frac{1}{\zeta}\left[\ln\left(\frac{G(0, T_j)}{G(0, T_i) + K}\right)\right.$$

$$+ \int_0^{T_i}\left(\sigma(s, T_j)^\top \overline{\eta}(s, T_i)\right.$$

$$+ \left.\frac{G(0, T_i)}{G(0, T_i) + K}\sigma(s, T_i)^\top \overline{\eta}(s, T_i)\right)ds \pm \frac{1}{2}\zeta^2\right],$$

$$\zeta^2 = \int_0^{T_i}\left\|\sigma(s, T_j) - \frac{G(0, T_i)}{G(0, T_i) + K}\sigma(s, T_i)\right\|^2 ds.$$

21.8 Conclusion

In the present paper, a joint model of commodity and interest rate dynamics based on the LIBOR Market Model approach was constructed. We have demonstrated how it can be effectively calibrated to market data, including at-the-money implied volatilities, and how less liquid instruments such as commodity spread options can be priced relative to the market using the model.

In closing, one should note that although neither seasonal cycles in commodity prices nor mean reversion of the commodity price process were explicitly considered in the model construction, both of these features (well documented in the empirical literature) can be captured by the model, to the extent that they are reflected in current market prices. Firstly, seasonal cycles in commodity prices are anticipated by the market and thus subsumed in the term structure of futures (or forward) prices, to which the model is calibrated by construction. Secondly, mean reversion is reflected in the term structure of volatility, to which the model is also calibrated, e.g. in the presence of mean reversion of the commodity spot price process, the volatility of the forward (or futures) prices increases as time to maturity decreases. Thus the martingale approach to the model construction does *not* preclude mean reversion simply because the objects considered explicitly are driftless—in fact, the corresponding processes under the appropriate probability measures are necessarily just as driftless in any other arbitrage-free model, including those that make mean reversion explicit in an Ornstein/Uhlenbeck process.

The main advantage of the LMM approach lies in its calibration to the market, and the model presented here opens up interesting avenues of further research (beyond the scope of the present paper) in terms of calibrating the model not just in the maturity dimension, but also in the strike dimension, along the lines of the 'smile'-fitting extensions of the basic LMM discussed by Brace (2007).

Acknowledgements

This research was supported as part of a project with SIRCA under the Australian Research Council's Linkage funding scheme (project No. LP0562616). The authors would like to thank the Editors of *Quantitative Finance* and two anonymous referees for helpful comments. The usual disclaimer applies.

References

Black, F., The pricing of commodity contracts. *J. Financ. Econometr.*, 1976, **3**, 167–179.

Brace, A., *Engineering BGM*, 2007 (Chapman & Hall/CRC: London).

Brace, A., Gatarek, D. and Musiela, M., The market model of interest rate dynamics. *Math. Finance*, 1997, **7**, 127–154.

Choy, B., Dun, T. and Schlögl, E., Correlating market models. *Risk*, 2004, **17**(9), 124–129.

Cortazar, G. and Schwartz, E.S., The valuation of commodity contingent claims. *J. Deriv.*, 1994, **1**, 27–39.

Cox, J.C., Ingersoll, J.E. and Ross, S.A., The relation between forward prices and futures prices. *J. Financ. Econometr.*, 1981, **9**, 321–346.

Gibson, R. and Schwartz, E.S., Stochastic convenience yield and the pricing of oil contingent claims. *J. Finance*, 1990, **45**, 959–976.

Heath, D., Jarrow, R. and Morton, A., Bond pricing and the term structure of interest rates: A new methodology for contingent claims valuation. *Econometrica*, 1992, **60**, 77–105.

Jamshidian, F., LIBOR and swap market models and measures. *Finance Stochast.*, 1997, **1**, 293–330.

Kirk, E., Correlation in the energy markets. In *Managing Energy Price Risk*, edited by R. Jameson, 1995 (Risk Publications and Enron: London).

Margrabe, W., The value of an option to exchange one asset for another. *J. Finance*, 1978, **33**, 177–186.

Miltersen, K., Commodity price modelling that matches current observables: A new approach. *Quantit. Finance*, 2003, **3**, 51–58.

Miltersen, K., Sandmann, K. and Sondermann, D., Closed form solutions for term structure derivatives with log-normal interest rates. *J. Finance*, 1997, **52**, 409–430.

Miltersen, K. and Schwartz, E., Pricing of options on commodity futures with stochastic term structures of convenience yield and interest rates. J. Financ. Quantit. Anal., 1998, **33**, 33–59.

Musiela, M. and Rutkowski, M., Continuous-time term structure models: Forward measure approach. *Finance Stochast.*, 1997, **1**, 261–291.

Pedersen, M.B., Calibrating Libor market models. SimCorp Financial Research Working Paper, 1998.

Pelsser, A., *Efficient Methods for Valuing Interest Rate Derivatives*, 2000 (Springer: Berlin).

Schlögl, E., Arbitrage-free interpolation in models of market observable interest rates. In *Advances in Finance and Stochastics*, edited by K. Sandmann and P. Schönbucher, 2002a (Springer: Heidelberg).

Schlögl, E., A multicurrency extension of the lognormal interest rate market models. *Finance Stochast.*, 2002b, **6**, 173–196.

Schwartz, E.S., The pricing of commodity-linked bonds. *J. Finance*, 1982, **37**, 525–539.

Schwartz, E.S., The stochastic behavior of commodity prices: Implications for valuation and hedging. *J. Finance*, 1997, **52**, 923–973.

Appendix: Volatility Integrals

This appendix describes the calculation of integrals as in (21.11) and (21.22) for piece-wise constant volatility matrices. We will focus on (21.22), since the integral in (21.11) can be obtained in the same way by setting $\bar{\eta} = \sigma$. The proposed method follows that used in Pedersen's LMM calibration (Pedersen 1998) to compute cap and swaption total variances.

In Section 21.3 a volatility matrix $U_C \in \mathbb{R}^{n_f \times d_C}$ was calculated for each calendar time $t_1, \ldots, t_{nc}$, and from the LMM calibration an analogous matrix $U_I \in \mathbb{R}^{m_f \times d_I}$ is available for each calendar time $s_1, \ldots, s_{mc}$. Merging over calendar times yields in each case three-dimensional arrays $\Lambda_C = \left(\lambda_{i,j,k}^C\right) \in \mathbb{R}^{n_c \times n_f \times d}$ and $\Lambda_I = \left(\lambda_{i,j,k}^I\right) \in \mathbb{R}^{m_c \times m_f \times d_I}$, respectively. Forward times are given by $x_1, \ldots, x_{nf}$ for commodity forwards and by $y_1, \ldots, y_m$ for interest forward rates. The calendar times $t_0 = s_0 = 0$ and forward times $x_0 = y_0 = 0$ are added for notational convenience. From these matrices of piecewise constant volatilities the integral in Equation (21.22) has to be calculated.

The first step illustrates the computation of $\bar{\eta}$ and can be omitted when calculating the integral in (21.11). The matrix Λ_I determines the forward interest rate volatilities $\lambda(s, s + y)$ as occurring in Equation (21.6). From these, a corresponding matrix $\Theta_I \in \mathbb{R}^{m_c \times m_f \times d_I}$ with entries $\bar{\eta}_{i,j,k}$ is calculated using relation (21.18). Unfortunately, the forward rates $L(t, T)$ in (21.18) are not known for $t > s_0$, but as a best guess we employ the actual forward rate curve in the sense that, for all future times $s_i > s_0$, an approximated forward rate curve $\bar{L}(s_i, T)$ is given by $L(s_0, T)$, essentially again making use of the standard 'frozen coefficient' approximation.

In the second step, the integral for a given time to maturity T in the upper integration limit is computed. We equalize the calendar time discretization of the volatility matrices for commodity forwards and forward interest rates by taking the union of $t_0, \ldots, t_{nc}$ and $s_0, \ldots, s_{mc}$. The new calendar times will be denoted by $t_0 = 0, t_1, \ldots, t_{p_c}$. The same is done for the discretization of times to maturity, which results in a discretization $x_0 = 0, x_1, \ldots, x_{p_f}$. The arrays Λ_C and Θ_I are extended accordingly (i.e. by simply inserting lines for different calendar times and columns for different times to maturity, respectively, but otherwise identical entries $\lambda_{i,j,k}$ and $\bar{\eta}_{i,j,k}$), such that $\Lambda_C, \Theta_I \in \mathbb{R}^{p_c \times p_f \times d}$. The purpose of this extension is to have a twofold sum in formula (A.1) instead of a fourfold sum.

Any tuple in $\mathcal{I} = \left\{(i, j) : 1 \leq i \leq p_c, 1 \leq j \leq p_f\right\}$ relates to a pair of $\sigma(t_i, t_i + x_j)$ and $\bar{\eta}(t_i, t_i + x_j)$, and in order to be relevant for the considered integral, the following inequalities need to be satisfied for some $0 \leq t \leq T$:

$$t_{i-1} \leq t < t_i, \quad x_{j-1} \leq T - t < x_j.$$

Equivalently, this can be written as

$$\kappa_{i,j} := \min\{t_i, T - x_{j-1}, T\} - \max\{t_{i-1}, T - x_j\} > 0.$$

Hence, the integral is given by

$$\int_t^T \sigma(u,T)\bar{\eta}(u,T)\mathrm{d}u = \sum_{(i,j)\in\mathcal{J}} \kappa_{i,j} \sum_{k=1}^d \lambda_{i,j,k}^{\mathrm{C}} \bar{\eta}_{i,j,k}, \qquad (A.1)$$

where $\mathcal{J} = \{(i,j) \in \mathcal{I} : \kappa_{i,j} > 0\}$.

22

Evaluation of Gas Sales Agreements with Indexation Using Tree and Least-Squares Monte Carlo Methods on Graphics Processing Units

W. Dong* and
B. Kang

22.1 Introduction*

A gas sales agreement (GSA, also known as a gas swing option) is an American-style option with daily exercise opportunities. The delivery of daily quantities is constrained between pre-determined minimum and maximum daily limits. Under a GSA, the gas buyer must take a minimum volume of gas, which is called the minimum bill; otherwise, a penalty applies. Typically, there is also a maximum annual quantity of gas that can be taken.

* Corresponding author. Email: wenfeng.dong@data61.csiro.au

DOI: 10.1201/9781003265399-25

Besides the above features, another feature, the indexation, introduces a further difficulty in the evaluation of a GSA. When the contractual price (strike price) in a GSA is a constant, the valuation of the GSA is a classical dynamic programming problem. This kind of problem has been fully investigated in the literature (see Thompson 1995, Dörr 2003, Jaillet *et al.* 2004, Barrera-Esteve *et al.* 2006, Thanawalla 2006, Bardou *et al.* 2009, Edoli *et al.* 2013, Chiarella *et al.* 2016, Dong and Kang 2019). In real contracts, however, the contractual price is set based on the indexation principle, under which the contractual price is called the index. For each month, the index value is determined by the weighted average price of certain energy products (e.g. crude oil) in the previous month (for details, see Asche *et al.* 2002). Under the indexation principle, the value of the GSA contract depends not only on the current state but also on the past values of certain other energy products. This feature links the valuation of the GSA to the moving average problem. Thus far, however, the literature has not presented an effective method to value GSAs embedded with moving average features.

In the literature, very few articles have discussed the valuation of options embedded with both the moving average feature and the early exercise feature. Among those that do, a common approach is the least-squares Monte Carlo simulation (LSMC), as in Longstaff and Schwartz (2001) and Broadie and Cao (2008). The idea is simply to find the continuation value by performing a regression on the polynomials of the current realized values of the underlying price (gas price) and the strike price (index). Based on the indexation principle, however, the index is determined by the average of past oil prices; hence, these oil prices should also affect the contract value. Grau (2008) and Dirnstorfer *et al.* (2013) suggested that, in the LSMC algorithm, the regression should be performed not only on the underlying price and strike price but also on all past values used to compute the strike price. This action leads to a very high-dimensional problem, especially in the evaluation of the GSAs, because the computation of the index requires 30 past oil values (we assume one month contains 30 days). Although both Grau (2008) and Dirnstorfer *et al.* (2013) proposed a technique based on the sparse grid to reduce the number of basis functions used in the LSMC algorithm, they can only solve a problem up to 10 dimensions and the computational time is very long. Bernhart *et al.* (2011) developed a method using a finite-dimensional approximation of the infinite-dimensional dynamics of the moving average process based on the weighted Laguerre polynomial expansion. They used several so-called Laguerre processes to approximate the moving average process, and they demonstrated that the accuracy of this approximation is better if more Laguerre processes are used. The regression in the LSMC algorithm is performed on all underlying prices and all realized values of these Laguerre processes. As reported in Warin (2012), there is no big improvement of the option value compared with the usual regression method by Broadie and Cao (2008). There are also works which price options with moving average features using the lattice-based method. Kao and Lyuu (2003) priced moving average lookback options with the help of a binomial tree. The binomial tree was also used by Dai *et al.* (2010) to price moving average barrier options. Furthermore, the willow tree method proposed by Curran (2001) was further developed by Xu *et al.* (2013) and Lu *et al.* (2017) to price path-dependent options.

Fewer studies have discussed the evaluation of GSAs with indexation. Bernhart (2011) used the methodology proposed by Bernhart *et al.* (2011) to price a GSA contract with only one exercise opportunity, which is an American option where the strike price is computed based on the indexation. In the LSMC, however, because all simulated paths of all underlying processes and Laguerre processes must be stored, one would have to store the path information on the hard drive due to the lack of random access memory (RAM). This would slow the computation speed significantly due to the need for file writing and/or reading operations. Edoli (2013) implemented the GSA evaluation by assuming that the moving average process is Markovian and implemented it using the finite difference method. Hanfeld and Schlüter (2017) investigated how the GSA operates on the gas market. With a focus on the short-term optimization of this contract, the authors compared the performance of the LSMC algorithm to a myopic approach* on the operational strategies in different periods of time (bullish, bearish, etc.).

---

* The contract is simply exercised if the current spot price is larger than the index.

Fanelli and Ryden (2018) priced the gas swing options traded on NetConnect Germany using a myopic approach with Monte Carlo simulations and provided some statistical tests.

In this paper, taking advantage of the rapid growth in GPU technology, we propose a new numerical method based on a two-dimensional (2-D) trinomial tree to fill the gap. Because the LSMC algorithm is well known to handle sophisticated models, such as multi-factor models, models with regime-switching, or models with jumps, we build this method for pricing GSAs with indexation and analyze its performance when implementing it on GPUs.

This paper is organized as follows. We suggest mean-reverting models for the gas price and oil price and formulate the details of the index price based on the indexation principle in Section 22.2. In addition, we generalize and formulate the key features of a GSA with indexation in Section 22.3. For pricing GSAs with indexation on a 2-D trinomial tree, we propose a detailed numerical method in Section 22.4. We build the LSMC algorithm together with an upper bound algorithm and a benchmark algorithm in Section 22.5. Section 22.6 presents a description of how we can implement both the tree algorithm and the LSMC algorithm on GPUs. Section 22.7 provides numerical examples that analyze the performance of the tree algorithm and LSMC algorithm. We conclude the work in Section 22.8.

22.2 The Pricing Framework and the Indexation Principle

Denote the spot prices of gas and oil at time t by $S(t)$ and $Z(t)$, respectively. This paper assumes that the log prices of the gas and the oil follow the mean-reverting processes

$$d\ln S(t) = \left[\theta^S(t) - \alpha_S \ln S(t)\right]dt + \sigma_S dB^S(t), \tag{22.1}$$

$$d\ln Z(t) = \left[\theta^Z(t) - \alpha_Z \ln Z(t)\right]dt + \sigma_Z dB^Z(t), \tag{22.2}$$

where $B^S(t)$ and $B^Z(t)$ are standard Brownian motions with correlation ρ, and α_S, σ_S, α_Z and σ_Z are constants. Let $F^S(t, T)$ and $F^Z(t, T)$ be the forward prices of gas and oil at time $t \in [0, T]$ with maturity T, respectively. $\theta^S(t)$ and $\theta^Z(t)$ are functions of time chosen to provide an exact fit to the observed forward curves $F^S(0, t)$ and $F^Z(0, t)$, respectively. These processes are standard models when it comes to the commodity price in the energy market (see Hull 2011, Edoli *et al.* 2013).

We have the time horizon $[0, T]$ equally spaced into L periods. Let δ be the interval of each period, $\delta = T/L$. Denote the beginning of the lth period by T_l, $l = 0, 1, \ldots, L - 1$, and $T_L = T$:

$$0 = T_0 < T_1 = \delta < T_2 = 2\delta < \cdots < T_l$$
$$= l\delta < \cdots < T_{L-1} = (L-1)\delta < T_L = L\delta = T.$$

Let the period $[T_l, T_{l+1}]$, $l = 0, \ldots, L-1$, be equally spaced into D pieces with interval $\Delta t = \delta/D$, where $T_l = l \cdot \delta$ and D is a positive integer. Then the time horizon $[0, T]$ is equally spaced into $L \cdot D$ pieces. Let $N = L \cdot D$ and $t_n = n \cdot \Delta t$, where $n = 0, 1, \ldots, N$. We have

$$T_l = t_{lD} < T_l + \Delta t = t_{lD+1} < T_l + 2\Delta t = t_{lD+2} < \cdots$$
$$< T_l + (D-1)\Delta t = t_{lD+(D-1)\Delta t}$$
$$< T_l + D\Delta t = t_{(l+1)D} = T_{l+1}.$$

Let $I(t_n)$ be the index at time t_n. For $t_n \in [T_0, T_1]$, the index is a fixed constant K which is specified in the contract. For $t_{lD+d} \in (T_l, T_{l+1}]$, $l = 1, \ldots, L-1$ and $d = 1, \ldots, D$, the index is the average value of the crude oil daily price realizations Z in the previous period $(T_{l-1}, T_l]$. That is,

$$I\left(t_{lD+d}\right) = \begin{cases} K & \text{for } l = 0, \\ \dfrac{1}{D}\displaystyle\sum_{j=1}^{D} Z\left(t_{(l-1)D+j}\right), & \text{for } l = 1,2,\ldots,L-1. \end{cases} \tag{22.3}$$

Since the average value of the oil prices in the current period is the index value for the next period, we need another variable, M, to track how this average moves. Thus, for $d = 1, \ldots, D$, we introduce the running average variable M, which is given by

$$M\left(t_{lD+d}\right) = \frac{1}{d}\sum_{j=1}^{d} Z\left(t_{lD+j}\right), \tag{22.4}$$

where $l = 0, 1, \ldots, L-2$, since we do not need to track the running average in the last month. Also, it is straightforward to observe that, for $t_n \in (T_l, T_{l+1}], l = 1, 2, \ldots, L-1$,

$$I\left(t_n\right) = M\left(T_l\right). \tag{22.5}$$

22.3 Gas Sales Agreements with Indexation

A gas sales agreement with indexation between the gas provider and the gas user could have many specific features to satisfy the needs of both parties to the agreement, but these contracts usually have a number of common features as set out below:

- T is the terminal date of the contract. Since we focus on the 1-year GSA, we let $T = 1$. L corresponds to the number of months and δ coincides with the length of each month. That is, $T = L \cdot \delta$. In the rest of this paper, we call the first month 'Month 0', the second month 'Month 1', ... , the last month 'Month $L - 1$'.
- We assume that the holder of a GSA has exactly one exercise right at each time t_n for $n = 1, \ldots, N$, which amounts to D rights for a whole month $(T_l, T_{l+1}]$ and N rights for a whole year $[0, T]$. Typical GSAs can usually be exercised daily (we assume that 1 year contains 360 days, that is, $L = 12$, $D = 30$ and $N = 360$).
- Let qt_n be the amount of gas taken (exercise decision) at time t_n, which is constrained by the minimum daily quantity q_{min} and the maximum daily quantity q_{max}, that is $q_{min} \le q_{tn} \le q_{max}$. This exercise decision qt_n is a random quantity depend on the observations of underlying prices. Let Q_{t_n} be the cumulative amount of gas taken before time t_n (also known as the period to date) which is given by

$$Q_{t_n} = \sum_{k=1}^{n-1} q_{t_k} \quad \text{for } n = 2,\ldots,N,$$

and $Qt_n = 0$ for $n = 0, 1$. In addition, we let Q_T be the total amount of gas taken. That is, $Q_T = Q_{tN} + q_{tN}$.

- Upon taking the volume q_{tn}, the payoff from the buyer's point of view at time $t_n = t_{lD+d} \in (T_l, T_{l+1}], l = 0, \ldots, L - 1$ and $d = 1, \ldots, D$, is given by

$$q_{t_{lD+d}}\left(S\left(t_{lD+d}\right) - I\left(t_{lD+d}\right)\right).$$

- There is a maximum quantity of gas the buyer can take, which is called the annual contract quantity (denoted by ACQ). Similarly, there is a minimum quantity of gas the buyer has to take, which is called the minimum bill (denoted by *MB*).Both the minimum bill and the annual contract quantity can be violated. If the total gas taken is below the minimum bill or above the annual contract quantity, the buyer has to pay penalties at the end of the contract. More precisely, if $Q_T <$ *MB* or $Q_T >$ *ACQ* at time *T*, there is an out cash flow generated by the penalty, in addition to the cash flow generated by the instant payoff. The possible penalty at time *T* is given by

$$
\begin{aligned}
\mathcal{P}(I, Q_T) = &-I \cdot \max\{MB - Q_T, 0\} \\
&- I \cdot \max\{Q_T - ACQ, 0\}.
\end{aligned}
\tag{22.6}
$$

- For t_{lD+d} within the period $(T_l, T_{l+1}]$, that is, $d = 1, 2, \dots, D - 1$, given the values of $M(t_{lD+d})$ and $I(t_{lD+d})$ (see (22.3) and (22.4)), we have the evaluations of the running average with respect to the oil price and the index, which can be described by

$$
M(t_{lD+d+1}) = \frac{d \cdot M(t_{lD+d}) + Z(t_{lD+d+1})}{d+1} \quad \text{and}
\tag{22.7}
$$
$$
I(t_{lD+d+1}) = I(t_{lD+d}),
$$

respectively.
- The risk-free rate is a fixed constant *r*.

From the buyer's point of view, the goal is to maximize the total expected discounted payoff of the contract, including the penalty. That is, to find the value $V(t_n, S, I, Q)$ where the gas price is *S*, the index is *I* and the period to date is *Q* of this contract at time t_n, which is given by

$$
\begin{aligned}
V(t_n, S, I, Q) = \sup_{q_{t_k}, n \leq k \leq N} \mathbb{E}\Bigg[&\sum_{k=n}^{N} e^{-r(t_k - t_n)} q_{t_k} \big(S(t_k) - I(t_k)\big) \\
&+ e^{-r(T - t_n)} \mathcal{P}\big(I(t_N), Q_T\big) \Big| S(t_n) = S, I(t_n) = I, Q_{t_n} = Q \Bigg].
\end{aligned}
\tag{22.8}
$$

Or, equivalently, for $t_n = t_{lD+d} \in (T_l, T_{l+1}]$, we have

$$
\begin{aligned}
V(t_n, S, I, Q) = \sup_{q_{tm}} \mathbb{E}\Bigg[&\underbrace{\sum_{k=d}^{D} e^{-r(t_{lD+k} - t_n)} q_{t_{lD+k}} \big(S(t_{lD+k}) - I(t_{lD+k})\big)}_{\text{part 1}} \\
&+ \underbrace{\sum_{g=l+1}^{L-1} \sum_{k=1}^{D} e^{-r(t_{gD+k} - t_n)} q_{t_{gD+k}} \big(S(t_{gD+k}) - I(t_{gD+k})\big)}_{\text{palt 2}} \\
&+ \underbrace{e^{-r(T - t_n)} \mathcal{P}\big(I(t_N), Q_T\big)}_{\text{part 3}} \Big| S(t_n) = S, \\
&I(t_n) = I, Q_{t_n} = Q \Bigg],
\end{aligned}
$$

where $m = lD\,d, lD\,d+1, \dots, N$, 'part 1' gives the value of all the other instant payoffs in the current month at t_n, 'part 2' gives the value of all instant payoffs in all future months at t_n, and 'part 3' gives the value of the penalty at t_n.

REMARK 1 In Bardou *et al.* (2009) (also see Bardou *et al.* 2010, Lari-Lavassani *et al.* 2001), the authors suggest a decomposition of the gas swing options. That is, a gas swing option can be decomposed into two parts: a swap part and a normalized part. We apply their idea to our GSA, and the time t_0 value of (22.8) can be written as follows:

$$
V\big(t_0, S(t_0), I(t_0), 0\big)
$$

$$
= \mathbb{E}\left[\underbrace{\sum_{n=1}^{N} e^{-t_n r} q_{\min} \cdot \big(S(t_n) - I(t_n)\big)}_{\text{The sway part}} \right]
$$

$$
+ \underbrace{\sup_{\overline{q}_{t_n} \in [0, \overline{q}]} \mathbb{E}\left[\sum_{n=1}^{N} e^{-t_n r} \overline{q}_{t_n} \cdot \big(S(t_n) - I(t_n)\big) + e^{-\mathrm{Tr}} \overline{\mathcal{P}}\big(I(t_N), \overline{Q}_T\big) \right]}_{\text{The normalized GSA}}.
$$

This normalized GSA has the following inputs: the daily minimum 0, the daily maximum $\overline{q} = q_{\max} - q_{\min}$. The penalty function of this GSA is given by

$$
\overline{\mathcal{P}}\big(I_N, \overline{Q}_T\big) = \mathcal{P}\big(I_N, \overline{Q}_T + N \cdot q_{\min}\big),
$$

where $\overline{Q}_T = \sum_{n=1}^{N} \overline{q} t_n$. Since most of the uncertainties lie in the normalized part, in this paper, we mainly focus on the normalized GSA contract. Without loss of generality, we usually let $q_{\min} = 0$ and $q_{\max} = \overline{q}$, where $\overline{q}$ is a constant.

22.4 The Evaluation Using Trinomial Trees

In this section, we build an algorithm that evaluates the price of the gas sales agreement using a 2-D trinomial tree. We first build two separate one-dimensional trinomial trees for the gas price (the gas tree) and the oil price (the oil tree), respectively. By combining these two one-dimensional trees together, we obtain a two-dimension trinomial tree. The tree building procedures are shown in Appendix 1.

Once the tree is built, each node on the 2-D tree can be referenced by a vector of integers (n, s, z), where n indicates that the current time is $n\Delta t$, s and z are the price indexes on the gas tree and the oil tree, respectively. Let $\Delta X = \sigma s \sqrt{3\Delta t}$ and $\Delta Y = \sigma z \sqrt{3\Delta t}$ be the space steps on the gas tree and the oil tree, respectively. Let a_n and b_n, n 0, 1, ... , N, be the amounts added on each node of the gas tree and the oil tree such that the gas tree and the oil tree are consistent with the observed forward curves (see (22.A4)), respectively. Then the gas price $S_{n,s}$ of level s on the gas tree and the oil price $Z_{n,z}$ of level z on the oil tree at time t_n are given by

$$
S_{n,s} = e^{s \cdot \Delta X + a_n} \quad \text{and} \quad Z_{n,z} = e^{z \cdot \Delta Y + b_n},
$$

respectively.

22.4.1 The Structure of a Trinomial Tree

As we can see in Figure 22.1, which shows a typical one-dimensional trinomial tree, there are three possible forms of the tree branching. Denote the highest indexes that can be reached by the gas tree and the oil tree by $s_{\max}$ and $z_{\max}$, respectively. Since both the gas tree and the oil tree are symmetrical, the lowest indexes that can be reached by the gas tree and the oil tree are $s_{\max}$ and $z_{\max}$, respectively. We can see all the forms of the tree branching of both the gas tree and the oil tree in Figure 22.2.

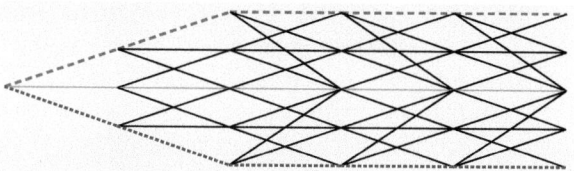

FIGURE 22.1   A one-dimensional trinomial tree.

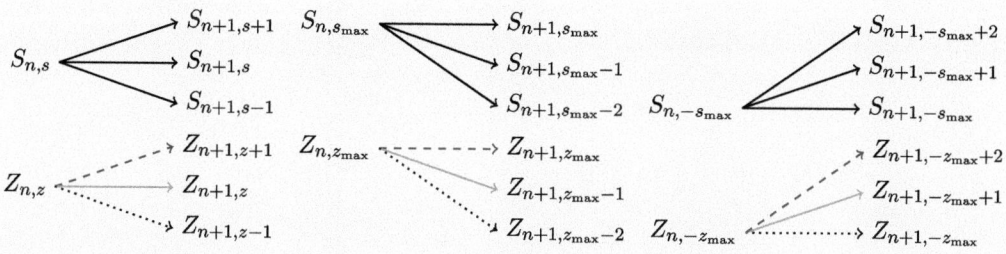

FIGURE 22.2   Possible forms of the tree branching for a mean-reverting trinomial tree.

Define the following functions:

$$\mathfrak{g}(s,b) = \begin{cases} s+b & \text{if } -s_{\max} < s < s_{\max} \\ s+b-1 & \text{if } s = s_{\max} \\ s+b+1 & \text{if } s = -s_{\max} \end{cases} \quad \text{and}$$

$$\mathfrak{h}(z,c) = \begin{cases} z+c & \text{if } -z_{\max} < z < z_{\max} \\ z+c-1 & \text{if } z = z_{\max} \\ z+c+1 & \text{if } z = -z_{\max} \end{cases}$$

where $b, c \in \{1, 0, 1\}$. Here we use integers -1, 0 and 1 to represent the lower, middle and upper branches on the trees, respectively. Then $(n+1, \mathfrak{g}(s, b), \mathrm{h}(z, c))$ gives the node associated with the b branch on the gas tree and the c branch on the oil tree emanating from the node (n, s, z). Now, we define one more function:

$$\mathrm{j}(n) = \min\{n, z_{\max}\}.$$

As we can see in Figure 22.1, the trinomial tree stops growing at some point. Before the oil tree stops growing, the size of the tree grows over time and n returns the highest level of the oil tree at time t_n. Once the tree stops growing, the size of the tree stops growing and the highest level of the oil tree is simply $z_{\max}$. Hence, $\mathrm{j}(n)$ gives the highest level of the oil tree at any $t_n \in [0, T]$. Since the oil tree is symmetrical, $-\mathrm{j}(n)$ gives the lowest level of the oil tree at any $t_n \in [0, T]$. It follows that the total number of nodes on the oil tree at time t_n is $2 \cdot \mathrm{j}(n)+1$. Similarly, we can define such a function $\mathfrak{k}(n)$ for the gas tree:

$$\mathfrak{k}(n) = \min\{n, s_{\max}\}.$$

In addition, we denote the probability associated with the b branch on the gas tree and the c branch on the oil tree emanating from node (n, s, z) by $p_{s,z,b,c}$. These probabilities can be computed using (22.A1) and (22.A2).

The discretization of the running average and the index. To perform the evaluation, at each time t_n, we generate two vectors, $\mathbf{M}_n$ and $\mathbf{I}_n$, in both the running average dimension and the index dimension by discretization, respectively. Since in the first $L - 1$ months (i.e. in Month l, l 0, 1, ... , $L - 2$), we have the running average variable M, then an integer m which represents the value of the running average should be introduced (we do not need to track the running average in the last month). Similarly, in the last $L - 1$ months (i.e. in Month l, $l = 1, 2, ... , L - 1$), we have the index price which is not a constant, thus we have to add one more integer i which represents the level of the value of the index (we do not need this integer i in the first month where the index is simply a constant K).

The running average vector $\mathbf{M}_n$. $\mathbf{M}_n$ is the vector containing the possible values of the running average M at time t_n. Let $M_{n,m}$ be the value of the running average at time t_n where the level in the running average vector $\mathbf{M}_n$ is m. Let H be the total number of values in $\mathbf{M}_n$, then

$$\mathbf{M}_n = \left(M_{n,0}, M_{n,1}, ..., M_{n,H-1} \right).$$

We also assume that $\mathbf{M}_n$ is constructed in such a way that

$$M_{n,0} < M_{n,1} < \cdots < M_{n,H-1}.$$

Once the 2-D tree is built, we assume that the level m value of the running average at each time t_n, $n = 1, 2, ... , N$, is referenced by a deterministic function $\partial_n(m)$ where the value solely depends on m. That is,

$$M_{n,m} = \partial_n(m)$$

Denote the lowest and highest values of the running average M that can be achieved at time t_n through the tree by $M_n^{\min}$ and $M_n^{\max}$, respectively. Referring back to Figure 22.1, $M_n^{\min}$ and $M_n^{\max}$ can be obtained by following the dotted branches and dashed branches, respectively. Then, for $t_{lD+d} \in (T_l , T_{l+1}]$, d = $1 , ... , D$ and $l = 0, 1, ... , L - 2$, we have

$$M_{lD+d}^{\min} = \frac{1}{d} \sum_{k=1}^{d} Z_{lD+d,-j(lD+d)} \quad \text{and}$$

$$M_{lD+d}^{\max} = \frac{1}{d} \sum_{k=1}^{d} Z_{lD+d,j(lD+d)}.$$

At time t_n, $n = lD + d$, $d = 1, ..., D - 1$ and $l = 0, 1, ..., L - 2$, for a given value of the running average $M_{n,m}$, we can get the evaluation of $M_{n,m}$ by following these three possible movements of the oil price Z in Figure 22.2. Note that the evaluation of the running average M only depends on the oil price Z, while the gas price S has no impact on M (see (22.4)). Denote the one time-step forward evaluation of $M_{n,m}$ by $\mathcal{M}_{n,m,c}$, together with (22.7), we have

$$\mathcal{M}_{n,m,c} = \frac{d \cdot M_{n,m} + Z_{n+1,h(z,c)}}{d+1}, \tag{22.9}$$

where $c = -1, 0, 1$. Figure 22.3 shows the evaluation of $M_{n,m}$ corresponding to the dashed, solid and dotted branches in Figure 22.2.

For the discretization in the running average dimension, we propose two possible approaches: further discretization of the space step on the oil fundamental tree and the non-uniform grid. We investigate the performance of these two approaches through numerical examples in Section 22.7.

Further discretization. We use the same discretization method as in Dai *et al.* (2010). Let the space step ΔM in the

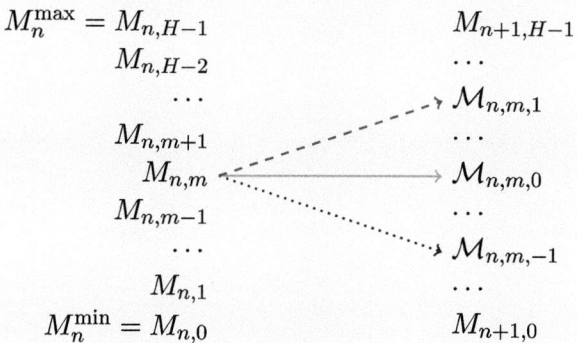

FIGURE 22.3 One step forward evaluation of the running average.

running average dimension be given by

$$\Delta M = \frac{\Delta Y}{F},$$

where ΔY is the space step on the oil tree and F is a positive integer. Then $\eth_n(m)$ is given by

$$\eth_n(m) = M_n^{\min}e^{m\Delta M}. \tag{22.10}$$

Recall that the total number of nodes on the oil tree at time t_n is $2\,\mathrm{j}(n)+1$, through the further discretization, the total number of values of the running average in the running average vector $\mathbf{M}_n$ is

$$H = 2 \cdot \mathrm{j}(n) \cdot F + 1. \tag{22.11}$$

Non-uniform grid. We adopt the non-uniform grid in in't Hout and Foulon (2010). Let M_n, $n = lD + d$, $d = 1, \dots, D$ and $l = 0, 1, \dots, L - 2$, be given by

$$\mathrm{M}_{lD+d} = \frac{1}{d}\sum_{k=1}^{d} Z_{lD+k,0}. \tag{22.12}$$

That is, M approximates the value of the running average M by assuming that the oil price moves on the oil tree by following the middle branches only (that is, we follow the green branches only in Figure 22.1). Let H be a positive integer and w_n, $n = 1, 2, \dots, N$, be a series of positive constants.[*] At each time t_n, we first construct a series of equidistant points

$$\xi_0 < \xi_1 < \xi_2 < \cdots < \xi_H,$$

where

$$\xi_m = \sinh^{-1}\left(\frac{M_n^{\min} - \mathrm{M}_n}{w_n \mathrm{M}_n}\right) + m\Delta\xi$$

and

---

[*] In our numerical example, a constant w_n for all $n = 1, 2, \dots, N$ is usually sufficient to get an accurate value. However, to add flexibility to this algorithm, we let w_n can be different constant numbers for different time steps.

$$\Delta\xi = \frac{1}{H}\left[\sinh^{-1}\left(\frac{M_n^{\max} - M_n}{w_n M_n}\right)\right.$$

$$\left. -\sinh^{-1}\left(\frac{M_n^{\min} - M_n}{w_n M_n}\right)\right].$$

Then a non-uniform grid at time t_n in the running average dimension can be obtained through the following transformation:

$$M_{n,m} = \eth_n(m)$$
$$= M_n + w_n M_n \sinh(\xi_m). \tag{22.13}$$

As reported in in't Hout and Foulon (2010), this grid is smooth, and these constants w_n control the fraction of points $M_{n,m}$ that lie in the neighborhood of M_n. A smaller w_n gives more points in the neighborhood of M_n. Intuitively speaking, the non-uniform grid could be a better approach than the further discretization if we let more points be in the neighborhood of M_n. Due to the mean-reverting nature of the oil price, the oil price is unlikely to be too high or too low for a long time period. It follows that value of the running average is even more unlikely to be too large or too small.

The index vector $\mathbf{I}_n$. $\mathbf{I}_n$ is the vector containing the possible values of the index I at time t_n. Let $I_{n,i}$ be the index at time t_n in Month l where the level in the index vector is i. Recall (22.5) and (22.3), the value of the index in Month l, $l = 1, 2, \ldots, L - 1$, equals the value of the running average on the last day of Month $l - 1$. Then we use the same vectors for $\mathbf{I}_{lD+d}$, $d = 1, \ldots, D$, as the vector for $\mathbf{M}_{lD}$. That is, for $t_{lD+d} \in (T_l, T_{l+1}]$, $d = 1, 2, \ldots, D$ and $l = 1, 2, \ldots, L - 1$, we have

$$I_{lD+d,i} = M_{lD,i} = \eth_{lD}(i). \tag{22.14}$$

In the rest of this paper, we use the word 'tree' to refer to the two-dimensional trinomial tree built in Appendix 1 together with the lattices constructed in this section.

22.4.2 Pricing Algorithm

In this section, we present an algorithm to price the GSA contract formulated in Section 22.3. This algorithm works backwards in time. Since in the last month there is no further month we do not need to track the running average, which means we also do not need to track the oil price either. It follows that the contract value is driven by three variables: the gas price, S, the index, I, and the period to date, Q. In the middle months, i.e. Month 1, Month 2, ..., Month $L - 2$, we need to track the running average and the oil price since they will be used to calculate the index in the coming month, which means the contract value is driven by five variables: S, I, Q, the running average, M, and the oil price, Z. In the first month, since the index is a constant K, the contract value is driven by just four variables: S, Q, M and Z.

Based on the above facts, the contract value at time t_n of Month l on the tree can be referenced by

$$\begin{cases} V^0\left(S_{n,s}, Z_{n,z}, M_{n,m}, Q\right) & \text{the first month} \\ V^l\left(S_{n,s}, Z_{n,z}, M_{n,m}, I_{n,i}, Q\right) & \text{the middle months, i.e.} \\ & l = 1, \ldots, L-2 \\ V^{L-1}\left(S_{n,s}, I_{n,i}, Q\right) & \text{the last month} \end{cases}$$

where the gas price is $S_{n,s}$, the oil price is $Z_{n,z}$, the value of the running average is $M_{n,m}$, the index is $I_{n,i}$ and the period to date is Q. Similarly, we have the optimal exercise decisions

$$\begin{cases} q^{0,*}\left(S_{n,s},Z_{n,z},M_{n,m},Q\right) & \text{the first month} \\ q^{l,*}\left(S_{n,s},Z_{n,z},M_{n,m},I_{n,i},Q\right) & \text{the middle months, i.e.} \\ & l=1,\ldots,L-2 \\ q^{L-1,*}\left(S_{n,s},I_{n,i},Q\right) & \text{the last month} \end{cases}$$

corresponding to values of those variables on the tree.

Terminal value. On the last day of the contract t_N, where $l=L-1$, by letting $q^* \; q^{L-1,*}(S_{N,s}, I_{N,i}, Q)$, we have the following scenarios:

- If $S_{N,s} \leq I_{N,i}$,
  - if $Q < MB$, that is the period to date at the terminal is less than the minimum bill. Upon taking a quantity of gas $\breve{q}$ up to that required to avoid the penalty, or the maximum possible, the actual payoff of this decision $\breve{q}$ is

  $$\breve{q}\cdot\left(S_{N,s}-I_{N,i}\right)+\breve{q}\cdot I_{N,i}=\breve{q}\cdot S_{N,s}>0,$$

  where $\breve{q}\cdot I_{N,i}$ is the reduced value of penalty by taking $\breve{q}$. Hence,

  $$q^* = \min\left\{\max\left\{MB-Q,q_{\min}\right\},q_{\max}\right\}.$$

  - if $Q \geq MB$, the buyer is under no pressure to meet the minimum bill. Since $S_{N,s} \leq I_{N,i}$, the optimal decision is to take the daily minimum so to reduce the loss from the instant payoff. The daily minimum also avoids or minimizes the penalty from taking more gas than the annual contract quantity. That is,

  $$q^* = q_{\min}.$$

- If $S_{N,s} > I_{N,i}$,
  - if $Q + q_{\max} \leq ACQ$, the buyer can safely take the daily maximum to maximize the profit of the instant payoff without worrying about the penalty arising from violating the annual contract quantity. Then,

  $$q^* = q_{\max}.$$

  - if $0 < ACQ - Q < q_{\max}$, it means the buyer can take a quantity of gas less than the daily maximum in order to gain the profit of the instant payoff without worrying about the penalty. In addition, he/she has the choice of further taking gas and making profit from the instant payoff and increasing the penalty at the same time. Suppose the buyer further takes a quantity of gas $\breve{q}$, then the actual payoff of this decision is

  $$\breve{q}\cdot\left(S_{N,s}-I_{N,i}\right)-\breve{q}\cdot I_{N,i}=\breve{q}\cdot\left(S_{N,s}-2I_{N,i}\right), \tag{22.15}$$

  where $\breve{q}\cdot I_{N,i}$ is the increased value of penalty by further taking $\breve{q}$. Then, we have the optimal decision in this scenario:

  $$q^* = \begin{cases} q_{\max} & \text{if } S_{N,s} > 2I_{N,i}, \\ \max\{\min\{ACQ \\ -Q,q_{\max}\},q_{\min}\} & \text{if } S_{N,s} \leq 2I_{N,i}. \end{cases} \tag{22.16}$$

- if $Q \geq ACQ$, this means that, by taking a quantity of gas, the buyer is profiting from the instant payoff and losing money from the penalty at the same time. That is, upon taking $\bar{q}$, the actual payoff of this decision is again given by (22.15), and the optimal decision is also given by (22.16).

In summary, we have the optimal decision at the last day of the contract which is given by

$$
q^* = \begin{cases} \min\{\max\{MB - Q, q_{\min}\}, q_{\max}\} & \text{if } S_{N,s} \leq I_{N,i} \\ \max\{\min\{ACQ - Q, q_{\max}\}, q_{\min}\} & \text{if } I_{N,i} < S_{N,s} \leq 2I_{N,i} \\ q_{\max} & \text{if } S_{N,s} > 2I_{N,i} \end{cases} \tag{22.17}
$$

Then the contract value on the last day is given by

$$
V^{L-1}\left(S_{N,s}, I_{N,i}, Q\right) = q^* \cdot \left(S_{N,s} - I_{N,i}\right) + \mathcal{P}\left(I_{N,i}, Q + q^*\right).
$$

REMARK 2 In this paper, we mainly focus on the GSAs introduced in Breslin *et al.* (2008) (also see Chiarella *et al.* 2016, Dong and Kang 2019) where $ACQ = N \cdot q_{\max}$. That is, we mainly focus on the penalty paid when the total volume of gas taken is below the minimum bill. However, for the purpose of completeness, we have also provided the optimal terminal decision for GSAs where $ACQ \leq N \cdot q_{\max}$.

Days within *the last month.* For $n = (L-1)D + d$, $d = 1, \dots, D-1$, we find the optimal exercise decision by

$$
\begin{aligned}
& q^{L-1,*}\left(S_{n,s}, I_{n,i}, Q\right) \\
& = \operatorname*{argmax}_{q_{t_n} \in \left[q_{\min}, q_{\max}\right]} \left\{ q_{t_n}\left(S_{n,s} - I_{n,i}\right) \right. \\
& \left. + \sum_{b=-1}^{1} p_{s,b} e^{-r\Delta t} V^{L-1}\left(S_{n+1, \mathfrak{g}(s,b)}, I_{n,i}, Q + q_{t_n}\right) \right\},
\end{aligned} \tag{22.18}
$$

where $p_{s,b}$ is the probability associated with the b branch emanating from the node (n, s) on the gas tree. $p_{s,b}$ can be computed by using Table A1. By letting $q^* = q^{L-1*}(S_{n,s}, I_{n,i}, Q)$, we have the contract value:

$$
\begin{aligned}
& V^{L-1}\left(S_{n,s}, I_{n,i}, Q\right) \\
& = q^*\left(S_{n,s} - I_{n,i}\right) \\
& + \sum_{b=-1}^{1} p_{s,b} e^{-r\Delta t} V^{L-1}\left(S_{n+1, g(s,b)}, I_{n,i}, Q + q^*\right).
\end{aligned} \tag{22.19}
$$

Days within *middle months.* Similarly, for $n = lD + d$, $d = 1, \dots, D-1$ and $l = 1, 2, \dots, L-2$, we find the optimal exercise decision by

$$
\begin{aligned}
& q^{l,*}\left(S_{n,S}, Z_{n,z}, M_{n,m}, I_{n,i}, Q\right) \\
& = \operatorname*{argmax}_{q_{t_n} \in \left[q_{\min}, q_{\max}\right]} \left\{ q_{t_n}\left(S_{n,s} - I_{n,i}\right) + \sum_{b,c=-1}^{1} p_{s,z,b,c} e^{-r\Delta t} \right. \\
& \left. \cdot V^{l}\left(S_{n+1, \mathfrak{g}(s,b)}, Z_{n+1, \mathrm{h}(z,c)}, M_{n,m,c}, I_{n,i}, Q + q_{t_n}\right) \right\}.
\end{aligned} \tag{22.20}
$$

Days within *the first month.* Again, for $n = d$, $d = 1, \dots, D-1$, we find the optimal exercise decision by

$$q^{0,*}\left(S_{n,s},Z_{n,z},M_{n,m},Q\right)$$

$$= \underset{q_{t_n}\in\left[q_{\min},q_{\max}\right]}{\text{argmax}}\left\{q_{t_n}\left(S_{n,s}-K\right)+\sum_{b,c=-1}^{1}p_{s,z,b,c}e^{-r\Delta t}\right.$$

$$\left.\cdot V^0\left(S_{n+1,g(s,b)},Z_{n+1,h(z,c)},M_{n,m,c},Q+q_{t_n}\right)\right\}. \tag{22.21}$$

Once we have the exercise decisions for days in the first and middle months, we can compute the contract value similarly as we did in (22.19).

Linear interpolation. Recall in Section 22.4.1, we have a discretized grid in the running average dimension. When

$$V^l\left(S_{n+1,g(s,b)},Z_{n+1,h(z,c)},M_{n,m,c},I_{n,i},Q+q_{t_n}\right)$$

or

$$V^0\left(S_{n+1,g(s,b)},Z_{n+1,h(z,c)},M_{n,m,c},Q+q_{t_n}\right)$$

in (22.20) and (22.21) (for simplicity, we use $V(M_{n,m,c})$) does not fall exactly on the discrete nodes of the grid, we use linear interpolation:

$$V\left(M_{n,m,c}\right)=V\left(M_{n+1,\bar{m}}\right)+\left(M_{n,m,c}-M_{n+1,\bar{m}}\right)$$

$$\times\frac{V\left(M_{n+1,\bar{m}+1}\right)-V\left(M_{n+1,\bar{m}}\right)}{M_{n+1,\bar{m}+1}-M_{n+1,\bar{m}}}, \tag{22.22}$$

where $\bar{m}$ is an integer such that $M_{n+1,\bar{m}}$ is the largest value in the running average vector $\mathbf{M}_{n+1}$ less than or equal to $M_{n,m,c}$.

Matching point. At time t_{lD}, $l=2,\dots,L-2$, we are on the last day of Month $l-1$. Given the current value of the running average $M(t_{lD})$, the index for time t_{lD+1}, which is the first day of Month l, is $I(t_{lD+1})=M(t_{lD})$ (see (22.3)). In addition, at time t_{lD+1}, $l=2,\dots,L-2$, by (22.4), the value of the running average is

$$M\left(t_{lD+1}\right)=\frac{1}{1}\sum_{j=1}^{1}Z\left(t_{lD+j}\right)=Z\left(t_{lD+1}\right).$$

That is, at time t_{lD}, if the contract value $V^l(S_{lD,s},Z_{lD,z},M_{lD,m},I_{n,i},Q)$ is given, upon taking the volume q_{tlD}, the contract value corresponding to the b branch on the gas tree and c branch on the oil tree is

$$V^l\left(S_{lD+1,g(s,b)},Z_{lD+1,h(z,c)},Z_{lD+1,h(z,c)},M_{lD,m},Q+q_{tlD}\right).$$

Then we can find the optimal decision $q^{l*}(S_{lD,s},Z_{lD,z},M_{lD,m},I_{lD,i},Q)$ by

$$q^{l,*}\left(S_{lD,s},Z_{lD,z},M_{lD,m},I_{lD,i},Q\right)$$

$$= \underset{q_{tlD}\in\left[q_{\min},q_{\max}\right]}{\text{argmax}}\left\{q_{tlD}\left(S_{lD,s}-I_{lD,i}\right)\right.$$

$$\left.+p_{s,z,b,c}e^{-r\Delta t}V^{l+1}\left(S_{lD+1,g(s,b)},Z_{lD+1,h(z,c)},Z_{lD+1,h(z,c)},M_{lD,m},Q+q_{tlD}\right)\right\}.$$

By letting $q^*=q^{l,*}(S_{lD,s},Z_{lD,z},M_{lD,m},I_{lD,i},Q)$, we have the contract value

$$V^l\left(S_{lD,s}, Z_{lD,z}, M_{lD,m}, I_{lD,i}, Q\right)$$

$$= q^*\left(S_{lD,s} - I_{lD,i}\right)$$

$$+ \sum_{b,c=-1}^{1} p_{s,z,b,c} e^{-r\Delta t} V^{l+1}\left(S_{lD+1,\mathfrak{g}(s,b)}, Z_{lD+1,h(z,c)}\right) \tag{22.23}$$

$$Z_{lD+1,h(z,c)}, M_{lD,m}, Q + q^*\Big).$$

We can have the optimal decision and contract value on the last day of Month 0 and on the last day of Month L–2 in a similar way.

The bang-bang consumption. It has been proved in Barrera-Esteve *et al.* (2006) that, if the penalty function (22.6) is continuously differentiable with respect to Q_T, then the GSA has so-called bang-bang consumption. That is, on each day except the last day, the optimal decision is either the daily maximum or the daily minimum. In our case, (22.6) is clearly not differentiable at some points with respect to Q_T. However, the bang-bang consumption is observed in our numerical examples. Based on our numerical observation, we make the following assumption: the GSAs in this paper have the so-called bang-bang consumption. For the discussion of the bang-bang consumption in the GSA, we refer interested readers to Barrera-Esteve *et al.* (2006), Bardou *et al.* (2010), Edoli and Vargiolu (2013) and Basei *et al.* (2014).

REMARK 3 In (22.20), we need to find the optimal decisions of all possible combinations of $(S_{n,s}, Z_{n,z}, M_{n,m}, I_{n,i}, Q)$. Recall that, at time t_n, the number of nodes on the gas tree and the oil tree are $2 \cdot k(n) + 1$ and $2 \cdot j(n)+1$, respectively. At the same time, the number of values in the running average vector $\mathbf{M}_n$ and the index vector $\mathbf{I}_n$ are both H. Under the bang-bang consumption, if we let $q_{max} = 1$ and $q_{min} = 0$, the possible periods to date at time t_n is $0, 1, \dots, n-1$, then it is not hard to see that (22.20) has complexity of the order of

$$\mathcal{O}\Big(n \cdot H^2 \cdot (2 \cdot \mathfrak{k}(n)+1)(2 \cdot j(n)+1)\Big).$$

Similarly, (22.21) has complexity of the order of

$$\mathcal{O}(n \cdot H \cdot (2 \cdot \mathfrak{k}(n)+1)(2 \cdot j(n)+1)),$$

and (22.18) has complexity of the order of

$$\mathcal{O}(n \cdot H \cdot (2 \cdot \mathfrak{k}(n)+1)).$$

22.5 The Evaluation Using LSMC

The least-squares Monte Carlo simulation (see Longstaff and Schwartz 2001) is a powerful tool when pricing options with early exercise features. It has been extended to accommodate options with multiple exercise opportunities in Mein-shausen and Hambly (2004), Aleksandrov and Hambly (2010) and Bender (2011). More specifically, the LSMC method has been used in Dörr (2003), Barrera-Esteve *et al.* (2006), Thanawalla (2006), Holden *et al.* (2011) and Bernhart (2011) to price GSA contracts.

To ease the notations in this section, we denote the gas price, the oil price, the value of the running average and the index at time t_n by S_n, Z_n, M_n and I_n, respectively. Recall (22.8), we have

$$V(t_n, S, I, Q)$$

$$= \sup_{q_{t_k} \in [q_{\min},\, q_{\mathrm{mur}}]} \mathbb{E}\left[\sum_{k=n}^{N} e^{-r(t_k - t_n)} q_{t_k} (S_k - I_k) \right. \tag{22.24}$$

$$\left. + e^{-r(T - t_n)} \mathcal{P}(I_N, Q_T) \,\Big|\, S_n = S, I_n = I, Q_{t_n} = Q \right].$$

where $Q_T = Q + \sum_{k=n}^{N} q_{t_k}$. Equation (22.24) can be rewritten as follows:

$$V(t_n, S, I, Q) = \sup_{q_{t_n} \in [q_{\min},\, q_{\max}]} \mathbb{E}\left[q_{t_n} (S - I) + C(t_n, S, I, Q, q) \right],$$

where $C(t_n, S, I, Q, q_{t_n})$ is called the continuation value which is given by

$$C(t_n, S, I, Q, q_{t_n})$$

$$= e^{-r\Delta t} \mathbb{E}\left[V\left(t_{n+1}, S_{n+1}, I_{n+1}, Q + q_{t_n} \right) \Big| S_n = S, I_n = I \right]. \tag{22.25}$$

The main idea of the LSMC is that the continuation value can be approximated by the linear combination of a set of basis functions of the current state. In our case, at each time t_n, $n = 1, 2, \dots , N - 1$, (22.25) can be approximated through

$$C(t_n, S, I, Q, q) \approx \sum_{k=1}^{K_b} \beta_{n,k}^{Q,q} \phi_k(\mathcal{E}_n),$$

where K_b is a positive integer. $\beta_{n,k}^{Q,q}$, $k = 1, \dots , K_b$, is the coefficient associated with the basis function $\phi_k(\varepsilon_n)$ where the period to date equals Q and the daily decision is q at time t_n. We call this the regression coefficient. $\mathcal{E}_n$, which is called the explanatory vector, contains some appropriate explanatory variables at time t_n. In the view of the value function (22.24), one straightforward option for the explanatory variables is that we use the gas price and the index. This is quite understandable since the gas price and the index are the variables in the calculation of everyday payoffs and hence should have a great impact on the continuation value. Recall (22.5) and (22.3), the index I is related to the running average M in the sense that the value of the index in the current month is the value of the running average at the end of the previous month. The running average can affect the contract value by influencing the index. Recall (22.4). The value of the running average is calculated by using oil prices. So, the oil price should also contribute to the contract value by influencing the running average and hence the index. Based on the above considerations, we let $\varepsilon = (S, Z, M, I)$.

In LSMC, the cross-paths information of many simulated paths is used to approximate the continuation value. Suppose we have G_1 simulated independent paths of (S, Z):

$$\left(\left(S_0^{(g)}, Z_0^{(g)} \right), \left(S_1^{(g)}, Z_1^{(g)} \right), \left(S_2^{(g)}, Z_2^{(g)} \right), \dots, \left(S_N^{(g)}, Z_N^{(g)} \right) \right),$$

where $g = 1, 2, \dots , G_1$. For each path g, we calculate the explanatory vector $\mathcal{E}^{(g)}$ and get G_1 paths of ε:

$$\left(\mathcal{E}_0^{(g)}, \mathcal{E}_1^{(g)}, \mathcal{E}_2^{(g)}, \dots, \mathcal{E}_N^{(g)} \right),$$

where $g = 1, 2, \dots , G_1$. Let $\hat{V}^{(g)}\left(t_n, S^{(g)}, I^{(g)}, Q \right)$ be the approximated contract value at time t_n of the gth realization which is given by

$$\hat{V}^{(g)}\left(t_n, S^{(g)}, I^{(g)}, Q\right)$$

$$= \sup_{q \in \left[q_{\min}, q_{\max}\right]} \left[q \cdot \left(S^{(g)} - I^{(g)}\right) + \sum_{k=1}^{K_b} \beta_{n,k}^{Q,q,*} \phi_k\left(\mathcal{E}_n^{(g)}\right) \right], \tag{22.26}$$

where $\beta_{n,k}^{Q,q,*}$ is the optimal regression coefficient associated with $\phi_k(\varepsilon_n)$. Write $\beta_n^{Q,q,*} = \left(\beta_{n,1}^{Q,q,*}, \beta_{n,1}^{Q,q,*}, \dots, \beta_{n,k_b}^{Q,q,*}\right)^{\text{trans}}$, where 'trans' gives the transpose of a matrix. The optimal regression coefficients $\beta_n^{Q,q,*}$ are found by solving the following over-determined system using the least-squares method:

$$\phi_n \beta_n^{Q,q,*} = V_n^q, \tag{22.27}$$

where ϕ_n and V_n are given by

$$\phi_n = \begin{pmatrix} \phi_1\left(\mathcal{E}_n^{(1)}\right) & \phi_2\left(\mathcal{E}_n^{(1)}\right) & \cdots & \phi_{K_b}\left(\mathcal{E}_n^{(1)}\right) \\ \phi_1\left(\mathcal{E}_n^{(2)}\right) & \phi_2\left(\mathcal{E}_n^{(2)}\right) & \cdots & \phi_{K_b}\left(\mathcal{E}_n^{(2)}\right) \\ \cdots & \cdots & \cdots & \cdots \\ \phi_1\left(\mathcal{E}_n^{(G_1)}\right) & \phi_2\left(\mathcal{E}_n^{(G_1)}\right) & \cdots & \phi_{K_b}\left(\mathcal{E}_n^{(G_1)}\right) \end{pmatrix} \text{ and}$$

$$V_n^q = \begin{pmatrix} e^{-r\Delta t}\hat{V}^{(1)}\left(t_{n+1}, S^{(1)}, I^{(1)}, Q+q\right) \\ e^{-r\Delta t}\hat{V}^{(2)}\left(t_{n+1}, S^{(2)}, I^{(2)}, Q+q\right) \\ e^{-r\Delta t}\hat{V}^{(G_1)}\left(t_{n+1}, S^{(G_1)}, I^{(G_1)}, Q+q\right) \end{pmatrix}, \tag{22.28}$$

respectively. Note that, the optimal $\beta_{n,k}^{Q,q,*}$ depends on both the period to date Q and the daily decision q. That is, at each time t_n, $n = 1, \dots, N-1$, we find the optimal coefficients for all possible combinations of Q and q. Once we have $\beta_{n,k}^{Q,q,*}$, we compute the approximated contract value of each path g using these optimal coefficients. In addition, at time t_N, we can compute the contract value of each path g through

$$\hat{V}^{(g)}\left(t_N, S^{(g)}, I^{(g)}, Q\right) = q^* \cdot \left(S^{(g)} - I^{(g)}\right) + \mathcal{P}\left(I^{(g)}, Q + q^*\right),$$

where the optimal decision $q^* = q^*(t_N, S^{(g)}, I^{(g)}, Q)$ is found by using (22.17).

The regression coefficients $\beta_{n,k}^{Q,q,*}$ provide an exercising rule. At each time t_n, for a given period to date Q and any realization of the underlying (S, Z), we can find the optimal decision $q^* = q^*(t_n, S, I, Q)$ through

$$q^*\left(t_n, S, I, Q\right) = \operatorname*{argmax}_{q \in \left[q_{\min}, q_{\max}\right]} \left[q(S - I) + \sum_{k=1}^{K_b} \beta_{n,k}^{Q,q,*} \phi_k\left(\mathcal{E}_n\right) \right]. \tag{22.29}$$

As we can see in (22.29), once we have the period to date Q and the realization of the underlying (S, Z), the exercising rule solely depends on the optimal coefficients $\beta_{n,k}^{Q,q,*}$, $k = 1, \dots, K_b$, since the continuation value is calculated through these coefficients.

When we compute the optimal coefficients by solving the over-determined system (22.27), at time t_n, we have assumed the knowledge of $\hat{V}^{(g)}\left(t_{n+1}, S^{(g)}, I^{(g)}, Q+q\right)$. At the same time, we are applying this exercising rule to the same set of paths in order to get the contract values, and this causes a high bias of the contract value (see, for instance, Broadie and Glasser-man 2004 and Section 22.8 in Glasserman 2003). To tackle this issue, one could simulate a second independent set of independent paths and apply the exercising rule obtained through the first set of paths to this newly generated set of paths through a forward induction.

The LSMC contains two steps. In the first step, given a set of K_b basis functions, at each time t_n, $n = 1$, $2, \ldots , N - 1$, we find the optimal coefficients $\beta_{n,k}^{Q,q,*}$, $k = 1, \ldots , K_b$, for all possible Q and q. In the second step, we generate a second set of paths and we find the approximated contract value of each path by applying the exercising rule (see (22.29)) on each path through a forward induction. Then the contract value is obtained by taking the average of all simulated paths. In the rest of this paper, the first step and the second step are called the backward scheme and the forward scheme, respectively.

REMARK 4 After building the 2-D trinomial tree, the model we use in the tree algorithm is actually given by

$$
\begin{cases}
dX(t) = -\alpha_S X(t)dt + \sigma_S dB^S(t) \\
S(t) = e^{X(t)+a(t)} \\
dY(t) = -\alpha_Z Y(t)dt + \sigma_Z dB^Z(t) \\
Z(t) = e^{Y(t)+b(t)}
\end{cases}
\tag{22.30}
$$

where $a(t)$ and $b(t)$ are functions of time which provide an exact fit to the observed forward curves. $a(t)$ and $b(t)$ are approximated in discrete time by using (22.A4). To provide a fair comparison between the tree algorithm and the LSMC algorithm, we use (22.30) as the underlying prices which are simulated in the LSMC algorithm. That is, the underlying processes S and Z are simulated using the Euler scheme (see Maruyama 1955):

$$
X_n = X_{n-1} - \alpha_S X_{n-1}\Delta t + \sigma_S \Delta B_n^S,
$$

$$
Y_n = Y_{n-1} - \alpha_Z Y_{n-1}\Delta t + \sigma_Z \Delta B_n^Z,
$$

$n = 1, 2, \ldots , N$, where $X_0 = Y_0 = 0$, $\Delta B_n^S = B^S(t_n) - B^S(t_{n-1})$ and $\Delta B_n^Z = B^Z(t_n) - B^Z(t_{n-1})$ are the Brownian increments. Then the realizations of S and Z at each time t_n, $n = 0, 1, \ldots , N$, are given by

$$
S_n = e^{X_n + a_n}, \quad Z_n = e^{Y_n + b_n},
$$

where a_n and b_n are computed using the tree building procedures. Once we have the realizations of S and Z, the running average M and the index I can be computed by (22.4) and (22.3), respectively. Due to the flexible nature of the Monte Carlo simulation, however, the LSMC algorithm can surely accommodate other models.

REMARK 5 In this section, we again assume that our GSAs have the bang-bang consumption. Under the bang-bang consumption, and letting $q_{min} = 0$ and $q_{max} = 1$, at each time t_n, for each possible period to date Q, we only need to find $\beta_n^{Q,0,*}$ and $\beta_n^{Q,1,*}$ Furthermore, based on the same reasons, at time t_n, the possible periods to date Q are integers from 0 to $n - 1$. That is, (22.26) becomes

$$
\hat{V}^{(g)}\left(t_n, S^{(g)}, I^{(g)}, Q\right)
$$

$$
= \max\left\{\sum_{k=1}^{K_b} \beta_{n,k}^{Q,0,*}\phi_k\left(\mathcal{E}_n^{(g)}\right), \left(S^{(g)} - I^{(g)}\right)\right.
$$

$$
\left. + \sum_{k=1}^{K_b} \beta_{n,k}^{Q,1,*}\phi_k\left(\mathcal{E}_n^{(g)}\right)\right\}.
$$

In addition, $\beta_n^{Q,0,*}$ and $\beta_n^{Q,1,*}$ for all possible periods to date Q at time t_n are found via (22.27). If the current period to date is Q, we find $\beta_n^{Q,0,*}$ by $\phi_n\beta_n^{Q,0,*} = V_n^0$. If the current period to date is $Q-1$, we find $\beta_n^{Q-1,1,*}$ by $\phi_n\beta_n^{Q-1,1,*} = V_n^1$. Recall the notations in (22.28), it is not hard to see that

$$\beta_n^{Q,0,*} = \beta_n^{Q-1,1,*}. \tag{22.31}$$

(22.31) gives us the opportunity to save a lot of computing time since it reduces nearly half of the least-squares fitting problems (22.27), although it needs to be processed with extra care when we use the parallel computing technique. Without (22.31), at each time t_n, the fitting problem (22.27) can be solved independently for each possible period to date Q. When we apply (22.31), this is not possible since we cannot update $\beta_{n,k}^{Q,0,*}$ until we find $\beta_{n,k}^{Q-1,1,*}$. Nevertheless, we can always use the parallel computing technique in the construction of ϕ_n and V_n^q in (22.27).

An upper bound of the GSA. Edoli (2013) proposes an algorithm for the upper bound of the GSA. The idea is quite straightforward. Instead of using an exercising rule, this upper bound algorithm maximizes the contract value of each path by linear programming using a deterministic algorithm. Then, this upper bound is obtained via the average of all maximized contract values of all paths. The value for each path g is computed by solving the following maximization problem:

$$\underset{q_{t_1},\ldots,q_{t_N},x_1,x_2}{\text{maximize}} \quad V^{(g)}(0) = \sum_{n=1}^{N} e^{-rt_n} q_{t_n}\left(S_n^{(g)} - I_n^{(g)}\right)$$

$$- I_N^{(g)} \cdot x_1 - I_N^{(g)} \cdot x_2$$

$$\text{subject to} \quad Q_T = \sum_{n=1}^{N} q_{t_n},$$

$$MB - Q_T \le x_1 \le MB,$$

$$Q_T - ACQ \le x_2 \le N - ACQ,$$

$$x_1 \ge 0,$$

$$x_2 \ge 0,$$

$$q_{t_n} \in \{0,1\}, \quad n = 1,2,\ldots,N.$$

This upper bound algorithm is fast and easy to implement using a deterministic algorithm (one suggestion is to use the built-in Matlab function intlinprog()). In addition, the above algorithm gives the upper bound of the GSA value in the sense that it has the so-called perfect foresight of the future movements of the underlying prices. That is, at time t_0, the contract value $V^{(g)}(0)$ of each path g is calculated via the maximization with the knowledge of all price realizations in all future time steps t_n, $n = 1, 2, \ldots, N$. Then, the value $V^{(g)}(0)$ is obtained in a deterministic environment instead of an uncertain environment. The perfect foresight can significantly overprice the contract value and hence gives an upper bound.

Benchmark values. Suppose the non-trivial condition does not hold. That is,

$$MB = 0 \quad \text{and} \quad N \le ACQ. \tag{22.32}$$

The buyer can freely take gas at any time t_n, $n = 1, 2, \ldots, N$, without worrying about the possible penalties as long as the daily constraints are satisfied. Then, when (22.32) holds, the GSA is equivalent to a strip of N European options covering the same exercisable dates. The nth European option has the terminal date t_n and the strike price I_n. The payoff of the nth European option is given by

$$\max\{S_n - I_n, 0\}.$$

Due to the unclear distribution of the index I, however, we still need to find the value of those European options via numerical approaches, such as Monte Carlo simulations. This can be easily done by finding the contract value of each path and then taking the average. Similarly to the upper bound algorithm, this benchmark algorithm also gives full knowledge of all future spot price realizations. It does not suffer from the perfect foresight, however, because that it takes the expectation over all the simulated paths. If the number of paths used is sufficiently large, based on the strong law of large numbers, value obtained by this benchmark algorithm should converge to the true contract value.

In this paper, this benchmark algorithm can only give benchmark values for the tree algorithm and cannot provide benchmarks for the LSMC algorithm. This is because, under (22.32), the LSMC algorithm degenerates to the benchmark algorithm and these two algorithms are equivalent. Since (22.32) holds, the terminal values of all paths are irrelevant to the period to date. When we solve the over-determined system, we get same coefficient vectors for all period to date. It follows that, in the forward scheme, the buyer will purchase gas if the instant payoff is positive and not purchase otherwise, which means that, for a fixed set of paths, the forward scheme and the benchmark algorithm return the same result.

The choice of basis functions. In this paper, we use functions in the complete set of polynomials (see Section 22.6.12 in Judd 1998) as basis functions. Given the explanatory vector (S, Z, M, I), the complete set of polynomials of total degree h is given by

$$\mathcal{P}_h = \left\{ S^{h_1} Z^{h_2} M^{h_3} I^{h_4} \mid h_n \geq 0, n = 1, 2, 3, 4 \text{ and } \sum_{n=1}^{4} h_n \leq h \right\}.$$

22.6 Evaluation on GPUs

22.6.1 General-Purpose Computing on GPUs

General-purpose computing on GPUs (GPGPU) is the use of GPUs in fields not related to graphics processing. One commonly used and successfully built GPGPU platform and programming model is the Compute Unified Device Architecture (CUDA), which was introduced by NVIDIA in 2006 (see NVIDIA 2018c). CUDA can be treated as an extension of the traditional programming language C/C++, and hence ease the coding effort for C/C ++ programmers. Next, we give some basic details about CUDA. We refer interested readers to NVIDIA (2018b, 2018c) for more information.

CUDA programming. On the GPU, the thread is the smallest unit which performs computing. It is similar to the thread when we use openMP on a multi-core CPU. Threads are grouped into a one-dimensional, two-dimensional or three-dimensional block. Furthermore, blocks are grouped into a one-dimensional, two-dimensional or three-dimensional grid. In the CUDA C/C ++ programme, a kernel is a function which is called from the CPU and executed on the GPU. The kernel is defined with the declaration specifier__global__. A kernel must be called together with triple angle brackets <<< dimGrid,dimBlock >>> , where dimGrid and dimBlock are one-dimensional, two-dimensional or three-dimensional vectors which specify the size of the grid and block, respectively.

The memory on the GPU side has a hierarchy which contains several types of memory. Different memory types offer different characteristics and it is important to understand these in order to optimize the CUDA C/C ++ program. Here we present some commonly used memory types. When launching a thread, this thread gets its own registers and these registers cannot be shared with other threads. The register has the fastest speed and, at the same time, is the smallest memory on the GPU. Each block has access to the shared memory and this memory is only shared among all threads in this single block. The shared memory is on-chip and also very fast. When a kernel is called, all threads in all blocks can access the global memory. The speed of the global memory is relatively slow since it is off chip and has high latency. The global memory is mainly used to store large data which needs to be accessed from all threads. The memory allocation is very important in CUDA, mainly because the reading and writing speed of the global memory on the GPU is not as fast as the speed of the RAM on the CPU. Taking advantages of the shared memory is the key to achieve better efficiency.

22.6.2 LSMC on GPUs

The LSMC algorithm can also be performed on GPUs. Intuitively speaking, the most time-consuming part of the LSMC algorithm is the backward scheme. This is because the back-ward scheme involves solving the least-squares fitting problem (22.27). Obviously, this fitting problem cannot be paralleled since it requires information from all paths. In Fatica and Phillips (2013), the authors proposed a way to price American options on GPUs, which also involves solving a fitting problem. In this section, we apply their methodology to our problem. Recall the following over-determined system:

$$\phi_n \beta_n^{Q,q} = V_n^q, \tag{22.33}$$

where ϕ_n is a $G_1 \times K_b$ matrix, $\beta_n^{Q,q}$ is a $K_b \times 1$ vector and V_n is a $G_1 \times 1$ vector. From (22.33), it can be easily seen that

$$\left(\phi_n^{\text{trans}} \phi_n\right) \beta_n^{Q,q} = \left(\phi_n^{\text{tans}} V_n^q\right), \tag{22.34}$$

where ϕ_n^{trans} is the transpose of ϕ_n. Now, we can find the optimal coefficient vector by solving (22.34). Since $\phi_n^{\text{trans}}\phi_n$ is a $K_b \times K_b$ matrix and $\phi_n^{\text{trans}}V_n^q$ is a $K_b \times 1$ vector, (22.34) is easy and fast to solve. Since solving (22.34) also requires all the information in $\phi_n^{\text{trans}}\phi_n$ and $\phi_n^{\text{trans}}V_n^q$, it can only be done on a single core (or thread). The matrix multiplication $\phi_n^{\text{trans}}\phi_n$ and $\phi_n^{\text{trans}}V_n^q$ can by paralleled on GPUs, however. The matrix multiplication is a classic CUDA programming problem, NVIDIA also provides the CUBLAS library (see NVIDIA 2018a) to ease the programming effort. In addition, since all the paths in both the backward scheme and the forward scheme are generated independently, the generations of all paths can be paralleled on the GPU. This is implemented in a similar manner as the paths generation on the CPU, except we assign one path to one thread on the GPU and then move forward in time to generate the full path.

However, computing $\beta_n^{Q,q}$ through (22.34) has a downside. Although (22.34) is theoretically correct, it causes problems when implemented on a computer. Recall that each row in the matrix ϕ_n contains the values of the basis functions with respect to the realization of the explanatory vector $\mathcal{E}$. Recall notations in (22.28), after the matrix multiplication, the entry in the k_1th row and k_2th column of the matrix $\phi_n^{\text{trans}}\phi_n$ is given by

$$\sum_{g=1}^{G_1} \phi_{k_1}\left(\mathcal{E}_n^{(g)}\right)\phi_{k_2}\left(\mathcal{E}_n^{(g)}\right), \tag{22.35}$$

$k_1, k_2 = 1, \dots, K_b$. When we use many simulated paths, a not very high degree h of the complete set of polynomials gives a very high value of (22.35). The information of all simulated paths at time t_n is suppressed into (22.35). When implemented the calculation on a digit computer, the matrix $\phi_n^{\text{trans}}\phi_n$ can be represented with only limited precision. That is, the LSMC algorithms that make use of (22.34) can usually obtain less accurate estimates of $\beta_n^{Q,q,*}$, compared with LSMC algorithms that use (22.33). We demonstrate the effect of this problem in Section 22.7.

When we implement the LSMC algorithm on the CPU, we can apply (22.31) to save computing time. The advantage of using (22.31) weakens if we have many parallel cores. This issue does not concern the CPU since the CPU usually contains four or eight cores. When we implement the LSMC algorithm on the GPU, we have hundreds of cores available. However, as we mentioned before, we mainly use cores on the GPU to compute the matrix multiplications $\phi_n^{\text{trans}}\phi_n$ and $\phi_n^{\text{trans}}V_n^q$ in (22.34). The matrix multiplications for different periods to date Q are not paralleled due to the limited storage on the GPU. In addition, (22.34) is solved by using a single core. Even using a single core, (22.34) can be solved very fast, since the sizes of $\phi_n^{\text{trans}}\phi_n$ and $\phi_n^{\text{trans}}V_n^q$ are small. Hence, the advantage of using (22.31) is much less significant on the GPU compared with it on the CPU.

Besides the LSMC on GPUs, we also provide notes on the implementation of the tree algorithms on GPUs in Appendix 2.

22.7 Numerical Examples

In this section, we provide numerical examples which evaluate the GSAs using the different algorithms that we have built in this paper. The GPU that we work with is a GeForce GTX 1060 on an Acer Helios 300 laptop. The CPU on the same laptop is an Intel(R) Core(TM) i7-7700HQ @ 2.80 GHz.

We use the parameter values in Table 22.1. We assume the flat forward curves are given by $F^S(0, t) = 100$ and $F^Z(0, t) = 100$ for $t \in [0, T]$. In terms of the non-uniform grid, when not variable, we use the constants $w_n = \dfrac{1}{3}$, $n = 1, 2, \dots, N$ (see (22.13)). Unless indicated otherwise, we use the contract values obtained using the LSMC algorithm on the CPU as pre-computed benchmark values.

22.7.1 Contract Values

In this section, we investigate the performance of the algorithms in terms of the contract values. In the rest of this section, the tree FD values refer to the contract values obtained by the tree algorithm with further discretization (22.10), whereas the tree NUG values refer to the contract values obtained by the tree algorithm with the non-uniform grid (22.13), and the LSMC values refer to the contract values obtained by the LSMC algorithm.

Figure 22.4 presents a comparison of the contract values obtained using different algorithms. In the left panel of Figure 22.4, F is the positive integer that we use to compute the further discretization (see (22.10)). A larger value of F results in a smaller value of ΔM, which further leads to a denser grid in the running average direction. In the right panel of Figure 22.4, H is the number of values in the running average vector $\mathbf{M}$ and a larger value of H also results in a denser grid in the running average direction. According to (22.14), a denser grid in the running average direction indicates a denser grid in the index direction. Both the tree FD value and the tree NUG value converge to just above the LSMC values. Furthermore, it seems that the tree FD value converges faster to the LSMC values than the tree NUG value. By (22.11), however, the running average vector $\mathbf{M}$ already contains 73 values at $F = 3$. At $F = 19$, the number of values in the running average vector $\mathbf{M}$ reaches 457. Figure 22.4 illustrates that the tree NUG value at $H = 320$ already achieved the same value as the tree FD value at $F = 19$. This becomes clear in the

TABLE 22.1   Parameters Set for Numerical Examples

$\alpha_S = 5$	$\sigma_S = 0.5$	$\alpha_Z = 6$	$\sigma_Z = 0.6$	$r = 0.05$
$\rho = 0.5$	$q_{\min} = 0$	$q_{\max} = 1$	$MB = 270$	$ACQ = 360$
$T = 1$	$N = 360$	$D = 30$	$L = 12$	

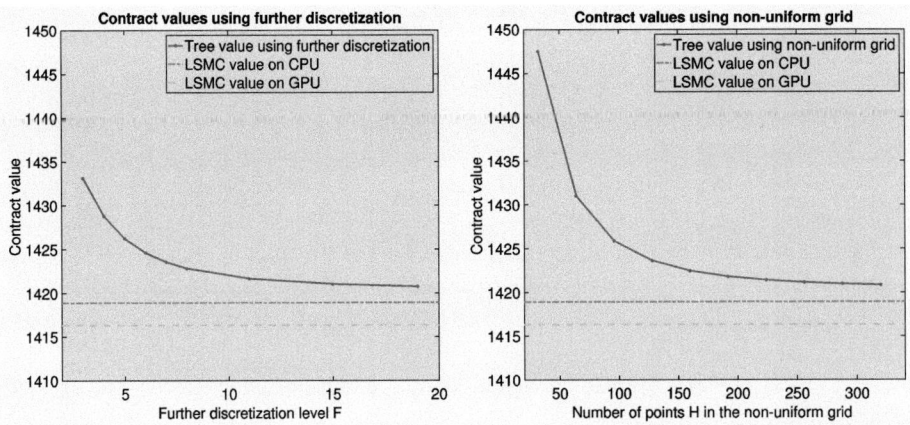

FIGURE 22.4 Contract values w.r.t. the average discretization. (The standard error of the LSMC values on the CPU and GPU are 2.317 and 2.332, respectively. The upper bound obtained by the upper bound algorithm is 1611.71 where 100,000 paths have been used.)

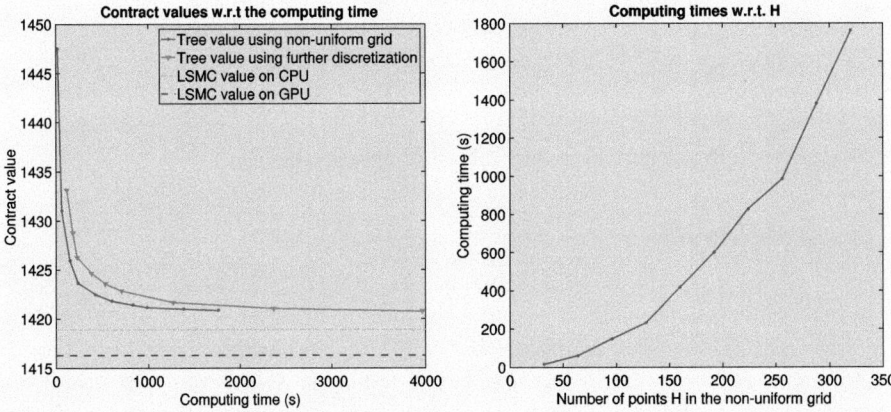

FIGURE 22.5 The convergence and computing time.

left plot of Figure 22.5, which displays the contract values with respect to the computing time using these two discretization methods. The tree NUG value converges much faster to the LSMC values than the tree FD value. This demonstrates that the non-uniform grid is a better discretization method than the further discretization method. In the remainder of this section, when we implement the tree algorithm, we only use the non-uniform grid.

In terms of the LSMC values, Figure 22.4 demonstrates that the LSMC value on the GPU is below the LSMC value on the CPU. This is primarily because of the limited precision while implementing it on a digital computer, as we mentioned in Section 22.6.2. However, the LSMC on the GPU has a great advantage: it is much more efficient than the LSMC on the CPU. In our numerical example, with 100,000 paths in the backward scheme, 1,000,000 paths in the forward scheme, and the complete set of fourth-degree polynomials, the LSMC on the CPU takes nearly 20 h to price our GSA contract. Even if we parallel the LSMC algorithm on four cores (the CPU we use contains four cores), it still requires more than 5 h. When the LSMC is implemented on the GPU, however, it takes only 244 s. Recall Remark 3, when we have a larger H, the computational cost of the tree algorithm is higher. The computing time with respect to H is shown in the right plot of Figure 22.5. The computing time grows rapidly as H increases. Compared with the LSMC on the CPU, however, the tree algorithm is still very fast.

Moreover, the contract values obtained by both the tree algorithm and LSMC algorithm are below the upper bound obtained by the upper bound algorithm (see the footnote for Figure 22.4). It demonstrates that our tree algorithm and LSMC algorithm produce reasonable results in a reasonable time frame and that the upper bound algorithm overestimates the contract by about 14%.

22.7.2 The LSMC Algorithm on the GPU

In Figure 22.5, the LSMC value on the GPU is not as accurate as the LSMC value on the CPU. In this section, we seek a method to obtain a better contract value using the LSMC algorithm on the GPU. Recall Section 22.6.2, when we apply (22.34) with many simulated paths and a complete set of high-degree polynomials, (22.35) returns a very large value. Due to the storage limit, some information is lost. Three methods can overcome this issue, we can use less simulated paths, use a complete set of lower-degree polynomials, or both. The Monte Carlo algorithms are based on the convergence of the sampled average to the mathematical one, which is why we can obtain better and more stable contract values using larger sets of simulated paths. That is, we can obtain very different contract values if we do not use sufficient large sets of paths in the LSMC algorithm. Therefore, we aim to use a complete set of lower-degree polynomials instead of decreasing the number of simulated paths. In the left panel of Figure 22.6, from the first to third degree, the contract value increases as the degree of the polynomials increases. If we further increase the degree of the polynomials, we should obtain a higher contract value. However, the left panel of Figure 22.6 shows that the contract value at the fourth degree is even lower than the contract value at the third degree because, at this point, the LSMC algorithm on the GPU starts to suffer from the problem mentioned in Section 22.6.2. At the third degree, the LSMC value on the GPU is only slightly lower than the LSMC value on the CPU at the fourth degree. The difference is primarily because the LSMC algorithm on the CPU does not suffer from reduced precision and can attain a better exercising rule using the complete set of fourth degree polynomials. The right panel of Figure 22.6 presents the computation time with respect to the degree of the polynomials. At the third degree, the LSMC algorithm on the GPU only requires 153 s to evaluate such a contract. That is, we can use a lower degree of polynomials to overcome the problem mentioned in Section 22.6.2 while achieving a faster implementation.

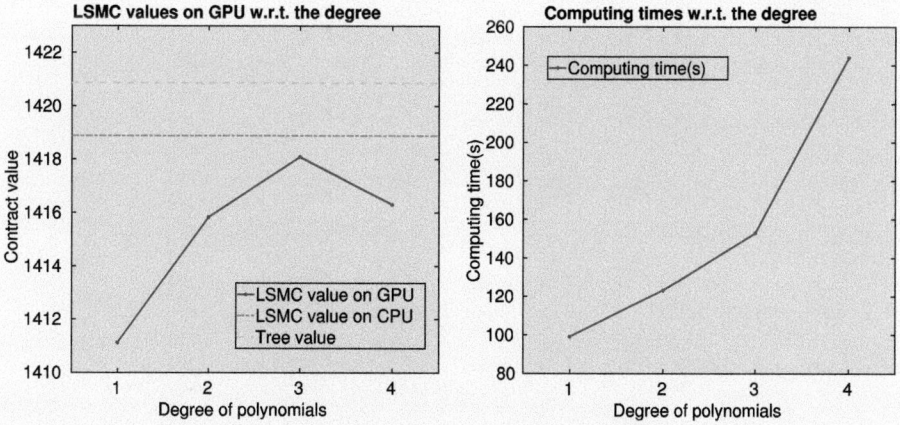

FIGURE 22.6 The LSMC value on GPU[b]. (100,000 paths and 1,000,000 paths have been used in the backward scheme and the forward scheme, respectively. From the degree $h = 1$ to $h = 4$, the standard errors of the LSMC values on the GPU are 2.324, 2.319, 2.321 and 2.561, respectively. The standard error of the LSMC value on the CPU with the degree $h = 4$ is 2.543. The tree value is obtained using a non-uniform grid with $H = 320$. In addition, the computing time in this figure refers to the LSMC algorithm on the GPU.)

22.7.3 Accuracy of the Tree Algorithm

Comparison between the tree algorithm and benchmark algorithm. We investigate the accuracy of the tree algorithm using a comparison between the tree algorithm and bench-mark algorithm. That is, we let the minimum bill MB in Table 22.1 be 0, meaning that no penalty is involved in our GSA. Figure 22.7 reports the results. At $H = 320$, the tree value is only 0.03% above the benchmark value.

The impact of w_n. Next, we investigate the effect of w_n on the tree algorithm. Recall that w_n, where $n = 1, 2, \ldots, N$, is the constant that controls the fraction of points $M_{n,m}$ in the running average vector $\mathbf{M}_n$ that lies in the neighborhood of M_n. In addition, a smaller w_n provides more points in the neighbor-hood of M_n. In our examples, we let w_n be a constant w for all $n = 1, 2, \ldots, N$. Figure 22.8 illustrates the effect of w. Both the convergence speed when using $w = \dfrac{1}{6}$ and the convergence speed when using $w = \dfrac{1}{10}$ out-performs the convergence speed using $w = \dfrac{1}{2}$. That is, when more points in the non-uniform grid lie in the neighborhood of M_n, the convergence speed of our tree algorithm is faster, which is understandable. Recall that M_n equals the value of the running average M, if the oil price moves on the oil tree by follow-ing the middle branches only (see (22.12)). Due to the mean-reverting nature of the oil price, the oil price is unlikely to be too high or too low for a long period. It follows that the value of the running average is even more unlikely to be too large or too small. Indeed, the reason the indexation of the GSA exists is to smooth the undesired volatility effects. In addition, Figure 22.8 demonstrates that the performance of our tree algorithm is worse when using $w = \dfrac{1}{20}$ than when using $w = \dfrac{1}{10}$. This indicates that, although the movement of the running average is relatively stable, assigning too many points in the neighborhood of M_n does not benefit the tree algorithm. No matter which value of w we use in Figure 22.8, the tree values for different values of w all essentially converge to the same value, at $H = 320$. This demonstrates that, when many points exists in the non-uniform grid, the tree algorithm is unlikely to be affected by the value of w. Figure 22.8 also demonstrates that, if the non-uniform grid contains fewer points, we can use a smaller w to obtain a better contract value. When using $w = \dfrac{1}{10}$, the contract value has only changed 0.08% from $H = 160$ to $H = 320$. Moreover, the computing time at $H = 160$ is only 417 s, which

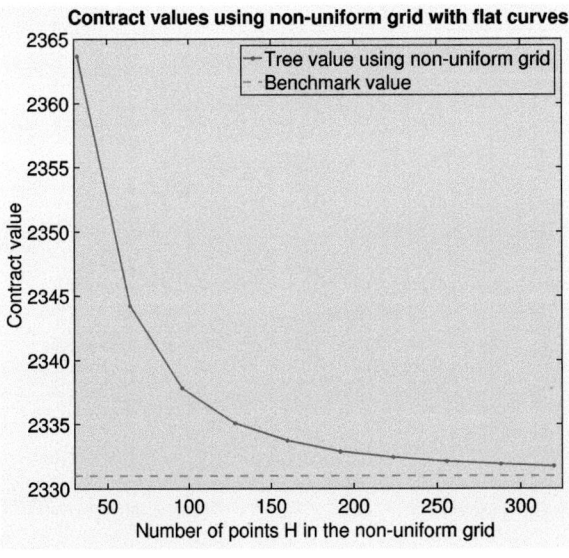

FIGURE 22.7   Contract values when $MB = 0$.

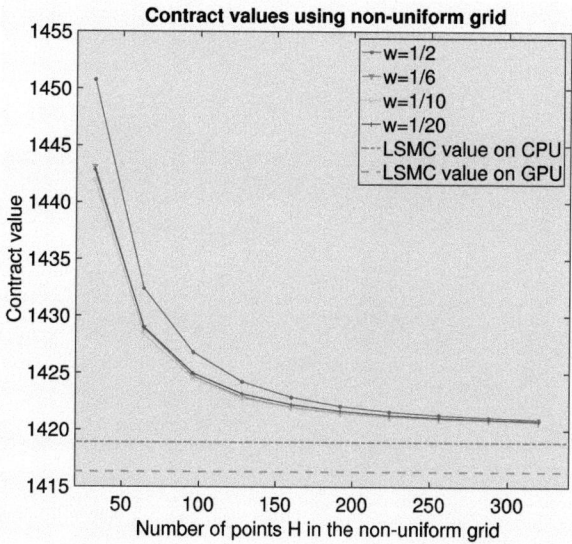

FIGURE 22.8   Contract values w.r.t. *w*.

is very competitive with the LSMC algorithm on the GPU. Even with $w = \dfrac{1}{2}$, the contract value has only changed by 0.13% from $H = 160$ to $H = 320$. This demonstrates that we can still attain a satisfactory contract value with fewer points in the non-uniform grid, even with a relatively large *w*.

22.7.4 Value Surfaces and Decision Surfaces from the Tree Algorithm

In this section, we investigate the value surfaces and decision surfaces by evaluating the GSA with flat forward curves using the tree algorithm.

Value surfaces. Figure 22.9 illustrates the value surfaces at time $t105$, which is the 15th day of Month 3. The value surfaces with respect to the running average, index and oil price have similar patterns. This is because the running average and oil price influence the contract values by affecting the index in the coming month. In addition, the contract value is more sensitive to the oil price than the running average and index, especially when the oil price is low. This is because, if the current oil price is low, then the oil price for the rest for this month is possibly in the low price regime. This leads to a low index in the coming month. In addition, in the surfaces with respect to the running average, index and oil price, for a fixed period to date Q, the contract value is stable when these corresponding values are high. This is because the running average, index and oil price are large enough to ensure the buyers avoid exercising gas rights under the contract. The bottom-left surface of Figure 22.9 shows that the contract value increases as the gas price increases because the buyers are willing to take gas under the GSA contract when the gas price is high. This action not only profits the buyers from the instant payoffs but also reduces or possibly avoids the penalties because more gas has been taken under the contract.

Decision surfaces. We define the exercise threshold as the gas price equal to or above which the buyer would better benefit from taking the daily maximum $q_{\max}$. A low exercise threshold means that the buyer is willing to take gas under the contract, whereas a high exercise threshold indicates the buyer is reluctant to take gas. For the buyers, the most important aspect is to make the optimal daily exercise decision at the beginning of each gas day, based on the current gas price, oil price, index price and running average value. We present the decision surfaces obtained using the tree algorithm at time t_{225} in Figure 22.10. In these surfaces, a big jump in the exercise threshold exists when the period to date is 135.

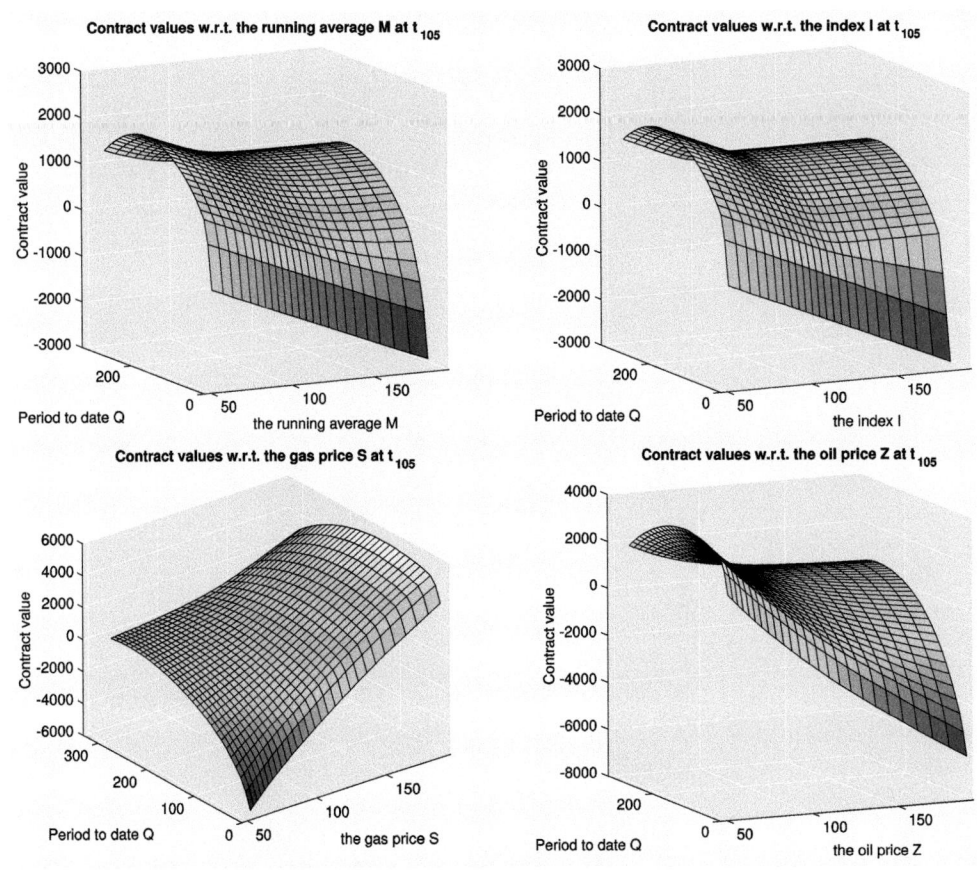

FIGURE 22.9 Value surfaces at time t_{105} by the tree algorithm[c]. (When not variable, these surfaces occur when the gas price is 98.815, the oil price is 98.543, the value of the running average is 108.835 and the index is 108.909.)

This is due to the requirement to meet the minimum bill. At day 225, $360 - 225 = 135$ exercise opportunities are left before the end of the contract. If the period to date is $270 - 135 = 135$, at day 225, then the buyer can only avoid the penalty by taking the daily maximum q_{max} for the remaining days. That is, the buyer must take gas under the contract even when the instant payoff is negative. In the middle plot of Figure 22.10, the exercise threshold is higher when the index is very large because the index is so large that the buyer loses more profits from the instant payoff than the penalty. Again, Figure 22.10 shows that the exercise threshold is more sensitive to the oil price than the running average. In addition, the decision surfaces with respect to the running average and oil price have different surface patterns with the surface with respect to the index. The exercise threshold increases as the running average and the oil price decrease because if the current oil price and the running average are low, then the index in the coming month is possibly very small. Hence, buyers are willing to save exercise opportunities to take advantage of the small index in the coming month. The exercise threshold increases as the index increases because the buyers do not benefit from a high index.

22.7.5 Contract Values with Respect to Parameters

For the same GSA contract with a fixed H and w, the tree algorithm provides the same contract value through separate valuations. Hence, the tree algorithm provides a powerful tool to analyze how the parameters of a GSA contract affect the contract value.

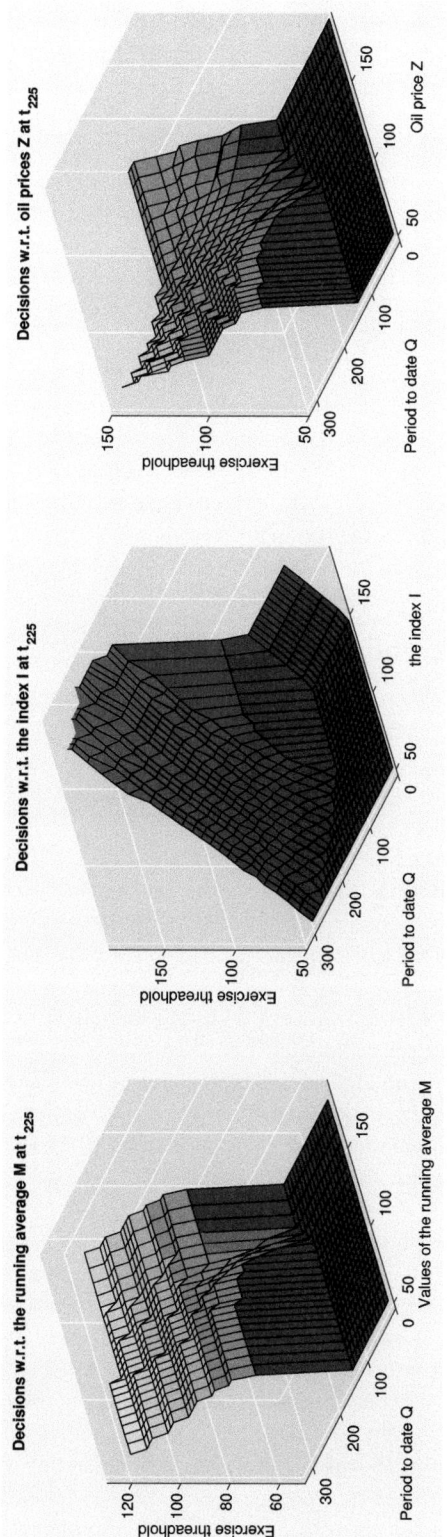

FIGURE 22.10 Decision surfaces at time t_{225} by the tree algorithm. (When not variable, these surfaces occur when the oil price is 98.5, the value of the running average is 108.774 and the index is 108.775.)

In this section, we investigate the effect of these parameters using the tree algorithm. Figure 22.11 shows how the contract value changes when the values of the parameters change. The contract value decreases when the interest rate increases because a high interest rate reduces the value of future cash flows. Furthermore, the contract value decreases as the correlation increases because buyers have fewer opportunities to make high profits when the gas price and oil price move in the same direction. In addition, the contract value increases with both the gas price volatility and the oil price volatility because the buyers could implement a more flexible trading strategy when the gas price or oil price fluctuates more. Moreover, the volatility of the gas price contributes more to the contract value than the oil price volatility because the volatility of the oil price is smoothed under the indexation principle. Furthermore, generally, the contract value decreases in both the mean-reverting rates of the gas price and oil price.

In the middle-bottom plot of Figure 22.11, the contract value decreases rapidly from $\alpha_S = 2$ to $\alpha_S = 5$, before becoming quite stable as the mean-reverting rate of the gas price increases. This demonstrates that α_S loses its influence on the contract if it is large enough. In the bottom-right plot of Figure 22.11, compared with α_S, the mean-reverting rate of the oil price α_Z has a much greater influence on the contract value because, when we have a small α_Z, the value of the index can be quite different in different months. This provides buyers with opportunities to take advantages of a low index and avoid taking gas under a high index. Recall Remark 3, different mean-reverting rates result in different sizes of the trinomial tree. Hence, the computing time of the tree algorithm should be related to α_S and α_Z. Figure 22.12 illustrates the computing time with respect to these two mean-reverting rates.

22.7.6 Self-Indexed Swing Options

With the development of liquified natural gas in the market place, fewer long-term contracts are indexed to oil on the market, but the indexation is based on the gas price itself, so that the contract provides flexibility in trading the price spread on the gas forward curves. To reflect this trend on the market, we introduce and evaluate the self-indexed swing options, where the value of the index is determined by the weighted average price of the gas prices in the previous month.

Technically, the calculation of the index can be performed using (22.3) by replacing $Z(t_{(l-1)D+j})$ with $S(t_{(l-1)D+j})$. The algorithms introduced in this paper can be easily modified to accommodate the self-indexed swing option. For the tree algorithm, we build the gas tree analogously to the oil tree, as described in Section 22.4.1 and use the gas tree only. An alternative and brute-force method is to simply set the correlation at $\rho = 1$.

Figure 22.13(a) displays the Dutch Title Transfer Facility gas forward curve observed on 1st February 2020. Using this forward curve, we price several self-indexed gas swing options to show how contract changes against different features. From Figure 22.13(a), we observe that a huge increase occurred in the forward price from the 9th month to the 10th month. Intuitively, the gas price in the 9th month would be a lot lower than that in the 10th month. Hence, in the 10th month, the index would be smaller than the gas price. That is, we should expect an increase in value when extending the maturity of a contract from 9 months to 10 months.

Figure 22.13(b) depicts the values of a self-indexed gas swing contract with $ACQ = 370$ and $MB = 210$ against the maturity T, supporting our expectations. The contract value almost doubled from a 9-month contract to a 10-month contract. Then, the contract value increases more slowly because of the indexation. The gap between the gas price and index is rarely very large. Additionally, we should always expect the contract value to not decrease when extending the maturity while holding the remaining inputs constant because we can either obtain better payoffs if the newly added days are favorable or continue with our previous decision if they are not favorable.

Figure 22.13(c) displays the value of a 1-year self-indexed gas swing option against the annual contract quantity ACQ. Overall, given an MB, the contract value increases as ACQ increases because a larger gap between ACQ and MB provides the buyer with more flexibility and increases the contract value. The

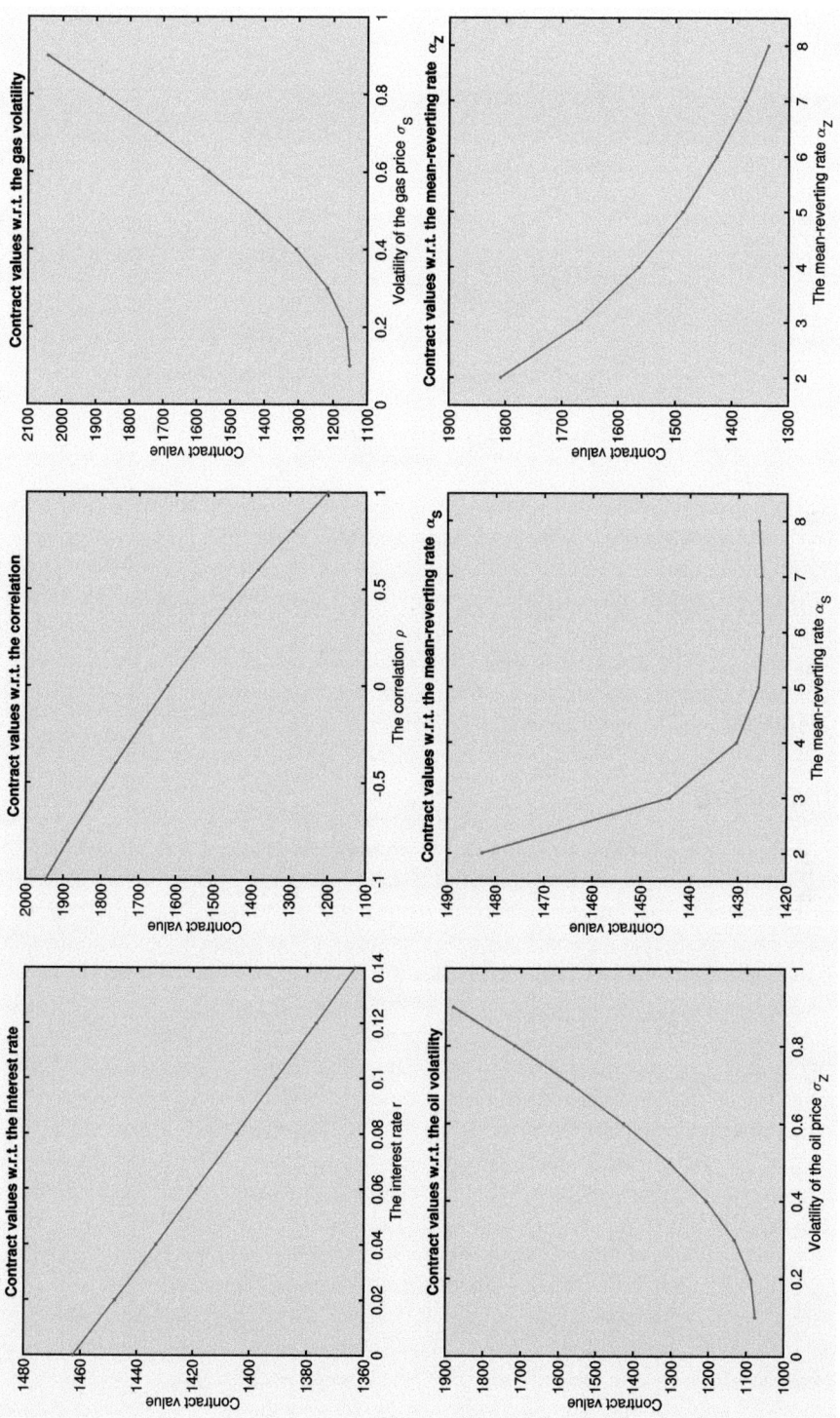

FIGURE 22.11 Contract values with respect to parameters.

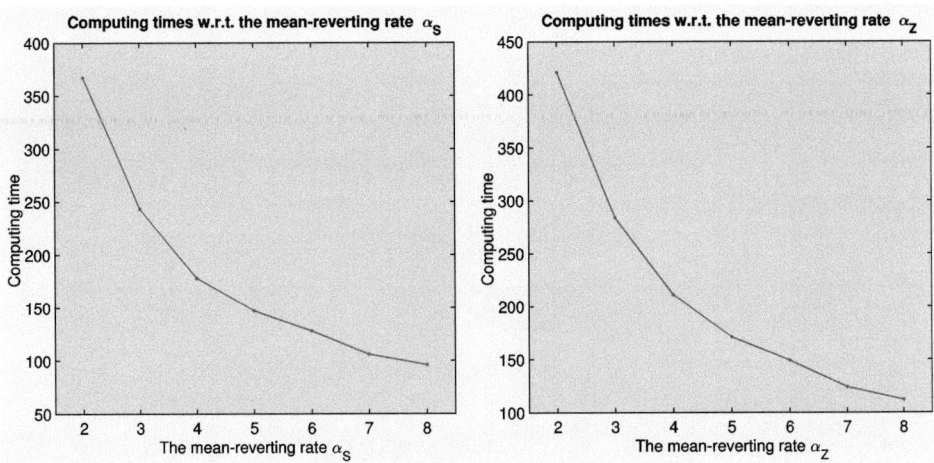

FIGURE 22.12   Computing times with respect to mean-reverting rates.

contract value stops growing when the gap between *ACQ* and *MB* is large enough to fully exploit the contract given the observed forward curve, as shown in Figure 22.13(c).

Figure 22.13(d) depicts the value of a 1-year self-indexed gas swing option against *MB*. In general, the contract value decreases as *MB* increases because narrowing the gap between *ACQ* and *MB* decreases the value. The contract values are the same when *ACQ* = 360 and *ACQ* = 315 because the annual contract quantity is large enough to be irrelevant to the contract value, and we observe this in Figure 22.13(c) as well. When *ACQ* does affect the contract, the contract value decreases faster for contracts with a smaller *ACQ* (e.g. the '*ACQ* 270' plot in Figure 22.13(d)).

22.8 Conclusion

In this paper, we build two algorithms for pricing the GSA with indexation: one lattice-based method that uses a 2-D trinomial tree and one LSMC algorithm that performs regression on four state variables. To achieve the best efficiency, we implement both algorithms on GPUs using CUDA programming. The numerical examples indicate that the LSMC algorithm on the GPU is more efficient than the tree algorithm and LSMC algorithm on the CPU. However, the contract value we obtained using the LSMC algorithm on the GPU is less accurate due to the limited precision on the computer. Although one can use fewer basis functions and simulated paths to ease this problem, based on our numerical analysis, it still provides a slightly less accurate estimation of the contract value compared with the other approaches. The LSMC algorithm on the CPU is time-consuming, requiring hours to evaluate a 1-year contract. The tree algorithm is implemented efficiently on the GPU using CUDA programming. The contract value we obtained using the tree algorithm is sufficiently accurate when compared with the benchmark value. In addition, our numerical examples indicate that the tree algorithm provides satisfactory values when we avoid using too many points in the non-uniform grid. The tree algorithm also has the advantage that, for fixed inputs, it returns a fixed contract value. Hence, it is a powerful tool to analyze the contract features of GSAs with indexation. Using the tree algorithm and several numerical studies, we demonstrate various features of this complex contract. For example, the contract value and the exercise threshold are more sensitive with respect to the oil price than the value of the running average. The running average and oil price have less influence on the contract value when their values are large.

As a lattice-based method, the tree algorithm can be very inefficient when introducing further dimensions. One can easily modify the LSMC algorithm to accommodate sophisticated models, such as multi-factor models, but it is difficult to duplicate this for a lattice-based method. Thus the advantage of the LSMC algorithm is its flexibility. When switching between different models, the only modification

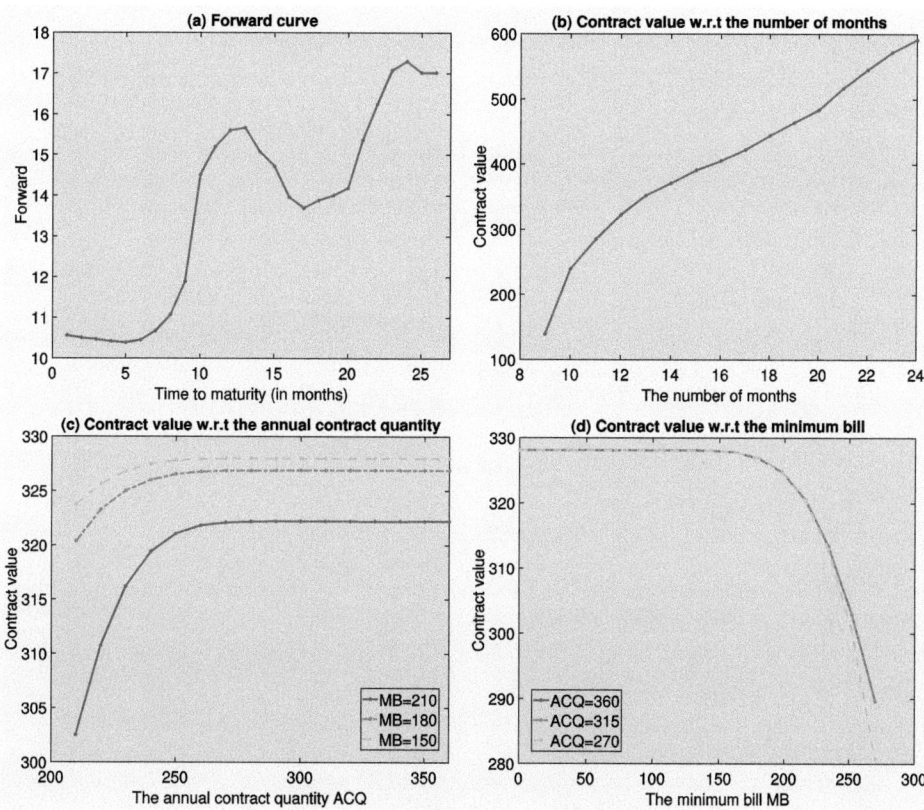

FIGURE 22.13   Value of self-indexed swing options.

that needs to be addressed is the simulation of the underlying prices. Therefore, the LSMC algorithm is popular among practitioners. To our knowledge, however, no lattice-based method has been built in the literature that addresses the real GSA contract with indexation. Hence, the tree algorithm provides a trustworthy benchmark for practitioners to test their results.

Other types of indexation also exist in the market. For example, a lag can exist between the month in which the index is used and the month in which the index is calculated. Moreover, the index can also be calculated on more than one energy product. These properties make the evaluation of a real contract much more challenging than the scenario envisaged in this paper, but we leave those to future research.

Acknowledgements

We would like to thank the Managing Editor, Publishing Editor and the anonymous referees for their useful comments to have improved the previous version of the paper.

Disclosure Statement

No potential conflict of interest was reported by the author(s).

ORCID

W. Dong http://orcid.org/0000-0001-7226-9533
B. Kang http://orcid.org/0000-0002-0012-0964

References

Aleksandrov, N. and Hambly, B.M., A dual approach to multiple exercise option problems under constraints. *Math. Methods Oper. Res.*, 2010, **71**, 503–533.

Asche, F., Osmundsen, P. and Tveterås, R., European market integration for gas? Volume flexibility and political risk. *Energy Econ.*, 2002, **24**, 249–265.

Bardou, O., Bouthemy, S. and Pagès, G., Optimal quantization for the pricing of swing options. *Appl. Math. Finance*, 2009, **16**, 183–217.

Bardou, O., Bouthemy, S. and Pagès, G., When are swing options bang-bang? *Int. J. Theor. Appl. Finance*, 2010, **13**, 867–899.

Barrera-Esteve, C., Bergeret, F., Dossal, C., Gobet, E., Meziou, A., Munos, R. and Reboul-Salze, D., Numerical methods for the pricing of swing options: A stochastic control approach. *Methodol. Comput. Appl. Probab.*, 2006, **8**, 517–540.

Basei, M., Cesaroni, A. and Vargiolu, T., Optimal exercise of swing contracts in energy markets: An integral constrained stochastic optimal control problem. *SIAM J. Financ. Math.*, 2014, **5**, 581–608.

Bender, C., Dual pricing of multi-exercise options under volume constraints. *Financ. Stoch.*, 2011, **15**, 1–26.

Bernhart, M., Modelization and valuation methods of gas contracts: Stochastic control approaches. PhD Thesis, Université Paris-Diderot-Paris VII, 2011.

Bernhart, M., Tankov, P. and Warin, X., A finite-dimensional approximation for pricing moving average options. *SIAM J. Financ. Math.*, 2011, **2**, 989–1013.

Breslin, J., Clewlow, L., Strickland, C. and van der Zee, D., Swing contracts: Take it or leave it. *Energy Risk*, 2008, February, 64–68.

Brigo, D. and Mercurio, F., *Interest Rate Models—Theory and Practice: With Smile, Inflation and Credit*, 2nd ed., 2006 (Springer-Verlag: Berlin).

Broadie, M. and Cao, M., Improved lower and upper bound algorithms for pricing American options by simulation. *Quant. Finance*, 2008, **8**, 845–861.

Broadie, M. and Glasserman, P., A stochastic mesh method for pricing high-dimensional American options. *J. Comput. Finance*, 2004, **7**, 35–72.

Chiarella, C., Clewlow, L. and Kang, B., The evaluation of multiple year gas sales agreement with regime switching. *Int. J. Theor. Appl. Finance*, 2016, **19**, 1–25.

Clewlow, L. and Strickland, C., Valuing energy options in a one factor model fitted to forward prices, 1999. Available online at: https://ssrn.com/abstract 160608 (accessed 14 November 2018).

Curran, M., Willow power: Optimizing derivative pricing trees. *Algo Res. Q.*, 2001, **4**, 15–23.

Dai, M., Li, P. and Zhang, J.E., A lattice algorithm for pricing moving average barrier options. *J. Econ. Dyn. Control*, 2010, **34**, 542–554.

Dirnstorfer, S., Grau, A.J. and Zagst, R., High-dimensional regression on sparse grids applied to pricing moving window Asian options. *Open. J. Stat.*, 2013, **3**, 427–440.

Dong, W. and Kang, B., Analysis of a multiple year gas sales agreement with make-up, carry-forward and indexation. *Energy Econ.*, 2019, **79**, 76–96.

Dörr, U., Valuation of swing options and examination of exercise strategies by Monte Carlo techniques. Master's Thesis, University of Oxford, 2003.

Edoli, E., Pricing of gas swing contracts with indexed strike: A viscosity solution approach with applications. PhD Thesis, University of Padova, 2013.

Edoli, E. and Vargiolu, T., Stochastic optimization for the pricing of structured contracts in energy markets. *ARGO Magazine*, 2013, **2**, 35–44.

Edoli, E., Fiorenzani, S., Ravelli, S. and Vargiolu, T., Modeling and valuing make-up clauses in gas swing contracts. *Energy Econ.*, 2013, **35**, 58–73.

Fanelli, V. and Ryden, A.K., Pricing a swing contract in a gas sale company. *Econ. Manage. Financ. Mark.*, 2018, **13**, 40–55.

Fatica, M. and Phillips, E., Pricing American options with least squares Monte Carlo on GPUs. In *Proceedings of the 6th Workshop on High Performance Computational Finance*, Denver, CO, 2013. Available online at: https://dl.acm.org/citation.cfm?doid 2535557.2535564 (accessed 14 November 2018).

Glasserman, P., *Monte Carlo Methods in Financial Engineering*, 2003 (Springer-Verlag: New York).

Grau, A.J., Applications of least-squares regressions to pricing and hedging of financial derivatives. PhD Thesis, Technical University of Munich, 2008.

Hanfeld, M. and Schlüter, S., Operating a swing option on today's gas markets – How least squares Monte Carlo works and why it is beneficial. *Zeitschrift für Energiewirtschaft*, 2017, **41**, 137–150.

Holden, L., Løland, A. and Lindqvist, O., Valuation of long-term flexible gas contracts. *J. Deriv.*, 2011, **18**, 75–85.

Hull, J.C., *Option, Futures and Other Derivatives*, 8th ed., 2011 (Prentice Hall: Upper Saddle River, NJ).

Hull, J.C. and White, A.D., Numerical procedures for implementing term structure models I: Single-factor models. *J. Deriv.*, 1994a, **2**, 7–16.

Hull, J.C. and White, A.D., Numerical procedures for implementing term structure models II: Two-factor models. *J. Deriv.*, 1994b, **2**, 37–48.

in't Hout, K.J. and Foulon, S., ADI finite difference schemes for option pricing in the Heston model with correlation. *Int. J. Numer. Anal. Model.*, 2010, **7**, 303–320.

Jaillet, P., Ronn, E.I. and Tompaidis, S., Valuation of commodity-based swing options. *Manage. Sci.*, 2004, **50**, 909–921.

Judd, K.L., *Numerical Methods in Economics*, 1998 (MIT Press: Cambridge, MA).

Kao, C.H. and Lyuu, Y.D., Pricing of moving-average-type options with applications. *J. Futures Mark.*, 2003, **23**, 415–440.

Lari-Lavassani, A., Simchi, M. and Ware, A., A discrete valuation of swing options. *Can. Appl. Math. Q.*, 2001, **9**, 35–73.

Longstaff, F.A. and Schwartz, E.S., Valuing American options by simulation: A simple least-squares approach. *Rev. Financ. Stud.*, 2001, **14**, 113–147.

Lu, L., Xu, W. and Qian, Z., Efficient willow tree method for European-style and American-style moving average barrier options pricing. *Quant. Finance*, 2017, **17**, 889–906.

Maruyama, G., Continuous Markov processes and stochastic equations. *Rend. Circ. Matemat.* Palermo, 1955, **4**, 48–90.

Meinshausen, N. and Hambly, B.M., Monte Carlo methods for the valuation of multiple-exercise options. *Mathe. Finance*, 2004, **14**, 557–583.

NVIDIA, CUBLAS library, 2018a. Available online at: https://docs. nvidia.com/cuda/cublas/index.html (accessed 14 November 2018).

NVIDIA, CUDA compiler driver NVCC, 2018b. Available online at: https://docs.nvidia.com/cuda/cuda -compiler-driver-nvcc/index. html (accessed 14 November 2018).

NVIDIA, NVIDIA CUDA C programming guide, 2018c. Available online at: https://docs.nvidia.com/ cuda/cuda-c-programming-guide/index.html (accessed 14 November 2018).

Thanawalla, R.K., Valuation of gas swing options using an extended least squares Monte Carlo algorithm. PhD Thesis, Heriot-Watt University, 2006.

Thompson, A.C., Valuation of path-dependent contingent claims with multiple exercise decisions over time: The case of take-or-pay. *J. Financ. Q. Anal.*, 1995, **30**, 271–293.

Warin, X., Hedging swing contract on gas markets, 2012. Available online at: https://arxiv.org/abs/1208 .5303 (accessed 14 November 2018).

Xu, W., Hong, Z. and Qin, C., A new sampling strategy willow tree method with application to path-dependent option pricing. *Quant. Finance*, 2013, 13, 861–872.

Appendix A: Construction of the Two-Dimensional Trinomial Tree

Denote the log prices of gas and oil at time $t \in [0, T]$ are denoted by $X(t)$ and $Y(t)$, respectively. To build a 2-D tree for (X, Y), we first build a simplified 2-D fundamental tree for a 2-D Markov process (x, y) below, which is obtained by assuming $\theta^S(t)$ $\theta^Z(t)$ 0 in (22.1) and (22.2):

$$dx(t) = -\alpha_S x(t)dt + \sigma_S dB^S(t),$$

$$dy(t) = -\alpha Zy(t)dt + \sigma_Z dB^Z(t),$$

where $x(0) = y(0) = 0$. Then we shift the nodes on the simplified tree by adding the integrated drift in order to be consistent with the observed forward curve. In the rest of this section, we summarize the tree building procedures in Clewlow and Strickland (1999), Hull and White (1994a, 1994b) and Brigo and Mercurio (2006).

Step one We build two separate fundamental one-dimensional trees for both x and y. Taking the log-normal price x as an example, the trinomial tree for y will be built in a very similar manner.

We discretize the time interval $[0, T]$ into N equally spaced pieces with the time step $\Delta t = T/N$. The node xn,s at time tn on the tree can is referenced by a pair of integers (n, s), where $t_n = n\Delta t$ and $x_{n,s} = s\Delta X$. In addition, due to convergence and stability considerations, it is suggested in Hull and White (1994a) that $\Delta X = \sigma s\sqrt{3\Delta t}$.

Because of the mean-reverting nature of this model, the trinomial tree will reach its maximum level,[*] $s_{\max}$, at some point (the minimal level, $s_{\min}$, is also reached at the same time, $s_{\min} = -s_{\max}$). Depending on the position of the node (n, s), branches emanating from node (n, s) have three forms as in Figure A1. Denote p_u, p_m and p_d the probabilities associated with the upper, middle and lower branches emanating from node (n, s), respectively. By matching the drift and variance of the underlying process, these probabilities for these three forms are given in table A1 (for detailed derivation, we refer to Dong and Kang 2019 and Appendix C in Clewlow and Strickland 1999).

It is shown in Hull and White (1994a) that, in order to ensure that p_u, p_m and p_d are all non-negative, $s_{\max}$ should be an integer between $0.184/\alpha_S\Delta t$ and $0.816/\alpha_S\Delta t$. To achieve the best efficiency, it is also suggested to set $s_{\max}$ at the smallest integer greater than $0.184/\alpha_S\Delta t$. In the rest of this paper, we will call this tree the fundamental tree.

Step two Once we have built two fundamental trinomial trees for both x and y, we combine these two trees together. Let node (n, z) be the node $y_{n,z}$ on the oil tree at time t_n where $y_{n,z} = z\Delta Y$ and $\Delta Y = \sigma Z\sqrt{3\Delta t}$. Any node on the 2-D tree can be referenced by a triplet of integers (n, s, z). There are three possible movements (an up movement, a middle movement and a down movement) at each node on both the gas tree and the oil tree, which gives a total of nine possible movements at each node on the 2-D tree. Let P_{uu} be the probability associated with the movement where the gas price follows the upper branch and the oil price follows the upper branch. Similar rule applies to the notations P_{um}, P_{ud}, P_{mu}, P_{mm}, P_{md}, P_{du}, P_{dm} and P_{dd}. Denote the probabilities associated with the upper, middle and lower branches on the oil tree by q_u, q_m and q_d, respectively. By assuming the correlation $\rho = 0$, the probabilities on the 2-D tree are the product of the corresponding probabilities associated with the branches on the gas tree and the oil tree. When the correlation is non-zero, according to Appendix F in Brigo and Mercurio (2006), each probability is shifted to maintain the marginal distribution and the covariance structure of x and y. The probabilities are given by

---

[*] The distribution of $x(t)$ converges to the normal distribution $N(0, \sigma^2/2\alpha)$ for $\alpha > 0$ as $t \to \infty\infty$. That is, in the long run, there is no natural maximum or minimum. However, because of the mean-reverting natural of the commodity prices and $T = 1$ considered in this paper, the two-dimensional trinomial tree provides a good approximation.

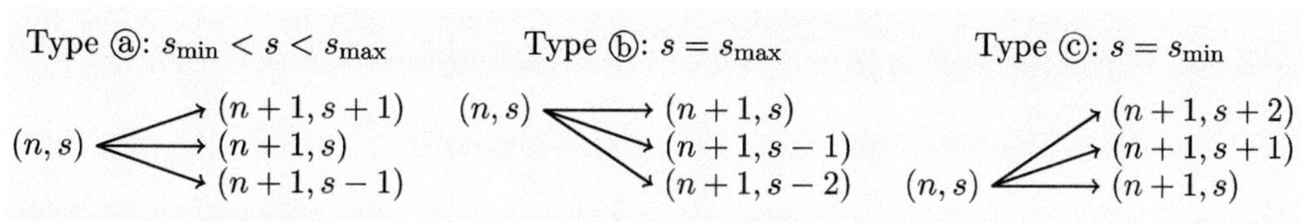

Type ⓐ: $s_{\min} < s < s_{\max}$ Type ⓑ: $s = s_{\max}$ Type ⓒ: $s = s_{\min}$

(n,s) ⟵ $(n+1,s+1)$ / $(n+1,s)$ / $(n+1,s-1)$

(n,s) ⟶ $(n+1,s)$ / $(n+1,s-1)$ / $(n+1,s-2)$

(n,s) ⟶ $(n+1,s+2)$ / $(n+1,s+1)$ / $(n+1,s)$

FIGURE A1 Possible forms of the tree branching.

TABLE A1 Probabilities

	Type ⓐ	Type ⓑ	Type ⓒ
p_u	$\dfrac{1}{6} + \dfrac{\alpha_S^2 s^2 \Delta t^2 - \alpha_S s \Delta t}{2}$	$\dfrac{7}{6} + \dfrac{\alpha_S^2 s^2 \Delta t^2 - 3\alpha_S s \Delta t}{2}$	$\dfrac{1}{6} + \dfrac{\alpha_S^2 s^2 \Delta t^2 + \alpha_S s \Delta t}{2}$
p_m	$\dfrac{2}{3} - \alpha_S^2 S^2 \Delta t^2$	$-\dfrac{1}{3} - \alpha_S^2 S^2 \Delta t^2$	$-\dfrac{1}{3} - \alpha_S^2 S^2 \Delta t^2 - 2\alpha_S S \Delta t$
p_d	$\dfrac{1}{6} + \dfrac{\alpha_S^2 s^2 \Delta t^2 + \alpha_S s \Delta t}{2}$	$\dfrac{1}{6} + \dfrac{\alpha_S^2 s^2 \Delta t^2 - \alpha_S s \Delta t}{2}$	$\dfrac{7}{6} + \dfrac{\alpha_S^2 s^2 \Delta t^2 + 3\alpha_S s \Delta t}{2}$

$$
\Pi_{\rho \geq 0} = \begin{pmatrix} P_{uu} & P_{um} & P_{ud} \\ P_{mu} & P_{mm} & P_{md} \\ P_{du} & P_{dm} & P_{dd} \end{pmatrix}
$$

$$
= \begin{pmatrix} p_u q_u + 5\varepsilon & p_u q_m - 4\varepsilon & p_u q_d - \varepsilon \\ p_m q_u - 4\varepsilon & p_m q_m + 8\varepsilon & p_m q_d - 4\varepsilon \\ p_d q_u - \varepsilon & p_d q_m - 4\varepsilon & p_d q_d + 5\varepsilon \end{pmatrix}
\tag{22.A1}
$$

and

$$
\Pi_{\rho < 0} = \begin{pmatrix} P_{uu} & P_{um} & P_{ud} \\ P_{mu} & P_{mm} & P_{md} \\ P_{du} & P_{dm} & P_{dd} \end{pmatrix}
$$

$$
= \begin{pmatrix} p_u q_u - 1\varepsilon & p_u q_m - 4\varepsilon & p_u q_d + 5\varepsilon \\ p_m q_u - 4\varepsilon & p_m q_m + 8\varepsilon & p_m q_d - 4\varepsilon \\ p_d q_u + 5\varepsilon & p_d q_m - 4\varepsilon & p_d q_d - \varepsilon \end{pmatrix},
\tag{22.A2}
$$

where $\varepsilon = \rho/36$ if $\rho \geq 0$ and $\varepsilon = -\rho/36$ if $\rho < 0$.

Step three Returning to the notations in Step 1, we now make both the gas tree to be consistent with the observed forward curve by adding an amount, a_n, at each node of each time step t_n. Again, the oil tree can be adjusted in a very similar manner.

We define the state price, $G_{n,s}$, as the value of a security that pays 1 if the node (n, s) is reached, and 0 otherwise, at time 0. Then the state price at each node can be obtained by forward induction:

$$
G_{n+1,s} = \sum_{s'} G_{n,s'} p_{s',s} e^{-r\Delta t},
$$

where (n, s') represents the node at time t_n that has a branch leading to node $(n+1, s)$ and $p_{s',s}$ is the probability of moving from node (n, s') to node $(n+1, s)$, r is the risk-free rate. Denote the gas price by S, that is $S(t)=e^{X(t)}$. In addition, let $S_{n,s}$ be the gas price where the time equals t_n and the level of the gas price on the adjusted gas tree is s. Thus the price at time 0 of any European claim with payoff function $C(S_{n,s})$ at time t_n on the tree is given by

$$C_0 = \sum_s G_{n,s} C\left(S_{n,s}\right),$$

(22.A3)

where the summation takes place through all nodes (n, s) at time tn. Now consider a case where $C(S_{n,s})$ $S_{n,s}$, then according to (22.A3), we have

$$e^{-rt_n} F^S\left(0, t_n\right) = \sum_s G_{n,s} S_{n,s}.$$

By letting $S_{n,s} = e^{xn,s+an}$, we obtain

$$a_n = \ln\left(\frac{e^{-rt_n} F^S\left(0, t_n\right)}{\sum_s G_{n,s} e^{xn,s}}\right).$$

(22.A4)

Therefore, by adding a_n at each node at each time t_n, we obtain the trinomial tree we need.

Appendix B: Notes on the Implementation of the Tree Algorithm on GPUs

In this appendix, we mainly focus on the evaluation in the middle months since it is this that requires the most computing effort (see Remark 3). The evaluations in other months can be conducted in a similar way, however. First recall the tree algorithm in the middle months:

$$V^l\left(S_{n,s}, Z_{n,z}, M_{n,m}, I_{n,i}, Q\right)$$

$$= \max_{q \in [0,q]}\left\{q \cdot \left(S_{n,s} - I_{n,i}\right) + \sum_{b,c=-1}^{1} p_{s,z,b,c} e^{-r\Delta t}\right.$$

$$\left. \cdot V^l\left(S_{n+1,g(s,b)}, Z_{n+1,h(z,c)}, \mathcal{M}_{n,m,c}, I_{n,i}, Q+q\right)\right\}.$$

(22.A5)

Note that not everything is paralleled on the GPU. We only let the GPU do the heavy work, that is equation (A5). Some easy work, such as the building of the 2-D trinomial tree, are implemented on the CPU. To ease the burden on the GPU, we can do some preparation before launching kernels. Recall that, when $\mathcal{M}_{n,m,c}$ in (A5) is not one of the values in the running average vector $\mathbf{M}_{n+1}$, we use linear interpolation (22.22). This involves finding $M_{n+1,\bar{m}}$ and $M_{n+1,\bar{m}+1}$. At any node (n, s, z) on the 2-D tree, the part

$$\frac{\mathcal{M}_{n,m,c} - M_{n+1,\bar{m}}}{M_{n+1,\bar{m}+1} - M_{n+1,\bar{m}}}$$

(22.A6)

in (22.22) only depends on the value of the running average $M_{n,m}$ at time t_n and the oil price $Z_{n+1,h(z,c)}$, $c = -1, 0, 1$, at time t_{n+1} (see (22.9)). At time t_n, for each combination of $M_{n,m}$ and $Z_{n,z}$, we can find $\bar{m}$ and the value of (A6) by following the up, middle and down movements on the oil tree before launching the kernels. In addition, at any node (n, s, z), for each combination of $S_{n,s}$ and $I_{n,i}$, we can also compute the instant payoff $(S_{n,s} - I_{n,i})$ in advance. The reason for these two actions is that, if we do these calculations inside a kernel, this kernel has to do the same calculations repeatedly, which wastes a lot of time.

Listing 1 gives the thread hierarchy of our CUDA programme. In Listing 1, H is the number of values in the running average vector $\mathbf{M}_n$, Q is the number of possible periods to date at time t_n, N_s and N_z are the total number of nodes on the gas tree and the oil tree at time t_n, respectively. dim3 is an integer vector type in CUDA which defines the dimensions of the block and the grid.

```
Listing 1. Block and grid dimensions
        dim3 dimBlock (H);
        dim3 dimGrid (N_s, N_z, Q);
```

Listing 2 gives the code on how we launch our kernel. The kernel which implements our algorithm (A5) is GAS _ Middle _ Months ().

```
    Listing 2. Call the kernel
 for (int i = 0; i < H; i++)
 {
 GAS_Middle_Months <<<dim Grid, dimBlock>>> ();
 }
```

Listings 1 and 2 mean that, if we have a strong enough GPU, at any time t_n, for each index In,i in the index vector $\mathbf{I}n$, the calculations of the values $V^l(S_{n,s}, Z_{n,z}, M_{n,m}, I_{n,i}, Q)$ for all combinations of $S_{n,s}, Z_{n,z}$, $M_{n,m}$ and Q are paralleled on the GPU. Listing 3 shows how we use shared memory to accelerate the reading and writing speed on the GPU.

```
Listing 3. Shared memory allocation
 __shared__ float linp_m[H][3];
     __shared__ int m_bar[H][3];
     __shared__ int n_1[6];
     __shared__ float prob[9];
     __shared__ float i_p;
     __shared__ float n_1_value[3][3][H];
```

In Listing 3, linp _ m[H][3] stores the values of (A6), m _ bar[H][3] stores the values of $\overline{m}$, n _ 1 [6] stores $g(s, b)$ and $\mathfrak{h}(z,c)$, prob[9] stores the probabilities associated with the nine possible movements on the 2-D tree, i _ p stores the instant payoff and n _ 1 _ value[3][3][H] stores the value of the GSA at time $t_n + 1$. As we can see, n _ 1[6], prob [9] and i _ p consume only a very small amount of memory. linp _ m[H][3], m _ bar[H][3] and n _ 1 _ value[3][3][H] may consume more memory since the sizes of these arrays depend on an integer H, which is the number of values in the running average vector $\mathbf{M}_n$. If we increase H to get a better approximation of the running average M, these arrays consume more memory, and this may reduce the number of blocks which can be paralleled at the same time. The advantage of the memory allocation in Listing 3, however, is that, if we increase H, the number of threads in each block is also increased. That is, even with less blocks paralleled, we still have a lot of threads running at the same time.

In each month, we evaluate the GSA for each value of index in the index vector $\mathbf{I}$. It means that, for each i in Listing 2, the kernel GAS _ Middle _ Months can also be paralleled. We do not do that mainly because of the shortage of memory on the GPU. Although the global memory is the largest memory on the GPU, its size is usually significantly smaller than the RAM on the CPU. Even when i in Listing 2 is not paralleled, we show in Section 22.7 that the computing time is satisfactory.

The data transfer between the CPU and the GPU can significantly slow the CUDA programme. In CUDA programming, memory transfers usually happen when the data is too large to store on the GPU. The advantage of our algorithm is that, under normal circumstances, we only need to do the data transfer when we want to retrieve specific data from the GPU. The whole evaluation of the GSA can be done on the GPU. This is because of the matching point condition (22.23). At the matching point between two consecutive months, we only need to store data when the oil price equals the value of the running average, hence, one dimension has been reduced.

4

Electricity Markets

23

Modeling the Distribution of Day-Ahead Electricity Returns: A Comparison

Sandro Sapio

23.1 Introduction

The 'problem of price variation', as Mandelbrot (1963) called it, has been one of the most debated issues in financial economics. Providing a correct description of the empirical distribution of returns is essential for the theory and practice of trading and investment. Indeed, portfolio selection theory based upon variance-based measures of risk only works under the assumption that returns have finite central moments. Furthermore, Value- at-Risk calculations, the pricing formulas for contingent claims, the accuracy of price forecasting and the appropriateness of econometric methods all depend on the distribution of returns.

The relevance of these issues is by no means confined to stock market analysis. In markets for non-storable commodities, such as electricity, trading mechanisms must continuously ensure market-clearing. In electricity, imbalances between demand and supply would cause blackouts. The prices quoted on wholesale power exchanges undergo sudden and short-lived excursions, caused by strategic behavior and accidental plant failures, while production and consumption smoothing are not feasible. A well-known empirical fact is that the heavy tails observed in electricity return distributions cannot be accounted for by the Gaussian law. Forecasting and risk management are therefore even more crucial than in stock markets, and a 'solution' to the problem of price variation in the context of power exchanges is even more urgent.

It is the aim of this paper to investigate the distributional nature of the day-ahead electricity price returns. More specifically, we ask whether electricity returns display heavy tails and skewness, and whether the central moments diverge; we also investigate the time scaling of risk and possible intra-daily differences in distributional shapes. Answers to these research questions are sought by comparing the goodness-of-fit performances of the α-stable, the Normal Inverse Gaussian (NIG), the Exponential Power (EP), and the Asymmetric Exponential Power (AEP) distribution laws. The data are drawn from

DOI: 10.1201/9781003265399-27

major European power exchanges, such as the Scandinavian NordPool, the Dutch APX, and the French Powernext. We focus on one-day returns computed on prices of individual hours, which allow us to obtain a grasp on the intra-day risk patterns.

The main findings of the study are the following. First, and expectedly, electricity returns display heavy tails. This fact is robust across markets and hourly auctions, and holds for various definitions of returns (log-returns, percentage returns, price changes) and regardless of the deseasonalization methodology. The tails in the APX market and in day-time auctions are fatter than in the other markets and in night-time auctions, and tend to be dampened by the logarithmic transform. Second, the NIG and the α-stable laws systematically outperform the EP and AEP distributions according to goodness-of-fit criteria, although no clear ranking can be established between the two 'winners'. Because only the α-stable and the NIG distributions are closed under convolution, fat tails are also expected to characterize returns computed over longer horizons. The estimated characteristic exponent of the α-stable distribution is always between 1 and 2, indicating that the expected value of the electricity returns converges, but the second moment does not. However, this is true only to the extent that the α-stable outperforms the NIG. Third, the skewness is an essential feature of the returns distributions. Some interesting cross-market variance emerges when Cholesky/scaling log-returns are considered (negative skewness in Powernext, positive in APX). Yet, using other definitions of returns and deseasonalization methods yields mixed results.

A number of previous works are closely related to the present paper. The α-stable model was shown to outperform the Hyperbolic and NIG distributions by Rachev *et al.* (2004, cited by Weron 2009) for EEX daily price differences, and by Weron (2005) for EEX and NordPool data. Mugele *et al.* (2005) found that NordPool and EEX daily price differences were also best described by stable laws, while the performance of the stable distribution for PolPX data was less successful. Further evidence of power-law tails was found by Bellini (2002), Bystro¨m (2005) and Chan and Gray (2006), who fitted generalized extreme-value distributions by means of peaks-over-threshold and block maxima methods, and by Deng and Jiang (2005), who followed a quantile function approach to model the distribution of CalPX and PJM returns. The Generalized Hyperbolic distribution, which includes the NIG as a special case, was estimated on NordPool data by Eberlein and Stahl (2003). Recent work by Weron (2009) reports estimates of the α-stable, Hyperbolic and NIG models on data from several markets (EEX, Omel, PJM, NEPOOL) and using various measures of price returns. The results vary across countries and are affected by how the returns are defined (see also Weron 2006). The probability density function of daily log-returns was modeled as a symmetric EP distribution by Bottazzi *et al.* (2005) and Bottazzi and Sapio (2007), whose estimates hinted at Laplacian or even heavier tails. Robinson and Baniak (2002) also fitted a Laplace distribution, whereas Bosco *et al.* (2007) used EP-distributed shocks in a PARMA-GARCH model. Deng *et al.* (2002) fitted a Cauchy–Laplace mixture to PJM and CalPX price differences. The present study covers three markets, analyses relatively large samples, and considers three definitions of price returns (log-returns, percentage returns, price differences). In this respect, the paper seeks to overcome the main limitations of the previous works, as highlighted, for instance, by Weron (2009, p. 460).

The paper is organized as follows. Section 23.2 describes the datasets and the deseasonalization methods, and provides summary statistics. The distributional models fitted in the paper are described in section 23.3, whereas section 23.4 illustrates the baseline estimation results concerning log-returns. In section 23.5 the robustness of the results is assessed with respect to other definitions of returns and deseasonalization methods. Section 23.6 concludes.

23.2 Data and Preliminary Analysis

For the purposes of this study, data on day-ahead electricity prices were collected for three major European power exchanges: NordPool (Denmark, Finland, Norway, Sweden), 2191 days from 1 January 1997 to 31 December 2002; APX (the Netherlands), 1457 days from 6 January 2001 to 31 December 2004; and Powernext (France), 1826 days from 1 February 2002 to 31 January 2007.* In these markets, each day,

---

* Data sources: NordPool FTP server, www.apx.nl and www.powernext.fr, respectively.

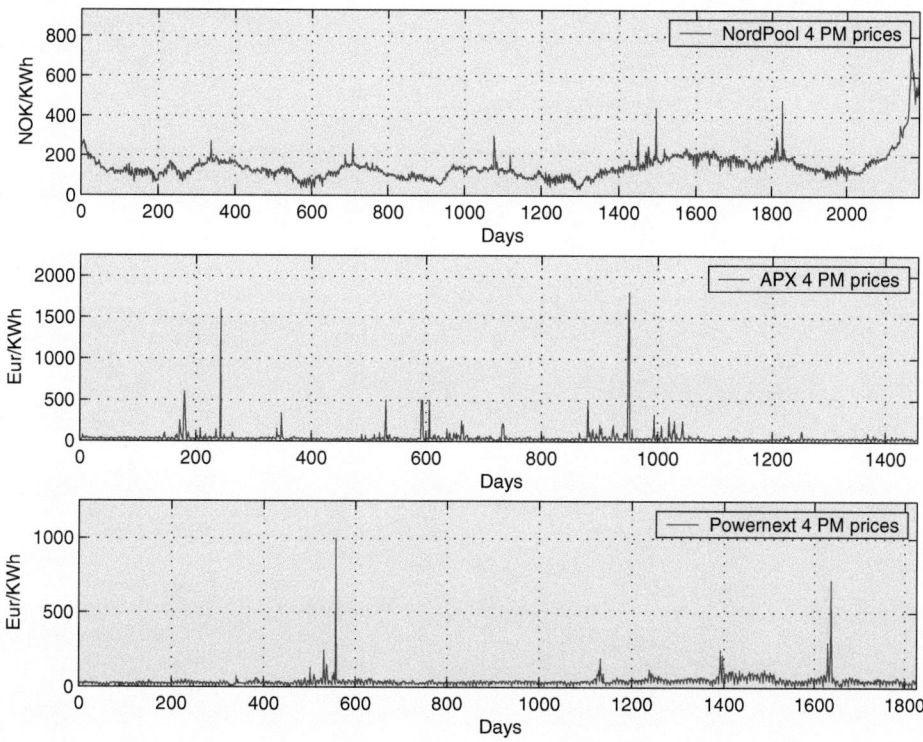

FIGURE 23.1 Plots of NordPool, APX and Powernext day-ahead prices for the 4 p.m. auctions. NOK, Norwegian kroner; Eur, Euro; KWh, Kilo Watts per hour.

24 auctions are run simultaneously in order to determine prices and quantities for each hour of the following day. The day-ahead prices are determined by means of uniform price auctions, so that all power is sold and purchased at the market-clearing price. The time series of day-ahead prices are depicted in Figure 23.1 for the 4 p.m. auctions, when demand is typically near its daily peak, and in Figure 23.2 for a night delivery session (4 a.m.), when average prices and demand are relatively low.

In finance, the price returns are usually defined as logarithmic price differences or log-returns $x_{ht} = \log p_{ht} - \log p_{h,t-1}$, where p_{ht} is the price at day t for the hour-h auction.* Table 23.1 provides summary statistics of the log-returns for selected hours, along with the outcomes of Shapiro–Wilk normality tests and autocorrelation coefficients. This table shows that, while drifts in power prices are rather weak, the standard deviations are largest at the beginning of the working day (8 a.m.) and decay there- after. The skewness is positive and stronger during the day, and the excess kurtosis is always positive and large, albeit without any clear intra-day pattern. Hence, the probability of observing large positive or negative fluctuations is greater than in a Gaussian process. The Shapiro– Wilk normality tests strongly reject the null of a Gaussian distribution ($SW = 1$) for all markets and all hourly auctions: the test statistics are always significantly below 1 (p values, not reported here, are all below 0.0001).

The serial correlations over a daily horizon are always negative, more so in the night-time auctions; the auto- correlations at lag 7 days are strong, up to 0.5–0.6 in some day-time auctions. The weekly pattern of economic activity is an obvious determinant of these patterns. Further time dependencies appear at lower frequencies, due to the seasonal patterns of economic activity and weather conditions. In addition to linear dependencies, the width of the power price fluctuations vary across hours, because

---

* Later in the paper (section 23.5) we shall discuss some drawbacks when using log-returns and assess the robustness of the results with respect to other definitions of returns.

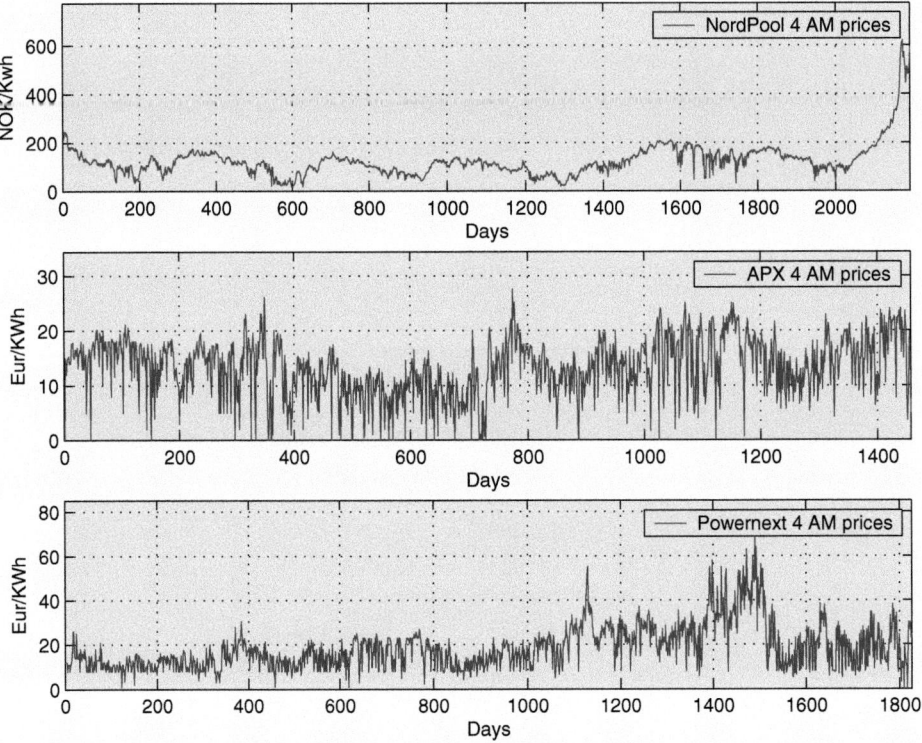

FIGURE 23.2   Plots of NordPool, APX and Powernext day-ahead prices for the 4 a.m. auctions. NOK, Norwegian kroner; Eur, Euro; KWh, KiloWatts per hour.

bidding strategies change under different relative scarcity (Karakatsani and Bunn 2004, Bottazzi *et al.* 2005, Simonsen 2005). All of this justifies the use of filters in order to remove the linear and higher-order autocorrelations in such a way that what remains is presumably the outcome of random shocks to market fundamentals.

The data are filtered in two steps. First, we remove all the linear autocorrelations by means of the semi- parametric Cholesky factor algorithm introduced by Diebold *et al.* (1997). The algorithm works as follows.

(1) Estimate the covariance matrix Σ: of the vector x_{ht} as the Toeplitz matrix built upon the autoco-variance vector γ.*

(2) Calculate C as the Cholesky factor of Σ, i.e. C: $CC'=\Sigma$.

(3) Extract the linearly uncorrelated, standardized residuals $\tilde{x}_{ht}$ as follows:

$$\tilde{x}_{ht} = C^{-1} x_{ht}. \tag{23.1}$$

Let us call $\tilde{x}$ the *Cholesky-filtered* log-returns. As a second step, we model the standard deviation of the filtered returns as a power function of the lagged price level (which is a proxy for market scarcity) or, in logs,

$$\begin{aligned} \log V\left[\tilde{x}_{ht} \mid p_{h,t-1}\right] = \chi + \chi' d_{ht} + \rho \log p_{h,t-1} \\ + \rho'\left(\log p_{h,t-1}\right) d_{ht} + \in_{ht}, \end{aligned} \tag{23.2}$$

---

* A Toeplitz matrix is a matrix that has constant values along all negative-sloping diagonals.

TABLE 23.1 Summary Statistics of Log-Returns in the NordPool, APX, and Powernext Markets, along with Shapiro–Wilk Statistics and Autocorrelation Coefficients

Auction	Mean	Std. dev.	Skewness	Kurtosis	SW	acf(1)	acf(7)
			NordPool				
4 a.m.	−0.0000	0.1380	0.4668	28.2410	0.7167	−0.2715	0.0371
8 a.m.	−0.0000	0.2107	1.0293	19.9273	0.7829	−0.1508	0.5278
12 (noon)	−0.0000	0.1384	1.0630	20.5106	0.8204	−0.1233	0.4812
4 p.m.	−0.0000	0.1193	1.2205	14.4644	0.8291	−0.0384	0.5035
8 p.m.	0.0000	0.0936	0.8641	43.0378	0.7597	−0.1489	0.1823
12 (midnight)	0.0000	0.0623	−0.3141	13.1957	0.8584	−0.0684	0.1075
			APX				
4 a.m.	−0.0009	1.2534	0.2439	23.7528	0.5273	−0.4427	0.0204
8 a.m.	−0.0002	1.6831	0.3925	15.1495	0.6517	−0.3893	0.3665
12 (noon)	0.0003	0.6921	0.8037	6.5865	0.9355	−0.2268	0.5297
4 p.m.	−0.0004	0.6345	0.8864	9.8323	0.8842	−0.1865	0.4836
8 p.m.	−0.0003	0.3410	−0.1048	9.8369	0.8879	−0.3801	0.1472
12 (midnight)	−0.0002	0.4831	0.6416	169.4452	0.4380	−0.4664	0.0193
			Powernext				
4 a.m.	−0.0000	0.5119	0.1385	83.7746	0.6818	−0.3742	0.3586
8 a.m.	−0.0000	0.6526	0.7423	6.8751	0.9071	−0.2713	0.6393
12 (noon)	−0.0000	0.4424	1.0837	11.1308	0.8926	−0.2023	0.4539
4 p.m.	0.0000	0.4484	1.0351	9.3050	0.9049	−0.2118	0.5202
8 p.m.	−0.0000	0.2924	0.6311	6.7971	0.9283	−0.1523	0.3808
12 (midnight)	0.0000	0.2103	0.4558	13.4018	0.8779	−0.2977	0.1724

and rescale the filtered returns in order to obtain homoskedastic samples[*]:

$$x^*_{ht} = \frac{\tilde{x}_{ht}}{e^{\hat{X} + \hat{x}' d_{hr} + \hat{\rho}\log p_{t,t-1} + \hat{\rho}'(\log p_{t,t-1})d_{ht}}}. \quad (23.3)$$

Finally, the hourly averages are subtracted. These returns will be referred to as *Cholesky-filtered and rescaled* log-returns. In the above equations, $V[\cdot]$ is the variance operator, χ, χ', ρ and ρ' are constant coefficients (their estimated values are indicated with a hat), $\tilde{x}_{ht}$ is the Cholesky-filtered series of log-returns for the hour-h auction at day t, $p_{h,t-1}$ is the price for the hour-h auction at day $t-1$, and ε_{ht} is an i.i.d. error term. The dummy variable d_{ht} allows both the slope and the intercept of the scaling regression to vary as the price reaches particularly high levels. This accounts for the possibility that the price dynamics is characterized by switching regimes (see De Jong (2006) and Weron (2009) and references therein). In order to estimate the power-law scaling coefficients, the data of each time series are grouped into equi-population bins. Next, the sample standard deviations of the log-returns in each bin are computed, and the logarithm of the sample standard deviations is OLS-regressed on a constant and on the logarithm of the mean price level within the corresponding bins.[†]

---

[*] A scaling relationship between log-return variance and volume levels was also considered, but it is seldom significant.

[†] Estimation of the scaling coefficients was performed for numbers of bins between 8 and 40. One finds that R^2 values are decreasing in the number of bins, and that the point estimates of $\hat{\rho}$ tend to decrease slightly in absolute value. A decision was made to focus on scaling based on 40 bins (NordPool), 28 bins (APX), and 34 bins (Powernext), corresponding to between 52 and 55 observations per bin. Indeed, Monte Carlo simulations performed by the author show that the profile of the scaling exponent estimates, with respect to the number of bins, is characterized by a flat region around the mentioned values. A larger number of bins implies more degrees of freedom in the regression, but the volatility estimates within each bin are more noisy, because they are based on a smaller number of observations, resulting in smaller R^2 values. With respect to the use of the mean prices, choosing the median prices does not affect the results significantly.

The estimates of the variance–price scaling relationship for NordPool and Powernext suggest that the standard deviations of the filtered returns are negatively correlated with the lagged price levels, but the dummy coefficients are large and positive, implying that the variance–price relationship is increasing at high price levels. The APX scaling exponents, instead, are quite variable across hours. Ljung–Box tests performed using 28 lags (4 weeks) cannot reject the null of zero serial correlation, while normality tests still reject the null of a Gaussian distribution.* Even after filtering and rescaling, the log- returns still display skewness and excess kurtosis.

This filtering procedure departs from what Weron (2009) calls the 'industry standard', wherein the electricity price is envisaged as the sum or the product of a (deterministic) trend/cycle/seasonal component and a stochastic component. The goal of the deseasonalization techniques used by De Jong (2006) and Weron (2009) is to isolate the stochastic component, which is then used to compute the price returns. De Jong (2006) regresses the log-prices on daily dummies, an annual sinusoidal, and an exponentially weighted moving average, while Weron (2009) applies a wavelet smoothing technique to deal with the annual cycle, plus a moving-average filter to remove the average weekly pattern. A problem with these approaches is that they may not yield the desirable i.i.d. samples. The approach followed in this paper is a way to solve this problem: the outcome of the Cholesky filter is a serially uncorrelated time series by construction, while the variance–price power-law scaling takes care of the remaining heteroskedasticity.

23.3 Distributions of Electricity Returns

When characterizing the probability density of the electricity price returns, it is desirable to select classes of probability distributions that are general and flexible enough so as to yield different implications concerning the decay of the tails, the skewness, the convergence of the central moments and the time scaling of risk. In this work, we focus on the α-stable, Normal Inverse Gaussian, Exponential Power and Asymmetric Exponential Power distribution families. These have frequently been analysed in the relevant literature, as mentioned in the Introduction.

The first class of probability distributions is relevant if one views the electricity returns as resulting from the sum of n i.i.d. shocks u_j, not restricted to having finite moments, with $j=1, \dots, n$. The Generalized Central Limit Theorem states that the distribution of $\left(1/\sqrt{n}\right)\sum_{j=1}^{n} u_{jt}$ converges to an α-stable distribution as $n \to \infty$ (Samorodnitsky and Taqqu 1994, Borak *et al.* 2005). A random variable is α-stable if and only if its characteristic function reads

$$\phi(v) = e^{-\sigma^{\alpha}|v|^{\alpha}\left\{1+i\beta(\text{sign }v)\tan(\pi v/2)\left((\sigma|v|)^{1-\alpha}-1\right)\right\}+i\mu v} \tag{23.4}$$

if $\alpha \neq 1$, or

$$\phi(v) = e^{-\sigma|v|\{1+i\beta(\text{sign }v)(2/\pi)\log(\sigma|v|)\}+i\mu v} \tag{23.5}$$

if $\alpha = 1$. An α-stable distribution is defined by four parameters: a characteristic exponent or stability index $\alpha \in (0, 2]$, a skewness parameter $\beta \in [-1, 1]$, a scale parameter $\sigma > 0$, and a location parameter $\mu \in \Re$. The stable distribution corresponds to a Normal when $\alpha = 2$, whereas $\alpha < 2$ implies that the variance is infinite and the tails asymptotically decay as power laws. When $\alpha = 1$, the Cauchy distribution results, but if $\alpha < 1$ even the first central moment diverges.

---

* The summary statistics for the Cholesky-filtered and rescaled variables and detailed information on the variance–price scaling estimates are available upon request.

A second model assumes that the electricity returns are variance–mean mixtures of Gaussian random variables. If the mixing distribution is a generalized inverse Gaussian law with $\lambda = -1/2$, the Normal Inverse Gaussian (NIG) law obtains. The probability density function reads (Barndorff-Nielsen 1997)

$$f_{\mathrm{NIG}}(x;\alpha,\beta,\sigma,\mu)$$

$$= \frac{\alpha\sigma}{\pi} e^{\sigma\sqrt{\alpha^2-\beta^2}+\beta(x-\mu)} \frac{K_1\left(\alpha\sqrt{\sigma^2+(x-\mu)^2}\right)}{\sqrt{\sigma^2+(x-\mu)^2}}. \tag{23.6}$$

The parameters are α (steepness), β (skewness), $\alpha > 0$ (scale), and $\mu \in \mathfrak{R}$ (location). The constant K_1 is the modified Bessel function of the third kind with index 1, also known as the MacDonald function. The NIG exhibits semi-heavy tails, i.e. heavier than Gaussian, but lighter than power law. The Cauchy distribution is the special case $f_{\mathrm{NIG}}(x; 0, 0, 1, 0)$. The tails of the NIG distribution taper off according to the following asymptotic formula:

$$f_{\mathrm{NIG}}(x) \approx |x|^{-3/2} e^{(\mp\alpha+\beta)x}, \tag{23.7}$$

for $x \to \pm\infty$. Note that both the α-stable and the NIG distribution are closed to convolution. This feature is particularly useful in the time scaling of risk, e.g. in deriving long-term risk from daily risk (Weron 2004).

The third class of distributions obtains if one assumes $x \sim$ i.i.d. $N(0, h^\psi)$, where $h \sim$ i.i.d. Exponential. As shown by Fu *et al.* (2005), if $\psi \geq 0$, this model yields an Exponential Power distribution with shape parameter $b > 0$ (inversely related to), scale parameter a, and position parameter μ:

$$f_{\mathrm{EP}}(x;a,b,\mu) = \frac{1}{2ab^{1/b}\Gamma[1+(1/b)]} e^{-(1/b)|(x-\mu)/a|^b}, \tag{23.8}$$

where $\Gamma(\cdot)$ is the gamma function.[*] The EP distribution reduces to a Laplace if $b = 1$ and to a Normal if $b = 2$. As b becomes smaller, the density becomes heavier-tailed and more sharply peaked.[†]

The EP family only includes symmetric probability distributions. However, it may be desirable to allow for more flexibility in modeling the skewness, in order to yield a more punctual comparison with the α-stable and NIG laws. The EP family has been generalized in this direction by Bottazzi and Secchi (2007), who introduced the Asymmetric Exponential Power (AEP) family:

$$f_{\mathrm{AEP}}\left(x;a_l,a_r,b_l,b_r,\mu\right)$$

$$= \frac{1}{a_l A_0\left(b_l\right) + a_r A_0\left(b_r\right)} \tag{23.9}$$

$$\times e^{-\left((1/b_l)|(x-\mu)/a_l|^{b_l}\theta(\mu-x)+(1/b_r)|(x-\mu)/a_r|^{b_r}\theta(x-\mu)\right)},$$

where $\theta(y)$ (for a generic variable y) is the Heaviside theta function, and

$$A_k(y) = y^{[(k+1)/y]-1}\Gamma\left(\frac{k+1}{y}\right).$$

---

[*] This distribution was first used in economics by Bottazzi and Secchi (2003), and is also known as the Subbotin distribution (Subbotin 1923).

[†] West (1987) represented the Exponential Power distribution as a scale mixture of Normals with an α-stable mixing distribution whose stability index is equal to $b/2$, but his result is limited to $b \geq 1$. See also Andrews and Mellows (1974) and Choy and Walker (2003).

The AEP density is characterized by two positive shape parameters (b_l, b_r), two positive scale parameters (a_l, a_r), and one position parameter (μ). The magnitudes of the shape parameters tune the behavior of the upper and lower tail, respectively. The AEP reduces to the EP distribution when $a_l = a_r$ and $b_l = b_r$. Unlike the α-stable and NIG distributions, the EP and AEP distributions are not closed under convolution. Hence, the distribution of returns computed over longer time horizons tends to converge to the Gaussian law.

The parameters of the stable distribution are estimated here by means of the characteristic function regression method (Koutrouvelis 1980, Kogon and Williams 1998). This method was shown by Weron (2004) and Scalas and Kim (2007) to be more accurate than alternative methods. We use Maximum Likelihood to estimate the parameters of the NIG, EP, and AEP distributions. The MFE Toolbox (Weron 2006) is exploited for the estimation of stable and NIG laws, whereas the EP and AEP distributions were fitted by making use of the Subbotools package (see also Bottazzi 2004 and Bottazzi and Secchi 2007).*

The goodness of fit of the estimated distribution models is assessed by means of the Kolmogorov–Smirnov and Cramer–von Mises statistics (D'Agostino and Stephens 1986). The Kolmogorov–Smirnov D statistic is defined as the maximum absolute deviation between the theoretical and empirical CDFs:

$$D = \max\left(\left|\frac{i}{N} - z_i\right|\right), \tag{23.10}$$

where z_i is the ith ordinate of the theoretical cumulative distribution function under test, and N the sample size. The Cramer–von Mises W^2 test statistics is based on the quadratic deviations between theoretical and empirical CDFs:

$$W^2 = \frac{1}{12N} + \sum_{i=1}^{N}\left[z_i - \frac{2i-1}{2N}\right]^2. \tag{23.11}$$

The asymptotic 5% limiting value of D is $1.36/\sqrt{N}$, that is 0.0291 (NordPool), 0.0356 (APX) and 0.0318 (Powernext). These are the Monte Carlo asymptotic values, under the assumption that sample sizes are large enough as to rule out the need for distribution-specific small sample corrections. The 5% limiting value for W^2 is 0.443—an exact result valid for any sample size greater than or equal to 5 (Stephens 1974).

23.4 Fitting the Empirical Probability Densities

The estimation results for the α-stable, NIG, EP and AEP distributions are reported in Tables 23.2–23.4 for selected hourly auctions in NordPool, APX, and Powernext, respectively, along with goodness-of-fit statistics.

The estimated stability index α for the stable distribution is always below the Normal value (i.e. 2). NordPool and Powernext point estimates of α lie within the range 1.70–1.85; APX estimates are slightly lower, more so in the 4 a.m. and 8 a.m. auctions: therefore, the tails of APX log-returns decay more slowly. The skewness parameter β in APX is positive for all hours, whereas $\beta < 0$ for all NordPool and Powernext hourly auctions. These patterns are confirmed, at least qualitatively, by the estimated NIG parameters: the log-returns in all markets display heavy tails, the steepness parameter is lower in the APX market, and the distribution of Powernext log-returns is characterized by a negative skew in all hours. This time the parameter β in the NordPool assumes negative values only for some hours (4 a.m., midnight), and is below zero in some APX auctions (4 a.m., 8 a.m.).

As to the EP distribution, the point estimates of the shape parameter are systematically below the Normal value, a sign of heavy tails. The shape coefficients are often around the Laplace value (namely 1), with some deviations (most frequently above 1 in the Powernext and below 1 in the APX

---

* Subbotools is available at http://cafim.sssup.it/~giulio/software/subbotools/.

TABLE 23.2 Parameter Estimates and Goodness-of-Fit Statistics for α-Stable, NIG, EP, and AEP Distributions Fitted to the Filtered and Rescaled Log-Returns: NordPool

Auction	Distribution	Tail	Parameter			Test value	
			Skewness	Scale	Location	D	W^2
4 a.m.	α-Stable	1.7750	−0.1536	0.5910	0.0263	**0.0133**	0.1908
	NIG	1.0299	−0.1707	1.0252	0.1723	0.0185	**0.1461**
	EP	1.1238		0.7803	0.0000	0.0380	0.6938
	AEP	1.0261, 1.4123		0.8033, 0.7961	0.0000	0.0300	0.3132
8 a.m.	α-Stable	1.6997	−0.0989	0.6101	−0.0177	**0.0135**	0.1066
	NIG	0.7107	0.0030	0.9164	−0.0039	0.0144	**0.0655**
	EP	0.9979		0.7992	0.0000	0.0211	0.2575
	AEP	1.0434, 0.9763		0.8014, 0.8047	0.0000	0.0249	0.2751
12 (noon)	α-Stable	1.7494	−0.0027	0.6447	−0.0162	**0.0112**	0.0991
	NIG	0.8139	0.0410	1.0472	−0.0528	0.0118	**0.0479**
	EP	1.0845		0.8471	0.0000	0.0263	0.2472
	AEP	1.1774, 1.0393		0.8512, 0.8562	0.0000	0.0220	0.2098
4 p.m.	α-Stable	1.7553	−0.1158	0.6350	−0.0213	0.0133	0.1711
	NIG	0.8327	0.0284	1.0324	−0.0352	**0.0116**	**0.0470**
	EP	1.0872		0.8324	0.0000	0.0256	0.2483
	AEP	1.1647, 1.0458		0.8350, 0.8391	0.0000	0.0250	0.2214
8 p.m.	α-Stable	1.7608	−0.1242	0.6446	−0.0285	**0.0110**	0.1276
	NIG	0.8325	0.0168	1.0559	−0.0213	0.0141	**0.0583**
	EP	1.0831		0.8411	0.0000	0.0235	0.2328
	AEP	1.1424, 1.0513		0.8432, 0.8467	0.0000	0.0234	0.2424
12 (midnight)	α-Stable	1.7946	−0.3804	0.6329	−0.0135	0.0202	0.2138
	NIG	1.0763	−0.1499	1.1719	0.1648	**0.0120**	**0.0264**
	EP	1.2292		0.8454	0.0000	0.0266	0.3865
	AEP	1.1209, 1.3770		0.8501, 0.8458	0.0000	0.0118	0.0487

Tail: α (stable and NIG), b (EP), b_l and b_r (AEP). Skewness: β (stable and NIG). Scale: σ (stable and NIG), a (EP), a_l and a_r (AEP). Location: μ. Asymptotic 5% limiting values: 0.0291 (D), 0.443 (W^2). Data in bold type indicate the minimum goodness-of-fit test values.

auctions). The AEP results show that the left tails are heavier in the NordPool and APX night auctions, and lighter during the day ($b_l < b_r$ by night, the opposite in the other hours). Consistent with the α-stable and NIG estimates of the skewness parameter, the Powernext left tails are longer in all auctions. Note that the point estimates of a_r and a_l are very similar to each other in all markets; as an implication, the skewness in log-returns is mainly due to tail asymmetries, not so much to asymmetries in scale.

The reported goodness-of-fit criteria are evidence in support of the α-stable and NIG models, which outperform the EP and AEP distributions. Both the α-stable and NIG laws provide excellent fits, but neither clearly prevails over the other, as both have some success in a certain number of hourly auctions. In fact, in some hours the NIG fits better according to one goodness-of-fit criterion, while the α-stable fits better according to the other criterion. Note further that the (symmetric) EP distribution virtually always provides the worst fit, presumably because of its inability to capture the skewness. Finally, although the AEP is outperformed by the stable and NIG laws, the goodness-of-fit statistics for the AEP distribution are often below the 5% critical values. The fitting performances of the α-stable, NIG, EP and AEP distributions in a selected hourly auction (4 p.m.) can be seen in Figure 23.3.

TABLE 23.3 Parameter Estimates and Goodness-of-Fit Statistics for α-Stable, NIG, EP, and AEP Distributions Fitted to the Filtered and Rescaled Log-Returns: APX

Auction	Distribution	Parameter				Test value	
		Tail	Skewness	Scale	Location	D	W^2
4 a.m.	α-Stable	1.5158	0.3067	0.5181	0.2085	**0.0219**	**0.0355**
	NIG	0.4264	−0.0869	0.6303	0.1312	0.0348	0.3584
	EP	0.7562		0.6841	0.0000	0.0659	1.9361
	AEP	0.6904, 1.0266		0.7324, 0.7045	0.0000	0.0377	0.4536
8 a.m.	α-Stable	1.3876	0.3903	0.5054	0.3332	**0.0341**	**0.0750**
	NIG	0.3093	−0.0649	0.5851	0.1255	0.0355	0.3809
	EP	0.6893		0.6893	0.0000	0.0735	2.4395
	AEP	0.6231, 0.9073		0.7363, 0.7028	0.0000	0.0406	0.3575
12 (noon)	α-Stable	1.7749	0.6614	0.6317	0.0215	0.0215	0.2227
	NIG	1.1050	0.2664	1.1675	−0.2900	**0.0132**	**0.0434**
	EP	1.2244		0.8559	0.0000	0.0480	0.6441
	AEP	1.5558, 1.0607		0.8654, 0.8717	0.0000	0.0149	0.0565
4 p.m.	α-Stable	1.6632	0.1908	0.5824	−0.0119	0.0202	0.1229
	NIG	0.7513	0.1063	0.8877	−0.1269	**0.0186**	**0.0812**
	EP	1.0219		0.7840	0.0000	0.0429	0.4586
	AEP	1.1819, 0.9421		0.7880, 0.7993	0.0000	0.0213	0.1735
8 p.m.	α-Stable	1.6520	0.1309	0.5986	−0.0158	**0.0161**	0.4738
	NIG	0.7178	0.0784	0.9181	−0.1010	0.0204	**0.0898**
	EP	1.0484		0.8218	0.0000	0.0412	0.3888
	AEP	1.1942, 0.9730		0.8256, 0.8353	0.0000	0.0269	0.2098
12 (midnight)	α-Stable	1.6831	0.1040	0.5425	0.0156	0.0258	**0.0346**
	NIG	0.5396	0.0012	0.6859	−0.0015	**0.0216**	0.1137
	EP	0.8146		0.6737	0.0000	0.0277	0.3375
	AEP	0.7958, 0.8584		0.6795, 0.6768	0.0000	0.0333	0.3597

Tail: α (stable and NIG), b (EP), b_l and b_r (AEP). Skewness: β (stable and NIG). Scale: α (stable and NIG), a (EP), a_l and a_r (AEP). Location: μ. Asymptotic 5% limiting values: 0.0356 (D), 0.443 (W^2). Data in bold type indicate the minimum goodness-of-fit test values.

23.5 Robustness

Unlike stock prices and exchange rates, the electricity price process cannot be approximated by a geometric random walk. Indeed, there is no widespread support in favor of either multiplicative or additive representations of electricity price processes. It is also quite clear that the price dynamics is driven by multiple seasonal factors and is possibly subject to regime shifts. As major implications, there is no unambiguous way to define price returns, neither can one determine beforehand what method is most appropriate to deseasonalize the data. Getting the tails and skewness 'right', a major issue in risk management, is not easy, as the estimated distributional parameters may change depending on the definition of returns and on the filtering methodology. These issues are addressed in this section, where the robustness of the results is assessed along two lines: (i) changing the definition of returns while keeping the same deseasonalization method as before (Cholesky filter plus volatility– price rescaling); and (ii) keep working on log-returns while changing the deseasonalization method.

TABLE 23.4 Parameter Estimates and Goodness-of-Fit Statistics for α-Stable, NIG, EP, and AEP Distributions Fitted to the Filtered and Rescaled Log-Returns: Powernext

Auction	Distribution	Parameter				Test value	
		Tail	Skewness	Scale	Location	D	W^2
4 a.m.	α-Stable	1.8312	−0.4342	0.6038	0.0082	**0.0120**	**0.0510**
	NIG	1.2566	−0.2787	1.1539	0.2624	0.0181	0.0828
	EP	1.1797		0.7905	0.0000	0.0480	0.7975
	AEP	1.0566, 1.6008		0.8201, 0.8150	0.0000	0.0288	0.2274
8 a.m.	α-Stable	1.7688	−0.6255	0.5946	−0.0235	0.0137	0.0577
	NIG	1.0771	−0.2594	1.0354	0.2569	**0.0108**	**0.0224**
	EP	1.1690		0.7963	0.0000	0.0490	0.9730
	AEP	1.0135, 1.4289		0.8072, 0.7973	0.0000	0.0229	0.1169
12 (noon)	α-Stable	1.7186	−0.2653	0.5860	−0.0295	**0.0123**	0.2397
	NIG	0.8389	−0.0525	0.9327	0.0585	0.0128	**0.0568**
	EP	1.0801		0.7820	0.0000	0.0300	0.2743
	AEP	1.0479, 1.1127		0.7836, 0.7808	0.0000	0.0238	0.2124
4 p.m.	α-Stable	1.7744	−0.5740	0.6182	−0.0558	0.0140	0.3900
	NIG	1.0057	−0.1229	1.0977	0.1351	**0.0117**	**0.0513**
	EP	1.2018		0.8338	0.0000	0.0354	0.4203
	AEP	1.1266, 1.2595		0.8331, 0.8281	0.0000	0.0261	0.2295
8 p.m.	α-Stable	1.7473	−0.2874	0.6255	−0.0241	0.0141	0.1306
	NIG	0.8779	−0.0660	1.0430	0.0786	**0.0098**	**0.0255**
	EP	1.1305		0.8325	0.0000	0.0263	0.2063
	AEP	1.0786, 1.1721		0.8319, 0.8280	0.0000	0.0185	0.1200
12 (midnight)	α-Stable	1.8102	−0.6648	0.6420	−0.0294	**0.0122**	0.1364
	NIG	1.1885	−0.2262	1.2658	0.2454	0.0138	**0.0673**
	EP	1.3072		0.8767	0.0000	0.0348	0.5336
	AEP	1.1684, 1.5158		0.8817, 0.8790	0.0000	0.0175	0.1478

Tail: α (stable and NIG), b (EP), b_l and b_r (AEP). *Skewness:* β (stable and NIG). *Scale:* σ (stable and NIG), a (EP), a_l and a_r (AEP). *Location:* μ. Asymptotic 5% limiting values: 0.0318 (D), 0.443 (W^2). Data in bold type indicate the minimum goodness-of-fit test values.

There are convincing reasons to expect that the shape of the returns distribution may be sensitive to the very definition of returns. Log-returns are usually considered as approximations of the percentage returns, defined as $(p_t - p_{t-1})/p_{t-1}$. The difference between a log-return x_t and a percentage return is of the order of $\frac{1}{2}x_t^2 + \frac{1}{6}x_t^3 + \cdots$ (Eberlein and Keller 1995). While this difference is negligible in financial markets, the approximation may be quite poor in power exchanges, due to the extremely large magnitudes of electricity price fluctuations. A second problem is that the logarithmic transformation dampens the extreme returns and makes the distribution of returns more symmetric, thereby affecting the estimated shape and skewness parameters. Empirically, this effect has been verified by Weron (2009), who finds that the distribution of price changes $p_t - p_{t-1}$ displays heavier tails than the distribution of log-returns. For the above reasons, the robustness of the foregoing results needs to be checked using alternative definitions of price returns.

Table 23.5 reports the estimation results when the α-stable, NIG, EP and AEP models are fitted on the empirical probability densities of percentage returns and price changes, Cholesky-filtered and rescaled as in section 23.2. The first result is that, in most hours, the distributions of price changes and percentage returns have heavier tails than the distributions of logarithmic returns: indeed, the point estimates of the shape parameters are lower. There are some exceptions concerning APX and Powernext: in those markets, the percentage returns are less fat-tailed than the log-returns from 4 a.m. to 4 p.m., and the

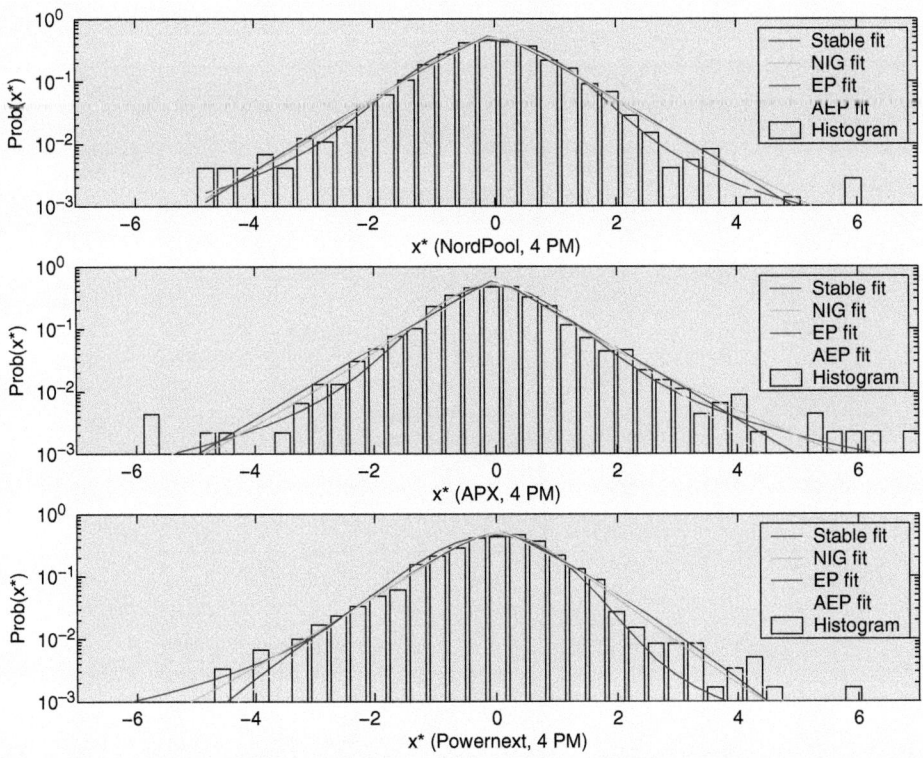

FIGURE 23.3   Density fit of filtered and rescaled log-returns: NordPool, APX, and Powernext 4 p.m. auctions.

tails of the price changes decay faster during the night. Second, the skewness patterns observed for log-returns are not robust when one considers percentage returns and price changes. For instance, the NordPool and Powernext price change distributions are closer to symmetric, but there is a negative skew in the APX (APX log-returns, on the contrary, display a positive skew). Less clear are the results for percentage returns, with alternating signs, although the magnitudes of the skewness parameters tend to be mild. Different distributions give contrasting results: the Powernext AEP left shape parameter b_l is now often slightly larger than the right shape parameter b_r, hinting at a positive skew, but α-stable skewness parameters for the same market are often negative. These findings are only partly in line with the evidence of Weron (2009), and show that analysing returns of individual hours can shed light on intricate intra-day patterns. Finally, the comparative performance of the α-stable law improves for the percentage returns and for the NordPool price changes (i.e. it is the best-fitting distribution in a greater number of hourly auctions), while the NIG prevails as a description of APX and Powernext price changes. The AEP law is now the best-fitting distribution in two instances (APX 4 a.m. percentage returns, Powernext 4 a.m. price changes).

The results presented in section 23.4 may also be sensitive to the filtering methodology. On the one hand, if power-law volatility–price scaling is an effective way of controlling for heteroskedasticity, the time series of log-returns filtered using only the Cholesky algorithm should be the superposition of underlying homoskedastic (and possibly non-heavy tailed) series. It should therefore display longer tails. One can therefore assess the impact of heteroskedasticity on the distributional shapes by fitting the empirical density functions of the log-returns after applying only the Cholesky filter (i.e. $\tilde{x}$), without considering volatility–price scaling. On the other hand, the Cholesky filter is built upon the sample autocovariance function, hence it may suffer from the same limitations as Fourier- based filters if the time series is non-stationary. The outcome of applying a filter that is less suited to deal with non-stationarities

TABLE 23.5 Parameter Estimates and Goodness-of-Fit Statistics for α-Stable, NIG, EP, and AEP Distributions. *Percentage Returns:* Cholesky-Filtered and Rescaled Percentage Returns. *Price Changes:* Cholesky-Filtered and Rescaled Price Changes

Auction	Distribution	NordPool				APX				Powernext			
		Tail	Skewness	D	W²	Tail	Skewness	D	W²	Tail	Skewness	D	W²
						Percentage returns							
4 a.m.	α-Stable	1.7601	0.0357	**0.0082**	**0.0207**	1.3853	−0.7461	0.0477	**0.5850**	1.6046	−0.5874	**0.0229**	**0.1364**
	NIG	0.9289	0.0098	0.0119	0.0526	0.1848	0.0043	0.0473	1.0026	0.5143	0.0694	0.0306	0.2866
	EP	1.1555		0.0184	0.1873	0.5968		0.0535	1.2675	0.7526		0.0454	0.9482
	AEP	1.1667, 1.1465		0.0178	0.1859	0.8469, 1.7175		**0.0469**	0.6978	0.7423, 0.7555		0.0486	0.9994
8 a.m.	α-Stable	1.6227	−0.0139	0.0176	0.1590	1.2095	−0.1256	**0.0221**	**0.1497**	1.6464	0.1672	0.0209	**0.1280**
	NIG	0.5102	0.0523	**0.0166**	**0.0713**	0.0425	0.0122	0.0388	0.6097	0.5762	0.0147	**0.0199**	0.2110
	EP	0.8235		0.0411	0.7424	0.4581	0.0000	0.1233	6.8636	0.8996		0.0349	0.5283
	AEP	0.9461, 0.7715		0.0296	0.2735	0.7217, 1.0243		0.0431	0.8278	0.9089, 0.8938		0.0326	0.5096
12 (noon)	α-Stable	1.7699	0.1106	**0.0147**	**0.0624**	1.1537	0.2139	0.0171	0.0672	1.4329	−0.1575	**0.0139**	**0.0766**
	NIG	0.7739	0.1199	0.0191	0.1656	0.1066	0.0229	**0.0159**	**0.0535**	0.4352	−0.0256	0.0218	0.1382
	EP	0.9735		0.0408	0.9430	0.5422		0.0894	2.5538	0.7390		0.0329	0.4407
	AEP	1.2500, 0.9019		0.0283	0.4420	0.5328, 0.5818		0.0384	0.5274	0.7288, 0.7464		0.0314	0.4142
4 p.m.	α-Stable	1.7506	0.1628	**0.0139**	0.0886	1.2728	0.1556	**0.0162**	**0.0568**	1.5934	−0.1482	**0.0127**	**0.0651**
	NIG	0.7985	0.1359	0.0143	**0.0403**	0.1701	0.0462	0.0235	0.1966	0.4663	0.0021	0.0199	0.1148
	EP	1.0205		0.0430	0.9258	0.5650		0.0996	3.7912	0.7762		0.0286	0.4574
	AEP	1.2837, 0.9251		0.0228	0.2031	0.6743, 0.5689		0.0436	0.7330	0.8105, 0.7597		0.0330	0.4611
8 p.m.	α-Stable	1.7344	0.0949	**0.0120**	0.0909	1.6523	0.3390	**0.0185**	**0.0585**	1.8009	0.2397	**0.0113**	**0.0235**
	NIG	0.7287	0.0928	0.0137	**0.0728**	0.7538	0.2542	0.0274	0.1576	0.9521	0.1711	0.0165	0.0958
	EP	1.0007		0.0376	0.5930	0.9065		0.0781	2.4277	1.0947		0.0385	0.6816
	AEP	1.1944, 0.9284		0.0232	0.2478	1.6815, 0.8225		0.0299	0.2811	1.3845, 0.9950		0.0255	0.2554

(Continued)

TABLE 23.5 (CONTINUED) Parameter Estimates and Goodness-of-Fit Statistics for α-Stable, NIG, EP, and AEP Distributions. *Percentage Returns*: Cholesky-Filtered and Rescaled Percentage Returns. *Price Changes*: Cholesky-Filtered and Rescaled Price Changes

Auction	Distribution	NordPool				APX				Powernext			
		Tail	Skewness	D	W²	Tail	Skewness	D	W²	Tail	Skewness	D	W²
12 (midnight)	α-Stable	1.8025	−0.2344	0.0157	0.1124	1.7001	−1.0000	0.1330	6.3918	1.8899	−0.2964	0.0149	0.0334
	NIG	1.0913	−0.0772	**0.0128**	**0.0392**	0.4942	0.1024	**0.0221**	**0.1155**	1.4407	−0.0210	**0.0113**	**0.0285**
	EP	1.2521		0.0245	0.1600	0.4621		0.4084	99.4721	1.4470		0.0155	0.0711
	AEP	1.1997, 1.3100		0.0197	0.0674	0.6949, 3.3029		0.1279	3.5804	1.4595, 1.4376		0.0168	0.0793
4 a.m.	α-Stable	1.6971	−0.2596	0.0183	0.1204	1.7924	−0.3495	0.0162	0.0487	1.8861	0.7159	0.0172	0.1173
	NIG	0.7944	−0.0678	**0.0132**	**0.0872**	1.0989	−0.1570	**0.0117**	**0.0263**	1.6595	0.2079	0.0152	0.0807
	EP	0.6237		0.0322	0.4607	1.2815		0.0266	0.2532	1.4724		0.0162	0.1359
	AEP	0.9799, 1.0960		0.0221	0.3037	1.0236, 1.1981		0.0153	0.0616	1.6274, 1.3686		**0.0190**	**0.0680**
8 a.m.	α-Stable	1.6136	−0.0794	**0.0124**	**0.0397**	1.7186	−0.2948	0.0374	**0.1449**	1.7706	−0.2979	0.0314	0.1547
	NIG	0.4924	0.0186	0.0208	0.2354	0.7468	−0.0124	**0.0224**	0.1915	0.9708	−0.0093	**0.0110**	**0.0302**
	EP	0.7959		0.0375	0.8018	0.9950		0.0352	0.4555	1.1434		0.0180	0.1221
	AEP	0.8523, 0.7687		0.0379	0.6920	0.1031, 0.3799		0.0417	0.4929	1.1850, 1.1193		0.0209	0.1504
12 (noon)	α-Stable	1.7239	0.1265	**0.0167**	**0.0891**	1.3951	−0.0885	0.0368	**0.1598**	1.5186	−0.0255	**0.0324**	**0.0892**
	NIG	0.6638	0.0758	0.0191	0.1708	0.3059	0.0743	**0.0239**	0.1747	0.4351	0.0440	**0.0231**	0.1789
	EP	0.8814		0.0469	0.9318	0.6739		0.0781	2.6601	0.7208		0.0494	1.0317
	AEP	1.0399, 0.8474		0.0352	0.5409	0.4287, 0.1061		0.0289	0.2943	0.8093, 0.6770		0.0315	0.4685
4 p.m.	α-Stable	1.7237	0.1361	0.0278	**0.1118**	1.3386	−0.1266	0.0307	0.2424	1.5252	−0.1906	0.0453	0.8678
	NIG	0.7389	0.1173	**0.0169**	0.1295	0.2013	0.0478	**0.0300**	**0.1919**	0.3569	0.0079	**0.0311**	**0.2986**
	EP	0.2201		0.0464	1.0360	0.5801		0.0953	3.3317	0.7068		0.0392	0.7774
	AEP	1.1818, 0.8735		0.0320	0.3599	0.4019, 0.1754		0.0336	0.4289	0.7435, 0.6884		0.0366	0.6522

Price changes

(Continued)

TABLE 23.5 (CONTINUED) Parameter Estimates and Goodness-of-Fit Statistics for α-Stable, NIG, EP, and AEP Distributions. *Percentage Returns*: Cholesky-Filtered and Rescaled Percentage Returns. *Price Changes*: Cholesky-Filtered and Rescaled Price Changes

Auction	Distribution	NordPool				APX				Powernext			
		Tail	Skewness	D	W²	Tail	Skewness	D	W²	Tail	Skewness	D	W²
8 p.m.	α-Stable	1.7509	0.0752	0.0320	**0.1361**	1.1560	−0.2662	**0.0176**	**0.0514**	1.7136	−0.0556	**0.0143**	**0.0701**
	NIG	0.6660	0.0400	**0.0189**	0.1407	0.1504	0.0283	0.0199	0.1369	0.7078	0.0730	0.0201	0.1237
	EP	0.4473		0.0326	0.4684	0.5909		0.0789	2.2140	0.9454		0.0420	0.6012
	AEP	0.9859, 0.8856		0.0234	0.3973	0.9745, 0.8882		0.0308	0.2658	1.0777, 0.8835		0.0272	0.3693
12 (midnight)	α-Stable	1.7473	−0.1336	0.0299	0.1708	1.6631	0.2094	0.0398	0.1232	1.8196	−0.1083	0.0310	0.1257
	NIG	0.9149	−0.0403	**0.0087**	**0.0261**	0.8023	0.2030	**0.0187**	**0.1034**	1.1197	0.0230	**0.0114**	**0.0388**
	EP	0.4005		0.0192	0.1206	0.9733		0.0596	1.2475	1.2703		0.0172	0.0776
	AEP	1.1200, 1.1702		0.0161	0.0817	0.0764, 0.2028		0.0240	0.2088	1.3464, 1.2336		0.0181	0.1131

Tail: α (stable and NIG), b (EP), b_l and b_r (AEP). *Skewness:* β (stable and NIG). Asymptotic 5% limiting values: 0.0291 (D, NordPool), 0.0356 (D, APX), 0.0318 (D, Powernext), 0.443 (W²). Data in bold type indicate the minimum goodness-of-fit test values.

would probably be a mixture of random variables, leading to spurious estimates of the asymmetry and tail parameters. Wavelet filters should be immune to these problems. Following Weron (2006, 2009), one approximates the long-term seasonal component of the time series of electricity prices by means of an S8 approximation, based on a Daubechies-20 wavelet filter; then the long-term seasonal component is subtracted from the time series and the weekly pattern is removed using a moving-average filter. The minimum of the resulting time series is aligned with the minimum of the original price series, and then the log-returns are computed, ready for the α-stable, NIG, EP, and AEP fit.

The estimates in Table 23.6 refer to Cholesky-filtered and wavelet-filtered log-returns. These estimates show that, compared with the distributions of Cholesky-filtered and rescaled log-returns, the tails are at least as fat. One noteworthy exception is the NIG distribution, whose shape parameter estimates are unusually high. The skewness estimates in the case of Cholesky-filtered log- returns are basically confirmed for APX (positive) and Powernext (negative), although the magnitudes are milder; NordPool log-returns are now approximately symmetric. For the wavelet-filtered log-returns, the skewness is now positive in the NordPool (it was often negative for Cholesky-filtered and rescaled log-returns), mildly negative in APX (it was positive), and the negative sign in the Powernext is confirmed in most hours, but the magnitudes are smaller. Finally, the α-stable law now displays an improved fitting performance and dominates the other distributions in a greater number of hours. The AEP provides the best fit twice (4 a.m. and 8 p.m.) in the Powernext market.

23.6 Conclusion

This paper contributes to the characterization of the probability density of the price returns in some European day-ahead electricity markets (NordPool, APX, Powernext) by fitting flexible and general families of distributions, such as the α-stable, Normal Inverse Gaussian (NIG), Exponential Power (EP), and Asymmetric Exponential Power (AEP), and comparing their goodness of fit. One finds that the probability of observing extremely large (positive or negative) returns is larger than for the Gaussian, confirming a very robust finding in the literature. The goodness-of-fit tests suggest that the α-stable and the NIG systematically outperform the EP and AEP models, although no clear ranking can be established between the two 'winners'. The evidence of heavy tails is robust to changing the definition of returns (from log-returns to percentage returns and price changes) and the deseasonalization methodology. Yet, the point estimates differ across cases; in particular, the logarithmic transform and volatility rescaling tend to dampen the extreme returns. Both the skewness and the tail behavior vary across markets, and one also observes interesting intra-daily patterns which were not detected in previous studies that focused on average daily returns.

The evidence of heavy tails is in accordance with the intuition behind the regime-switching and jump-diffusion models. If the returns can be represented as scale–location mixtures—as the good performance of the NIG distribution seems to suggest—the market is characterized by low volatility most of the time, but instances of extreme volatility are not negligibly rare. This is very much consistent with the idea that power exchanges undergo transitions between quiet and turbulent price regimes. The better fitting performance of the α-stable and NIG laws compared with the EP and AEP models suggests that there is a time scaling between daily and longer-term market risk: both the α-stable and the NIG distributions are closed under convolution, therefore their shapes are preserved under time aggregation. Moreover, to the extent that the α-stable distribution outperforms the NIG, the results signal the non-convergence of the second moment. This is bad news vis-a` -vis the use of price volatility as a measure of price risk in power exchanges. The results also suggest that the skewness and the kurtosis are at least as important for risk management as volatility, in line with the theoretical results obtained by Bessembinder and Lemmon (2002) concerning the pricing of power forwards. Still, in interpreting the results one has to acknowledge the relative scarcity of data for the tails of electricity returns distributions, compared with the wealth of high-frequency financial data. This poses estimation and goodness-of-fit problems, as pointed out by Weron *et al.* (2004). Such problems are going to persist unless we

TABLE 23.6 Parameter Estimates and Goodness-of-Fit Statistics for α-Stable, NIG, EP, and AEP Distributions. *Cholesky-Filtered*: Cholesky-Filtered Log-Returns. *Wavelet-Filtered*: Wavelet-Filtered Log-Returns

Auction	Distribution	NordPool				APX				Powernext			
		Tail	Skewness	D	W²	Tail	Skewness	D	W²	Tail	Skewness	D	W²
						Cholesky-filtered							
4 a.m.	α-Stable	1.5964	0.0079	**0.0167**	**0.0860**	1.5781	0.2587	**0.0284**	**0.2567**	1.8050	−0.4188	**0.0125**	**0.0435**
	NIG	0.7127	−0.0578	0.0230	0.2309	0.7232	−0.1685	0.0396	0.3625	1.1979	−0.2509	0.0164	0.0904
	EP	0.9215		0.0355	0.7323	0.8370		0.0666	1.9110	1.1230		0.0491	0.9247
	AEP	0.8728, 0.9904		0.0304	0.5458	0.7553, 1.2110		0.0445	0.5167	1.0020, 1.4278		0.0259	0.2578
8 a.m.	α-Stable	1.6432	0.0686	0.0147	**0.0549**	1.5114	0.3584	0.0355	0.4784	1.7001	−0.4300	**0.0119**	**0.0496**
	NIG	0.7614	0.0052	0.0133	0.0575	0.5966	−0.1339	**0.0348**	**0.3671**	0.9606	−0.2285	0.0136	0.0538
	EP	0.9515		0.0213	0.2579	0.7781		0.0697	2.1414	1.0416		0.0483	1.1732
	AEP	0.9602, 0.9422		0.0232	0.2574	0.6994, 1.1055		0.0399	0.4188	0.9146, 1.3346		0.0190	0.1437
12 (noon)	α-Stable	1.7258	0.0659	**0.0103**	**0.0351**	1.7843	0.6160	0.0194	0.0802	1.6686	−0.1333	**0.0101**	**0.0229**
	NIG	0.9168	0.0559	0.0129	0.0577	1.2039	0.2783	**0.0137**	**0.0418**	0.8066	−0.0267	0.0120	0.0532
	EP	1.0639		0.0268	0.2820	1.2328		0.0466	0.5844	1.0165		0.0257	0.2312
	AEP	1.1297, 1.0181		0.0187	0.2149	1.5485, 1.0719		0.0150	0.0544	1.0065, 1.0216		0.0242	0.2242
4 p.m.	α-Stable	1.6814	0.0106	**0.0107**	**0.0363**	1.6672	0.1706	0.0191	**0.0519**	1.6853	−0.3301	**0.0111**	**0.0240**
	NIG	0.8287	0.0436	0.0153	0.1184	0.8550	0.1176	**0.0184**	0.0775	0.8761	−0.1011	0.0126	0.0542
	EP	1.0130		0.0272	0.4278	1.0258		0.0411	0.4397	1.0752		0.0352	0.4411
	AEP	1.0499, 0.9746		0.0221	0.3581	1.1896, 0.9489		0.0216	0.1692	1.0056, 1.1325		0.0238	0.2382
8 p.m.	α-Stable	1.6941	0.0087	**0.0091**	**0.0186**	1.6988	0.1062	**0.0148**	**0.0604**	1.6705	−0.0576	**0.0118**	0.0530
	NIG	0.8317	0.0280	0.0150	0.1060	0.9226	0.0907	0.0193	0.0950	0.8416	−0.0813	0.0147	**0.0516**
	EP	0.9868		0.0260	0.3963	1.1021		0.0393	0.3406	1.0505		0.0285	0.3461
	AEP	1.0166, 0.9593		0.0251	0.3643	1.2435, 1.0287		0.0256	0.2134	0.9805, 1.1437		0.0233	0.1708

(*Continued*)

TABLE 23.6 (CONTINUED) Parameter Estimates and Goodness-of-Fit Statistics for α-Stable, NIG, EP, and AEP Distributions. *Cholesky-Filtered*: Cholesky-Filtered Log-Returns. *Wavelet-Filtered*: Wavelet- Filtered Log-Returns.

Auction	Distribution	NordPool				APX				Powernext			
		Tail	Skewness	D	W²	Tail	Skewness	D	W²	Tail	Skewness	D	W²
12 (midnight)	α-Stable	1.6232	−0.0442	**0.0099**	**0.0285**	1.7510	0.2557	**0.0232**	**0.0718**	1.7033	−0.3023	**0.0126**	**0.0531**
	NIG	0.7315	−0.0469	0.0133	0.0627	0.9172	0.0250	**0.0227**	0.1257	0.9383	−0.1440	0.0160	0.0831
	EP	0.9572		0.0296	0.3633	0.8640		0.0305	0.3945	1.1021		0.0408	0.5833
	AEP	0.9181, 1.0044		0.0208	0.2555	0.8475, 0.9016		0.0357	0.4419	1.0050, 1.2632		0.0206	0.2059
							Wavelet-filtered							
4 a.m.	α-Stable	1.3899	0.1277	**0.0193**	**0.0740**	1.6652	−0.1434	**0.0124**	**0.0224**	1.6328	−0.0079	0.0253	0.1164
	NIG	9.2782	−0.0463	0.0201	0.0904	3.3436	−0.1840	0.0272	0.1785	4.9146	−0.0844	0.0152	**0.0249**
	EP	0.7769		0.0312	0.2811	0.6982		0.0498	0.5507	1.0240		0.0191	0.0339
	AEP	0.7640, 0.7859		0.0305	0.2781	0.6850, 0.7121		0.0442	0.5128	1.0026, 1.0458		**0.0147**	0.0304
8 a.m.	α-Stable	1.3670	0.0655	0.0230	**0.0638**	1.6838	−0.1257	**0.0253**	**0.1012**	1.7980	−0.1765	**0.0237**	**0.0575**
	NIG	5.7047	0.3095	**0.0203**	0.0873	3.4372	−0.2307	0.0375	0.4312	7.8396	−0.4915	0.0274	0.1315
	EP	0.7312		0.0347	0.3582	0.7567		0.0576	0.9541	1.2164		0.0352	0.2467
	AEP	0.7485, 0.7111		0.0300	0.3124	0.7379, 0.7779		0.0511	0.9002	1.1703, 1.2707		0.0308	0.2260
12 (noon)	α-Stable	1.5447	0.2860	0.0200	**0.0515**	1.0698	−0.1326	0.0241	0.1144	1.5319	0.0961	**0.0238**	**0.0880**
	NIG	11.6727	0.8902	**0.0199**	0.0584	1.6736	−0.0037	**0.0168**	**0.0372**	3.9648	−0.1671	0.0319	0.2900
	EP	0.9010		0.0393	0.2715	0.5230		0.0298	0.2460	0.7398		0.0467	0.6475
	AEP	0.9381, 0.8586		0.0295	0.1976	0.5221, 0.5238		0.0291	0.2458	0.7200, 0.7638		0.0420	0.6102
4 p.m.	α-Stable	1.5365	0.1359	**0.0167**	**0.0329**	1.0541	−0.0543	0.0268	0.1109	1.5584	−0.2774	**0.0208**	**0.0863**
	NIG	12.9554	1.5886	0.0201	0.0705	0.5572	0.0024	**0.0215**	**0.1004**	5.3607	−0.0180	0.0374	0.2960
	EP	0.8617		0.0474	0.4613	0.4238		0.0462	0.6175	0.7932		0.0462	0.6001
	AEP	0.9379, 0.7957		0.0273	0.2256	0.4355, 0.4120		0.0473	0.5795	0.7867, 0.7990		0.0460	0.5902

(Continued)

TABLE 23.6 (CONTINUED) Parameter Estimates and Goodness-of-Fit Statistics for α-Stable, NIG, EP, and AEP Distributions. *Cholesky-Filtered:* Cholesky-Filtered Log-Returns. *Wavelet-Filtered:* Wavelet- Filtered Log-Returns.

Auction	Distribution	NordPool				APX				Powernext			
		Tail	Skewness	D	W²	Tail	Skewness	D	W²	Tail	Skewness	D	W²
8 p.m.	α-Stable	1.6042	0.2205	**0.0177**	**0.0345**	1.2354	0.0548	**0.0244**	**0.1109**	1.7631	−0.3230	0.0207	**0.0440**
	NIG	16.8065	−0.0877	0.0341	0.1956	2.7449	−0.0124	0.0331	0.1757	12.0221	−0.0761	0.0295	0.2032
	EP	0.8731		0.0455	0.3815	0.6965		0.0459	0.3761	1.1684		0.0347	0.2634
	AEP	0.9081, 0.8376		0.0359	0.3130	0.6622, 0.7336		0.0295	0.2206	1.0527, 1.3768		**0.0184**	0.0562
12 (midnight)	α-Stable	1.5117	0.2004	**0.0196**	**0.0673**	1.4497	0.1648	**0.0274**	**0.0868**	1.7528	−0.3078	0.0247	0.0734
	NIG	15.4975	0.4670	0.0261	0.1362	2.3668	−0.0098	0.0321	0.2113	21.0111	−2.3941	**0.0180**	**0.0359**
	EP	0.8135		0.0342	0.3425	0.6598		0.0468	0.5054	1.1671		0.0330	0.2026
	AEP	0.8190, 0.8061		0.0341	0.3371	0.6696, 0.6480		0.0410	0.4851	1.0810, 1.2551		0.0184	0.0649

Tail: α (stable and NIG), b (EP), b_l (AEP). Skewness: β (stable and NIG). Asymptotic 5% limiting values: 0.0291 (D, NordPool), 0.0356 (D, APX), 0.0318 (D, Powernext), 0.443 (W^2). Data in bold type indicate the minimum goodness-of-fit test values.

wait long enough to have many years of data available. Thus, estimation on data from power exchanges established more recently is not expected to yield better results, at least not in the near future.

The results reported in this paper can be seen as the starting point for further work. One could extend the policy oriented analysis performed by Robinson and Baniak (2002) on the impact of Contracts for Differences (CfDs) and test the effects of further policy measures, such as the introduction of the EU ETS scheme for carbon emissions and the liberalization of retail trading.

References

Andrews, D.F. and Mallows, C.L., Scale mixtures of Normal distributions. J. R. Statist. Soc., Ser. B, 1974, 36(1), 99–102.

Barndorff-Nielsen, O.E., Normal Inverse Gaussian distributions and stochastic volatility modelling. Scand. J. Statist., 1997, 24(1), 1–13.

Bellini, F., Empirical analysis of electricity spot prices in European deregulated markets, Quaderni Ref. 7/2002, 2002.

Bessembinder, H. and Lemmon, M.L., Equilibrium pricing and optimal hedging in electricity forward markets. J. Finance, 2002, 57(3), 1347–1382.

Borak, S., Ha¨rdle, W. and Weron, R., Stable distributions. SFB 649 Discussion Paper 2005–008, Humboldt Universita¨t zu Berlin, 2005.

Bosco, F., Parisio, L. and Pelagatti, M., Deregulated wholesale electricity prices in Italy: An empirical analysis. Int. Adv. Econ. Res., 2007, 13, 415–432.

Bottazzi, G., Subbotools: A reference manual. LEM Working Paper 2004/14, S. Anna School of Advanced Studies, Pisa, 2004.

Bottazzi, G. and Sapio, S., Power exponential price returns in day-ahead power exchanges. In Econophysics and Sociophysics of Markets and Networks, edited by A. Chatterjee and B.K. Chakrabarti, 2007 (Springer: Berlin).

Bottazzi, G., Sapio, S. and Secchi, A., Some statistical investigations on the nature and dynamics of electricity prices. Physica A, 2005, 355(1), 54–61.

Bottazzi, G. and Secchi, A., Why are distributions of firm growth rates tent-shaped? Econ. Lett., 2003, 80, 415–420.

Bottazzi, G. and Secchi, A., Maximum Likelihood estimation of the symmetric and Asymmetric Exponential Power distribution. LEM, S. Anna School of Advanced Studies, Pisa, 2007.

Bystro¨m, H., Extreme value theory and extremely large electricity price changes. Int. Rev. Econ. Finance, 2005, 14(1), 41–55.

Chan, K.F. and Gray, P., Using extreme value theory to measure value-at-risk for daily electricity spot prices. Int. J. Forecast., 2006, 22, 283–300.

Choy, S.T.B. and Walker, S.G., The extended exponential power distribution and Bayesian robustness. Statist. Probab. Lett., 2003, 65, 227–232.

D'Agostino, R.B. and Stephens, M.A., Goodness-of-fit Techniques, 1986 (CRC Press: Boca Raton, FL).

De Jong, C., The nature of power spikes: A regime-switch approach. Stud. Nonlinear Dynam. Econometr., 2006, 10(3), (article 3).

Deng, S.-J. and Jiang, W., Levy process-driven mean-reverting electricity price model: The marginal distribution analysis. Decis. Supp. Syst., 2005, 40(3/4), 483–494.

Deng, S.-J., Jiang, W. and Xia, Z., Alternative statistical specifications of commodity price distribution with fat tails. Adv. Model. Optimiz., 2002, 4, 1–8.

Diebold, F.X., Ohanian, L.E. and Berkovitz, J., Dynamic equilibrium economies: A framework for comparing models and data. Working Paper No. 97-7, Federal Reserve Bank of Philadelphia, 1997.

Eberlein, E. and Keller, U., Hyperbolic distributions in finance. Bernoulli, 1995, 1(3), 281–299.

Eberlein, E. and Stahl, G., Both sides of the fence: A statistical and regulatory view of electricity risk. Energy Power Risk Mgmt, 2003, 8, 34–38.

Fu, D., Pammolli, F., Buldyrev, S., Riccaboni, M., Matia, K., Yamasaki, K. and Stanley, H.E., The growth of business firms: Theoretical framework and empirical evidence. Proc. Natn Acad. Sci., 2005, 102, 18801–18806.

Karakatsani, N.V. and Bunn, D., Modelling stochastic volatility in high-frequency spot electricity prices. Department of Decision Sciences, London Business School, 2004.

Kogon, S.M. and Williams, D.B., Characteristic function based estimation of stable distribution parameters. In A Practical Guide to Heavy Tails: Statistical Techniques and Applications, edited by R.J. Adler, R.E. Feldman, and M. Taqqu, 1998 (Springer: Berlin).

Koutrouvelis, I.A., Regression-type estimation of the parameters of stable laws. J. Am. Statist. Assoc., 1980, 75(372), 918–928.

Mandelbrot, B., The variation of certain speculative prices. J. Business, 1963, 34(4), 394–419.

Mugele, C., Rachev, S.T. and Trück, S., Stable modeling of different European power markets. Invest. Mgmt Financial Innov., 2005, 3, 65–85.

Robinson, T. and Baniak, A., The volatility of prices in the English and Welsh electricity pool. Appl. Econ., 2002, 34, 1487–1495.

Samorodnitsky, G. and Taqqu, M.S., Stable NonGaussian Random Processes, 1994 (Chapman & Hall: London).

Scalas, E. and Kim, K., The art of fitting financial time series with Levy stable distributions. Korean J. Phys., 2007, 50, 105–111.

Simonsen, I., Volatility of power markets. Physica A, 2005, 355, 10–20.

Stephens, M.A., EDF statistics for goodness of fit and some comparisons. J. Am. Statist. Assoc., 1974, 69(347), 730–737.

Subbotin, M.F., On the law of frequency of errors. Matematicheskii Sbornik, 1923, 31, 296–301.

Weron, R., Computationally intensive value at risk calculations. In Handbook of Computational Statistics, edited by J.E. Gentle, W. Härdle, and Y. Mori, pp. 911–950, 2004 (Springer: Berlin).

Weron, R., Heavy tails and electricity prices. In The Deutsche Bundesbank's 2005 Annual Fall Conference on Heavy Tails and Stable Paretian Distributions in Finance and Macroeconomics, Eltville, 10–12 November 2005.

Weron, R., Modeling and Forecasting Electricity Loads and Prices: A Statistical Approach, 2006 (Wiley: Chichester).

Weron, R., Heavy-tails and regime-switching in electricity prices. Math. Meth. Oper. Res., 2009, 69, 457–473.

Weron, R., Bierbrauer, M. and Trück, S., Modeling electricity prices: Jump diffusion and regime switching. Physica A, 2004, 336, 39–48.

West, M., On scale mixtures of Normal distributions. Biometrika, 1987, 74, 646–648.

24

Stochastic Spot Price Multi-Period Model and Option Valuation for Electrical Markets

Eivind Helland,
Timur Aka and
Eric Winnington

24.1 Introduction

The deregulation of electricity markets has resulted in competitive prices, but also in higher fluctuations of electricity price development since electricity is scarcely storable. Most markets exhibit high price volatility and intermittent price spikes. However, the prices have a strong daily, weekly and yearly periodicity that is explained by looking at the market price as an equilibrium price based on supply and demand curves. Since the demand is very inelastic, the marginal costs of the supply side determine the price to a large extent. If the total load is low, plants with the lowest variable production costs are used, and if the total load is high, gas- or oil-fired plants with high fuel costs are run additionally. The periodicity of the total load is responsible for the periodicity of the electricity prices. The total load has a random component, depending on short-term weather conditions and other uncertain parameters, but it also has a clearly predictable part, and so do electricity prices (Müller *et al.* 2004). This has led to a demand for derivative products that hedge the owner against high prices. A thorough understanding of the stochastic price dynamics is thus necessary for the purposes of risk management and derivative pricing. At Axpo AG, the largest energy company in Switzerland, we have developed a stochastic spot price multi-period model for optimising hydro power plant production, pricing complex options and structured products, and hedging the asset portfolio.

A stochastic spot price model has been developed in order to evaluate standard and off-standard products in the electrical market and for the optimisation of power production in pumped storage hydro

DOI: 10.1201/9781003265399-28

parks. The spot price model is built up in several steps and calibration is carried out based on the price distribution curves of spot prices and the log returns of forward contracts as well as standard option contracts over several delivery periods.

The base price level of the stochastic spot price model is an hourly price forward curve that has been modelled using a regression model based on past European Exchange (EEX) price data. The main input parameters were the calendar dates (weekdays, holidays, etc.) to obtain seasonality effects (intraday, weekly and yearly), historical meteorological data (daily temperature, wind and precipitation data) and long-term market expectations. The price forward curve (PFC) was then scaled such that the average price of the hourly PFC was equal to the future power contract at the corresponding time interval to obtain the arbitrage-free condition. Secondly, the daily changes of the PFC have been modelled based on a log-normal, mean-reverting (weekly) process of the front-year forward base contract in order to reflect the shift in spot prices due to changes in long-term forward contracts. Thirdly, a lognormal, mean-reverting, short-term (hourly) process was modelled to represent the intra-day and day-ahead perturbations in spot prices. Additionally, two jump process are admixed to model the positive and negative jumps observed on the German market.

Based on the most recent PFC, and a calibrated set of weekly, hourly and jump processes, a set of price scenarios is generated using the spot simulation model. These scenarios are used to evaluate the price of future options contracts available on the EEX. These contracts are created and evaluated in TS-Energy (software developed by Time-Steps). The distribution of the scenarios' end values are used to evaluate the fair price of a call and a put option based on selected strike prices. These values are compared with the observed prices for call options on the EEX or broker data available for the German market. The stochastic spot price multi-period models are calibrated so that calculated option prices for monthly, quarterly and annual contracts fall within the quoted bid–ask spread.

24.2 Spot Price Multi-Period Model

Our spot price simulation model captures the typical features observed in the electricity market as described above in the Introduction. In this paper we calibrate it to data for German electricity prices on the European Energy Exchange (EEX), but it can readily be adapted to any other market. Therefore, we describe the spot market price by a discrete-time stochastic process with an hourly granularity. The full model can be considered as a three-factor model in order to fully capture the different features observed in electricity markets.

- The base price level of the spot price model is an hourly price forward curve that has been modelled using a regression model accounting for the seasonal patterns and periodicities and scaled with liquid future power contracts.
- A short-term process with positive and negative jumps is used to represent the intraday and day-ahead perturbations in spot prices which account for unexpected variations in supply and demand due to weather conditions and production outages.
- Along-term process including a volatility term structure, which accounts for the uncertainty in supply and demand, fuel costs and macroeconomic variables in the long term.

The fundamental equation of our model can be written as

$$S_t = \left(X_t + L_{p,t} + L_{op,t} \right) \frac{Y_t}{Y_0}. \tag{24.1}$$

The factors X_t, $L_{p,t}$ and $L_{op,t}$ produce the short-term variations in price behaviour, whereas the factor Y_t is responsible for its long-term variation. More precisely, the process X_t describes the deviation of spot prices from the hourly price forward curve θ_t, which reflects the current market situation, and is given by

$$X_t = \theta_t \left(1 + \eta \frac{X_{t-1} - \theta_{t-1}}{\theta_{t-1}} + \sigma_{X,t}\, \varepsilon_{X,t} + a \cdot \ln\left(\frac{\theta_{t+24}}{\theta_t} \right) \Delta t_h \right), \tag{24.2}$$

where ε_t are standard normally distributed random numbers, i.e. $\varepsilon_{X,t} \sim \mathcal{N}(0,1)$. η is the autocorrelation parameter of the previous hourly shift of the intraday spot price versus the price forward curve. If the spot price is higher than the expected price forward curve level then higher prices will also be experienced in the following hours. The jump processes $L_{k,t}, k \in \{p,op\}$ take into account occasionally observed spikes in the hourly spot prices. Introducing two independent jump processes allows us to properly distinguish the jump intensities in peak and off-peak hours. The discrete jump processes have the form

$$L_k = \mathbf{1}_{\{u_{k,t} < \lambda_{k,t}\}} \cdot |\varepsilon_{k,t}| \cdot h_k \cdot \mathbf{1}_{\{t \in T_k\}} + L_{k,t-1} \cdot d_k, k \in \{\text{p,op}\}, \tag{24.3}$$

where $\mathbf{1}_{\{\cdot\}}$ denotes the usual indicator function, the disjoint sets T_p and T_{op} subdivide hourly time grid nodes into peak (p) and off-peak (op) hours, $u_{k,t}$ are uniformly distributed random numbers, i.e. $u_{k,t} \sim \mathcal{U}(0,1)$ and $\varepsilon_{k,t} \sim \mathcal{N}(0,1)$. The volatility $\sigma_{X,t}$ in Equation (24.2) and the jump intensities $\lambda_{k,t}$ in (24.3) are allowed to be time dependent in order to consider seasonality effects. Finally, the evolution of the long-term variation process is given by

$$Y_{t+1} = \begin{cases} Y_t \cdot \left(1 + 1/7 \left(\ln\left(\frac{K}{Y_t} \right) \Delta t_d + \sigma_{Y,t} \varepsilon_{Y,t} \right) \right), & \text{if } t \in \{0,7,14,21,28,\ldots\} \\ Y_t + \Delta Y_t, & \text{else,} \end{cases} \tag{24.4}$$

where K is a user-defined long-term reference price. Note that the time-step length Δt_d of one day in (24.4) differs from Δt_h in (24.2), which is based on an hourly time grid. Therefore, the long-term process exhibits a random change in its value once a week based on $\varepsilon_{Y,t} \sim \mathcal{N}(0,1)$ and is updated by a constant value $\Delta Y_t := Y_t - Y_{t-1}$ in the remaining time steps. Again, the daily volatility $\sigma_{Y,t}$ in Equation (24.4) is allowed to be a function of time. We point out that all random variables in the above equations are assumed to be stochastically independent.

24.3 Model Calibration

The model is calibrated separately for each factor. The short-term and long-term factor calibrations are described in detail below. For the statistical analysis in the following sections, we used EEX prices from January 1, 2011, to January 1, 2012.

24.3.1 The Price Forward Curve

The price forward curve (PFC) reflects the expectation of hourly energy prices for the coming years. Two basic elements are required in order to create a PFC, (1) an estimate of the relative price structure, with an hourly resolution for the entire time period, and (2) current market values of forward products (EEX). It is important to emphasise that the PFC is not a forecast in the sense of an independent market evaluation, but rather a market-consistent breakdown of the market conditions to an hourly granularity. These two elements are supported by regression analysis, which represents the core elements in the creation of the PFC. The main input parameters are the calendar dates (weekdays, holidays, etc.) to obtain seasonality effects (intraday, weekly and yearly), historical meteorological data (daily temperature, wind and precipitation data) and the long-term market expectations. The PFC is then scaled so that the average price of the hourly PFC is equal to the future power contract at its corresponding time interval to obtain arbitrage-free conditions on weekly, monthly, quarterly and yearly peak and off-peak

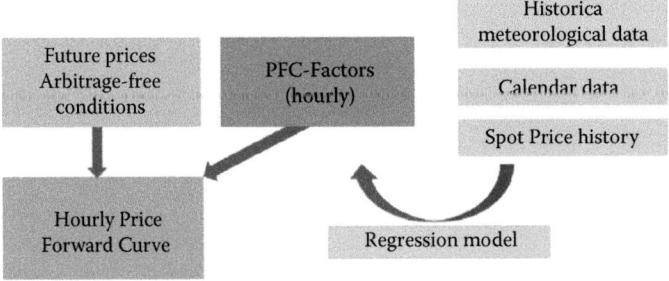

FIGURE 24.1   Price forward curve modelling approach.

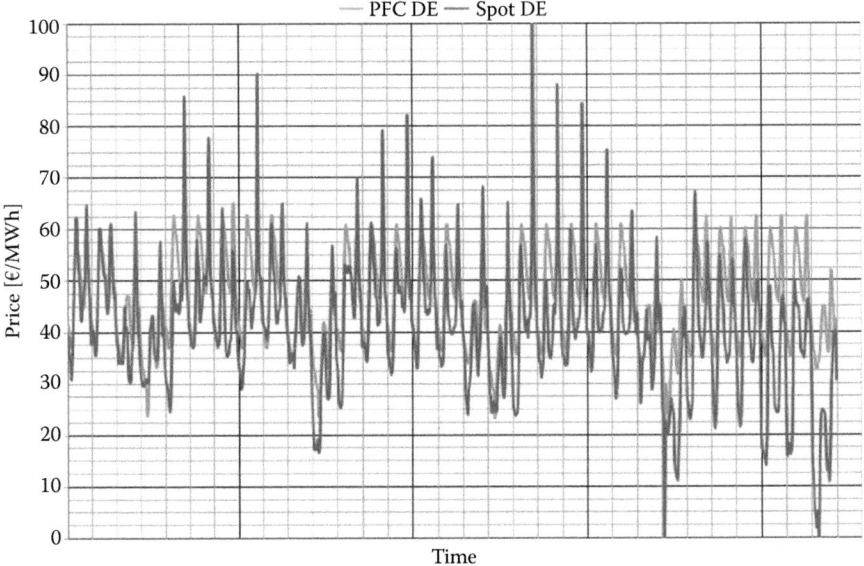

FIGURE 24.2   A generated price forward curve (light grey) and the historically realised spot price (dark grey) compared.

forward products. The calibration process is depicted in Figure 24.1. Figure 24.2 shows the hourly price Forward Curve created *a priori* over a monthly period together with the realised spot prices. The PFC is far smoother and less volatile than the realised spot prices since it is based on averaged historical data.

24.3.2 The Short-Term Process

The short-term process X_t is described by Equation (24.2). We have calibrated the volatility and the mean reversion to fit the distribution curve of the price difference between the Price Forward Curve and the realised spot price with EEX prices from January 1, 2011, until January 1, 2012. We have extracted the effect of price variations of the forward contracts by using daily updated PFCs to only include unexpected spot price variations due to short-term weather fluctuations and unexpected outages of power plants (PFC*). The empirical autocorrelation function η of the difference of the realized spot price and the PFC* can be approximated by the function η^τ, with τ the time lag in hours, where η was calibrated to be 0.8. Then, with least-squares regression modelling, we estimate the mean reversion and the volatility parameters. Figure 24.3 depicts the comparison of the realised and the modelled price differences, i.e. (PFC* − realized spot price) and (PFC* − short-term spot price model).

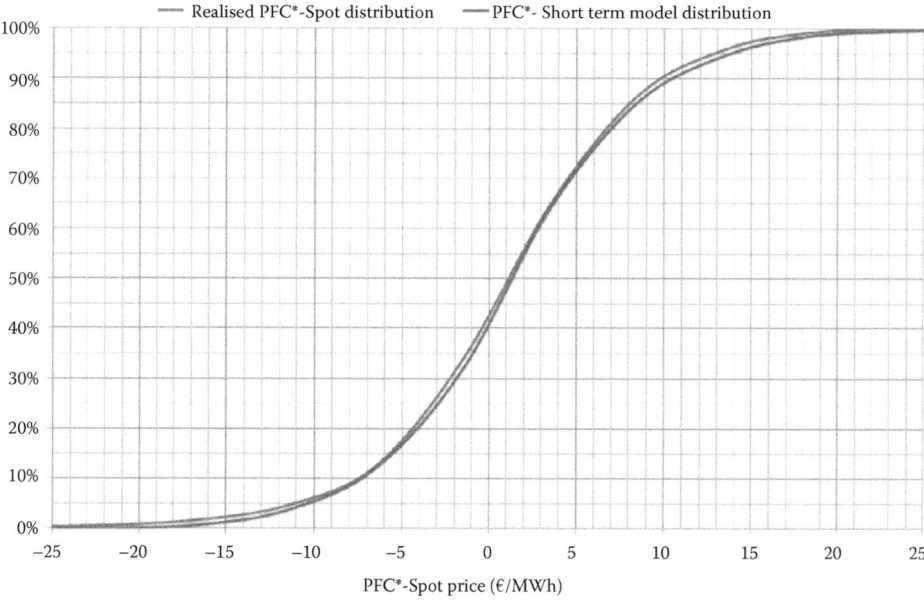

FIGURE 24.3 Comparison of the realised and modelled price differences: (PFC* − realised spot price) vs. (PFC* − short-term spot price model).

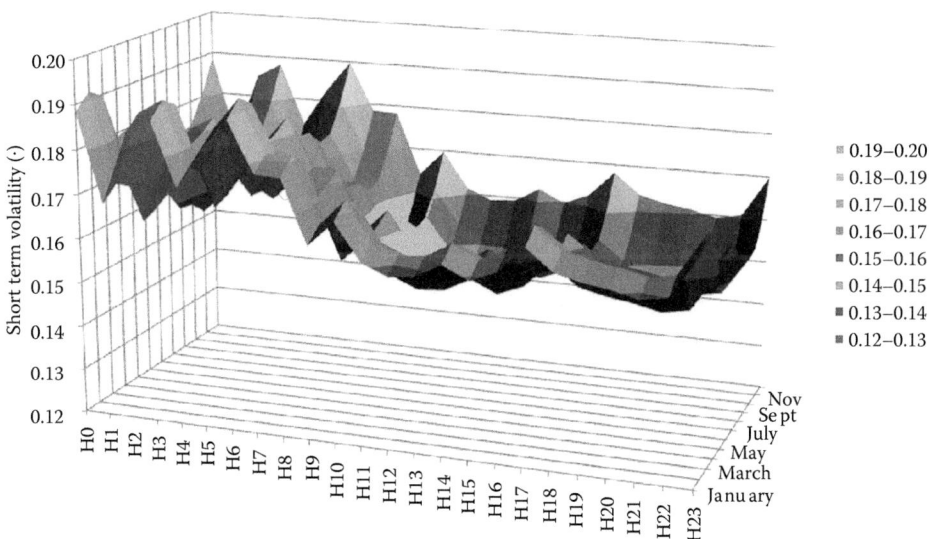

FIGURE 24.4 Realised short-term volatility as a function of hour and month.

We also investigated the variation of the short-term volatility as a function of peak and off-peak hours, week days and months (Figure 24.4). We found significant differences in the hourly and monthly structures, thus developing an empirical expression to model the short-term volatility as a function of the hour and the month. The parameters in the discrete jump process (24.3) were estimated by analysing the frequency and magnitude of the outliers. For this purpose a realised spot price outside the range of two standard deviations from the adjusted PFC* was defined as an outlier.

24.3.3 The Long-Term Process

In order to reflect shifts in the level of spot prices due to changes in long-term forward contracts, we included daily shifts of the PFC based on a log-normal mean-reverting process of the front-year forward base contract. The parameters of the model are calibrated with 5 years of front-year forward contract data. The long-term process is expressed by Equation (24.4). Figure 24.5 shows a qq-plot of the realised front-year forward price and the forward model. The qq-plot suggests that forward model prices follow a distribution that fits the forward price cumulative distribution. Figure 24.6 depicts the realised

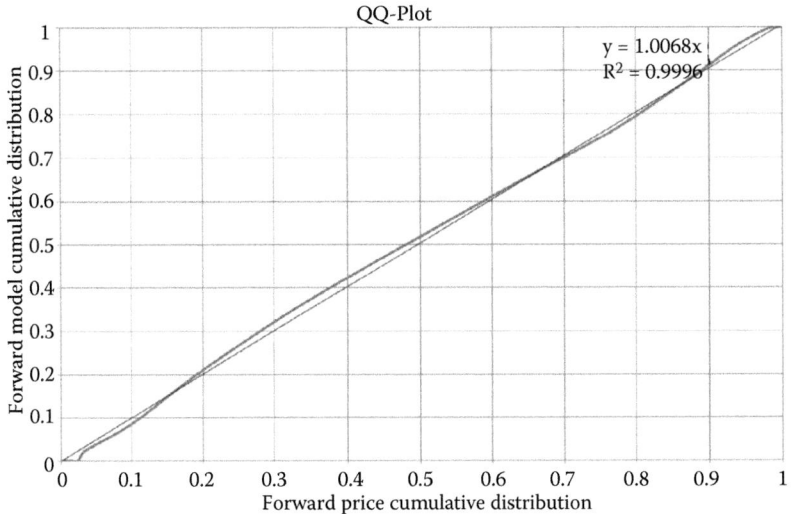

FIGURE 24.5   qq-Plot of the realised front-year forward and forward model prices.

FIGURE 24.6   Realised front-year forward price and one generated scenario with the forward price model.

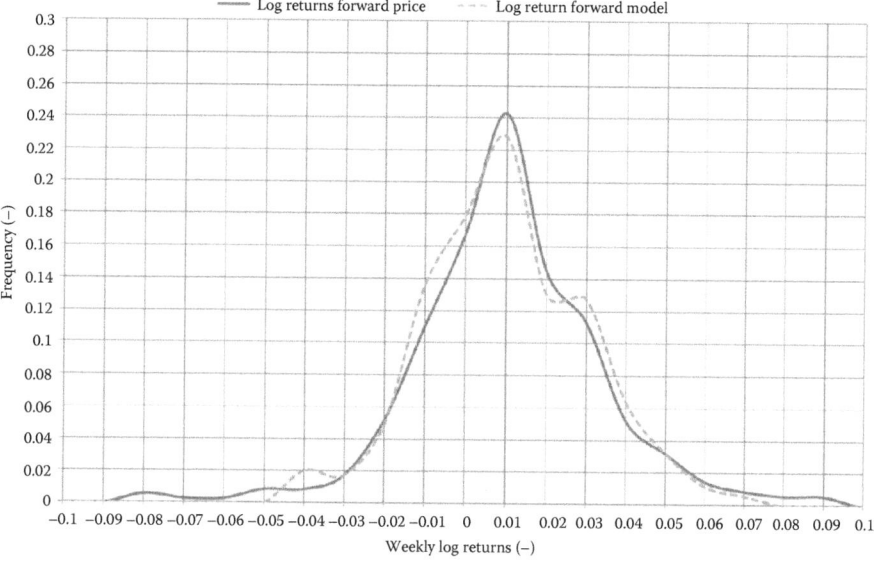

FIGURE 24.7   Weekly log returns of realised forward prices and one simulated path of the forward price model.

front-year forward price and one scenario generated with the forward price model. The model seems to mimic the price behaviour of the realised prices. Figure 24.7 shows the corresponding comparison of the weekly log returns of realised forward prices and the path simulated by the forward price model.

24.3.4 The Term Structure of Volatility

A set of generated spot price scenarios is used to evaluate the price of future options contracts available on the EEX. The distribution of the averaged spot price values for different delivery periods is used to evaluate the fair price of call and put options based on selected strike prices. These values are compared with the available quoted market prices. The long-term volatility in Equation (24.4) is calibrated so that calculated option prices for monthly, quarterly and annual contracts fall within the quoted bid–ask spread. A simplified scaling factor of the form $a \cdot t_d^{-\beta}$ is used to include the volatility term structure in the long-term volatility. The parameters are adjusted on a daily basis in order to be consistent with the market quotes from the brokers for the different delivery periods (see Figure 24.8). For comparison with the implied volatility of yearly options, we note that the annualised volatility of the forward price model is of the order of 13%–14%.

24.4 Scenario Generation and Valuation of Structured Products

The model defined above is used to generate multiple spot price paths. Figure 24.9 shows one path in comparison with the price forward curve, and Figure 24.10 shows a selection of five simulated paths. The model defined above is also used to generate multiple price paths for forward prices. Figure 24.11 shows the cumulative forward price distribution curve of 200 paths of a simulated front-year product for a day, week and monthly time period. Based on this information we can estimate the value at risk for different time periods.

24.4.1 Estimation of the Volatility Surface

Generated spot price scenarios are used to evaluate the price of option contracts available on the EEX. The distributions of the averaged spot price values for different delivery periods are used to evaluate the

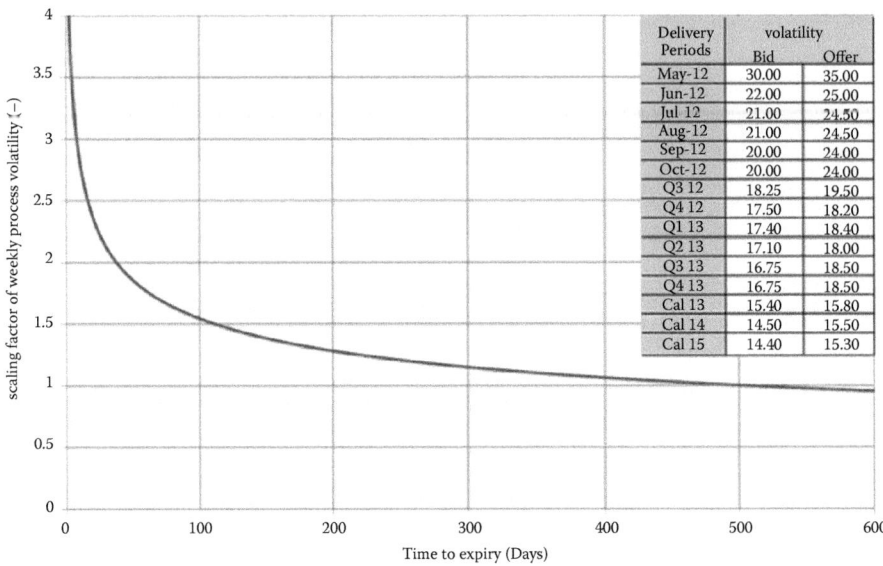

Delivery Periods	volatility	
	Bid	Offer
May-12	30.00	35.00
Jun-12	22.00	25.00
Jul 12	21.00	24.50
Aug-12	21.00	24.50
Sep-12	20.00	24.00
Oct-12	20.00	24.00
Q3 12	18.25	19.50
Q4 12	17.50	18.20
Q1 13	17.40	18.40
Q2 13	17.10	18.00
Q3 13	16.75	18.50
Q4 13	16.75	18.50
Cal 13	15.40	15.80
Cal 14	14.50	15.50
Cal 15	14.40	15.30

FIGURE 24.8   Term structure of volatility.

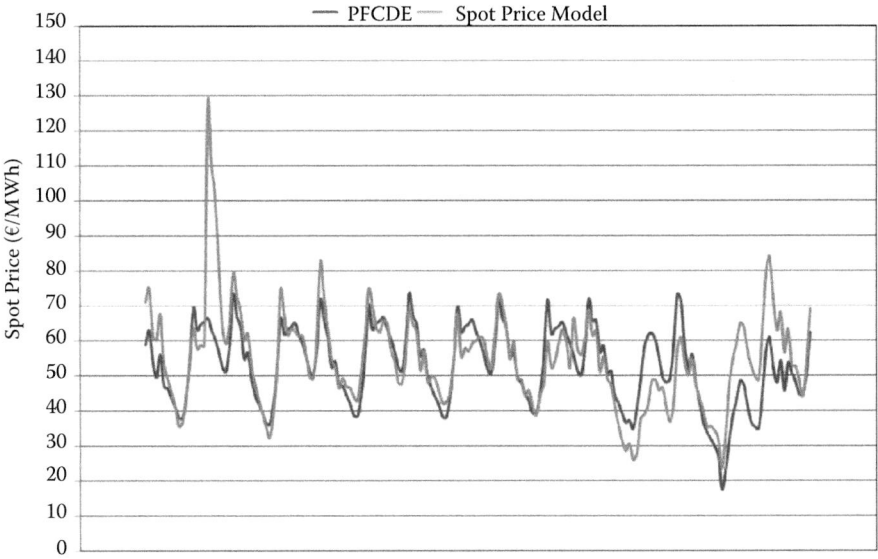

FIGURE 24.9   Price forward curve (dark grey) and spot price model (light grey).

fair price of call and put options for different strike prices and times to expiry. Figure 24.12 shows the volatility surface of a call option on the yearly forward contract Cal13 based on quotes from 20.04.2012, when the underlying price was 51 Euro/MWh. Figure 24.13 shows that the calculated call option premiums are within the range of the bid–ask spread for an out-of-the-money option.

24.5 Structured Product Valuation

Non-standard financial contracts are products tailored to the needs of specific clients, but not sold in sufficient quantity for a standard price model to have been established. They are also referred to as exotic

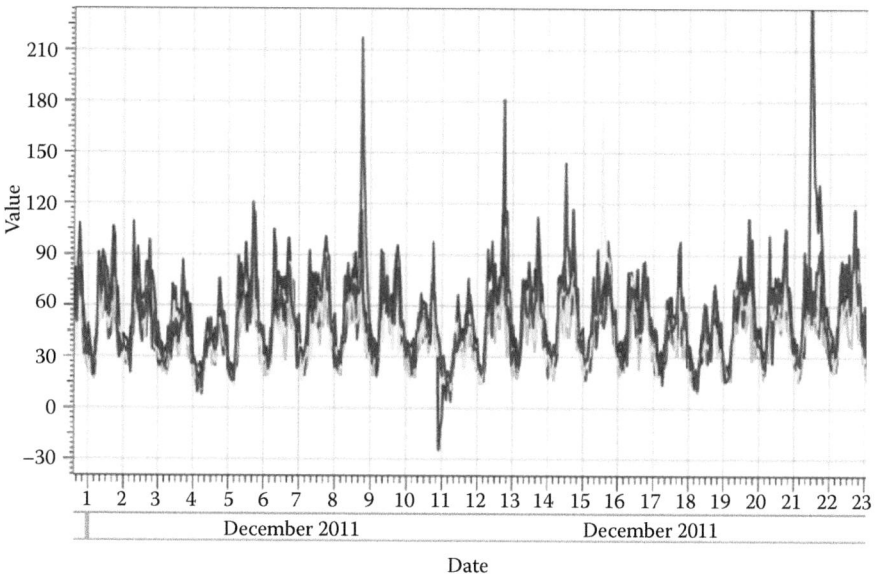

FIGURE 24.10 A sample set of spot price scenarios.

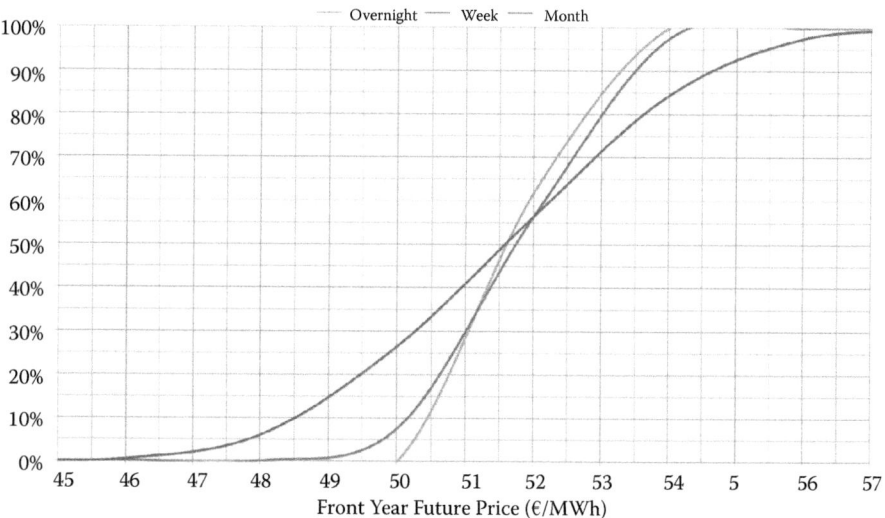

FIGURE 24.11 Cumulative forward price distribution.

or structured products. A standard product usually trades actively and there is little uncertainty about its price so that the choice of model primarily affects how hedging is done. In the case of structured products, model risk is much greater because there is the potential for both pricing and hedging being impacted. Examples of structured products are swing options and virtual power plants (VPPs), see Figure 24.14, which address the need of market participants to have flexibility in both the timing and volume of the energy delivered. The increasing variability of both generation (from solar and wind) and loads will also require more sophisticated and decentralised decision making. As a result of all these factors, interest in virtual power plants is gaining significant momentum within the energy industry. The price forward curve and spot price scenarios described in this paper are used to evaluate a fair price

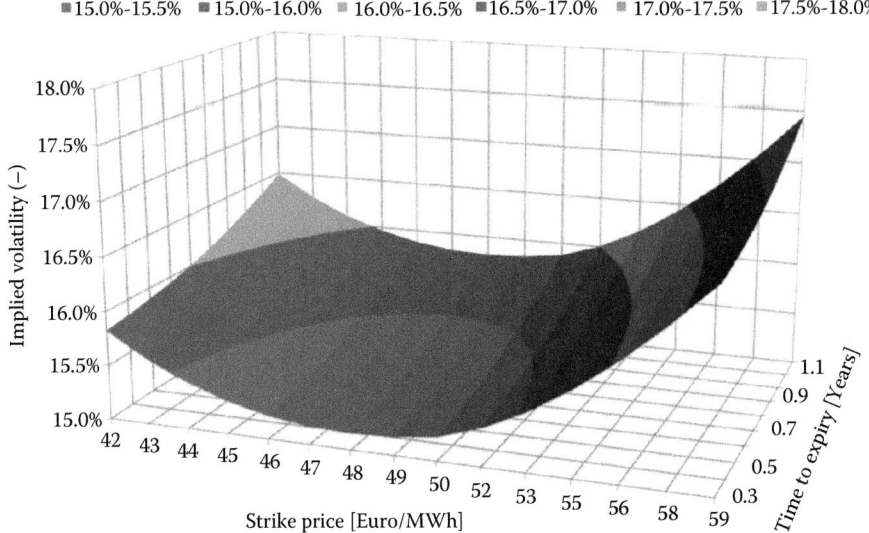

FIGURE 24.12   Volatility surface for a yearly call option, Cal13 (quoted 20.04.2012).

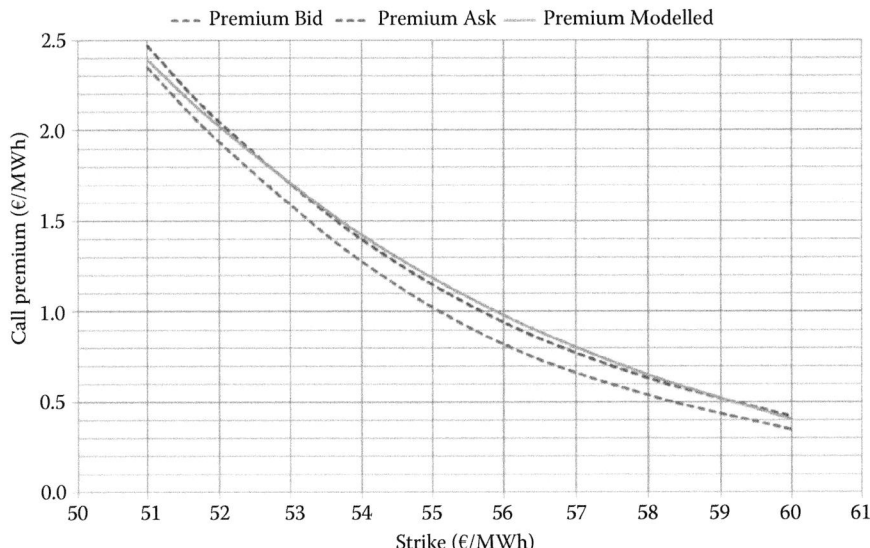

FIGURE 24.13   Estimated call option premiums for a yearly call option, Cal13 (quoted 20.04.2012).

and value at risk for these non-standard products using the stochastic optimisation software TS-Energy (Zuur 2004, 2008, 2012).

24.5.1 Valuation and Risk Analysis: TS-Energy

TS-Energy is a valuation and risk analysis application for complex power assets and contracts. Its functionalities are as follows.

- Optimal use of storage capacity (e.g. hydro power, gas, wind capacity): When to sell, hold or buy (e.g. wind power at night); optimal operation, fair price, value at risk and bidding information.

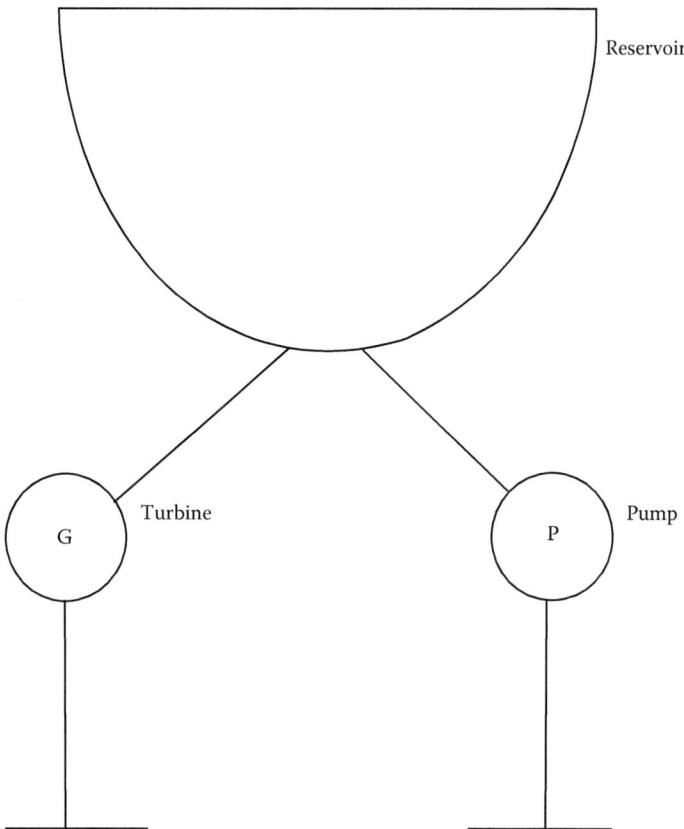

FIGURE 24.14 Simplified setup of a virtual power plant (VPP).

- Optimal use and pricing of mid-term power contracts/VPPs.
- Pricing of contracts between suppliers of storage power and their customers.
- Arbitrage—taking advantage of today's forward pricing inefficiencies (brokers).
- Portfolio analysis and optimisation.
- Efficient planning and valuation of power generation plants and storage capacity. Expected net present value, value at risk and payback period of a business model.

The approach uses stochastic processes to describe uncertain inputs such as electricity prices and water influx to optimise assets without the requirement of perfect foresight imposed by standard linear programming software. TS-Energy uses a two-step process, a backwards integration step that determines an optimal operation strategy for an asset and a forwards integration step that applies this strategy to a set of scenarios to determine fair value and risk information for the asset (Zuur 2004, 2008, 2012, Helland and Winnington 2012). Hydropower assets are described by basins and operating units (generators and pumps) (see Figure 24.14), as well as by stochastic processes describing the net water influx for each basin. During the valuation step, the backward integration method is carried out in order to derive the action grid which holds the information about the best action at each settle date (e.g. to hold or exercise the option) given the market state (e.g. the underlying price) and the product state (e.g. the number exercises left of the swing contract or the volume of a basin). The modelling of the optimal dispatch strategy implies the determination of the value-maximising decisions for all states at each point in time in order to determine the right time for generation or pumping (see Figure 24.15). In the forwards integration step, this optimal dispatch strategy is applied to spot price scenarios generated by means of Monte-Carlo simulations. An adequate number of scenarios for convergence is in the range of 100 to

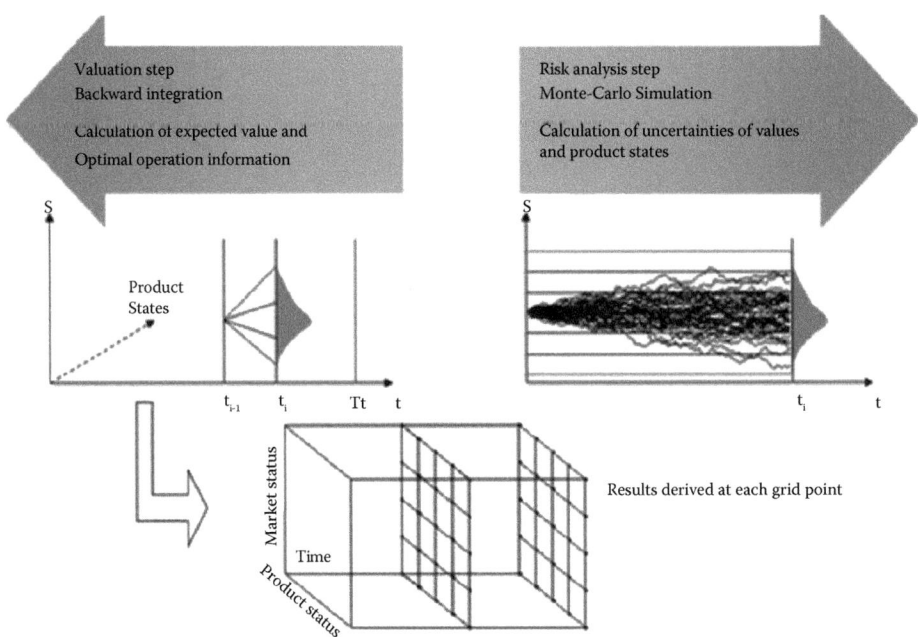

FIGURE 24.15   Backward and forward process (TS-Energy 2010).

10,000, which can be used to derive the expected earnings, the probability distribution and the value at risk of the asset.

24.5.2 Virtual Power Plant Valuation

Based on the most recent PFC, and a calibrated set of weekly, hourly and jump processes, a set of price scenarios is generated using the spot price simulation model. These scenarios are used to evaluate the price of a VPP (virtual power plant) contract. These contracts are created and evaluated in TS-Energy. The distribution of the scenarios' end values is used to evaluate the fair price of the contract. Since there is no inflow to the VPP, the only stochastic variable is the spot price.

The calculations give the plant operator the hourly price boundaries for generation, holding and pumping in order to maximise net income until maturity by providing the optimal price-dependent hourly bids (see Figure 24.16). Figure 24.17 shows a typical result report of a plant simulation. It provides information about the fair price, the expected income from generation, the expected cost for pumping, the standard deviation and the Value at Risk of operating a hydropower plant for a given time period. Figure 24.18 depicts the cumulative distribution of the fair value of the VPP contract.

24.6 Conclusion

A stochastic spot price model has been developed in order to evaluate standard and off-standard products in the electrical market and for the optimisation of the power production in pump storage hydro parks. The spot price model is built up in several steps and calibration is carried out based on the price distribution curves of spot prices and the log returns of forward contracts, as well as standard option contracts over several time periods. The price model is used to evaluate non-standard contracts, to optimise the scheduling of a hydro power pumped-storage plant, and for dynamical hedging of production.

Price in €/MWh

	0.0	6.5	9.6	14.1	20.7	30.5	44.8	66.0	97.1	142.8	210.1	309.0	3000.0
1	−180	−180	−180	−180	−180	0	24	24	172	172	172	172	172
2	−180	−180	−180	−180	−180	−180	−156	24	24	204	204	204	204
3	−180	−180	−180	−180	−180	−180	−156	−156	24	24	204	204	204
4	−138	−138	−138	−138	−138	−138	−138	−114	−114	24	24	204	204
5	0	0	0	0	0	0	24	24	24	24	24	204	204
6	0	0	0	0	0	0	24	24	24	204	204	204	204
7	0	0	0	0	0	24	24	204	204	204	204	204	204
8	0	0	0	0	0	204	204	204	204	204	204	204	204
9	−180	−180	−180	−180	−180	24	24	204	204	204	204	204	204
10	−180	−180	−180	−180	−180	24	24	204	204	204	204	204	204
11	−180	−180	−180	−180	−180	24	204	204	204	204	204	204	204
12	−180	−180	−180	−180	0	24	204	204	204	204	204	204	204
13	−180	−180	−180	−180	−180	0	204	204	204	204	204	204	204
14	−180	−180	−180	−180	−180	0	24	204	204	204	204	204	204
15	−180	−180	−180	−180	−180	0	24	204	204	204	204	204	204
16	−180	−180	−180	−180	−180	−180	24	24	204	204	204	204	204
17	−180	−180	−180	−180	−180	−180	0	24	24	204	204	204	204
18	−180	−180	−180	−180	−180	−180	0	24	24	204	204	204	204
19	−180	−180	−180	−180	−180	0	24	24	24	204	204	204	204
20	−180	−180	−180	−180	0	0	24	204	204	204	204	204	204
21	−180	−180	0	0	180	180	204	204	204	204	204	204	204
22	−180	−180	−180	0	0	100	100	157	157	157	157	157	157
23	−180	−180	−180	0	0	0	0	150	150	150	150	150	150
24	−180	−180	−180	−180	0	0	0	86	143	143	143	143	143

Delivery hour

Volume in MWh

FIGURE 24.16 Price-dependent hourly bids (TS-Energy 2010).

	01.06.2012	31.08.2012	30.11.2012	01.03.2013	01.06.2013
Expected value	723'131	723'131	723'131	723'131	723'131
Expected price	723'131	669'815	497'454	286'380	0
Expected income (cumulated)	0	53'316	225'678	436'751	723'131
Expected income	0	53'316	172'362	211'073	286'380
Expected income by generation	0	250'967	812'522	1'534'952	2'465'884
Expected costs of pumping	0	197'651	586'845	1'098'201	1'742'753
Expected penalties	0	0	0	0	0
Expected costs of switching on/off	0	0	0	0	0
Standard deviation of value	0	31'253	64'642	102'054	129'977
Value at Risk (95%)	0	56'109	123'980	208'839	269'170
Generated electricity (MWh)	0	4'429	13'435	21'910	34'786
Consumed electricity (MWh)	0	6'667	18'821	30'030	46'381
Days generated	0	12	37	61	97
Days pumped	0	14	39	63	97
Expected volume (m³)	0	571	680	612	0
Expected volume used for generation (m³)	0	4'429	13'435	21'910	34'786
Expected volume used for pumping (m³)	0	5'000	14'115	22'522	34'786

FIGURE 24.17 Valuation report (TS-Energy 2010).

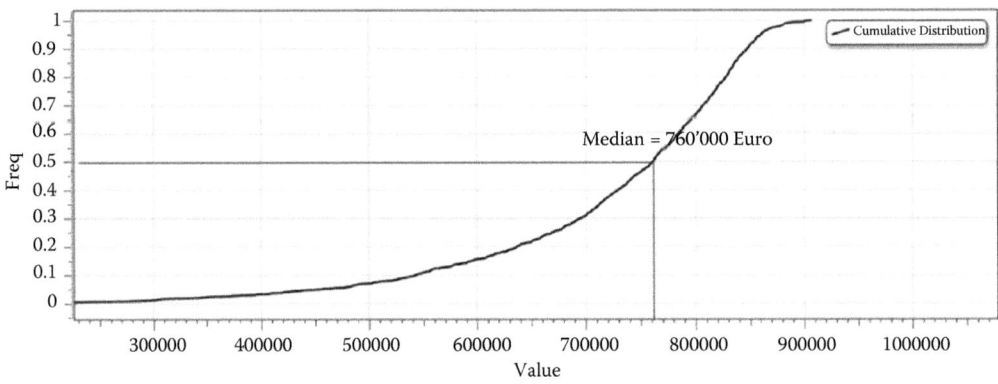

FIGURE 24.18   Cumulative distribution of the VPP valuation.

Acknowledgements

The results of this paper were obtained in a joint development project of Axpo AG and Time-steps AG. The project was supported financially by the Axpo Trading Division. The authors wish to express their gratitude to Roman von Siebenthal, Ed Zuur, André Röthig and Jens Wartenburger for their valuable contributions.

References

Helland, E. and Winnington, E., Stochastic optimization of the scheduling of a multi-stage pumped-storage hydro power plant. Report, Hydro scheduling in competitive markets, Bergen, Norway 2012.

Müller, A., Schindmayr, G., Burger, M. and Klar, B., A spot market model for pricing derivatives in electricity markets. *Quant. Finance*, 2004, **4**, 109–122.

Zuur, E., Ts-hydropower, approach document. Report, Time-steps AG, Switzerland, 2004.

Zuur, E., Hydro 2008: calibration the mean reversion process. Report, Time-steps AG, Switzerland, 2008.

Zuur, E., Ts-energy 2012, technical description. Report, Time-steps AG, Switzerland, 2012.

Modelling Spikes and Pricing Swing Options in Electricity Markets

Ben Hambly,
Sam Howison
and Tino Kluge

25.1 Introduction

A distinctive feature of electricity markets is the formation of price spikes which are caused when the maximum supply and current demand are close, often when a generator or part of the distribution network fails unexpectedly. The occurrence of spikes has far reaching consequences for risk management and pricing purposes. In this context a parsimonious model with some degree of analytic tractability has clear advantages, and in this paper we propose and examine in detail a simple mean-reverting spot price process exhibiting spikes.

In our model the spot price process S is defined to be the exponential of the sum of three components: a deterministic periodic function f characterizing seasonality, an Ornstein–Uhlenbeck (OU) process X and a mean-reverting process with a jump component to incorporate spikes Y. We set this up formally. Let $(\Omega, \mathcal{F}, \mathbb{P})$ be a probability space equipped with a filtration $(\mathcal{F}_t)$ satisfying the usual conditions. We let

$$
\begin{aligned}
S_t &= \exp\left(f(t) + X_t + Y_t \right), \\
dX_t &= -\alpha X_t dt + \sigma dW_t, \\
dY_t Y_1 &= -\beta Y_{t-} dt + J_t dN_t,
\end{aligned}
\tag{25.1}
$$

where N is a Poisson process with intensity λ and J is an independent identically distributed (i.i.d.) process representing the jump size, W is a Brownian motion and alpha, beta, sigma are constants. We assume W, N and J to be mutually independent processes adapted to the filtration.

Our model generalizes a number of earlier models. For example, the commonly used model of Lucia and Schwartz (2002) is obtained by setting $\beta = 0$ and taking $J = 0$. In this model, S_t is log-normally

DOI: 10.1201/9781003265399-29

distributed giving analytic option price formulae very similar to those in the Black–Scholes model. To allow for a stochastic seasonality, a further component can be inserted into the model and, as long as this process has a normal distribution, analytic tractability is maintained. The main disadvantage of the Lucia and Schwartz models is their inability to mimic price spikes. To overcome this, jumps can be added to the model; for example, the case $\beta = 0$ in our model,

$$dX_t = -\alpha X_t dt + \sigma dW_t + J_t dN_t, \quad S_t = \exp\left(f(t) + X_t\right). \tag{25.2}$$

This model is briefly mentioned by Clewlow and Strickland (2000, Section 2.8). Analytic results are given by Deng (2000) based on transform analysis described by Duffie *et al.* (2000). Calibration to historical data and the observed forward curve is discussed by Cartea and Figueroa (2005) where practical results for the UK electricity market are given. More general versions are discussed by Benth *et al.* (2003) and applied to data from the Nordpool market. For these models to exhibit typical spikes the mean-reversion rate α must be extremely high, otherwise the jumps do not revert quickly enough.

Benth *et al.* (2007) introduce a set of independent pure mean-reverting jump processes of the form

$$S_t = \sum_i w_i Y_t^{(i)}, dY_t^{(i)} = -\alpha_i Y_t^{(i)} dt + \sigma_i \, dL_t^{(i)},$$

where w_i are some positive weights and the $L(i)$ are independent increasing càdlàg pure jump processes. The spot price process is a linear combination of the pure jump processes and, as there is no exponential function involved, positivity of the spot is achieved by allowing positive jumps only. An advantage of this formulation is that semi-analytic formulae for option prices on forwards with a delivery period can be derived. However, a full analysis of this class of models still seems to be in its early stages; there is some empirical work on fitting this model in the work of Klüppelberg *et al.* (2008).

There are a number of papers which discuss the stylized facts of electricity markets and seek to model them empirically. An empirical comparison of a number of models for the Californian market can be found in the work of Knittel and Roberts (2005). Other empirical approaches try to fit more advanced time series models (e.g. Koopman *et al.* 2007), or alternative mathematical models, such as jump diffusion and regime switching (e.g. Weron *et al.* 2004). It is clear that there is a need to capture seasonality and rapid price changes.

The model (25.1) we consider in this paper is an extension of (25.2), in which we allow for two different mean-reversion rates, one for the diffusive part and one for the jump part. The introduction of a mean-reverting spike process Y allows us to choose a higher mean-reversion rate β in order for the jump to revert much more quickly and so mimic a price spike. This is useful for modelling the NordPool market where estimates of the mean-reversion speed are typically small, but this might not be needed in markets where the speed of mean reversion α is generally very high, like in the UKPX or EEX, where a jump diffusion with mean reversion may also be appropriate.

Returning to our model (25.1) and solving for X_t and Y_t, we have

$$X_t = X_0 e^{-\alpha t} + \sigma \int_0^t e^{-\alpha(t-s)} dW_s,$$

$$Y_t = Y_0 e^{-\beta t} + \sum_{i=1}^{N_t} e^{-\beta(t-\tau_i)} J_{\tau_i}, \tag{25.3}$$

where τ_i indicates the random time of the occurrence of the ith jump. Thus, given $X_0 = x_0, X_t \sim N\left(x_0 e^{-\alpha t}, \left(\sigma^2/2\alpha\right)\left(1 - e^{2\alpha t}\right)\right)$. Properties of the spike process Y are not as obvious and will be examined in the following section. At this point we make no assumption on the jump size J but will later give results for exponentially and normally distributed jump sizes.

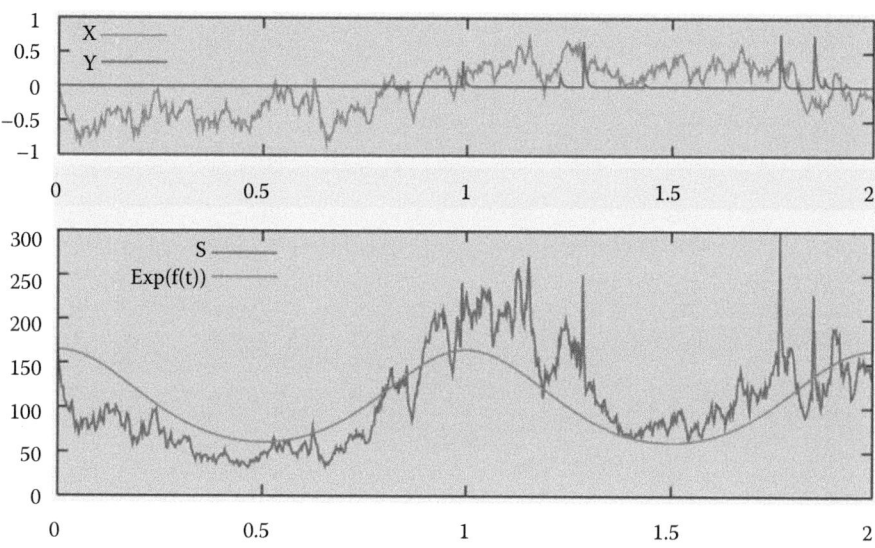

FIGURE 25.1 Simulated sample paths of X, Y and S of the spot model (25.1). We use the following parameters which are of the same order of magnitude as calibrated values of the NordPool spot market, except the seasonality function $f(\cdot)$ which has been chosen arbitrarily: $f(t) = \ln(100) + 0.5\cos(2\pi t)$, $\alpha = 7$, $\sigma = 1.4$, $\beta = 200$, $J_t \sim \exp(1/\mu_J)$, $\mu_J = 0.4$, $\lambda = 4$.

Note also that, although X and Y are both Markov processes, the price process S is not. We will therefore assume that all three components of the price process, i.e. the time t values $f(t)$, X_t and Y_t, are individually observable, implying we expect jumps not to be small.

Figure 25.1 shows a simulated sample path of the processes X, Y and the composed process S.

In Section 25.2 we derive important stochastic properties of the process, including the moment generating function and various approximations to its probability density function. Pricing of a variety of derivative contracts will be discussed in Sections 25.3 and 25.4, using the results obtained in Section 25.2.

25.2 Properties of the Model for Spot Prices

25.2.1 The Spike Process

The following result is known and given by Duffie *et al.* (2000) in a more general framework.

Lemma 25.2.1: Moment Generating Function of the Spike Process, Y_t

Let $\{J_1, J_2, \ldots\}$ be a series of i.i.d. random variables. We assume that there is a $\theta_0 > 0$ such that the moment generating function $\Phi_J(\theta) := \mathbb{E}\left[e^{\theta J}\right]$ exists for $\theta < \theta_0$. Let $\{\tau_1, \tau_2, \ldots\}$ be the random jump times of a Poisson process N with intensity λ. Then the process Y with initial condition $Y_0 = 0$ has, for all $\theta < \theta_0$, $t \geq 0$, the moment generating function

$$\Phi_Y(\theta, t) := \mathbb{E}\left[e^{\theta Y_t}\right] = \exp\left(\lambda \int_0^t \Phi_J\left(\theta e^{-\beta s}\right) - 1 ds\right). \tag{25.4}$$

Furthermore, the first two moments of Y_t are given by

$$\mathbb{E}\left[Y_t\right] = \Phi_Y'(0, t) = \frac{\lambda}{\beta} \mathbb{E}[J]\left(1 - e^{-\beta t}\right),$$

$$\mathbb{E}\left[Y_t^2\right] = \Phi_Y''(0, t) = \mathbb{E}\left[Y_t\right]^2 + \frac{\lambda}{2\beta} \mathbb{E}\left[J^2\right]\left(1 - e^{-2\beta t}\right),$$

and in particular we have

$$\mathrm{var}\big[Y_t\big] = \frac{\lambda}{2\beta}\mathbb{E}\big[J^2\big]\big(1 - e^{-2\beta t}\big).$$

Remark 1 (asymptotics for $\beta \to \infty$): As remarked above, in practice the timescale $1/\beta$ for mean reversion of spikes is much shorter than any of: the contract lifetime T; the diffusive mean-reversion time $1/\alpha$; the volatility timescale $1/\sigma^2$; and the mean arrival time of spikes $1/\lambda$. We therefore calculate approximations for the moment generating function of the spike process, and for its distribution, as $\beta \to \infty$.

To analyse the behaviour of the moment generating function for large β we make the substitution $u = \theta e^{-\beta s}$ in the integrand to obtain

$$\Phi_{Y_t}(\theta) = \exp\left(\frac{\lambda}{\beta}\int_{\theta e^{-\beta t}}^{\theta}\frac{\Phi_J(u)-1}{u}du\right).$$

For fixed θ, t, as $\beta \to \infty$ we have

$$\int_0^{\theta e^{-\beta t}}\frac{\Phi_J(u)-1}{u}du = \theta e^{-\beta t}\mathbb{E}[J] + \mathrm{O}\big(e^{-2\beta t}\big),$$

because $\Phi_J(u) = 1 + \mathbb{E}[J]u + \mathrm{O}(u^2), u \to 0$, and so

$$\Phi_{Y_t}(\theta) = \exp\left(\frac{\lambda}{\beta}\left(\int_0^{\theta}\frac{\Phi_J(u)-1}{u}du - \theta e^{-\beta t}\mathbb{E}[J] + \mathrm{O}\big(e^{-2\beta t}\big)\right)\right). \tag{25.5}$$

Example 25.2.2: Exponentially Distributed Jump Size

If $J \sim \mathrm{Exp}(1/\mu_J)$ with mean jump size μ_J, then $\Phi_J(\theta) = 1/(1 - \mu_J)$ exists for $\theta < \theta_0 = 1/\mu_J$. We obtain

$$\Phi_Y(\theta, t) = \left(\frac{1 - \theta\mu_J e^{-\beta t}}{1 - \theta\mu_J}\right)^{\lambda/\beta}, \quad \theta < 1/\mu_J$$

As $t \to \infty$, we have $\Phi_Y(\theta, t) \to (1 - \theta\mu_J)^{-\lambda/\beta}$ for $\theta < 1/\mu_J$ so the stationary distribution for Y is the Gamma distribution $Gamma(\lambda/\beta, 1/\mu_J)$. As $\beta \to \infty$, we also have $\Phi_Y(\theta, t) = 1 + \theta\mu_J\lambda/\beta + O(\beta^{-2})$ $\approx (1-\theta\mu_J)^{-\lambda/\beta}$ for $\theta < 1/\mu_J$. Thus Y_t is distributed approximately as $Gamma(\lambda/\beta, 1/\mu_J)$ for large β.

The mean and variance of the spike process Y_t with $Y_0 = 0$ are

$$\mathbb{E}\big[Y_t\big] = \frac{\lambda\mu_J}{\beta}\big(1 - e^{-\beta t}\big), \quad \mathrm{var}\big[Y_t\big] = \frac{\lambda\mu_J^2}{\beta}\big(1 - e^{-2\beta t}\big).$$

25.2.2 The Combined Process

Having examined the properties of the spike process Y we conclude properties of the sum $X_t + Y_t$ and consequently of the price $S_t = \exp(f(t) + X_t + Y_t)$.

Theorem 25.2.3:

Let the spot process S be defined by (25.1) and let $Z_t := \ln S_t = f(t) + X_t + Y_t$ with X_0 and Y_0 given. The moment generating function of Z_t exists for $\theta < \theta_0$ and is given by

$$\mathbb{E}e^{\theta Z t} = \exp\left(\theta f(t) + \theta X_0 e^{-\alpha t} + \theta^2 \frac{\sigma^2}{4\alpha}\left(1 - e^{-2\alpha t}\right) \right.$$
$$\left. + \theta Y_0 e^{-tt} + \lambda \int_0^t \Phi_J\left(\theta e^{-\beta s}\right) - 1 ds \right) \tag{25.6}$$

Proof: The processes X and Y are independent so the expectation of the product is the product of the expectations. The moment generating function of Y_t is given in Lemma 25.2.1 which yields the result.

If $\theta_0 > 1$, the expectation value of the spot process at time t, S_t, immediately follows by setting $\theta = 1$. We note that, in general, the moments for the spot price $\mathbb{E}S_t^\theta$ only exist for $\theta < \theta_0$. For instance, if the jump distribution is exponential with mean μ_J, the spot price has less than $1/\mu_J$ moments.

25.2.3 Approximations

We now derive approximations to the density functions of the spike process at maturity T for large mean-reversion values β of the spike process. Although we can always compute the density by Laplace inversion of the moment generating function, an explicit expression for the density allows for more efficient algorithms and more explicit option pricing formulae. Here we only provide an expression for the density of a 'truncated' spike process $\tilde{Y}_T$ as defined below. However, knowledge of the density of $\tilde{Y}_T$ alone will help us to efficiently construct a grid to price swing options. We start by defining the truncated spike process $\tilde{Y}$, showing that $\tilde{Y}_T$ provides a good approximation to the value $\tilde{Y}_T$ for large values of β, deriving a general formula for the density function of $\tilde{Y}_T$ and finally making it more explicit by considering an exponential jump size distribution.

For very high mean-reversion rates β and small jump intensities λ, the dominant contribution to the density of the spike process comes from the last jump. We therefore introduce the truncated spike process

$$\tilde{Y}_t := \begin{cases} J_{N_t} e^{-\beta(t - \tau_{N_t})}, & N_t > 0, \\ 0, & N_t = 0. \end{cases} \tag{25.7}$$

Note that we only consider Y starting from 0; any other starting point can be incorporated by adding the initial value.

Lemma 25.2.4:

$\tilde{Y}_T$ is identically distributed as

$$Z_t := \begin{cases} J_1 e^{-\beta \tau_1}, & \tau_1 \leq t, \\ 0, & \tau_1 > t. \end{cases}$$

Proof: We use the reversibility property that if $N = \{N_t ; t \in \mathbb{R}_+\}$ is a Poisson process, then $\hat{N} = \{-N_{-t} ; t \in \mathbb{R}_+\}$ is also a Poisson process. As τ_{N_t} is the jump time of the last jump before t, this translates into the first jump of the reversed process and hence $t - \tau_{N_t}$ and τ_1 are identically distributed, given $N_t > 0$. If $N_t = 0$, then there has been no jump in $[0, t]$ and the same applies for the reversed process and so this is equivalent to $\tau_1 > t$.

Lemma 25.2.5 (moment generating function of the truncated spike process):

The random variable $\tilde{Y}_t$ of the truncated spike process at time t with initial condition $\tilde{Y}_0 = 0$ has a moment generating function for $\theta \prec \theta_0$ which is given by

$$\Phi_{Y_t}(\theta) = 1 + \lambda \int_0^t \left(\Phi_J\left(\theta e^{-\beta s}\right) - 1\right) e^{-\lambda s} ds.$$

The first two moments are given by

$$\mathbb{E}\left[\tilde{Y}_t\right] = \frac{\lambda}{\beta + \lambda} \mathbb{E}[J]\left(1 - e^{-(\beta + \lambda)t}\right),$$

$$\mathbb{E}\left[\tilde{Y}_t^2\right] = \frac{\lambda}{2\beta + \lambda} \mathbb{E}\left[J^2\right]\left(1 - e^{-(2\beta + \lambda)t}\right).$$

Proof: By Lemma 25.2.4 we only need to determine the moment generating function of

$$Z_t := J e^{-\beta \tau} I_{\tau \le t}, \quad \tau \sim \text{Exp}(\lambda),$$

where I_A denotes the indicator function for the event A. Given the jump time τ we have

$$\mathbb{E}\left[e^{\theta Z_t}|\tau = s\right] = \Phi_J\left(\theta e^{-\beta s} I_{s \le t}\right),$$

and so

$$
\begin{aligned}
\mathbb{E}\left[e^{\theta Z_t}\right] &= \mathbb{E}\left[\mathbb{E}\left[e^{\theta Z_t}|\tau\right]\right] \\
&= \int_0^\infty \Phi_J\left(\theta e^{-\beta s} I_{S \le t}\right) \lambda e^{-\lambda s} ds \\
&= \int_0^t \Phi_J\left(\theta e^{-\beta s}\right) \lambda e^{-\lambda s} ds + e^{-\lambda t}.
\end{aligned}
$$

The first two moments are given by $\mathbb{E}\left[\tilde{Y}_t\right] = \Phi'_{\tilde{Y}_t}(0)$ and $\mathbb{E}\left[\tilde{Y}_t^2\right] = \Phi''_{\tilde{Y}_t}(0)$.

Remark 2 (pointwise convergence of the moment generating functions): The moment generating function of the truncated spike process converges pointwise to the moment generating function of the spike process for either $\lambda \to 0$ or $\beta \to \infty$ with t and θ fixed. First consider $\lambda \to 0$. Fix all other parameters and set $g(s; \beta, \theta) := \Phi_J(\theta e^{-\beta s}) - 1$, then

$$
\begin{aligned}
\Phi_{Y_t}(\theta) &= \exp\left(\lambda \int_0^t g(s; \beta, \theta) ds\right) \\
&= 1 + \lambda \int_0^t g(s; \beta, \theta) ds + O\left(\lambda^2\right), \\
\Phi_{\tilde{Y}_t}(\theta) &= 1 + \lambda \int_0^t g(s; \beta, \theta) e^{-\lambda s} ds \\
&= 1 + \lambda \int_0^t g(s; \beta, \theta) ds + O\left(\lambda^2\right).
\end{aligned}
$$

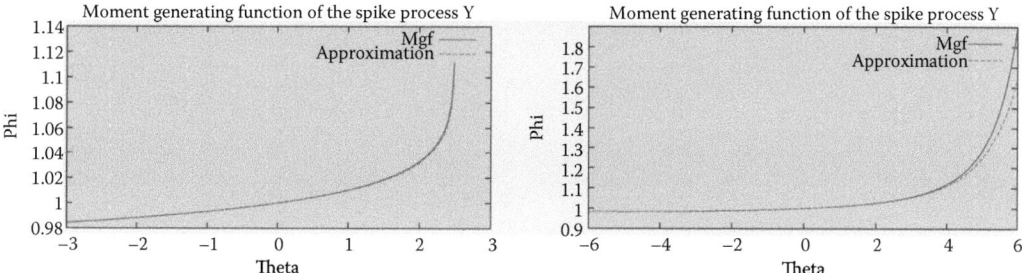

FIGURE 25.2 Moment generating function of Y_t and $\tilde{Y}_T$ denoted by mgf and approximation, respectively. On the left we use $J \sim \text{Exp}(1/\mu_J)$ and on the right $J \sim N\left(\mu_J, \mu_J^2\right)$. The same parameters as in Figure 25.1 are used and we set $t = 1$.

To see the convergence for $\beta \to \infty$ with λ, t and θ fixed, first note that from (25.5) we have

$$\Phi_{Y_t}(\theta) = 1 + \frac{\lambda}{\beta} \int_0^\theta \frac{\Phi_J(u) - 1}{u} du + O\left(1/\beta^2\right).$$

Also from Lemma 25.2.5,

$$\Phi_{\tilde{Y}_t}(\theta) = 1 + \lambda \int_0^t \left(\Phi_J\left(\theta e^{-\beta s}\right) - 1\right) e^{-\lambda s} ds$$

$$= 1 + \frac{\lambda}{\beta} \int_{\theta_e - \beta t}^\theta \frac{\Phi_J(u) - 1}{u} \left(\frac{u}{\theta}\right)^{\lambda/\beta} du$$

by setting $\theta e^{-\beta s} = u$. Now as $\beta \to \infty$, $(u/\theta)^{\lambda/\beta} \to 1$ except in a small region $u = O(\theta e^{-\beta/\lambda})$, which makes a negligible (exponentially small) contribution to the integral. Likewise, we may replace the lower limit of integration by 0 and incur a similarly small error. Hence,

$$\Phi_{\tilde{Y}_t}(\theta) = 1 + \frac{\lambda}{\beta} \int_0^\theta \frac{\Phi_J(u) - 1)}{u} du + o\left(\frac{1}{\beta}\right)$$

$$= \Phi_{Y_t}(\theta) + o\left(\frac{1}{\beta}\right).$$

Two examples of the approximated and exact moment generating function using our standard parameters can be seen in Figure 25.2.

Lemma 25.2.6 (distribution of the truncated spike process):

Let the jump size distribution have density function f_J. Then the truncated spike process $\tilde{Y}_t$ as defined above has the cumulative distribution function

$$F_{\tilde{Y}_t}(x) = e^{-\lambda t} I_{x \geq 0} + \int_{-\infty}^x f_{\tilde{Y}_t}(y) dy \quad t \geq 0,$$

with probability density function

$$f_{\tilde{Y}_t}(x) = \frac{\lambda}{\beta} \frac{1}{|x|^{1-\lambda/\beta}} \left| \int_x^{x e^{\beta t}} f_{J_t}(y) |y|^{-\lambda/\beta} dy \right|, \quad x \neq 0. \tag{25.8}$$

Proof: Based on Lemma 25.2.4 it suffices to determine the distribution of

$$\tilde{Y}_t = JZI_{\tau \le t}, \quad Z := e^{-\beta\tau}, \quad \tau \sim \text{Exp}(\lambda).$$

It follows that Z is the (β/λ)th power of a uniformly distributed random variable on $[0, 1]$ and its density is given by

$$f_z(x) = \frac{\lambda}{\beta} x^{-(1-(\lambda/\beta))} I_{x \in [0,1]}.$$

As $\mathbb{P}(\tau > t) = e^{-\lambda t}$ we obtain the cdf of $ZI_{\tau \le t}$ as

$$F_{ZI_{\tau \le t}}(x) = e^{-\lambda t} I_{x \ge 0} + \int_{-\infty}^{\infty} f_{ZI_{\tau \le t}}(y) dy,$$

$$f_{ZI_{\tau \le t}}(x) = \frac{\lambda}{\beta} x^{-(1-(\lambda/\beta))} I_{x \in \left[e^{-\beta t}, 1\right]},$$

and the distribution of the product of two independent random variables J and $ZI_{\tau \le t}$ is then given by

$$F_{JZI_{\tau \le t}}(c) = e^{-\lambda t} I_{c \ge 0} + \int_{-\infty}^{c} f_{JZI_{\tau \le t}}(x) dx,$$

$$f_{JZI_{TSI}}(c) = \int_{-\infty}^{\infty} f_{ZI_{\tau \le t}}(c/x) \frac{f_J(x)}{|x|} dx.$$

With

$$f_{ZI_{\tau \le t}}(c/x) = \frac{\lambda}{\beta} \frac{1}{c^{1-(\lambda-\beta)}} I_{x \in \left[c, ce^{\beta t}\right]} x^{1-(\lambda/\beta)}, \quad c > 0,$$

$$f_{ZI_{\tau \le t}}(c/x) = \frac{\lambda}{\beta} \frac{1}{|c|^{1-(\lambda/\beta)}} I_{x \in \left[ce^{\beta t}, c\right]} |x|^{1-(\lambda/\beta)}, \quad c < 0,$$

the desired result follows.

Example 25.2.7 (exponential jump size distribution):

Let $J \sim \text{Exp}(1/\mu_J)$ be exponentially distributed. The probability density function for the truncated spike process $\tilde{Y}_T$ is

$$f_{Y_t}(x) = \frac{\lambda}{\beta\mu_J^{\lambda/\beta}} \frac{\Gamma\left(1 - \lambda/\beta, x/\mu_J\right) - \Gamma\left(1 - \lambda/\beta, xe^{\beta t}/\mu_J\right)}{x^{1-\lambda/\beta}}, \quad x > 0, \tag{25.9}$$

where $\Gamma(a, x)$ is the incomplete Gamma function. The approximation is a good fit to the exact density for typical market parameters as can be seen in Figure 25.3. The only discrepancy occurs at $Y_t = 0$ where the density has a singularity. We use this approximation in Section 25.4.1 to efficiently generate a grid to price swing options.

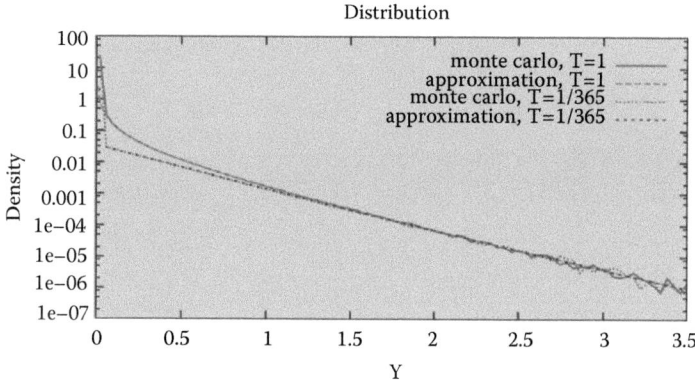

FIGURE 25.3 Distribution of the spike process (Y_t) at T with a jump size of $J \sim \mathrm{Exp}(1/\mu_J)$. We use approximation (25.9) and compare it with the exact density as produced by a Monte-Carlo simulation. We use the same parameters as in Figure 25.1.

25.3 Option Pricing

The electricity market with the model presented is obviously incomplete. Not only are we faced with a non-hedgeable jump risk but also we cannot use the underlying process (S_t) to hedge derivatives due to inefficiencies in storing electricity. From now on we assume the model is specified in the risk-neutral measure $\mathbb{Q}$ as

$$
\begin{aligned}
S_t &= \exp\big(f(t) + X_t + Y_t\big), \\
dX_t &= -\alpha X_t dt + \sigma dW_t, \\
dY_t &= -\beta Y_{t-} dt + J_t dN_t,
\end{aligned}
\tag{25.10}
$$

where W is a Brownian motion under $\mathbb{Q}$ and N a Poisson process with intensity λ under $\mathbb{Q}$. For simplicity of notation we use the same parameters as in (25.1) but note that they might differ from the parameters under the real-world measure. A common modelling assumption is that the risk-neutral model has a jump structure that is similar to that observed under $\mathbb{P}$. This is certainly true for a discrete jump size distribution, so we will assume that it is the case here.

Lemma 25.3.1 (seasonal function consistent with the forward curve):

Let $t = 0$ and $F_0^{[T]}$ be the forward at time 0 maturing at time T; then the risk-neutral seasonality function is given by

$$
f(T) = \ln F_0^{[T]} - X_0 e^{-\alpha T} - Y_0 e^{-\beta T} - \frac{\sigma^2}{4\alpha}\left(1 - e^{-2\alpha T}\right)
$$

$$
- \lambda \int_0^T \Phi_J\left(e^{-\beta s}\right) - 1 ds.
\tag{25.11}
$$

Proof: The forward price is $F_0^{[T]} = \mathbb{E}^{\mathbb{Q}}\big[S_T\big]$ and so the result follows from (25.6).

This result forms an important part of the calibration of the model. As the model is incomplete the calibration procedure depends on the set of liquid derivatives used. Here we assume a continuous

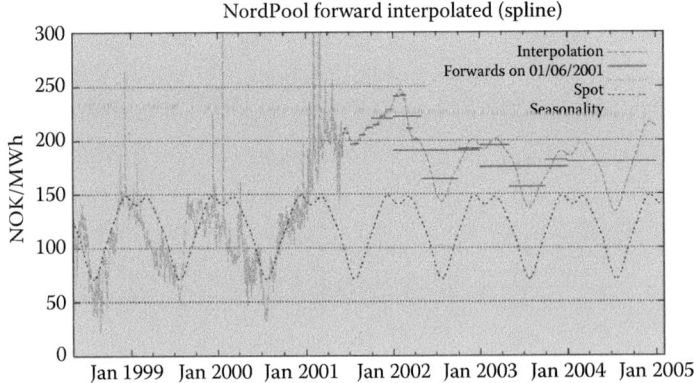

FIGURE 25.4 Interpolation of the forward curve by a seasonal function and spline correction. Three years, worth of spot history data has been used to calibrate a seasonality function which is then used as a first approximation of the forward curve. The difference between the seasonality function and the observed forward prices is then corrected by a piecewise quadratic polynomial.

for- ward curve is observable in the market, i.e. values of $F_0^{[T]}$ are given. This is not a realistic assumption but there are ways to generate a continuous curve consistent with discretely observed prices; see Figure 25.4 and Kluge (2006, Section 2.2.2.5) for more details.

For the sake of simplicity we adopt a policy of choosing parameters from the real-world measure $\mathbb{P}$ if they are not uniquely determined by the set of observed derivative prices. This is equivalent to saying we choose a risk-neutral measure $\mathbb{Q}$ which changes as few parameters of the model as possible. As there are at least five parameters (α, β, σ, λ, μ_J) and seasonality to fit we need to trade-off those calculated using historical data and the observed derivative prices. The volatility parameter σ remains unchanged by any equivalent measure change so we can always determine it from historical data. If we only see forward prices in the market we can also calibrate all other parameters to historical data except the seasonal function. In brief, this can be done by de-seasonalizing the data using the real-world seasonality and then making a first estimation of α from which we can start filtering suspected spikes. From the reduced dataset, α can be re-estimated and suspected spikes filtered recursively. Having determined all parameters from historical data we finally calculate the risk-neutral seasonality function f from the observed forward curve based on (25.11).

25.3.1 Pricing Path-Independent Options

If the pay-off of an option on the spot at maturity T is given by $g(S_T)$ then its arbitrage free price at time t is given by

$$V(x,y,t) = e^{-r(T-t)} \mathbb{E}^Q \Big[g(S_T) \,\big|\, X_t = x, Y_t = y \Big].$$

Although we do not have an expression for the density of S_T we know its moment generating function and so can apply Laplace transform methods to calculate the expectation value. For an overview, see Cont and Tankov (2004, Section 11.1.3)* or Carr and Madan (1998) and Lewis (2001). Consider, for example, put or call options. Let $Z_t = \ln S_t$ and let $\Phi_{Z_t}(\theta)$ be its moment generating function, as given in (25.6). Now define its truncated moment generating function by

---

* They describe the method in terms of a complex valued characteristic function and Fourier inversion, but by allowing complex values the method can also be written in terms of Laplace transforms.

$$G_v(x,t) := \mathbb{E}\left[e^{vZ_t} I_{[Z_t \leq x]}\right] = \int_{-\infty}^{x} e^{vy} \, dF_{Z_t}(y),$$

which can be computed using a generalization of Lévy's inversion theorem:

$$G_v(x) = \frac{\Phi_{Z_t}(v)}{2} - \frac{1}{\pi}\int_0^{\infty} \frac{\Im\left(\Phi_{Z_t}(v + 1\theta)e^{-1\theta x}\right)}{\theta} d\theta.$$

The price of a put option is then

$$\mathbb{E}\left[\left(K - S_T\right)^+\right] = K\mathbb{E}\left[I_{S_T \leq K}\right] - \mathbb{E}\left[S_T I_{S_T \leq K}\right]$$

$$= KG_0(\ln K) - G_1(\ln K),$$

and by put–call parity we obtain the price of a call option.

25.3.2 Pricing Options on Forwards

For a forward contract at time t, understood to be today, maturing at T the strike of a zero-cost forward is given by

$$F_t^{[T]} = \mathbb{E}^Q\left[S_T \mid \mathcal{F}_t\right].$$

The most common options on forwards are puts or calls maturing at the same time as the underlying forward, i.e. the pay-off is given by $\left(F_T^{[T]} - K\right)^+$, which is equivalent to $(S_T - K)^+$. We can price these contracts based on the dynamics of the spot and using methods developed above. However, by analysing the dynamics of the forward curve implied by the spot price model we will gain further insights and be able to relate the price of an option to the Black-76 formula (Black 1976), which is still widely used in practice.

Recall that the expectation value of S_T is equal to the moment generating function given in (25.6) at $\theta = 1$. For $F_t^{[T]} = \mathbb{E}^Q\left[S_T \mid X_t \cdot Y_t\right]$ we obtain

$$F_t^{[T]} = \exp\left(f(T) + X_t e^{-\alpha(T-t)} + Y_t e^{-\beta(T-t)} + \frac{\sigma^2}{4\alpha}\left(1 - e^{-2\alpha(T-t)}\right) + \lambda\int_0^{T-t} \Phi_J\left(e^{-\beta s}\right) - 1 ds\right). \tag{25.12}$$

For fixed T, the dynamics of the forward maturing at T is then

$$\frac{dF_t^{[T]}}{F_t^{[T]}} = -\lambda\left(\Phi_J\left(e^{-\beta(T-t)}\right) - 1 dt + \sigma e^{-\alpha(T-t)}\right)dW_t$$

$$+ \left(\exp\left(J_t e^{-\beta(T-t)}\right) - 1\right)dN_t. \tag{25.13}$$

The forward is a martingale under Q by definition, and so the drift term compensates the jump process. For large time to maturities $T-t$, a jump in the underlying process has only very limited effect on the forward. More precisely, if the relative change in the underlying is $\exp(J_t)-1$ the forward changes relatively by $\exp(J_t e^{-\beta(T-t)})-1$. In addition to the jump component the dynamics follows a deterministic volatility process starting with a low volatility $\sigma e^{-\alpha T}$ at $t = 0$ and increasing to σ at maturity. Without the jump component there are clear similarities with the Black-76 model.

For pricing purposes we need to find the distribution of $F_T^{[T]}$ in terms of its initial condition $F_t^{[T]}$. We have

$$\ln F_T^{[T]} = f(T) + X_T + Y_T,$$

$$\ln F_T^{[T]} = f(T) + X_t e^{-\alpha(T-t)} + Y_t e^{-\beta(T-t)} + \frac{\sigma^2}{4\alpha}\left(1 - e^{-2\alpha(T-t)}\right) \tag{25.14}$$

$$+\lambda \int_0^{T-t} \Phi_J\left(e^{-\beta s}\right) - 1 \, ds.$$

Eliminating the seasonality component $f(T)$, and using the relations

$$X_T - X_t e^{-\alpha(T-t)} = \sigma \int_t^T e^{-\alpha(T-s)} \, dW_s,$$

$$Y_T - Y_t e^{-\beta(T-t)} = \sum_{i=N_t}^{N_T} J_{\tau_i} e^{-\beta(T-\tau_i)},$$

we finally get

$$\ln F_T^{[T]} = \ln F_t^{[T]} + \sigma \int_t^T e^{-\alpha(T-s)} \, dW_s + \sum_{i=N_i}^{N_t} J_{\tau_i} e^{-\beta(T-\tau_i)}$$

$$+\frac{\sigma^2}{4\alpha}\left(1 - e^{-2\alpha(T-t)}\right) + \lambda \int_0^{T-t} \Phi_J\left(e^{-\beta s}\right) - 1 \, ds.$$

Without the jump component, $F_T^{[T]}$ would be log-normally distributed. In order to relate the pricing of options to the Black-76 formula even in the presence of spike risks, we assume that $F_T^{[T]}$ is log-normally distributed in a first approximation. We ignore the heavy tails caused by the spike risk and so expect to underestimate prices of far out-of-the-money calls but should do well with at-the-money calls.

We define the approximation by matching the first two moments but take into account that by definition $F_T^{[T]}$ is a martingale for a fixed maturity T and in order to keep the same property we set

$$\ln F_T^{[T]} \approx \ln F_t^{[T]} + \xi, \quad \xi \sim N\left(-\frac{1}{2}\breve{\sigma}^2(T-t), \breve{\sigma}^2(T-t)\right),$$

and $\hat{\sigma}^2(T-t) := \operatorname{var}\left[\ln F_T^{[T]} \mid F_t\right]$, i.e.

$$\hat{\sigma}^2(T-t) = \operatorname{var}\left[\sigma \int_t^T e^{-\alpha(T-s)} \, dW_s + \sum_{i=N_{it}}^{N_T} J_{t_{i=n_t}} e^{-\beta(T-t_i)}\right]$$

$$= \frac{\sigma^2}{2\alpha}\left(1 - e^{-2\alpha(T-t)}\right) + \frac{\lambda}{2\beta} \mathbb{E}\left[J^2\right]\left(1 - e^{-2\beta(T-t)}\right).$$

Remark 1 (term structure of implied volatility):

Comparing this result with the setting of Black-76 (Black 1976) where $dF = F\sigma \, dW$ and so $F_T = F_t \exp(\xi)$ with $\xi \sim N\left(-\frac{1}{2}\sigma^2(T-t), \sigma^2(T-t)\right)$, we conclude that $\hat{\sigma}$ is the implied Black-76 volatility and in a first approximation given by

$$\hat{\sigma}^2 \approx \frac{\left(\sigma^2/2\alpha\right)\left(1 - e^{-2\alpha(T-t)}\right) + (\lambda/2\beta)\mathbb{E}\left[J^2\right]\left(1 - e^{-2\beta(T-t)}\right)}{T-t}, \tag{25.15}$$

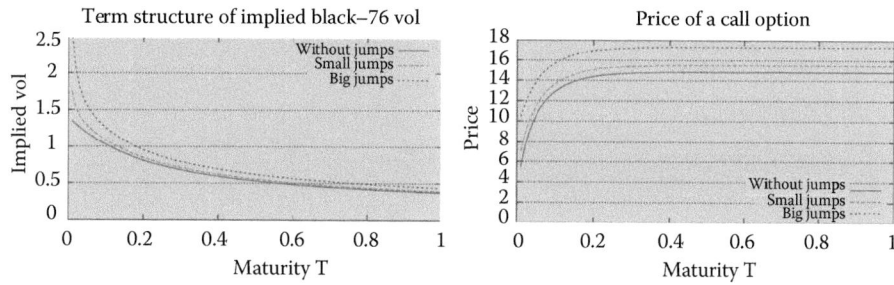

FIGURE 25.5   Implied volatilities and prices. The left graph shows implied volatilities with respect to time to maturity where approximation (25.15) is used. The three lines correspond to no jumps ($\mu_J = 0$), small jumps ($\mu_J = 0.4$) and big jumps ($\mu_J = 0.8$). In the right graph the corresponding prices of an at-the-money call are plotted. Parameters are $r = \ln(1.05)$, $\alpha = 7$, $\beta = 200$, $\sigma = 1.4$, $\lambda = 4.0$, $F_0^{[T]} = 100$, $K = 100$. Note, for an exponential distributed jump size J we have $\mathbb{E}\left[J^2\right] = 2\mu_j^2$.

which is shown in Figure 25.5. It can be seen that the spike process has a much more significant impact on the implied volatility for short maturities rather than for long-term maturities. As far as the price of an at-the-money call is concerned, the additional jump risk adds an almost constant premium to the price to be paid without any jump risk.

Remark 2 (implied volatility across strikes):

The approximation does not predict a change of implied volatility across strikes. However, the jump risk introduces a skew as can be seen in Figure 25.6 where the exact solution based on Section 25.3.1 has been used to calculate implied volatilities. The bigger the mean jump size and hence the bigger $\mathbb{E}\left[J_2\right]$, the more profound is the skew.

25.3.3 Pricing Options on Forwards with a Delivery Period

As electricity is a flow variable, forwards always specify a delivery period. The results of the previous section can therefore only be seen as an approximation to option prices on forwards with short delivery periods, like one day. Here we only consider options on forwards maturing at the beginning of the delivery period, i.e. the pay-off is given by some function of $F_{T_1}^{[T_1,T_2]}$ at time T_1. An option on such a forward is conceptually similar to an Asian option in the Black–Scholes world. One method of pricing Asian options is to approximate the distribution of the integral by a log-normal distribution and this can be done by matching the first two moments; see Turnbull and Wakeman (1991), for example. Once the parameters of the approximate log-normal distribution have been determined, pricing options comes down to pricing in the Black–Scholes or Black-76 setting.

The strike price of a zero cost forward with a delivery period is generally given by a weighted average of instantaneous forwards of the form

$$F_t^{[T_1,T_2]} = \int_{T_1}^{T_2} w\left(T;T_1,T_2\right)F_t^{[T]}\mathrm{d}T,$$

where for a settlement at maturity T_2 the weighting factor is given by $w(T; T_1, T_2) = 1/(T_2-T_1)$ and for instantaneous settlement the discounting alters the weighting to $w(T;T_1,T_2) = re^{-rT}/(e^{-rT_1} - e^{-rT_2})$.

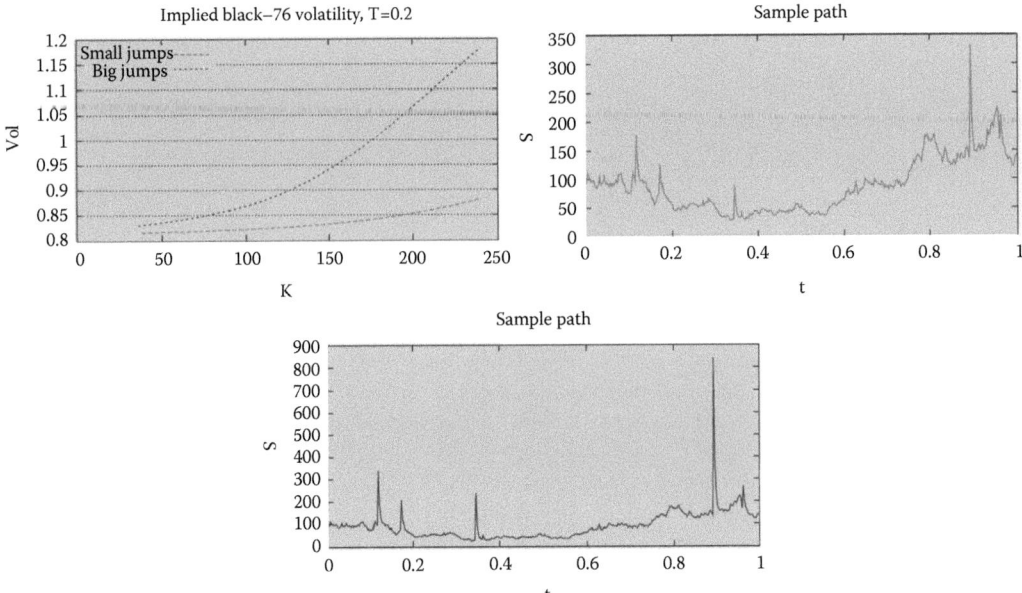

FIGURE 25.6 Implied volatilities across strikes and sample paths. The upper graph shows the implied volatility for one maturity $T = 0.2$ based on the exact solution. The approximate solution (25.15) yields 0.82 and 0.85 for the small and big jumps, respectively. Sample paths of the model with the same parameters are drawn in the lower two graphs, where the left path is generated with a low mean jump size ($\mu_J = 0.4$) and the right with a high mean jump size ($\mu_J = 0.8$). All the other parameters are the same as in Figure 25.5.

The second moment of $F_{T_1}^{[T_1, T_2]}$ is given by

$$\mathbb{E}^Q\left[\left(\int_{T_1}^{T_2} w(T) F_{T_1}^{[T]} dT\right)^2 \middle| \mathcal{F}_t\right]$$

$$= \int_{T_1}^{T_2}\int_{T_1}^{T_2} w(T) w\left(T^*\right) \mathbb{E}^Q\left[F_{T_1}^{[T]} F_{T_1}^{[T^*]} \middle| \mathcal{F}_t\right] dT\, dT^*,$$

and the expectation of the product of two individual forwards $\mathbb{E}^Q\left[F_{T_i}^{[T]} F_{T_1}^{[T^*]} \middle| \mathcal{F}_t\right]$ can be derived using the solution of the forward (25.12) as follows:

$$\ln F_{T_1}^{[T]} = \ln F_t^{[T]} + e^{-\alpha(T-T_1)}\sigma\int_t^{T_1} e^{-\alpha(T_1-s)} dW_s + e^{=\beta(T-T_1)}$$

$$\times \sum_{i=N_t}^{N_{T_1}} J_{\tau_i} e^{-\beta(T_1-\tau_i)} - \frac{\sigma^2}{4\alpha}\left(e^{-2\alpha(T-T_1)} - e^{-2\alpha(T-t)}\right)$$

$$+ \lambda\int_0^{T-T_1} \Phi_J\left(e^{-\beta s}\right) - 1\, ds - \lambda\int_0^{T-t} \Phi_J\left(e^{-\beta s}\right) - 1\, ds$$

$$= \ln F_t^{[T]} + e^{-\alpha(T-T_1)}\sigma\int_t^{T_1} e^{-\alpha(T_1-s)} dW_s + e^{=\beta(T-T_1)}$$

$$\times \sum_{i=N_t}^{N_1} J_{\cdot_i} e^{-\beta(T_1-\tau_i)} - \frac{\sigma^2}{4\alpha}\left(e^{-2\alpha(T-T_1)} - e^{-2\alpha(T-t)}\right)$$

$$- \lambda\int_0^{T_1-t} \Phi_J\left(e^{-\beta(T-T_1)}e^{-\beta s}\right) - 1\, ds,$$

and so

$$\ln F^{[T]}_{T_1} = \ln F^{\left[T^*\right]}_t$$

$$= \ln F^{[T]}_t + \ln F^{[T^*]}_t \left(e^{-\alpha(T-T_1)} + e^{-\alpha\left(T^*-T_1\right)} \right) \sigma$$

$$\times \int_t^{T_1} e^{-\alpha(T_1-s)} dW_s + \left(e^{-\beta(T-T_1)} + e^{-\beta\left(T^*-T_1\right)} \right)$$

$$\times \sum_{i=N_t}^{N_{T_1}} J_{\tau_i} e^{-\beta(T_1-\tau_i)} - \frac{\sigma^2}{4\alpha} \left(1 + e^{-2\alpha\left(T^*-T\right)} \right)$$

$$\times \left(e^{-2\alpha(T-T_1)} - e^{-2\alpha(T-t)} \right) - \ln\Phi_Y \left(e^{-\beta(T-T_1)} \right)$$

$$- \ln\Phi_Y \left(e^{-\beta\left(T^*-T_1\right)} \right),$$

which gives

$$\mathbb{E}^Q \left[F^{[T]}_{T_1} F^{\left[T^*\right]}_{T_1} \middle| \mathcal{F}_t \right]$$

$$= \mathbb{E}^Q \left[\exp\left(\ln F^{[T]}_{T_1} + F^{\left[T^*\right]}_{T_1} \middle| \mathcal{F}_t \right. \right.$$

$$= F^{[T]}_t F^{[T^*]}_t \right] \frac{\Phi_Y \left(e^{-\beta(T-T_1)} + e^{-\beta\left(T^*-T_1\right)} \right)}{\Phi_Y \left(e^{-\beta(T-T_1)} \right) \Phi_Y \left(e^{-\beta\left(T^*-T_1\right)} \right)}$$

$$\times \exp\left(-\frac{\sigma^2}{4\alpha} \left(1 + e^{-2\alpha\left(T^*-T\right)} \right) \left(e^{-2\alpha(T-T_1)} \right) - e^{-2\alpha(T-t)} \right)$$

$$\times \exp\left(\frac{\sigma^2}{4\alpha} \left(1 + e^{-\alpha\left(T^*-T\right)} \right)^2 \left(e^{-2\alpha(T-T_1)} - e^{-2\alpha(T-t)} \right) \right).$$

How well the moment matching procedure works is shown in Figure 25.7 where the density of a forward $F^{[T_1,T_2]}_{T_1}$ is compared with the density obtain by the approximation. The shapes of both densities are similar but differences in values are clearly visible. As a result, one might not expect call option prices based on the approximate distribution to be very close to the exact prices for all strikes K but still close enough to be useful. For an at-the-money strike call option, prices for varying maturities are shown in Figures 25.8 and 25.9. As it turns out, the approximation gives very good results for short delivery periods and is still within a 5% range for delivery periods of one year.

25.4 Pricing Swing Options

Swing contracts are a broad class of path-dependent options allowing the holder to exercise a certain right multiple times over a specified period but only one right at a time* or per time interval like a day.

---

* This will also involve a 'refraction period' in which no further right can be exercised.

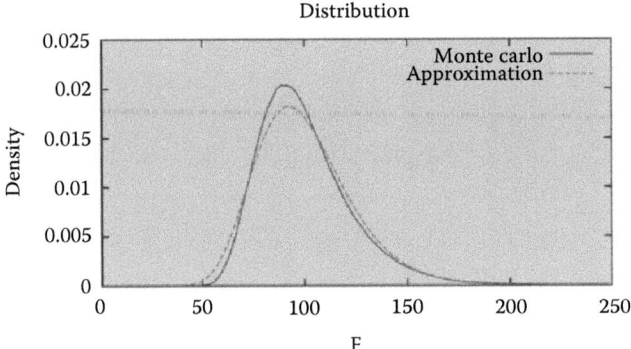

FIGURE 25.7 Distribution of $F_{T_1}^{[T_1,T_2]}$. Parameters are as before, see Figure 25.5, and $\mu_J = 0.4$, $T_1 = 1$, $T_2 = 1.25$. The red curve is based on a Monte-Carlo simulation with 10^6 sample paths; the green curve is the log-normal approximation matching the first two moments.

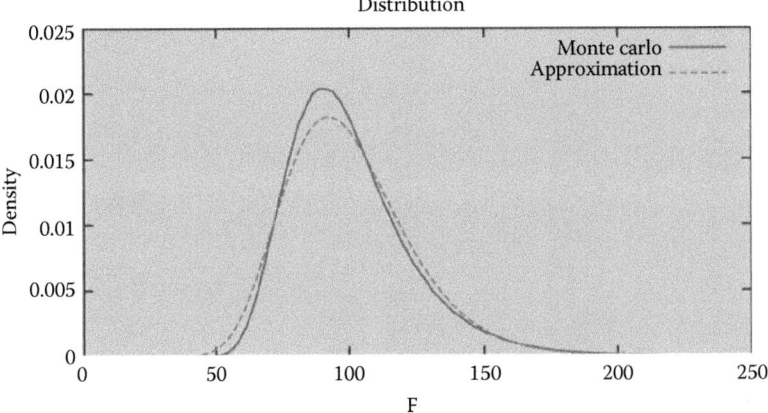

FIGURE 25.8 Value of an at-the-money call option on a forward $F_{T_1}^{[T_1,T_2]}$ depending on the delivery period $T_2 - T_1$. Parameters are as before, see Figure 25.5, and $\mu_J = 0.4$, $T_1 = 1$, $K = 100$. The Monte-Carlo result is based on 100,000 sample paths for each duration. Note, any forward $F_{T_1}^{[T_1,T_2]}$ delivers exactly 1 MWh over the delivery period $[T_1, T_2]$.

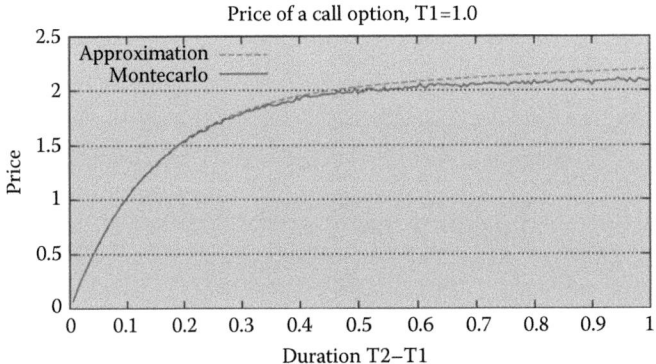

FIGURE 25.9 Same as in Figure 25.8 but where the volume of the forward is proportional to the delivery period, i.e. we assume a constant consumption of 1 MWh per year, which is about 114 W. Here, the price is simply $T_2 - T_1$ times the price of a standard call option on $F_{T_1}^{[T_1,T_2]}$.

Such a right could be the right to receive the pay-off of a call option. Other possibilities include a mixture of different pay-off functions like calls and puts or calls with different strikes. Another very common feature is to allow the holder to exercise a multiple of a call or put option at once, where the multiple is called volume. This generally involves further restriction on the volume, like upper and lower bounds for each right and for the sum of all trades.

Swing contracts can be seen as an insurance for the holder against excessive rises in electricity prices. Assuming the prices generally revert to a long-term mean, even a small number N of exercise opportunities suffices to cover the main risks and hence make the premium of the contract cheaper. Sometimes, swing contracts are bundled with forward contracts. The forward contract then supplies the holder with a constant stream of energy at a fixed pre-determined price. If the strike price of the call options of the swing contract is set to the forward price, the swing contract will allow for flexibility in the volume the customer receives for the fixed price. They can either 'swing up' or 'swing down' the volume of energy and hence the name swing contract. One cannot assume that the holder always exercises the contract in an optimal way to maximize expected profit but they might also exercise according to their own internal energy demands.

It is only very recently that articles on numerical pricing methods for swing options have appeared in the literature. We can identify a few main approaches, all based on dynamic programming principles. A Monte-Carlo method and ideas of duality theory are utilized by Meinshausen and Hambly (2004) to derive lower and upper bounds for swing option prices. The main advantages of the method being its flexibility, as it can be easily adapted to any stochastic model of the underlying, and its ability to produce confidence intervals of the price. Monte-Carlo techniques are also used by Ibanez (2004) and Carmona and Touzi (2008), where the latter uses the theory of the Snell envelope to determine the optimal exercise boundaries and also utilizes the Malliavin calculus for the computation of greeks. A constructive solution to the perpetual swing case for exponential Brownian motion is also given by Carmona and Touzi (2008). Unfortunately, these methods only work for the most basic versions of swing contracts where, at each time, only one unit of an option can be exercised.

More general swing contracts with a variable volume per exercise and an overall constraint can be priced with a tree-based method introduced by Jaillet *et al.* (2004).

In the above papers a discrete-time model for the underlying is used where one time step corresponds to the time frame in which no more than one right can be exercised, i.e. one day in most of the traded contracts. A special case where the number of exercise opportunities is equal to the number of exercise dates is considered by Howison and Rasmussen (2002) and a continuous optimal exercise strategy derived which yields a partial integro-differential equation for the option price.

Our method is based on the tree approach of Jaillet *et al.* (2004) with some slight modifications to adapt it to the peculiarities of our model for the underlying electricity price process.

25.4.1 The Grid Approach

The tree method of Jaillet *et al.* (2004) requires a discrete time model of the underlying. This is due to the fact that their swing contracts allow the holder to exercise at most one option within a specified time interval, say a day, and this is best modelled if the underlying process has the same time discretization. Assuming (S_t) is some continuous stochastic process for the spot price we obtain a discrete model by observing it on discrete points in time only, i.e.

$$S_{t_0}, S_{t_1}, S_{t_2}, \ldots, S_{t_m},$$

with $t_0 = 0$, $t_{i+1} = t_i + \Delta t$, $t_m = T$ and $\Delta t = \dfrac{1}{365}$, indicating we can exercise on a daily basis.

Let the maturity date T be fixed and the pay-off at time t for simplicity* be given by $(S_t - K)^+$ for some strike price K and we assume only one unit of the underlying can be exercised in any time period. Let

---

* We could assume any general payoff function.

$V(n, s, t)$ denote the price of such a swing option at time t and spot price s which has n out of N exercise rights left. The dynamic programming principle allows us to write

$$V(n,s,t) = \max \left\{ \begin{array}{l} e^{-r\Delta t}\mathbb{E}^Q\left[V\left(n,S_{t+\Delta t},t+\Delta t\right)|S_t = s\right], \\ e^{-r\Delta t}\mathbb{E}^Q\left[V\left(n-1,S_{t+\Delta t},t+\Delta t\right)|S_t = s\right], \\ \qquad +(s-K)^+ \end{array} \right\}, \tag{25.16}$$

$$n < N,$$

and $V(n, s, T) = (S - K)^+, 0 < n \le N$ and $V(0, s, t) = 0$. The conditional expectations can be written

$$\mathbb{E}^Q\left[V\left(n,S_{t+\Delta t},t+\Delta t\right)|S_t = s_i\right]$$

$$= \int_{-\infty}^{\infty} V(n,x,t+\Delta t)f_S(x;s)\mathrm{d}x$$

where $f_S(x; s)$ is the density of $S_{t+\Delta t}$ given $S_t = s$. Discretizing the spot variable we approximate

$$\mathbb{E}^Q\left[V\left(n,S_{t+\Delta t},t+\Delta t\right)|S_t = s_i\right]$$

$$\cong \sum_j V\left(n,s_j,t+\Delta t\right)f_S\left(s_j;s_i\right)\Delta s_j$$

This is only one possible approximation; others might be to use higher-order integration rules or using only a few grid points in the sum based on the fact that $f_S(x; s) \to 0$ for $|s - x|$ large. For a trinomial tree, one only uses three grid points, i.e.

$$\mathbb{E}^Q\left[V\left(n,S_{t+\Delta t},t+\Delta t\right)|S_t = s_i\right] \approx \sum_{j=-1}^{1} V\left(n,s_{i+j},t+\Delta t\right)p_{i,i+j},$$

$p_{i,j}$ being the probability of going from node i to node j. However, such a tree approach is not well suited to our case for two reasons. First, the time step size is determined by the shortest time between two possible exercise dates, which is mainly one day for swing contracts. This limits the accuracy of the algorithm as a refinement of the grid in the spot direction will not improve the result. Second, in the presence of jumps, a three-point approximation for the conditional density is insufficient due to the heavy tails in the distribution. As a result, we keep our method general and say

$$\mathbb{E}^Q\left[V\left(n,S_{t+\Delta t},t+\Delta t\right)|S_t = s\right] \approx \sum_j V\left(n,s_j,t+\Delta t\right)p_{i,j},$$

where $p_{i}j$ is an approximation to the density $f_S(s_j; s_i)\Delta s_j$ (it can accommodate higher-order integration rules and boundary approximations). With the notation $V_{i,k}^n := V\left(n,s_i,t_k\right)$ we can then write the method as

$$V_{i,k}^n = \max\left\{ e^{-r\Delta t}\sum_j V_{j,k+1}^n p_{i,j}, e^{-r\Delta t}\sum_j V_{j,k+1}^{n-1} p_{i,j} + \left(s_i - K\right)^+ \right\}, \tag{25.17}$$

$$V_{i,k}^0 = 0, \quad V_{i,m}^n = 0.$$

25.4.2 Numerical Results

We now turn to the model of interest, (25.10), which exhibits spikes. Assume that the mean-reversion process (X_t) and the spike process (Y_t) are individually observable and so the value function V of a swing option depends on both variables and the general pricing principle (25.16) becomes

$$V(n,x,y,t) = \max \begin{cases} e^{-r\Delta t}\mathbb{E}^Q\left[V\left(n, X_{t+\Delta t}, Y_{t+\Delta t}, t+\Delta t\right) \mid X_t = x, Y_t = y\right], \\ e^{-r\Delta t}\mathbb{E}^Q\left[V\left(n-1, X_{t+\Delta t}, Y_{t+\Delta t}, t+\Delta t\right) \mid X_t = x, Y_t = y\right] + \left(e^{f(t)+x+y} - K\right)^+. \end{cases}$$

In order to calculate conditional expectations we need to define transition probabilities. Given one starts at node $(X_t, Y_t) = (x_i, y_j)$ the probability to arrive at node $(X_{t+\Delta t}, Y_{t+\Delta t}) = (x_k, y_l)$ is approximately given by

$$p_{i,j,k,l} \approx f_{X_{t+\Delta t}|X_t=x_i}\left(x_k\right)f_{Y_{t+\Delta t}|Y_t=y_j}\left(y_l\right)\Delta x\Delta y,$$

because X_t and Y_t are independent. The conditional density of the mean-reverting process (X_t) is known as $X_{t+\Phi t}$ given $X_t = x$ is normally distributed with N $(xe^{-\alpha\Delta t}, (\sigma^2/2\alpha)(1-e^{-2\alpha\Delta t}))$. As we do not have a closed-form expression for the density of the spike process we use approximations developed in Section 25.2.1. For an exponential jump size distribution $J \sim \text{Exp}(1/\mu_J)$, for example, we use approximation (25.9) for the spike process at time t given zero initial conditions.

The introduction of a second space dimension increases the complexity of the algorithm considerably, essentially by a factor proportional to the square of the number of grid points in the y direction. To price the swing contract shown in Figure 25.10 which has 365 exercise dates and up to 100 exercise opportunities, our C++ implementation requires about 10 minutes to complete the calculation on an Intel P4, 3.4 GHz, and for a grid of 120×60 points in the x and y direction, respectively. The same computation but with no spikes and a grid of 120×1 points only takes about one second.

Based on Figure 25.10 we make two observations. First, the price per exercise right decreases with the number of exercise rights. This is the correct qualitative behaviour one would expect because n swing options each with one exercise right[*] only, offer more flexibility than one swing option with n exercise

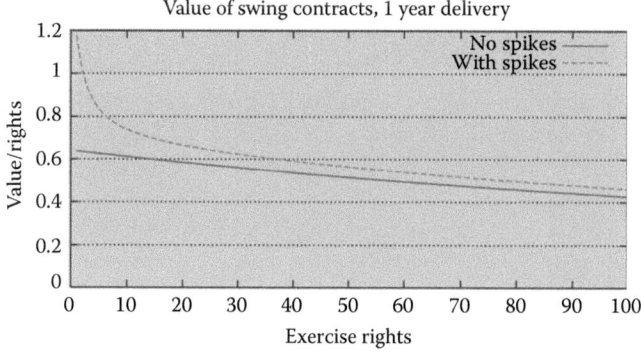

FIGURE 25.10 Value of a one year swing option per exercise right. Market parameters of the underlying are as before, see Figure 25.1, i.e. $\alpha = 7$, $\beta = 200$, $\sigma = 1.4$, $\lambda = 4$, $J \sim \text{Exp}(1/\mu_J)$ with $\mu_J = 0.4$, $f(t) = 0$, $r = 0$, and initial conditions $X_0 = 0$ and $Y_0 = 0$. The swing contract delivers over a time period of one year $T \in [0, 1]$ with up to 100 call rights and a strike price of $K = 1$, where a right can be exercised on any day. As a comparison the price of the same swing option is plotted but where the underlying does not exhibit spikes, i.e. $\lambda = 0$.

---

[*] This is actually an American option.

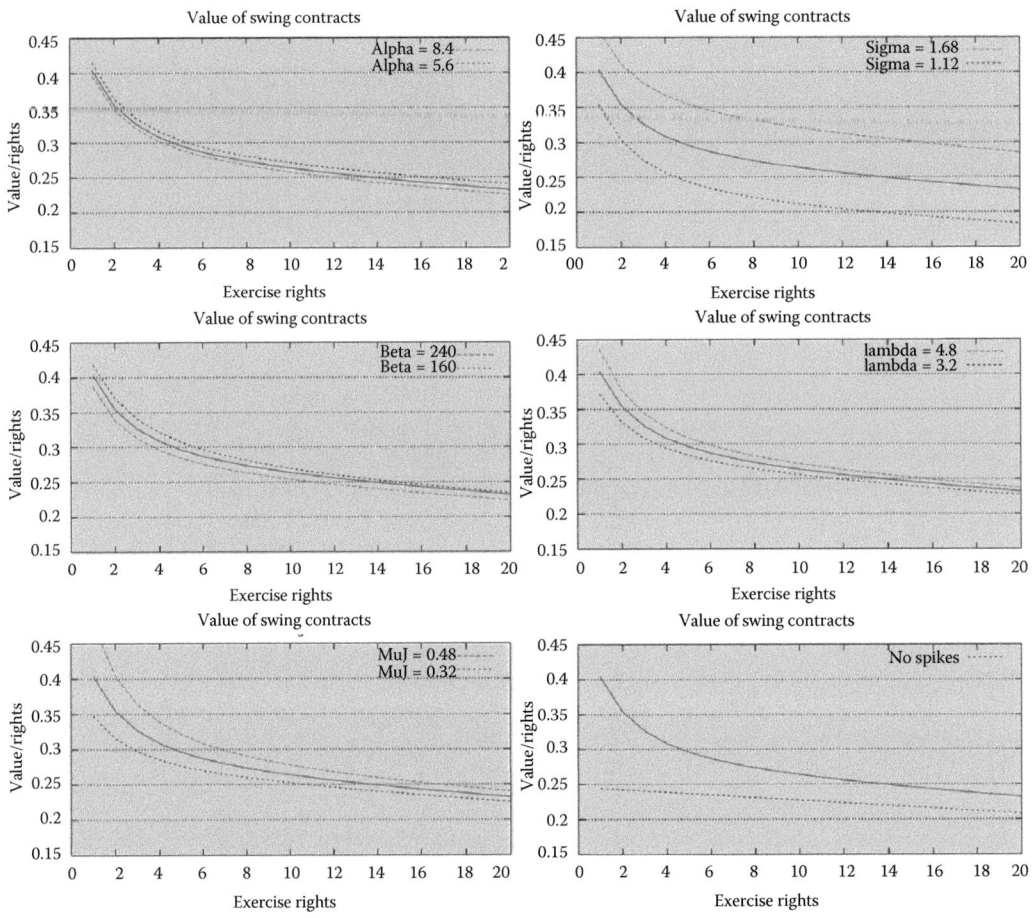

FIGURE 25.11 Sensitivity of swing option prices with respect to model parameters. A swing option with 60 exercise dates and up to 20 rights is considered, where the red curve is based on the parameters of Figure 25.10. In each graph, one market parameter is shifted up by 20% (green line) and down by 20% (blue line). We always plot the option price divided by the number of exercise rights.

rights.* Second, the premium added due to the spike risk is much more significant for options with small numbers of exercise rights than for a large number. This is also intuitively clear, as an option with say 100 exercise rights will mainly be used against high prices caused by the diffusive part and only occasionally against spiky price explosions.

In Figure 25.11 we show how sensitive swing option prices are to changes in market parameters. There we consider a swing option with a duration of 60 days and up to 20 exercise opportunities. In each graph we change one parameter by 20% up and down. The most significant change is caused by a change in the volatility parameter σ. Note, the long-term variance of the mean-reverting process (X_t) is $\sigma^2/2\alpha$ and we expect some direct relationship between the long-term variance and the option price. Hence, a change in the mean-reversion parameter α is inversely proportional to the price and quantitatively changes the price less than the volatility σ. The mean-reversion parameter β of the spike process has a similar effect on the option price as α has, but where the influence slightly decreases with the number of options. This is consistent with previous observations of the impact of jumps on option prices as seen in Figure 25.10.

---

* The rights of a swing option can only be exercised one at a time.

This effect is much more clearly visible for the other jump parameters λ and μ_J which have the greatest impact on options with only a few exercise rights. For one exercise right, a 20% change in the jump size μ_J has an even greater effect on the price than a 20% change in volatility σ. A possible explanation is that we deal with the exponential of an exponentially distributed jump size which is very heavy tailed.

References

Benth, F., Ekeland, L., Hauge, R. and Neilsen, B., A note on arbitrage-free pricing of forward contracts in energy markets. Appl. Math. Finan., 2003, **10**, 325–336.

Benth, F., Kallsen, J. and Meyer-Brandis, T., A non-Gaussian Ornstein-Uhlenbeck process for electricity spot price modeling and derivative pricing. Appl. Math. Finan., 2007, **14**, 153–169.

Black, F., The pricing of commodity contracts. J. Finan. Econ., 1976, **3**, 167–179.

Carmona, R. and Touzi, N., Optimal multiple stopping and valuation of swing options. Math. Finan., 2008, **18**, 239–268.

Carr, P. and Madan, D., Option valuation using the fast Fourier transform. J. Comput. Finan., 1998, **2**, 61–73.

Cartea, A. and Figueroa, M., Pricing in electricity markets: a mean reverting jump diffusion model with seasonality. Appl. Math. Finan., 2005, **12**, 313–335.

Clewlow, L. and Strickland, C., Energy Derivatives: Pricing and Risk Management, 2000 (Lacima Publications: London).

Cont, R. and Tankov, P., Financial Modelling with Jump Processes, 2004 (Chapman & Hall/CRC: New York).

Deng, S., Stochastic models of energy commodity prices and their applications: mean-reversion with jumps and spikes. Working paper PWP-073, 2000.

Duffie, D., Pan, J. and Singleton, K., Transform analysis and asset pricing for affine jump-diffusions. Econometrica, 2000, **68**, 1343–1376.

Howison, S. and Rasmussen, H., Continuous swing options. Working paper, 2002.

Ibanez, A., Valuation by simulation of contingent claims with multiple early exercise opportunities. Math. Finan., 2004, **14**, 223–248.

Jaillet, P., Ronn, E. and Tompadis, S., Valuation of commodity based swing options. Mgmt. Sci., 2004, **50**, 909–921.

Kluge, T., Pricing swing options and other electricity derivatives. PhD thesis, University of Oxford, 2006.

Klüppelberg, C., Meyer-Brandis, T. and Schmidt, A., Electricity spot price modelling with a view towards extreme spike risk. Preprint, 2008.

Knittel, C.R. and Roberts, M.R., An empirical examination of restructured electricity prices. Energy Econ., 2005, **27**, 791–817.

Koopman, S.J., Ooms, M. and Carnero, M.A., Periodic seasonal Reg-ARFIMA-GARCH models for daily electricity spot prices. J. Am. Statist. Assoc., 2007, **102**, 16–27.

Lewis, A., A simple option formula for general jump diffusion and other exponential Lévy processes. Working paper, 2001.

Lucia, J. and Schwartz, E., Electricity prices and power derivatives: evidence from the Nordic power exchange. Rev. Deriv. Res., 2002, **5**, 5–50.

Meinshausen, N. and Hambly, B., Monte Carlo methods for the valuation of options with multiple exercise opportunities. Math. Finan., 2004, **14**, 557–583.

Turnbull, S.M. and Wakeman, L.M., A quick algorithm for pricing European average options. J. Finan. Quant. Anal., 1991, **26**, 377–389.

Weron, R., Bierbrauer, M. and Trück, S., Modeling electricity prices: jump diffusion and regime switching. Physica A, 2004, **33**, 39–48.

26

Efficient Pricing of Swing Options in Lévy-Driven Models

Oleg Kudryavtsev
and
Antonino Zanette

26.1 Introduction

The motivation for this work comes from energy markets, where financial instruments are becoming increasingly important for risk management. In a deregulated market, energy contracts will need to be priced according to their financial risk. Due to the uncertainty of consumption and the limited fungibility of energy, new financial contracts such as swing options have been introduced in the commodity market. Swing options are an American-type option with many exercise rights. Their owner can exercise them many times under the condition that they respect the refracting time that separates two successive exercises. Swing options are also widely used in the gas and oil markets. Thus, pricing swing options will become increasingly important.

Many numerical methods, essentially based on the solution to the dynamic programming equation, have been introduced recently in the financial literature. In the context of swing options, two different probabilistic strategies have been developed. In the first, swing options are priced using an extension of the binomial tree algorithm, leading to the so-called forest tree (Lari-Lavassani *et al.* 2001, Jaillet *et al.* 2004). In the second, Monte Carlo methods are used, in which conditional expectations are computed using either regression techniques (Barrera-Esteve *et al.* 2006) or Malliavin calculus (Mnif and Zhegal 2006, Carmona and Touzi 2008). In particular, Carmona and Touzi (2008) propose a Monte Carlo approach to the problem of pricing American put options, in a finite time horizon, with multiple exercise rights in the case of geometric Brownian motion. In that paper, they introduce the inductive hierarchy of Snell envelopes needed in the multiple exercise case.

Energy expenditure increases sharply with higher daily temperature variation, and, consequently, the price varies. Although these spikes of power consumption are infrequent, they have a large financial

DOI: 10.1201/9781003265399-30

impact, and therefore, many authors propose pricing swing options in a model with jumps (Mnif and Zhegal 2006, Wilhelm and Winter 2008). Mnif and Zhegal (2006) extend the results of Carmona and Touzi (2008) to a market with jumps. In fact, the multiple stopping time problem for swing options can be reduced to a cascade of single stopping time problems in a Lévy market where jumps are permitted. With regards to deterministic methods, Wilhelm and Winter (2008) develop a finite element algorithm for pricing swing options in different models, including jump models.

Boyarchenko and Levendorskiĭ (2006) apply the Wiener–Hopf method to similar but distinct multistage investment or disinvestment problems (sequences of embedded perpetual American (real) options) under uncertainty, modelled as a monotone function of a Lévy process; in the case of the Kou model, closed-form solutions are given.

In this paper, we propose two approaches to solving multiple parabolic partial integro-differential equations (PIDEs) for pricing swing options in jump models. The first method, which is very simple and is introduced for comparison purposes, uses a finite difference scheme to solve the system of variational inequalities associated with the swing option problem approached through the splitting method proposed by Barles *et al.* (1995).

The second method uses fast Wiener–Hopf factorization (FWHF), introduced by Kudryavtsev and Levendorskiĭ (2009), where a fast and accurate numerical method for pricing barrier options for a wide class of Lévy processes was constructed. The FWHF method is based on an efficient approximation of the Wiener–Hopf factors and the Fast Fourier Transform algorithm. The advantage of the Wiener–Hopf approach over finite difference schemes in terms of accuracy and convergence was demonstrated by Kudryavtsev and Levendorskiĭ (2009). The FWHF-method was developed further in a series of papers by Kudryavtsev (2011, 2013). We will propose here a new efficient pricing algorithm for swing options that involves dynamic programming and the solving of multiple PIDEs by the FWHF method. We apply this algorithm for pricing swing options where the spot electricity price is a Lévy process that allows the consideration of jump risk. Numerical results, developed as in Wilhelm and Winter (2008) in the Black–Scholes and CGMY models, show the efficiency and accuracy of the proposed algorithms. The results presented in Chapter 28 were developed by Kudryavtsev and Zanette (2013).

The rest of the paper is organized as follows. Section 26.2 is devoted to the basic facts on Lévy processes. In Section 26.3, we present the multiple optimal stopping problem for swing options. In Sections 26.4 and 26.5, we propose, respectively, a finite difference and a Wiener–Hopf approach for pricing swing options. The numerical results are presented in Section 26.6.

26.2 Lévy Processes: Basic Facts

26.2.1 General Definitions

A Lévy process is a stochastically continuous process with stationary independent increments (for general definitions, see, e.g. Sato (1999)). A Lévy process may have a Gaussian component and/or a pure jump component. The latter is characterized by the density of jumps, which is called the Lévy density. A Lévy process X_t can be completely specified by its characteristic exponent, ψ, definable from the equality $E\left[e^{i\xi X(t)}\right] = e^{-t\psi(\xi)}$ (we confine ourselves to the one-dimensional case).

The characteristic exponent is given by the Lévy–Khintchine formula:

$$\psi(\xi) = \frac{\sigma^2}{2}\xi^2 - i\mu\xi + \int_{-\infty}^{+\infty}\left(1 - e^{i\xi y} + i\xi y \mathbf{1}_{|y|\leq 1}\right)\nu(dy), \tag{26.1}$$

where $\sigma^2 \geq 0$ is the variance of the Gaussian component, and the Lévy measure $\nu(dy)$ satisfies

$$\int_{\mathbb{R}\setminus\{0\}} \min\left\{1, y^2\right\}\nu(dy) < +\infty \tag{26.2}$$

Assume that, under a risk-neutral measure chosen by the market, the price process has the dynamics $S_t = e^{X_t}$, where X_t is a certain Lévy process. Then we must have $E[e^{X_t}] < +\infty$, and, therefore, ψ must admit analytic continuation into a strip $\Im\xi \in (-1,0)$ and continuous continuation into the closed strip $\Im\xi \in [-1,0]$.

The infinitesimal generator of X, denoted by L, is an integro-differential operator that acts as follows:

$$
\begin{aligned}
Lu(x) \quad = \quad & \frac{\sigma^2}{2}\frac{\partial^2 u}{\partial x^2}(x) + \mu\frac{\partial u}{\partial x}(x) \\
& + \int_{-\infty}^{+\infty}\left(u(x+y) - u(x) - y\mathbf{1}_{|y|\le1}\frac{\partial u}{\partial x}(x)\right)\nu(dy).
\end{aligned} \tag{26.3}
$$

The infinitesimal generator L can also be represented as a pseudo-differential operator (PDO) with the symbol $-\Psi(\xi)$, i.e. $L = -\psi(D)$, where $D = i\partial_x$. Recall that a PDO $A = a(D)$ acts as follows:

$$
Au(x) = (2\pi)^{-1}\int_{-\infty}^{+\infty}e^{ix\xi}a(\xi)\breve{u}(\xi)d\xi \tag{26.4}
$$

where $\hat{u}$ is the Fourier transform of a function u,

$$
\hat{u}(\xi) = \int_{-\infty}^{+\infty}e^{-ix\xi}u(x)dx.
$$

Note that the inverse Fourier transform in (26.4) is defined in the classical sense only if the symbol $a(\xi)$ and function $\hat{u}(\xi)$ are sufficiently nice. In general, one defines the (inverse) Fourier transform by duality.

Further, if the riskless rate, r, is constant, and if the stock does not pay dividends, then the discounted price process must be a martingale. Equivalently, the following condition (the EMM requirement) must hold (see, e.g. Boyarchenko and Levendorskiĭ (2002)):

$$
r + \psi(-i) = 0, \tag{26.5}
$$

which can be used to express μ via the other parameters of the Lévy process:

$$
\mu = r - \frac{\sigma^2}{2} + \int_{-\infty}^{+\infty}\left(1 - e^y + y\mathbf{1}_{|y|\le1}\right)\nu(dy) \tag{26.6}
$$

Hence, the infinitesimal generator may be rewritten as follows:

$$
\begin{aligned}
Lu(x) = \quad & \frac{\sigma^2}{2}\frac{\partial^2 u}{\partial x^2}(x) + \left(r - \frac{\sigma^2}{2}\right)\frac{\partial u}{\partial x}(x) \\
& + \int_{\mathbb{R}}\left[u(x+y) - u(x) - \left(e^y - 1\right)\frac{\partial u}{\partial x}(x)\right]\nu(dy)
\end{aligned} \tag{26.7}
$$

26.2.2 Regular Lévy Processes of Exponential Type

Loosely speaking, a Lévy process X is called a *Regular Lévy Process of Exponential type* (RLPE) if its Lévy density has a polynomial singularity at the origin and decays exponentially at infinity (Boyarchenko and Levendorskiĭ 2002). An almost equivalent definition is as follows: the characteristic exponent is analytic in a strip $\Im\xi \in (\lambda_-,\lambda_+), \lambda_- < -1 < 0 < \lambda_+$, is continuous up to the boundary of the strip, and admits the representation

$$
\psi(\xi) = -i\mu\xi + \phi(\xi), \tag{26.8}
$$

where $\phi(\xi)$ stabilizes to a positively homogeneous function at infinity:

$$\phi(\xi) \sim c_{\pm} |\xi|^{\nu}, \quad \text{as } \Re\xi \to \pm\infty, \text{ in the strip } \Im\xi \in (\lambda_-, \lambda_+), \tag{26.9}$$

where $c_{\pm} > 0$. 'Almost' means that the majority of classes of Lévy processes used in empirical studies of financial markets satisfy the conditions of both definitions. These classes are as follows: Brownian motion, Kou's model (Kou 2002), Hyperbolic processes (Eberlein and Keller 1995), Normal Inverse Gaussian processes and their generalization (Barndorff-Nielsen 1998, Barndorff-Nielsen and Levendorskiĭ 2001), and the extended Koponen's family. Koponen (1995) introduced a symmetric version; Boyarchenko and Levendorskiĭ (2000) gave a non-symmetric generalization; later, a subclass of this model appeared under the name the CGMY model in Carr *et al.* (2002), and Boyarchenko and Levendorskiĭ (2002) used the name KoBoL family.

The important exception is Variance Gamma Processes (VGP; see, e.g. Madan *et al.* (1998)). VGP satisfy the conditions of the first definition but not the second, as the characteristic exponent behaves like const $\cdot \ln |\xi|$ as $\xi \to \infty$.

Example 26.2.1

The characteristic exponent of a pure jump CGMY model is given by

$$\psi(\xi) = -i\mu\xi + C\Gamma(-Y)\left[G^Y - (G + i\xi)^Y + M^Y - (M - i\xi)^Y \right], \tag{26.10}$$

where $C > 0, \mu \in \mathbb{R}, Y \in (0,2), Y \neq 1$, and $-M < -1 < 0 < G$.

Example 26.2.2

If the Lévy measure of a jump diffusion process is given by a normal distribution

$$v(dx) = \frac{\lambda}{\delta\sqrt{2\pi}} \exp\left(-\frac{(x-\gamma)^2}{2\delta^2} \right) dx,$$

then we obtain the Merton model. The parameter λ characterizes the intensity of the jumps. The characteristic exponent of the process is of the form

$$\psi(\xi) = \frac{\sigma^2}{2}\xi^2 - i\mu\xi + \lambda\left(1 - \exp\left(-\frac{\delta^2\xi^2}{2} + i\gamma\xi \right) \right), \tag{26.11}$$

where $\sigma, \delta, \lambda \geq 0, \mu, \gamma \in \mathbb{R}$.

There are two important degenerate cases.

- If the intensity of jumps $\lambda = 0$, then we obtain the Black–Scholes model with $\mu = r - (\sigma^2/2)$ fixed by the EMM requirement.
- If the intensity of jumps $\lambda > 0$, but $\delta = 0$, then we obtain a jump diffusion process with a constant jump size γ; the drift term $\mu = r - (\sigma^2/2) + \lambda(1 - e^\gamma)$ is fixed by the EMM requirement.

26.2.3 The Wiener–Hopf Factorization

There are several forms of the Wiener–Hopf factorization. The Wiener–Hopf factorization formula used in probability reads as follows:

$$E\left[e^{i\xi X_T}\right] = E\left[e^{i\xi\bar{X}_T}\right]E\left[e^{i\xi\underline{X}_T}\right], \forall \xi \in \mathbb{R} \tag{26.12}$$

where $T \sim \text{Exp } q$, and $\bar{X}_t = \sup_{0 \le s \le t} X_s$ and $\underline{X}_t = \inf_{0 \le s \le t} X_s$ are the supremum and infimum processes. Introducing the notation

$$\varphi_q^+(\xi) = qE\left[\int_0^\infty e^{-qt}e^{i\xi\bar{X}_t}dt\right] = E\left[e^{i\xi\bar{X}_T}\right], \tag{26.13}$$

$$\varphi_q^-(\xi) = qE\left[\int_0^\infty e^{-qt}e^{i\xi\underline{X}_t}dt\right] = E\left[e^{i\xi\underline{X}_T}\right], \tag{26.14}$$

we can write (26.12) as

$$\frac{q}{q+\psi(\xi)} = \varphi_q^+(\xi)\varphi_q^-(\xi). \tag{26.15}$$

Equation (26.15) is a special case of the Wiener–Hopf factorization of the symbol of a PDO. In applications to Lévy processes, the symbol is $q/q+\psi(\xi))$, and the PDO is $\mathcal{E}_q := q/(q-L) = q(q+\psi(D))^{-1}$: the normalized resolvent of the process X_t or, using the terminology of Boyarchenko and Levendorskiĭ (2005), the expected present value operator (EPV operator) of the process X_t. The name is due to the observation that, for a stream $g(X_t)$,

$$\mathcal{E}_q g(x) = E\left[\int_0^{+\infty} qe^{-qt}g(X_t)dt \mid X_0 = x\right].$$

We introduce the following operators:

$$\mathcal{E}_q^\pm := \varphi_q^\pm(D), \tag{26.16}$$

which also admit interpretation as the EPV operators under supremum and infimum processes. One of the basic observations in the theory of PDO is that the product of symbols corresponds to the product of operators. In our case, it follows from (26.15) that

$$\mathcal{E}_q = \mathcal{E}_q^+\mathcal{E}_q^- = \mathcal{E}_q^-\mathcal{E}_q^+ \tag{26.17}$$

as operators in appropriate function spaces.

For a wide class of Lévy models, $\mathcal{E}$ and $\mathcal{E}^\pm$ admit interpretation as expectation operators:

$$\mathcal{E}_q g(x) = \int_{-\infty}^{+\infty} g(x+y)P_q(y)dy,$$

$$\mathcal{E}_q^\pm g(x) = \int_{-\infty}^{+\infty} g(x+y)P_q^\pm(y)dy,$$

where $P_q(y)$ and $P_q^\pm(y)$ are certain probability densities with

$$P_q^\pm(y) = 0, \quad \forall \pm y < 0.$$

Moreover, the characteristic functions of the distributions $P_q(y)$ and $P_q^\pm(y)$ are $q(q+\psi(\xi))^{-1}$ and $\varphi_q^\pm(\xi)$, respectively.

The general results in this paper are based on simple properties of the EPV operators, which follow immediately from the interpretation of $\mathcal{E}^{\pm}$ as expectation operators. For details, see Boyarchenko and Levendorskiĭ (2005).

Proposition 26.2.3

EPV operators $\varepsilon_q^{\pm}$ have the following properties.

(1) *If $g(x) = 0\ \forall x \geq h$, then $\forall x \geq h, \left(\varepsilon_q^+ g\right)(x) = 0$ and $\left(\left(\varepsilon_q^+\right)^{-1} g\right)(x) = 0$.*

(2) *If $g(x) = 0\ \forall x \geq h$, then $\forall x \geq h, \left(\varepsilon_q^- g\right)(x) = 0$ and $\left(\left(\varepsilon_q^-\right)^{-1} g\right)(x) = 0$.*

(3) *If $g(x) \geq 0\ \forall x$, then $\left(\varepsilon_q^+ g\right)(x) \geq 0\ \forall x$. If, in addition, there exists x_0 such that $g(x) > 0\ \forall x > x_0$, then $\left(\varepsilon_q^+ g\right)(x) > 0\ \forall x$.*

(4) *If $g(x) \geq 0\ \forall x$, then $\left(\varepsilon_q^- g\right)(x) \geq 0\ \forall x$. If, in addition, there exists x_0 such that $g(x) > 0\ \forall x < x_0$, then $\left(\varepsilon_q^- g\right)(x) > 0\ \forall x$.*

(5) *If g is monotone, then $\varepsilon_q^+ g$ and $\varepsilon_q^- g$ are also monotone.*

(6) *If g is continuous and satisfies*

$$|g(x)| \leq C\left(e^{\sigma_- x} + e^{\sigma_+ x}\right), \quad \forall x \in \mathbb{R} \tag{26.18}$$

where $\sigma_- \geq 0 \geq \sigma_+$ and C are all independent of x, then $\varepsilon_q^+ g$ and $\varepsilon_q^- g$ are continuous.

26.3 The Multiple Optimal Stopping Problem for Swing Options

We consider a price process that evolves according to the formula

$$S_t = e^{X_t},$$

where $\{X\}_{t \geq 0}$, the driving process, is an adapted Lévy process defined on the filtered probability space $\left(\Omega, \mathcal{F}, \mathbb{F} = \{\mathcal{F}_t\}_{t \geq 0}, \mathbb{P}\right)$, satisfying the usual conditions.

Let T be the option's maturity time, and let $\mathcal{T}_{t,T}$ be the set of $\mathbb{F}$-stopping times with values in $[t, T]$. Consider a swing option that gives the right to multiple exercise with a $\delta > 0$ refracting period that separates two successive exercises (the number of possible exercises is fixed). We consider the possibility of n put exercises. We shall denote by $\mathcal{T}^n$ the collection of all vectors of stopping times $(\tau_1, \tau_{1, \ldots}, \tau_n)$, such that

- $\tau_i \geq T$ a.s.;
- $\tau_i - \tau_{i-1} \geq \delta$ on $\{\tau_{i-1} \leq T\}$ a.s. for all $i = 2, \ldots, n$.

Denote by $v^{(i)}(t, x)$ the swing option value with the possibility of i exercises at spot level $S = e^x$ and time $t \geq T$.

Following Carmona and Touzi (2008), the multiple exercise problem can be solved by computing

$$v^{(n)}(0, x) = \sup_{(\tau_1, \ldots, \tau_n) \in T^n} \sum_{i=1}^{n} E\left[e^{-r\tau_i} \phi\left(X_{\tau_i}\right)\right] \tag{26.19}$$

where

$$\phi(x) = \left(K - e^x\right)_+$$

is the pay-off function.

To solve the multiple optimal stopping problem, Carmona and Touzi (2008) introduce the idea of an inductive hierarchy. In fact, they reduce the multiple stopping problem to a cascade of n optimal single stopping problems. Define the value function for $i = 1, \ldots, n$

$$v^{(i)}(t,x) = \sup_{\tau \in \mathcal{T}_{t,T}} E\left[e^{-r\tau} \phi^{(i)}\left(\tau, X_\tau^{t,x}\right) \right], \tag{26.20}$$

where the reward function $\phi^{(i)}$ is now defined as

$$\phi^{(i)}(t,x) = \phi(x) + E\left[e^{-r\delta} v^{(i-1)}\left(t + \delta, X_{t+\delta}^{t,x}\right) \right], \quad t \le T - \delta, \tag{26.21}$$

$$\phi^{(i)}(t,x) = \phi(x), \quad t > T - \delta. \tag{26.22}$$

The problem can be solved using a Monte Carlo algorithm. Let $t_0 = 0 < t_1 < t_2 < \cdots < t_N = T$ be a time discretization grid. The price of a swing option can be computed by the backward induction procedure

$$\begin{cases} v^{(i)}\left(t_N, x\right) = \phi(x) \\ v^{(i)}\left(t_{k-1}, x\right) = \max\left\{ \phi^{(i)}\left(t_{k-1}, x\right); \quad k = N, \ldots, 1. \right. \\ \left. e^{-r(t_k - t_{k-1})} E\left[v^{(i)}\left(t_k, X_{t_k}^{t_{k-1}, x}\right) \right] \right\} \end{cases}$$

Carmona and Touzi (2008) and Mnif and Zeghal (2006), respectively, considered a Monte Carlo Malliavin-based algorithm to compute the price in the Black–Scholes and jump models frameworks. Barrera-Esteve *et al.* (2006) used a regression-based method to approximate conditional expectations. In the following sections, we propose two PIDE-based approaches.

26.4 The Finite Difference Scheme for Pricing Swing Options

We can compute the swing option price using the formulation given in (26.20) with an analytical approach. In fact, we propose to solve the following system of variational inequalities associated with the swing options formulation:

$$\begin{cases} \max\left(\phi^{(i)}(t,x) - v^{(i)}(t,x), \dfrac{\partial v^{(i)}}{\partial t} + Lv^{(i)} - rv^{(i)} \right) = 0 \\ (t,x) \in [0,T[\times \mathbb{R}, \\ v^{(i)}(T,x) = \phi^{(i)}\left(T, e^x\right), \end{cases} \tag{26.23}$$

with $i = 1, \ldots, n$, where the integro-differential operator L is defined in (26.7).
Now recall that, for $t \ge T - \delta$,

$$\phi^{(i)}(t,x) = \left(K - e^x\right)_+ + E\left[e^{-r\delta} v^{(i-1)}\left(t + \delta, X_{t+\delta}^{t,x}\right) \right].$$

Let us define for $t \le T - \delta$

$$u^{(i)}(t,x) = E\left[e^{-r\delta} v^{(i)}\left(t + \delta, X_{t+\delta}^{t,x}\right) \right].$$

By the Feyman–Kac theorem, $u^{(i)}(t, x) = z(0, x)$, where $z(t, x)$ is the solution of the following partial integro-differential equation (PIDE):

$$\begin{cases} \dfrac{\partial z}{\partial t} + Lz - rz = 0, & (t,x) \in [0,\delta[\times \mathbb{R}, \\ z(\delta,x) = v^{(i)}(t+\delta,x), \end{cases} \qquad (26.24)$$

which can be computed numerically using a finite difference approach. To price a swing option, therefore, we can solve the system of variational inequalities (26.23) computing the reward pay-off function $\phi^{(i)}(t, x)$ in the following way:

$$\phi^{(i)}(t,x) = \phi(x)$$

for $T - \delta < t \leq T$, and

$$\phi^{(i)}(t,x) = \left(K - e^x \right)_+ + u^{(i-1)}(t,x),$$

for $t \leq T - \delta$. As stated above, the reward pay-off function can be computed numerically using a finite difference scheme. The numerical solution of the variational inequalities (26.23) requires numerically solving each PIDE problem (26.24). To solve (26.23) and (26.24), we perform the following steps.

- *Localization.* We choose a spatial bounded computational domain Ω_l, which implies that we must choose some artificial boundary conditions.
- *Truncation of large jumps.* This step corresponds to truncating the integration domain in the integral part.
- *Discretization.* The derivatives of the solution are replaced by finite differences, and the integral terms are approximated using the trapezoidal rule. Then the problem is solved by using an explicit–implicit scheme (see Briani *et al.* (2004) and Cont and Voltchkova (2005) and the program implementation in the PREMIA software (www.premia.fr)). In particular, we introduce a time grid $t = s\Delta t$, $s = 0, \dots, N$, where $\Delta t = T/N$ is the time step. At each time step, it is necessary to solve a linear system for the linear problem (26.24) and a linear complementarity problem for the nonlinear problem (26.23). The idea of the explicit–implicit method is based on an asymmetric treatment of the differential and integral parts of L. The operator L in (26.24) is split into two parts,

$$Lz = Dz + Jz,$$

- where D and J are the differential and integral parts of L, respectively. We replace Dz with a finite difference approximation $D_\Delta z$ and Jz with the trapezoidal quadrature approximation $J_\Delta z$ and use the following explicit–implicit time-stepping:

$$\frac{z^{s+1} - z^s}{\Delta t} + D_{\Delta z^s} + + J_\Delta z^{s+1} - rz^s = 0.$$

The integral part is treated in an explicit way to avoid a dense matrix, while the differential part is treated in an implicit way. Details of the algorithms are given in Cont and Voltchkova (2005).

- *Treatment of the variational inequalities.* We solve each of the variational inequalities (26.23) using the splitting method of Barles *et al.* (1995). The splitting methods can be viewed as an analytical version of dynamic programming. The idea of this scheme is to split the American problem into two steps: we construct recursively the approximate solution $v^{(i)}(s\Delta t, x)$ at each time step $s\Delta t$ by starting from $v^{(i)}(N\Delta t, x) = \phi(x)$ and computing at each time step the values of $v^{(i)}(s\Delta t, x)$ for $s = N - 1, \dots, 0$ as follows:

- compute the solution of the following linear Cauchy problem on $[s\Delta t, (s + 1)\Delta t[\times\Omega_l$ using an explicit–implicit scheme:

$$\begin{cases} \dfrac{\partial w^{(i)}(s\Delta t, x)}{\partial t} + Lw^{(i)}(s\Delta t, x) - rw^{(i)}(s\Delta t, x) = 0 \\ \text{in } \left[s\Delta t, (s + 1)\Delta t \right[\times\Omega_l, \\ w^{(i)}((s + 1)\Delta t, x) = v^{(i)}((s + 1)\Delta t, x); \end{cases}$$

- apply the early exercise $v^{(i)}(s\Delta t, x) = \max(w^{(i)}(s\Delta t, x), \phi^{(i)}(s\Delta t, x))$, where the reward function $\phi^{(i)}(s\Delta t, x)$ is obtained by solving the linear problem (26.24) with an explicit–implicit finite difference method.

One could also apply the method of horizontal lines or Carr's randomization to (26.20), then use the explicit–implicit finite difference scheme to solve the corresponding sequence of free boundary problems. The analytical method of lines was introduced to finance by Carr and Faguet (1994); Carr (1998) suggested an important new probability interpretation of the method, which we call Carr's randomization. In the case of American options, the convergence of Carr's randomization algorithm was proved by Bouchard *et al.* (2005) for a wide class of strong Markov processes. In the next section, we will start with Carr's randomization procedure.

26.5 Pricing Swing Options Using the Wiener–Hopf Approach

In this section, we apply the Wiener–Hopf approach to pricing swing options. The first step is to discretize the time $(0 =)t_0 < t_1 < \dots < t_N (= T)$, but not the space variable. Set $v_N^i(x) = \left(K - e^x \right)_+$. For $s = N - 1, N - 2, \dots, 0$, set $\Delta_s = t_{s+1} - t_s$, $q^s = r + (\Delta_s)^{-1}$, and denote by $v_s^i(x)$ Carr's randomized approximation to $v^i(t_s, x)$.

The early exercise boundary h_s^i for an interval (t_s, t_{s+1}) and $v_s^i(x)$ can be found using backward induction. For $s = N - 1, N - 2, \dots$, the boundary h_s^i is chosen to maximize

$$v_s^i(x) = E\left[\int_0^{\tau_s^i} e^{-q^s t} v_{s+1}^i\left(X_t^{0,x} \right) dt \right] + E\left[e^{-q^s \tau_s^i} \phi_s\left(X_{\tau_s^i}^{0,x} \right) \right], \tag{26.25}$$

where τ_s^i is the hitting time of the interval of the form $\left(-\infty, h_s^i \right]$, and

$$\phi_s^{(i)}(x) = \left(K - e^x \right) + E\left[e^{-r\delta} v^{(i-1)}\left(t_s + \delta, X_{t_s+\delta}^{t_s,x} \right) \right], \quad t_s \le T - \delta,$$

and

$$\phi_s^{(i)}(x) = \left(K - e^x \right), \quad t_s > T - \delta.$$

As in Boyarchenko and Levendorskiĭ (2009), where the case of American options was considered, to derive (26.25), we replace $\phi(x) = (K - e^x)_+$ in (26.19) with $(K - e^x)$. This replacement is justified by a simple consideration that it is non-optimal to exercise the option when $(K - e^x) \le 0$.

In this paper, we use uniform spacing; therefore, q^s and Δ^s are independent of s and denoted q and Δt, respectively. For the case of put swing options, v_s^i given by (26.25) is a unique solution of the boundary problem

$$(q - L)v_s^i(x) = (\Delta t)^{-1} v_{s+1}^i(x), \quad x > h_s^i, \tag{26.26}$$

$$v_s^i(x) = \phi_s^{(1)}(x), \quad x \le h_s^i. \tag{26.27}$$

Note that the problem (26.26) and (26.27) can be obtained by discretization of the time derivative in the generalized Black–Scholes equation (see details in Boyarchenko and Levendorskiĭ (2009) and the bibliography therein).

Let the refracting period δ be equal to $k\Delta t$, where k is a certain positive integer. Then, for $i = 1, \ldots, n$,

$$\phi_s^{(i)}(x) = \left(K - e^x\right) + u_s^{i-1}(x), \tag{26.28}$$

where

$$u_s^0(x) = 0, \tag{26.29}$$

$$u_s^i(x) = 0, \quad t_s > T - \delta. \tag{26.30}$$

$$u_s^i(x) = E\left[e^{-r\delta} v_{s+k}^{(i)}\left(X_{t_{s+k}}^{t_s, x}\right)\right], \quad t_s \le T - \delta. \tag{26.31}$$

Introduce $\tilde{v}_s^i(x) = v_s^i(x) - \phi_s^{(i)}(x)$, and substitute $v_s^i(x) = \tilde{v}_s^i(x) + \phi_s^{(i)}(x)$ into (26.26)–(26.27) as follows:

$$(q - L)\tilde{v}_s^i(x) = (\Delta t)^{-1} G_s^i(x), \quad x > h_s^i, \tag{26.32}$$

$$\tilde{v}_s^i(x) = 0, \quad x \le h_s^i, \tag{26.33}$$

where $G_s^i = \tilde{v}_{s+1}^i + \phi_{s+1}^{(i)} - \Delta t(q - L)\phi_s^{(i)}$.

Using arguments similar to those of Boyarchenko and Levendorskiĭ (2009), it can be shown that, for $s = n - 1, n - 2, \ldots, 0$, the function G_s^i is a non-decreasing continuous function satisfying bound (26.18) with $\sigma_+ = 1, \sigma_- = 0$; in addition,

$$G_s^i(-\infty) < 0 < G_s^i(+\infty) = +\infty. \tag{26.34}$$

Then $G_s^i(x)$ satisfies the conditions of theorem 26.2.6 of Boyarchenko and Levendorskiĭ (2009). Due to this theorem and proposition 26.2.3, the following statements hold.

1. The function

$$\tilde{w}_s^i := \varepsilon_q^+ G_s^i \tag{26.35}$$

   is continuous; it increases and satisfies (26.34).

2. The equation

$$\tilde{w}_s^i(h) = 0 \tag{26.36}$$

has a unique solution, denoted by h_s^i.

3. The hitting time of $\left(-\infty, h_s^i\right], \tau\left(h_s^i\right)$, is a unique optimal stopping time.

4. (Carr's approximation to) the swing option value with i exercise rights at the moment s is given by

$$v_s^i = (q\Delta t)^{-1}\varepsilon_q^-\mathbf{1}_{\left(h_s^i,+\infty\right)}\tilde{w}_s^i + \phi_s^{(i)}. \tag{26.37}$$

Equivalently,

$$\tilde{v}_s^i = (q\Delta t)^{-1}\varepsilon_q^-\mathbf{1}_{\left(h_s^i,+\infty\right)}\tilde{w}_s^i. \tag{26.38}$$

5. $\tilde{v}_s^i = v_s^i - \phi_s^{(i)}$ is a positive non-decreasing function that admits bound (26.18) with $\sigma_+ = 1, \sigma_- = 0$ and satisfies $\tilde{v}_s^i(+\infty) = +\infty$; it vanishes below h_s^i and increases on $\left[h_s^i,+\infty\right)$.

Because functions G_s^i and $\tilde{w}_s^i$ tend to plus infinity as $x \to +\infty$, the numerical calculation of the integrals in (26.35) and (26.37) may face certain difficulties. To improve the convergence, we reformulate the algorithm in terms of the bounded functions v_s^i. Taking into account (26.28) and (26.5), G_s^i can be rewritten as follows:

$$\begin{aligned} G_s^i(x) &= v_{s+1}^i(x) - \Delta t(q-L)u_s^{(i-1)}(x) - \Delta t(q-L)\left(K - e^x\right) \\ &= v_{s+1}^i(x) - \tilde{u}_s^{(i-1)}(x) - \left(\Delta t Kq - e^x\right), \end{aligned} \tag{26.39}$$

where $\tilde{u}_s^i(x)$ can be approximated by the formulae

$$\tilde{u}_s^0(x) = 0, \tag{26.40}$$

$$\tilde{u}_s^i(x) = 0, \quad t_s > T - \delta, \tag{26.41}$$

$$\tilde{u}_s^i(x) = E\left[e^{-r(\delta-\Delta t)}v_{s+k}^{(i)}\left(X_{t_{s+k}}^{t_{s+1},x}\right)\right] + o(\Delta t), \quad t_s \leq T - \delta. \tag{26.42}$$

Note that we can easily compute the expectation on the RHS of (26.42) using the Fourier transform technique (see, e.g. Carr and Madan (1999) or Boyarchenko and Levendorskiĭ (2002)) as follows:

$$\tilde{u}_s^i(x) \approx (2\pi)^{-1}e^{-\rho x}\int_{-\infty}^{+\infty} e^{ix\xi-(\delta-\Delta t)(r+\psi(\xi+i\rho))}\hat{v}_{s+k}^{(i),\rho}(\xi)d\xi \tag{26.43}$$

where $\hat{v}_s^{(i),\rho}(\xi)$, is the Fourier transform of the price $v_s^{(i)}(x)$ multiplied by an appropriate damping exponential factor $e^{\rho x}$; in our case, $\rho > 0$. Numerically, formula (26.43) can be efficiently realized by means of the FFT technique (Carr and Madan 1999).

However, for very short refracting periods δ, the integrand in (26.43) may decay slowly at infinity (see, e.g. Lord et al. (2008)). Hence, the numerical implementation of the Fourier transform may not be sufficiently accurate. To circumvent the potential numerical pricing difficulties when dealing with the case $\delta = k\Delta t, k > 1$, ($k$ is not too large and Δt is small), the finite difference approach proposed in Section 26.4 can be used efficiently to find $\tilde{u}_s^i(x)$. The expectation in (26.42) can be interpreted as the solution at time Δt to the problem (26.24), with t_s instead of t. Finally, if $\delta = \Delta t$, then $\tilde{u}_s^i(x) \approx v_{s+k}^{(i)}(x)$.

We can now rewrite (26.37) as follows:

$$v_s^i = (q\Delta t)^{-1}\varepsilon_q^-\left(\mathbf{1}_{\left(h_s^i,+\infty\right)}w_s^i - \mathbf{1}_{\left(-\infty,h_s^i\right]}w_{s,0}^i\right), \tag{26.44}$$

where

$$w_s^i = \varepsilon_q^+ v_{s+1}^i, \tag{26.45}$$

$$
\begin{aligned}
w_{s,0}^i &= \varepsilon_q^+ \left(\tilde{u}_s^{(i-1)}(x) + \Delta t K q - e^x \right) \\
&= \varepsilon_q^+ \tilde{u}_s^{(i-1)}(x) + \Delta t K q - \varphi_q^+(-i)e^x,
\end{aligned}
\tag{26.46}
$$

and h_s^i is a solution to the equation

$$w_s^i = w_{s,0}^i. \tag{26.47}$$

Note that, in (26.44) and (26.45), the functions in the arguments of the operators ε_q^- and ε_q^+ are bounded. The algorithm can be efficiently realized by using the Fast Wiener–Hopf factorization method (see details in Section 2 of Kudryavtsev and Levendorskiĭ (2009)).

26.6 Numerical Results

In this section, we illustrate the efficiency and the robustness of the proposed methods numerically using the parameters of the numerical examples for pricing swing options in the Black–Scholes and CGMY models provided by Wilhelm and Winter (2008).

In the case of the Wiener–Hopf approach, we use the adaptive method from Kudryavtsev and Levendorskii (2009). For a fixed number of time steps, N, and a step in x-space, Δx, we increase the domain in x-space twofold to ensure that the price does not change significantly. In the dual space it corresponds to increasing the number of points M. Fix the space step $\Delta x > 0$ and the number of space points (and dual space points) $M = 2^m$. Define the partitions of the normalized log-price domain $[-M\Delta x/2; M\Delta x/2]$ by points $x_k = -(M\Delta x/2) + k\Delta x$, $k = 0, \ldots, M - 1$, and the frequency domain $[-\pi/\Delta x; \pi/\Delta x]$ by points $\xi_l = 2\pi l/\Delta x$, $l = -M/2, \ldots, M/2$.

We consider a put swing option with $n = 1, 2, 3$ exercise numbers and a refracting period $\delta = 0.1$. We assume that the initial value of the stock prices is $S = 100$, the exercise price $K = 100$, the maturity $T = 1$ and the force of the interest rate $r = 0.05$.

In order to solve the PIDE numerically using the finite difference scheme, we first localize the variables and the integral term to bounded domains. We use for this purpose the estimates for the localization domain and the truncation of large jumps given by Voltchkova and Tankov (2008).

In our examples, both methods use a spatial discretization step $\Delta x = 0.001$ and a varying number of time steps $N = 50, 100, 200$. In the Wiener–Hopf approach, the optimal choice for the number of space points is $M = 2^{12}$ (doubling the number M changes the option prices by 0.0001% or less).

We propose first to assess the numerical robustness of our algorithm in the Black–Scholes case, using the volatility $\sigma = 0.3$. Table 26.1 reports the prices (with time in seconds in parentheses) with the relative errors in a Black–Scholes framework, using the finite difference method (FD) proposed in Section 26.4 and the Wiener–Hopf approach (FWHF) proposed in Section 26.5. As benchmark solutions, we take those provided by Wilhelm and Winter (2008) (B-WW).

Furthermore, we provide numerical results in a Lévy market model. To be exact, we use the CGMY model (Carr *et al.* 2002) with $C = 1$, $G = 10$, $M = 10$ and $Y = 0.5$. No comparison results are available in the paper of Wilhelm and Winter (2008), and thus we use as the benchmark value the FWHF method with a very fine mesh grid ($\Delta x = 0.0002$, $N = 800$ and $M = 2^{15}$). In order to justify the 'benchmark' values, we observed that with a finer mesh grid ($\Delta x = 0.0001$, $N = 1600$ and $M = 2^{17}$) the option prices change only by 0.02% or less. Table 26.2 reports the numerical results for the CGMY model.

Table 26.3 reports the prices of swing options in the CGMY model with decreasing values of refracting periods $\delta = 0.1, 0.01, 0.001, 0$. The prices are calculated using the FWHF method with a spatial

TABLE 26.1   Swing Option Prices in the Black–Scholes Model

	N	Prices			Relative errors (%)		
		$n = 1$	$n = 2$	$n = 3$	$n = 1$	$n = 2$	$n = 3$
FWHF	50	9.786 (0.22)	19.130 (0.7)	27.968 (1.17)	−0.85	−0.65	−0.56
	100	9.826 (0.42)	19.190 (1.38)	28.045 (2.33)	−0.45	−0.34	−0.29
	200	9.848 (0.83)	19.222 (2.72)	28.085 (4.59)	−0.22	−0.17	−0.15
B-WW		9.8700	19.2550	28.1265			
FD	50	9.834 (0.12)	19.096 (5.16)	27.711 (8.10)	−0.36	−0.83	−1.48
	100	9.864 (0.27)	19.184 (8.64)	27.925 (16.47)	−0.06	−0.37	−0.72
	200	9.867 (0.57)	19.220 (16.68)	28.027 (33.0)	−0.03	−0.18	−0.35

TABLE 26.2   Swing Option Prices in the CGMY Model

	N	Prices			Relative errors (%)		
		$n = 1$	$n = 2$	$n = 3$	$n = 1$	$n = 2$	$n = 3$
FWHF	50	7.100 (0.22)	13.859 (0.7)	20.228 (1.17)	−0.80	−0.61	−0.54
	100	7.131 (0.42)	13.905 (1.38)	20.287 (2.33)	−0.36	−0.28	−0.25
	200	7.147 (0.83)	13.928 (2.72)	20.317 (4.59)	−0.14	−0.11	−0.10
B-FWHF		7.157	13.944	20.337			
FD	50	7.173 (1.20)	13.887 (37.1)	20.102 (76.1)	0.22	−0.41	−1.16
	100	7.172 (2.31)	13.928 (146)	20.238 (286.5)	0.21	−0.11	−0.49
	200	7.171 (4.56)	13.948 (751)	20.306 (1398)	0.20	0.03	−0.15

TABLE 26.3   Convergence of Swing Option Prices in the CGMY Model

	δ	$n = 3$
FWHF	0.1	20.3416
	0.01	21.3701
	0.001	21.4699
	0	21.4760

discretization step $\Delta x = 0.001$ and number of time steps $N = 1000$; the other parameters remain the same. We see that the sequence of prices increases to the limit value as the refracting period goes to 0. The limit value is the solution to problem (26.20) with the reward function in (26.21) defined for zero refracting period. Hence, in the limit case, we obtain a simplified problem, because we do not need to calculate the expectation in (26.21).

All computations were performed in double precision on an Eee PC with the following characteristics: CPU Atom N450, 1.67 GHz, 2 Gb of RAM.

The numerical results confirm the reliability of both approaches, demonstrating the robustness of the methods. In particular, the Wiener–Hopf approach is undoubtedly a very precise and efficient method for pricing swing options in the presence of multiple jumps.

Acknowledgements

The first author gratefully acknowledges financial support from the European Science Foundation (ESF) through the Short Visit Grant number 3404 of the program 'Advanced Mathematical Methods for Finance' (AMaMeF) and the Russian Foundation for Humanities, the project number 15–32–01390.

References

Barles, G., Daher, C. and Romano, M., Convergence of numerical schemes for problems arising in finance theory. *Math. Models Meth. Appl. Sci.*, 1995, **5**(1), 125–143.

Barndorff-Nielsen, O.E., Processes of normal inverse Gaussian type. *Finance Stochast.*, 1998, **2**, 41–68.

Barndorff-Nielsen, O.E. and Levendorskiĭ, S., Feller processes of normal inverse Gaussian type. *Quant. Finance*, 2001, **1**, 318–331.

Barrera-Esteve, C., Bergeret, F., Dossal, C., Gobet, E., Meziou, A., Munos, R. and Reboul-Salze, D., Numerical methods for the pricing of swing options: A stochastic control approach. *Methodol. Comput. Appl. Probab.*, 2006, **8**(4), 517–540.

Bouchard, B., El Karoui, N. and Touzi, N., Maturity randomization for stochastic control problems. *Ann. Appl. Probab.*, 2005, **15**(4), 2575–2605.

Boyarchenko, S.I. and Levendorskiĭ, S.Z., Option pricing for truncated Lévy processes. *Int. J. Theor. Appl. Finance*, 2000, **3**, 549–552.

Boyarchenko,S.I.andLevendorskiĭ, S.Z.,*Non-GaussianMerton–Black–ScholesTheory*, 2002 (World Scientific: Singapore).

Boyarchenko, S.I. and Levendorskiĭ, S.Z., American options: The EPV pricing model. *Ann. Finance*, 2005, **1**(3), 267–292.

Boyarchenko, S.I. and Levendorskiĭ, S.Z., General option exercise rules, with applications to embedded options and monopolistic expansion. *Contrib. Theor. Econ.*, 2006, **6**(1), article 2.

Boyarchenko, S.I. and Levendorskiĭ, S.Z., Pricing American options in regime-switching models. *SIAM J. Control Optimiz.*, 2009, **48**, 1353–1376.

Briani, M., La Chioma, C. and Natalini, R., Convergence of numerical schemes for viscosity solutions to integro-differential degenerate parabolic problems arising in financial theory. *Numer. Math.*, 2004, **98**(4), 607–646.

Carmona, C. and Touzi, N., Optimal multiple stopping and valuation of swing options. *Math. Finance*, 2008, **18**(2), 239–268.

Carr, P., Randomization and the American put. *Rev. Financ. Stud.*, 1998, **11**, 597–626.

Carr, P. and Faguet, D., Fast accurate valuation of American options. Working Paper, Cornell University, 1994.

Carr, P., Geman, H., Madan, D.B. and Yor, M., The fine structure of assets return: An empirical investigation. *J. Business*, 2002, **75**(3), 305–332.

Carr, P. and Madan, D., Option valuation using the Fast Fourier Transform. *J. Comput. Finance*, 1999, **3**, 463–520.

Cont, R. and Voltchkova, E., A finite difference scheme for option pricing in jump-diffusion and exponential Lévy models. *SIAM J. Numer. Anal.*, 2005, **43**(4), 1596–1626.

Eberlein, E. and Keller, U., Hyperbolic distributions in finance. *Bernoulli*, 1995, **1**, 281–299.

Jaillet, P., Ronn, E.I. and Tompaidis, S., Valuation of commodity-based swing options. *Mgmt Sci.*, 2004, **50**, 909–921.

Koponen, I., Analytic approach to the problem of convergence of truncated Lévy flights towards the Gaussian stochastic process. *Phys. Rev. E*, 1995, **52**, 1197–1199.

Kou, S.G., A jump-diffusion model for option pricing. *Mgmt Sci.*, 2002, **48**, 1086–1101.

Kudryavtsev, O.E. and Levendorskiĭ, S.Z., Fast and accurate pricing of barrier options under Levy processes. *Finance Stochast.*, 2009, **13**(4), 531–562.

Kudryavtsev, O.Ye., An efficient numerical method to solve a special cass of integro-differential equations relating to the Levy models. *Mathematical Models and Computer Simulations*, 2011, **3**(6), 706–711.

Kudryavtsev, O., Finite difference methods for option pricing under Levy processes: Wiener–Hopf factorization approach. *The Scientific World Journal*, 2013, **2013**, Article ID 963625, 12 pages

Kudryavtsev, O. and Zanette, A., Efficient pricing of swing options in Levy-driven models. Quantitative Finance, 2013, **13**(4), 627–635.

Lari-Lavassani, A., Simchi, M. and Ware, A., A discrete valuation of swing options. *Can. Appl. Math. Q.*, 2001, **9**(1), 35–73.

Lord, R., Fang, F., Bervoets, F. and Oosterlee, C.W., A fast and accurate FFT-based method for pricing early-exercise options under Levy processes. *SIAM J. Sci. Comput.*, 2008, **30**(4), 1678–1705.

Madan, D.B., Carr, P. and Chang, E.C., The variance Gamma process and option pricing. *Eur. Finance Rev.*, 1998, **2**, 79–105.

Mnif, M. and Zhegal, A.B., Optimal multiple stopping and valuation of swing options in Lévy models. *Int. J. Theor. Appl. Finance*, 2006, **9**(8), 1267–1298.

Sato, K., *Lévy Processes and Infinitely Divisible Distributions*, 1999 (Cambridge University Press: Cambridge).

Voltchkova, E. and Tankov, P., Deterministic methods for option pricing in exponential Lévy models. *PREMIA documentation*, 2008. Available online at: http://www.premia.fr

Wilhelm, M. and Winter, C., Finite element valuation of swing options. *J. Comput. Finance*, 2008, **11**(3), 107–132.

27

The Valuation of Clean Spread Options: Linking Electricity, Emissions and Fuels

René Carmona,
Michael Coulon and
Daniel Schwarz

27.1 Introduction

Spread options are most often used in the commodity and energy markets to encapsulate the profitability of a production process by comparing the price of a refined product to the costs of production, including, but not limited to, the prices of the inputs to the production process. When the output commodity is electric power, such spread options are called *spark spreads* when the electricity is produced from natural gas, and *dark spreads* when the electricity is produced from coal. Both processes are the sources of Greenhouse Gas (GHG) emissions, in higher quantities for the latter than the former. In this paper we concentrate on the production of electricity and CO_2 emissions and the resulting dependence structure between prices.

Market mechanisms aimed at controlling CO_2 emissions have been implemented throughout the world, and whether they are mandatory or voluntary, cap-and-trade schemes have helped to put a price

DOI: 10.1201/9781003265399-31

on carbon in Europe, the US, and around the world. In the academic literature, equilibrium models have been used to show what practitioners have known all along, namely that the price put on CO_2 by the regulation should be included in the costs of production to set the price of electricity (Carmona *et al.* 2010).

Strings of spark spread options (options on the spread between the price of 1 MWh of electricity and the cost of the amount of natural gas needed to produce such a MWh) with maturities covering a given period are most frequently used to value the optionality of a gas power plant which can be run when it is profitable to do so (namely when the price of electricity is greater than the cost of producing it), and shut down otherwise. In a nutshell, if an economic agent takes control on day t of a gas power plant for a period $[T_1,T_2]$, then for every day $\tau \in [T_1,T_2]$ of this period, he or she can decide to run the power plant when $P_\tau > h_g S_\tau^g + K$, booking a profit $P_\tau - h_g S_\tau^g - K$ for each unit of power produced, and shut the plant down if $P_\tau \le h_g S_\tau^g + K$. Here P_τ denotes the price at which one unit (1 MWh) of power can be sold on day τ, S_τ^g the price of one unit of natural gas (typically one MMBtu), h_g the efficiency or heat rate of the plant (i.e. the number of units of natural gas needed to produce one unit of electricity) and K the daily (fixed) costs of operations and maintenance of the plant. Ignoring constraints such as ramp-up rates and start-up costs, this scheduling is also automatically induced when generators bid at the level of their production costs in the day-ahead auction for power. So in this somewhat oversimplified analysis of the optionality of the plant, the value at time t of the control of the plant operation on day τ can be expressed as $e^{-r(\tau-t)}\mathbb{E}\left[\left(P_\tau - h_g S_\tau^g - K\right)^+ | \mathcal{F}_t\right]$, where, as usual, the exponent + stands for the positive part, i.e. $x^+=x$ when $x \ge 0$ and $x+ = 0$ otherwise, r for the constant interest rate used as discount factor to compute the present values of future cash flows, and F_t denotes the information available on day t when the conditional expectation is actually computed. So the operational control (for example, as afforded by a tolling contract) of the plant over the period $[T_1,T_2]$ could be valued on day t as

$$V_t^{PP} = \sum_{\tau=T_1}^{T_2} e^{-r(\tau-t)}\mathbb{E}\left[\left(P_\tau - h_g S_\tau^g - K\right)^+ | \mathcal{F}_t\right].$$

This rather simplistic way of valuing a power generation asset in the spirit of the theory of real options had far-reaching implications in the developments of the energy markets, and is the main reason why spread options are of the utmost importance. However, such a valuation procedure is flawed in the presence of emission regulation as the costs of production also have to include the costs specific to the regulation. To be more specific, the day-τ potential profit $\left(P_\tau - h_g S_\tau^g - K\right)^+$ of the spark spread has to be modified to $\left(P_\tau - h_g S_\tau^g - e_g A_\tau - K\right)^+$ in order to accommodate the cost of the regulation. Here A_τ is the price of one allowance certificate worth one ton of CO_2 equivalent, and e_g is the emission coefficient of the plant, namely the number of tons of CO_2 emitted by the plant during the production of one unit of electricity. Such a spread is often called a *clean spread* to emphasize the fact that the externality is being paid for, and the real option approach to power plant valuation leads to the following *clean price*:

$$V_t^{CPP} = \sum_{\tau=T_1}^{T_2} e^{-r(\tau-t)}\mathbb{E}\left[\left(P_\tau - h_g S_\tau^g - e_g A_\tau - K\right)^+ | \mathcal{F}_t\right]$$

for the control of the plant over the period $[T_1, T_2]$ in the presence of the regulation.

In order to price such cross-commodity derivatives, a joint model is clearly required for fuel prices, electricity prices and carbon allowance prices. Various studies have analysed the strong links between these price series (De Jong and Schneider 2009, Koenig 2011). Many reduced-form price models have been proposed for electricity (see Eydeland and Wolyniec (2003) and Benth *et al.* (2008) reviews) with a focus on capturing its stylized features such as seasonality, high volatility, spikes, mean-reversion and fuel price correlation. On the other hand, many authors have argued that these same features are better captured via a structural approach, modelling the dynamics of underlying factors such as demand (load), capacity and fuel prices (early examples include Barlow (2002), Cartea and Villaplana (2008), Pirrong and Jermakyan (2008) and Coulon and Howison (2009)).

Similarly, for carbon emission allowances, exogenously specified processes that model prices directly have been proposed by some (Carmona and Hinz 2011). Others have instead treated the emission process as the exogenously specified underlying factor; in this case the allowance certificate becomes a derivative on cumulative emissions (Seifert *et al.* 2008, Chesney and Taschini 2012). However, these models do not take into account the important feedback from the allowance price to the rate at which emissions are produced in the electricity sector—a feature that is crucial for the justification of any implementation of a cap-and-trade scheme. In a discrete-time framework this feedback mechanism has been addressed, for example, by Coulon (2009) and Carmona *et al.* (2010). In continuous time the problem has been treated by Carmona *et al.* (2012b) and Howison and Schwarz (2012), whereby the former models the accumulation of emissions as a function of an exogenously specified electricity price process, while the latter uses the bid stack mechanism to infer the emission rate.

The literature on spread options is extensive. In industry, Margrabe's classical spread option formula (Margrabe 1978) is still widely used, and has been extended by various authors (see Carmona and Durrleman (2003) for an overview), including to the three commodity case, as required for the pricing of clean spreads (Alos *et al.* 2011). Carmona and Sun (2012) analyse the pricing of two-asset spread options in a multiscale stochastic volatility model. For electricity markets, pricing formulae for dirty spreads based on structural models have been proposed by Carmona *et al.* (2012a), who derive a closed-form formula in the case of $K = 0$, and by Aïd *et al.* (2012), who derive semi-closed form formulae for $K \neq 0$ at the expense of a fixed merit order.

The original contributions of the paper are twofold. First, we express the value of clean spread options in a formulation where demand for power and fuel prices are the only factors whose stochastic dynamics are given exogenously, and where the prices of power and emission allowances are derived from a bid-stack-based structural model and a forward backward stochastic differential system, respectively. The second contribution is the development of a numerical code for the computation of the solution of the pricing problem. First, we solve a 4 + 1-dimensional semilinear partial differential equation to compute the price of an emission allowance, then we use Monte Carlo techniques to compute the price of the spread option. These computational tools are used to produce the numerical results for the case studies presented in Section 27.6 of the paper for the purpose of illustrating the impact of a carbon regulation on the price of spread options. In this final section we first compare the price of spark and dark spread options in two different markets, one with no emission regulation and the other governed by an increasingly strict cap-and-trade system. Second, we analyse the impact that different merit order scenarios have on the option prices. Third, we demonstrate the difference between the structural and the reduced-form approach by comparing the option prices produced by our model with those produced by two candidate reduced-form models. Fourth and last, we contrast two competing policy instruments: cap-and-trade, represented by the model we propose, and a fixed carbon tax.

27.2 The Bid Stack: Price Setting in Electricity Markets

In order to capture the dependency of the electricity price on production costs and fundamental factors in a realistic manner, we use a structural model in the spirit of those reviewed in the recent survey of Carmona and Coulon (2012). The premises of structural models for electricity prices depend upon an explicit construction of the supply curve. Since electricity is sold at its marginal cost, the electricity spot price is given by evaluating the supply function for the appropriate values of factors used to describe the costs of production in the model.

In practice, electricity producers submit day-ahead bids to a central market operator, whose task it is to allocate the production of electricity amongst them. Typically, firms' bids have the form of price–quantity pairs, with each pair comprising the amount of electricity the firm is willing to produce, and the price at which the firm is willing to sell this quantity. Given the large number of generators in most markets, it is common in structural models to approximate the resulting step function of market bids by a continuous increasing curve. Firms' bid levels are determined by their costs of production. An important feature of our model, distinguishing it from most of the commonly used structural models, is the inclusion, as part of the production costs, of the costs incurred because of the existence of emission regulation.

We assume that, when deciding which firms to call upon to produce electricity, the market operator adheres to the merit order, a rule by which cheaper production units are called upon before more expensive ones. For simplicity, operational and transmission constraints are not considered.

Assumption 27.2.1

The market operator arranges bids according to the merit order, in increasing order of production costs.

The map resulting from ordering market supply in increasing order of electricity costs of production is what is called the bid stack. As it is one of the important building blocks of our model, we define it in a formal way for later convenience.

Definition 27.2.2: The *bid stack* is given by a measurable function

$$b : [0, \overline{x}] \times \mathbb{R} \times \mathbb{R}^n \ni (x, a, s) \to b(x, a, s) \in \mathbb{R},$$

with the property that, for each fixed $(a, s) \in \mathbb{R} \times \mathbb{R}^n$, the function $[0, \overline{x}] \ni x \to b(x, a, s)$ is strictly increasing.

In this definition, $\overline{x} \in \mathbb{R}_{++}$ represents the *market capacity* (measured in MWh) and the variable x the *supply of electricity*. The integer $n \in \mathbb{N} \setminus \{0\}$ gives the number of economic factors (typically the prices, in € say, of the fuels used in the production of electricity), and $s \in \mathbb{R}^n$ the numeric values of these factors. Here and throughout the rest of the paper the *cost of carbon emissions* (measured in € per metric ton of CO_2) is denoted by a. So for a given allowance price, say a, and fuel prices, say s, the market is able to supply x units of electricity at price level $b = b(x, a, s)$ (measured in € per MWh). In other words, $b(x, a, s)$ represents the bid level of the marginal production unit in the event that demand equals x.

The choice of a function b which captures the subtle dependence of the electricity price upon the level of supply and the production costs, is far from trivial, and different approaches have been considered in the literature, as reviewed recently by Carmona and Coulon (2012). In Section 27.5.1 we extend the model proposed by Carmona *et al.* (2012a) to include the cost of carbon as part of the variable costs driving bid levels.

27.3 Risk-Neutral Pricing of Allowance Certificates

As the inclusion of the cost of emission regulation in the valuation of spread options is the main thrust of the paper, we explain how emission allowances are priced in our model. The model we introduce is close to Howison and Schwarz (2012). However, we extend the results found therein to allow the equilibrium bids of generators to be stochastic and driven by fuel prices, a generalization that is vital for our purpose.

We assume that carbon emissions in the economy are subject to cap-and-trade regulation structured as follows: at the end of the compliance period, each registered firm needs to offset its cumulative emissions with emission allowances or incur a penalty for each excess ton of CO_2 not covered by a redeemed allowance certificate. Initially, firms acquire allowance certificates through free allocation, e.g. through National Allocation Plans (NAP) like in the initial phase of the European Union (EU) Emissions Trading Scheme (ETS), or by purchasing them at auctions as in the Regional Greenhouse Gas Initiative (RGGI) in the North East of the US. Allowances change hands throughout the compliance period. Typically, a firm which thinks that its initial endowment will not suffice to cover its emissions will buy allowances, while firms expecting a surplus will sell them. Adding to these *naturals*, speculators enter the market providing liquidity. Allowances are typically traded in the form of forward contracts and options. In this paper, we denote by A_t the spot price of an allowance certificate maturing at the end of the compliance period. Because their cost of carry is negligible, we treat them as financial products liquidly traded in a market without frictions, and in which long and short positions can be taken.

In a competitive equilibrium, the level of cumulative emissions relative to the cap (i.e. the number of allowance certificates issued by the regulation authority) determines whether—at the end of the compliance period—firms will be subjected to a penalty payment and create a demand for allowance certificates (see Carmona *et al.* (2010) for details). For this reason, allowance certificates should be regarded as derivatives on the emissions accumulated throughout the regulation period. This type of option written on a non-tradable underlying interest is rather frequent in the energy markets: temperature options are a case in point.

27.3.1 The Market Emission Rate

As evidenced by the above discussion, the rate at which CO_2 is emitted into the atmosphere as a result of electricity production has to be another important building block of our model. Clearly, at any given time, this rate is a function of the amount of electricity produced and because of their impact on the merit order, the variable costs of production, including fuel prices, and, notably, the carbon allowance price itself.

Definition 27.3.1: The *market emission rate* is given by a measurable function

$$\mu_e : [0, \bar{x}] \times \mathbb{R} \times \mathbb{R}^n \ni (x, a, s) \to \mu_e(x, a, s) \in \mathbb{R}_+,$$

on which we will impose technical assumptions later on.

With the above definition, for a given level of electricity supply and for given allowance and fuel prices, $\mu_e = \mu_e(x, a, s)$ represents the rate at which the market emits, measured in tons of CO_2 per hour. Cumulative emissions are then computed by integrating the market emission rate over time. Since any increase in supply can only increase the emission rate, it is of course reasonable from a modelling point of view to expect μ_e to be increasing as a function of x. Similarly, as the cost of carbon increases, the variable costs (and hence the bids) of pollution-intensive generators increase by more than those of environmentally friendlier ones. Dirtier technologies become relatively more expensive and are likely to be scheduled further down in the merit order. As a result, cleaner technologies are brought online earlier and we expect μ_e to be decreasing as a function of a.

In Section 27.5.2 we propose a specific functional form for μ_e consistent with the bid stack model introduced in Section 27.5.1.

27.3.2 The Pricing Problem

We shall use the following notation. For a fixed time horizon $T \in \mathbb{R}_+$, let $\left(W_t^0, W_t\right)_{t \in [0,T]}$ be[*] a $(n + 1)$-dimensional standard Wiener process on a probability space $(\Omega, \mathcal{F}, \mathbb{P})$, $\mathcal{F}^0 := \left(\mathcal{F}_t^0\right)$ the filtration generated by W^0, $\mathcal{F}^W := \left(\mathcal{F}_t^W\right)$ the filtration generated by W, and $\mathcal{F} := \mathcal{F}^0 \vee \mathcal{F}^W$ the market filtration. All relationships between random variables are to be understood in the almost sure sense.

---

[*] To simplify the presentation, from now on we drop the subscript $t \in [0, T]$, which specifies the time interval on which a stochastic process is defined.

Consumers' *demand for electricity* is given by an $\mathcal{F}^0$-adapted stochastic process (D_t) taking values in $[0,\bar{x}]$. In response to this demand, producers supply electricity, and we assume that demand and supply are always in equilibrium, so that at any time $t \in [0,T]$ an amount D_t of electricity is supplied. The *prices of fuels* are observed $\mathcal{F}^W$-adapted stochastic processes (S_t) taking values in $\mathbb{R}^n$, where $S_t := (S_t^1,\ldots,S_t^n)$. As we will see in Section 27.3.3, the price of an *allowance certificate* at time t, say A_t, is now constructed as a $\mathcal{F}$-adapted stochastic process solving a Forward Backward Stochastic Differential Equation (FBSDE). The rate of emission $\mu_s(D_t,A_t,S_t)$ can then be evaluated and the cumulative emissions computed by integrating over time, resulting in a $\mathcal{F}$-adapted process (E_t).

Since we do not present a calibration of the model to any particular electricity or emissions market, we avoid the difficulties of estimating market prices of risk (see, for example, Eydeland and Wolyniec (2003) and Carmona and Coulon (2012) for discussions of some possible ways to approach this thorny issue), and instead choose to specify the dynamics of the processes (D_t) and (S_t) directly under a risk-neutral measure $\mathbb{Q} \sim \mathbb{P}$ chosen by the market for pricing purposes. In practice, various alternative approaches to parameter estimation and calibration could be used to identify a risk-neutral measure which is consistent with liquidly traded products such as power and fuel forward contracts, thereby also estimating the magnitude of risk premia. However, our focus here is on the overall nature of energy price correlations, their structural origins and their important impact on spread option prices, rather than carrying out a study of any one market. For this reason, we also ignore any questions of market incompleteness (i.e. the uniqueness of $\mathbb{Q}$), transaction costs, illiquidity, inelasticity of demand (and also any risk that demand could ever exceed supply, for example causing a black-out), and finally the role of non-power-sector emissions. While interesting details of the markets, we argue that the inclusion of such effects should not cause any substantial change to the important qualitative conclusions drawn from the model in Section 27.6.

27.3.3 An FBSDE for the Allowance Price

We assume that, at time $t = 0$, demand for electricity is known. Thereafter, it evolves according to an Itô diffusion. Specifically, for $t \in [0, T]$, demand for electricity D_t is the unique strong solution of a stochastic differential equation of the form

$$dD_t = \mu_d(t,D_t)dt + \sigma_d(D_t)d\tilde{W}_t^0, \quad D_0 = d_0 \in (0,\bar{x}), \tag{27.1}$$

where $(\tilde{W}_t^0)$ is an $\mathcal{F}_t^0$-adapted is $\mathbb{Q}$-Brownian motion. The time dependence of the drift allows us to capture the seasonality observed in electricity demand, and the resulting seasonality in prices.

Similarly to demand, the prices of the fuels used in the production process satisfy a system of stochastic differential equations written in vector form as follows:

$$dS_t = \mu_s(S_t)dt + \sigma_s(S_t)d\tilde{W}_t, \quad S_0 = s_0 \in \mathbb{R}^n, t \in [0,T], \tag{27.2}$$

where $(\tilde{W}_t)$ is an $\mathcal{F}_t^W$-adapted $\mathbb{Q}$-Brownian motion. We note that, in some cases, it may be appropriate to also include time-dependence in the drift or volatility above, in order to capture seasonal patterns in some fuels such as natural gas.

Cumulative emissions are measured from the beginning of the compliance period at time $t = 0$, so that $E_0 = 0$. Subsequently, they are determined by integrating the market emission rate μ_e introduced in definition 27.3.1. So assuming that the price A_t of an allowance certificate is known, the cumulative emissions process is represented by an absolutely continuous process, i.e. for $t \in [0, T]$,

$$dE_t = \mu_e(D_t,A_t,S_t)dt, E_0 = 0. \tag{27.3}$$

Note that, with this definition, the process (E_t) is non-decreasing, which makes intuitive sense considering that it represents a cumulative quantity.

To complete the formulation of the pricing model, it remains to characterize the allowance certificate price process $(A_t)_{t\in[0,T]}$. If our model is to apply to a one compliance period scheme, in a competitive equilibrium, at the end of the compliance period $t = T$, its value is given by a deterministic function of the cumulative emissions:

$$A_T = \phi(E_T), \tag{27.4}$$

where $\phi : \mathbb{R} \hookrightarrow \mathbb{R}$ is bounded, measurable and non-decreasing. Usually, $\phi(\cdot) := \pi \mathbb{1}_{[\Gamma,\infty)}(\cdot)$, where $\pi \in \mathbb{R}_+$ denotes the penalty paid in the event of non-compliance and $\Gamma \in \mathbb{R}_+$ the cap chosen by the regulator as the aggregate allocation of certificates (see Carmona *et al.* (2010) for details). Since the discounted allowance price is a martingale under $\mathbb{Q}$, it is equal to the conditional expectation of its terminal value, i.e.

$$A_t = \exp(-r(T-t))\mathbb{E}^{\mathbb{Q}}\big[\phi(E_T) \mid \mathcal{F}_t\big], \quad \text{for } t \in [0,T], \tag{27.5}$$

which implies, in particular, that the allowance price process (A_t) is bounded. Since the filtration $(\mathcal{F}_t)$ is being generated by the Wiener process, it is a consequence of the Martingale Representation Theorem (Karatzas and Shreve 1999) that the allowance price can be represented as an Itô integral with respect to the Brownian motion $(\tilde{W}_t^0, \tilde{W}_t)$. It follows that

$$dA_t = rA_t dt + z_t^0 d\tilde{W}_t^0 + Z_t \cdot d\tilde{W}_t, \quad \text{for } t \in [0,T], \tag{27.6}$$

for some $\mathcal{F}_t$-adapted, square integrable process (Z_t^0, Z_t).

Combining equations (27.1), (27.2), (27.3), (27.4) and (27.6), the pricing problem can be reformulated as the solution of the FBSDE

$$\begin{cases} dD_t = \mu_d(t,D_t)dt + \sigma_d(D_t)d\tilde{W}_t^0, & D_0 = d_0 \in (0,\bar{x}), \\ dS_t = \mu_s(S_t)dt + \sigma_s(S_t)d\tilde{W}_t, & S_0 = s_0 \in \mathbb{R}^n, \\ dE_t = \mu_e(D_t,A_t,S_t)dt, & E_0 = 0, \\ dA_t = rA_t dt + Z_t^0 d\tilde{W}_t^0 + Z_t \cdot d\tilde{W}_t, & A_T = \phi(E_T). \end{cases} \tag{27.7}$$

Note that the first two equations are standard stochastic differential equations (in the forward direction of time) which do not depend upon the cumulative emissions and the allowance price. We will choose their coefficients so that existence and uniqueness of solutions hold. Also, for the sake of convenience, we implicitly assume that the function μ_e, whose first argument was originally restricted to the interval $[0,\bar{x}]$, is defined on the whole $\mathbb{R} \times \mathbb{R} \times \mathbb{R}^n$ by setting $\mu_e(x,a,s) = \mu_e(0,a,s)$ for $x < 0$ and $\mu_e(x,a,s) = \mu_e(\bar{x},a,s)$ for $x > \bar{x}$. Finally, we make the following assumptions on the coefficients of (27.7).

Assumption 27.3.2

The functions $\mu_d : [0,T] \times [0,\bar{x}] \to \mathbb{R}, \sigma_d : [0,\bar{x}] \to \mathbb{R}, \mu_s : \mathbb{R}^n \to \mathbb{R}^n, \sigma_s : \mathbb{R}^n \to \mathbb{R}^n \times \mathbb{R}^n$ are such that the first two equations in (27.7) have a unique strong solution.

27.3.4 Existence of a Solution to the Pricing Problem

Theorem 27.3.3:

If Assumption 27.3.2 holds, the function μ_e giving the emission rate is Lipschitz with respect to the variable a uniformly in x and s, and $\mu_e(x,0,s)$ is uniformly bounded in x and s, and the function ϕ giving the terminal condition is bounded, non-decreasing and Lipschitz, then the FBSDE (27.7) has a unique square integrable solution.

Proof: Let (D_t) and (S_t) represent the strong solutions of the first two equations of (27.7) whose existence is guaranteed by Assumption 27.3.2. These equations being decoupled from the remaining ones, the latter can be treated as a FBSDE with random coefficients and one-dimensional forward and backward components. We claim that existence and uniqueness hold because of Theorem 7.1 of Ma *et al.* (2011).* Strictly speaking, we cannot apply directly Theorem 7.1 of Ma *et al.* (2011) because our Wiener process is $(n + 1)$-dimensional. However, a close look at the proof shows that what is really needed is to prove the well-posedness of what the authors call the characteristic BSDE, and the boundedness of its solution and the solutions of the dominating Ordinary Differential Equations (ODE). In the present situation, these equations are rather simple due to the fact that (E_t) has bounded variation, and, as a consequence, its volatility vanishes. The two dominating ODEs can be solved explicitly and one can check that the solutions are bounded by inspection. Moreover, the function ϕ giving the terminal condition being uniformly Lipschitz, the characteristic BSDE is one-dimensional, and although driven by a multidimensional Brownian motion, its terminal condition is bounded, and Kobylanski's comparison results (see the original contribution (Kobylanski 2000)) can be used to conclude the proof.

The above result is proven for a terminal condition given by a smooth function ϕ. However, as already mentioned earlier, competitive equilibrium arguments for single compliance period models suggest that the function ϕ should be singular (see, for example, Carmona *et al.* (2010)). Indeed, in the event of non-compliance, that is, when the cumulative emissions strictly exceed the cap at the end of the compliance period, i.e. when $E_T > \Gamma$, the penalty and the allowance certificate are perfect substitutes; therefore, they ought to have the same monetary value and one should have $A_t = \pi$, which suggests $\phi(e) = \pi$ whenever $e > \Gamma$. Similarly, in the event of compliance, that is, when the cumulative emissions are strictly below the cap, there will be spare certificates in the market; these certificates will be in zero demand and will therefore expire worthless, so, in this case, $A_T = 0$, which suggests $\phi(e) = 0$ whenever $e > \Gamma$. This economic interpretation of the function ϕ giving the terminal condition gives the whole story when the event $\{E_T = \Gamma\}$ has zero probability since we do not have to worry about the definition of $\phi(e)$ when $e = \Gamma$. Hence the importance of knowing if the random variable E_T is continuous (e.g. has a density). Again, see an early discussion of this property in Carmona *et al.* (2010), and a systematic analysis in Carmona *et al.* (2012b) and Carmona and Delarue (2012). We conjecture that a proof in the spirit of the one given by Carmona and Delarue (2012) should work in the setting of this paper if μ_e is strictly decreasing in a, providing existence and uniqueness of a solution of the FBSDE when the binary terminal condition is weakened. Furthermore, Carmona and Delarue also proved that, still under strict monotonicity of μ_e, the aggregate emissions are equal to the cap with positive probability at the end of the compliance period. This shows that the competitive equilibrium argument given earlier is enough to specify a unique emission process (E_T) and a unique price process (A_T) for the allowance, even though the terminal price of an allowance A_T at the end of the compliance period cannot be prescribed *ex ante* on a set of scenarios of positive probability. We suspect that this is also the case in the present situation.

Note added in proof: The conjectured existence and uniqueness of a solution to the FBSDE (27.7) in our setting was recently proved by Schwarz (2012).

27.4 Valuing Clean Spread Options

In this section we consider the problem of spread option pricing as described in the introduction. Whether the goal is to value a physical asset or to manage the risk associated with financial positions, one needs to compute the price of a European call option on the difference between the price of electricity and the costs of production for a particular power plant. The costs that we take into account are the

---

* We thank Francois Delarue for suggesting this approach.

fixed operation and maintenance costs, the cost of the fuel needed to generate one MWh of electricity and the cost of the ensuing emissions. Letting the $\mathcal{F}_t$-adapted process (P_t) denote the spot price of electricity, and recasting the informal discussion in the introduction with the notation we chose to allow for several input fuels, a clean spread option with maturity $\tau \in [0, T]$ is characterized by the terminal pay-off

$$\left(P_\tau - h_i S_\tau^i - e_i A_\tau - K\right)^+,$$

where K represents the value of the fixed operation and maintenance costs and, for $i \in 1,...,n, h_i \in \mathbb{R}_{++}$ and $e_i \in \mathbb{R}_+$ denote the specific heat and emission rates of the power plant under consideration, and S^i is the price at time τ of the fuel used in the production of electricity. In the special case when S^i is the price of coal (gas) the option is known as a *clean dark (spark) spread* option.

Since we are pricing by expectation, for $i \in \{1,...,n\}$, the value V_t^i of the clean spread is given by the conditional expectation under the pricing measure of the discounted pay-off, i.e.

$$V_t^i = \exp(-r(\tau - t))\mathbb{E}^{\mathbb{Q}}\left[\left(P_\tau - h_i S_\tau^i - e_j A_\tau - K\right)^+ | \mathcal{F}_t\right],$$

for $t \in [0, \tau]$.

27.5 A Concrete Two-Fuel Model

We now turn to the special case of two fuels, coal and gas, which differ significantly in their level of emissions per MWh of power generated.

27.5.1 The Bid Stack

Our bid stack model is a slight variation of the one we proposed in Carmona *et al.* (2012a). Here we extend it to include the cost of emissions as part of the variable costs driving firms' bids.

We assume that the coal and gas generators have aggregate capacities $\bar{x}_c$ and $\bar{x}_g$, respectively, so that the market capacity is $\bar{x} = \bar{x}_c + \bar{x}_g$, and their bid levels are given by linear functions of the allowance price and the price of the fuel used for the generation of electricity. We denote these bid functions by b_c and b_g, respectively. The coefficients appearing in these linear functions correspond to the marginal emission rate (measured in ton equivalent of CO_2 per MWh) and the heat rate (measured in MMBtu per MWh) of the technology in question. Specifically, for $i \in \{c,g\}$, we assume that

$$b_i(x,a,s) := e_i(x)a + h_i(x)s, \qquad \text{for}(x,a,s) \in \left[0, \bar{x}_i\right] \times \mathbb{R} \times \mathbb{R}, \tag{27.8}$$

where the *marginal emission rate* e_i and the *heat rate* h_i are given by

$$e_i(x) := \hat{e}_i \exp\left(m_i x\right),$$

$$\text{for } x \in [0, \bar{x}_i].$$

$$h_i(x) := \hat{h}_i \exp\left(m_i x\right),$$

Here $\hat{e}_i$, $\hat{h}_i$ and m_i are strictly positive constants. We allow the marginal emission rate and the heat rate of each technology to vary to reflect differences in efficiencies within the fleet of coal and gas generators. Less efficient plants with higher heat rates have correspondingly higher emission rates. We assume that, for each technology, the ratio h_i/e_i is fixed, a reasonable approximation which implies that the emissions rate of any coal (gas) plant is simply a fixed multiple of the quantity of coal (gas) burned.

Proposition 27.5.1: *With b_c and b_g as above and $I = \{c, g\}$, the market bid stack b is given by*

$$b(x,a,s) = \begin{cases} \left(\hat{e}_i a + \hat{h}_i s_i\right)\exp\left(m_i x\right), & \begin{array}{l} \text{if } b_i\left(x,a,s_i\right) \leq b_j\left(0,a,s_j\right) \\ \text{for } i, j \in I, i \neq j, \end{array} \\[1em] \left(\hat{e}_i a + \hat{h}_i s_i\right)\exp\left(m_i\left(x - \bar{x}_j\right)\right), & \begin{array}{l} \text{if } b_i\left(x - \bar{x}_j, a, s_i\right) > b_j\left(0,a,s_j\right) \\ \text{for } i, j \in I, i \neq j \end{array} \\[1em] \prod_{i \in I}\left(\hat{e}_i a + \hat{h}_i s_i\right)^{\beta_i}\exp\left(\gamma x\right), & \text{otherwise,} \end{cases}$$

for $(x,a,s) \in [0,\bar{x}] \times \mathbb{R} \times \mathbb{R}^2$, where $\beta_i = m_{I\setminus\{i\}}/\left(m_c + m_g\right)$ and $\gamma = m_c m_g/\left(m_c + m_g\right)$.

Proof: The proof is a straightforward extension of corollary 1 of Carmona *et al.* (2012a).

27.5.2 The Emission Stack

In order to determine the rate at which the market emits we need to know which generators are supplying electricity at any time. By the merit order assumption the market operator calls upon firms in increasing order of their bid levels. Therefore, given electricity, allowance and fuel prices $(p,a,s) \in \mathbb{R} \times \mathbb{R} \times \mathbb{R}^2$, for $i \in \{c,g\}$, the set of active generators of fuel type i is in one-to-one correspondence with the set $\{x \in [0,\bar{x}_i] : b_i(x,a,s) \leq p\}$.

Proposition 27.5.2: *Assuming that the market bid stack is of the form specified in proposition 5.1, the market emission rate μ_e is given by*

$$\mu_e(x,a,s) := \sum_{i \in \{c,g\}} \frac{\hat{e}_i}{m_i}\left(\exp\left(m_i \hat{b}_i^{-1}\left(b(x,a,s),a,s_i\right)\right) - 1\right), \tag{27.9}$$

for $(x,a,s) \in [0,\bar{x}] \times \mathbb{R} \times \mathbb{R}^2$, where, for $i \in \{c,g\}$, we define

$$\hat{b}_i^{-1}\left(p,a,s_i\right) := 0 \vee \left(\bar{x}_i \wedge \frac{1}{m_i}\log\left(\frac{p}{\hat{e}_i a + \hat{h}_i s_i}\right)\right),$$

for $(p,a,s) \in \mathbb{R} \times \mathbb{R} \times \mathbb{R}^2$, and, as usual, $a \wedge b = \min(a,b)$ and $a \vee b = \max(a,b)$.

Proof: The market emission rate follows from integrating the marginal emission rate e_i for each technology over the corresponding set of active generators and then summing the two. Given the monotonicity of b_i in x and its range $[0,\bar{x}_i]$, the function $\hat{b}_i^{-1}$ describes the quantity of electricity supplied by fuel $i \in \{c,g\}$, and hence the required upper limit of integration.

27.5.3 Specifying the Exogenous Stochastic Factors

The Demand Process. We posit that, under $\mathbb{Q}$, the process (D_t) satisfies for $t \in [0, T]$ the stochastic differential equation

$$\begin{aligned} dD_t &= -\eta\left(D_t - \bar{D}(t)\right)dt + \sqrt{2\eta\hat{\sigma}D_t\left(\bar{x} - D_t\right)}d\tilde{W}_t, \\ D_0 &= d_0 \in (0,\bar{x}), \end{aligned}$$

where $[0,T] \ni t \to \bar{D}(t) \in (0,\bar{x})$ is a deterministic function giving the level of mean reversion of the demand and $\eta,\hat{\sigma} \in \mathbb{R}_{++}$ are constants. With this definition (D_t) is a Jacobi diffusion process; it has a linear, mean-reverting drift component and degenerates on the boundary. Moreover, subject to min $(\bar{D}(t),\bar{x} - \bar{D}(t))\bar{x}\hat{\sigma}$, for $t \in [0,T]$, the process remains within the interval $(0,\bar{x})$ at all times (Forman and Sørensen 2008). To capture the seasonal character of demand, we choose a function $\bar{D}(t)$ of the form

$$\bar{D}(t) := \varphi_0 + \varphi_1 \sin(2\pi\vartheta t),$$

where the values of the coefficients will be specified in the next section.

The Fuel Price Processes. We assume that the prices of coal (S_t^c) and gas $\left(S_t^g\right)$ follow correlated exponential (or geometric) Ornstein–Uhlenbeck processes under the measure $\mathbb{Q}$, i.e. for $i \in \{c,g\}$ and $t \in [0,T]$,

$$\begin{aligned}
dS_t^i &= -\eta_i\left(\log S_t^i - \bar{s}_i - \frac{\hat{\sigma}_i^2}{2\eta_i}\right)S_t^i dt + \hat{\sigma}_i S_t^i d\tilde{W}_t^i, \\
S_0^i &= s_0^i \in \mathbb{R}_{++},
\end{aligned}$$

where $d\left\langle W^c, W^g \right\rangle_t = \rho dt$.

27.6 Numerical Analysis

We now turn to the detailed analysis of the model we propose. For this purpose we consider a number of case studies in Sections 27.6.2 to 27.6.5. To produce the following results we used the numerical schemes explained in Appendices A and B.

27.6.1 Choice of Parameters

The tables in this section specify the values of the parameters used for the numerical analysis of our model that follows below. We refer to the parameter values specified in Tables 27.1.–27.5 as the 'base case' and indicate whenever we depart from this choice. Note that our choices do not correspond to a particular electricity market, but that all values are within a realistic realm.

TABLE 27.1 Parameters Relating to the Bid and Emission Stacks

$\hat{h}$	$\hat{e}_c$	m_c	$\bar{x}_c$	$\hat{h}_g$	$\hat{e}_g$	m_g	$\hat{x}_g$	$\bar{x}$
3	0.9	0.00005	12,000	7	0.4	0.00003	18,000	30,000

TABLE 27.2 Parameters Relating to the Demand Process

η	φ_0	φ_1	ϑ	$\hat{\sigma}$	d_0
50	21,000	3000	1	0.1	21,000

TABLE 27.3 Parameters Relating to the Fuel Price Processes

η_c	$\bar{s}_c$	$\hat{\sigma}_c$	s_0^c	η_g	$\bar{s}_g$	$\hat{\sigma}_g$	s_0^g	ρ
1.5	2	0.5	exp(2)	1.5	2	0.5	exp(2)	0.3

TABLE 27.4   Parameters Relating to the Cap-and-Trade Scheme

π	Γ	T	r
100	1.4×10^8	1	0.05

TABLE 27.5   Parameters Relating to the Spread Options

High eff. coal		Low eff. coal		High eff. gas		Low eff. gas	
h_c	e_c	h_c	e_c	h_g	e_g	h_g	e_g
3.5	1.05	5.0	1.5	7.5	0.43	1.5	0.66

Table 27.1 provides the parameter values specifying the bid curves. We consider a medium-sized electricity market served by coal and gas generators and with gas being the dominant technology. For the marginal emission rates, Table 27.1 implies that $e_c \in [0.9, 1.64]$ and $e_g \in [0.4, 0.69]$ (both measured in t CO_2 per MWh), so that all gas plants are 'cleaner' than all coal plants. For the heat rates, we observe that $h_c \in [3, 5.5]$ and $h_g \in [7, 12]$ (both measured in MMBtu per MWh). Using (27.9) now with $D_t = \bar{x}$, for $0 \le t \le T$, and the assumption that there are 8760 production hours in the year, we find, denoting the maximum cumulative emissions by $\bar{e}$, that $\bar{e} = 2.13 \times 10^8$.

Table 27.2 contains the parameter values for the demand process (D_t). We model periodicities on an annual and a weekly time scale and the chosen rate of mean-reversion assumes that demand reverts to its (time-dependent) mean over the course of one week.

In Table 27.3 we give the parameter values that specify the behaviour of the prices of coal and gas. Both are chosen to be slowly mean-reverting, at least in comparison to demand. To ease the analysis, we assume that all parameters are identical for the two fuels, including mean price levels, both measured in MMBtu.*

Table 27.4 defines the cap-and-trade scheme that we assume to be in place. The duration of the compliance period T is measured in years and we set the cap at 70% of the upper bound $\bar{e}$ for the cumulative emissions, in order to incentivise a reduction in emissions. This choice of parameter values results in A_0 being approximately equal to $\pi/2$, a value for which there is significant initial overlap between gas and coal bids in the stack. Furthermore, the values imply a bid stack structure such that, at mean levels of coal and gas prices, $A_t=0$ pushes all coal bids below gas bids, while for $A_t=\pi$ almost all coal bids are above all gas bids.

Finally, in Table 27.5 we specify the four spread option contracts used in the base case scenario to represent high and low efficiency coal plants, and high and low efficiency gas plants (note that low (high) efficiency means dirtier (cleaner) and corresponds to high (low) h_i and e_i).

We now consider a series of case studies to investigate various features of the model's results in turn. As the model captures many different factors and effects, this allows us to isolate some of the most important implications. In Case study I, we investigate the impact on coal and gas plants of different efficiencies of creating an increasingly strict carbon emissions market. In Case study II, we assess the impact on these plants of changes in initial fuel prices. In Case study III, we compare spread option prices in our model with two simple reduced-form approaches for A_t, which allows us to better understand the role of key model features such as bid-stack-driven abatement. Finally, in case study IV, we

---

* We note that gas and coal prices are typically quoted in different units, and can often differ by a factor of 10 or more. However, in our analysis, as we are not fitting to data, coal and gas only play the role of common representative fuel types (and other possibilities include lignite, oil, etc.). Therefore, our parameter choices simply reflect typical characteristics of energy price behavior. Much more importantly for our analysis, we require that one fuel be significantly 'cleaner' than the other, and that the relative price levels allow for merit order changes driven by the cap-and-trade market.

consider the overall impact of cap-and-trade markets in the electricity sector, by comparing them with a well-known alternative, a fixed carbon tax.

27.6.2 Case Study I: Impact of the Emission Market

The first effect that we are interested in studying is the impact of the cap-and-trade market on clean spread option prices, for increasingly strict levels of the cap Γ. At one extreme (when the cap is so generous that $A_t \approx 0$, for all $t \in [0,T]$), the results correspond to the case of a market without a cap-and-trade system, while at the other extreme (when the cap is so strict that $A_t \approx \pi \exp(-r(T-t))$, for all $t \in [0,T]$), there is essentially a very high carbon tax which tends to push most coal generators above gas generators in the stack. It is intuitively clear that higher carbon prices typically lead to higher spark spread option prices and lower dark spread option prices, thus favouring gas plants over coal plants, but the relationships can be more involved as we vary between low and high efficiency plants.

In Figure 27.1, we compare spread option prices corresponding to different efficiency generators (i.e. to different h_i, e_i in the spread pay-off) as a function of maturity τ. 'High' and 'low' efficiency plant indicates values of h_i, e_i chosen to be near the lowest and highest, respectively, in the stack, as given by Table 27.5. Within each of the four subplots, the five lines correspond to five different values of the cap Γ, ranging from very lenient to very strict. We immediately observe from Figure 27.1 the seasonality in spread prices caused by the seasonality in power demand. This is most striking for the low efficiency cases (high h_i, e_i), as such plants would rarely be used in shoulder months, particularly in the case of gas.

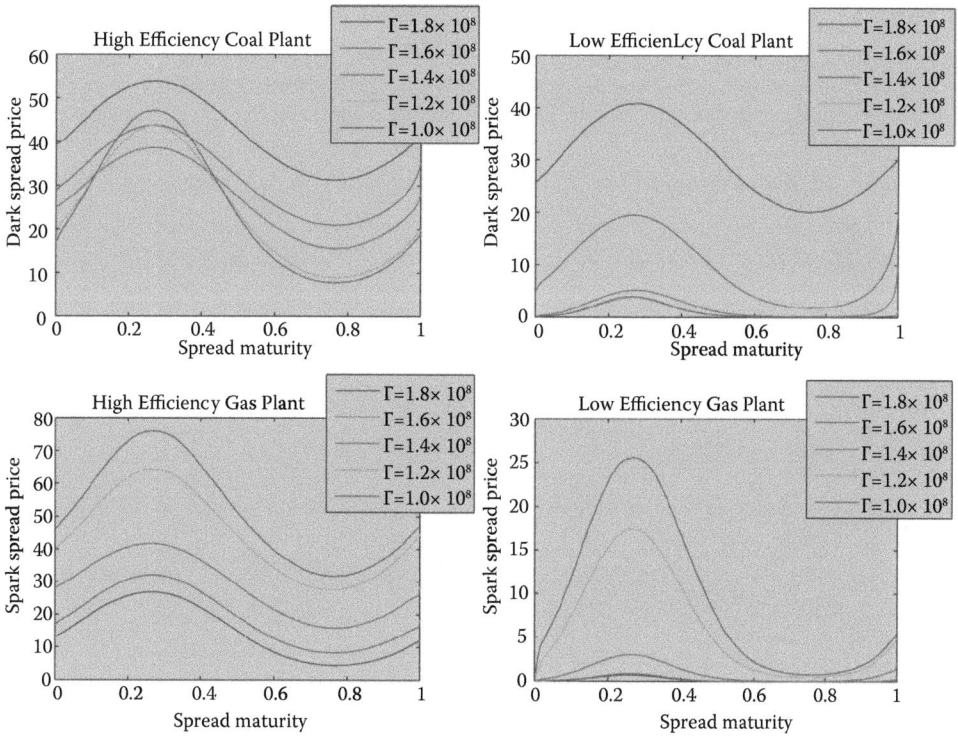

FIGURE 27.1 Cap strictness analysis for high efficiency coal (top left), low efficiency coal (top right), high efficiency gas (bottom left) and low efficiency gas (bottom right): spark and dark spread option values plotted against maturity, for varying levels of the cap Γ. Note that the five equally spaced cap values from 1.8×10^8 to 1.0×10^8 tons of CO_2 imply initial allowance prices of \$5, \$28, \$52, \$80 and \$94.

For low efficiency plants, the relationship with cap level (and corresponding initial allowance price) is as one would expect: a stricter cap greatly increases the value of gas plants and greatly decreases the value of the dirtier coal plants. This is also true for high efficiency gas plants, although the price difference (in percentage terms) for different Γ is less, since these are effectively 'in-the-money' options, unlike those discussed above. However, the analysis becomes more complicated for high efficiency coal plants, which tend to be chosen to run in most market conditions, irrespective of emission markets. Interestingly, we find that, for these options, the relationship with Γ (and hence A_0) can be non-monotonic under certain conditions, particularly for high levels of demand, when the price is set near the very top of the stack. In such cases a stricter cap provides extra benefit for the cleaner coal plants via higher power prices (typically set by the dirtier coal plants on the margin) which outweighs the disadvantage of coal plants being replaced by gas plants in the merit order.

27.6.3 Case Study II: Impact of Fuel Price Changes

Note that, in Table 27.3, the initial conditions of both gas and coal have been set to be equal to their long-term median levels. We now consider the case of the gas price s_0^g being either above or below its long-term level, thus inducing a change in the initial merit order. Given the record low prices of under \$2 recently witnessed in the US natural gas market (due primarily to shale gas discoveries), it is natural to ask how such fuel price variations affect our spread option results. Note, however, that since $\eta_c = \eta_g = 1.5$ (implying a typical mean-reversion time of 8 months), by the end of the trading period the simulated fuel price distributions will again be centred near their mean-reversion levels. Thus in this case study, we capture a temporary, not permanent, shift in fuel prices.

In Figure 27.2, we plot the value of coal and gas power plants, as given by the sum of spread options of all maturities $\tau \in [0, T]$. In the first plot, we consider low efficiency (high h_i and e_i) plants, while in the second we consider high efficiencies. The latter are much more likely to operate each day and to generate profits, and are hence much more valuable than the former. However, they also show different relationships with s_0^g, as illustrated for several different cap levels Γ (like in Case study I above) which correspond to high, low or medium (base case) values of A_0.* Firstly, for low efficiency plants (left plot), we observe that the gas plant value is typically decreasing in s_0^g, as we expect, since higher gas prices tend

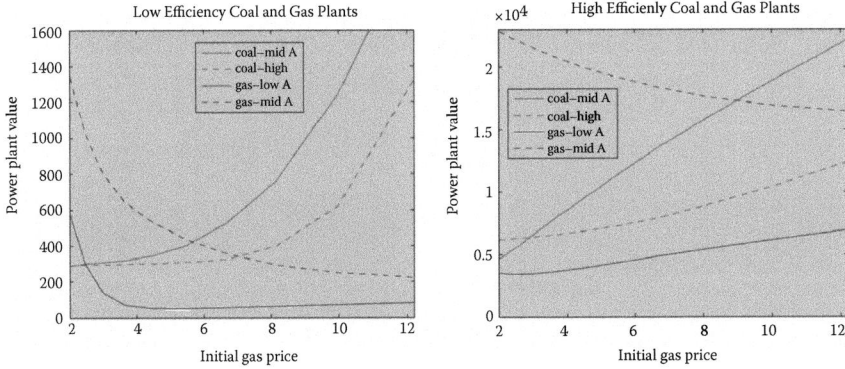

FIGURE 27.2 Power Plant Value (sum of spreads over τ) versus s_0^g for low efficiency (left) and high efficiency (right). 'High A' corresponds to $\Gamma = 1 \times 10^8$, 'mid A' to $\Gamma = 1.4 \times 10^8$ (base case) and 'low A' to $\Gamma = 1.8 \times 10^8$, with corresponding values $A_0 = 94$, $A_0 = 52$ and $A_0 = 5$.

---

* In the first plot, the cases 'coal – low A' and 'gas – high A' would produce values much higher than the other cases, and hence we instead choose 'coal – mid A' and 'gas – mid A' in order to illustrate the effects on a single plot.

to push the bids from gas above those from coal, meaning there is less chance that the gas plant will be used for electricity generation. Similarly, coal plant values are typically increasing in s_0^g, as more coal plants will be used. Note however that, for some cases, the curves flatten out, as no more merit order changes are possible. This is particularly true for the coal plant when A_0 is very high (and hence once gas drops below a certain point, the coal plant is almost certainly going to remain more expensive to run than all gas plants) and for the gas plant when A_0 is very low (and hence once coal increases above a certain point, the gas plant is almost certainly going to remain more expensive to run than all coal plants).

We now turn our attention to the high efficiency case (right plot), meaning the relatively cheap and clean plants for each technology. As expected, the coal plant benefits from low values of A_0 (as implied by a lenient cap) and gas from high values of A_0 (i.e. a strict cap). On the other hand, the relationship with s_0^g is now increasing for all cases plotted except that of a gas plant with high A_0. While it may seem surprising that, for low values of A_0 (or medium though not plotted), the gas plant value increases with s_0^g, this is quite intuitive when one considers that the range of bids from gas generators widens as s_0^g increases, implying that the efficient plants can make a larger profit when the inefficient plants set the power price. Indeed, as demand is quite high on average, and gas is 60% of the market, it is likely that these efficient gas plants will almost always be 'in-the-money' even if coal is lower in the stack. Only in the case where coal is typically above gas and now marginal (i.e. the high A_0 case) is the value of the gas plant decreasing in s_0^g since the plant's profit margins shrink as gas and coal bids converge.

27.6.4 Case Study III: Comparison with Reduced-Form

The second analysis we consider is to compare the results of our structural model for the allowance price, with two other simpler models, both of which belong to the class of 'reduced-form' models. The first of these treats the allowance price itself as a simple Geometric Brownian Motion (with drift r under $\mathbb{Q}$), and hence A_τ is log-normal at spread maturity, like S_τ^c and S_τ^g. The second comparison treats the emission process as a Geometric Brownian Motion (GBM), and retains the digital terminal condition $A_T = \pi^{\lceil}_{\{ET \geq \Gamma\}}$. As the drift of (E_t) is then simply a constant (chosen to match the initial value A_0 in the full model), there is no feedback from (A_t) to (E_t), or in other words, no abatement induced by the allowance price. For any time t, A_t is then given in closed form by a formula resembling the Black–Scholes digital option price. In order to fully specify the two reduced-form models, we need to choose a volatility parameter σ_a or σ_e for each of the GBMs, as well as correlations ρ_{ac}, ρ_{ag} or ρ_{ec}, ρ_{eg} with the Brownian Motions driving the other exogenous factors, coal and gas prices. All of these parameters are chosen to approximately match the levels of volatility and correlation produced by simulations in the full structural model, and are given in Table 27.6. Finally, note that, in all three models that we compare, the power price is given by the same bid stack function as usual, so our aim is to isolate and evaluate the effect of our more sophisticated framework for the allowance price, in comparison with simpler approaches. The cap throughout is $\Gamma = 1.4 \times 10^8$, the base case.

Figure 27.3 reveals that the difference between the reduced-form models and the full structural model is relatively small for high efficiency gas and coal plants which are typically 'in-the-money'. In contrast, a larger gap appears for low efficiency cases, where the reduced-form models significantly overprice spread options relative to the stack model. In particular, the case of log-normal emissions produces much higher prices, especially for dark spreads. The intuition is as follows. In the full model, the bid

TABLE 27.6 Parameters for Reduced-Form Comparisons, Treating A_t and E_t as GBMs

σ_a	ρ_{ac}	ρ_{ag}	σ_e	ρ_{ec}	ρ_{eg}
0.6	−0.2	0.4	0.006	−0.2	0.2

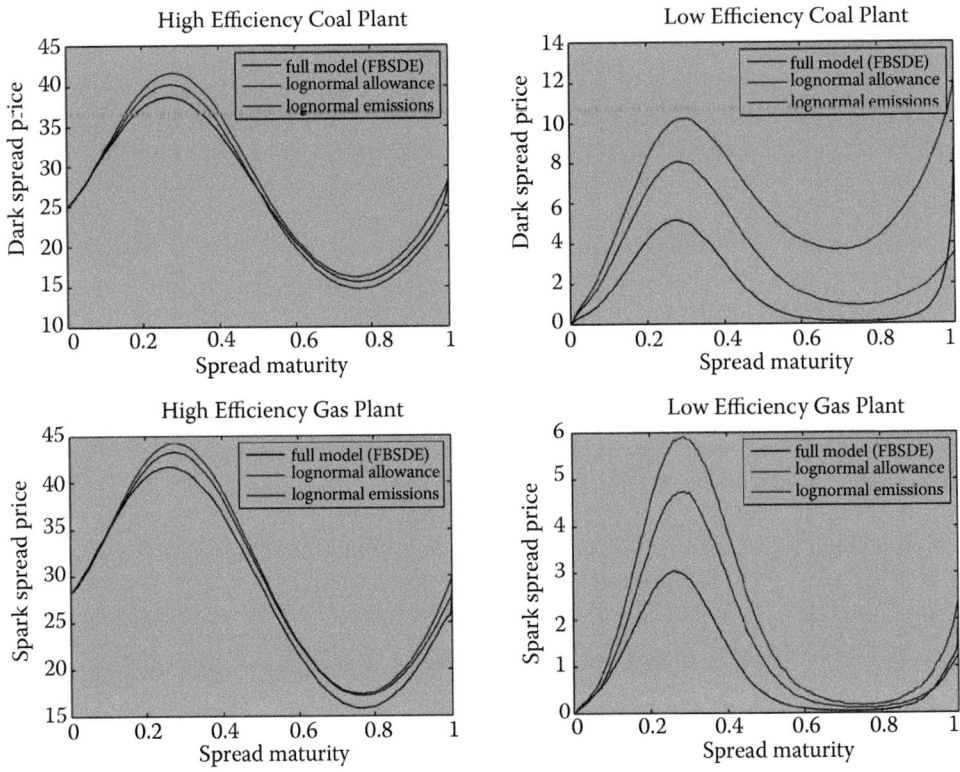

FIGURE 27.3 Model comparison against reduced-form: spark and dark spread option values for varying heat rates, emission rates and maturities.

stack structure automatically leads to lower emissions when the allowance price is high, and higher emissions when the allowance price is low, producing a mean-reversion-like effect on the cumulative emissions, keeping the process moving roughly towards the cap, with the final outcome (compliance or not) in many simulations only becoming clear very close to maturity. In contrast, if (E_t) is a GBM, much of the uncertainty is often resolved early in the trading period, with (A_t) then sticking near zero or π for much of the period. In such cases, there is a much larger benefit for deep OTM options (low efficiency plants), for which the tails of the allowance price distribution provide great value either for coal (when the price is near zero) or for gas (when the price is near the penalty). We observe that, in some of the subplots (particularly low efficiency coal), this extra benefit is indeed realized in the full model, but only very near the end of the trading period when the volatility of (A_t) often spikes, and the process either rises or falls sharply. In contrast, for the other reduced-form model with log-normally distributed allowance price, the volatility of the allowance price is constant throughout and (A_t) never moves rapidly towards zero or the penalty. However, the overall link with fuel and power prices is much weaker when simply using correlated Brownian Motions, which serves to widen the spread distribution in most cases relative to the full structural model. This result is somewhat similar to the observation of Carmona *et al.* (2012a) that a stack model generally produces lower spread option prices than Margrabe's formula for correlated log-normals.

27.6.5 Case Study IV: Cap-and-Trade vs. Carbon Tax

Finally, we wish to investigate the implications of the model for cap-and-trade systems, as compared with fixed carbon taxes. This question has been much debated by policy makers as well as academics, and can be roughly summarized as fixing quantity versus fixing price. Carmona *et al.* (2010) compare several different designs for cap-and-trade systems with a carbon tax, using criteria such as cost to society and windfall profits to power generators. Here we follow a related approach by analysing the power sector as a whole, but we build on our previous case studies by using spread option prices as a starting point. Firstly, we observe that the total expected discounted profits of the power sector are equal to the value of all the power plants implied by the bid stack structure, which in turn equals a portfolio of (or integral over) sums of spread option prices with varying h_i and e_i. Therefore, for each simulation over the period $[0, T]$, total profits (total revenues minus total costs) are*

$$\text{Total profits} \quad = \quad \sum_{\tau \in [0,T]} \left(P_\tau D_\tau - \int_0^{D_\tau} b(x, A_\tau, S_\tau) \, dx \right)$$

$$= \quad \sum_{\tau \in [0,T]} \int_0^{\bar{x}} \left(P_\tau - b(x, A_\tau, S_\tau) \right)^+ dx$$

$$= \quad \sum_{\tau \in [0,T]} \left(\int_0^{\bar{x}_c} \left(P_\tau - h_c(x) S_\tau^c - e_c(x) A_\tau \right)^+ dx \right.$$

$$+ \int_0^{\bar{x}_g} \left(P_\tau - h_g(x) S_\tau^g - e_g(x) A_\tau \right)^+ dx \Bigg),$$

where the second line follows from the fact that the events $\{P_\tau \geq b(x, A_\tau, S_\tau)\}$ and $\{D_\tau \geq x\}$ are equal.

Hence, instead of picking particular coal and gas plants with efficiencies specified by the parameters in Table 27.5, we now integrate power plant value over all the efficiencies of plants in the stack, as defined by the parameters in Table 27.1. In the case of the carbon tax, we simply force $A_t = A_0 \exp(rt)$ for all $t \in [0, T]$, including the exponential function in order to match the mean of the process in the cap-and-trade model. This is equivalent to setting the volatility σ_a equal to zero in the GBM model for the allowance price in Case study III.

In Figure 27.4, we first plot the expected total market profits in the base case as a function of time. It is interesting to observe that two important effects occur, pulling the profits in opposite directions, but varying in strength over the trading period. In particular, although the profits must be equal at time zero, a gap quickly appears in the early part of the trading period, with expected profits to power generators

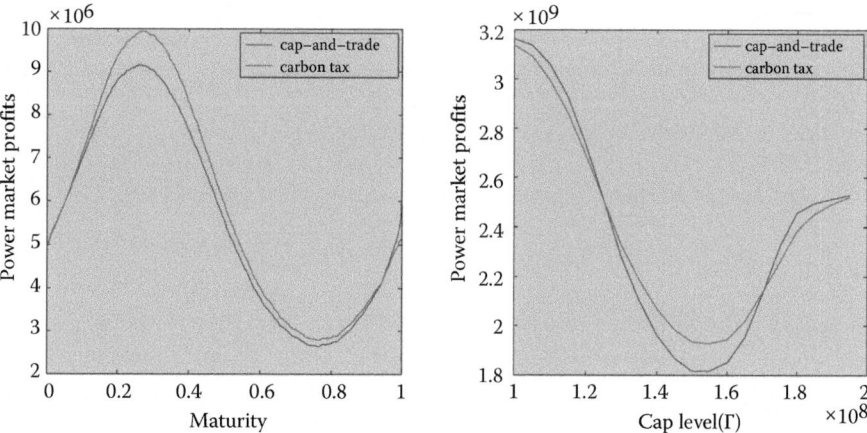

FIGURE 27.4 Cap-and-trade vs. carbon tax: power sector profits versus time for the 'base case' (left); total profits over one year for equally spaced cap values from 1×10^8 to 1.95×10^8 tons of CO_2 (right).

---

* Note that we do not consider here additional issues such as whether allowances are auctioned or freely allocated to generators. Instead, we assume that allowances are bought on the market by generators as and when they need them.

significantly higher under a carbon tax than cap-and-trade. However, as maturity approaches, the gap narrows and the order reverses over the final days, as cap-and-trade generates higher expected profits. We reason as follows: firstly, as $A_0=52$ in the base case, the bids of coal and gas begin the period at very similar levels, a state which generally keeps profits low, since the variance of electricity prices is low and the profit margins of both coal and gas generators are quite low. As time progresses and fuel prices move, the coal and gas bids will tend to drift apart in most simulations, for example with gas sometimes moving above coal, say. However, in our structural model for the cap-and-trade scheme, in such a case the higher emissions will induce a higher allowance price, and in turn a feedback effect due to the coupling in (27.7), which acts to keep coal and gas bids closer together. A similar argument can be made for the case of gas bids tending to move below coal bids but then being counteracted by lower allowance prices. Again we see that the power market structure induces mean-reversion on (E_t), which in this scenario (of a mid-range cap level) corresponds to keeping coal and gas bids close together. On the other hand, under a carbon tax with fixed (or deterministic) allowance price, there is of course no feedback mechanism (no price-sensitive abatement), and bids tend to wander apart. However, as the end of the trading period approaches, in the cap-and-trade system the allowance price eventually gets pulled to either zero or the penalty, which will separate the bids in one way or the other, either leading to very large profits for coal plants (if $A_T=0$) or for gas plants (if $A_T=\pi$). This is a similar effect to that discussed when comparing with a log-normal allowance price in Case study III, as neither a carbon tax nor a log-normal allowance price model sees the extra volatility near maturity caused by the terminal condition.

Finally, in the second plot of Figure 27.4, we consider how these conclusions change if the cap is made stricter or more lenient. Instead of plotting against maturity, we consider the total profits of the power sector over the entire period $[0, T]$. Firstly, we observe that, under both forms of emission regulation, power sector profits are lowest if the cap is chosen close to the base case, under which the bids from coal and gas generators are more tightly clustered together. Secondly, it is important to notice that the conclusion in the previous discussion that a carbon tax provides more profits to the power sector does not hold for all scenarios of the cap. In particular, for either very high or very low values of the cap, the cap-and-trade scheme provides more profits than a tax. The explanation here is that for the automatic abatement mechanism in the stack to have its largest impact (keeping bids together, and emissions heading towards the cap), there needs to be significant uncertainty at time zero as to whether the cap will be reached. The feedback mechanism of a cap-and-trade system then allows this uncertainty to be prolonged through the period. On the other hand, for an overly strict or overly lenient cap (or similarly for a merit order which does not allow for much abatement), the second effect discussed above dominates over the first. In other words, the terminal condition which guarantees large profits to either coal or gas at maturity begins to take precedence earlier in the trading period, instead of just before maturity as in the base case. Although in practice there are many other details to consider when comparing different forms of emission legislation, our stylized single-period model sheds some light on the differences between cap-and-trade and carbon tax, as well as the clear importance of choosing an appropriate cap level.

27.7 Conclusion

As policy makers debate the future of global carbon emission legislation, the existing cap-and-trade schemes around the world have already significantly impacted the dynamics of electricity prices and the valuation of real assets, such as power plants, particularly under the well-known European Union Emissions Trading Scheme. Together with the recent volatile behaviour of all energy prices (e.g. gas, coal, oil), the introduction of carbon markets has increased the risk of changes in the merit order of fuel types, known to be a crucial factor in the price-setting mechanism of electricity markets. In the US, the recent sharp drop in natural gas prices is already causing changes in the merit order, which would be further magnified by any new emission regulation, such as the upcoming cap-and-trade market in California. Such considerations are vital for describing the complex dependence structure between electricity, its input fuels, and emission allowances, and thus highly relevant for both market participants and policy makers designing emission trading schemes. In this paper, we derived the equilibrium carbon allowance price as the solution of an FBSDE, in which feedback from allowance price on market emission rates is linked to the electricity stack structure. The resulting model specifies simultaneously both electricity and allowance price dynamics as a function of fuel prices, demand and accumulated emissions; in this way, it captures consistently the highly state-dependent correlations between all the energy prices, which would not be achievable in a typical reduced-form approach. We used a PDE representation for the solution of the pricing FBSDE and implemented a finite difference scheme to solve for the price of carbon allowances. Finally, we compared our model for allowance prices with other reduced-form approaches and analysed its important implications on price behaviour, spread option pricing and the valuation of physical assets in electricity markets covered by emission regulation. The four case studies illustrate the many important considerations needed to understand the complex joint dynamics of electricity, emissions and fuels, as well as the additional insight that can be provided by our structural approach.

Acknowledgements

R.C. is partially supported by NSF-DMS 0806591. M.C. is partially supported by NSF-DMS-0739195.

References

Aïd, R., Campi, L. and Langrené, N., A structural risk-neutral model for pricing and hedging power derivatives. *Math. Finance*, 2012, to appear.

Alos, E., Eydeland, A. and Laurence, P., A Kirks and a Bacheliers formula for three-asset spread options. *Energy Risk*, 2011, September, 52–57.

Barlow, M., A diffusion model for electricity prices. *Math. Finance*, 2002, **12**, 287–298.

Benth, F.E., Benth, J.S. and Koekebakker, S., *Stochastic Modeling of Electricity and Related Markets*, 2008 (World Scientific: Singapore).

Carmona, R. and Coulon, M., A survey of commodity markets and structural models for electricity prices. In *Financial Engineering for Energy Asset Management and Hedging in Commodity Markets. Proceedings from the Special Thematic Year at the Wolfgang Pauli Institute, Vienna*, edited by F.E. Benth, 2012 (Springer).

Carmona, R. and Delarue, F., Singular FBSDEs and scalar conservation laws driven by diffusion processes. *Probab. Theory Relat. Fields*, 2012, to appear.

Carmona, R. and Durrleman, V., Pricing and hedging spread options. *SIAM Rev.*, 2003, **45**(4), 627–685.

Carmona, R. and Hinz, J., Risk neutral modeling of emission allowance prices and option valuation. *Mgmt Sci.*, 2011, **57**(8), 1453–1468.

Carmona, R. and Sun, Y., Implied and local correlations from spread options. *Appl. Math. Finance*, 2012.

Carmona, R., Coulon, M. and Schwarz, D., Electricity price modelling and asset valuation: A multi-fuel structural approach. *Math. Financ. Econ.*, 2012a, to appear.

Carmona, R., Delarue, F., Espinoza, G.E. and Touzi, N., Singular forward backward stochastic differential equations and emission derivatives. *Ann. Appl. Probab.*, 2012b.

Carmona, R., Fehr, F., Hinz, J. and Porchet, A., Market designs for emissions trading schemes. *SIAM Rev.*, 2010, **52**, 403–452.

Cartea, Á. and Villaplana, P., Spot price modeling and the valuation of electricity forward contracts: The role of demand and capacity. *J. Bank. Finance*, 2008, **32**, 2501–2519.

Chesney, M. and Taschini, L., The endogenous price dynamics of the emission allowances: An application to CO_2 option pricing. *Appl. Math. Finance*, 2012, 447–475.

Coulon, M., Modelling price dynamics through fundamental relationships in electricity and other energy markets. PhD thesis, University of Oxford, 2009.

Coulon, M. and Howison, S., Stochastic behaviour of the electricity bid stack: From fundamental drivers to power prices. *J. Energy Mkts*, 2009, **2**, 29–69.

De Jong, C. and Schneider, S., Cointegration between gas and power spot prices. *J. Energy Mkts*, 2009, **2**(3), 27–46.

Eydeland, A. and Wolyniec, K., *Energy and Power Risk Management: New Developments in Modeling, Pricing and Hedging*, 2003 (Wiley: New York).

Forman, J.l. and Sørensen, M., The Pearson diffusions: A class of statistically tractable diffusion processes. *Scand. J. Statist.*, 2008, **35**, 438–465.

Glasserman, P., *Monte Carlo Methods in Financial Engineering*, 2004 (Springer: Berlin).

Howison, S. and Schwarz, D., Risk-neutral pricing of financial instruments in emission markets: A structural approach. *SIAM J. Financ. Math.*, 2012, to appear.

Karatzas, I. and Shreve, S.E., *Brownian Motion and Stochastic Calculus*, 1999 (Springer: Berlin).

Kobylanski, M., Backward stochastic differential equations and partial differential equations with quadratic growth. *Ann. Probab.*, 2000, **28**, 558–602.

Koenig, P., Modelling correlation in carbon and energy markets. Working Paper, Department of Economics, University of Cambridge, 2011.

LeVeque, R.J., *Numerical Methods for Conservation Laws*, 1990 (Birkhäuser).

Ma, J., Wu, Z., Zhang, D. and Zhang, J., On well-posedness of forward-backward SDEs|A unified approach. Working Paper, University of Southern California, 2011.

Margrabe, W., The value of an option to exchange one asset for another. *J. Finance*, 1978, **33**, 177–186.

Oleinik, O.A. and Radkevic, E.V., *Second Order Equations with Nonnegative Characteristic Form*, 1973 (AMS).

Pirrong, C. and Jermakyan, M., The price of power: The valuation of power and weather derivatives. *J. Bank. Finance*, 2008, **32**, 2520–2529.

Schwarz, D., Price modelling and asset valuation in carbon emission and electricity markets. PhD thesis, University of Oxford, 2012.

Seifert, J., Uhrig-Homburg, M. and Wagner, M.W., Dynamic behaviour of CO_2 spot prices. *J. Environ. Econ. Mgmt*, 2008, **56**(2), 180–194.

Appendix A: Numerical Solution of the FBSDE

A.1 Candidate Pricing PDE

In Theorem 27.3.3 we addressed the existence and uniqueness of a solution to the FBSDE (27.7). Given the Markov nature of the equation, we conjecture that there exists a deterministic function $\alpha:[0,T]\times[0,\bar{x}]\times\mathbb{R}_+[0,\bar{e}]\to 0,\pi]$ such that $A_t=\alpha(t,D_t,E_t,S_t^c,S_t^g)$, and sufficiently smooth to be a classical solution to the semilinear PDE

$$\mathcal{L}\alpha+\mathcal{N}\alpha=0,\quad \text{on}\,U_T,\tag{A.1}$$

$$\alpha=\phi(e),\quad \text{on}\{t=T\}\times U,\tag{A.2}$$

where $U:=(0,\bar{x})\times\mathbb{R}_{+}\times\mathbb{R}_{+}\times(0,\bar{e})$ and $U_T:=[0,T)\times U$ and the operators $\mathcal{L}$ and $\mathcal{N}$ are defined by

$$\mathcal{L}:=\frac{\partial}{\partial t}+\frac{1}{2}\sigma_d(d)^2\frac{\partial^2}{\partial d^2}+\frac{1}{2}\sigma_c\left(s_c\right)^2\frac{\partial^2}{\partial s_c^2}+\frac{1}{2}\sigma_g\left(s_g\right)^2\frac{\partial^2}{\partial s_g^2}$$

$$+\rho\sigma_c\left(s_c\right)\sigma_g\left(s_g\right)\frac{\partial^2}{\partial s_c\partial s_g}$$

$$+\mu_d(t,d)\frac{\partial}{\partial d}+\mu_c\left(s_c\right)\frac{\partial}{\partial s_c}+\mu_g\left(s_g\right)\frac{\partial}{\partial s_g}-r,$$

and $\mathcal{N} := \mu_e\left(d, \mathbf{s}\left(s_c, s_g\right)\right)(\partial / \partial e)$. As previously, we specify for our purposes that $\phi(e) = \pi I_{[\Gamma,\infty)}(e)$, for $e \in \mathbb{R}$.

With regards to the problem (A.1), the question arises at which parts of the boundary we need to specify boundary conditions and, given the original stochastic problem (27.7), of what form these conditions should be. To answer the former question we consider the Fichera function f at points of the boundary where one or more of the diffusion coefficients disappear (Oleinik and Radkevic 1973). Defining $n := (n_d, n_c, n_g, n_e)$ to be the inward normal vector to the boundary, Fichera's function for the operator $(\mathcal{N} + \mathcal{L})$ reads

$$
\begin{aligned}
f\left(t, d, s_c, s_g, e\right) \quad := \quad & \left(\mu_d - \frac{1}{2}\frac{\partial}{\partial d}\sigma_d^2\right)n_d \\
& + \left(\mu_c - \frac{1}{2}\frac{\partial}{\partial s_c}\sigma_c^2 - \frac{\partial}{\partial s_c}\rho\sigma_c\sigma_g\right)n_c \\
& + \left(\mu_g - \frac{1}{2}\frac{\partial}{\partial s_g}\sigma_g^2 - \frac{\partial}{\partial s_g}\rho\sigma_c\sigma_g\right)n_g \\
& + \mu_e n_e, \quad on\ \partial U_T.
\end{aligned}
$$ (A.3)

At points of the boundary where $f \geq 0$ the direction of information propagation is outward and we do not need to specify any boundary conditions; at points where $f < 0$ information is inward flowing and boundary conditions have to be specified. We evaluate (A.3) for the choice of coefficients presented in Section 27.5.3.

Considering the parts of the boundary corresponding to $d = 0$ and $d = \bar{x}$, we find that $f \geq 0$ if and only if $\min(\overline{D}(t), \bar{x} - \overline{D}(t))\bar{x}\breve{\sigma}$, which is the same condition prescribed in Section 27.5.3 to guarantee that the Jacobi diffusion stays within the interval $(0, \bar{x})$. At points of the boundary corresponding to $e = 0$, we find that $f \geq 0$ always. On the part of the boundary on which $e = \bar{e}$, $f < 0$ except at the point $(d, \cdot, \cdot, e) = (0, \cdot, \cdot, \bar{e})$, where $f = 0$, an ambiguity which could be resolved by smoothing the domain. Similarly, we find that $f \geq 0$ on parts of the boundary where $s_c = 0$ or $s_g = 0$. Therefore, no boundary conditions are necessary except when $e = \bar{e}$, where we prescribe

$$
\alpha = \exp(-r(T-t))\pi, \quad on\ U_T\big|_{e=\bar{e}}.
$$ (A.4)

In addition, we need to specify an asymptotic condition for large values of s_c and s_g. We choose to consider solutions that, for $i \in \{c, g\}$, satisfy

$$
\frac{\partial \alpha}{\partial s_i} \sim 0, \quad on\ U_T\big|_{s_i \to \infty}.
$$ (A.5)

A.2 An Implicit–Explicit Finite Difference Scheme

We approximate the domain U_T by a finite grid spanning $\left[0, T\right] \times \left[0, \bar{x}\right] \times \left[0, \bar{s}_c\right] \times \left[0, \bar{s}_g\right] \times \left[0, \bar{e}\right]$. For the discretization we choose mesh widths Δd, Δs_c, Δs_g, Δe and a time step Δt. The discrete mesh points $(t_k, d_m, s_{c_i}, s_{g_j}, e_n)$ are then defined by

$$
\begin{aligned}
t_k &:= k\Delta t, \qquad d_m := m\Delta d, \\
s_{c_i} &:= i\Delta s_c, \qquad s_{g_j} := j\Delta s_g, \quad e_n := n\Delta e.
\end{aligned}
$$

The finite difference scheme we employ produces approximations $\alpha_{m,i,j,n}^k$, which are assumed to converge to the true solution α as the mesh width tends to zero.

Since the partial differential equation (A.1) is posed backwards in time with a terminal condition, we choose a backward finite difference for the time derivative. In order to achieve better stability properties we make the part of the scheme relating to the linear operator $\mathcal{L}$ implicit; the part relating to the operator $\mathcal{N}$ is made explicit in order to handle the nonlinearity.

In the e-direction we are approximating a conservation law PDE with discontinuous terminal condition (for an in-depth discussion of numerical schemes for this type of equation, see LeVeque (1990)). The first derivative in the s-direction, relating to the nonlinear part of the partial differential equation, is discretized against the drift direction using a one-sided upwind difference. Because characteristic information is propagating in the direction of decreasing e, this one-sided difference is also used to calculate the value of the approximation on the part of the boundary corresponding to $e = 0$. On the part of the boundary corresponding to $e = \bar{e}$ we apply the condition (A.4).

In the d-direction the equation is elliptic everywhere except on the boundary, where it degenerates. Therefore, we expect the convection coefficient to be much larger than the diffusion coefficient near the boundaries. In order to keep the discrete maximum principle we again use a one-sided upwind difference for the first-order derivative. Thereby we have to pay attention that, due to the mean-reverting nature of (D_t), the direction of information propagation and therefore the upwind direction changes as the sign of μ_d changes. The same upwind difference is also used to calculate the value of the approximation at the boundaries $d = 0$ and $d = \bar{x}$. To discretize the second-order derivative we use central differences.

The s_c- and s_g-directions are treated similarly to the d-direction. We use one-sided upwind differences for the first-order derivatives, thereby taking care of the boundaries corresponding to $s_c = 0$ and $s_g = 0$. The second-order derivatives are discretized using central differences. At the boundary corresponding to $s_c = \bar{s}_c$ and $s_g = \bar{s}_g$ we apply the asymptotic condition (A.5) as a boundary condition.

With smooth boundary data, on a smooth domain, the scheme described above can be expected to exhibit first-order convergence. In our setting, we expect the discontinuous terminal condition to have adverse effects on the convergence rate. We refer the interested reader to Howison and Schwarz (2012) for a numerical estimation and a detailed analysis of the convergence rate of the numerical scheme described in this section, as applied to a very similar case.[*]

Appendix B: Numerical Calculation of Spread Prices

B.1 Time Discretization of SDEs

Let $(D_k, S_k^c, S_k^g, E_k, A_k)$ denote the discrete-time approximation to the FBSDE solution $(D_t, S_t^c, S_t^g, E_t, A_t)$ on the time grid $0 < \Delta t < 2\Delta t < \cdots < n_k \Delta t = \tau$. At each time step we calculate A_k by interpolating the discrete approximation $\alpha_{m,i,j,n}^k$ at (D_k, S_k^c, S_k^g, E_k) beginning with the initial values $D_0 = d_0, S_0^c = s_0^c, S_0^g = s_0^g$, $E_0 = 0$. The approximations (D_k, S_k^c, S_k^g, E_k) are obtained using a simple Euler scheme (Glasserman 2004) for the forward components of (27.7). The discretized version of (D_t) is forced to be instantaneously reflecting at the boundaries $D_k = 0$ and $D_k = \bar{x}$; similarly, the discretized versions of (S_t^c) and (S_t^g) are made instantaneously reflecting at $S_k^c = 0$ and $S_k^g = 0$.

B.2 Monte Carlo Calculation of Option Prices

Using this discretization we simulate n_{mc} paths and, as usual, for $t \in [0, \tau)$, calculate the mean clean spread price $\check{V}_t^j$, given by

---

[*] Although the setting of Howison and Schwarz (2012) leads to a 2 + 1-dimensional model instead of the 4 + 1-dimensional model we are considering here, we emphasize that structurally the two models are identical. Therefore, we believe that the convergence properties of the two numerical schemes are also similar.

$$\check{V}_t^j := \exp(-r(\tau-t))\frac{1}{n_{mc}}\sum_{i=1}^{n_{mc}}\left(b\left(D_{n_k}^i, S_{n_k}^{c,i}, S_{n_k}^{g,i}, A_{n_k}^i\right)\right.$$
$$\left. -h_j S_{n_k}^{j,i} - e_j A_{n_k}^i\right)^+,$$

where the index i refers to the simulation scenario and $j \in \{c, g\}$. The corresponding standard error $\hat{\sigma}_v$ is obtained by

$$\check{\sigma}_v := \sqrt{\frac{1}{n_{mc}\left(n_{mc}-1\right)}\sum_{i=1}^{n_{mc}}\left(V_{n_k}^i - \check{V}_\tau^j\right)^2}.$$

28

Is the EUA a New Asset Class?

Vicente Medina
and Angel Pardo

28.1 Introduction

In order to achieve the reduction objectives for greenhouse gas emissions, the European Union decided that a big part of that reduction would have to be directly assumed by the companies of the most polluting sectors. Since 2005, the companies of these sectors, included in the 2003/87/EC Directive, receive each year entitlements to emit one tonne of carbon dioxide equivalent gas, which are denominated European Union Allowances (EUAs).* At the end of the control period, the firms covered by the environmental regulations have to give a sufficient number of allowances to cover their verified real emissions. If these companies emit more CO_2 than the allowances they own, they would have to go to the European Union Emission Trading Scheme (EU ETS) and buy the difference. It is important to highlight that not only the polluting companies participate in the EU ETS, but also external agents can trade in it. Therefore, knowledge of the statistical properties of EUAs is of interest not only for hedging operations, but also for speculative or risk management purposes.

In spite of the youth of the EU ETS, the academic literature has analysed the CO_2 market from diverse perspectives. Mansanet-Bataller and Pardo (2008) study the market at an institutional level. Daskalakis and Markellos (2008), Miclaus *et al.* (2008), Daskalakis *et al.* (2009) and Mansanet-Bataller and Pardo (2009) study different aspects of the market efficiency. Benz and Hengelbrock (2008), Rotfuss (2009), Uhrig-Homburg and Wagner (2009) and Rittler (2012) analysed the lead–lag relationship between the spot and futures markets, this last one being the market that leads the price discovery process. Finally, Borak *et al.* (2006), Benz and Trück (2009), Daskalakis *et al.* (2009) and Chesney and Taschini (2012) proposed different alternatives to model the price dynamics of CO_2 emission allowances.[†]

---

* The sectors included in the 2003/87/CE Directive are: energy (electricity, co-generation), refining petroleum, iron and steel industry, mineral products, cement, lime, glass, ceramics (roofing tiles, bricks, floor tiles, etc.), cardboard, pulp and paper.

[†] An excellent review of this kind of literature can be found in Convery (2009).

DOI: 10.1201/9781003265399-32

The papers mentioned above analyse short periods of time and the statistical properties of EUA returns appear as a secondary objective. This is precisely the main purpose of our study. Specifically, starting from the stylised facts described by Cont (2001) for financial assets, and from the statistical properties analysed by Gorton and Rouwenhorst (2006) and Gorton *et al.* (2012) for commodities, we study the stylised empirical facts of EUA returns. By doing so, we will determine whether EUA behaves like a financial asset, a commodity, or a new asset. The knowledge of its statistical properties is of interest for both policy makers and portfolio managers, in terms of regulation and portfolio diversification, respectively. The remainder of the paper is organised as follows. Section 28.2 presents the stylised facts observed in assets grouped into four categories. Section 28.3 describes the data we used in the study. The results of the analysis are presented and discussed in Section 28.4. The last section summarises with some concluding remarks.

28.2 Stylised Facts of Asset Returns

Following Cont (2001, p. 233), we define the stylised facts of an asset as a set of statistical facts which emerge from the empirical study of asset returns and which are common to a large set of assets and markets. Some papers have overviewed the stylised facts that are characteristics of the financial assets (see among others, Pagan 1996, Cont 2001, Morone 2008, and Sewell 2011) and those that are common in the commodities (see Gorton and Rouwenhorst 2006 and Gorton *et al.* 2012). Although some stylised facts are related to each other, the statistical characteristics of asset returns can be grouped into four large sets of phenomena that have to do with the distribution of returns, correlation facts, volatility-related properties and commodity-related facts.

In the first group, the characteristics observed in the historical returns distribution are analysed. The first step is to look into the existence of normality in the distribution of returns. In the case of not rejecting normality, we could not reject both symmetry and the absence of heavy tails. However, the rejection of the hypothesis of normality would make necessary a more exhaustive analysis to determine if the reason for the rejection comes from the existence of asymmetry or because the frequency curve is more or less peaked than the mesokurtic curve. The second aspect we analyse is the *intermittency*, which refers to the phenomenon that returns present, at any time scale, high variability that is translated in the appearance of *outliers* throughout the asset life.* Thirdly, we investigate the *aggregational gaussianity*, a third aspect that is detected in historical returns distributions and which makes reference to the fact that the aggregation of data in bigger time intervals approaches the Gaussian data distribution. While the two last phenomena have been observed both in financial assets and commodity futures, the empirical evidence regarding asymmetry usually indicates that the skewness is negative in financial assets and positive in commodities. Finally, we have analysed if EUA returns are stationary. When series are non-stationary and follow a unit root process, persistence of shocks will be infinite. If EUA prices follow a trend stationary process, then there exists a tendency for the price level to return to its trend path over time and investors may be able to forecast futures returns by using information on past EUA returns. However, the majority of economic and financial time series exhibit trending behaviour or non-stationarity in the mean. If EUA price series were non-stationary, any shock to EUA price would be permanent, implying that EUA futures returns would be unpredictable based on historical observations.

The second group refers to some correlation facts observed in the returns of whatever asset and to their consequences. Firstly, we study the autocorrelation of the returns. This has been the classical way to test the weak form of the efficient hypothesis in financial markets. The absence of significant linear correlations in returns has been widely documented and it is usually not detected except for very small intraday time scales. Secondly, given that the correlation test may be influenced by the presence of extreme returns, we carry out a test run in order to check the existence of randomness in EUA returns generation. With this test, we get robustness on a possible predictability of EUA returns over the short

---

* Outliers refer to those returns which, by their magnitude, are considered as unusual and infrequent.

term. Thirdly, we analyse the presence of a slow decay of autocorrelation in absolute returns, also known as the *Taylor effect*. Following Taylor (2007), this effect makes reference to the fact that a big return in absolute terms is more probably followed by another big one, rather than a small one. This phenomenon, unlike the two previous ones, is interpreted as a sign of long-range dependence.

The third group of facts investigates the specific characteristics observed in volatility. The first well-known property about volatility is the positive autocorrelation observed in different measures of volatility over several days. This fact is generally detected through the existence of autocorrelation in squared returns and it is known as *volatility clustering*. Secondly, we study three volatility-related cross-correlation facts. The first one looks at whether volatility responds differently to positive and negative shocks of the same magnitude, the second one analyses the correlation between volume and volatility, and the last one examines the correlation between the change in the open interest and the volatility.* Although studies on the relationship between trading measures and the underlying price volatility provide mixed evidence, following Bhargava and Malhotra (2007) a number of other researchers report a positive correlation between trading activity and volatility. The significance of these two relationships would indicate that the volume and/or the change in the open interest, in EUA markets, could be used as explanatory variables of volatility.

The last group alludes to specific features extracted from commodity behaviours. The first aspect we investigate is the presence of a non-trading effect observed in the volatility of weather-sensitive assets. Following Fleming *et al.* (2006), trading versus non-trading period variance ratios in weather-sensitive markets are lower than those in the equity markets. Given that Mansanet-Bataller *et al.* (2007) and Alberola *et al.* (2008) have shown empirical evidence about the influence of weather on carbon returns, we study the non-trading effect on EUAs by testing whether the information flow on EUA markets is evenly distributed around the clock. Another stylised fact characteristic of commodities is the negative correlation with stocks and bonds. Generally, commodity futures exhibit a certain negative correlation, mainly in the early part of falling periods, as noted in Gorton and Rouwenhorst (2006). Furthermore, the negative correlation becomes greater as we increase the time lag in which we hold the positions. Finally, we test the property of *inflation hedge*. Assets hedge against inflation when they correlate positively and significantly against it. Following Gorton and Rouwenhorst (2006), commodity futures usually show better behaviour against unexpected inflation than stocks or bonds do, and therefore they can be used for this kind of hedging.

28.3 Data

To carry out the analysis of the stylised facts of EUA returns, we have to select the most representative EUA asset in the market and the time frame. EUAs can be traded in several organised markets such as spot, futures and options markets. From among these markets, most of the trading volume is concentrated in the futures markets, especially in the futures contracts listed for electronic trading at the European Climate Exchange (ECX).† Furthermore, empirical evidence cited in Section 28.1 supports the notion that the price discovery process is led by the futures markets. For these reasons, we have chosen futures prices in order to obtain the most informative EUA return.

Related to the time frame used in this study, it is convenient to clarify some aspects of the EU ETS. Trading in EUAs has been organised into different phases. Phase I covers the years from 2005 to 2007 and was considered as a *pilot* or *learning* phase, characterised by an excess of EUAs that provoked a sharp

---

* Open interest is the total sum of all outstanding long and short positions of futures contracts that have not been closed.

† The unit of trading of one contract is one lot of 1000 CO_2 EU allowances. See the User Guide of ICE ECX Contracts: EUAs and CERs at https://www.theice.com/productguide/ProductGroupHierarchy.shtml?groupDetail=&group.groupId=19 for further information about the contract specifications of ECX EUA futures contract (last accessed on 20 March 2012).

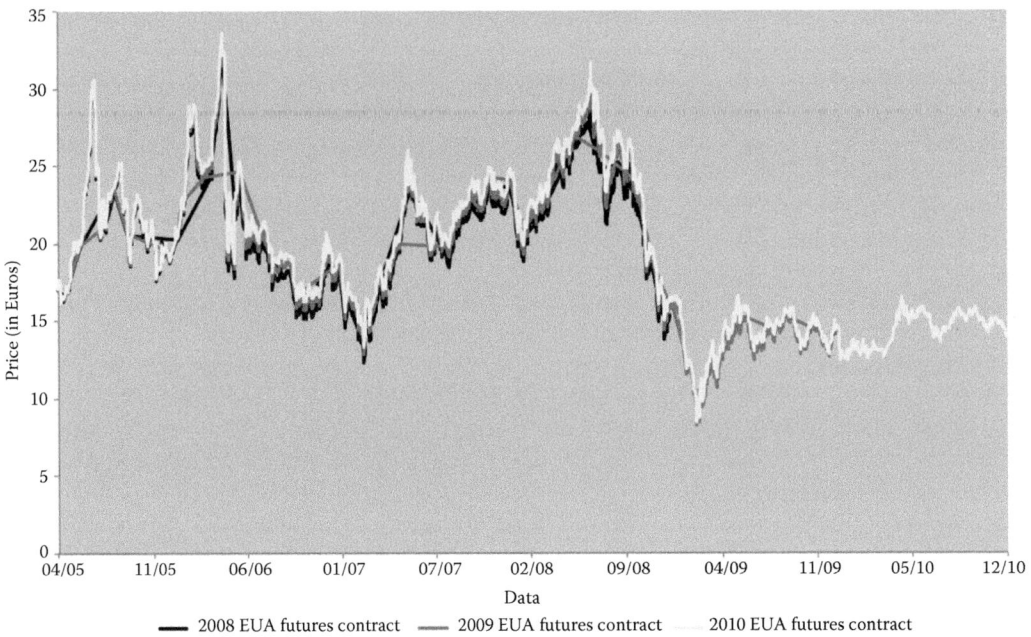

FIGURE 28.1 Prices evolution: the prices evolution is shown for the 2008, 2009 and 2010 EUA futures contracts. The sample period goes from 22 April 2005 to 15 December 2008 for the first contract, from 22 April 2005 to 14 December 2009 for the second one, and from 22 April 2005 to 20 December 2010 for the last one.

decline in prices.* Phase II matches the Kyoto protocol fulfilment period, which goes from January 2008 to December 2012, while Phase III will include the period 2013 to 2020.† Another important aspect to consider is that Phase I EUAs could not be used in Phase II, meaning that *banking* was not allowed between Phase I and Phase II. Nevertheless, banking is allowed between Phases II and III, a reason that Phase II EUAs might be used during the 2013 to 2020 period. For these reasons, we limit our sample to Phase II EUA futures returns. In particular, we focus on 2008, 2009 and 2010 EUA ECX futures contracts. The period sample goes from 22 April 2005 to 15 December 2008 for the first contract, from 22 April 2005 to 14 December 2009 for the second one, and from 22 April 2005 to 20 December 2010 for the third one, collecting 936, 1190 and 1446 daily returns for 2008, 2009 and 2010 EUA ECX futures contracts, respectively. Figure 28.1 shows the price evolution for the three EUA futures contracts. In general, all three contracts share a common behaviour. Until 24 April 2006, futures prices had a positive trend, over-passing the 30 Euros price per contract, but continuous rumours of over-allocation crashed Phase I prices, negatively affecting Phase II prices too. From then, prices declined until reaching their first minimum of 14 Euros in February 2007. In July 2008, EUA prices approached their maximum quote of over 30 Euros, following which prices started a new bearish period because of the financial crisis. By February 2009, EUA prices reached their historical minimum of around eight Euros. Finally, from April 2009 to the end of 2010, EUA prices have been fluctuating around 14 Euros.

---

* The special features observed in this period are treated in detail by, among others, Miclaus *et al.* (2008), Paolella and Taschini (2008), Benz and Trück (2009), Daskalakis *et al.* (2009) and Mansanet-Bataller and Pardo (2009).

† Given that the EU ETS is organised in phases, the member states must elaborate a National Allocation Plan (NAP) for the first two phases, in which the member states attribute their emission allowances to the different companies, included in the sectors involved in the 2003/87/EC Directive, and establish the emission limits for the different sectors, as well as for each one of the facilities covered by EU ETS. Therefore, these NAPs establish the EUA supply available in the market until 2012. Contrary to Phase I and Phase II, in Phase III only one EU-wide emissions allocation under ETS Phase III will be elaborated, where each installation's individual allocation will be decided by the European Commission.

Additionally, in order to test some correlation facts, other series of data have been used: a stock index, a stock index futures contract, bond futures contracts, a commodity futures contract, a risk-free asset series and the observed inflation. We have chosen for all of these series the European benchmarks, using the same time frame as the one we have selected for each EUA contract. Specifically, we have chosen the Euro Stoxx 50 and its futures contract, traded at the EUREX market, as indicative of a European stock index and its future. Regarding interest rates, the fixed income futures have also been obtained from the EUREX market. The three references chosen for this case are Euro Schatz futures, Euro Bolb futures, and Euro Bund futures, as benchmarks for European short-, mid- and long-term bond prices, respectively. As a representative commodity futures contract, we have chosen the Brent futures contract traded at the International Petroleum Exchange (IPE). The risk-free asset has been approached through the one-month Euribor rate and, as a proxy of the inflation rate, we have used the European Harmonized Index of Consumer Prices (HICP).

We have carried out our study using returns defined as $r_t = \log(p_t / p_{t-1})$, where p_t is the closing price of the futures contract on day t. Finally, when it has been needed, we have switched futures contracts on the last trading day in order to create a continuous series in futures contracts.

28.4 Results

In this section, we test the presence of the stylised facts described in Section 28.2: the distribution of returns, correlation facts, volatility-related features and commodity-related facts. For each case, we present the methodology used to test the empirical property, the results and the financial implications.

28.4.1 Data Distribution

From a financial point of view, the *gaussianity* of returns data is of interest for portfolio theory, derivatives pricing and risk management. However, one of the most well known properties in the distributions of asset returns is precisely the non-normality of the returns series and the fact that they present a greater number of extreme values than those observed in series with Gaussian distributions.[*] As a background, Table 28.1 presents the summary statistics of the EUA returns for 2008, 2009 and 2010 futures contracts, and Figure 28.2(a), (b) and (c) show the histogram of the historical daily returns distribution. All three daily series are leptokurtic and present negative skewness. The Jarque–Bera statistic rejects the null hypothesis of normality for all three series at the 5% level. The histogram shows heavy tails and a high number of extreme returns, both positive and negative.

The Jarque–Bera statistic is based on the assumption of normality; therefore the daily returns series can be symmetric but non-normal due to the excess of kurtosis. To check this fact, we test the distribution symmetry by applying the method proposed by Peiró (2004). Firstly, we divide each sample into two subsamples, where the first subsample contains the excesses of positive returns with respect to the mean $r^+ = \{r_t - \bar{r} \mid r_t > \bar{r}\}$, and the second one, the excesses of negative returns with respect to the mean, in absolute terms $r^- = \{\bar{r} - r_t \mid r_t < \bar{r}\}$. Secondly, we use the Wilcoxon rank test to check whether the two subsamples come from the same distribution.

Given the negative skewness coefficients reported in Table 28.1, the asymmetry tests, non-reported, show a statistic of 2.9854 (*p*-value 0.0028) for futures with maturity in 2008, 2.0044 (*p*-value 0.0450) for 2009 futures and 1.8161 (*p*-value 0.0694) for 2010 futures. Therefore, we observe a significant negative asymmetry for the first two contracts, something common in financial assets such as stocks, while we cannot reject symmetry at the 5% level for 2010 futures. This characteristic has a direct implication in margins requirements asked by clearing houses in EUA derivatives markets. A negative asymmetry

---

[*] In Catalán and Trívez (2007), some reasons concerning the importance of extreme values are presented.

TABLE 28.1 Summary Statistics for EUA Futures Returns

	Daily	Monday to Monday	Wednesday to Wednesday	Monthly Based Day 1	Monthly Based Day 15	Residuals EGARCH
			Panel A: 2008 EUA futures contract			
Mean	−0.0001	−0.0007	−0.0007	−0.0011	−0.0025	−0.0140
Median	0.0013	0.0072	0.0033	−0.0026	0.0022	0.0209
Maximum	0.1865	0.2076	0.1441	0.3189	0.2786	3.9433
Minimum	−0.2882	−0.5146	−0.3757	−0.3780	−0.3405	−7.5034
Std. Dev.	0.0286	0.0767	0.0646	0.1532	0.1317	1.0003
Skewness	−1.3166	−1.8022	−1.3332	−0.2364	−0.2879	−0.5897
Kurtosis	17.9459	13.1149	9.5885	2.7371	3.0530	7.6354
Jarque–Bera	8982.2510	912.8140	397.8340	0.5240	0.5990	891.3000
p-value	0.0000	0.0000	0.0000	0.7690	0.7410	0.0000
K–S stat	2.9130	1.2370	1.1710	0.5710	0.5210	2.6360
p-value	0.0000	0.0940	0.1280	0.9000	0.9490	0.0000
Observations	936	190	189	43	43	935

	Daily	Monday to Monday	Wednesday to Wednesday	Monthly Based Day 1	Monthly Based Day 14	Residuals EGARCH
			Panel B: 2009 EUA futures contract			
Mean	−0.0001	−0.0009	−0.0006	−0.0035	−0.0031	−0.0085
Median	0.0007	0.0050	0.0032	−0.0013	0.0132	0.0283
Maximum	0.1932	0.2069	0.1782	0.3219	0.4039	3.6552
Minimum	−0.2811	−0.5087	−0.3667	−0.3697	−0.4580	−7.8527
Std. Dev.	0.0286	0.0757	0.0649	0.1498	0.1487	1.0003
Skewness	−0.8873	−1.4620	−1.0739	−0.2569	−0.3153	−0.5654
Kurtosis	14.4420	10.9417	8.0011	2.7422	4.1379	7.1086
Jarque–Bera	6647.5940	722.1760	2.9740	0.7570	3.8780	899.6650
p-value	0.0000	0.0000	0.0000	0.6850	0.1440	0.0000
K–S stat	2.9180	1.2680	1.3740	0.6240	0.6300	2.9260
p-value	0.0000	0.0800	0.0460	0.8310	0.8220	0.0000
Observations	1190	242	241	55	55	1189

	Daily	Monday to Monday	Wednesday to Wednesday	Monthly Based Day 1	Monthly Based Day 20	Residuals EGARCH
			Panel C: 2010 EUA futures contract			
Mean	−0.0001	−0.0008	−0.0005	−0.0015	−0.0030	−0.0003
Median	0.0002	0.0019	0.0018	−0.0020	−0.0020	0.0005
Maximum	0.1912	0.2078	0.1830	0.3249	0.2801	0.1979
Minimum	−0.2743	−0.5016	−0.3582	−0.3654	−0.3710	−0.2711
Std. Dev.	0.0268	0.0708	0.0597	0.1392	0.1254	0.0267
Skewness	−0.8500	−1.4094	−1.0625	−0.2087	−0.2978	−0.8110
Kurtosis	14.5575	11.3714	8.6198	3.1254	3.5760	14.4896
Jarque–Bera	8222.0183	959.0620	442.2060	0.5304	1.9451	8106.6428
p-value	0.0000	0.0000	0.0000	0.7670	0.3781	0.0000
K–S stat	3.0339	1.5029	1.7856	0.4845	0.5570	3.0453
p-value	0.0000	0.0218	0.0034	0.9730	0.9157	0.0000
Observations	1446	295	294	67	68	1445

Notes: This table presents the descriptive statistics for 2008, 2009 and 2010 EUA futures contracts traded on the ICE ECX market. Panel A, B and C present the statistics for the 2008, 2009 and 2010 EUA futures contracts, respectively. The 'Monday to Monday' ('Wednesday to Wednesday') column shows the results for weekly returns calculated from Monday (Wednesday) close to Monday (Wednesday) close. 'Monthly based day 1' presents the results for monthly returns, calculated from the first trading day of each month to the last trading day in the same month. 'Monthly based day 15', 'Monthly based day 14' and 'Monthly based day 20' show the results for monthly returns, calculated from one day in mid-month to the same day in the next month, being the number for that specific day. 'Residuals EGARCH' stands for the statistics of the residuals obtained after fitting an AR(1)-EGARCH(1,1) model for each contract. 'K–S stat' stands for the Kolmogorov–Smirnov statistic, which tests the null hypothesis of the normality of the distribution. '2008 EUA futures contract' refers to the futures contract maturing on 15 December 2008, '2009 EUA futures contract' refers to the futures contract maturing on 14 December 2009 and '2010 EUA futures contract' refers to the futures contract maturing on 20 December 2010.

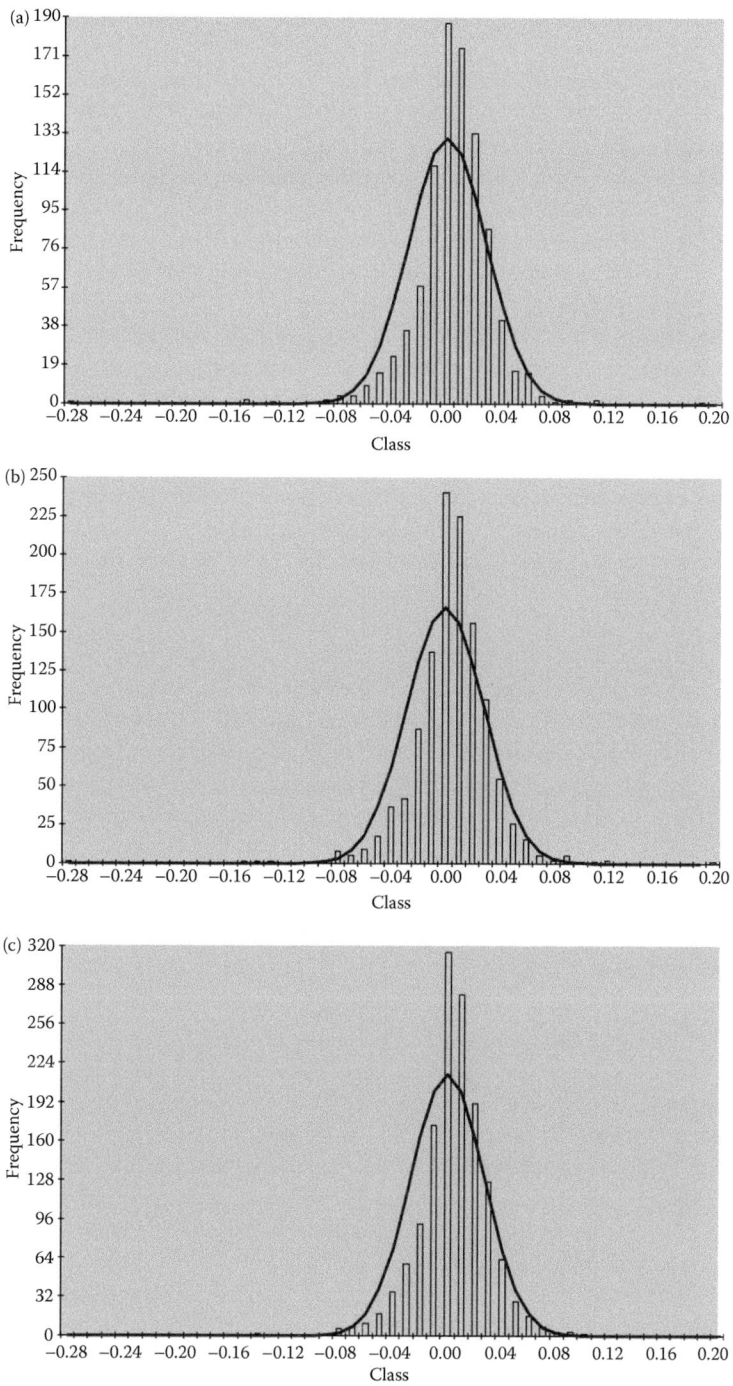

FIGURE 28.2 Returns histograms: (a), (b) and (c) depict the histograms of the daily returns for the 2008, 2009 and 2010 EUA futures contracts, respectively, in comparison with the histogram of a normal distribution. The period sample goes from 22 April 2005 to 15 December 2008 for the first contract, from 22 April 2005 to 14 December 2009 for the second one and from 22 April 2005 to 20 December 2010 for the last one, collecting 936, 1190 and 1446 daily returns for 2008, 2009 and 2010 EUA ECX futures contracts, respectively.

implies more risk for long positions and, as a consequence, bigger margins should be asked for those positions.

The second aspect we have analysed is the *intermittency* of the series. To study the existence of this empirical fact, we have used the VaR approach and, following Galai *et al.* (2008), the procedure based on the calculation of the Huber M-estimator to distinguish between outliers and the body of the distribution. Table 28.2 presents the outlier detection based on daily returns for the EUA 2008, 2009 and 2010 futures contracts. Five percent VaR (95% VaR) provides outlier detection through the Value at Risk method, considering as extreme positive (negative) outliers those returns that are over (under) the 5% one-day VaR (the 95% one-day VaR).* The 5% one-day VaR measure (3.89, 3.98 and 3.78% for 2008, 2009 and 2010 futures, respectively) is of interest for estimating the risk of loss in short positions in futures markets, while the 95% one-day VaR (4.53, 4.56 and 4.24% for 2008, 2009 and 2010 futures, respectively) is relevant for measuring the same risk but for long positions. Obviously, the number of positive and negative extreme returns is the same, but the 95% VaR measure in absolute terms is higher than the 5% VaR measure, confirming the negative asymmetry previously observed.

To get robustness in the outlier detection analysis, we have obtained the Huber M-estimator. The fifth, ninth and thirteenth columns in Table 28.2 present the results following this procedure.† Based on the M-estimator, over 17% of returns in all three samples are classified as outliers. Furthermore, in all three samples, the number of negative outliers (90, 110 and 126, respectively) far exceeds the number of positive ones (66, 89 and 105, respectively). Note that in both cases the number of outliers has increased over time.

Panel B of Table 28.2 presents the size of events measured by the number of standard deviations (σ) that an outlier return deviates from the mean. Independently of the approach, the majority of the outliers have a size between one to four standard deviations. This could be considered as usual for normal distributions. However, the empirical VaR and the Huber M-estimator approaches also detect diverse extreme large movements, in all three futures contracts, that diverge from the mean and the median, respectively, from 5 σs to 11 σs. Additionally, Panel C of Table 28.2 presents the clusters of outliers. The Huber M-estimator shows concentrations of up to six consecutive outliers while the VaR approach detects clusters of up to three outliers. It is important to note that the number of clusters with this procedure is higher when the outliers are considered independently of their sign, indicating that extreme returns are followed by themselves. On the whole, the results of Table 28.2 indicate a high probability of occurrence of large movements in the EUA market and prove the existence of *intermittency* in EUA returns.

Thirdly, we have studied another aspect that is usually detected in historical returns distribution, which is the fact that the aggregation of data in bigger time intervals approaches a Gaussian data distribution. In order to test the *aggregational gaussianity*, we have generated two subsamples both with weekly and monthly returns. The weekly returns have been calculated taking the close price from Monday close to the following Monday close, while the monthly returns take the closing prices of the last trading day of two consecutive months. Table 28.1 presents the statistics for all three sub-samples. Given the absence of normality, we have applied the non-parametric Kolmogorov–Smirnov test. Moving from daily returns to lower frequency returns, the normality of the distributions cannot be rejected at the conventional levels of significance. To confirm our results, we have generated two additional sub-samples. In the case of weekly returns from Wednesday closing to the following Wednesday closing and

---

* Malevergne and Sornette (2004) summarise the virtues of VaR measures in three characteristics: simplicity, relevance in addressing the ubiquitous large risks often inadequately accounted for by the standard volatility, and their prominent role in the recommendations of the international banking authorities.

† To obtain outliers by applying the M-Huber estimator, it is necessary to transform the data until convergence is attained. Furthermore, for the generation of the M-Huber estimator, we have to define the term k, which will be used as a limit during the detection of outliers. Following Galai *et al.* (2008), we have chosen $k = 4.496$. With this selection, the iterative process ends with an M-Huber estimator of 0.000879 for 2008 futures, 0.000561 for 2009 futures and 0.000465 for 2010 futures. A detailed description of this procedure may be found in Hoaglin *et al.* (1983, chapter 11).

TABLE 28.2 Outlier Detection

	2008 EUA Futures Contract				2009 EUA Futures Contract				2010 EUA Futures Contract			
	5% VaR	95% VaR	Total VaR	Huber	5% VaR	95% VaR	Total VaR	Huber	5% VaR	95% VaR	Total VaR	Huber
Number of observations	936				1190				1446			
Panel A: Outliers summary												
Number of outliers	47	46	93	156	60	59	119	199	73	72	145	231
Number of outliers	5.02%	4.91%	9.94%	16.67%	5.04%	4.96%	10.00%	16.72%	5.05%	4.98%	10.03%	15.98%
VaR return	3.89%	−4.53%	—	—	3.98%	−4.56%	—	—	3.78%	−4.24%	—	—
Number of positive outliers	47	0	47	66	60	0	60	89	73	0	73	105
Number of negative outliers	0	46	46	90	0	59	59	110	0	72	72	126
From 0 to 1	0	0	0	0	0	0	0	0	0	0	0	0
From 1 to 2	34	20	54	118	42	28	70	150	48	35	83	170
From 2 to 3	8	18	26	24	11	22	33	33	17	25	42	41
From 3 to 4	4	4	8	9	6	5	11	11	5	8	13	13
From 4 to 5	0	1	1	1	0	1	1	1	2	0	2	2
From 5 to 6	0	2	2	2	0	2	2	2	0	3	3	3
From 6 to 7	1	0	1	1	1	0	1	1	0	0	0	0
From 7 to 8	0	0	0	0	0	0	0	0	1	1	1	1
From 8 to 9	0	0	0	0	0	0	0	0	0	0	0	0
From 9 to 10	0	0	0	0	0	1	1	1	0	0	0	0
From 10 to 11	0	1	1	1	0	0	0	0	0	1	1	1
Panel C: Clusters of outliers												
1 value	37	36	63	84	44	45	73	106	57	54	94	129
2 values	5	5	12	24	8	7	17	31	8	9	18	30
3 values	0	0	2	5	0	0	4	4	0	0	5	6
4 values	0	0	0	1	1	0	0	2	0	0	0	2
5 values	0	0	0	1	0	0	0	1	0	0	0	2
6 values	0	0	0	0	0	0	0	1	0	0	0	1

Notes: This table presents the outlier detection based on daily returns for the EUA 2008, 2009 and 2010 futures contracts. '5% VaR' ('95% VaR') provides the outlier detection through the Value at Risk (VaR) method, considering as extreme positive (negative) outliers those returns which are over (under) the 5% one-day VaR (the 95% one-day VaR). Total VaR considers the aggregation of extreme returns. 'Huber' provides the outlier detection through the M-Huber estimator. The table is divided into three panels. Panel A summarises the outlier detection. Panel B presents the magnitude of the outliers detected in terms of standard deviations of returns (σ). Panel C shows the number of clusters of outliers that have been detected. '2008 EUA futures contract' refers to the futures contract maturing on 15 December 2008, '2009 EUA futures contract' refers to the futures contract maturing on 14 December 2009 and '2010 EUA futures contract' refers to the futures contract maturing on 20 December 2010.

for monthly returns calculating returns from two consecutive mid-months days, for all three futures contracts. We obtain similar results, presented in Table 28.1, that confirm the fact that the empirical distribution of EUA returns tends to normality as the frequency of observation decreases.

Finally, we have analysed if EUA returns are stationary by applying the Kwiatkowski *et al.* (1992) unit root test. We have tested the null of stationarity with intercept and deterministic time trend for the price series in levels. The critical value (at the 1% level) is 0.216 (see Table 1 in Kwiatkowski *et al.*'s paper, p. 166). If rejected, we have then tested the null of stationarity plus intercept for the time series in first differences. The results indicate that EUA prices should first be differenced to render the data stationary in all three contracts analysed (see Table 28.3). Therefore EUA prices are difference-stationary and EUA futures returns would be unpredictable based on past observations. This fact should be taken into account for cointegration analysis.

28.4.2 Correlation Facts

The second group of empirical aspects is related to some correlation facts observed in the returns and to their consequences. On the one hand, we study the significance of the autocorrelation coefficients. The study of the presence/absence of autocorrelation is a typical way to detect *short-term predictability of returns* from past information. Given the non-normality of daily distributions, we have decided to apply the non-parametric Spearman's rank correlation coefficient. Table 28.4 presents the Spearman's autocorrelation coefficients and their associated *p*-values. Autocorrelation is significant and positive only for the first lag at the 1% level.

Note that the square of the autocorrelation coefficient could be interpreted as the fraction of the variation of return on day *t* explained by a lagged return in simple linear regression. In the case of one lagged, the variation of today's return is $(0.124)^2 = 1.53$, $(0.109)^2 = 1.10$ and $(0.0779)^2 = 0.61\%$ for 2008, 2009 and 2010 futures contracts, respectively. Therefore, although the influence of the past return is positive and significant, the short-term predictability is very weak. Additionally, and given that there exist a large number of outliers, we have eliminated their effects in the autocorrelation analysis by applying a test run. This test assumes an ordered sequence of n_1 returns above the median and n_2 returns below the median. A run of type 1 (type 2) is defined as a sequence of one or more returns above (below) the median that are followed and preceded by returns below (above) the median. r_1 and r_2 are the number of runs of type 1 and 2, respectively, and *r* is the total number of observed runs. By comparing the observed number of runs and the expected number of runs, we test the hypothesis that returns follow a random sequence. Too few or too many runs would suggest non-randomness in the

TABLE 28.3 Unit Root Tests

	Level	Differences
2008 EUA futures contract	0.3281	0.0949*
2009 EUA futures contract	0.3867	0.0783*
2010 EUA futures contract	0.3392	0.0733*

Notes: This table summarises the Kwiatkowski *et al.* (1992) unit root tests for the time series of EUAs future prices both in levels and in first differences. We test the null of stationarity with intercept and deterministic time trend for the price series in levels. The critical value (at the 1% level) is 0.216 (see Table 1 in Kwiatkowski *et al.*'s paper, p. 166). If rejected, we then test the null of stationarity plus intercept for the time series in first differences. The critical value is 0.739 (see Table 1 in Kwiatkowski *et al.*'s paper, p. 166). '2008 EUA futures contract' refers to the futures contract maturing on 15 December 2008, '2009 EUA futures contract' refers to the futures contract maturing on 14 December 2009 and '2010 EUA futures contract' refers to the futures contract maturing on 20 December 2010.

* The null of stationarity cannot be rejected (at the 1% level).

TABLE 28.4   Autocorrelation Tests

	2008 EUA Futures Contract		2009 EUA Futures Contract		2010 EUA Futures Contract	
Lag	ρ	*p*-Value	ρ	*p*-Value	ρ	*p*-Value
1	0.1240	0.0001	0.1090	0.0002	0.0779	0.0031
2	−0.0095	0.7714	−0.0160	0.5814	−0.0250	0.3421
3	0.0444	0.1748	0.0480	0.0979	0.0395	0.1339
4	0.0386	0.2381	0.0484	0.0949	0.0385	0.1434
5	0.0083	0.7996	0.0066	0.8198	0.0131	0.6183
6	−0.0003	0.9936	0.0044	0.8798	0.0016	0.9526
7	−0.0280	0.3922	−0.0170	0.5577	−0.0143	0.5875
8	0.0221	0.4987	0.0156	0.5911	0.0144	0.5858
9	−0.0083	0.7998	0.0073	0.8023	0.0259	0.3257
10	0.0660	0.0435	0.0548	0.0587	0.0398	0.1315

Notes: This table provides the first 10 lagged sample autocorrelation coefficients for the 2008, 2009 and 2010 EUA futures contracts returns. ρ stands for the Spearman's rank correlation coefficient. '2008 EUA futures contract' refers to the futures contract maturing on 15 December 2008, '2009 EUA futures contract' refers to the futures contract maturing on 14 December 2009 and '2010 EUA futures contract' refers to the futures con- tract maturing on 20 December 2010.

TABLE 28.5   Run Tests

	Observed Number of Runs	Expected Number of Runs	Statistic	*p*-Value
2008 EUA futures contract	429	469	2.4855	0.0129
2009 EUA futures contract	555	596	1.6820	0.0926
2010 EUA futures contract	698	724	1.3679	0.1797

Notes: This table reports the test run statistics for all three EUA futures contracts where the null hypothesis is that returns are generated in a random way. The table shows the observed and the expected numbers of runs, the statistic and its *p*-value. '2008 EUA futures contract' refers to the futures contract maturing on 15 December 2008, '2009 EUA futures contract' refers to the futures contract maturing on 14 December 2009 and '2010 EUA futures contract' refers to the futures contract maturing on 20 December 2010.

distribution. Table 28.5 shows results that indicate that the null of randomness is rejected at the 5% level for 2008 futures, but not for 2009 and 2010 futures contracts. In summary, correlation analysis and test runs indicate a positive but weak short-run predictability of past returns. Furthermore, this small predictability has diminished over time.

Finally, we analyse the presence of a slow decay of autocorrelation in absolute returns, also known as the Taylor effect, which would indicate that large returns in absolute terms are more probably followed by another large return, rather than a small one. Note that we have already obtained preliminary evidence of dependence among outliers; however, the Taylor effect makes reference to a feature that is present in the entire sample. To test this effect, we have obtained the Spearman's rank autocorrelation coefficient taking into account absolute returns. Table 28.6 presents results that indicate a significant slow decay of sample autocorrelation coefficients at the 5% level, suggesting a long-range dependence in absolute returns.

28.4.3 Volatility-Related Features

The third group of facts investigates some aspects of EUA returns related to some phenomena of the time series volatility. The first is related to the positive autocorrelation observed in different measures of volatility; the second makes reference to the fact that volatility responds differently to positive and

TABLE 28.6   Taylor Effect

Lag	2008 EUA Futures Contract		2009 EUA Futures Contract		2010 EUA Futures Contract	
	ρ	p-Value	ρ	p-Value	ρ	p-Value
1	0.1071	0.0010	0.1322	0.0000	0.1301	0.0000
2	0.0498	0.1280	0.0848	0.0034	0.1106	0.0000
3	0.1202	0.0002	0.1533	0.0000	0.1583	0.0000
4	0.0792	0.0154	0.1012	0.0005	0.0964	0.0002
5	0.0873	0.0075	0.1156	0.0001	0.1233	0.0000
6	0.0907	0.0055	0.1331	0.0000	0.1398	0.0000
7	0.0639	0.0507	0.0722	0.0127	0.0776	0.0032
8	0.0740	0.0235	0.1153	0.0001	0.1268	0.0000
9	0.0861	0.0084	0.1102	0.0001	0.0988	0.0002
10	0.0466	0.1541	0.0758	0.0089	0.0874	0.0009

Notes: This table provides the first 10 lagged sample autocorrelation coefficients for the 2008, 2009 and 2010 EUA futures contracts absolute returns. ρ stands for the Spearman's rank correlation coefficient. '2008 EUA futures contract' refers to the futures contract maturing on 15 December 2008, '2009 EUA futures contract' refers to the futures contract maturing on 14 December 2009 and '2010 EUA futures contract' refers to the futures contract maturing on 20 December 2010.

negative shocks of the same magnitude; and the last one analyses the correlation between trading-related measures and volatility.

Firstly, we have depicted the squared returns of all three futures contracts as a standard approach to detect volatility clusters. Figure 28.3(a), (c) and (e) exhibit this fact, that is very common both in financial and commodities assets series.

Additionally, we have obtained a high–low daily volatility measure. We have chosen the volatility measure proposed by Garman and Klass (1980) due to its higher relative efficiency than the standard estimators. The Garman and Klass volatility measure (hereinafter, the Garman–Klass volatility) is obtained in the following way:

$$\sigma_{GK} = 0.511(u-d)^2 - 0.019\{c(u+d)-2ud\} - 0.383c^2,$$

where u=log H_t–log O_t is the normalised high (H_t is today's high price; O_t is today's opening price); d=log L_t–log O_t is the normalised low (L_t is today's low price); and c=log C_t–log O_t is the normalised close (C_t is today's close price).

Figure 28.3(b), and (f) represent the evolution of the Garman–Klass volatility over the trading days of all three futures. Both figures show *volatility clustering*. There are two striking clusters in July 2005 and April 2006. Benz and Hengelbrock (2008) relate the first one to unexpected selling by some Eastern European countries and to the insecurity perceived in the market as a consequence of the terrorist attacks in London in July 2005. The second cluster took place in April 2006 and is explained by the fact that in April 2006 the market understood that there was a great excess of allowances in Phase I, which also affected Phase II EUA futures.

Panel A of Table 28.7 presents the positive and significant Spearman's cross autocorrelation coefficients of Garman–Klass volatility for the first 10 lags. The results confirm the existence of volatility clusters that decay very slowly.

Secondly, we study the *asymmetric volatility*, which is the fact that volatility responds differently to positive and negative shocks of the same magnitude.[*] Cont (2001), Bouchaud *et al.* (2007) and Bouchaud

---

[*] A survey of the empirical literature about this fact can be found in Taylor (2005).

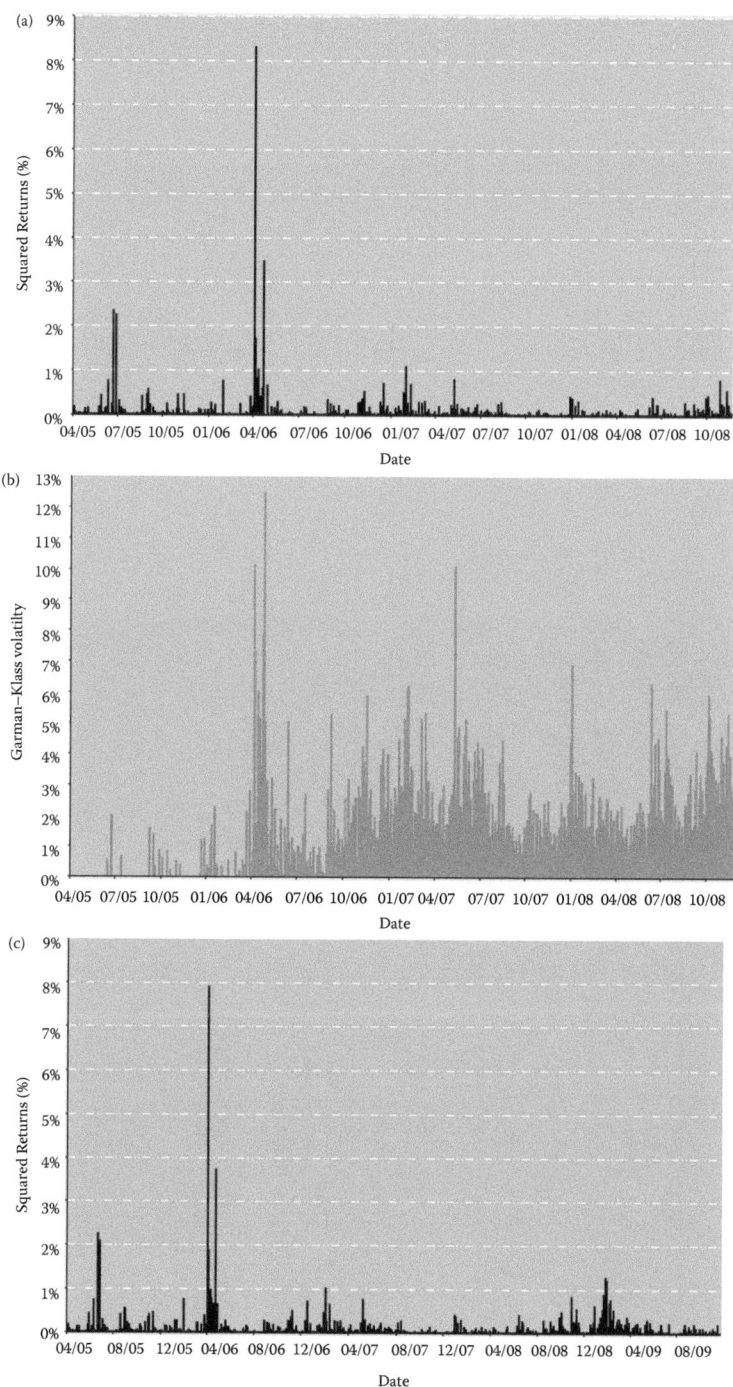

FIGURE 28.3 Volatility clustering: (a), (c) and (e) show the squared returns representation for the 2008, 2009 and 2010 EUA futures contracts, respectively; (b), (d) and (f) show the Garman–Klass volatility representation for the 2008, 2009 and 2010 EUA futures contracts, respectively. Volatility clustering is observed in all figures. '2008 EUA futures contract' refers to the futures contract maturing on 15 December 2008, '2009 EUA futures contract' refers to the futures contract maturing on 14 December 2009 and '2010 EUA futures contract' refers to the futures contract maturing on 20 December 2010.

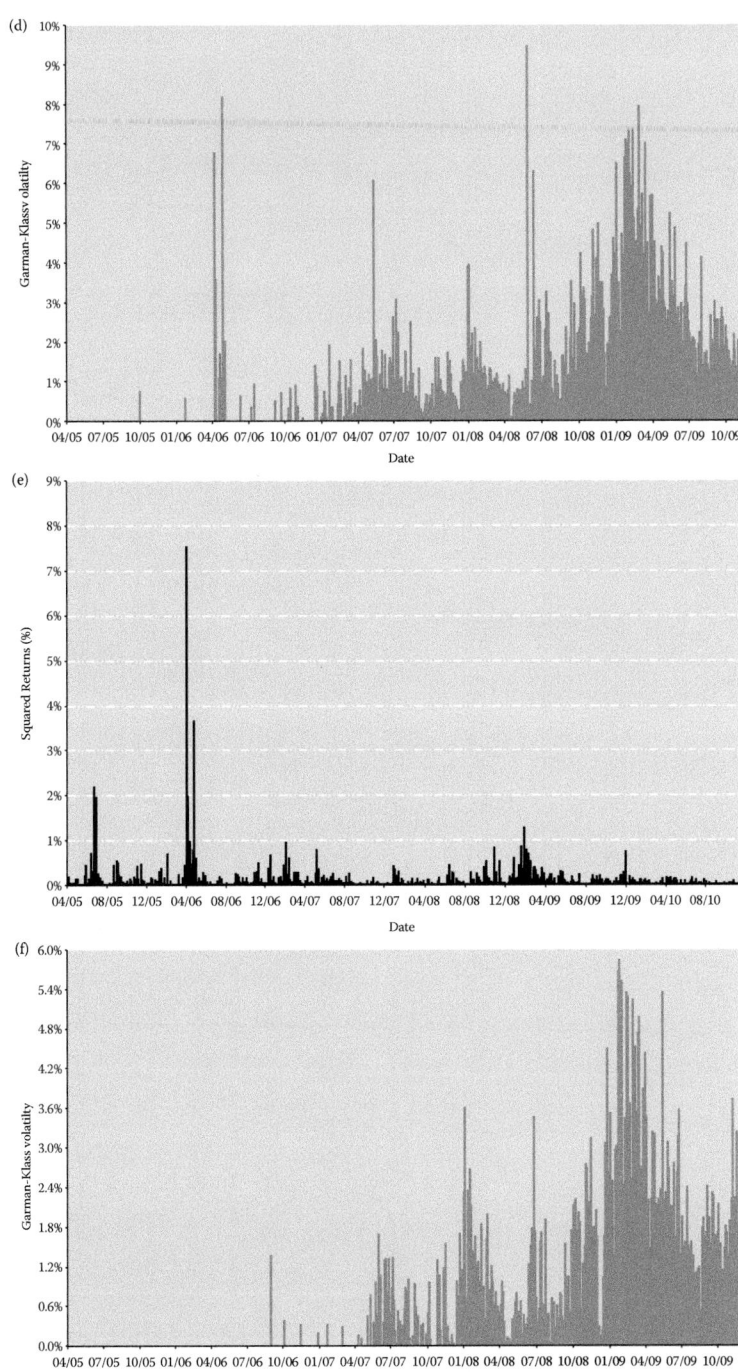

FIGURE 28.3   (CONTINUED)

TABLE 28.7 Volatility-Related Features

Lag	2008 EUA Futures Contract		2009 EUA Futures Contract		2010 EUA Futures Contract	
	Correlation	*p*-Value	Correlation	*p*-Value	Correlation	*p*-Value
	Panel A: $\rho_s(\sigma, \sigma_{-\tau})$					
1	0.7634	0.0000	0.8428	0.0000	0.8641	0.0000
2	0.7132	0.0000	0.8384	0.0000	0.8514	0.0000
3	0.6968	0.0000	0.8267	0.0000	0.8400	0.0000
4	0.7207	0.0000	0.8304	0.0000	0.8287	0.0000
5	0.7189	0.0000	0.8266	0.0000	0.8377	0.0000
6	0.7036	0.0000	0.8090	0.0000	0.8249	0.0000
7	0.6796	0.0000	0.8128	0.0000	0.8243	0.0000
8	0.6515	0.0000	0.8034	0.0000	0.8218	0.0000
9	0.6651	0.0000	0.8025	0.0000	0.8181	0.0000
10	0.6592	0.0000	0.8156	0.0000	0.8226	0.0000
	Panel B: $\rho_s(\sigma, r_{-\tau})$					
0	−0.1008	0.0020	−0.0931	0.0013	−0.0666	0.0113
1	−0.0915	0.0051	−0.0781	0.0070	−0.0564	0.0321
2	−0.0777	0.0174	−0.0608	0.0361	−0.0485	0.0652
3	−0.0626	0.0557	−0.0777	0.0074	−0.0637	0.0154
4	−0.0655	0.0453	−0.0614	0.0341	−0.0572	0.0298
5	−0.0613	0.0609	−0.0387	0.1821	−0.0580	0.0278
6	−0.0692	0.0342	−0.0630	0.0298	−0.0571	0.0304
7	−0.0506	0.1222	−0.0510	0.0787	−0.0341	0.1956
8	−0.0554	0.0901	−0.0497	0.0867	−0.0541	0.0402
9	−0.0572	0.0803	−0.0312	0.2815	−0.0497	0.0594
10	0.0789	0.0158	−0.0676	0.0197	−0.0589	0.0256

Notes: Panel A presents the Spearman's autocorrelation coefficients (ρ) between the daily Garman–Klass volatility (σ) and the daily Garman–Klass volatility lagged τ periods. Panel B presents the Spearman's cross correlations coefficients between the daily Garman–Klass volatility and the daily futures return (r) lagged τ periods. '2008 EUA futures contract' refers to the futures contract maturing on 15 December 2008, '2009 EUA futures contract' refers to the futures contract maturing on 14 December 2009 and '2010 EUA futures contract' refers to the futures contract maturing on 20 December 2010.

and Potters (2001) comment on the existence of a negative correlation between volatility and returns in financial assets, and particularly in stocks, which tends to zero as we increase the time lag. Panel B in Table 28.7 present the Spearman's cross correlation coefficients between Garman–Klass volatility and returns. Both the contemporaneous and the one-period lagged correlation coefficient are significant and negative at the 1% level, confirming that EUA volatility responds differently to positive and negative returns.

Although volatility modelling is not an objective of this study, to get additional insights about volatility asymmetry we have tested standard GARCH models that take into account this stylised fact in Phase II.[*] Following Rodríguez and Ruiz (2009), we have estimated EGARCH models as the most flexible GARCH models that capture the asymmetric effect. In all three series of data, the asymmetric parameter is significant at the 1% significance level; however, the residuals of the estimated models still show persistence of heavy tails (see the last column in Table 28.1). Therefore, when modelling EUA volatility,

---

[*] Borak *et al.* (2006), Miclaus *et al.* (2008), Paolella and Taschini (2008) and Benz and Trück (2009), among others, present GARCH models to analyse the conditional volatility of Phase I EUAs returns.

TABLE 28.8   Trading-Related Features

Lag	2008 EUA Futures Contract		2009 EUA Futures Contract		2010 EUA Futures Contract	
	Correlation	p Value	Correlation	p-Value	Correlation	p-Value
Panel A: $\rho_s(\sigma, VOL_{-\tau})$						
0	0.7644	0.0000	0.8832	0.0000	0.8403	0.0000
1	0.7070	0.0000	0.8559	0.0000	0.8210	0.0000
2	0.6875	0.0000	0.8492	0.0000	0.8122	0.0000
3	0.6865	0.0000	0.8512	0.0000	0.8132	0.0000
4	0.6980	0.0000	0.8454	0.0000	0.8082	0.0000
5	0.6956	0.0000	0.8444	0.0000	0.8082	0.0000
6	0.6831	0.0000	0.8377	0.0000	0.8049	0.0000
7	0.6758	0.0000	0.8354	0.0000	0.8045	0.0000
8	0.6667	0.0000	0.8298	0.0000	0.7981	0.0000
9	0.6710	0.0000	0.8352	0.0000	0.7918	0.0000
10	0.6690	0.0000	0.8390	0.0000	0.7946	0.0000
Panel B: $\rho_s(\sigma, \Delta OI_{-\tau})$						
0	0.1720	0.0000	0.3186	0.0000	0.2705	0.0000
1	0.1146	0.0006	0.2767	0.0000	0.2658	0.0000
2	0.1025	0.0021	0.2817	0.0000	0.2460	0.0000
3	0.1319	0.0001	0.2934	0.0000	0.2536	0.0000
4	0.1286	0.0001	0.2977	0.0000	0.2554	0.0000
5	0.1087	0.0011	0.2793	0.0000	0.2570	0.0000
6	0.1019	0.0022	0.2581	0.0000	0.2641	0.0000
7	0.1148	0.0006	0.2581	0.0000	0.2504	0.0000
8	0.0841	0.0117	0.2588	0.0000	0.2419	0.0000
9	0.1203	0.0003	0.2838	0.0000	0.2325	0.0000
10	0.1394	0.0000	0.2649	0.0000	0.2334	0.0000

Notes: Panel A presents the Spearman's cross correlation coefficients (ρ) between the daily Garman–Klass volatility (σ) and the daily trading volume (VOL) lagged τ periods. Panel B presents the Spearman's cross correlation coefficients between the daily Garman–Klass volatility (σ) and the change in the daily open interest (ΔOI) lagged τ periods. '2008 EUA futures contract' refers to the futures contract maturing on 15 December 2008, '2009 EUA futures contract' refers to the futures contract maturing on 14 December 2009 and '2010 EUA futures contract' refers to the futures contract maturing on 20 December 2010.

we suggest assuming fat-tailed unconditional distributions in order to capture all the volatility features of EUA return data.*

Finally, we have tested the relationship between trading activity and volatility. Given that our measure of volatility is an intraday volatility, we have chosen two flow variables, the level of daily trading volume (VOL) and the change in the open interest (ΔOI), to calculate the cross correlations between these trading-related measures and volatility.

Panels A and B in Table 28.8 present results for the cases of volume and change in open interest, respectively. The contemporaneous cross correlation coefficients in the case of volume are positive and significant (76, 88 and 84% for 2008, 2009 and 2010 futures contract, respectively). The same occurs in the case of open interest (17, 32 and 27% for 2008, 2009 and 2010 futures contracts, respectively). Furthermore, the cross correlations for lags from one to ten decay very slowly in both measures and in

---

* See Rachev and Mittnik (2000, chapter 4) for a comparison of conditional and unconditional distributional models for returns on the Nikkei 225 stock market index.

all three futures contracts. All in all, these facts indicate that both trading-related measures should be considered as explanatory variables in volatility models. Additionally, following the classical approach in the financial literature, which considers volume as a proxy for day trading and speculative activity, and open interest as a proxy to measure hedging activity, we could say that both speculators and hedgers destabilise the EUA market.

28.4.4 Commodity Features

The last group of relevant stylised facts is focused on specific features observed in commodities series. The first we study is the non-trading effect in EUA volatility in order to test whether the information flow is more evenly distributed around the clock in weather-sensitive markets than in the equity market. Firstly, we have tested a possible daily seasonality in returns or variances. Neither non-parametric Kruskal–Wallis nor Brown–Forsythe tests can reject the equality of medians and variances.* Therefore, returns and variances are evenly distributed over the trading days. Secondly, two subsamples have been generated for each maturity. We have separated the weekend from the rest of the week, with the returns from Friday closing to Monday closing standing for weekend returns. Finally, we have split each subsample into two, separating between trading periods (open-to-close returns, OC) and non-trading periods (close-to-open returns, CO).

The results are presented in Table 28.9. All three futures contracts show similar results. We can reject the equality in variances between weekend and weekdays neither for trading nor non-trading periods, with the exception of non-trading periods for 2009 futures contract. However, when comparing volatility between trading and non-trading periods conditioned on weekend or weekdays, we reject the null hypothesis at the 1% level of significance (except for weekend periods for 2009 futures contract), showing that the trading volatility (OC) is higher than non-trading volatility (CO). Taking into account that the number of trading hours (7:00 to 17:00, London local time) is lower than the number of non-trading hours, our results clearly indicate that information flow is concentrated during the business day, similar to the equity markets, and does not evolve randomly around the clock.

A second phenomenon characteristic of commodities futures is the negative correlation with stocks and bonds (see Gorton and Rouwenhorst 2006). We have analysed this aspect by calculating the existing correlation between EUA futures and the main European benchmarks for bond and stock markets. Specifically, we have chosen Schatz bond futures, Bolb bond futures and Bund bond futures as representatives of short-term, medium-term and long-term bonds, and the Euro Stoxx 50 futures contract and its underlying asset as a benchmark for the stock market. Table 28.10 presents Spearman's rank correlation coefficients between the different described assets and the three EUA futures contracts.

The significance of the correlation with all the assets can be seen. Nevertheless, the correlation is negative with bond assets and positive with stocks and Brent, mainly in the 2009 and 2010 futures contracts. A possible explanation of these results can be found in the fact that, in business cycle expansions (contractions), when the stock market is normally in a bullish (bearish) moment, companies produce more (less) and as a consequence they pollute more (less) and need higher (lower) numbers of EUAs. Expansions (contractions) are normally accompanied by an increase (decrease) in interest rates and a decrease (increase) in bond prices. Note that the whole significant correlation picture has appealing implications for portfolio diversification purposes.

Finally, we test the property of *inflation hedge*. If the correlation between EUAs and inflation were significant and positive, they would let us hedge against inflation. To study this fact, we have carried out a monthly analysis. Furthermore, inflation has been divided into three types: the observed, the expected and the unexpected. It is interesting to extract the unexpected component of inflation because this is

---

* The Kruskal–Wallis statistic is 3.849 (p-value = 0.427) for 2008 futures, 4.210 (p-value = 0.378) for 2009 futures and 6.4956 (p-value = 0.1651) for 2010 futures. The Brown–Forsythe statistic takes the values 1.591 (p-value = 0.174), 1.817 (p-value = 0.123) and 1.9257 (p-value = 0.1037) for 2008, 2009 and 2010 futures contracts, respectively.

TABLE 28.9   Information Flows

	No. observations	σ_{OC}	σ_{CO}	BF-Statistic	p-Value
Panel A: 2008 EUA futures contract					
Weekend	144	47.43	25.17	11.3814	0.0008
Weekdays	574	41.01	28.24	30.0853	0.0000
BF-statistic		0.0144	0.7702		
p-value		0.9044	0.3805		
Total	718	42.36	27.66	41.2757	0.0000
Panel B: 2009 EUA futures contract					
Weekend	142	42.14	34.53	2.0256	0.1558
Weekdays	572	39.88	28.14	25.2061	0.0000
BF-statistic		0.6069	6.3929		
p-value		0.4362	0.0117		
Total	714	40.32	29.57	25.5893	0.0000
Panel C: 2010 EUA futures contract					
Weekend	161	34.21	25.17	10.7123	0.0012
Weekdays	656	30.76	24.91	32.4802	0.0000
BF-statistic		1.1193	0.0713		
p-value		0.2904	0.7895		
Total	817	31.49	24.95	43.1415	0.0000

Notes: This table reports the Brown–Forsythe statistic (BF-statistic), which tests the null of equality of variances among trading and non-trading periods. Panels A, B and C present the results for weekends and for weekdays for 2008, 2009 and 2010 futures contracts, respectively. Weekend results are based on returns from Friday close to Monday close and weekday results are based on returns from Monday close to Friday close. The table also reports the number of observations and the annualised volatility as a percentage for trading (σ_{OC}) and non-trading (σ_{CO}) periods. '2008 EUA futures contract' refers to the futures contract maturing on 15 December 2008, '2009 EUA futures contract' refers to the futures contract maturing on 14 December 2009 and '2010 EUA futures contract' refers to the futures contract maturing on 20 December 2010.

TABLE 28.10   Correlation with Bonds, Stocks and Brent

	2008 EUA Futures Contract		2009 EUA Futures Contract		2010 EUA Futures Contract	
Asset	ρ	p-Value	ρ	p-Value	ρ	p-Value
Schatz bond futures	−0.0950	0.0038	−0.1311	0.0000	−0.1145	0.0000
Bolb bond futures	−0.0971	0.0031	−0.1442	0.0000	−0.1248	0.0000
Bund bond futures	−0.0951	0.0038	−0.1465	0.0000	−0.1268	0.0000
Euro Stoxx 50 futures	0.0585	0.0751	0.1476	0.0000	0.1412	0.0000
Euro Stoxx 50 index	0.0577	0.0787	0.1434	0.0000	0.1362	0.0000
Brent oil futures	0.2413	0.0000	0.2499	0.0000	0.2302	0.0000

Notes: The table gives the Spearman's rank correlation coefficients and their p-values between the EUA 2008, 2009 and 2010 futures contracts and Schatz bond futures (short-term fixed rent), Bolb bond futures (medium-term fixed rent), Bund bond futures (long-term fixed rent), Euro Stoxx 50 futures, the Euro Stoxx 50 index, and the Brent futures contract. '2008 EUA futures contract' refers to the futures contract maturing on 15 December 2008, '2009 EUA futures contract' refers to the futures contract maturing on 14 December 2009 and '2010 EUA futures contract' refers to the futures contract maturing on 20 December 2010.

TABLE 28.11 Inflation Hedge

	Correlation		
	Observed Inflation	Expected Inflation	Non-Expected Inflation
2008 EUA futures contract			
Correlation coefficient	0.2895	−0.0908	0.3241
Sig. (bilateral)	0.0601	0.5627	0.0345
2009 EUA futures contract			
Correlation coefficient	0.2953	−0.0618	0.2863
Sig. (bilateral)	0.0290	0.6542	0.0345
2010 EUA futures contract			
Correlation coefficient	0.3379	−0.0316	0.2973
Sig. (bilateral)	0.0054	0.7998	0.0149

Notes: The table shows the Spearman's rank correlation coefficients and their p-values between the EUA futures contracts returns and the series of the observed, expected and non-expected inflation. The number of monthly observations is 43, 55 and 67 for the 2008, 2009 and 2010 EUA futures contracts, respectively. '2008 EUA futures contract' refers to the futures contract maturing on 15 December 2008, '2009 EUA futures contract' refers to the futures contract maturing on 14 December 2009 and '2010 EUA futures contract' refers to the futures contract maturing on 20 December 2010.

the part we are interested in hedging against since the expected component is anticipated by the interest rate, and, with fixed income assets we might hedge against it. The Harmonized Index of Consumer Prices (HICP) monthly series has been chosen for the observed inflation.[*] The expected inflation is the free risk interest rate, which is the reason we use the one month Euribor rate as a proxy, leaving the non-expected inflation as the difference between the observed and the expected inflation.

Table 28.11 shows the correlation coefficients between our asset and the different definitions of inflation. Note that the Spearman's correlation coefficients between the futures and the unexpected inflation are significant in all the futures contracts; therefore, as commodity futures, EUA assets could help us to hedge against inflation.

28.5 Conclusions

The objective of our paper is to study the stylised facts of EUA returns. Given that Phase I (2005–2007) is generally considered as a pilot and learning phase, we have chosen 2008, 2009 and 2010 futures contracts to test the accomplishment of several statistical features that usually appear in both financial and commodities assets.

We have found evidence of intermittency, *aggregational gaussianity*, short-term predictability, the Taylor effect, volatility clustering, asymmetric volatility, and a positive relation between volatility and volume and between volatility and change in open interest. Our findings suggest that temporal dependences should be modelled with ARMA-GARCH structures. However, the persistence of conditional heavy-tails indicates that, in order to capture the volatility features of EUA return data; fat-tailed unconditional distributions should be assumed. Furthermore, unlike commodities properties, we have observed negative asymmetry, positive correlation with stocks indexes and higher volatility levels

---

[*] Eurostat defines the Harmonized Index of Consumer Prices (HICP) as an economic indicator constructed to measure the changes over time in the prices of consumer goods and services acquired by households. The HICP gives comparable measures of inflation in the euro-zone, the EU, the European Economic Area and for other countries including accession and candidate countries. They are calculated according to a harmonised approach and a single set of definitions. They provide the official measure of consumer price inflation in the euro-zone for the purposes of monetary policy in the euro area and assessing inflation convergence as required under the Maastricht criteria. The elected series is the European Union HICP with 25 countries. More details about this index can be found at http://ec.europa.eu/eurostat (last accessed on 20 March 2012).

during the trading session—an indication that EUA information flow is concentrated during the trading day and does not evolve randomly around the clock. The property of inflation hedge and the positive correlation with bonds, both characteristics being typical of commodity futures, are also detected. Therefore, our results indicate that EUAs do not behave like either common commodities futures or financial assets, and suggest that the EUA is a new asset class. The entirety of these facts, robust over time, has appealing implications for portfolio analysis, volatility modelling, hedging activities and cointegration analysis.

Acknowledgements

A preliminary version of this paper was published by Instituto Valenciano de Investigaciones Econó micas (IVIE) as Working Paper WP-AD 2012-14. The authors are grateful for the financial support of the Spanish Ministry of Education and Science (projects ECO2009- 14457-C04-04 and CGL200909604), the Fonds Européen de Développement Régional (FEDER), and the Cátedra de Finanzas Internacionales—Banco Santander. They are also indebted to the European Climate Exchange (ECX) market for providing the database. The usual caveats apply.

References

Alberola, E., Chevallier, J. and Chèze, B., Price drivers and structural breaks in European carbon prices 2005–2007. *Energy Policy*, 2008, **36**, 787–797.

Benz, E. and Hengelbrock, J., Liquidity and price discovery in the European CO_2 futures market: An intraday analysis. Working Paper, Bonn Graduate School of Business, 2008.

Benz, E. and Trück, S., Modelling the price dynamics of CO_2 emission allowances. *Energy Econ.*, 2009, **31**, 4–15.

Bhargava, V. and Malhotra, D.K., The relationship between futures trading activity and exchange rate volatility, revisited. *J. Multinat. Finan. Mgmt*, 2007, **17**, 95–111.

Borak, S., Härdle, W., Trück, S. and Weron R., Convenience yields for CO_2 emission allowances futures contracts. Discussion Paper 2006-076, SFB 649 Economic Risk, Humboldt-University of Berlin, 2006.

Bouchaud, J.P., Matacz, A. and Potters, M., The leverage effect in financial markets: Retarded volatility and market panic. *Phys. Rev. Lett.*, 2007, **87**, 228701.

Bouchaud, J.P. and Potters, M., More stylized facts of financial markets: Leverage effect and downside correlations. *Physica A*, 2001, **299**, 60–70.

Catalán, B. and Trívez, F.J., Forecasting volatility in GARCH models with additive outliers. *Quant. Finan.*, 2007, **7**, 591–596.

Chesney, M. and Taschini, L., The endogenous price dynamics of emission allowances: An application to CO_2 option pricing. *Appl. Math. Finan.*, 2012, 447–475.

Cont, R., Empirical properties of asset returns: Stylized facts and statistical issues. *Quant. Finan.*, 2001, **1**, 223–236.

Convery, F.J., Reflections—the emerging literature on emission trading in Europe. *Rev. Environ. Econ. & Policy*, 2009, **3**, 121–137.

Daskalakis, G. and Markellos, R.N., Are the European carbon markets efficient? Rev. *Futures Mkts*, 2008, **17**, 103–128.

Daskalakis, G., Psychoyios, D. and Markellos, R.N., Modelling CO_2 emission allowances prices and derivatives: Evidence from the European trading scheme. *J. Banking & Finance*, 2009, **33**, 1230–1241.

Directive 2003/87/EC of the European Parliament and of the Council of 13 October 2003 establishing a scheme for greenhouse gas emission allowance trading within the Community and amending Council Directive 96/61/EC. Available online at: http://eur-lex.europa.eu/ LexUriServ/LexUriServ. do?uri=OJ:L:2003:275:0032:0032:EN: PD (accessed 1 December 2011).

Fleming, J., Kirby, C. and Ostdiek, B., Information, trading, and volatility: Evidence from weather-sensitive markets. *J. Finan.*, 2006, **61**, 2899–2930.

Galai, D., Kedar-Levy, H. and Schreiber, B.Z., Seasonality in outliers of daily stock returns: A tail that wags the dog? *Int. Rev. Finan. Anal.*, 2008, **17**, 784–792.

Garman, M.B. and Klass, M.J., On the estimation of security price volatilities from historical data. *J. Bus.*, 1980, **53**, 67–78.

Gorton, G.B., Hayashi, F. and Rouwenhorst K.G., The fundamentals of commodity futures returns. *Rev. Finan.*, 2012, **17**, 35–105.

Gorton, G.B. and Rouwenhorst, K.G., Facts and fantasies about commodity futures. *Finan. Analysts J.*, 2006, **62**, 47–68.

Hoaglin, D.C., Mosteller, F. and Tukey, J.W., *Understanding Robust and Exploratory Data Analysis*, 1983 (Wiley Classics Library: New York).

Kwiatkowski, D., Phillips, P., Schmidt, P. and Shin, Y., Testing the null hypothesis of stationarity against the alternative of a unit root: How sure are we that economic time series have a unit root? *J. Econometr.*, 1992, **54**, 1–3, 159–178.

Malevergne, Y. and Sornette, D., Value-at-Risk-efficient portfolios for a class of super- and subexponentially decaying assets return distributions. *Quant. Finan.*, 2004, **4**, 17–36.

Mansanet-Bataller, M. and Pardo, Á., What you should know about carbon markets. *Energies*, 2008, **1**, 120–153.

Mansanet-Bataller, M. and Pardo, Á., Impacts of regulatory announcements on CO_2 prices. *J. Energy Mkts*, 2009, **2**, 77–109.

Mansanet-Bataller, M., Pardo, Á. and Valor, E., CO_2 prices, energy and weather. *Energy J.*, 2007, **28**, 73–92.

Miclaus, P.G., Lupu, R., Dumitrescu, S.A. and Bobirca, A., Testing the efficiency of the European carbon futures market using event-study methodology. *Int. J. Energy & Environ.*, 2008, **2**, 121–128.

Morone, A., Financial markets in the laboratory: An experimental analysis of some stylized facts. *Quant. Finan.*, 2008, **8**, 513–532.

Pagan, A., The econometrics of financial markets. *J. Emp. Finan.*, 1996, **3**, 15–102.

Paolella, M.S. and Taschini, L., An econometric analysis of emission allowances prices. *J. Banking & Finan.*, 2008, **32**, 2022–2032.

Peiró, A., Asymmetries and tails in stock index returns: Are their distributions really asymmetric? Quant. Finan., 2004, **4**, 37–44.

Rachev, S. and Mittnik, S., *Stable Paretian Models in Finance*, 2000 (Wiley: Chichester).

Rittler, D., Price discovery and volatility spillovers in the European Union Emissions Trading Scheme: A high-frequency analysis. *J. Banking & Finan.*, 2012, **36**, 774–785.

Rodríguez, M.J. and Ruiz, E., GARCH models with leverage effect: Differences and similarities. UC3M Working Paper, Statistics and Econometrics 09–02, Instituto Valenciano de Investigaciones Económicas, 2009.

Rotfuss, W., Intraday price formation and volatility in the European Union Trading Scheme: An introductory analysis. Discussion Paper 09-018, Centre for European Economic Research (ZEW), Mannheim, Germany, 2009.

Sewell, M., Characterization of financial time series. Research Note, University College London, 2011.

Taylor, S.J., *Asset Price Dynamics, Volatility and Prediction*, 2005 (Princeton University Press: Princeton, NJ).

Taylor, S.J., Modelling Financial Time Series, 2nd ed., 2007 (World Scientific: Hackensack, NJ).

Uhrig-Homburg, M. and Wagner, M., Futures price dynamics of CO_2 emission allowances: An empirical analysis of the trial period. *J. Derivatives*, 2009, **17**, 73–88.

Contemporary Topics

29

Volatility Is Rough

Jim Gatheral,
Thibault Jaisson and
Mathieu
Rosenbaum

29.1 Introduction

29.1.1 Volatility Modeling

In the derivatives world, log-prices are often modelled as continuous semi-martingales. For a given asset with log-price Y_t, such a process takes the form

$$dY_t = \mu_t dt + \sigma_t dW_t,$$

where μ_t is a drift term and W_t is a one-dimensional Brownian motion. The term σ_t denotes the volatility process and is the most important ingredient of the model. In the Black–Scholes framework, the volatility function is either constant or a deterministic function of time. In Dupire's local volatility model, see Dupire (1994), the local volatility $\sigma(Y_t, t)$ is a deterministic function of the underlying price and time, chosen to match observed European option prices exactly. Such a model is by definition time-inhomogeneous; its dynamics are highly unrealistic, typically generating future volatility surfaces (see section 29.1.3 below) completely unlike those we observe. A corollary of this is that prices of exotic options under local volatility can be substantially off-market. On the other hand, in so-called stochastic volatility models, the volatility σ_t is modelled as a continuous Brownian semi-martingale. Notable amongst such stochastic volatility models are the Hull and White model Hull and White (1993), the Heston model Heston (1993) and the SABR model Hagan *et al.* (2002). Whilst stochastic volatility dynamics are more realistic than local volatility dynamics, generated option prices are not consistent with observed European option prices. We refer to Gatheral (2006) and Musiela and Rutkowski (2006) for more detailed reviews of the different approaches to volatility modelling. More recent market practice is to use local stochastic volatility (LSV) models which both fit the market exactly and generate reasonable dynamics.

Consistent with our focus on derivatives, our goal in this work is to replicate features of the observed time series of volatility over timescales from 1 day to 10 years say. Indeed, volatility modelling can probably only be relevant at timescales of order one day or more. Below this, at the sub-second timescale for example, it is not even clear what the meaning of volatility is (independently of a specific model). Nevertheless, in order to get accurate volatility measurements, we will rely on high frequency methods in our estimation procedures.

29.1.2 Fractional Volatility

In terms of the smoothness of the volatility process, the preceding models offer two possibilities: very regular sample paths in the case of Black–Scholes, and volatility trajectories with regularity close to that of Brownian motion for the local and stochastic volatility models. Starting from the stylized fact that volatility is a long memory process, various authors have proposed models that allow for a wider range of regularity for the volatility. In a pioneering paper, Comte and Renault (1998) proposed to model log-volatility using fractional Brownian motion (fBm for short), ensuring long memory by choosing the Hurst parameter $H > 1/2$. A large literature has subsequently developed around such fractional volatility models, for example Cheridito *et al.* (2003), Comte *et al.* (2012), Rosenbaum (2008).

The fBm $\left(W_t^H\right)_{t\in\mathbb{R}}$ with Hurst parameter $H \in (0, 1)$, introduced in Mandelbrot and Van Ness (1968), is a centred self-similar Gaussian process with stationary increments satisfying for any $t \in \mathbb{R}, \Delta \geq 0, q > 0$:

$$\mathbb{E}\left[\left|W_{t+\Delta}^H - W_t^H\right|^q\right] = K_q \Delta^{qH}, \tag{29.1.1}$$

with K_q the moment of order q of the absolute value of a standard Gaussian variable. For $H = 1/2$, we retrieve the classical Brownian motion. The sample paths of W^H are Hölder-continuous with exponent r, for any $r < H$.* Finally, when $H > 1/2$, the increments of the fBm are positively correlated and exhibit long memory in the sense that

$$\sum_{k=0}^{+\infty} \text{Cov}\left[W_1^H, W_k^H - W_{k-1}^H\right] = +\infty.$$

Indeed, $\text{Cov}\left[W_1^H, W_k^H - W_{k-1}^H\right]$ is of order k^{2H-2} as $k \to \infty$. Note that in the case of the fBm, there is a one-to-one correspondence between regularity and long memory through the Hurst parameter H.

As mentioned earlier, the long memory property of the volatility process has been widely accepted as a stylized fact since the seminal analyses of Ding et al. (1993), Andersen and Bollerslev (1997) and Andersen *et al.* (2001b). Initially, it appears that the term *long memory* referred to the slow decay of the autocorrelation function (of absolute returns for example), anything slower than exponential. Over time however, it seems that this term has acquired the more precise meaning that the autocorrelation

---

* Actually H corresponds to the regularity of the process in a more accurate way: in terms of Besov smoothness spaces, see section 29.2.1.

function is not integrable, see Beran (1994), and even more precisely that it decays as a power law with exponent less than 1. Much of the more recent literature, for example Bentes and Cruz (2011), Chen *et al.* (2006), Chronopoulou (2011), assumes long memory in volatility in this more technical sense. Indeed, meaningful results can probably only be obtained under such a specification since it is not possible to estimate the asymptotic behaviour of the covariance function without assuming a specific form. Nevertheless, analyses such as that of Andersen *et al.* (2001b) use data that predate the advent of high-frequency electronic trading, and the evidence for long memory has never been sufficient to satisfy remaining doubters such as Mikosch and Stărică (2000). To quote Cont (2007):

> ... the econometric debate on the short range or long range nature of dependence in volatility still goes on (and may probably never be resolved)...

In the spirit of the above quote, in our view, the question as to whether the volatility time series exhibits long memory (in a technical sense) or not is not a very useful or fruitful one. Indeed, we believe that in practice, the concept of long memory is too fragile to be applicable to an analysis involving ultrahigh frequency data (for example any seasonality may give rise to spurious long memory). Therefore, we do not focus on long memory in this work. Still, we do show that the autocorrelation function of volatility does not behave as a power law, at least at usual timescales of observation. In particular, we are able to provide explicit expressions enabling us to analyse thoroughly the dependence structure of the volatility process.

29.1.3 The Shape of the Implied Volatility Surface

As is well known, the implied volatility $\sigma_{BS}(k,\tau)$ of an option (with log-moneyness k and time to expiration τ) is the value of the volatility parameter in the Black–Scholes formula required to match the market price of that option. Plotting implied volatility as a function of strike price and time to expiry generates the *volatility surface*, explored in detail in, for example, Gatheral (2006). A typical such volatility surface generated from a 'stochastic volatility inspired' (SVI) (Gatheral and Jacquier 2014) fit to closing SPX option prices as of 20 June 2013* is shown in figure 29.1. It is a stylized fact that, at least in equity markets, although the level and orientation of the volatility surface do change over time, the general overall shape of the volatility surface does not change, at least to a first approximation. This suggests that it is desirable to model volatility as a time-homogenous process, i.e. a process whose parameters are independent of price and time.

However, conventional time-homogenous models of volatility such as the Hull and White, Heston and SABR models do not fit the volatility surface. In particular, as shown in figure 29.2, the observed term structure of at-the-money ($k = 0$) volatility skew

$$\psi(\tau) := \left| \frac{\partial}{\partial k} \sigma_{BS}(k,\tau) \right|_{k=0}$$

is well approximated by a power law function of time to expiry τ. In contrast, conventional stochastic volatility models generate a term structure of at-the-money (ATM) skew that is *constant* for small τ and behaves as a sum of decaying exponentials for larger τ.

In section 29.3.3 of Fukasawa (2011), as an example of the application of his martingale expansion, Fukasawa shows that a stochastic volatility model where the volatility is driven by fractional Brownian motion with Hurst exponent H generating an ATM volatility skew of the form $\psi(\tau) \sim \tau^{H-1/2}$, at least for small τ. This is interesting in and of itself in that it provides a counterexample to the widespread belief that the explosion of the volatility smile as $\tau \to 0$ (as clearly seen in figures 29.1 and 29.2) implies the presence of jumps (Carr and Wu 2003). The main point here is that for a model of the sort analysed by

---

* Closing prices of SPX options for all available strikes and expirations as of 20 June 2013 were sourced from OptionMetrics (www. optionmetrics.com) via Wharton Research Data Services (WRDS).

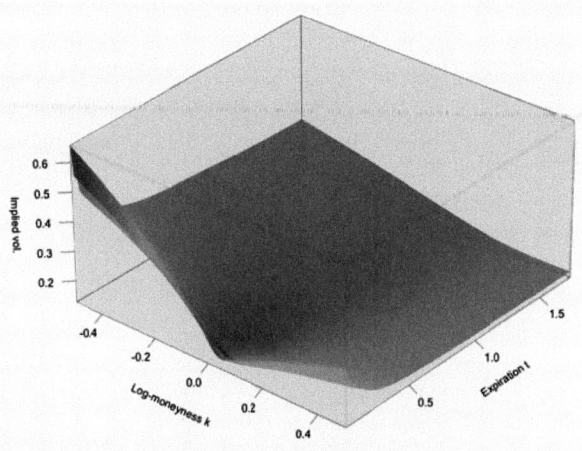

FIGURE 29.1 The S&P volatility surface as of 20 June 2013.

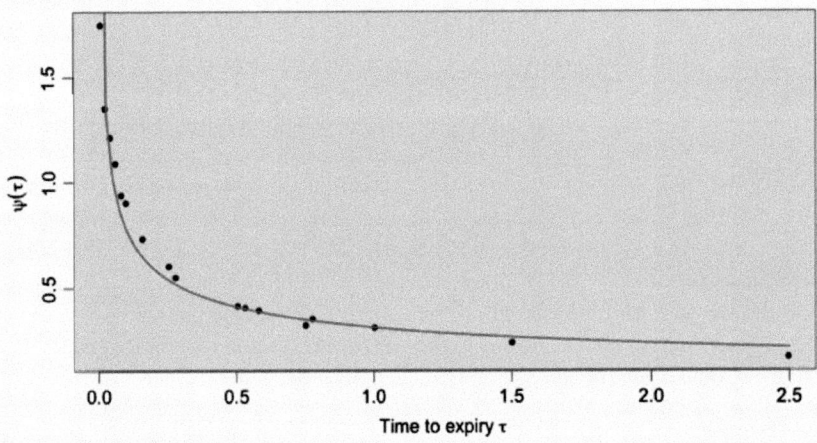

FIGURE 29.2 The black dots are non-parametric estimates of the S&P ATM volatility skews as of 20 June 2013; the red curve is the power law fit $\psi(\tau) = A \tau^{-0.4}$.

Fukasawa to generate a volatility surface with a reasonable shape, we would need to have a value of H close to zero. As we will see in section 29.2, our empirical estimates of H from time series data are in fact very small.

The volatility model that we will specify in section 29.3.1, driven by fBm with $H < 1/2$, therefore has the potential to be not only consistent with the empirically observed properties of the volatility time series but also consistent with the shape of the volatility surface. In this paper, we focus on the modelling of the volatility time series. A more detailed analysis of the consistency of our model with option prices is provided in Bayer *et al.* (2016).

29.1.4 Main Results and Organization of the Paper

In section 29.2, we report our estimates of the smoothness of the log-volatility for selected assets. This smoothness parameter lies systematically between 0.08 and 0.2 (in the sense of Hölder regularity for example). Furthermore, we find that increments of the log-volatility are approximately normally distributed and that their moments enjoy a remarkable monofractal scaling property. This leads us to model the log of volatility using a fBm with Hurst parameter $H < 1/2$ in section 29.3. Specifically we adopt the fractional stochastic volatility (FSV) model of Comte and Renault (1998). We call our model Rough FSV (RSFV) to underline that, in contrast to FSV, we take $H < 1/2$. We also show in the same section that the RFSV model is remarkably consistent with volatility time series data. The issue of volatility persistence is considered through the lens of the RFSV model in section 29.4. Our main finding is that although the RFSV model does not have any long memory property, classical statistical procedures aiming at detecting volatility persistence tend to conclude the presence of long memory in data generated from it. This sheds new light on the supposed long memory in the volatility of financial data. In section 29.5, we finally apply our model to volatility forecasting. In particular, we show that RFSV volatility forecasts outperform conventional AR and HAR volatility forecasts. Some proofs are relegated to the appendix.

29.2 Smoothness of the Volatility: Empirical Results

In this section, we report estimates of the smoothness of the volatility process for four assets:

- The DAX and Bund futures contracts, for which we estimate integrated variance directly from high frequency data using an estimator based on the model with uncertainty zones (see Robert and Rosenbaum 2011, Robert and Rosenbaum 2012). This model enables us to safely use all the ultra high frequency price data in order to perform our estimation, and thus to obtain accurate estimates over short time windows.
- The S&P and NASDAQ indices, for which we use precomputed realized variance estimates from the Oxford-Man Institute of Quantitative Finance Realized Library.*

29.2.1 Estimating the Smoothness of the Volatility Process

Let us first pretend that we have access to discrete observations of the volatility process, on a time grid with mesh Δ on $[0, T]$: $\sigma_0, \sigma_\Delta..., \sigma_{k\Delta}..., k \in \left\{0, \lfloor T / \Delta \rfloor\right\}$. Set $N = \lfloor T / \Delta \rfloor$, then for $q \geq 0$, we define

$$m(q,\Delta) = \frac{1}{N} \sum_{k=1}^{N} \left|\log\left(\sigma_{k\Delta}\right) - \log\left(\sigma_{(k-1)\Delta}\right)\right|^q.$$

In the spirit of Rosenbaum (2011), our main assumption is that for some $s_q > 0$ and $b_q > 0$, as Δ tends to zero,

$$N^{qs_q} m(q,\Delta) \rightarrow b_q. \tag{29.2.1}$$

Under additional technical conditions, equation (29.2.1) essentially says that the volatility process belongs to the Besov smoothness space $\mathcal{B}_{q,\infty}^{s_q}$ and does not belong to $\mathcal{B}_{q,\infty}^{s_q'}$, for $s_q' > s_q$ (see Rosenbaum 2009). Hence, s_q can really be viewed as the regularity of the volatility when measured in l_q norm. In particular, functions in $\mathcal{B}_{q,\infty}^{s}$ for every $q > 0$ enjoy the Hölder property with parameter h for any $h < s$. For example, if $\log(\sigma_t)$ is a fBm with Hurst parameter H, then for any $q \geq 0$, equation (29.2.1) holds in probability with $s_q = H$ and it can be shown that the sample paths of the process indeed belong to $\mathcal{B}_{q,\infty}^{H}$

---

* http://realized.oxford-man.ox.ac.uk/data/download. The Oxford-Man Institute's Realized Library contains a selection of daily non-parametric estimates of volatility of financial assets, including realized variance (rv) and realized kernel (rk) estimates. A selection of such estimators is described and their performances compared in, for example, Gatheral and Oomen (2010).

almost surely. Assuming the increments of the log-volatility process are stationary and that a law of large numbers can be applied, $m(q, \Delta)$ can also be seen as the empirical counterpart of

$$\mathbb{E}\left[\left|\log(\sigma_\Delta) - \log(\sigma_0)\right|^q\right].$$

Of course the volatility process is not directly observable, and an exact computation of $m(q, \Delta)$ is not possible in practice. We must therefore proxy spot volatility values by appropriate estimated values. Since the minimal Δ will be equal to one day in the sequel, we proxy the (true) spot volatility daily at a fixed given time of the day (11 am for example). Two daily spot volatility proxies will be considered:

- For our ultrahigh frequency intraday data (DAX future contracts and Bund future contracts,[*] 1248 days from 13 May 2010 to 01 August 2014[†]), we use the estimator of the integrated variance from 10 am to 11 am London time obtained from the model with uncertainty zones (see Robert and Rosenbaum 2011, 2012). After renormalization, the resulting estimates of integrated variance over very short time intervals can be considered as good proxies for the unobservable spot variance. In particular, the one-hour long window on which they are computed is small compared to the extra day timescales that will be of interest here.
- For the S&P and NASDAQ indices,[‡] we proxy daily spot variances by daily realized variance estimates from the Oxford-Man Institute of Quantitative Finance Realized Library (3,540 trading days from 3 January 2000 to 31 March 2014). Since these estimates of integrated variance are for the whole trading day, we expect estimates of the smoothness of the volatility process to be biased upwards, integration being a regularizing operation. We compute the extent of this bias by simulation in section 29.3.4 and more quantitatively in appendix 3.

In the following, we retain the notation $m(q, \Delta)$ with the understanding that we are only proxying the (true) spot volatility as explained above. We now proceed to estimate the smoothness parameter s_q for each q by computing the $m(q, \Delta)$ for different values of Δ and regressing $\log m(q, \Delta)$ against $\log \Delta$. Note that for a given Δ, several $m(q, \Delta)$ can be computed depending on the starting point. Our final measure of $m(q, \Delta)$ is the average of these values.

29.2.2 DAX and Bund Futures Contracts

DAX and Bund futures are amongst the most liquid assets in the world and moreover, the model with uncertainty zones used to estimate volatility is known to apply well to them (see Dayri and Rosenbaum 2015). So we can be confident in the reliability of our volatility proxy. Nevertheless, as an extra check, we will confirm the quality of our volatility proxy by Monte Carlo simulation in section 29.3.4.

Plots of $\log m(q, \Delta)$ vs $\log \Delta$ for different values of q are displayed for the DAX in figure 29.3, and for the Bund in figure 29.4.

For both DAX and Bund, for a given q, the points essentially lie on a straight line. Under stationarity assumptions, this implies that the log-volatility increments enjoy the following scaling property in expectation:

$$\mathbb{E}\left[\left|\log(\sigma_\Delta) - \log(\sigma_0)\right|^q\right] = b_q \Delta^{\zeta_q},$$

where $\zeta_q = q\, s_q > 0$ is the slope of the line associated to q. Moreover, the smoothness parameter s_q does not seem to depend on q. Indeed, plotting ζ_q against q, we obtain that $\zeta_q \sim H q$ with H equal to 0.125 for the DAX and to 0.082 for the Bund, see figure 29.5.

---

[*] For every day, we only consider the future contract corresponding to the most liquid maturity.
[†] Data kindly provided by QuantHouse EUROPE/ASIA, http://www.quanthouse.com.
[‡] And also the CAC40, Nikkei and FTSE indices in some specific parts of the paper.

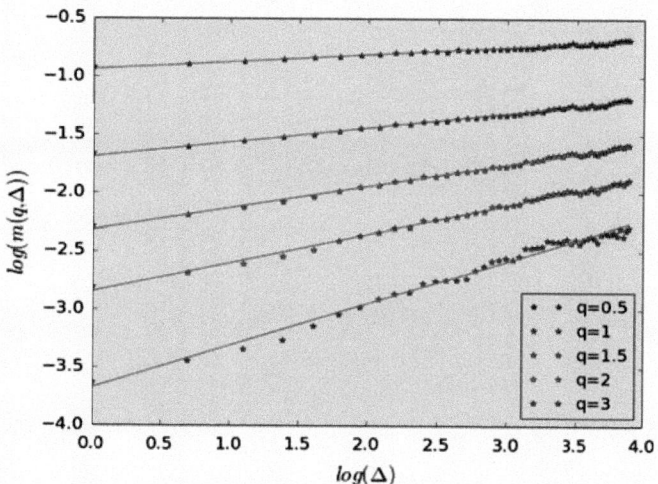

FIGURE 29.3   log $m(q, \Delta)$ as a function of log Δ, DAX.

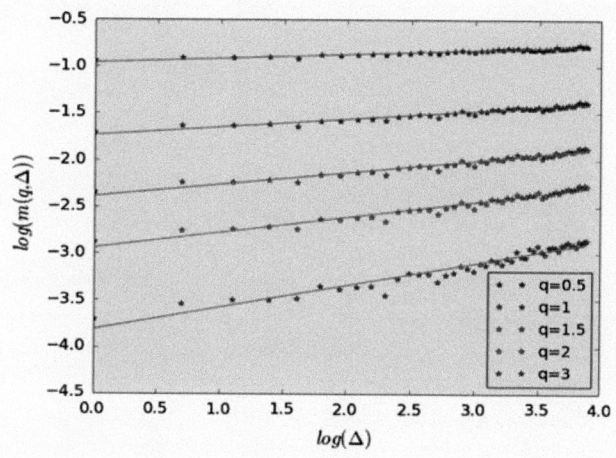

FIGURE 29.4   log $m(q, \Delta)$ as a function of log Δ, Bund.

We remark that the graphs for ζ_q are actually very slightly concave. However, we observe the same small concavity effect when we replace the log-volatility by simulations of a fBm with the same number of points. We conclude that this effect relates to finite sample size and is thus not significant.

29.2.3 S&P and NASDAQ Indices

We report in figures 29.6 and 29.7 similar results for the S&P and NASDAQ indices. The variance proxies used here are the precomputed 5-min realized variance estimates for the whole trading day made publicly available by the Oxford-Man Institute of Quantitative Finance.

We observe the same scaling property for the S&P and NASDAQ indices as we observed for DAX and Bund futures and again, the s_q do not depend on q. However, the estimated smoothnesses are slightly higher here: $H = 0.142$ for the S&P and $H = 0.139$ for the NASDAQ, see figure 29.8.

Once again, we do expect these smoothness estimates to be biased high because we are using whole-day realized variance estimates, as explained earlier in section 29.2. Finally, we remark that as for DAX and Bund futures, the graphs for ζ_q are slightly concave.

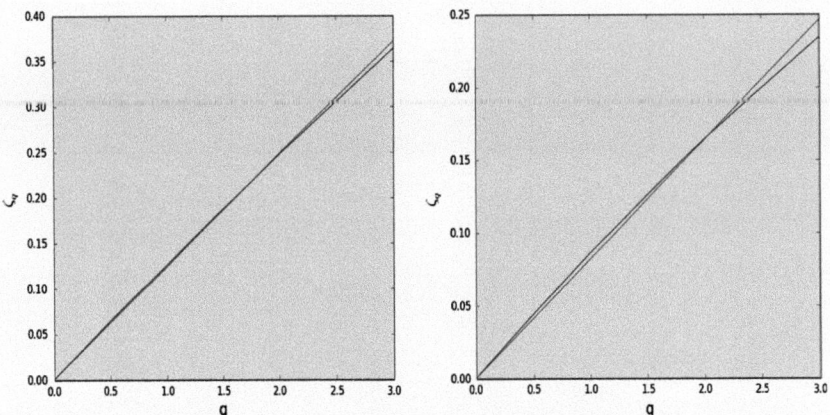

FIGURE 29.5 ζ_q (blue) and $0.125 \times q$ (green), DAX (left); ζ_q (blue) and $0.082 \times q$ (green), Bund (right).

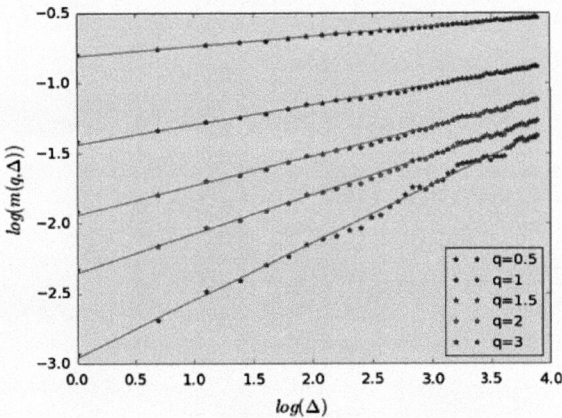

FIGURE 29.6 $\log m(q, \Delta)$ as a function of $\log \Delta$, S&P.

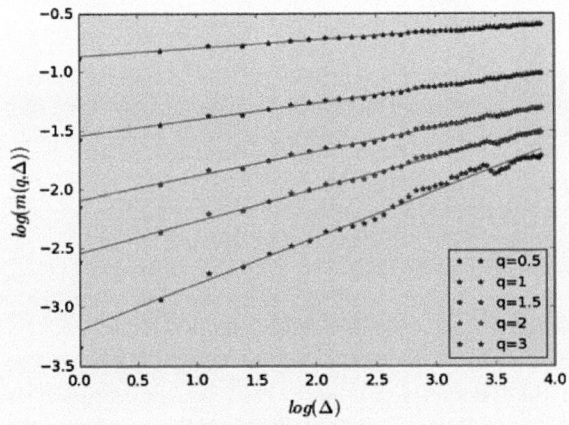

FIGURE 29.7 $\log m(q, \Delta)$ as a function of $\log(\Delta)$, NASDAQ.

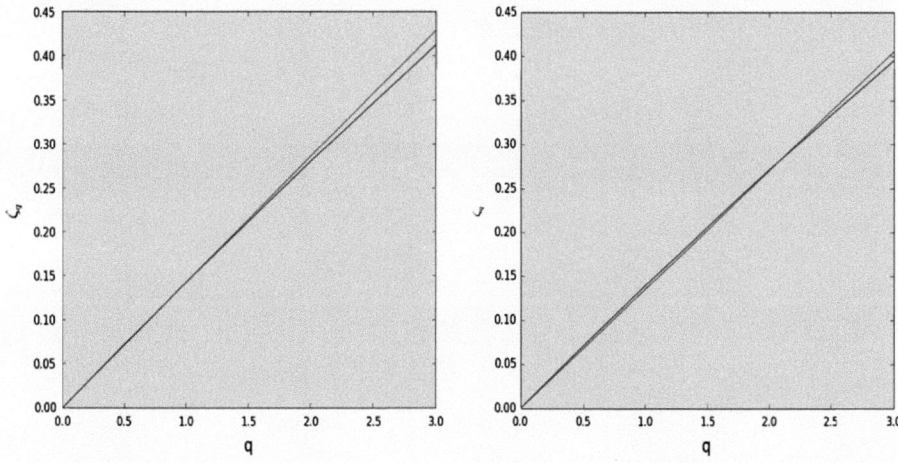

FIGURE 29.8 ζ_q (blue) and $0.142 \times q$ (green), S&P (left); ζq (blue) and $0.139 \times q$ (green), NASDAQ (right).

29.2.4 Other Indices

Repeating the analysis of section 29.2.3 for each index in the Oxford-Man data-set, we find the $m(q, \Delta)$ present a universal scaling behaviour. For each index and for $q = 0.5, 1, 1.5, 2, 3$, by doing a linear regression of $\log(m(q, \Delta))$ on $\log(\Delta)$ for $\Delta = 1, \dots, 30$, we obtain estimates of ζ_q that we summarize in table B1 in the appendix.

29.2.5 Distribution of the Increments of the Log-Volatility

Having established that all our underlying assets exhibit essentially the same scaling behavior,[*] we focus in the rest of the paper only on the S&P index, unless specified otherwise. That the distribution of increments of log-volatility is close to Gaussian is a well-established stylized fact reported for example in the papers Andersen *et al.* (2001a, 2001b). Looking now at the histograms of the increments of the log-volatility in figure 29.9, with the fitted normal density superimposed in red, we see that, for anyΔ, the empirical distribution of log-volatility increments is verified as being close to Gaussian. More impressive still is that rescaling the 1-day fit of the normal density by $^{\Delta H}$ generates (blue dashed) curves that are very close to the red fits of the normal density, consistent with the observed scaling.

The slight deviations from the Normal distribution observed in figure 29.9 are again consistent with the computation of the empirical distribution of the increments of a fractional Brownian motion on a similar number of points.

29.2.6 Does H Vary Over Time?

In order to check whether our estimations of H depend on the time interval, we split the Oxford-Man realized variance dataset into two halves and re-estimate H for each half separately. The results are presented in table B2 in the appendix. We note that although the estimated H all lie between 0.06 and 0.20, they seem to be higher in the second period which includes the financial crisis.

---

[*] We have also verified that this scaling relationship holds for Crude Oil and Gold futures, with similar smoothness estimates.

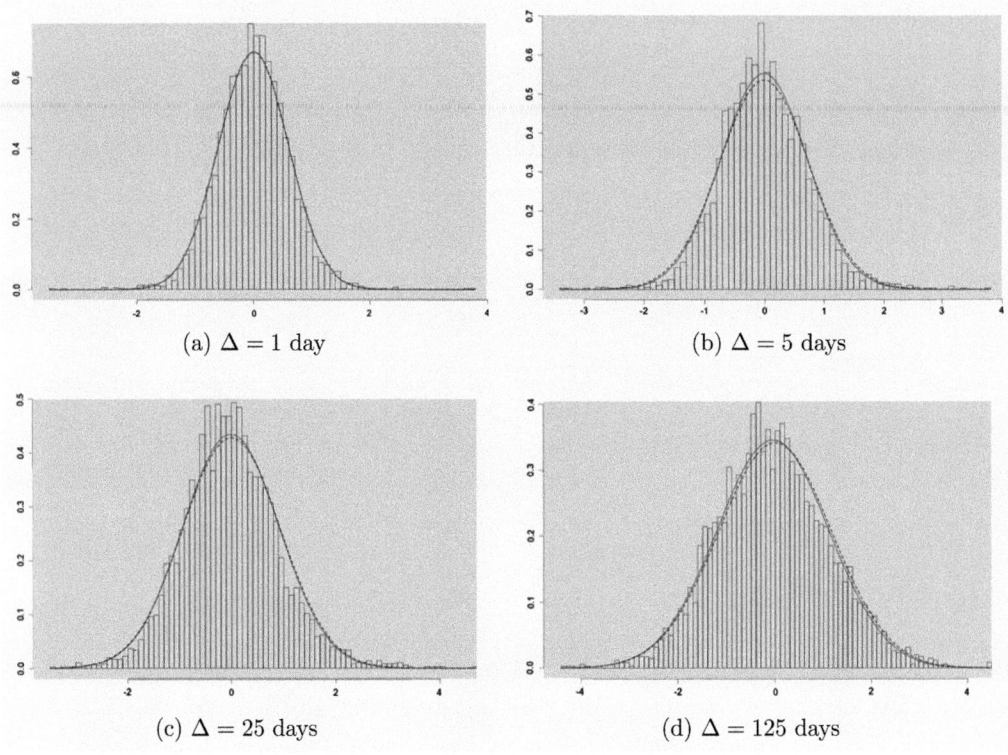

FIGURE 29.9 Histograms for various lags Δ of the (overlapping) increments $\log \sigma_{t+\Delta} - \log \sigma_t$ of the S&P log-volatility; normal fits in red; normal fit for $\Delta = 1$ day rescaled by $^{\Delta H}$ in blue.

29.3 A Simple Model Compatible with the Empirical Scaling of the Volatility

In this section, we specify the Rough FSV model and demonstrate that it reproduces the empirical facts presented in section 29.2.

29.3.1 Specification of the RFSV Model

In the previous section, we showed that, empirically, the increments of the log-volatility of various assets enjoy a scaling property with constant smoothness parameter and that their distribution is close to Gaussian. This naturally suggests the simple model:

$$\log \sigma_{t+\Delta} - \log \sigma_t = \nu \left(W_{t+\Delta}^H - W_t^H \right), \tag{29.3.1}$$

where W^H is a fractional Brownian motion with Hurst parameter equal to the measured smoothness of the volatility and ν is a positive constant. We may of course write (29.3.1) under the form

$$\sigma_t = \sigma \exp\left(\nu W_t^H \right), \tag{29.3.2}$$

where σ is another positive constant.

However, this model is not stationary, stationarity being desirable both for mathematical tractability and also to ensure reasonableness of the model at very large times. This leads us to impose stationarity by modelling the log-volatility as a fractional Ornstein–Uhlenbeck process with a very long reversion timescale.

A stationary fractional Ornstein–Uhlenbeck process (X_t) is defined as the stationary solution of the stochastic differential equation

$$dX_t = v dW_t^H - \alpha (X_t - m) dt,$$

where $m \in \mathbb{R}$ and v and α are positive parameters (see Cheridito *et al.* 2003). As for usual Ornstein–Uhlenbeck processes, there is an explicit form for the solution which is given by

$$X_t = v \int_{-\infty}^{t} e^{-\alpha(t-s)} dW_t^H + m. \tag{29.3.3}$$

Here, the stochastic integral with respect to fBm is simply a pathwise Riemann–Stieltjes integral, see again Cheridito *et al.* (2003).

We thus arrive at the final specification of our Rough Fractional Stochastic Volatility (RFSV) model for the volatility on the time interval of interest $[0, T]$:

$$\sigma_t = \exp(X_t), t \in [0, T], \tag{29.3.4}$$

where (X_t) satisfies equation (29.3.3) for some $v > 0$, $\alpha > 0$, $m \in \mathbb{R}$ and $H < 1/2$ the measured smoothness of the volatility. This model provides a very parsimonious description of the volatility process with only four parameters (and in practice only three, see below). Moreover, statistical inference methods for fractional Brownian motion and fractional Ornstein– Uhlenbeck process are well known (see Istas and Lang 1997, Coeurjolly 2001, Brouste and Iacus 2013).

Such a model is indeed stationary. However, if $\alpha \ll 1/T$, the log-volatility behaves locally (at timescales smaller than T) as a fBm. This observation is formalized in the following proposition.

Proposition 29.3.1 *Let W^H be a fBm and X^α defined by (29.3.3) for a given $\alpha > 0$. As α tends to zero,*

$$\mathbb{E} \left[\sup_{t \in [0,T]} \left| X_t^\alpha - X_0^\alpha - v W_t^H \right| \right] \to 0.$$

The proof is given in appendix A.1.

Proposition 29.3.1 implies that in the RFSV model, if $\alpha \ll 1/T$, and we confine ourselves to the interval $[0, T]$, we can proceed as if the log-volatility process were a fBm. Indeed, simply setting $\alpha = 0$ in (29.3.3) gives (at least formally) $X_t - X_s = v \left(W_t^H - W_s^H \right)$ and we immediately recover our simple non-stationary fBm model (29.3.1). Consequently, although the RFSV model is technically stationary, its ergodic behaviour is of no interest for us; for example, estimation of the mean of the volatility is not possible in practice. Indeed, at any timescale of practical interest (from one day to several years), we see no evidence of ergodicity in the data, see figure 29.13.

The following corollary implies that the (exact) scaling property of the fBm is approximately reproduced by the fractional Ornstein–Uhlenbeck process when α is small.

COROLLARY 29.3.1 *Let $q > 0$, $t > 0$, $\Delta > 0$. As α tends to zero, we have*

$$\mathbb{E} \left[\left| X_{t+\Delta}^\alpha - X_t^\alpha \right|^q \right] \to v^q K_q \Delta^{qH}.$$

The proof is given in appendix A.2.

29.3.1.1 RFSV vs. FSV

We recognize our RFSV model (29.3.4) as a particular case of the classical FSV model of Comte and Renault (1998). The key difference is that here we take $H < 1/2$ and $\alpha \ll 1/T$, whereas to accommodate the assumption of long memory, Comte and Renault have to choose $H > 1/2$. The analysis of Fukasawa referred to earlier in section 29.1.3 implies in particular that if $H > 1/2$, the volatility skew function $\psi(\tau)$ is *increasing* in time to expiration τ (at least for small τ), which is obviously completely inconsistent with the approximately $1/\sqrt{\tau}$ skew term structure that is observed. To generate a decreasing term structure of volatility skew for longer expirations, Comte and Renault are then forced to choose $\alpha \gg 1/T$. Consequently, for very short expirations ($\tau \ll 1/\alpha$), models of the Comte and Renault type with $H > 1/2$ still generate a term structure of volatility skew that is inconsistent with the observed one, as explained for example in section 29.4 of Comte *et al.* (2012).

In contrast, the choice $H < 1/2$ enables us to reproduce the observed smoothness and scaling of the volatility process and generate a term structure of volatility skew in agreement with the observed one. The choice $H < 1/2$ is also consistent with what is improperly called mean reversion by practitioners, which in fact corresponds to strong oscillations in the volatility process. Finally, taking α very small implies that the dynamics of our process is close to that of a fBm, see proposition 29.3.1. This last point is particularly important. Indeed, recall that at the timescales we are interested in, the important feature we have in mind is really this fBm like-behaviour of the log-volatility.

Finally, note that we could no doubt have considered other stationary models satisfying proposition 29.3.1 and corollary 29.3.1, where log-volatility behaves as a fBm at reasonable timescales; the choice of the fractional Ornstein–Uhlenbeck process is probably the simplest way to accommodate this local behavior together with the stationarity property.

29.3.2 RFSV Model Autocovariance Functions

From proposition 29.3.1 and corollary 29.3.1, we easily deduce the following corollary, where $o(1)$ tends to zero as α tends to zero.

COROLLARY 29.3.2 *Let $q > 0$, $t > 0$, $\Delta > 0$. As α tends to zero,*

$$\mathrm{Cov}\left[X_t^\alpha, X_{t+\Delta}^\alpha\right] = \mathrm{Var}\left[X_t^\alpha\right] - \frac{1}{2}v^2\Delta^{2H} + o(1).$$

Consequently, in the RFSV model, for fixed t, the covariance between X_t and $X_{t+\Delta}$ is linear with respect to Δ^{2H}. This result is very well satisfied empirically. For example, in figure 29.10, we see that for the S&P, the empirical autocovariance function of the log-volatility is indeed linear with respect to Δ^{2H}. Note in passing that at the timescales we consider, the term $\mathrm{Var}\left[X_t^\alpha\right]$ is higher than $\frac{1}{2}v^2\Delta^{2H}$ in the expression for $\mathrm{Cov}\left[X_t^\alpha, X_{t+\Delta}^\alpha\right]$.

Having computed the autocovariance function of the log-volatility, we now turn our attention to the volatility itself. We have

$$\mathbb{E}\left[\sigma_{t+\Delta}\sigma_t\right] = \mathbb{E}\left[e^{X_t^\alpha + X_{t+\Delta}^\alpha}\right],$$

with X^α defined by equation (29.3.3). Since X^α is a Gaussian process, we deduce that

$$\mathbb{E}\left[\sigma_{t+\Delta}\sigma_t\right]$$
$$= e^{\mathbb{E}\left[X_t^\alpha\right] + \mathbb{E}\left[X_{t+\Delta}^\alpha\right] + \mathrm{Var}\left[X_t^\alpha\right]/2 + \mathrm{Var}\left[X_{t+\Delta}^\alpha\right]/2 + \mathrm{Cov}\left[X_t^\alpha, X_{t+\Delta}^\alpha\right]}.$$

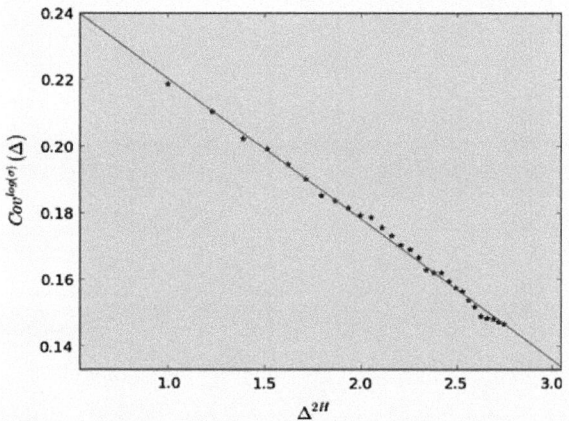

FIGURE 29.10 Autocovariance of the log-volatility as a function of Δ^{2H} for $H = 0.14$, S&P.

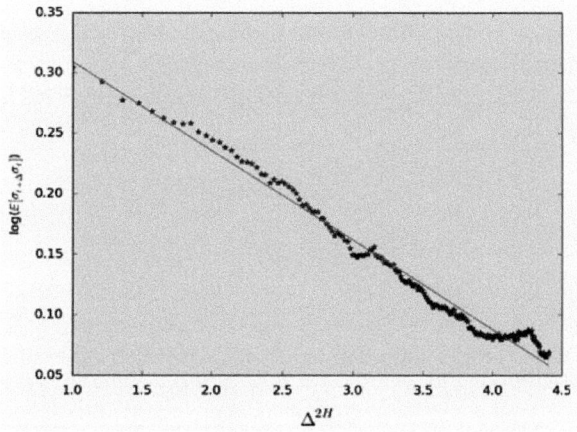

FIGURE 29.11 Empirical counterpart of $\log\left(\mathbb{E}\left[\sigma_{t+\Delta}\sigma_t\right]\right)$ as a function of Δ^{2H}, S&P.

Applying corollary 29.3.2, we obtain that when α is small, $\mathbb{E}\left[\sigma_{t+\Delta}\sigma_t\right]$ is approximately equal to

$$e^{2\mathbb{E}\left[X_t^\alpha\right]+2\operatorname{Var}\left[X_t^\alpha\right]}e^{-v^2\frac{\Delta^{2H}}{2}}. \tag{29.3.5}$$

It follows that in the RFSV model, $\log\left(\mathbb{E}\left[\sigma_{t+\Delta}\sigma_t\right]\right)$ is also linear in Δ^{2H}. This property is again very well satisfied on data, as shown by figure 29.11, where we plot the logarithm of the empirical counterpart of $\mathbb{E}\left[\sigma_{t+\Delta}\sigma_t\right]$ against Δ^{2H}, for the S&P with $H = 0.14$.

We note that putting Δ^{2H} on the x-axis of figure 29.11 is really crucial in order to retrieve linearity. In particular, a corollary of (29.3.5) is that the autocovariance function of the volatility does not decay as a power law as widely believed; see figure 29.12 where we show that a log–log plot of the autocovariance function does not yield a straight line.

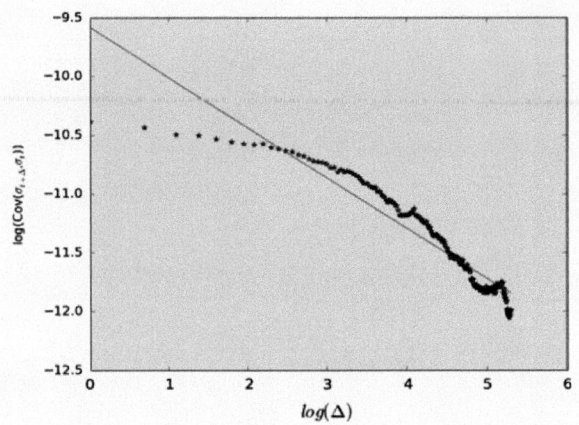

FIGURE 29.12   Empirical counterpart of $\log\left(\mathrm{Cov}\left[\sigma_{t+\Delta},\sigma_t\right]\right)$ as a function of log (Δ), S&P.

29.3.3 RFSV vs. FSV Again

To further demonstrate the incompatibility of the classical long memory FSV model with volatility data, consider the quantity $m(2, \Delta)$. Recall that in the data (see section 29.2) we observe the linear relationship $\log m(2, \Delta) \approx \zeta_2 \log \Delta + k$ for some constant k. Also, in both FSV and RFSV, we can consider

$$m(2,\Delta) = \mathbb{E}\left[\left(X_{t+\Delta} - X_t\right)^2\right]$$

$$= 2\left(\mathrm{Var}\left[X_t\right] - \mathrm{Cov}\left[X_t, X_{t+\Delta}\right]\right).$$

In figure 29.13, we plot $m(2, \Delta)$ with the parameters $H = 0.53$, corresponding to the FSV model parameter estimate of Chronopoulou and Viens (2012), and $\alpha = 0.5$ to ensure some visible decay of the volatility skew. The slope of $m(2, \Delta)$ in the FSV model for small lags is driven by the value of H; the lag at which $m(2, \Delta)$ begins to flatten and stationarity kicks in corresponds to a timescale of order $1/\alpha$. It is clear from the picture that to fit the data, we must have $\alpha \ll 1/T$ and the value of H must be set by the initial slope of the regression line, which as reported earlier in section 29.2 is $\zeta_2 = 2 \times 0.14$.

29.3.4 Simulation-Based Analysis of the RFSV Model

Our goal in this section is to show that in terms of smoothness measures, one obtains on simulated data from the RFSV model the same behaviours as those observed on empirical data. In particular, we would like to be able to quantify the positive bias associated with estimating H from whole-day realized variance data as in section 29.2.3, relative to using data from a one-hour window as in section 29.2.2.

   We simulate the RFSV model for 2000 days (chosen to be between the lengths of our two data-sets). In order to account for the overnight effect, we simulate the volatility $\sigma_t{}^*$ and efficient price $P_t{}^\dagger$ over the whole day. The parameters: $H = 0.14$, $\nu = 0.3$, $m = X_0 = -5$ and $\alpha = 5 \times 10^{-4}$ are chosen to be consistent with our empirical estimates from section 29.2. To model microstructure effects such as the discreteness of the price grid, we consider that the observed price process is generated from P_t using the uncertainty zones model of Robert and Rosenbaum (2011) with tick value 5×10^{-4} and parameter $\eta = 0.25$.

---

* To simulate the fBm, we use a spectral method with 40 000 000 points (20 000 points per day). We then simulate X taking
$$X_{(n+1)\cdot} - X_{n\cdot} = \nu\left(W_{(n+1)\cdot}^H - W_{n\delta}^H\right) + \alpha\delta\left(m - X_{n\delta}\right)\left(\text{with } \delta = 1/20000\right).$$
† $P_{(n+1)\delta} - P_{n\delta} = P_{n\delta}\sigma_{n\delta}\sqrt{\delta}U_n$ where the U_n are iid standard Gaussian variables.

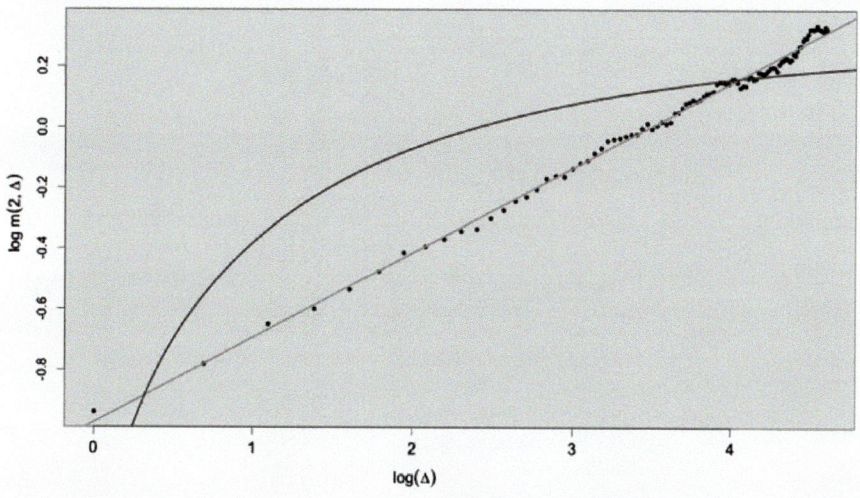

FIGURE 29.13 Long memory models such as the FSV model of Comte and Renault are not compatible with S&P volatility data. Black points are empirical estimates of $m(2, \Delta)$; the blue line is the FSV model with $\alpha = 0.5$ and $H = 0.53$; the orange line is the RFSV model with $\alpha = 0$ and $H = 0.14$.

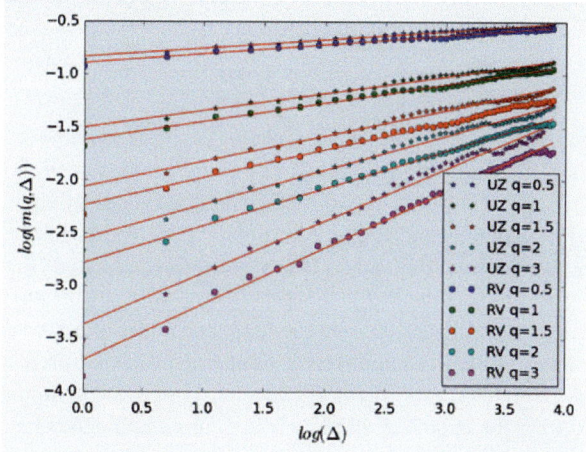

FIGURE 29.14 $\log(m(q, \Delta))$ as a function of $\log(\Delta)$, simulated data, with realized variance and uncertainty zones estimators.

Exactly as in section 29.2, for each of the 2000 days, we consider two volatility proxies obtained from the observed price and based on:

- The integrated variance estimator using the model with uncertainty zones over one-hour windows, from 10 am to 11 am.
- The 5-min realized variance estimator, over eight-hour windows (the trading day).

We now repeat our analysis of section 29.2, generating graphs analogous to figures 29.3, 29.4, 29.6 and 29.7 obtained on empirical data. Figure 29.14 compares smoothness measures obtained using the uncertainty zones estimator on one-hour windows with those obtained using the realized variance estimator on eight-hour windows.

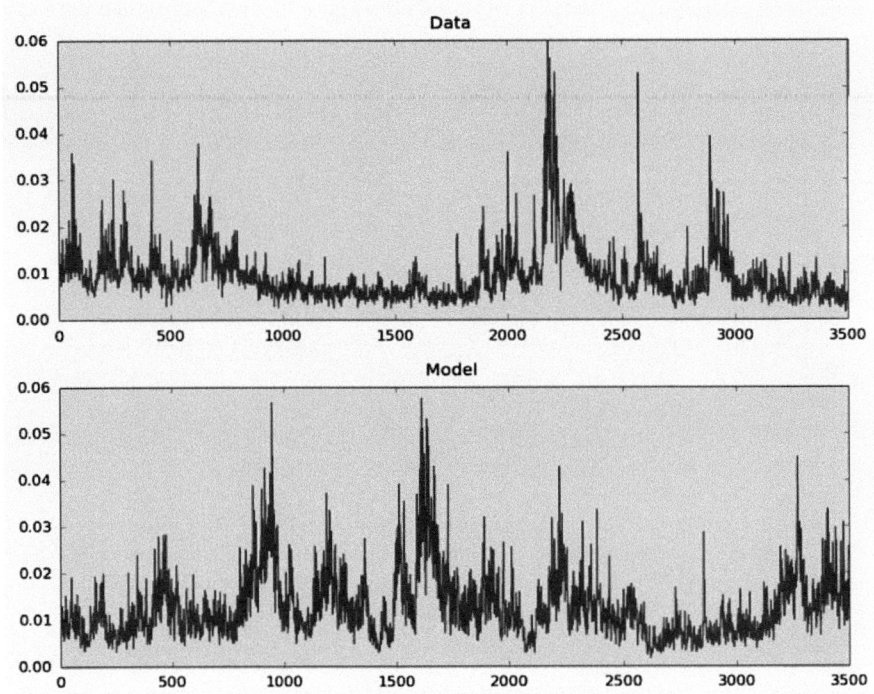

FIGURE 29.15   Volatility of the S&P (above) and generated by the model (below).

When the uncertainty zones estimator is applied on a one-hour window (1/24 of a simulated day) as in section 29.2.2, we estimate $H=0.16$, which is close to the true value $H=0.14$ used in the simulation. The results obtained with the realized variance estimator over daily eight-hour windows (1/3 of a simulated day) do exhibit the same scaling properties as those we see in the empirical data with a smoothness parameter that does not depend on q. However, the estimated H is biased slightly higher at around 0.18. As discussed in section 29.2.1, this extra positive bias is no surprise and is due to the regularizing effect of the integral operator over the longer window. We note also that the estimated values of ν ('volatility of volatility' in some sense), obtained from the intercepts of the regressions, are lower with the longer time windows, again as expected. A detailed computation of the bias in the estimated H associated with the choice of window length in an analogous but more tractable model is presented in appendix 3.

We end this section by presenting in figure 29.15 a sample path of the model-generated volatility (spot volatility, directly from the simulation rather than estimated from the simulated price series) together with a graph of S&P volatility over 3500 days.

A first reaction to figure 29.15 is that the simulated and actual graphs look very alike. In particular, in both of them, persistent periods of high volatility alternate with low volatility periods. On closer inspection of the empirical volatility series, we observe that the sample path of the volatility on a restricted time window seems to exhibit the same kind of qualitative properties as those of the global sample path (for example periods of high and low activity). This fractal-type behaviour of the volatility has been investigated both empirically and theoretically in, for example, Bacry and Muzy (2003), Bouchaud and Potters (2003), Mantegna and Stanley (2000).

At the visual level, we observe that this fractal-type behaviour is also reproduced in our model, as we now explain. Denote by $L^{x,H}$ the law of the geometric fractional Brownian motion with Hurst exponent H and volatility x on $[0, 1]$, that is $\left(e^{xW_t^H}\right)_{t\in[0,1]}$. Then, when α is very small, the rescaled volatility process on [0, Δ]: $(\sigma_t \Delta /\sigma_0)_{t \in [0,1]}$, has approximately the law $L^{v\Delta^H, H}$. Now remark that for H small, the function u^H increases very slowly. Thus, over a large range of observation scales Δ, the rescaled volatility processes on [0, Δ] have approximately the same law. For example, between an observation scale of one day and five years (1250 open days), the coefficient x characterizing the law of the volatility process is 'only' multiplied by $1250^{0.14} = 2.7$. It follows that in the RFSV model, the volatility process over one day resembles the volatility process over a decade.

29.4 Spurious Long Memory of Volatility?

We revisit in this section the issue of long memory of volatility through the lens of our model. Recall that a stationary time series is said to exhibit long memory if the autocovariance function Cov [log(σ_t), log($\sigma_{t+\Delta}$)] (or sometimes Cov [σ_t, $\sigma_{t+\Delta}$]) goes slowly to zero as $\Delta \to \infty$, and often even more precisely that it behaves as $\Delta^{-\gamma}$,[*] with $\gamma < 1$ as $\Delta \to \infty$.

Thus, the classical approach to long memory is to consider a parametric class of models and to estimate within this class the parameter γ, typically based on empirical autocovariances, see Andersen *et al.* (2003) and figure 29.12. As mentioned earlier in the introduction, the long memory of volatility is widely accepted as a stylized fact.

Specifically, in the RFSV model, we have from corollary 29.3.2 that

$$\text{Cov}\Big[\log\big(\sigma_t\big), \log\big(\sigma_{t+\Delta}\big)\Big] \approx A - B\Delta^{2H}$$

and from equation (29.3.5) that

$$\text{Cov}\Big[\sigma_t, \sigma_{t+\Delta}\Big] \approx Ce^{-B\Delta^{2H}} - D,$$

for some constants A, B, C and D. Moreover, we demonstrated in figures 29.10 and 29.11 that these relations are consistent with the data. Thus, the autocovariance function does not decay as a power law in the RFSV model nor does it appear to decay as a power law in the data.

Nevertheless, as an experiment, we can apply both to the data and to sample paths of the RFSV model some standard statistical procedures aimed at identifying long memory that have been used in the financial econometrics literature. Such procedures are of course designed to identify long memory under rather strict modelling assumptions. Consequently, spurious results may obviously then be obtained if the model underlying the estimation procedure is misspecified, which is the case with the RFSV model.[†]

With the same model parameters as in section 29.3.4, we simulate our model over 3500 days, which corresponds to the size of our data-set. Consider first the procedure in Andersen *et al.* (2001b), where in the context of a fractional Gaussian noise (FGN) model with Hurst parameter $\check{H}$, the authors test for long memory in the volatility by studying the scaling behaviour of the quantity

$$V(\Delta) = \text{Var}\left[\int_0^\Delta \sigma_s^2 ds\right]$$

with respect to Δ. In the FGN model, as $\Delta \to \infty$, the autocorrelation function $\rho(\Delta)$ behaves asymptotically as $\Delta^{2\check{H}-2}$ and $V(\Delta)$ behaves asymptotically as $\Delta^{2\check{H}}$ as $\Delta \to \infty$. Figure 29.16 presents the graph of the

---

[*] Indeed, the notion of empirical long memory does not make much sense outside the power law case; the empirical values of covariances at very large timescales are never measurable and thus one cannot conclude whether the series of covariances converges in general.

[†] Recall in particular that the RFSV model is only formally stationary.

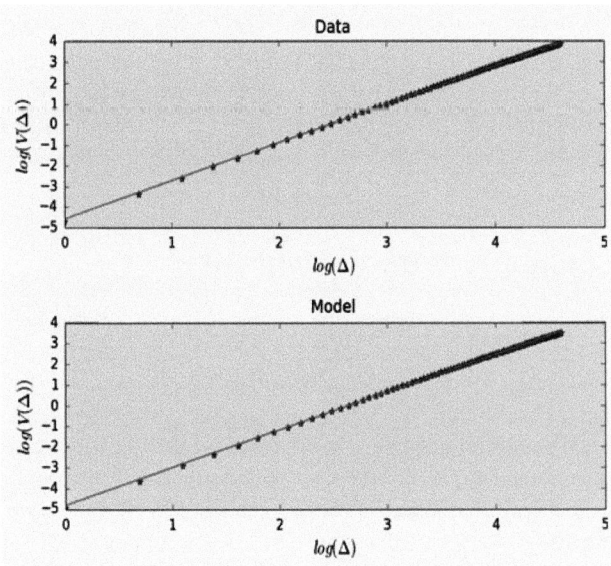

FIGURE 29.16   Empirical counterpart of log($V(\Delta)$) as a function of log(Δ) on S&P (above) and simulation (below).

logarithm of the empirical counterpart of $V(\Delta)$ against the logarithm of Δ, on the S&P data and within our simulation framework.

We note from figure 29.16 that both our simulated model and market data lead to very similar graphs, close to straight lines with slope 1.86, giving $\hat{H} = 0.93$.[*] Accordingly, in the setting of Andersen *et al.* (2001b), we would deduce power law behaviour of the autocorrelation function with exponent 0.14 and therefore long memory. Thus, if the data are generated by a model like the RFSV model, one can easily be wrongly convinced that the volatility time series exhibits long memory.

In Andersen *et al.* (2003), in the context of an ARFIMA (0, d, 0) model, the authors deduce long memory in the volatility by showing that the process ε_t obtained by fractional differentiation of the log-volatility $\varepsilon_t = (1 - L)^d \log(\sigma_t)$, with $d = 0.401$[†] (which is obtained by regression of the log-periodogram using the GPH estimator Geweke and Porter-Hudak (1983)) and L the lag operator behaves as a white noise. To check for this, they compute the autocorrelation function of ε_t. We give in figure 29.17 the autocorrelation functions of the logarithm of σ_t and ε_t, again both on the data and on the simulated path.

Once again, the data and the simulation generate very similar plots. We conclude that this procedure for estimating long memory is just as fragile as the first, and it is easy to wrongly deduce volatility long memory when applying it.

In conclusion, the RFSV model is yet another model in which classical estimation procedures identify spurious long memory; see Andersen and Bollerslev (1997), Diebold and Inoue (2001), Granger and Hyung (2004), Granger and Teräsvirta (1999) for various other such examples. Moreover, these procedures estimate the same long memory parameter from data generated from a suitably calibrated RFSV model as they estimate from empirical data. Once again, although the (near) non-stationarity of the RFSV model induces long swings in volatility, mirroring long-range dependence, it does not exhibit long memory in the classical power law sense.

---

[*] Note that there is no reason to expect that there should be any direct connection between $\hat{H}$ estimated for the FGN model and the H we estimated for the RFSV model.

[†] It is shown in Geweke and Porter-Hudak (1983) that the autocorrelation functions of the ARFIMA(0, d, 0) and the FGN model with Hurst parameter $\hat{H}$ have the same asymptotic behaviour as $\Delta \rightarrow \infty$ if $d = \hat{H} - \dfrac{1}{2}$.

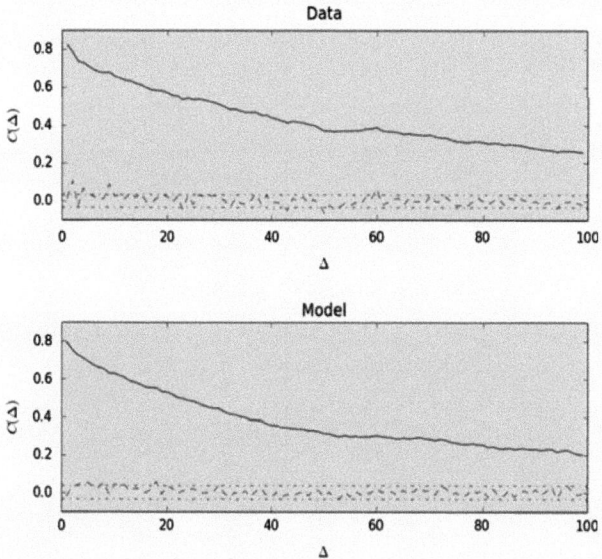

FIGURE 29.17 Autocorrelation functions of $\log(\sigma_t)$ (in blue) and ε_t (in green) and the Bartlett standard error bands (in red), for S&P data (above) and for simulated data (below).

29.5 Forecasting Using the RFSV Model

In this section, we present an application of our model: fore-casting the log-volatility and the variance.

29.5.1 Forecasting Log-Volatility

The key formula on which our prediction method is based is the following one:

$$
\mathbb{E}\left[W_{t+\Delta}^H \middle| \mathcal{F}_t\right]
$$
$$
= \frac{\cos(H\pi)}{\pi} \Delta^{H+1/2} \int_{-\infty}^{t} \frac{W_s^H}{(t-s+\Delta)(t-s)^{H+1/2}} ds,
$$

where W^H is a fBm with $H < 1/2$ and $\mathcal{F}_t$ the filtration it generates, see Theorem 29.4.2 of Nuzman and Poor (2000). By construction, over any reasonable timescale of interest, as formalized in corollary 29.3.1, we may approximate the fractional Ornstein–Uhlenbeck volatility process in the RFSV model as log $\log \sigma_t^2 \approx 2\nu W_t^H + C$, for some constants ν and C. Our prediction formula for log-variance then follows:[*]

$$
\mathbb{E}\left[\log \sigma_{t+\Delta}^2 \middle| \mathcal{F}_t\right]
$$
$$
= \frac{\cos(H\pi)}{\pi} \Delta^{H+1/2} \int_{-\infty}^{t} \frac{\log \sigma_s^2}{(t-s+\Delta)(t-s)^{H+1/2}} ds. \tag{29.5.1}
$$

This formula, or rather its approximation through a Riemann sum (we assume in this section that the volatilities are perfectly observed, although they are in fact estimated), is used to forecast the log-volatility 1, 5 and 20 days ahead ($\Delta = 1, 5, 20$).

---

[*] The constants 2ν and C cancel when deriving the expression.

We now compare the predictive power of Formula (29.5.1) with that of autoregressive (AR for short) and heterogeneous autoregressive (HAR for short) forecasts, in the spirit of Corsi (2009).* Recall that for a given integer $p > 0$, the AR(p) and HAR predictors take the following form (where the index i runs over the series of daily volatility estimates):

- AR(p):

$$\widehat{\log\left(\sigma_{t+\Delta}^2\right)} = K_0^\Delta + \sum_{i=0}^{p} C_i^\Delta \log\left(\sigma_{t-i}^2\right).$$

- HAR:

$$\widehat{\log\left(\sigma_{t+\Delta}^2\right)} = K_0^\Delta + C_0^\Delta \log\left(\sigma_t^2\right)$$

$$+ C_5^\Delta \frac{1}{5} \sum_{i=0}^{4} \log\left(\sigma_{t-i}^2\right)$$

$$+ C_{20}^\Delta \frac{1}{20} \sum_{i=0}^{19} \log\left(\sigma_{t-i}^2\right).$$

We estimate AR coefficients using the R stats library[†] on a rolling time window of 500 days. In the HAR case, we use standard linear regression to estimate the coefficients as explained in Corsi (2009). In the sequel, we consider $p = 5$ and $p = 10$ in the AR formula. Indeed, these parameters essentially give the best results for the horizons at which we wish to forecast the volatility (1, 5 and 20 days). For each day, we forecast volatility for five different indices.[‡]

We then assess the quality of the various forecasts by computing the ratio P between the mean squared error of our predictor and the (approximate) variance of the log-variance:

$$P = \frac{\sum_{k=500}^{N-\Delta} \left(\log\left(\sigma_{k+\Delta}^2\right) - \widehat{\log\left(\sigma_{k+\Delta}^2\right)}\right)^2}{\sum_{k=500}^{N-\Delta} \left(\log\left(\sigma_{k+\Delta}^2\right) - \mathbb{E}\left[\log\left(\sigma_{t+\Delta}^2\right)\right]\right)^2},$$

where $\mathbb{E}\left[\log\left(\sigma_{t+\Delta}^2\right)\right]$ denotes the empirical mean of the log-variance over the whole time period.

We note from table 29.1 that the RFSV forecast consistently outperforms the AR and HAR forecasts, especially at longer horizons. Moreover, our forecasting method is more parsimonious since it only requires the parameter H to forecast the log-variance. Compare this with the AR and HAR methods, for which coefficients depend on the forecast time horizon and must be recomputed if this horizon changes.

---

* Note that we do not consider GARCH models here since we have access to high frequency volatility estimates and not only to daily returns. Indeed, it is shown in Andersen *et al.* (2003) that forecasts based on the time series of realized variances outperform GARCH forecasts based on daily returns.

[†] More precisely, we use the default Yule-Walker method.

[‡] In addition to S&P and NASDAQ, we also investigate CAC40, FTSE and Nikkei, over the same time period as S&P and NASDAQ. For simplicity, the parameter H used in our predictor is computed only once for each asset, using the whole time period. This yields similar results to using a moving time window adapted in time.

TABLE 29.1 Ratio P for the AR, HAR and RFSV Predictors

	AR(5)		AR(10)	HAR(3)	RFSV
SPX2.rv $\Delta = 1$	0.317		0.318	0.314	**0.313**
SPX2.rv $\Delta = 5$	0.459		0.449	0.437	**0.426**
SPX2.rv $\Delta = 20$	0.764		0.694	0.656	**0.606**
FTSE2.rv $\Delta = 1$	0.230		0.229	0.225	**0.223**
FTSE2.rv $\Delta = 5$	0.357		0.344	0.337	**0.320**
FTSE2.rv $\Delta = 20$	0.651		0.571	0.541	**0.472**
N2252.rv $\Delta = 1$	0.357		0.358	0.351	**0.345**
N2252.rv $\Delta = 5$	0.553		0.533	0.513	**0.504**
N2252.rv $\Delta = 20$	0.875		0.795	0.746	**0.714**
GDAXI2.rv $\Delta = 1$	0.237		0.238	0.234	**0.231**
GDAXI2.rv $\Delta = 5$	0.372		0.362	0.350	**0.339**
GDAXI2.rv $\Delta = 20$	0.661		0.590	0.550	**0.498**
FCHI2.rv $\Delta = 1$	0.244		0.244	0.241	**0.238**
FCHI2.rv $\Delta = 5$	0.378		0.373	0.366	**0.350**
FCHI2.rv $\Delta = 20$	0.669		0.613	0.598	**0.522**

Note: Best predictions are in boldface.

Remark that our predictor can be linked to that of Duchon *et al.* (2012), where the issue of the prediction of the log-volatility in the multifractal random walk model of Bacry and Muzy (2003) is tackled. In this model,

$$\mathbb{E}\left[\log\left(\sigma_{t+\Delta}^2 \right) | \mathcal{F}_t \right] = \frac{1}{\pi} \sqrt{\Delta} \int_{-\infty}^{t} \frac{\log\left(\sigma_s^2 \right)}{(t-s+\Delta)\sqrt{t-s}} \, ds,$$

which is the limit of our predictor when H tends to zero.

Note also that our prediction formula may be rewritten as

$$\mathbb{E}\left[\log\left(\sigma_{t+\Delta}^2 \right) | \mathcal{F}_t \right] = \frac{\cos(H\pi)}{\pi} \int_{0}^{+\infty} \frac{\log\left(\sigma_{t-\Delta u}^2 \right)}{(u+1)u^{H+1/2}} \, du.$$

For a given small $\varepsilon > 0$, let r be the smallest real number such that

$$\int_{r}^{+\infty} \frac{1}{(u+1)u^{H+1/2}} \, du \leq \varepsilon.$$

Then, we have, with an error of order ε,

$$\mathbb{E}\left[\log\left(\sigma_{t+\Delta}^2 \right) | \mathcal{F}_t \right] \approx \frac{\cos(H\pi)}{\pi} \int_{0}^{r} \frac{\log\left(\sigma_{t-\Delta u}^2 \right)}{(u+1)u^{H+1/2}} \, du.$$

Consequently, the volatility process needs to be considered (roughly) down to time $t - \Delta r$ if one wants to forecast up to time Δ in the future. The relevant regression window is thus linear in the forecasting horizon. For example, for $r = 1$, $\varepsilon = 0.35$ which is not so unreasonable. In this case, as is well known to practitioners, to predict log-volatility one week ahead, one should essentially look at the volatility over the last week. If trying to predict log-volatility one month ahead, one should look at the volatility over the last month.

29.5.2 Variance Prediction

Recall that $\sigma_t^2 \approx 2vW_t^H + C$ for some constant C. In Nuz-man and Poor (2000), it is shown that $W_{t+\Delta}^H$ is conditionally Gaussian with conditional variance

$$\mathrm{Var}\left[W_{t+\Delta}^H | \mathcal{F}_t\right] = c\Delta^{2H}$$

where

$$c = \frac{\Gamma(3/2-H)}{\Gamma(H+1/2)\Gamma(2-2H)}.$$

Thus, we obtain the following form for the RFSV predictor of the variance:

$$\widehat{\sigma_{t+\Delta}^2} = \exp\left(\widehat{\log\sigma_{t+\Delta}^2} + 2cv^2\Delta^{2H}\right) \qquad (29.5.2)$$

where $\widehat{\log\left(\sigma_{t+\Delta}^2\right)}$ is the predictor from section 29.5.1 and v^2 is estimated as the exponential of the intercept in the linear regression of $\log(m(2, \Delta))$ on $\log(\Delta)$.

We compare in table 29.2 the performances of the RFSV fore-cast with those of AR and HAR forecasts (constructed on variance rather than log-variance this time).

We find again that the RFSV forecast typically outperforms AR and HAR, although it is worth noting that the HAR forecast is already visibly superior to the AR forecast.

Since our working paper first appeared, much work has been done to estimate the roughness of volatility of other assets, notably by Bennedsen, Lunde and Pakkanen in Bennedsen *et Hal.* (2016). In an analysis of E-mini S&P 500 futures data at all timescales over 15 min, the out-of-sample forecasting performance of the estimator (29.5.2) is shown to be very similar to the performance of the other more highly parameterized estimators proposed. Interestingly, at daily and higher timescales, the simple estimator (29.5.2) is actually shown to outperform the more complicated estimators. It is also notable that

TABLE 29.2   Ratio P for the AR, HAR and RFSV Predictors

	AR(5)		AR(10)	HAR(3)	RFSV
SPX2.rv $\Delta = 1$	0.520		0.566	0.489	**0.475**
SPX2.rv $\Delta = 5$	0.750		0.745	0.723	**0.672**
SPX2.rv $\Delta = 20$	1.070		1.010	1.036	**0.903**
FTSE2.rv $\Delta = 1$	0.612		0.621	0.582	**0.567**
FTSE2.rv $\Delta = 5$	0.797		0.770	0.756	**0.707**
FTSE2.rv $\Delta = 20$	1.046		0.984	0.935	**0.874**
N2252.rv $\Delta = 1$	0.554		0.579	**0.504**	0.505
N2252.rv $\Delta = 5$	0.857		0.807	0.761	**0.729**
N2252.rv $\Delta = 20$	1.097		1.046	1.011	**0.964**
GDAXI2.rv $\Delta = 1$	0.439		0.448	0.399	**0.386**
GDAXI2.rv $\Delta = 5$	0.675		0.650	0.616	**0.566**
GDAXI2.rv $\Delta = 20$	0.931		0.850	0.816	**0.746**
FCHI2.rv $\Delta = 1$	0.533		0.542	0.470	**0.465**
FCHI2.rv $\Delta = 5$	0.705		0.707	0.691	**0.631**
FCHI2.rv $\Delta = 20$	0.982		0.952	0.912	**0.828**

Note: Best predictions are in boldface.

rough volatility was confirmed for more than five thousand equities in Bennedsen *et al.* (2016); we are not yet aware of any asset for which rough volatility has not been confirmed.

We emphasize that our point in this forecasting exercise is not to show that our rough volatility-based approach is really superior to others. For example, we could have considered alternative competitors to the RFSV formula or refined the HAR procedure. We only wish to demonstrate that our forecast is probably at least as good as other predictors whilst being simpler, requiring only the estimation of *H*.

29.6 Conclusion

Using daily realized variance estimates as proxies for daily spot (squared) volatilities, we uncovered two startlingly simple regularities in the resulting time series. First, we found that the distributions of increments of log-volatility are approximately Gaussian, consistent with many prior studies. Secondly, we established the monofractal scaling relationship

$$\mathbb{E}\left[\left|\log\left(\sigma_\Delta\right) - \log\left(\sigma_0\right)\right|^q\right] = K_q \nu^q \Delta^{qH}, \tag{29.6.1}$$

where *H* can be seen as a measure of smoothness characteristic of the underlying volatility process; typically, $0.06 < H < 0.2$. The simple scaling relationship (29.6.1) naturally suggests that log-volatility may be modelled using fractional Brownian motion.

The resulting Rough Fractional Stochastic Volatility (RFSV) model turns out to be formally almost identical to the FSV model of Comte and Renault (1998), with one major difference: in the FSV model, $H > 1/2$ to ensure long memory, whereas in the RFSV model $H < 1/2$, typically, $H \approx 0.1$. Moreover, in the FSV model, the mean reversion coefficient α has to be large compared to $1/T$ to ensure a decaying volatility skew; in the RFSV model, the volatility skew decays naturally just like the observed volatility skew, $\alpha \ll 1/T$ and indeed for timescales of practical interest, we may proceed as if α were exactly zero.

We further showed that applying standard test procedures to volatility time series simulated with the RFSV model would lead us to erroneously deduce the presence of long memory, with parameters similar to those found in prior studies. Despite that volatility in the RFSV model (or in the data) is not a long memory process, we can therefore explain why long memory of volatility is widely accepted as a stylized fact.

Thus, the RFSV model is able to replicate the stylized facts of the time series, i.e. *volatility is rough*. More precisely, we have shown that within a specific class of models (which we strongly argue to be relevant), empirical daily realized variance values are much more likely to be sampled from a rough volatility process than from a smooth process. Whether or not instantaneous variance is rough cannot of course be determined ultimately since instantaneous variance is latent and not observable; as mentioned in the introduction, it is not even clear that such an object exists independently of a specific model.

It is of course plausible that other models are compatible with many of our observations. In fact, there are probably many ways to design a process so that most of our empirical results are reproduced (for example estimation errors when estimating volatility can be quite significant for some models, leading to downward biases in the measurement of the smoothness). However, what we show here is that we cannot find any evidence against the RFSV model. In statistical terms, the null hypothesis that the data generating process of the volatility is a RFSV model cannot be rejected based on our analysis. Even more, it is likely that the RFSV model is simpler, more parsimonious and more tractable than any other such model. In particular, we do not address the question of volatility jumps. Neither do we insist that there are no jumps in volatility. Rather, one of the main messages of our work is that our model is able to replicate the stylized facts of the time series without having to appeal to jumps, see Bibby *et al.* (2005) for another example where a continuous process can mimic the properties obtained from data generated by a process with jumps.

As an application of the RFSV model, we showed how to forecast volatility at various times-cales, at least as well as when using Fulvio Corsi's impressive HAR estimator, but with only one parameter – H!

We focus in this work on the statistical properties of the RFSV model. In Bayer *et al.* (2016), the authors explore the implications of the RFSV model (written under the physical measure $\mathbb{P}$) for option pricing (under the pricing measure $\mathbb{Q}$). In particular, following Mandelbrot and Van Ness, the fBm that appears in the definition (29.3.4) of the RFSV model may be represented as a fractional integral of a standard Brownian motion as follows Mandelbrot and Van Ness (1968):

$$W_t^H = \int_0^t \frac{dW_s}{(t-s)^\gamma} + \int_{-\infty}^0 \left[\frac{1}{(t-s)^\gamma} - \frac{1}{(-s)^\gamma} \right] dW_s, \qquad (29.6.2)$$

with $\gamma = \frac{1}{2} - H$. The observed anticorrelation between price moves and volatility moves may then be modelled naturally by anticorrelating the Brownian motion W that drives the volatility process with the Brownian motion driving the price process. As already proved by Fukasawa (2011), such a model with a small H reproduces the observed decay of at-the-money volatility skew with respect to time to expiry, asymptotically for short times. It is shown that an appropriate extension of Fukasawa's model, consistent with the RFSV model, fits the entire implied volatility surface remarkably well. In particular, this model accurately reproduces the extreme short dated smiles, with no jumps. Moreover, despite that it would seem from (29.6.2) that knowledge of the entire path $\{W_s : s < t\}$ of the Brownian motion would be required, it turns out that the statistics of this path necessary for option pricing are traded and thus easily observed. Remarkably, Heston-type formulas can also be obtained in the rough volatility framework (see El Euch and Rosenbaum 2016, 2017).

Finally, note that there are microstructural foundations to rough volatility models. Indeed, it is explained in El Euch *et al.* (2016) how rough volatility emerges as the scaling limit of a Hawkes process-based description of the order flow in the context of high frequency trading and metaorder splitting.

Acknowledgements

We are grateful to the referees of Econometrica, the Journal of the American Statistical Association, the Journal of Business and Economic Statistics and Quantitative Finance whose careful reading and valuable comments have helped us improve substantially our presentation of this work. We also thank Masaaki Fukasawa for several interesting discussions.

Disclosure Statement

No potential conflict of interest was reported by the authors.

Funding

Thibault Jaisson was financially supported by the chair 'Risques Financiers' of the Risk Foundation and the chair 'Marchés en Mutation' of the French Banking Federation. Mathieu Rosenbaum was financially supported by the ERC [679836 STAQAMOF] and the chair 'Analytics and Models for Regulation'.

References

Andersen, T.G. and Bollerslev, T., Heterogeneous information arrivals and return volatility dynamics: Uncovering the long-run in high frequency returns. *J. Finance*, 1997, **52**(3), 975–1005.

Andersen, T.G. and Bollerslev, T., Intraday periodicity and volatility persistence in financial markets. *J. Emp. Finance*, 1997, **4**(2), 115–158.

Andersen, T.G., Bollerslev, T., Diebold, F.X. and Ebens, H., The distribution of realized stock return volatility. *J. Financ. Econ.*, 2001a, **61**(1), 43–76.

Andersen, T.G., Bollerslev, T., Diebold, F.X. and Labys, P., The distribution of realized exchange rate volatility. *J. Amer. Stat. Assoc.*, 2001b, **96**(453), 42–55.

Andersen, T.G., Bollerslev, T., Diebold, F.X. and Labys, P., Modeling and forecasting realized volatility. *Econometrica*, 2003, **71**(2), 579–625.

Bacry, E. and Muzy, J.F., Log-infinitely divisible multifractal processes. *Commun. Math. Phys.*, 2003, **236**(3), 449–475.

Bayer, C., Friz, P. and Gatheral, J., Pricing under rough volatility. *Quant. Finance*, 2016, **16**(6), 887–904.

Bennedsen, M., Lunde, A. and Pakkanen, M.S., Decoupling the short- and long-term behavior of stochastic volatility. Available at SSRN 2846756, 2016.

Bentes, S.R. and Cruz, M.M.d., Is stock market volatility persistent? A fractionally integrated approach. Instituto Superior de Contabilidade e Administração de Lisboa, Comunicações, 2011.

Beran, J., *Statistics for Long-memory Processes*. Vol. 61, 1994 (CRC Press: Boca Raton, FL).

Bibby, B.M., Skovgaard, I.M., Sørensen, M., et al., Diffusion-type models with given marginal distribution and autocorrelation function. *Bernoulli*, 2005, **11**(2), 191–220.

Bouchaud, J.P. and Potters, M., *Theory of Financial Risk and Derivative Pricing: From Statistical Physics to Risk Management*, 2003 (Cambridge University Press: Cambridge).

Brouste, A. and Iacus, S.M., Parameter estimation for the discretely observed fractional Ornstein-Uhlenbeck process and the Yuima R package. *Comput. Stat.*, 2013, **28**(4), 1529–1547.

Carr, P. and Wu, L., What type of process underlies options? A simple robust test. *J. Finance*, 2003, **58**(6), 2581–2610.

Chen, Z., Daigler, R.T. and Parhizgari, A.M., Persistence of volatility in futures markets. *J. Futures Markets*, 2006, **26**(6), 571–594.

Cheridito, P., Kawaguchi, H. and Maejima, M., Fractional Ornstein- Uhlenbeck processes. *Electron. J. Probab*, 2003, **8**(3), 1–14.

Chronopoulou, A., Parameter estimation and calibration for long- memory stochastic volatility models. In *Handbook of Modeling High-frequency Data in Finance*, edited by F.G. Viens, M.C. Mariani and I. Florescu, pp. 219–231, 2011 (John Wiley & Sons: Hoboken, NJ).

Chronopoulou, A. and Viens, F.G., Estimation and pricing under long-memory stochastic volatility. *Ann. Finance*, 2012, **8**(2–3), 379–403.

Coeurjolly, J.-F., Estimating the parameters of a fractional Brownian motion by discrete variations of its sample paths. *Stat. Inference Stoch. Process.*, 2001, **4**(2), 199–227.

Comte, F., Coutin, L. and Renault, É., Affine fractional stochastic volatility models. *Ann. Finance*, 2012, **8**(2–3), 337–378.

Comte, F. and Renault, E., Long memory in continuous-time stochastic volatility models. *Math. Finance*, 1998, **8**(4), 291–323.

Cont, R., Volatility clustering in financial markets: Empirical facts and agent-based models. In *Long Memory in Economics*, edited by G. Teyssière and A.P. Kirman, pp. 289–309, 2007 (Springer: Berlin Heidelberg).

Corsi, F., A simple approximate long-memory model of realized volatility. *J. Financ. Econom.*, 2009, **7**(2), 174–196.

Dayri, K. and Rosenbaum, M., Large tick assets: Implicit spread and optimal tick size. *Microstruct. Liquidity*, 2015, **1**(1), 155003.

Diebold, F.X. and Inoue, A., Long memory and regime switching. *J. Econom.*, 2001, **105**(1), 131–159.

Ding, Z., Granger, C.W. and Engle, R.F., A long memory property of stock market returns and a new model. *J. Emp. Finance*, 1993, **1**(1), 83–106.

Duchon, J., Robert, R. and Vargas, V., Forecasting volatility with the multifractal random walk model. *Math. Finance*, 2012, **22**(1), 83–108.

Dupire, B., Pricing with a smile. *Risk Mag.*, 1994, **7**(1), 18–20.

El Euch, O., Fukasawa, M. and Rosenbaum, M., The microstructural foundations of leverage effect and rough volatility, preprint arXiv:1609.05177, 2016.

El Euch, O. and Rosenbaum, M., The characteristic function of rough heston models, arXiv preprint arXiv:1609.02108, 2016.

El Euch, O. and Rosenbaum, M., Perfect hedging in rough heston models, arXiv preprint arXiv:1703.05049, 2017.

Fukasawa, M., Asymptotic analysis for stochastic volatility: Martingale expansion. *Finance Stoch.*, 2011, **15**(4), 635–654.

Gatheral, J., *The Volatility Surface: A Practitioner's Guide*. Vol. 357, 2006 (John Wiley & Sons: Hoboken, NJ).

Gatheral, J. and Jacquier, A., Arbitrage-free SVI volatility surfaces. *Quant. Finance*, 2014, **14**(1), 59–71.

Gatheral, J. and Oomen, R.C., *Zero-intelligence realized variance estimation*. *Finance Stoch.*, 2010, **14**(2), 249–283.

Geweke, J. and Porter-Hudak, S., The estimation and application of long memory time series models. *J. Time Ser. Anal.*, 1983, **4**(4), 221–238.

Granger, C.W. and Hyung, N., Occasional structural breaks and long memory with an application to the s&p 500 absolute stock returns. *J. Emp. Finance*, 2004, **11**(3), 399–421.

Granger, C.W. and Teräsvirta, T., A simple nonlinear time series model with misleading linear properties. *Econ. Lett.*, 1999, **62**(2), 161–165.

Hagan, P.S., Kumar, D., Lesniewski, A.S. and Woodward, D.E. Managing smile risk. *Wilmott Mag.*, 2002, 84–108.

Heston, S.L., A closed-form solution for options with stochastic volatility with applications to bond and currency options. *Rev. Financ. Stud.*, 1993, **6**(2), 327–343.

Hull, J. and White, A., One-factor interest-rate models and the valuation of interest-rate derivative securities. *J. Financ. Quant. Anal.*, 1993, **28**(02), 235–254.

Istas, J. and Lang, G., Quadratic variations and estimation of the local Hölder index of a Gaussian process. *Ann. l'IHPProbab. Stat.*, 1997, **33**(4), 407–436.

Mandelbrot, B.B. and Van Ness, J.W., Fractional Brownian motions, fractional noises and applications. *SIAM Rev.*, 1968, **10**(4), 422–437.

Mantegna, R.N. and Stanley, H.E., *Introduction to Econophysics: Correlations and Complexity in Finance*, 2000 (Cambridge University Press: Cambridge).

Mikosch, T. and Stărică, C., Is it really long memory we see in financial returns. In *Extremes and Integrated Risk Management*, edited by P. Embrechts, pp. 149–168, 2000 (Risk Books: London).

Musiela, M. and Rutkowski, M., *Martingale Methods in Financial Modelling*. Vol. 36, 2006 (Springer: Heidelberg, Germany).

Novikov, A. and Valkeila, E., On some maximal inequalities for fractional Brownian motions. *Stat. Probab. Lett.*, 1999, **44**(1), 47–54.

Nuzman, C.J. and Poor, V.H., Linear estimation of self-similar processes via Lamperti's transformation. *J. Appl. Probab.*, 2000, **37**(2), 429–452.

Robert, C.Y. and Rosenbaum, M., A new approach for the dynamics of ultra-high-frequency data: The model with uncertainty zones. *J. Financ. Econom.*, 2011, **9**(2), 344–366.

Robert, C.Y. and Rosenbaum, M., Volatility and covariation estimation when microstructure noise and trading times are endogenous. *Math. Finance*, 2012, **22**(1), 133–164.

Rosenbaum, M., Estimation of the volatility persistence in a discretely observed diffusion model. *Stoch. Process. Appl.*, 2008, **118**(8), 1434–1462.

Rosenbaum, M., First order p-variations and Besov spaces. *Stat. Probab. Lett.*, 2009, **79**(1), 55–62.

Rosenbaum, M., A new microstructure noise index. *Quant. Finance*, 2011, **11**(6), 883–899.

Appendix A: Proofs

A.1 Proof of Proposition 29.3.1

Starting from equation (29.3.3) and applying integration by parts, we get

$$X_t^\alpha = v W_t^H - \int_{-\infty}^t v\alpha e^{-\alpha(t-s)} W_s^H ds + m.$$

Therefore,

$$\left(X_t^\alpha - X_0^\alpha\right) - v W_t^H = -\int_0^t v\alpha e^{-\alpha(t-s)} W_s^H ds$$
$$- \int_{-\infty}^0 v\alpha\left(e^{-\alpha(t-s)} - e^{\alpha s}\right) W_s^H ds.$$

Consequently,

$$\sup_{t\in[0,T]} \left|\left(X_t^\alpha - X_0^\alpha\right) - v W_t^H\right| \le v\alpha T \hat{W}_T^H$$
$$+ \int_{-\infty}^0 v\alpha\left(e^{\alpha s} - e^{-\alpha(T-s)}\right) \hat{W}_s^H ds,$$

where $\hat{W}_t^H = \sup_{s\in[0,t]} \left|W_s^H\right|$. Using the maximum inequality of Novikov and Valkeila (1999), we get

$$\mathbb{E}\left[\sup_{t\in[0,T]} \left|\left(X_t^\alpha - X_0^\alpha\right) - v W_t^H\right|\right]$$
$$\le c\left(v\alpha T T^H + \int_{-\infty}^0 v\alpha\left(T\alpha e^{\alpha s}\right)|s|^H \, ds\right),$$

with c some constant. The term on the right-hand side is easily seen to go to zero as α tends to zero.

A.2 Proof of Corollary 29.3.1

We first recall equation (29.2.2) in Cheridito *et al.* (2003) which writes:

$$\text{Cov}\left[X_{t+\Delta}^\alpha, X_t^\alpha\right] = K \int_{\mathbb{R}} e^{i\Delta x} \frac{|x|^{1-2H}}{\alpha^2 + x^2} \, dx,$$

with $K = v^2 \Gamma(2H+1)\sin(\pi H)/(2\pi)$.[*] Now remark that

$$\mathbb{E}\left[\left(X_{t+\Delta}^\alpha - X_t^\alpha\right)^2\right] = 2\text{Var}\left[X_t^\alpha\right] - 2\text{Cov}\left[X_{t+\Delta}^\alpha, X_t^\alpha\right].$$

Therefore,

$$\mathbb{E}\left[\left(X_{t+\Delta}^\alpha - X_t^\alpha\right)^2\right] = 2K \int_{\mathbb{R}} \left(1 - e^{i\Delta x}\right)\frac{|x|^{1-2H}}{\alpha^2 + x^2} \, dx.$$

---

[*] This covariance is real because it is the Fourier transform of an even function.

This implies that for fixed Δ, $\mathbb{E}\left[\left|X_{t+\Delta}^\alpha - X_t^\alpha\right|^2\right]$ is uniformly bounded by

$$2K\int_{\mathbb{R}}\left(1-e^{i\Delta x}\right)\frac{|x|^{1-2H}}{x^2}\,\mathrm{d}x.$$

Moreover, $X_{t+\Delta}^\alpha - X_t^\alpha$ is a Gaussian random variable and thus for every q, its moment of order $(q+1)$ is uniformly bounded (in α) so that the family $\left|X_{t+\Delta}^\alpha - X_t^\alpha\right|^q$ is uniformly integrable. Therefore, since by proposition 29.3.1,

$$\left|X_{t+\Delta}^\alpha - X_t^\alpha\right|^q \to v^q\left|W_{t+\Delta}^H - W_t^H\right|^q, \text{ in law,}$$

we get the convergence of the sequence of expectations.

Appendix B: Estimations of H

B.1 On Different Indices

Table B.1.

B.2 On Different Time Intervals*

Table B.2.

Appendix C: The Effect of Smoothing

Although we are really interested in the model

$$\log\sigma_{t+\Delta} - \log\sigma_t = v\left(W_{t+\Delta}^H - W_t^H\right),$$

consider the more tractable (fractional Stein and Stein or fSS) model:

$$v_{t+\Delta} - v_t = \alpha\left(W_{t+\Delta}^H - W_t^H\right),$$

where $v_t = \sigma^2$. We cannot observe v_t but suppose we can proxy it by the average

$$\hat{v}_t^\delta = \frac{1}{\delta}\int_0^\delta v_u\mathrm{d}u.$$

We would, for example, like to estimate $m(2,\Delta) = \mathbb{E}\left[\left(v_{t+\Delta} - v_t\right)^2\right]$. However, we need to proxy spot variance with integrated variance so instead we have the estimate

---

* Note that we used realized kernel rather than realized variance estimates to generate table B2. Results obtained using different estimators are almost indistinguishable.

TABLE B.1 Estimates of ζq for All Indices in the Oxford-Man Data-Set

Index	$\zeta 0.5/0.5$	$\zeta 1$	$\zeta 1.5/1.5$	$\zeta 2/2$	$\zeta 3/3$
SPX2.rv	0.128	0.126	0.125	0.124	0.124
FTSE2.rv	0.132	0.132	0.132	0.131	0.127
N2252.rv	0.131	0.131	0.132	0.132	0.133
GDAXI2.rv	0.141	0.139	0.138	0.136	0.132
RUT2.rv	0.117	0.115	0.113	0.111	0.108
AORD2.rv	0.072	0.073	0.074	0.075	0.077
DJI2.rv	0.117	0.116	0.115	0.114	0.113
IXIC2.rv	0.131	0.133	0.134	0.135	0.137
FCHI2.rv	0.143	0.143	0.142	0.141	0.138
HSI2.rv	0.079	0.079	0.079	0.080	0.082
KS11.rv	0.133	0.133	0.134	0.134	0.132
AEX.rv	0.145	0.147	0.149	0.149	0.149
SSMI.rv	0.149	0.153	0.156	0.158	0.158
IBEX2.rv	0.138	0.138	0.137	0.136	0.133
NSEI.rv	0.119	0.117	0.114	0.111	0.102
MXX.rv	0.077	0.077	0.076	0.075	0.071
BVSP.rv	0.118	0.118	0.119	0.120	0.120
GSPTSE.rv	0.106	0.104	0.103	0.102	0.101
STOXX50E.rv	0.139	0.135	0.130	0.123	0.101
FTSTI.rv	0.111	0.112	0.113	0.113	0.112
FTSEMIB.rv	0.130	0.132	0.133	0.134	0.134

$$
m^\delta(2,\Delta) = \mathbb{E}\left[\left(\hat{v}^\delta_{t+\Delta} - \hat{v}^\delta_t\right)^2\right]
$$

$$
= \frac{1}{\delta^2}\mathbb{E}\left[\left(\int_0^\delta \left(v_{u+\Delta} - v_u\right)du\right)^2\right]
$$

$$
= \frac{\alpha^2}{\delta^2}\int_0^\delta\int_0^\delta \mathbb{E}\left[\left(W^H_{u+\Delta} - W^H_u\right)\left(W^H_{s+\Delta} - W^H_s\right)\right]duds
$$

$$
= \int_0^\delta\int_0^\delta \left(|u - s + \Delta|^{2H} - |u - s|^{2H}\right)duds, \qquad (C1)
$$

where the last step uses that:

$$
\mathbb{E}\left[W^H_u W^H_s\right] = \frac{1}{2}\left(u^{2H} + s^{2H} - |u - s|^{2H}\right),
$$

and the symmetry of the integral.

We assume that the length δ of the smoothing window is less than one day so $\Delta > \delta$. Then, easy computations give

$$
\int_0^\delta\int_0^\delta |u - s + \Delta|^{2H}\,du\,ds
$$

$$
= \frac{1}{2H+1}\frac{1}{2H+2}\left((\Delta+\delta)^{2H+2} - 2\Delta^{2H+2} + (\Delta-\delta)^{2H+2}\right)
$$

TABLE B.2 Estimates of *H* over Two Different Time Intervals for All Indices in the Oxford-Man Data-Set

Index	H (first half)	H (second half)
SPX2.rk	0.115	0.158
FTSE2.rk	0.140	0.156
N2252.rk	0.083	0.134
GDAXI2.rk	0.154	0.168
RUT2.rk	0.098	0.149
AORD2.rk	0.059	0.114
DJI2.rk	0.123	0.151
IXIC2.rk	0.094	0.156
FCHI2.rk	0.140	0.146
HSI2.rk	0.072	0.129
KS11.rk	0.109	0.147
AEX.rk	0.168	0.151
SSMI.rk	0.206	0.183
IBEX2.rk	0.122	0.149
NSEI.rk	0.112	0.124
MXX.rk	0.068	0.118
BVSP.rk	0.074	0.134
GSPTSE.rk	0.075	0.147
STOXX50E.rk	0.138	0.132
FTSTI.rk	0.080	0.171
FTSEMIB.rk	0.133	0.140

and

$$\int_0^\delta \int_0^\delta |u - s|^{2H} \, du \, ds = \frac{2}{2H+1} \frac{1}{2H+2} \delta^{2H+2}.$$

Substituting back into (C1) gives

$$
\begin{aligned}
m^\delta(2,\Delta) \quad &= \alpha^2 \Delta^{2H} \frac{1}{2H+1} \\
&\times \frac{1}{2H+2} \frac{1}{\theta^2} \Big((1+\theta)^{2H+2} - 2 - 2\theta^{2H+2} \\
&\quad + (1-\theta)^{2H+2} \Big) \\
&=: \alpha^2 \Delta^{2H} f(\theta).
\end{aligned}
$$

where $\theta = \delta/\Delta$.

Figure C1 shows the effect of smoothing on the estimated variance in the fSS model. Keeping δ fixed, as Δ increases, $f(\theta) = f(\delta/\Delta)$ increases towards one. Thus, in a linear regression of $\log m^\delta(2, \Delta)$ against $\log \Delta$, we will obtain a higher effective H (from the higher slope) and a lower effective ('volatility of volatility') α, exactly as we observed in the RSFV model simulations in section 29.3.4.

C.1 Numerical Example

In the simulation of the RSFV model in section 29.3.4, we have $H = 0.14$, $\delta_1 = 1/24$ for the UZ estimate and $\delta 2 = 1/3$ for the RV estimate. We now reproduce a fSS analogue of the RFSV simulation plots of

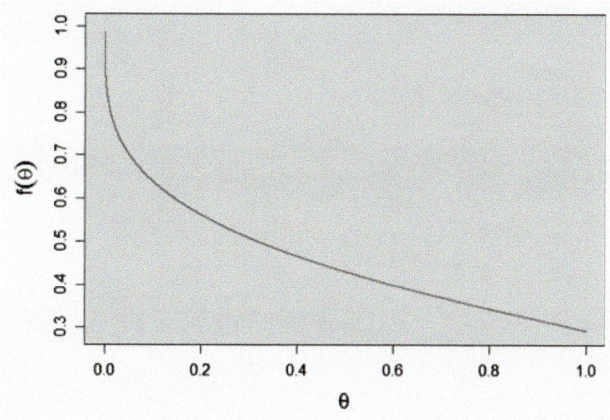

FIGURE C.1 $f(\theta)$ vs $\theta = \delta/\Delta$ with $H = 0.14$.

TABLE C3 Estimated Model Parameters from the Regressions Shown in Figure C2

Estimate	Est. α	Est. H
Exact ($\delta = 0$)	0.300	0.140
UZ ($\delta = 1/24$)	0.263	0.161
RV ($\delta = 1/3$)	0.230	0.184

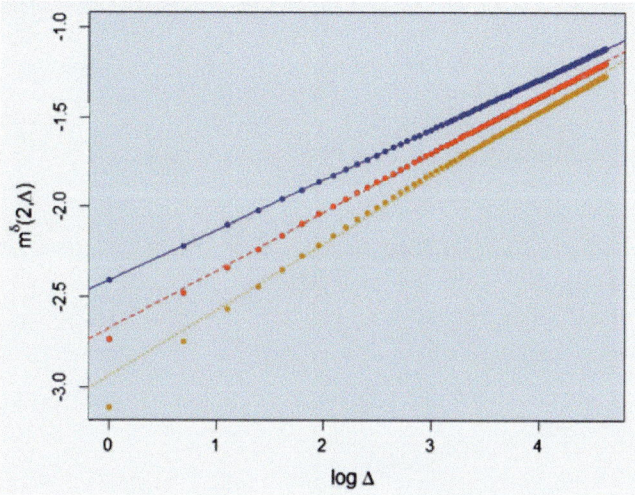

FIGURE C.2 Analogue of figure 14 in the fSS model: the blue solid line is the true $m(2, \Delta)$; the red long-dashed line is the UZ estimate $m^\delta_1 (2, \Delta)$; the orange short-dashed line is the RV estimate $m^\delta_2 (2, \Delta)$.

$m(2, \Delta)$ in figure 29.14. Specifically, for each $\Delta \in \{1, 2, ..., 100\}$, with $\alpha = 0.3$ and $\delta = \delta_1$ or $\delta = \delta_2$, we compute the $m^\delta(2, \Delta)$ and regress $\log m^\delta(2, \Delta)$ against $\log \Delta$. The regressions are shown in figure C2 and results tabulated in table C3.

In figure C2 and table C3, we observe similar qualitative and quantitative biases from our fSS model simulation as we observe in our simulation of the RSFV model with equivalent parameters in section 29.3.4.

Algorithmic Trading in a Microstructural Limit Order Book Model

Frédéric Abergel,
Côme Huré and
Huyên Pham

30.1 Introduction

Most of the markets use a limit order book (order book) mechanism to facilitate trade. Any market participant can interact with the order book by sending either market orders or limit orders. In such type of markets, the market makers play a fundamental role by providing liquidity to other market participants, typically to impatient agents who are willing to cross the bid-ask spread. The profit made by a market-making strategy comes from the alternation of buy and sell orders.

From the mathematical modeling point of view, the market- making problem corresponds to the choice of an optimal strategy for the placement of orders in the order book. Such a strategy should maximize the expected utility function of the wealth of the market maker up to a penalization of their inventory. In the recent litterature, several works focused on the problem of market-making through stochastic control methods. The seminal paper by Avellaneda and Stoikov (2007) inspired by the work of Ho and Stoll (1979) proposes a framework for trading in an order driven market. They modeled a reference price for the stock as a Wiener process, and the arrival of a buy or sell liquidity-consuming order at a distance δ from the reference price is described by a point process with an intensity in an exponential

DOI: 10.1201/9781003265399-35

form decreasing with δ. They characterized the optimal market-making strategies that maximize an exponential utility function of terminal wealth. Since this paper, other authors have worked on related market-making problems. Guéant *et al.* (2012) generalized the market-making problem of Avellaneda and Stoikov (2007) by dealing with the inventory risk. Cartea and Jaimungal (2013) also designed algorithms that manage inventory risk. Fodra and Pham (2015b) and Fodra and Pham (2015a) considered a model designed to be a good compromise between accuracy and tractability, where the stock price is driven by a Markov Renewal Process, and solved the market-making problem. Guilbaud and Pham (2013) also considered a model for the mid-price, modeled the spread as a discrete Markov chain that jumps according to a stochastic clock, and studied the performance of the market-making strategy both theoretically and numerically. Cartea and Jaimungal (2010) employed a hidden Markov model to examine the intra-day changes of dynamics of the order book. Very recently, Cartea *et al.* (2015) and Guéant (2016) published monographs in which they developped models for algorithmic trading in different contexts. El Aoud and Abergel (2015) extended the framework of Avellaneda and Stoikov to the options market-making. A common feature of all these works is that a model for the price or/and the spread is considered, and the order book is then built from these quantities. This approach leads to models that predict well the long-term behavior of the order book. The reason for this choice is that it is generally easier to solve the market-making problem when the controlled process is low- dimensional. Yet, some recent works have introduced accurate and sophisticated micro-structural order book models. These models reproduce accurately the short-term behavior of the market data. The focus is on conditional probabilities of events, given the state of the order book and the positions of the market maker. Abergel *et al.* (2016) proposed models of order book where the arrivals of orders in the order book are driven by Poisson processes or Hawkes processes. Cont *et al.* (2007) also modeled the orders arrivals with Poisson processes. Huang *et al.* (2015) proposed a queue-reactive model for the order book. In this model the arrivals of orders are driven by Cox point processes with intensities that only depend on the state of the order book (they are not time dependent). Other tractable dynamic models of order-driven market are available (see e.g. Cont *et al.* 2007, Rosu 2008, Cartea *et al.* 2014).

In this paper we adopt the micro-structural model of order book in Abergel *et al.* (2016), and solve the associated trading problem. The problem is formulated in the general frame- work of Piecewise Deterministic Markov Decision Process (PDMDP), see Bäuerle and Rieder (2011). Given the model of order book, the PDMDP formulation is natural. Indeed, between two jumps, the order book remains constant, so one can see the modeled order book as a point process where the time becomes a component of the state space. As for the control, the market maker fixes their strategy as a deterministic function of the time right after each jump time. We prove that the value function of the market-making problem is equal to the value function of an associated non-finite horizon Markov decision process (MDP). This provides a characterization of the value function in terms of a fixed point dynamic programming equation. Jacquier and Liu (2018) recently followed a similar idea to solve an optimal liquidation problem, while Baradel *et al.* (2018) and Lehalle *et al.* (2018) also tackled this problem of reward functional maximization in a micro-structure model of order book framework.

The second part of the paper deals with the numerical simulation of the value functions. The computation is challenging because the micro-structural model used to model the order book leads to a high-dimensional pure jump con- trolled process, so evaluating the value function is computationally intensive. We rely on control randomization and Markovian quantization methods to compute the value functions. Markovian quantization has been proved to be very efficient for solving control problems associated with high- dimensional Markov processes. We first quantize the jump times and then quantize the state space of the order book. See Pagès *et al.* (2004) for a general description of quantization applied to controlled processes. The projections are time-consuming in the algorithm, but Fast approximate nearest neighbors algorithms (see e.g. Muja and Lowe 2009) can be implemented to alleviate the procedure. We borrow the values of intensities of the arrivals of orders for the order book simulations from Huang *et al.* (2015) in order to test our optimal trading strategies.

The paper is organized as follows. The model setup is introduced in Section 30.2: we present the micro-structural model for the order book, and show how the market maker interacts with the market. In Section 30.3, we prove the existence and provide a characterization of the value function and optimal trading strategies. In Section 30.4, we introduce a quantization-based algorithm to numerically solve a general class of discrete-time control problem with finite horizon, and then apply it on our trading problem. We then present some results of numerical tests on simulated order book. Section 30.5 presents an extension of our model when order arrivals are driven by Hawkes processes, and finally the appendix collects some results used in the paper.

30.2 Model Setup

30.2.1 Order Book Representation

We consider a model of the order book inspired by the one introduced in chapter 6 of Abergel *et al.* (2016).

Let us fix $K \geq 0$. An order book is supposed to be fully described by K *limits** on the bid side and K limits on the ask side. Denote by pa_t the *best ask* at time t, which is the cheapest price a participant in the market is willing to sell a stock at time t, and by pb_t the *best bid* at time t, which is the highest price a participant in the market is willing to buy a stock at time t. We use the pair of vectors $\left(\underline{a_t}, \underline{b_t}\right) = \left(a_t^1, ..., a_t^K, b_t^1, ..., b_t^K\right)$ where

- a_t^i is the number of shares available i ticks away from pb_t,
- $-b_t^i$ is the number of shares available i ticks away from pa_t,

to describe the order book. The vectors $\underline{a_t}$ and $\underline{b_t}$ describe respectively the ask and the bid sides at time t. The quantities $a_t^i, 1 \leq i \leq K$, live in the discrete space $q\mathbb{N}$ where $q \in \mathbb{R}^*$ is the minimum order size on each specific market (*lot size*). The quantities $b_t^i, 1 \leq i \leq K$, live in the discrete space $-q\mathbb{N}$. By convention, the a^i are non-negative, and the b^i are non-positive for $0 \leq i \leq K$. The tick size ϵ represents the smallest intervall between different price levels. We assume in the sequel that the orders arrivals have the same size $q = 1$, and set the tick size to $\epsilon = 1$ for simplicity.

Constant boundary conditions are imposed outside the moving frame of size $2K$ in order to guarantee that both sides of the LOB are never empty: we assume that all the limits up to the Kth ones are equal to a_∞ in the ask side, and equal to b_∞ in the bid side, with $a_\infty, -b_\infty \in \mathbb{N}$.

The order book can receive at any time three different kinds of orders from general market participants: market orders, limit orders and cancel orders. The orders arrivals are modeled by the following point processes:

- M^+ stands for the buy market orders flow, and we denote by λ^{M+} its intensity,
- M^- stands for the sell market orders flow, and we denote by λ^{M-} its intensity,
- L_i^+, for $i \in \{1, ...K\}$, stands for the sell orders flow at the ith limit on the ask side, and we denote by λ_i^{L+} its intensity,
- L_i^-, for $i \in \{1, ...K\}$, stands for the buy orders flow at the ith limit on the bid side, and we denote by λ_i^{L-} its intensity,
- C_i^+, for $i \in \{1, ...K\}$, stands for the cancel orders flow at the ith limit on the ask side, and we denote by λ_i^{C+} its intensity,
- C_i^-, for $i \in \{1, ...K\}$, stands for the cancel orders flow at the ith limit on the bid side, and we denote by λ_i^{C-} its intensity.

---

* Limit is also referred to as *quote* or *price level* in the literature.

We assume in the sequel that

(Harrivals) The orders arrivals from general market participants (market orders, limit orders and cancel orders) occur according to Markov jump processes which intensities only depends on the couple (a, b). Moreover, we assume that the all the intensities are at most linear w.r.t. the couple (a, b) and are constant between two events.

Under **(Harrivals)**, Let $\lambda^L, \lambda^C, \lambda^M$ be positive real constants such that

$$\sum_{i=1}^{K} \lambda_i^{L^+}(\underline{a}, \underline{b}) + \sum_{i=1}^{K} \lambda_i^{L^-}(\underline{a}, \underline{b}) \le \lambda^L(|\underline{a}| + |\underline{b}|),$$

$$\sum_{i=1}^{K} \lambda_i^{C^+}(\underline{a}, \underline{b}) + \sum_{i=1}^{K} \lambda_i^{C^-}(\underline{a}, \underline{b}) \le \lambda^C(|\underline{a}| + |\underline{b}|),$$

$$\lambda^{M^+}(\underline{a}, \underline{b}) + \lambda^{M^-}(\underline{a}, \underline{b}) \le \lambda^M(|\underline{a}| + |\underline{b}|),$$

for all state (a, b) of the LOB, where $|a| := \sum_{k=1}^{K} -a^k$ and $|b| := \sum_{k=1}^{K} b^k$.

> REMARK 30.2.1 The linear conditions on the intensities are required to prove that the control problem is well-posed.
>
> REMARK 30.2.2 We assume the intensity to be constant between jumps in **(Harrivals)** for simplicity. All the results proposed in Section 30.3 can be extended to the case where the intensities of the jump processes are deterministic between jumps. Such an extension is considered in Section 30.5, where the arrivals are modeled using Hawkes processes with exponential kernel.
>
> REMARK 30.2.3 Some information can be integrated to the order book model by adding new processes. For example, some exogenous processes that send orders to the best-ask and best-bid limits can be added to model the predictions of the mid-price and its volatility that an agent may have. Doing so is critical to manage the risk-reward tradeoffs.

30.2.2 Market Maker Strategies

We assume that the order book matching is done on price/time priority, which means that each limit of the order book is a queue where the first order in the queue is the first one to be executed.[*]

We consider a market maker who stands ready to send buy and sell limit orders on a regular and continuous basis at quoted prices. A usual assumption in stochastic control theory to characterize the value function as solution to a HJB equation is to constrain the control space to be compact. In this spirit, we shall make the following assumption:

(Hcontrol) Assume that at any time, the total number of limit orders placed by the marker maker does not exceed a fixed (possibly large) integer $\bar{M}$.

30.2.2.1 Control of the Market Maker

The market maker can choose at any time to keep, cancel or take positions in the order book (as long as she does not hold more than $\bar{M}$ positions in the order book). Her positions are fully described by the following $\bar{M}$–dimensional vectors $\underline{ra}_t, \underline{rb}_t, \underline{na}_t, \underline{nb}_t$ where $\underline{ra}$ (resp. $\underline{rb}$) records the limits in which the market maker's sell (resp. buy) orders are located; and $\underline{na}$ (resp. $\underline{nb}$) records the ranks in the queues of each market maker's sell (resp. buy) orders. In order to guarantee that the strategy of the market maker

---

[*] Such an order book is sometimes referred to in the literature as an order book governed by a *FIFO* (*First In First Out*) rule.

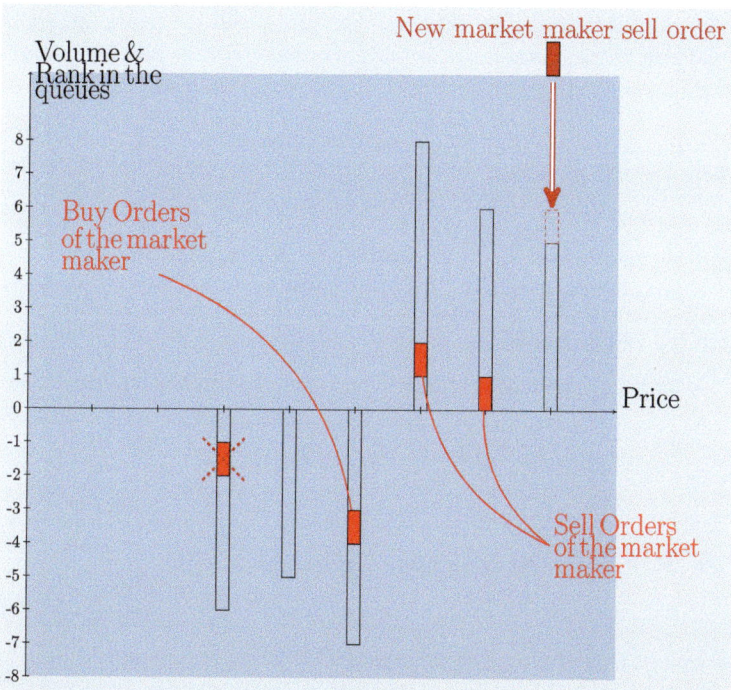

FIGURE 30.1 Example of market maker's placements and decisions she might make. In this example: the ask-side of the order book is described by $\underline{a} = (7, 5, 4)$; and the bid-side by $\underline{b} = (-6, -4, -5)$. The market maker positions are described by $\underline{ra} = (0, 1, -1)$ and $\underline{rb} = (0, 2, -1)$; and the associated ranks vectors are $\underline{na} = (2, 1, -1\ ...)$ and $\underline{nb} = (4, 2, -1)$. We have $i0 = 0$. After each order arrival, she can send new limit orders (see action on the top right), cancel some positions (see dashed cross on the bottom left), or just keep their orders unchanged.

is predictable w .r .t. the natural filtration generated by the orders arrivals processes, we shall make the following assumption.

(Harrivals2) The intensities do not depend on the control. Moreover, the market maker does not cross the spread.

To simplify the theoritical analysis, we also make the following assumption: **(Harrivals3)** Assume that the market maker does not change their strategy between two orders arrivals of the order book. In other words, the market maker makes a decision right after one of the order arrivals processes $L^\pm$, $C^\pm$, $M^\pm$ jumps, and keep it until the next the jump of an order arrival.

REMARK 30.2.4 Assumption **(Harrivals3)** is mild if the order book jumps frequently, since the market maker can change their decisions frequently in such a case. It can also easily be relaxed by considering piecewise constant controls between jumps (which seems well-adapted to most of the time-discretized control problems met in the industry) or any other parametric family of functions. The results and proofs can then be extended, relying mainly on the PDMDP.*

We provide in Figure 30.1 a graphical representation of the controlled LOB. Notice that the market maker inter- acts with the order book by placing orders at some limits. The latter have ranks that evolve after each orders arrivals. Denote by $(T_n)_{n\in\mathbb{N}}$ the sequence of jump times of the order book. We denote

---

* PDMDP stands for Piecewise Deterministic Markov Decision Process, and refers to the the control processes which have deterministic dynamics between (random) jumps.

by $\mathbb{A}$ the set of the admissible strategies, defined as the predictable processes $(\underline{ra}_t, \underline{rb}_t)_{t \leq T}$ such that the control is constant between two consecutive arrivals of orders from the participants, and such that the order of the market maker do not cross the spread. These conditions reads:

- for all $n \in \mathbb{N}$, $(\underline{ra}_t, \underline{rb}_t) \in \{1, \dots, K\}^{\bar{M}} \times \{1, \dots, K\}^{\bar{M}}$ are constant on $(T_n, T_{n+1}]$
- $ra_*, rb_* \geq i0$

where, for every vector $\underline{a} : a_* = \min_{1 \leq i \leq K} \{a_i \ s.t. \ \underline{a}_i \neq -1\}$; and: $i0 = \operatorname{argmin}_{1 \leq i \leq K} (a_i \ s.t. \ a_i > 0)$. The control is the double vector of the positions of the $\bar{M}$ market maker's orders in the order book.

By convention, we set in the sequel $ra_i(t) = -1$ if the ith market maker's order is not placed in the order book.

30.2.2.2 Controlled Order Book

The order book, controlled by the market maker, is fully described by the following state process Z:

$$Z_t := \left(X_t, Y_t, a_t, b_t, na_t, nb_t, pa_t, pb_t, ra_t, rb_t \right),$$

where, at time t:

- X_t is the cash held by the market maker on a zero interest account.
- Y_t is the inventory of the market maker, i.e. it is the (signed) number of shares held by the market maker.
- pa_t is the ask price, i.e. the cheapest price a general market participant is willing to sell stock.
- pb_t is the bid price, i.e. the highest price a general market participant is willing to buy stock.
- a_t $(a_1(t), \dots, a_K(t))$ (resp. $\underline{b}_t$ $(b_1(t), \dots, b_K(t))$) describes the ask (resp. bid) side: $i \in \{1, \dots, K\}, a_i(t)$ is the sum of all the general market participants' sell orders which are i ticks away from the bid (resp. ask) price.
- ra_t (resp. $\underline{ra}_t$) describes the market maker's orders in the ask (resp. bid) side: for $i \in \{1, \dots, \bar{M}\}, \underline{ra}_t(i)$ is the number of ticks between the ith market maker's sell (resp. bid) order and the bid (resp. ask) price. By convention, we set $\underline{ra}_t(i) = -1$ (resp. $\underline{rb}_t(i) = -1$) if the ith sell (resp. buy) order of the market maker is not placed in the order book. As a result $\underline{ra}_t(i), \underline{rb}_t(i) \in \{1, \dots, K\} \cup \{-1\}$.
- na_t (resp. $\underline{nb}_t$) describes the ranks of the market maker's orders in the ask (resp. bid) side. For $i \in \{1, \dots, \bar{M}\}, \underline{na}_t(i) \in \{-1, \dots, |a| + \bar{M}$ (resp. $\underline{nb}_t(i) \ 1, \dots, b \ \bar{M}\})$ is the rank of the ith sell (resp. buy) orders of the market maker in the queue. By convention, we assume that $\underline{na}_t(i) = -1$ (resp. $\underline{nb}_t(i) = -1$) if the ith sell (resp. buy) order of the market maker is not placed in the order book.

30.3 Presentation of the Market-Making Problem. Theoretical Resolution

30.3.1 Definition of the Market-Making Problem and Well-Posedness of the Value Function

We denote by V the value function for the following market-making problem:

$$V(t,z) = \sup_{\alpha \in \mathbb{A}} \mathbb{E}_{t,z}^{\alpha} \left[\int_t^T f\left(\alpha_s, Z_s\right) ds + g\left(Z_T\right) \right], \tag{30.1}$$

$$(t,z) \in [0,T] \times E,$$

where:

- $\mathbb{A}$ is the set of the admissible strategies, defined in Section 30.2.2.1.
- f and g are respectively the instantaneous and terminal reward functions.
- $\mathbb{E}_{t,z}^{\alpha}$ stands for the expectation conditioned by $Z_t = z$ and when strategy $\alpha = (\alpha_s)_{t \leq s < T}$ is followed on $[t, T]$ (Figure 30.2).

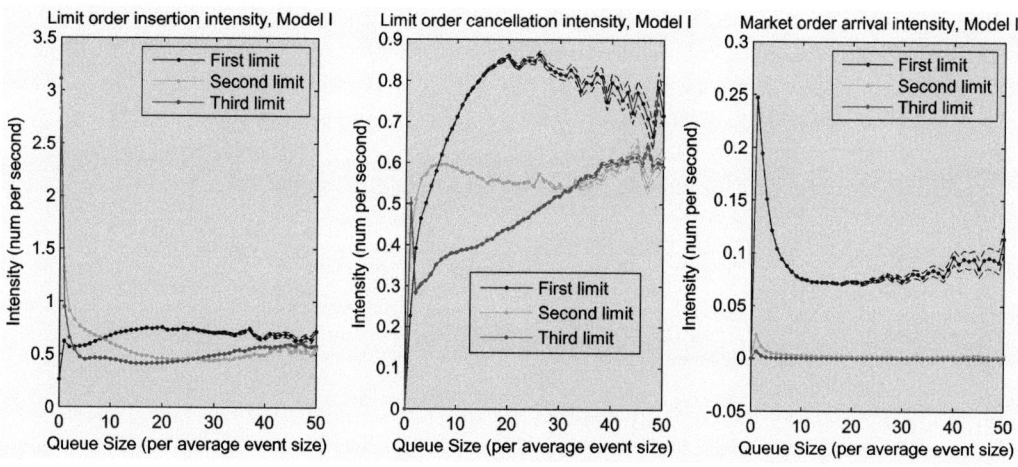

FIGURE 30.2 Values of the intensities w .r .t. the queue size used in the numerical tests of Section 30.4.4. The plot is taken from Huang *et al.* (2015) (see its Figure 13). The intensities are estimated by the authors using their data from Alcatel Lucent.

EXAMPLE 30.3.1 The terminal reward g can be defined as the sum of the market maker's terminal wealth function and an inventory penalization term, i.e. $g : z \mapsto x + L(y) - \eta y^2$ where L is the amount earned from the immediate liquidation of the inventory.* We remind that z stands for a state of order book; ε is the tick size of the LOB; x is the value of the risk-free account of the market maker; η is the penalization parameter of the market maker; and where we remind that y stands for the (signed) market maker's inventory.

The running reward f can stand for a penalization of inventory term: $f(z) := -\gamma y^2$, with $\gamma > 0$.

We shall assume the following conditions on the rewards to insure the well-posedness of the market-making problem.

(Hrewards) The expectation of the integrated running reward is uniformly upper-bounded w .r .t. the strategies in $\mathbb{A}$, i.e.

$$\sup_{\alpha \in \mathbb{A}} \mathbb{E}^{\alpha}_{t,z} \left[\int_t^T f^+ \left(Z_s, \alpha_s \right) \mathrm{d}s \right] < +\infty$$

holds; where for all state z and action a, we denote $f^+(z, a) := \max(f(z, a), 0)$. Moreover, the terminal reward $g(Z_T)$ is a.s. at most linear with respect to the number of events up to time T, denoted by N_T in the sequel, i.e. there exists a constant $c_1 > 0$ such as $g(Z_T) \le c_1 N_T$, a.s..

---

* L is defined as follows:

$$L(z) = \begin{cases} \sum_{k=1}^{J-1} \left[a_k(pa + k\epsilon) \right] \\ \quad + \left(y - a_0 - \cdots - a_{J-1} \right)(pa + J\epsilon) & \text{if } y < 0 \\ -\sum_{k=1}^{J-1} \left[b_k(pb - k\epsilon) \right] \\ \quad + \left(y + b_0 + \cdots + b_{J-1} \right)(pb - J\epsilon) & \text{if } y > 0 \\ 0 & \text{if } y = 0, \end{cases}$$

for all $z = (x, y, \underline{a}, \underline{b}, \underline{na}, \underline{nb}, \underline{pa}, \underline{pb}, \underline{ra}, \underline{rb})$, and where we define:

$$J := \begin{cases} \min\left\{ j \mid \sum_{i=1}^{j} a_i > -y \right\} & \text{if } y < 0 \\ \min\left\{ j \mid \sum_{i=1}^{j} |b_i| > y \right\} & \text{if } y > 0. \end{cases}$$

REMARK 30.3.1 Under Assumption **(Hcontrol)**, Assumption **(Hrewards)** holds when g is defined as the wealth of the market maker plus an inventory penalization. In particular, we have $g(Z_T) \leq N_T \bar{M}$, where $\bar{M}$ is the maximal number of orders that can be sent by the market maker, which holds a.s. since the best profit the market maker can make is when their buy (resp. sell) limit orders are all executed, and then the price keeps going to the right (resp. left) direction. Hence the second condition of **(Hrewards)** holds with $c_1 = \bar{M}$.

The following Lemma 30.3.1 tackles the well-posedness of the control problem.

LEMMA 30.3.1 *Under (Hrewards) and (Hcontrol), the value function is well-defined, i.e.*

$$\sup_{\alpha \in \mathbb{A}} \mathbb{E}_{t,z}^{\alpha} \left[g(Z_T) + \int_t^T f(\alpha_s, Z_s) ds \right] < +\infty,$$

where, as defined previously, $\mathbb{E}_{t,z}^{\alpha}[.]$ stands for the expectation conditioned by the event $\{Z_t = z\}$, assuming that strategy $\alpha \in \mathbb{A}$ is followed in $[t, T]$.

Proof Denote by $(N_t)_t$ the sum of all the arrivals of orders up to time t. Under **(Hrewards)**, we can bound $\mathbb{E}_{t,z}^{\alpha} \left[\int_t^T f(\alpha_s, Z_s) ds + g(Z_T) \right]$, the reward functional at time t associated to a strategy $\alpha \in \mathbb{A}$, as follows:

$$\mathbb{E}_{t,z}^{\alpha} \left[\int_t^T f(\alpha_s, Z_s) ds + g(Z_T) \right]$$

$$\leq \sup_{\alpha \in \mathbb{A}} \mathbb{E}_{t,z}^{\alpha} \left[g(Z_T) \right] + \sup_{\alpha \in \mathbb{A}} \mathbb{E}^{\alpha} \left[\int_t^T f^+(Z_s, \alpha_s) ds \right] \tag{30.2}$$

$$\leq c_1 \sup_{\alpha \in \mathbb{A}} \mathbb{E}_{t,0}^{\alpha} \left[N_T \right] + \sup_{\alpha \in \mathbb{A}} \mathbb{E}_{t,z}^{\alpha} \left[\int_t^T f^+(Z_s, \alpha_s) ds \right],$$

where once again, for all general process M and all $m \in E$, $\mathbb{E}_{t,m}^{\alpha} \left[M_T \right]$ stands for the expectation of M_T conditioned by $M_t = m$ and assuming that the market maker follows strategy $\alpha \in \mathbb{A}$ in $[t, T]$. Let us show that the first term in the r.h.s. of (30.2) is bounded. On one hand, we have:

$$\mathbb{E}_{t,0}^{\alpha} \left[N_T \right] \leq \| \lambda \|_\infty \int_0^T \mathbb{E} \left(|a|_t + |b|_t \right) dt, \tag{30.3}$$

where $\lambda_\infty := \lambda^L + \lambda^C + \lambda^M$ is a bound on the intensity rate of N_t. On the other hand, there exists a constant $c_2 > 0$ such that $d(|a| + |b|)_t \leq c_2 \, dL_t$ so that: $\mathbb{E}_{t,|a|_0 + |b|_0}^{\alpha} \left[|a|_t + |b|_t \right] \leq |a|_0 + |b|_0 + c_3 \int_0^t \mathbb{E} \left[|a|_s + |b|_s \right] ds$. Applying Gronwall's inequality, we then get:

$$\mathbb{E}_{t,|a|_0 + |b|_0}^{\alpha} \left[|a|_t + |b|_t \right] \leq \left(|a|_0 + |b|_0 \right) e^{c_3 t}. \tag{30.4}$$

Plugging (30.4) into (30.3) finally leads to:

$$\mathbb{E}_{t,0}^{\alpha} \left[N_T \right] \leq c_4 e^{c_3 T}$$

wit c_3 and $c_4 > 0$ that do not depends on α, which proves that the first term in the r. h. s. of (30.2) is bounded. Also, its second term in the r.h.s. of (30.2) is bounded under **(Hrewards)**. Hence, the reward functional is bounded uniformly in α, which proves that the value function of the considered market-making problem is well-defined.

30.3.2 Markov Decision Process Formulation of the Market-Making Problem

In this section, we first reformulate the market-making problem as a Markov Decision Process (MDP), and then characterize the value function as solution to a Bellman equation.

Let us denote (T_n) is the increasing sequence of the arrivals of market/limit/cancel order to the market; and let $Z_n := \phi^{a\left(Z_{T_n}\right)}\left(Z_{T_n}\right)$, where $\phi^a(z) \in E$ is the state of the order book at time t such that $T_n < t <$ *Tn*+1, given that $Z_{T_n} = z$ and given that the strategy a has been chosen by the market maker at time T_n.

Let us consider the Markov Decision Process $\left(T_n, Z_n\right)_{n\in\mathbb{N}}$, which is characterized by the following information

$$\underbrace{[0,T]\times E,}_{\text{state space}} \quad \underbrace{A_z}_{\text{market maker control}} \quad, \quad \underbrace{\lambda}_{\text{intensity of the jump}} \quad,$$

$$\underbrace{Q}_{\text{transitions kernel}} \quad, \quad \underbrace{r}_{\text{reward}}$$

where:

- $[0, T] \times E$ is the state space of the time-continuous controlled process $(T_n, Z_n)_{n\in\mathbb{N}}$; and $E := \mathbb{R} \times \mathbb{N} \times \mathbb{N}^K \times \mathbb{N}^K \times \mathbb{N}^{\bar{M}} \times \mathbb{N}^{\bar{M}} \times \mathbb{N}^{\bar{M}} \times \mathbb{N}^{\bar{M}} \times \mathbb{R} \times \mathbb{R}$ is the state space of (Z_t). For $z \in E$, $z = (x, y, \underline{a},$ $\underline{b}, \underline{na}, \underline{nb}, \underline{ra}, \underline{rb}, pa, pb)$ where: x is the cash held by the market maker, y their inventory; $\underline{a}$ and $\underline{b}$, introduced in Section 30.2.2.2. represent the orders in the ask and bid sides of the order book of all the participants except the market maker's; $\underline{na}$ (resp. $\underline{nb}$) is the $\bar{M}$-dimensional vector of the ranks of the market maker's sell (resp. buy) orders in the queues ; $\underline{ra}$ (resp. $\underline{rb}$) is the $\bar{M}$-dimensional vector of the number of ticks the $\bar{M}$ market maker's sell (resp. buy) orders are from the bid (resp. ask) price; pa (resp. pb) is the ask-price (resp. bid-price).
- A_z, for every state $z \in E$, is the set of the admissible actions (i.e. the actions the market maker can take) when the order book is at state z:

$$A_z = \left\{\underline{ra}, \underline{rb} \in \{1,\ldots,K\}^{\bar{M}}\right.$$

$$\left.\times\{1,\ldots,K\}^{\bar{M}} \mid rb_*, ra_* \geq i0\right\},$$

where we define $c_* = \min_{1\leq i\leq K}\left\{c_i | c_i \neq -1\right\}$ and $c0 = \mathrm{argmin}_{1\leq i\leq K}\left\{c_i > 0\right\}$ for $\underline{c} \in \mathbb{N}^{\bar{M}}$. We recall that this condition means that the market maker is not allowed to cross the spread.
- λ is the intensity of the controlled process (Z_t), and reads:

$$\lambda(z) := \lambda^{M^+}(z) + \lambda^{M^-}(z) + \sum_{1\leq j\leq K} \lambda^{L_j^+}(z)$$

$$+ \sum_{1\leq j\leq K} \lambda^{L_j^-}(z) + \sum_{1\leq j\leq K} \lambda^{C_j^+}(z) + \sum_{1\leq j\leq K} \lambda^{C_j^-}(z).$$

Observe that λ does not depend on the strategy α chosen by the market maker since we assumed that the general participants does not 'see' the market maker's orders in the order book. Although we wrote z as argument for he intensity of the order book process, it cannot depend on any controlled component variable of the latter. To simplify, the reader can assume that the intensities only depend on the vectors $\underline{a}$ and $\underline{b}$.

- Q is the transition kernel of the MDP, which is defined as follows:

$$Q(B \times C \,|\, t, z, \alpha)$$

$$:= \mathring{\lambda}(z) \int_0^{T-t} e^{-\lambda(z)s} \mathbf{1}_B(t+s) Q'\left(C \,|\, \phi^{\alpha}(z), \alpha\right) ds \qquad (30.5)$$

$$+ e^{-\lambda(z)(T-t)} \mathbf{1}_{T \in B, z \in C},$$

for all Borelian sets $B \subset \mathbb{R}_+$ and $C \subset E$, for all $(t, z) \in [0, T] \times E$, for all $\alpha \in A$, and where Q' is the transition kernel of (Z_t) defined for all state z as:

$$Q'\left(z' \,|\, z, u\right) = \begin{cases} \dfrac{\lambda^{M^+}(z)}{\lambda(z)} & \text{if } z' = e^{M^+}\left(\phi^u(z)\right) \\ \vdots & \\ \dfrac{\lambda^{C^+}(z)}{\lambda(z)} & \text{if } z' = e^{C_K^+}\left(\phi^u(z)\right), \end{cases}$$

where $\phi^u(z)$ is the new state of the controlled order book when decision u as been taken and when the order book was at state z before the decision; $e^{M^+}(z)$ is the new state of the order book right after it received a buy market order, given that it was at state z before the jump; and $e^{C_i^{\pm}}(z)$ is the new state of the order book right after it received a cancel order from a general market participant on its ith ask/bid limit, given that it was at state z.

- $r : [0, T] \times E^C \to \mathbb{R}$ is the running reward associated to the MDP with infinite horizon defined as follows:

$$r(t, z, a) := -c(z, a) e^{-\lambda(z)(T-t)}(T-t) \mathbf{1}_{t > T}$$

$$+ c(z, a) \left(\frac{1}{\lambda(z)} - \frac{e^{-\lambda(z)(T-t)}}{\lambda(z)} \right) \qquad (30.6)$$

$$+ e^{-\lambda(z)(T-t)} g(z) \mathbf{1}_{t \leq T},$$

and its definition is motivated by Proposition 30.3.1 below.

The cumulated reward functional associated to the MDP $(T_n, Z_n)_{n \in \mathbb{N}}$ for an admissible policy $(f_n)_{n=0}^{\infty}$ is defined as:

$$V_{\infty, (f_n)}(t, z) = \mathbb{E}_{t,z}^{(f_n)} \left[\sum_{n=0}^{\infty} r\left(T_n, Z_n, f_n\left(T_n, Z_n\right)\right) \right],$$

and the associated value function is the supremum of the cumulated reward functional over all the admissible controls in $\mathbb{A}$, i.e.

$$V_{\infty}(t, z) = \sup_{(f_n)_{n=0}^{\infty} \in \mathbb{A}} V_{\infty, \alpha}(t, z), \quad (t, z) \in [0, T] \times E, \qquad (30.7)$$

Notice that we used the same notation for admissible controls of the MDP and those of the continuous-time control problem.

REMARK 30.3.2 Q is defined as in (30.5) because

$$\mathbb{P}\left(T_{n+1} - T_n \leq t, Z_{n+1} \in B \mid T_0, Z_0, \ldots, T_n, Z_n\right)$$

$$= \lambda\left(Z_n\right)\int_0^t e^{-\lambda(Z_n)s} Q'\left(B \mid Z_{T_n}, \alpha_{T_n}\right) ds$$

$$= \lambda\left(Z_n\right)\int_0^t e^{-\lambda(Z_n)s} Q'\left(B \mid Z_{T_n}, f_n\left(Z_n\right)\right) ds,$$

holds for any admissible policy $\alpha = \left(f_n\right)_{n=0}^{\infty} \in \mathbb{A}$, for all Borelian $B \subset E$, and for all $t \in [0, T]$.

In the sequel, we denote $([0,T] \times E)^C := \{(t,z,a) \in E \times \{1,\ldots,K\}^{2M} \mid t \in [0,T], z \in E, a \in A_z\}$, and $E^C := \{(z,a) \in E \times \{1,\ldots,K\}^{2\bar{M}} \mid z \in E, a \in A_z\}$. Q' is the stochastic kernel from E^C to E that describes the distribution of the jump goals, i.e. $Q'(B|z, u)$ is the probability that the order book jumps in the set B given that it was at state $z \in E$ right before the jump, and the control action $u \in A_z$ has been chosen right after the jump time.

REMARK 30.3.3 The MDP is defined in such a way that the control is feedback and constant between two consecutive arrivals of market/limit/cancel orders in the market, i.e. in the time- continuous setting: we restrict ourselves to the control $\alpha = (\alpha_t)$ which are entirely characterized by the decision functions $f_n : [0, T] \times E \to A$, and such that

$$\alpha_t = f_n\left(T_n, Z_n\right) \text{ for } t \in \left(T_n, T_{n+1}\right]$$

By abuse of notation, we denote in the sequel by α the sequence of controls $\left(f_n\right)_{n=0}^{\infty}$.

The following Proposition 30.3.1 motivates the special choice of the running reward r as defined in (30.6):

PROPOSITION 30.3.1 *The value function of the MDP defined by* **30.7)** *coincides with* **(30.1)**, *i.e. we have for all* $(t, z) \in E^C$:

$$V_\infty(t,z) = V(t,z). \tag{30.8}$$

Proof Let us show that for all $\alpha = (f_n) \in \mathbb{A}$ and all $(t, z) \in E^C$

$$V_\alpha(t,z) = V_\infty^{(f_n)}(t,z). \tag{30.9}$$

Let us first denote by $H_n := (T_0, Z_0, \ldots, T_n, Z_n)$. Notice then that for all admissible strategy α:

$$V_\alpha(t,z) = \mathbb{E}_{t,z}^\alpha\left[\sum_{n=0}^{\infty} \mathbf{1}_{T > T_{n+1}}\left(T_{n+1} - T_n\right)c\left(Z_n, \alpha_n\right)\right.$$

$$\left. + \mathbf{1}_{\left[T_n \leq T < T_{n+1}\right]}\left(g\left(Z_T\right) - \eta Y_T^2 + \left(T - T_n\right)c\left(Z_n, \alpha_n\right)\right)\right] \tag{30.10}$$

$$= \sum_{n=0}^{\infty} \mathbb{E}_{t,z}^{(f_n)}\left[r\left(T_n, Z_n, f_n\left(T_n, Z_n\right)\right)\right],$$

where we conditioned by H_n between the first and the second line. We recognize $V_\infty^{(f_n)}$ in the r. h. s. of (30.10), so that the proof of (30.9) is completed.

It remains to take the supremum over all the admissible strategies $\mathbb{A}$ in (30.9) to get (30.8).

From Proposition 30.3.1, we deduce that the value function of the market-making problem is the same as the value function V_∞ of the discrete-time MDP with infinite horizon. We now aim at solving the MDP control problem. To proceed, we first define the maximal reward mapping for the infinite horizon MDP:

$$
\begin{aligned}
(\mathcal{T}v)(t,z) &:= \sup_{a \in A_z} \left\{ r(t,z,a) + \int v\left(t',z'\right) Q\left(t',z' \mid t, \phi^a(z), a\right) \right\} \\
&= \sup_{a \in A_z} \left\{ r(t,z,a) + \lambda(z) \int_0^{T-t} e^{-\lambda(z)s} \right. \\
&\qquad \times \left. \int v\left(t+s,z'\right) Q'\left(dz' \mid \phi^a(z), a\right) ds \right\},
\end{aligned}
\tag{30.11}
$$

where we recall that:

- $\phi^a(z)$ is the new state of the order book when the market maker follows the strategy α and the order book is at state z before the decision is taken.
- $\lambda(z)$ is the intensity of the order book process given that the order book is at state z.

We shall tighten assumption **(Hrewards)** in order to guarantee existence and uniqueness of a solution to (30.1), as well as characterizing the latter.

(HrewardsBis): The running and terminal rewards are at most quadratic w. r. t. the state variable, uniformly w .r .t. the control variable, i.e.

(i) The running reward f is such that $|c|$ is uniformly bounded by a quadratic in z function, i.e. there exists $c_5 > 0$ such that:

$$
\forall (z,a) \in E \times A, \quad |f(z,a)| \le c_5\left(1 + |z|^2\right).
$$

(ii) The terminal reward g has no more than a quadratic growth, i.e. there exists $c_6 > 0$ such that:

$$
\forall z \in E, \quad |g(z)| \le c_6\left(1 + |z|^2\right).
$$

REMARK 30.3.4 Assumption **(HrewardsBis)** holds in the case where g is the terminal wealth of the market maker plus a penalization of their inventory, and where with no running reward, i.e. $f = 0$.

The main result of this section is the following theorem that gives existence and uniqueness of a solution to (30.1), and moreover characterizes the latter as fixed point of the maximal reward operator defined in (30.11).

THEOREM 30.3.1 $\mathcal{T}$ *admits a unique fixed point v which coincides with the value function of the MDP. Moreover we have:*

$$
v = V_\infty = V.
$$

Denote by f^* the maximizer of the operator $\mathcal{T}$. Then $(f^*, f^*, \ldots)$ is an optimal stationary (in the MDP sense) policy.

REMARK 30.3.5 Theorem 30.3.1 states that the optimal strategy is stationary in the MDP formulation of the problem, but of course, it is not stationary for the original time-continuous trading problem with finite horizon (30.1), since the time component is not a state variable anymore in the original formulation. Actually, given $n \in \mathbb{N}$ and the state of order book z at that time, the optimal decision to take at time T_n is given by $f^*(T_n, z)$.

We devote the next section to the proof of Theorem 30.3.1.

30.3.3 Proof of Theorem 30.3.1

Remind first that we defined in the previous section $E^C := \left\{ (z,a) \in E \times \{1, \ldots, K\}^{2\bar{M}} \mid z \in E, a \in A_z \right\}$ and $([0,T] \times E)^C := \left\{ (t,z,a) \in [0,T] \times E \times \{1, \ldots, K\}^{2\bar{M}} \mid t \in [0,T], z \in E, a \in A_z \right\}$.

DEFINITION 30.3.1 A measurable function $b : E \to \mathbb{R}_+$ is called a *bounding function* for the controlled process (Z_t) if there exists positive constants c_c, c_g, c_Q, c_ϕ such that:

(i) $|f(z, a)| \le c_c b(z)$ for all $(z, a) \in E^C$.
(ii) $|g(z)| \le c_g b(z)$ for all z in E.
(iii) $\int b(z') Q'(dz' | z, a) \le c_Q b(z)$ for all $(z, a) \in E^C$.
(iv) $b(\phi^\alpha(z)) \le c_\phi b(z)$ for all $(t, z, \alpha) \in ([0, T] \times E)^C$.

PROPOSITION 30.3.2 *Let b be such that*:

$$\forall z \in E, b(z) := 1 + |z|^2 .$$

*Then, b is a bounding function for the controlled process (Z_t), under Assumption **(HrewardsBis)**.*

Proof Let us check that b defined in Proposition 30.3.2 satisfies the four assertions in Definition 3.1.

- Assertion 1 and 2 of Definition 3.1 holds under **(HrewardsBis)**.

- First notice that $\underline{ra}$, $\underline{rb}$ are bounded by $\sqrt{\bar{M}}K$ (where we recall that K is the number of limits in each side of the order book, and $\bar{M}$ is the biggest number of limit orders that the market maker is allowed to send in the market). Secondly, $pa' \in B(pa, K)$, $pb' \in B(pb, K)$, where $B(x, r)$ is the ball centered in x with radius $r > 0$, because of the limit conditions that we imposed in our LOB model. And last, we can see that $|\underline{a}'| \le |\underline{a}| + a_\infty K$. These three bounds are linear w .r. t. z so that assertion 3 holds.

- $\phi^\alpha(z) = z^\alpha$ only differs from z by its $\underline{na}$, $\underline{nb}$, and $\underline{ra}$, $\underline{rb}$ components. But $|\underline{na}| \le \sqrt{\bar{M}}(|\underline{a}| + \bar{M})$ and $|\underline{nb}| \le \sqrt{\bar{M}}(|\underline{b}| + \bar{M})$ are bounded by a linear function of $(\underline{a}, \underline{b})$, also $|\underline{ra}|$ and $|\underline{rb}|$ are bounded by the universal constant Definition 3.1 holds.

Let us define

$$\Lambda := (4K + 2) \sup \left\{ \frac{\lambda^{M^\pm}}{|\underline{a}| + |\underline{b}|}, \frac{\lambda^{L^\pm}}{|\underline{a}| + |\underline{b}|}, \frac{\lambda^{C^\pm}}{|\underline{a}(z)| + |\underline{b}(z)|} \right\},$$

which is well-defined under **(Harrivals)**.

PROPOSITION 30.3.3 *If b is a bounding function for (Z_t), then*

$$b(t, z) := b(z) e^{\gamma(z)(T - t)},$$

$$\text{with } \gamma(z) = \gamma_0 (4K + 2) \Lambda (1 + |\underline{a}| + |\underline{b}|) \quad \text{and} \quad \gamma_0 > 0$$

is a bounding function for the MDP, i.e. for all $t \in [0, T]$, $z \in E$, $a \in A_z$, we have:

$$|r(t,z,a)| \leq c_g b(t,z),$$

$$\int b(s,z')Q(ds,dz'|t,z,a) \leq c_\phi c_Q e^{C(T-t)} \frac{1}{1+\gamma_0} b(t,z),$$

with $C = \gamma_0 \Lambda K(4K+2)(|a|_\infty + |b|_\infty)$.

Proof Let z' $(x', y', \underline{a}', \underline{b}', \underline{na}', \underline{nb}', \underline{ra}', \underline{rb}')$ be the state of the order book after an exogenous jump occurs given that it was in state z before the jump. Since $|a'| \leq |a| + a_\infty K$ and $|b'| \leq |b| + b_\infty$, where a_∞ and b_∞ are defined as the border conditions of the order book, we have:

$$\gamma(z') \leq \gamma(z) + C, \tag{30.12}$$

with $C = \gamma_0 \Lambda K(4K+2)(a_\infty + b_\infty)$. Then, we get:

$$\int b(s,z')Q(ds,dz'|t,\phi^\alpha(z),\alpha)$$

$$= \lambda(z) \int_0^{T-t} e^{-\lambda(z)s} \int b(t+s,z')Q'(dz'|\phi_s^\alpha(z),\alpha)ds$$

$$= \lambda(z) \int_0^{T-t} e^{-\lambda(z)s} \int b(z')e^{\gamma(z')(T-(t+s))}Q'(dz'|\phi_s^\alpha(z),\alpha)ds$$

$$\leq \lambda(z) \int_0^{T-t} e^{-\lambda(z)s} \int b(z')e^{(\gamma(z)+C)(T-(t+s))}Q'(dz'|\phi_s^\alpha(z),\alpha)ds$$

$$\leq \lambda(z) \int_0^{T-t} e^{-\lambda(z)s} e^{(\gamma(z)+C)(T-(t+s))} \int b(z')Q'(dz'|\phi_s^\alpha(z),\alpha)ds$$

$$\leq \lambda(z) \int_0^{T-t} e^{-\lambda(z)s} e^{(\gamma(z)+C)(T-(t+s))} c_Q c_\phi b(z)ds$$

$$\leq \frac{\lambda(z)c_Q c_\phi}{\lambda(z)+\gamma(z)+C} e^{(\gamma(z)+C)(T-t)}\left(1 - e^{-(T-t)(\lambda(z)+\gamma(z)+C)}\right)b(z)$$

$$\leq c_Q c_\phi \frac{\lambda(z)}{\lambda(z)+\gamma(z)+C} e^{C(T-t)}\left(1 - e^{-(T-t)(\lambda(z)+\gamma(z)+C)}\right)b(t,z),$$

where we applied (30.12) at the third line. It remains to notice that

$$\frac{\lambda(z)}{\lambda(z)+\gamma(z)+C} = \frac{\lambda(z)}{\lambda(z)(1+\gamma_0)+\gamma_0\underbrace{[\Lambda(|a|+|b|)-\lambda(z)]}_{\geq 0}}$$

$$\leq \frac{1}{1+\gamma_0},$$

to complete the proof of the proposition.

Let us denote by $\|\cdot\|_b$ the *weighted supremum norm* such that for all measurable function $v : E' \to \mathbb{R}$,

$$\|v\|_b := \sup_{(t,z)\in E'} \frac{|v(t,z)|}{b(t,z)},$$

and define the set:

$$\mathbb{B}_b := \left\{ v : E' \to \mathbb{R} \mid v \text{ is measurable and } \|v\|_b < \infty \right\}.$$

Moreover let us define

$$\alpha_b := \sup_{(t,z,a) \in E' \times \mathcal{R}} \frac{\int b\left(s,z'\right) Q\left(ds, dz' \mid t, \phi^\alpha(z), \alpha\right)}{b(t,z)}.$$

From the preceding estimations we can bound α_b as follows:

$$\alpha_b \le c_Q c_\phi \frac{1}{1+\gamma_0} e^{CT},$$

so that, by taking: $\gamma_0 = c_Q c_\phi e^{CT}$, we get: $\alpha_b < 1$. In the sequel, we then assume w. l. o. g. that $\alpha_b < 1$. Recall that the maximal reward mapping for the MDP has been defined as:

$$\mathcal{T}v : (t,z) \mapsto \sup_{a \in A_z} \left\{ r(t,z,a) + \lambda(z) \int_0^{T-t} e^{-\lambda(z)s} \right.$$

$$\left. \times \int v\left(t+s,z'\right) Q'\left(dz' \mid \phi^a(z), a\right) ds \right\}$$

It is straightforward to see that:

$$\|\mathcal{T}v - \mathcal{T}w\|_b \le \alpha_b \|v - w\|_b, \tag{30.13}$$

which implies that $\mathcal{T}$ is contracting, since $\alpha_b < 1$.

Let $\mathcal{M}$ be the set of all the continuous function in $\mathbb{B}_b$. Since b is continuous, $\left(\mathcal{M}, \|\cdot\|_b\right)$ is a Banach space.

$\mathcal{T}$ sends $\mathcal{M}$ to $\mathcal{M}$. Indeed, for all continuous function v in $\mathbb{B}_b$, $(t,z,a) \mapsto r(t,z,a) + \lambda(z) \int_0^{T-t} e^{-\lambda(z)s}$
$\int v(t+s,z') Q'\left(dz' \mid \phi^a(z), a\right) ds$ is continuous on $[0, T] \times E^C$. A_z is finite, so we get the continuity of the application:

$$\mathcal{T}v : (t,z) \mapsto \sup_{a \in A_z} \left\{ r(t,z,a) + \lambda(z) \int_0^{T-t} e^{-\lambda(z)s} \right.$$

$$\left. \times \int v\left(t+s,z'\right) Q'\left(dz' \mid \phi^a(z), a\right) ds \right\}.$$

PROPOSITION 30.3.4 *There exists a maximizer for $\mathcal{T}$, i.e. let $v \in \mathcal{M}$, then there exists a Borelian function $f : [0, T] \times E \to A$ such that for all $(t, z) \in E'$:*

$$\mathcal{T}v(t,z,f(t,z)) = \sup_{a \in A} \left\{ r(t,z,a) + \lambda(z) \int_0^{T-t} e^{-\lambda(z)s} \right.$$

$$\left. \times \int v\left(t+s,z'\right) Q'\left(dz' \mid \phi^a(z), a\right) ds \right\}$$

Proof $D^*(t,z) = \left\{ a \in A \mid \mathcal{T}_a v(t,z) = \mathcal{T}v(t,z) \right\}$ is finite, so it is compact. So $(t,z) \mapsto D^*(t,z)$ is a compact-valued mapping. Since the application $(t,z,a) \mapsto \mathcal{T}_a(t,z) - \mathcal{T}(t,z)$ is continuous, we get that

$D^* = \left\{ (t,z,a) \in E'^C \mid \mathcal{T}_a v(t,z) = \mathcal{T}v(t,z) \right\}$ is borelian. Applying the measurable selection theorem yields to the existence of the maximizer. (see Bäuerle and Rieder 2011, p. 352)

LEMMA 30.3.2 *The following holds:*

$$\sup_{\alpha \in \mathcal{A}} \mathbb{E}_{t,z}^{\alpha} \left[\sum_{k=n}^{\infty} \left| r(T_k, Z_k) \right| \right] \leq \frac{\alpha_b^n}{1 - \alpha_b} b(t,z),$$

and in particular, we have:

$$\limsup_{n \to \infty} \mathbb{E}_{t,z}^{\alpha} \left[\sum_{k=n}^{\infty} \left| r(t_k, Z_k) \right| \right] = 0.$$

Proof By conditioning we get $\mathbb{E}_{t,z}^{\alpha} \left[\left| r(T_k, Z_k) \right| \right] \leq c_g \alpha_b{}^k b(t,z)$ for $k \in \mathbb{N}$, and for all $\alpha \in \mathbb{A}$. It remains to sum this inequality to complete the proof of Lemma 30.3.2.

We can now prove Theorem 30.3.1.

Proof We divided the proof of Theorem 30.3.1 into four steps.

Step 1: Inequality (30.13) and Proposition 30.3.3 imply that $\mathcal{T}$ is a stable and contracting operator defined on the Banach space $\mathcal{M}$. Banach's fixed point theorem states that $\mathcal{T}$ admits a fixed point, i.e. there exists a function $v \in \mathbb{M}$ such that $v = \mathcal{T}v$, and moreover we have $v = \lim_{n \to \infty} \mathcal{T}^n 0$. Notice that $\mathcal{T}^N 0$ coincides with v_0 defined recursively by the following Bellman equation:

$$\begin{aligned} v_N &= 0 \\ v_n &= \mathcal{T}v_{n+1} \quad \text{for } n = N-1,\dots,0. \end{aligned} \tag{30.14}$$

The solution of the Bellman equation is always larger than the value function of the MDP associated (see e.g. Theorem 2.3.7 p. 22 in Bäuerle and Rieder 2011). Then we have: $\mathcal{T}^n 0 \geq \sup_{(f_k)} \mathbb{E}_n^{(f_k)} \left[\sum_{k=0}^{n-1} r(t_k, X_k) \right] =: J_n$, where J_n is the value function of the MDP with finite horizon n and terminal reward 0, associated to (30.14). Moreover, by Lemma 7.1.4 p. 197 in Bäuerle and Rieder (2011), we know that $(J_n)_n$ converges as $n \to \infty$ to a limit that we denote by J. Passing at the limit in the previous inequality we get: $\lim_{n \to \infty} T^n 0 \geq J$, i.e.

$$v \geq J. \tag{30.15}$$

Step 2: Let us fix a strategy $\alpha \in \mathbb{A}$, and take $n \in \mathbb{N}$. We denote $J_n(\alpha) := \mathbb{E}_0^{(\alpha_k)} \left[\sum_{k=0}^{n-1} r(t_k, X_k) \right]$, the reward functional associated to the control α on the discrete finite time horizon $\{0, \dots, n\}$. By definition, we have $J_n(\alpha) \leq J_n$. We get by letting $n \to \infty$: $\lim_{n \to +\infty} J_n(\alpha) : J_{\infty}(\alpha)$ J. Taking the supremum over all the admissible strategies α finally leads to:

$$V_{\infty} \leq J. \tag{30.16}$$

Step 3: Let us denote by f a maximizer of $\mathcal{T}$ associated to v, which exists, as stated in Proposition 30.3.4. v is the fixed point of $\mathcal{T}$ so that $v = \mathcal{T}_f^n(v)$, for $n \in \mathbb{N}$. Moreover $v \leq \delta$ where $\delta := \sup_{\alpha \in \mathcal{A}} \mathbb{E} \left[\sum_{k=0}^{\infty} r^+(Z_k, \alpha_k) \right]$, so that $\mathcal{T}_f^n(v) \leq \mathcal{T}_f^n 0 + \mathcal{T}_o^n \delta$, where $\mathcal{T}_o^n \delta = \sup_{\alpha} \mathbb{E}_n^{\alpha} \left[\sum_{k=n}^{\infty} r^+(t_k, Z_k) \right]$. Lemma 30.3.2 implies that $\mathcal{T}_o^n \delta \to 0$ as $n \to \infty$. Hence, we get:

$$v \leq J_f. \tag{30.17}$$

Step 4: Conclusion. Since

$$J_f \le V_\infty \tag{30.18}$$

holds, we get by combining (30.15)–(30.18):

$$V_\infty \le J \le v \le J_f \le V_\infty. \tag{30.19}$$

All the inequalities in (30.19) are then equalities, which completes the proof of Theorem 30.3.1.

30.4 Numerical Resolution of the Market-Making Control Problem

In this section, we first introduce an algorithm to numerically solve a general class of discrete-time control problem with finite horizon, and then apply it on the trading problem (30.1).

30.4.1 Framework

Let us consider a general discrete-time stochastic control problem over a finite horizon $N \in \mathbb{N} \setminus \{0\}$. The dynamics of the controlled state process $Z^\alpha = (Z^\alpha)_n$ valued in $\mathbb{R}^d$ is given by

$$Z_{n+1}^\alpha = F\left(Z_n^\alpha, \alpha_n, \varepsilon_{n+1}\right), \quad n = 0,\dots,N-1, Z_0^\alpha = z \in \mathbb{R}^d,$$

with $(\varepsilon_n)_n$ is a sequence of i.i.d. random variables valued in some Borel space $(E, \mathcal{B}(E))$, and defined on some probability space $(\Omega, \mathbb{F}, \mathbb{P})$ equipped with the filtration $\mathbb{F} = \left(\mathcal{F}_n\right)_n$ generated by the noise $(\varepsilon_n)_n$ ($\mathcal{F}_0$ is the trivial σ-algebra), the control α $(\alpha_n)_n$ is an $\mathbb{F}$-adapted process valued in $A \subset \mathbb{R}^q$, and F is a measurable function from $\mathbb{R}^d \times \mathbb{R}^q \times E$ into $\mathbb{R}^d$.

Given a running cost function f defined on $\mathbb{R}^d \times \mathbb{R}^q$, a terminal cost function g defined on $\mathbb{R}^d$, the cost functional associated to a control process α is

$$J(\alpha) = \mathbb{E}\left[\sum_{n=0}^{N-1} f\left(Z_n^\alpha, \alpha_n\right) + g\left(Z_N^\alpha\right)\right].$$

The set $\mathbb{A}$ of admissible controls is the set of control processes α satisfying some integrability conditions ensuring that the cost functional $J(\alpha)$ is well-defined and finite. The con- trol problem, also called Markov decision process (MDP), is formulated as

$$V_0\left(x_0\right) := \sup_{\alpha \in \mathcal{A}} J(\alpha),$$

and the goal is to find an optimal control $\alpha^* \in \mathcal{A}$, i.e. attaining the optimal value: $V_0(z) J(\alpha^*)$. Notice that problem (30.20) may also be viewed as the time discretization of a continuous time stochastic control problem, in which case, F is typically the Euler scheme for a controlled diffusion process.

Problem (30.20) is tackled by the dynamic programming approach. For $n = N, \dots, 0$, the value function V_n at time n is characterized as solution of the following backward (Bellman) equation:

$$V_N(z) = g(z)$$

$$V_n(z) = \sup_{a \in A}\left\{f(z,a) + \mathbb{E}_{n,z}^a\left[V_{n+1}\left(Z_{n+1}\right)\right]\right\}, \quad z \in \mathbb{R}^d, \tag{30.20}$$

Moreover, when the supremum is attained in the DP formula at any time n by $a_n^*(z)$, we get an optimal control in feedback form given by: $\alpha^* = \left(a_n^*\left(z_n^*\right)\right)_n$ where $Z^* = Z^{\alpha*}$ is the Markov process defined by

$$Z_{n+1}^* = F\left(Z_n^*, a_n^*\left(Z_n^*\right), \varepsilon_{n+1}\right), \quad n = 0, \ldots, N-1, Z_0^* = z.$$

There are two usual ways that have been studied in the literature, to solve numerically (30.20): one way is to use local regression methods, relying e.g. on quantization, k-nearest neighbors or kernel ideas to approximate the conditional expectations by cubature methods; another way is to rely on MC regress-now or later methods to regress the value functions V_{n+1} at time n for $n = 0, \ldots, N-1$ on basis functions or neural networks. See e.g. Kharroubi *et al.* (2014) for the regress-now and Balata and Palczewski (2017) for the regress- later methods for algorithms using basis functions, and e.g. Huré *et al.* (2018) for regression on neural networks based on regress-now or regress-later techniques.

30.4.2 Presentation and Rate of Convergence of the Qknn Algorithm

In this section, we present an algorithm based on k-nn estimates for local non-parametric regression of the value function, and optimal quantization to quantize the exogenous noise, in order to numerically solve (30.20).

Let us first introduce some ingredients of the quantization approximation:

- We denote by $\hat{\varepsilon}$ a K-quantizer of the E-valued random variable $\varepsilon_{n+1} \sim \varepsilon_1$, that is a discrete random variable on a grid $\Gamma = \{e_1, \ldots, e_K\} \subset E^K$ defined by

$$\hat{\varepsilon} = \text{Proj}_\Gamma\left(\varepsilon_1\right) := \sum_{\ell=1}^K e_\ell 1_{\varepsilon_1 \in C_i(\Gamma)},$$

where $C_1(\Gamma), \ldots, C_K(\Gamma)$ are Voronoi tesselations of Γ, i.e. Borel partitions of the Euclidian space $(E, |\cdot|)$ satisfying

$$C_\ell(\Gamma) \subset \left\{e \in E : |e - e_\ell| = \min_{j=1,\ldots,K}|e - e_j|\right\}.$$

The discrete law of $\hat{\varepsilon}$ is then characterized by

$$\hat{p}_\ell := \mathbb{P}\left[\hat{\varepsilon} = e_\ell\right] = \mathbb{P}\left[\varepsilon_1 \in C_\ell(\Gamma)\right], \quad \ell = 1, \ldots, K.$$

The grid points (e_ℓ) which minimize the L^2- quantization error $\left\|\varepsilon_1 - \hat{\varepsilon}\right\|_2$ lead to the so-called optimal L-quantizer, and can be obtained by a stochastic gradient descent method, known as Kohonen algorithm or competitive learning vector quantization (CLVQ) algorithm, which also provides as a byproduct an estimation of the associated weights $\left(\hat{p}_\ell\right)$. We refer to Pagès *et al.* (2004) for a description of the algorithm, and mention that for the normal distribution, the optimal grids and the weights of the Voronoi tesselations are precomputed and available on the website http://www.quantize.maths-fi.com

- Recalling the dynamics (30.20), the conditional expectation operator is equal to

$$P^{\hat{a}_n^M(z)}W(x) = \mathbb{E}\left[W\left(Z_{n+1}^{\hat{a}_n^M}\right) \mid Z_n = x\right]$$

$$= \mathbb{E}\left[W\left(F\left(z, \hat{a}_n^M(z), \varepsilon_1\right)\right)\right], \quad z \in \mathcal{R}^d,$$

that we shall approximate analytically by quantization via:

$$\hat{P}^{\hat{a}_n^M(z)} W(z) := \mathbb{E}\left[W\left(F\left(z, \hat{a}_n^M(z), \hat{\varepsilon}\right)\right)\right]$$

$$= \sum_{\ell=1}^{K} \hat{p}_\ell W\left(F\left(z, \hat{a}_n^M(z), e_\ell\right)\right).$$

Let us secondly introduce the notion of training distribution that will be used to build the estimators of value functions at time n, for $n = 0, \dots, N - 1$. Let us consider a measure μ on the state space E. We refer to it in the sequel as the training measure. Let us take a large integer M, and for $n = 0, \dots, N$, introduce $\Gamma_n = \{Z^{(1)}, \dots, Z^{(M)}\}$, where $\left(Z_n^{(m)}\right)_{m=1}^{M}$ is a i.i.d. sequence of r.v. following law μ. Γ_n should be seen as a training sampling to estimate the value function V_n at time n.

The proposed algorithm reads as:

$$\hat{V}_N^Q(z) = g(z), \quad \text{for } z \in \Gamma_N,$$

$$\hat{Q}_n(z,a) = \sum_{\ell=1}^{K} p_\ell \left[f(z,a) + \hat{V}_{n+1}^Q\left(\text{Proj}_{n+1}\left(F(z, e_\ell, a)\right)\right)\right], \quad (30.21)$$

$$\hat{V}_n^Q(z) = \sup_{a \in A} \hat{Q}_n(z,a), \quad \text{for } z \in \Gamma_n, n = 0, \dots, N-1.$$

where, for $n\ 0, \dots, N$, $\text{Proj}_n(z)$ stands for the closest neighbor of $z \in E$ in the grid Γ_n, i.e. the operator $z \mapsto \text{Proj}_n(z)$ is actually the euclidean projection on the grid Γ_n.

In the sequel, we refer to (30.21) as the Qknn algorithm.

We shall make the following assumption on the transition probability of $(Z_n)_{0 \le n \le N}$, to guarantee the convergence of the Qknn algorithm.

(Htrans) Assume that the transition probability $\mathbb{P}(Z_{n+1} \in A | Z_n = z, a)$ conditioned by $Z_n = z$ when control a is followed at time n admits a density r w.r.t. the training measure μ, which is uniformly bounded and lipschitz w.r.t. the state variable z, i.e. there exists $\|r\|_\infty > 0$ such that for all $z \in E$ and control u taken at time n:

$$\left| r(y; n, x, a)\right| \le \|r\|_\infty \quad \text{and}$$

$$\left| r(y; n, x, a) - r\left(y; n, x', a\right)\right| \le [r]_L \left| x - x'\right|$$

and r is defined as follows:

$$\mathbb{P}\left(Z_{n+1} \in O \,|\, Z_n = z, u\right) = \int_O r(y; n, x, a) \mathrm{d}\mu(y).$$

and where we denoted by $[r]_L$ the Lipschitz constant of r w.r.t. x.

Denote by $\text{Supp}(\mu)$ the support of μ. We shall assume smoothness conditions on μ and F to provide a bound on the projection error.

(Hμ) We assume $\text{Supp}(\mu)$ to be bounded, and denote by $\|\mu\|_\infty$ the smallest real such that $\text{Supp}(\mu) \subset B(0, \|\mu\|_\infty)$. Moreover, we assume $x \in E \mapsto \mu(B(x, \eta))$ to be Lipschitz, uniformly w.r.t. η, and we denote by $[\mu]_L$ its Lipschitz constant.

(HF) For $x \in E$ and $a \in A$, assume F to be $\mathbb{L}^1$-Lipschitz w.r.t. the noise component ε, i.e. there exists $[F]_L > 0$ such that for all $x \in E$ and $a \in A$, for all r.v. ε and ε', we have:

$$\mathbb{E}\left[\left| F(x,a,\varepsilon) - F\left(x,a,\varepsilon'\right)\right|\right] \le [F]_L \mathbb{E}\left[\left|\varepsilon - \varepsilon'\right|\right]$$

We now state the main result of this section whose proof is postponed in Appendix 1.

THEOREM 30.4.1 *Take $K = M^{2+d}$ points for the optimal quan- tization of the exogenous noise ε_n, $n = 1$, ... , N. There exist constants $\left[\hat{V}_n^Q\right]_L > 0$, that only depend on the Lipschitz coefficients of f, g and F, such that, under **(IIIruns)**, it holds for*

$n = 0, \ldots, N-1$, as $M \to +\infty$:

$$\left\| \hat{V}_n^Q(X_n) - V_n(X_n) \right\|_2 \leq \sum_{k=n+1}^{N} r^{N-k}_\infty \left[\hat{V}_k^Q\right]_L \left(\varepsilon_k^{proj} + [F]_L \varepsilon_k^Q\right)$$

$$+ \mathcal{O}\left(\frac{1}{M^{1/d}}\right),$$

(30.22)

where $\varepsilon_k^Q := \left\| \hat{\varepsilon}_k - \varepsilon_k \right\|_2$ stands for the quantization error, and

$$\varepsilon_n^{proj} := \sup_{a \in A} \left\| \text{Proj}_{n+1}\left(F(X_n, a, \hat{\varepsilon}_n)\right) - F(X_n, a, \hat{\varepsilon}_n) \right\|_2$$

stands for the projection error, at time n.

REMARK 30.4.1 The constants $\left[\hat{V}_n^Q\right]_L > 0$ are defined in (A8).

From Theorem 30.4.1, we can deduce consistency and provide a rate of convergence for the estimator $\hat{V}_n^Q, n = 0, \ldots, N-1$, under some rather tough yet usual compactness conditions on the state space.

COROLLARY 30.4.1 *Under **(Hμ)** and **(HF)**, the Qknn-estimator $\hat{V}_n^Q$ is consistent for $n = 0, \ldots, N-1$, when taking M^{d+1} points for the quantization; and moreover, we have for $n = 0, \ldots, N-1$, as $M \to +\infty$:*

$$\left\| \hat{V}_n^Q(X_n) - V_n(X_n) \right\|_2 = \mathcal{O}\left(\frac{1}{M^{1/d}}\right).$$

Proof We postpone the proof of Theorem 30.4.1 to Appendix 1.

30.4.3 Qknn Algorithm Applied to the Order Book Control Problem (30.1)

We recall the expression of the time-continuous controlled order book process

$$Z_t = \left(X_t, Y_t, \underline{a}_t, \underline{b}_t, \underline{na}_t, \underline{nb}_t, pa_t, pb_t, \underline{ra}_t, \underline{rb}_t\right)$$

admits a representation as a MDP as shown in Section 30.3.2. In Section 30.3.3, we proved that the value function associated to the MDP is characterized as limit as $N \to \infty$ of the the Bellman equation (30.14). In this section, some implementation details on the Qknn algorithm are presented in order to numerically solve (30.14).

30.4.3.1 Training Set Design

Inspired by Fiorin *et al.* (2019), we use product-quantization method and randomization techniques to build the training set Γ_n on which we project (T_n, Z_n) that lies on $[0, T] \times E$, where T_n and Z_n stands for the *n*th jump of *Z* and the state of *Z* at time t_n, i.e. $Z_n = Z_{T_n}$ for $n \geq 0$. This basic idea of Control Randomization consists in replacing in the dynamics of *Z* the endogenous control by an exogenous control $\left(I_{T_n}\right)_{n \geq 0}$, as introduced in Kharroubi *et al.* (2014). In order to alleviate the notations, we denote by I_n the control taken at time T_n, for $n \geq 0$.

Initialization. Set: $\Gamma_0^E = \{z\}$ and $\Gamma_0^T = \{0\}$.

Randomize the control, using e.g. uniform distribution on A at each time step, and then simulate D randomized processes to generate $\left(T_n^k, Z_n^k\right)_{n=0,k=1}^{N,D}$.

For all $n = 1, \ldots, N$, set $\Gamma_n^T = \left\{T_n^k, 1 \le k \le D\right\}$, which stands for the grid associated to the quantization of the nth jump time T_n, and set $\Gamma_n^E = \left\{Z_n^k, 1 \le k \le D\right\}$ which stands for the grid associated to the quantization of the state Z_n of Z at time T_n.

REMARK 30.4.2 The way we chose our training sets is often referred to as an *exploration strategy* in the reinforcement learning literature. Of course, if one has ideas or good guess of where to optimally drive the controlled process, she should not follow an exploration-type strategy to build the training set, but should rather use the guess to build it, which is referred to as the *exploitation strategy* in the reinforcement learning and the stochastic bandits literature. We refer to Balata *et al.* (2019) for several other applications of the *exploration strategy* to build training sets. Note that this idea is the root of all the Q-learning based algorithms. See Sutton and Barto (2011) for more details on Q-learning.

Let F and G be the Borelian functions such that $Z_n = F(Z_{n-1}, d_n, I_n)$ and $T_n = G(T_{n-1}, \epsilon_n, I_n)$, where $\epsilon_n \sim \mathcal{E}(1)$ stands for the temporal noise, and d_n is the state noise, for $n \ge 0$.

Let us fix $N \ge 1$ and consider $\left(\hat{T}_n, \hat{Z}_n\right)_{n=0}^N$, the dimensionwise projection of $\left(T_n, Z_n\right)_{n=0}^N$ on the grids $\Gamma_n^T \times \Gamma_n^E, n = 0, \ldots, N$, i.e. $\hat{T}_0 = 0, \hat{Z}_0 = z$, and

$$\hat{T}_n = \mathrm{Proj}\left(G\left(\hat{T}_{n-1}, \epsilon_n, I_n\right), \Gamma_n^T\right),$$

$$\hat{Z}_n = \mathrm{Proj}\left(F\left(\hat{Z}_{n-1}, d_n, I_n\right), \Gamma_n^E\right), \quad \text{for } n = 1, \ldots, N.$$

$\left(\hat{T}_n, \hat{Z}_n, I_n\right)_{n \in \{0, N\}}$ is a Markov chain.

Define then $\left(\hat{T}_n^Q, \hat{Z}_n^Q\right)_{n=0}^N$ as temporal noise-quantized version of $\left(\hat{T}_n, \hat{Z}_n, I_n\right)_{n=0}^N$. Note that we do not need to quantize the spacial noise since this noise already takes a finite number of states. Let $\hat{\epsilon}_n$ be the quantized process associated to ϵ_n. The process $\left(\hat{T}_n^Q, \hat{Z}_n^Q\right)_{n=0}^N$ is then defined as follows: $\hat{Z}_0^Q = z, \hat{T}_0^Q = 0$ and $\forall 1 \le n \le N$:

$$\hat{T}_n^Q = \mathrm{Proj}\left(G\left(\hat{t}_{n-1}, \hat{\epsilon}_n, I_n\right), \Gamma_n^T\right),$$

$$\hat{Z}_n^Q = \mathrm{Proj}\left(F\left(\hat{Z}_{n-1}, d_n, I_n\right), \Gamma_n^E\right).$$

Denote by $\left(\hat{V}_n^{Q,(N,D)}\right)_{n=0}^N$ the solution of the Bellman equation associated to $\left(\hat{T}_n^Q, \hat{Z}_n^Q\right)_{n=0}^N$:

$$\left(\hat{B}_{N,D}^Q\right): \begin{cases} \hat{V}_N^{Q,(N,D)} = 0 \\ \hat{V}_n^{Q,(N,D)}(t,z) = r(t,z,a) \\ \quad + \sup_{a \in A}\left\{\mathbb{E}_{t,z}^a\left[\hat{V}_{n+1}^{Q,(N,D)}\left(\hat{T}_{n+1}^Q, \hat{Z}_{n+1}^Q\right)\right]\right\}, \\ \text{for } n = 0, \ldots, N, \end{cases}$$

where $\mathbb{E}_{t,z}^a[.]$ stands for the expectation conditioned by the events $\hat{T}_n^Q = t, \hat{Z}_n^Q = z$ and when decision $I_n = a$ is taken at time t.

We wrote the pseudo-code of the Qknn algorithm to compute $\left(\hat{B}_{N,D}^Q\right)$ in Algorithm 1.

We discuss in Remark 30.4.3 the reasons why we can apply Theorem 30.4.1.

Algorithm 1 Generic Qknn Algorithm
Inputs:

- N : number of time steps
- z: state in E at time $T_0 = 0$
- $\Gamma^\varepsilon = \{e_1, \ldots, e_L\}$ and $(p_\ell)_{\ell=1}^L$: the grid and the weights for the optimal quantization of $(\varepsilon_n)_{n=1}^N$.
- Γ_n and Γ_n^E the grids for the projection of respectively the time and the state components at time n, for $n = 0, \ldots, N$.
1: **for** $n = N - 1, \ldots, 0$ **do**
2: Compute the approximated $Qknn$-value at time n:

$$\hat{Q}_n(z,a) = r\left(T_n,z,a\right) + \sum_{\ell=1}^L p_\ell \hat{V}_{n+1}^Q\left(\text{Proj}\left(G\left(z,e_\ell,a\right),\Gamma_{n+1}^T\right),\text{Proj}\left(F\left(z,e_\ell,a\right),\Gamma_{n+1}^E\right)\right),$$

for $(z,a) \in \Gamma_n \times A_z$;
3: Compute the optimal control at time n

$$\hat{A}_n(z) \in \underset{a \in A_z}{\arg\min}\, \hat{Q}_n(z,a), \quad \text{for } z \in \Gamma_n,$$

where the argmin is easy to compute since A_z is finite for all $z \in E$;
4: Estimate analytically by quantization the value function:

$$\hat{V}_n^Q(z) = \hat{Q}_n\left(z,\hat{A}_n(z)\right), \quad \forall z \in \Gamma_n;$$

5: **end for**

Output:

- $\left(\hat{V}_0^Q\right)$: Estimate of $V(0, z)$;

REMARK 30.4.3 When the number of jumps of the LOB $N \geq 1$ is fixed, the set of all the states that can take the controlled order book by jumping less than N times, denoted by $\mathcal{K}$ in the sequel, is finite. Hence, the reward function r, defined in (30.6), is bounded and Lipschitz on $\mathcal{K}$.

The following proposition states that $\hat{V}_n^{Q,(N,D)}$, built from the combination of time-discretization, k-nearest neighbors and optimal quantization methods, is a consistent estimator of the value function at time T_n, for $n = 0, \ldots, N - 1$. It provides a rate of convergence for the Qknn-estimations of the value functions.

PROPOSITION 30.4.1 *The estimators of the value functions provided by Qknn algorithm are consistent. Moreover, it holds as $M \to +\infty$:*

$$\left\|\hat{V}_n^{Q,(N,M)}\left(\hat{T}_n,\hat{Z}_n\right) - V_n\left(T_n,Z_n\right)\right\|_{M,2} = \mathcal{O}\left(\alpha^N + \frac{1}{M^{2/d}}\right),$$

for $n = 0,\ldots,N-1,$

where we denote by $\|\cdot\|_{M,2}$ the $\mathbb{L}^2(\mu)$ norm conditioned by the training sets that have been used to build the estimator $\hat{V}_{n+1}^{Q,(N,M)}$.

Proof Splitting the error of time cutting and quantization, we get:

$$\left\| V_n\left(T_n, Z_n\right) - \hat{V}_n^{(N,M)}\left(\hat{T}_n, \hat{Z}_n\right) \right\|_{M,2}$$

$$\leq \left\| V_n\left(T_n, Z_n\right) - V_n^{(N)}\left(T_n, Z_n\right) \right\|_{M,2} \tag{30.23}$$

$$+ \left\| V_n^{(N)}\left(T_n, Z_n\right) - \hat{V}_n^{(N,M)}\left(\hat{T}_n, \hat{Z}_n\right) \right\|_{M,2}.$$

Step 1: Applying Lemma 30.3.2, we get the following bound on the first term in the r.h.s. of (30.23):

$$\left\| V_n\left(T_n, Z_n\right) - V_n^{(N)}\left(T_n, Z_n\right) \right\|_{M,2} \leq \frac{\alpha^N}{1-\alpha} \|b\|_\infty, \tag{30.24}$$

where $\|b\|_\infty$ stands for the supremum of b over $[0, T] \times E$.

Step 2: Note that the assumptions of Theorem 30.4.1 are met as noticed in Remark 30.4.3, so that the latter provides the following bound for the second term in the r.h.s. of (30.23):

$$\left\| V_n^{(N)}\left(T_n, Z_n\right) - \hat{V}_n^{Q,(N,M)}\left(\hat{T}_n, \hat{Z}_n\right) \right\|_{M,2} \underset{M\to\infty}{=} \mathcal{O}\left(\frac{1}{M^{2/d}}\right). \tag{30.25}$$

It remains to plug (30.24) and (30.25) into (30.23) to complete the proof of Proposition 30.4.1.

30.4.4 Numerical Results

In this section, we propose several settings to test the efficiency of Qknn on simulated order books. We take no running reward, i.e. $f = 0$, and take the wealth of the market maker after liquidating their inventory as terminal reward, i.e. $g(z) = x + L(y)$. The intensities are taken constant in the first tests, and state dependent in the second ones. The values of the state dependent intensities are similar to the ones in Huang *et al.* (2015). Although the intensities are assumed to be uncontrolled in Section 30.3 for predictability reasons, the latter are controlled processes in this section, i.e. the intensities of the order arrivals depends on the orders in the order book from all the participant plus the ones of the market maker. The optimal trading strategies have been computed among two different classes of strategies: in Section 30.4.4.1, we tested the algorithm to approximate the optimal strategy among those where the market maker is only allowed to place orders only at the best bid and the best ask; in Section 30.4.4.2, we computed the optimal trading strategy among the class of the strategies where the market maker allows herself to place orders on the two best limits on each side of the order book. Note that the second class of controls is more general than the first one. The code is available on https://github.com/comeh.

The search of the k nearest neighbors, that arises when estimating the conditional expectations using the Qknn algorithm, is very time-consuming; especially in the considered market- making problem which is of dimension more than 10. The efficiency of Qknn then highly depends on the algorithm used to find the k nearest neighbors in high-dimension. We implemented the method using the Fast Library for Approximate Nearest Neighbors algorithm (FLANN), introduced in Muja and Lowe (2009) and already available as a library of C + +, Python, Julia and many other languages. This algorithm is based on tree methods. Note that recent algorithms based on graph also proved to perform well and can also be used.

30.4.4.1 Case 1: The Market Maker only Place Orders at the Best Ask and Best Bid

Denote by $\mathcal{A}$llim the class of controls where the placements of orders are allowed on the best ask and best bid exclusively. We implement the Qknn algorithm to compute the optimal strategy among those in A1lim. We then compared the optimal strategy with a naive strategy which consists in always placing one order at the best bid and one order at the best ask. The naive strategy is called 11 in the plots, and can be seen as a benchmark. The naive strategy is a good benchmark when the model for the intensities of order arrivals is symmetrical, i.e. the intensities for the bid and the ask sides are the same. Indeed, in this case, the market maker can expect to earn the spread in average.

In Figure 30.3(a), we take constant intensities to model the limit and market orders arrivals, and linear intensity to model the cancel orders. In this setting, as we can see in the figure, the strategy computed using Qknn algorithm performs as well as the naive strategy. Note that, obviously, the market maker has to take enough points for the state quantization in order for Qknn algorithm to perform well. In Figure 30.3(b), we plotted the P& L of the market maker when the latter compute the optimal strategy using only 6000 points for the state space discretization, and for such a low number of points for the grid, Qknn algorithm performs poorly. In this setting, notice that the naive strategy seems to perform well.

In Figure 30.4, we plotted the empirical histogram of the P& L of the market maker using the Qknn-estimated optimal strategy, computed with grids of size $N = 10^3, 10^4, 10^5, 10^6$ for the state space discretization; and the empirical histogram of the P &L of the market maker using the naive strategy. We took intensities that are state dependent. One can see that the larger the size of the grids are, the better the Qknn-estimation of the optimal strategy is.

In Figure 30.5, we plot the P &L of the market maker following the Qknn-estimated optimal strategy and the naive strategy. We took the same parameters as in Figure 30.4 to run the simulations except from the terminal time that we set to be equal to $T = 10$. In this setting, the Qknn-estimated optimal strategy performs much better than the naive strategy, which highlights the fact that the naive strategy is not optimal.

We plotted in Figure 30.6 the reaction of Qknn when a trend is added in the dynamic of the market. In this example, we took a higher intensity for the sell market order than the one for the buy market order, which creates an artificial positive trend in the dynamic of the price. Observe that Qknn understood correctly that it is better not to sell when the price goes up.

30.4.4.2 Case 2: The Market Maker Place Orders on the First Two Limits of the Orders Book

We extend the class of admissible controls to the ones where the market maker places order on the first two limits on the bid and ask sides of the order book. Denote by *A2lim* the latter. We run simulations to test the Qknn algorithm on A2lim. In Figures 30.7 and 30.8, we plot the empirical distributions of the P& L when the market maker follows the three different strategies:

- Qknn-estimated optimal strategy among those in A2lim (PLOpt2lim).
- Qknn-estimated optimal strategy among those in A1lim (PLOpt1lim).
- naive strategy, i.e. always place orders on the best bid and best ask queues (PL11).

Note that the P& L of the market maker is always better when the class of admissible controls is extended, see Figure 30.7; but in some numerical tests, the extended set of controls does not seem to improve the P &L. Indeed, we observed that the terminal P& L estimated using Qknn among A2lim and A1lim have the similar empirical distribution in the tests whose results are presented in Figure 30.8.

30.5 Model Extension to Hawkes Processes

We consider in this section a market maker who aims at maximizing a function of their terminal wealth, penalizing their inventory at terminal time T in the case where the orders arrivals are driven by Hawkes processes. Let us first present the model with Hawkes processes for the LOB.

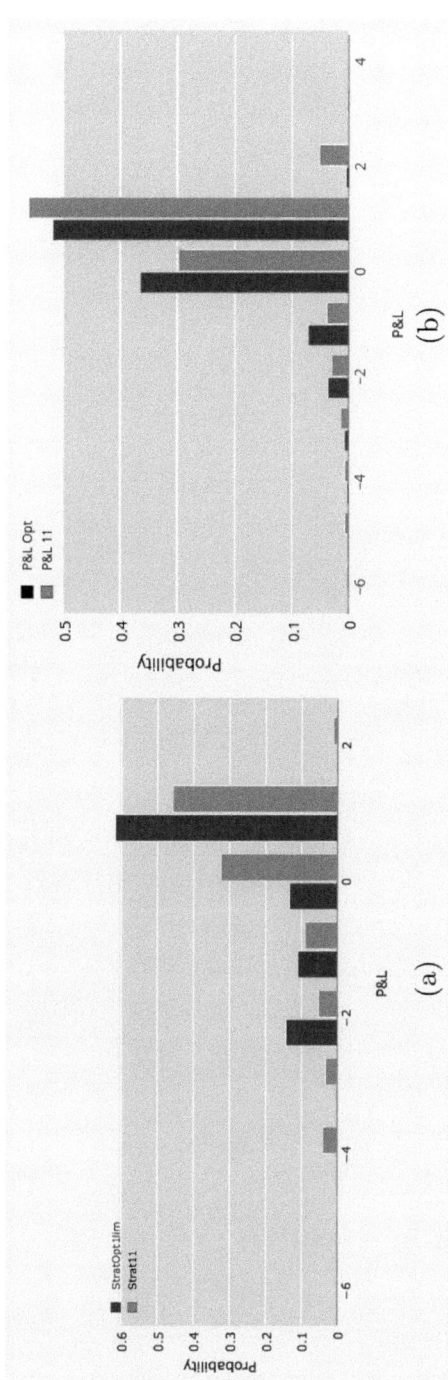

FIGURE 30.3 Histogram of the P&L when following the Qknn estimated optimal strategy (Opt) and the naive strategy (30.11). We took symmetrical and constant intensities, and a short terminal time $T = 1$. Notice that the Qknn strategy looks to improve the P&L by reducing the losses when enough points are taken to build the grids (see Figure 3(a)), and that its performance is worse if less points are taken to build the grids (see Figure 3(b)). (a) 10^5 points for the grids and (b) 6000 points for the grids.

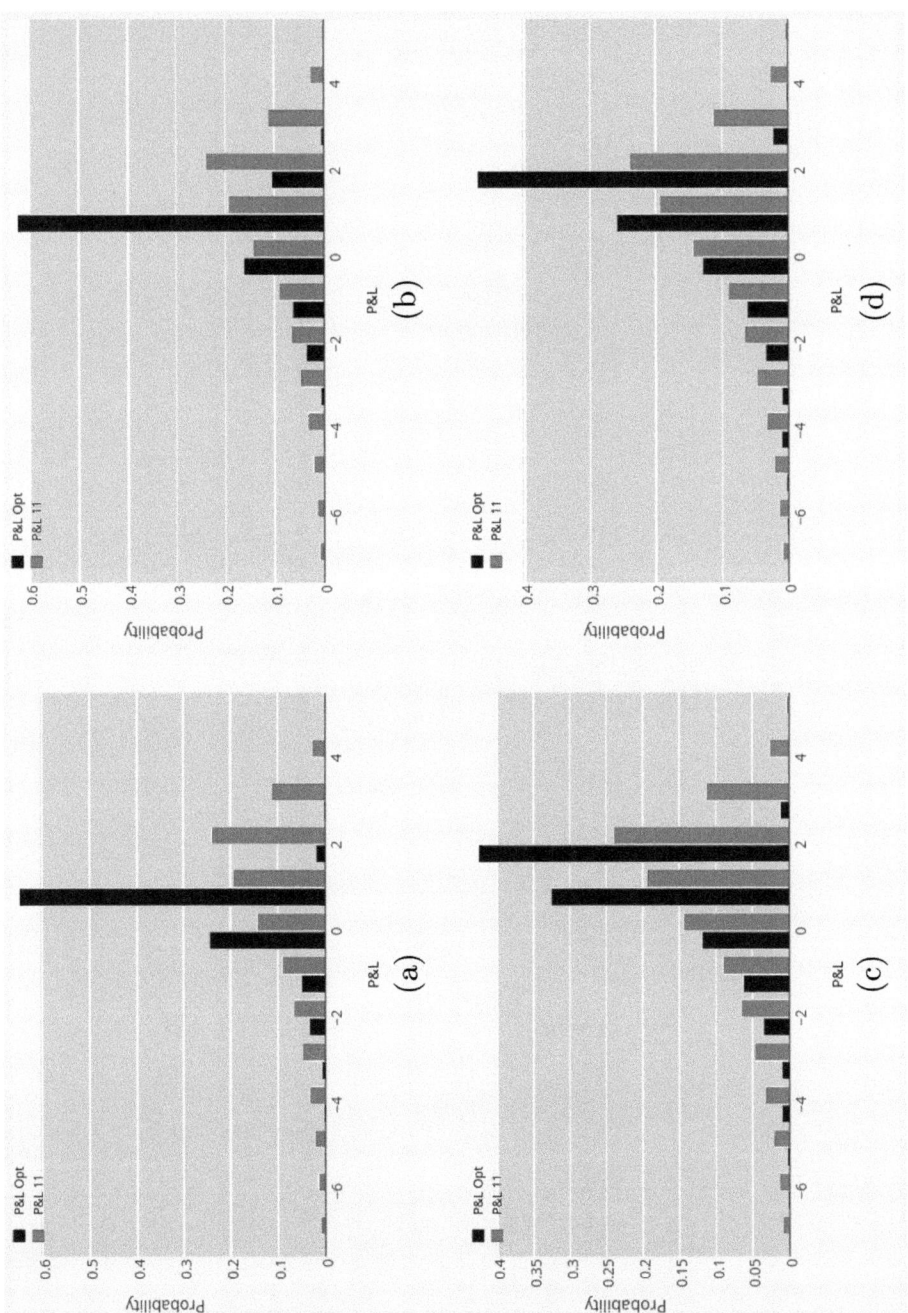

FIGURE 30.4   Histogram of the P&L, when the market maker follow the Qknn estimated optimal strategy (Opt) and the naive strategy (30.11). The intensities λ^M, λ_i^L and λ^C depend on the state of the order book. The P&L of the market maker when following the Qknn-estimated optimal strategy is computed using 10^3, 10^4, 10^5, and 10^6 points for the grids: see respectively Figure 4(a–d). The reader can see that the market maker increases their expected terminal wealth (after liquidation) by taking more and more points for the state space discretization. Also, the naive strategy is beaten when the intensities are state dependent.

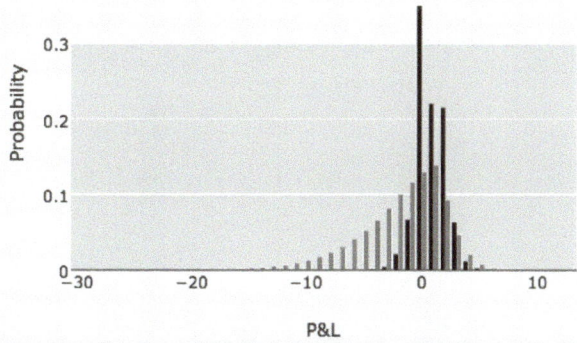

FIGURE 30.5 Distribution of the P&L when the market maker follows the optimal strategy (dark blue) and the naive strategy (light blue). We took symmetrical and state dependent intensities; and long terminal time: $T = 10$. Notice that the Qknn strategy does better than the naive strategy when the intensities are state dependent.

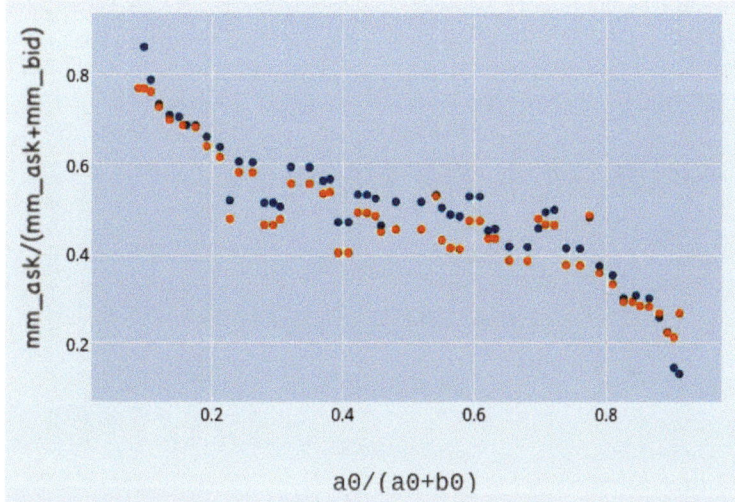

FIGURE 30.6 Reaction of Qknn to an artificial positive trend in the LOB. The y-axis represents the ratio of orders sent on the ask side by the market maker to their orders sent in the ask and bid sides in the order book. The x-axis is represents the ratio of the size of the best-ask limit to the sum of the sizes of the best-bid and best-ask limits. The blue points are those without trend in the market. The orange points are those with a positive trend in the market. We can see that Qknn took correctly the trend into account in its decisions: for two same order books, Qknn is less willing to sell when the price is expected to increase.

Model for the LOB: We assume that the order book receives limit, cancel, and market orders. We denote by L^+ (resp. L^-) the limit order arrivals process the ask (resp. bid) side; by C^+

(resp. C^-) the cancel order on the ask (resp. bid) side; and by M^+ (resp. M^-) the buy (resp. sell) market order arrivals processes. In this section, the limit orders arrivals are assumed to follow Hawkes processes dynamics, and moreover we assume the kernel to be exponential. The order arrivals are then modeled by a $(4K + 2)$-variate Hawkes process (N_t) with a vector of exogenous intensities λ_0 and exponential kernel ϕ, i.e. $\phi^{ij}(t) = \alpha_{ij}\beta e^{\beta t} \mathbf{1}_{t\geq 0}$.

Let $\alpha = (\alpha^{ij})_{i,j}$. We assume: $(\alpha)_{i,j}$ to have spectral radius strictly smaller than 1, which is a sufficient condition to guarantee stationarity of the model and convergence of $\mathbb{E}[\lambda_u \mid \mathcal{F}_t]$ as $u \to +\infty$, as shown e.g. in Hawkes (1971).

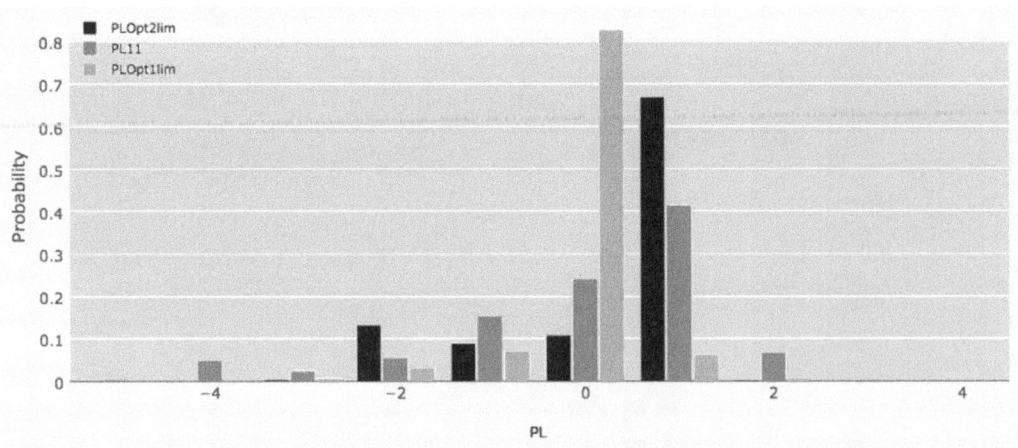

FIGURE 30.7   P& L of the market maker who follows optimal strategies and the naive strategy (PL11). Short Terminal Time. asymmetrical intensities for the market order arrivals: the intensity for the buying market order process is taken higher than the one for the selling market order process. The wealth of the market maker is greater when she places orders on the two first limits of each sides of the order book, rather than when she places orders only on the best limits at the bid and ask sides.

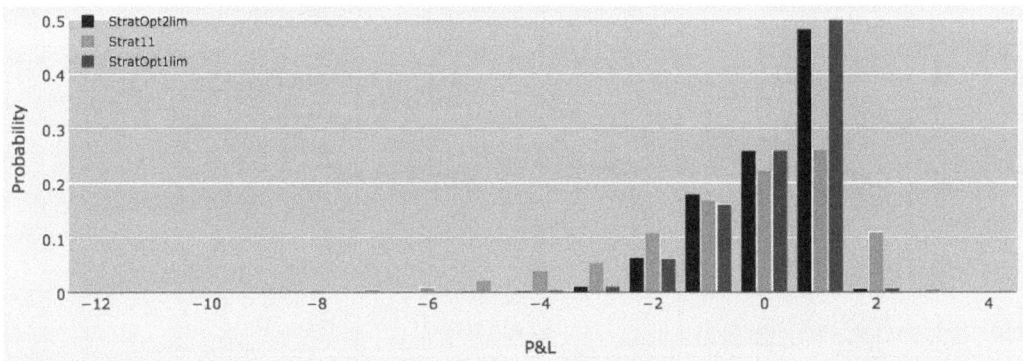

FIGURE 30.8   P& L when following the optimal strategy or the naive strategy (PL11). Long Terminal Time. Symmetrical intensities for the arrival of market orders. 4.10^4 points for the quantization. Notice that the Qknn strategy computed on the extended class of controls, i.e. order placements on the two first limits (StratOpt2lim), performs as well as the one computed on the original class of controls, i.e. order placements on the best-bid and best-ask (StratOpt1lim).

Note that in the presented model, the following holds:

(Hλ) λ is assumed to be independent of the control.

Denoting by $D = 4K + 2$ the dimension of (N_t), the mth component of the intensity λ of N_t writes, under **(Hλ)**:

$$\lambda_t^m = \lambda_0^m + \sum_{j=1}^{D} \alpha_{mj} \int_0^t e^{-\beta(t-s)} \mathrm{d}N_s^j, \quad \text{for } m = 1,\dots,D,$$

It is well-known that for this choice of intensity, the couple $(N_t, \lambda_t)_{t \geq 0}$ becomes Markovian, see e.g. Lemma 6 in Mas-soulié (1998) for a proof of this result, and moreover we have:

$$d\lambda_t^m = -\beta D\left(\lambda_t^m - \lambda_0^m\right)dt + \sum_{j=1}^{D} \alpha_{mj} dN_t^j, \quad \text{for } m = 1, \ldots, D,$$

with given initial conditions: $\lambda_0^m \in \mathbb{R}_+^*$ for $m = 1, \ldots, D$.

We can now rewrite the control problem (1) in the particular case where the order book is driven by Hawkes processes, there is no running reward, i.e. $f = 0$, and where the terminal reward G stands for the terminal wealth of the market maker penalized by their inventory. We then consider the following problem in this section:

$$V(t, \lambda, z) := \sup_{\alpha \in \mathbb{A}} \mathbb{E}_{t,z,\lambda}^{\alpha}\left[G\left(Z_T\right)\right], \tag{30.26}$$

where $G(z)$ denotes the wealth of the market maker when the controlled order book is at state z, plus a term of penalization of their inventory; and where $\mathbb{A}$ is the set of the admissible controls, i.e. the predictable decisions taken by the market maker until a terminal time $T > 0$.

We now present the main result of this section.

THEOREM 30.5.1 *V is characterized as the unique solution of the following HJB equation:*

$$f(T, z, \lambda) = G(z), \quad \text{for } z \in E$$

$$0 = \frac{\partial f}{\partial t}(t, z, \lambda) - D\beta \sum_{m=1}^{D}\left[\left(\lambda^m - \lambda_0^m\right)\frac{\partial f}{\partial \lambda^m}(t, z, \lambda)\right.$$

$$\left. + \lambda^m \sup_{a \in A_z}\left[f\left(t, e_m^a(z), \lambda + \alpha_m\right) - f(t, z, \lambda)\right]\right], \tag{30.27}$$

for $0 \leq t < T$, and $(t, z, \lambda) \in \mathbb{R}_+ \times E \times \mathbb{R}_+^*$.

where $\alpha_m = (\alpha_{1m}, \ldots, \alpha_{Dm})$. Moreover, V admits the following representation:

$$V(t, z, \lambda) = \sup_{\alpha \in \mathbb{A}} \sum_{n=0}^{\infty} \mathbb{E}_{t,z,\lambda}^{\alpha}\left[1_{T_n \leq T} G(Z_{T_n}^{\alpha}) \exp\left\{-\left|\lambda_0\right|\left(T - T_n\right)\right.\right.$$

$$\left.\left. + \sum_{m=1}^{D} \frac{\lambda_{T_n}^m - \lambda_0^m}{D\beta}\left(e^{-\sum_{j=1}^{D} \beta_{mj}(T - T_n)} - 1\right)\right\}\right], \tag{30.28}$$

where, for $n \geq 0$, T_n stands for the nth jump time of Z after time t, and $\left(Z_{T_n}^{\alpha}, \lambda_{T_n}\right)_{n=0}^{\infty}$ is as a MDP controlled by $\alpha \in \mathbb{A}$; and where $\mathbb{E}_{t,z,\lambda}^{\alpha}[.]$ stands for the expectation conditioned by $Z_t = z$, $\lambda_t = \lambda$ when the control α is followed.

REMARK 30.5.1 *V is characterized in (30.28) as the value function associated to an MDP with infinite horizon, where the running reward reads:*

$$r(t, z, \lambda) = 1_{t \leq T} G(z) \exp\left\{-\left|\lambda_0\right|_1 (T - t)\right.$$

$$\left. + \sum_{m=1}^{D} \frac{\lambda^m - \lambda_0^m}{D\beta}\left(e^{-D\beta(T-t)} - 1\right)\right\},$$

where $|\cdot|_1$ denotes the $\mathbb{L}^1\left(\mathbb{R}^D\right)$ norm.

Proof of Theorem 30.5.1. Step 1: Let us check that (30.28) holds, where V is defined as solution of (30.26).

First notice that $\left(\lambda_t, Z_t\right)_t$ is a PDMDP,[*] i.e. $\left(\lambda_t, Z_t\right)_t$ is deterministic between two jumping times. We then aim at rewriting the expression of the value function defined in (30.26) as the value function associated to a infinite horizon control problem of the PDMDP $\left(\lambda_t, Z_t\right)_t$. To do so, we first notice that by conditioning on the time jumps we get:

$$
\begin{aligned}
V(t,z,\lambda) &= \sup_{\alpha \in \mathbb{A}} \mathbb{E}_{t,z,\lambda}^{\alpha}\left[G\left(Z_T^{\alpha}\right)\right] \\
&= \sup_{\alpha \in \mathbb{A}} \mathbb{E}_{t,z,\lambda}^{\alpha}\left[\sum_{n=0}^{\infty} 1_{T_n \leq T < T_{n+1}} G\left(Z_{T_n}^{\alpha}\right)\right] \\
&= \sup_{\alpha \in \mathbb{A}} \sum_{n=0}^{\infty} \mathbb{E}_{t,z,\lambda}^{\alpha}\left[1_{T_n \leq T} G\left(Z_{T_n}^{\alpha}\right) \mathbb{P} \right. \\
&\qquad \left. \times \left(T - T_n \leq T_{n+1} - T_n \mid T_n\right)\right],
\end{aligned}
\tag{30.29}
$$

where $(T_n)_n$ is the sequence of jump times of N. This process is a jump process with intensity $\mu_s = \sum_{m=1}^{D} \lambda_s^m$. Since it holds, conditioned on $\mathcal{F}_{T_n}$:

$$
\mu_s = \sum_{m=1}^{D} \lambda_0^m + \left(\lambda_{T_n}^m - \lambda_0^m\right)e^{-D\beta\left(s-T_n\right)}, \quad \text{for} s \in \left[T_n, T_{n+1}\right),
$$

then, we have:

$$
\begin{aligned}
&\mathbb{P}\left(T_{n+1} - T_n \geq T - T_n \mid T_n\right) \\
&= \int_{T-T_n}^{\infty} \mu_s e^{-\int_0^s \mu_u du} ds \\
&= \exp\left\{-\left|\lambda_0\right|\left(T - T_n\right) + \sum_{m=1}^{D} \frac{\lambda_{T_n}^m - \lambda_0^m}{D\beta}\left(e^{-D\beta(T-T_n)} - 1\right)\right\}.
\end{aligned}
\tag{30.30}
$$

Plugging (30.30) into (30.29), the value function rewrites:

$$
\begin{aligned}
V(t,z,\lambda) &= \sup_{\alpha \in \mathbb{A}} \sum_{n=0}^{\infty} \mathbb{E}_{t,z,\lambda}^{\alpha}\left[1_{T_n \leq T} G(Z_{T_n}^{\alpha}) \exp\left\{-\left|\lambda_0\right|\left(T - T_n\right)\right.\right. \\
&\qquad \left.\left. + \sum_{m=1}^{D} \frac{\lambda_{T_n}^m - \lambda_0^m}{D\beta}\left(e^{-D\beta(T-T_n)} - 1\right)\right\}\right],
\end{aligned}
\tag{30.31}
$$

which completes the step 1. The r.h.s of (30.31) can be seen as the value function of an infinite horizon control problem associated to the PDMDP.

Step 2: Let us show that V is the unique solution to (30.27). Notice first that the solutions to the following HJB equation

---

[*] PDMDP stands for Piecewise Deterministic Markov Decision Process, which is a MDP whose dynamic is deterministic between jumps.

$$G(z) = f(T, z, \lambda)$$

$$0 = \frac{\partial f}{\partial t} - \sum_{m=1}^{D} D\beta \left(\lambda^m - \lambda_0^m \right) \frac{\partial f}{\partial \lambda^m}$$

$$+ \lambda^m \sup_{a \in A_z} \left[f\left(t, e_m^a(z), \lambda + \alpha_m \right) - f(t, z, \lambda) \right],$$

for $0 \leq t < T$.

are the fixed points of the operator $\mathcal{T} = \mathcal{T}_1 \circ \mathcal{T}_2$ where $\mathcal{T}_1$ and $\mathcal{T}_2$ are defined as follows:

$$\mathcal{T}_1 : F \mapsto f \text{ solution of } \begin{cases} \dfrac{\partial f}{\partial t} - D\beta \sum_{m=1}^{D} \left(\lambda^m - \lambda_0^m \right) \dfrac{\partial f}{\partial \lambda^m} \\ \qquad = F(t, z, \lambda) \\ f(T, z, \lambda) = G(z), \end{cases}$$

and:

$$\mathcal{T}_2 : f \mapsto -\sum_{m=1}^{D} \lambda^m \sup_{a \in A_z} \left[f\left(t, e_m^a(z), \lambda + \alpha_m \right) - f(t, z, \lambda) \right].$$

We now use the characteristic method to rewrite the image of $\mathcal{T}_1$.

Let us take function F, and define $f = \mathcal{T}_1(F)$. Let us fix $t \in [0, T]$ and $\lambda \in (\mathbb{R}_+)^D$, and denote by g the function $g(s, z) = f\left(s, z, \lambda_s^1, \ldots, \lambda_s^D \right)$ where, for $m = 1, \ldots, D, s \mapsto \lambda_s^m$ is a differentiable function defined on $[t, T]$ as solution to the following ODE:

$$\frac{d\lambda_s^m}{ds} = -D\beta \left(\lambda_s^m - \lambda_0^m \right), \quad \text{for all } t < s \leq T, \tag{30.32}$$

$$\lambda_t^m = \lambda^m.$$

For $m = 1, \ldots, D$, basic theory on ODE provides existence and uniqueness of a solution to (30.32), which is given by:

$$\lambda_s^m = \lambda_0^m + \left(\lambda^m - \lambda_0^m \right) e^{-D\beta(s-t)}, \quad \text{for } s \in [t, T],$$

and $m = 1, \ldots, D$.

Since

$$\frac{\partial g}{\partial s} = \frac{\partial f}{\partial s} + \sum_{m=1}^{D} \frac{d\lambda_s^m}{ds} \frac{\partial f}{\partial \lambda^m},$$

then $g(t, z) = G(z) - \displaystyle\int_t^T F\left(s, z, \lambda_s \right) ds$, which finally leads to the following expression of $\mathcal{T}_1(F)$:

$$\mathcal{T}_1(F) = f(t, z, \lambda) = G(z) - \int_t^T F\left(s, z, \lambda_s \right) ds. \tag{30.33}$$

Replacing F by $\mathcal{T}_2(f)$ in (30.33), we get that f is fixed point of $\mathcal{T}_1 \circ \mathcal{T}_2$ if and only if:

$$f(t,\lambda,z) + \sum_{m=1}^{D} \int_t^T \lambda_s^m f(s,z,\lambda_s) ds$$

$$= G(z) - \sum_{m=1}^{D} \int_t^T \lambda_s^m \sup_{a\in A_z} f(s,e_m^a(z),\lambda_s + \alpha_m) ds.$$

Notice

$$\frac{\partial f(s,\lambda_s,z) e^{-\sum_{j=1}^{D}\int_t^s \lambda_u^j du}}{\partial s}$$

$$= -\sum_{m=1}^{D} \lambda_s^m e^{-\sum_{j=1}^{D}\int_t^s \lambda_u^j du} \sup_{a\in A_z} f(s,e_m^a(z),\lambda_s + \alpha_m),$$

so that:

$$f(t,\lambda,z) = G(z) e^{-\sum_{m=1}^{D}\int_t^T \lambda_s^m ds}$$

$$+ \sum_{m=1}^{D} \int_t^T \lambda_s^m e^{-\int_t^s \lambda_u du} \sup_{a\in A_z} f(s,e_m^a(z),\lambda_s + \alpha_m) ds \tag{30.34}$$

$$= G(z) e^{-\sum_{m=1}^{D}\int_t^T \lambda_s^m ds}$$

$$+ \sup_{a\in A_z} \mathbb{E}_{t,\lambda,z}^a \left[f(T_1,Z_1,\lambda_{T_1} + \alpha_m) \right],$$

where T_1 is the first jump time of N larger than t, we denote $Z_1 = Z_{T_1}$. Equation (30.34) shows that the fixed point of $\mathcal{T}_1 \circ \mathcal{T}_2$ is characterized as the fixed point of the operator $\mathcal{T}$ defined for any smooth enough function f by:

$$\mathcal{T}(f) = G(z) e^{-\sum_{m=1}^{D}\int_t^T \lambda_s^m ds} + \sup_{a\in A_z} \mathbb{E}_{t,\lambda,z}^a \left[f(T_1,Z_1,\lambda_{T_1} + \alpha_m) \right],$$

where $\mathbb{E}_{t,\lambda,z}^a[.]$ stands for the expectation conditioned by the events $\lambda_t = \lambda$ and $Z_t = z$, when decision a is taken at time t. We recognize here the maximal reward operator of the value function defined in (30.31). Basic theory on PDMDP shows that the maximal reward operator $\mathcal{T}$ admits V as unique fixed point, which completes step 2.

30.6 Conclusion

In this paper, we solved theoretically and numerically a general market-making problem with different microstructural models of order books, by rewriting the problem as a Markov decision process with infinite horizon. This new representation offers a nice and simple characterization of the optimal strategy of market-making, which is implementable relying for example on some quantization and control randomization ideas as proposed in this paper. Others algorithms, like those based on reinforcement learning (see e.g. Chapter 6.5 of Sutton and Barto 2011 for an introduction to (deep) Q-learning, and/or Guéant and Manziuk 2019 for its application to market-making), look particularly well-adapted to solve the market-making problem using its MDP representation, especially in the context of high- dimension.

The proposed methodology can be adapted to solve theoretically and numerically control problems associated to a general class of controlled point processes.

Acknowledgments

We would like to thank the anonymous referees for their valuable comments on the first version of the paper.

Disclosure Statement

No potential conflict of interest was reported by the authors.

ORCID

Côme Huré http://orcid.org/0000-0002-8196-7445

References

Abergel, F., Jedidi, A. and Muni Toke, I., *Limit Order Books*, 2016 (Cambridge University Press: Cambridge).

Avellaneda, M. and Stoikov, S., High-frequency trading in a limit order book. *Quant. Finance*, 2007, **8**, 217–224.

Balata, A. and Palczewski, J., Regress-Later Monte Carlo for optimal control of Markov processes. eprint arXiv:1712.09705, 2017.

Balata, A., Huré, C., Laurière, M., Pham, H. and Pimentel, I., A class of finite-dimensional numerically solvable McKean-Vlasov control problems. *ESAIM Proc. Surv.*, 2019, **65**, 114–144.

Baradel, N., Bouchard, B., Evangelista, D. and Mounjid, O., Optimal inventory management and order book modeling. arXiv:1802.08135, 2018.

Bäuerle, N. and Rieder, U., *Markov Decision Processes with Applications to Finance*, 2011 (Springer: Berlin).

Cartea, A. and Jaimungal, S., Modeling asset prices for algorithmic and high frequency trading. *Appl. Math. Finance*, 2010, **20**(6), 512–547.

Cartea, A. and Jaimungal, S., Risk metrics and fine tuning of high-frequency trading strategies. *Math. Finance*, 2013, **25**(3), 576–611.

Cartea, A., Jaimungal, S. and Ricci, J., Buy low sell high: A high frequency trading perspective. *SIAM J. Financ. Math.*, 2014, **5**(1), 415–444.

Cartea, A., Jaimungal, S. and Penalva, J., *Algorithmic and High-Frequency Trading*, 2015 (Cambridge University Press: Cam- bridge).

Cont, R., Stoikov, S. and Talreja, R., A stochastic model for order book dynamics. *Oper. Res.*, 2007, **58**, 549–563.

El Aoud, S. and Abergel, F., A stochastic control approach to option market making. *Mark. Microstruct. Liquidity*, 2015, **1**(1), 1550006.

Fiorin, L., Pagès, G. and Sagna, A., Product Markovian quantiza- tion of a diffusion process with appli- cations to finance. *Methodol. Comput. Appl. Probab.*, 2019, **21**(4), 1087–1118.

Fodra, P. and Pham, H., High frequency trading and asymptotics for small risk aversion in a Markov renewal model. *SIAM J. Financ. Math.*, 2015a, **6**(1), 656–684.

Fodra, P. and Pham, H., Semi Markov model for market microstructure. *Appl. Math. Financ.*, 2015b, **22**(3), 261–295.

Graf, S. and Luschgy, H., *Foundations of Quantization for Probability Distributions*, Vol. 1730, 2000 (Springer-Verlag: Berlin Heidelberg).

Guéant, O., *The Financial Mathematics of Market Liquidity*, 2016 (Chapman and Hall/CRC: New York).

Guéant, O. and Manziuk, I., Deep reinforcement learning for market making in corporate bonds: Beating the curse of dimensionality. ar Xiv preprint arXiv:1910.13205, 2019.

Guéant, O., Lahalle, C.A. and Fernandez-Tapia, J., Dealing with the inventory risk. *Math. Financ. Econ.*, 2012, **7**, 477–507.

Guilbaud, F. and Pham, H., Optimal high frequency trading with limit and market orders. *Quant. Finance*, 2013, **13**(1), 79–94.

Gyorfi, L., Kohler, M., Krzyzak, A. and Walk, H., *A Distribution-Free Theory of Nonparametric Regression*, 2002 (Springer- Verlag: New York).

Hawkes, A.G., Spectra of some self-exciting and mutually exciting point processes. *Biometrika*, 1971, **58**, 83–90.

Ho, T. and Stoll, H., Optimal dealer pricing under transactions and return uncertainty. *J. Financ. Econ.*, 1979, **9**, 47–73.

Huang, W., Lehalle, C.A. and Rosenbaum, M., Simulating and analyzing order book data: The queue-reactive model. *J. Am. Stat. Assoc.*, 2015, **110**(509), 107–122.

Huré, C., Pham, H., Bachouch, A. and Langrené, N., Deep neural networks algorithms for stochastic control problems on finite horizon: Convergence analysis. arXiv:1812.04300, 2018.

Jacquier, A. and Liu, H., Optimal liquidation in a level-I limit order book for large tick stocks. *SIFIN*, 2018, **9**(1), 875–906.

Kharroubi, I., Langrené, N. and Pham, H., A numerical algorithm for fully nonlinear HJB equations: An approach by control randomization. *Monte Carlo Method. Appl.*, 2014, **20**, 145–165.

Lehalle, C.A., Othmane, M. and Rosenbaum, M., Optimal liquidity- based trading tactics. arXiv:1803.05690, 2018.

Massoulié, L., Stability results for a general class of interacting point processes dynamics, and applications. *Stoch. Process. Appl.*, 1998, **75**(1), 1–30.

Muja, M. and Lowe, D., Fast approximate nearest neighbors with automatic algorithm configuration. International Conference on Computer Vision Theory and Applications (VISAPP), 2009.

Pagès, G., Pham, H. and Printems, J., Optimal quantization methods and applications to numerical problems in finance. In *Handbook on Numerical Methods in Finance*, edited by S.T. Rachev and G.A. Anastassiou, chap. 7, pp. 253–298, 2004 (Birkhäuser: Boston).

Rosu, I., A dynamic model of the limit order book. Rev. Financ. Stud., 2008, 22, 4601–4641.

Sutton, R.S. and Barto, A.G., Reinforcement Learning: An Introduc- tion, 2011 (The MIT Press: Cambridge, MA).

Appendices

Appendix A: Proof of Theorem 30.4.1 and Corollary 30.4.1

We divided the proofs of Theorem 30.4.1 and Corollary 30.4.1 into the following Lemmas.

Lemma A.1 aims at bounding the projection error. It relies on Gyorfi *et al.* (2002), see p.93, as well as Zador's theorem, stated in Section B for the sake of completeness.

LEMMA A.1 *Assume $d \geq 3$, and take $K = M^{d+2}$ points for the optimal quantization of ε_n, then it holds under (**Hμ**) and (**HF**), as $M \to +\infty$,*

$$\varepsilon_n^{proj} = \mathcal{O}\left(\frac{1}{M^{1/d}}\right), \tag{A1}$$

where we remind that $\varepsilon_n^{proj} := \sup_{a \in A}||\mathrm{Proj}_{n+1}\left(F\left(X_n, a, \hat{\varepsilon}_n\right)\right) - F\left(X_n, a, \hat{\varepsilon}_n\right)||_2$ stands for the average projection error.

Proof Let us take $\eta > 0$, and observe that

$$\mathbb{P}\left(\left| \text{Proj}_{n+1}\left[F\left(X_n, a, \hat{\varepsilon}_{n+1} \right) \right] - F\left(X_n, a, \hat{\varepsilon}_{n+1} \right) \right|^2 > \eta \right)$$

$$= \mathbb{E}\left[\prod_{m=1}^{M} \mathbb{E}\left[\mathbf{1}_{\left| X_{n+1}^{t,(m)} - F\left(X_n, a, \hat{\varepsilon}_{n+1} \right) \right| > \sqrt{\eta}} \mid X_n, \hat{\varepsilon}_{n+1} \right] \right]$$

$$= \mathbb{E}\left[\left(1 - \mu\left[B\left(F\left(X_n, a, \hat{\varepsilon}_{n+1} \right), \sqrt{\eta} \right) \right] \right)^M \right],$$

where for all $x \in E$ and $\eta > 0$, $B(x, \eta)$ denote the ball of center x and radius η. Since $x \mapsto (1 - x)^M$ is M-Lipschitz, we get by application of Zador's theorem:

$$\mathbb{P}\left(\left| \text{Proj}_{n+1}\left[F\left(X_n, a, \hat{\varepsilon}_{n+1} \right) \right] - F\left(X_n, a, \hat{\varepsilon}_{n+1} \right) \right|^2 > \eta \right)$$

$$\leq M[F]_L[\mu]_L \left\| \hat{\varepsilon}_{n+1} - \varepsilon_{n+1} \right\|_2$$

$$+ \mathbb{E}\left[\left(1 - \mu\left(B\left(F\left(X_n, a, \varepsilon_{n+1} \right), \sqrt{\eta} \right) \right) \right)^M \right]$$

$$= \frac{M[F]_L[\mu]_L}{K^{1/d}} + \mathbb{E}\left[\left(1 - \mu\left(B\left(F\left(X_n, a, \varepsilon_{n+1} \right), \sqrt{\eta} \right) \right) \right)^M \right]$$

$$+ \mathcal{O}\left(\frac{M}{K^{1/d}} \right),$$

as the number of points for the quantization of the exogenous noise K goes to $+\infty$, and where M stands for the size of the grids Γ_n.

Let us introduce $A_1, \dots, A_{N(\eta)}$, a cubic partition of $\text{Supp}(\mu)$, which is bounded under $(\mathbf{H}\mu)$, such that for all $j = 1, \dots, N(\eta)$, A_j has diameter η. Also, Notice that there exists $c > 0$, which only depends on $\text{Supp}(\mu)$, such as

$$N(\eta) \leq \frac{c}{\eta^d}. \tag{A2}$$

If $x \in Aj$, then $Aj \subset B(x, \eta)$, therefore:

$$\mathbb{E}\left[\left(1 - \mu\left(B\left(X_n, \eta \right) \right) \right)^M \right] = \sum_{j=1}^{N(\eta)} \int_{A_j} (1 - \mu(B(x, \eta)))^M \mu(dx)$$

$$\leq \sum_{j=1}^{N(\eta)} \int_{A_j} \left(1 - \mu\left(A_j \right) \right)^M \mu(dx). \tag{A3}$$

Also notice that:

$$\sum_{j=1}^{N(\eta)} \mu\left(A_j \right) \left(1 - \mu\left(A_j \right) \right)^M \leq \sum_{j=1}^{N(\eta)} \max_z z(1-z)^M \leq \frac{e^{-1}N(\eta)}{M}. \tag{A4}$$

Combining (A3) and (A4) leads to

$$\mathbb{E}\left[\left(1-\mu\left(B\left(X_n,\eta\right)\right)\right)^M\right] \leq \frac{e^{-1}N(\eta)}{M}. \tag{A5}$$

Let $L = 2\|\mu\|_\infty$ stands for the diameter of the support of μ. We then get, as $M \to +\infty$,

$$\mathbb{E}\left[\left|\mathrm{Proj}_{n+1}\left[F\left(X_n,a,\hat{\varepsilon}_{n+1}\right)\right]-F\left(X_n,a,\hat{\varepsilon}_{n+1}\right)\right|^2\right]$$

$$= \int_0^\infty \mathbb{P}\left(\left|\mathrm{Proj}_{n+1}\left[F\left(X_n,a,\hat{\varepsilon}_{n+1}\right)\right]-F\left(X_n,a,\hat{\varepsilon}_{n+1}\right)\right|^2 > \eta\right)d\eta$$

$$\leq \int_0^{L^2} \frac{M[F]_L[\mu]_L}{K^{2/d}} + \mathbb{P}\left(\left|\mathrm{Proj}_{n+1}\left[F\left(X_n,a,\hat{\varepsilon}_{n+1}\right)\right]\right.\right.$$

$$\left.\left. - F\left(X_n,a,\varepsilon_{n+1}\right)\right| > \sqrt{\eta}\right)d\eta$$

$$= \int_0^{L^2} \min\left(1,\frac{e^{-1}N(\sqrt{\eta})}{M}\right)d\eta + \mathcal{O}\left(\frac{M}{K^{1/d}}\right) \tag{A6}$$

$$= \int_0^{L^2} \min\left(1,\frac{c\eta^{-d/2}}{eM}\right)d\eta + \mathcal{O}\left(\frac{M}{K^{1/d}}\right)$$

$$= \int_0^{(c/(eM))^{(2/d)}} 1 d\eta + \int_{(c/(eM))^{(2/d)}}^{L^2} \frac{c\eta^{-d/2}}{eM}d\eta + \mathcal{O}\left(\frac{M}{K^{1/d}}\right)$$

$$= \frac{\tilde{c}^2}{M^{2/d}} + \mathcal{O}\left(\frac{M}{K^{1/d}}\right),$$

where $\tilde{c}$ is defined as $\tilde{c} := \sqrt{\frac{d}{d-2}}\left(\frac{c}{e}\right)^{1/d}$, and where we used (A5) and (A2) to go from the second to the third line. It remains to take $K = M^{d+1}$ points for the optimal quantization of the exogenous noise, and then take square root of equality (A6), in order to derive (A1).

LEMMA A.2 *Assume $d \geq 3$, take $K = M^{d+2}$ points for the optimal quantization of ε_n, and let $x \in E$. Then it holds under (Hμ) and (HF), as $M \to +\infty$:*

$$\varepsilon_n^{proj}(x) = \mathcal{O}\left(\frac{1}{M^{1/d}}\right),$$

where $\varepsilon_n^{proj}(x)$, defined as $\varepsilon_n^{proj}(x) := \sup_{a \in A}\|\mathrm{Proj}_{n+1}\left(F\left(x,a,\hat{\varepsilon}_n\right)\right)-F\left(x,a,\hat{\varepsilon}_n\right)\|_2$, stands for the later-projection error at state x.

Proof Following the same steps as those used to prove Lemma A.1, we show that:

$$\mathbb{P}\left(\left|\mathrm{Proj}_{n+1}\left[F\left(x,a,\hat{\varepsilon}_{n+1}\right)\right]-F\left(x,a,\hat{\varepsilon}_{n+1}\right)\right|^2 > \eta\right)$$

$$= \frac{M[F]_L[\mu]_L}{K^{1/d}} + \mathbb{E}\left[\left(1-\mu\left(B\left(F\left(x,a,\varepsilon_{n+1}\right),\sqrt{\eta}\right)\right)\right)^M\right]$$

$$+ \mathcal{O}\left(\frac{M}{K^{1/d}}\right),$$

as $K \to +\infty$, and moreover,

$$\mathbb{E}\left[\left(1 - \mu\left(B\left(F\left(x, a, \varepsilon_{n+1}\right), \sqrt{\eta}\right)\right)\right)^M\right] \leq \frac{e^{-1}N(\eta)}{M}.$$

holds, which is enough to complete the proof of Lemma A.2.

LEMMA A.3 *Under **(HF)**, for $n = 0, \dots, N$ there exists constant $\left[\hat{V}_n^Q\right]_L > 0$ such that for $x, x' \in E$, it holds as $M \to \infty$:*

$$\left|\hat{V}_n^Q(x) - \hat{V}_n^Q\left(x'\right)\right| \leq \left[\hat{V}_n^Q\right]_L \left|x - x'\right| + \mathcal{O}\left(\frac{1}{M^{1/d}}\right). \tag{A7}$$

Moreover, following bounds holds on $\left[\hat{V}_n^Q\right]_L$, for $n = 0, \dots, N$:

$$\left[\hat{V}_N^Q\right]_L \leq [g]_L$$

$$\left[\hat{V}_n^Q\right]_L \leq [f]_L + [F]_L \left[\hat{V}_{n+1}^Q\right]_L, \quad \text{for } n = 0, \dots, N-1. \tag{A8}$$

Proof Let us show that by induction that $\hat{V}_N^Q$ is Lipschitz. First, notice that (A7) holds at terminal time $n = N$, if one define $\left[\hat{V}_N^Q\right]_L$ as $\left[\hat{V}_N^{Q_N}\right]_L = [g]_L$. Let us take $x, x' \in E$. Assume $\left|\hat{V}_{n+1}^Q(x) - \hat{V}_{n+1}^Q\left(x'\right)\right| \leq \left[\hat{V}_{n+1}^Q\right]_L \left|x - x'\right| + \mathcal{O}\left(\frac{1}{M^{1/d}}\right)$ holds for some $n = 0, \dots, N-1$. Let us show that

$$\left|\hat{V}_n^Q(x) - \hat{V}_n^Q\left(x'\right)\right| \leq \left[\hat{V}_n^Q\right]_L \left|x - x'\right| + \mathcal{O}\left(\frac{1}{M^{1/d}}\right),$$

where $\left[\hat{V}_N^Q\right]_L$ is defined in (A8). Notice that, by the dynamic programming principle and the triangular inequality, it holds:

$$|\hat{V}_n^Q(x) - \hat{V}_n^Q\left(x'\right)|$$

$$\leq [f]_L \left|x - x'\right| + \sup_a \mathbb{E}_n^a\left[\hat{V}_{n+1}^Q\left(\text{Proj}_{n+1}\left(F\left(x, a, \hat{\varepsilon}_{n+1}\right)\right)\right)\right.$$

$$\left. - \hat{V}_{n+1}^Q\left(\text{Proj}_{n+1}\left(F\left(x', a, \hat{\varepsilon}_{n+1}\right)\right)\right)|\right]$$

$$\leq [f]_L \left|x - x'\right| + \left[\hat{V}_{n+1}^Q\right]_L \sup_a \mathbb{E}\left[|\text{Proj}_{n+1}\left(F\left(x, a, \hat{\varepsilon}_{n+1}\right)\right)\right.$$

$$\left. - F\left(x, a, \hat{\varepsilon}_{n+1}\right)|\right] + \mathcal{O}\left(\frac{1}{M^{1/d}}\right)$$

$$\leq \left([f]_L + \left[\hat{V}_{n+1}^Q\right]_L [F]_L\right)\left|x - x'\right| + \mathcal{O}\left(\frac{1}{M^{1/d}}\right)$$

$$\leq \left[\hat{V}_n^Q\right]_L \left|x - x'\right| + \mathcal{O}\left(\frac{1}{M^{1/d}}\right),$$

which completes the proof of (A7).

We now proceed to the proof of Theorem 30.4.1.

Proof of Theorem 30.4.1. Combining inequality $|u_1 + u_2 + u_3|^2 \leq 3(|u_1|^2 + |u_2|^2 + |u_3|^2)$ that holds for all $u_1, u_2, u_3 \in \mathbb{R}$ with inequality $|\sup_{i \in I} a_i - \sup_{i \in I} b_i| \leq \sup_{i \in I} |a_i - b_i|$ that holds for all families $(a_i)_{i \in I}$ and $(b_i)_{i \in I}$ of reals, and all set I, we have:

$$\| \hat{V}_n^Q (X_n) - V_n (X_n) \|_2^2$$

$$\leq 3 \mathbb{E} \Bigg[\sup_{a \in A} \mathbb{E}_{n,X_n} | \hat{V}_{n+1}^Q \big(\mathrm{Proj}_{n+1} \big(F(X_n, a, \hat{\varepsilon}_{n+1}) \big) \big)$$

$$- \hat{V}_{n+1}^Q \big(F(X_n, a, \hat{\varepsilon}_{n+1}) \big) \Big|^2$$

$$+ \sup_{a \in A} \mathbb{E}_{n,X_n} \Big| \hat{V}_{n+1}^Q \big(F(X_n, a, \hat{\varepsilon}_{n+1}) \big) - \hat{V}_{n+1}^Q \big(F(X_n, a, \varepsilon_{n+1}) \big) \Big|^2$$

$$+ \sup_{a \in A} \mathbb{E}_{n,X_n} \Big| \hat{V}_{n+1}^Q \big(F(X_n, a, \varepsilon_{n+1}) \big) - V_{n+1} \big(F(X_n, a, \varepsilon_{n+1}) \big) \Big|^2 \Bigg]$$

where $\mathbb{E}_{n,X_n}$ stands for the expectation conditioned by the state X_n at time n. It holds as $M \to +\infty$, using Lemma A.3:

$$\left\| \hat{V}_n^Q (X_n) - V_n (X_n) \right\|_2^2$$

$$\leq 3 \Big[\hat{V}_n^Q \Big]_L \mathbb{E} \Bigg[\sup_a \mathbb{E}_{n,X_n} \Big[\big| \mathrm{Proj}_{n+1} \big(F(X_n, a, \hat{\varepsilon}_{n+1}) \big) \right.$$

$$- F(X_n, a, \hat{\varepsilon}_{n+1}) \big|^2 \Big] \tag{A9}$$

$$+ \sup_a \mathbb{E}_{n,X_n} \Big[\big| F(X_n, a, \hat{\varepsilon}_{n+1}) - F(X_n, a, \varepsilon_{n+1}) \big|^2 \Big] \Bigg]$$

$$+ 3 \| r \|_\infty \mathbb{E} \Big[| \hat{V}_{n+1}^Q (X_{n+1})) - V_{n+1} (X_{n+1})) |^2 \Big] + \mathcal{O} \left(\frac{1}{M^{1/d}} \right)$$

Under **(HF)**, (A9) can then be rewritten as:

$$\left\| \hat{V}_n^Q (X_n) - V_n (X_n) \right\|_2^2$$

$$\leq 3 \Big[\hat{V}_n^Q \Big]_L \left([F]_L^2 \big(\epsilon_n^Q \big)^2 + \big(\epsilon_n^{proj} \big)^2 \right)$$

$$+ 3 \ r \ _\infty \left\| \hat{V}_{n+1}^Q (X_{n+1}) - V_{n+1} (X_{n+1}) \right\|_2^2 + \mathcal{O} \left(\frac{1}{M^{1/d}} \right).$$

(30.22) then follows by induction, which completes the proof of Theorem 30.4.1.

Proof of Corollary 4.1. Corollary 4.1 is straightforward by plugging the bound for the projection error provided by Lemma A.1 and the one of the quantization error provided by the Zador's Theorem into (30.22).

Appendix B: Zador's Theorem

THEOREM A.1 Zador's theorem *Let us take $n = 0, \ldots, N$, and denote by K the number of points for the quantization of the exogenous noise ε_n.*

Assume that $\mathbb{E}\left[\left| \varepsilon_n \right|^{2+\eta} \right] < +\infty$ for some $\eta > 0$. Then, there exists a universal constant $C > 0$ such that:

$$\lim_{M \to +\infty} \left(M^{\frac{1}{d}} \left\| \hat{\varepsilon}_n - \varepsilon_n \right\|_2 \right) = C$$

Proof We refer to Graf and Luschgy (2000) for a proof of Theorem A.1.

Cryptocurrency Liquidity During Extreme Price Movements: Is There a Problem with Virtual Money?

Viktor Manahov

31.1 Introduction

Cryptocurrencies emerged as a new class of tradable assets several years ago. Historically, bitcoin was the first and most popular cryptocurrency and was created by following a manuscript published in 2008 by an unknown author under the pseudonym 'Nakamoto'. The ecosystem of cryptocurrencies has experienced rapid expansion in the last two years. As of September 2017, there were over 1100 digital currencies with a total market capitalisation of $60 billion compared to 500 cryptocurrencies with a capitalisation of only $10 million in January 2015 (Bariviera *et al.* 2018). Unlike traditional fiat currencies issued by central banks, some cryptocurrencies are characterised with clearly defined programmatic sets of rules that prevent an excess in supply.* Cryptocurrencies have no tangible existence and they are decentralised,† with no central authority accountable for maintaining the ledger or responsible for

---

* For instance, the sum of all bitcoin balances is set to 21 million bitcoins (Velde 2013).
† This decentralisation is achieved by using the blockchain—a comprehensive ledger that enables the recording of transactions and keeps track of individual accounts.

DOI: 10.1201/9781003265399-36

sustaining the algorithm used to implement the ledger system* (Kumar and Smith 2017). Therefore, cryptocurrency security is based on the security of the underlying cryptographic mechanism, which raises concerns regarding consumer protection, money laundering and tax evasion.

While academic research on cryptocurrencies is almost exclusively focused on their fair value and their bubble existence† during normal conditions, no attention has been given to liquidity‡ provision during periods of market stress such as extreme price movements (EPMs). This is important because EPMs are by far the biggest issue in cryptocurrency markets. For instance, the largest cryptocurrency—bitcoin—experienced extreme growth during 2017, from less than $1000 to almost $20,000, but its price dramatically decreased to approximately $6000 between December 2017 and February 2018 (Hattori and Ishida 2019). Moreover, while much attention has been paid to the significant increase in price and volume of cryptocurrencies, there has not been a study related to the profitability of digital currencies and herding behaviour under conditions of EPMs.

In this study, we deploy a unique dataset, stamped at the millisecond timeframe, of the most traded cryptocurrencies—bitcoin, ethereum, ripple, litecoin and dash—and two major cryptocurrency indices—Crypto Index (CRIX) and CCI30 Crypto Currencies Index—to investigate the behaviour of cryptocurrency traders (CTs) during periods of EPMs in digital markets. Our dataset contains the user identification (ID), which is the internal number associated with exchange users. The user ID is critical because it enables us to analyse cryptocurrency transactions by particular users. The information of the trader identifier and the actual direction of trade (buyer- or seller-initiated trades) included in our dataset are key components in our empirical analysis. Thus, we are able to directly identify liquidity supply by individual CTs and develop a set of methods to investigate the relationship between cryptocurrency liquidity, herding behaviour and profitability during periods of stressful market conditions. To the best of our knowledge this is the first such study related to the cryptocurrency markets.

The contributions of our study are threefold. First, our empirical findings suggest that CTs demand liquidity even during the utmost EPMs. We also observe that CTs facilitate EPMs and are likely to continue this process in the future.

Second, in contrast with the equity markets, there is no investment consensus on how to evaluate digital currencies and therefore analysts' opinions vary significantly from viewing cryptocurrencies as fraud, to seeing them as the currency of the future. This disagreement amongst investors has enabled cryptocurrency markets to be highly dependent on socially established views.§ Moreover, the very recent introduction of cryptocurrencies to the general public has required some investors to have a

---

* Cryptocurrency transactions are validated by competing market participants from different markets known as *miners*, who solve highly sophisticated cryptologic algorithms. The winner in this validation process receives a fraction of the validated cryptocurrency in return.

† Corbet *et al.* (2018) investigate the existence of bubbles in Bitcoin and Ethereum and conclude that there are periods of clear bubble behaviour, while Cheah and Fry (2015) observe that Bitcoin exhibits speculative bubbles and the fundamental price of Bitcoin is zero.

‡ In the last two years, cryptocurrency liquidity dramatically increased because most cryptocurrencies are gaining fast acceptance as a form of payment, mainly in online shops. This acceptance increased by nearly 800% in September 2017, compared to the beginning of 2015. Now, more than 370,000 merchants in 182 different countries around the world accept cryptocurrencies, including some of the big US corporations (such as Microsoft, Apple, Amazon, IBM PayPal and Home Depot).

§ Most cryptocurrency market participants are individual investors who mainly depend on information from social media and online blogs for research and, therefore, are more likely to follow the investment decisions of others. The online community of heterogeneous users equipped with different information has a significant impact on how investors process information and make buy and sell decisions in cryptocurrencies. Some cryptocurrency investors observe and mimic the trading strategies of big cryptocurrency holders known as 'whales', using designated whale-following websites and mobile applications. Reddit is a social news website where users can discuss different cryptocurrency topics with a community of over 600,000 digital currency subscribers. Menkhoff, Schmidt and Brozynski (2006) argue that herding behaviour increases with investors' inexperience.

modest degree of technical and fundamental understanding about the topic and to rely on other market participants, regardless of their own view, to develop and adjust their investment behaviour.*

Stavroyiannis and Babalos (2019) obtain data of the largest cryptocurrencies for a period between August 2015 and February 2018 to examine herding behaviour in the cryptocurrencies' market. Their results derived from the standard testing procedure using ordinary least squares demonstrate the existence of herding behaviour in the cryptocurrencies' market. However, the authors report that herding behaviour disappears when a more robust time-varying regression model is introduced. Kallinterakis and Wang (2019) deploy different econometric models and a dataset that range between December 2013 and July 2018 to examine herding behaviour and its possible determinants. This study shows that herding behaviour is substantial and strongly asymmetric (stronger appearance during up-markets, low volatility and high volume days), with evidence of smaller cryptocurrencies intensifying its size. Bouri *et al.* (2019) perform a rolling-window analysis with the daily closing prices of 14 leading cryptocurrencies and observe significant timevarying herding behaviour, mostly driven by uncertainty of economic policy. The sensitivity of the rolling-window size, however, represents the main limitation of this study. This is because short rollingwindows are appropriate for fast timescales, whereas low timescales demand longer rollingwindows. Vidal-Tomás *et al.* (2019) obtain daily data from 65 digital currencies from January 2015 to December 2017 and cross-sectional standard deviation of returns as a measure of herding dispersion to analyse the existence of herding behaviour in cryptocurrency markets. Opposite to our findings, the authors highlight the possible presence of herding in down markets, implying cryptocurrency market inefficiency. Da Gama Silva *et al.* (2019) observe extreme periods of adverse herding behaviour in the most liquid 50 cryptocurrencies between March 2015 and November 2018. Calderón (2018) analyses a sample of the leading 100 cryptocurrencies and suggests that cryptocurrency investors frequently depart from the rational asset pricing model and instead follow the crowd during periods of market stress. While Kaiser and Stöckl (2019) find evidence of statistically significant herding behaviour in a large cross-section of digital coins, Haryanto *et al.* (2020) document the same behaviour during bearish and bullish periods in a large dataset consisting of 21.2 million trade records from April 2011 to November 2013. Under similar market activity conditions and using data from six major digital coins, Ajaz and Kumar (2018) capture the presence of herding behaviour, which is found to be dependent on market activity rather than market volatility. Consistent with our findings, Ballis and Drakos (2019) report that investors follow the herd in a dataset of six major cryptocurrencies, where market dispersion follows market movements at a faster pace in an up-events market compared to a down-events one. Opposite to the findings of the above studies, the empirical results of Gümüş, *et al.* (2019) indicate no evidence of investors following the herd in bitcoin, litecoin, stellar, monero, dogecoin, dash, and the CCI30 index.

In this study, we directly measure herding by observing the buying and selling behaviour of cryptocurrency users trading on 28 different exchanges. We find significant evidence of herding behaviour in the bitcoin, ethereum, ripple, litecoin, dash, CRIX and CCI30 markets across the entire dataset. We observe that the decrease in dispersion (presence of herding behaviour) in cryptocurrency markets is more evident during periods of significant increase in cryptocurrency prices. In other words, cryptocurrency market participants tend to follow the market consensus when prices are increasing. Our robustness checks obtained from a more dynamic time-varying approach indicate that herding behaviour follows a dynamic pattern that varies over time with decreasing magnitude.

In our third and final contribution, we estimate the profits of individual CTs by taking into account the appropriate transaction costs, including software and hardware expenses. Our profitability analysis suggests that between 1 October 2016, and 1 October 2017, the average gross daily profit is $1,874,312 for bitcoin, $709,005 for ethereum, $244,762 for ripple, $108,338 for litecoin, $100,990 for dash, $71,122 for

---

* Bouri *et al.* (2018) suggest that when some of the leading cryptocurrency developers—like Vitalik Buterin (the cofounder of ethereum) and Charlie Lee (the creator of litecoin)—express their own views on specific digital currency topics, they affect the prices of the related cryptocurrencies.

CRIX and $54,806 for CCI30. Our examination of profit persistency suggests that CTs are capable of extending their strong profitability-generating record in the future.

Cryptocurrency prices are characterised by EPMs, even by the standards of other financial instruments that have been subject to speculative trading, such as dot-com shares in the late 1990s. Zimmerman (2018) suggests that this is a result of the limited capacity of blockchain to accommodate speculative trading. In the cryptocurrency markets, speculators offer fees to miners to execute their transactions first. Most speculators obtain noisy signals related to the long-term value of any particular digital currency, before executing trades on an exchange. Speculators who are convinced that short-term cryptocurrency prices will increase, will offer higher fees to miners to have their transactions processed before their information becomes stale. In cases when a small number of blocks are created in the blockchain, it is the transactions with the highest fees paid—and thus the most EPMs—that are incorporated in the cryptocurrency prices. This creates EPMs that are comparable to a benchmark in which there is no overcrowding on the blockchain, implying that EPMs are essential to any financial instrument that is transacted through a public blockchain.

A central principle of cryptocurrency markets is endogenous liquidity provision, in which CTs supply liquidity by submitting limit orders. Nevertheless, liquidity provision by cryptocurrency CTs is endogenous because they are not obligated to facilitate trading and stabilise markets in periods of EPMs. A growing number of studies find that endogenous liquidity providers (ELPs) usually withdraw from the equities markets during periods of market stress. Raman *et al.* (2014) obtain proprietary intra-day data of all WTI crude oil futures from 2006 and 2011 and find strong evidence that electronic market makers diminish their participation and liquidity provision in periods of substantially high and persistent volatility, trading order imbalances and bid-ask spreads. The authors also suggest that electronic market-makers with longer trading horizons are less likely to withdraw from the market during periods of market stress.

Cespa and Vives (2015) deploy a two-period model in which two types of traders—full and standard—execute short-term trading orders and observe that an increase in the mass of traders with a continuous presence in the market is leading to a decrease in liquidity and welfare. Bongaerts and Van Achter (2015) design a novel model of high-frequency trading liquidity provision with informed trading and endogenous entry and exit in order to examine the implications of liquidity provision on market stability. This study finds that ELPs—such as high-frequency traders—can make market liquidity less stable over time. Anand and Vankataraman (2016) use audittrail data of 1286 stocks from the Toronto Stock Exchange in the calendar year 2006, and observe that ELPs withdraw from the market in a synchronised way when market conditions are unfavourable. As noted by the authors, this withdrawal has the potential to facilitate periodic illiquidity in large and small stocks. The findings of these studies are particularly important because a systematic withdrawal of liquidity during conditions of market stress has the potential to destabilise markets and lower investor confidence (Anand and Vankataraman 2016).

The remainder of this paper is organised as follows: Section 31.2 comprises cryptocurrency and data description; Section 31.3 examines the relationship between cryptocurrency liquidity and EPMs, herding behaviour and profitability; Section 31.4 presents various robustness checks; and Section 31.5 concludes the paper.

31.2 Cryptocurrency Data Description

We obtain millisecond level data for bitcoin, ethereum, ripple, litecoin, dash, CRIX and CCI30 Crypto Currencies Index from Kaiko,* ranging from 1 October 2016, to 1 October 2017. Our choice was based on liquidity, which is one of the most important factors in cryptocurrency trading. In the context of cryptocurrency trading, liquidity can be defined as the ability of a coin to be converted into cash or other

---

* We use proper traded prices to execute the topics under our investigation. Alexander and Dakos (2019) suggest that it is very important to use traded data from crypto trading venues rather than data from coin-ranking sites when examining market efficiency, hedging and portfolio optimisation and trading in cryptocurrency markets.

coins easily. However, cryptocurrency liquidity has represented a significant issue since the beginning of cryptocurrency trading. As of March 2020, there are more than 1600 cryptocurrencies traded on 200 cryptocurrency exchanges and 21 decentralised exchanges, but not every virtual currency has the capacity to attract potential investors.

Asolo (2019) suggests that investors prefer trading more liquid cryptocurrencies because there will be no negative impact on cryptocurrency prices when they attempt to exit a position. Furthermore, more liquid cryptocurrencies like bitcoin, ethereum, ripple, litecoin, and dash are significantly harder to manipulate, making them less likely to participate in pump and dump strategies that are popular in illiquid cryptocurrency markets. Pump and dump schemes entered the cryptocurrency world from the stock market. This strategy suggests that the developer selects an illiquid cryptocurrency of stable low value and artificially inflates its cost, obtaining profits from the price increase.

Moreover, illiquid cryptocurrency markets provide the opportunity for 'whales' to manipulate prices. 'Whales' represent people or a group of people working together to hold a large fraction of any particular cryptocurrency, and they can use this to their advantage to manipulate the price of that coin. Whales often deploy a strategy known as 'rinse and repeat', which is very profitable to a whale if executed at the right time. The whale usually begins by selling lower than the market rate, which in turn causes cryptocurrency market participants to start selling off their digital money in panic. Then the whale will re-purchase when the price of the coin reaches a new low level. This process is repeated to accumulate more wealth, more coins and more control over that particular coin. Another strategy that whales use to manipulate cryptocurrency prices is by deploying buy and sell walls. If the cryptocurrency price decreases, investors will usually purchase at a lower price and sell when it reaches a higher price. A whale can impose either buy or sell walls and closely monitors the price until it hits exactly the anticipated price. Subsequently, the wall vanishes because the whale has cancelled their large buy or sell order.

Kaiko also provides data for approximately 5000 crypto-to-crypto and crypto-to-fiat currency pairs traded on 28 different exchanges: Btcbox, BTCC, Bittrex, Bitstamp, BTCe, Bitfinex, Bithumb, Bit-Z, bit-Flyer, BTC38, BitMEX, Binance, Coinbase, CEX.io, Gatecoin, Gemini, Ethfinex, HitBTC, Huobi, Itbit, Kraken, MtGox, OKEx, OkCoin, Poloniex, Quoine, Yobit and Zaif. Kaiko receives the data by using querying Application Programming Interfaces (APIs) provided by the above trading venues. The data is stamped at the millisecond timeframe (reported in Universal Coordinated Time) and presented in the form of a different series of comma-separated files with each row recording the trade and user ID, transaction type (buy or sell), cryptocurrency volume, transaction fees and a time stamp. In addition, the dataset specifies whether trading orders are initiated by the buyer or the seller and, more specifically, if they are aggressive bids or asks. This particular information is important for our profitability analysis.

We perform a data cleaning procedure by aggregating trades in accordance to their trade IDs. We remove trades where bid or ask orders are missing, those that are duplicated and those that have the same identifier for buy and sell orders. Also, we only consider US dollar-denominated trades.

31.2.1 Cryptocurrency Data Management

The very high presence of no price changes is one of the statistical properties of cryptocurrency data stamped at the millisecond interval. We modified the Student's t distribution associated with the standardised residuals to account for this apparent market inactivity:

$$f\left\langle \frac{\epsilon_t}{\sigma_t} \mid \delta_t \right\rangle = \left\{ p_0 \quad \text{If } \delta_t = 1 \right. \tag{31.1}$$

Or

$$f\left\langle \frac{\epsilon_t}{\sigma_t} \mid \delta_t \right\rangle = \left\{ \frac{g_{v(\epsilon_t/\sigma_t)}}{1 - p_0} \quad \text{If } \delta_t = 0 \right. \tag{31.2}$$

where $g_v(\cdot)$ represents the Student's density function and δ_t measures market inactivity as follows:

$$\delta_t = \begin{cases} 1 & \text{If } |x_t| + |x_{t-1}| = 0, \text{ otherwise} \\ 0 \end{cases} \qquad (31.3)$$

If δ_t=1, the forecast $\hat{x}_{t+i/t} = 0$ for i=1, 2 ... (Meade 2002).

Considering the large dataset size, another important statistical issue that occurs is the Lindley's paradox. This paradox could lead to the overstatement of statistical significance and a tendency to reject the null hypothesis, even when the posterior odds favour the null. We follow Connolly (1989) to estimate sample size adjusted critical values for t-statistics:

$$t_* = \left[(T - k)\left(T^{1/T} - 1 \right) \right]^{1/2} \qquad (31.4)$$

where T represents the sample size and k measures the number of estimated parameters.

We conduct large-sample adjustments to the critical t-values in order to eliminate the statistical significance overstatement. When the absolute value of a regression t-statistic is greater than the t^* value of equation (4), its absolute value is reduced by the adjustment t^*. When the estimated standard statistic is greater than the critical value of t^*, the null hypothesis is rejected.

31.3 Empirical Results

We compute the daily returns of the five cryptocurrencies and the two indices as:

$$R_t = Ln\left[(P_t) / (P_{t-1}) \right] \times 100 \qquad (31.5)$$

where R_t measures the return of bitcoin, ethereum, ripple, litecoin, dash, CRIX and the CCI30 Index and $Ln(P_t)$ and $Ln(P_{t-1})$ represent each financial instrument's natural log at time t and t 1 respectively.

Table 31.1 reports the descriptive statistics of the five cryptocurrencies and the two indices. The distribution of the returns for each digital financial instrument is positively skewed, indicating that large positive price changes are more likely than large negative price changes. The kurtosis is significantly higher than three for all five cryptocurrencies and the two indices, implying a fat-tailed distribution of returns. The Jarque-Bera statistics suggest that the null hypothesis of the normally distributed returns is rejected in all periods.

TABLE 31.1   The Descriptive Statistics for Daily Returns for Bitcoin, Ethereum, Ripple, Litecoin, Dash, CRIX and CCI30 from 1 October 2016–1 October 2017

Cryptocurrency	Mean	Median	Min	Max	SD	SK	K	J-B
Bitcoin	0.81	0.62	−77.12	82.34	12.15	0.81	10.10	0.00
Ethereum	0.76	0.33	−63.77	72.76	11.80	0.77	9.16	0.00
Ripple	0.72	0.24	−58.06	66.99	9.36	0.63	8.73	0.00
Litecoin	0.79	0.08	−45.25	55.35	6.47	0.42	5.48	0.00
Dash	0.62	0.05	−35.38	50.11	5.98	0.40	4.84	0.00
CRIX	0.56	0.04	−29.46	42.48	4.18	0.38	3.73	0.00
CCI30	0.44	0.01	−22.88	36.75	3.33	0.29	2.94	0.00

'SD' refers to the standard deviation, while 'Max' and 'Min' refers to the maximum and minimum returns. 'SK' denotes the skewness, while 'K' refers to the kurtosis. 'JB', refers to the Jarque-Bera test.

31.3.1 The Relationship between Cryptocurrency Liquidity and EPM

We would first like to investigate the implications of increased cryptocurrency liquidity on EPMs by deploying the Amihud's measure of liquidity (AM), the Amivest liquidity ratio (AR) and the Corwin and Schultz bid-ask spread estimator (CS). We select the three measures of liquidity after considering the fact that all trading volume numbers are self-reported by cryptocurrency exchanges. Gandal *et al.* (2018) suggest that cryptocurrency exchanges inflate volumes or report unreal trades to make their exchange seem more active than they actually are. Therefore, cryptocurrency exchanges have a strong incentive to manipulate activity by moving coins back and forth between some of their own accounts.

Amihud (2002) introduced a liquidity measure that detects the daily price response related to one dollar of trading volume:

$$\text{Illiquidity} = \text{Average}\left(\frac{|r_t|}{\text{Volume}_t}\right) \tag{31.6}$$

where r_t measures the cryptocurrency and index return on day t and Volume_t represents the dollar volume on day t. The liquidity ratio is estimated over all positive-volume days.

The Amivest liquidity ratio measures the price impact as:

$$\text{Liquidity} = \text{Average}\left(\frac{\text{Volume}_t}{|r_t|}\right) \tag{31.7}$$

This ratio is estimated over all non-zero return days (Goyenko *et al.* 2009).

Corwin and Schultz (2012) developed a bid-ask spread estimator based on daily high and low prices:

$$S = \frac{2\left(e^\alpha - 1\right)}{1 + e^\alpha} \tag{31.8}$$

where S is the bid-ask spread for the five cryptocurrencies and the two indices and α measures the difference between the adjustments of one-day and two-day periods:$\alpha = \frac{\sqrt{2\beta} - \sqrt{\beta}}{3 - 2\sqrt{2}} - \sqrt{\frac{\gamma}{3 - 2\sqrt{2}}}$ where γ is the high and low price adjustments of a two-day period:$\gamma = E\left\{\sum_{j=0}^{1}\left[\ln\left(\frac{H_{t,t+1}^0}{L_{t,t+1}^0}\right)\right]^2\right\}$ and β is the daily high (H) and low (L) price adjustments to the high price: $\beta = E\left\{\sum_{j=0}^{1}\left[\ln\left(\frac{H_{t+j}^0}{L_{t+j}^0}\right)\right]^2\right\}$

The Corwin and Schultz (2012) estimator is based on the assumption that high cryptocurrency prices are typically buyer-initiated, while low prices are seller-initiated and therefore, the ratio takes into account both the fundamental volatility of the seven financial instruments and their bid-ask spreads.

To estimate the EPMs we use the following regression and two additional criteria:

$$D_t = \alpha + \beta_1 CMR_t^L + \beta_2 CMR_t^U + \epsilon_t \tag{31.9}$$

where D_t measure the level of dispersion in the extreme tails of the distribution of cryptocurrency market and index returns; CMR_t^L is a dummy variable representing the cryptocurrency market and index return positioned in the lower tail of the distribution on day t ($CMR_t^L = 1$ when the cryptocurrency market and index return is positioned in the extreme lower tail of the return distribution and $CMR_t^L = 0$ otherwise); CMR_t^U is a dummy variable that measures the cryptocurrency market and index return positioned in the upper tail of the distribution on day t ($CMR_t^U = 1$ when the cryptocurrency market and index return

TABLE 31.2 Feasible Generalised Least Squares Regression (FGLS) of Market Efficiency on a Set of Three Explanatory Variables in Different Specifications of the Regression

Model	(1)	(2)	(3)
AM	248	152**	128**
AR	8.07**	6.57**	9.56**
CS	−267**	−209	−372**
Adj. R^2	0.90	0.91	0.96
F	3.43**	4.18**	6.12**

'AM' denotes the Amihud's measure of liquidity, while 'AR' refers to the Amivest liquidity ratio and 'CS' measure the Corwin and Schultz bid-ask spread. **indicates statistical significance at the 5% level.

is positioned in the extreme upper tail of the return distribution and $CMR_t^U = 0$ otherwise) and α is the average dispersion of the sample. We compute equation (9) by adopting the following two main criteria to define EPMs: the one per cent (ten per cent) criterion limit CMR_t^L and CMR_t^U to one per cent (ten per cent) of the lower tail distribution and one per cent (ten per cent) of the upper tail distribution.

We adopt the Lewis and Linzer (2005) feasible generalised least squares regression (FGLS) because the explained variable (market efficiency) is based on several estimates.* Table 31.2 describes the results for three different multivariate specifications (Model 1, Model 2 and Model 3) in addition to the descriptive properties of regressors. AM_i, AR_i and CS_i serve as explanatory variables in our regression models. Overall, we find that Model 3 exhibits statistical significance for the Amihud's measure of liquidity, the Amivest liquidity ratio and the Corwin and Schultz bid-ask spread estimator and, therefore, it is the most appropriate statistical model This finding implies a positive relationship between increased liquidity and EPMs.

31.3.2 The Implications of Cryptocurrency Traders' Activity on EPM

To examine the cryptocurrency traders (CTs) activity during periods of EPMs, we implement directional trade imbalances estimated as the difference between the actual trading activity in the direction of the EPM and the trading activity orientated in the opposite direction: $CTs^D = CTs^{D+} - CTs^{D-}$ and $CTs^S = CTs^{S+} - CTs^{S-}$, where CTs^D measure cryptocurrency liquidity demand, CsT^S measures cryptocurrency liquidity supply and the superscripts + and – denote trading activity in the same or opposite direction of the EMP return, respectively. For instance, if CTs request five digital tokens of liquidity in the direction of the price movement and request two digital tokens against the direction of the EPM, CTs^D will be + 3. If CTs supply five digital tokens of liquidity in the opposite direction of the EPM and supply three digital tokens in the direction of the price movement, then CTs^S will be - 2. We also design a net imbalance metric, CTs^NET, estimated as the sum of CTs^D and CTs^S. The net imbalance metric is an important indicator that highlights the direction of trading activity by CTs in comparison with the EPM direction. Therefore, a positive net CTs imbalance suggests that trading is orientated in the direction of the EPM, whilst a negative net CTs imbalance suggests that trading is orientated in the opposite direction of the EPM.

Table 31.3 shows the activity of CTs during periods of 20 s before and after the EPM events and reports the liquidity demand and supply findings for CTs. Our choice of 20 s intervals is based on the findings

---

* Lewis and Linzer (2005) find a superior performance of the feasible generalised least squares regression (FGLS) compared to ordinary least squares (OLS) in cases where the explained variable is based on estimates.

TABLE 31.3 Cryptocurrency Traders (CTs) Liquidity Supply and Demand Around Extreme Price Movements (EPMs)

Panel A	CTs^{NET}		
EPMs period	t-20 seconds	t	t + 20 seconds
Bitcoin	85.8***	452.8***	214.7***
	(0.00)	(0.00)	(0.00)
Ethereum	73.9***	316.2***	210.5***
	(0.00)	(0.00)	(0.00)
Ripple	68.2**	223.9***	204.2***
	(0.01)	(0.00)	(0.00)
Litecoin	46.1***	202.1***	109.7**
	(0.00)	(0.00)	(0.01)
Dash	36.0***	151.5**	101.7***
	(0.00)	(0.02)	(0.00)
CRIX	33.9***	107.9***	92.6***
	(0.00)	(0.00)	(0.00)
CCI30	28.1***	88.2***	64.3***
	(0.00)	(0.00)	(0.00)
Panel B	CTs^{D}		
Bitcoin	1322.4***	14273.0***	3711.2***
	(0.00)	(0.00)	(0.00)
Ethereum	1002.5***	9967.1***	2660.0***
	(0.00)	(0.00)	(0.00)
Ripple	983.2***	8003.4***	2158.4***
	(0.00)	(0.00)	(0.00)
Litecoin	405.5***	6161.0**	1995.2***
	(0.00)	(0.02)	(0.00)
Dash	400.1***	4934.3***	1142.0**
	(0.00)	(0.00)	(0.02)
CRIX	244.9***	1958.2***	442.0***
	(0.00)	(0.00)	(0.00)
CCI30	171.0**	1154.1***	381.1***
	(0.03)	(0.00)	(0.00)
Panel C	CTs^{S}		
Bitcoin	−800.5***	−12985.0***	−1968.1***
	(0.00)	(0.00)	(0.00)
Ethereum	−712.3***	−8023.4***	−1021.0**
	(0.00)	(0.00)	(0.01)
Ripple	−510.8***	−6353.3***	−999.1***
	(0.00)	(0.00)	(0.00)
Litecoin	−418.2***	−4994.2***	−201.5***
	(0.00)	(0.00)	(0.00)
Dash	−399.0***	−3008.1**	−195.7***
	(0.00)	(0.01)	(0.00)

(*Continued*)

TABLE 31.3 (CONTINUED)   Cryptocurrency Traders (CTs) Liquidity Supply and
Demand Around Extreme Price Movements (EPMs)

Panel C		CTs^S	
CRIX	−89.9***	−1563.0***	−108.0***
	(0.00)	(0.00)	(0.00)
CCI30	−70.4**	−1008.2***	−72.1***
	(0.02)	(0.00)	(0.00)

This table shows the directional trading volume around EPMs for bitcoin, ethereum, ripple, litecoin, dash, CRIX and CCI30 from 1 October 2016, to 1 October 2017. We report the results for time prior to an EPM (t-0.3milliseconds) and after an EPM (t 0.3milliseconds). CTs^D measure the liquidity difference in liquidity-demanding CT volume in the direction of the EPMs and liquidity-demanding volume in the opposite direction of the EPMs. CTs^S measure the liquidity difference in liquidity-providing volume in the opposite direction of the EPMs and liquidity-providing volume in the direction of the EMPs. CTs^{NET} represents the difference between CTs^D and CTs^S. We present p-values in parentheses.*** indicates significance at the 1% level. ** indicates significance at the 5% level.

of Li *et al.* (2018) who report that cryptocurrency market participants generate positive returns if they purchase tokens within the first 20 s after pump-and-dump schemes occurence.* Therefore, we implement event window periods of 20 s before and after the EPM interval to examine the liquidity demand and supply of CTs for the seven financial instruments. Panel A of Table 31.3 reveals several important initial findings. The CTs^{NET} imbalance is positive, which implies that liquidity demand exceeds CTs liquidity supply for the five cryptocurrencies and the two indices. Moreover, the positive CTs^{NET} imbalance indicates that trading is orientated in the direction of the EPMs. We find that CTs^{NET} imbalance is statistically significant in the direction of the EPMs (interval t) and the two other intervals—20 s prior to an EPM (t–20 s) and 20 s after an EPM (t 20seconds)—mostly at the one per cent level of significance. When we break CTs activity to demand and supply, we observe that CTs trade in the direction of the EPMs based on their liquidity demanding trading orders and in the opposite direction with their liquidity supplying trading orders for the five cryptocurrencies and the two indices. For instance, the CTs^D for bitcoin is 14,273 tokens (Table 31.3, Panel B), while the liquidity supply (CTs^S) for the same cryptocurrency is 12,985 tokens. There-fore, CTs trading bitcoin demand 1288 tokens (the difference between CT_S^D and CT_S^S values) of net liquidity in the direction of the EPMs. We observe the same trend for the other four cryptocurrencies and the two indices. This empirical finding is consistent with the results of similar studies in the equity and derivative markets.

Raman *et al.* (2014), Cespa and Vives (2015), Bongaerts and Van Achter (2015) and Anand and Vankataraman (2016) all report a decline in liquidity during EPMs. On the other hand, Brogaard *et al.* (2017) documented that high-frequency traders are providers of net liquidity in NASDAQ stocks during the EPMs between 2008 and 2009. However, in contrast to our study, the authors were unable to observe the trading activities of individual market participants and they only examined trading on one exchange—NASDAQ. This is an important limitation because trades on NASDAQ represent between 30% and 40% of the total trading in US equities, hence, there is a possibility that market participants provide liquidity on NASDAQ during EPMs while at the same time demanding it from other trading venues.

Table 31.3, Panel A, also shows that in the interval beginning 20 s prior to an EPM (t–20 s) CTs execute trades in the direction of the EPMs, suggesting that CTs facilitate EPMs. Our findings are consistent

---

* Pump-and-dump schemes in the cryptocurrency markets represents price manipulation that involves artificially increasing crytocurrency prices before selling the cheaply purchased tokens at higher price levels. When the tokens are 'dumped,' their price decreases and traders accumulate losses. These schemes are usually performed in microcap stocks trading but have recently become common in the cryptocurrency markets (Shifflet and Vigna 2018).

with the results of several studies investigating equity markets. Golub *et al.* (2013) argue that mini-crashes in individual stocks dramatically increased in the last few years and suggest a link between these crashes and the activities of high-frequency traders. Leal *et al.* (2014) used computational techniques to replicate a stock market in which market participants play a significant role in generating flash crashes.

To further investigate whether the activity of CTs trigger EPMs, we need to capture the trading direction relative to permanent price changes and transitory pricing errors. When CTs trading is orientated in the direction of the pricing error, we can conclude that the activity of CTs facilitates EPMs. In contrast, CTs will not trigger EPMs when their activity is oriented in the opposite direction to the transitory pricing error. The state space model of the cryptocurrencies can be decomposed into two distinguished parts—permanent and transitory constituents:

$$p_{i,t} = m_{i,t} + s_{i,t} \tag{31.10}$$

where $p_{i,t}$ represents the (log) midquote (the average of the bid and ask quote) at time t for financial instrument i; $m_{i,t}$ is the permanent component of a martingale type—$m_{i,t} = m_{i,t-1} w_{i,t}$ with an innovative element $w_{i,t}$ included in the permanent price component; and $s_{i,t}$ represents the transitory price component.

We develop two different state space models to examine the implications of CTs activity on EPMs. The first model examines all trading activity denoted $CT s^{NET}$ while the second model analyses the demand and supply components of CTs denoted as $CT s^D$ and $CT s^S$. We compute the aggregate model as:

$$w_{i,t} = k_i^{all} \ \widetilde{CT}s_{i,t}^{NET} + \mu_{i,t} \tag{31.11}$$

where $\widetilde{CT}_s^{NET}$ is the surprise innovation factor in CTs^{NET}, which is the residual of an autoregressive model used to eliminate autocorrelation. We implement the Kalman filter where cryptocurrency price changes, CTs^{NET}, $CT s^D$ and $CT s^S$ are non-zero in order to estimate the state space model for each digital currency in a trading day for the entire sample period.

Table 31.4 shows the empirical results of the $CT s^{NET}$ state space model for each financial instrument and the overall space model. We observe that all space models are negatively correlated with the permanent price component and positively correlated with the transitory price component, suggesting that the activity of CTs facilitates EPMs. This particular finding can be explained by the limited capacity of the blockchain to handle speculative trading generated by CTs and that CTs therefore offer fees to miners to prioritise their transactions. Most CTs obtain noisy signals related to the long-term value of the five cryptocurrencies and the two indices before executing trades on any of the 28 trading venues in our study. CTs who believe that short-term cryptocurrency prices will increase are able to offer higher fees to miners to have their transactions processed before their information becomes stale. In cases when fewer blocks are created in the blockchain, it is the transactions with the highest fees paid—and thus the most EPMs—that are incorporated in the cryptocurrency prices leading to the creation of EPMs.

The values of k and ψ are estimated in basis points per $1,000,000 traded. The value of -0.69 for the overall k coefficient listed in the last column of Panel A in Table 31.4, indicates that $1,000,000 of negative surprise order flow (bid minus ask orders) corresponds to a -0.69 basis points decrease in the permanent price component. The aggregate proportion of permanent price variance $(k^{NET} \sigma (CT s^{NET}))^2$ correlated with overall $CT s^{NET}$ order flow is -17.92 basis points squared out of the millisecond permanent price variance of -305.78 basis points squared. The positive values of ψ coefficients in the transitory price component suggest that CTs trade in the direction of the pricing errors and therefore the activity of CTs creates EPMs.

Furthermore, we employ a disaggregated state space model to examine the individual implications of bid and ask cryptocurrency trading orders on EMPs. Table 31.5, Panel A, shows that CTs bid and ask orders are negatively correlated (k^{bid} and k^{ask} values are negative in all cryptocurrencies and indices)

TABLE 31.4 Sate Space Model for Bitcoin, Ethereum, Ripple, Litecoin, Dash, CRIX and CCI30 from 1 October 2016, to 1 October 2017

	Measures	Bitcoin	Ethereum	Ripple	Litecoin	Dash	CRIX	CCI30	All
Panel A	Permanent price component								
k^{NET}	bps/\$1,000,000	-0.135	-0.81	-0.64	-0.41	-0.36	-0.29	-0.21	-0.69
(t-stat)		(-9.19)	(-7.09)	(-6.25)	(-3.38)	(-2.99)	(-3.14)	(-1.96)	(-5.64)
$\sigma^2(\widetilde{CT_s}^{NET})$	\$1,000,000	-30.03	-25.20	-21.26	-19.88	-15.70	-14.52	-12.08	-24.03
$(k^{NET} * \sigma(\widetilde{CT_s}^{NET}))^2$	bps.2²	-18.11	-14.88	-13.54	-11.90	-11.62	-10.43	-9.08	-17.92
(t-stat)		(-20.61)	(-17.03)	(-11.99)	(-8.74)	(-12.53)	(-7.55)	(-6.87)	(-14.26)
$\sigma^2(w_{i,t})$	bps.2²	-357.01	-344.95	-312.90	-217.77	-201.12	-112.43	-99.15	-305.78
Panel B	Transitory price component								
ϕ	bps/\$1,000,000	0.34	0.33	0.24	0.20	0.19	0.16	0.09	0.21
ψ^{NET}		0.20	0.16	0.11	0.08	0.06	0.07	0.05	0.14
(t-stat)		(6.12)	(8.44)	(10.15)	(4.11)	(5.01)	(2.19)	(3.01)	(2.98)
$\sigma^2(\widetilde{CT_s}^{NET})$	\$1,000,000	1.37	1.24	1.20	0.99	0.83	0.64	0.51	1.11
$(\psi^{NET} * \sigma(\widetilde{CT_s}^{NET}))^2$	bps.2²	8.59	7.70	6.99	5.03	4.76	2.97	2.67	6.18
(t-stat)		(12.91)	(10.14)	(10.03)	(5.87)	(4.55)	(1.99)	(2.01)	(8.28)
$\sigma^2(s_{i,t})$	bps.2²	114.36	102.50	94.83	88.09	78.43	61.63	50.24	96.85

$CTS_{i,t}^{NET}$ is the overall net order flow and $\widetilde{CTs}_{i,t}^{NET}$ represent the surprise component of the order flow. We estimate the t-statistics using standard errors double-clustered on the seven financial instruments and millisecond data.

TABLE 31.5 Disaggregated Sate Space Model for Bitcoin, Ethereum, Ripple, Litecoin, Dash, CRIX and CCI30 from 1 October 2016, to 1 October 2017

Measures		Bitcoin	Ethereum	Ripple	Litecoin	Dash	CRIX	CCI30	All
Panel A	Permanent price component								
k^{bid}	bps/$1,000,000	30.61	27.90	23.22	16.90	13.88	15.84	15.77	21.99
(t-stat)		(18.11)	(16.02)	(15.99)	(13.82)	(8.28)	(9.98)	(6.10)	(11.62)
k^{ask}	bps/$1,000,000	28.75	26.12	19.71	20.95	12.16	9.45	8.32	17.97
(t-stat)		(16.03)	(12.83)	(10.60)	(9.78)	(9.87)	(4.17)	(3.85)	(8.89)
$\sigma^2(\widetilde{CTs}^{bid})$	$1,000,000	1.92	1.11	1.03	0.92	0.85	0.73	0.61	1.01
$\sigma^2(\widetilde{CTs}^{ask})$	$1,000,000	0.87	1.02	1.11	0.71	0.84	0.63	0.75	0.99
$(k^{bid} * \sigma^2(\widetilde{CTs}^{bid}))^2$	bps.2²	42.02	40.55	38.99	30.07	19.97	13.65	11.72	26.53
(t-stat)		(23.91)	(20.18)	(13.71)	(9.82)	(8.04)	(6.87)	(5.08)	(16.35)
$(k^{ask} * \sigma^2(\widetilde{CTs}^{ask}))^2$	bps.2²	30.54	38.12	46.89	16.16	21.26	8.89	12.90	19.37
(t-stat)		(20.11)	(17.22)	(14.18)	(9.53)	(8.13)	(4.76)	(3.78)	(5.58)
$\sigma^2(w_{i,t})$	bps.2²	487.12	403.01	322.98	299.91	162.17	87.34	71.62	404.43
Panel B	Transitory price component								
φ		0.61	0.58	0.47	0.33	0.28	0.18	0.10	0.29
ψ^{bid}	bps/$1,000,000	19.94	13.72	10.44	7.66	6.23	5.19	4.61	10.83
(t-stat)		(5.63)	(5.09)	(4.19)	(3.99)	(3.35)	(2.01)	(2.19)	(4.77)
ψ^{ask}	bps/$1,000,000	-10.88	-6.88	-5.90	-4.55	-3.67	-3.11	-2.99	-9.85
(t-stat)		(-3.11)	(-2.91)	(-2.44)	(-1.89)	(-1.02)	(-0.99)	(-0.82)	(-2.17)
$\sigma^2(\widetilde{CTs}^{bid})$	$1,000,000	0.84	0.91	0.98	0.62	0.74	0.57	0.65	0.80
$\sigma^2(\widetilde{CTs}^{ask})$	$1,000,000	0.91	0.82	0.72	0.60	0.54	0.42	0.38	0.71
$(\psi^{bid} * \sigma^2(\widetilde{CTs}^{bid}))^2$	bps.2²	24.72	20.18	15.15	10.68	9.09	6.59	5.43	13.62
(t-stat)		(11.90)	(10.80)	(11.44)	(9.71)	(8.54)	(3.17)	(2.81)	(12.63)
$(\psi^{ask} * \sigma^2(\widetilde{CTs}^{ask}))^2$	bps.2²	19.64	12.99	15.37	10.81	9.66	7.42	5.75	11.04
(t-stat)		(7.19)	(6.82)	(5.10)	(4.11)	(4.01)	(3.23)	(3.01)	(4.24)
$\sigma^2(s_{i,t})$	bps.2²	411.02	321.16	284.09	254.17	205.81	155.10	72.90	395.45

$CTs_{i,t}^{bid}$ and $CTs_{i,t}^{ask}$ represent bid and ask order flows; $\widetilde{CTs}_{i,t}^{bid}$ and $\widetilde{CTs}_{i,t}^{ask}$ are the surprise components in those order flows. We estimate the t-statistics using standard errors double-clustered on the seven financial instruments and millisecond data.

with price changes in the permanent price component of the state space model. The positive values of k^{ask} and k^{bid} coefficients suggest that trading is conducted in the direction of the permanent pricing errors and therefore causes EPMs.

We also observe a positive relation between ψ^{bid} and the transitory price component across all cryptocurrencies and indices (Table 31.5, Panel B), implying that CTs^{bid} trading orders follow the direction of the transitory component of the state space model. This finding indicates that when cryptocurrency prices diverge from their fundamental values, CTs do not facilitate trading activities in order to restore prices to their fundamental levels. As a consequence, the distance between the quoted prices of the seven financial instruments and the efficient price levels increases confirming the presence of speculative trading in the 28 cryptocurrency markets under investigation.

The values of ψ^{ask} coefficients are all negative, however, suggesting the adverse selection of coefficients based on the uninformed transitory component. Therefore, CTs^{bid} trading orders are related to significantly less information being incorporated into the prices of the five cryptocurrencies and the two indices, and bigger pricing errors in comparison with CTs^{ask} trading orders. This finding provides evidence that CTs play a detrimental role in the process of price discovery both in terms of bigger pricing errors and less information being incorporated into the prices of the seven digital assets. Moreover, increased pricing errors indicate cryptocurrency price inefficiency and deprived allocation of resources. This finding is consistent with the majority of studies on cryptocurrency market efficiency. The initial studies on this topic employed a battery of empirical tests to examine mainly bitcoin markets. Urquhart (2016) used daily bitcoin data and a wide variety of different testing procedures to observe market inefficiency, though there has been a tendency towards efficient markets since August 2013. Nadarajah and Chu (2017) adopted similar econometric tests and bitcoin data to report market inefficiency. Bariviera (2017) investigated bitcoin market efficiency by estimating the Hurst exponent and overlapping sliding windows technique. The author captured initial market inefficiency, but there was a trend towards efficient markets after 2014.

The first cryptocurrency market efficiency studies are examined bitcoin; the tremendous growth of trading in digital money has led to a number of studies concerning alternative virtual financial instruments. Zhang *et al.* (2018) collected daily data for bitcoin, ethereum, ripple, litecoin, stellar, dash monero, and nem to obtain mixed market efficiency results. Brauneis and Mestel (2018) adopted a much bigger dataset, consisting of 73 different cryptocurrencies and nine different econometric models, in order to look into market efficiency. The results indicate that the higher the liquidity of a digital currency, the more inefficient this coin is. Charfeddine and Maouchi (2018) examined the daily closing prices of bitcoin, ethereum, ripple, and litecoin to find evidence of market inefficiency in three out of the four digital coins, with ethereum being the most efficient market. In a much larger study, Wei (2018) investigated 458 different cryptocurrencies, using a series of efficiency tests. The results show that while bitcoin returns indicate efficient markets, numerous other digital coins exhibit signs of market inefficiency. The markets of bitcoin, litecoin, ripple, and dash were also found to be inefficient, but with steps towards efficiency in the study of Caporale *et al.* (2018). In a somewhat similar vein, Bouri *et al.* (2018) failed to detect efficiency in seven major cryptocurrency markets. More recently, Al-Yahyaee *et al.* (2020) have studied bitcoin, ethereum, monero, dash, litecoin, and ripple, using the time-rolling MF-DFA approach to demonstrate time-varying market inefficiency, with litecoin being the most inefficient market.

31.3.3 The Magnitude of EPM

In this section, we investigate whether CTs provide liquidity to the market during different EPMs between 1 October 2016, and 1 October 2017. For this purpose, we divide the EPMs into three different magnitude categories by return magnitude—small (C1), medium (C2) and large (C3). Table 31.6, Panel A, reports the descriptive statistics for the three categories. Trading volume and the quoted and the relative spreads increase in return magnitude (Panel A). The liquidity provision of CTs also increases from 451 cryptocurrency tokens in C1 to 728 tokens in C2 (Panel B). However, we observe that liquidity provision

TABLE 31.6 The Magnitude of EPMs for Bitcoin, Ethereum, Ripple, Litecoin, Dash, CRIX and CCI30 from 1 October 2016, to 1 October 2017

| | Descriptive statistics | | | | | |
| | C1 (small) | | C2 (medium) | | C3 (large) | |
Panel A	mean	std.dev.	mean	std.dev.	mean	std.dev.
Cryptocurrency return	0.518	0.109	0.726	0.135	0.858	0.172
Trading orders	62.92	73.84	69.99	80.15	77.04	92.10
Volume ($)	2,726,623	3,103,357	2,994,037	3,346,282	3,665,833	3,889,255
Asset volume	31,828	46,901	47,221	54,180	50,189	61,273
Quoted spread ($)	0.095	0.123	0.141	0.667	0.169	0.828
Relative spread (%)	0.117	0.181	0.308	0.726	0.572	0.898
Panel B	t-0.3		CTs_t^{NET}		$t + 0.3$	
C1	97.6**		451.2***		273.4***	
C2	114.8***		728.1***		408.1**	
C3	80.5***		363.3***		225.9***	

In Panel A, we divide the EPMs into three different categories by return magnitude—C1 (small), C2 (medium) and C3 (large). Panel B shows the CTs^{NET} statistics. .*** indicates significance at the 1% level. ** indicates significance at the 5% level.

decreases during the largest EPMs (C3) to 363 tokens. This finding is in line with Kirilenko *et al.* (2017), who report that market participants withdrew liquidity provision when equity prices reached very low levels during the flash crash in 2010.

31.3.4 The Activity of CTs during Normal Trading and in EPM

This section examines the behaviour of CTs under conditions of normal trading and in cases of EPMs. This is important because CTs may demand less liquidity than normal when several cryptocurrencies experience EPMs. We study this issue by developing the following regression:

$$CTs_{i,t}^{NET} = \alpha + \beta_1 1_{EPM_{i,t}} + \beta_2 RT_{i,t} + \beta_3 \text{Vol}_{i,t}$$

$$+ \beta_4 QS_{i,t} + vLags_{kit - \sigma\gamma k\sigma} + \varepsilon_{i,t}$$

(31.12)

where CTs^{NET} represent the difference between CTs^D and CTs^S, 1_{EPM} represents a dummy variable, that is equal to one when the 0.3 millisecond interval t in cryptocurrency and index i is captured as EPM and is equal to zero otherwise, $RT_{i,t}$ measures the absolute return, $Vol_{i,t}$ measures the traded cryptocurrency volume, $QS_{i,t}$ is the percentage of quoted spread, and $vLags_{kit - \sigma\gamma k\sigma}$ is a vector of different lags related to each of the dependent and independent variables, with $\sigma \in \{1, 2, 3...., 10\}$. Both the dependent and the independent variables in the vector are indexed with a subscript k. We jointly elucidate the coefficients of the $1_{EPMi,t}$ dummy and the $RT_{i,t}$ variable because they are both related to cryptocurrency and index returns. The values of the $RT_{i,t}$ variable in Column 1 of Table 31.7 suggest that CTs usually demand liquidity in the direction of cryptocurrency and index returns. The $1_{EPMi,t}$ coefficients also show that CTs increase liquidity demand during EPMs, with 0.826 standard deviations.

Column 2 of Table 31.7 illustrates that CTs also increase liquidity demand during periods of both permanent and transitory EPMs. We observe a similar trend of increased liquidity demand in different levels of EPMs magnitude (Column 3 of Table 31.7). CTs seem to intensify liquidity demand during the largest EPMs incorporated in C3. We also find that CTs demand liquidity during periods of EPMs and their demand is significantly more than they would usually demand under conditions of normal trading. The positive and statistically significant values of CTs^{NET} (Table 31.6, Panel B) imply that CTs

TABLE 31.7 The Activity of CTs During Normal Trading and in EPMs for Bitcoin, Ethereum, Ripple, Litecoin, Dash, CRIX and CCI30 from 1 October 2016, to 1 October 2017

Model	(1)	(2)	(3)
$1_{EPMi,t}$ (t-stat)	0.826***		
	(0.00)		
$1_{EPMi,t-PERMANENT}$ (t-stat)		0.684***	
		(0.00)	
$1_{EPMi,t-TRANSITORY}$		0.811***	
(t-stat)		(0.00)	
$1_{EPMi,t-C1}$			0.599***
(t-stat)			(0.00)
$1_{EPMi,t-C2}$			0.747***
(t-stat)			(0.00)
$1_{EPMi,t-C3}$			0.912***
(t-stat)			(0.00)
$RT_{i,t}$ (t-stat)	0.072***	0.091***	0.098***
	(0.00)	(0.00)	(0.00)
$VOL_{i,t}$	0.080***	0.083***	0.091***
(t-stat)	(0.00)	(0.00)	(0.00)
$QS_{i,t}$ (t-stat)	0.014**	0.018***	0.023***
	(0.01)	(0.00)	(0.00)
Adj. R^2	0.92	0.94	0.95

$1_{EPMi,t}$ represent a dummy variable, which is equal to one when the 20 s interval t in cryptocurrency and index i is captured as EPM and is equal to zero otherwise, $RT_{i,t}$ measure the absolute return, $Vol_{i,t}$ measure the traded cryptocurrency volume, $QS_{i,t}$ is the percentage of quoted spread. *** indicates significance at the 1% level. ** indicates significance at the 5% level.

consistently demand liquidity in the direction returns. Brogaard *et al.* (2014) documented the same findings for high-frequency traders operating in the US equity markets, while Brogaard *et al.* (2017) reported the opposite pattern for standalone EPMs on NASDAQ between 2008 and 2009.

31.3.5 Herding Behaviour during EPM

Some cryptocurrency investors often observe and replicate the trading strategies of big cryptocurrency holders known as 'whales', using designated whale-following websites and mobile applications. Whales are market participants with large cryptocurrency holdings and are capable of significantly changing the market by influencing cryptocurrency prices. The activities of whales are probably the biggest single contributor to herding behaviour in the cryptocurrency markets. Whales use 'buy and sell walls' to exploit digital currency prices. A 'buy wall' is when whales generate and place a very large buy order on a cryptocurrency exchange. CTs operating with small trading orders are likely to be attracted by this very large buy position and accept it as forthcoming increase in price and therefore initiate cryptocurrency purchase. When this happens, the cryptocurrency price will increase. The major issue in this herding behaviour situation is that the whales are able to increase the price without actually investing in the cryptocurrency. The price increase in this case is facilitated by the activities of the regular CTs, instead of the whales. When the cryptocurrency price is high enough, the whales will alter their buy and sell walls and profit from the price increase. As a result the price of the cryptocurrency will decrease. The lack of position price limits in cryptocurrency exchanges is the main motivator for this type of unethical activity by the whales. The presence of position price limits will prevent the creation of large buy and sell trading orders and therefore mitigate the implications of herding behaviour in cryptocurrency markets.

Christie and Huang (1995) suggest that herding behaviour occurs during periods of exaggerated market movements,* where the market participants abandon their own beliefs in favour of the market consensus. Under such a description of herding behaviour, dispersion—defined as a cross-sectional standard deviation of returns—appears as a natural measure of herding in our study. Dispersions evaluate the average proximity of cryptocurrency returns to the mean. The cryptocurrency returns are usually below zero in cases when returns fluctuate simultaneously with the market. The level of dispersion substantially increases when cryptocurrency returns of individual digital currencies begin to differ from the whole market return.

During intervals of extreme market fluctuations, the Capital Asset Pricing Model (CAPM) predicts that large market changes are increasing dispersion because individual cryptocurrencies differ in their sensitivity to the market return. In economic words, this dispersion in sensitivities will force individual digital currency returns away from the market. Alternatively, the herding behaviour of individual cryptocurrency returns around the market corresponds to a decreased level of dispersion. Therefore, the CAPM and herding behaviour models provide diverging assumptions for the conduct of dispersion intervals of extreme market movements.

We investigate whether the levels of dispersion are significantly lower than average during periods of extreme cryptocurrency market fluctuations. Since our dataset specifies whether trading orders are initiated by the buyer or the seller, we are able to measure herding behaviour directly from the cryptocurrency markets. We estimate the level of dispersion, D_t by using equation (9).

CAPM postulates that the β_1 and β_2 coefficients in equation (9) should be positive, while in cases of herding behaviour the same two coefficients should be negative. In Table 31.8, Panels A and B show the regression results for the five cryptocurrencies and the two indices using the one and ten-per cent criterion for extreme cryptocurrency market fluctuations. Both β_1 and β_2 for the seven financial instruments (under one and ten per cent criterion) are negative, which implies the presence of herding behaviour in digital markets. Additionally, β_2 estimates possess considerable consistency across the cryptocurrency and index markets, while a much wider distribution is present in β_1 estimates. The heteroscedasticity t-statistics presented in parentheses confirm the reliability of β_1 and β_2 coefficients. These findings contradict the assumptions of CAPM and are consistent with the predictions of herding behaviour.

Table 31.8 (Panels A and B) illustrate the asymmetric dispersion behaviour in response to significant fluctuations in average cryptocurrency and index returns. For instance, the β_2 estimates under the ten per cent criterion in Table 31.8 (Panel B) are between one and five times smaller than those for the β_1 coefficients. We observe that this ratio is even more pronounced under the one per cent criterion, where the β_2 coefficients (Table 31.8, Panel A) are between one and eight times smaller than those for the β_1 estimates. This important finding implies that the decrease in dispersion (the presence of herding behaviour) in cryptocurrency and index markets is more evident during periods of significant increase in cryptocurrency and index prices. This observation is in line with the findings of Kallinterakis and Wang (2019) and Ballis and Drakos (2019) who report stronger herding behaviour appearance during up-markets but opposite to Vidal-Tomás *et al.* (2019) who highlight the possible presence of herding in down markets.

In other words, CTs tend to follow the market consensus when prices are increasing. Hence, digital currency markets are more likely to respond to positive news than to negative news, i.e. cryptocurrency investors respond to positive events more quickly.

Evidence of herding behaviour in the seven digital markets suggests that the trading decisions of market participants are not developed in isolation. Instead, CTs tend to disregard the individual characteristics of digital currencies and herd on the overall performance of the cryptocurrency markets. Evidence of herding is also indicative of cryptocurrency market inefficiency, which makes the occurrence of

---

* Cryptocurrency markets are characterised by excessive price fluctuations, providing conditions for the creation of herding behaviour.

TABLE 31.8   Regression Coefficients for Bitcoin, Ethereum, Ripple, Litecoin, Dash, CRIX and CCI30 Daily Returns from 1 October 2016, to 1 October 2017 during EPMs

Panel A	1% criterion		
Cryptocurrency	α	$\beta 1$	$\beta 2$
Bitcoin	7.82	−0.22	−1.75
(*t*-stat)		(8.14)	(10.14)
Ethereum	7.05	−0.76	−1.88
(*t*-stat)		(10.13)	(11.09)
Ripple	5.89	−1.15	−2.01
(*t*-stat)		(21.37)	(20.43)
Litecoin	3.90	−0.12	−0.24
(*t*-stat)		(4.69)	(4.88)
Dash	3.37	−0.63	−0.65
(*t*-stat)		(7.90)	(6.36)
CRIX	2.12	−0.09	−0.14
(*t*-stat)		(2.14)	(1.99)
CCI30	1.98	−0.12	−0.18
(*t*-stat)		(2.31)	(2.43)
Panel B	10% criterion		
Bitcoin	8.23	−0.38	−1.81
(*t*-stat)		(9.02)	(9.99)
Ethereum	7.99	−0.99	−2.10
(*t*-stat)		(12.12)	(18.08)
Ripple	6.31	−1.66	−2.26
(*t*-stat)		(14.63)	(22.24)
Litecoin	4.46	−0.20	−0.37
(*t*-stat)		(4.99)	(6.25)
Dash	3.77	−0.88	−0.80
(*t*-stat)		(8.18)	(7.88)
CRIX	2.80	−0.15	−0.22
(*t*-stat)		(2.24)	(3.14)
CCI30	2.44	−0.26	−0.27
(*t*-stat)		(3.87)	(3.55)

α is measure the average dispersion of the sample. We report the heteroscedasticity *t*-statistics in parentheses.

systematic risk more likely. Herding behaviour could result in price instability, excess volatility, the formation of bubbles and even market crashes (Hwang and Salmon 2004).

31.3.6 Profitability of CTs during EPM

We quantify the profits from cryptocurrency trading by taking into account the appropriate transaction costs. Our dataset specifies whether trading orders are initiated by the buyer or the seller with the user ID, enabling us to estimate the profits obtained by individual traders. We estimate the daily gross profit after transaction costs* as:

---

* Kraken (www.kraken.com), one of the biggest cryptocurrency exchanges, suggests transaction costs of 26 basis points per trade. We implement slightly higher transaction costs of 30 basis points in order to take into account any software and hardware trading-related expenses.

$$\bar{P} = \frac{1}{T} \sum_{t=1}^{T} n_{t-1} \Delta m_t - \left| \Delta n_t \right| m_t \left(s_t + \tau \right) \| \tag{31.13}$$

where $\bar{P}$ measures the average gross profit for bitcoin, ethereum, ripple, litecoin, dash, CRIX and CCI30 after accounting for transaction costs of 30 basis points per transaction; T denotes time; n_t measures the net inventory position of each trader; m_t is the midquote price (the average of the bid and ask quote) for the five cryptocurrencies and the two indices; s_t represents the half-spread; and τ denotes the sum of the transaction fees per transaction.

Table 31.9 (Panel A) shows that between 1 October 2016, and 1 October 2017, the average gross daily profits are \$1,874,312 for bitcoin; \$709,005 for ethereum; \$244,762 for ripple; \$108,338 for litecoin; \$100,990 for dash; \$71,122 for CRIX and \$54,806 for CCI30.

While all values reported in the second column of Table 31.9 (Panel A) indicate the magnitude of cryptocurrency and index profitability, risk-adjusted trading performance plays a significant role in profit estimations. Hence, we estimate the Sharpe ratio for the five cryptocurrencies and the two indices on a daily basis:

$$SR_{i,t} = \frac{r_{i,t} - r_f}{\sigma_t} * \sqrt{365} \tag{31.14}$$

where $r_{i,t}$ measures the daily return based on the daily profit for bitcoin, ethereum, ripple, litecoin, dash, CRIX and CCI30; r_f represents the risk-free rate measured by the estimated value of the daily compounded rate that is obtained from the annual investment yield of monthly US Treasury bills (we obtained related data up to 1 October 2017, from the Federal Reserve https://www.federalreserve.gov

TABLE 31.9 The Average Gross Daily Profits, Daily Sharpe Ratios and Daily Probabilistic Sharpe Ratios for Bitcoin, Ethereum, Ripple, Litecoin, Dash, CRIX and CCI30 from 1 October 2016, to 1 October 2017

	Panel A—SR and PSR		
Cryptocurrency	$\bar{P}$	SR	PSR
Bitcoin	\$1,874,312	2.94	0.997**
Ethereum	\$709,005	2.01	0.991**
Ripple	\$244,762	1.92	0.966**
Litecoin	\$108,338	1.06	0.951**
Dash	\$100,990	1.01	0.923**
CRIX	\$71,122	0.68	0.917**
CCI30	\$54,806	0.51	0.910**
	Panel B—CSR		
Test statistics	Avg.CSR	Std error	Sig level
Subperiod 1	0.1972	0.8372	0.0000
Subperiod 2	0.1644	0.7004	0.0000
Subperiod 3	0.1107	0.9289	0.0001
Subperiod 4	0.0795	0.6618	0.0000

Chi-squared(4) = 15.3622 with significance P denotes the daily gross profits; SR is the Sharpe ratio; PSR is the observed Sharpe ratio; CSR denotes the conditional Sharpe ratio **indicates significance at the 5% level. Subperiod 1 (1st October 2016 to 31st December, 2016); subperiod 2 (1st January, 2017 to 31st March, 2017); subperiod 3 (1st April, 2017 to 31st June, 2017); subperiod 4 (1st July, 2017 to 31st, September, 2017).

/releases/h15/); σ denotes the standard deviation of market participant i's return within the sample period; and 365 represents the typical number of trading days per year in cryptocurrency markets.

Table 31.9 (Panel A) shows high daily Sharpe ratios of 2.94 for bitcoin, 2.01 for ethereum, 1.92 for ripple, 1.06 for litecoin, 1.01 for dash, 0.68 for CRIX and 0.51 for CCI30. Fama and French (2002) estimate an average Sharpe ratio of 0.31 for the S&P 500, using daily and annual data observations. Direct comparison with the Sharpe ratio values in Table 31.9 (Panel A) suggests that the risk-adjusted performance of cryptocurrency investors trading the five digital coins and the two indices between 1st October 2016 and 1st October 2017 is between one and nine times higher than the Sharpe ratio of the S&P 500. Hence, we conclude that cryptocurrency traders' risk-adjusted returns at the millisecond time frame are much higher than that of institutional investors operating in the equity markets at lower frequencies.

Moreover, we test the significance of our Sharpe ratio estimations by adopting the observed Sharpe ratio $(\widehat{SR})$ introduced by Bailey and Lopez de Prado (2012). The testing procedure can be represented in the following probabilistic terms:

$$\widehat{PSR}(SR^*) = Z \left[\frac{(\widehat{SR} - SR^*)\sqrt{n-1}}{\sqrt{1 - \widehat{y_3}\widehat{SR} + \frac{\widehat{y_4}}{4}\widehat{SR}^2}} \right] \tag{31.15}$$

where Z denotes the *cdf* of the standard normal distribution and, SR^* represents the Sharpe's ratio value under the null hypothesis. For a given SR^*, $\widehat{PSR}$ increases with greater $\widehat{SR}$ in the millisecond sample, or positively skewed returns $(\widehat{y_3})$, or longer track records (n). $\widehat{PSR}$ decreases when tails are fatter $(\widehat{y_4})$. The last column of Table 31.9 (Panel A) illustrates that all Sharpe ratios for all five cryptocurrencies and the two indices are statistically significant at the five per cent level.

We follow the rolling regressions procedure of Tang and Whitelaw (1997) and Da Fonseca (2010) to determine if there is significant time-variation in estimated Sharpe ratios. Tang and Whitelaw's procedure requires choosing a fixed sample size and running rolling regressions. That is, a fixed number of data observations are used to compute Sharpe ratio values, and the estimation window is moved forward by one month at a time. Tang and Whitelaw (1997) suggests that the advantage of this approach is that if Sharpe ratio values vary over time, either because of empirical model misspecification or structural shifts, then the values from the rolling regressions will 'adapt' to these changes. In order to run the rolling regressions we divide our data sample into four equal subsamples: subperiod 1 (1 October 2016 to 31 December 2016), subperiod 2 (1 January 2017 to 31 March 2017), subperiod 3 (1 April 2017 to 31 June 2017) and subperiod 4 (1 July 2017 to 31 September 2017). We follow Da Fonseca (2010) to estimate the conditional Sharpe ratio:

$$S_{i,t} = -r_{f,t}\sigma_{M,t} \operatorname{Corr}_t \left(M_{t+1}, R_{i,t+1} \right) \tag{31.16}$$

where $r_{f,t}$ denotes the risk-free return; $\sigma_{M,t}$ represents the conditional standard deviation of the stochastic discount factor (or pricing kernel) of the Harrison and Kreps (1979) type, M_{t+1}; Corr_t measures the conditional correlation between the pricing kernel and digital financial instrument i and $R_{i,t+1}$ denotes the expected return of digital financial instrument i at time t.

Table 31.9 (Panel B) shows that the average of the time-varying Sharpe ratios is positive across the four different subperiods. We observe that the highest conditional Sharpe ratios is subperiod 1, with values decreasing in subperiods 2, 3, and 4. These variations in the values—if the conditional Sharpe ratios are inconsistent with the conditional Capital Asset Pricing Model (CAPM) and related models—imply a close to constant Sharpe ratio. One possible explanation is that the results are due to market inefficiency that we observe in our herding behaviour investigations. The reported statistics in Table 31.9 (Panel B) are complemented by a test of equality across the four subperiods. The results of the test-related chi-squared statistics and the respective level of significance confirm that the behaviour of the Sharpe ratios was not equal across the four subperiods under investigation. Our findings are broadly in line with Tang and Whitelaw (1997) and Da Fonseca (2010), with the notable difference that Tang and Whitelaw (1997) reports low Sharpe ratios at the peak of the business cycle and high at the trough.

Next, we investigate the persistence of cryptocurrency and index profits by examining whether the profits from the previous day's trading are a good predictor of the profits of the current day. This robust examination is important because the potential distribution of profits over time indicates whether cryptocurrency and index profits will extend their future strong performance. To test the persistence of cryptocurrency and index profits, we run the following OLS regression:

$$\text{Pr ofit}_{i,t} = \alpha + \beta_1 \text{Pr ofit}_{i,t-1} + \beta_2 \text{Volume}_t$$
$$+\beta_3 \text{Volatility}_t + \varepsilon_{i,t}$$

(31.17)

where $Profit_{i,t}$ represents modified log profits $sign(profits) * \log(1 + |profits|)$ to incorporate any negative trading profits; $Profit_{i,t-1}$ denotes the signed logged profits for bitcoin, ethereum, ripple, litecoin, dash, CRIX and CCI30 on day $t-1$; $Volume_t$ is the log of each cryptocurrency and index trading volume for day t; $Volatility_t$ measures the price volatility for day t computed as the volume-weighted standard deviation of the price process for the same day and $\varepsilon_{i,t}$ represents the noise in each order i on day t.

Table 31.10 reports the persistency of cryptocurrency and index profits for bitcoin, ethereum, ripple, litecoin, dash, CRIX and CCI30 between 1 October 2016, and 1 October 2017. Although the coefficient

TABLE 31.10 Persistency of Profits for Bitcoin, Ethereum, Ripple, Litecoin, Dash, CRIX and CCI30 from 1 October 2016–1 October 2017

Variables	$Profit_{t-1}$	Intercept	$Volume_t$	$Volatility_t$	Adj-R^2	F-test
Bitcoin						
Univariate regressions	0.48***	5.034**			0.91	7.133***
Multivariate regressions	0.021***	6.223***	0.954***	0.668**	0.96	22.009***
Ethereum						
Univariate regressions	0.55***	6.987***			0.94	8.663**
Multivariate regressions	0.044***	7.004**	0.989***	0.704***	0.96	30.115***
Ripple						
Univariate regressions	0.79***	8.272***			0.95	11.257***
Multivariate regressions	0.088***	9.102***	0.999***	0.897**	0.97	35.803***
Litecoin						
Univariate regressions	0.21***	2.174***			0.92	4.772***
Multivariate regressions	0.009***	3.818***	0.739***	0.277***	0.93	16.373***
Dash						
Univariate regressions	0.24***	3.990***			0.95	5.990***
Multivariate regressions	0.012***	4.264**	0.889***	0.371***	0.94	25.002***
CRIX						
Univariate regressions	0.11***	2.007***			0.97	3.289**
Multivariate regressions	0.008***	2.983**	0.632**	0.291***	0.94	14.039***
CCI30						
Univariate regressions	0.007***	1.996***			0.91	3.006***
Multivariate regressions	0.004***	2.120***	0.582***	0.203**	0.93	12.601***

Profit$i,t-1$ denotes the signed logged profits for bitcoin, ethereum, ripple, litecoin, dash, CRIX and CCI30 on day $t-1$; t Volume is the log of each cryptocurrency and index trading volume for day t; t Volatility measures the price volatility for day t computed as the volume-weighted standard deviation of the price process for the same day and i, $t \epsilon$ represents the noise in each order i on day t.

***indicates significance at the 1% level of significance. **indicates significance at the 10% level.

of the lagged profitability variable (*Profit$_{t-1}$*) and its statistical significance is the main indicator of persistence in cryptocurrency and index profits, we include two control variables (*Volume$_t$* and *Volatility$_t$*) in order to determine whether profitability persistence is due to the lagged profitability being a proxy for the two control variables. The first column of Table 31.10 shows both the univariate and multivariate regressions with the two control variables. The lagged profitability variable (*Profit$_{t-1}$*) coefficients demonstrate statistical significance at the one per cent level in the seven financial instruments. This finding indicates that one-day lagged trading performance is a good predictor of the performance for the current day.

In addition, the multivariate specification with control variables maintains the statistical significance at different levels of significance. This finding is consistent with Momtaz (2019) who obtained data of 1403 cryptocurrencies sorted in four different portfolios to report abnormal returns between 1000 per cent and 1200 per cent for three of the portfolios within three years of continuous trading. Lintilhac and Tourin (2019) deploy stochastic control models and bitcoin data from 2014 to 2016 to estimate positive out-of-sample profits at different sampling frequencies. However, in our study, we observe that the size of the lagged profitability variable decreases significantly after the introduction of the control variables. This decline in the coefficient of the lagged profitability variable indicates a lower persistence of cryptocurrency and index profits when we introduce control variables.

31.3.7 CTs Activity during Future EPM

We investigate the CTs activity during future EPMs by developing probit regressions. We compute the probability of EPMs as a function of lagged coefficients of CTs^{NET}, volume, return and spread:

$$\text{Prob}(EPM = 1)_{i,t} = \alpha + \beta_1 CTs_{i,t-1}^{NET} + \beta_2 RT_{i,t-1} + \beta_3 \text{Vol}_{i,t-1}$$
$$+ \beta_4 QS_{i,t-1} + \varepsilon_{i,t}$$

(31.18)

We define all variables in equation (12), while we lag all variables by one interval in equation (18).

Table 31.11 reports significant evidence of CTs being related to higher probability of future EPMs. The marginal effect of the CTs^{NET} variable indicates that the probability of an EPM increases by 7.4 per cent of the unconditional probability with every standard deviation in the pre-EPM CTs^{NET} (Column 3 of Table 31.11). We observe the same pattern of increased probability of future EPM in both the permanent and transitory components. The probability of an EPM increases by 2.9 per cent in the permanent component, while the same probability in the transitory component increases by 2.1 per cent at the one per cent level of significance (Columns 1 and 2 of Table 31.11).

TABLE 31.11   CTs Activity and the Probability of Future Occurrence of EPMs for Bitcoin, Ethereum, Ripple, Litecoin, Dash, CRIX and CCI30 from 1 October 2016, to 1 October 2017 during EPMs

Model	Permanent (1)	Transitory (2)	All (3)
Intercept	4.727***	6.023***	8.547***
(*t*-stat)	(0.00)	(0.00)	(0.00)
CTs_{t-1}^{NET}	0.014***	0.037***	0.059***
Marginal Effect	0.021***	0.029***	0.074***
(*t*-stat)	(0.00)	(0.00)	(0.00)
Controls	Yes	Yes	Yes
Pseudo-R^2	0.89	0.90	0.94

CTsNET measure the trading volume of all seven financial instruments traded against the direction of the price movement for all CT trades. *** indicates significance at the 1% level.

31.4 Robustness Checks

Most EPMs are the result of very brief liquidity dislocations and, therefore, we implement a short sampling interval of 20 s in order to capture them. However, a short sampling interval could divide the EPMs into various price changes that are not sufficiently large enough to be detected by the identification procedure. To check the robustness of the results, we re-run the main analysis for two alternative time intervals: one minute and ten minutes. We also run regression coefficients for the five cryptocurrencies and the two indices during internals of extreme market fluctuations measured using five-per cent criterion. Finally, we adopted a more dynamic time-varying approach to test the robustness of our herding behaviour results under one, five and ten-per cent criterion.

Tables 31.12 and 31.13 report the activity of CTs during periods of one minute and ten minutes before and after an EPM event and illustrate the liquidity demand and supply findings for CTs. Panel A of

TABLE 31.12 Cryptocurrency Traders (CTs) Liquidity Supply and Demand Around Extreme Price Movements (EPMs)

Panel A	CTs^{NET}		
EPMs period	$t{-}1$ minute	t	$t+1$ minute
Bitcoin	166.1** (0.03)	648.8*** (0.00)	304.2*** (0.00)
Ethereum	135.2*** (0.00)	496.4** (0.01)	219.9*** (0.00)
Ripple	129.3***	344.0***	204.7***
	(0.00)	(0.00)	(0.00)
Litecoin	98.8*** (0.00)	232.1*** (0.00)	199.6** (0.01)
Dash	90.3*** (0.00)	201.9** (0.02)	121.7*** (0.00)
CRIX	55.9** (0.01)	188.1*** (0.00)	84.6*** (0.00)
CCI30	34.04** (0.03)	168.2*** (0.00)	73.9* (0.11)
Panel B	CTs^{D}		
Bitcoin	1518.2*** (0.00)	16048.2*** (0.00)	4711.2*** (0.00)
Ethereum	1254.9*** (0.00)	13095.4*** (0.00)	2660.5*** (0.00)
Ripple	999.5***	9463.4***	2008.2***
	(0.00)	(0.00)	(0.00)
Litecoin	701.5*** (0.00)	6222.5*** (0.00)	1893.2*** (0.00)
Dash	670.0*** (0.00)	4994.3*** (0.00)	1442.0** (0.02)
CRIX	273.8* (0.07)	1898.9*** (0.00)	877.4** (0.03)
CCI30	191.9** (0.03)	1154.1*** (0.00)	595.8*** (0.00)
Panel C	CTs^{S}		
Bitcoin	−1101.5*** (0.00)	−13899.0*** (0.00)	−2028.1*** (0.00)
Ethereum	−799.4*** (0.00)	−9003.2*** (0.00)	−1327.9*** (0.01)
Ripple	−584.6***	−5115.0***	−1002.8***
	(0.00)	(0.00)	(0.00)
Litecoin	−346.1** (0.02)	−3374.7*** (0.00)	−1101.3** (0.03)
Dash	−299.9*** (0.00)	−2469.2*** (0.00)	−717.7*** (0.00)
CRIX	−106.0*** (0.00)	−963.7*** (0.00)	−408.1*** (0.00)
CCI30	−98.8*** (0.02)	−723.2* (0.01)	−293.7*** (0.00)

This table shows the directional trading volume around EPMs for bitcoin, ethereum, ripple, litecoin, dash, CRIX and CCI30 from 1 October 2016, to 1 October 2017. We report the results for time prior to an EPM ($t − 1$ minute) and after an EPM ($t + 1$ minute). CTs^{D} measure the liquidity difference in liquidity-demanding CT volume in the direction of the EPMs and liquidity-demanding volume in the opposite direction of the EPMs. CTs^{S} measure the liquidity difference in liquidity-providing volume in the opposite direction of the EPMs and liquidity-providing volume in the direction of the EMPs. CTs^{NET} represents the difference between CTs^{D} and CTs^{S}. We present p-values in parentheses.*** indicates significance at the 1% level. ** indicates significance at the 5% level.*indicates significance at the 10% level.

TABLE 31.13   Cryptocurrency Traders (CTs) Liquidity Supply and Demand Around Extreme Price Movements (EPMs)

Panel A	CTs^{NET}		
EPMs period	t–10 minutes	t	t + 10 minutes
Bitcoin	319.5*** (0.00)	883.1*** (0.00)	580.3*** (0.00)
Ethereum	228.2*** (0.00)	592.8*** (0.00)	411.4** (0.02)
Ripple	287.7***	510.6**	385.0***
	(0.00)	(0.01)	(0.00)
Litecoin	134.1** (0.02)	414.5*** (0.00)	297.9*** (0.00)
Dash	103.0*** (0.00)	383.8*** (0.02)	221.5*** (0.00)
CRIX	90.7*** (0.00)	207.9*** (0.00)	170.0*** (0.00)
CCI30	71.7* (0.12)	188.7*** (0.00)	101.3* (0.11)
Panel B	CTs^{D}		
Bitcoin	3492.6** (0.02)	18715.1*** (0.00)	7561.5*** (0.00)
Ethereum	2352.0** (0.03)	14883.4*** (0.00)	4099.9*** (0.00)
Ripple	1733.1***	11772.3***	3608.7***
	(0.00)	(0.00)	(0.00)
Litecoin	1300.9*** (0.00)	8277.0** (0.02)	2505.2*** (0.00)
Dash	1152.5*** (0.00)	6434.1*** (0.00)	2002.6*** (0.02)
CRIX	663.5*** (0.00)	1393.2*** (0.00)	1163.0** (0.01)
CCI30	459.0*** (0.03)	1201.1*** (0.00)	1101.8*** (0.00)
Panel C	CTs^{S}		
Bitcoin	−2231.3*** (0.00)	−15246.3*** (0.00)	−5478.4*** (0.00)
Ethereum	−1090.4** (0.01)	−11189.8*** (0.00)	−3066.9*** (0.00)
Ripple	−919.8***	−9555.3***	−1452.5***
	(0.00)	(0.00)	(0.00)
Litecoin	−800.7*** (0.00)	−7006.0*** (0.00)	−1291.0*** (0.00)
Dash	−563.0*** (0.00)	−4218.4*** (0.01)	−999.1*** (0.00)
CRIX	−389.3*** (0.00)	−1008.1*** (0.00)	−688.0*** (0.00)
CCI30	−246.5*** (0.00)	−647.4** (0.01)	−344.8*** (0.00)

This table shows the directional trading volume around EPMs for bitcoin, ethereum, ripple, litecoin, dash, CRIX and CCI30 from 1 October 2016, to 1 October 2017. We report the results for time prior to an EPM (t –10 minutes) and after an EPM (t + 10 minutes). CTs^{D} measure the liquidity difference in liquidity-demanding CT volume in the direction of the EPMs and liquidity-demanding volume in the opposite direction of the EPMs. CTs^{S} measure the liquidity difference in liquidity-providing volume in the opposite direction of the EPMs and liquidity-providing volume in the direction of the EMPs. CTs^{NET} represents the difference between CTs^{D} and CTs^{S}. We present p-values in parentheses.*** indicates significance at the 1% level. ** indicates significance at the 5% level.

Tables 31.12 and 31.13 show that the CTs^{NET} imbalance is positive, suggesting that liquidity demand exceeds CTs liquidity supply for the seven digital instruments. Furthermore, the positive CTs^{NET} imbalance indicates that trading is orientated in the direction of the EPMs. We observe that CTs^{NET} imbalance is statistically significant in the direction of the EPMs (interval t). The CTs^{NET} imbalance is also statistically significant in the same direction when we consider the other four intervals—one minute (t–1 min) and ten minutes (t–10 minutes) prior to an EPM and 1 min (t + 1 minute) and ten minutes (t + 10 min) after an EPM. Panel A of Tables 31.12 and 31.13 also show that in the intervals beginning one minute and ten minutes prior to an EPM, CTs execute trades in the direction of the EPMs—indicating that CTs facilitate EPMs.

We find that CTs trade in the direction of the EPMs based on their liquidity demanding trading orders and in the opposite direction with their liquidity supplying trading orders for the five cryptocurrencies and the two indices (Panels B and C of Tables 31.12 and 31.13). For example, CTs^D for bitcoin is 16,048 tokens for one minute interval (Table 31.12, Panel B), and 18,715 tokens for ten minutes time interval. The liquidity supply (CTs^S) for the same cryptocurrency is 13,899 tokens for one minute interval and 15,246 tokens for ten minutes interval. Therefore, CTs trading bitcoin for the one minute interval demand 2149 tokens of net liquidity in the direction of the EPMs and 3469 tokens in the same direction for the ten minutes interval. Although our robustness checks results are consistent with the findings of the main analysis, we observe that liquidity demand for the seven financial instruments increases when the frequency of trading decreases. CTs executing bitcoin trading orders demand 1288 tokens of net liquidity for the 20 s interval. However, this demand increases to 2149 tokens for the one second interval and 3469 tokens for the ten minutes interval. One possible explanation for this finding is that in cryptocurrency markets there are fewer high-frequency traders executing orders at the microsecond and even nanosecond timeframes.

Table 31.14 shows the regression results for the five cryptocurrencies and the two indices using the fiveper cent criterion for extreme cryptocurrency market fluctuations. Consistent with our main findings under the one and ten-per cent criterion, both β_1 and β_2 coefficients are negative, suggesting the presence of herding behaviour. The heteroscedasticity t-statistics presented in parentheses confirm the reliability of β_1 and β_2 coefficients. In addition, the β_2 estimates under the five per cent criterion are between one and three times smaller than those for the β_1 coefficients. We observe that this ratio is even more pronounced under the one per cent criterion, where the β_2 coefficients are between one and two times smaller than those for the β_1 estimates. This finding confirms our main results where the presence of herding behaviour in cryptocurrency and index markets is more evident during periods of significant increase in cryptocurrency and index prices. However, direct comparison with the one and ten-per cent criterion results in Table 31.8 (Panels A and B) reveals that the magnitude of herding during periods of significant increase in cryptocurrency and index prices is less prevalent under the five-per cent criterion. This is because the ten per cent criterion β_2 coefficients in Table 31.8 (Panel B) are between one and

TABLE 31.14 Regression Coefficients under the 5% Criterion for Bitcoin, Ethereum, Ripple, Litecoin, Dash, CRIX and CCI30 Daily Returns from 1 October 2016, to 1 October 2017 during EPMs

Panel A		5% criterion	
Cryptocurrency	a	P 1	P 2
Bitcoin	7.91	−0.29	−0.33
(t-stat)		(8.77)	(9.01)
Ethereum	7.62	−0.81	−1.94
(t-stat)		(11.09)	(10.55)
Ripple	6.03	−1.24	−2.11
(t-stat)		(16.88)	(21.76)
Litecoin	4.28	−0.15	−0.31
(t-stat)		(4.71)	(5.19)
Dash	3.59	−0.70	−0.76
(t-stat)		(6.83)	(6.50)
CRIX	2.41	−0.11	−0.20
(t-stat)		(2.21)	(2.56)
CCI30	2.16	−0.14	−0.25
(t-stat)		(2.46)	(2.62)

a is measure the average dispersion of the sample. We report the heteroscedasticity /-statistics in parentheses.

five times smaller than those for the β_1 coefficients, while the β_2 coefficients under the one per cent criterion in Table 31.8 (Panel A) are between one and eight times smaller than those for the β_1 coefficients.

Any investment approach, including following the herd, could change over time (Ge.bka and Wohar 2013). For example, Christoffersen and Tang (2010) demonstrate that herding behaviour is more persistent during periods of low returns, and Chiang and Zheng (2010) and Kremer and Nautz (2013) document herding to be more pronounced during periods of market stress. Numerous studies such as Andreu *et al.* (2009); Pierdzioch *et al.* (2010); Ge.bka and Wohar (2013); Sensoy (2013); Stavroyiannis and Babalos (2013); Babalos and Stavroyiannis (2015); Babalos *et al.* (2015); Sharma *et al.* (2015); Clements *et al.* (2017); Arjoon and Bhatnagar (2017); Curto *et al.* (2017); Yasir and Önder (2019) and Stavroyiannis *et al.* (2019) implement time-varying and rolling-window models to demonstrate the dynamic nature of herding behaviour in equity and commodity markets. However, less than a handful of studies, such as Stavroyiannis and Babalos (2019) and Bouri *et al.* (2019), investigate this phenomenon in cryptocurrency markets using time-varying techniques.

To formally examine the dynamic nature of herding behaviour in cryptocurrency markets, we follow the time-varying testing procedure of Ge.bka and Wohar (2013). We divide our data sample into four equal subsamples: subperiod 1 (1 October 2016 to 31 December 2016), subperiod 2 (1 January 2017 to 31 March 2017), subperiod 3 (1 April 2017 to 31 June 2017) and subperiod 4 (1 July 2017 to 31 September 2017).

Robustness checks estimations of equation (9) under the one, five and ten per cent criteria are reported in Table 31.15. In general, they confirm our previous findings: both and coefficients are negative, implying the presence of herding behaviour in the seven virtual currency markets under the three criteria. However, it is interesting to highlight that most β_1 and β_2 coefficients decline over time. The two herding parameters, β_1 and β_2, decline in value in the four subperiods, with evidence of significantly lowered values between subperiods 3 and 4. This finding indicates that herding behaviour follows a dynamic pattern that varies over time with decreasing magnitude. The evidence of time-varying herding behaviour in cryptocurrency markets is in line with the findings of Bouri *et al.* (2019) but opposite to Stavroyiannis and Babalos (2019). Bouri *et al.* (2019) obtained daily data from the leading 14 cryptocurrencies from April 2013 to May 2018 and used a rolling-window approach to investigate whether cryptocurrency traders mimic the investment decisions of others. The authors demonstrate that the cryptocurrency market experiences the presence of the type of herding behaviour that seems to vary over time. In contrast, Stavroyiannis and Babalos (2019) collected daily data from eight digital coins between August 2015 and February 2018 and used two testing procedures: the standard testing procedure based on ordinary least squares and a more robust time-varying regression model to revisit the topic of cryptocurrency herding behaviour. This study reveals that the presence of herding behaviour under the static testing procedure is applied, but this type of behaviour diminishes when a more dynamic testing model is introduced.

31.5 Conclusion

Since the introduction of bitcoin in 2009, there have been more than 1100 projects that have implemented similar ideas to bitcoin and these digital projects are collectively known as 'cryptocurrencies'. Despite the technological advances in the ever growing field of cryptocurrencies, there a number of major issues that concern the market and policy makers around the world. These issues are based on different reasons such as the infant nature of the digital market, the lack of understanding of cryptocurrency trading and some of the abnormal economic properties of the cryptocurrencies. Furthermore, so far there is little academic research related to the entire field of cryptocurrencies.

In this study, we deploy a unique dataset, stamped at the millisecond timeframe, of the most traded cryptocurrencies—bitcoin, ethereum, ripple, litecoin and dash—and two major cryptocurrency indices—Crypto Index (CRIX) and CCI30 Crypto Currencies Index—to investigate the relationship between cryptocurrency liquidity, herding behaviour and profitability during periods of stressful market conditions.

TABLE 31.15 Rolling Regressions of the Conditional Sharpe Ratio For Bitcoin, Ethereum, Ripple, Litecoin, Dash, CRIX and CCI30

Cryptocurrency	1%			5%			10%		
	α	β_1	β_2	α	β_1	β_2	α	β_1	β_2
Bitcoin									
Subperiod 1	4.13	−0.1636***	−1.0272***	4.28	−0.1737***	−1.1885***	5.08	−0.2733***	−1.2991***
Subperiod 2	3.80	−0.1982***	−1.0788**	4.03	−0.1993***	−1.1936***	4.86	−0.2982**	−1.5775**
Subperiod 3	2.99	−0.8220***	−1.9828***	3.56	−1.0388***	−2.4822***	3.16	−1.4074***	−2.7260***
Subperiod 4	2.16	−1.3738***	−2.0282***	3.11	−1.6835***	−3.6673***	3.09	−1.9091***	−4.8009***
Ethereum									
Subperiod 1	4.01	−0.1403**	−1.0046**	4.09	−0.1603***	−1.1002***	4.76	−0.2084***	−1.1274***
Subperiod 2	3.87	−0.1711***	−1.0993***	3.47	−0.1900***	−1.1416***	4.18	−0.2233***	−1.2900***
Subperiod 3	3.15	−0.7363***	−1.7222***	2.80	−0.9967***	−2.1289**	3.72	−1 3729***	−2.2157***
Subperiod 4	2.99	−1.2180***	−1.9945***	2.22	−1.5174***	−3.0182***	3.34	−1.8366***	−4.4255***
Ripple									
Subperiod 1	3.90	−0.1077***	−0.8815***	3.99	−0.1212*	−1.0568*	4.17	−0 1499***	−1.0982***
Subperiod 2	3.81	−0.1338***	−0.9863***	3.50	−0.1395***	−1.1104***	4.02	−0.1788**	−1.1062**
Subperiod 3	3.72	−0.5409***	−1.6377**	2.88	−0.7790***	−2.0043***	3.70	−1.2165***	−2.0556**
Subperiod 4	3.25	−1.1400***	−1.8882***	2.34	−1.3214***	−2.8477***	3.21	−1.7351***	−3.9434***
Litecoin									
Subperiod 1	2.80	−0.1004***	−0.6938***	2.91	−0.1188***	−0 9995***	3.37	−0.1402***	−1.0043***
Subperiod 2	2.66	−0.1255***	−0.7704***	2.63	−0.1209**	−1.0384**	3.11	−0.1699***	−1.0472***
Subperiod 3	2.20	−0.3117**	−1.4662***	2.05	−0.6367***	−1.8744***	2.99	−1.1330**	−2.0002***
Subperiod 4	2.04	−1.0500***	−1.7812***	1.34	−1.1766***	−2.3220***	2.75	−1.5290***	−3.4210***
Dash									
Subperiod 1	1.77	−0.0822*	−0.4033**	1.89	−0.0973***	−0.6522**	2.12	−0.1039*	−0.9052*
Subperiod 2	1.61	−0.1010***	−0.6637***	1.78	−0.1200**	−0.9035**	2.00	−0.1189***	−0.9233***
Subperiod 3	1.49	−0.1279**	−1.2747**	1.01	−0.2491***	−1.2336***	1.85	−1.0466**	−1.8771**
Subperiod 4	1.12	−1.0030***	−1.4466***	0.96	−1.0835***	−2.1078***	1.59	−1.2790***	−2 9993***
CRIX									
Subperiod 1	1.10	−0.0480***	−0.2366**	1.15	−0.0681***	−0.4003**	1.64	−0.0884***	−0.7355***
Subperiod 2	1.03	−0.0773**	−0.4383**	1.00	−0.0943**	−0.7556**	1.52	−0.1010***	−0.8884***
Subperiod 3	0.78	−0.0999*	−1.1121*	0.65	−0.1288***	−1.1409***	1.39	−0.6755*	−1.4431*
Subperiod 4	0.52	−0.8352***	−1.2004***	0.23	−0.9812***	−1.7200***	1.26	−1.0409**	−2.2660**
CCI30									
Subperiod 1	0.83	−0.0296**	−0.1537**	0.99	−0.0457*	−0.2371**	1.18	−0.0401***	−0.5882***
Subperiod 2	0.70	−0.0309***	−0.3854***	0.86	−0.0582**	−0.6004***	0.98	−0.0743*	−0.6993*
Subperiod 3	0.57	−0.0614**	−1.0007**	0.53	−0.0823***	−1.1152***	0.74	−0.0936**	−1.2201**
Subperiod 4	0.29	−0.5580**	−1.1003**	0.44	−0.8830**	−1.2909**	0.53	−0.9320***	−1.6401***

Subperiod 1 (1st October 2016 to 31st December, 2016); subperiod 2 (1st January, 2017 to 31st March, 2017); subperiod 3 (1st April, 2017 to 31st June, 2017); subperiod 4 (1st July, 2017 to 31st, September, 2017).*** indicates significance at the 1% level. ** indicates significance at the 5% level. * indicates significance at the 10% level.

Our empirical findings suggest that CTs demand liquidity even during the utmost EPMs. We also observe that CTs facilitate EPMs and are likely to continue this process in the future. We find significant evidence of herding behaviour in the seven financial instruments and we observe that the presence of herding behaviour in cryptocurrency markets is more evident during periods of significant increase in cryptocurrency prices. In other words, cryptocurrency market participants tend to follow the market consensus when prices are increasing. Our robustness checks obtained from a more dynamic time-varying approach indicate that herding behaviour follows a dynamic pattern that varies over time with decreasing magnitude.

We also estimate the profits of individual CTs by taking into account the appropriate transaction costs, including software and hardware expenses. Our profitability analysis suggests that between 1October 2016, and 1October 2017 the average gross daily profit is $1,874,312 for bitcoin; $709,005 for ethereum; $244,762 for ripple; $108,338 for litecoin; $100,990 for dash; $71,122 for CRIX and $54,806 for CCI30. Our examination of profit persistency suggests that CTs are capable of extending their strong profitability-generating record in the future.

Although beyond the scope of this study, more research can shed light on some significant cryptocurrency market issues. For instance, it will be important to investigate the implications of position price limits in cryptocurrency exchanges on the development of herding behaviour. Will the introduction of position price limits help mitigate the creation of such behaviour?

Disclosure Statement

No potential conflict of interest was reported by the author(s).

References

Ajaz, T. and Kumar, A. S., Herding in crypto-currency markets. *Ann. Financ. Econ.*, 2018, **13**(02), 1850006. doi:10.1142/S2010495218500069.

Al-Yahyaee, K. H., Mensi, W., Ko, H.-U., Yoon, S.-M. and Kang, S. H., Why cryptocurrency markets are inefficient: The impact of liquidity and volatility. *N. Am. J. Econ. Finance*, 2020, **52**, 101–168. doi:10.1016/j.najef.2020.101168.

Alexander, C. and Dakos, M., A critical investigation of cryptocurrency data and analysis. *Quant. Finance*, 2019, **20**(2), 173–188. doi:10.1080/14697688.2019.1641347.

Amihud, Y., Illiquidity and stock returns: Cross-section and time series effects. *J. Financ. Mark.*, 2002, **5**, 31–56. doi:10.1016/S1386-4181(01)00024-6.

Anand, A. and Vankataraman, K., Market conditions, fragility, and the economics of market making. *J. Financ. Econ.*, 2016, **8**(1), 31–53. doi:10.1016/j.jfineco.2016.03.006.

Andreu, L., Ortiz, C. and Sarto, J. L., Herding behaviour in strategic asset allocations: New approaches on quantitative and intertemporal imitation. *Appl. Financ. Econ.*, 2009, **19**, 1649–1659. doi:10.1080/09603100903018786.

Arjoon, V. and Bhatnagar, C., Dynamic herding analysis in a frontier market. *Res. Int. Bus. Finance*, 2017, **42**, 496–508. doi:10.1016/j.ribaf.2017.01.006

Asolo, B. The ultimate beginner's guide to cryptocurrency trading, 2019. Available online at: https://www.mycryptopedia.com/theultimate-beginners-guide-to-trading-cryptocurrency/ (accessed 1 March 2020).

Babalos, V. and Stavroyiannis, S., Herding, anti-herding behaviour in metal commodities futures: A novel portfolio-based approach. *Appl. Econ.*, 2015, **47**(46), 4952–4966. doi:10.1080/00036846.2015.1039702.

Babalos, V., Stavroyiannis, S. and Gupta, R., Do commodity investors herd? Evidence from a timevarying stochastic volatility model. *Resour. Policy*, 2015, **46**, 281–287. doi:10.1016/j.resourpol.2015.10.011.

Bailey, D. H. and Lopez de Prado, M. M., The Sharpe ratio efficient frontier. *J. Risk*, 2012, **15**(2), 34–57.

Ballis, A. and Drakos, K., Testing for herding in the cryptocurrency market. *Finance Res. Lett.*, 2020, **33**, 101210.

Bariviera, A. F., The inefficiency of bitcoin revisited: A dynamic approach. *Econ. Lett.*, 2017, **161**, 1–4. doi:10.1016/j.econlet.2017.09.013.

Bariviera, A. F., Zunino, L. and Rosso, O. A., An analysis of high-frequency cryptocurrencies prices dynamics using permutation-information-theory quantifiers. *Chaos: Interdisc. J. Nonlinear Sci.*, 2018, **28**, 34–61.

Bhosale, J. and Mavale, S., Volatility of select crypto-currencies: A comparison of bitcoin, ethereum and litecoin. *Annu. Res. J. SCMS Pune*, 2018, **6**, 132–141.

Bongaerts, D. and Van Achter, M. High-frequency trading and market stability. Working Paper, Erasmus University Rotterdam (EUR)—Finance, 2015.

Bouri, E., Gupta, R. and Roubaud, D., Network causality structures among bitcoin and other financial assets: A directed acyclic graph approach. *Q. Rev. Econ. Financ.*, 2018, **70**, 203–213.

Bouri, E., Gupta, R. and Roubaud, D., Herding behaviour in cryptocurrencies. *Finance Res. Lett.*, 2019, **29**, 216–221. doi:10.1016/j.frl.2018.07.008.

Bouri, E., Lau, C. K. M., Lucey, B. and Roubaud, D., Trading volume and the predictability of return and volatility in the cryptocurrency market. *Finance Res. Lett.*, 2018, **29**, 340–346. doi:10.1016/j.frl.2018.08.015.

Brauneis, A. and Mestel, R., Price discovery of cryptocurrencies: Bitcoin and beyond. *Econ. Lett.*, 2018, **165**, 58–61. doi:10.1016/j.econlet.2018.02.001.

Brogaard, J., Carrion, A., Mayaert, T., Riordan, R., Shkilko, A. and Sokolov, K., High frequency trading and extreme price movements. *J. Financ. Econ.*, 2017, **128**(2), 253–265.

Brogaard, J., Hendershott, T., and Riordan, R., High-frequency trading and price discovery. *Rev. Financ. Stud.*, 2014, **27**(8), 2267–2306.

Calderón, P. O. Herding behavior in cryptocurrency markets, 2018. Available online at: https://www.researchgate.net/profile/Obryan_ Poyser/publication/324993841_Herding_behavior_in_crypto curr ency_market/links/5c0506e092851cf05c6712e6/Herding-behavi or-incryptocurrency-mar-ket.pdf (accessed 15 March 2020).

Caporale, G. M., Gil-Alana, L. and Plastun, A., Persistence in the cryptocurrency market. *Res. Int. Bus. Finance*, 2018, **46**, 141–148.

Cespa, J. and Vives, X. The welfare impact of high-frequency trading. Working Paper, 2015.

Charfeddine, L. and Maouchi, Y., Are shocks on the returns and volatility of cryptocurrencies really persistent? *Finance Res. Lett.*, 2018, **28**, 423–430. doi:10.1016/j.frl.2018.06.017.

Cheah, E. T. and Fry, J., Speculative bubbles in bitcoin markets? An empirical investigation into the fundamental value of bitcoin. *Econ. Lett.*, 2015, **130**, 32–36. doi:10.1016/j.econlet.2015.02.029.

Chiang, T. C. and Zheng, D., An empirical analysis of herd behavior in global stock markets. *J. Bank. Financ.*, 2010, **34**(8), 1911–1921.

Christie, W. G. and Huang, R. D., Following the pied piper: Do individual returns herd around the market? *Financ. Anal. J.*, 1995, **51**(4), 31–37. doi:10.2469/faj.v51.n4.1918.

Christoffersen, S. R. and Tang, Y., Institutional herding and information cascades: Evidence from daily trades. Working paper, 2010, https://papers.ssrn.com/sol3/papers.cfm?abstract_id = 1572726.

Clements, A., Hurn, S. and Shi, S., An empirical investigation of herding in the U.S. stock market. *Econ. Model.*, 2017, **67**, 184–192. doi:10.1016/j.econmod.2016.12.015.

Connolly, R. A., An examination of the robustness of the weekend effect. *J. Financ. Quant. Anal.*, 1989, **24**, 133–169. doi:10.2307/2330769.

Corbet, S., Lucey, B. and Yarovaya, L., Datestamping the bitcoin and ethereum bubbles. *Finance Res. Lett.*, 2018, **26**, 81–88. doi:10.1016/j.frl.2017.12.006.

Corwin, S. A. and Schultz, P., A simple way to estimate bid and ask spreads from daily high and low prices. *J. Finance*, 2012, **67**(2), 719–759. doi:10.1111/j.1540-6261.2012.01729.x.

Curto, J. D., Falcão, P. F. and Braga, A. A., Herd behaviour and market efficiency: Evidence from the Iberian stock exchanges. *J. Adv. Stud. Finance*, 2017, **2**(16), 81–94. doi:10.14505/jasf.v8.2(16).01.

Da Fonseca, J. S., The performance of the European stock markets: A time-varying Sharpe ratio approach. *Eur. J. Finance*, 2010, **16**(7), 727–741. doi:10.1080/1351847X.2010.495479.

Da Gama Silva, P. V. J., Klotzle, M. C., Pinto, A. C. F. and Gomes, L. L., Herding behavior and contagion in the cryptocurrency market. *J. Behav. Exp. Finance.*, 2019, **22**, 41–50. doi:10.1016/j.jbef.2019.01.006.

Empirica. Different types of cryptocurrency, 2018. Available online at: http://empirica.io/blog/different -types-cryptocurrency/ (accessed 19 January 2018).

Fama, E. and French, K., The equity premium. *J. Finance*, 2002, **57**(2), 637–659. doi:10.1111/1540-6261.00437.

Gandal, N., Hamrick, J. T., Moore, T. and Oberman, T., Price manipulation in the bitcoin ecosystem. *J. Monet. Econ.*, 2018, **95**, 86–96. doi:10.1016/j.jmoneco.2017.12.004.

Gębka, B. and Wohar, M. E., International herding: Does it differ across sectors? *J. Int. Financ. Mark. Inst. Money*, 2013, **23**, 55–84. doi:10.1016/j.intfin.2012.09.003.

Golub, A., Keane, J. and Poon, S.-H. High-frequency trading and mini-flash crashes. Working Paper, 2013.

Goyenko, R. Y., Holder, C. W. and Trzcinka, C. A., Do liquidity measure liquidity? *J. Financ. Econ.*, 2009, **92**, 153–181. doi:10.1016/j.jfineco.2008.06.002.

Gümüs, G. K., Gümüs, Y. and Çimen, A., Herding behaviour in cryptocurrency market: CSSD and CSAD analysis. In *Blockchain Economics and Financial Market Innovation*, pp. 103–114, 2019 (Springer).

Harrison, M. and Kreps, D., Martingales and arbitrage in multiperiod security markets. *J. Econ. Theory.*, 1979, **20**(3), 381–408.

Haryanto, S., Subroto, A. and Ulpah, M., Disposition effect and herding behavior in the cryptocurrency market. *J. Ind. Bus. Econ.*, 2020, **47**, 1–18.

Hattori, T. and Ishida, R. Did the introduction of bitcoin futures crash the bitcoin market at the end of 2017? 2019, Working paper.

Hwang, S. and Salmon, M., Market stress and herding. *J. Empir. Finance*, 2004, **11**(4), 585–616. doi:10.1016/j.jempfin.2004.04.003.

Kaiser, L. and Stöckl, S., Cryptocurrencies: Herding and the transfer currency. *Finance Res. Lett.*, 2020, **33**, 101214.

Kallinterakis, V. and Wang, Y., Do investors herd in cryptocurrencies– and why? *Res. Int. Bus. Finance*, 2019, **50**, 240–245. doi:10.1016/j.ribaf.2019.05.005.

Kirilenko, A., Kyle, A. S., Samadi, M. and Tuzun, T., The flash crash: The impact of high-frequency trading on an electronic market. *J. Finance*, 2017, **72**(3), 967–998. doi:10.1111/jofi.12498

Kremer, S., and Nautz, D., Causes and consequences of short-term institutional herding. *J. Bank. Financ.*, 2013, **37**(5), 1676–1686.

Kumar, A. and Smith, C. Cryptocurrencies—An introduction to not-so-funny moneys, 2017, Reserve Bank of New Zealand Analytical Notes series AN2017/07, Reserve Bank of New Zealand.

Leal, S. J., Napoletano, M., Roventini, A. and Fagiolo, G., Rock around the clock: An agent-based model of low- and high-frequency trading. *J. Evol. Econ.*, 2014, **26**(1), 49–76.

Lewis, J. B. and Linzer, D. A., Estimating regression models in which the dependent variable is based on estimates. *Polit. Anal.*, 2005, **13**, 345–364. doi:10.1093/pan/mpi026.

Li, T., Shin, D. and Wang, B. Cryptocurrency pump-and-dump schemes. Working Paper, 2018.

Lintilhac, P. S. and Tourin, A., Model-based pairs trading in the bitcoin markets. *Quant. Finance*, 2019, **17**(5), 703–716. doi:10.1080/14697688.2016.1231928.

Meade, N., A comparison of short term foreign exchange forecasting methods. *Int. J. Forecast.*, 2002, **18**, 67–83. doi:10.1016/S0169-2070(01)00111-X.

Menkhoff, L., Schmidt, U. and Brozynski, T., The impact of experience on risk taking, overconfidence, and herding of fund managers: Complementary survey evidence. *Eur. Econ. Rev.*, 2006, **50**(7), 1753–1766.

Momtaz, P. P., The pricing and performance of cryptocurrency. *Eur. J. Finance*, 2019, 367–380. https://doi .org/10.1080/1351847X. 2019.1647259.

Nadarajah, S. and Chu, J., On the inefficiency of bitcoin. *Econ. Lett.*, 2017, **150**, 6–9. doi:10.1016/j. econlet.2016.10.033.

Nakamoto, S. Bitcoin: A peer-to-peer electronic cash system, 2008. Available online at: https://bitcoin .org/bitcoin.pdf (accessed 14 December 2017).

Pierdzioch, C., Rülke, J. C. and Stadmann, G., New evidence of anti-herding of oil-price forecasters. *Energy Econ.*, 2010, **32**, 1456–1459. doi:10.1016/j.eneco.2010.05.014.

Phillip, A., Chan, J. S. K. and Peiris, S., A new look at cryptocurrencies. *Econ. Lett.*, 2018, **163**, 6–9.

Raman, V., Robe, M. and Yadav, P. Electronic market makers, trader anonymity and market fragility, Working Paper, 2014.

Sensoy, A., Generalized Hurst exponent approach to efficiency in MENA markets. *Phys. A*, 2013, **392**, 5019–5026. doi:10.1016/j.physa.2013.06.041.

Sharma, S. S., Narayan, P. K. and Thuraisamy, K., Time-varying herding behaviour, global financial crisis, and the Chinese stock market. *Rev. Pac. Basin Financ. Mark. Policies*, 2015, **18**(2):1550009. doi:10.1142/S0219091515500095

Shifflet, S. and Vigna, P. Traders are talking up cryptocurrencies, then dumping them, costing others millions, 2018. Available online at: https://www.wsj.com/graphics/cryptocurrency-schemesgener-ate-big-coin/ (accessed 15 April 2018).

Stavroyiannis, S. and Babalos, V. On the time varying nature of herding behaviour: Evidence from major European indices. Working Paper, University of Peloponnese, 2013.

Stavroyiannis, S. and Babalos, V., Herding behavior in cryptocurrencies revisited: Novel evidence from a TVP model. *J. Behav. Exp. Finance.*, 2019, **22**, 57–63. doi:10.1016/j.jbef.2019.02.007.

Stavroyiannis, S., Babalos, V., Bekiros, S. and Lahmiri, S., Is anti-herding behaviour spurious? *Finance Res. Lett.*, 2019, **29**, 379–383. doi:10.1016/j.frl.2018.09.003.

Tang, Y. and Whitelaw, R. F., Time-varying Sharpe ratios and market timing. *Q. J. Finance*, 1997, **1**(3), 465–493. doi:10.1142/S2010139211000122.

Urquhart, A., The inefficiency of bitcoin. *Econ. Lett.*, 2016, **148**, 80–82. doi:10.1016/j.econlet.2016.09.019.

Velde, F. R. Bitcoin: A primer. Chicago Federal Letter 317, 2013, Federal Reserve Bank of Chicago.

Vidal-Tomás, D., Ibánes, A. M. and Farinós, J. E., Herding in the cryptocurrency market: CSSD and CSAD approaches. *Finance Res. Lett.*, 2019, **30**, 181–186. doi:10.1016/j.frl.2018.09.008.

Wei, W. C., Liquidity and market efficiency in cryptocurrencies. *Econ. Lett.*, 2018, **168**, 21–24. doi:10.1016/j.econlet.2018.04.003.

Yasir, M. and Önder, A. O. E. Time varying herding behaviour in US stock market. International Congress of Management, Economy and Policy 2019, Istanbul, Turkey, 2019.

Zhang, W., Wang, P., Li, X. and Shen, D., The inefficiency of cryptocurrency and its cross-correlation with Dow Jones Industrial average. *Phys. A*, 2018, **510**, 658–670. doi:10.1016/j.physa.2018.07.032.

Zimmerman, P. Blockchain and price volatility. Working Paper, University of Cambridge, 2018.

Appendix

Cryptocurrency Description

In this study, we consider the five largest-capped cryptocurrencies—bitcoin, ethereum, ripple, litecoin and dash—and two cryptocurrency indices—Crypto Index (CRIX) and CCI30 Crypto Currencies Index.

Bitcoin is the first cryptocurrency and it appeared in 2009, providing a solution to the issue of double spending* (Nakamoto 2008). The network is peer-to-peer and all transactions are conducted between users directly, therefore, there are no third-party entities or financial institutions. Bitcoin transactions are validated by network nodes using cryptography (the SHA-256 algorithm) and are stored in a publicly distributed ledger known as blockchain. Bitcoin is separable to around eight decimal places, but

---

* In other words, a bitcoin can be sent securely and one should not be able to spend the same bitcoin again without anyone else being able to facilitate a transaction and without one being able to chargeback the same bitcoin.

this could be increased further if needed. In economic terms, a single bitcoin can be used at a fractional increment, which can be as small as 0.0000001 bitcoins per single transaction. This particular fractional increment is known as *Satoshi*, named after the developer. As of January, 2018, the current market capitalisation of bitcoin is $191 billion (Bhosale and Mavale 2018).

Ethereum was first introduced in 2013 by Vitalik Buterin, providing a decentralised platform for smart contracts and distributed applications (DApps) to operate without any fraud, downtime or intervention from an intermediary. Similar to bitcoin, this cryptocurrency represents a public platform with open source, blockchain computing and smart scripting features. *Ether* is the token run on the platform, although the Turing-complete programming language is able to trade, secure and codify financial derivatives, insurance contracts and many other types of transactions. As of January, 2018, ethereum has a market capitalisation of approximately $105 billion (Bhosale and Mavale 2018).

Ripple was introduced in 2012 by Chris Larsen and his company, OpenCoin, with the aim to provide instant and cheap international payments. Opposite to bitcoin, this cryptocurrency possesses a consensus ledger that does not require mining from other network users, leading to less computing power and lower network latency. The payment mechanism allows payments to another network user in a matter of five seconds, compared to between one and ten minutes in mining-based protocols. Therefore, ripple has a much better likelihood of competing with the conventional debit and credit cards' point-of-sale transactions. Moreover, this particular cryptocurrency has been used by some financial institutions as their main settlement infrastructure technology, due the lack of counterparty credit risk (Phillip *et al.* 2018). As of January, 2018, the market capitalisation of ripple is $48 billion (Bhosale and Mavale 2018).

Litecoin was launched in 2011 by Charles Lie to provide much faster transaction confirmation times compared to bitcoin. In contrast to the SHA-256 encryption algorithm of bitcoin, litecoin's Application-Specific Integrated Circuit Chips (ASICs) and Proof-of- Work (PoW) algorithm makes the protocol four times faster (two and a half minutes versus ten minutes) and it also has a lower transaction fee. This is due to the heavy mining process involved in bitcoin compared to the fast litecoin mining process that can be performed by a normal desktop computer with substantially less computing power. As of January, 2018, the market capitalisation of litecoin stood at $10 billion (Bhosale and Mavale 2018).

The darkcoin project was launched in 2014 by Evans Duffield and was recently renamed 'dash'. Dash provides more anonymity than most cryptocurrencies by implementing a decentralised master code network that makes it very difficult to detect transactions. Dash is the only digital currency that uses an instant transactions feature called InstandSend. This digital currency can be mined using either a Central Processing Unit (CPU) or a Graphics Processing Unit (GPU), and as of January, 2018, has a market capitalisation of approximately $9.6 billion (Empirica 2018).

Since its introduction in 2016, CRIX serves as a benchmark for the crypto market. CRIX is a real-time index developed jointly by Humboldt University in Germany, the Singapore Management University and CoinGeco, a company that still supplies the data for computation. CRIX currently consists of 20 members determined by the Akaike Information Criterion (AIC) and the Schwartz Information Criterion (BIC). Since the cryptocurrency market is very fast, the reallocation period for the index is one month to ensure that CRIX is always up-to-date (http://thecrix.de/).

The CCI30 Crypto Currencies Index was developed by an independent team of fund managers, mathematicians and quantitative analysts and was launched in the beginning of 2015. The index is calculated in real-time and consists of the 30 cryptocurrencies with the largest market capitalisation. The index is re-computed on a quarterly basis, but the rebalancing of the constituents takes place every month. The developers adopted the exponentially weighted moving average of the market capitalisation technique to proportionally weight each index component to the square root of its smoothed market capitalisation (https://cci30.com/).

32

Identifying the Influential Factors of Commodity Futures Prices through a New Text Mining Approach

Jianping Li,
Guowen Li,
Xiaoqian Zhu and
Yanzhen Yao

32.1 Introduction

Along with the deepening of economic globalization and financial integration, the dynamics of futures prices is of great interest to practitioners, researchers, and regulators as it plays an increasingly prominent role in the financial markets (Chang *et al.* 2018). It is noteworthy that the futures market is full of uncertainty and vulnerable to numerous influential factors. How to effectively identify these influential factors is key to understanding the dynamics of futures prices (Heath 2019).

From the perspective of the underlying assets of a contract, there are broadly two categories of futures at present, namely, financial futures and commodity futures. Financial futures are based on intangible financial assets, such as interest rates and stock indices, while commodity futures are based on physical commodities, such as gold and corn. Compared with financial futures, there are several types of commodity futures, mainly divided into agricultural commodity futures, metal futures, and energy futures.

DOI: 10.1201/9781003265399-37

In addition, the factors that affect the dynamics of these commodity futures prices are more extensive and differ from each other due to the intrinsic properties of the underlying assets (Natanelov *et al.* 2011). Thus, the systematic identification of the influential factors of various commodity futures prices is a complex and tough issue.

The existing literature has made significant contributions to identifying influential factors. A flood of literature has demonstrated that the movements of the spot market and the commodity futures market themselves affect the dynamics of commodity futures markets (Wang *et al.* 2017, Ready 2018). Certain other studies have attempted to investigate the effects of macroeconomic factors on commodity futures markets, such as inflation and economic policies (Szymanowska *et al.* 2014, Liu *et al.* 2018). Several studies have also illustrated the potential effects of climate change on agricultural commodity futures markets (Algieri 2014). However, these studies have mainly identified the relevant influential factors based on researchers' subjective analysis or on summaries of previous studies. This usually makes the derived influence factors relatively limited and subjective.

More recently, increasing amounts of financial textual data, such as online financial news and financial analysis reports, have emerged owing to the rapid development of the Internet, which provides us with a new source of data for more in-depth analyses (Li *et al.* 2020a, 2020b, Wei *et al.* 2019b). We have noticed that the news about commodity futures usually incorporates valuable information regarding the influential factors of various commodity futures markets. The amount of online news is tremendous, so these articles contain a wealth of information about influential factors. By using proper text mining approaches, the influence factors of commodity futures prices can be identified relatively comprehensively based on this news data.

Therefore, the objective of this paper is to identify the influence factors of commodity futures prices from massive online news sources. A new text mining approach called Dependency Parsing-Sentence-Latent Dirichlet Allocation (hereafter DP-Sent-LDA for short) is proposed to effectively analyse the news headlines about commodity futures markets. As concise summaries of the news reports, the massive number of news headlines contain abundant relevant information about the influence factors of commodity futures prices. Moreover, the analysis of news headlines with DP-Sent-LDA is an unsupervised process, which does not rely on a human's subjective judgement. Thus, the discovery of the influence factors about commodity futures prices in this paper is relatively comprehensive and objective compared with previous studies.

In the empirical analysis, the proposed DP-Sent-LDA model is used to analyse 49 501 news headlines about commodity futures, covering the period 2011.07.05–2018.07.18. They are collected from *Hexun.com* (http://www.hexun.com/), a leading vertical financial portal website in China with massive, high-end, and prime-quality financial information. In total, 104 specific influential factors that affect the dynamics of commodity futures are identified. Then, a four-layer influential factor system for commodity futures is developed based on these factors. The relative importance of these influential factors is also provided according to their occurrence frequency in the news headlines. Finally, the effectiveness of these identified influential factors is tested by conducting regression analysis.

The rest of this paper is organized as follows. In Section 32.2, two strands of related literature are reviewed. Section 32.3 introduces the proposed text mining approach named DP-Sent-LDA. Section 32.4 describes the data in detail. The empirical analysis is conducted, and the empirical results are analysed in Section 32.5. Section 32.6 concludes with the major conclusions.

32.2 Literature Review

This paper is related to two strands of the literature: (1) influential factors of commodity futures markets and (2) topic models in financial text mining.

32.2.1 Influential Factors of Commodity Futures Markets

It is generally perceived that the dynamics of commodity futures markets (such as risk premium, futures prices, and market volatility) is determined by market factors stemming from the commodity futures market itself and the spot market (Wang *et al.* 2017). However, as an important element in both national and worldwide financial markets, the commodity futures market is also deeply affected by macroeconomic variables, such as inflation and interest rates. Additionally, the potential effects of certain factors related to the production process of the underlying assets on the dynamics of particular commodity futures markets cannot be ignored. Therefore, based on the existing relevant research, the various influential factors are classified into three major categories, called market factors, macro-level factors, and production-related factors.

Market factors refer to variables that are extracted from the commodity futures market itself and the corresponding spot market, such as spot prices (Gibson and Schwartz 1990, Hamilton and Wu 2014, Wang *et al.* 2017). For example, Gibson and Schwartz (1990) illustrated the close relationship between spot prices and oil futures pricing. Moreover, the dynamics of commodity futures volatility is affected by the equilibrium of supply and demand. Ready (2018) showed that a change in the dynamics of the oil supply may provide an explanation for the changes in the term structure of oil futures. Other market factors that have been identified in the literature also incorporate the hedging pressure of commodities (Basu and Miffre 2013), convenience yields of commodities (Dempster *et al.* 2012), investor sentiment (Smales 2014a), and so on.

Macro-level factors refer to variables that are related to macroeconomics, state politics, and public policies. It has been well-recognized in many studies that macro-level factors have a great impact on commodity markets (Christie–David *et al.* 2004, Mo *et al.* 2018). However, research regarding the reactions of commodity futures markets to macro-level variables is surprisingly scarce. Recent studies have identified the effects of inflation (Szymanowska *et al.* 2014), inventories (Heath 2019), exchange rates (Smales *et al.* 2014b), and GDP growth (Tsvetanov *et al.* 2016) on the dynamics of commodity futures markets.

Production-related factors refer to factors that are highly related to the production process of the underlying assets, including internal factors such as manufacturing techniques and external factors such as climate changes and seasonality. For instance, Sørensen (2002) found strong evidence for seasonality in pricing corn futures, soybean futures, and wheat futures based on weekly panel-data observations of futures prices from the Chicago Board of Trade. The work implemented by Algieri (2014) stated that climate change mitigation had been a frequent rationale behind biofuel policies as biofuel production could increase the demand for feed-stocks and bring about a consequent increase in agricultural commodity prices.

As reviewed above, the existing identification of the influential factors of commodity futures markets greatly relies on researchers' subjective judgements or on summarizing of previous studies. A relatively comprehensive and objective list of the influential factors is difficult to obtain due to the wide variety of commodities futures. Numerous news reports generally contain a great deal of valuable information regarding the influential factors. Thus, identifying the factors that affect the dynamics of commodity futures prices from a large collection of news reports is a feasible way to address this issue.

32.2.2 Topic Models in Financial Text Mining

As one of the major types of text mining methods, the topic model is a statistical model for the discovery of hidden semantic structures in a text body in an unsupervised learning manner. It can discover the topics of the documents and cluster these documents according to their similarities in topic space, which helps to organize and extract the key information for us to understand large collections of unstructured text bodies. Heretofore, topic models have been widely applied in a large body of finance-related literature.

The most common topic model is the Latent Dirichlet Allocation (LDA) model proposed by Blei *et al.* (2003). The application of LDA has received substantial attention in the quantitative analysis of long-length texts in the financial domain, such as financial research reports (Yan and Bai 2016), financial news (Hagenau *et al.* 2013, Pröllochs *et al.* 2015) and financial literature (Moro *et al.* 2015). For instance, Yan and Bai (2016) efficiently predicted the most correlated industries of financial news among the 24 first-level industries of the Chinese market based on 2.3 GB of financial research reports from Chinese securities companies. Based on 14 463 German financial news announcements, Pröllochs *et al.* (2015) examined how detecting negation scopes could improve the accuracy of sentiment analysis in financial news. Moro *et al.* (2015) analysed 219 published articles in the search for current research topics and future trends in business intelligence applications for the banking industry.

Furthermore, there is also a recent increasing interest in the use of topic models in the text analysis of financial statements, especially the Form 10-K reports filed with the U.S. Securities Exchange Commission (SEC). For example, using LDA-based analysis on the whole Form 10-K reports of 10 452 firms over the period of 1996–2013, Dyer *et al.* (2017) quantified a variety of 10-K disclosure attributes and found that the majority of the increase in length is the result of disclosure associated with three new regulatory requirements. In addition, some studies focus on certain sections of corporate 10-K reports and found that these short textual disclosures, which are easier to retrieve, contain substantial significant financial information as well (Campbell *et al.* 2014). Recently, a topic model called Sentence Latent Dirichlet Allocation (Sent-LDA) has been put forward to discover and quantify risk types from the textual risk disclosures (Section 32.1A) of Form 10-K reports. Having been demonstrated to be effective in dealing with the summary headings instead of the whole textual risk disclosure section, Sent-LDA has gained popularity within the text analysis literature (Wei *et al.* 2019a).

However, there still exists a drawback for the Sent-LDA model. When dealing with the summary headings of documents, all words are assumed to be meaningful and closely related to the topic of the sentence. As a matter of fact, only certain keywords convey the relevant information pertinent to the topic. Incorporating irrelevant or meaningless words into the topic identification process might bring about inaccurate results and thus reduce the accuracy of Sent-LDA. In other words, if a method can be developed to extract the keywords in sentences in advance, the most valuable information can be figured out and quantified, and the accuracy of sentence-wide topic allocation will be improved substantially. Thus, a new topic model called DP-Sent-LDA is developed in our paper to address this problem.

32.3 Proposed DP-Sent-LDA Model

In this section, we elaborate on the proposed DP-Sent-LDA model. DP-Sent-LDA is developed based on the rationale of the Sent-LDA model. First, the rationale and the graphical representation of Sent-LDA are described. Then, the concept, generative process, and learning algorithm of DP-Sent-LDA are given.

32.3.1 Sent-LDA

As the most common topic model currently in use, LDA is an unsupervised Bayesian machine-learning model to identify the topics contained in a set of documents (Dyer *et al.* 2017). More specifically, LDA generates automatic summaries of topics in terms of a discrete probability distribution over the words for each topic and further infers per-document discrete distributions over topics. The interaction between the set of documents and latent topic structures can be manifested in a probabilistic generative process associated with LDA. Thus, this generative process can be considered as a random process that produces the observed documents.

LDA is based on the 'bag-of-words' assumption that the order of words in a document can be neglected (Blei *et al.* 2003). Under this assumption, the boundaries between sentences conveying the information about which words should be classified into the same topic are ignored. This might result in scenarios where each word in a sentence is sampled from different topics. However, in some cases, each sentence in a document is only regarding one topic. Therefore, the Sent-LDA model is proposed to take the boundaries between sentences into account and assume that all words in a sentence are sampled from the same topic (Bao and Datta 2014). Under this 'one topic per sentence' assumption, the words in different sentences are no longer interchangeable, and the sampling of each word in the same sentence is dependent on each other word.

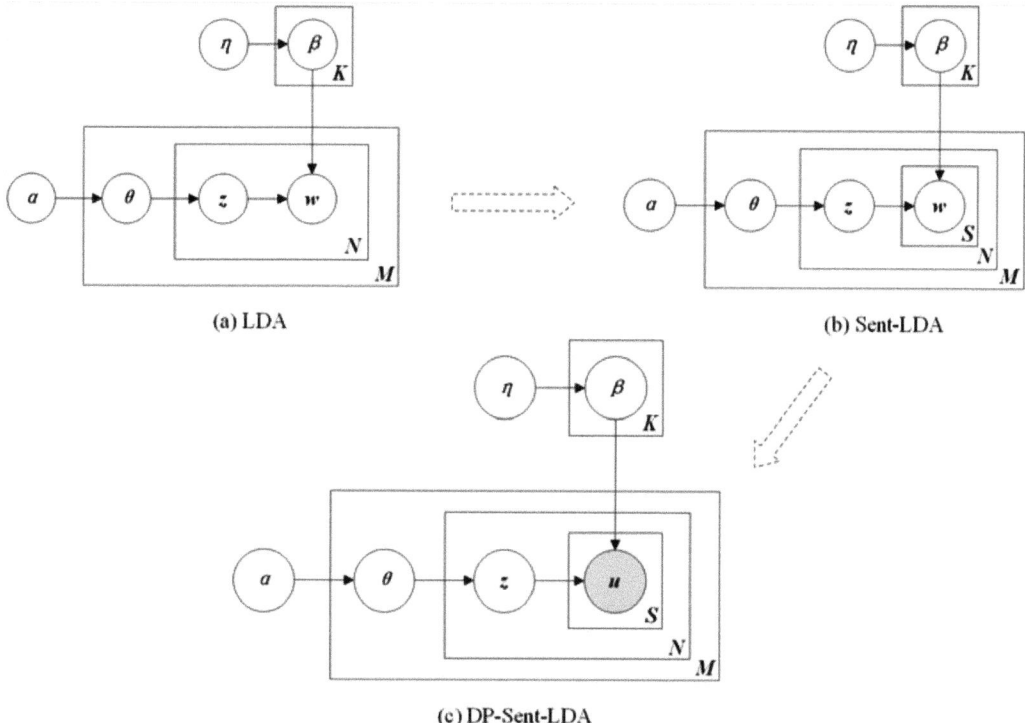

(a) LDA

(b) Sent-LDA

(c) DP-Sent-LDA

FIGURE 32.1   Graphical model representations of LDA, Sent-LDA, and the proposed DP-Sent-LDA.

Given a corpus of texts, let M, S, N, K, and V denote the number of documents, the number of sentences in a document, the number of words in a document, the number of topics, and the vocabulary size, respectively. *Dirichlet*($\cdot$) and *Multinomial*($\cdot$) represent the Dirichlet and multinomial distributions with parameter ($\cdot$), respectively. βk denotes the V-dimensional word distribution for the topic k, and $\boldsymbol{\theta}_d$ denotes the K-dimensional topic proportion for the document d. η and $\boldsymbol{\alpha}$ denote the hyper-parameters of the corresponding Dirichlet distribution. w denotes the set of words from each sentence s. u denotes the set of keywords from each sentence s. Based on these notations, the graphical of Sent-LDA is presented in Figure 32.1 (b), which adds a sentence layer S to the hierarchy of the original LDA shown in Figure 32.1 (a). The graphical of DP-Sent-LDA in Figure 32.1 (c) uses the keyword set u to replace the original word set w, which will be illustrated in detail in Section 32.3.2.

32.3.2 DP-Sent-LDA

32.3.2.1 Concept of DP-Sent-LDA

DP-Sent-LDA is a further improved topic model based on Sent-LDA. As stated in Section 32.2, the redundant information incorporated in the irrelevant words of sentences might disturb the judgement for the number and details of topics when using Sent-LDA. Therefore, a dependency parsing (DP) process is introduced in our model to tackle this problem. The capital letters 'DP' in DP-Sent-LDA is the abbreviation of dependency parsing.

Dependency parsing refers to a process that works out the grammatical structure of sentences by using dependency representations over more informative lexicalized phrase structures (Mcdonald 2006). There has been a surge of interest in dependency parsing for applications such as relation extraction and synonym generation. By using the dependency parsing process, a sentence could be parsed into an intuitive tree or graph structure that represents words and their relationship to syntactic modifiers using directed edges (Mcdonald 2006). Specifically, this tree structure consists of a unique root node and several other nodes, among which directed edges indicate their relationship. The word in the root node usually conveys the most critical information, and as the tree grows deeper, the word of its node becomes less important. For example, if the sentence '*The weather is abnormal this year, so the corn production will be affected*' is parsed, the parsing result in the form of a tree structure is shown as Figure 32.2. The detailed explanation of the relationship labels for this example is provided in Table 32.1. As Figure 32.2 makes clear, the word *abnormal* is the root node, which carries the most important information; as the tree grows deeper, the words *the, this*, and *so* carry less information and become less important. Thus, using the dependency parsing process, the keywords from each sentence can be effectively extracted and a collection of pre-processed financial textual data is generated.

In simple terms, LDA is a basic model to extract topics from the general type of text. The Sent-LDA adds a 'one topic per sentence' assumption to LDA to make it fit for some special types of short text, such as the news headlines and the risk factor headings in the Form 10-K of U.S. companies. The DP-Sent-LDA further improves the effectiveness of Sent-LDA by removing the redundant words in the short text using the dependency parsing process.

32.3.2.2 Generative Process of DP-Sent-LDA

DP-Sent-LDA defines a generative process that demonstrates how documents are generated with a known document-topic distribution (i.e. multinomial distribution with parameter θ) and topic-word distribution (i.e. multinomial distribution with parameter β). Then, by applying the learning algorithm of DP-Sent-LDA presented in section 32.3.2.3, the optimal parameters with observed documents can be estimated. The words extracted from the DP process are defined as keywords and the remaining

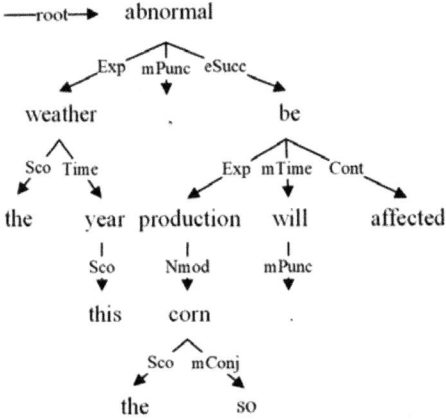

FIGURE 32.2 The tree structure of the sentence sample.

TABLE 32.1 Detailed Explanations of Relationship Labels for the Sentence Sample

No.	Tag	Relationship	No.	Tag	Relationship
1	Root	Root Word	6	Time	Time
2	Exp	Experiencer	7	mTime	Time
3	mPunc	Punctuation Marker	8	Cont	Content
4	eSucc	Event Successor	9	Nmod	Name-modifier
5	Sco	Scope	10	mConj	Recount Marker

redundant words are defined as supplementary words. Using the same notations as in Section 32.3.1, the generative process of DP-Sent-LDA is as follows.

(1) For each topic $k \in \{1, \cdots, K\}$,
    (a) draw a distribution over keywords $\boldsymbol{\beta}_k \sim Dirichlet\,(\boldsymbol{\eta})$

(2) For each document d,
    (a) draw a vector of topic proportions $\boldsymbol{\theta}_d \sim Dirichlet\,(\boldsymbol{\alpha})$
    (b) draw each sentence s in the document d,
        (i) draw a topic assignment $z_{d,s} \sim Multinomial\,(\boldsymbol{\theta}_d)$.
        (ii) for each keyword $\boldsymbol{u}_{d,s,n}$ in sentences,
            (ii-a) draw a keyword $\boldsymbol{u}_{d,s,n} \sim Multinomial\,(\beta_{z_d,s})$.
        (iii) generate complete sentences using supplementary words.

The graphical representation of DP-Sent-LDA is presented in Figure 32.1 (c), in which the keyword set $\boldsymbol{u}$ replaces the original word set $\boldsymbol{w}$ in the hierarchy of Sent-LDA presented in Figure 32.1 (b).

32.3.2.3 Learning Algorithm of DP-sent-LDA

The premise of DP-Sent-LDA is the dependency parsing process. Dependency parsing is generally translated into that of finding the maximum spanning tree (MST) y for a given sentence s (Mcdonald 2006). By defining the score of an edge in the tree to be the dot product between a high-dimensional feature representation of edge $f(i, j)$ and a weight vector $\boldsymbol{w}$, the MST problem is also equivalent to finding the dependency tree with the highest score for a given sentence. To deal with this problem, a high-order Eisner parsing algorithm put forward by Che *et al.* (2009) based on the first-order parsing algorithm of Eisner (1996) is employed in our work. The basic idea of the high-order Eisner parsing algorithm is to parse the left and the right dependence of a word independently and combine them in the later stage. The detailed steps of the high-order Eisner parsing algorithm can be found in Che *et al.* (2009) for parsimony.

Then, to use DP-Sent-LDA to discover topics from the sentences composed of keywords after dependency parsing process, the second key sub-problem is to draw the posterior distribution of the hidden variables $\boldsymbol{\theta}$ (topic proportions) and $\boldsymbol{z}$ (topic assignments) given the model parameters and the observed set of keywords $\boldsymbol{u}$ from sentences. That is:

$$p(\theta, z | u, \alpha, \beta) = \frac{p(\theta, z, u | \alpha, \beta)}{p(u)} \tag{32.1}$$

Since this distribution is generally intractable to compute, topic model learning algorithms are employed to approximate the above equation by forming an alternative distribution over the latent topic structure that is adapted to be close to the true posterior. According to Bao and Datta (2014), the learning algorithms fall into two categories: sampling-based algorithms and variational algorithms. The Collapsed Gibbs Sampling (CGS) algorithm (Griffiths and Steyvers 2004) and the Variational Expectation Maximization (VEM) algorithm (Blei *et al.* 2003, Wei *et al.* 2019b) are the most commonly used sampling-based and variational methods, respectively. Many studies have discussed the advantages and disadvantages of these two types of learning algorithms and compared their performances. Bao and Datta (2014) demonstrated that the VEM algorithm performs better than the CGS algorithm when the Sent-LDA model is used. Therefore, in this paper, the VEM learning algorithm is employed to approximate the intractable distribution.

The basic idea of VEM is to introduce variational parameters to estimate the posterior parameters of the topic model. The algorithm is divided into two stages, which are the variational inference stage and the Expectation Maximization (EM) stage. In the variational inference stage, a probability distribution is obtained by Kullback-Leibler divergence to approximate the required posterior distribution. The variational parameters need to be solved in this stage. The EM stage is divided into two steps. The E-step is intended to determine the variational parameters in variational inference, and the goal of M-step is to solve the posterior parameters in the topic model using the determined variational parameters. For more details about the VEM algorithm, please refer to Blei *et al* (2003).

32.4 Data

In our empirical analysis, six representative types of commodity futures markets in China are chosen as the sample, covering three major categories of commodity futures including agricultural commodity futures, metal futures, and energy futures. The details of the futures samples are shown in Table 32.2. Crude oil is by far the most actively traded commodity around the world. The evolution of its price and volatility is of utmost importance for almost all economies. Among more than 20 kinds of agricultural products, soybean, corn, and wheat are three primary agricultural commodities with the highest attention from investors (Sørensen 2002). For gold, there is always a widespread interest and demand for investors since it plays an important role as a store of value and as a monetary cushion. Rebar is an important raw material for the industrial and real estate markets. Thus, six types of commodity futures based on these underlying assets constitute our sample set.

The textual data is collected from *Hexun.com*, which is a leading vertical financial portal website in China devoted to providing professional, high-end, and prime-quality financial information. In *Hexun .com*, news reports on different commodity futures markets are organized into separate sections and updated frequently. Specifically, each news report consists of a summary headline and the corresponding detailed content. News headlines, instead of full reports, are used in our study because condensed headlines are easier to retrieve and contain sufficient key information on the factors that affect the dynamics of commodity futures prices. Eventually, a total of 49 501 news headlines of the six types of commodity futures covering the period 2011.07.05–2018.07.18 are extracted. Examples of our data are presented in Table 32.3 to provide an intuitive view of Chinese sentence structures.

Furthermore, the regression analysis is conducted to validate the effectiveness of the identified influential factors. Thus, the time series data of proxy variables regarding the futures prices are also collected. The period for the time series is 2011.06–2018.07 monthly. The proxy variables used in the regression analysis are from the WIND database, International Country Risk Guide (ICRG), and the Economic Policy Uncertainty (EPU) Website (http://www.policyuncertainty.com/).

Among the three data sources, WIND is an integrated Chinese financial data provider. As one of the world's best commercial sources of country risk analysis and rating, ICRG provides the financial, political, and economic risk information and forecasts of 140 countries. The EPU Website develops indices of economic policy uncertainty for countries around the world, a geopolitical risk index and other economic and political indicators, which are publicly accessible.

32.5 Empirical Analysis

In this section, the proposed DP-Sent-LDA model is used to identify the influential factors of six types of Chinese commodity futures prices. First, the data preparation and parameter settings are described. Then,

TABLE 32.2 The Six Types of Commodity Futures in Our Sample Set

Futures category	Underlying assets	Sample period	Number of headlines
Energy futures	Crude oil	2011.07.05–2018.07.18	5282
Agricultural	Wheat	2013.06.13–2018.07.05	4991
commodity	Corn	2011.06.21–2018.07.12	12 069
futures	Soybean	2012.07.19–2018.07.09	9715
Metal futures	Rebar	2011.07.11–2018.07.05	12 503
	Gold	2011.12.27–2018.07.12	4941

TABLE 32.3 Examples of News Headlines for the Six Types of Futures

Commodity futures	Date	New headlines	
		In Chinese	In English
Crude oil	2018.07.17	油价大跌！美油跌 4.2% 布油跌 4.7%	Oil prices have fallen! WTI fell by 4.2% while Brent by 4.7%.
	2018.03.23	贸易战忧虑波及油市，油价收跌止步两连阳	Sino-US trade war worries spread to the oil market, and oil prices decline after two consecutive rises.
	2018.03.19	伊朗核协议岌岌可危，对石油市场影响多大？	The Iran nuclear deal is in jeopardy. What is the impact on the oil market?
Wheat	2018.06.04	黄淮、江淮天气晴好 小麦收获进度加快	Sunny weather in Huang-Huai and Jiang-Huai areas speeds up the harvest of wheat.
	2017.06.17	小麦赤霉病有了"克星"	Fusarium Head Blight (FHB) in wheat has a "buster".
	2016.04.15	小麦穗期重大病虫害形势严峻	Prevention and control of the situation that wheat ears suffer from insect pests and diseases face a grim challenge.
Corn	2018.06.25	玉米市场供应过剩 不为贸易摩擦"所动"	Excessive supply in the corn market is not "moved" by the trade friction.
	2018.06.13	气候变暖可增加全球玉米歉收概率	Climate warming can increase the probability of a failure in global corn.
	2017.02.07	"一号文件"点燃市场热情 玉米期货成交和持仓齐创新高	"No. 1 Document" ignited the market enthusiasm, and both trade volume and open interest of corn futures hit new highs.
Soybean	2018.07.09	贸易战对商品市场影响逐步减弱 豆类品种表现不一	Impact of trade wars on commodity markets has gradually weakened, and different bean varieties have different performances.
	2017.06.12	天气升水暂时主导美豆反弹	Weather premiums temporarily dominate the rebound of the US soybean.
	2017.09.20	原油上涨，美元指数下跌，大豆期货收涨	Crude oil rose, dollar index fell, and soybean futures went up.
Steel rebar	2018.04.02	螺纹钢行情关键是需求 后续或震荡整理	The key to the rebar market is demand, which will be dominated by shocks later.
	2018.01.15	现货价格持续下跌 螺纹钢或迎方向选择的重要时点	Spot prices continue to fall, and steel might usher an important point in the market trend.
	2017.03.20	楼市分类调控政策压制 螺纹高位承压波动	Suppressed by housing market classification control polices, the rebar steel market fluctuates under high pressure.
Gold	2018.07.03	为何通胀上升 黄金暴跌？	Why has the inflation risen and the gold price plummeted?
	2017.09.26	朝美紧张局势升温 推动金价上涨	Tension between the DPRK and the US is heating up, pushing the gold price up.
	2016.04.14	欧美股市、美元指数双双走强 金价承压下行	Stock markets across Europe and U.S. and U.S. dollar index both strengthen, and gold prices go down under pressure.

our proposed method is compared with the Sent-LDA model in terms of predictive power. Next, the empirical results regarding the influential factors of commodity futures prices are analysed in detail. Finally, a further regression analysis is conducted to validate the effectiveness of the identified influential factors.

32.5.1 Implementation of DP-Sent-LDA

As noted above, the implementation of DP-Sent-LDA consists of two parts, namely, the dependency parsing process and the topic discovery process of DP-Sent-LDA. Primarily, to conduct the dependency parsing process on the dataset of Chinese news headlines, an integrated Chinese processing platform called Language Technology Platform (LTP) is employed. Designed by Che *et al.* (2010), LTP has been proven to be well at handling Chinese textual data by providing a suite of high-performance natural language processing modules. Based on the high-order Eisner parsing algorithm embedded in LTP, the 49 501 news headlines are processed by extracting the keywords sentence by sentence.

Table 32.4 compares the descriptive statistics of the original data and the processed data in terms of six types of commodity futures, including the values of the mean and standard deviation. More specifically, the mean value (or standard deviation) denotes the average (or square root of the variance) of the number of words of each sentence in the original or the processed dataset. For each type of commodity futures price, the values of the mean and standard deviation of the original data are significantly larger than that of the processed data. This result demonstrates that the dependency parsing process can greatly remove redundant words in the original headlines and effectively keep the key and valuable information pertinent to the sentence-level topic.

During the learning process of the DP-Sent-LDA model, the specific parameter settings of the VEM and DP-Sent-LDA are summarized in Table 32.5. Specifically, the maximum number of iterations of the coordinate ascent variational inference for a single document is set as 1000. The convergence criteria for variational inference is 1×10^{-8}. The maximum number of iterations of VEM is set as 1000. The convergence criteria for VEM is 1×10^{-5}. For the hyperparameters of the corresponding Dirichlet distributions, every element of α is set as 50/K, where K is the number of risk factors, and every element of η is set as 0.1.

32.5.2 Comparison of Sent-LDA and DP-Sent-LDA on Predictive Power

To examine the performance of DP-Sent-LDA, Sent-LDA with the same parameter settings as those in Table 32.5 is used as the benchmark model. The most typical evaluation criterion of topic models involves measuring how well a topic model performs when predicting the topic clusterings of new documents. Specifically, when estimating the probability of unseen held-out documents given a set of training documents, a 'good' model should give rise to a higher probability of held-out documents. Therefore,

TABLE 32.4 Comparison of the Original News Headline Data and the Processed Data Obtained by the Dependency Parsing Process

Descriptive statistics			Crude oil	Wheat	Corn	Soybean	Rebar	Gold
Number of words in a sentence	Mean	Original data	8.92	8.16	8.50	9.22	7.78	8.36
		Processed data	5.47	5.70	6.48	5.93	5.47	5.40
	Std. dev.	Original data	2.88	2.26	2.38	2.39	1.97	2.18
		Processed data	1.89	1.80	1.70	1.55	1.42	1.28

TABLE 32.5 The Parameter Settings of DP-Sent-LDA and its Learning Algorithm VEM

Parameters	Specific settings
The maximum number of iterations of coordinate ascent variational inference for a single document	1000
The convergence criteria for variational inference	1×10^{-8}
The maximum number of iterations of VEM	1000
The convergence criteria for VEM	1×10^{-5}
The number of topics K	5–100
The hyperparameter of the corresponding Dirichlet distributions α	50/K
The hyperparameter of the corresponding Dirichlet distributions η	0.1

to measure the predictive power of the two models, a conventional metric in language modelling, which is called 'perplexity', is adopted. The perplexity can be understood as the predicted number of equally likely words for a word position on average and is monotonically decreasing the function of the log-likelihood. Thus, a lower perplexity over a held-out document is equivalent to a higher log-likelihood, which indicates a better predictive performance. For a test set D_{test} of M documents, the formula of the per-word perplexity is defined as follows:

$$\text{perplexity}\left(D_{test}\right) = \exp\left(-\sum_{d=1}^{M} \log p\left(u_d\right) / \sum_{d=1}^{M} N_d\right) \qquad (32.2)$$

where N_s is the number of words in the document d.

In this paper, the per-word perplexity is obtained via a ten-fold cross-validation, as in Blei and Lafferty (2007). Figure 32.3 presents how the predictive power of the two models changes with the varying of the number of topics in terms of six types of commodity futures. The horizontal axis represents the number of topics, while the vertical axis denotes the perplexity. The dotted line is the perplexity of Sent-LDA, while the solid line is that of DP-Sent-LDA. The number of topics is set from 5 to 200 with a step of 5. The lower the perplexity is, the better the predictive performance of the model. As shown in Figure 32.3, regardless of the type of commodity futures, the perplexity of each of the two models first decreases

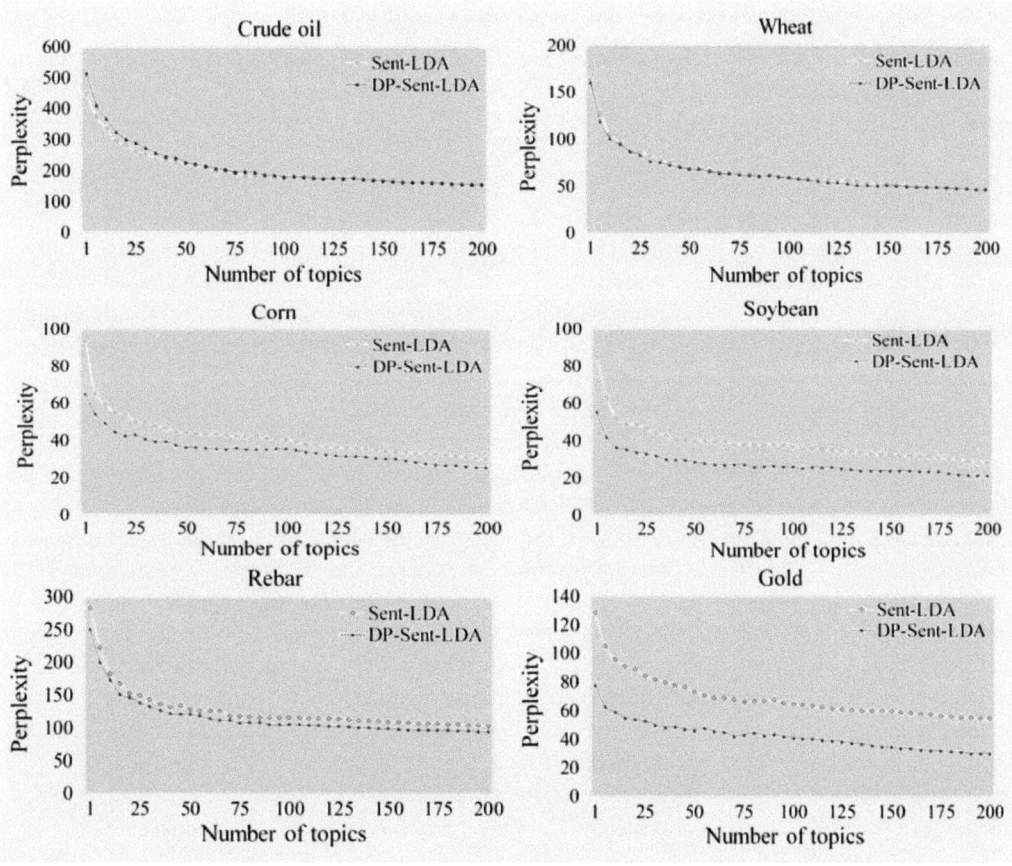

FIGURE 32.3 Comparison of the predictive power of DP-Sent-LDA and Sent-LDA for the six types of commodity futures.

with the increase in the number of topics but then slowly converges to a fixed value eventually. For the six types of commodity futures except for crude oil futures and wheat futures, DP-Sent-LDA obviously outperforms Sent-LDA with a lower perplexity for all the number of topics. For crude oil and wheat, the difference between the perplexities of the two models is trivial. Therefore, it can be concluded that the proposed DP-Sent-LDA is more effective than Sent-LDA in topic modelling because it holds a better predictive power.

The number of topics is an important parameter for LDA models, which are also usually determined on the basis of perplexity. Therefore, Figure 32.3 can help to decide the optimal topic number of DP-Sent-LDA. It can be roughly seen from Figure 32.3 that the perplexities for six types of commodity futures begin to converge when the topic number ranges from 100 to 200. Thus, the number of topics is set as 150 in learning the DP-Sent-LDA model for the crude oil futures, wheat futures, corn futures, soybean futures, rebar futures, and gold futures. In this case, the number of influential factors that can be identified will reach the maximum. There are 900 topics in total for the six types of commodity futures. Then, we merge similar topics and eliminate redundant topics. At last, 104 topics that represent 104 specific influential factors of the six commodity futures are uncovered, which will be illustrated in Section 32.5.3.1. Based on the analysis of these 104 specific factors, a general influential factor system for the commodity futures prices is constructed, which will be presented in Section 32.5.3.2.

32.5.3 The Identified Influential Factors of Commodity Futures Prices

32.5.3.1 The Influential Factor System of Commodity Futures Prices

By applying DP-Sent-LDA to analyse the news headlines, 104 specific factors that affect the crude oil, wheat, corn, soybean, rebar, and gold commodity futures markets are eventually identified in total. Figure 32.4 presents the word clouds of the 104 specific factors for the six types of commodity futures markets.

In Figure 32.4, each word cloud consists of all the features words about a certain topic. The topic here refers to a specific factor affecting these commodity futures prices. In each word cloud, the bigger the font size of the Chinese word is, the larger the probability of the corresponding text occurring in the news headlines. According to these words, the factors are labelled manually and put above the word clouds. Using the first word cloud of crude oil in Figure 32.4 as an example, two Chinese words that have the largest font size are '油价' (oil price) and '价格' (price). This indicates that oil price is closely related to the dynamics of the crude oil futures price. Thus, this word cloud is labelled as the oil price (spot price). The second word cloud of crude oil is labelled as supply and demand because the most obvious words are '产量' (yield) and '供应' (supply). In the same way, the specific factors that affect the dynamics of the six individual commodity futures prices are all determined. The numbers of the identified factors for crude oil, corn, wheat, soybean, rebar, and gold futures are 19, 18, 14, 20, 19 and 14, respectively.

After determining these specific factors, an influential factor system for commodity futures prices is developed. In total, 30 influential factors are summarized from the 104 specific factors. Then, the 30 influential factors can be further classified into 9 sub-categories based on their similarities. The 9 sub-categories of influential factors can be finally merged into 3 major categories, which are market factors, macrolevel factors, and production-based factors. Table 32.6 presents the four-layer hierarchical system of identified influential factors in detail. The market factors include the spot market, the futures market, the international market, and other highly related markets; macro-level factors incorporate policies, politics, and macroeconomics; and production-based factors include internal factors and external factors. The internal factors specifically refer to variables that can be controlled by the firm itself, while the external factors are related to uncontrollable external variables, such as climate changes.

From the influential factor system in Table 32.6, it can be seen that almost all of the influential factors mentioned in the existing literature (reviewed in Section 32.2.1) are covered. Furthermore, some additional influential factors that are rarely discussed are identified. To be specific, the three major categories

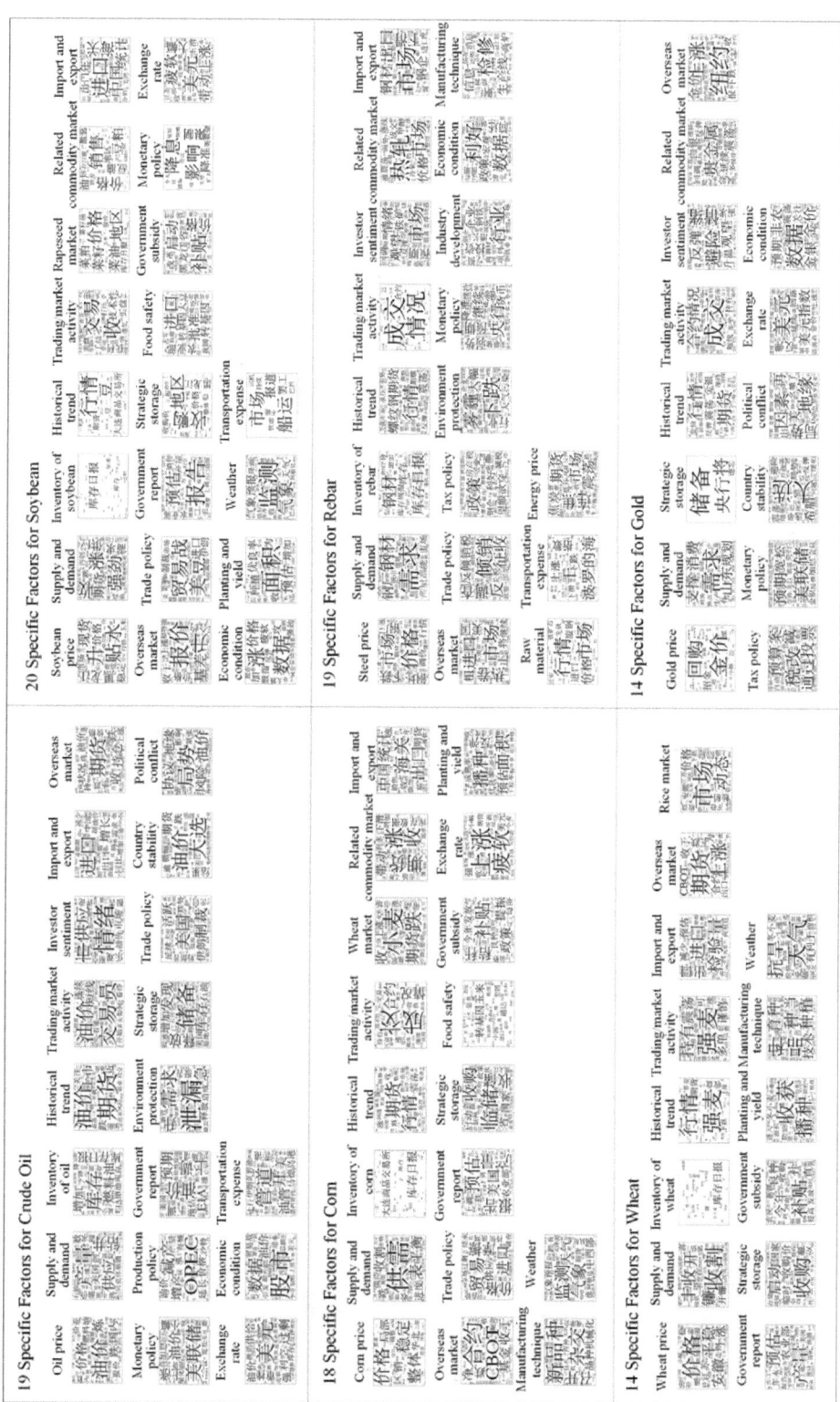

FIGURE 32.4 Word clouds of the 104 specific factors that affect the six types of commodity futures prices.

TABLE 32.6 The Hierarchical System of the Identified Influential Factors for Commodity Futures Prices

Major category	Sub-category	Influential factor	Specific factors
Market factors	Spot market	Spot price	Commodity spot price
		Supply and demand Inventory	Supply and demand for commodity spot
			Commodity inventory of futures exchanges; central bank gold reserves
	Futures market	Historical trend Trading/market activity	Historical price trend of futures
			Trading and holding behaviour of investors in the futures markets
		Investor sentiment	Level of investor confidence in investing in crude oil, rebar, and gold
	International market	Import and export	Import and export volume between China and other countries except for gold
		Overseas market	Futures and spot markets in countries other than China
	Other related markets	Substitute market	Wheat and rice market for corn; Corn and rice for Wheat; Rapeseed market for soybean.
		Related commodity market	Upstream and downstream commodity market of corn and soybean, such as ethanol, soybean oil and soybean meal; Commodity market with similar properties to rebar and gold, such as hot rolling and sliver
Macro-level factors	Policies	Trade policy	Tariffs policy on corn, soybeans, and rebar between China and other countries
		Production policy Tax policy	OPEC's oil production policy
			Rebar-related tax policies, such as property taxes; Tax policy reform for gold
		Government report	OPEC reports on oil production, import and export; USDA reports on wheat, corn and soybean production and demand
		Food safety	Food safety policy on genetically modified corn and soybeans
		Environment protection	Environmental protection policies related to oil and steel production
		Strategic storage	National or local government strategic reserve policies for wheat, corn, and soybeans
		Government subsidy	National or local government subsidy policies for wheat, corn, and soybeans
	Politics	Country stability	Internal stability of major countries in the world, such as government changes, leader elections
		Political conflict	Geopolitical conflicts between countries, such as the North Korean nuclear issue
	Macroeconomic	Exchange rate	The exchange rate between RMB and the currency of major corn and soybean producing and consuming countries; the U.S. dollar index
		Monetary policy	China and the United States monetary policy, such as interest rates and reserve ratios
		Economic condition	Economic conditions of China and other major countries in the world, such as inflation, stock market conditions, and the unemployment rate
		Industry development	Investment, profitability as well as the capacity transformation of the steel industry
Production-based factors	Internal factors	Planting and yield	Growth of wheat, corn, and soybeans before harvest, such as planting area, pest disease
		Manufacturing technique	Status of hybridization technologies related to wheat and corn; equipment maintenance and renewal related to rebar production
	External factors	Raw material Weather	Iron ore and scrap prices related to rebar production
			Weather conditions related to wheat, corn, and soybeans
		Transportation expense	Expenses related to the transportation of soybeans, steel, and iron ore, such as the Baltic Dry Index; oil pipeline laying and repair costs
		Energy price	Energy prices related to rebar production, such as coking coal prices

and the nine sub-categories can be determined based on a thorough literature review. This means that the existing studies have conducted in-depth research on this issue, and a general understanding of the factors affecting commodity futures has already formed. However, as the analysis granularity is further refined, some newly discovered influential factors can be observed. To be more specific, regarding macro-level factors, although many works in the literature discuss the impact of policy releases on the futures market (Anderson and Nelgen 2012; Anderson *et al.* 2013; Gouel 2016; Stephen and Hillard 2017), few of them have focused on tax policies, food safety policies and environmental protection policies. For production-based factors, manufacturing technique is rarely mentioned in previous studies. Therefore, it can be concluded that the proposed DP-Sent-LDA model can effectively find the influential factors for commodity futures prices from a massive number of news headlines.

32.5.3.2 The Relative Importance of the Influential Factors

The relative importance of the influential factors for the commodity futures prices is also a concern, which can be given by the proposed DP-Sent-LDA model. In this model, it is assumed that the more news headlines classified into an influential factor there are, the more attention is paid by the public to this influential factor, and so, the more important this influential factor is for the commodity futures prices. Thus, the relative importance ratio can be computed by dividing the number of news headlines classified into an influential factor by the total number of news headlines regarding the individual commodity futures. Figures 32.5–32.7 present the relative importance of the 3 major categories, 9 sub-categories and 30 influential factors for each commodity futures, respectively.

From Figure 32.5 it can be clearly seen that the relative importance ratio of the market factors almost exceeds 70% for all six commodity futures. This indicates that more than 70% of the news discusses the market factors. The market factors are significantly more important than the macro-level factors and production-based factors in influencing the commodity futures market dynamics. Figure 32.6 further shows that for the sub-categories, the spot market factor and futures market factor are generally more important than other factors. Furthermore, the prominent factors of different commodity futures markets are different. In Figure 32.7, the differences among these commodity futures' prominent factors are more evident. For example, for the corn futures, the relative importance ratio of the spot price is the largest, indicating that it is the most important factor that affects the fluctuation of corn futures. For the gold futures, there are four prominent factors, i.e. historical trend, trading/market activity, economic condition, and monetary policy. For the wheat futures, the substitute market is the most important. For the soybean futures, the related commodity market dominates all the influential factors.

32.5.4 Regression Analysis of the Effectiveness of the Influential Factors

In Section 32.5.3 the possible influential factors about six types of futures are derived from the massive news headings. This section further employs the regression analysis to explore whether these factors indeed have an influence on the dynamics of these futures in the Chinese market.

Among the six types of commodity futures, the transaction data regarding the Chinese oil futures is lacking since it was not launched until March 26, 2018. Thus, the regression analysis of the other five types of commodity futures in China, i.e. the gold futures, the rebar futures, the corn futures, the soybean futures, and the wheat futures are used to validate the effectiveness of the identified influential factors.

32.5.4.1 Proxy Variable Selection Results

How to choose proper proxy variables for the influential factors are always controversial. The word clouds derived by the proposed DP-Sent-LDA can provide an easier way to solve this problem. Based on the feature words in these word clouds, most of the proxy variables for the corresponding influential factors can be intuitively found out.

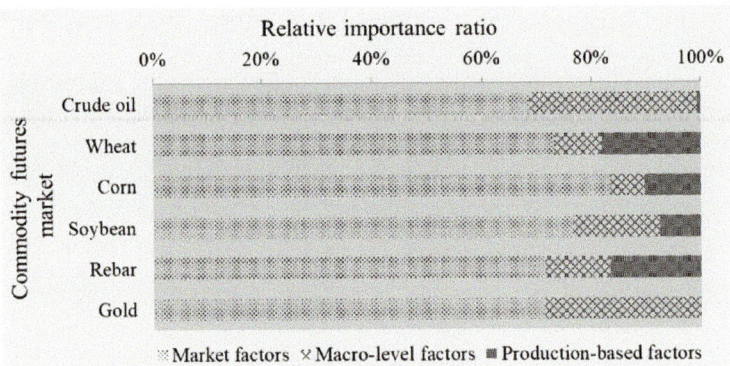

FIGURE 32.5 The relative importance of the three major categories of influential factors.

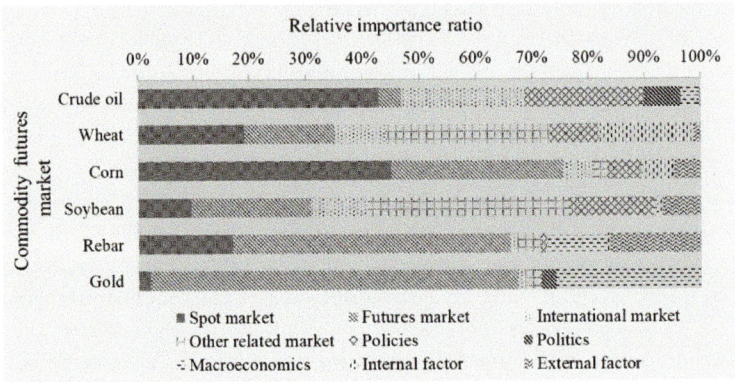

FIGURE 32.6 The relative importance of the nine sub-categories of influential factors.

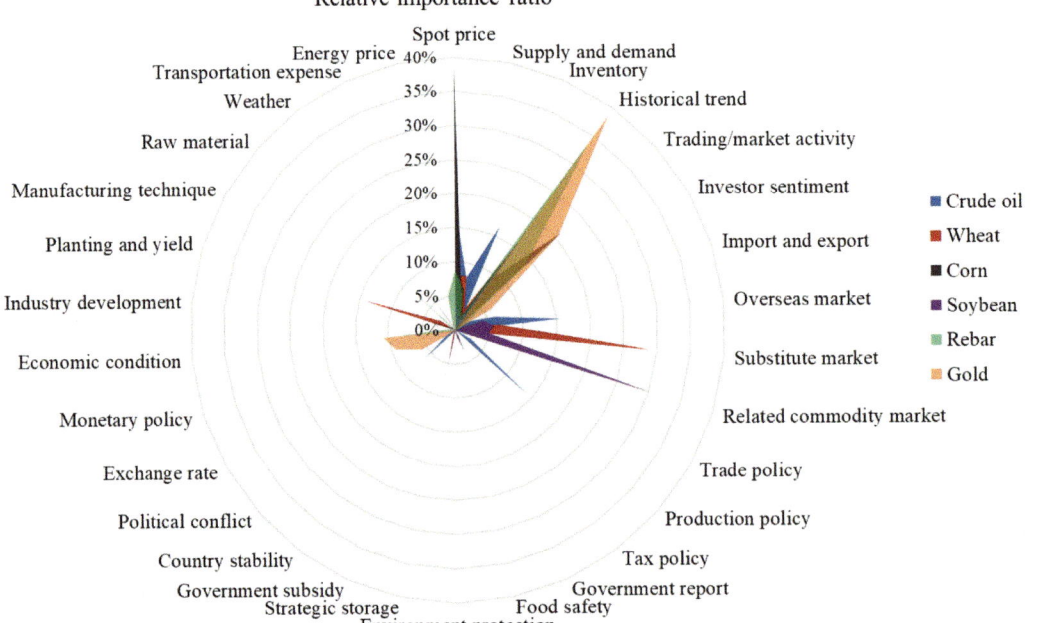

FIGURE 32.7 The relative importance of the thirty influential factors.

The word clouds in Figure 32.4 show specific influential factors that can affect the dynamic of five futures prices except for the crude oil futures. The bigger the font size of the word in the word clouds, the more important this word is. Then the proxy variables are selected based on these feature words and related researches. Taking the gold futures for example, in the word cloud labelled as exchange rate, the three most frequent feature words are '美元' (dollar), '美元指数' (dollar index) and '反弹' (rebound). Therefore, in line with Wang and Chueh (2013) and Aye *et al.* (2015), the Dollar index is chosen as the proxy variable for this influential factor. In the same way, all the proxy variables for all five commodity futures can be determined. Table 32.6 shows that there are 30 different influential factors for all six commodity futures. Among these influential factors, production policy only affects crude oil, so it is not included in the regression analysis. All proxy variables for the remaining 29 influential factors are determined according to their word clouds and related researches, which are shown in Table 32.7.

The data of these proxy variables are in either monthly frequency or daily frequency. Monthly data is used directly for regression. For the trade volume data of commodity futures in daily frequency, the monthly volatility of this variable is calculated for regression. For other types of daily data, such as the Dollar index and substitute commodity futures price, the monthly volatilities of their logarithmic returns are calculated for regression. Unless otherwise specified, the independent variable data are all from the Chinese market. The dependent variable is the volatility of the logarithmic return of the settlement price of commodity futures in the Chinese market, where the logarithmic return is enlarged by multiplying by 1000 to ensure that the regression coefficients are of suitable magnitude. The volatility here is measured by using standard deviation.

32.5.4.2 Regression Analysis Results

In general, the most commonly used regression model is the ordinary least squares (OLS) regression. However, if the multicollinearity, heteroscedasticity, and autocorrelation problems exist in the OLS regression model, the accuracy of the estimated results will be seriously affected. In line with other studies such as Chu and Lin (2001), Stanko and Henard (2017), and Born and Breitung (2016), this paper employs the condition number, White test and Durbin-Watson (DW) test to examine whether the OLS regression model has the three problems. The results of the White test and DW test reject that there are heteroscedasticity and autocorrelation. However, the conditional numbers of the independent variables for gold futures, rebar futures, corn futures, soybean futures, and wheat futures are 45 201, 59 626, 225 796, 105 320, and 37 461, which are all greatly exceeding the threshold 1000, meaning that there are strong multicollinearities between these variables.

In order to better validate the effectiveness of the influential factors, the ridge regression model is used in this paper. It is a popular improved OLS regression model that can eliminate the effects of multicollinearity on the results. Compared with the OLS regression model, the ridge regression adds a penalty term $\lambda \beta^T \beta$ to the objective function of least squares to limit the regression coefficients β to a reasonable range (Assaf *et al.* 2019). The coefficient λ in the penalty term is called the penalty factor. The greater the penalty factor, the stronger the constraint on β. The penalty factor λ here is set to 0.1.

The estimation results of the ridge regression model for the five commodity futures are shown in Table 32.8. From the table it can be seen that the R^2 of the five regression models for gold futures, rebar futures, corn futures, soybean futures, and wheat futures are 0.37, 0.50, 0.38, 0.47 and 0.07 respectively, which means, on the whole, these independent variables explain the dependent variable relatively well except for wheat futures. The results of F-statistics support the same conclusion. The regression model fits the data better, the F-statistics tends to be greater and its estimated value tends to be more significant. The values of the F-statistics are significant at different levels except for the wheat futures. The possible reason why the regression model of wheat futures is not statistically significant may be that its price is very stable compared with other futures. Because wheat is one of the staple food of the Chinese people, the government keeps a tight price control on it. In terms of the significance of the variables, it can be seen that the coefficients of the great majority of variables are significant in regression models, which further validates the effectiveness of the identified influential factors. Although R^2 of the regression

TABLE 32.7   The Proxy Variables for Influential Factors based on the Word Clouds and Related Researches

No.	Influential factor	Proxy variable	No.	Influential factor	Proxy variable
1	Spot price	Commodity futures related spot prices	16	Strategic storage	Ratio of national spot inventory in previous to current period
2	Supply and demand	Ratio of spot supply to demand	17	Government subsidy	Ratio of government agricultural subsidies in previous to the current period
3	Inventory	Ratio of futures inventory in previous to the current period	18	Country stability	The average value of the national stability index of China, the U.S. and Greece
4	Historical trend	The commodity futures price for one period earlier	19	Political conflict	Geopolitical Index
5	Trading/market activity	Trading volume of commodity futures	20	Exchange rate	Dollar index
6	Investor sentiment	China Stock Market Investment Confidence Index	21	Monetary policy	China 3-month government bond yield*
7	Import and export	Export volume to import volume of related spot	22	Economic condition	The Shanghai Composite Index*
8	Overseas market	U.S. corresponding commodity futures prices	23	Industry development	Industry fixed asset investment, YoY
9	Substitute market	Substitute commodity futures price	24	Planting and yield	Planting area of related agricultural commodity
10	Related commodity market	Related commodity futures price	25	Manufacturing technique	(1) Operating rate of steel company for rebar; (2) Crop yield per unit area for agricultural commodities
11	Trade policy	U.S. Trade Uncertainty Index	26	Raw material	Iron ore price
12	Tax policy	China's tax burden index	27	Weather	El Niño Index
13	Government report	Ratio of production expectation released by the government in previous to the current period	28	Transportation expense	Baltic Dry Index
14	Food safety	Relevant genetically modified food market proportion	29	Energy price	Coke futures price
15	Environment protection	Government energy conservation and environmental protection expenditure, MoM			

Note: For the proxy variables of the spot price of agricultural products, the monthly price indices are used because their spot prices are not always available; For gold futures, because it is more affected by the U.S. economy, the data of some variables from the U.S. market, which is the U.S. 3-month government bond yield and the S&P 500 index; Only one typical proxy variable is selected for each influential factor in this study.

model for wheat futures is relatively low, most of its variables are still significant, such as supply and demand, historical trends, and so on.

At last, a notable finding is that the influential factors that appear very frequently in the news are also very significant in the regressions. From Figure 32.7, it is easy to find the influential factors with the largest relative importance ratios, which means they appear most often in news headlines. For all the five

TABLE 32.8 Regression Analysis Results for the Five Types of Commodity Futures

Influential factor	Gold β	Gold t-stat	Rebar β	Rebar t-stat	Corn β	Corn t-stat	Soybean β	Soybean t-stat	Wheat β	Wheat t-stat
Spot price	5.72***	75.32	5.22***	44.96	0.75	0.99	−1.37*	−1.92	−0.25	−0.28
Supply and demand	−6.01***	−4.33	8.36***	5.31	0.95	0.32	2.17***	14.00	−12.23***	−9.81
Inventory			−0.03	−0.54	−0.49	−0.86	0.07	0.32	−0.49	−0.28
Historical trend	0.67***	8.04	7 88***	40.94	23.80***	19.36	|0.45***	7.03	−2.10***	−9.15
Trading/market activity	0.25***	2.88	0.34***	4.30	−0.08	−0.07	1.54***	3.27	0.83	0.39
Investor sentiment	0.02	0.36	−0.07	−1.20						
Import and export			−0.03	−0.96	0.08	0.18	−0.12***	−3.92	−2.85***	−5.54
Overseas market	5.17***	63.86	2.02***	11.07	6.13***	9.25	0.66***	4.79	0.69***	3.93
Substitute market					−2.04***	−4.26	−0.06	−0.48	0.50**	2.36
Related commodity market	6.96***	61.14	6.55***	50.67	5.51	1.65	1.23***	15.24		
Trade policy			0.01*	1.93	0.00	0.36	0.01***	3.40		
Tax policy	5.01***	5.96	−7.06***	−5.95						
Government report					12.95***	5.90	10.55***	14.24	6.10***	9.45
Food safety					29.32***	−5.28	−6.38***	−10.14		
Environment protection			0.00	0.27						
Strategic storage	5.24***	5.59			−11.71**	−2.02	0.20	0.70	5.16***	3.20
Government subsidy					−0.07***	−2.76	−0.02***	−5.30	0.02*	1.86
Country stability	1.31	1.32								
Political conflict	−0.02***	−2.80								
Exchange rate	0.24***	8.17			1.25***	8.98	−0.13***	−4.90		
Monetary policy	−2.59***	−7.21	−3.88***	−8.29			1.66***	6.53		
Economic condition	1.72***	20.74	−6.72***	−24.51			0.53***	4.47		
Industry development			−0.26**	−2.46						
Planting and yield					1.74	0.77	0.51	1.63	0.56	0.81
Manufacturing technique			−0.02	−0.18	9.49***	3.72			3.72***	9.13
Raw material			12.23***	49.24						
Weather					5.81***	3.45	−0.33	−0.85	0.17	0.18
Transportation expense			−1.48***	−5.24			2.3***	15.68		
Energy price			12.55***	49.97						
Constant	5.96***	6.39	7.84***	4.91	12.96***	6.43	9.48***	14.00	1.52**	2.22
R^2	0.37		0.50		0.38		0.47		0.071	
F–statistics	2.80**		3.41***		2.26***		2.30***		0.25	

Note:

1. By convention, *, **, *** respectively represents the significance at 10%, 5% and 1% level (see e.g. Bao and Datta 2014, Li *et al.* 2020c).

2. The significant variables are in bold.

3. The top influential factors with the highest importance ratios are shadowed for each commodity futures.

futures, the historical trend is one of the most frequently mentioned influential factors. Other frequent influential factors for the two metal futures include trading/market activity and economic condition, etc. For the three agricultural commodity futures, the influential factors such as supply and demand, related commodity market, and substitute market account for a large proportion. These influential factors are shadowed in Table 32.8 for the ease of observation. It can be seen that the variables of these influential factors are almost all significant at 1% level. This indicates that the public is fully recognized about what impacts the fluctuation of the commodity futures price most.

32.6 Conclusions

Previous studies have mainly identified influential factors of commodity futures prices based on researchers' judgements or on summarizing previous studies. The major contribution of this paper is to propose a new text mining method named DP-Sent-LDA to identify the influential factors of commodity futures from the massive number of news headlines. The proposed method has two notable advantages compared with previous studies. Firstly, since DP-Sent-LDA is an unsupervised algorithm, the influential factor identification process does not completely rely on human's subjective judgment. Secondly, based on a large number of news headlines, the identified influential factors are usually more comprehensive.

The proposed method is empirically tested on the commodity futures in the Chinese market based on 49 501 news headlines regarding six types of commodity futures collected from a leading financial portal website in China. First and foremost, by using the DP-Sent-LDA model, a total of 104 specific influential factors that affect the dynamics of six commodity futures prices are automatically identified from these news headlines. Then an in-depth analysis shows that the 104 factors cover almost all the influential factors that have been mentioned in the existing literature. More importantly, some macro-level factors and production-based factors that rarely appear in previous studies have been found, such as food safety policy, tax policy, and manufacturing technique. Last but not least, a regression analysis is conducted to test whether these identified influential factors indeed have impacts on the commodity futures prices. The results show that they explain the dynamics of the commodity futures prices well and that their coefficients are significant at different levels in the regression models. All of these results demonstrate that the proposed method can objectively, comprehensively, and effectively extract the influential factors of commodity futures prices from the news headlines.

This study offers an important supplement to traditional qualitative analysis methods. It can quickly give a comprehensive list of influential factors for commodity futures prices without experts' experience or reviewing a large amount of literature, which makes it much easier to understand the movement of commodity futures prices. These factors lay the foundation for the commodity futures pricing and forecasting, helping the decision-making about hedging and investment strategies. The last thing worth noting is that the proposed method can also be applied to other financial products or assets as long as the news contains information about their influential factors.

Disclosure Statement

No potential conflict of interest was reported by the authors.

Funding

This study was supported by grants from the National Natural Science Foundation of China (71425002, 71971207, and 71601178) and the Youth Innovation Promotion Association of Chinese Academy of Sciences (2012137 and 2017200).

ORCID

Jianping Li http://orcid.org/0000-0003-4976-4119

References

Algieri, B., The influence of biofuels, economic and financial factors on daily returns of commodity futures prices. *Energy Policy*, 2014, **69**, 227–247.

Anderson, K. and Nelgen, S., Trade barrier volatility and agricultural price stabilization. *World. Dev.*, 2012, **40**(1), 36–48.

Anderson, K., Rausser, G. and Swinnen, J., Political economy of public policies: Insights from distortions to agricultural and food markets. *J. Econ. Lit.*, 2013, **51**(2), 423–477.

Assaf, A.G., Tsionas, M. and Tasiopoulos, A., Diagnosing and correcting the effects of multicollinearity: Bayesian implications of ridge regression. *Tour. Manag.*, 2019, **71**, 1–8.

Aye, G., Gupta, R., Hammoudeh, S. and Kim, W.J., Forecasting the price of gold using dynamic model averaging. *Int. Rev. Financial Anal.*, 2015, **41**, 257–266.

Bao, Y. and Datta, A., Simultaneously discovering and quantifying risk types from textual risk disclosures. *Manage. Sci.*, 2014, **60**(6), 1371–1391.

Basu, D. and Miffre, J., Capturing the risk premium of commodity futures: The role of hedging pressure. *J. Bank. Financ.*, 2013, **37**(7), 2652–2664.

Blei, D.M. and Lafferty, J.D., A correlated topic model of science. *Annal. Appl. Stat.*, 2007, **1**(1), 17–35.

Blei, D.M., Ng, A.Y. and Jordan, M.I., Latent Dirichlet allocation. *J. Mach. Learn. Res.*, 2003, **3**, 993–1022.

Born, B. and Breitung, J., Testing for serial correlation in fixed-effects panel data models. *Econom. Rev.*, 2016, **35**(7), 1290–1316.

Campbell, J.L., Chen, H., Dhaliwal, D.S., Lu, H. and Steele, L.B., The information content of mandatory risk factor disclosures in corporate filings. *Rev. Account. Studies*, 2014, **19**(1), 396–455.

Chang, M.C., Tsai, C.-L., Wu, R.C.-F. and Zhu, N., Market uncertainty and market orders in futures markets. *J. Futures Mark.*, 2018, **38**(8), 865–880.

Che, W., Li, Z. and Liu, T., LTP: A Chinese language technology platform. The 23rd International Conference on Computational Linguistics: Demonstrations, Beijing, China, August 23–27, 2010.

Che, W., Li, Z., Li, Y., Guo, Y., Qin, B. and Liu, T., Multilingual dependency-based syntactic and semantic parsing. The 13th Conference on Computational Natural Language Learning: Shared Task, Boulder, Colorado, June 4, 2009.

Christie–David, R., Chaudhry, M. and Koch, T.W., Do macroeconomics news releases affect gold and silver prices? *J. Econ. Bus.*, 2004, **52**(5), 405–421.

Chu, Q.C. and Lin, Y.Y., Determinants of the dollar value of default risk: A put option perspective. *Rev. Quant. Financ. Account.*, 2001, **16**(2), 131–148.

Dempster, M.A.H., Medova, E. and Tang, K., Determinants of oil futures prices and convenience yields. *Quant. Finance*, 2012, **12**(12), 1795–1809.

Dyer, T., Lang, M. and Stice-Lawrence, L., The evolution of 10-K textual disclosure: Evidence from Latent Dirichlet allocation. *J. Account. Econ.*, 2017, **64**(2–3), 221–245.

Eisner, J., The new probabilistic models for dependency parsing: An exploration. Proceedings of the 16th International Conference on Computational Linguistics, Copenhagen, Denmark, August 5–9, 1996.

Gibson, R. and Schwartz, E.S., Stochastic convenience yield and the pricing of oil contingent claims. *J. Financ.*, 1990, **45**(3), 959–976.

Gouel, C., Trade policy coordination and food price volatility. *Am. J. Agric. Econ.*, 2016, **98**(4), 1018–1037.

Griffiths, T.L. and Steyvers, M., Finding scientific topics. *Proc. Natl. Acad. Sci. U. S. A.*, 2004, **101**(suppl 1), 5228–5235.

Hagenau, M., Liebmann, M. and Neumann, D., Automated news reading: Stock price prediction based on financial news using context-capturing features. *Decis. Support. Syst.*, 2013, **55**(3), 685–697.

Hamilton, J.D. and Wu, J.C., Risk premia in crude oil futures prices. *J. Int. Money. Finance.*, 2014, **42**, 9–37.

Heath, D., Macroeconomic factors in oil futures markets. *Manage. Sci.*, 2019, **65**(9), 4407–4421.

Li, J., Li, G., Liu, M., Zhu, X. and Wei, L., A novel text-based framework for forecasting agricultural futures using massive online news headlines. *Int. J. Forecast.*, 2020a. doi:10.1016/j.ijforecast.2020.02.002.

Li, J., Li, J. and Zhu, X., Risk dependence between energy corporations: A text-based measurement approach. *Int. Rev. Econ. Financ.*, 2020b, **68**, 33–46.

Li, J., Li, J., Zhu, X., Yao, Y. and Casu, B., Risk spillovers between FinTech and traditional financial institutions: Evidence from the U.S. *Int. Rev. Financial Anal.*, 2020c, **71**, 101544.

Liu, Y., Han, L. and Yin, L., Does news uncertainty matter for commodity futures markets? Heterogeneity in energy and non-energy sectors. *J. Futures Mark.*, 2018, **38**, 1–2.

Mcdonald, R., Discriminative learning and spanning tree algorithms for dependency parsing. Ph.D. Thesis, University of Pennsylvania, Pennsylvania, USA, 2006.

Mo, D., Gupta, R., Li, B. and Singh, T., The macroeconomic determinants of commodity futures volatility: Evidence from Chinese and Indian markets. *Econ. Model.*, 2018, **70**, 543–560.

Moro, S., Cortez, P. and Rita, P., Business intelligence in banking: A literature analysis from 2002 to 2013 using text mining and latent Dirichlet allocation. *Expert. Syst. Appl.*, 2015, **42**(3), 1314–1324.

Natanelov, V., Alam, M.J., Mckenzie, A.M. and Van Uylenbroeck, G., Is there co-movement of agricultural commodities futures prices and crude oil? *Energy Policy*, 2011, **39**(9), 4971–4984.

Pröllochs, N., Feuerriegel, S. and Neumann, D., Enhancing sentiment analysis of financial news by detecting negation scopes. 48th Hawaii International Conference on System Sciences, Hawaii, USA, January 5–8, 2015.

Ready, R.C., Oil consumption, economic growth, and oil futures: The impact of long-run oil supply uncertainty on asset prices. *J. Monet. Econ.*, 2018, **94**, 1–26.

Smales, L.A., News sentiment in the gold futures market. *J. Bank. Financ.*, 2014a, **49**, 275–286.

Smales, L.A., O'Grady, B. and Yang, Y., Examining the impact of macroeconomic announcements on gold futures in a VAR-GARCH framework. *Quant. Finance*, 2014b, **22**(9), 710–716.

Sørensen, C., Modeling seasonality in agricultural commodity futures. *J. Futures Mark.*, 2002, **22**(5), 393–426.

Stanko, M.A. and Henard, D.H., Toward a better understanding of crowdfunding, openness and the consequences for innovation. *Res. Policy.*, 2017, **46**(4), 784–798.

Stephen, P.A. and Hillard, G., OPEC and world oil security. *Energy Policy*, 2017, **108**(1), 512–523.

Szymanowska, M., De Roon, F., Nijman, T. and Goorbergh, R.V.D., An anatomy of commodity futures risk premia. *J. Financ.*, 2014, **69**(1), 453–482.

Tsvetanov, D., Coakley, J. and Kellard, N., Is news related to GDP growth a risk factor for commodity futures returns? *Quant. Finance*, 2016, **16**(12), 1887–1899.

Wang, D., Tu, J., Chang, X. and Li, S., The lead-lag relationship between the spot and futures markets in China. *Quant. Finance*, 2017, **17**(9), 1447–1456.

Wang, Y.S. and Chueh, Y.L., Dynamic transmission effects between the interest rate, the US dollar, and gold and crude oil prices. *Econ. Model.*, 2013, **30**, 792–798.

Wei, L., Li, G., Zhu, X. and Li, J., Discovering bank risk factors from financial statements based on a new semi-supervised text mining algorithm. *Account. Financ.*, 2019a, **59**(3), 1519–1552.

Wei, L., Li, G., Zhu, X., Sun, X. and Li, J., Developing a hierarchical system for energy corporate risk factors based on textual risk disclosures. *Energy Econ.*, 2019b, **80**, 452–460.

Yan, L. and Bai, B., Correlated industries mining for Chinese financial news based on LDA trained with research reports. International Symposium on Communications and Information Technologies, Qingdao, China, September 26–28, 2016.

33

Classification of Flash Crashes Using the Hawkes (p,q) Framework[*]

Alexander Wehrli
and Didier Sornette

33.1 Introduction

On 6 May 2010, a then unique intraday price drop in U.S. equity markets to lows of 9–10% below the market's opening price within a timespan of just over four minutes shook financial markets. The event was accompanied by significant spikes in trading activity in US equity, equity derivatives and equity futures markets. In fact, it was later concluded that the crash could be attributed to a large, aggressive sell order executed in the E-mini S&P futures market, which upset liquidity conditions to the point

[*] The views, opinions, findings, and conclusions or recommendations expressed in this paper are strictly those of the authors. They do not necessarily reflect the views of the Swiss National Bank. The Swiss National Bank takes no responsibility for any errors or omissions in, or for the correctness of, the information contained in this paper.

of cascading into so-called 'hot-potato trading* (Commodity Futures Trading Commission-Securities Exchange Commission (CFTC-SEC) 2010a, 2010b). The order flows at this stage apparently grew ever more toxic (Easley *et al.* 2011, Andersen and Bondarenko 2014), undermining the willingness of market makers to hold inventory (Kirilenko *et al.* 2017) and prices collapsed as a result. The event, post-mortem coined as a 'flash crash', would become the first of its kind, followed by (supposedly) similar events in other financial instruments since then. For example on 15 October 2014, the price for U.S. Treasury futures contracts spiked sharply, resulting in the benchmark 10-year contract's yield experiencing a 16-basis-point drop and then rebound within only a few minutes (USDT, BGFRS, FRBNY, SEC and CFTC 2015). Also in over-the-counter traded financial instruments, like foreign exchange, such events have occurred (Bank for International Settlements 2017a), making clear that they are a feature of modern electronic financial markets, independent of the underlying asset. These large, precipitous and transitory price drops generated significant concern among legislators, regulatory bodies and consequently spurred interest in the phenomenon by both the investing public and academia (Madhavan 2012, Cespa and Foucault 2014, McInish *et al.* 2014, Kyle and Obizhaeva 2019).

A common culprit for the incidence of flash crashes—in particular also in the investing public[†]—is HFT activity (Bouveret *et al.* 2015). While there is however little empirical evidence that these market participants significantly alter their inventory dynamics when faced with large liquidity imbalances, they are however found to amplify volatility in such situations (Kirilenko *et al.* 2017, Schroeder *et al.* 2020). Other aspects that are cited to encompass, amplify or even cause flash crashes are procyclical trading strategies (von der Becke and Sornette 2011), the exhaustion of risk constraints and an evaporation of market depth (Anderson *et al.* 2015, Bank for International Settlements 2017b), a breakdown of crossmarket arbitrage activity (Menkveld and Yueshen 2019), data feed anomalies (Aldrich *et al.* 2017) and other mechanistic amplifiers like circuit breakers in related markets or hedging flows (Bank for International Settlements 2017a). Clearly, all of the aforementioned mechanisms can lead to a sudden deterioration of liquidity.[‡] Furthermore, any significant shock to liquidity has the potential to become self-reinforcing. For example, as liquidity dries up, the expectations of market makers may change to the point where finding trading partners to clear their exposures may seem increasingly unlikely, which will cause them to withdraw or extremely skew their prices, thus further exacerbating the liquidity crisis and potentially destabilizing the price formation process to the point of breakdown. Such endogenous emergence of market crashes[§] through imbalances between liquidity demand and risk-bearing capacities of market makers is also theoretically well understood (Huang and Wang 2009).

The picture that emerges is that the dynamics during such an event are resulting from a highly complex interaction between the behaviors of different market participants and a growing number of technical factors—owing to an ever-increasing technological sophistication and structural fragmentation of financial markets—each of which could have unforeseeably precipitated or amplified a critical event and a cascade of price changes. The involved feedback loops can in turn materialize in highly endogenous price process dynamics where, when pushed to a critical point, rapid movements from the minimum tick size all the way to fluctuations of the order of unity in the price are possible. Due to the increased connectedness between marketplaces, such extreme events are also more likely to be of systemic nature (Calcagnile *et al.* 2018). In this spirit, this paper refrains from trying to identify potential causes of flash crashes (in fact we hypothesize, and evidence suggests, that the possible triggers for such extreme price dynamics are manifold), but attempts to characterize the activity dynamics around such events in

---

[*] Referring to the repeated passing of inventory imbalances between market participants.

[†] A survey by BlackRock (2010) for example reports that 46% of surveyed professionals believed high frequency trading (HFT) was the biggest contributor to the E-mini crash.

[‡] We use the term liquidity quite casually here, without a strict definition and simply aim to describe an 'ease of trading'.

[§] We will use the term crash generically for the resulting price dynamics in such events. While most price moves during such events indeed were negative—mostly by construction, as the asset under consideration is traded against cash—there is however no reason why upward price moves should not be a consequence of such a deterioration in liquidity. And in fact, in our empirical study, we will also look at examples where the considered price spiked up.

detail. In particular, we are interested in investigating potential instantaneous characteristic signatures of anomalous market regimes, which could foreshadow significant short-term market instabilities.

Clearly, a very natural and rich description of the computerized and fragmented markets under consideration here could be gained by adopting a complex systems perspective and employing some sort of agent-based model (Samanidou *et al.* 2007, Huang 2015, Dieci and He 2018). Indeed, agent-based models have been found to reproduce a range of phenomena that are associated with flash crashes. In minority games for example, the crowding phase provides an example of how hot-potato trading emerges as a result of the dependence between recent individual actions and aggregate behavior (Coolen 2005, Challet *et al.* 2013). Also, an amplification of volatility through adapting behavior of agents has been successfully documented in an agent-based setup (Harras and Sornette 2011). The approach we propose in the sequel can be viewed as being between an agent-based model and a purely stochastic model, as it captures the collective activity process resulting from the interactions of all agents. In this representation, each event caused by one market participant is a new signal to other market participants, which react themselves in a causal manner. The time-dependence of the relevant feedback parameter of the model can hereby be interpreted as the varying nature of the extent of that response—either due to changes in the behavior of agents or their composition in the market. Our framework is an extension of the classical Hawkes process (Hawkes 1971b, 1971a), which has proven to be a very useful framework to statistically measure and describe the interaction between events at the lowest time scale in a variety of scientific fields. In particular, this self-excited point process—where applicable—allows one to quantify and disentangle the relative contributions of endogenous (endo) and exogenous (exo) activities within a given system. Thanks to its mapping to a branching process (Ogata 1981), the classical Hawkes model provides a single parameter—the branching ratio η—which directly measures the degree of endogenous feedback effects in the arrival of events. In the context of fluctuations in high frequency financial prices, Filimonov and Sornette (2012) have opened the road toward the full utilization of the dynamics of the branching ratio in order to diagnose different regimes and to possibly forecast impending crises associated with the approach to criticality, which is determined by $\eta = 1$. Here, we leverage recent methodological developments in Wheatley *et al.* (2019) and Wehrli *et al.* (2020) and extend the framework proposed therein to allow estimation of the full dynamics of $\eta(t)$ in a structural manner.

The study of feedback parameters, such as the branching ratio, in order to characterize interactions and the stability of systems, in the same spirit as we do here, has a rich history in many scientific fields. Sornette (2006) provides a study of general precursory and recovery signatures accompanying endogenous and exogenous shocks in complex networks. Theoretical predictions for signatures of precursors and responses to shocks in very general processes were developed in Sornette and Helmstetter (2003), also providing a mean-field theory for a variant of the stochastic process we will use for the purposes of this study. A key insight is hereby that responses to shocks depend on the nature and extent of the feedback mechanisms present in the system dynamics. In Crane and Sornette (2008) for example, these differences are leveraged to classify dynamics in social systems as endogenous or exogenous. Other applications encompass book sales (Deschatres and Sornette 2005), seismicity (Helmstetter *et al.* 2003), or financial crashes (Sornette 2017). Naturally, the study of financial volatility time series also lends itself to such considerations (Sornette *et al.* 2004b, Joulin *et al.* 2008). For example in the recently proposed jump-diffusion model of Bates (2019), large market movements are the result of the accumulation of rapid and self-exciting intraday increases in conditional volatility, a quantity closely related to the conditional intensity of price changes used to describe the price process in the sequel of this paper. A recent study in Fosset *et al.* (2020) further links the occurrence of liquidity crises to such endogenous feedback effects and finds that, for these events to occur, financial markets must either constantly operate just below criticality, or exhibit time-dependent endogenous dynamics that occasionally visit the critical regime in order to produce the events in question. Resonating with the latter scenario and tying back to the argument of approximating aggregate agent behavior using Hawkes processes, also in agent-based models, the parameter controlling the social coupling/herding strength of traders must be allowed to intermittently reach the critical regime in order to reproduce stylized facts of financial markets like volatility clustering, realistic bubbles and corrections (Kaizoji *et al.* 2015, Westphal and Sornette 2020). In this spirit, our present study in fact proposes a framework to empirically distinguish a wide range of such dynamics on a case-by-case basis.

To test our approach as broadly as possible, we investigate events in equity and bond futures, as well as in the interbank foreign exchange market, all of which are among the most liquid financial markets traded electronically today. In particular, we investigate the E-mini flash crash of 6 May 2010, the U.S. Treasury yield flash crash of 15 October 2014 and several crashes in interbank currency spot markets. To complement our study of flash crashes in these markets, we also look at an event in the Ethereum cryptocurrency market. Since their creation in 2008, cryptocurrencies have significantly gained in attention and trading activity, with in excess of 100 cryptocurrencies currently traded in the market. In particular, Bitcoin and Ethereum (ETH) have assumed a leading role, making up more than half of the market capitalization of the top 1000 cryptocurrencies, even after the drastic decline in Bitcoin's market share in early 2017 (Gerlach *et al.* 2019). We propose the study of a flash crash in one of these markets to provide an important complement to the study of traditional financial assets for two reasons. First, the notion, that crashes appear to occur also in the absence of any particular exogenous event, suggests that—as discussed—destabilising feedback loops of behavioral origin may be present. The lack of a way to reliably assess the fundamental value of a cryptocurrency, as opposed to the other asset classes considered here, makes such a behavioral root of crashes a plausible hypothesis. Second, while equities, bonds and foreign exchange are highly mature markets, cryptocurrencies have only been around for a good decade, and their marketplaces have a much more contemporary technological foundation. As such, any commonalities arising from the joint study of mature and adolescent markets will point towards mechanisms that are indeed more fundamental than simple differences in asset characteristics or trading technologies. Furthermore, to our knowledge, this is also the first study to provide a comprehensive cross-asset analysis of flash crash dynamics and their exogenous and endogenous constituents. Finally, it is important to note that we select events without a systematic definition of what exactly constitutes a flash crash and simply look at events that were designated as such by the investing public and the financial press.

The rest of the paper is organized as follows. Section 33.2 describes the data sets used in the study and provides some institutional details of the various financial markets investigated here. Section 33.3 introduces the point process model used to characterize price formation dynamics and Section 33.4 summarizes the methodology—described in detail in a technical appendix—used to estimate this model on high frequency price data. Section 33.5 surveys a series of flash crashes and provides estimated dynamics of exogenous and endogenous market activity around the events. Section 33.6 explores the goodness of the fitted processes and tests the proposed model against an alternative specification with a refined treatment of exogenous clustering. Section 33.7 concludes.

33.2 Data Sets

Since the financial instruments we are studying might be traded in different places/on different platforms simultaneously—foreign exchange for example being an OTC market with a highly fragmented market structure (Bank for International Settlements 2017, Schrimpf and Shushko 2019)—we aim to base our analysis on data from the most liquid marketplace available for each instrument. In the sequel, we provide a description of the data sets we rely on—with a particular focus on technical aspects that could influence observed activity patterns—and discuss some institutional details of the various markets.

33.2.1 The 10-Year U.S. Treasury Note Futures Contract

Treasury futures are standardized contracts to trade U.S. government debt obligations for future delivery or settlement. They were introduced in the 1970s at the Chicago Board of Trade (CBOT), now part of the Chicago Mercantile Exchange (CME), as a tool to hedge short-term risks on U.S. treasury yields.

One contract for the 10-year futures contract has a face value at maturity of USD 100 000. That is, each contract affords the buyer the right to buy an underlying Treasury note with a face value of USD 100 000. The prices of these contracts are quoted as percentages of their par value, in the case of the 10-year contract the minimum price increment is 1/64%. In USD terms, this amounts to USD 15.625. Treasury futures are traded on an electronic limit orderbook, the CME Globex trading platform, but there are also still open outcry trading pits, which we however exclude from the analysis as they only account for a negligible share of trading. Overall, the Treasury futures market is a highly computerized financial market, where high frequency trading is estimated to account for well more than half of the trading activity on benchmark contracts (Bouveret *et al.* 2015). Trading on the Globex platform happens continuously from 5 pm (Chicago Time; CT) to 4 pm (CT) of the next day, Sunday to Friday. Between 4 pm and 5 pm CT, the exchange imposes a maintenance window where no trading is possible. Treasury futures contracts have a quarterly expiration cycle in March, June, September and December. At any given point in time, several contracts written on the same underlying security (differing only in their expiration dates) may trade side by side. The contract closest to expiry, the so-called front-month contract, offers the most liquidity and is the contract we focus our analysis on. In October 2014, when the flash crash occurred, the front-month contract was the one expiring in December (symbol ZNZ4). In principle, the data feed also contains quotes from a so-called implied orderbook, which originate from some order book related to the futures contract under consideration. For our study, we ignore quotes from the implied order book and compute mid-prices from the direct order book exclusively. Our data set is obtained from the CME DataMine and contains all market data messages required to reconstruct the limit order book tick-by-tick, 10 price levels deep. Each market data message is timestamped to the millisecond at the time of market data publication, but can contain multiple modifications to the book. Each modification is additionally given a timestamp when the respective change to the book actually occurred on the Globex exchange, but only timestamped to the second.

33.2.2 The E-Mini S&P500 Futures Contract

The E-mini S&P 500 futures contract was introduced in 1997 by the CME as a complement to the regular S&P 500 futures contract and has become one of the most actively traded securities in the world. It trades exclusively electronically on the CME Globex platform. As for U.S. Treasury futures, the share of high frequency trading is typically estimated to be 50% or higher (Sussman *et al.* 2009, Aite Group, New World Order 2009)—even though there are some issues surrounding estimates of the size of high frequency trading volumes (SEC 2010). Trading is possible from 5 pm CT to 4 pm CT of the next day, Sunday to Friday. In addition to the one hour maintenance period starting at 4 pm, there is an additional temporary trading halt of 15 minutes, starting at 3:15 pm CT. The size of S&P 500 E-mini contracts is one-fifth of the regular S&P 500 futures contracts—which are still partially floor-traded—and it has a notional value of USD 50 times the value of the S&P 500 stock index, denominated in index points. The contract has a tick size of 0.25 index points (USD 12.50) and four expiration months per year. We use data for the front-month contract to ensure that we use the most actively traded contract for our analysis, which in May 2010 was the contract with expiry in June. Our data set for the E-mini is obtained from TickData,* which offers research-quality, historical tick-by-tick prices of futures and index markets. The data set contains entries for the best bid and ask quotes whenever either the respective price or amount changed, timestamped with millisecond resolution.

33.2.3 Foreign Exchange Spot Interbank Market

For the events in foreign exchange markets, we rely on data from the trading venue which is considered as the respective primary interbank trading place for the currency pair. In particular, this is the

---

* www.tickdata.com

Electronic Broking Services (EBS) interbank trading platform for the EUR/USD and USD/JPY currency pairs and Thomson Reuters Matching, now a service of Refinitiv, for the other currency pairs in the present study. All currency pairs analyzed are also traded on a plethora of other trading venues simultaneously, but EBS and Refinitiv are still considered primary marketplaces and important points of price discovery, despite their declining market share (Bank for International Settlements 2017). The EBS dataset provides a snapshot of the limit order book, whenever a change occurs in the 10 best price levels, within the last 100 milliseconds, with corresponding timestamp resolution of 100 milliseconds. The data for the Refinitiv currency pairs is obtained from Thomson Reuters Tick History (TRTH). An entry encompassing the best quoted price levels on each side of the market is available when either the best available price or respective quoted amount has changed. The data set contains two timestamps for each modification, one representing the time when the change happened on the trading platform with millisecond resolution and one taken at collection time by TRTH with nanosecond resolution. Both foreign exchange trading platforms operate continuously from Sunday to Friday and their minimum order sizes are one million of the base currency. Prices on EBS are quoted in half-pips,* whereas on Reuters the tick size is typically one pip.[†]

33.2.4 The Ethereum Cryptocurrency Market

For the study of the Ethereum cryptocurrency—or rather the Ethereum platform's native cryptocurrency 'Ether'—we look at data from GDAX (Global Digital Asset Exchange), the first licensed U.S. bitcoin exchange and largest market to trade Ether against the US-Dollar in the world, launched in 2016. Even though technically speaking, the Ethereum platform itself is not a cryptocurrency, we will use the names Ethereum and Ether interchangeably. We obtain data on quote and trade activity on the GDAX order book from coinapi.io, a provider of unified data APIs to cryptocurrency markets. The data set contains tick-by-tick information on the top-of-book quoted prices and amounts, as well as all trading activity on the GDAX exchange. For each update, two timestamps are available, one provided by the GDAX exchange with millisecond resolution and one with even higher resolution, taken when the market data update was recorded by the data provider. The ETH/USD pair is traded with price increments of 0.01 USD. As opposed to the other financial markets considered here, ETH/USD can be traded on the GDAX 24/7.

33.2.5 Sample Selection and Summary

From each of the data sets, we compute mid-prices, i.e. the average of the best prevailing bid and ask price at any given point in time. To fit the mid-price data for the various events, we collect the times where the mid-price changed, in a window of one day centered around the trough/peak of each event. Since not all of our data sets encompass 12 hours of pre- and post-event data—or include a weekend/non-trading period within the desired window—not all samples will be 24 hours long. We provide a short summary of the events under consideration and the respective sample sizes in table 33.1. Additionally, we illustrate the distribution of the events across the day in figure 33.1, which shows that, while there are indeed some events that happen during traditionally illiquid times of the day, there are also crashes that occurred during the more liquid times when trading centres in both Europe and the United States are active.

---

* One pip being 0.0001 USD for EUR/USD and 0.01 JPY for USD/JPY.
[†] Being 0.0001 units of the price currency for the considered currency pairs, except for USD/ZAR, where the tick size is 0.0005 South African Rand.

TABLE 33.1 Overview of Data Sets Used in the Study

#	Market	Date	T	N_t	$\langle\lambda\rangle$	$\bar{\lambda}$
(1)	E-mini	6 May 2010	13 h 30min	86 486	1.78	172
(2)	U.S. Treasury	15 October 2014	19 h 20 min	108 833	1.56	270
(3)	EUR/USD	18 March 2015	24 h	68 832	0.80	10
(4)	NZD/USD	24 August 2015	24 h	12 861	0.15	2
(5)	USD/ZAR	11 January 2016	16 h 18 min	5427	0.09	2
(6)	GBP/USD	7 October 2016	24 h	27 169	0.31	4
(7)	ETH/USD	21 June 2017	12 h 23 min	110 856	2.49	31
(8)	EUR/USD	25 December 2017	24 h	869	0.01	5
(9)	USD/JPY	2 January 2019	24 h	44 782	0.52	10

Notes: The fitting window T of each event is the result of taking one day of data around the trough/peak of the event. Where the sample duration is less than one day, this means that either: the resulting sample window covers a weekend/non-trading period; or is limited by the size of the available data set. The number of events within each window is denoted N_T. We provide average unconditional event intensities $\langle\lambda\rangle$, as well as the respective maximum $\bar{\lambda}$ in the sample. The units of the intensities are sec^{-1}.

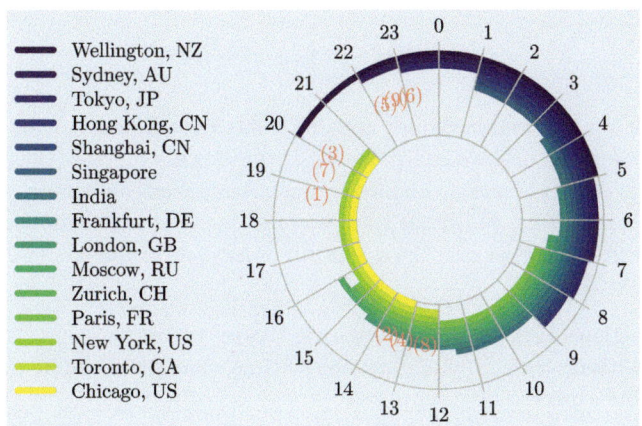

FIGURE 33.1 Shows a Coordinated Universal Time (UTC) clock of the trading hours of major financial centers, where we have used the representation of Chen and Stindl (2018). The events from table 33.1 are indicated with their number at the respective time of day.

33.3 The Hawkes Model with Time-Dependent Endogenous Dynamics

The classical Hawkes process (Hawkes 1971a, 1971b) has provided an important tool in the study of financial price fluctuations (see e.g. Bowsher 2007, Bacry *et al.* 2013, Bacry and Muzy 2014, Rambaldi *et al.* 2019 and reviews in Bacry *et al.* 2015 and Hawkes 2018). In particular, it parsimoniously allows one to distinguish endogenous responses from exogenous influences (Filimonov and Sornette 2012, Hardiman *et al.* 2013, Filimonov *et al.* 2014, Hardiman and Bouchaud 2014, Filimonov and Sornette 2015b)—a task that we call 'solving the endo-exo problem' (Wheatley *et al.* 2019, Wehrli *et al.* 2020). In the standard Hawkes model, the parameter η—being defined most naturally in the language of branching processes (Harris 1963, Daley and Vere-Jones 2008) as the average number of offspring of first generation triggered per observed point—characterizes also the degree of self-excitation inherent to the dynamics of the system. At the same time, η characterizes the fraction of triggered events in the observed population (Helmstetter and Sornette 2003, Saichev and Sornette 2006). When $\eta < 1$, then the process is *subcritical* and, starting from a single immigrant point at some time, dies out with probability one. When the process is *super-critical* ($\eta > 1$), the process has finite probability to explode to an infinite number of points (Saichev et al. 2005). The critical regime ($\eta - 1$) delineates the transition between the two other regimes and can still produce stationary dynamics under some conditions for the intensity of exogenously occurring events (Brémaud and Massoulié 2001, Saichev and Sornette 2014, Jaisson and Rosenbaum 2015).

Here, we consider a generalization of the standard Hawkes process, which features both exogenous non-stationarity (in the form of a time-dependent, deterministic arrival intensity for new exogenous points) as well as time-dependent endogenous dynamics in the form of a deterministically vary-ing branching ratio. The point process N we work with is defined as a random measure on $\mathbb{R}$ and counts the random number of points in some Borel set $A \subseteq \mathbb{R}$, denoted $N(A)$. For ease of notation, we write $N_t := ((0,t])$ for $t \in [0,\infty)$. A single realization of the process consists of all the times $s \in [0,t]$, where $dN_s = N_s - \lim_{u \to s^-} N_u = 1$ and we assume that the process is orderly. The history of the process is given by $\sigma(N) = \left(\mathcal{F}_t \right)_{t \in (-\infty,\infty)}$, where $\mathcal{F}_t$ is the σ-algebra generated by N up to and including t, i.e. $\mathcal{F}_t = \sigma \left(N_s, s \leq t \right)$. The $\mathcal{F}$-conditional intensity function, which fully defines our process, has the form

$$
\begin{aligned}
\lambda(t) &= \mu(t) + \int_{-\infty}^{t} \eta(s) h(t-s) dN_s \\
&= \mu(t) + \sum_{i:t_i < t} \eta\left(t_i\right) h\left(t - t_i\right),
\end{aligned}
\tag{33.1}
$$

where $\mu(t) \geq 0 \ \forall t$ is the intensity of exogenously arriving points and can be called the immigration inten-sity in view of the branching process interpretation of the Hawkes process (Ogata 1981). The endog-enous feedback effects of the process are defined by the instantaneous branching ratio $(t) \geq 0 \ \forall t$ and an offspring density $h:[0,\infty) \to [0,\infty)$, which has an L^1 norm of one. It is worth considering some possible specifications for the instantaneous branching ratio, in order to delineate properly the class of models we are considering here.

(i)   A constant function $\eta(t) \equiv \eta$ recovers the classical Hawkes process.

(ii) If $\eta(t)$ is some fixed, deterministic function of time, then there can be time-varying trends and shifts in the expected number of first generation offspring that are triggered by a given point.

(iii) An often considered general specification defines $\left\{ \eta\left(t_i\right) \right\}_{i \geq 1}$ to be a sequence of independent and identically distributed random variables, e.g. also encompassing the dynamic contagion process of Dassios and Zhao (2011).

(iv) If $\eta(t)$ is allowed to evolve according to a stochastic differential equation, then we recover the Hawkes process with stochastic excitations from Lee et al. (2016).

(v)   The instantaneous branching ratio can also be envisaged to depend on some other observed quantity beyond their occurrence time, which would cast the setting into a marked point process specification (Daley and Vere-Jones 2003). The marks m_i associated with some point t_i relate to the instantaneous branching ratio through a 'productivity law', given by some function f, as $\eta(t_i) = f(m_i)$. This specification for example includes the well-known epidemic-type aftershock (ETAS) model of Ogata (1988) developed in statistical seismology to describe complex sequences of earth-quake aftershocks, where the productivity is an exponential function of the magnitude mi of the main shock.

(vi) The specification $\eta(t) = f(\lambda(t))$, where the function $f(\cdot)$ encodes the dependence of the branching ratio on the contemporaneous conditional intensity, corresponds to the recursive point process model of Schoenberg et al. (2017), where the branching ratio depends on the conditional intensity, which itself is a function of the instantaneous branching ratio of preceding points.

(vii) A renewal-type version of the instantaneous branching ratio can be obtained by choosing $\eta(t_i)=f(t_i-t_{i-1})$, e.g. applied in Harte (2014).

In our setup here, $\eta(t)$ is a *deterministic* function of time as in (ii), which is sampled at random jump times t_i. We will hereby model the function describing the time-evolution of η in a flexible, non-parametric fashion.

Also in this specific setup, we can interpret λ as a conditional hazard function, i.e. $\lambda(t) = \lim_{\Delta \to 0} \Delta^{-1} \mathbb{E}\left[N_{t+\Delta} - N_t \mid \mathcal{F}_{t-} \right]$. The branching representation of this point process is furthermore very similar to the standard Hawkes process in that (t) induces immigrants according to an inhomogeneous Poisson process. All existing points t_i then trigger a first generation of offspring with intensity η $(t_i)h(t-t_i)$, which in turn trigger their own generation of offspring in the same manner, and so on. The collection of these independent Poisson processes forms the process N and the $\eta(t)$ are the expected number of immediate offspring of a single point occurring at time t. This branching representation also provides the basis for efficient simulation* of the process. Our model for the endogenous feedback mechanism—i.e. the memory of the process—is an approximate Pareto distribution (Hardiman *et al.* 2013) with tail index $\varepsilon > 0$,

$$h(t) = Z^{-1} \left[\sum_{i=0}^{M-1} \xi_i^{-(1+\varepsilon)} e^{-\frac{t}{\xi_i}} - S e^{-\frac{t}{\xi_{-1}}} \right]. \tag{33.2}$$

The $\xi_i = \tau_0 b^i$ are power law weights, $\tau_0 > 0$ a short-lag cutoff and S and Z normalizing factors. The range over which the power law is approximated can be controlled by M and b. If one is, e.g. interested in computing the log-likelihood of the process (Daley and Vere-Jones 2008) for an observed sequence of events at times $t_i \varepsilon (0,T]$,

$$\log L\left(t_1, \dots, t_{N_T}\right) = \sum_{i=1}^{N_T} \log \lambda\left(t_i\right) - \int_0^T \lambda(t) dt, \tag{33.3}$$

then this approximate power law form for the offspring density allows a reduction of computational complexity from $\mathcal{O}\left(N^2\right)$ to $\mathcal{O}(N)$ (Ozaki 1979).

33.4 Calibration to Mid-Price Data—the Hawkes(p,q) Model

Our goal is to estimate the model (33.1) based on realizations $\{t_i\}_{i=1}^{N_T}$ of the random sequence of event times induced by the process of high frequency mid-price changes. The predominant approach to create such time-varying estimates is to numerically maximize the log-likelihood (33.3) of a time- invariant parameter model on windows of some fixed length (i.e. on sub-samples of the jump times t_i) and rolling them forward in time to trace out a time-variant estimate of the parameters (e.g. Filimonov and Sornette 2012, Hardiman *et al.* 2013, Filimonov *et al.* 2014, Mark *et al.* 2020). This approach of rectangular windowing brings several practical difficulties, like the selection of window size in a way that does not neglect relevant memory, or treatment of edge and truncation effects, which when neglected can bias parameter estimates. Similar issues are for example well-known in the signal processing literature, e.g. in the form of the 'Gibbs phenomenon'. Resembling approaches in signal processing to cope with

---

* The process can be simulated recursively, generation-bygeneration. The immigrant points from $\mu(t)$ are simulated first—either by thinning Lewis and Shedler (**1979**) or by drawing the number of points on [0, T] from the Poisson distribution with mean $\int_0^T \mu(t)dt$ and choosing their time-location by inverse-transform sampling (Cinlar 2013) from a density $\propto \mu(t)$. The number of points for each offspring process of a given mother $\mathbf{t}_i$ are then each drawn from a Poisson distribution with mean $\boldsymbol{\eta}(\mathbf{t}_i)$ and dispersed on the time axis according to random inter-event times drawn from $\mathbf{h}(t)$.

these issues, also for models such as (33.1), the rectangular windowing can be replaced by more general approaches like *local maximum likelihood* (Tibshirani and Hastie 1987, Fan *et al.* 1998), which introduces kernel-weights into the log-likelihood in order to localize its parameters. However, bandwidth selection for the kernel is a highly involved procedure when done systematically. Furthermore, the optimal bandwidth in general also depends on bias and variance estimates of the non-parametric estimator, which might not be easily available. Additionally, independent control over the flexibility of the various parameters in the model is also non-trivial.

For the present application, where adaptive and time-dependent estimators for both $\mu(t)$ and $\eta(t)$ are required, we prefer an approach where the flexibility of the individual models can be controlled and traded off more explicitly. Our approach is to rely on the Expectation Maximizaton (EM) algorithm (Dempster *et al.* 1977) to provide a maximum likelihood estimate (MLE). The EM algorithm is a widely applicable computational approach for the iterative construction of MLE in incomplete data or latent variable problems. In the context of point processes with affine conditional intensities like (33.1), the unobserved information can be chosen to be the corresponding branching structure of the point process. In this case, the EM decomposes the estimation into sub-problems, where at each iteration of the algorithm, separate models can be fitted for $\mu(t)$ and $\eta(t)$. Here, we will use adaptive spline models for both functions, which allow one to control their flexibility by choosing a number of degrees of freedom (i.e. knots of the splines). We denote these degrees of freedom as p for $\mu(t)$, and q for $\eta(t)$, labeling the resulting full model as a *Hawkes(p,q)* model. To perform model selection, one can then simply apply the estimation procedure for a range of values for p and q and take the model which minimizes some information criterion. Based on past results (Wheatley *et al.* 2019), we choose the Bayesian Information Criterion (BIC) (Schwarz 1978) for our purposes.

A detailed description of the EM estimation procedure is given in appendix 1. The ability of the approach to estimate the dynamics of interest in the presence of time-varying exogenous and endogenous excitation—in particular during a regime where the process approaches criticality temporarily—is furthermore demonstrated in appendix 2.

33.5 Endogeneity of Financial Markets around Flash Crashes

To characterize exogenous and endogenous dynamics around flash crashes, we now apply the Hawkes(p,q) methodology to our data sets from section 33.2. In the sequel, we provide a short description of each event—in chronological order—and discuss how endogenous and exogenous activity evolved through the lens of our Hawkes(p,q) fits. We begin by a short discussion on the treatment of timestamps used for estimation and provide the specific settings of the EM algorithm used in estimation.

33.5.1 Preprocessing of Timestamps and Calibration Specifics

As was discussed in section 33.2, the data sets in our study exhibit different limitations regarding the resolution of the timestamps associated with the mid-prices. This brings several practical difficulties. First, any limitation in time resolution introduces a spurious discretization of the duration data, which might affect estimation of the offspring probability density during the M-step of the EM algorithm. Second, while for our foreign exchange and cryptocurrency data sets, even where timestamps are available with millisecond precision, the timestamps are in fact unique, for the E-mini and U.S. Treasury data set, around 45% of the events have identical timestamps. For our point process model to be well-defined, we need to assume that the associated counting process is *orderly*—i.e. $\mathbb{P}\left[N_{t+\Delta} - N_t > 1 \mid \mathcal{F}_{t-}\right] = o(\Delta)$. As such, contemporaneously occurring events are incompatible with the definition of the process. The principled approach to deal with such limited timestamp resolution and/or duplicate timestamps is to randomize them within the interval of uncertainty (Filimonov and Sornette 2012, Hardiman *et al.* 2013, Lallouache and Challet 2016). Additionally, depending on the timestamp which is used as a base for the calibration, an additional source of uncertainty is introduced by a potential bundling of updates to the limit order book during data collection/processing, which when combined with inadequate assumptions on the related timestamp uncertainty can induce spurious clustering in the data (Filimonov *et al.* 2014, Filimonov and Sornette 2015b).

Here, we follow along the lines suggested in Filimonov and Sornette (2015b) and whenever available, consider timestamps provided by the exchange as a reliable source of data, despite their typically inferior resolution as compared to potential additional timestamps taken by the provider of the data sets. For all data sets except the E-mini, we have the exchange timestamps available and simply smooth the resulting duration data by randomizing their timestamps within the resolution of the respective timestamps, which is—as elaborated in section 33.2—$\Delta = 1$ second for the U.S. Treasury futures data, $\Delta = 100$ milliseconds for the EBS data and $\Delta = 1$ millisecond for the TRTH and Ethereum data. Additionally, for these data sets, we will also produce fits on the durations resulting from raw timestamps (denoted $\Delta = 0$). For the U.S. Treasury futures data, we will use for this purpose the higher resolution timestamps taken at the time the market data update was published, but remove all duplicates. Additionally, we can look at the difference between the time a market data message was published (with millisecond precision) and the timestamp of each modification at the exchange contained within such a message (with second precision). This difference Δ_i might in fact give a more narrow definition of the uncertainty around the true timestamp of a mid-price jump than the one-second window assumed in the first set of fits. It turns out that the distribution of these time differences follows a uniform distribution quite well. As such, we produce an additional set of fits for the U.S. Treasury futures event, applying a random time shift $\Delta \sim U(0, \Delta_i)$ to each second-precision timestamp t_i, contained within a market data message used to derive Δ_i. For the E-mini data, we do not have any exchange timestamps available and thus rely on the timestamps with millisecond precision from the data provider, but perform a search over different assumptions on Δ. As suggested in Filimonov *et al.* (2014), when relying on such enriched timestamps, the choice of Δ can be informed by the distribution of durations. As can be seen from table 33.2, the median duration of the E-mini data is approximately 40 milliseconds, while its average is around one second. This suggests that a reasonable value for Δ is of the order of 100–1000 milliseconds.* In order to be as close as possible to the analysis in Filimonov and Sornette (2012), we will focus on $\Delta = 1$ second, but

TABLE 33.2 Overview of Timestamp Sources (EXCH = Exchange, DP = Data Provider) and Durations per Event

#	Market	Source	Median duration (sec)	Mean duration (sec)
(1)	E-mini	DP	0.043	1.090
(2)	U.S. Treasury	EXCH	0.006	1.128
(3)	EUR/USD	EXCH	6.7	99.480
(4)	NZD/USD	EXCH	2	6.716
(5)	USD/ZAR	EXCH	1	10.701
(6)	GBP/USD	EXCH	0.751	3.180
(7)	ETH/USD	EXCH	0.101	0.402
(8)	EUR/USD	EXCH	0.3	1.255
(9)	USD/JPY	EXCH	0.5	1.930

Note: For the U.S. Treasury data, the timestamps are chosen to be the ones taken at the time of publication of the market data.

---

* We remark that Filimonov *et al.* (2014) find consistent uncertainties (based on data from a different data provider) of the timestamps for the E-mini in the year 2010 and additionally show that constant branching ratio estimates for different assumptions in the range 100 ms < Δ < 1 second are practically indistinguishable, while $\Delta = 1$ second results in slightly higher branching ratios, but similar dynamics.

also consider alternative assumptions of $\Delta = 0$ (fitting on unique timestamps only, without additional randomization), $\Delta = 1$ millisecond and $\Delta = 100$ milliseconds and assess the effect of a particular choice for Δ on the branching ratio and in particular the resulting goodness of the fit.

Finally, in our EM estimation, we vary the allowed degrees of freedom for the immigration and branching ratio functions $p,q \in [1,25]$ and, as in our simulation study before, truncate the interaction length of $h(\cdot)$ at 1000 points. The algorithm is run for 500 iterations, or early stopped as soon as convergence is achieved. Our criterion for convergence will be that the supremum norm of the change in the parameter vector from one iteration to the next stays below a threshold of 10^{-4} for a couple of consecutive iterations.

33.5.2 The E-Mini S&P 500 Futures Crash in May 2010

We begin our study by looking at the E-mini flash crash of 6 May 2010. In figure 33.2, we show the evolution of the mid-price of the futures contract and the surrounding activity in quoting and trading activity. The spike in quoting activity is indeed very impressive with over 4000 mid-price changes per minute at the peak. Regarding the particular sequence of events that encompassed the crash, we refrain from repeating them here, as they are covered in great detail in Commodity Futures Trading Commission-Securities Exchange Commission (CFTC-SEC) (2010a) and Commodity Futures Trading Commission-Securities Exchange Commission (CFTC-SEC) (2010b) and several other studies. We want to highlight however the many technical frictions that accompanied the event. For example the fact that there was a significant dislocation between ETF prices on the NYSE and the E-mini futures on the CME is well-known, as is the fact that there was a five second trading halt on the CME due to its stop logic functionality during the event. What is less well covered and was recently highlighted in great detail in Aldrich *et al.* (2017) is that there were multiple message-level issues on the data feeds just a few minutes before the trading halt was triggered on the CME, leading to compromised market data integrity and market participants' withdrawal of liquidity. We believe that such noise introduced into the data feeds indeed has the potential to amplify feedback effects, much like other mechanical amplifiers do.

Looking at the resulting estimates in figure 33.2, we see that the optimal Hawkes(19,7) fit using $\Delta = 1$ second recovers a branching ratio that is rather stable around 0.7 before the crash and then rises to criticality as the mid-price diverges. The dynamics of the branching ratio are in very good agreement with what was recovered in Filimonov and Sornette (2012), using the same randomization window, even though their peak branching ratio of around 0.95 is slightly lower than what we find here. A transient rise in the branching ratio is also found when fitting on raw/unique timestamps ($\Delta = 0$). There, the optimal Hawkes(17,18) fit recovers a branching ratio that rises from values around 0.5 to ≈ 0.8 as the mid-price diverges. Unsurprisingly, removing almost half of the points to achieve uniqueness in the timestamps results in a downward bias of the average level of self-excitation of ≈ 0.2, but to a large extent preserves its dynamics. The branching ratio stops its rise and drops noticeably around the time the CME halted trading. Once trading resumed, the branching ratio relaxes towards pre-crash levels. The exogenous intensity on the other hand also explodes sharply as the crash materializes, but stays elevated for a much longer period of time than the branching ratio. Comparing the goodness of the fits for different assumptions on Δ (see also section 33.6) reveals that for all values except $\Delta = 0$, a significant amount of autocorrelation persists in the residuals of the process, even though the average estimates for $\eta(t)$ are comparable across the different assumptions and in line with other previous studies at around 0.7–0.8. Fits using $\Delta = 1$ second exhibit the least residual autocorrelation on randomized timestamps, thus produces the best goodness of fit among the candidates for Δ tested, and we use this value as the optimal choice for the purpose of this study.

33.5.3 The U.S. Treasury Bond Yield Crash in October 2014

Between 13:33 and 13:45 UTC, on October 15, 2014, the U.S. Treasury bond market experienced a significant spike in prices, against the backdrop of falling yields that had already set in with the U.S. stock

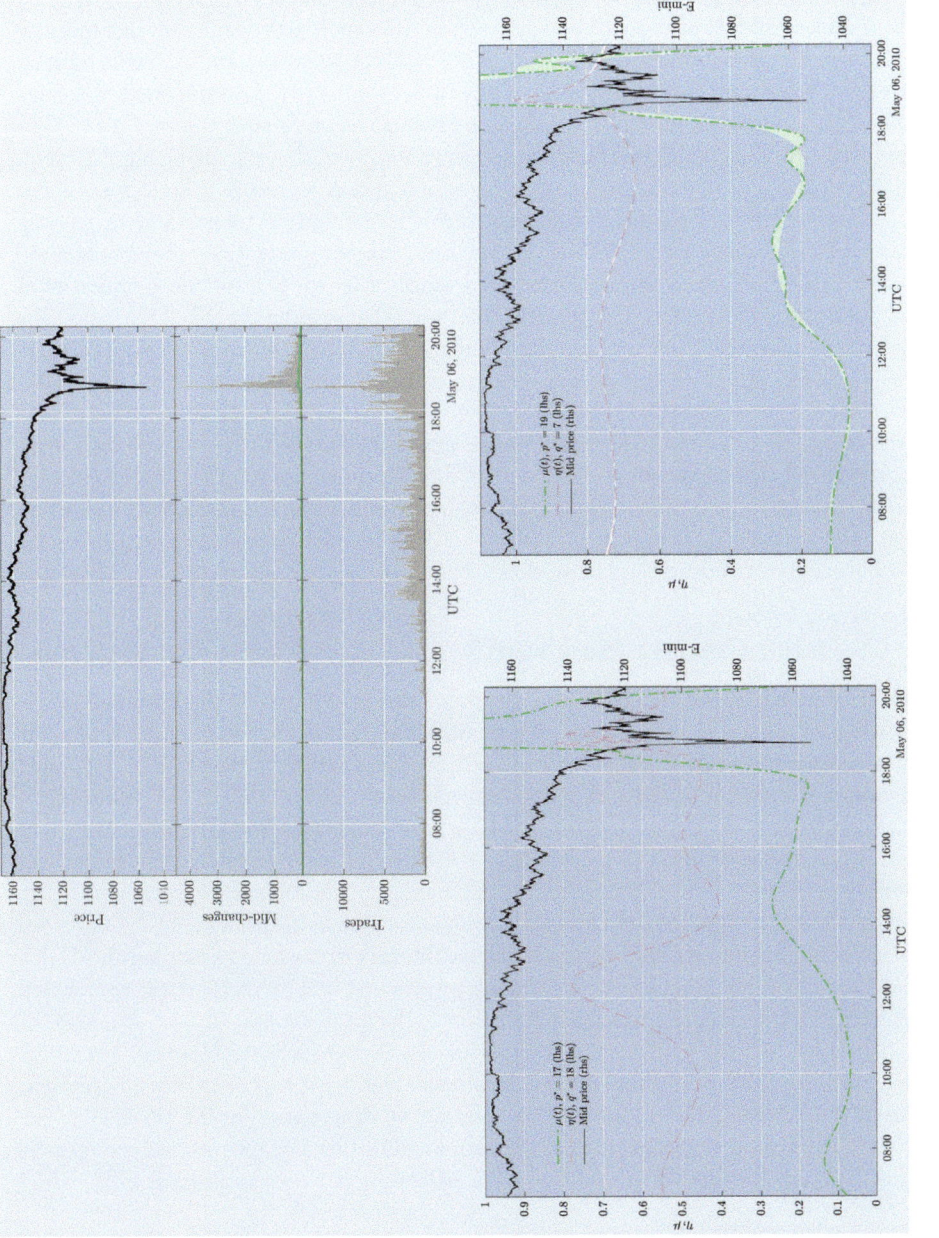

FIGURE 33.2 The top row shows the evolution of the mid-price of the S&P 500 E-mini futures contract around the event on 6 May 2010. Additionally shown below the price are the number of mid-price changes and trades per minute. In the bottom row, the estimated immigration intensities $\mu(t)$, in sec^{-1}, and branching ratio $\eta(t)$ are shown in green and purple for fits using $\Delta = 0$ (left) and $\Delta = 1$ second (right), respectively. The lines indicate the BIC-optimal functions and the shaded regions indicate 95% credible intervals, obtained by bootstrap sampling 1000 times according to the approximate posterior probabilities of the models, see Appendix 1. The BIC-optimal degrees of freedom for the immigration ($p^\star$) and the branching ratio ($q^\star$) are indicated in the legend. Furthermore, the optimal immigration estimate from fits using $\Delta = 1$ second is also indicated in green in the upper panel, scaled by integrating $\mu(t)$ within the respective bin of the mid-change counts.

market opening one hour earlier. During the 12-minute period of the crash, the yield on the 10-year U.S. Treasury note traded in an extraordinary range of 37 basis points. Market activity was indeed extremely elevated at almost 3000 mid-price updates per minute at the peak (see figure 33.3). Despite these movements, there was no trading halt on the CME, bid ask spreads remained fairly tight and prices showed very few gaps. USDT, BGFRS, FRBNY, SEC and CFTC (2015) highlights a number of structural factors relevant to the event, including the increasingly electronic nature of the U.S. treasury market and the prominent role of automated trading strategies. The report also documents that message traffic increased so dramatically just before the event that the average latency on the exchange built up from sub-milliseconds to around 60 milliseconds for a short time. Similar to the E-mini event, there was a substantial amount of hot-potato trading during the U.S. Treasury event. As such, many of the mechanisms that we suspect to result in highly endogenous price dynamics were also present during this event.

And indeed, the selected Hawkes(21,12) fit under the assumption of $\Delta = 1$ second identifies a branching ratio that peaks at ≈ 0.96 just as the mid-price also reaches its most extreme value. Assuming $\Delta \sim U(0, \Delta_i)$ recovers a Hawkes(19,12) model, which identifies a similar gradual rise of the branching ratio as a precursor to the crash, even rising up to $\eta(t) \approx 1$. Since, as for the E-mini, the goodness of fit under the assumption of $\Delta = 1$ second is superior, we will focus on that value in the rest of the paper. The endogenous dynamics we identify, irrespective of the assumption on Δ, calm down as the price recovers and then hours later in fact $\eta(t)$ rises again to near-critical levels without triggering a similarly drastic price response—the price in fact stabilizes further during this period. As we will see in other events subsequently, the branching ratio can indeed also rise in times when market dynamics recover from perturbations—a behavior that we think can be understood in the context of adaptive control dynamics as we will discuss below. The optimal Hawkes(18,12) on raw/unique timestamps ($\Delta = 0$) recovers very similar dynamics for the endogenous dynamics, with a slightly lower peak value of ≈ 0.93 for the branching ratio and some more uncertainty in model selection.

33.5.4 The US-Dollar Crash in March 2015

Within minutes on 18 March 2015, the US-Dollar depreciated around 2% against the Euro and recovered most of this loss within a very short period of time thereafter. The depreciation of the US-Dollar had already started around two hours earlier after the Federal Open Market Committee (FOMC) had released a statement that was not aligned with market expectations. Simultaneous to the depreciation, U.S. stock markets traded sharply higher. The flash crash then occurred just a few minutes after the U.S. stock markets closed, visible at 20:00 UTC in figure 33.4. The event was accompanied by highly elevated trading and quoting activity on the foreign exchange spot market. In addition, Euro foreign exchange futures contracts on the CME triggered a circuit breaker which resulted in the price being stuck at some level for several minutes,[*] leading the Euro foreign exchange futures price to being artificially held back relative to other currency futures.[†] Since foreign exchange futures and spot markets are tightly coupled through arbitrage relations, we can expect such dislocations in the futures market to also propagate to spot markets if they are persistent enough. Looking at the estimates of the selected Hawkes(23,3) model in figure 33.4, we see first that the estimator for the exogenous intensity spikes in excellent agreement with the release of the FOMC statement and the incident of the depreciation of the US-Dollar. The branching ratio however rises only weakly during the period, indicating a slightly higher endogenous excitation during the crash and subsequent recovery. The estimator recovers a peak value of $(t) \approx 0.75$

---

[*] The CME 'Special Price Fluctuation Limits', or Rule 589 that was triggered, states that, if certain lead-month primary futures contracts trade more than some pre-determined amount outside the previous day's settlement price, a 5-minute monitoring period will be triggered. During this period, trading is only allowed at or above (below) the price limit if the price is falling (rising).

[†] See e.g. http://www.nanex.net/aqck2/4689.html for a detailed overview of the event on the futures market.

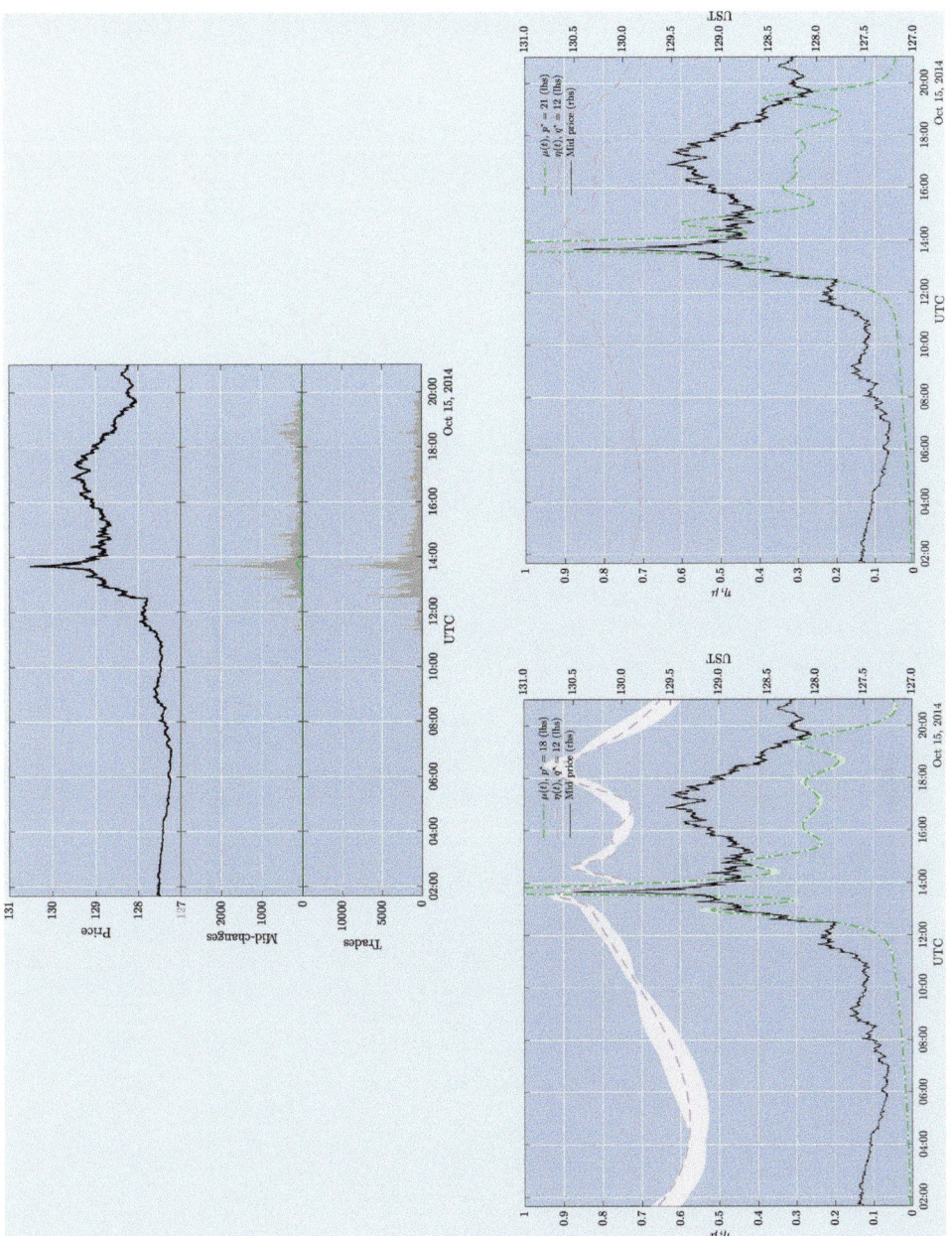

FIGURE 33.3　Same as figure 33.2 for the 10-year U.S. Treasury Note Futures contract event on 15 October 2014. The bottom panels show the estimated immigration intensities $\mu(t)$, in sec^{-1} and branching ratio $\eta(t)$, in green and purple, for fits using unique/raw timestamps ($\Delta = 0$; left) and using timestamps that are randomized assuming $\Delta = 1$ second (right), respectively.

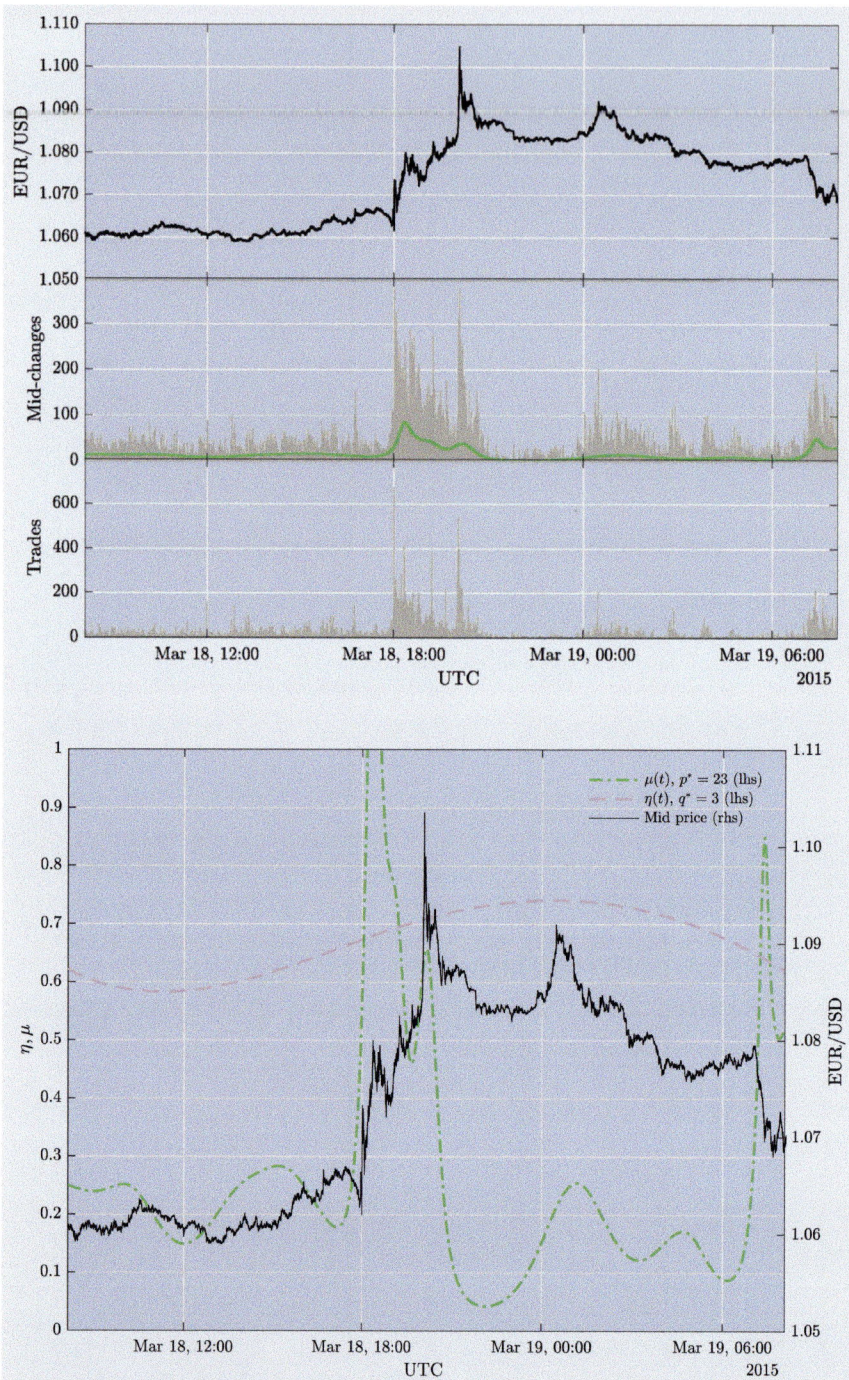

FIGURE 33.4 Same as figure 33.2 for the EUR/USD event on 18 March 2015. The bottom panel shows estimates based on randomized timestamps ($\Delta = 100$ milliseconds).

at that time, rising from around 0.6, but remaining far from criticality. We find no difference in the dynamics when fitting on raw timestamps.

33.5.5 The New Zealand Dollar Crash in August 2015

On 25 August 2015, the NZD/USD cross rate dropped almost 8% over the course of just a couple of minutes. Within half an hour, the move was largely reverted. During the episode, both quoting and trading activity increased significantly (see figure 33.5). Similar price dynamics could also be observed in the NZD/JPY pair. Several reports suggested that a sell-off in the US equity market had triggered a wave of risk reduction and closing of carry trades, the following increase in correlation led to dealers failing to hedge cross-asset exposures and consequently withdrawing from two-sided market making, which resulted in the excessive price reaction in the NZD (Callaghan 2017, Bank for International Settlements 2017a, 2017b). The description of this event—having its source even in another asset altogether—already suggests that there were a lot of mechanisms at play that are exogenous to the price dynamics of the NZD/USD currency pair at that time. As such, it comes as little surprise that we find no changes in the branching ratio over the event (see figure 33.5). The optimal Hawkes(8,1) model on randomized timestamps has constant $\eta(t) \approx 0.6$ and attests the spike in the mid-price intensity around the event to a rise in exogenous activity. Also when fitting the dynamics on the raw timestamps, the resulting optimal model is a Hawkes(8,1) process with constant endogenous dynamics of identical magnitude. We stress that our calibration finding $q^* = 1$ is encouraging evidence for the fact that—despite the significant flexibility allowed by the Hawkes(p,q) model—our approach does not bias towards always identifying some changes in the instantaneous branching ratio. The observation that a constant branching ratio is supported by the narratives of the event and is also recovered by the fits, validates our approach nicely.

33.5.6 The South African Rand Crash in January 2016

The South African Rand—the currency with the 18th highest turnover in OTC foreign exchange according to Bank for International, Settlements (2019) and thus one of the most liquid emerging markets currencies—depreciated around 8% within just 10 minutes on 10 January 2016. Bank for International Settlements (2017a) finds that the move was triggered by retail investors unwinding carry trades, which was subsequently exacerbated by hedging of barrier options and the execution of stop-loss orders in illiquid Asian trading hours. The price dynamics during this event are reminiscent of typical dynamics in complex systems, where nonequilibrium abrupt transitions, due to exogenous shocks, are generally followed by power law relaxations (Sornette and Helmstetter 2003, Sornette *et al.* 2004a, Crane and Sornette 2008, Crane *et al.* 2010)—in particular also observed in financial time series (Lillo and Mantegna 2003, Sornette *et al.* 2004b, Weber *et al.* 2007, Petersen *et al.* 2010). The resulting Hawkes(10,4) fits* corroborate this story insofar as that the initial price drop is accompanied by a spike in the exogenous intensity, while the branching ratio only later starts to rise (see figure 33.6). Interestingly enough, during the subsequent relaxation/stabilization of the price move, the estimated branching ratio peaks just below criticality ($\eta(t) = 1$ is in fact within the credible interval of [0.93, 1.01] at the peak) at a time when price dynamics finally seem to have calmed down—similar to what we recovered around the event in the U.S. Treasury futures market. The fact that, for these events, a rise in the endogenous dynamics towards criticality seems to have acted as a stabilizing mechanism, is reminiscent of a concept from the study of adaptive control systems, where locally adaptive stabilization is found to have the potential to move a system to critical points. In Patzelt and Pawelzik (2011, 2013) for example, it is shown that rapidly adaptive stabilization creates attractive critical points under quite general assumptions about the system dynamics. As such, we could interpret the observations around the event in the South African Rand and

---

* On raw timestamps, the resulting Hawkes(9,4) fit shows almost identical dynamics, however the critical branching ratio is not within the resulting credible interval of [0.92,0.94].

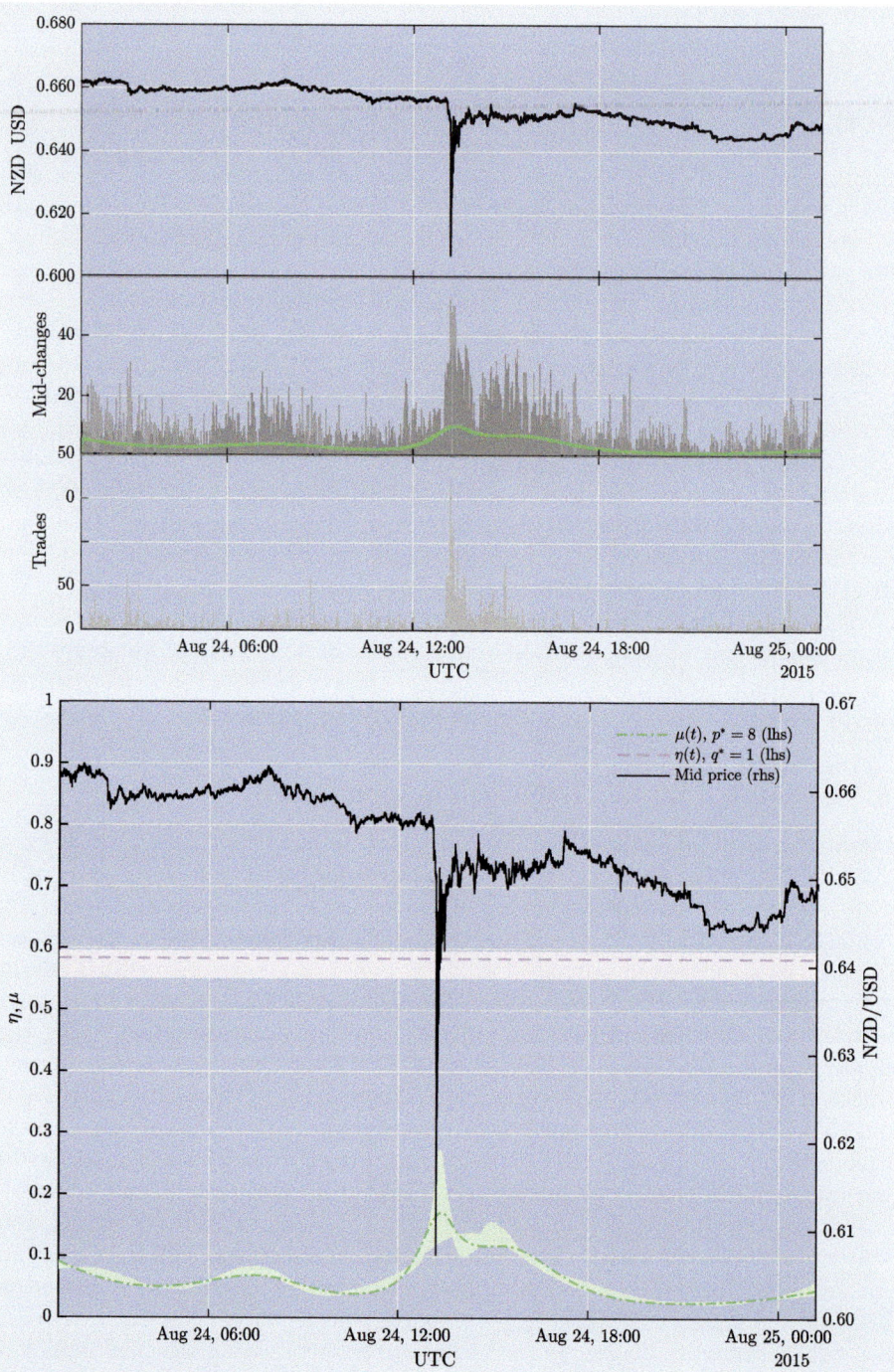

FIGURE 33.5 Same as figure 33.2 for the NZD/USD event on 24 August 2015. The bottom panel shows estimates based on randomized timestamps $(\Delta = 1$ millisecond$)$.

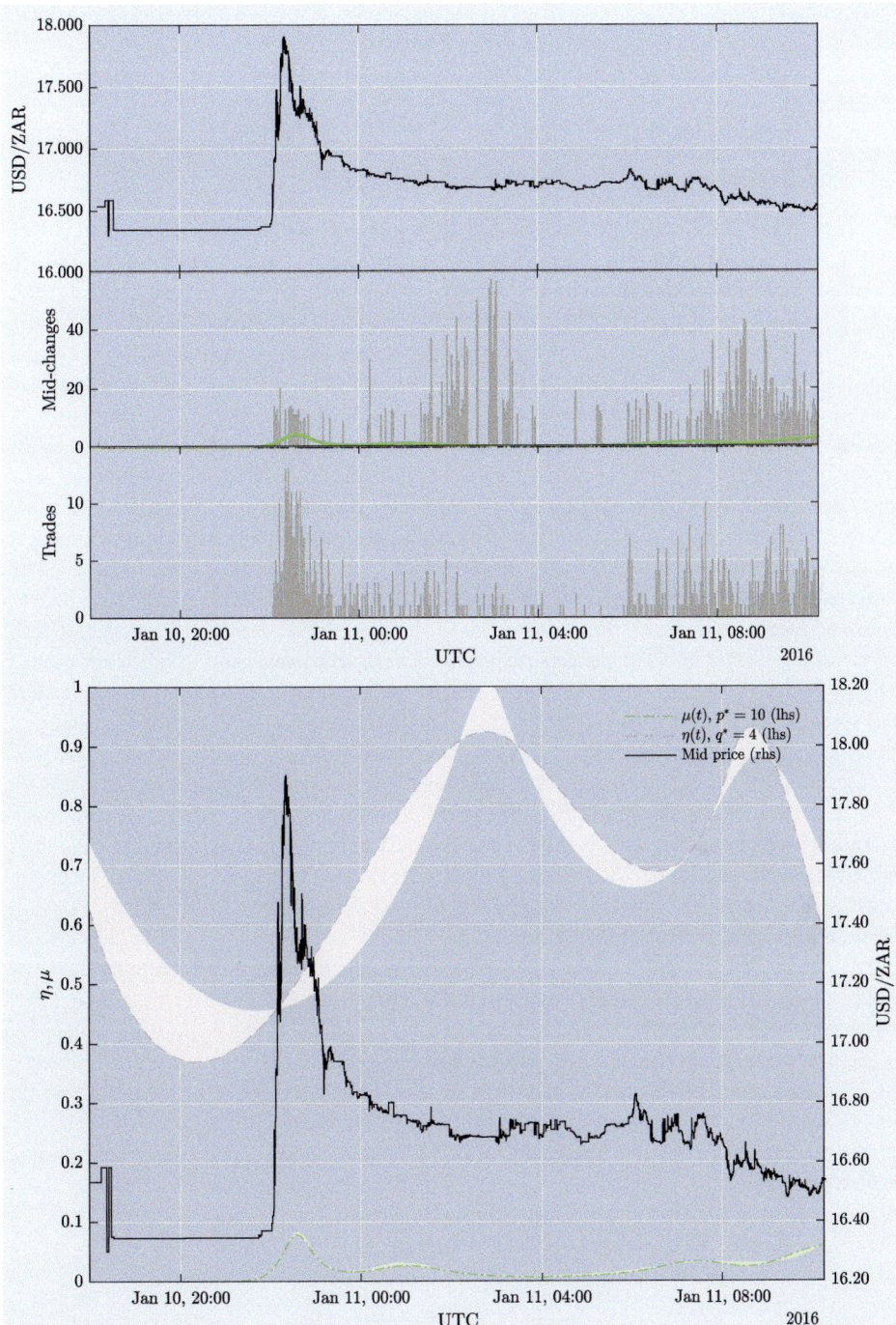

FIGURE 33.6   Same as figure 33.2 for the USD/ZAR event on 11 January 2016. The bottom panel shows estimates based on randomized timestamps (Δ = 1 millisecond).

the U.S. Treasury futures as market participants collectively adapting to and removing predictabilities (e.g. local trends) from the price process. In the process of doing so, the market indeed self-organized temporarily towards a locally critical point, which we supposedly recover with our dynamic branching ratio estimator.

33.5.7 The British Pound Crash in October 2016

Within less than a minute, on 7 October 2016, the British Pound depreciated against the US-Dollar by almost 10% and reverted most of the drop within the subsequent 10 minutes. The event remains one of the best covered ones in foreign exchange markets, with official institutions providing a meticulous account of the incident (Noss *et al.* 2016, Bank for International Settlements 2017a). The reports find that the excessive movements in one of the most liquid currency pairs in the world have resulted from a confluence of factors. First they mention the execution of a large sell order during typically illiquid trading hours, which the market apparently still absorbed in a relatively orderly manner. Several amplifying factors are then found to have contributed to a severe deterioration in liquidity and subsequent price drop. In particular, significant demand to sell Pound Sterling to hedge options positions, as the currency depreciated, appears to have played an important amplifying role. The execution of stoploss orders and the closing-out of positions as the currency traded through key levels may also have had an impact. The reports also note the potentially amplifying role of circuit breakers being triggered in GBP/USD futures contracts on the CME,* which may have created larger price pressure on other trading platforms and increased the price impact of trades in the spot market. Studies of transaction level data around the event further find that dealers (primarily investment banks) were significant contributors to the liquidity dry-up during the crash. They reduced their market-making activity significantly and widened spreads, while other financial firms (such as high frequency traders and hedge funds) stepped in to provide some liquidity in their place (Schroeder *et al.* 2020).

The description of the event suggests that many of the previously discussed mechanisms that are suspected to amplify endogenous price dynamics were present during the crash—from stop-loss execution to trading halts in related markets and increased HFT participation, almost all of the ingredients for high endogeneity were apparently present. And indeed, if we look at the estimates of the optimal Hawkes(11,8) model around the event in figure 33.7, we find that the branching ratio rises significantly from around 0.5 to peak values of ≈ 0.98 during the price drop, reaching near-critical levels and then relaxing to average levels of 0.6 once the price dynamics stabilize. Interestingly enough, the branching ratio seems to start to increase already well before the actual drop. We suspect this leading effect to at least partly also be influenced by the fact that, at the overnight time of the day during which the event took place, trading is predominantly implemented by automated systems, typically exhibiting reflexive activity.† The large sell transaction may then have acted as a trigger, which through mechanistic amplifiers like stop-loss orders or trading halts evolved to the observed price drop—all of which can be expected to manifest in an elevated branching ratio. Fitting on raw timestamps yields similar dynamics represented by a Hawkes(11,6) model, albeit with a slightly lower point estimate for the peak fraction of endogenous activity of 0.91, but wider credible intervals, which in fact contain 0.98.

---

* Trading on the CME was halted multiple times during the event. First, for 10 seconds due to a so-called 'velocity-logic' event, then for several minutes due to reaching its daily lower price limit and later again due to a velocity-logic event (see Bank for International Settlements 2017a for more details).

† We have indeed found evidence for such a diurnal overnight rise in the branching ratio also in other currency pairs, see e.g. Wehrli *et al.* (2020).

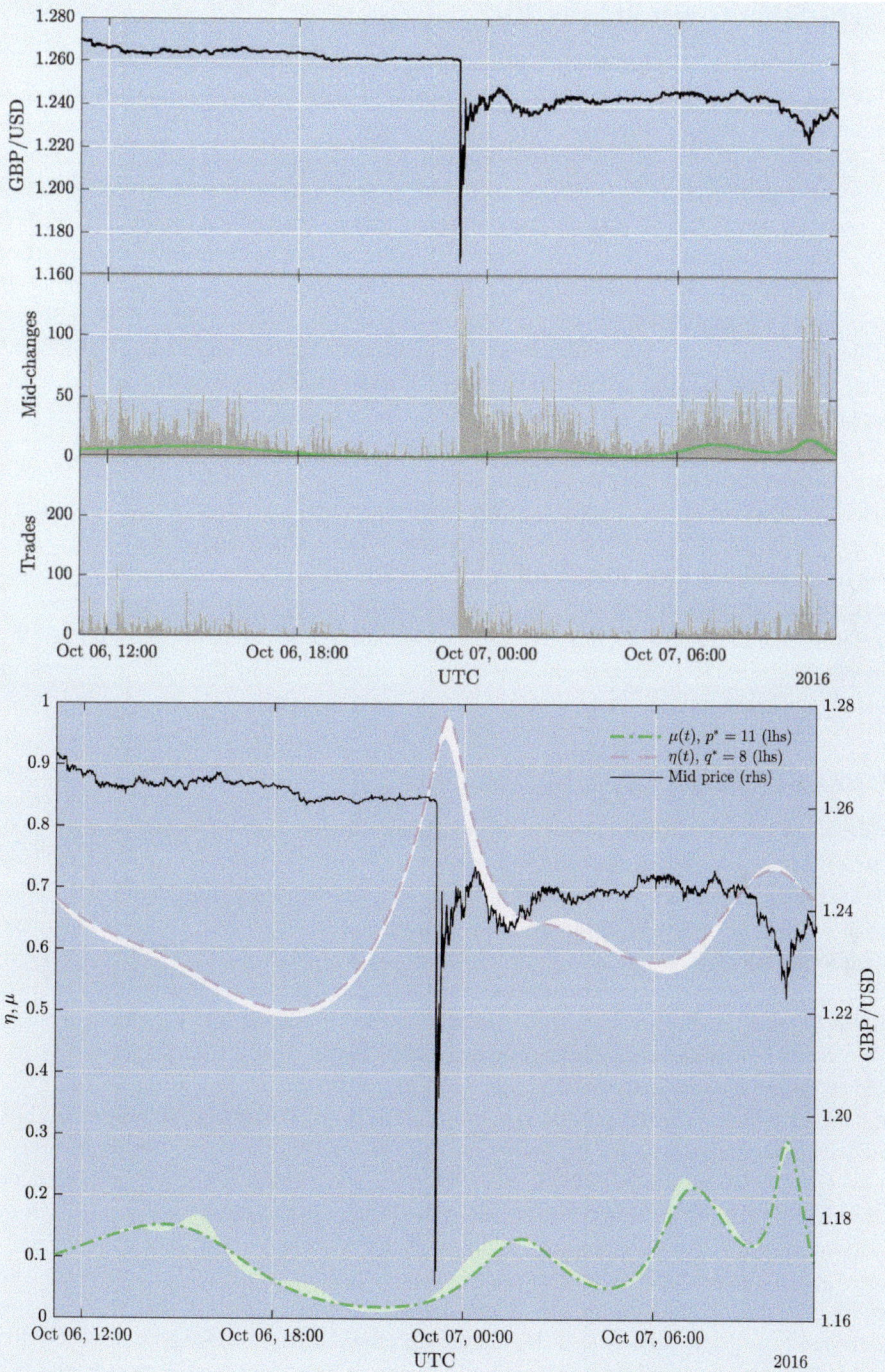

FIGURE 33.7 Same as figure 33.2 for the GBP/USD event on 7 October 2016. The bottom panel shows estimates based on randomized timestamps ($\Delta = 1$ millisecond).

33.5.8 The Ether Crash in June 2017

At 19:30 UTC on 21 June 2017 a multimillion dollar market sell order was placed on the GDAX ETH/USD order book. The extraordinary trading activity resulted in a price drop of around 30%, see figure 33.8. The series of market orders occurred as a single atomic operation on the matching engine, completing in less than 45 milliseconds.[*] As a result, hundreds of stop loss orders and margin calls were automatically executed by the matching engine of GDAX, causing Ether to momentarily trade as low as 0.10 USD. As a precautionary measure, GDAX disabled new margin funding and trading on the ETH/USD book at 20:10 UTC. Trading resumed at 23:10 UTC. Even before the trading halt was implemented, the GDAX webcast experienced technical issues and latency due to an extreme surge in customer activity. To avoid data inconsistencies resulting from these problems, we cut off our sample at 19:53 UTC, just before the data feed became unavailable. On randomized timestamps, the optimal model is a Hawkes(19,5)[†] and reveals that the branching ratio is constantly above 0.8 and visits the critical regime twice before the occurrence of the crash, whereas it is only around 0.85 at the actual time of the price drop. To gauge how the overall level of self-excitation on the day of the crash compares to 'typical' dynamics of cryptocurrencies, we can look at the recent study of Mark *et al.* (2020), done for the Bitcoin market. They estimate a standard Hawkes on mid-price changes that were induced by market orders. At one hour sample sizes, they find $\eta \in [0.7, 0.9]$, slightly lower, but compatible with our results for Ether, which yield an average branching ratio of 0.9. The exogenous intensity in our estimates furthermore also only spikes after the price drop, which suggests that the extreme price move in fact most likely can simply be traced back to the market impact of the excessively large sell transaction. This is also corroborated by the fact that trade intensity spikes significantly, whereas quoting intensity as measured by the mid-price jump intensity remains fairly stable around the crash.

33.5.9 The EUR/USD Crash in December 2017

As an example of the fact that high levels of endogeneity in the price process can also be apparent around irregular price moves in situations when overall activity is extremely subdued, we look at the EUR/USD exchange rate on 25 December 2017. As can be seen from figure 33.9, the EUR/USD exchange rate gapped lower by around 3% instantaneously and recovered hours later to unchanged levels with very little quoting and trading activity. Even though many markets are closed during Christmas holidays, the foreign exchange market remains open for trading and prices are in fact streamed even when most staff at financial institutions are not present. Clearly, we can expect that this will lead to a lot of the live pricing decisions being made in an automated fashion—and with appropriately restrictive automated safeguards—similar to more illiquid times of the day during the European overnight times. Following this rationale, it comes as little surprise that the optimal Hawkes(3,1) model exhibits a constant and even *supercritical* level for the branching ratio at ≈ 1.5. On raw timestamps, the dynamics of the optimal Hawkes(5,1) model are comparable, with $\eta \approx 1.3$. We however also see for both fits that the credible intervals for the branching ratio, as well as the immigration are very wide—apart from model uncertainty most likely also reflecting the highly irregular distribution of inter-arrival times in this sample, which can also be expected to bias estimates of η upward[‡], especially if heavy-tailed offspring densities, like the approximate power law we rely on here, are used. Overall, the activity patterns and estimates around the event suggest that the supposed flash crash in this instance was more likely just some market makers stopping from quoting on EBS for some time or widening spreads significantly in subdued holiday trading, to the point where the mid-price dislocated downward.

---

[*] https://medium.com/@AdamLWhite/eth-usd-trading-update-3- 69e623335e00.

[†] On raw timestamps, the optimal Hawkes(16,4) produces almost identical dynamics for the branching ratio.

[‡] For a discussion regarding the effect of outliers on branching ratio estimates, see Filimonov and Sornette (2015b).

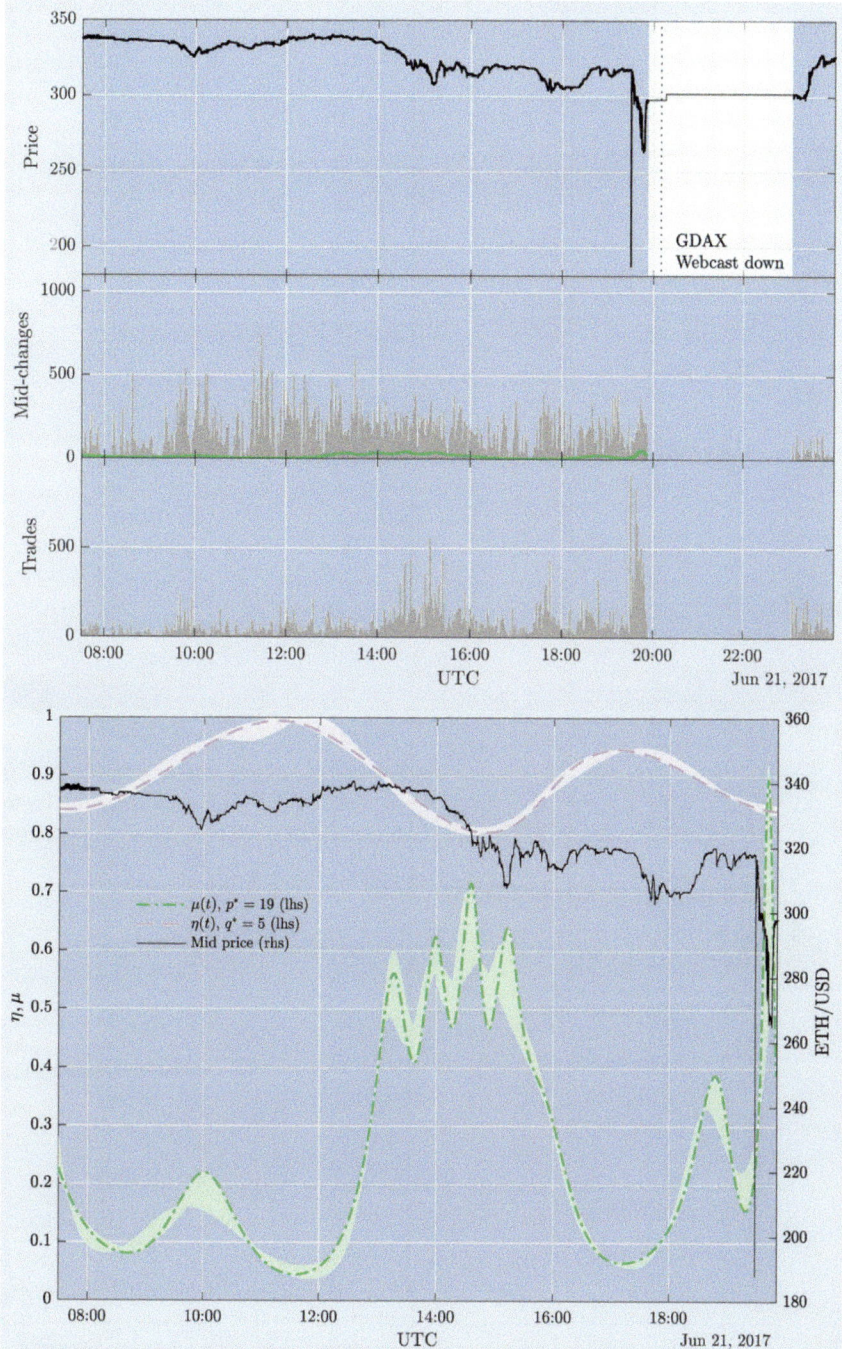

FIGURE 33.8 Same as figure 33.2 for the ETH/USD event on 21 June 2017. In the top panel, the time during which the GDAX data feed was down is shaded in gray, and the vertical dotted line indicates the start of the trading halt (see main text for details). The bottom panel shows estimates based on randomized timestamps ($\Delta = 1$ millisecond).

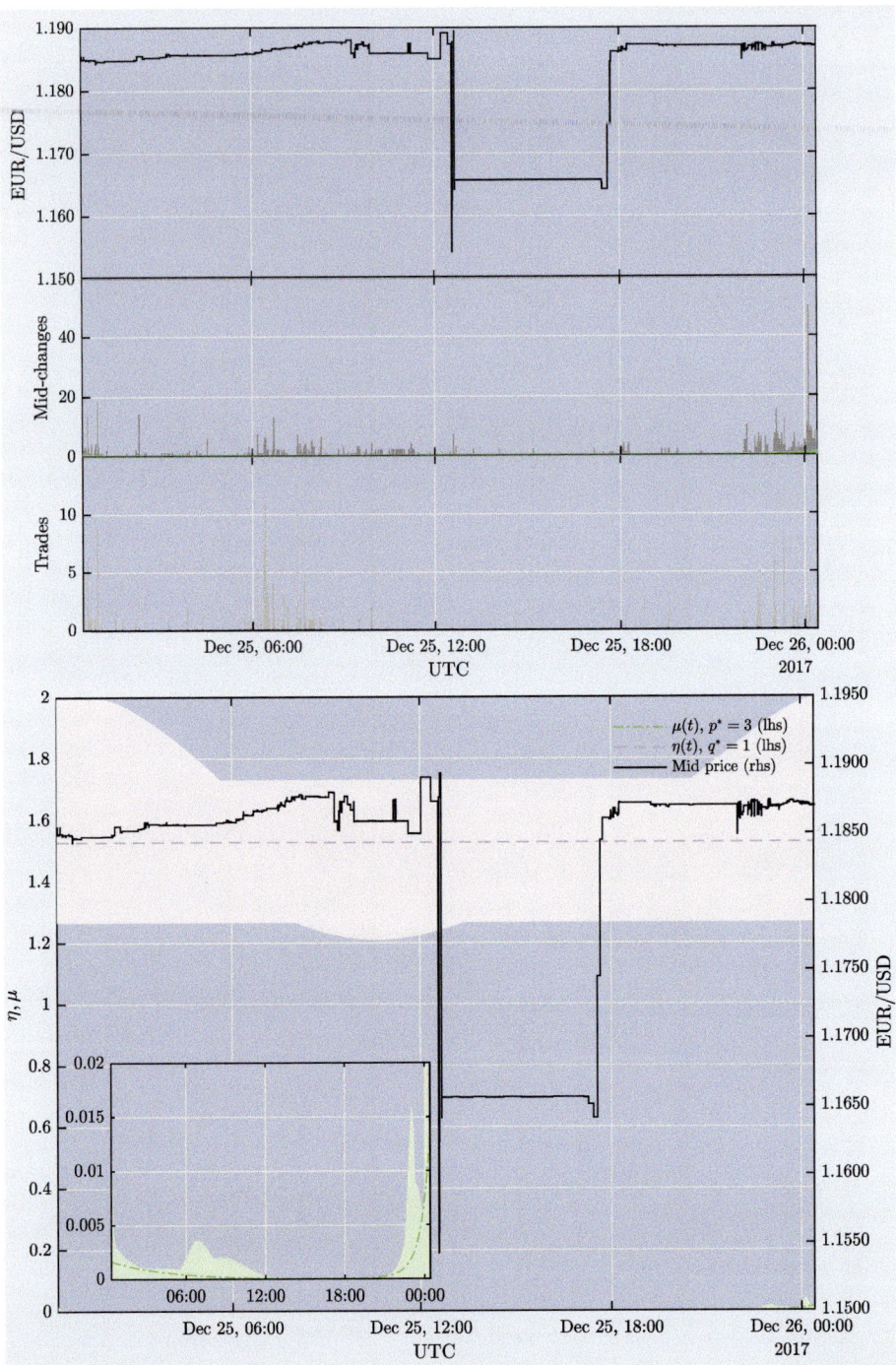

FIGURE 33.9 Same as figure 33.2 for the EUR/USD event on 25 December 2017. The bottom panel shows estimates based on randomized timestamps (Δ = 100 milliseconds).

33.5.10 The USD/JPY Crash in January 2019

One of the most recent events in foreign exchange concerns the Japanese Yen on 2 January 2019. In just 30 seconds, the Yen appreciated by 3% against the US-Dollar, a comprehensive description of the event can be found in Reserve Bank of Australia (2019). Three key factors are found to have contributed to the deterioration in market conditions. First, there were significant liquidations of carry positions by Japanese retail investors (see also Han and Westelius 2019). Because of daily system halts at retail aggregators, clients were not able to post additional margin or close positions to avoid triggering automatic stop-losses. Failure to post additional margin resulted in an automatic liquidation by brokers of the positions. Second, liquidity was seasonally reduced at the time, due to the time of year and day, but also since Japanese banks were additionally closed for a public holiday. Third, the safety mechanisms of automated market making systems may have acted as amplifiers during the event as they automatically switched off when market conditions turned unusual. Further exacerbating all these facts might be that CME futures markets were closed at the time, making it more difficult to hedge positions, particularly if they were one-sided.

Looking at our model estimates for the event in figure 33.10, we can see that the branching ratio indeed rises leading up to and during the crash, in line with the endogeneity-inducing factors like stop-loss executions reported to play a role in the crash. However, even at the peak, the branching ratio from the optimal* Hawkes(25,4) model stays well away from criticality, with a maximum of ≈ 0.75. The exogenous intensity also spikes during the event, but only after the trough of the crash, during the initial rebound of the price.

33.6 Residual Analysis and Refined Treatment of Exogenous Clustering

In order to address the basic adequacy of our Hawkes(p,q) model to describe the univariate mid-price microstructure around flash crashes, we perform residual analysis and a comparison with alternative models.

33.6.1 Residual Analysis

The standard way to assess the quality of point process fits also applies to the Hawkes(p,q) model and is based on the standardized residual process $\xi_i = \int_0^{t_i} \lambda(t \mid \hat{\theta})dt$ with corresponding increments $r_i = \xi_i - \xi_{i-1}$. Under the null hypothesis that the t_i were generated by the Hawkes(p,q) process, the residual process ξ_i should be a Poisson process with unit intensity (Cinlar 2013). Given the fact that we attempt to fit days with particularly extreme realizations of non-stationary price behaviors, with tens of thousands of points, and with a randomization procedure to correct for limited timestamp resolution and coinciding points, it is not reasonable to assume that our still fairly simple model (33.1) will be able to fully explain a complete day of complex and diverse trading activity in a period around a flash crash. As such, it is unsurprising that, e.g. a Kolmogorov-Smirnov test confidently rejects the null hypothesis of unit exponentiality of r, except for the event in USD/ZAR. The deviations from the null hypothesis of the unit exponential distribution hereby occur predominantly for very short durations. For our purpose of explaining dynamics in exogenous and endogenous activity, the precise distribution of the residuals is however not essential. Rather, we are interested in (i) trends and (ii) autocorrelation being described

---

* The optimal model on raw timestamps has (20,8) degrees of freedom and a peak branching ratio of 0.78 at 00:00 UTC. Since $p^* = 25$ on pre-processed timestamps is at the border of the explored range, we have additionally extended the model search up to $p = 35$. Using the extended search, the optimal p^* is 27 and q^* remains unchanged at 4. Also the estimates of the branching ratio around the flash crash remain identical, only the model uncertainty is reduced.

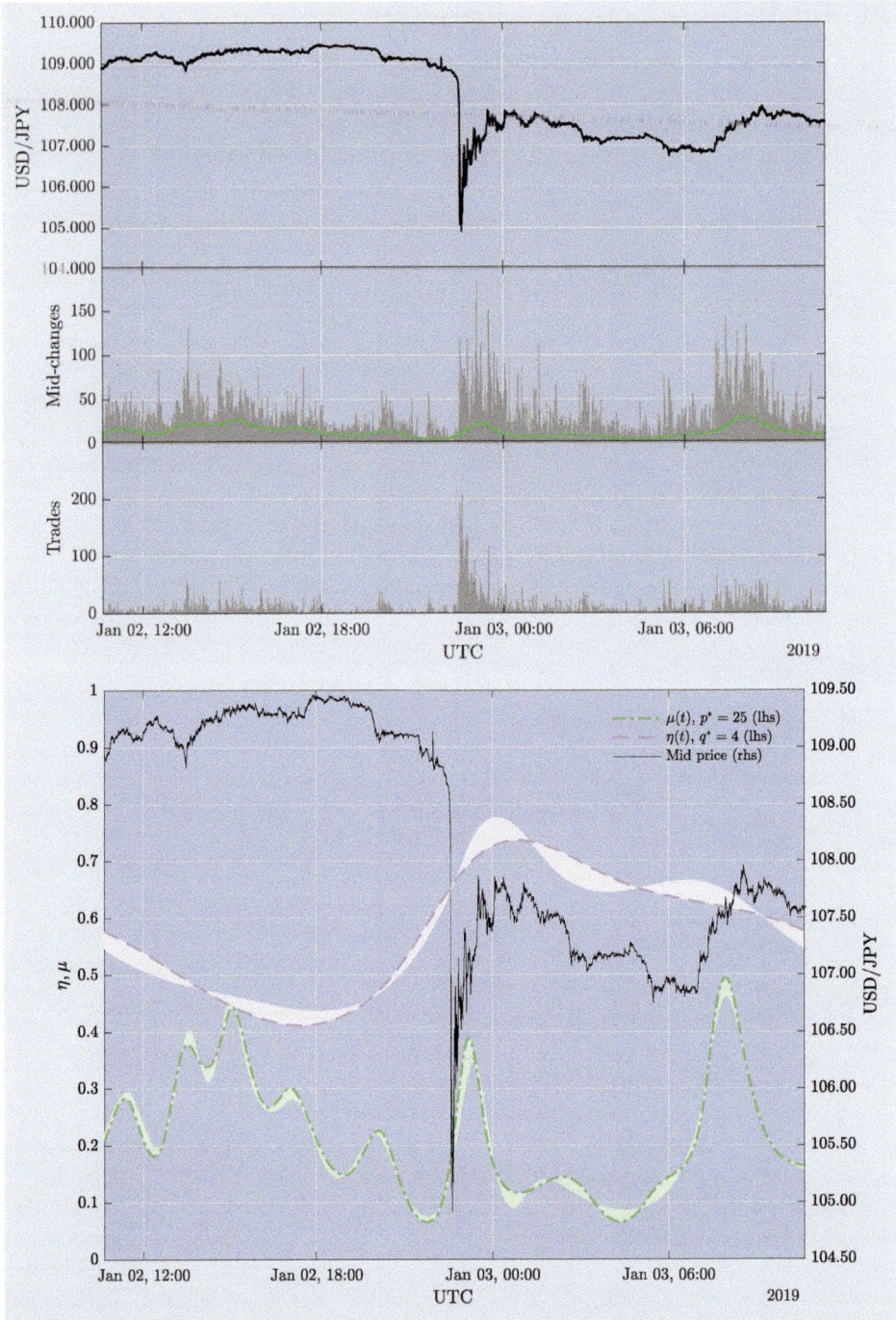

FIGURE 33.10 Same as figure 33.2 for the USD/JPY event on 2 January 2019. The bottom panel shows estimates based on randomized timestamps ($\Delta = 100$ milliseconds).

sufficiently well by our model. For (i) we can check the uniformity of the ξ_i on $(0, N_T]$, e.g. by plotting a histogram. (ii) can be evaluated with standard tests for significant autocorrelations, here we choose to complement a visual check of the ACF with a Ljung-Box test (Ljung and Box 1978).

Looking at the summary of the residual checks in figures 33.11 ($\Delta = 0$) and 33.12 (randomized time-stamps), we can see that the Hawkes(p,q) model does a good job in achieving approximate uniformity of the residuals. Thus, trends are successfully removed by our flexible treatment of $\mu(t)$. Regarding remaining significant autocorrelations, we find that on raw/unique timestamps ($\Delta = 0$) the model indeed captures a wide variance of the event rate and in general very little residual correlation remains. But, there are some significant autocorrelations present at short lags, leading to confident rejection of uncorrelated residuals r and squared residuals r^2 for the majority of events—even when the broadly present, first lag negative autocorrelation is excluded from the Ljung-Box test. This first-order feature of the residual ACF is indeed particularly noticeable and warrants a more detailed investigation below. On random-ized timestamps, we notice a significant amount of autocorrelation remaining for the E-mini and U.S. Treasury data sets, where duplicates were present before randomization. While the magnitudes of the remaining autocorrelations are rather small, the goodness of fit for the two futures data sets, even for the best fitting assumption of $\Delta = 1$ second presented here, is clearly inferior to the ones on the foreign exchange and cryptocurrency data—where after randomization the goodness of fit is comparable to when using raw timestamps. These results clearly highlight the importance of high-quality timestamps in the application of financial point processes. In fact, we find that, when duplicates are present, the assumptions on Δ have a significant effect on the goodness of fit as measured by the residual autocor-relation. In line with this result, the choice of Δ also has a clear effect on the estimates of the offspring density parameters (see table 33.3). In particular, we observe that both the mode and tail index of the approximate power law tend to increase with larger randomization windows, the triggering strength according to $\langle \eta \rangle$ is however largely comparable. We note that some of these issues with data quality could be remedied by working with aggregated point processes in the form of integer-valued time series models, where the spurious microstructure induced by timestamp inaccuracies are naturally aggregated out. The integer-valued autoregressive (INAR) process representation of the Hawkes process for exam-ple and its estimation using EM algorithms, as developed in Wehrli *et al.* (2020), can be extended to time-varying branching ratios like we use here in a straightforward manner. This however goes beyond the scope of this paper and is left for future work (Figure 33.12).

33.6.2 Comparison to Alternative Models

Here, we return to the broadly present, negative first-order autocorrelation of the r_i uncovered in the residual analysis above and take it as a motivation to compare the Hawkes(p,q) to alternative models. A first conjecture could be that this observed residual behavior is spuriously introduced by the random-ization procedure. However, fits on raw timestamps also exhibit comparable residual ACF and we thus must assume that it is a genuine feature of the data set, which the process (33.1) is not able to capture. As an alternative hypothesis, we could think that the residual first-lag dependence is an indication that the Hawkes(p,q) process—or more particularly our estimator for $\mu(t)$—does not capture a moving- aver-age type feature of the data, which in the context of point processes corresponds to a shot-noise type of exogenous clustering. To support this intuition, we simulate from a shot-noise Cox process (Møller 2003) with conditional intensity

$$\lambda^*(t) = \mu + \int_{-\infty}^{t} \Theta(t-s)\mathrm{d}N_s^{\mu}, \tag{33.4}$$

which is the intensity of a process N^*, driven by a homogenous Poisson process N^{μ} with intensity $\mu > 0$. The exogenous clustering in this process is described by the kernel $\Theta:[0,\infty) \rightarrow [0,\infty)$ with $\gamma = \int_0^{\infty} \Theta(t)\mathrm{d}t \in [0,\infty)$,

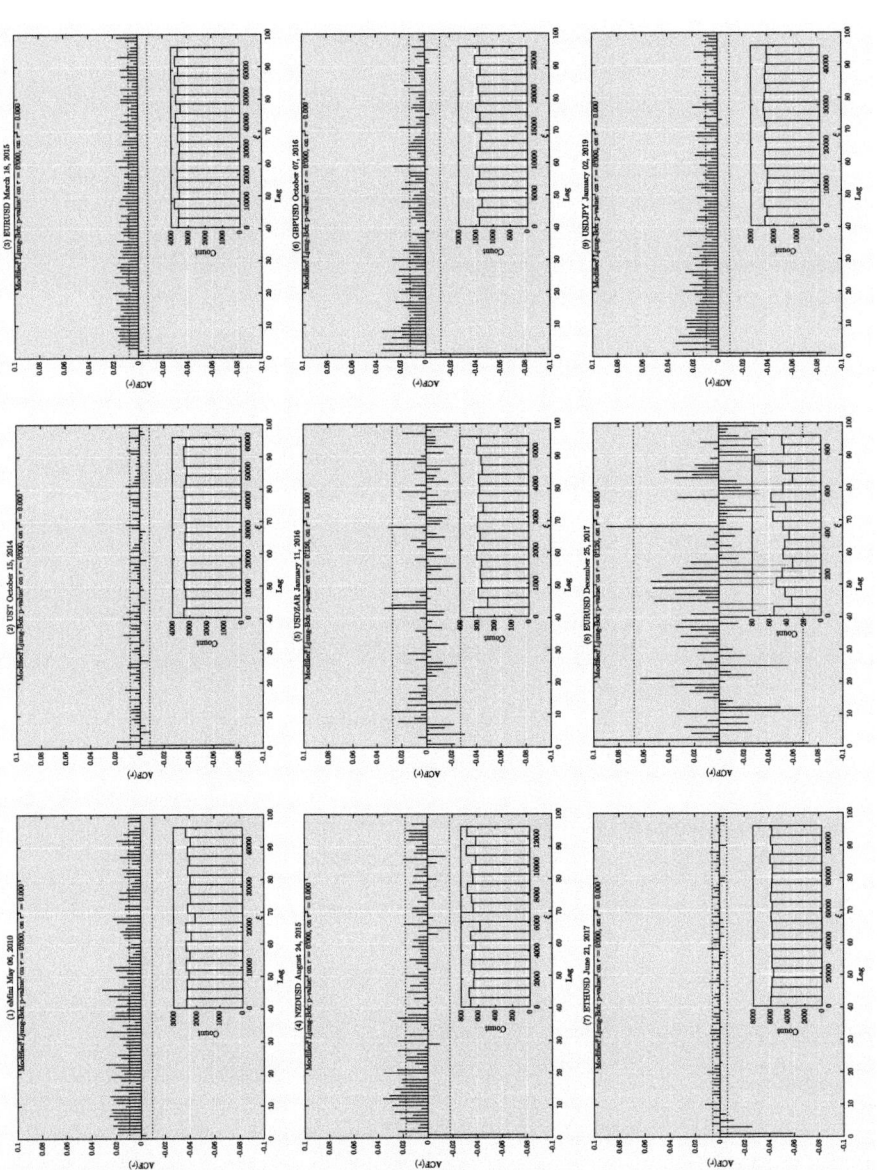

FIGURE 33.11 Goodess of fits on raw/unique timestamps. Shows the sample autocorrelation function $\check{\rho}_k$ of the residuals $r_i = \xi_i - \xi_{i-1}$ from Hawkes(p,q) fits on timestamps without randomization ($\Delta = 0$). The horizontal dashed lines are 95% approximate confidence intervals for a white noise process. The ξ_i are obtained from integrating the conditional intensity of the optimal fit up to t_i. Each plot includes the p-values from a modified Ljung-Box test, where the first-order autocorrelation is excluded from the test, i.e. the test statistic is $S = N_T(N_T + 2)\sum_{k=2}^{h+1} \dfrac{\widehat{\rho}_k^2}{N_T - k} \sim \chi^2(h)$. In the insets, we additionally plot histograms of the standardized residuals ξ on $(0, N_T]$.

TABLE 33.3　Summary of the Approximate Power Law Offspring Density Parameter Estimates (The Mode τ_0 in Seconds and the Tail Index ε) from the Optimal Hawkes(p,q) fit for Each Event and Different Randomization Windows Δ

	Event	$\Delta = 0$		$\Delta = 1\text{ms}$		$\Delta = 100\text{ms}$		$\Delta = 1\text{sec}$		$U(0, \Delta_v)$	
		τ_0	ε	τ_0	ε	τ_0	ε	τ_0	ε	τ_0	ε
(1)	E-mini	0.014	0.49	10^{-4}	0.28	0.022	0.98	0.185	1.29		
(2)	U.S. Treas.	0.002	0.38					0.127	1.45	0.020	0.59
(3)	EUR/USD	0.280	0.65			0.282	0.65				
(4)	NZD/USD	3.079	0.71	3.077	0.71						
(5)	USD/ZAR	3.047	0.71	2.996	0.70						
(6)	GBP/USD	0.652	0.51	0.650	0.52						
(7)	ETH/USD	0.145	0.53	0.145	0.53						
(8)	EUR/USD	0.142	0.06			0.122	0.03				
(9)	USD/JPY	0.227	0.55			0.219	0.54				

so that $f(t)=\Theta(t)/\gamma$ is a density with primitive $F(t) = \int_0^t f(s)\mathrm{d}s$. The resulting ACF of the inter-arrival times is given in figure 33.13 for the particular case of an exponential density.

The negative first-order dependence we are looking for is apparent, suggesting that an extension of the Hawkes(p,q) model in the spirit of the autoregressive moving-average (ARMA) point process introduced in Wheatley et al. (2018) to

$$\lambda(t) = \mu(t) + \int_{-\infty}^{t} \Theta(t-s)\mathrm{d}N_s^{\mu} + \int_{-\infty}^{t} \eta(s)h(t-s)\mathrm{d}N_s \tag{33.5}$$

might provide a more comprehensive treatment of exogenous clustering than simply relying on flexible $\mu(t)$ to capture such bursts. In addition, this allows us to test whether the point clusters in the data are better explained by Hawkes type (endogenous) or shot-noise Cox type* (exogenous) clustering, which has particular relevance when assessing a hypothesis of criticality.

Because the process with conditional intensity (33.5) is an extension of the ARMA point process, it can be estimated using a Monte Carlo Expectation Maximization (MCEM) algorithm (Wei and Tanner 1990), where Markov Chain Monte Carlo (MCMC) sampling through a Metropolis- Hastings algorithm is used to perform conditional simulation of the immigrant process in the E-step (Wheatley *et al.* 2018). A description of the algorithm to estimate (33.5) is given in Appendix 3.

A significant practical difficulty of the process (33.5) is that, because the immigrants N^{μ} are unobserved, both residuals and unconditional likelihood are inaccessible for testing and model selection. We solve the latter here by assuming that the optimal number of degrees of freedom from the Hawkes(p,q) fits are also optimal for the extended 'MA-Hawkes(p,q)' model (33.5) and fix them for estimation. Since we can expect that the inclusion of additional structure for exogenous clustering through $\Theta(t)$ will—if anything—reduce the required flexibility in $\mu(t)$, it seems reasonable to assume that the optimal extended model would require $p \leq p^*, q \leq q^*$. As making the models for μ too flexible usually does not introduce significant bias in the estimated average levels of endogenous excitation (see also Wheatley *et al.* 2019 and Wheatley *et al.* 2018), we propose that this approach can be used to circumvent the inaccessibility of the unconditional likelihood for the present purpose. To assess the significance of the improvement from including $\Theta(t)$, we estimate likelihood ratios using MCMC as discussed in Møller and Waagepetersen

---

* Note that the particular specification of the dynamics for the shot- noise contributions $\Theta(\cdot)$—having unit marks—as in (5) specializes the shot-noise Cox process to a modification of the well-known Neyman-Scott process (Neyman and Scott **1958**).

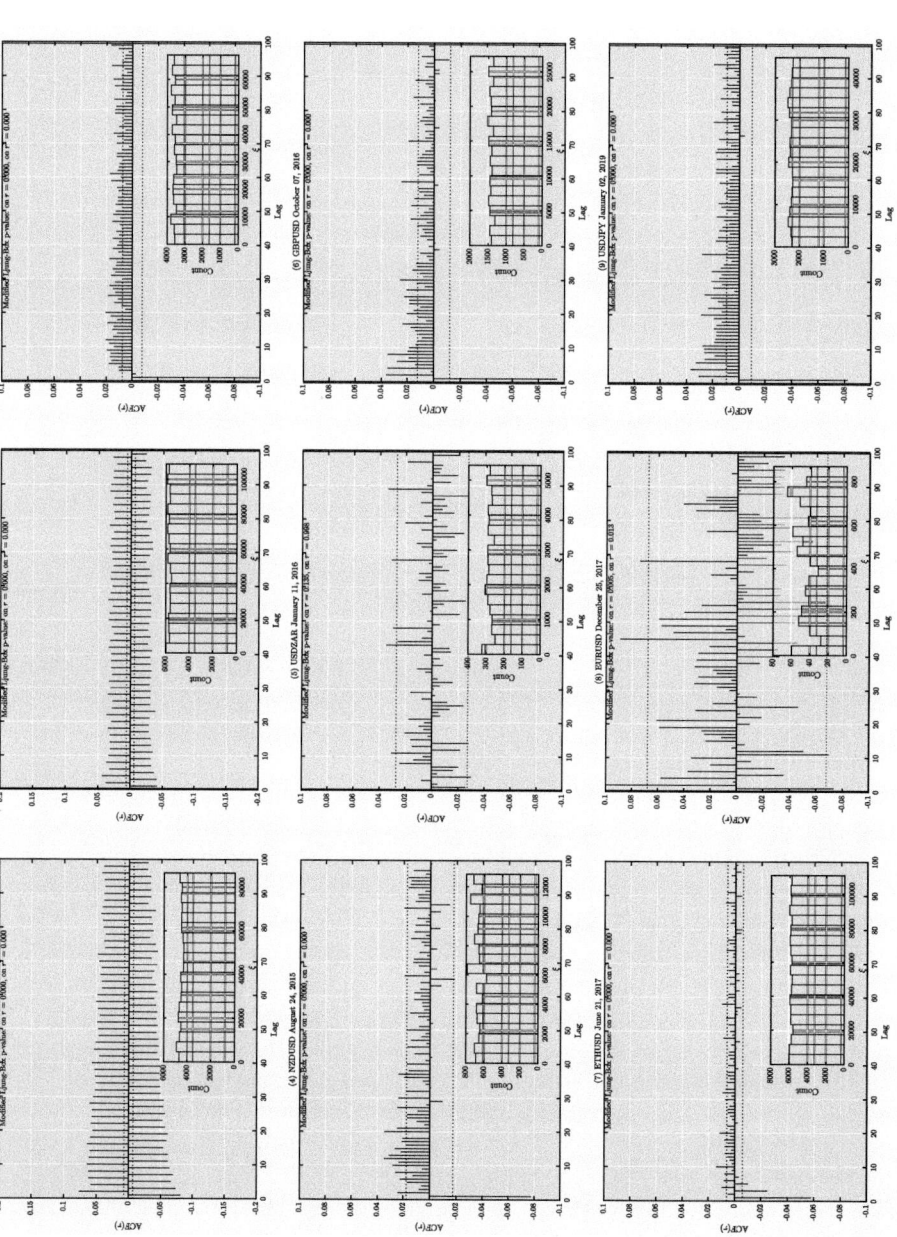

FIGURE 33.12 Goodness of fit on randomized timestamps. Shows the sample autocorrelation function $\check{\rho}_k$ of the residuals $r_i = \xi_i - \xi_{i-1}$ from Hawkes(p,q) fits using the best fitting assumption on the timestamp uncertainty Δ. The horizontal dashed lines are 95% approximate confidence intervals for a white noise process. The ξ_i are obtained from integrating the conditional intensity of the optimal fit up to t_i. For the E-mini data and the U.S. Treasury data, the fits were performed using $\Delta = 1$ second, the remaining fits are on randomized exchange timestamps as described in the main text. Each plot includes the p-values from a modified Ljung-Box test, where the first-order autocorrelation is excluded from the test, i.e. the test statistic is $S = N_T (N_T + 2) \sum_{k=2}^{h+1} \dfrac{\hat{\rho}_k^2}{N_T - k} \sim \chi^2(h)$. In the insets, we additionally plot histograms of the standardized residuals ξ on $(0, N_T)$.

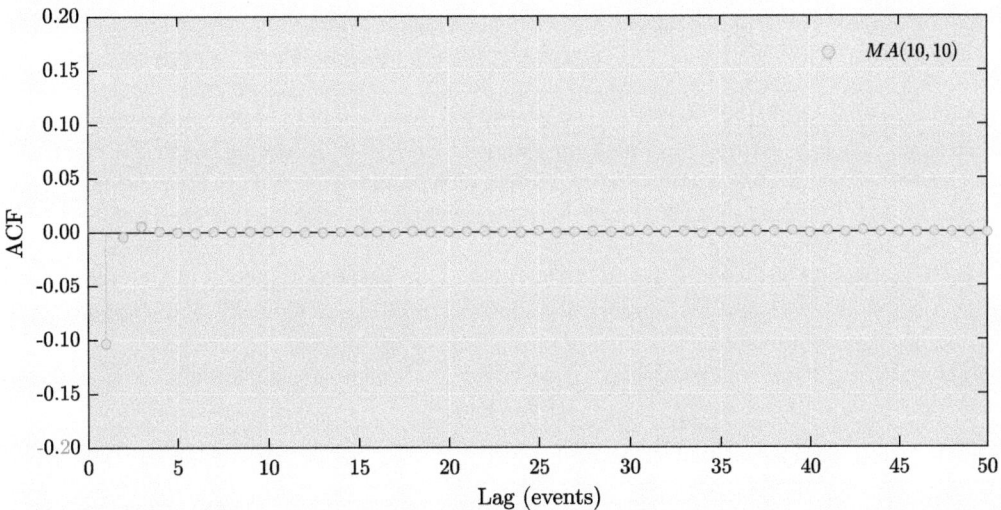

FIGURE 33.13 Shows the average empirical autocorrelation function of the inter-arrival times from 100 simulations of a moving-average point process (a shot-noise or Neyman-Scott process) with conditional intensity (33.4). The kernel is of the form $\Theta(t) = \alpha e^{-\beta t}$ and the process thus denoted MA(α,β). The standard errors of the average autocorrelations across the simulations are extremely small and not shown here.

(2003). In particular, the likelihood ratio for some Hawkes(p,q) parameter estimate $\hat{\theta}$ and a corresponding MA-Hawkes(p,q) process with estimates $\hat{\theta}_1$ is approximated by

$$r\left(\hat{\theta},\hat{\theta}_1\right) := \frac{p(\omega|\hat{\theta})}{p(\omega|\hat{\theta}_1)} \approx \frac{1}{K}\sum_{i=1}^{K}\frac{p\left(\omega,c^{(i)}|\hat{\theta}\right)}{p\left(\omega,c^{(i)}|\hat{\theta}_1\right)}, \tag{33.6}$$

where $c^{(1)},c^{(2)},\ldots,c^{(K)}$ is an MCMC sample of immigrant realizations with density $\propto p\left(c\,|\,\omega,\hat{\theta}_1\right)$-just like in the E-step of the MCEM algorithm (see Appendix 3) -being the density of an immigrant realization c, conditional on the parameter estimates $\hat{\theta}_1$ and the data ω. Once the approximate likelihood ratio is computed on a sufficiently large MCMC sample, standard hypothesis tests are available since $-2\log r\left(\hat{\theta},\hat{\theta}_1\right) \dot{\sim} \chi^2(1)$, and we can thus also directly test the null of the Hawkes(p,q) model against the alternative of the MA-Hawkes(p,q).

To further assess the goodness of fit of the extended model without having access to the true residual process, we again resort to conditional MCMC simulation. Each draw $c^{(K)}, k=1,\ldots,K$ from a simulated Markov chain of immigrants is used to compute a 'conditional' residual series

$$\xi_i^k = \int_0^{t_i}\mu(t)dt + \sum_{j:c_j^k<t_i}\gamma F\left(t_i-c_j^k\right) + \sum_{j:t_j<t_i}\eta\left(t_j\right)H\left(t_i-t_j\right), \tag{33.7}$$

where $F(t) = \int_0^t f(s)ds$ and $H(t) = \int_0^t h(s)ds$ are, as before, the primitives of the exogenous and endogenous offspring densities. On these conditional residuals, the standard tests mentioned above apply. One can then perform these tests on each draw from the Markov chain and look at distributions of the test results over the ensemble of performed draws.

We present results from fitting the MA-Hawkes(p,q) in table 33.4, where we have used $\Theta(t)=\alpha e^{-\beta t}$, so that the exogenous branching ratio is given by $\gamma=\alpha/\beta$, and as before use an approximate power law offspring density for the Hawkes kernel. Overall, the estimates suggest only very weak exogenous clustering effects apart from what our flexible estimator for $\mu(t)$ already covers. At most, we find an exogenous clustering coefficient of around 1.5, but mostly γ is in fact close to zero and the corresponding characteristic times are rather short. With the estimated exogenous clustering coefficient γ being less than 2, the shot-noise process is providing near-Poissonian immigration to drive the Hawkes processes and thus the MA-Hawkes(p,q) is not an obvious replacement for the standard Hawkes(p,q) model. This is also formally confirmed by p-values of the performed likelihood ratio tests failing to reject the null of the Hawkes(p,q) for all events. In line with the Hawkes(p,q) prevailing over the MA-Hawkes (p,q) in a formal test, the average autocorrelation of the simulated residuals in figure 33.14 also shows that the short-lag negative autocorrelation remains in most cases (except where $\gamma > 1$ it is visibly reduced, e.g. for ETH/USD), and the overall goodness of fit according to this measure is not improved significantly.

TABLE 33.4   MA-Hawkes(p,q) Fits with $\Theta(t)=\gamma\beta e^{-\beta t}$ and Approximate Power Law Offspring for the Hawkes Kernel

#	Event	Δ	p^*	q^*	$\langle\mu\rangle$	γ	β^{-1}	$\langle\eta\rangle$	p-value	$\eta(t)$
(1)	E-mini (06/05/2010)	0	17	18	0.299 (0.342)	0.01	0.018	0.56 (0.54)	0.854	
		1 sec	19	7	0.246 (0.299)	1.26	0.409	0.50 (0.74)	0.990	
(2)	U.S. Treasury (15/10/2014)	0	18	12	0.149 (0.153)	0.39	0.247	0.63 (0.71)	0.979	
		1 sec	21	12	0.148 (0.178)	1.55	0.493	0.58 (0.80)	0.990	
(3)	EUR/USD (18/03/2015)	100 ms	23	3	0.220 (0.262)	0.16	0.297	0.67 (0.67)	0.958	
(4)	NZD/USD (24/08/2015)	1 ms	8	1	0.058 (0.062)	0.00	17.259	0.61 (0.58)	1.000	
(5)	USD/ZAR (11/01/2016)	1 ms	10	4	0.018 (0.019)	0.01	32.822	0.68 (0.68)	0.978	
(6)	GBP/USD (7/10/2016)	1 ms	11	8	0.090 (0.103)	0.01	1.204	0.68 (0.64)	0.990	
(7)	ETH/USD (21/06/2017)	1 ms	19	5	0.179 (0.247)	1.50	0.085	0.82 (0.90)	0.991	
(8)	EUR/USD (25/12/2017)	100 ms	3	1	<0.001 (0.001)	0.41	0.173	1.69 (1.53)	0.622	
(9)	USD/JPY (02/01/2019)	100 ms	25	4	0.198 (0.218)	0.04	0.251	0.60 (0.57)	0.881	

Notes: The MCEM estimation is done with $K = 50$ samples taken uniformly from the second half of a chain of length 100 000, generated by the Metropolis-Hastings algorithm. The p-values are from a likelihood ratio test for the null of the Hawkes(p,q) model against the alternative of the corresponding MA-Hawkes(p,q), performed on an MCMC sample of $K = 100$ draws from a Markov chain of length 300 000. The values in parentheses denote the respective estimates from the Hawkes(p,q) model and the units of the average immigration $\langle\mu\rangle$ are sec^{-1} and the MA kernel time scale β^{-1} is expressed in seconds. The column $\eta(t)$ provides sparklines for the branching ratio function estimates of the respective MA-Hawkes(p,q) fit in solid green and the corresponding Hawkes(p,q) fit in dashed purple, where the functions are plotted on $[0, T]$ and the three horizontal grid lines indicate levels for $\eta(t)$ of 0, 0.5 and 1, respectively. The vertical black line indicates the trough/peak of the flash crash.

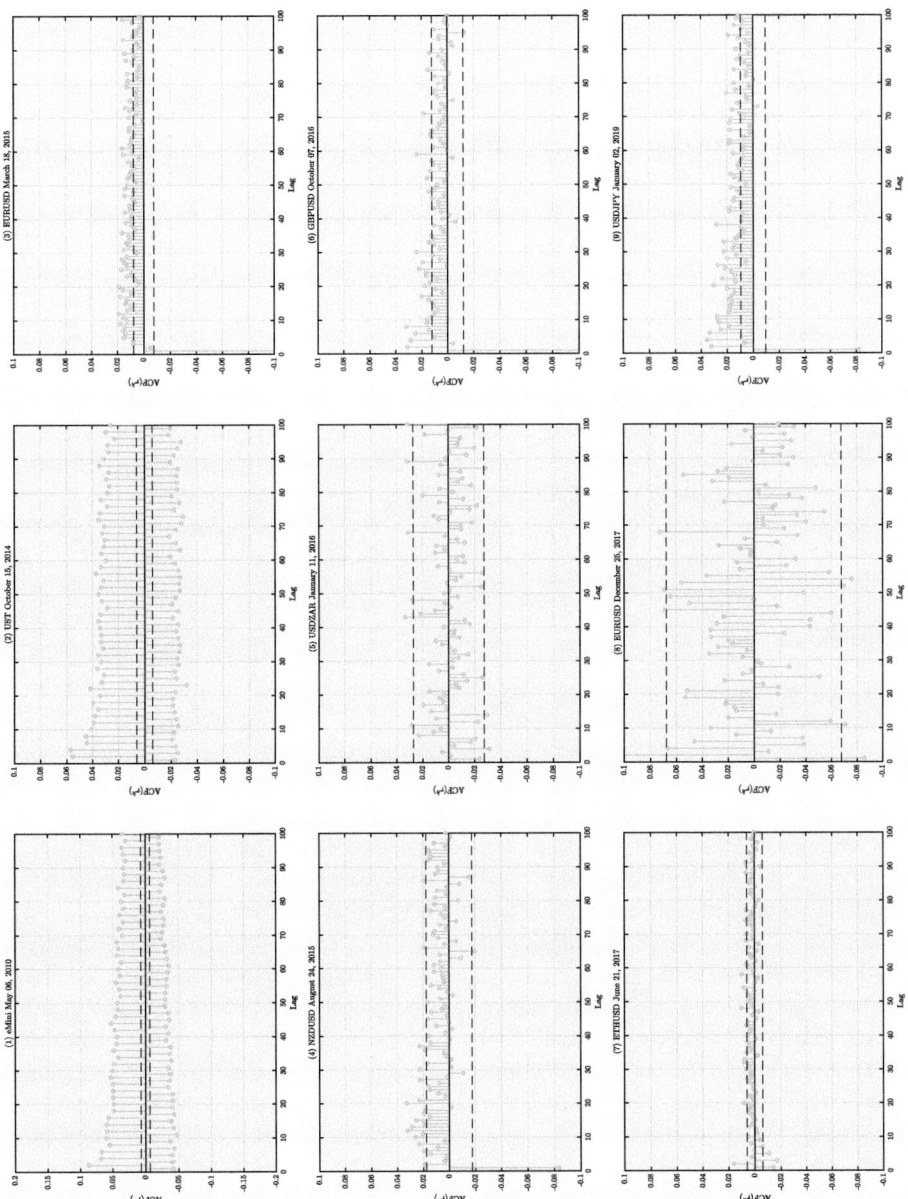

FIGURE 33.14 Shows the average sample autocorrelation function of the residuals r^k obtained by MCMC simulation of the immigrant process, conditional on the parameter estimates of the process (33.5). The average is constructed by drawing a realization of the immigrant process $K = 100$ times from a Markov chain of length 300 000, generated by the Metropolis-Hastings algorithm, and then computing the empirical ACF of the resulting residuals $r_i^k = \xi_i^k - \xi_{i-1}^k$, according to (33.7). The horizontal dashed lines indicate 95% approximate confidence intervals for a white noise process. The standard errors of the mean autocorrelations at each lag across the MCMC draws are extremely small (< 0.001) and not shown here. The MA-Hawkes(p,q) processes were produced using the best fitting assumption on the timestamp uncertainty Δ, see main text for details.

Regarding the dynamics of $\eta(t)$ recovered from the MA-Hawkes(p,q) fits, we observe that, while the average level of η is lower for larger γ, its peak values at crash times remain largely unchanged when additional shot-noise effects are modeled.

A third hypothesis for the negative short-lag residual pattern in our Hawkes(p,q) fits is that the estimated approximate power law kernel fails to describe the true offspring kernel dynamics in some respect, which then manifests in the observed residual autocorrelation. To test this hypothesis, we evaluate the effect of a bias in the parameter estimates of the approximate power offspring density on the residuals in a simulation study, for the simplified case of a Hawkes(1,1) model. We simulate realizations for a set of kernel parameters and then compute the residual process for these synthetic realizations, based on a parameter set where one focal parameter is increasingly biased away from the true value. From figure 33.15, we can see that underestimating the branching ratio—as expected—leads to increasingly positively autocorrelated residuals, where the autocorrelation is more pronounced and persistent, the more downward bias is introduced. Clearly, this is not the pattern we observe in our empirical fits and we thus conclude that a bias in the estimates for $\eta(t)$ can be ruled out. However, when the tail index of the approximate power law is overestimated, we indeed observe a pattern in the residual autocorrelation that is very similar to what can be observed for the real data sets in Figures 33.12 and 33.14, with negative autocorrelation at very short lags that increases in magnitude as the bias in the tail index grows.[*] In fact, the presence of this upward bias for the the tail index in our estimator was already documented in the simulation study in Section 33.appendix 2. Naturally, such an effect can also be expected to arise, if the true empirical kernel shape simply does not exactly follow the power law dynamics we impose here.

As such, we conclude that, while our Hawkes(p,q) estimator with approximate power law memory does not perfectly describe the shape of the empirical offspring density $h(t)$, it seems to satisfactorily encode both time-varying exogenous activity $\mu(t)$ and endogenous triggering strength $\eta(t)$, thus capturing the main features of the data, which is indeed what our objective was for the present study.

33.7 Conclusion

We have presented a cross-asset analysis of self-excited dynamics around events with highly irregular and cascading price dynamics, commonly referred to as flash crashes. Using a novel extension of the class of self-excited conditional Poisson processes with time-varying endogenous feedback effects—the Hawkes(p,q) process—we document instantaneous characteristic signatures of potentially anomalous market regimes around these flash crashes. Systematically balancing the flexibility of estimators for exogenous (p) and endogenous (q) activity sources, we find that, out of the nine events studied here, five exhibit a precursory rise of the branching ratio, of which three reached (near-)critical levels of self-excitation. On the basis of our exemplary selection of events, we find that the events in equity and bond futures are both to be classified to be of endogenous nature, whereas only few of the foreign exchange crashes are found to fall in the endogenous category and many are in fact triggered exogenously. The fact that flash crashes in foreign exchange more often appear to be of exogenous nature is in line with the fact that, at time scales up to a few days, we generally find higher levels of endogeneity in equity futures than foreign exchange markets (Wehrli *et al.* 2020), which are naturally more exposed to shocks from international news and macroeconomic releases. In some cases, the abnormal price moves furthermore seem to be a simple consequence of a large transaction instantaneously exhausting available liquidity. To improve identification in such situations where trading activity might have acted as a trigger, quote and trade arrivals could also be modeled jointly. For this purpose, the Hawkes(p,q) model—just like

---

[*] We remark that the fact that several different mechanisms (bias in offspring density estimates and additional exogenous clustering) can produce similar patterns in second-order statistics is additional evidence for the superiority of EM-based estimation over moment- based estimators for Hawkes-type point processes. While the EM, basing its estimation on the full conditional intensity, is able to discern between true exogenous clustering (here of Neyman-Scott type) and Hawkes-type effects, an estimator on the basis of second-order statistics would find them indistinguishable.

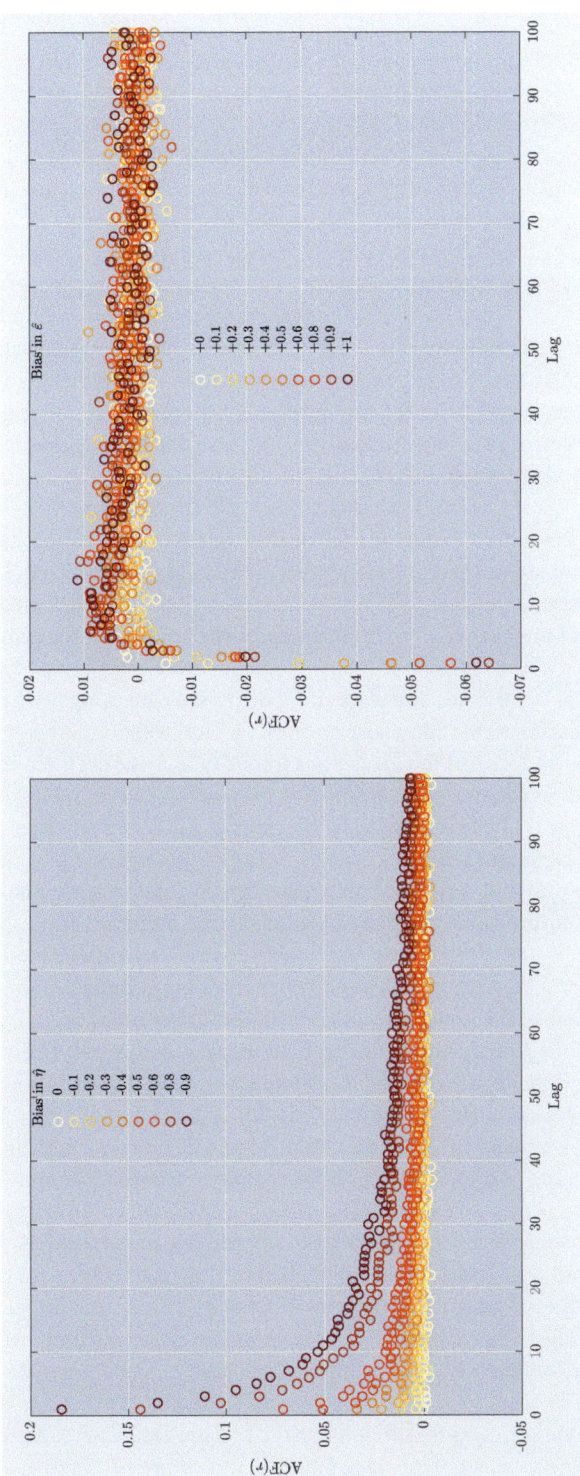

FIGURE 33.15 Shows the average autocorrelation function of the residuals $r_i = \xi_i - \xi_{i-1}$, depending on the bias of the parameter estimates for the branching ratio (left panel) and the tail index of the approximate power law kernel (right panel). The residuals are obtained by simulating a Hawkes(1,1) process and computing the corresponding residual process using parameters from a model where all but one parameter are kept at their true values, but the focal parameter is shifted according to the bias indicated in the legends of the plots. The immigration is fixed to a constant of $\mu = 0.1$ throughout and the kernel is parametrized as an approximate power law with $\tau_0 = 0.1$. In the left panel, the tail index is fixed at a value of $\varepsilon = \check{\varepsilon} = 0.5$, the true branching ratio is set to $\eta = 0.9$ and the branching ratio of the artificial estimator is varied from $\check{\eta} = 0.9$ to $\check{\eta} = 0$. In the right panel, the same procedure is performed for the tail index of the kernel. Now the branching ratio is fixed at $\eta = \check{\eta} = 0.8$, the true tail index is still $\varepsilon = 0.5$ and the artificial estimator of the tail index is varied from $\check{\varepsilon} = 0.5$ to $\check{\varepsilon} = 1.5$. Each combination is simulated 50 times.

the standard Hawkes process—can be extended to a multivariate form, where all the self- and cross-triggering strengths are time-dependent. Clearly, this will result in a highly complex model and can be expected to be computationally demanding to apply to large data sets. In the multivariate setting, taking the necessary effort, the EM algorithm may however be even more useful than in the univariate case, as it decomposes and renders tractable estimation that would otherwise become highly problematic, with far too many parameters for reliable direct numerical MLE.

Regarding broader methodological aspects of financial point process modeling, we also highlight that by systematically selecting models for (t) and $\eta(t)$, we consistently recover more exogenous than endogenous heterogeneity in the mid-price dynamics. This means that, even though q^* is frequently bigger than one, we overwhelmingly obtain $p^* > q^*$ in our estimates, providing additional evidence for the paramount necessity to control for exogenous heterogeneity in financial point process estimation, as argued in Wheatley *et al.* (2019) and Wehrli *et al.* (2020). Furthermore, our extensions in the spirit of the ARMA point process have indicated that, when allowing for more rich exogenous clustering mechanisms, the branching ratio in regular times might in fact even be lower than typically documented values of 0.6-0.8, whereas peak values of $\eta(t) > 0.9$ remain in the case of supposedly endogenously triggered events. For the broader question of reflexivity in financial markets, the moving average extension that we made here is only the simplest one possible. The shot-noise term can be extended to have general shot-noise Cox bursts, with a distribution of marks that allows for the co-existence of a wide range of impacts of external shocks—in particular, so that large shocks are relatively rare. This produces a similar effect to the models from insurance risk theory that have been analyzed for example in Dassios and Jang (2003), Albrecher and Asmussen (2006), Klüppelberg and Mikosch (1995), Boumezoued (2016) and Brémaud and Massoulié (2002) and might shed even more comprehensive light on the general level of market endogeneity.

Apart from these general points, we also see some specific potential avenues for further improvement in the study of extreme financial events like flash crashes. In fact, when looking at our (MA-) Hawkes(p,q) fits, we can observe a tendency of the exogenous intensity to spike even more drastically than the branching ratio during the events. As such, through the lenses of the Hawkes(p,q) model, the explosive behavior of the empirical activity during these crashes is impossible to reproduce just by self-excitation and thus—by design—the exogenous intensity has to explain a large share of the observed dynamics. These limitations of the present formulation in explaining the level of self-excitation during flash crashes can be addressed in future work by either (i) allowing for point fertilities that follow a power law distribution (Saichev *et al.* 2005, Saichev and Sornette 2013) or (ii) extend the model to a non-linear process akin to Filimonov and Sornette (2011). Both approaches are complementary to a time-varying $\eta(t)$ and might alleviate some of the limitations of the current formulation.

Finally, in the spirit of cross-pollination of scientific disciplines, our method of adaptively estimating a time-dependent triggering strength might also be useful to fit nonstandard earthquake series that, e.g. appear under transient stress changes caused by aseismic forces—for instance by fluid injection—where stationary epidemic-type aftershock sequence (ETAS) models fit poorly (Kumazawa and Ogata 2014).

So why do flash crashes occur? The (perhaps unsatisfactory) answer according to our efforts here is, it depends. Our survey of flash crashes revealed that there seems to be a 'zoo' of triggers for flash crashes to occur. However, one possible theme is indeed that the presence of mechanistic factors like level-dependent hedging, automated stop-losses or circuit breakers in related markets can produce the necessary amplifying effects which result in cascading price change dynamics that are accompanied by sharply rising measures of endogenous excitation. However, we have also documented that flash crashes can happen in the *absence* of a critical price process. The fact, that there is no universality in the crash dynamics and many causes seem to exist for flash crashes to occur, is similar to what we see for big market crashes on longer time scales, as found in Johansen and Sornette (2010). Also Filimonov and Sornette (2015a)—based on an assessment of intraday financial drawdowns—find that some events can be attributed to an internal mutual-excitation between market participants, while others are pure responses to external shocks. While we document here that flash crashes seem to be accompanied by precursory rises in estimates of self-exciting activity with some empirical regularity, there is no apparent universality to the dynamics of $\eta(t)$ around the events and a flexible estimator is thus required to gain insight into the price dynamics. As such, we believe that the endo-exo classification framework we propose here can be useful post-mortem in developing remedies and better future processes, such as circuit breakers, latency floor mechanisms and other market design choices. Furthermore, it can potentially be used ex-ante for short-term forecasts in the case of the endogenously triggered class of crashes, by monitoring the feedback parameter in real-time. The design of such a surveillance system would have to consider several important points that were not addressed by the present study, such as rolling window calibration, sensitivity to window size and the elaboration of a signal detection algorithm. One particular difficulty concerning the latter, which such a potential early-warning system would have to solve based on the findings presented here, is filtering local critical points without diverging price movements. The fact that the branching ratio at times goes close to criticality intermittently without the occurrence of abnormal returns means that relying solely on the branching ratio to predict flash crashes necessarily produces false positives. A potential avenue to address these false positives could be to combine monitoring the branching ratio with some measure of liquidity risk, e.g. in the spirit of Donier and Bouchaud (2015). To make progress and validate such ideas, we think that, even though our Hawkes(p,q) framework cannot solve the problem of forecasting flash crashes on its own, it can provide a valuable forensic tool that allows building a catalog of dynamics and assessing similarities between events. This in turn can lead to the identification and validation of additional precursory measures, which when combined with the branching ratio can help bring down the type I error rate of the signal.

Finally, on a general level, our model might help understand how changes at the micro-level may affect macro-level properties of financial markets, such as the occurrence of flash crashes. If it is understood which behaviors of market participants lead to changes in the endogeneity of the price process, there will also be more clarity about which constraints can affect the overall market ecology beneficially (and adversely!). We hope that our (MA-)Hawkes(p,q) framework can provide a tool to inform such issues. A natural way to gain traction on the behavioral roots of specific endogenous activity dynamics as uncovered by our model is to construct an agent-based model of limit order book markets, where different classes of traders and their behavior can be modeled and controlled explicitly. In this theoretical/simulated setup, the contributions of the actions of each agent to subsequently measured endogenous activity dynamics could be quantified objectively and the effect of policies and market design choices explored in an experimental fashion.

Disclosure Statement

No potential conflict of interest was reported by the author(s).

References

Aite Group, New World Order, The high frequency trading community and its impact on market structure. The Aite Group report, 2009, pp. 1–39 (unpublished).

Albrecher, H. and Asmussen, S., Ruin probabilities and aggregate claims distributions for shot noise Cox processes. *Scand. Actuar. J.*, 2006, 2006, 86–110.

Aldrich, E.M., Grundfest, J.A., and Laughlin, G., The flash crash: A new deconstruction (March 26, 2017). Available at SSRN: https://ssrn.com/abstract = 2721922 or http://dx.doi.org/10.2139/ ssrn.2721922

Andersen, T.G. and Bondarenko, O., Reflecting on the VPIN dispute. *J. Financ. Markets*, 2014, 17, 53–64.

Anderson, N., Webber, L., Noss, J., Beale, D. and Crowley-Reidy, L., The resilience of financial market liquidity. Bank of England Financial Stability Paper No. 34, October 2015.

Bacry, E. and Muzy, J.F., Hawkes model for price and trades high-frequency dynamics. *Quant. Finance*, 2014, 14, 1147–1166.

Bacry, E., Delattre, S., Hoffmann, M. and Muzy, J.F., Some limit theorems for Hawkes processes and application to financial statistics. *Stoch. Process. Their. Appl.*, 2013, 123, 2475–2499.

Bacry, E., Mastromatteo, I. and Muzy, J.F., Hawkes processes in finance. *Market Microstruct. Liquid.*, 2015, 1, 1550005.

Bank for International Settlements, The sterling 'flash event' of 7 October 2016, BIS Markets Committee Papers No. 9, 2017a.

Bank for International Settlements, Foreign exchange liquidity in the Americas, BIS Papers No. 90, 2017b.

Bank for International Settlements, Monitoring of fast-paced electronic markets, BIS Report submitted by a Study Group established by the Markets Committee, 2018.

Bank for International Settlements, Triennial central bank surevey—Foreign exchange turnover in April 2019, 2019.

Bates, D.S., How crashes develop: Intraday volatility and crash evolution. *J. Finance*, 2019, 74, 193–238.

BlackRock, 'Flash crash' perceptions study, 2010. Available online at: https://www.sec.gov/comments /265-26/265-26-33.pdf.

Boumezoued, A., Population viewpoint on Hawkes processes. *Adv. Appl. Probab.*, 2016, 48, 463–480.

Bouveret, A., Breuer, P., Chen, Y., Jones, D. and Sasaki, T., Fragilities in the U.S. Treasury Market: Lessons from the 'Flash Rally' of October 15, 2015. IMF Working Paper, 2015, 15/222.

Bowsher, C., Modelling security market events in continuous time: Intensity based, multivariate point process models. *J. Econom.*, 2007, **141**, 876–912.

Brémaud, P. and Massoulié, L., Hawkes branching point processes without ancestors. *J. Appl. Probab.*, 2001, **38**, 122–135.

Brémaud, P. and Massoulié, L., Power spectra of general shot noises and Hawkes point processes with a random excitation. *Adv. Appl. Probab.*, 2002, 34, 205–222.

Calcagnile, L.M., Bormetti, G., Treccani, M., Marmi, S. and Lillo, F., Collective synchronization and high frequency systemic instabilities in financial markets. *Quant. Finance*, 2018, 18, 237–247.

Callaghan, M., The New Zealand dollar in global markets. *Res. Bank N.Z. Bull.*, 2017, **80**.

Cespa, G. and Foucault, T., Illiquidity contagion and liquidity crashes. *Rev. Financ. Stud.*, 2014, 27, 1615–1660.

Challet, D., Marsili, M. and Zhang, Y.C., *Minority Games: Interacting Agents in Financial Markets*, 2013 (Oxford University Press: New York, NY).

Chen, F. and Stindl, T., Direct likelihood evaluation for the renewal Hawkes process. *J. Comput. Graph. Stat.*, 2018, 27, 119–131.

Cinlar, E., *Introduction to Stochastic Processes*, 2013 (Dover: New York, NY).

Commodity Futures Trading Commission-Securities Exchange Commission (CFTC-SEC), Findings regarding the market events of May 6. Staff report, 2010a. Available online at: https://www.sec.gov /news/studies/2010/marketevents-report.pdf.

Commodity Futures Trading Commission-Securities Exchange Commission (CFTC-SEC), Preliminary findings regarding the market events of May 6. Staff report, 2010b. Available online at: https://www .sec.gov/sec-cftc-prelimreport.pdf.

Coolen, A., The Mathematical Theory of Minority Games—Statistical Mechanics of Interacting Agents, 2005 (Oxford University Press: New York, NY).

Crane, R. and Sornette, D., Robust dynamic classes revealed by measuring the response function of a social system. *Proc. Nat. Acad. Sci. USA*, 2008, **105**, 15649–15653.

Crane, R., Schweitzer, F. and Sornette, D., Power law signature of media exposure in human response waiting time distributions. *Phys. Rev. E*, 2010, **81**, 056101.

Daley, D. and Vere-Jones, D., *An Introduction to the Theory of Point Processes. Volume I: Elementary Theory and Methods*, 2nd ed., Probability and Its Applications Vol. 1, 2003 (Springer Verlag: New York).

Daley, D. and Vere-Jones, D., *An Introduction to the Theory of Point Processes. Volume II: General Theory and Structure*, 2nd ed., Probability and Its Applications Vol. 2, 2008 (Springer Verlag: New York).

Dassios, A. and Jang, J.W., Pricing of catastrophe reinsurance and derivatives using the Cox process with shot noise intensity. *Finance Stoch.*, 2003, 7, 73–95.

Dassios, A. and Zhao, H., A dynamic contagion process. *Adv. Appl. Probab.*, 2011, 43, 814–846.

Dempster, A.P., Laird, N.M. and Rubin, D.B., Maximum likelihood from incomplete data via the EM algorithm. *J. R. Statist. Soc. Ser. B (Methodol.)*, 1977, 39, 1–38.

Deschatres, F. and Sornette, D., Dynamics of book sales: Endogenous versus exogenous shocks in complex networks. *Phys. Rev. E.*, 2005, 71, 016112.

Dieci, R. and He, X.Z., Chapter 5 - Heterogeneous agent models in finance. In *Handbook of Computational Economics*, edited by C. Hommes and B. LeBaron, Vol. 4, pp. 257–328, 2018 (Elsevier: Amsterdam).

Donier, J. and Bouchaud, J.P., Why do markets crash? Bitcoin data offers unprecedented insights. *PLoS. ONE.*, 2015, 10, 1–11.

Duan, N., Smearing estimate: A nonparametric retransformation method. *J. Am. Stat. Assoc.*, 1983, 78, 605–610.

Easley, D., Lopez de Prado, M. and O'Hara, M., The microstructure of the 'flash crash': Flow toxicity, liquidity crashes, and the probability of informed trading. *J. Portfolio Manage.*, 2011, 37, 118–128.

Fan, J., Farmen, M. and Gijbels, I., Local maximum likelihood estimation and inference. *J. R. Statist. Soc. Ser. B (Stat. Methodol.)*, 1998, 60, 591–608.

Filimonov, V. and Sornette, D., Self-excited multifractal dynamics. *EPL (Europhys. Lett.)*, 2011, 94, 46003.

Filimonov, V. and Sornette, D., Quantifying reflexivity in financial markets: Toward a prediction of flash crashes. *Phys. Rev. E.*, 2012, 85, 056108.

Filimonov, V. and Sornette, D., Power law scaling and 'Dragon-Kings' in distributions of intraday financial drawdowns. *Chaos, Solitons Fract.*, 2015a, 74, 27–45.

Filimonov, V. and Sornette, D., Apparent criticality and calibration issues in the Hawkes self-excited point process model: Application to high-frequency financial data. *Quant. Finance*, 2015b, 15, 1293–1314.

Filimonov, V., Bicchetti, D., Maystre, N. and Sornette, D., Quantification of the high level of endogeneity and of structural regime shifts in commodity prices. *J. Int. Money Finance*, 2014, 42, 174–192.

Fosset, A., Bouchaud, J.P. and Benzaquen, M., Endogenous liquidity crises. arXiv preprint arXiv:1912.00359, 2020.

Gerlach, J.C., Demos, G. and Sornette, D., Dissection of Bitcoin's multiscale bubble history from January 2012 to February 2018. *R. Soc. Open. Sci.*, 2019, 8, 180643.

Han, F. and Westelius, N., Anatomy of sudden yen appreciations. IMF Working Paper, 2019, **19/136**.

Hardiman, S.J. and Bouchaud, J.P., Branching-ratio approximation for the self-exciting Hawkes process. *Phys. Rev. E—Statist. Nonlinear Soft Matter Phys.*, 2014, 90, 1–7.

Hardiman, S.J., Bercot, N. and Bouchaud, J.P., Critical reflexivity in financial markets: A Hawkes process analysis. *Euro. Phys. J. B*, 2013, 86, 442.

Harris, T.E., *The Theory of Branching Processes*, Die Grundlehren der Mathematischen Wissenschaften, Bd. 119, 1963 (Springer: Berlin).

Harras, G. and Sornette, D., How to grow a bubble: A model of myopic adapting agents. *J. Econ. Behav. Organ.*, 2011, 80, 137–152.

Harte, D., An ETAS model with varying productivity rates. *Geophys. J. Int.*, 2014, 198, 270–284.

Hawkes, A., Point spectra of some mutually exciting point processes. *J. R. Statist. Soc. Ser. B (Methodol.)*, 1971a, 33, 438–443.

Hawkes, A., Spectra of some self-exciting and mutually exciting point processes. *Biometrika*, 1971b, 58, 83–90.

Hawkes, A.G., Hawkes processes and their applications to finance: A review. *Quant. Finance*, 2018, 18, 193–198.

Helmstetter, A. and Sornette, D., Importance of direct and indirect triggered seismicity in the ETAS model of seismicity. *Geophys. Res. Lett.*, 2003, 30, 1576. doi:10.1029/2003GL017670

Helmstetter, A., Sornette, D. and Grasso, J.R., Mainshocks are aftershocks of conditional foreshocks: How do foreshock statistical properties emerge from aftershock laws. *J. Geophys. Res. Solid Earth*, 2003, 108, 2046.

Huang, J., Experimental econophysics: Complexity, self-organization, and emergent properties. *Phys. Rep.*, 2015, 564, 1–55.

Huang, J. and Wang, J., Liquidity and market crashes. *Rev. Financ. Stud.*, 2009, 22, 2607–2643.

Jaisson, T. and Rosenbaum, M., Limit theorems for nearly unstable Hawkes processes. *Ann. Appl. Probab.*, 2015, 25, 600–631.

Johansen, A. and Sornette, D., Shocks, crashes and bubbles in financial markets. *Brussels Econom. Rev.*, 2010, 53, 201–253.

Joulin, A., Lefevre, A., Grunberg, D. and Bouchaud, J.P., Stock price jumps: News and volume play a minor role. *Wilmott Mag.*, 2008, 46, 1–7.

Kaizoji, T., Leiss, M., Saichev, A. and Sornette, D., Super-exponential endogenous bubbles in an equilibrium model of rational and noise traders. *J. Econom. Behavior Org.*, 2015, 112, 289–310.

Kass, R.E. and Wasserman, L., A reference Bayesian test for nested hypotheses and its relationship to the Schwarz criterion. *J. Amer. Statist. Assoc.*, 1995, 90, 928–934.

Kirilenko, A., Kyle, A.S., Samadi, M. and Tuzun, T., The flash crash: High-frequency trading in an electronic market. *J. Finance*, 2017, 72, 967–998.

Klüppelberg, C. and Mikosch, T., Explosive Poisson shot noise processes with applications to risk reserves. *Bernoulli*, 1995, 1, 125–147.

Kooperberg, C. and Stone, C.J., A study of logspline density estimation. *Comput. Stat. Data. Anal.*, 1991, 12, 327–347.

Kooperberg, C. and Stone, C.J., Logspline density estimation for censored data. *J. Comput. Graph. Stat.*, 1992, 1, 301–328.

Kumazawa, T. and Ogata, Y., Nonstationary ETAS models for nonstandard earthquakes. *Ann. Appl. Stat.*, 2014, 8, 1825–1852.

Kyle, A.S. and Obizhaeva, A.A., Large bets and stock market crashes (March 22, 2019). Available at SSRN: https://ssrn.com/abstract = 2023776 or http://dx.doi.org/10.2139/ssrn.2023776

Lallouache, M. and Challet, D., The limits of statistical significance of Hawkes processes fitted to financial data. *Quant. Finance*, 2016, 16, 1–11.

Lee, Y., Lim, K.W. and Ong, C.S., Hawkes processes with stochastic excitations. In *Proceedings of the 33rd International Conference on International Conference on Machine Learning—Volume 48*, ICML'16: New York, NY, USA, pp. 79–88, 2016, JMLR.org.

Lewis, E. and Mohler, G., A nonparametric EM algorithm for multiscale Hawkes processes. *J. Nonparametr. Stat.*, 2011, 1, 1–16.

Lewis, P.A. and Shedler, G.S., Simulation of nonhomogeneous Poisson processes by thinning. *Naval Res. Logist. Quart.*, 1979, 26, 403–413.

Lillo, F. and Mantegna, R.N., Power-law relaxation in a complex system: Omori law after a financial market crash. *Phys. Rev. E*, 2003, 68, 016119.

Ljung, G. and Box, G., On a measure of a lack of fit in time series models. *Biometrika*, 1978, 65, 297–303.

Madhavan, A., Exchange-traded funds, market structure, and the flash crash. *Financ. Anal. J.*, 2012, 68, 20–35.

Mark, M., Sila, J. and Weber, T.A., Quantifying endogeneity of cryptocurrency markets. *Euro. J. Finance*, 2020. https://doi.org/10.1080/1351847X.2020.1791925

Marsan, D. and Lengliné, O., Extending earthquakes' reach through cascading. *Science*, 2008, 319, 1076–1079.

McInish, T., Upson, J. and Wood, R.A., The flash crash: Trading aggressiveness, liquidity supply, and the impact of intermarket sweep orders. *Financ. Rev.*, 2014, 49, 481–509.

McLachlan, G. and Krishnan, T., *The EM Algorithm and Extensions*, 2nd ed., Wiley series in probability and statistics, 2008 (John Wiley & Sons, Inc: Hoboken, NJ).

Menkveld, A.J. and Yueshen, B.Z., The flash crash: A cautionary tale about highly fragmented markets. *Manage. Sci.*, 2019, 65, 4470–4488.

Møller, J., Shot noise Cox processes. *Adv. Appl. Probab.*, 2003, 35, 614–640.

Møller, J. and Waagepetersen, R., *Statistical Inference and Simulation for Spatial Point Processes*, Monographs on Statistics and Applied Probability 100, 2003 (Chapman & Hall/CRC: Boca Raton, FL).

Neyman, J. and Scott, E., Statistical approach to problems of cosmology. *J. R. Statist. Soc. Ser. B.*, 1958, 20, 1–43.

Noss, J., Pedace, L., Tobek, O., Linton, O.B. and Crowley-Reidy, L., The October 2016 sterling flash episode: When liquidity disappeared from one of the world's most liquid markets. Bank of England Working Paper, 2017.

Ogata, Y., On Lewis' simulation method for point processes. *IEEE Trans. Inform. Theory*, 1981, 27, 23–31.

Ogata, Y., Statistical models for earthquake occurrences and residual analysis for point processes. *J. Am. Stat. Assoc.*, 1988, 83, 9–27.

Ozaki, T., Maximum likelihood estimation of Hawkes' self-exciting point processes. *Ann. Inst. Stat. Math.*, 1979, 31, 145–155.

Patzelt, F. and Pawelzik, K., Criticality of adaptive control systems. *Phys. Rev. Lett.*, 2011, 00, 238103.

Patzelt, F. and Pawelzik, K., An inherent instability of efficient markets. Sci. Rep., 2013, 3, 2784.

Petersen, A.M., Wang, F., Havlin, S. and Stanley, H.E., Quantitative law describing market dynamics before and after interest-rate change. *Phys. Rev. E*, 2010, 81, 066121.

Rambaldi, M., Bacry, E. and Muzy, J.F., Disentangling and quantifying market participant volatility contributions. *Quant. Finance*, 2019, 19, 1613–1625.

Reserve Bank of Australia, Statement on monetary policy—February 2019. Monetary Policy Statements, 2019.

Saichev, A., Helmstetter, A. and Sornette, D., Power-law distributions of offspring and generation numbers in branching models of earthquake triggering. *Pure Appl. Geophys.*, 2005, 162, 1113–1134.

Saichev, A. and Sornette, D., Renormalization of branching models of triggered seismicity from total to observable seismicity. *Euro. Phys. J. B-Condensed Matter Complex Syst.*, 2006, 51, 443–459.

Saichev, A. and Sornette, D., Fertility heterogeneity as a mechanism for power law distributions of recurrence times. *Phys. Rev. E*, 2013, 87, 022815.

Saichev, A. and Sornette, D., Super-linear scaling of offsprings at criticality in branching processes. *Phys. Rev. E.*, 2014, 89, 012104.

Salakhutdinov, R., Roweis, S. and Ghahramani, Z., Optimization with EM and expectation-conjugate-gradient. Proceedings of the ICML, pp. 672–679, 2003.

Samanidou, E., Zschischang, E., Stauffer, D. and Lux, T., Agent-based models of financial markets. *Rep. Prog. Phys.*, 2007, 70, 409–450.

Schoenberg, F., Hoffmann, M. and Harrigan, R., A recursive point process model for infectious diseases. *Ann. Inst. Stat. Math.*, 2017, 71, 1271–1287.

Schrimpf, A. and Shushko, V., FX trade execution: Complex and highly fragmented. *BIS Quartlery Review*, 2019.

Schroeder, F., Lepone, A., Leung, H. and Satchell, S., Flash crash in an OTC market: Trading behaviour of agents in times of market stress. *Euro. J. Finance*, 2020, 1–21.

Schwarz, G., Estimating the dimension of a model. *Ann. Stat.*, 1978, 6, 461–464.

SEC, Concept release on equity market structure. Exchange Act Release No. 34-61358, 17 CFR 242, 2010.

Sornette, D., Endogenous versus exogenous origins of crises. In *Extreme Events in Nature and Society*, edited by S. Albeverio, V. Jentsch and H. Kantz, pp. 95–119, 2006 (Springer: Berlin, Heidelberg).

Sornette, D., *Why Stock Markets Crash: Critical Events in Complex Financial Systems*, Princeton Science Library, 2017 (Princeton University Press).

Sornette, D. and Helmstetter, A., Endogeneous versus exogeneous shocks in systems with memory. *Phys. A.*, 2003, 318, 577–591.

Sornette, D., Deschâtres, F., Gilbert, T. and Ageon, Y., Endogenous versus exogenous shocks in complex networks: An empirical test using book sale rankings. *Phys. Rev. Lett.*, 2004a, 93, 228701.

Sornette, D., Malevergne, Y. and Muzy, J.F., Volatility fingerprints of large shocks: Endogenous versus exogenous. In *Proceedings of the The Application of Econophysics*, edited by H. Takayasu, pp. 91–102, 2004b (Springer Japan: Tokyo).

Sussman, A., Tabb, L. and Iati, R., US equity high frequency trading: Strategies, sizing and market structure. TABB Group report, 2009, pp. 1–32 (unpublished).

Tibshirani, R. and Hastie, T., Local likelihood estimation. *J. Am. Stat. Assoc.*, 1987, 82, 559–567.

USDT, BGFRS, FRBNY, SEC and CFTC, The US Treasury Market on October 15, 2014. Joint Staff Report, 2015.

Veen, A. and Schoenberg, F., Estimation of space-time branching process models in seismology using an EM-type algorithm. *J. Amer. Statist. Assoc.*, 2008, 103, 614–624.

von der Becke, S. and Sornette, D., Crashes and high frequency trading. An evaluation of risks posed by high-speed algorithmic trading. In *Report for the UK Government project entitled 'The Future of Computer Trading in Financial Markets'*, pp. 1–26, 2011 (Foresight Driver Review—DR7: London, UK).

Weber, P., Wang, F., Vodenska-Chitkushev, I., Havlin, S. and Stanley, H.E., Relation between volatility correlations in financial markets and Omori processes occurring on all scales. *Phys. Rev. E*, 2007, 76, 016109.

Wehrli, A., Wheatley, S. and Sornette, D., Scale-, time- and asset- dependence of Hawkes process estimates on high frequency price changes. Swiss Finance Institute Research Paper No. 20-39, 2020.

Wei, G.C.G. and Tanner, M.A., A Monte Carlo implementation of the EM algorithm and the poor man's data augmentation algorithms. *J. Am. Stat. Assoc.*, 1990, 85, 699–704.

Westphal, R. and Sornette, D., Market impact and performance of arbitrageurs of financial bubbles in an agent-based model. *J. Econom. Behav. Org.*, 2020, 171, 1–23.

Wheatley, S., Filimonov, V. and Sornette, D., The Hawkes process with renewal immigration & its estimation with an EM algorithm. *Comput. Stat. Data. Anal.*, 2016, 94, 120–135.

Wheatley, S., Schatz, M. and Sornette, D., The ARMA point process and its estimation. arXiv preprint arXiv:1806.09948, 2018.

Wheatley, S., Wehrli, A. and Sornette, D., The endo-exo problem in high frequency financial price fluctuations and rejecting criticality. *Quant. Finance*, 2019, 19, 1165–1178.

Wu, C., On the convergence properties of the EM algorithm. *Ann. Statist.*, 1983, 11(1), 95–103.

Appendices

Appendix A: The EM Algorithm for the Estimation of the Hawkes(p,q) Process

Generally, the EM algorithm can be used to iteratively build an MLE using the complete data likelihood function $L^c(\theta/\omega,Z)$ (McLachlan and Krishnan 2008), where $\omega := \{t_i\}_{i=1}^{N_T}$ denotes a realization of the process N, θ is a vector of model parameters and Z introduces an additional set of unobserved data, so that the estimation can be decomposed into more manageable sub-problems. For point processes with linear conditional intensities, if we were to know the branching structure of the underlying process, estimation would reduce to separate probabilistic estimations of the immigration and offspring processes (Marsan and Lengliné 2008, Veen and Schoenberg 2008, Lewis and Mohler 2011, Wheatley *et al.* 2016). As such, we define the missing data to be the branching structure of the process, represented as a lower-triangular indicator-matrix $Z_{n \times n} \in \{0,1\}^{n \times n}$, where $Z_{i,j}=1$ if point t_i is an offspring of point t_j and $Z_{i,i}=1$ if t_j is an immigrant.

Provided an initial parameter estimate $\hat{\theta}^{(0)}$, the EM algorithm iterates the expectation step (E-step) and maximization step (M-step) in each iteration l until $\hat{\theta}^{(l)}$ (and the log-likelihood of the process) converges (Wu 1983, Salakhutdinov et al. 2003). In the E-step, the algorithm computes the expected value of the complete data log-likelihood with respect to the conditional distribution of the missing data, given the observed data and the current parameter estimates

$$Q\left(\theta \mid \omega, \hat{\theta}^{(l)}\right) = \mathbb{E}_{Z \mid \omega, \hat{\theta}^{(l)}}\left[\log L^c\left(\theta \mid \omega, Z\right)\right]. \tag{A1}$$

The subsequent M-step maximizes the expected complete data log-likelihood (A1) to obtain new parameter estimates

$$\hat{\theta}^{(l+1)} = \arg\max_{\theta} Q\left(\theta \mid \omega, \hat{\theta}^{(l)}\right). \tag{A2}$$

The expected complete data log-likelihood of the process with conditional intensity (33.1) has the form

$$Q\left(\theta \mid \omega, \hat{\theta}^{(l)}\right) = \sum_{i=1}^{N_T}\left(\pi_{i,i}\log\mu\left(t_i\right) + \sum_{j:t_j < t_i}\pi_{i,j}\log\left(\eta\left(t_j\right)h\left(t_i - t_j\right)\right)\right) \\ - \int_0^T \lambda\left(t\right)\mathrm{d}t, \tag{A3}$$

where the branching probabilities are $\pi_{i,j} = \mathbb{P}\left[Z_{i,j}=1 \mid \omega, \hat{\theta}^{(l)}\right]$ and can be estimated in the E-step using random thinning (Ogata 1981) as

$$\pi_{i,j} = \frac{\mu\left(t_i \mid \hat{\theta}^{(l)}\right)}{\lambda\left(t_i \mid \hat{\theta}^{(l)}\right)} \text{ and } \pi_{i,j} = \frac{\eta\left(t_j \mid \hat{\theta}^{(l)}\right)h\left(t_i - t_j \mid \hat{\theta}^{(l)}\right)}{\lambda\left(t_i \mid \hat{\theta}^{(l)}\right)}, \quad t_i > t_j. \tag{A4}$$

Using this probabilistic branching structure and, assuming that μ does not share any parameters with η and h, the M-step of the algorithm can estimate the various parameters of the immigration and offspring processes separately. First, the exogenous intensity $\mu(t)$ is estimated based on the weighted sample composed of point and weight pairs $\left\{\left(t_i, \pi_{i,i}\right)\right\}_{i=1}^{N_T}$. In the simplest case of a constant immigration estimator, this is simply $\hat{\mu} = T^{-1}\sum_i \pi_{i,i}$. Regarding a particular model for $\mu(t)$, we follow the approach in Wheatley *et al.* (2019) and Wehrli *et al.* (2020) and use adaptive *logspline* models (Kooperberg and Stone 1991, 1992) with p degrees of freedom as a flexible technique to capture nuisance structures such as shocks and trends. Next, an analytical maximization of (A3) gives the objective function to maximize for the offspring density h as

$$\ell_h^c\left(\theta \mid, \pi\right) = \sum_{i=1}^{N_T}\sum_{j:t_j < t_i}\pi_{i,j}\log\left(h\left(t_i - t_j\right)\right) - \sum_{j=1}^{N_T}\eta\left(t_j\right)H\left(T - t_j\right), \tag{A5}$$

where $H(t) = \int_0^t h(s)\mathrm{d}s$ is the cumulative distribution function corresponding to the density h. Finally, to estimate $\eta(t)$, we consider the solutions $\hat{\eta}_j = \arg\max_{\eta(t_j)} Q\left(\theta \mid \omega, \hat{\theta}^{(l)}\right)$ as (potentially noisy) estimates for the random number of points that are triggered by events at times t_j, which are given by

$$\hat{\eta}_j = \frac{\sum_{i:t_i>t_j} \pi_{i,j}}{H(T-t_j)}. \tag{A6}$$

Estimating the full function $\eta(t)$, being the expected instantaneous number of points triggered by an event at time t, can then be done from pairs $(t_j, \hat{\eta}_j)$ using traditional regression/smoothing approaches. A straightforward approach would be to form a piecewise constant estimator by aggregating the $\hat{\eta}_j$ in some given intervals. A special case of this estimator is the constant branching ratio estimate $\check{\eta} = \sum_i \sum_{j<1} \pi_{i,j} / \sum_i H(T-t_i)$. This approach however lacks adaptivity and a more flexible estimator is desirable to properly evaluate the dynamics of η. A first natural candidate is to perform a kernel regression using some kernel $K(\cdot)$ and estimate the branching ratio as

$$\hat{\eta}(t \mid \tau) = \frac{\sum_{i=1}^{N_T} \hat{\eta}_i K\left((t-t_i)/\tau\right)}{\sum_{i=1}^{N_T} K\left((t-t_i)/\tau\right)} \tag{A7}$$

for some bandwidth τ. While this approach indeed allows flexible estimation of the branching ratio, model selection (i.e. determining the bandwidth τ) is a notoriously difficult issue in nonparametric statistics and we prefer a more straightforward way to control the flexibility of our estimator. Nevertheless, this kernel estimator can be a valuable tool when doing exploratory work, for example when the influence of the flexibility of the immigration or the type of offspring density used on the estimated dynamics of η are of interest.

An alternative approach we propose here, in analogy to the treatment of $\mu(t)$, is to use some form of adaptive spline, where the degrees of freedom permitted can be controlled explicitly and balanced with the flexibility of μ. To this end, at each iteration of the M-step, we draw a Monte Carlo sample from the t_i with weights $\propto \hat{\eta}_i$. On this sample, we fit a logspline density with q degrees of freedom (knots) to obtain an optimal allocation of the knot sequence $\tilde{\mathbf{t}} := \{\tilde{t}_n\}_{n=1,\ldots,q}$. We then use these logspline-optimal knot locations to fit a piecewise cubic spline to the pairs $(t_i, \hat{\eta}_i)$ with knots at the $\tilde{t}_n$ and periodic boundary conditions, so that the spline has the desired number of degrees of freedom. This procedure has the advantage that we adaptively place knots at locations where the probability of endogenous points occurring is high and simultaneously retain the possibility to control the flexibility of the resulting function by choosing the allowed number of knots q in a similar fashion to the degrees of freedom p of the immigration. Additionally, since the logspline estimate for the immigration (log-)density is also based on cubic splines, the estimators for both endogenous and exogenous dynamics have identical flexibility and can thus be compared more easily. A final practical consideration concerns the fact that we require $\eta(t) \geq 0 \forall t$, which is not a priori guaranteed even though $\eta_i \geq 0$ by construction. A cubic spline which interpolates positive function values is not necessarily positive for all arguments—in particular this can arise when the spline needs to interpolate over regions with abrupt changes. In order to preserve non-negativity of our regression splines for $\eta(t)$, we thus choose to estimate the respective spline $\hat{s}(t \mid \tilde{t})$ on a transformed scale, performing the regression

$$\log(1+\eta_i) = \hat{s}\left(t_i \mid \tilde{\mathbf{t}}\right) + \epsilon_i, \tag{A8}$$

for some residual ϵ_i. A well-known issue in regression analysis is that retransforming a problem like (A8) back to the original metric for inference purposes can yield a biased estimator of the mean response—in our case the instantaneous branching ratio $\eta(t)$. In particular, we are interested in $\eta(t_i) = \mathbb{E}[\eta_i] = \exp\left(\hat{s}(t_i \mid \tilde{t})\right) \mathbb{E}[e^{\epsilon_i}] - 1$ without knowledge of the distribution of e^{ϵ_i}. Our approach here is to apply a simple 'smearing factor' (Duan 1983) to approximate $\mathbb{E}\left[e^{\epsilon_i}\right]$ using $\tilde{S} := N_T^{-1} \sum_{i=1}^{N_T} e^{\epsilon_i}$ and characterize any potentially remaining bias of the resulting estimator in a simulation study presented in the sequel. The final estimator for the instantaneous branching ratio is then

$$\hat{\eta}(t) = \left(\tilde{S} e^{\hat{s}(t|\tilde{t})} - 1 \right)^+ \tag{A9}$$

and $(x)^+ = \max(x,0)$ denotes the positive part, applied to remove any potentially remaining negative branching ratio values arising from the smearing estimate.

The resulting overall model candidate finally has $p + q$ degrees of freedom and we call it a Hawkes(p,q) model. To select an appropriate model, we propose in the spirit of Wheatley *et al.* (**2019**) and Wehrli *et al.* (**2020**) to use the Bayesian Information Criterion (BIC) (Schwarz **1978**). Furthermore, to characterize the uncertainty inherent in the selection of p and q, we can approximate the posterior probability of some model defined by the tuple (p, q), in a Bayesian fashion (Kass and Wasserman **1995**), given the data ω using

$$P((p,q) \mid \omega) \approx \frac{\exp\left(-\frac{1}{2} \Delta \mathrm{BIC}_{(p,q)} \right)}{\sum_{m \in \mathcal{M}} \exp\left(-\frac{1}{2} \Delta \mathrm{BIC}_m \right)}, \tag{A10}$$

where $\Delta \mathrm{BIC}_{(p,q)}$ is the difference between the BIC of some specific model defined by (p,q) and the minimum BIC over a set of models $\mathcal{M}$ we are evaluating. Having computed the posterior probabilities for a set of models then also allows us to form credible intervals for $\mu(t)$ and $\eta(t)$ by bootstrapping percentiles of the posterior distribution over the functions by sampling with replacement from the models in $\mathcal{M}$ according to their approximate posterior probabilities (A10).

Appendix B: Estimating Transient Criticality—A Simulation Study

We evaluate the EM-based Hawkes(p,q) estimator in a simulation study. For this purpose, we draw synthetic realizations of a process with bimodal immigration intensity $\mu(t) = 0.1 + 0.05\sin(3\pi t / T)$ and a branching ratio $\eta(t) = 0.6 + \bar{S} \cdot \Phi(t; T/4, 500)$, being constant most of the time, but with a peak described by the Gaussian function $\Phi(\cdot)$ centered on location $T/4$ and with width (standard deviation) 500. We perform two types of simulations, one where the branching ratio just touches criticality instantaneously and one where it visits the critical regime for around 600 seconds and thus the divergence of the process has some time to develop. For this purpose, the Gaussian peak is scaled with a factor $\bar{S}$, so that its value at the mode is {0.4, 0.45} and the resulting peak branching ratios are thus {1, 1.05}, resembling the hypothesized temporary rise to (or above) criticality during flash crashes. Finally, we use an offspring density like in (33.2) with $\tau_0 = 0.1$, $\varepsilon = 0.5$ and set $M = 15$ and $b = 5$.

Simulation and fitting is then done on 50 realizations for both types of processes considered, each on a window of length $T = 10\,000$ seconds. The degrees of freedom for the immigration and the branching ratio estimator are scanned in $p,q \in [1,10]$, where $p = 1$ and $q = 1$ correspond to the constant estimators of the immigration intensity and branching ratio, respectively. For performance, we truncate the interaction length of the memory kernel to 1000 points (Wheatley *et al.* **2019**). We summarize the results of the simulation study in figure 33.A1. Overall, we find that the Hawkes(p,q) estimator recovers the dynamics of the branching ratio quite well—in particular its mode, which will be of central interest in empirical studies. The maximum of $\eta(t)$ is underestimated marginally, where the peak branching ratio recovered on average is 0.98 when its true value is 1 and 1.01 when it is 1.05. The downward bias also extends to the the average level of $\eta(t)$, an error of ≈ 0.05 is hereby observable uniformly across [0, T]. The variance of $\eta(t)$ is quite small across the simulations.

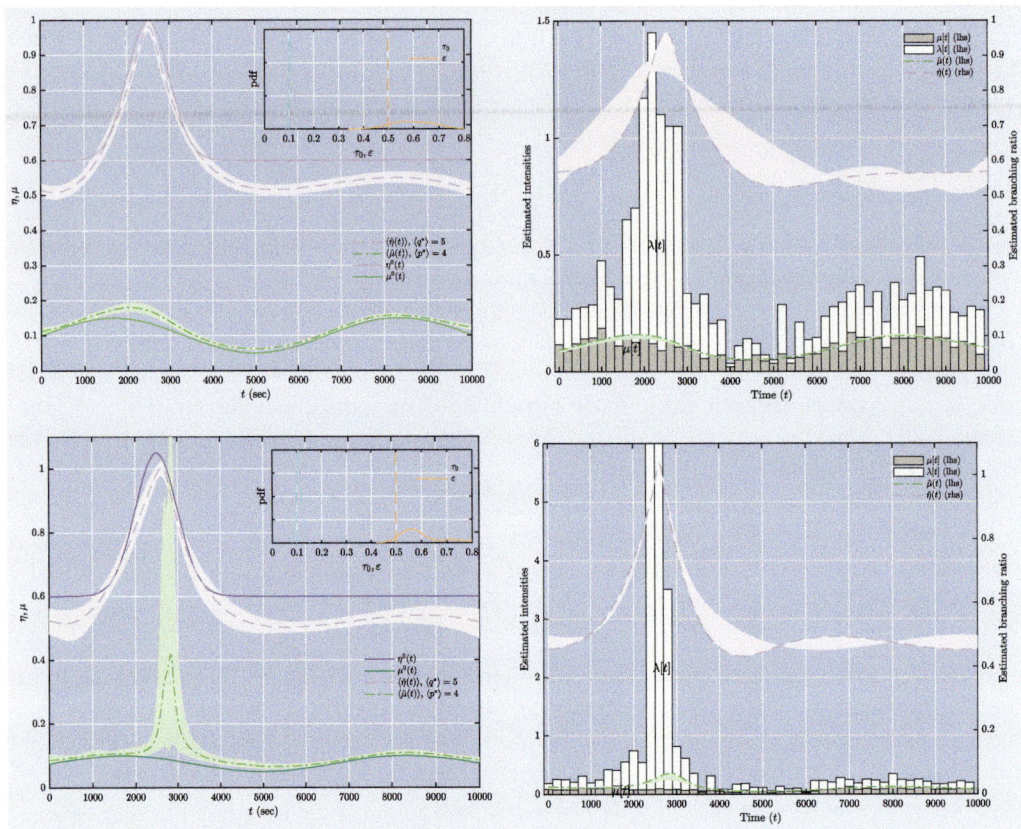

FIGURE 33.A1   Simulation results for processes with peak branching ratio of $\eta(t) = 1$ (top row) and $\eta(t) = 1.05$ (bottom row). The *left column* shows the true immigration $\mu^0(t)$ and branching ratio $\eta^0(t)$ in solid lines, and their estimates $\mu(t)$, $\eta(t)$ based on a Hawkes(p,q) obtained from the EM algorithm in dashed lines. For each of the 50 simulations performed, the model with the highest posterior probability (A10) was chosen and the dashed lines represent the average over these estimates. The shaded regions indicate corresponding 95% confidence intervals for the mean, obtained by bootstrapping 1000 times from the resulting function estimates. We also provide the average selected degrees of freedom for the immigration (p^*) and branching ratio (q^*) in the legend. In the inset, the distributions of the optimal offspring density parameter estimates τ_0 and ε are shown. Their true values are indicated with vertical dashed lines. The *right column*, for a single representative simulation, counts the number of immigrants and overall points, and normalizes them into histogram unconditional intensity estimates, labeled $\mu[t]$ and $\lambda[t]$, respectively. The best fit of the immigration intensity $\breve{\mu}(t)$ is given by the green line. Additionally, on the secondary y-axis, the best estimate of the branching ratio $\breve{\eta}(t)$ is shown. The shaded regions indicate 95% credible intervals obtained from bootstrapping 1000 times from all the fitted functions for this realization, with weights proportional to their posterior probability (A10).

Regarding the immigration intensity estimate, we observe that it is overestimated locally where the transient approach to criticality occurs, indicating more difficult separability of the dynamics when both exogenous and endogenous activity are high. This effect is naturally more pronounced when $\eta(t)$ remains in the critical regime for some time. The confidence interval also widens accordingly in the region where the endogenous dynamics spike, as endogenous and exogenous time scales become more difficult to disentangle by the EM. In the transient supercritical case, the uncertainty in estimating (t) is very high in this region. For single realizations, we typically observe that the uncertainty in model selection as indicated by the credible intervals around the estimate of $\eta(t)$ is higher than for the corresponding estimate of (t). Furthermore, we observe that the bootstrapped credible intervals for the branching ratio can be as high as 0.1, which highlights the importance to also look at the uncertainty in model selection as proposed here. Regarding the offspring density parameter estimates, we find that the high frequency cutoff $_{\tau 0}$ is estimated very accurately and with little variance, whereas the tail index ϵ is overestimated on average by ≈ 0.1—in line with the slight overall underestimation of the average level of self excitation and the applied kernel truncation—and its variance is also larger than what we observe for τ_0.

Appendix C: The MCEM Algorithm for the Estimation of the MA-Hawkes(*p,q*) Process

Here, we give a short practical description of the extended MCEM algorithm required to estimate the MA-Hawkes(*p,q*) process. The algorithm synthesizes the principles described in section A for the estimation of the Hawkes(*p,q*) model, with the MCEM algorithm as derived in Wheatley *et al.* (2018) for the inclusion of the additional shot-noise/moving-average effects in the spirit of the ARMA point process.

For the estimation of a model like (33.5), the missing data has to be extended to $(\mathbf{Z}, C)$, where C denotes the immigrant process induced by N^μ, and the complete data log-likelihood thus takes the form

$$Q\left(\theta \mid \omega, \hat{\theta}^{(l)}\right) = \mathbb{E}_{Z,C|\omega,\hat{\theta}^{(l)}}\left[\log L^c(\theta \mid \omega, Z, C)\right], \tag{A11}$$

which can be approximated by averaging over a sample $c^{(k)}$, $k = 1, \ldots, K$ of immigrant realizations from a density$\propto p\left(c \mid \omega, \hat{\theta}^{(l)}\right)$ as

$$Q\left(\theta \mid \omega, \hat{\theta}^{(l)}\right) \approx \frac{1}{K}\sum_{k=1}^{K} \mathbb{E}_{Z|\omega,\hat{\theta}^{(l)},c^{(k)}}\left[\log L^c\left(\theta \mid \omega, Z, c^{(k)}\right)\right]. \tag{A12}$$

To obtain a conditional MCMC sample from the immigrant process, a Metropolis-Hastings algorithm is employed. Given some state $c^{(k)}$ of the Markov chain, each Metropolis-Hastings iteration starts with randomly proposing the birth or death of an immigrant. In case we choose birth, we uniformly select some $c^* \in \omega \setminus c^{(k)}$. The Metropolis-Hastings birth ratio to decide whether c^* is accepted, given some immigrant vector c, is obtained from

$$r_b\left(c,c^*\right)$$
$$= \frac{N_T - |c|}{|c|+1}\mu\left(c^*\right)e^{-\gamma F\left(T - c^*\right)} \tag{A13}$$
$$\prod_{t_i \in \omega \setminus \left(c \cup c^*\right)}\left(1 + \frac{\Theta\left(t_i - c^*\right)}{\lambda\left(t_i \mid c\right) - \mu\left(t_i\right)}\right) / \left(\lambda\left(c^* \mid c\right) - \mu\left(c^*\right)\right).$$

and we accept $c^{(k+1)} = c^{(k)} \cup c^*$ with acceptance probability $\min\left\{r_b\left(c^{(k)}, c^*\right), 1\right\}$. A derivation of this result can be found in Wheatley *et al.* (**2018**). In case we chose death, $c^* \in c^{(k)}$ is removed from the current state of the Markov chain with acceptance probability $\min\left\{1/r_b\left(c^{(k)} \setminus c^*, c^*\right), 1\right\}$. The Metropolis-Hastings iterations can then be repeated until a sufficiently long chain is available, and the desired K samples be drawn from this chain.

Then, given some $c^{(k)} = \left(c_1^k, \ldots, c_{n_\mu^k}^k\right)$ and due to the fact that the ARMA point process, as well as its generalization in the form of the MA-Hawkes(*p,q*), maps to a branching process, the inner expectation in (A12) decouples analogously to (A3), using branching probabilities which can be evaluated due to random thinning (Ogata **1981**) in a similar fashion like (A4) as

$$
\pi_{i,j}^{\Theta,k} = \begin{cases} \dfrac{\Theta\left(t_i - c_j^k \,|\, \hat{\theta}^{(l)}\right)}{\lambda\left(t_i \,|\, \hat{\theta}^{(l)}, c^{(k)}\right) - \mu\left(t_i \,|\, \hat{\theta}^{(l)}\right)} & \text{for } t_i \in \omega \setminus c^{(k)}, t_j \in \omega \\[2ex] 0 & \text{otherwise} \end{cases}
$$

(A14)

$$
\pi_{i,j}^{h,k} = \begin{cases} \dfrac{\eta\left(t_j \,|\, \hat{\theta}^{(l)}\right) h\left(t_i - t_j \,|\, \hat{\theta}^{(l)}\right)}{\lambda\left(t_i \,|\, \hat{\theta}^{(l)}, c^{(k)}\right) - \mu\left(t_i \,|\, \hat{\theta}^{(l)}\right)} & \text{for } t_i \in \omega \setminus c^{(k)}, t_j \in \omega \\[2ex] 0 & \text{otherwise.} \end{cases}
$$

In the M-step of the algorithm, the objective (A12) is maximized, which-as for estimation without shot-noise effects-has the structure of decoupled density estimation. First, $\hat{\mu}(t)$ is estimated on the MCMC sample of immigration times $\{c^{(1)}, \ldots, c^{(K)}\}$. For a time-varying estimate, a PDF $m(t)$ can be estimated by a density estimator of choice and scaled up by $\hat{N}_T^\mu = K^{-1} \sum_{k=1}^K n_\mu^k$. As for the Hawkes($p$ q) estimation presented in the main text, we resort to adaptive logspline density estimates with p degrees of freedom here. A constant immigration intensity estimate is simply obtained from $\hat{N}_T^\mu / T$. Next, the offspring density $h(\cdot)$ is estimated as in the Hawkes(p,q) case with objective (A5), with the difference that the offspring probabilities used in the objective function need to be averaged over the MCMC samples as

$$
\pi_{i,j}^h := \frac{1}{K} \sum_{k=1}^K \pi_{i,j}^{h,k}.
$$

(A15)

Estimation of the time-dependent branching ratio is then identical to the Hawkes(p,q) case, only using the above average offspring probabilities in the computation of the $\hat{\eta}_j$ in (A6). Estimation of $\Theta(\cdot)$ finally requires to pool inter-event times and weights over all MCMC samples $\pi^\Theta := \left\{\left(\pi_{i,j}^{\Theta,k}, t_i - c_j^k\right)\right\}_{k=1,\ldots,K}$ and maximize

$$
\ell_\Theta^c\left(\theta \,|\, \pi^\Theta\right) = \frac{1}{K} \sum_{k=1}^K \sum_{i=1}^{N_T} \sum_{j=1}^{n_\mu^k} \pi_{i,j}^{\Theta,k} \log\left(\Theta\left(t_i - c_j^k\right)\right)
$$

$$
- \frac{\gamma}{K} \sum_{k=1}^K \sum_{j=1}^{n_\mu^k} F\left(T - c_j^k\right),
$$

(A16)

where $F(t)$ is the cumulative distribution corresponding to the exogenous offspring density $f(t) = \Theta(t) / \gamma$. The resulting full estimation procedure is sketched in the following box.

    1: Start from some initial parameter vector $\theta^{(0)} = (\mu(t), \Theta(t), \eta(t), h(t))$
    2: Choose an initial set of immigrants $c^{(0)}$ from ω
    3: Define some convergence threshold ϵ
    4: $l \leftarrow 0$
    5: **do**
    6:    **E-step:**
    7:    **Generate Markov chain of immigrants:**

8:        $u \leftarrow U(0,1)$

9:        **if** $u > 0.5$ **then**

10:          Choose some $c^* \in \omega \setminus c^{(k-1)}$

11:          Compute the birth-ratio $r_b\left(c^{(k-1)}, c^*\right)$ as in (A13)

12:          Accept $c^{(k)} \leftarrow c^{(k-1)} \cup c^*$ with probability $\min\left\{r_b\left(c^{(k-1)}, c^*\right), 1\right\}$

13:        **else**

14:          Choose some $c^* \in c^{(k-1)}$

15:          Compute the death-ratio $1/r_b\left(c^{(k-1)} \setminus c^*, c^*\right)$ using (A13)

16:          Accept $c^{(k)} \leftarrow c^{(k-1)} \setminus c^*$ with probability $\min\{1/r_b\left(c^{(k-1)} \setminus c^*, c^*\right), c^*), 1\}$

17:        **end if**

18:        If no modification was accepted in 12 or 16, $c^{(k)} \leftarrow c^{(k-1)}$

19:        $k \leftarrow k+1$ and go to 8, until the chain has the desired length

20:

21:        Choose K states of the generated Markov chain as samples for the immigrant process

22:        $\hat{N}_T^\mu \leftarrow K^{-1} \sum_{k=1}^{K} n_\mu^k$

23:        Compute the probabilistic branching structures $\pi_{i,j}^{\Theta,(k)}$ and $\pi_{i,j}^{h,(k)}$ from (A14)

24:        Compute the offspring probabilities $\pi_{i,j}^h$ as in (A15)

25:

26: **M-step:**

27:        Estimate the density $m(t)$ on $\left\{c^{(k)}\right\}_{k=1,\dots,K}$

28:        $\hat{\mu}(t) \leftarrow \hat{N}_T^\mu \cdot m(t)$

29:        Estimate the offspring density $\hat{h}(t)$ by maximizing (A5) given the current $\pi_{i,j}^h$

30:        Estimate $\hat{\eta}(t)$ as in (A9) using $\hat{\eta}_j \leftarrow \sum_{i:t_i > t_j} \pi_{i,j}^h / \hat{H}\left(T - t_j \mid \hat{\theta}^{(l)}\right)$

31:        Estimate $\hat{\Theta}(t)$ by maximizing (A16), given $\pi^\Theta \leftarrow \left\{\left(\pi_{i,j}^{\Theta,k}, t_i - c_j^k\right)\right\}_{k=1,\dots,K}$

32:

33: $l \leftarrow l+1$

34:        $\hat{\theta}^{(l)} \leftarrow (\hat{\mu}(t), \hat{\Theta}(t), \hat{\eta}(t), \hat{h}(t))$

35: **while** $\left\| \hat{\theta}^{(l)} - \hat{\theta}^{(l-1)} \right\|_\infty > \epsilon$

36: $\hat{\theta}^{(l)}$ is the MCEM estimate for the parameters of (33.5)

Epilogue

We hope that the contents of this book will contribute to the contemporary heated debates about commodity markets amongst investors, practitioners, academics and policymakers. Recent regulation regarding commodity trading is currently reducing the involvement of investment banks in commodity markets. In the medium term this may be expected to decrease commodity investment return volatility. We feel that only a deeper understanding of commodity futures markets and related derivative products will lead to financial practitioners creating truly innovative commodity-based products which are beneficial to investors.

Michael Dempster, Cambridge Ke Tang, Beijing

August 2021

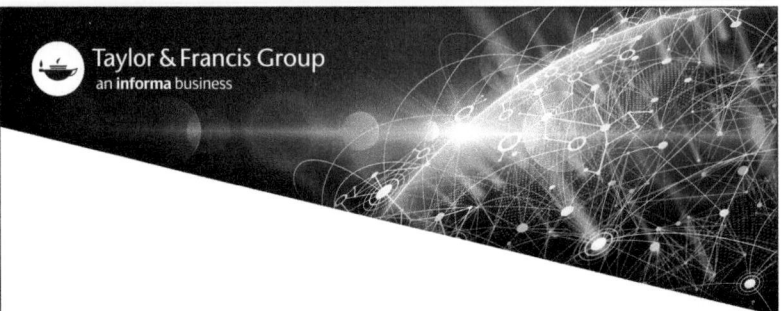